2nd International Symposium on Advanced Magnetic Materials and Applications 2010

(ISAMMA 2010)

Journal of Physics: Conference Series Volume 266

Sendai, Japan
12 – 16 July 2010

ISBN: 978-1-61782-300-8
ISSN: 1742-6588

Printed from e-media with permission by:

Curran Associates, Inc.
57 Morehouse Lane
Red Hook, NY 12571

Some format issues inherent in the e-media version may also appear in this print version.

Printed by Curran Associates, Inc. (2011)

For permission requests, please contact the Institute of Physics
at the address below.

Institute of Physics
Dirac House, Temple Back
Bristol BS1 6BE UK

Phone: 44 1 17 929 7481
Fax: 44 1 17 920 0979

techtracking@iop.org

Additional copies of this publication are available from:

Curran Associates, Inc.
57 Morehouse Lane
Red Hook, NY 12571 USA
Phone: 845-758-0400
Fax: 845-758-2634
Email: curran@proceedings.com
Web: www.proceedings.com

TABLE OF CONTENTS

PREFACE

The 2nd International Symposium on Advanced Magnetic Materials and Applications 2010 (ISAMMA 2010) was held in Sendai, Japan, from 12–16, July 2010. ISAMMA is the first consolidated symposium of three independent symposia held in the Asian region: ISPMM (International Symposium on Physics of Magnetic Materials) of Japan which was first held in 1987 in Sendai, and was subsequently held five times, Beijing (1992), Seoul (1995), Sendai (1998), Taipei (2001), and Singapore (2005); ISAMT (International Symposium of Advanced Magnetic Technology) of Taiwan, and SOMMA (International Symposium on Magnetic Materials and Applications) of Korea, both of which were started in 1999, and were held five times up to 2005. ISAMMA was established as a new international symposium which will be held every 3 years in Asia. The concept of this unified international symposium was mainly developed by Prof. M. Takahashi, Conference Chair of this conference, ISAMMA 2010. The first memorial symposium, ISAMMA 2007, was held on Jeju Island, Korea, from 28 May to 1 June 2007. The main purpose and scope of the ISAMMA conferences are to provide an opportunity for scientists and engineers from all over the world to meet in Asia to discuss recent advances in the study of magnetic materials and their physics, and spin related phenomena and materials.

ISAMMA 2010,Sendai Japan 2010.7.12~16

The categories of ISAMMA 2010 were:

- Fundamental Properties of Magnetic Materials;
- Hard/Soft Magnetic Materials and Applications;
- Spintronics Materials and Devices;
- Structured Materials;
- Multi Functional Magnetic Materials;
- Spin Dynamics and Micromagnetics;
- Magnetic Storage;
- Materials for Applications (Sensors, High Frequency, Power, and Bio/Medical devices);
- Magnetic Imaging and Characterization.

The scientific program commenced on Tuesday 13 July 2010 with opening remarks by the Symposium Chairman and the plenary talks were presented by T Rasing, P Fischer, H Yoda and S Sugimoto.

The conference was attended by 511 participants from 23 countries, with about 40 percent of participants attending from overseas (see figure). The program involved 4 plenary talks (45 minutes each), 37 invited talks (30 minutes), 85 contributed talks (15 minutes), and 352 posters.

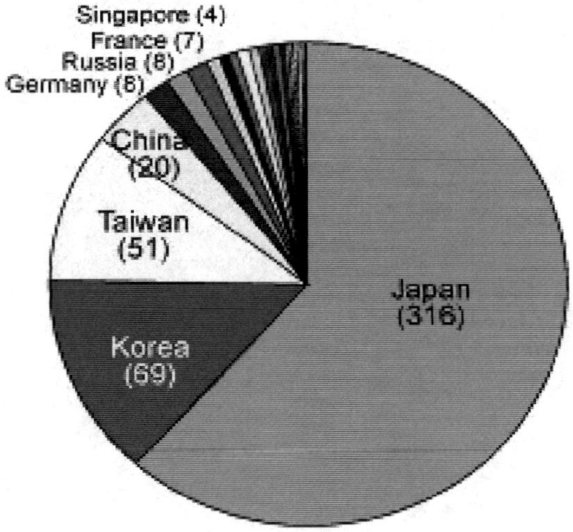

Organizing Committee of ISAMMA 2010

M Takahashi	*Tohoku Univ., Japan, Chairman*
K Takanashi	*Tohoku Univ., Japan, Chair of the Program Committee*

H Saito	*Akita Univ., Japan, Chair of the Publication Committee*
M Sahashi	*Tohoku Univ., Japan, Chair of the Treasury Committee*
M Tsunoda	*Tohoku Univ., Japan, General Secretary*
H Akinaga	*AIST, Japan*
H Fukunaga	*Nagasaki Univ., Japan*
K Hono	*NIMS, Japan*
S Ishio	*Akita Univ., Japan*
S Iwata	*Nagoya Univ., Japan*
K Nakagawa	*Nihon Univ., Japan*
S Nakagawa	*Tokyo Inst. of Tech., Japan*
T Ono	*Kyoto Univ., Japan*
Y Suzuki	*Osaka Univ., Japan*
M Tanaka	*Ehime Univ., Japan*
T Tanaka	*Univ. of Tokyo, Japan*

Program Committee of ISAMMA 2010

K Takanashi	*Tohoku Univ., Japan, Chair*
M Mizuguchi	*Tohoku Univ., Japan, Vice-chair*
M Doi	*Tohoku Univ., Japan*
A Fujita	*Tohoku Univ., Japan*
K Ishiyama	*Tohoku Univ., Japan*
T Kato	*Nagoya Univ., Japan*
T Kawagoe	*Osaka Pref. Univ.of Edu., Japan*
O Kitakami	*Tohoku Univ., Japan*
Y Kitamoto	*Tokyo Inst. of Tech., Japan*
F Matsukura	*Tohoku Univ., Japan*
C Mitsumata	*Hitachi Metals, Japan*
S Mizukami	*Tohoku Univ., Japan*
H Naganuma	*Tohoku Univ., Japan*
S Nakagawa	*Tokyo Inst. of Tech., Japan*
K Nakamura	*Tohoku Univ., Japan*
K Ono	*KEK, Japan*
T Ono	*Kyoto Univ., Japan*
F Sato	*Tohoku Univ., Japan*
M Shirai	*Tohoku Univ., Japan*
S Sugimoto	*Tohoku Univ., Japan*
M Yamaguchi	*Tohoku Univ., Japan*

Publication Committee of ISAMMA 2010

H Saito	*Akita Univ., Japan, Chair*
S Yoshimura	*Akita Univ., Japan, Vice-chair*
Y Ando	*Tohoku Univ., Japan*
J Ariake	*AIT, Japan*
S Mitani	*NIMS, Japan*
H Muraoka	*Tohoku, Japan*
M Nakano	*Nagasaki Univ., Japan*
R Nakatani	*Osaka Univ., Japan*

H Asano	*Nagoya Univ., Japan*
M Futamoto	*Chuo Univ., Japan*
J Hayakawa	*Hitachi, ARL, Japan*
T Honda	*Kyushu Inst. of Tech., Japan*
M Igarashi	*Hitachi, CRL, Japan*
H Ito	*Kansai Univ., Japan*
H Iwasaki	*Toshiba, Japan*
H Kato	*Yamagata Univ., Japan*
M Konoto	*AIST, Japan*
H Kubota	*AIST, Japan*
K O'Grady	*Univ. of York, UK*
A Sakuma	*Tohoku Univ., Japan*
T Sato	*Keio Univ., Japan*
T Sato	*Shinshu Univ., Japan*
K Tajima	*Akita Univ., Japan*
M Takeda	*JAEA, Japan*
Y Takemura	*Yokohama Nat'l Univ., Japan*
M Tanaka	*Univ. of Tokyo, Japan*
A Tsukamoto	*Nihon Univ., Japan*
S Yabukami	*Tohoku Gakuin Univ., Japan*

Treasury Committee of ISAMMA 2010

M Sahashi	*Tohoku Univ., Japan, Chair*
K Ishiyama	*Tohoku Univ., Japan*
K Nakagawa	*Nihon Univ., Japan*
T Ogawa	*Tohoku Univ., Japan*
S Saito	*Tohoku Univ., Japan*
T Tanaka	*Ehime Univ., Japan*
N Tezuka	*Tohoku Univ., Japan*

Executive Committee of ISAMMA 2010

M Takahashi	*Tohoku Univ., Japan, Chair*
K Takanashi	*Tohoku Univ., Japan, Vice-chair*
K Miyake	*Tohoku Univ., Japan*
T Ogawa	*Tohoku Univ., Japan*
S Okamoto	*Tohoku Univ., Japan*
M Oogane	*Tohoku Univ., Japan*
S Saito	*Tohoku Univ., Japan*
Y Sakuraba	*Tohoku Univ., Japan*
T Shima	*Tohoku Gakuin Univ., Japan*
N Tezuka	*Tohoku Univ., Japan*
M Tsunoda	*Tohoku Univ., Japan*

We are grateful to all the participants for their valuable contributions and active discussions. We gratefully acknowledge the financial support from 17 Japanese companies (ASAKA RIKEN CO., LTD, Fujikin Incorporated, Furukawa Electric Co., Ltd, Hitachi Metals, Ltd, IZUMI-TEC CO., LTD, Miwa Electric Industrial CO., LTD, MIWA MFG CO., LTD, NEOARK Corporation, Optima Corporation, PRESTO CO., LTD, SHOWA DENKO K.K., TAIYO YUDEN CO., LTD, TDK Corporation, TEIJIN LIMITED, Ube Material Industries, Ltd, ULVAC, Inc, and V TEX Corporation) and 7 foundations (SENDAI TOURISM & CONVENTION BUREAU, The Iwatani Naoji Foundation, Tohoku University Electro-Related Departments Global COE Program 'Center of Education and Research for

Information Electronics Systems', The Murata Science Foundation, Research Foundation for Materials Science, Nippon Sheet Glass Foundation for Materials Science and Engineering, and Aoba Foundation for The Promotion of Engineering).

Co-Publication

This is a co-publication with *Journal of Physics D: Applied Physics*. The bulk of the papers, after peer review, are published in *Journal of Physics: Conference Series*. However a selection of papers, are published separately in a special issue of *Journal of Physics D: Applied Physics*.

Magnetodielectric effect in $CoCr_{2-X}Fe_XO_4$

Huixin Bao[1,2], Sen Yang[1], Xiaobing Ren[1,2]
[1]Multi-disciplinary Materials Research Center, Frontier Institute of Science and Technology, and State Key Laboratory for Mechanical Behavior of Materials, Xi'an Jiaotong University, Xi'an 710049, China;
[2]Ferroic Physics Group, National Institute for Materials Science, Tsukuba, 305-0047, Ibaraki, Japan

E-mail: baohx2006@gmail.com

Abstract. Conventional multiferroics show the multiferroic behaviors below two successive transitions (a ferroelectric transition and a ferromagnetic transition). In the present study, we report the magneto-dielectric (MD) effect (both in-plane and out-of-plane) in $CoCr_{2-X}Fe_XO_4$ which undergoes only a ferromagnetic transition. We found the MD effect in the pure ferromagnetic phase ($T_S<T<T_C$) of $CoCr_2O_4$. And such MD effect is temperature dependent. The magnitude of the MD effect gradually decreases with approaching Curie temperature (T_C) and finally becomes zero at T_C. With doping Fe into $CoCr_2O_4$, we found the magnitude of MD effect increases and the MD effect exists in a wider temperature region because T_C increases and T_S decrease with doping Fe. Our results show the coupling between the ferromagnetic and ferroelectric orders. It suggests that both the ferromagnetic and ferroelectric properties originate from the same transition.

1. Introduction

Multiferroics are materials in which ferromagnetism, ferroelectricity and ferroelasticity coexist in the same phase [1-3]. This implies that they possess spontaneous magnetization which can be reoriented by external magnetic field, spontaneous polarization which can be reoriented by an applied electric field and spontaneous deformation which can be reoriented by an applied stress. However, it is customary to exclude ferroelasticity and only consider ferromagnetic and ferroelectric characteristics (magnetoelectrics) [4]. Multiferroics have attracted great interest in the past few years since their application potential on the dual read-write memory devices and spintronic devices [1, 2]. Besides the applications, the science of these materials is truly fascinating because the coupling between ferromagnetism and ferroelectricity challenges the conventional mechanism which shows ferroelectric ordering is incompatible with the ferromagnetic ordering [5].

Due to the incompatibility of ferromagnetism and ferroelectricity, there are rare single phase multiferroics [5]. Most of the single phase multiferroic materials (such as $BiFeO_3$, $YMnO_3$ and $BiMnO_3$) have a ferroelectric transition at high temperature and a ferromagnetic (or ferrimagnetic/antiferromagnetic) transition at low temperature [6]. Very recently, based on the spiral spin model [7, 8], a number of new multiferroic systems ($TbMnO_3$, $DyMnO_3$ and MCr_2O_4 with M=Mn, Fe and Co) were discovered [6, 9, 10]. With decreasing temperature, they undergo a magnetic transition to the ferromagnetic (or ferrimagnetic/antiferromagnetic) state firstly. Then, spiral spin transition occurs at lower temperature, the materials get to multiferroic state with the emerging of ferroelectricity induced by the spiral spins. All the above multiferroics show the magnetoelectric coupling below two successive transitions. Here we report a magnetoelectric coupling phenomenon which undergoes only one transition: the ferromagnetic transition, which may imply the ferromagnetic and ferroelectric properties have the same origin.

In the present work, we studied magneto-dielectric effect [11-14] (MD effect: the external magnetic field dependence of dielectric permittivity (ε)) in the pure ferrimagnetic phase of $CoCr_{2-X}Fe_XO_4$. When the magnetic field is parallel with the AC electric field (in-plane), we found that ε gradually decreases and finally saturates with increasing the magnetic field. Such a MD effect exist in the pure ferrimagnetic phase ($T_S<T<T_C$) of $CoCr_2O_4$, which was known the multiferroic properties only exist below the spiral transition ($T<T_S$) [9]. The magnitude of the MD effect gradually decreases with approaching Curie temperature (T_C) and finally becomes to zero at T_C, which is quite similar with the change of magnetization with temperature (M-T curve). With doping Fe into $CoCr_2O_4$, we found the magnitude of MD effect increases and the MD effect exists in a wider temperature region because T_C increases and Ts decrease with doping Fe. When the magnetic field is perpendicular with the AC electric field (out-of-plane), it show similar results, the only difference is that ε increases with increasing the magnetic field and the magnitude of the MD effect under the same field is smaller than the parallel case. Our results suggest the coupling between the ferromagnetism (or ferrimagnetism /antiferromagnetism) and ferroelectricity exist in ferromagnetic insulators.

2. Experimental

The polycrystalline samples $CoCr_{2-X}Fe_XO_4$ (X=0 and 0.5) were prepared from CoO, Cr_2O_3, and Fe_2O_3 by the conventional solid state reaction method. The diameter and thickness of the cylinder is 8mm and 1mm respectively. The samples with good spinel structure were initially characterized by X-ray diffraction. The temperature dependence of the DC magnetization of $CoCr_{2-X}Fe_XO_4$ was measured by a Quantum Design SQUID. Silver paste was painted as electrodes. The temperature dependence of dielectric permittivity was measured with 3 kHZ by a dielectric measurement system (including a LCR 3532 meter and a cryostat for temperature changing). The change of dielectric permittivity with magnetic field up to 5T both in-plane and out-of-plane (in-plane: the direction of the magnetic field is parallel with the AC electric field; out-of-plane: the direction of the magnetic field is perpendicular with the AC electric field) was measured with 3 kHZ by the combination of the dielectric measurement system and a super-conducting magnet.

3. Results

Fig. 1 (a) and (b) show the temperature dependence of magnetization (M) and dielectric permittivity (ε) for a pure polycrystalline $CoCr_2O_4$. The onset of the magnetization (Fig 1(a)) at T_C=96K is due to the ferrimagnetic transition. The tiny anomalies in both M-T (Fig 1(a)) and ε-T (Fig 1(b)) which are discerned at 28 K, correspond to the spiral transition (T_S). These results keep good consistence with the reported results of $CoCr_2O_4$ single crystal [9] and polycrystal [15]. It was reported [9] that the multiferroic properties only exist in the multiferroic phase (Fig 1(a) region B) which is below the spiral transition temperature T_S. In the following, we shall show the multiferroic properties in the pure ferromagnetic phase (Fig 1(b) region A: $T_S<T<T_C$).

Here it should be mentioned that direct measurement of ferroelectricity (or polarization P) caused by magnetic field becomes difficult when ferroelectricity is weak. In such a situation, measuring the change of dielectric permittivity ($\varepsilon = dP/dE$) by magnetic field (i.e., magnetodielectric effect) becomes a much more sensitive tool to detect the existence of coupling between P and M. [11, 12] Magnetodielectric effect can be described by, $MD = \varepsilon(H)/\varepsilon(0) - 1$ where $\varepsilon(H)$ is the dielectric permittivity under field H and $\varepsilon(0)$ is the dielectric permittivity without field" line 41 of page 2 of our paper.

Fig.2 shows the magnetodielectric effect (in-plane) at various temperatures in pure ferromagnetic phase of $CoCr_2O_4$ polycrystal. The magnetodielectric effect is defined as $MD = \varepsilon(H)/\varepsilon(0) - 1$, where $\varepsilon(H)$ is the dielectric permittivity under magnetic field and $\varepsilon(0)$ is the dielectric permittivity without field. As shown in Fig 2a at 30K, MD gradually decreases with the increase of the external magnetic field along the positive direction. After removing the magnetic field, MD can go back to the original value. MD has the same variance while applied a magnetic field with opposite direction. At a relative higher temperature 84K, Fig 2b shows similar results as Fig 2a, but the magnitude of MD effect under the same magnetic field is smaller than that of 30K. MD finally becomes zero at 114K above T_C (Fig 2c). Fig 2d shows the change of the magnitude of magnetodielectric effect (MD) with temperature. It obviously shows the MD effect only exist below T_C become to zero at T_C, which is much similar with the change of the magnetism with temperature (M-T curve) of the common ferromagnetic materials, indicating that the ferromagnetic properties (M) must couple with the dielectric properties (ε). The switching of M under magnetic field makes the change of ε. It may suggest a small ferroelectric polarization (P) changes with the switching of the ferromagnetic domains.

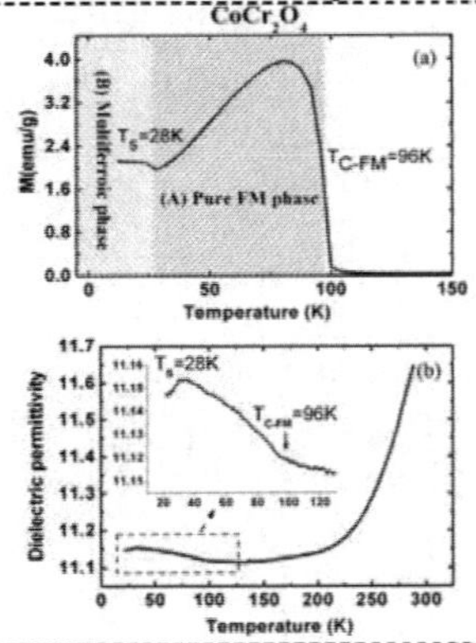

Figure 1. The temperature dependence of magnetization (M) and dielectric permittivity (ε) for a polycrystalline $CoCr_2O_4$.

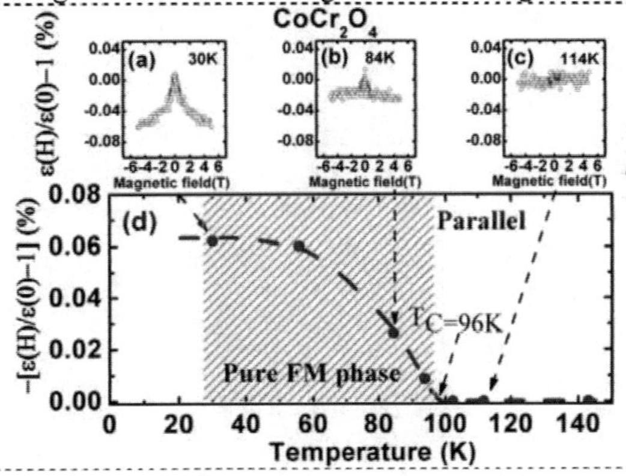

Figure 2. The temperature dependence of parallel (in-plane) MD effect for $CoCr_2O_4$. (a), (b) and (c) show perpendicular MD results at 30K, 84K, and 114K respectively.

Fig.3 (a) and (b) show the temperature dependence of M and ε of $CoCr_{1.5}Fe_{0.5}O_4$ polycrystalline. Comparing with the pure $CoCr_2O_4$, the T_C of $CoCr_{1.5}Fe_{0.5}O_4$ increases to 175K, while the spiral transition temperature (T_S) decreases to lower temperature (<4K, beyond the temperature range of the Squid). Therefore the pure ferrimagnetic phase of $CoCr_{1.5}Fe_{0.5}O_4$ is wider than that of pure $CoCr_2O_4$. We can study the MD effect in a wider temperature and closer to the room temperature.

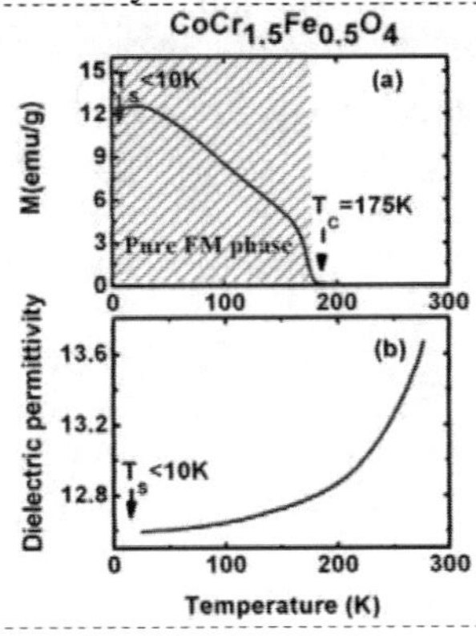

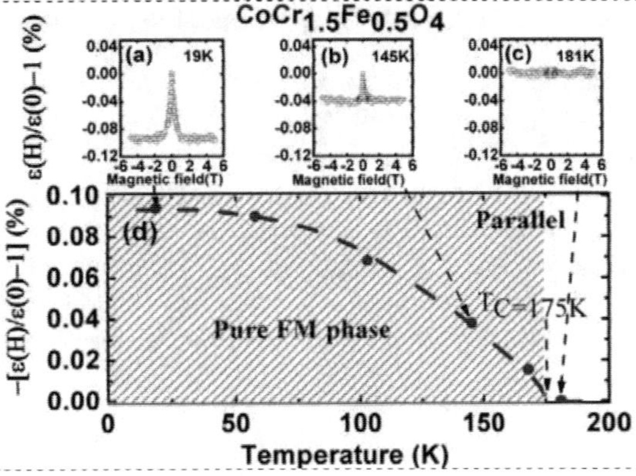

Figure 3. The temperature dependence of magnetization

Figure 4. The temperature dependence of parallel (in-plane) MD effect for $CoCr_{1.5}Fe_{0.5}O_4$. (a), (b) and (c)

(M) and dielectric permittivity (ε) for a polycrystalline $CoCr_{1.5}Fe_{0.5}O_4$.

show perpendicular MD results at 19K, 145K, and 181K respectively.

Fig 4 shows magneto-dielectric effect (in-plane) of $CoCr_{1.5}Fe_{0.5}O_4$ at various temperatures. It is similar with that (Fig 2) of pure $CoCr_2O_4$. The magnitude of MD effect under the same magnetic field gradually decreases with the increase of temperature and becomes to zero at T_C (175K). The only difference is that the magnitude of MD of $CoCr_{1.5}Fe_{0.5}O_4$ is larger than that of pure $CoCr_2O_4$ at the same temperature, which is due to the increase of coupling between M and ε by doping some Fe (increasing of T_C). It indicates that doping Fe can enhance the coupling between ferromagnetic and ferroelectric properties, and widen the multiferroic region.

Fig 5 shows the magneto-dielectric effect (out-of-plane) of $CoCr_{1.5}Fe_{0.5}O_4$ at various temperatures. Interestingly, the dielectric permittivity increases with the increase of magnetic field, which is opposite with the in-plane case. However, with the increase on temperature, the magneto-dielectric effect also becomes smaller and finally disappears at T_C. Comparing Fig 5 with Fig 4, we found that the magnitude of MD effect out-of plane is smaller than that of MD effect in-plane. It indicates the anisotropy of the coupling between M and ε.

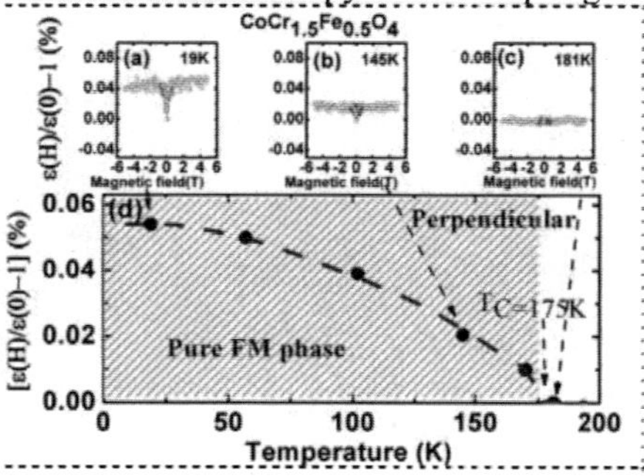

Figure 5. The temperature dependence of perpendicular (out-of-plane) MD effect for $CoCr_{1.5}Fe_{0.5}O_4$. (a), (b) and (c) show perpendicular MD results at 19K, 145K, and 181K respectively.

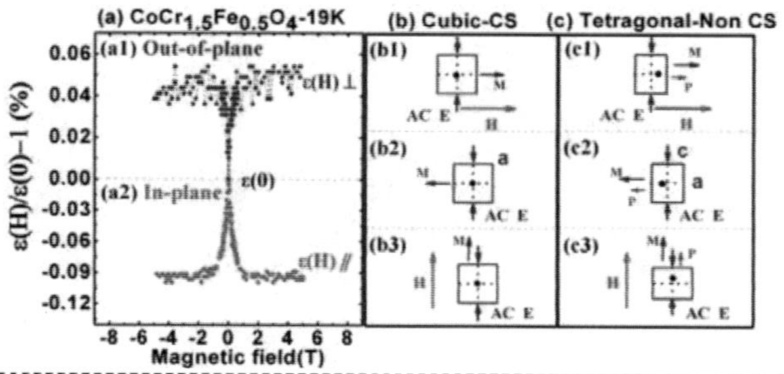

Figure 6. The schematic explanation for the MD behavior in both paramagnetic ((b): $T>T_C$) and ferrimagnetic phase ((c): $T<T_C$).

4. Discussion

The above magneto-dielectric results (Fig 2, Fig 4 and Fig 5) in $CoCr_{2-x}Fe_xO_4$ show there must be coupling between the magnetization (M) and the dielectric permittivity (ε). As we know the definition of the dielectric permittivity is that $\varepsilon = dP/dE$. When we measure the MD effect with an external magnetic field, the small AC electric field is a constant field, so dE is unchangeable during the measurement. Therefore the change of the dielectric permittivity must be ascribed that dP is changed by the external magnetic field H. This may poses a potential challenge to the convention view [9] that the ferrimagnetic state hosts a cubic structure with central symmetry.

Fig 6 shows the analysis of the MD effect with different structure view. Fig 6a shows the MD effect in-plane and out-of-plane of $CoCr_{1.5}Fe_{0.5}O_4$ at 19K. Obviously, it is easy to get $\varepsilon(H)\perp > \varepsilon(0) > \varepsilon(H)//$. In the following, we will analyze the MD results (Fig 6a) with a cubic ferrimagnetic state with central symmetry and a tetragonal ferromagnetic state with off-center symmetry respectively.

Generally, the ferrimagnetic state is considered as a cubic state with central symmetry. ε(0) is measured in such a multi-domain state (as shown in Fig 6b2) with an AC electric field. After applied a magnetic field which is perpendicular with the electric field, M switches to the direction of the magnetic field (Fig 6b1). Similarly, Fig 6b3 show the state while applied a magnetic field parallel with the AC electric field. ε(H)□ and ε(H)□of the single domain state are got. Comparing the Fig 6b1, b2 and b3, all the three states have a cubic structure with central symmetry, only the directions of M are different, which has no influence on the dP. So, the dielectric permittivity of the three states should be the same ε(H)□=ε(0)= ε(H)□. It is different from the experimental results as shown in Fig 6a. Therefore, it is difficult to understand the MD results with a cubic structure with central symmetry.

Recent results [16, 17] with a high-resolution synchrotron XRD clearly show that the ferromagnetic transition is a structural phase transition. The structure of the ferromagnetic state is non-cubic. The lattice distortion and the spontaneous magnetization (M) couples each other. Therefore, we assume the ferrimagnetic $CoCr_{2-X}Fe_XO_4$ state host a tetragonal structure with off centre symmetry and the direction of P is conforming to the M (as shown in Fig 6c2) because the M of the ferrimagnetic $CoCr_{2-X}Fe_XO_4$ is along <001> direction [9]. ε(0) is measured in such a multi-domain state with an AC electric field. After applied a magnetic field which is perpendicular with the electric field, both M and P switches to the direction of the magnetic field (Fig 6c1). Similarly, Fig 6c3 show the state while applied a magnetic field parallel with the AC electric field. ε(H)□ and ε(H)□of the single domain state are got respectively. In this case, it is can be considered as ferroelectrics (such as $BaTiO_3$)[18, 19]. Since the dielectric permittivity of BaTiO3 along ***a*** axis (ε□) is larger than dielectric permittivity along ***c*** axis (ε□) [20]. Then, it is easy to understand why ε(H)□>ε(0)>ε(H)□.

5. Conclusion

Our magneto-dielectric results show the coupling between the ferromagnetic and ferroelectric orders. The disappearance of the MD effect at T_C implies that both the ferromagnetic and ferroelectric properties originate from the same transition and the ferromagnetic state may posses a non-central symmetry. The ferromagnetic insulators are a kind of weak ferroelectrics. Doping Fe instead of Cr can enhance the coupling between M and P and make the MD effect exist in a wider temperature range.

Acknowledgement

We thank Y. Wang, D. Z. Xue, Chao. Zhou, J. H. Gao and X. D. Ding for technical support and helpful discussions. The authors acknowledge the support of National Basic Research Program of China (Grant No. 2010CB613003), National Science Foundation of China (50720145101 and 50702042), as well as 111 project of China.

References

[1] W. Eerenstein, N. D. Mathur, and J. F. Scott, Nature **442**, 759 (2006).
[2] M. Fiebig, J. Phys. D: Appl. Phys. **38**, R123 (2005).
[3] N. A. Spaldin, and M. Fiebig, Science **309**, 391 (2005).
[4] C. N. R. Rao, and C. R. Serrao, J. Mater. Chem. **17**, 4931 (2007).
[5] N. A. Hill, J. Phys. Chem. B **104**, 6694 (2000).
[6] W. Prellier, M. P. Singh, and P. Murugavel, J. Phys.-Condes. Matter **17**, R803 (2005).
[7] H. Katsura, N. Nagaosa, and A. V. Balatsky, Phys. Rev. Lett. **95**, 4 (2005).
[8] Y. Tokura, Science **312**, 1481 (2006).
[9] Y. Yamasaki *et al.*, Phys. Rev. Lett. **96**, 207204 (2006).
[10] Y. Tokura, J. Magn. Magn. Mater. **310**, 1145 (2007).
[11] T. Kimura *et al.*, Phys. Rev. B **67**, 180401(R) (2003).
[12] T. Suzuki, and T. Katsufuji, Phys. Rev. B **77**, 220402(R) (2008).
[13] G. Catalan, Appl. Phys. Lett. **88**, 102902 (2006).
[14] N. Mufti, G. R. Blake, and T. T. M. Palstra, J. Magn. Magn. Mater. **321**, 1767 (2009).
[15] G. Lawes *et al.*, Phys. Rev. B **74**, 024413 (2006).
[16] S. Yang, and X. B. Ren, Phys. Rev. B **77**, 014407 (2008).
[17] S. Yang *et al.*, Phys. Rev. Lett. **104**, 197201 (2010) .
[18] H. X. Bao *et al.*, Appl. Phys. Lett. **91**, 142903 (2007).

[19] X. B. Ren, Nat. Mater. **3**, 91 (2004).
[20] W. J. Merz, Phys. Rev. **76**, 1221 (1949).

Journal of Physics: Conference Series **266** (2011) 012002 doi:10.1088/1742-6596/266/1/012002

Magnetocaloric effect of the Gd-*LRE*-Al-(Co, Fe) (*LRE*=La, Ce, Pr, Nd) glassy ribbons in the intermediate temperature

Y K Fang[1,2], C H Lai[1], C C Hsieh[1], X G Zhao[3], H W Chang[4], W C Chang[1,5] and W Li[2]

[1] Department of Physics, National Chung Cheng University, Ming Hsiung, Chia-Yi, 62102 Taiwan

[2] Division of Functional Materials Research, Central Iron & Steel Research Institute, Beijing, 100081 China

[3] Shenyang National Laboratory for Materials Science, Institute of Metal Research, and International Centre for Materials Physics, Chinese Academy of Sciences, Shenyang 110016, China

[4] Department of Physics, Tunghai University, Taichung, Taiwan.

E-mail: ykfang@126.com, phywcc@ccu.edu.tw

Abstract. The magnetic properties and mgnetocaloric effect (MCE) of the $Gd_{65}Al_{15}Co_{20-x}Fe_x$ (x = 0 - 20) and $Gd_{65-y}LRE_yAl_{15}Fe_{20}$ (*LRE*=La, Ce, Pr, Nd; y = 0 - 20) glassy ribbons have been investigated. The Curie temperature (T_c) of the Gd-based $Gd_{65}Al_{15}Co_{20-x}Fe_x$ glassy ribbons can be greatly increased by moderate Fe substitution. Furthermore, it is found that Ce atom among the other light-rare-earth elements is more prefer one for adjusting the T_c of the $Gd_{65-y}LRE_yAl_{15}Fe_{20}$ glassy ribbons. The low materials cost, improved MCE properties make $Gd_{65-y}Ce_yAl_{15}Fe_{20}$ glassy ribbons attractive candidates for magnetic refrigeration application in the intermediate temperatures of 80-220 K.

1. Introduction

Magnetic refrigeration based on magnetocaloric effect (MCE) attracts very much attention because it is an environmentally friendly and high energy efficiency cooling technology [1, 2]. Generally, the working temperature range of the studied magnetic refrigeration materials is divided into four regions, i.e., so-called low temperature (<80 K), intermediate temperature (80-220 K), near room temperature (220-350 K), and relative high temperature (350-500 K). To date, large or giant magnetic entropy changes have been achieved in polycrystalline magnetic materials, $Gd_5(Si_{1-x}Ge_x)_4$ [3], $MnFeP_{1-x}As_x$ [4], $La(Fe_{1-x}Si_x)_{13}$ [5], $MnAs_{1-x}Sb_x$ [6], Ni-Mn-Ga [7], etc. These materials have potential applications for magnetic refrigeration near room temperature. In the so-called intermediate temperature range, MCE of rare-earth (RE) intermetallics REM_2 (M=Al, Co, and Ni) Laves phases [8] and manganites [9] have been extensively studied.

Recent studies [10-15] verified that amorphous magnetic materials/ bulk metallic glasses have advantages over the crystalline refrigerants because they may have a larger temperature range of the half maximum peak ΔS_M, as well as larger magnetic entropy change, giving rise to larger refrigeration

[5] To whom any correspondence should be addressed.

capacity (RC). It has been reported that the working temperature for heavy-rare-earth (*HRE*)-Al-Co (*HRE* = Gd, Dy, Er, and Ho) bulk metallic glasses is usually less than ~100 K [13-15]. And there has been less report on the HRE-based bulk metallic glasses/ glassy ribbons for working temperature beyond 100 K [16, 17].

In this paper, the working temperature of $Gd_{65}Al_{15}Co_{20-x}Fe_x$ (x = 0 - 20) glassy ribbons is increased to be ~ 200 K by substituting Co with Fe. Furthermore, for lowering the material cost, the Gd metal in these materials is substituted by low-cost light-rare-earth (LRE) elements (such as La, Ce, Pr, and Nd) in part, and the corresponding magnetic properties and MCE properties have been investigated for $Gd_{65-y}LRE_yAl_{15}Fe_{20}$ (*LRE* = La, Ce, Pr, Nd; y = 0 - 20) using in the intermediate temperature.

2. Experimental

Glassy ribbons with the nominal compositions (in at. %) of $Gd_{65}Al_{15}Co_{20-x}Fe_x$ (x = 0- 20) and $Gd_{65-y}LRE_yAl_{15}Fe_{20}$ (*LRE* = La, Ce, Pr, and Nd; y = 0 -20) have been prepared by rapid quenching technology on a single copper wheel with a speed of 40 m/s. The ribbons had the width of 2-3 mm and the thickness of about 25μm. The amorphous character of the as-quenched ribbons was checked by x-ray diffraction (XRD) using Cu-K_α radiation. All the magnetic measurements were performed on a vibrating sample magnetometer. The thermomagnetic analysis was carried out under a magnetic field of 100 Oe. The field dependence of magnetization was measured using a maximum applied field H = 50 kOe for selected temperatures in the vicinity of the Curie temperature T_c.

3. Results and discussion

The amorphous character for the studied ribbons in this work has been ascertained by XRD analysis. For increasing the working temperature of Gd-based glassy ribbons, the influence of Fe substitution on the Curie temperature T_c of the $Gd_{65}Al_{15}Co_{20-x}Fe_x$ (x = 0- 20) glassy ribbons has been investigated. Figure 1 shows the dependence of the magnetization M of these glassy ribbons on temperature in the magnetic field of 100 Oe. The values of T_c of the specimens are defined as the temperature where $|dM/dT|$ is maximal. The insert in figure 1 is the T_c versus x, the Fe contents. It is found that the T_c increases from 129 K at x = 0 to 193 K when x reaches 15. For further increasing Fe content, the T_c of the ribbons decreases. It has been reported that the Gd-Gd interactions are ferromagnetic and predominant in Gd-base amorphous alloys [18]. These interactions are influenced by the Fe substitution and then determine the T_c for these amorphous alloys. When x is about 15, the Gd-Gd interactions are strongest. Therefore, the T_c of the Gd-based $Gd_{65}Al_{15}Co_{20-x}Fe_x$ amorphous ribbons can be greatly increased by proper Fe substitution. For the $Gd_{65}Al_{15}Fe_{20}$ amorphous ribbon, the value of T_c (~ 181 K) is good consistent with the reported value (~180 K) in Ref. [17]. For searching for the magnetic refrigeration candidates using in the intermediate temperature range (80-220 K), and further lowering the material cost, the focus in the following researches is on the $Gd_{65-y}LRE_yAl_{15}Fe_{20}$ (*LRE* = La, Ce, Pr, Nd; y = 0 -20) glassy ribbons by substituting Gd element with LRE elements in part.

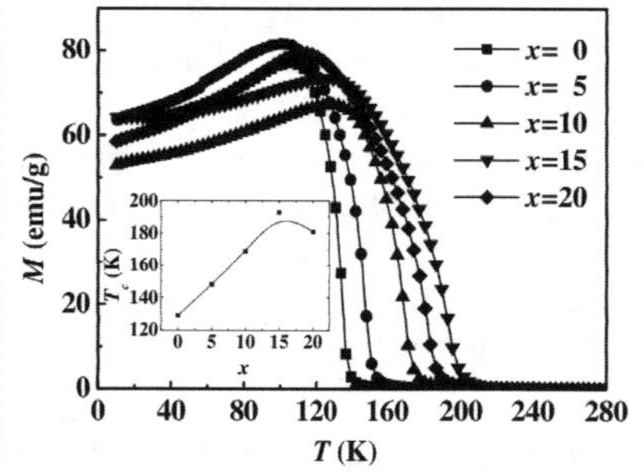

Figure 1. *M-T* curves of the $Gd_{65}Al_{15}Co_{20-x}Fe_x$ glassy ribbons. The insert is the T_c versus x, the Fe content.

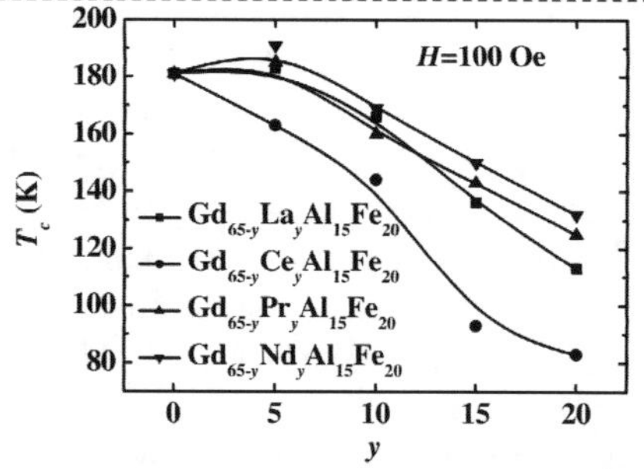

Figure 2. Dependence of T_c of the $Gd_{65-y}LRE_yAl_{15}Fe_{20}$ glassy ribbons on the contents of the light rare earth elements.

Figure 2 shows the dependence of T_c of the $Gd_{65-y}LRE_yAl_{15}Fe_{20}$ (*LRE* = La, Ce, Pr, and Nd; y = 0 -20) glassy ribbons on the contents of the light-rare-earth elements. Obviously, the T_c values of these ribbons decrease with increasing the light-rare-earth elements contents. This reduction of T_c can be attributed to not only the lower transition temperatures of La, Ce, Pr, and Nd metals but also the weaker magnetic interactions between them with Gd atom [19]". From figure 2 it can also be seen that the influence of La, Pr, and Nd elements on the T_c values is similar. However, a large T_c variation from 80 K to about 180 K has been obtained for Ce-substituted $Gd_{65-y}Ce_yAl_{15}Fe_{20}$ glassy ribbons, indicating that Ce element is much more effective for decreasing the T_c of the $Gd_{65-y}LRE_yAl_{15}Fe_{20}$ glassy ribbons comparing with the others. In 1970, J. D. Speight [20] reported that the magnetic properties of the rapidly quenched Gd-La (30 at. %), Gd-Ce (30%), Gd-Pr (35%) and Gd-Nd (50%) Sm-type alloys. The author pointed out that the Gd-Gd and the Gd-*LRE* interactions were ferromagnetic and anti-ferromagnetic (AFM), and AFM ordering of the Gd moments occurs for Sm-type Gd-Ce alloy. Therefore, Gd-Ce alloy exhibited the lower T_c relative to the others with the same amount of *LRE* substitution. Accordingly, this is the reason of the variation of T_c of the $Gd_{65-y}Ce_yAl_{15}Fe_{20}$ glassy ribbons.

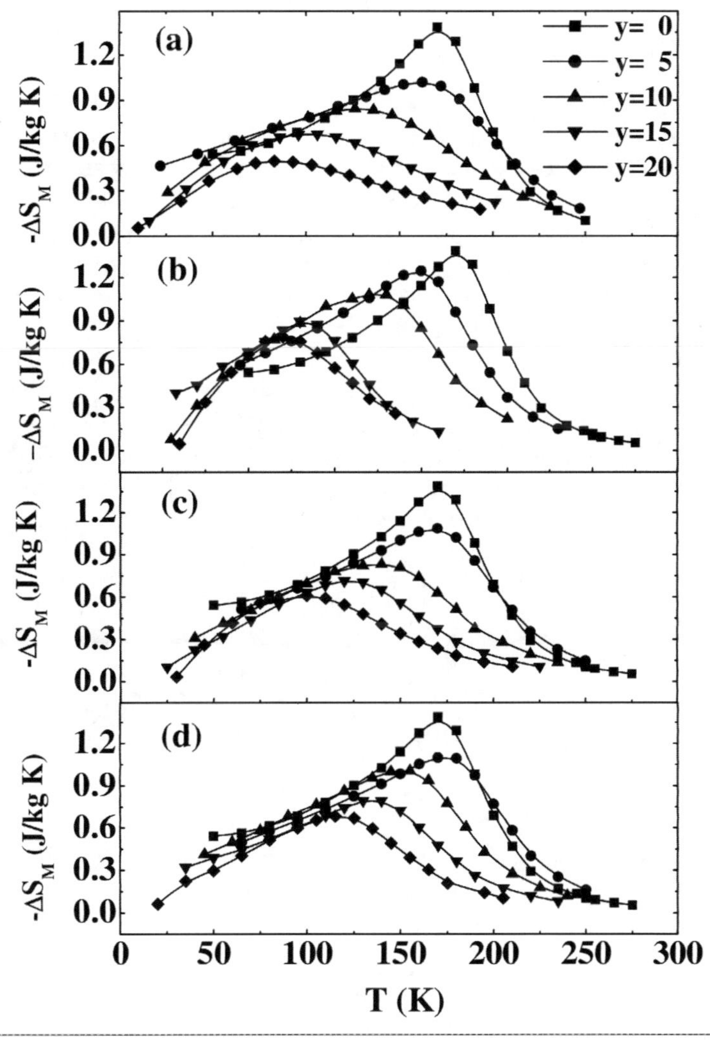

Figure 3. Magnetic entropy changes, $-\Delta S_M$, of the $Gd_{65-y}LRE_yAl_{15}Fe_{20}$ (*LRE* = La, Ce, Pr, Nd; y = 0-20) glassy ribbons under 10 kOe. (a) $Gd_{65-y}La_yAl_{15}Fe_{20}$, (b) $Gd_{65-y}Ce_yAl_{15}Fe_{20}$, (c) $Gd_{65-y}Pr_yAl_{15}Fe_{20}$, (d) $Gd_{65-y}Nd_yAl_{15}Fe_{20}$.

The isothermal magnetization curves for the $Gd_{65-y}LRE_yAl_{15}Fe_{20}$ (*LRE* = La, Ce, Pr, Nd; y = 0 - 20) glassy ribbons are measured with an increasing magnetic field. The sweeping rate of the field (~100 Oe/sec.) was slow enough to ensure that the data were recorded in an isothermal process. The magnetic entropy changes (ΔS_M) of the specimens as a function of temperature and magnetic field was calculated from the isothermal magnetization curves using the Maxwell relation

$$\Delta S_M(T,\Delta H)=\int_0^H\left(\frac{\partial M}{\partial T}\right)_H dH \tag{1}$$

which, for magnetization measurements made at constant temperature and at discrete H intervals, can be approximated by [8]

$$\Delta S_M\approx\frac{1}{\Delta T}\left[\int_0^H M(T+\Delta T,H)dH-\int_0^H M(T,H)dH\right]. \tag{2}$$

The temperature dependence of $-\Delta S_M$ of the $Gd_{65-y}LRE_yAl_{15}Fe_{20}$ (*LRE* = La, Ce, Pr, Nd; y = 0-20) glassy ribbons at the magnetic field changes of 10 kOe is shown in figure 3. All the specimens present a similar behavior, with a peak at the Curie temperature of the amorphous alloy and a progressive fall to zero as the measuring temperature increases or decreases. The maximal magnetic entropy changes, $-\Delta S_M^{max}$, of these $Gd_{65-y}LRE_yAl_{15}Fe_{20}$ specimens under 10 kOe are shown in figure 4. These values are located in the range of 0.5-1.4 J/ kg K under 10 kOe. It is found that the $-\Delta S_M^{max}$ decreases with increasing the substitution contents, which is due to the lower saturation moment of the light-rare-earth atoms than Gd. It was reported that the magnetic moment of La, Ce, Pr, Nd, and Gd are 0, 2.54, 3.68, 3.52, 7.94 μ_B/atom, respectively [19]. Therefore, the magnetization of Gd-based alloy is decreased to be about 10-20% when twenty percent Gd atoms are substituted by La, Ce, Pr, Nd atoms, respectively. However, it is interesting that the $-\Delta S_M^{max}$ of $Gd_{65-y}Ce_yAl_{15}Fe_{20}$ glassy ribbons is larger than those of the other light rare earth containing specimens under the same substitution content y. Typically, figure 5 shows the $-\Delta S_M$ curves of the $Gd_{55}LRE_{10}Al_{15}Fe_{20}$ (*LRE* = La, Ce, Pr, and Nd) glassy ribbons with y =10 at the magnetic field change of 50 kOe. As seen in figure 3, the $-\Delta S_M^{max}$ values of these specimens are around 4J/ kg K under 50 kOe. These values are comparable with or even higher than that of $Gd_{55}Co_{20}Fe_5Al_{20}$ (2.24 J/kg K at 20 kOe) [14] BMGs alloy, $Gd_{60}Al_{10}Mn_{30}$ nanocomposite (3.3 J/kg K at 50 kOe) [11]. Moreover, it can also be found that the $-\Delta S_M^{max}$ of the $Gd_{55}Ce_{10}Al_{15}Fe_{20}$ glassy ribbons is the largest among these ribbons studied. As-mentioned above, the relative large T_c variation of $Gd_{65-y}Ce_yAl_{15}Fe_{20}$ glassy ribbons is about 100K when y changes from 0 to 20. Therefore, for the application of MCE refrigeration at the intermediate temperature (80-220 K), the $Gd_{65-y}Ce_yAl_{15}Fe_{20}$ (y =0-20) glassy ribbons are preferred.

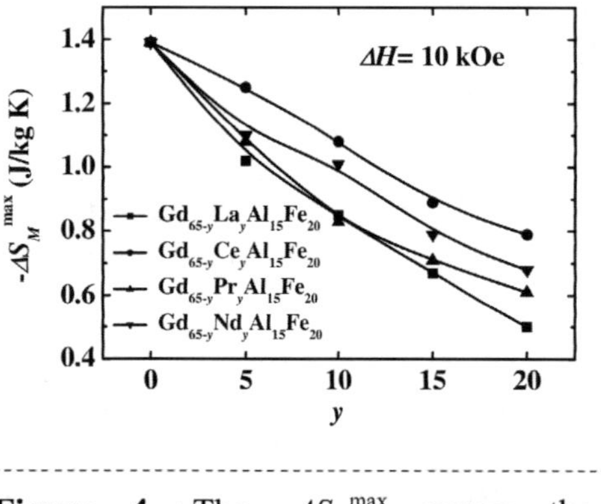

Figure 4. The $-\Delta S_M^{max}$ versus the different contents of the light rare earth elements for the $Gd_{65-y}LRE_yAl_{15}Fe_{20}$ glassy ribbons under 10 kOe.

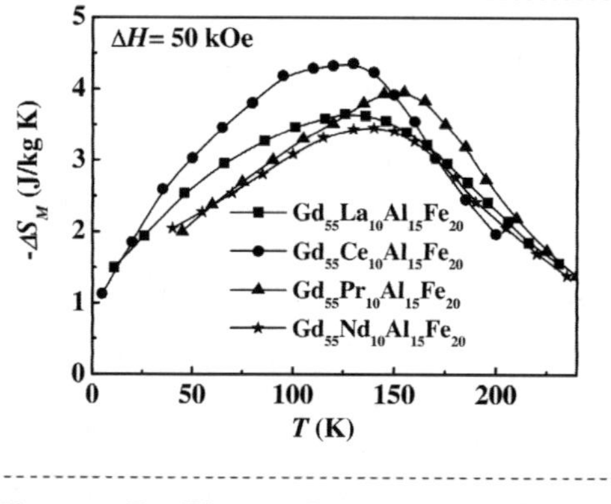

Figure 5. The $-\Delta S_M$ curves of the $Gd_{55}LRE_{10}Al_{15}Fe_{20}$ (*LRE* = La, Ce, Pr, Nd) glassy ribbons under 50 kOe.

Besides the magnetic entropy changes, the refrigerant capacity (RC) values of these ribbons have been investigated. In the present work the RC values are estimated by numerically integrating the area under the $-\Delta S_M$-T curves using the full width at half maximum of the peaks as the integration limit. Typically, the RC values of the $Gd_{55}LRE_{10}Al_{15}Fe_{20}$ (*LRE* = La, Ce, Pr, and Nd) glassy ribbons are determined to be 540-600 J/kg at 50 kOe, which is comparable with or even higher than those of $Gd_{36}Y_{20}Al_{24}Co_{20}$(459 J/kg)[14], $Gd_{33}Er_{22}Al_{25}Co_{20}$ (574 J/kg)[22], $Gd_5Si_2Ge_2$ (305 J/kg) [3], and $Dy_{50}Gd_7Al_{23}Co_{20}$(290 J/kg)[13].

4. Conclusion

With moderate Co substitution by Fe the Curie temperature of the Gd-based $Gd_{65}Al_{15}Co_{20-x}Fe_x$ glassy ribbons is increased by about 200K. Furthermore, for lowering the material cost, the Gd metal in these materials is substituted by low-cost light rare earth elements (such as La, Ce, Pr, and Nd) in part. The maximum magnetic entropy changes, $-\Delta S_M^{max}$, of these specimens are around 4J/kg K under 50 kOe in the temperature range of 80-200K. Relative large refrigeration capacity is reached to 540-600 J/kg for $Gd_{55}LRE_{10}Al_{15}Fe_{20}$ (*LRE* = La, Ce, Pr, and Nd) glassy ribbons. Therefore, for the application of MCE refrigeration at intermediate temperature (80-220 K), the $Gd_{65-y}Ce_yAl_{15}Fe_{20}$ (y =0-20) glassy ribbons are preferred.

Acknowledgements

This work was supported by the National Science Council, Taiwan, under grant No. 98-2112-M194-005-MY3, the National Natural Science Foundation of China, the National High Technology Research and Development Program of China, and the National Program on Key Basic Research Project.

References

[1] Tishin A 1999 *Handbook of Magnetic Materials* Vol 12, ed K. H. J. Buschow (Amsterdam: Elsevier) pp 395-524
[2] Brück E 2005 *J. Phys. D* **38** R381
[3] Pecharsky V K and Gschneidner K A 1997 *Phys. Rev. Lett.* **78** 4494
[4] Tegus O, Brück E, Buschow K H J and de Boer F R 2002 *Nature* (London) **415** 150
[5] Hu F X, Shen B G, Sun J R, Cheng Z H, Rao G H and Zhang X X 2001 *Appl. Phys. Lett.* **78** 3675
[6] Wada H and Tanabe Y 2001 *Appl. Phys. Lett.* **79** 3302
[7] Hu F X, Shen B G, Sun J R 2000 *Appl. Phys. Lett.* **76** 3460
[8] Gschneidner K A, Pecharsky V K Jr andTsokol A O 2005 *Rep. Prog. Phys.* **68** 1479
[9] Phana M H and Yu S C, 2007 *J. Magn. Magn. Mater.* **308** 325
[10] Franco V, Conde C F, Blázquez J S, Millán M and Conde A 2007 *J. Appl. Phys.* **102** 013908
[11] Gorsse S, Chevalier B and Orveilon G 2008 *Appl. Phys. Lett.* **92** 122501
[12] Fang Y K, Yeh C C, Hsieh C C, Chang C W, Chang H W, Chang W C, Li X M and Li W 2009 *J. Appl. Phys.* **105** 07A910
[13] Luo Q, Zhao D Q, Pan M X and Wang W H 2007 *Appl. Phys. Lett.* **90** 211903
[14] Liang L, Hui X, Wu Y and Chen G L 2008 *J. Alloys Compd.* **457** 541
[15] Du J, Zheng Q, Li Y B, Zhang Q, Li D and Zhang Z D 2009 *J. Appl. Phys.* **103** 023918
[16] Jo C L, Xia L, Ding D, Dong Y D and Ekoko G 2008 *J. Alloys Compd.* **458** 18
[17] Dong Q Y, Shen B G, Chen J, Shen J, Wang F, Zhang H W and Sun J R 2009 *J. Appl. Phys.* **105** 053908
[18] Uhlmann D R 1972 *J Non-Crystal Solids* **7** 337
[19] Legvold S 1980 *Ferromagnetic Materials* Vol 1, ed E P Wohlfarth (Amsterdam: Elsevier) pp 184-295
[20] Speight J D 1970 *J. Less-Common Metals* **20** 251
[21] Luo Q, Zhao D Q, Pan M X and Wang W H 2006 *Appl. Phys. Lett.* **89** 081914

Journal of Physics: Conference Series **266** (2011) 012003 doi:10.1088/1742-6596/266/1/012003

Linearity of Giant Hall resistivity versus magnetic field in Fe/Pt/[CoFe/Pt]$_n$/Fe/Pt multilayers

M F Liu[1], X C Song[1], H Lv[1], L R Zhang[1], J M Bai[1,*], F L Wei[1], S Yoshimura[2], H Saito[2]

[1]Key Laboratory for Magnetism and Magnetic Materials of the Ministry of Education, Lanzhou University,

TianShui Road, S 222, LanZhou, GanSu, 730000, P.R. CHINA

[2] Center for Geo-environmental Science, Faculty of Engineering & Resource Science, Akita University,

Tegatagakuen-machi 1-1, Akita, 010-8502, JAPAN

Email: jmbai@lzu.edu.cn

Abstract: The Giant Hall effect (GHE) material is one of potential effective materials for magnetic sensors. Meanwhile, because of the magnetic hysteresis of the magnetic materials, the linearity of Giant Hall resistivity could be an extremely important factor for industrial application. The Fe(*a* Å)/Pt(*b* Å)/[$Co_{0.9}Fe_{0.1}$(3 Å)]/Pt(12 Å)]$_n$/Fe(*a* Å)/Pt(*b* Å) magnetic multilayers were fabricated and their linearity of Giant Hall resistivity versus magnetic field was investigated in out-of-plane direction. By choosing the different structures and adjusting the thickness of the sublayers, the perpendicular interface anisotropy and in-plane shape anisotropy in the multilayers can be fine tuned to control the Giant Hall effect behaviors of the films. As an example of middle-magnetic-field sensor application, the Fe(4 Å)/Pt(6 Å)/[$Co_{0.9}Fe_{0.1}$(3 Å)/Pt(12 Å)]$_3$/Fe(4 Å)/Pt(6 Å) structure was fabricated to show a good linearity of Hall resistivity of 2.2% with a Hall slope of 16 $\mu\Omega\cdot$cm/T and a magnetic field range of [-160, 160] Oe at room temperature in out-of-plane direction.

1. Introduction

Semiconductor hall effect devices are being widely used for magnetic sensor application, which

[*] To whom any correspondence should be addressed.

accounts for more than one billion dollars of market supply in the 2009. But there are still factors which limits its further application, such as the high resistivity, low frequency response, high power consumption, poor linearity, and poor thermal robustness. Interestingly, the semiconductor hall devices are rarely used to measure just a magnetic field for its poor linearity[1]. To overcome these demerits, we can use metallic materials instead of semiconductor. But the Hall effect for common metals and alloys is too weak for sensor using. The Giant Hall Effect (GHE) in magnetic metals and alloys, which arises from spin-orbit scattering, is obviously larger than the ordinary one. Some magnetic materials exhibit very large GHE, which makes it almost suitable for hall detectors. Most of the recently research focuses on improving the hall sensitivity of the magnetic materials. The hall slope of $Co_{0.9}Fe_{0.1}$/Pt multilayers reached up to 545 μΩ·cm/T at room temperature with a low saturation field of about 10 Oe[2]. The hall slope of Cu(Co) nanoparticle film reached 40mΩ cm/T in the field below 40 Oe by aging-induced method[3]. The Hongkong University reported that the saturated hall resistivity of as large as 200 μΩcm at 4 kG in Ni-SiO_2 granular films[4]. These make the Giant Hall effect materials a competitive and potential effective material for the magnetic sensors application.

Because of the magnetic hysteresis of the magnetic materials, the linearity could be an extremely important factor for production. In order to get a good linearity of the Giant Hall effect in magnetic multilayers, it is required that the multilayers exhibit a small negative effective perpendicular magnetic anisotropy (K_{eff})[2,5]. It has been demonstrated that we can change the perpendicular magnetic anisotropy (K_{eff}) by adjusting the thickness of the sublayers in the magnetic multilayers[6]. By choosing different structures and adjusting the thickness of the sublayers can fine tune the perpendicular interface anisotropy and the shape anisotropy in the multilayers[5,6]. In this article, our attention focus on the linearity of Giant Hall resistivity dependence on the applied field of the magnetic multilayers, especially, the Fe/Pt/$[CoFe/Pt]_n$/Fe/Pt multilayers.

2. Experimental

The films with the structure of Fe(*a* Å)/Pt(*b* Å)/[$Co_{0.9}Fe_{0.1}$(3 Å)]/Pt(12 Å)$]_n$/Fe(*a* Å)/Pt(*b* Å) were deposited on 4 Inch silicon wafers at room temperature by DC magnetron sputtering by a Lesker CMS-18 Sputtering system. Before deposition, the background pressure was better than 6×10^{-8} Torr, the sputtering pressure was 5 mTorr. The $Co_{0.9}Fe_{0.1}$ sublayers were deposited by co-sputtering with adjustable sputtering power for Co and Fe targets. To avoid the offset hall voltage at zero applied field, a five-pin mask has been used to format a 5-pad magnetic films.

After the disposition, the hall resistivity behavior was tested by a 5-pin method, and the offset hall voltage was eliminated by adjusting the variable résistance parallel to the hall voltage direction. The applied magnetic field was perpendicular to the film plane. Then 1mA current was input to one direction of the films by using a Keithley 2400 source meter, and hall voltage is acquired by a 16-bits differential analog Data-Acquisition (DAQ) card at the direction perpendicular to the current in plane. To filter the random noise, an average algorithm has been introduced here. The root mean square (RMS) noise of the DAQ card is about 20 μV. As is known that, the GHE of magnetic multilayers is characterized by a parameter of hall resistivity (ρ_{xy}):

$$\rho_{xy} = \frac{V_y}{I_x} \times t \qquad (1)$$

where V_y is the hall voltage, I_x the current, t the thickness of the film.

3. Results and Discussions

The hall resistivity behavior of the structure Fe(a Å)/Pt(b Å)/[$Co_{0.9}Fe_{0.1}$(3 Å)]/Pt(12 Å)]$_3$/Fe(a Å)/Pt(b Å) with a = 4 Å, b = 16 Å, 12 Å, 7 Å, 6 Å, 4 Å and b = 12 Å, a = 3 Å, 4 Å, 5 Å, 6 Å, 8 Å are shown in the figure 1. The Giant Hall resistivity showed approximately linear dependence on the magnetic field applied perpendicular to the film before saturation, but with different hysteresis for the different structures. Then the hall resistivity dependence on the applied field over the best linearity range was fitted by using the least square method to obtain the response of the GHE films on the applied field. Also the linearity Δ is calculated as below:

$$\Delta = \Delta Y_{max} / Y_{FS} \qquad (2)$$

ΔY_{max} is the maximum error between the test data and the best fitting line. Y_{FS} is the full scale of the best fitting line of Y.

To analysis the data in detail, we can get the linearity, hall sensitivity, best linearity range, dynamic range and the saturated hall resistivity. The hall sensitivity K equals to the slope of the best fitting line. The dynamic range is defined as the maximum field before saturation, or the saturation field, the best linearity range is a range over which the hall resistivity vs. applied field has a best linearity for industrial application. The main characteristics of the different samples are listed in tab.1.

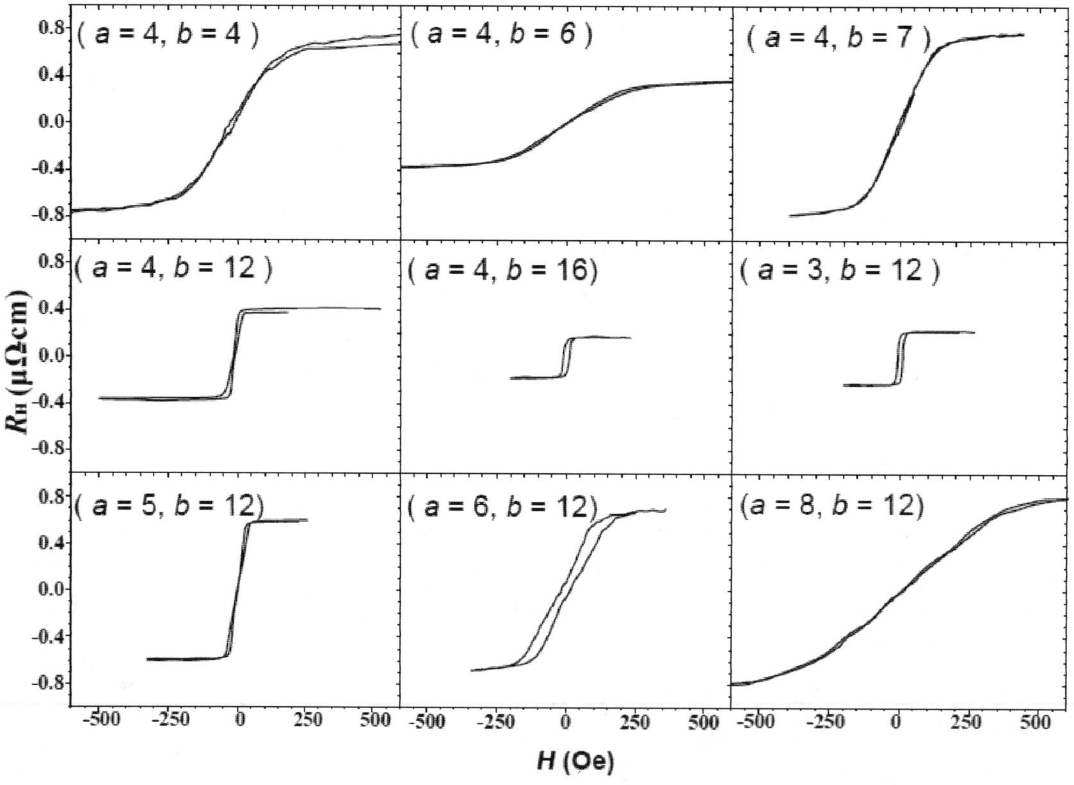

Journal of Physics: Conference Series **266** (2011) 012003 doi:10.1088/1742-6596/266/1/012003

Figure 1.The Giant Hall Resistivity behaviors of the structure of Fe(*a* Å)/Pt(*b* Å)/[$Co_{0.9}Fe_{0.1}$(3 Å)]/Pt(12 Å)]$_3$/Fe(*a* Å)/Pt(*b* Å) in out-of-plane direction.

From figure 1 and table 1, we can find that:

1. Keep the Fe sublayer thickness a = 4 Å, increase the Pt sublayers thickness b, or keep the Pt sublayer thickness b = 12 Å, decrease the Fe sublayers thickness a, the saturation field of hall resistivity decreased. At the same time, the hall slope increase similarly. This comes from the increasing ratio of magnetic metal of Fe.
2. Keep the a as 4 Å, the samples with smaller b have better linearity, such as the b = 4 Å, 6 Å, 7 Å have better linearity than the 12 Å, 16 Å samples. While keep the b as 12 Å , the samples with larger a have better linearity, such as a = 5 Å, 6 Å, 8 Å have better linearity than a = 3 Å, 4 Å samples.
3. The Fe(4 Å)/Pt(6 Å)/[$Co_{0.9}Fe_{0.1}$(3 Å)/Pt(12 Å)]$_3$/Fe(4 Å)/Pt(6 Å) multilayers has a best linearity, which reaches to 2.2% over the applied field range of [-160, 160] Oe without any compensation. It makes the material attractive for industrial application on magnetic sensors.
4. Keep the Pt sublayer thickness b as 12 Å, the saturated hall resistivity increased obviously with the increasing Fe sublayers thickness from 3 Å to 8 Å, which may come from the increasing ratio of magnetic metal of Fe.

Table 1. The main characteristics of the structure Fe(*a* Å)/Pt(*b* Å)/[CoFe(3 Å)/Pt(12 Å)]3/Fe(*a* Å)/Pt(*b* Å) with different sublayer thickness of *a* and *b*.

Structure: *a*, *b* (Å)		Linearity Δ (%)	Sensitivity *K* (μΩcm/T)	Best Linearity Range (Oe)	Dynamic range (Oe)	Saturated Hall Resistivity (μΩcm)
a = 4	b = 16	Nonlinearity	91	Nonlinearity	[-30, 30]	[-0.17, 0.17]
	b = 12	15.5%	178	[-15,15]	[-40, 40]	[-0.38, 0.38]
	b = 7	3.8%	55	[-100,100]	[-200, 200]	[-0.75, 0.75]
	b = 6	2.2%	16	[-160,160]	[-300, 300]	[-0.32, 0.32]
	b = 4	6.0%	35	[-140,140]	[-300, 300]	[-0.75, 0.75]
b = 12	a = 3	Nonlinearity	133	Nonlinearity	[-20, 20]	[-0.22, 0.22]
	a = 4	15.5	178	[-15,15]	[-40, 40]	[-0.38, 0.38]
	a = 5	9.7%	164	[-30,30]	[-50, 50]	[-0.55, 0.55]
	a = 6	10.9%	46	[-120,120]	[-200, 200]	[-0.65, 0.65]
	a = 8	5.1%	19	[-350,350]	[-600, 600]	[-0.91, 0.91]

4. Conclusion

The effective perpendicular magnetic anisotropy of the Fe(*a* Å)/Pt(*b* Å)/[CoFe(3 Å)/Pt(12 Å)]$_n$/Fe(*a* Å)/Pt(*b* Å) multilayers was tuned by adjusting the thickness of the Fe and Pt sublayers to improve the linearity of Hall resistivity versus applied magnetic field along out-of-plane direction. Meanwhile, the hall reisitivity behavior, including the saturation field, the saturation hall resistivity, and the linearity, can also be controlled to fullfill the specified requirements of certain applications. As an example, the Fe(4 Å)/Pt(6 Å)/[$Co_{0.9}Fe_{0.1}$(3 Å)/Pt(12 Å)]$_3$/Fe(4 Å)/Pt(6 Å) multilayers showed best linearity of Hall resistivity vs applied magnetic field in perpendicular direction, which reached to 2.2% over the applied field range of [-160, 160] Oe without any compensation. This structure has high possibility on industrial application as a middle-magnetic-field sensor.

Acknowledgements

This work was partially supported by National Natural Science Foundation of China, NSFC #60803035.

References

[1] R.S. Popovic *et al.*, *Sensors and Actuators*, A **91**(2001),46-50.
[2] Y. Zhu and JW. Cai , *Appl. Phys. Lett*,**90**, 012104 (2007) .
[3] Weiping Cai *et al.*, *J. Phys. Chem. C* 2008, **112**, 1837-1841.
[4] A. B. Pakhomov *et al.*, *Appl. Phys. Lett.* **67** (23), 4 December 1995.
[5] C. L. Canedy, X. W. Li, and G. Xiao, *J. Appl. Phys*, **81**, 5367 (1997).
[6] M T Johnson, *Rep. Prog. Phys.* **59** (1996) 1409–1458.

Journal of Physics: Conference Series **266** (2011) 012004 doi:10.1088/1742-6596/266/1/012004

Heat capacity of the $Ni_{50}Mn_{37}(In_{0.2}Sn_{0.8})_{13}$ alloy

S M Podgornykh[1,3], **E G Gerasimov**[1], **N V Mushnikov**[1] **and T Kanomata**[2]
[1]Institute of Metal Physics, Ural Division of RAS, 620990 Ekaterinburg, Russia,
[2]Department of Applied Physics, Tohoku Gakuin University, Tagajo, Miyagi 985, Japan

E-mail: sp@imp.uran.ru

Abstract. Heat capacity of the Heusler-type $Ni_{50}Mn_{37}(In_{0.2}Sn_{0.8})_{13}$ alloy has been studied in external magnetic fields up to 3 T. The alloy has the austenite Curie temperature $T_{CA}\approx 290$ K, the martensitic phase transformation temperature $T_M\approx 220$-240 K, and the martensite Curie temperature $T_{CM}\approx 185$ K. The martensitic phase transition is accompanied by a considerable change of both magnetization and resistivity of the sample. It was found that upon the first-order martensitic transformation the heat capacity demonstrate hysteresis, both temperature and peak values being different. The maximum value of the heat capacity near T_M on cooling strongly differs from that on heating. In magnetic field the peaks of the heat capacity shift to lower temperatures with the rate $\Delta T_M/\Delta B\approx 2.5$ K/T. The difference in the peak values of ΔC is caused by either overheating or overcooling the sample during the measurement of each point and the release of the latent heat of the martensitic phase transformation. The estimations of magnetocaloric effect (MCE) from the magnetic field effect on the heat capacity and direct measurements clearly demonstrate that the transition is accompained by a negative (MCE) at T_M. The hysteresis observed leads to a decrease of the (MCE) in magnetic refrigeration cycle.

1. Introduction

Ferromagnetic shape memory alloys have attracted much attention during the two past decades, because they are promising materials for application in magnetic refrigerators. In particular, a large inverse magnetocaloric effect (MCE) has been predicted from magnetic measurements of the Heusler-type $Ni_{0.50}Mn_{0.50-x}Sn_x$ alloys with $0.13 \le x \le 0.15$ [1]. Recently, it has been reported on the ferromagnetic shape-memory effect in the NiMnIn and NiMnSn alloy systems, in which the magnetization in martensitic phase is very low in comparison with the parent phase. Magnetic properties of the $Ni_{0.45}Co_{0.5}Mn_{0.50-x}In_x$ alloys [2] are similar to those of the $Ni_{0.50}Mn_{0.37}(Sn_{1-x}In_x)_{0.13}$ compound showing the first - order spontaneous and field-induced martensitic phase transition [3]. Temperature dependence of the magnetization shows three critical temperatures: the Curie temperature of the austenite phase $T_{CA}\approx 290$ K, the Curie temperature of the martensite phase $T_{CM}\approx 185$ K, and the martensitic transformation temperature T_M=220-240 K [3,4]. Temperature dependences of the resistivity and magnetoresistivity have been studied in various magnetic fields [4]. In present paper, temperature dependence of the heat capacity for the $Ni_{0.50}Mn_{0.37}(Sn_{0.8}In_{0.2})_{0.13}$ alloy has been studied in comparison with the behaviour of the magnetization and electrical resistivity. Additionally, we have examined the sign of the MCE by direct method.

[3] To whom any correspondence should be addressed.

2. Hysteresis of the heat capacity of the $Ni_{50}Mn_{37}(In_{0.2}Sn_{0.8})_{13}$

We confirmed by X-ray diffraction that the parent austenite phase has the $L2_1$ Heusler-type ordered cubic structure with the lattice parameter a = 0.5993 nm, while the martensite phase has a complex modulated orthorombic structure. The heat capacity measurements were performed using a PPMS-9 (Quantum Design) between 1.8 and 320 K in applied magnetic fields up to 3T. It has been found that the low-temperature heat capacity can be well described by the relation $C = \gamma T + \beta T^3$ with γ = 4.5 mJ/(g-atom·K) and Θ_D = 310 K. As seen from figure 1, two distinct anomalies which correspond to the critical temperatures T_M and T_{CA} are observed for the $C(T)$ dependence. No any anomalies can be resolved at T_{CM}.

The $C(T)$ dependences were measured on heating and cooling the sample. As seen from figure 2, the thermal hysteresis in C is observed near the T_M. The excess in C near the transition is equal to

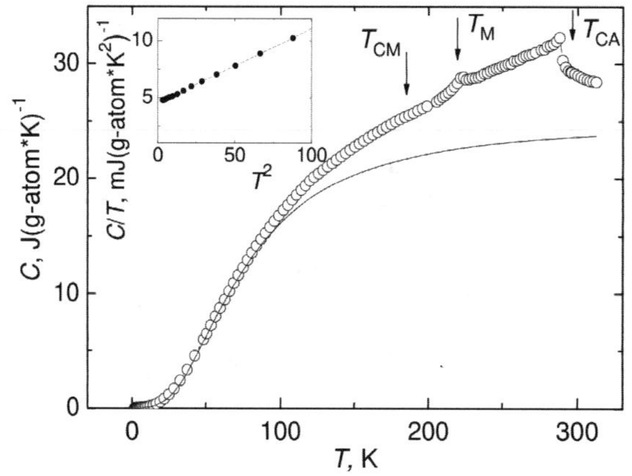

Figure 1. Temperature dependence of the heat capacity C for the $Ni_{50}Mn_{37}(In_{0.2}Sn_{0.8})_{13}$. The solid line shows the Debye function with Θ_D =310 K. The inset shows dependence of C/T vs T^2.

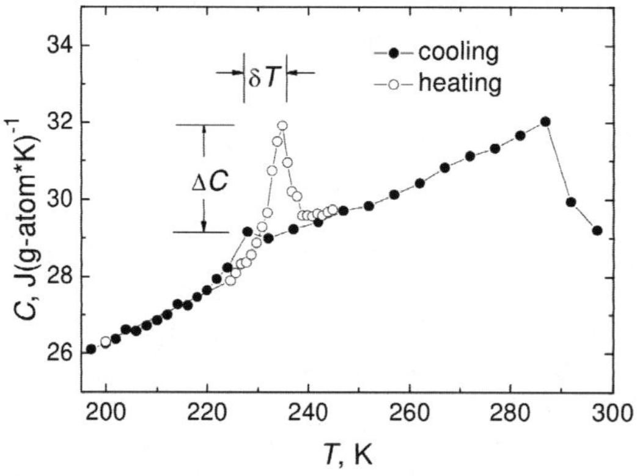

Figure 2. Temperature dependence of the heat capacity C for the $Ni_{50}Mn_{37}(In_{0.2}Sn_{0.8})_{13}$ upon cooling (closed circles) and heating (open circles) in zero external magnetic field.

Journal of Physics: Conference Series **266** (2011) 012004 doi:10.1088/1742-6596/266/1/012004

$\Delta C_{down} \approx 1$ J/(g-atom·K) at T=228 K and $\Delta C_{up} \approx$ 4-10 J/(g-atom·K) at T=235 K upon cooling and heating, respectively. The difference between the values of ΔC_{up} and ΔC_{down} may be caused by overheating or overcooling the sample during the measurement of the each point and release of the latent heat of the martensitic phase transformation. In magnetic field, the position of the peaks in C shifts to lower temperatures with the rate $\Delta T_M/\Delta B$=2.5 K/T at T_M. Figure 3 shows that the transition broadens and the excess of the heat capacity becomes lower when the measurements are performed in high magnetic fields. Earlier the significant thermal hysteresis in C near the T_M was observed for the $Ni_{2+x}Mn_{1-x}Ga$ with x=0.18 [5]. The thermal hysteresis δT (figure 2) is due to the martensitic phase transformation, as it was discussed earlier for $Ni_{2.19}Mn_{0.81}Ga$ [6]. Another notable feature is the reduction in the height of the peak of C measured on the cooling. The value of T_M determined from cooling is found to be always several degrees lower than that from heating.

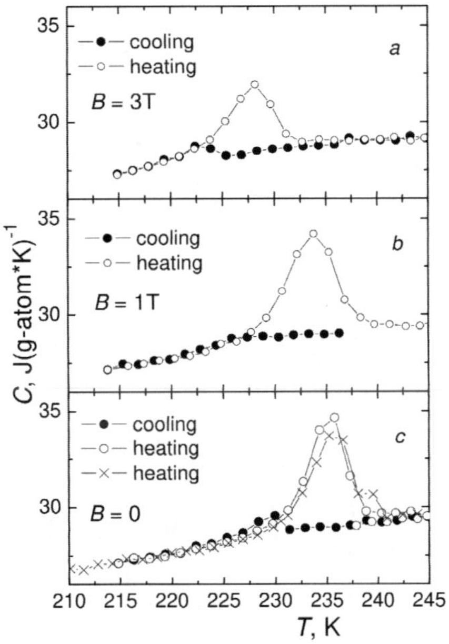

Figure 3. Temperature dependence of the heat capacity C for the $Ni_{50}Mn_{37}(In_{0.2}Sn_{0.8})_{13}$ upon cooling (closed circles) and heating (open circles and ×) in various external magnetic fields: a – 3 T; b – 1 T; c – B = 0.

The peak value of $C(T)$ depends on both the temperature steps and the sweep rate during measurement. This feature may be attributed to a relaxation process associated with the first-order type phase transition, which might give some contribution to the heat capacity.

3. Hysteresis of magnetization, resistivity and magnetocaloric effect

The specific magnetization σ is different for the austenitic and martensitic phases. It decreases as the temperature decreases near the T_M, as shown in figure 4 [3]. The estimations of the magnetocaloric effect using Maxwell relation gives the inverse (negative) MCE near the T_M.

Figure 5 shows temperature dependence of the electrical resistivity ρ for the $Ni_{50}Mn_{37}(In_{0.2}Sn_{0.8})_{13}$. Distinct anomalies in $\rho(T)$ at both T_{CA} (a slope change) and T_M (a jump) can be clearly seen. The temperature T = 220 K was chosen for direct measurement of the MCE since the magnetoresistance at this temperature possesses its maximum value. We studied the adiabatic temperature change using the resistivity thermometer attached to the sample surface. In order to monitor the martensitic transformation in magnetic field, we performed simultaneous measurement of the electrical resistivity.

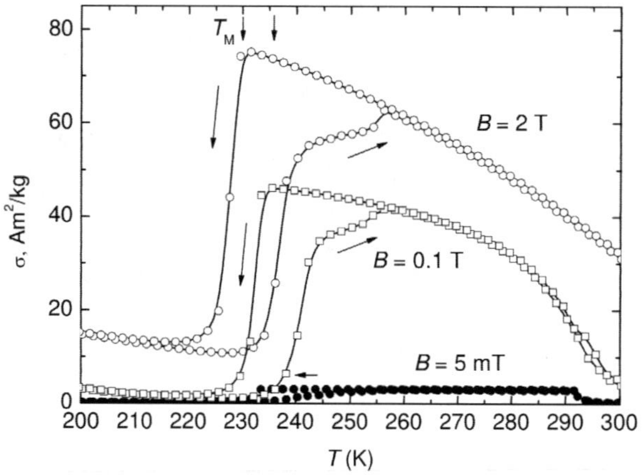

Figure 4. Temperature dependencies of the magnetization σ for the $Ni_{50}Mn_{37}(In_{0.2}Sn_{0.8})_{13}$ in various external magnetic fields upon cooling and heating [3]. The vertical arrows correspond to the temperatures of the heat capacity peaks in Figure 3c.

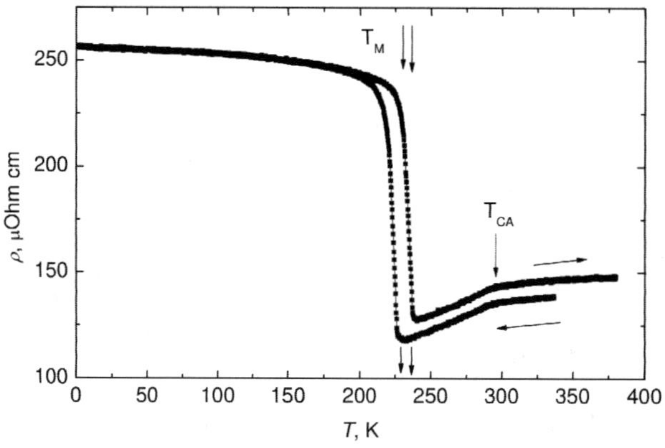

Figure 5. Temperature dependence of the electrical resistivity ρ for the $Ni_{50}Mn_{37}(In_{0.2}Sn_{0.8})_{13}$ upon cooling and heating [4]. The vertical arrows correspond to the temperatures of the heat capacity peaks in Figure 3c.

Figure 6 presents the simultaneously measured time dependence of the resistivity, adiabatic temperature change of the sample and magnetic field. Since the resistance of the austenite and martensite phases substantially differs, the magnetoresistance is a very sensitive tool to control the martensitic transformation. It is clearly seen that application of magnetic field leads to the field-induced martensitic phase transition which is accompanied by the decrease of the sample temperature (ΔT is negative) and the resistance. When magnetic field decreases one can observe the growth of both temperature and resistivity in a certain magnetic field. By cycling magnetic field we found that the

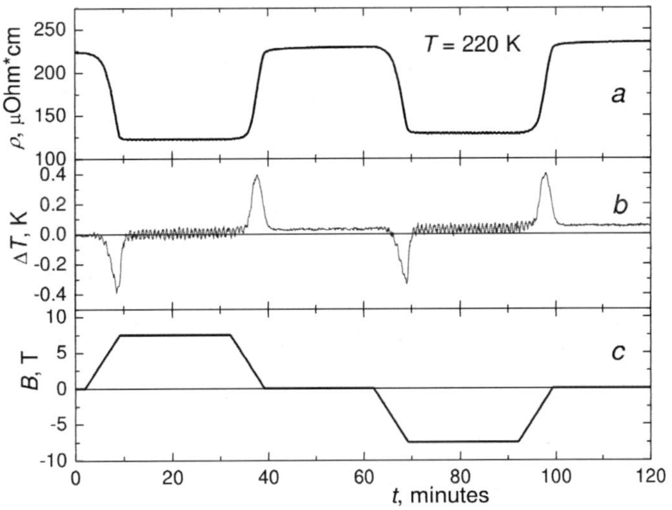

Figure 6. Time dependencies of the electrical resistivity - ρ (*a*), the temperature change of the sample - ΔT (*b*) received from the thermometer of resistance during to increasing and decreasing magnetic field -*B* (*c*) for the $Ni_{50}Mn_{37}(In_{0.2}Sn_{0.8})_{13}$ at the temperature $T = 220$ K.

value of ΔT gradually decreases. Perhaps, cycling is accompanied by the change of the sample microstructure which gives rise to an increase of the field hysteresis around the first-order phase transition.

4. Conclusion

The heat capacity hysteresis may result in an erroneous estimation of the value of magnetocaloric effect determined from the heat capacity data. Hovewer, both the estimations from the magnetic field effect on the heat capacity and direct measurements of the adiabatic temperature change ΔT clearly demonstrate that for $Ni_{50}Mn_{37}(In_{0.2}Sn_{0.8})_{13}$ the transition at T_M is accompained by a inverse MCE. The hysteresis observed leads to a decrease of the MCE in magnetic refrigeration cycle.

Acknowledgments
This work was partially supported by Russian foundation for basic research, Project 10-02-96019.

References

[1] Krenke T, Duman E, Acet M, Wassermann E F, Moya X, Mañosa L and Planes A 2005 *Nature materials* **4** 450
[2] Kainuma R, Imano Y, Ito W, Sutou Y, Morito H, Okamoto S, Kitakami O, Oikawa K, Fujita A, Kanomata T and Ishida K 2006 *Nature* **4493** 957
[3] Gerasimov E G, Mushnikov N V, Kanomata T and Nishihara H 2009 *News in the Magnetism and Magnetic Materials:Proc. MSU (Moscow, MSU, 28 June-2 July 2009)* (Moscow, Moscow State University, Physics Departament) pp 538-540
[4] Gerasimov E G, Podgornykh S M Mushnikov N V, Dykina V P, and Kanomata T 2010 *Euro-Asian Symposium: Trends in Magnetism EastMag2010. Program and Abstracts* (Ekaterinburg, Institute of Metal Physics) p 162
[5] Kuo Y K, Sivakumar K M, Chen H C, Su J H and Lue C S 2005 *Phys. Rev.* B **72** 054116
[6] Khovaylo V V, Skokov K P, Loshkid'ko Yu S, Koledov V V, Shavrov V G, Buchelnikov V D, Taskaev S V, Miki H, Takagi T and Vasiliev A N 2008 *Phys. Rev.* B **78** 060403(R)

Journal of Physics: Conference Series **266** (2011) 012005 doi:10.1088/1742-6596/266/1/012005

Study of inter-dot coupling in nano-patterned permalloy dots array

Chien-Tu Chao, Cheng-Yi Kuo, Yu-Ching Tsai, C K Chang, J F Wang, Lance Horng, and Jong-Ching Wu [1]
Department of Physics, National Changhua University of Education, Taiwan

E-mail: phjcwu@cc.ncue.edu.tw

Abstract. We present a series of studies on interdot coupling in the nanometer-scaled permalloy dots array. A standard electron beam lithography in conjunction with lift-off process was employed for patterning 30 nm thick of millions of permalloy dots array with diameter (D) of 500 nm and the aspect ratio S/D of spacing (S) to diameter ranging from 0.2 to 1.25. The magnetization reversal processes were identified to evolve through the vortex nucleation, movement, and annihilation based on magnetic force microscopy (MFM) imaging in the presence of external magnetic fields. The nucleation field, annihilation field, and moving rate of vortex core were analyzed using M-H loops measured by the alternating gradient magnetometer (AGM). These behaviors are associated with the dipole-dipole interaction in dots array with various interdot spacings.

1. Introduction

Magnetic thin films and multilayer in the small lateral length scales have provided a wealth of scientific interest and potential applications, such as magnetic field sensor and magnetic memory [1], [2]. For the memory application, ultra-high density and thus lager memory capacity is one of the main concerns [3]-[7], of which the challenges in the development of high density magnetic data storage and magnetoresistive random access memory devices are reducing the lateral size of magnetic elements and cross-talk between the adjacent magnetic elements. Until now, advanced nanofabrication techniques have enabled the realization of magnetic dots in few tens of nanometers. As a result, the practical application for memory relies on how well one can control every single dot in the array. It is therefore that the influence on the magnetization reversal of each dot from adjacent dots needs to be understood [8]-[10]. Namely, the interdot interactions may strongly restrict the memory density in nature. Herein, millions of identical permalloy ($Ni_{80}Fe_{20}$) dots with various interdot spacing have been fabricated by electron beam lithography for the study of interdot interactions. The advantage of using e-beam lithography is that one can precisely define the lateral size and shape of designed patterns. Alternating Gradient Magnetometer (AGM) and Magnetic Force Microscopy (MFM) were adopted to investigate the interdot interaction through the magnetic hysteresis loops and magnetic imaging, respectively. In this study, we found that the vortex state reveal a significant transition during vortex core movement in hysteresis loops, which was not observed previously.

[1] To whom any correspondence should be addressed.

2. Experiments

The samples were manufactured by a standard electron beam lithography through a lift-off technique. A commercial scanning electron microscope (SEM) modified for direct writing system with auto-focusing technique and step-motorized movement was employed, thus millions of identical dots were arranged in a square lattice with a total area of 3 mm×3 mm. The huge amount of permalloy dots can give rise to a total magnetic moment strength up to 10^{-5} emu, which is large enough to be resolved by AGM. AGM was performed to measure the net magnetization of samples in the presence of external magnetic field in room temperature. We were able to make the permalloy dot arrays with the size uniformity within few percent of variation, as shown in Fig. 1. Millions of dot with individual dot diameter (D) of 500 nm and the aspect ratio S/D of spacing (S) to diameter 0.2, 0.35, 0.5, 1.0, and 1.25 were delineated. The permalloy film was thermally evaporated in the absence of any external magnetic fields and the film thickness was controlled to be 30 nm, monitored by a quartz crystal. Eventually, AGM and MFM were used to investigate the magnetic configurations of these peramlloy dot arrays due to their excellent field sensitivity and high spatial-resolution, respectively.

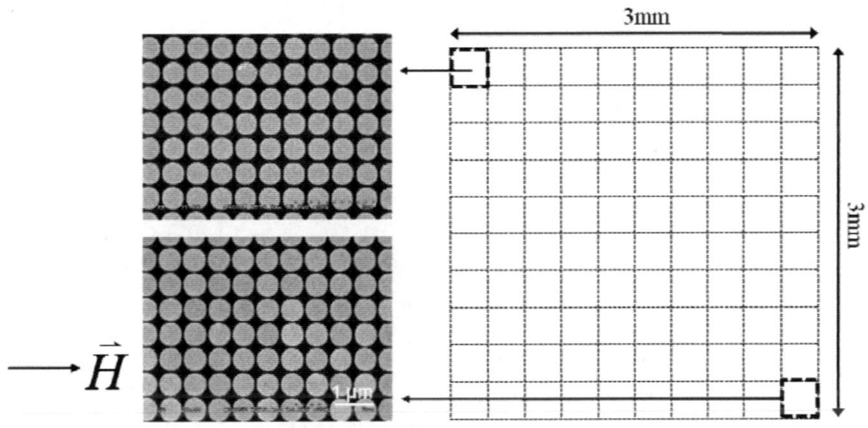

Figure 1. SEM image of dot array with D=500 nm and S=100 nm. Notice that the direct writing system equipped with an auto-focusing technique and step-motorized movement enables millions of identical dots arranged in a square lattice and a total area of 3mm×3mm, in which the size uniformity is within few percent of variation.

3. Results and discussion

Figure 2 shows a hysteresis loop of dots array with D=500 nm and S=250 nm, which was measured by AGM. In addition, the magnetization behaviours in the central area of whole dots array were also imaged by MFM technique and some of the images taken at certain fields are shown in the inset of Fig. 2. Notice that the external magnetic field was applied in the film plane and parallel to the edge of the square lattice for AGM measurement as well as MFM imaging. The magnetization reversal processes are mainly through the vortex nucleation, movement, and annihilation, respectively. A comparison is made between the normalized magnetization strength and the contrast of MFM images, as shown in the inset of Fig. 2. In saturated states, the magnetization reveals a maximum value and MFM image represents a dark to bright contrast from one end to another. Namely, as the magnetization saturates under external magnetic field, the orientation of magnetic moment has uniform direction and displays a single domain state. The magnetization maintains in maximum from saturation until the external magnetic field reduces to a so called nucleation field.

Within vortex state region, as shown in Fig. 3, we find out that obvious transitions were related to vortex core position. Under the external magnetic field sweeping, the vortex core nucleates in the border of dot and moves toward center with a direction that is perpendicular to field. The net magnetization of dots array in the presence of certain magnetic field can be used to estimate the vortex core position directly, i.e., the net magnetization is zero while the core moves to the center of dot. The slope of hysteresis loop in specific point indicates the moving rate of vortex core during external field sweeping. The corresponding MFM images and sketches in point (a)-(e) in hysteresis loop are also shown in Fig. 3 (a)-(e), respectively. The MFM images show clear contrast to indicate the position of vortex core in various magnetic fields, the corresponding configuration as shown in the sketches of Fig. 3. In Fig.3 (a), vortex core have nucleated near the top edge of dot, and we also detected other area to make sure that the vortex nucleations are completed. Whereas the vortex core positions of Fig.3 (a) and (b) are almost at the same place, the slop of M-H curve from “a” to “b” is slack. Noted that the vortex core position of Fig.3 (b) and (d) are almost at a quarter and three quarter of diameter, respectively. The quantitative of M/Ms at “b” and “d” is approximately ±0.5 which are correspond with the position of vortex core. Comparing with core position in MFM images and net magnetization in the hysteresis loop, the vortex core moves fast when centrifugal distance is less than half of radius. Within the central area, less pinning potential makes the vortex core moving easily with slight variation of magnetic field.

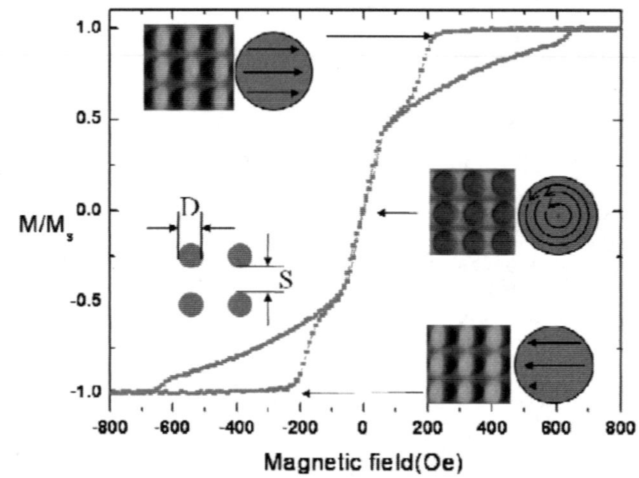

Figure 2. Hysteresis loop of dot array with D=500 nm and S=250 nm, the inset shows the corresponding MFM images and sketches in certain external magnetic fields. The dimension arrangement is also indicated in the inset.

Regarding to the interaction strength between adjacent dots in the dots array with various interdot spacings, the M-H loops measured by AGM are used for analysis and comparison. The hysteresis loops with various interdot spacing are shown in Fig. 4. Significant differences in magnetization reversal process are present. The two specific points, nucleation and annihilation fields, are defined by taking the derivative of the hysteresis loop, in which the significant peaks indicate that transitions were occurred at these fields. In addition, those specific points were taken by drawing two straight from the linear parts before and after the turning points, and the crossed points are the nucleation/annihilation fields. We found out that these two approaches reveal the same results. In fact, same way was used in the Ref. 5. Nucleation and annihilation fields decrease with decreasing the dot spacing; the plateau in vortex state region is not manifest in dense dots array. A stronger stray field interaction, which is induced by magnetic moments of one dot, is expected in dense dots array. The stray field prompts the magnetic moment of adjacent dots to align in the same direction. Accordingly, the nucleation field decreases in dense dots array due to stronger stray field interaction. After vortex core nucleates, a significant plateau appears and then terminates in the same external magnetic field with various dots spacing. As mentioned above, the plateau length is determined only by nucleation field and becomes apparent in larger spacing array. With interdot spacing increasing, the magnetization reversal process of dots array tends to behave without interaction, the same as single dot. Similarly, the nucleation field also saturates until the spacing reaches a critical value, as shown in inset of Fig. 4. These results show the same tendency with previous study [11]-[12], we can ensure that the sparse dots array reveal the behavior the same as the one of single dot.

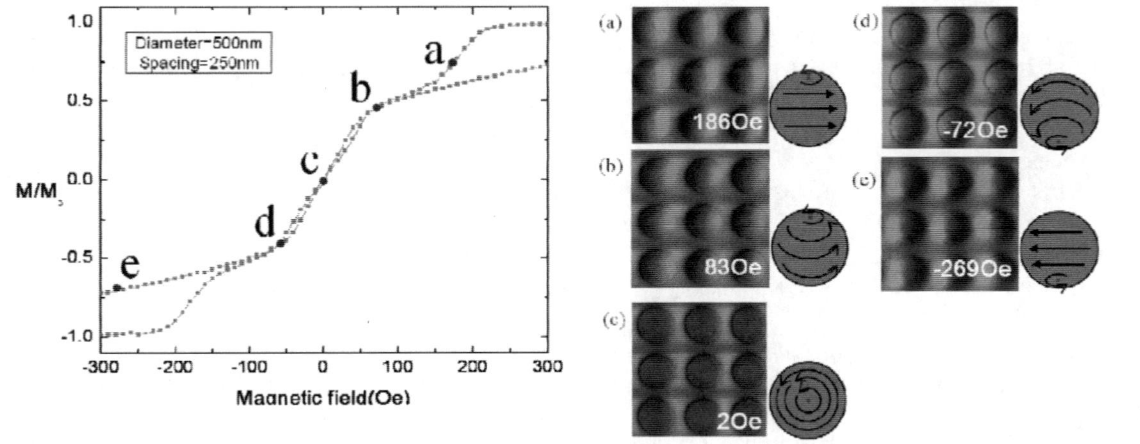

Figure 3. The hysteresis loop focused within the vortex state region. The corresponding MFM images and schematic drawings in certain points of field are shown to the right.

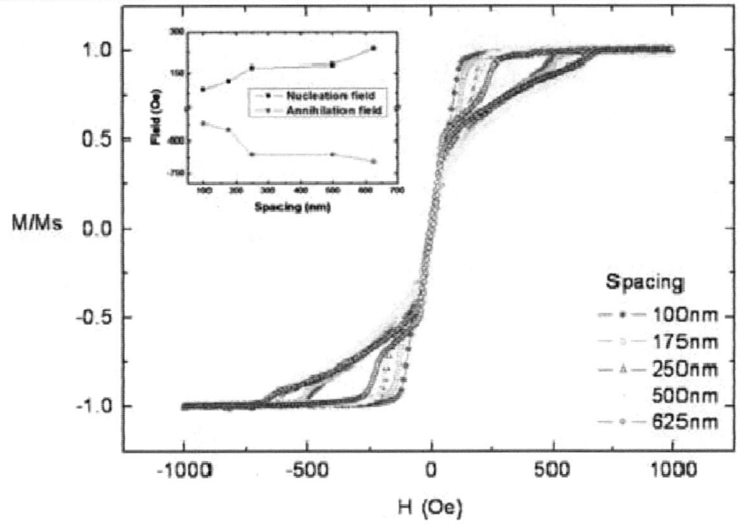

Figure 4. M-H loops of dots array with D=500 nm and spacing of 100, 175, 250, 500, and 625 nm, respectively. The inset shows the curves of nucleation and annihilation fields versus interdot spacing during magnetic field sweeping down.

Journal of Physics: Conference Series **266** (2011) 012005 doi:10.1088/1742-6596/266/1/012005

4. Conclusions

Huge amounts of identical permalloy dots array with size uniformity within few percent of variation were fabricated to identify the interaction effect. In the presence of external magnetic field, AGM and MFM techniques were employed to explore the magnetic configurations and interaction strength between adjacent dots. A stronger interdot coupling due to stray field were resolved evidently through magnetization reversal process. As dot spacing increases, the nucleation field rises and the plateau becomes distinct. The behavior is believed to be associated with interdot coupling in dot array with various interdot spacings.

Acknowledgments

This work was supported in part by the Ministry of Economic Affairs of Taiwan under Grant 97-EC-17-A-01-S1-026 and in part by the National Science Council of Taiwan under Grant NSC 98-2112-M-018-004-MY3.

References

[1] Ross C A, Smith H I, Sacas T, Schattenburg M, Farhoud M, Hwang M, Walsh M, Abraham M C, and Ram R J 1999 *J. Vac. Sci. Technol.* B **17** 3168-76
[2] Pokhil Taras, Song Dian, and Nowak Janusz 2000 *J. Appl. Phys.* **87** 6319-21
[3] Cowburn R P, Koltsov D K, Adeyeye A O, Welland M E, and Tricker D M 1999 *Phys. Rev. Lett.* **83** 1042-5
[4] Lebib A, Li S P, Natali M, and Chen Y 2001 *J. Appl. Phys.* **89** 3892-6
[5] Novosad V, Guslienko K Yu, Shima H, Otani Y, Kim S G, Fukamichi K, Nikuchi N, Kitakami O, and Shimada Y 2002 *Phys. Rev. B* **65** 060402(R)
[6] Natali M, Prejbeanu I L, Lebib A, Buda L D, Ounadjela K, and Chen Y 2002 *Phys. Rev. Lett.* **88** 157203-6
[7] Zhu Xiaobin, Grütter P, Metlushko V, and Ilic B 2002 *Appl. Phys. Lett.* **80** 4789-91
[8] Guslienko K Yu 1999 *Appl. Phys. Lett.* **75** 394-6
[9] Schneider M,and Hoffmann Horst 1999 *J. Appl. Phys.* **86** 4539-43
[10] Guslienko K Yu, Novosad V, Otani Y, and Fukamichi K 2001 *Appl. Phys. Lett.* **78** 3848-50
[11] K. Y. Guslienko, V. Novosad, Y. Otani, H. Shima, and K. Fukamichi 2001 *Physical Review B.* **65**, 024414
[12] V. Novosad, K. Yu. Guslienko, H. Shima, Y. Otani, S. G. Kim, K. Fukamichi, N. Kikuchi, O. Kitakami, and Y. Shimada 2002 *Physical Review B.* **65**, 060402

Journal of Physics: Conference Series **266** (2011) 012006 doi:10.1088/1742-6596/266/1/012006

Electrical and Magnetic Properties of the Ni based Ternary Compounds R_2NiGe_3 (R = Rare Earth Ions)

J. W. Chen[1], S. Y. Guan, C. H. Wang
Department of Physics, National Taiwan University, Taipei, Taiwan, R. O. C.

E-mail: jwchen@phys.ntu.edu.tw

Abstract. The crystal structure, electrical, and magnetic properties of the Ni based ternary compounds R_2NiGe_3 (R = rare earth ions) have been investigated using powder X-ray diffraction, electrical resistivity, and magnetic susceptibility measurements. X-ray diffraction patterns reveal that all the samples studied crystallize in the AlB_2-derived hexagonal structure with space P6/mmm and the obtained values of the lattice constants a and c are a = 0.4188(1) nm and c = 0.4339(3) nm for La_2NiGe_3 and the lattice constants are found to decrease monotonically with increasing atomic number of R. A sudden drop to zero value in the $\rho(T)$ curve reveals that La_2NiGe_3 becomes superconducting with a midpoint transition temperature T_C of ~ 0.45 K. The $\chi(T)$ curves for all samples studied follow the Curie-Weiss behavior for T > 100 K, with the obtained value of effective moment μ_{eff} close to the moment of the corresponding R^{3+} ions. The occurrence of peak in the $\chi(T)$ curve and a drop off in the $\rho(T)$ curve at the corresponding temperatures indicate that antiferromagnetic transition occurs below 10 K in Gd_2NiGe_3 and Dy_2NiGe_3. In addition, Pr_2NiGe_3 and Ho_2NiGe_3 also become magnetically ordered at low temperatures as inferred from the drop off in the $\rho(T)$ curve and the anomaly in the $\chi(T)$ curves in these compounds. The existence of minimun at 18 K followed by a $-\ln T$ dependence below 18 K in the ρ(T) curve and a $T^{1/2}$ dependence in the $\chi(T)$ curve observed for Ce_2NiGe_3 reveal characteristic of a Kondo lattice in this compound.

[1] To whom any correspondence should be addressed.

1. Introduction

The ternary rare-earth (R) and actinide compounds crystallizing in AlB_2-derived hexagonal structure have attracted a lot of interest in the recent years because of their interesting properties [1]. For example, the Y_2PdGe_3 is a superconductor with a superconducting transition temperature (T_c) of 3 K [1,2], the Nd_2PdGe_3 compound exhibits magnetic ordering of a ferromagnetic type below 6 K [2], and Gd_2PdGe_3 orders antiferromagnetically below 10 K [3]. Recent report revealed that the AlB_2-derived hexagonal type ternary compound Ce_2NiGe_3 is a Kondo lattice compound that exhibits spin-glass behaviour due to the frustration and random distribution of Ni and Ge atoms on the crystallographic sites low temperatures [4, 5]. Recently, compounds with AlB_2-derived hexagonal structure and the Ni-based compounds have attracted a lot of research interest due to the observation of superconductivity in these systems. In this report, we present our results of the crystal structure, electrical, and magnetic properties of the R_2NiGe_3 (R = rare earth ions) compounds.

2. Experimental details

The polycrystalline samples of R_2NiGe_3 (R = rare earth ions) were prepared by arc-melting the stoichiometric amount of high purity elements (R: 99.99 %, Ni: 99.99 %, Ge: 99.999 %) together in a water cooled copper hearth in a Zr-gettered argon atmosphere. The alloy buttons were flipped over and re-melted carefully several times. A two-stage procedure was utilized with negligible weight loss (< 0.5 %). To improve the sample homogeneity, the as melted samples were subsequently wrapped in Ta foils, sealed in the evacuated quartz tubes and annealed at 800 °C for one week. The crystallographic data for all samples were obtained with powder X-ray diffractometer utilizing Cu Kα radiation. Ac electrical resistivity of the bar-shaped samples has been measured between 4.2 K and 300 K in a He^4 cryostat using a four probe technique. Dc magnetic susceptibility measurements were performed in a commercial superconducting quantum interference device (SQUID) from 2 K to 300 K in an applied magnetic of 5000 Oe. In addition, a He^3 refrigerator was used for measuring the resistivity data of La_2NiGe_3 for 0.3 K < T < 4 K.

3. *Results and discussion*

Powder X-ray diffraction patterns reveal that all the samples studied are of the of the hexagonal AlB_2-derived type structure. This is shown in Fig. 1(a), where the X-ray diffraction pattern and the Miller indices of the diffraction peaks for La_2NiGe_3 are plotted. Except for a small extra peak at $2\theta = 33.691°$ (indicated by the star symbol) due to the existence of a tiny amount of the 122 phase, the X-ray pattern can be indexed with the AlB_2-derived hexagonal structure and the obtained values of the lattice parameters are $a = 4.188(1)$ Å and $c = 4.339(3)$ Å for La_2NiGe_3. In Fig. 1(b), the lattice parameters a and c, and the unit cell volume V for the R_2NiGe_3 compounds are plotted with respect to the rare earth

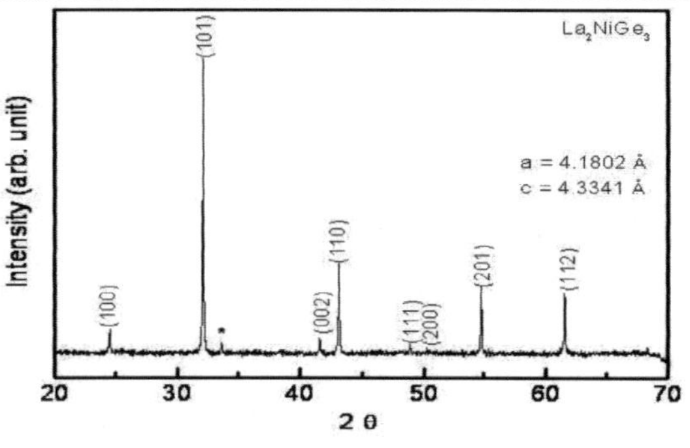

Figure 1(a). X-ray diffraction pattern of La_2NiGe_3 with the Miller indices of the peaks labeled..

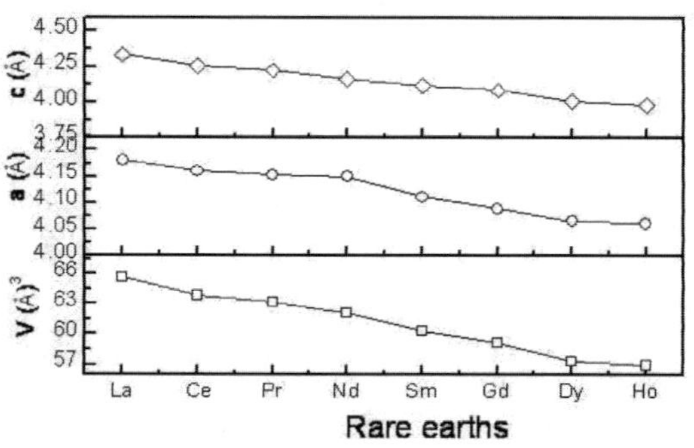

Figure 1(b). Lattice parameters a, c, and the unit cell volume V for the R_2NiGe_3.

R. The values of *a*, *c* and *V* are found to decrease monotonically with increasing atomic number R. The lanthanum contraction (monotonic decrease of *a* and *c*) observed here indicates that the rare earth ions in the R_2NiGe_3 compounds are trivalent at room temperature.

The normalized electrical resistance *R(T)/R(300 K)* versus *T* curves for the R_2NiGe_3 (R = La, Pr, Nd, and Gd) compounds are plotted in Fig. 2(a) for 0 K ≤ *T* ≤ 300 K. The *R(T)/R(300 K)* curves for these compounds exhibit typical characteristics of a common metal and decrease monotonically with decreasing temperature *T*. In addition, a change of the slope in the *R(T)/R(300 K)* curve is observed at ~ 12 K and 5.5 K for Pr_2NiGe_3 and Gd_2NiGe_3, respectively (Fig. 2(b)), indicating the occurrence of magnetic ordering in these compounds. The low temperature $\rho(T)$ curve for La_2NiGe_3 is plotted in the inset of Fig. 2(a). A sudden drop of the $\rho(T)$ to zero value at T = 0.55 K indicates that La_2NiGe_3 becomes superconducting with a mid-point transition temperature of 0.4 K. The drop off in the *R(T)/R(300 K)* curve reveals that Ho_2NiGe_3 become magnetically ordered at ~ 8 K (the $\chi(T)$ curve is not shown here)

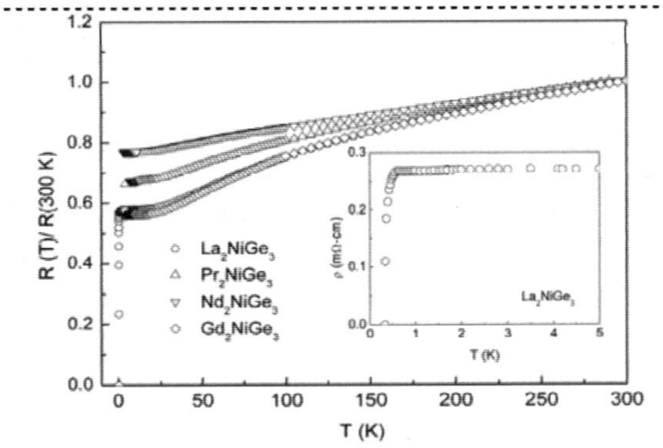

Figure 2(a). *R(T)/R(300 K)* versus *T* curves for R_2NiGe_3 (R = La, Pr, Nd, and Gd). The inset depicts the $\rho(T)$ curve for La_2NiGe_3 for T < 5 K.

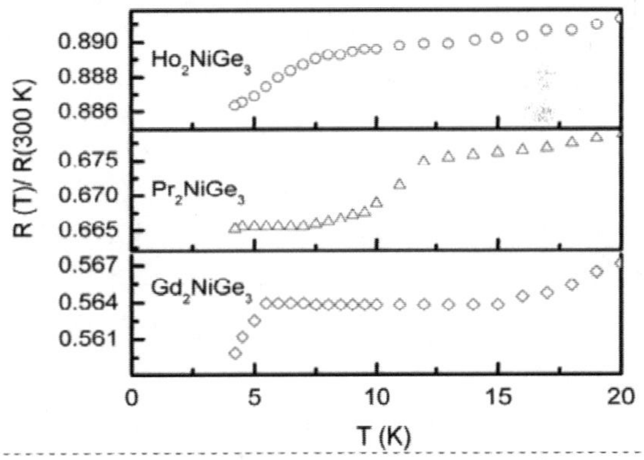

Figure 2(b). Low temperature *R(T)/R(300 K)* versus *T* curves for Ho_2NiGe_3, Pr_2NiGe_3 and Gd_2NiGe_3.

The magnetic susceptibility χ and inverse susceptibility χ^{-1} as a function of temperature *T* curves for R_2NiGe_3 (R = Pr, Nd, Gd and Dy) measured in an applied field of 5 kOe for 0 K ≤ *T* ≤ 300 K are

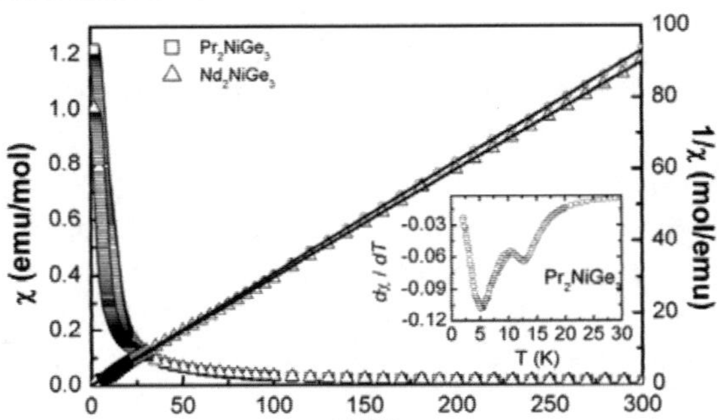

Figure 3(a). χ vs *T* and $1/\chi$ vs *T* of Pr_2NiGe_3 and Nd_2NiGe_3 measured in an applied field of 5 kOe. The inset depicts the $d\chi/dT$ of Pr_2NiGe_3.

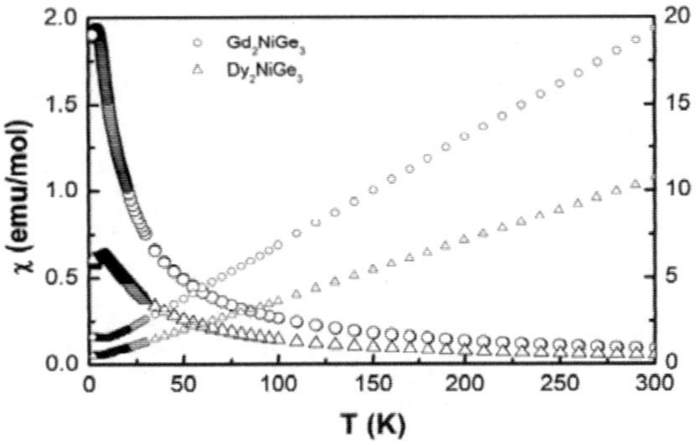

Figure 3(b). χ vs *T* and $1/\chi$ vs *T* of Gd_2NiGe_3 and Dy_2NiGe_3 measured in an applied field of 5 kOe.

plotted in Fig. 3. The susceptibility data reveal that Nd_2NiGe_3 are nonmagnetic down to 2 K. The $d\chi/dT$ curve (inset of Fig.3(a)) for Pr_2NiGe_3 reveals the occurrence of two magnetic transitions at 12 K and 6.9 K, respectively. For $T > 50$ K, χ follows a Curie-Weiss behavior $\chi = C/(T - \Theta_p)$, with Θ_p = -0.61 K and -0.58 K for Pr_2NiGe_3 and Nd_2NiGe_3, respectively. The effective moment obtained from the Curie constant $C = N\mu_{eff}^2/3k_B$ yield the values of $\mu_{eff} = 3.59\ \mu_B$ per Pr ion and $\mu_{eff} = 3.65\ \mu_B$ per Nd ion. The obtained values are close to that of the Pr^{3+} ($\mu_{eff} = 3.58\ \mu_B$) and Nd^{3+} ($\mu_{eff} = 3.62\ \mu_B$) ions. The $\chi(T)$ curve for Gd_2NiGe_3 reveals the occurrence of an antiferromagnetic transition at 8 K, near the temperature at which a change of the slope in the *R(T)/R(300 K)* curve was observed for this compound. The magnetic susceptibility curve for Dy_2NiGe_3 exhibits a peak at T = 4.2 K, which indicates the occurrence of antiferromagnetic ordering in this compound. The $\chi(T)$ curves for both compounds can also be described with the Curie-Weiss law for $T > 100$ K and the obtained values of the effective moment are $\mu_{eff} = 8.00\ \mu_B$ per Gd ion and $\mu_{eff} = 10.68\ \mu_B$ per Dy ion, respectively.

The susceptibility data (Fig. 4) reveal that Ce_2NiGe_3 are nonmagnetic down to 2 K and the susceptibility saturates gradually at low temperatures. The $\chi(T)$ curve for Ce_2NiGe_3 can be fitted with the Curie-Weiss law for $T > 100$ K and the obtained values of the effective moment are $\mu_{eff} = 2.57\ \mu_B$ per Ce ion, a value that is close to the theoretical value of Ce^{3+} ion (2.54 μ_B). The $\rho(T)$ curve for Ce_2NiGe_3 reveals characteristics of a Kondo system. As shown in the inset of fig. 4, where the low temperature $\rho(T)$ curve for Ce_2NiGe_3 is plotted. The $\rho(T)$ curve exhibit a minimum at 18 K and followed by a -ln$T$ dependence from 18 K to 6 K and drop off rapidly below 6 K. Such behaviour is typical of a Kondo lattice in the presence of crystal field effects [6, 7].

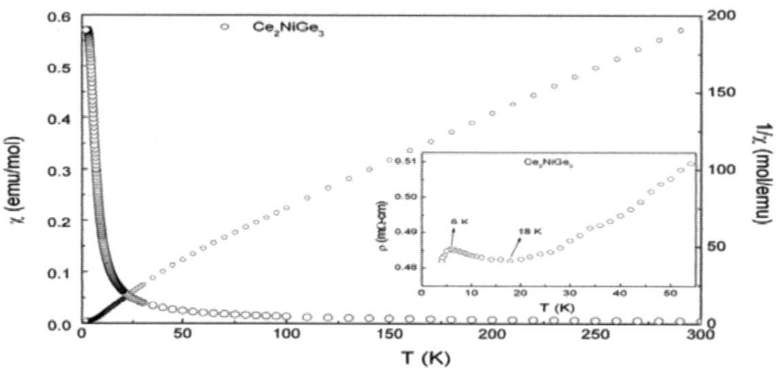

Figure 4. χ vs T and $1/\chi$ vs T of Ce_2NiGe_3 measured in an applied field of 5 kOe. The inset depicts the $\rho(T)$ curve for Ce_2NiGe_3 for T < 50 K.

In summary, we have studied the crystal structure, electrical, and magnetic properties of the R_2NiGe_3 (R = rare earth ions) compounds. We found that all the samples studied are of the AlB_2-derived hexagonal structure and the lattice constants *a* and *c* decrease monotonically with increasing atomic number of R. La_2NiGe_3 becomes superconducting with T_C = 0.4 K, Ce_2NiGe_3 is a Kondo lattice as inferred from the $\rho(T)$ data. Both $\chi(T)$ and $\rho(T)$ data indicate that antiferromagnetic transition occurs in Gd_2NiGe_3 and Dy_2NiGe_3 with transition temperature T_N of 8 K and 4.2 K, respectively. In addition, Ho_2NiGe_3 also become magnetically ordered at ~ 8 K and Pr_2NiGe_3 exhibits two magnetic transitions at ~ 12 K and 5.5 K, respectively.

Acknowledgements
This work was supported by the ROC National Science Council under Grant No. NSC 98-2112-M-002-017-MY2.

References:

[1] Majumdar S and Sampathathkumaran E V 2001 *Phys. Rev.* B **63** 3360
[2] Sampathathkumaran E V, Majumdar S, Schneider W, Molodtsov S L and Laubschat C 2002 *Physica* B **312-313** 61
[3] Majumdar S, Kumar M, Sampathathkumaran E V, 1999 *J. of Alloys and comp.*, **61-64,** 288
[4] Contardi V, Ferro R, Marazza R and Rossi D 1972 *J. Less-Common Met.* **51**, 277
[5] Huo D, Sakurai J, Kuwai T, Isikawa Y and Lu Q 2001 *Phys. Rev.* B **64**, 224405
[6] Hossain Z, Gupta L C and Geibel C 2002 *J. Phys. : Condens. Matter* **14** 9687
[7] Hossain Z, Hamashima S, Umeo K, Takabatake T, Geibel C and Steglich F 2000 *Phys. Rev. B* **62** 8905

Journal of Physics: Conference Series **266** (2011) 012007 doi:10.1088/1742-6596/266/1/012007

Sm_2Co_{17} Alloy Fabricated by Copper Mould Casting

Haibo Feng[a], Wei Pan[a], Anhua Li[a], Ronghai Yu[b], Wei Li[a]

[(a)] *Division of Functional Materials, Central Iron & Steel Research Institute, Beijing, 100081, P.R.China*

[(b)] *Department of Materials Science and Engineering, Tsinghua University, Beijing, 100084, P.R.China*

Abstract

The ingot and fast-cooling Sm_2Co_{17} alloys were fabricated by arc-melting and suction casting methods, respectively. Magnetic properties of the fast-cooling Sm_2Co_{17} alloy are enhanced than the ingot samples. The fast-cooling alloy with a diameter of 6mm has a high coercivity of 16.47kOe at room temperature. Moreover, the fast-cooling samples show a better performance than the ingot and commercial alloy, a high coercovity can be maintained below 500 °C and remain 70% of the value when temperature increased to 600°C.

Keywords: SmCo alloy; fast-cooling; Coercivity; high temperature;

It's well known that Sm_2Co_{17} magnets owe their outstanding magnetic properties exclusively to the complicated three-phase nanostructure which developed in a self-organized process inside the micrometer-scaled grains during the three-step heat-treatment procedure consisting of homogenization, isothermal aging and slow cooling. The Sm_2Co_{17} type alloys are the best candidate materials for high temperature (above 400°C) applications [1,2]. A high maximum magnetic energy product value of 30MGOe can be obtained in $Sm(Co_{bal}Fe_xCu_{0.128}Zr_{0.02})_{7.0}$ magnet with 0.2 at% Fe addition [3]. Magnets with high Fe content have higher saturation magnetization, but larger temperature coefficient of coercivity. The ideal high temperature magnets should have a moderately high coercivity at room temperature and a smaller temperature coefficient of intrinsic coercivity β. G. C.Hadjipanayis et al. reported the $Sm(Co_{bal}Fe_{0.1}Cu_{0.08}Zr_{0.033})_{7.0}$ magnet having a small temperature dependence of Hci, the coercivity is 11.6 kOe at room temperature and 10 kOe at 500°C[4]. Chen et al. found that high temperature magnet has smaller and more uniform cells comparing with the commercial SmCo type magnets, which was related to the better temperature dependence of intrinsic coercivity. This indicated that one can change the intrinsic coercivity and its temperature dependence by changing the processing parameter and microstructure. In this paper, a fast-cooling Sm_2Co_{17} alloy was fabricated by copper mould casting method, and show a better high temperature performance than the ingot and commercial alloy.

The magnets with nominal composition of $Sm(Co_{bal}Fe_{0.08}Cu_{0.10}Zr_{0.025})_{7.45}$ were prepared by arc melting the mixture of pure metals. Rod samples with a diameter of 6 mm and a length of about 10 mm were produced by suction casting of the molten alloy into a water-cooled copper mold in argon atmosphere. The suction casted alloy rod has a bright and clean surface as shown in Fig.1. The arc melting ingots and suction casting rods were soluted at 1180°C for 2 hours, and then isothermally aged at about 830 °C for 2-12 hours, followed by a slow cooling at 0.6°C/min to 400°C. The room temperature magnetic properties were measured using a NIM-2000H closed-circuit hysteresigraph. High temperature magnetic properties were tested using a vibrating sample magnetometer (VSM) with a maximum applied field of 50 kOe.

X-ray diffraction spectra of the aged $Sm(Co_{bal}Fe_{0.08}Cu_{0.10}Zr_{0.025})_{7.45}$ ingot and rod sample with 6mm diameter are shown in Fig.2. Both the two samples consist of 2:17R and 1:5H phases. The intensities of $SmCo_5$ diffraction peaks of the suction casted Φ6mm sample are higher than that of the ingot. This related to the phase orientation in the suction casted samples. Moreover, diffraction peaks of 2:17R and 1:5H phase in the suction casted Φ6mm samples are wider than that of the ingot sample, which indicated that smaller grains were obtained in the suction casted samples. An optimized cellular/lamellar microstructure and a higher coercivity can be obtained in the suction samples with a smaller grain size.

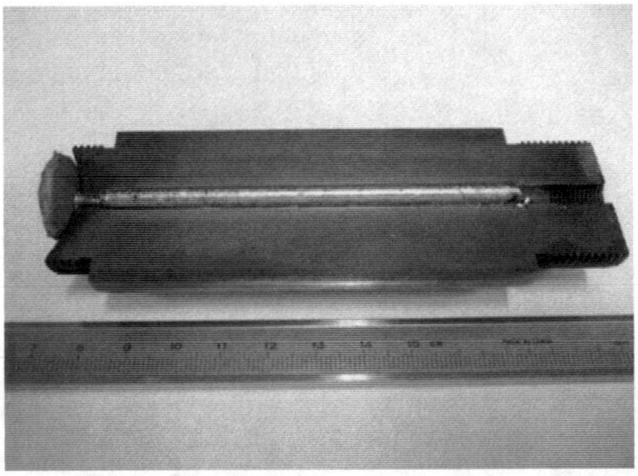

Fig.1 Photograph of the suction casted Sm_2Co_{17} alloy with a 6mm diameter.

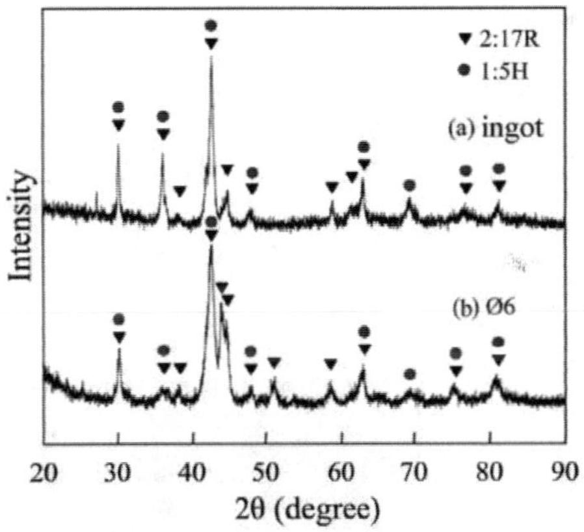

Fig.2 XRD spectra of the magnets: (a) ingot and (b)Φ6mm.

High performance of the Sm(Co,Fe,Cu,Zr)z magnets is attributed to the unique fine cellular/lamellar microstructure formed after the solution and isothermal aging heat-treatment. The suction casted rod magnet exhibits a high intrinsic coercivity (iHc) of 16.47 kOe after aging at 830°C for 12 hours. While, a low intrinsic coercivity of 5~6 kOe was obtained in the aged ingot sample. Typical demagnetization curve of the aged $Sm(Co_{bal}Fe_{0.08}Cu_{0.10}Zr_{0.025})_{7.45}$ rod magnet at room temperature are shown in Fig3. High temperature magnetic properties of the suction casted rod samples were investigated. As show in Fig.4, the Φ6mm rod sample exhibits a much smaller coercivity loss below 500 °C, and a 30% decrease on coercivity at 600°C. This indicates that a high intrinsic coercivity of 10~11kOe can be obtained at 550 °C. The coercivity of ingot sample with the same composition dropped slowly with the temperature increasing. For commercial Sm_2Co_{17} magnets (z=8.0), the coercivity value decreases 40% when temperature increased to 400 °C. This confirmed that the lower the ratio z, the smaller the temperature coefficient of coercivity [5]. Moreover, a better temperature dependence of intrinsic coercivity can be obtained by microstructure optimization in magnets with the same composition. Optical microscopy images of the suction casted and ingot magnets are shown in Fig.5. The grain size of suction casted rod is about 50~80 um, which is much smaller and more uniform than that (100~200um) in the ingot sample. So, the smallest coercivity loss may be obtained by optimizing the grain size and microstructure in Sm_2Co_{17} type magnetic alloys.

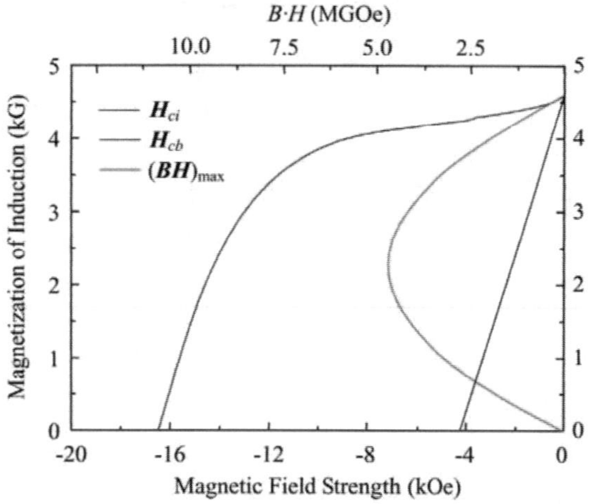

Fig.3 Demagnetization curves of the suction casted Φ6mm magnets.

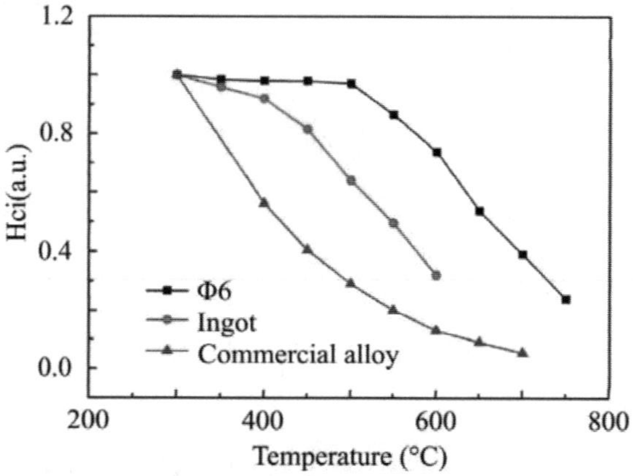

Fig.4 Temperature dependence of coercivity for Φ6mm, ingot and commercial magnets.

Fig.5 OM images of the magnets: (a) suction casted Φ6mm sample and (b) ingot.

[1] X.Chen, J.F.Liu, C.Ni, G.C.Hadjipanayis. *J.Appl.Phys.* **83**, 7139 (1998).
[2] C.H.Chen, M.S.Walmer, M.H.Walmer, et al. *J.Appl.Phys.* **83**, 6706 (1998).
[3] W.Tang, Y.Zhang, G. C.Hadjipanayis. *J. Magn. Magn. Mat.* **221**,268 (2000).
[4] G.C.Hadjipanayis, W.Tang, Y.Zhang, et al. *IEEE Trans. Magn.* **36**, 3382 (2000).
[5] J.F.Liu, Y.Zhang, D.Dimitrov, G. C.Hadjipanayis. *J.Appl.Phys.* **85**, 2800 (1999).

Journal of Physics: Conference Series **266** (2011) 012008 doi:10.1088/1742-6596/266/1/012008

Large Isotropic Volume Change due to Thermal-induced First-order Transition in $La_{1-z}Pr_z(Fe_{0.88}Si_{0.12})_{13}$

S. Fujieda,[1] **A. Fujita**[2] **and K. Fukamichi**[1]
[1] Institute of Multidisciplinary Research for Advanced Materials, Tohoku University, Katahira 2-1-1, Sendai 980-8577, Japan
[2] Department of Materials Science, Graduate School of Engineering, Tohoku University, Aoba-yama 6-6-02, Sendai 980-8579, Japan

E-mail: fujieda@tagen.tohoku.ac.jp

Abstract $NaZn_{13}$-type $La_{0.5}Pr_{0.5}(Fe_{0.88}Si_{0.12})_{13}$ exhibits an isotropic volume change associated with the thermal-induced first-order transition at the Curie temperature T_C = 186 K. The magnitude of the isotropic volume change at T_C of $La_{0.5}Pr_{0.5}(Fe_{0.88}Si_{0.12})_{13}$ is about 1.3 %, larger than that of $La(Fe_{0.88}Si_{0.12})_{13}$. As a result, the enhancement in the local magnetic moment of Fe at T_C is caused by the partial substitution of Pr. The volume change around T_C of $La_{1-z}Pr_z(Fe_{0.88}Si_{0.12})_{13}$ is comparable in magnitude to that of $La_{1-z'}Ce_{z'}(Fe_{0.88}Si_{0.12})_{13}$ having the same T_C. Accordingly, the enhancement in the local magnetic moment of Fe at T_C for $La_{1-z}Pr_z(Fe_{0.88}Si_{0.12})_{13}$ is closely related with the decrease of T_C without the marked decrease of thermal stability of Fe moment.

1. Introduction

$NaZn_{13}$-type $La(Fe_xSi_{1-x})_{13}$ compounds with $0.86 \leq x \leq 0.90$ exhibit the thermal-induced first-order transition between the ferromagnetic (F) to the paramagnetic (P) states at the Curie temperature T_C [1]. The thermal-induced first-order transition at T_C is accompanied by a marked volume change [2, 3]. Since the cubic $NaZn_{13}$-type structure of the $Fm\bar{3}c$ space group is kept after the thermal-induced first-order transition, the volume change at T_C is isotropic [2, 3]. Furthermore, T_C is sensibly decreased with decreasing the unit cell volume by applying the hydrostatic pressure, though the decrease of saturation magnetization M_s is not so significant [2]. Such magnetovolume effects in $La(Fe_xSi_{1-x})_{13}$ are discussed in terms of Landau-type expansion of magnetic free energy by taking the effects of spin fluctuations and the magnetoelastic energy into account [2, 4-6].

Recently, it has been reported that La in $La(Fe_xSi_{1-x})_{13}$ is replaced by other rare earth elements such as Ce [7], Pr [8] and Nd [9, 10]. The $NaZn_{13}$-type single phase is obtained in $La_{1-z}Pr_z(Fe_{0.88}Si_{0.12})_{13}$ with $z \leq 0.5$ [8]. Since the unit cell volume decreases with increasing z because of the lanthanide contraction, T_C is decreased from 195 K to 186 K by a partial substitution of z = 0.5 with keeping the thermal-induced first-order transition. In addition, M_s increases with increasing z, and hence the magnetic moment per Pr atom is estimated to be about 3.3 μ_B, very close to that of a free Pr^{3+} ion, indicating that the magnetic moment of Pr is regarded as the parallel to that of Fe. As a result, the magnetization change at T_C in the thermomagnetization curve becomes larger with increasing z. The amplitude of local magnetic moment is closely related with the volume change [11, 12]. In the present study, therefore, the volume change associated with the thermal-induced first-order transition at T_C of

$La_{0.5}Pr_{0.5}(Fe_{0.88}Si_{0.12})_{13}$ has been evaluated in order to investigate the influence of the partial substitution of Pr on the local magnetic moment change of Fe .

2. Experimental

$NaZn_{13}$-type $La_{0.5}Pr_{0.5}(Fe_{0.88}Si_{0.12})_{13}$ and $La(Fe_{0.88}Si_{0.12})_{13}$ compounds were prepared by arc-melted in an argon gas atmosphere and the heat-treatments for homogenization were carried out in a vacuum quartz tube. The annealing temperature and duration were 1173 K and 10 days for $La_{0.5}Pr_{0.5}(Fe_{0.88}Si_{0.12})_{13}$ and 1323 K for 10 days for $La(Fe_{0.88}Si_{0.12})_{13}$. The powder x-ray diffraction measurements were made by using CuKα radiation. The magnetization was measured with a SQUID magnetometer.

3. Results and discussion

Figure 1 presents the temperature dependence of x-ray diffraction patterns in the 2θ region between 99 and 105° with a step interval of 0.02° for $La_{0.5}Pr_{0.5}(Fe_{0.88}Si_{0.12})_{13}$ with T_C = 186 K. Since the diffraction patterns below T_C is almost the same as that at 220 K, the cubic $NaZn_{13}$-type structure is kept after the thermal-induced first-order transition. A significant shift of the pattern toward a lower angle side is observed with increasing temperature form 180 to 185 K. Accordingly, $La_{0.5}Pr_{0.5}(Fe_{0.88}Si_{0.12})_{13}$ exhibits an isotropic volume change associated with the thermal-induced first-order transition at T_C in analogy with $La(Fe_{0.88}Si_{0.12})_{13}$ [2].

Figure 2 shows the temperature dependence of the relative volume change $\Delta V/V$ evaluated from x-ray diffraction data of $La_{0.5}Pr_{0.5}(Fe_{0.88}Si_{0.12})_{13}$ and $La(Fe_{0.88}Si_{0.12})_{13}$. The value of T_C is indicated by the arrow. A marked change of $\Delta V/V$ in the vicinity of T_C is observed in $La_{0.5}Pr_{0.5}(Fe_{0.88}Si_{0.12})_{13}$. By using the local magnetic moment M and the mean-squared amplitude of thermal spin fluctuations ξ^2, the contribution of the magnetism to the volume as a function of the temperature $\omega(T)$ is given as [12]

$$\omega(T) = \kappa C_{mv}\{M^2(T) + \xi^2(T)\}, \tag{1}$$

where κ and C_{mv} are the compressibility and the magnetovolume coupling constant, respectively. The value of ξ^2 increases with increasing temperature [12]. In contrast, M decreases with increasing temperature and becomes zero in the P state. Therefore, the volume change associated with the thermal-induced first-order transition at T_C, $\Delta\omega(T_C)$, is expressed as [12];

$$\Delta\omega(T_C) = \kappa C_{mv}\{M^2(T_C)_F - \xi^2(T_C)_P\}, \tag{2}$$

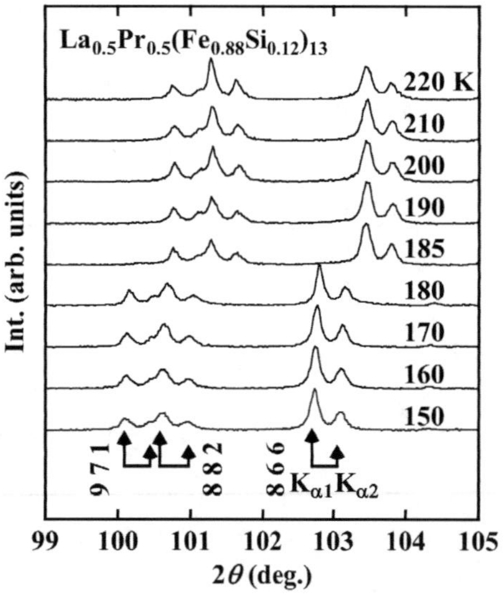

Fig.1 Temperature dependence of X-ray diffraction patterns in the 2θ region between 99 and 105° with a step interval of 0.02° for $La_{0.5}Pr_{0.5}(Fe_{0.88}Si_{0.12})_{13}$ with T_C = 185 K.

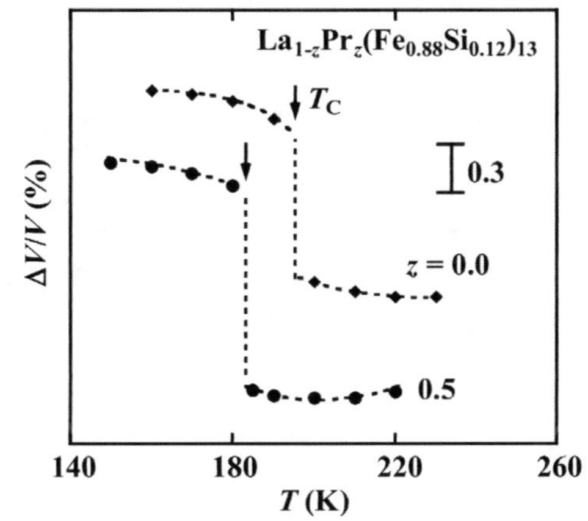

Fig.2 Temperature dependence of the relative volume change $\Delta V/V$ evaluated from X-ray diffraction data of $La_{1-z}Pr_z(Fe_{0.88}Si_{0.12})_{13}$.

where $M(T_C)_F$ is M at T_C in the F state and $\xi^2(T_C)_P$ is ξ^2 at T_C in the P state. Thus the large value of M of Fe results in the large volume change at T_C. Note that the magnitude of the volume change around T_C of $La_{0.5}Pr_{0.5}(Fe_{0.88}Si_{0.12})_{13}$ is about 1.3 %, clearly larger than that of $La(Fe_{0.88}Si_{0.12})_{13}$. Accordingly, the enhancement of isotropic volume change associated with the thermal-induced first-order transition of $La_{1-z}Pr_z(Fe_{0.88}Si_{0.12})_{13}$ is mainly attributed to the increase in M of Fe.

Recently, we have investigated the temperature dependence of $\Delta V/V$ for $La_{1-z'}Ce_{z'}(Fe_{0.88}Si_{0.12})_{13}$ with $z' \leq 0.3$ [13]. The magnitude of the volume change in the vicinity of T_C, $\Delta V/V(T_C)$, as a function of T_C for $La_{1-z'}Ce_{z'}(Fe_{0.88}Si_{0.12})_{13}$ is given in Fig. 3, together with the data of $La_{0.5}Pr_{0.5}(Fe_{0.88}Si_{0.12})_{13}$. It has been revealed that T_C of $La_{1-z'}Ce_{z'}(Fe_{0.88}Si_{0.12})_{13}$ decreases with increasing the Ce concentration z' because of the decrease of the unit cell volume [14]. The volume dependence of T_C of $La_{1-z}Ce_z(Fe_{0.88}Si_{0.12})_{13}$ is similar to that of $La(Fe_{0.86}Si_{0.14})_{13}$ under hydrostatic pressure. However, the value of the spin-wave dispersion coefficient is slightly increased by the partial substitution of Ce, although the value of $La(Fe_{0.86}Si_{0.14})_{13}$ under hydrostatic pressure decreases with decreasing T_C [13]. Therefore, T_C of $La_{1-z'}Ce_{z'}(Fe_{0.88}Si_{0.12})_{13}$ is decreased without the marked decrease of thermal stability of Fe moment. As a result, the magnetization change at T_C is enhanced by the partial substitution of Ce, although the saturation magnetization at 5 K is hardly changed [14]. In other words, the partial substitution of Ce brings about the increase in M of Fe at T_C, resulting in the enhancement of volume change associated with the thermal-induced first-order transition. It should be noted that the relation between $\Delta V/V(T_C)$ and T_C of $La_{1-z}Pr_z(Fe_{0.88}Si_{0.12})_{13}$ is similar to that of $La_{1-z'}Ce_{z'}(Fe_{0.88}Si_{0.12})_{13}$, indicating that M of Fe in the former is almost the same as that in the latter having the same T_C. Therefore, it is concluded that the increase in M of Fe due to the partial substitution of Pr is caused by

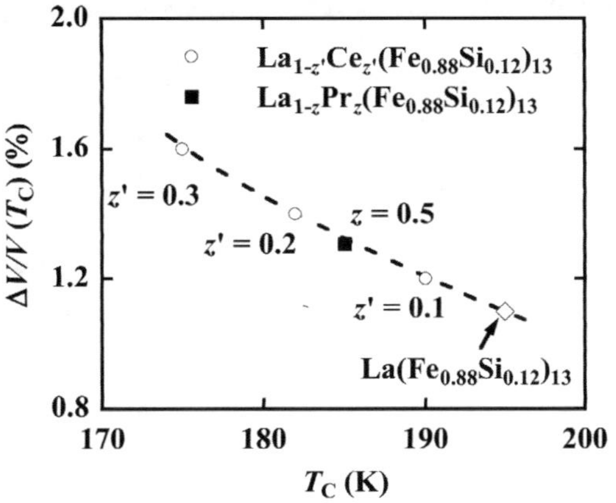

Fig.3 The value of $\Delta V/V(T_C)$ associated with the thermal-induced first-order transition as a function of T_C for $La_{1-z}Pr_z(Fe_{0.88}Si_{0.12})_{13}$ and $La_{1-z'}Ce_{z'}(Fe_{0.88}Si_{0.12})_{13}$.

the decrease of T_C without the marked decrease of thermal stability of Fe moment.

4. Conclusion

The relative volume $\Delta V/V$ change associated with the thermal-induced first-order transition at the Curie temperature T_C of $La_{0.5}Pr_{0.5}(Fe_{0.88}Si_{0.12})_{13}$ has been investigated. $La_{0.5}Pr_{0.5}(Fe_{0.88}Si_{0.12})_{13}$ exhibits an isotropic volume change at T_C = 186 K in a similar way of $La(Fe_{0.88}Si_{0.12})_{13}$. The magnitude of $\Delta V/V$ at T_C of $La_{0.5}Pr_{0.5}(Fe_{0.88}Si_{0.12})_{13}$ is about 1.3 %, which is larger than that for $La(Fe_{0.88}Si_{0.12})_{13}$. The value of $\Delta V/V$ associated with the thermal-induced first-order transition as a function T_C of $La_{1-z}Pr_z(Fe_{0.88}Si_{0.12})_{13}$ is comparable with that of $La_{1-z'}Ce_{z'}(Fe_{0.88}Si_{0.12})_{13}$. Therefore, it is concluded that the increase in the local magnetic moment of Fe at T_C is connected with the decrease of T_C caused by the partial substitution of Pr.

References

[1] Fujita A, Akamatu K and Fukamichi K 1999 *J. Appl. Phys* **85** 4756
[2] Fujita A, Fujieda S, Fukamichi K, Mitamura H and Goto T 2002 *Phys. Rev. B* **65** 014410
[3] Fukamichi K and Fujita A 2000 *J. Mater. Sci. Technol.* **16** 167
[4] Fujita A, Fukamichi K, Wang JT and Kawazoe Y 2003 *Phys. Rev. B* **68** 104431
[5] Fujita A, Fukamichi K, Yamada M and Goto T 2006 *Phys. Rev. B* **73** 104420
[6] Fujieda S, Fujita A, Fukamichi K, Yamaguchi Y and Ohoyama K 2008 *J. Phys. Soc. Jpn.* **77** 074722
[7] Fujieda S, Fujita A and Fukamichi K 2004 *Mater. Trans.* **45** 3228
[8] Fujieda S, Fujita A and Fukamichi K 2007 *J. Appl. Phys.* **102** 023907
[9] Kim Anh DT, Thuy NP, Duc NH, Nhien TT, Nong NV, 2003 *J. Magn. Magn. Mater.* **263** 427
[10] Fujieda S, Fujita A and Fukamichi K 2007 *Mater. Sci. Forum* **561-565** 1093
[11] Morita T and Usami K 1980 *Solid State Commun.* 34 95
[12] Yamada H and Terao K 1994 *J. Phys.: Condens. Matter* **6** 10805
[13] Fujieda S, Fujita A and Fukamichi K to be submitted.
[14] Fujieda S, Fujita A, Fukamichi K, Hirano N and Nagaya S 2006 *J. Alloys Compd.* **408-412** 1165

Journal of Physics: Conference Series **266** (2011) 012009 doi:10.1088/1742-6596/266/1/012009

Dual-cavity magnetophotonic crystals when applying bounding: an analysis by the matrix approach

Taichi Goto, Yoji Haga, Alexander V. Baryshev, Mitsuteru Inoue
Toyohashi University of Technology, Hibari-ga-oka 1-1, Tempaku,
441-8580 Toyohashi, Aichi, Japan

E-mail: inoue@ee.tut.ac.jp

Abstract. Optical responses of one-dimensional magnetophotonic crystals (1D MPCs) with a dual-cavity structure were analyzed by the transfer matrix method. The analysis implied that such dual cavities are fabricated when bounding two identical single microcavities. While keeping sameness of the cavities, we analyzed the influence of a synchronous layers' thickness deviation in both cavities. High transmittance and large Faraday rotation of the dual cavities were shown to be robust to the thickness deviation of the layers within the cavities and that of their interface layer.

1. Introduction

One-dimensional magnetophotonic crystals (MPCs) comprising magneto-optical materials are shown to provide large enhancement of magneto-optical (MO) responses due to strong light coupling to the nonreciprocal magnetic subsystem of MPCs [1]. This is attractive to several applications including thin film isolator [2], magneto-optic spatial light modulator [3], and Q-switch for lasers [4]. Bismuth and dysprosium-substituted yttrium iron aluminum garnet $(BiDyY)_3(FeAl)_5O_{12}$ (garnet hereafter) is a well-known MO material for the visible spectral range. Garnet films have high transmissivity over a wide spectral range; however, their MO responses weaken for near infrared light. Concepts of MPC having the microcavity structure can be fruitful for the enhancement of MO responses of garnet. However for such single-cavity MPCs, the transmission peak located within the photonic band gap splits into two peaks due to the difference of resonant wavelengths corresponding to the right and left circularly polarized light (eigen modes of MO materials). This causes an unwanted effect – the reduction of the transmitted signal, and, when designing responses of single-cavity MPCs, the magnitudes of their transmissivity and MO responses is a fundamental trade-off. In order to solve such an issue, 1D MPCs containing two defects, the so-called dual-cavity, are suggested and expected to exhibit high transmissivity and the Faraday rotation up to 45 degrees in the near-infrared range [5].

The present work deals with an analysis of properties of dual-cavity MPCs by the transfer matrix method [6]. The analysis takes into account that samples are fabricated with the help of adhesive bonding, where two identical single-cavity MPCs are bounded through an interface layer (see figure 1 (b)). Robustness of dual-cavity MPC's spectra to different structural parameters was studied and large optical and magneto-optical responses were demonstrated. Experimental spectra of fabricated samples are also discussed.

2. Result of modeling and discussion

Let us first recall our recent experimental data. We demonstrated MO responses of MPCs with a dual-cavity structure fabricated by ion beam sputtering with in-situ controlling thicknesses of successively sputtered layers [7]; a SEM image of the sample is shown in figure 1(a). Structure of this MPC was $Q/(T/S)^5/G/(S/T)^5/S/(T/S)^5/G/(S/T)^5$, where Q was a quartz substrate, T was a Ta_2O_5, S was a SiO_2, G was a garnet which composition was $Bi_{0.5}Dy_{0.7}Y_{1.8}Fe_{3.3}Al_{1.7}O_{12}$. Ta_2O_5 and SiO_2 layers had optical thickness (*nd*) of $\lambda/4$ and that for garnet was $\lambda/2$ [8]. Experimental spectra of the samples were fitted with calculated ones and showed an intrinsic feature of dual-cavities. However, the magnitudes of transmissivity and the Faraday rotation angle were up to 5 times smaller than that in the calculated spectra of the sample. This was addressed to the layers' thickness deviation from the designed thicknesses.

Figure 2 shows transmission (thick line) and Faraday rotation (triangles) spectra of the successively sputtered MPC (shown in figure 1(a)) and also a transmission spectrum of a bounded dual cavity with a structure of $Q/(T/S)^3/T/(S/T)^3/S/(T/S)^3/T/(S/T)^3/Q$ (thin line). (The sample was assembled from two half-inch fragments of a single cavity sputtered on top of a one-inch substrate; the bonded area was approximately 1 cm^2.) Here, one can see a striking difference in transmission spectra – the bounded sample had a spectrally narrow intensive peak (localized state) inside the photonic band gap. Though the transmission spectrum of the MPC had a similar structure, the intensity of the resonance was low. The limitation of the successive sputtering method is that garnet layers have to be annealed in air for crystallization. Since the double cavity comprises two garnet layers, the sample is annealed twice. The second annealing contributes destructively to structural properties of the sample, i.e. breaks sameness of single cavities [7]. Characteristic feature of dual cavities – double structure of the resonant peak – was observed in the magneto-optical spectra together with the enhancement of the Faraday rotation at λ = 1064 nm (see figure 2 (b)). This shows that dual-cavity modes were excited.

The model dual-cavity MPC had a structure of $SG/(T/S)^8/G/(S/T)^8/S/(T/S)^8/G/(S/T)^8/SG$, where SG is a SGGG substrate (see figure 1 (b)). Resonant wavelength of λ = 1064 nm – the working wavelength of giant pulse lasers [4] – was selected as the microcavity mode. Optical thickness (nd_0) of Ta_2O_5 and SiO_2 layers was $\lambda/4$, and that for the garnet layers was $\lambda/2$. Transmission spectra and that of the Faraday rotation angle were calculated using the transfer matrix method taking into account the dielectric parameters of the real sputtered materials (SiO_2, Ta_2O_5, $Bi_{0.5}Dy_{0.7}Y_{1.8}Fe_{3.3}Al_{1.7}O_{12}$) and the SGGG substrate [10]. For demonstrating advantages of bounding against successive sputtering, two sets of spectra were calculated – spectra of sample 1 comprising nonidentical single cavities, and those of sample 2 comprising identical twins; a deviation of the garnet layer's thickness (α) from d_0 was introduced to both samples.

Figure 3 illustrates spectra of sample 1 where one cavity had a detuned structure: the thickness of garnet layer d varied in a range of $d_0 \pm \alpha$ (α = 0, 0.2, 0.4,..., 1 nm). This consideration describes a realistic MPC which can be fabricated using successive sputtering. One can see that light transmission for the resonant wavelength of 1064 nm decreased as d departed from d_0. Note that the transmission dropped down to 75 % even for a negligibly small α = 0.2 nm, whereas the Faraday rotation angle insignificantly reduced (only short (or long) wavelength spectral shift of the peak happened in accord with variation of d). This indicates that breaking sameness of two microcavities to a very little degree substantially worsens the figure of merit of dual cavities. With further rise of α, transmissivity decreases to < 10 % for $\alpha = \pm 0.2$ nm. Since the thickness deviation tolerance is extremely low, fabrication (by successive sputtering) of the dual cavity with good performance becomes extremely difficult.

Calculated spectra of sample 2 are discussed in figures 4 and 5. Transmission and Faraday rotation angle spectra were calculated where the thickness of garnet layer in both single cavities varied in the range discussed above. Figure 4 shows that, if sameness of single cavities was kept, any synchronous variation of structural properties of the single cavities in the range of α = 0.1 nm had a negligible influence on the magnitudes of the transmissivity and the Faraday rotation of sample 2. The only destructive effect observed in this modelling is that, for fabrication of the dual cavity with desired spectral position of Fabri-Perot resonance, thicknesses of layers has to be controlled with extremely high precision: zero transmissivity is observed for λ = 1064 nm at $\alpha = \pm 0.6$ nm. Figure 5 illustrates changes in the spectra where the thickness of the interface layer was changed in the range of $d_0 \pm 50$ nm. One can see that, even if the optical thickness of the bounding interface SiO_2 layer in the dual cavity considerably deviates from $nd_0 = \lambda/4$, these deviations bring an insignificant alteration of the transmissivity and Faraday rotation angle for light with a desired wavelength.

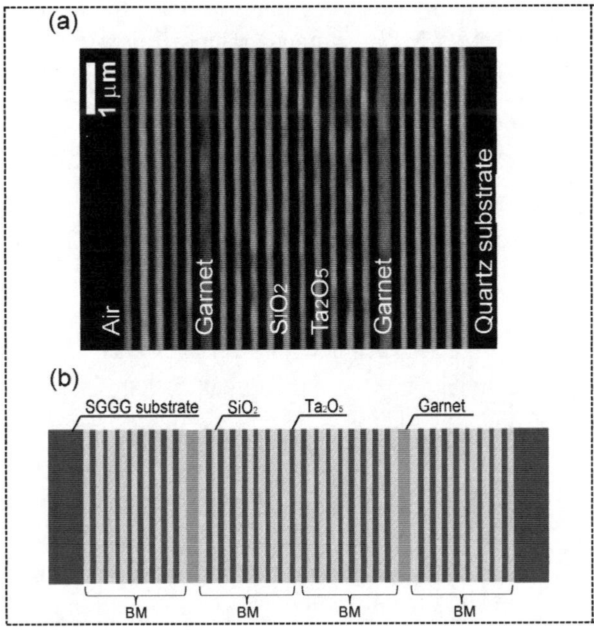

Figure 1. (a) Cross sectional SEM image of an MPC with a structure of $Q/(T/S)^5/G/(S/T)^5/S/\ (T/S)^5/G/\ (S/T)^5$. (b) Sketch of a dual cavity, $SG/S/(T/S)^8/G/(S/T)^8/S/(T/S)^8/G/(S/T)^8/S/SG$.

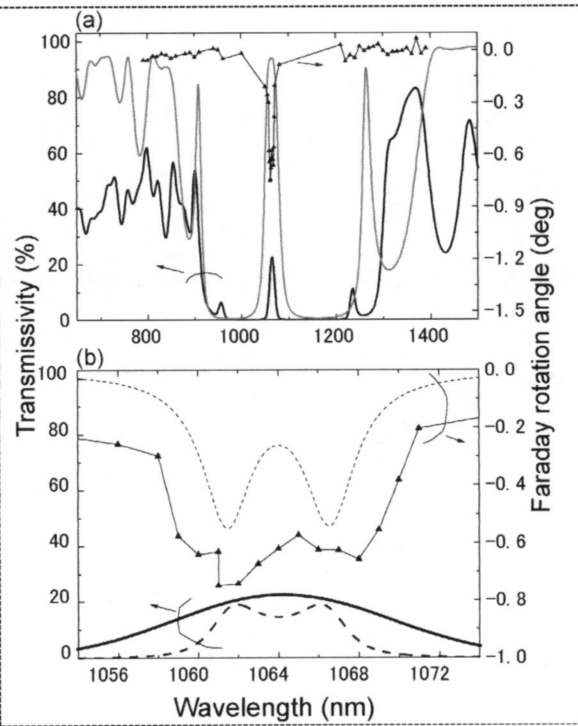

Figure 2. Experimental transmission (bold line) and Faraday rotation angle (triangles) spectra of the MPC shown in figure 1 (a). Calculated transmission (bold dashed line) and Faraday rotation angle (thin dashed line) of the MPC; magnitudes in the spectra were scaled by a factor of 0.2. Gray line in plot (a) shows a dual cavity fabricated by the adhesive bonding method; the sample had a structure of $Q/(T/S)^3/G/(S/T)^3/S/(T/S)^3/G/(S/T)^3/Q$.

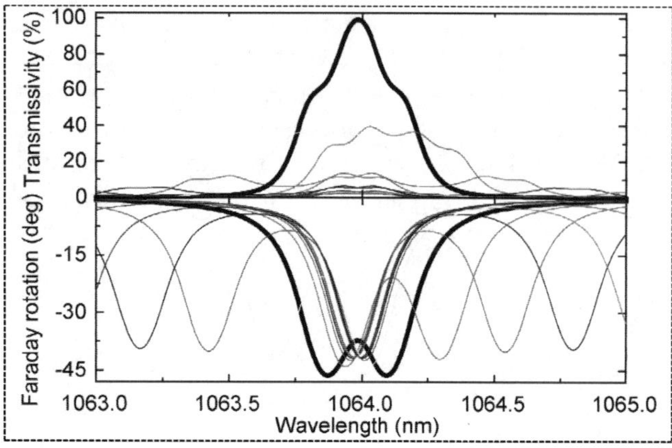

Figure 3. Calculated transmission and Faraday rotation angle spectra of the dual-cavity MPC, sample 1. Here a thickness deviation ($\lambda/2 \pm 0.2$ nm) was made for one of garnet layers. Bold line shows a spectrum of the ideal structure.

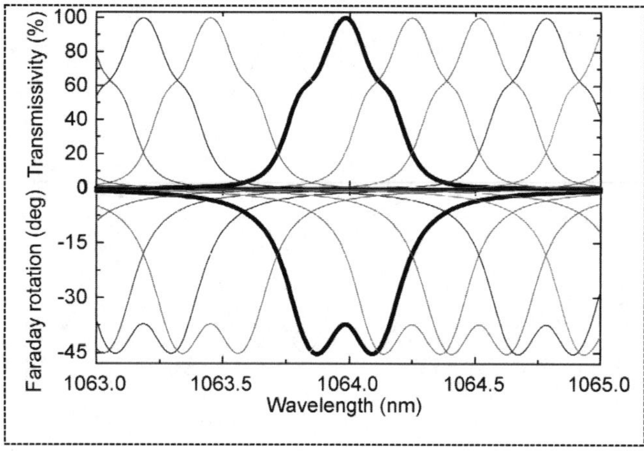

Figure 4. Calculated transmission and Faraday rotation angle spectra of the dual-cavity MPC, sample 2. Here a thickness deviation ($\lambda/2 \pm 0.2$ nm) was made for both garnet layers. Bold line shows a spectrum of the ideal structure.

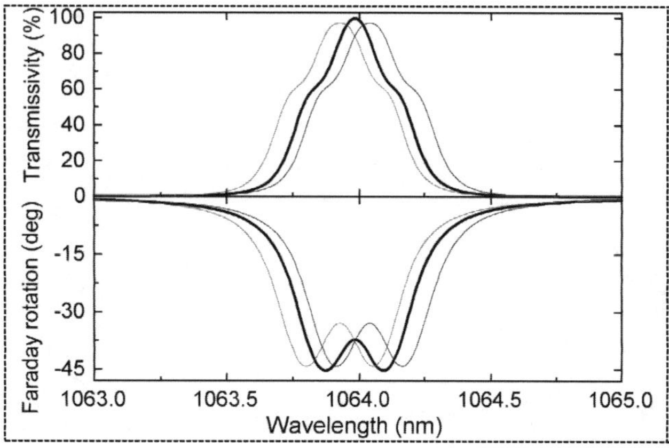

Figure 5. Calculated transmission and Faraday rotation angle spectra of the dual-cavity MPC, sample 2. Here a thickness deviation ($\lambda/4 \pm 50$ nm) was made for the bounding interface SiO_2 layer.

3. Conclusion

We have demonstrated dual cavities fabricated utilizing successive sputtering and adhesive bounding and compared their properties. Optical responses of experimental samples and an analysis of spectra by the transfer matrix approach showed that sameness of two single cavities is essential requirement for realizing dual-cavity MPCs. Calculated spectra of a dual cavity were shown to be robust to a synchronous deviation of layers' thickness in the single cavities, and lacking symmetry of dual cavities to a very little degree results in strong deterioration of their responses. Bounding technique is believed to be the promising approach for fabrication of dual-cavity MPCs. However, our up-to-date results show that bonding single cavities comprising garnet layers is still challenging task because garnet annealing brings a number of restricting factors; main of them are appearance of grains and microcracks on bonding surfaces, and bending of a single cavity as a whole.

Acknowledgments

This work was supported in part by Research Fellowships of Japan Society for the Promotion of Science (JSPS) for Young Scientists Grant No. 21-7394, Regional Innovation Cluster Program (2nd stage) and Global COE Program "Frontiers of Intelligent Sensing" from the Ministry of Education, Culture, Sports, Science and Technology of Japan (MEXT).

References

[1] Inoue M, Fujikawa R, Baryshev A V, Khanikaev A, Lim P B, Uchida H, Aktsipetrov O, Fedyanin A, Murzina T and Granovsky A 2006 *J. Phys. D* **39** R151-R161

[2] Kato H, Matsushita T, Takayama A, Egawa M, Nishimura K and M. I 2003 *Opt. Commun.* **219** 271-276

[3] Mito S, Kim J, Chung K H, Yamada K, Kato T, Takagi H, LIM P B and Inoue M 2010 *J. Appl. Phys.* **107** 09A948

[4] Tsunekane M and Taira T 2006 *Opt. Lett.* **31** 2003-2005

[5] Levy M, Yang H C, Steel M J and Fujita J 2001 *J. Lightwave Technol.* **19** 1964-1969

[6] Kato H, Matsushita T, Takayama A, Egawa M, Nishimura K and Inoue M 2003 *J. Appl. Phys.* **93** 3906-3911

[7] Goto T, Baryshev A V, Tobinaga K and Inoue M 2010 *J. Appl. Phys.* **107** 09A 946-943

[8] For the sample shown in figure 1 (a), the first Bragg mirror was purchased from Ito Kogaku Co. Ltd.: the refractive indices were 2.1814 (Ta_2O_5) and 1.4474 (SiO_2), the thicknesses of Ta_2O_5 and SiO_2 were 121 nm and 183 nm, respectively. Refractive indices in the second Bragg mirror were 2.1108 (Ta_2O_5), 1.5139 (SiO_2) and 2.101 (garnet), and thicknesses were 126 nm (Ta_2O_5), 175 nm (SiO_2) and 253 nm (garnet). The thicknesses of the constituent layers were such that the resultant dual cavity had a resonant mode at λ = 1064 nm. The refractive index of the SGGG substrate was 1.935.

Journal of Physics: Conference Series **266** (2011) 012010 doi:10.1088/1742-6596/266/1/012010

Magnetic properties of nanocrystalline FeCuNbSiB with huge creep induced anisotropy

Giselher Herzer[1], Viktoria Budinsky and Christian Polak
Development Rapid Solidification Technology,
Vacuumschmelze GmbH & Co. KG, D-63450 Hanau, Germany

E-mail: gherzer@googlemail.com

Abstract. Ribbons of originally amorphous $Fe_{73.5}Cu_1Nb_3Si_{15.5}B_7$ have been annealed under tensile stress for about 4 seconds at temperatures between 450°C and 750°C. Under such short time conditions the optimum nanocrystalline state is achieved for annealing temperatures between 600°C and 720°C. This is about 100°C higher than in more conventional heat treatments with annealing times in the order of an hour. The stress annealed ribbons reveal an almost perfectly linear hysteresis loop and a transverse domain pattern with zigzag domain walls characteristic for a magnetic easy plane perpendicular to the stress axis. The induced magnetic anisotropy increases proportional to the annealing stress and is comparable to that obtained after more prolonged annealing. We succeeded to induce anisotropy constants as high as $K_u \approx 12$ kJ/m^3 which even exceed the local magneto-crystalline anisotropy ($K_1 \approx 8$ kJ/m^3) of the crystalline Fe-Si phase. Nonetheless, the coercivity is still as small as $H_c \approx 8$ A/m. This is to be compared to $H_c \approx 0.5$ A/m for field annealed samples where K_u (≈ 20 J/m^3) is smaller by more than two orders of magnitude. The behaviour of H_c can be understood within the framework of the random anisotropy model.

1. Introduction

Nanocrystalline Fe-Cu-Nb-Si-B alloys reveal a homogeneous ultrafine grain structure of bcc-FeSi with grain sizes of typically 10–15 nm and random orientation, embedded in an amorphous minority matrix [1], [2]. This particular microstructure is obtained by devitrification from the amorphous state and enables excellent soft magnetic properties comparable to those of permalloys and Co-based amorphous alloys. Optimised alloy compositions like $Fe_{73.5}Cu_1Nb_3Si_{15.5}B_7$ moreover reveal near-zero saturation magnetostriction in the nanocrystalline state which makes them particularly attractive for application [2].

Similar to other soft magnetic alloys, the hysteresis loop can be tailored by uniaxial magnetic anisotropies induced by annealing in a magnetic field [3], [4] or under tensile stress [5], [6]. For Si-contents larger than about 10 at% stress annealing of nanocrystalline Fe-Cu-Nb-Si-B yields a creep induced anisotropy characterised by an easy magnetic plane perpendicular to the stress axis. This macroscopic anisotropy originates from the *local* magneto-elastic anisotropy of the FeSi crystallites associated with their elongation induced by stress annealing [6]–[8].

The induced magnetic anisotropy energy is proportional to the stress applied during annealing and can amount to be several thousands of J/m^3 which is two orders of magnitude higher than the typical

[1] To whom any correspondence should be addressed.

magnitude of field induced anisotropies. Such strong and well-controlled anisotropies are of interest, for example, for magnetic energy storage cores which require low a permeability of typically a few hundreds being constant over a wide magnetic field range [9], [10].

The practical use of stress annealing for application ultimately requires a continuous reel-to-reel annealing technique with possibly short annealing times [10]. The objective of the present work is to investigate in more detail the magnetic properties of stress annealed $Fe_{73.5}Cu_1Nb_3Si_{15.5}B_7$ after such short time annealing treatments lasting a few seconds only. A particular emphasis hereby was to increase the creep induced anisotropy as far as possible.

2. Experimental

Amorphous ribbons of $Fe_{73.5}Cu_1Nb_3Si_{15.5}B_7$ were prepared by rapidly quenching from the melt in about 20 µm thickness. The nanocrystalline state was achieved by annealing under a tensile stress σ_a along the ribbon axis. For this purpose the ribbon was continuously transported through a furnace with an approximately 10 cm long homogeneous temperature zone kept under nitrogen atmosphere. The ribbon was guided in close contact to an annealing fixture made of stainless steel in order to ensure a good heat transfer. The annealing temperature T_a was measured by a thermocouple attached to the fixture in the centre of the oven. The annealing speed was 1.6 m/min which corresponds to an annealing time of 4 seconds at T_a with heating and cooling rates in the range of 100-200 K/s. The set-up allows inline recording of the ac (60 Hz) hysteresis loop after annealing. The investigated ribbons were 1.6 mm and 6 mm wide. The narrower width was chosen to achieve highest tensile stress.

The saturation magnetostriction λ_s was measured by the small angle magnetisation rotation (SAMR) method [11], [12]. The magnetic anisotropy was determined from the dc hysteresis loop recorded on straight ribbons. The samples were 0.5 m long which ensured negligibly small demagnetisation effects. Coercivity was determined either from the hysteresis loops or, in particular for high anisotropies, with a Förster Koerzimat 1.096 [13]. The domain structure was determined by magneto-optical Kerr microscopy using digital image processing to enhance the contrast [14]. The images shown are difference images between the demagnetised and the magnetically saturated state.

3. Results and Discussion

Figure 1 shows the saturation magnetostriction λ_s, the anisotropy field H_a and the coercivity H_c as a function of the annealing temperature T_a after annealing for 4 seconds under a tensile stress of 45 MPa. The decrease of λ_s and the simultaneous increase of H_a, both starting at about 550°C, indicate the onset of the first stage of crystallisation which is largely completed at annealing temperatures of about

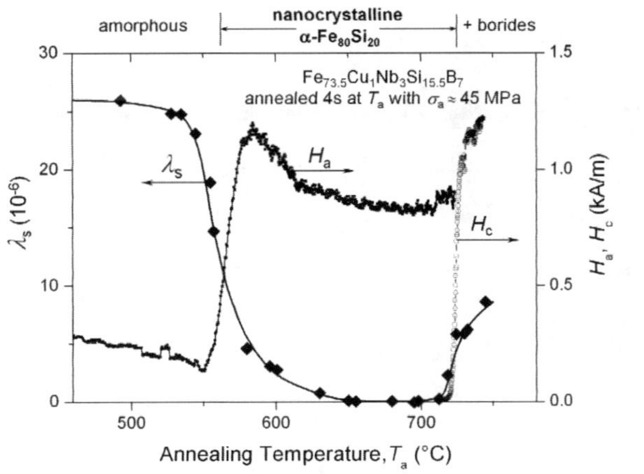

Figure 1. Saturation magnetostriction λ_s, creep induced anisotropy field H_a and coercivity H_c of $Fe_{73.5}Cu_1Nb_3Si_{15.5}B_7$ stress annealed for 4 sec at T_a with a tensile stress of 45 MPa. H_a and H_c were recorded inline during heating up of the furnace.

Journal of Physics: Conference Series **266** (2011) 012010 doi:10.1088/1742-6596/266/1/012010

600°C. As a result we obtain the typical nanocrystalline microstructure consisting of bcc FeSi grains embedded in an amorphous matrix [1], [2]. X-ray defraction revealed an average grain size of about 12 nm. This structure remains stable up to about 720°C where the precipitation of Fe-B compounds leads to a severe degradation of coercivity and to an increase of the saturation magnetostriction. The optimum nanocrystalline state is thus achieved at annealing temperatures which are about 100°C higher than in more conventional heat treatments with annealing times in the order of an hour [1], [2]. Due to the very short annealing time the onset of crystallisation even occurs at higher temperatures than in typical DSC (differential scanning calorimetry) scans where the corresponding onset temperatures are T_{x1} = 508°C and T_{x2} = 694°C for a heating rate of 10 K/min.

Figure 2 shows the typical hysteresis loops obtained in almost the whole range of annealing temperatures T_a. Only above $T_a \approx$ 720°C the coercivity and the remanence start to increase (cf figure 1) indicating that the hysteresis loop is more and more controlled by the high magneto-crystalline anisotropy of Fe-B precipitates (e.g. $K_1 \approx$ -400 kJ/m^3 for Fe_2B [15]). At $T_a \approx$ 745°C, for example, where H_c approaches its maximum value, the hysteresis loop reveals a high remanence to saturation ratio of $J_r/J_s \approx$ 80% ($\sigma_a \approx$ 20 MPa). The remanence slightly decreases to about 75% when the annealing stress σ_a increases to about 200 MPa indicating the formation of a still transversely induced anisotropy.

Figure 3 depicts the induced anisotropy in the nanocrystalline state as a function of the annealing

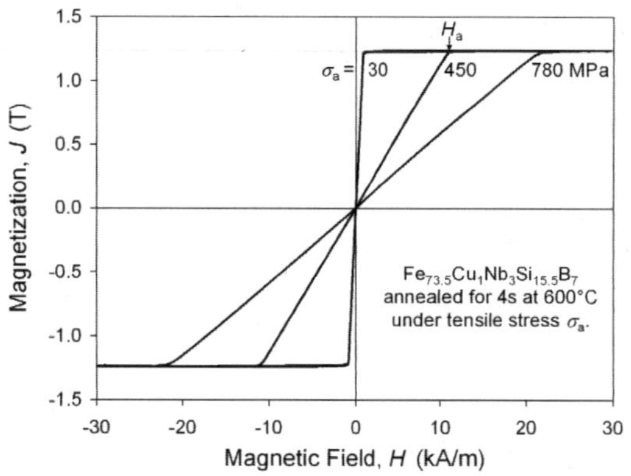

Figure 2. Typical hysteresis loops obtained after stress annealing. The arrow indicates the definition of the anisotropy field H_a.

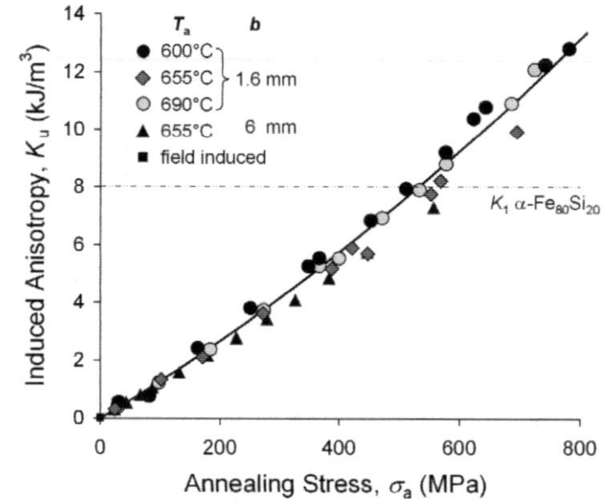

Figure 3. Induced anisotropy K_u of nanocrystalline $Fe_{73.5}Cu_1Nb_3Si_{15.5}B_7$ as a function of the annealing stress.

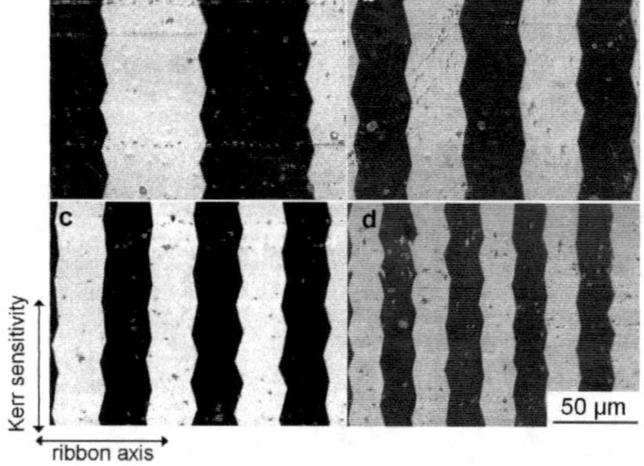

Figure 4. Domain patterns of 1.6 mm wide ribbons annealed for 4 s at 600°C with a tensile stress of (a) 30 MPa, (b) 160 MPa, (c) 430 MPa and (d) 750 MPa.

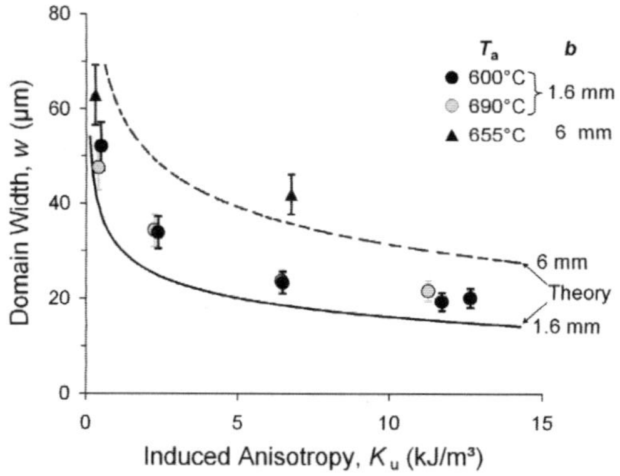

Figure 5. Domain width w as a function of the magnetic anisotropy K_u induced by stress annealing 1.6 mm and 6 mm wide ribbons for 4 s at T_a.

stress. The anisotropy constant K_u was evaluated by $K_u = H_a J_s/2$ where J_s denotes the saturation polarisation. Although the annealing time is only a few seconds, the magnitude of the induced anisotropy is comparable to previous results obtained for more prolonged annealing times in the order of an hour [5]-[8]. Indeed, we achieved induced anisotropies up to $K_u \approx 12$ kJ/m^3 resulting in a relative permeability of $\mu \approx 50$ constant up to a magnetic field of about 20 kA/m (cf figures 2 and 3). Such record high values of K_u even exceed the local magneto-crystalline anisotropy constant of the FeSi crystallites ($K_1 \approx 8$ kJ/m^3 [2]) indicated by the horizontal, dash-dotted line in figure 3. The achievable maximum value of K_u was limited by the circumstance that the ribbons broke when we tried to apply annealing stresses larger than about 800 MPa. The plastic elongation of the ribbons annealed under the maximum load close to 800 MPa was already as large as 50% relative to the original length.

Figures 4 and 5 supplement the characteristic features of the domain structure. The transverse domains separated by zigzag walls indicate a well developed magnetic easy plane perpendicular to the direction of the annealing stress. The domain width w can be well described theoretically by the Landau-Lifshitz domain model [14], i.e.

$$w = \sqrt{8b\sqrt{A/K_u}} \tag{1}$$

where b denotes the ribbon width and A (= 6×10^{-12} J/m) is the exchange stiffness. In comparison field annealed samples reveal straight domain walls and significantly wider domains ($w \approx 100 - 200$ μm) due to their comparatively small anisotropy [16].

Figure 6, finally, shows the coercivity H_c as a function of the induced anisotropy K_u. For comparison the figure also contains the data obtained after magnetic field annealing. The latter yields considerably smaller induced anisotropies of $K_u \approx 5 - 25$ J/m^3 and correspondingly small coercivities of $H_c \approx 0.3 - 0.6$ A/m [2]–[4]. Yet even for the huge creep induced anisotropies exceeding the local magneto-crystalline anisotropy of the crystallites, H_c remains well below 10 A/m.

The coercivity mechanism is schematically sketched in figure 7. Thus, the sample reveals a small remanent magnetisation J_r which basically is determined by anisotropy fluctuations δK [16], [17]. At H_c the average magnetisation is slightly rotated out of the magnetic easy axis thus compensating the remanent magnetisation. Coercivity and remanence, hence, are related by $\mu_0 \chi H_c = J_r$, where μ_0 is the vacuum permeability and χ is the susceptibility associated with K_u. Assuming $J_r/J_s \propto \delta K/K_u$ [17], this results in

$$H_c = p_c \, \delta K / J_s \tag{2}$$

where p_c denotes a dimensionless pre-factor related to the proportionality factor between J_r/J_s and

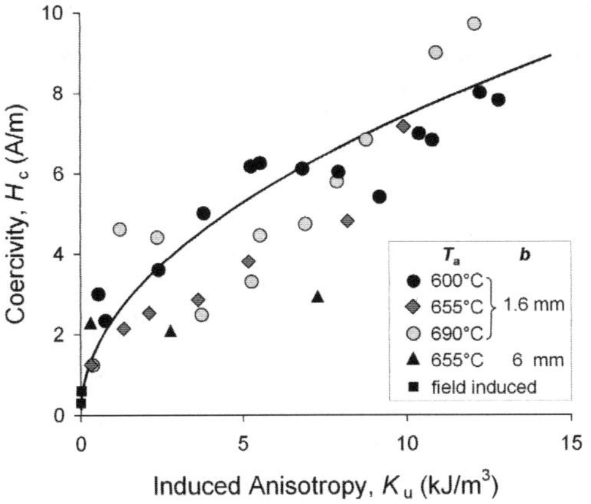

Figure 6. Coercive field H_c as a function of the magnetic anisotropy K_u.

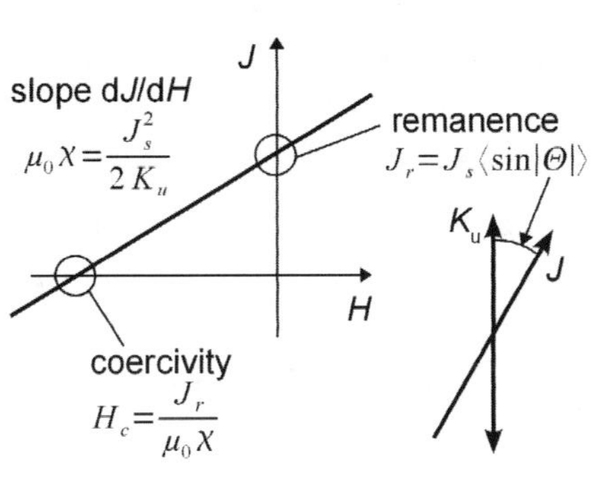

Figure 7. Schematic sketch of the coercivity mechanism.

$\delta K/K_u$. The anisotropy fluctuations δK include a significant contribution arising from the randomly oriented magneto-crystalline anisotropy of the nanocrystals [16], [17]. The magnitude of these fluctuations can been evaluated within the framework of the random anisotropy model for mixed uniform and random anisotropy which yields for $K_u > \delta K$ [17]

$$\delta K \approx \tfrac{1}{2} x \sqrt{\beta K_1 K_u}\,(D/L_0)^3 . \tag{3}$$

Here x is the crystalline volume fraction, β a factor taking into account the cubic symmetry of the crystals, D is the grain size and $L_0 = \varphi_0\,(A/K_1)^{1/2}$ the basic exchange length related to the local magneto-crystalline anisotropy K_1. For a theoretical estimate we assumed the following material parameters: $K_1 = 8$ kJ/m^3, $\beta = 0.4$, $L_0 = 40$ nm, $x = 0.75$ [2], [17] and $D = 12$ nm as determined by X-ray diffraction. The experimental data can then be reasonably matched assuming $p_c = 0.16$. The numerical result is represented by the full line shown in figure 6.

It should be finally noted that the determination of H_c is a non-trivial experimental task, in particular for very high anisotropies. Owing to the extreme loop shape where both J_r/J_s and H_c/H_a are less than 0.1%, the accuracy needed to reliably resolve H_c is at the limit of the experimental resolution. The most reproducible results could still be obtained by using a commercial fluxgate magnetometer (Förster Koerzimat CS 1.096 [13]). In this case we used 60 mm long samples where 8 ribbons were stapled on each other in order to increase the signal to noise ratio. Yet the experimental error for H_c is still relatively large ranging in the order of about ± 10 – 20%.

References

[1] Yoshizawa Y, Oguma S and Yamauchi K 1988 *J. Appl. Phys.*, **64** 6044
[2] Herzer G 1997 *Handbook of magnetic materials* vol 10, ed K. H. Buschow (Elsevier Science B.V) chapter 3 pp 415–462.
[3] Yoshizawa Y and Yamauchi K 1990 *IEEE Trans. Magn.* **5** 1070
[4] Herzer G 1994 *Mater. Sci. Eng.* A **181-182** 876
[5] Kraus L, Záveta K, Heczko O, Duhaj P, Vlasák G and Schneider J 1992 *J. Magn. Magn. Mat.* **112** 275
[6] Herzer G 1994 *IEEE Trans. Magn.* **30** 4800
[7] Ohnuma M, Hono K, Yanai T, Fukunaga F and Yoshizawa Y 2003 *Appl. Phys. Letters* **83** 2859
[8] Ohnuma M, Hono K, Yanai T, Nakano M, Fukunaga F and Yoshizawa Y 2005 *Appl. Phys. Letters* **86** 152513
[9] Fukunaga H, Furukawa N, Tanaka H and M. Nakano M 2000 *J. Appl. Phys.* **87** 7103
[10] Alves F, Simon F, Kane S N, Mazaleyrat F, Waeckerle T, Save T and Gupta A 2005 *J. Magn. Magn. Mat.* **294** e141
[11] Narita N, Yamasaki J and Fukunaga H 1980 *IEEE Trans. Magn.* **16** 435
[12] Herzer G, Flohrer S and Polak Ch 2010 *IEEE Trans. Magn.* **46** 341
[13] http://www.foerstergroup.com/KOERZIMAT-CS-coercivity-field-strength.157.0.html
[14] Hubert A and Schäfer R 1998 *Magnetic Domains* (Springer)
[15] Iga A, Tawara Y and Yanase A 1966 *J. Phys. Soc. Jpn.* **21** 404
[16] Flohrer S, Schäfer R, Polak Ch and Herzer G 2005 *Acta Materialia* **53** 2937; Flohrer S and Herzer G 2010 *J. Magn. Magn. Mat.* **322** 1511
[17] Herzer G 2005 *J. Magn. Magn. Mat.* **294** 99; Herzer G 2005 *Properties and Applications of Nanocrystalline Alloys from Amorphous Precursors*, ed B Idzikowski, P Svec and M Miglierini (Kluwer Academic Publishers) pp 15-34

Journal of Physics: Conference Series **266** (2011) 012011 doi:10.1088/1742-6596/266/1/012011

Transport Properties of Heusler Compound $Ru_{2-x}Fe_xCrSi$ under Pressure

Masakazu Ito[1], Toru Hisamatsu[1], Shingo Nakashima[1], Iduru Shigeta[1], Kazuyosi Matsubayashi[2], Yoshiya Uwatoko[2] and Masahiko Hiroi[1]
[1]Graduate School of Science and Engineering, Kagoshima University, Korimoto 1-21-35, Kagoshima 890-0065, Japan
[2]Institute for Solid State Physics, The University of Tokyo, Kashiwanoha, Kashiwa, Chiba 277-8581, Japan

E-mail: showa@sci.kagoshima-u.ac.jp

Abstract. We investigated transport properties of the full-Heusler compound $Ru_{2-x}Fe_xCrSi$ (x = 0.1), which is an insulator, under pressure P up to 8 GPa. The temperature dependence of electric resistance $R(T)$ for $Ru_{1.9}Fe_{0.1}CrSi$ is insensitive to pressure. $R(T)$ is not described by the simple activation-type but -log$T$ in the wide T and all P range we measured. This might be due to atomic disorder in the crystalline.

1. Introduction

Heusler compounds have been studied for long time because they show various magnetic properties. So-called full-Heusler compounds have the generic chemical formula X_2YZ, where X and Y are transition metals and Z is a nonmagnetic *sp* element. The crystal structure of the full-Heusler compound is the cubic $L2_1$-type structure as shown in Fig. 1. Recently, the full-Heusler compound with half-metallicity is very attractive because it has potential of technological applications, a tunnelling magneto-resistance (TMR) device, for example. The half-metal material is a ferromagnetic and has the 100% spin polarized conduction electrons at the Fermi level (E_F), because only the majority spin band has the finite density of states at E_F. Ishida *et al.* predicted that the full-Heusler compound Co_2MnZ (Z= Si or Ge) has half-metallicity, by a first-principle band structure calculations [1, 2]. In the experimental study, however, Co_2MnZ and other candidate half-metals show insufficient polarization of the conduction electrons [3]. It is considered that this insufficiency come from disorder in the crystalline structure. The full-Heusler compound $Ru_{2-x}Fe_xCrSi$ is one of the candidate materials with half-metallicity. The theoretical study for $Ru_{2-x}Fe_xCrSi$ predicted that half-metallicity is preserved in the presence of the disorder in the crystalline structure [4].

So far, one of the authors (M. H) synthesized the $Ru_{2-x}Fe_xCrSi$ compounds and investigated the transport and magnetic properties for the wide x range [5, 6]. For x > 1.4, the system is the ferromagnetic with the relative high Curie temperature T_C ~ 500 K, and T_C decreases linearly with decreasing x from 1.4 to 0.3. At around x ~ 0.2, the magnetic property changes to an antiferromagnetic. On the other hand, the transport property of this system changes from the metal to insulator around x ~ 0.7 with decreasing x.

In this study, we investigated the pressure effects on the transport property of $Ru_{2-x}Fe_xCrSi$ (x=0.1).

2. Experimental

Journal of Physics: Conference Series **266** (2011) 012011 doi:10.1088/1742-6596/266/1/012011

The polycrystalline samples of $Ru_{2-x}Fe_xCrSi$ were prepared employing an arc melting technique under Ar atmosphere. The crystal structure of $Ru_{2-x}Fe_xCrSi$ was confirmed by an X-ray powder diffraction with Cu Kα radiation. The electric resistance was measured by a conventional four-probe DC technique. The measuring of the electric resistance under hydrostatic pressure from 2.0 GPa to 8.0 GPa employing a cubic anvil device.

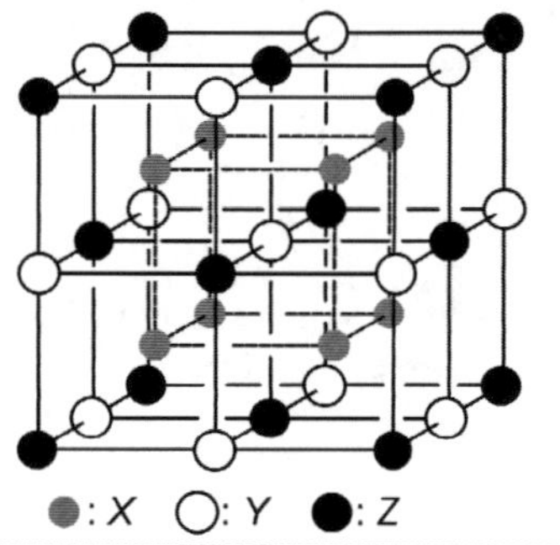

Figure 1. Crystal structure of full-Heusler compound. The space group is *Fm3m*.

3. Results and discussion

Figure 2 shows the X-ray diffraction patterns of $Ru_{1.9}Fe_{0.1}CrSi$ as a representation. This profile shows that the sample is a single phase and has the $L2_1$-type crystal structure. The space group is *Fm3m*. The Fe concentration x dependence of the lattice constant a is shown in Fig. 3. With increasing x, the value of a decreases monotonically. This may suggest that the Fe atoms are homogenously mixed in the sample [7].

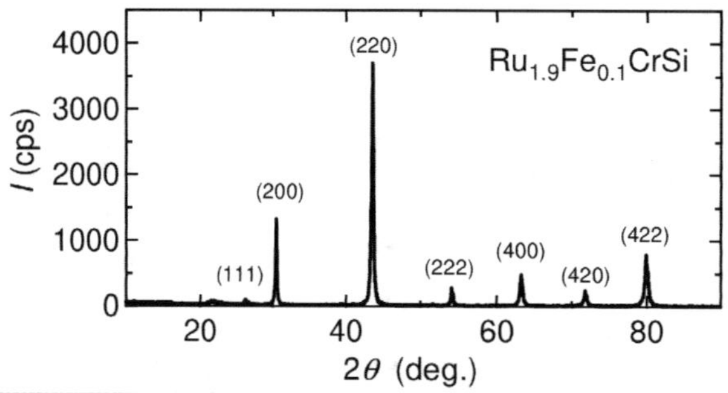

Figure 2. X-ray diffraction patterns of $Ru_{1.9}Fe_{0.1}CrSi$ at room temperature.

The temperature T dependence of electric resistivity, $\rho(T)$, for $Ru_{2-x}Fe_xCrSi$ with various x values is shown in Fig. 4 . The samples with the higher x value have a metallic conductivity defined as $\partial\rho/\partial T > 0$. On the other hand, the transport property in the lower x range changes to an insulator defined as $\partial\rho/\partial T < 0$. Namely, an insulator to metal transition occurs by substituting the Fe atoms for the Ru atoms, as was reported previously [5]. As mentioned above, the lattice constant of the Ru-rich phase of this system is larger than that of the Fe-rich phase. This makes us expect the occurrence of a pressure-induced transition from the insulator to the metal, because pressure makes the lattice shrink. From this motive, we have investigated the transport properties of insulating $Ru_{1.9}Fe_{0.1}CrSi$ under pressure P up to 8 GPa. Figure 5 (a) shows the temperature dependence of electric resistance divided by the value at 300 K, $R(T)/R(300K)$ for $Ru_{1.9}Fe_{0.1}CrSi$ under pressure. Unfortunately, for $Ru_{1.9}Fe_{0.1}CrSi$, we did not observe the pressure induced metallisation. The value of $R(T)/R(300K)$ seems to be insensitive for pressure. From the previous our specific heat measurements for $Ru_{2-x}Fe_xCrSi$, the Debye temperature θ

of this system is estimated to be 490 K [8]. This value is close to that of an iron, thus, it is considered that $Ru_{2-x}Fe_xCrSi$ is not so 'hard'.

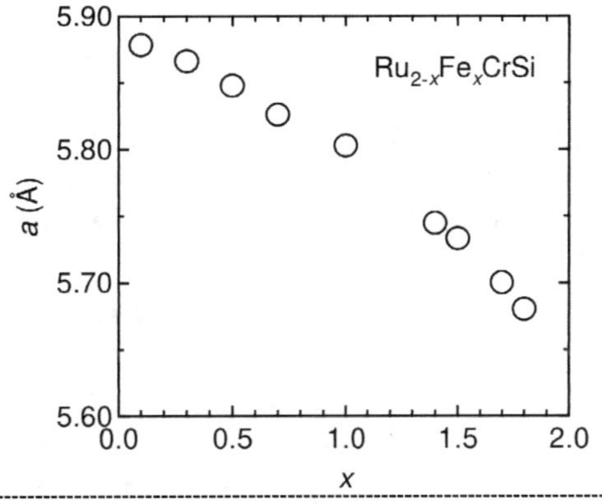

Figure 3. Fe concentration x dependence of the lattice constant a for $Ru_{2-x}Fe_xCrSi$.

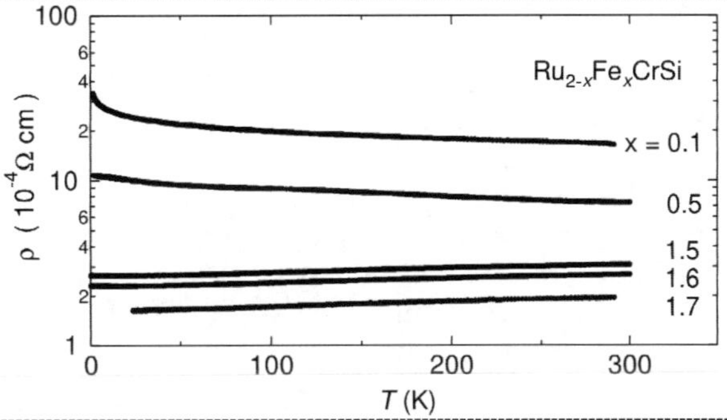

Figure 4. Temperature dependence of electric resistivity $\rho(T)$ for $Ru_{2-x}Fe_xCrSi$ at ambient pressure. The vertical axis is a logarithmic scale.

As shown in Fig. 5 (b), the most characteristic feature is that $R(T)/R(300K)$ does not show a simple activation-type temperature dependence, instead of that, a $-\log(T)$ dependence is observed in the relative wide T and all P range we measured. Some of the transition metal oxides with the site doping such as $Ca_{1-x}La_xRuO_3$ and $La_{1-x}Pr_xSr_2MnO_7$, and of the strongly disordered crystalline such as (La, $Pd_3)_8Mn$ also show the $-\log T$ dependence [9, 10, 11]. For these compounds, it is considered that the electron scattering by a random potential in the crystalline causes the $-\log T$ dependence of $R(T)$. In the theoretical study, Lee and Ramakrishnan have predicted that an electron scattering by random potential gives rise to the logarithmic temperature dependence [12].

Recently, the neutron scattering measurements were carried out on the analogue compound system, $Ru_{2-x}Fe_xCrGe$ [13]. Ru_2CrGe is highly ordered in the $L2_1$-type crystal structure. By substitution of Ru atoms by Fe atoms, the X site is occupied with Ru, Fe, and Cr atoms, and the Y site is occupied by Fe and Cr atoms. Namely, atomic substitution induces atomic disorder. Also in the crystalline of $Ru_{1.9}Fe_{0.1}CrSi$, atomic disorder such as $Ru_{2-x}Fe_xCrGe$ might occur owing to element substitution.

For $Ru_{1.9}Fe_{0.1}CrSi$, the existence of atomic disorder may cause random potential in the system and $-\log T$ dependence of $R(T)$. If the $-\log T$ dependence of $R(T)$ comes from atomic disorder, pressure insensitivity of $R(T)/R(300K)$ might be reflected in that the level of atomic disorder does not change by applying *P*. To confirm this assumption, the detailed crystal structural analysis of $Ru_{1.9}Fe_{0.1}CrSi$ under pressure is needed.

4. Conclusion

To investigate the relation between the lattice constant and transport properties on the full-Heusler insulating compound $Ru_{1.9}Fe_{0.1}CrSi$, we measured electric resistance under pressure. The temperature dependence of electric resistance *R*(*T*) for $Ru_{1.9}Fe_{0.1}CrSi$ is insensitive to *P* up to 8.0 GPa. *R*(*T*) does not show the activation-type *T* dependence which is observed in the conventional band gap insulator. Instead of that, the –log(*T*) dependence are observed in the relative wide *T* and all *P* range. This might be come from atomic disorder in the crystalline structure.

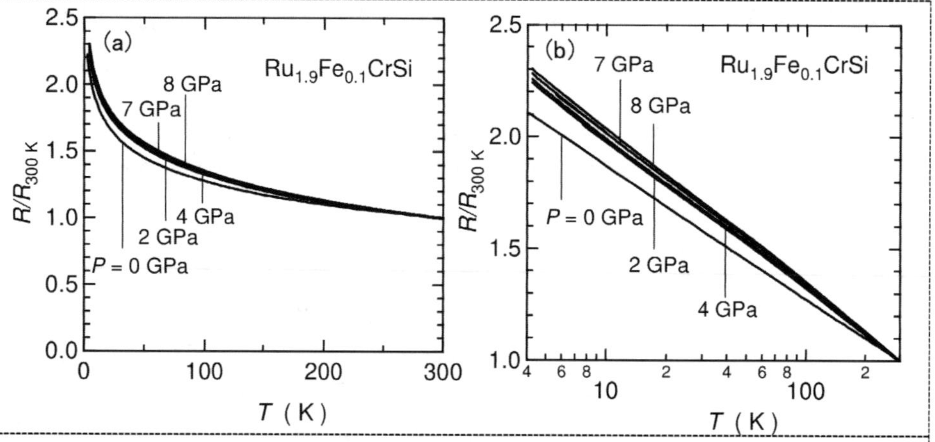

Figure 5. Temperature dependence of electric resistance divided by the value at 300 K, $R(T)/R_{300K}$, for $Ru_{1.9}Fe_{0.1}CrSi$ under pressure. The plots with the (a) linear scales and the (b) semi-logarithmic scale.

[1] Ishida S, Fujii S, Kashiwagi S, and Asano S, 1995 *J. Phys. Soc. Jpn.* **64** 2152
[2] Ishida S, Masaki T, Fujii S, and Asano S, 1998 *Physica B.* **245** 1
[3] Ravel B, Raphael M P, Harris V G, and Huang Q, 2002 *Phys. Rev.* B **65** 184431
[4] Ishida S, Mizutani S, Fujii S, and Asano S, 2006 *Mater. Trans.* **47** 31
[5] Hiroi M, Matsuda K, and Rokkaku T, 2007 *Phys. Rev.* B **76** 132401
[6] Hiroi M, Rokkaku T, Matsuda K, Hisamatsu T, Shigeta I, Ito M, Sakon T, Koyama K, Watanabe K, Nakamura S, Nojima T, Nakano T, Chen L, Fujiwara T, Uwatoko Y, Manaka H, and Terada N, 2009 *Phys. Rev.* B **79** 224423
[7] Kusakari Y, Kanomata T, Fukushima K, and Nishihara H, 2007 *J. Magn.Magn. Mater.* **310** e607
[8] Ito M, Hisamatsu T, Rokkaku T, Shigeta T, Manaka H, Terada N, and Hiroi M, 2010 *Phys. Rev.* B **82** 024406
[9] Sugiyama T and Tsuda N, 1999 *J. Phys. Soc. Jpn.* **68** 3980.
[10] Bhatia S N, Mohapatra N, Kundaliya D, and Malik S K, 2006 *J. Phys.:Condens. Matter.* **18** 2179
[11] Moorjani K and Coey J M D (ed), 1984 *Magnetic Glasses* (Amsterdam: Elsevier)
[12] Lee P A and Ramakrishnan T V, 1985 *Rev. Mod. Phys.* **57** 287
[13] Brown P J, Gandy A P, Kanomata T, Kusakari Y, Sheikh A, Neumann K-U, Ouladdisf, and Ziebeck R A, 2008 *J. Phys.: Condens. Matter* **17** 1583

Structural and Magnetic Properties of Glass Doped with Iron Oxide

J Kaewkhao[1] W Siriprom[2] S Insiripong[3] T Ratana[4] T Ratana[5]
C Kedkaew[4] and P Limsuwan[4]

[1]Center of Excellence in Glass Technology and Materials Science (CEGM), Nakhon Pathom Rajabhat University, Nakorn Pathom, Thailand, 73000
[2]Faculty of Resources and Environment, Kasetsart University, Sriracha Campus, Chonburi Thailand, 20110
[3]Physics Program, Faculty of Science and Technology, Muban Chombueng Rajabhat University Thailand, 70150
[4]Department of Physics, King Mongkut's University of Technology Thonburi, Bangkok Thailand, 10140
[5]Industrial Chemistry Department, Faculty of Applied Science, King Mongkut'sInstitute of Technology North Bangkok, Bangkok, Thailand, 10800

e-mail: mink110@gmail.com

Abstract. Glasses based on borosilicate system were prepared by melt-quenching method in the composition range of (mol%) $20Na_2O : 1.0Al_2O_3 : 13B_2O_3 : 6.3CaO : 0.2Sb_2O_3 : 4.5BaO : (55-X)SiO_2 : xFe_2O_3$ (where x = 0, 5, 10, 15, 20 and 25) and the structural, optical and magnetic properties were investigated. The result showed that the density and molar volume of glasses increased with increasing of Fe_2O_3 concentration. This might be due to an increase of Fe_2O_3 concentration at the expense of SiO_2 caused the opened glass network structure. Fe_2O_3 also affected the change of B_2O_3 structure detected by FTIR technique. The magnetic measurement revealed that the glass sample exhibited a ferrimagnetic behavior when doping with 25 % mol Fe_2O_3.

Introduction

Borosilicate glass is a well known prevalent type of glass that can be used in many applications, for examples, laboratory equipment and glass cookware because of its excellent chemical and physical properties such as a low thermal expansion coefficient and a high chemical corrosion resistance. In the recent years, glasses doped with transition metal oxides have attracted a great deal of attention due to their important physical and chemical properties **[1-3]**. In particular, Fe_2O_3 has been considered as one of the most promising candidate for improving physical properties of glasses such as electrical, optical and magnetic properties **[4-6]**. In this work, glasses based on borosilicate system doped with Fe_2O_3 in a range of 0-25 % mol were prepared through a melt-quenching method. The effect of Fe_2O_3 concentrations on structural, optical and magnetic properties of the samples was examined by Fourier transformed infrared spectroscopy (FTIR) and vibrating sample magnetometer (VSM), respectively.

Materials and Method

The glass samples were prepared by using high purity Na_2CO_3, Al_2O_3, H_3BO_3, CaO, Sb_2O_3, $BaCO_3$, SiO_2 and Fe_2O_3 in a composition range of (mol%) $20Na_2O : 1.0Al_2O_3 : 13B_2O_3 : 6.3CaO : 0.2Sb_2O_3 : 4.5BaO : (55-x)SiO_2 : xFe_2O_3$ (where x = 0, 5, 10, 15, 20 and 25). The starting powders were thoroughly mixed by grinding in a high purity alumina crucible for 30 minutes and then melted at 1,100 °C in an electrical muffle furnace for 3 h. After melting completely, the melted powders were then poured quickly into a preheated stainless steel mould and annealed at 500 °C for 3 h before cooled to room temperature. Finally, the as-prepared glass samples were cut and finely polished to obtain a dimension of 1.0cm x 2.0cm x 0.3cm. By applying Archimedes principle, the weight of the prepared glass samples was measured in air and in xylene using a 4-digit sensitive microbalance (Denver, Pb214). Density, ρ, was determined by the relation as follow.

$$\rho = \frac{W_a}{W_a - W_b} \times \rho_b \tag{1}$$

where W_a is the weight in air, W_b is the weight in xylene, and ρ_b is the density of xylene ($\rho_b = 0.863$ g/cm^3). The corresponding molar volume (V_M) was calculated using the relationship, $V_M = M_T/\rho$, where M_T is the total molecular weight of the multi-component glass system. The FT-IR transmission spectra of Fe_2O_3 doped glass were recorded with Elmer FTIR Spectrometer Spectrum 2000, using the KBr pellet technique. Magnetic property measurements for all samples were carried out at room temperature with a vibrating sample magnetometer (VSM, Lakeshore 7404)

Results and Discussion

The glass with no Fe_2O_3 composition was shows colorless and transparency. When 5 % of Fe_2O_3 add in glass, the high intensity yellow-brown color was appeared and decrease transmission in visible region. The glasses doped with Fe_2O_3 was turned into an opaque when Fe_2O_3 concentration at 25% mol. The variation of the density and molar volume with Fe_2O_3 concentration is presented in figure 1 and figure 2 respectively. As can be seen in figure 1, the density of glasses show linear increase (with $R^2 = 0.99$) with the content of Fe_2O_3 incorporated in the network. This indicated that the substitution of SiO_2 by Fe_2O_3 resulted in an increase of the average molecular weight of oxide ions in the glass due to Fe_2O_3 has a higher relative molecular mass than that of SiO_2. The molar volume depended on both the changing rate of density and molecular weight. However, when the Fe_2O_3 concentration was increased, the increasing rate of molecular weight is greater than the increasing rate of density. As shown in figure 2, the molar volume increased with increasing of Fe_2O_3 content, this is attributed to an increase in the number of non-bridging oxygen (NBOs). The obtained results indicated that the Fe_2O_3 could enter in the glass network and acted as a modifier by occupying the interstitial space in the network and generating the NBOs to the structure. It was observed that the addition of Fe_2O_3 may accordingly result in an extension of glass network [7]. The following result shows a best fit was polynomial trend line to illustrate the relationship between molar volume and Fe_2O_3 content. Notice that the R^2 value is 0.9858, which is a best fit of the line to the data.

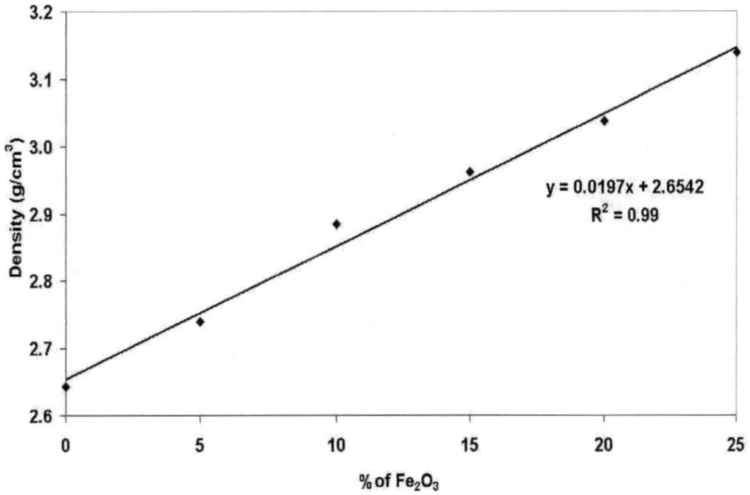

Figure 1. The densities of Fe_2O_3-doped glass

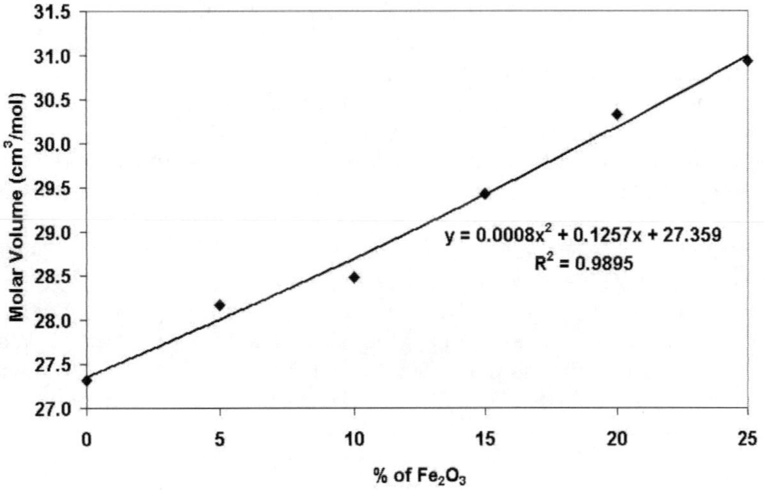

Figure 2. Molar volumes of Fe_2O_3-doped glass

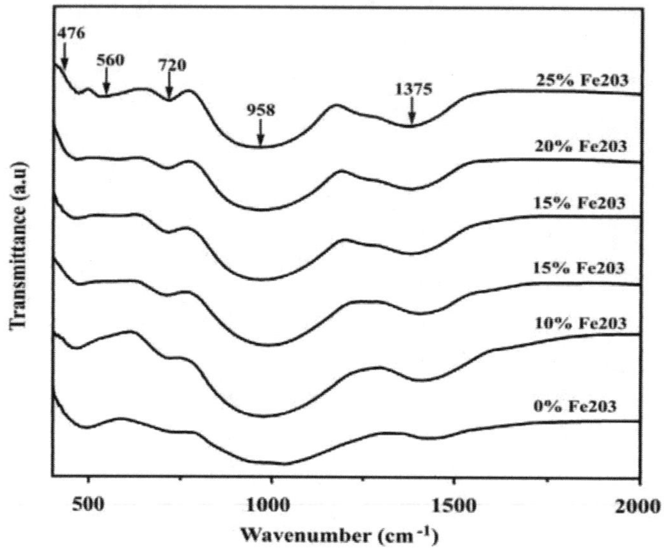

Figure 3. The FTIR spectra of Fe_2O_3-doped glass

Figure 3 shows FTIR spectra of $20Na_2O$: $1.0Al_2O_3$: $13B_2O_3$: $6.3CaO$: $0.2Sb_2O_3$: $4.5BaO$: (55-x)SiO_2 : xFe_2O_3 (where x = 0, 5, 10, 15, 20 and 25) glass samples in a range of 400-2000 cm^{-1}. It is clearly seen that the main absorption band in a region of 900–1100 cm^{-1} is attributed to the combined stretching vibrations of Si–O–Si bonds and the stretching vibrations of B–O bonds in BO_4 structural units **[8-9]**. The absorption band in a range of 1200-1500 cm^{-1} can be assigned to the B–O stretching vibrations of BO_3 structural units **[9, 10]**. The intensity of this band increased as a function of Fe_2O_3 concentration. This is due to the structural changes involving the conversions of the BO_4 into BO_3 structural units with an increase of the Fe_2O_3 glass modifier in good agreement with molar volume result. The absorption peaks at about 700 cm^{-1} and 460 cm^{-1} are assigned to the bending vibration of B-O-B and Si–O–Si bonds, respectively **[8,9]**. The IR spectrum of 25% mol Fe_2O_3-doped glasses show the appearance of absorption band in a range of 550-580 cm^{-1} which is ascribed to stretching vibrations of Fe–O bonds **[8]**. This peak might come from the formation of iron oxide or ferrite phase in the glass network.

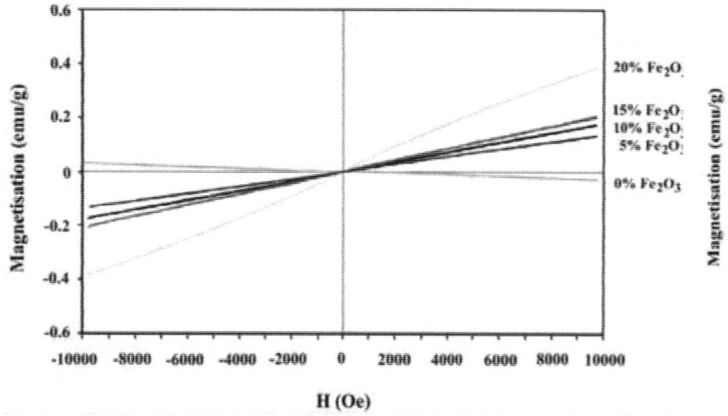

Figure 4. Magnetization for 0-20 % mol Fe_2O_3-doped glass

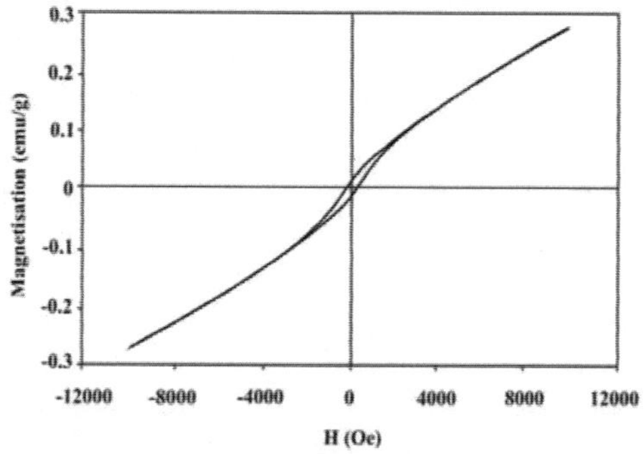

Figure 5. Magnetization for the 25% mol Fe_2O_3-doped glass

The magnetization (M) as a function of magnetic field (H) of $20Na_2O : 1.0Al_2O_3 : 13B_2O_3 : 6.3CaO : 0.2Sb_2O_3 : 4.5BaO : (55-x)SiO_2 : xFe_2O_3$ (where x = 0, 5, 10, 15, 20 and 25) glass measured by a vibrating sample magnetometer (VSM) at room temperature are presented in figure 4 and 5. It was observed that undoped glass exhibited a diamagnetic behaviour while the glass doped with Fe_2O_3 at 5-20 % mol, showed a paramagnetic behaviour and the magnetization increased with increasing Fe_2O_3 concentration. The paramagnetism originates from the presence of permanent magnetic dipole moments of iron in the glass (see in figure 4). The magnetization curve of the 25% mol Fe_2O_3 -doped glass was shown in figure 5. It is obviously seen that the hysteretic loop indicated a ferromagnetic behavior. The coercive field of hysteretic loop is about 240 Oe. The magnetization and FTIR results of 25% mol Fe_2O_3-doped glass could confirm that the formation of iron oxide or ferrite phase formed in the glass sample. For high concentration (25 % Fe_2O_3), each Fe^{3+} ion in glass network become closer and Fe^{3+} ion is not fully compensated by NBOs which tend to aggregate with each other and finally, cluster or particle of Fe_2O_3 was formed.

Conclusions

In this present work, glasses based on borosilicate system doped with Fe_2O_3 in the range of 0-25 % mol were prepared by a melt-quenching method. The structural, optical and magnetic properties of glass samples were studied. The result showed that the density and molar volume of glasses increased with increasing of Fe_2O_3 concentration. This might be due to the increase of Fe_2O_3 concentration at the expense of SiO_2 caused the opened glass network structure. The FTIR results showed that the BO_3 structural units increased with increasing of Fe_2O_3 concentration. Herein, The FTIR and magnetic results could confirm the presence of iron oxide or ferrite phase in the glass sample when doping with 25 % mol Fe_2O_3.

Acknowledgement

The authors would like to thank the Center of Excellence in Glass Technology and Materials Science (CEGM), Faculty of Science and Technology, Nakhon Pathom Rajabhat University for instrument and facilities. J. Kaewkhao and C. Kedkaew would like to thank the National Research Council of Thailand (NRCT) for funding this research. Thanks are also due to Research and Development Institute, NPRU for facilities.

5. References

[1] Mekki A Khatak G D Holland D Chinkhota M and Wenger L E 2003 *J. Non-Cryst. Solids* **318** 193
[2] Bergo P Pontuschka W M Prison J M Motta C C and Martinelli J R 2004 *J. Non-Cryst. Solids* **348** 84
[3] Zotov N Kirfel A Beuneu B Delaplane R Hohlwein D Reinauer F and Glaum R 2004 *Physica* B **350** 1071
[4] Licina V Mogus-Milankovic A Reis S T Day D E 2007 *J. Non-Cryst. Solids* **353** 4395

[5] Singh R K Srinivasan A and Kothiyal G P 2009 *J Mater Sci* **20** 147
[6] Ardelean L Lungu R and Pacuta P 2007 *J Mater Sci* **18** 837
[7] Abdel-Baki M Abdel-Wahab F A Radi A and El-Diasty F 2007 *J. Phys. Chem. Solids* **68** (8) 1457
[8] Battisha L K Afify H H Ibrahim M 2006. *J. Magn. Mag. Mater.* **306** 211
[9] Pascuta P Pop L Rada S Bosca M and Culea E 2008. *J Mater Sci* **19** 424
[10] Pascuta P 2009. *J Mater Sci* **21** 338

Journal of Physics: Conference Series **266** (2011) 012013 doi:10.1088/1742-6596/266/1/012013

Detection of ferromagnetic resonance in a single-crystalline Fe wire using a rectifying effect

Y. Kasatani[1*], **A. Yamaguchi**[1,2*], **H. Miyajima**[1] **and Y. Nozaki**[1,3]
[1] Department of Physics, Keio University, Hiyoshi 3-14-1, Yokohama, Kanagawa, 223-8522, Japan
[2] PRESTO, JST, Honcho 4-1-8, Kawaguchi, Saitama, 332-0012, Japan
[3] JST, CREST, Sanbancho, Chiyoda-ku, Tokyo, 102-0075, Japan

E-mail: ykasatan@phys.keio.ac.jp

Abstract. The broadband ferromagnetic resonance (FMR) of a single-crystalline Fe microwire was measured using a rectifying effect. The experimental result indicates the uniform mode dominates the system. The effective Gilbert damping factor of the Fe wire was estimated by the rectifying effect. A rectifying spectrum is broadened because of the spin wave excitation and the thermal fluctuation as the input power increases.

1. Introduction

The revealing of magnetization dynamics in nanoscale and micronscale artificial ferromagnets is one of the most interesting issues. With the development of the radio frequency (rf) measurement technique, the investigation of the magnetization dynamics in nanoscale and micron-scale patterned magnets has been proceeded [1-10]. For example, the rectifying effect induced by a rf current has been recently applied to detect ferromagnetic resonance (FMR) in the artificial ferromagnetic wire [8].

Because high-quality epitaxial growth method has been established, low-dimensional single-crystalline ferromagnets are available for studying on the magnetization dynamics. Therefore, the Gilbert damping factor has been estimated in the single-crystalline Fe ultrathin film [1,2] and in the single-crystalline Fe strip line [5].

In this paper, the measurement technique for the broadband FMR of a single-crystalline Fe microwire using a rectifying effect is established. We reveal the effect of the thermal fluctuation on the magnetization relaxation caused by Joule heating effect due to a rf current injection.

2. Experimental Preparation

High-quality epitaxial $Fe(100)_{bcc}$ films of 50 nm in thickness were grown in ultrahigh vacuum (UHV) at 100 °C by molecular beam epitaxy (MBE) on MgO(100) substrates, followed by 10 nm-thick protective Au cap at room temperature using MBE in UHV. A Fe wire of 1 μm width was fabricated using e-beam lithography and Ar ion milling. The longitudinal axis of wire was aligned along $Fe(100)[011]_{bcc}$ direction in the plane. A coplanar waveguide (CPW) constructed with Cr (5 nm)/Au (80 nm) were continuously made by the e-beam lithography and the lift-off method. Optical

* corresponding author YK: ykasatan@phys.keio.ac.jp
AY: yamaguch@phys.keio.ac.jp

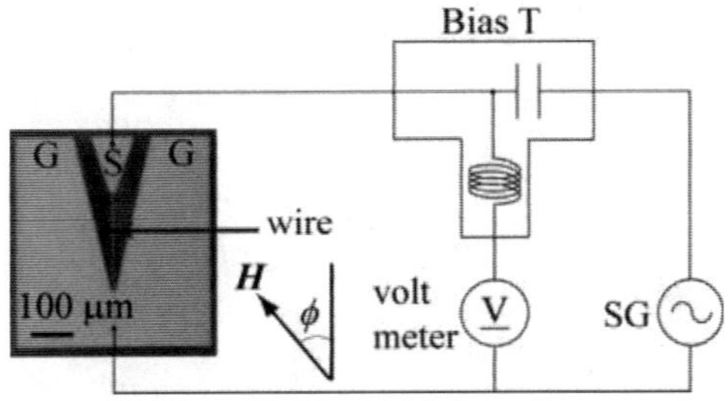

Figure 1. Optical micrograph of a sample and schematic of an electrical measurement circuit.

micrograph of a typical system is shown in figure 1. A ground-signal-ground (GSG)-type microwave probe was connected to the CPW, and the dc voltage difference induced by the rf current was detected via a bias-tee (Bias T) circuit. The rf current was injected into the wire by a signal generator with the frequency range from 500 MHz to 40 GHz. The structural characteristic of the Fe layer was determined by reflection high-energy electron diffraction (RHEED) and out-of-plane thin-film X-ray diffraction (TF-XRD) measurements. The RHEED image of Fe deposited on an MgO(100) substrate (figure 2(a)) showed some spots and streaks from $Fe(100)_{bcc}$. The out-of-plane TF-XRD result is shown in figure 2(b). These results indicate that the Fe film is epitaxially grown on an MgO(100) substrate with a little strain.

3. Results and Discussion

The external static magnetic field $|\boldsymbol{H}|$ sweeping from 0 to 2023 Oe was applied in the substrate plane at a tilting angle ϕ from the longitudinal axis of the wire. All FMR spectra were obtained by using the rectifying effect at room temperature.

The rf power dependence of rectifying voltage V spectra for the magnetic field of $|\boldsymbol{H}| = 2023$ Oe at $\phi = 30^\circ$ are shown in figure 3(a). The resonance frequency ω_k is almost constant against increasing the microwave power. Figure 3(b) shows the differential spectra of rf current power -5 dBm and +17 dBm. The differential rectifying spectra can provide highly-sensitive detection of spin-wave modes excited by the rf current. As seen in figure 3(b), the almost only uniform mode was excited even when the input rf power changed. Therefore, we describe the system with the macrospin model. The macrospin will be treated as an ellipsoid with principal axes along x, y and z and corresponding demagnetizing factors N_x, N_y and N_z. Therefore, the frequency of the uniform-mode spin wave is represented by

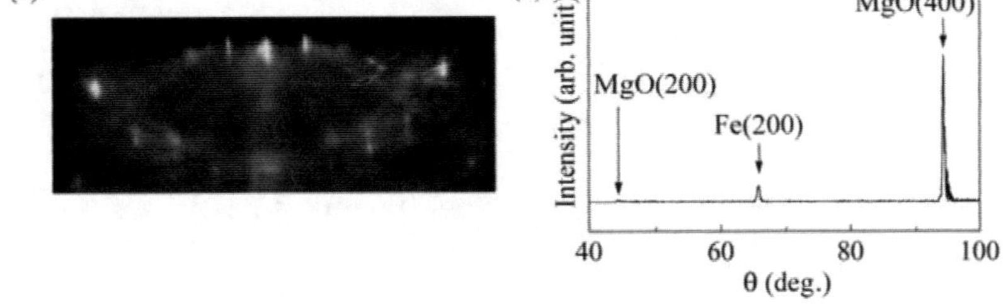

Figure 2. (a) RHEED image of the annealed Fe layer. (b) TF-XRD spectra of a Fe film deposited on an MgO(100) substrate.

$$\omega_k^{\,2} = \gamma^2 H_b' H_c' \tag{1}$$

with

$$H_b' \approx H_{\text{ext}} + M_S \left(N_y - N_x\right) \cos 2\phi \tag{2a}$$

and

$$H_c' \approx M_S \left(N_z - N_x \cos^2 \phi\right). \tag{2b}$$

Here, H_{ext}, γ and M_S represent the applied external magnetic field, the gyromagnetic ratio and the saturation magnetization, respectively. The rectifying spectrum V is described by

$$V(f) = \frac{C_1\left(f^2 - f_k^{\,2}\right) + C_2 f^2}{\left(f^2 - f_k^{\,2}\right)^2 + f^2 \alpha_{\text{eff}}^{\,2} \Delta^2} + V_0, \tag{3}$$

where $f = \omega/2\pi$, $f_k = \omega_k/2\pi$, V_0, C_1, C_2, and α_{eff} are the rf current frequency, resonance frequency, offset voltage, asymmetric and symmetric amplitude coefficients and the effective Gilbert damping factor, respectively[8]. The full width half maximum (FWHM) of FMR spectrum Δ is written by

$$\Delta = \frac{\gamma\left(H_b' + H_c'\right)}{2\pi}. \tag{4}$$

When we apply Eq. (3) to the measured spectra, we estimate the effective damping constant α_{eff} and the asymmetric and symmetric amplitude coefficients C_1 and C_2.

The effective Gilbert damping factor α_{eff} is plotted as a function of the rf current power in figure 4. This result indicates that α_{eff} increases with increasing the input rf current power. This is attributable to the heat fluctuation of the magnetization induced by the Joule heating due to the rf current.

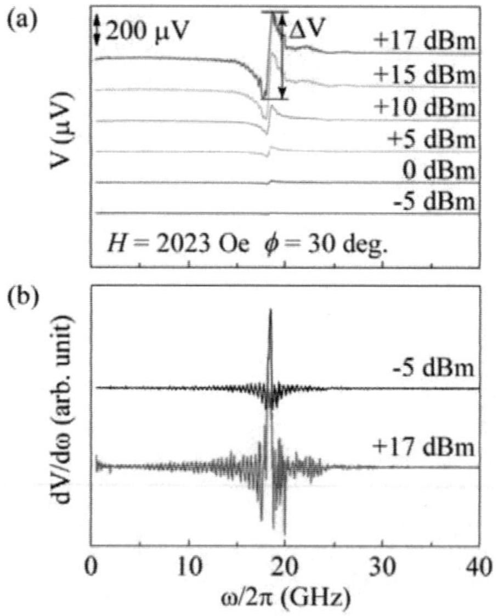

Figure 3. (a) Typical spectra of rectifying voltage V induced by intensity changed rf current under $|\boldsymbol{H}| = 2023$ Oe at $\phi = 30^\circ$. (b) Differential voltage $\mathrm{d}V/\mathrm{d}\omega$ plotted as a function of the rf frequency.

Meanwhile, it is reported that the damping factor of the single-crystalline Fe[110] wires of 45-nm in thickness and 80-μm in width is $\alpha = 0.014$ □ 0.016 in Ref. 5, the damping factor for the single-crystalline Fe(001) film of 30-nm in thickness is $\alpha = 0.003$ in Ref. 2, and in Ref. 11, the damping factor for the epitaxial Fe(001) square dot with 50 nm on a side and 10 ML in thickness is $\alpha \approx 0.02$ □ 0.09 , respectively. It is indicated that the effective damping factor of the single-crystalline Fe decreases with growing the sample size because the boundary condition is extremely important in the magnetization damping process. As the sample size becomes small, as several modes of spin waves come to be excited and the energy dissipation via the excited spin waves increases.

4. Summary

The rectifying measurement in the single-crystalline Fe wire was established. The effective Gilbert damping factor was affected by the size of the sample in terms of the spin wave excitation and influenced by the rf power in terms of the thermal fluctuation.

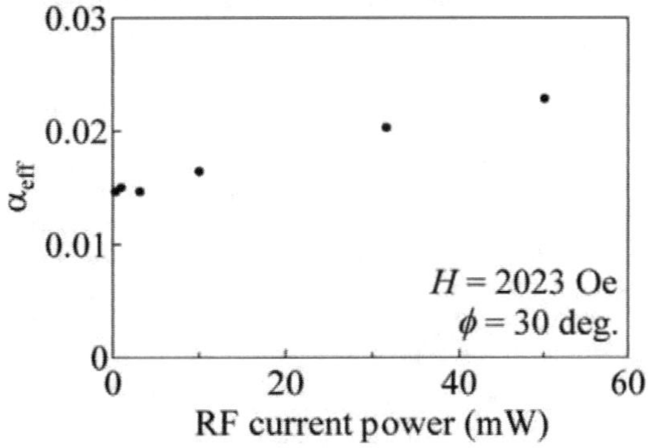

Figure 4. Dependence of the effective Gilbert damping constant on the rf current power under the application of the field of 2023 Oe at 30 degree.

5. Acknowledgements

The present study was partly supported by MEXT Grants-in-Aid for Scientific Research in a Priority Area, a JSPS Grant-in-Aid for Scientific Research, the JST PRESTO program and the JST, CREST.

6. References

[1] Fermin J R, Azevedo A, Aguiar F M, Li B and Rezende S M 1999 *J. Appl. Phys.* **85** 7316
[2] Kuanr B K, Camley R E, Celinski Z 2004 *J. Appl. Phys.* **95** 6610
[3] Oogane M, Wakitani T, Yakata S, Yilgin R, Ando Y, Sakuma A and Miyazaki T 2006 *Jpn. J. Appl. Phys.* **45** 3889
[4] Hui X, Wirthmann A, Gui Y S, Tian Y, Jin X F, Chen Z H, Shen S C and Hu C-M 2008 *Appl. Phys. Lett.* **93** 232502
[5] Su J, Tsai C S and Lee C C 2000 *J. Appl. Phys.* **87** 5968
[6] Demand M, Encinas-Oropesa A, Kenane S, Ebels U, Huynen I and Piraux L 2002 *J. Magn. Magn. Mater.* **249** 228
[7] Kohmoto O 2007 *Mat. Sci. Eng. A* **449-451** 394
[8] Yamaguchi A, Motoi K, Hirohata A, Miyajima H, Miyashita Y and Sanada Y 2008 *Phys. Rev. B* **78** 104401
[9] Medina J D L T, Darques M, Piraux L and Encinas A 2009 *J. Appl. Phys.* **105** 023909
[10] Ruzmetov D and Chandrasekhar V 2008 *J. Magn. Magn. Mater.* **320** 47
[11] Lepadatu S, Wu J, Bunce C, Zou X, Niu D, Xu Y B, Chantrell R amd Ju G P 2007 *J. Appl. Phys.* **101** 09C111

Journal of Physics: Conference Series **266** (2011) 012014 doi:10.1088/1742-6596/266/1/012014

Four fold in-plane magnetic anisotropy in sputtered (100)MgO/Fe epitaxial films

K. Noda, M. Higuchi, Y. Komaki, T. Tanaka[1], Y. Nozaki and K. Matsuyama
Department of Electronics, ISEE, Kyushu University
Motooka 744, Nishi-ku, Fukuoka 819-0395, Japan

E-mail: t-tanaka@ed.kyushu-u.ac.jp

Abstract. We report structural and magnetic properties of Fe thin films, deposited onto (100)MgO substrates by rf magnetron sputtering. The successful epitaxial growth of Fe<100>(100) parallel to MgO<110>(100) was achieved at a practical growth rate of 0.73 nm/s. The *M-H* hysteresis measured along Fe<100> directions exhibits a superior squareness with a low coercivity (H_c = 5 Oe). The saturation field along <110> directions, exceeding 500 Oe, demonstrates the crystalline cubic anisotropy comparable to the bulk Fe. The anisotropic properties of the initial susceptibility against an oscillation field of 1 mOe amplitude were also confirmed with inductive measurements.

1. Introduction

The epitaxial growth of magnetic thin films is a key issue for integrating magnetic elements into various planar electronic devices. Representative ferromagnetic material of Fe attracts renewed research interests because of the highest saturation magnetization and the considerable cubic magnetic anisotropy, which can be utilized for wide-band microwave devices [1,2], non-volatile memory elements [3], and quaternary memory devices [4]. Extensive efforts have been devoted for the epitaxial growth of Fe thin films on GaAs substrates, because of the low lattice mismatch of ~1.4 % and the potential of semiconductor/magnetic-material hybrid devices [5-7]. The successful epitaxial growth was also reported for another hetero-epitaxial system of Fe/MgO-substrate with the reasonable lattice mismatch of 3.8 % [8-10].

As for the inductive application at a high frequency region, the superior susceptibility under well defined magnetic anisotropy are essential characteristics. Novel device performances of a multi-bit-state or a field programmable function can be realized with the fourfold cubic anisotropy. In the present study, characterization of Fe thin films, sputtered onto MgO(100) substrates, are performed taking various deposition conditions as parameters. The structural and magnetic properties are systematically investigated, focussing the anisotropic susceptibility originated from the four fold anisotropy.

2. Experimental

Fe films were deposited onto (100)MgO substrates by rf magnetron sputtering. Prior to the film deposition, Ar ion etching of the substrate surface was performed in order to eliminate contaminations.

1 t-tanaka@ed.kyushu-u.ac.jp

The deposition was carried out in a 5 mTorr Ar gass of 99.9999 % purity with a base pressure of 5×10^{-7} Torr before sputtering. Fe target size is 3 inch in diameter and with 99.99 % purity. The crystallographic and magnetic properties of the sample were studied with X-ray diffractometery (XRD) and vibrating sample magnetometer (VSM). The anisotropic property in the susceptibility (χ) was investigated by using the inductive measurements. The variation of χ was sensitively detected from the shift of the LC resonant frequency f_r (~10 MHz), measured for a flat pick-up coil inserted with the Fe-film [11]. Although the samples with various thickness were prepared, the results for 100-nm thick film are focused in the present manuscript.

3. Result and discussion

Fig. 1(a) shows XRD patterns of Fe films, fabricated with various substrate temperatures (T_s) and a fixed sputtering power P_s = 150 W. The growth rate GR is 0.73 nm/s. The results demonstrate that the Fe(100) is preferentially aligned parallel to MgO(100) at all of the sputtering temperature T_s (28 – 600 °C), although iron oxide phase is observed. Similar preferential growth was also observed for a sample sputtered at lower P_s of 20 W. However, the XRD peaks of iron oxides become conspicuous in the low power sputtering, probably due to the low GR = 0.09 nm (P = 20 W). The evaluated lattice constant c values decreases with T_s, excepting T_s = 28 °C. The c value of 0.2864 nm for T_s = 600 °C is in good agreement with that for bulk Fe (0.2866 nm). Fig. 1(b) presents the values of grain diameter d, estimated from the Scherrer's equation, as a function of T_s. The diameter d increases from 3.73 nm (T_s = 28 °C) to 27.0 nm (T_s = 600 °C) with increasing T_s.

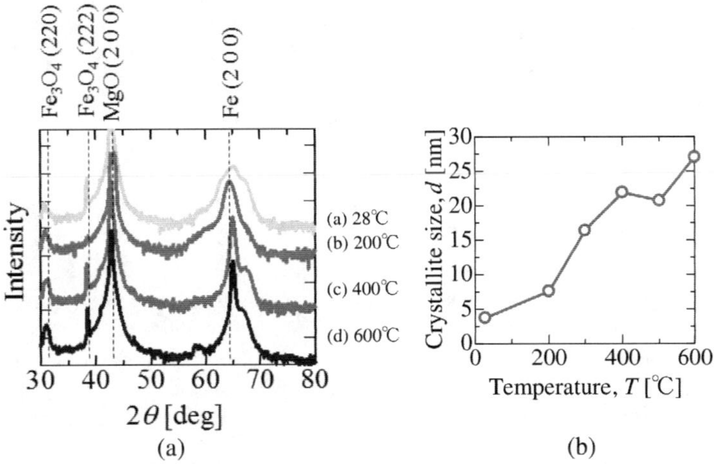

Fig. 1. (a) XRD patterns for Fe films deposited at various sputtering temperatures. (b) Dependence of crystallite size d on sputtering temperature T_s.

Fig. 2 presents a comparison of magnetic properties for samples prepared with different sputtering power P_s ((a) 150 W, (b) 20 W) and a fixed T_s of 600 °C. As for a sample sputtered at P_s = 150W, the square hysteresis curve with a coercive field H_c = 12.65 Oe is measured along the MgO[100] direction, noted as 45° in the figure. The gradual demagnetization curve from the magnetization saturation along the [110] direction implies the magnetization rotation in the individual domain. The normalized remanent magnetization of 0.7 indicates that the magnetization relaxes to one of the <100> directions. The measured saturation field, exceeding 600 Oe, is comparable to that for the cubic anisotropy field of bulk Fe (H_k = 570 Oe). The squareness ratio S measured along various directions presents the four fold angular symmetry, as shown in the right hand side panel. The anisotropic magnetic properties demonstrate the successful epitaxial growth of the 100-nm thick Fe films onto MgO substrate with the crystalline orientation of Fe<100>(100)//MgO<110>(100). The MgO<110>(100) direction is,

Journal of Physics: Conference Series **266** (2011) 012014 doi:10.1088/1742-6596/266/1/012014

hereafter, referred as the <100> direction, indicating the easy axis of the fabricated Fe films. Contrarily, the sample sputtered at P_s = 20 W presents the isotropic hysteresis curves with an order of enlarged coercive H_c = 110 Oe for the both of two directions, as shown in Fig. 2(b). The similar isotropic magnetic properties were also reported for Fe films, sputtered at relatively low temperatures (315 °C, 400 °C)[12], and reasoned to be the homogeneous distribution of morphological features. Probable explanation for the degraded magnetic anisotropy in our samples would be a dispersion of anisotropy axis due to the crystallographic disorientation in the film plane.

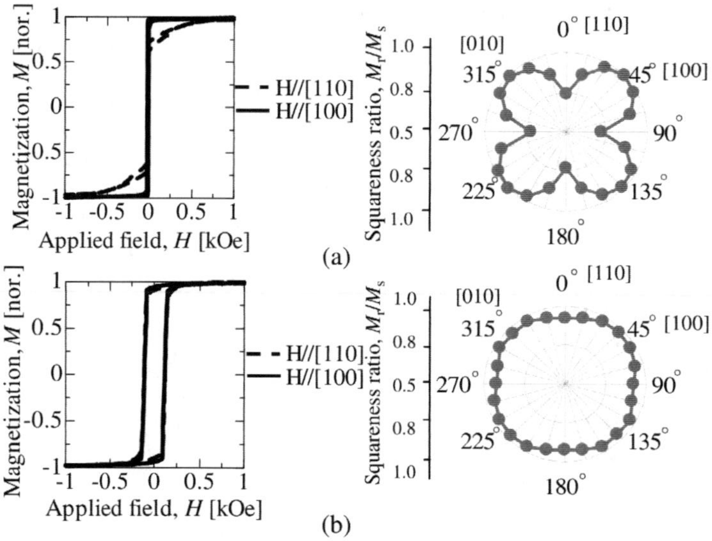

Fig. 2. Comparison of hysteresis curve and angular dependence of squareness ratio for Fe films, deposited at different sputtering power, (a) P = 150 W, (b) P = 20 W.

Fig. 3 shows the ratio of the values of S, measured along the [110] ($S_{[110]}$) and the [100] ($S_{[100]}$) directions. Most of the results present practically identical values of $S_{[110]}/S_{[100]}$ ranged from 0.71 to 0.75, which indicate that the four fold magnetic anisotropy originated from the epitaxial growth. Extraordinary large $S_{[110]}/S_{[100]}$ of 0.95 is observed for a sample deposited with P_s = 20 W and T_s = 600 °C, probably due to the significant in-plane dispersion of the crystalline axis as mentioned before.

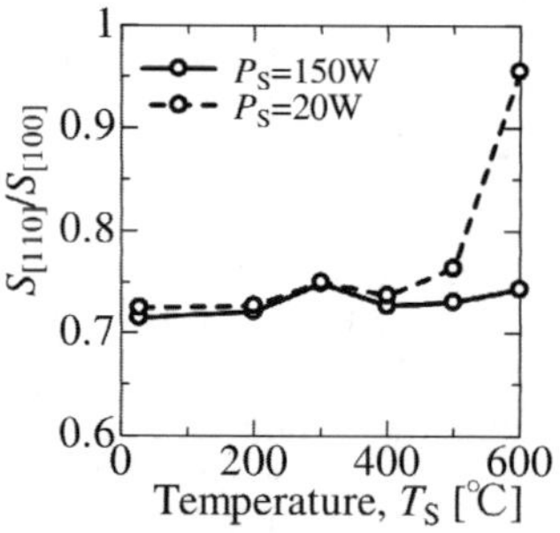

Fig. 3. Dependence of relative squareness ratio $S_{[110]}/S_{[100]}$ on sputtering temperature T_s

Dependence of the coercive field H_c and the coercivity squareness S^* on the sputtering temperature T_s is plotted in Fig 4(a), taking the P_s as a parameter. The H_c for samples sputtered with P_s = 150 W presents relatively low coercive fields < 13 Oe. The minimum value of 4.5 Oe is obtained at T_s = 300 °C. The H_c for samples sputtered with P_s = 20 W drastically increases at T_s higher than 500 °C. Fig. 4(b) plots the coercive squareness S^*, defined as H_n/H_c (H_n: nucleation field), as a function of T_s. As shown in the figure, the S^* for samples fabricated with P_s = 150 W exhibits superior coercivity squareness. The value of S^* significantly decreases at T_s lower than 28 °C (P_s = 150 W), 200 °C (P_s = 20 W). Fig. 5 shows the angular dependence of hysteresis curves and the H_c values for the sample sputtered with P_s = 150 W and T_s = 500 °C. The applied field angle is measured from the [110] direction. As can be seen in the figure, no significant angular dependence is observed for the magnetization reversal process. This cannot be predicted from the simple domain nucleation and propagation model, and the further detailed study is required for an explanation. Somewhat gradual magnetization reversal can be attributable to the inhomogeneous nucleation fields in the film plane.

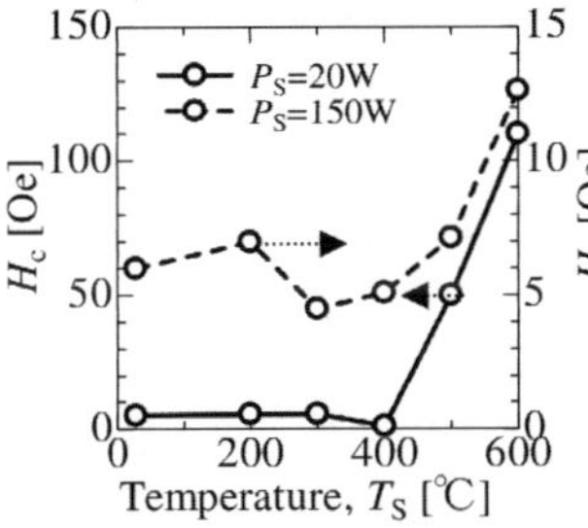

Fig. 4(a). Sputtering temperature dependence of coercive field H_c.

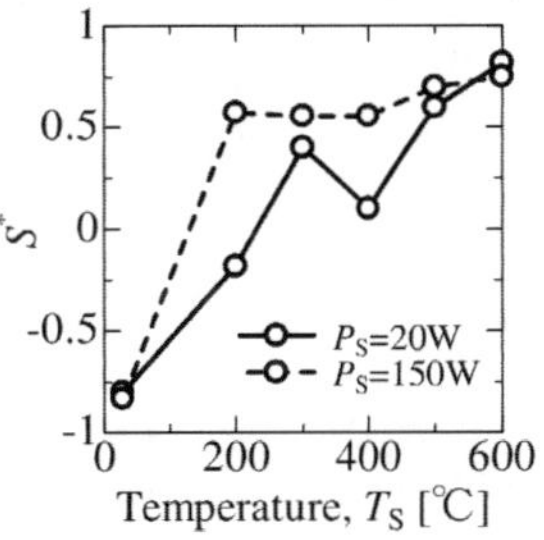

Fig. 4(b). Sputtering temperature dependence of coercivity squareness S^*.

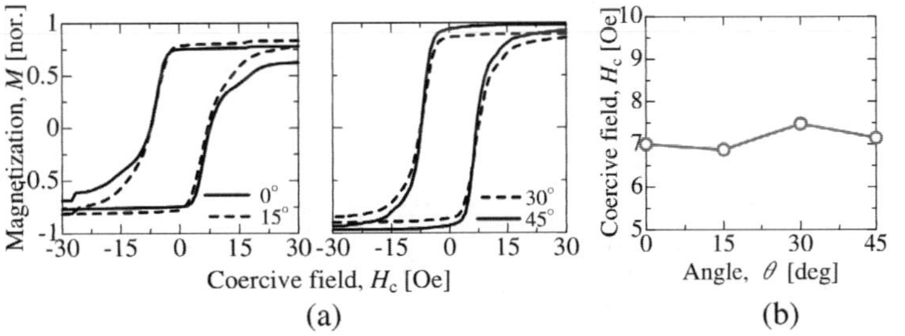

Fig 5. Angular dependence of (a) *M-H* hysteresis properties and (b) coercive filed H_c for a sample deposited with P_s =150 W and T_s = 500 °C.

Resonance curves for a pick-up coil inserted with the Fe film were measured for different exciting field ($H_{exc.}$: 1 mOe amplitude) directions, as shown in Fig. 6. The magnetization was initially saturated along the [100] or [110] direction by applying an external field H_s. The sample saturated along the [100] direction exhibits marked shift of the resonance curve depending on the $H_{exc.}$ direction, as shown in Fig. 6(a). The result indicates the anisotropic susceptibility owing to the superior squareness along the easy axis. While, no distinguished difference of the resonance frequency is observed for the sample saturated along the [110] direction, as shown in Fig. 6(b). This means that the magnetization is equally rotated to the easy axes orthogonal to each other ([100] and [010]), which results in the isotropic susceptibility for the excitation along the orthogonal two directions. The results also demonstrate the magnetic property with the four fold symmetry, that is, the absence of the additional uniaxial anisotropy, as reported in the other hetero epitaxial system of Fe/GaAs[13,14].

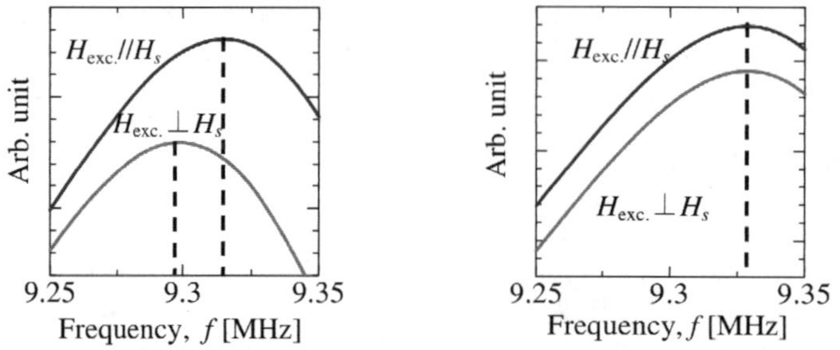

(a) H_s//[100] (b) H_s//[110]

Fig. 6. Comparison of resonance curves of pick-up coil inserted with Fe film. The magnetization is initially saturated with an external field H_s applied along (a) [100], (b) [110] directions. Measurements are performed for orthogonal exciting field directions of parallel ($H_{exc.}//H_s$) and perpendicular ($H_{exc.} \perp H_s$) to the saturation direction.

4. Conclusion

The structural and magnetic properties were systematically studied for Fe thin films, deposited onto MgO(100) substrates by rf magnetron sputtering. The successful epitaxial growth of Fe<100>(100) parallel to MgO<110>(100) was achieved at a practical growth rate of 0.73 nm/s. A well defined in-plane four fold symmetry is observed in the angular dependence of *M-H* hysteresis profile for samples deposited with various sputtering temperature (28 ~ 600 °C) and power (20 ~ 150 W). The saturation filed of exceeding 500 Oe, measured along the hard axis, demonstrates the crystalline cubic anisotropy comparable to that for a bulk value. The *M-H* hysteresis measured along the hard axis exhibits a superior squareness with a low coercivity (H_c = 5 Oe). Noticeable difference in the initial susceptibility, reflecting the four fold magnetic anisotropy, was observed for the different exciting field directions by using inductive measurements.

References

[1] Camley R E and Mills D L 1997 *J. Appl. Phys.* **82** 3058
[2] Tsai C S, Su J and Lee C C 1999 *IEEE Trans. Magn.* 35 3178
[3] Schloemann E, Tustison R, Weissman J, Hook V H and Varitimos T 1988 *J. Appl. Phys.* **63** 3140
[4] Yoo T, Khym S, Yea S-y, Chung S, Lee S, Liu X and Furdyna J K 2009 *Appl. Phys. Lett.* **95** 201505
[5] Prinz G A, Rado G T and Krebs J J 1982 *J. Appl. Phys.* **53** 2087
[6] Tustison R W, Varitimos T, Hook V J and Schloemann E F *Appl. Phys. Lett.* **51** 285
[7] Ding Y, Alexander, Jr. C and Klemmer T J 2003 *J. Appl. Phys.* **93** 6674
[8] Fuke H, Sawabe A and Mizoguchi T 1993 *Jpn. J. Appl. Phys.* **32** L1137
[9] Subagyo A, Sueoka K, Mukasa K and Hayakawa K 1999 *J. Appl. Phys.* **38** 38020
[10] Shikada K, Tabuchi K, Ohtake M, Fumiyoshi K and Futamoto M 2008 *J. Magn. Soc. Jpn.* **32** 296-303
[11] Zhang M, Shimizu K, Nozaki Y and Matsuyama K 2007 *J. Magn. Magn. Mater.* **310** 2746
[12] Cebollada F, H-Maneru A, Hernando A, M-Boubeta C, Cebollada A and Gonzalez J M 2002 *Phys. Rev. B* **66** 174410
[13] Krebs J J, Jonker B T and Prinz G A 1987 *J. Appl. Phys.* **67** 2596
[14] Florczak J M and Dahlberg E D 1991 *Phys. Rev. B* **44** 9338

Journal of Physics: Conference Series **266** (2011) 012015 doi:10.1088/1742-6596/266/1/012015

Domain-Wall Pinning in Er and Dy Studied by Minor-Loop Scaling Laws

S Kobayashi, S Takahashi, H Kikuchi and Y Kamada
NDE&Science Research Center, Iwate University, Morioka 020-8551, Japan

E-mail: koba@iwate-u.ac.jp

Abstract. Scaling laws of magnetic minor hysteresis loops for heavy rare-earth metals Er and Dy were studied. We found that there exists a scaling power law between hysteresis loss and remanence of minor loops with a constant exponent of 1.2-1.3 in the low-temperature ferromagnetic phase below T_c as was commonly observed in soft ferromagnetic materials. However, the scaling law persists even in the higher temperature incommensurate antiferromagnetic phase with no net spontaneous magnetization. These observations reflect the irreversible movement of Bloch walls, which separate ferromagnetic domains below T_c or small ferromagnetic regions in the antiferromagnetic phase.

1. Introduction

Magnetic hysteresis loops of bulk ferromagnetic materials have been extensively investigated both in the scientific and technological areas. The hysteresis properties including coercive force, remanence, and hysteresis loss, reflect the various internal features of these ferromagnetic materials, such as chemical composition, crystal structure, anisotropy, and lattice defects, and are affected by temperature as well as stress. These properties, which are averaged over a system with many domains, can be interpreted in view of the statistical behaviour of the Bloch wall movement and domain rotation.

Scaling laws of magnetic minor hysteresis loops have been extensively studied for evaluating the magnetic quality of bulk ferromagnetic materials and thin ferromagnetic films. In particular, the power-law relation between hysteresis loss W_F* and maximum flux density B_a* of minor hysteresis loops with an exponent of n_F=1.6, given below, was found about one century ago and is well known as the Steinmetz law [1];

$$W_F^* = W_F^0 \left(\frac{B_a^*}{B_s} \right)^{n_F} , \qquad (1)$$

where B_s is the maximum flux density of the major loop and W_F^0 is a coefficient which offers useful information about intrinsic properties such as magnetocrystalline anisotropy, magnetostriction, and defect density. The Steinmetz law holds true in the medium magnetization range where irreversible motion of Bloch walls dominates magnetization mechanism. However, the applicability of the law is limited to soft ferromagnetic materials with low coercive force. On the other hand, a scaling power law between W_F* and remanence B_R* of minor loops, recently proposed for transition-metal ferromagnets [2] as well as strain-induced ferromagnetic α'-martensite embedded in a paramagnetic austenitic matrix [3], given below, was found to selectively reflect irreversible motion of Bloch walls even in the presence of rotation mechanism.

$$W_F^* = W_m^0 \left(\frac{B_R^*}{B_R} \right)^{n_m} , \qquad (2)$$

where B_R is remanence of the major loop, W_m^0 is a coefficient, and n_m is an exponent of 1.3-1.4. While the Steinmetz law generally fails in the low-field regime due to a significant contribution of domain rotation, this scaling law holds true from very low to medium range of magnetization. This unique character of the law opens up a possibility of examining domain-wall kinetics in various magnetic materials where the Steinmetz law fails.

In this paper, we have examined the scaling law between hysteresis loss and remanence of minor loops for Er and Dy which show a rich H-T phase diagram associated with various types of magnetic ordering [4,5]. Below a Néel temperature of $T_N \sim 80$ K, Er exhibits a sinusoidally modulated structure with magnetic moments confined along the hexagonal *c* axis. As the temperature decreases, the periodicity of magnetic moments along the *c* axis increases, associated with a development of the basal plane components, and finally below $T_c \sim 20$ K a low-temperature conical structure with ferromagnetic *c*-axis components appears. On the other hand, Dy exhibits an antiferromagnetic helical structure below $T_N \sim 180$ K where magnetic moments confined on the hexagonal basal planes rotate from basal plane to basal plane. With decreasing temperature, the periodicity along the *c* axis increases and below $T_c \sim 90$ K all the magnetic moments align ferromagnetically along the *a* axis. Detailed temperature dependence of the scaling law and the coefficient of magnetic minor hysteresis loops in these complex magnetic structures has been investigated in the present study.

2. Experimental

The toroidal polycrystalline samples of Er and Dy with a purity of 99.9% were used for magnetic measurements. A 125-turn exciting and 100-turn detecting coils were wound around the samples. A triangular current was applied to the exciting coil to generate a cyclic magnetic field and magnetize the sample. The magnetic field within the sample was obtained from the voltage across a 1-Ω resistance connected to the exciting coil in series. The induced voltage of the pickup coil was integrated in order to obtain a magnetic flux within the sample.

A set of magnetic minor hysteresis loops with various field amplitudes H_a up to 30 kA/m was measured by gradually increasing the amplitude of the wave current, being the speed of applied field dH/dt to be constant. In the present study, we tested various dH/dt conditions; 32 - 192 kA/m/s. Before measuring a set of minor loops, the sample was demagnetized with a decaying alternating magnetic field with a frequency of 10 Hz. The samples were placed in a He-gas closed-cycle refrigerator by which the temperature was varied from 5 to 300 K.

Magnetic phase transition temperatures of our samples were accurately determined from the temperature dependence of dc magnetization under a magnetic field of 50 Oe, measured with a superconducting quantum interference device (SQUID) magnetometer.

3. Results and discussions

Figure 1 shows a set of minor hysteresis loops for Er, taken at various temperatures on heating. At T = 5K below T_c, hysteresis loss is very small and minor loops are dominated by reversible process of domain wall motion as well as domain rotation as shown in Fig. 1(a). As the temperature increases, the hysteresis loss of each minor loop gradually develops and maximizes at around $T \sim 20$ K, followed by the sudden decrease with the onset of a higher temperature incommensurate antiferromagnetic phase at T_c^{heat} (= 23 K). As shown in Fig. 1(d), however, the hysteretic behaviour persists even above T_c^{heat}, although it almost disappears at $T \sim 30$ K as shown in Fig. 1(e). This behaviour was also seen for data taken on cooling above T_N.

In order to investigate whether existing scaling laws of minor loops hold true below and above T_c, parameters of minor loops were defined as denoted in Fig. 1(f) and the relations were examined.

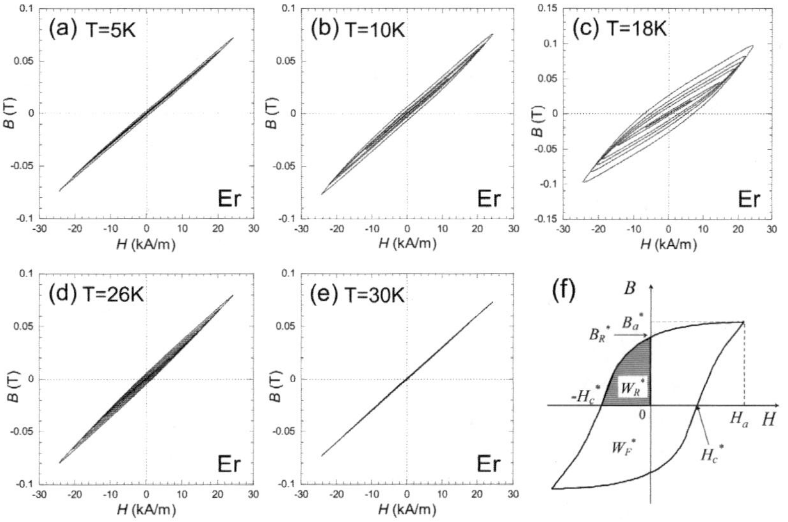

Figure 1. A set of magnetic minor hysteresis loops for Er, taken at various temperatures on heating; (a) T=5 K, (b) T=10 K, (c) T=18 K, (d) T=26 K, and (e) T=30 K. The speed of applied dH/dt was 192 kA/m/s. (f) Parameters of a minor loop with a field amplitude H_a.

Figure 2(a) shows a double logarithmic plot of W_F*-B_a* curves for Er, taken at various temperatures on heating. At all measurement temperatures, no Steinmetz law with a power-law exponent of 1.5 was confirmed. This is attributed to the dominant contribution of reversible motion of domain walls and domain rotation to magnetization, because the Steinmetz law is generally valid when irreversible motion of Bloch walls dominates [2].

On the other hand, as is seen in a linear relationship between W_F* and B_R* in a double logarithmic plot shown in Fig. 2(b), W_F*-B_R* curves show a power law in a wide magnetization range below and above T_c and the slope is almost independent of temperature. Least-squares fits to Eq. (2) yielded $n_m = 1.17 \pm 0.06$ for data obtained on heating and cooling.

Figure 3(a) shows W_m^0 as a function of temperature for Er, obtained on heating and cooling. Though W_m^0 shows a temperature hysteresis of ~ 2K at the first-order phase transition temperature of T_c, W_m^0 obtained on heating and cooling exhibits a similar behaviour to each other; as the temperature increases, W_m^0 decreases below T_c and then exhibits a sudden increase above T_c.

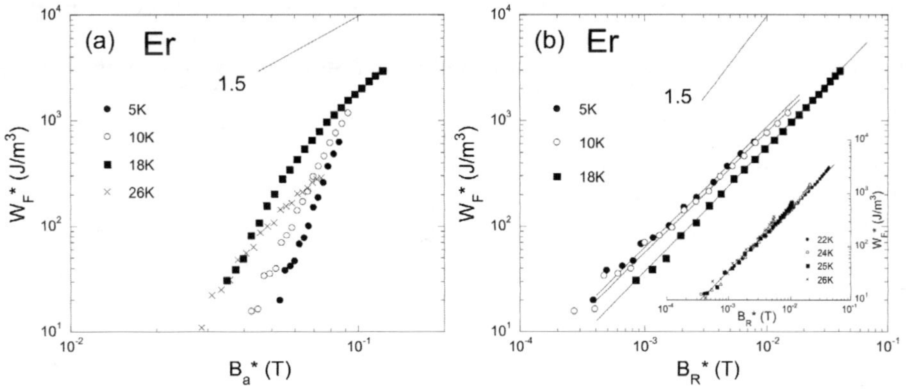

Figure 2. Relations (a) between W_F* and B_a* and (b) between W_F* and B_R* for Er, taken at various temperatures on heating. The inset in (b) shows W_F*-B_R* curves at higher temperatures.

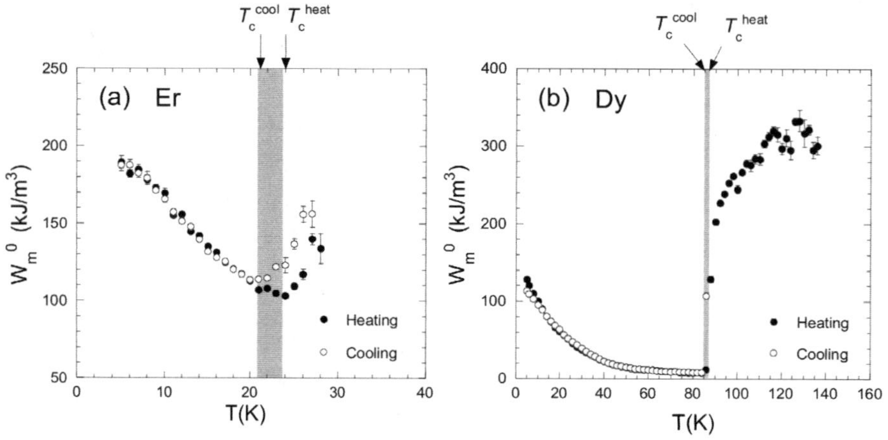

Figure 3. W_m^0 as a function of temperature, obtained on heating and cooling for (a) Er and (b) Dy. The shade represents the temperature hysteresis.

Figure 3(b) shows W_m^0 as a function of temperature for Dy, obtained on heating and cooling. Also in the case of Dy, the scaling law of Eq. (2) with $n_m = 1.33 \pm 0.06$ was found both below and above T_c and the temperature dependence of W_m^0, which is similar to that for Er, was observed. However, the hysteresis behaviour of minor loops above T_c was only observed on heating from below T_c and persists up to a rather high temperature of ~ 130 K. Note that the Steinmetz law of Eq. (1) with $n_F = 1.43 \pm 0.04$ was confirmed below T_c for Dy because of the formation of the simple ferromagnetic structure with magnetic moments aligned along the *a* axis; the coefficient W_F^0 shows a monotonic decrease with increasing temperature below T_c (not shown).

In the low temperature ferromagnetic phase below T_c, Er exhibits a conical structure with ferromagnetic components of magnetic moments along the *c* axis, whereas Dy exhibits ferromagnetic ordering with magnetic moments confined along the *a* axis. In both cases, irreversible motion of Bloch walls would contribute to the hysteretic behaviour of minor loops [4,5], although reversible process due to domain wall motion and domain rotation also plays a significant role in the magnetization. This irreversible process would yield the scaling power law between hysteresis loss and remanence of minor loops for both metals as was observed in soft ferromagnetic materials [2].

As the temperature increases from below T_c, antiphase domains are nucleated within ferromagnetic domains at T_c to form the antiferromagnetic structure with incommensurate modulation along the *c* axis. This would produce a large number of energetically unstable domain walls separating small ferromagnetic regions. These walls in the antiferromagnetic phase are different from those formed during cooling from above T_N and would give a significant contribution to irreversible magnetization process and therefore the scaling law above T_c. Actually, such domain walls in the antiferromagnetic phase were confirmed for Dy only after heating from below T_c and were found to persist up to T ~ 130K well above T_c [4,5,6]. A further investigation of the scaling law for both metals is in progress in order to elucidate the magnetization mechanism due to domain wall motion and qualitatively interpret the behaviour of the coefficient in both the ferromagnetic and antiferromagnetic phases.

4. References

[1] Steinmetz C P 1984 *Proc. IEEE* **72** 197
[2] Kobayashi S, Takahashi S, Shishido T, Kamada Y and Kikuchi H 2010 *J. Appl. Phys.* **107** 023908
[3] Kobayashi S, Saito A, Takahashi S, Kamada Y and Kikuchi H 2008 *Appl. Phys Lett.* **92** 182508
[4] Palmer S B 1975 *J. Phys. F: Metal Phys.* **5** 2370
[5] Loffler E and Rauch H 1969 *J. Phys. Chem. Solids* **30** 2175
[6] McKenna T J, Campbell S J, Chaplin D H and Wilson G V H 1983 *phys. stat. sol.* **75** 421

Journal of Physics: Conference Series **266** (2011) 012016 doi:10.1088/1742-6596/266/1/012016

Magnetic properties of mechanically alloyed Mn-Al-C powders

O Kohmoto, N Kageyama, Y Kageyama, H Haji, M Uchida, Y Matsushima

Division of Mathematics and Physics, Graduate school of Natural Science and Technology, Okayama University, Okayama 700-8530, Japan

E-mail: kohmoto@cc.okayama-u.ac.jp

Abstract. We have prepared supersaturated-solution Mn-Al-C alloy powders by mechanical alloying using a planetary high-energy mill. The starting materials were pure Mn, Al and C powers. The mechanically-alloyed powders were subjected to a two-step heating. Although starting particles are Al and Mn with additive C, the Al peak disappears with MA time. With increasing MA time, transition from α-Mn to β-Mn does not occur; the α-Mn structure maintains. At 100 h, a single phase of supersaturated-solution α-Mn is obtained. The lattice constant of α-Mn decreases with increasing MA time. From the Scherrer formula, the crystallite size at 500 h is obtained as 200Å, which does not mean amorphous state. By two-step heating, high magnetization (66 emu/g) was obtained from short-time-milled powders (t=10 h). The precursor of the as-milled powder is not a single phase α-Mn but contains small amount of fcc Al. After two-step heating, the powder changes to τ-phase. Although the saturation magnetization increases, the value is less than that by conventional bulk MnAl (88 emu/g). Meanwhile, long-time-milled powder of single α-Mn phase results in low magnetization (5.2 emu/g) after two-step heating.

1. Introduction

Hard magnetic Mn-Al alloys were first obtained using conventional powder-metallurgy method by Kono in 1958 [1]. The alloys are known to be obtained by two methods. The one method is quenching form the high temperature ε-phase (hcp structure [2]) followed by heating at around 770 K. The other method is cooling from the high temperature ε-phase at a controlled cooling rate. The obtained hard-magnetic alloys have τ-phase (CuAu structure). Typical properties are saturation magnetization (σ_s) of 88-96 emu/g [1,3] and coercieve force of 1.5 kOe [4]. The coercieve force is increased by additional cold working: coercive force of 4.6 kOe is obtained [3]. Furthermore, carbon addition is effective to cold working [5].

On the other hand, mechanically alloying (MA) has been applied to this alloy system [6-9]. As-milled alloys does not show τ-phase: it is not clear whether the phase of the milled alloys is α-Mn or β-Mn structure. Furthermore, the as-milled alloys have low saturation magnetization and low coercivity. By heating, high coercivity is obtained.

In this study, we prepare Mn-Al-C powders by mechanical alloying. The alloys were followed by two-step heating to obtain high saturation magnetization as well as high coercivity. Main parameters for comparing magnetic properties of the two-step-heated alloys are C addition (0-6 at %), milling time (10 h, 200 h) and second-step heating temperature (500-700°C).

2. Experimental

MA was carried out in a planetary high-energy ball mill (Fritsch Pulverisette P-5) starting from pure element powders of Mn (75 μm), Al (53-106 μm) and C (75 μm) with 99.9 % purity. All the alloys studied were prepared from mixtures of Mn, Al and C. The nominal compositions are two series: (1) 60, 55, 50 at %Mn-bal. Al and (2) *x* at %C- (55-0.5*x*) at %Mn-bal. Al (*x*=0, 2, 4, 6). The powder samples were loaded into a cylindrical stainless-steel vial together with balls of the same material in a nitrogen atmosphere [10]. The ball-to-powder weight ratio was 12:1. The milling process was performed at a speed of 150 r.p.m. (revolutions per minute) up to 500 h. The mechanically-alloyed powders were subjected to a two-step heating. The homogenization is done at 1050°C for 3 hours, and the heating is done at 500-700 °C for 10 to 60 min.

The characterization was carried out by X-ray diffraction (XRD) using CuKα radiation, by thermal analysis with differential thermal analyzer (DTA) under an argon atmosphere at a heating rate of 0.33 K/s, magnetic measurements obtained from a vibrating-sample magnetometer (VSM). Also, the mean particle sizes were measured by laser diffraction particle size analyzer.

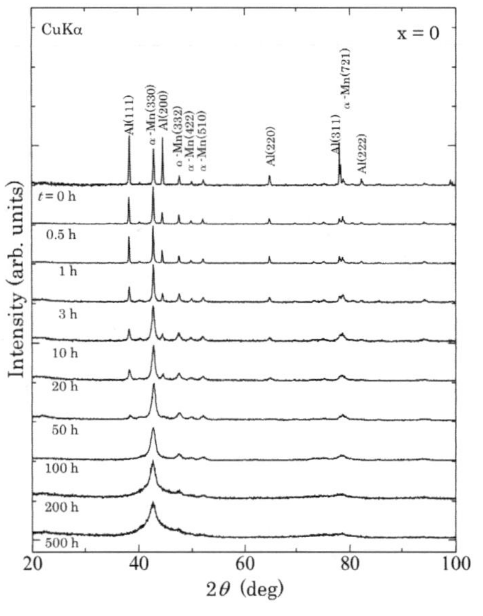

Figure 1. Change in X-ray diffraction for 55 at %Mn-bal. Al alloys with alloying time (*t*).

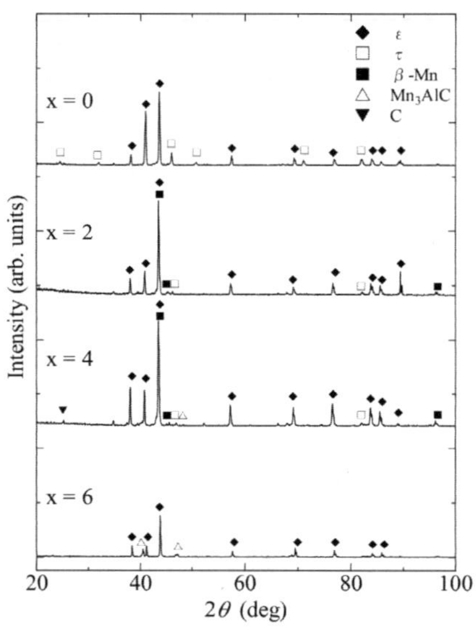

Figure 2. X-ray diffraction for *x* at.%C-(55-0.5*x*) at %Mn-bal. Al alloys being milled for 10 h and homogenized at 1150 °C.

3. Results and discussion

3.1 As-milled alloy

Figure 1 shows change in X-ray diffractions for Mn-Al alloys with alloying time (*t*). At *t*=0 h, strong Al (111) peak is seen and is decreases with milling time. In equilibrium Mn-rich Mn-Al system, pure Mn is α-Mn structure and transform to β-Mn with Al addition. Both α-Mn and β-Mn are cubic and the XRD spectra are similar. Almost all peaks appear at the nearly-same angles except for α-Mn(422) and β-Mn(310). Since CuKα target is used, a peak of β-Mn(310) with an intensity of 62% is expected at 2θ=45.4 deg. However, no peak at the angle is observed: we conclude that the milled powders have supersaturated α-Mn structure. This analysis is different from the previous reports of β-Mn [6, 9].

Since the XRD data is not shown [6,9], the changing process from α-Mn to β-Mn in the structure is not clear. In the equilibrium state, the solubility of Al into α-Mn in only 2 at % at room temperature [1]. Replacement of more than 2 at% results in transition to β-Mn and the β-Mn keeps up to 40 % Al. At t =500 h, all the fcc-Al peaks disappear and only α-Mn(330) and α-Mn(721) peaks remain as shown in Fig. 1. The lattice constant of α-Mn solution decreases with MA time, i.e. Al inclusion. This is partly due to included stress in the alloys. We confirmed a decrease in lattice constant of pure α-Mn by mechanical grinding. The crystallite size is calculated using Sherrer formula neglecting included stress. The crystallite size obtained from the α-Mn(330) decreases from 800 Å to 150-200 Å at t =500 h. The apparent mean-particle sizes observed by laser diffraction particle size analyzer increase once from 40 μm of pure Al to 150 μm which depends on the composition, and then decrease to 10 μm at 100-h milling. The size keeps constant after 100 h. Effect of C addition on the crystallographic change is not observed in XRD spectra.

The supersaturated-solution α-Mn is ferromagnetic with a low saturation magnetization (σ_s) of 0.8 emu/g and the maximum coercivity (H_c) of 0.3 kOe. This is due to imbalance in the number of anti-ferromagnetic Mn atom.

3.2 Homogenized alloys

Figure 2 shows change in XRD spectrum with C content (x=0-6) for the alloys with x at %C-(55-0.5x) at %Mn-bal. Al after 1150°C heating. The milling time is only 10 h; therefore the precursor is a mixture of main phase α-Mn with a small amount of fcc-Al (see Fig. 1).

By the homogenizing heating, both saturation magnetization and coercivity increase. The x=0 alloy includes small amount of τ-phase and have higher saturation magnetization of 25 emu/g. Meanwhile, C-containing alloys (x=2-6) has almost single ε-phase. The x=6 alloy is almost ε-phase except for a small amount of Mn_3AlC intermetallic compound. Mn_3AlC is paramagnetic at room temperature having ferromagnetism at low temperature and Curie temperature of 15 °C [11]. Therefore, we find that the ε-phase has the magnetization at 16 kOe (σ_{16}) of 5.5 emu/g. It should be noted that ε-phase has ferromagnetism. This is contrasted to the previous report [1]. A typical hysteresis loop of the ε-phase alloy (x=6) is shown in Fig. 3. From the XRD, the alloy has a single phase (see Fig. 2). This shows the C-addition stabilizes as reported by Zeng *et al.* [12]. The change in H_c with C-addition is slightly different from σ_s: the largest H_c is for x=0 alloy and decreases with C-addition.

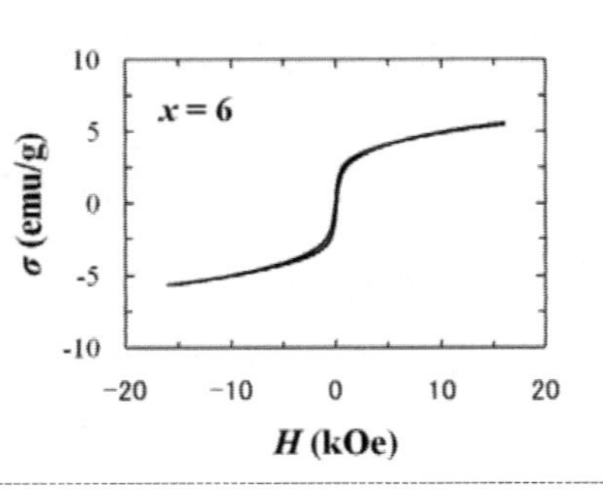

Figure 3 Hysteresis of 6%C-52%Mn-bal.Al alloy (x=6) having the ε-phase.

3.3 Second-step heated alloys

After the second-step heating, main phase of the alloys is changed to τ-phase. XRD spectra for x at %C-(55-0.5x) at %Mn-bal. Al (x=0, 6) is shown in Fig. 4. However, small amount of γ_2, ε, and β-

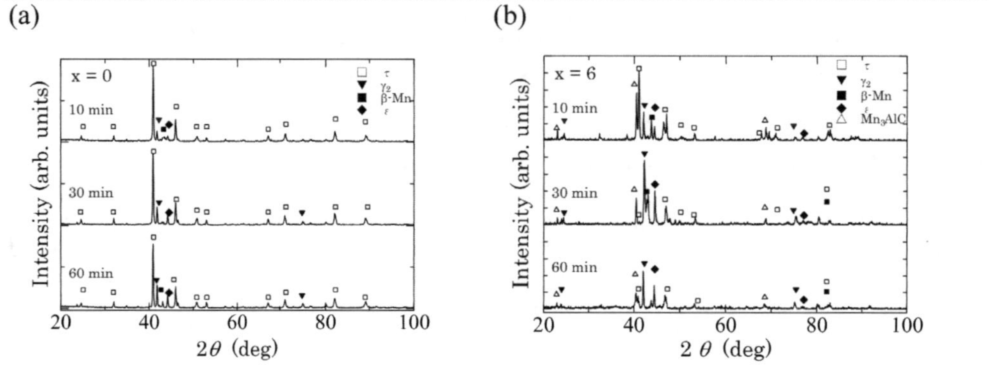

Figure 4. Change in X-ray diffraction for x at.%C-(55-0.5x) at %Mn-bal. Al alloys with increasing second-step heating time (10, 30, 60 min) at 500°C. (a): x=0, (b) x=6.

Mn are observed. The milling time is only 10 h and homogenising is 1150°C. By carbon addition, intermetallic Mn_3AlC (perovskite structure) is observed.

The largest σ_s is obtained for C-free alloys. However, H_c is as low as 1 kOe. The hysteresis loop is shown in Fig. 5(a). Since the magnetization is not saturated at 16 kOe, saturation magnetization (σ_s) is obtained by extrapolation to $1/H$=0 using H^{-1} law as shown in Fig. 5 (b). In an exact calculation, we must consider an H^{-2} term [13], however, we used only a main H^{-1} term. A σ_s value of 66 emu/g is obtained. High H_c of 2.6 kOe is obtained for x=4 alloy, however, σ_{16} is decreased to 40 emu/g and σ_s is 52 emu/g.

These magnetic and crystallographic properties for the short-time milled alloys (t=10 h) are contrasted to long-time milled powder (t=200 h). After homogenization, the main phase of 200 h milled powder is ε-phase. However, after second–step heating, the main phase of powder is β-Mn with small amount of τ-phase, which has as small as 5.2 emu/g. The reason is thought as follows. After long time milling, a large amount of stress is stored in the supersaturated solution α-Mn. Although the crystal structure is changed to ε-phase after the homogenizing heat treatment, the following transition to τ-phase does not sufficiently occur by the second heating due to residual stress.

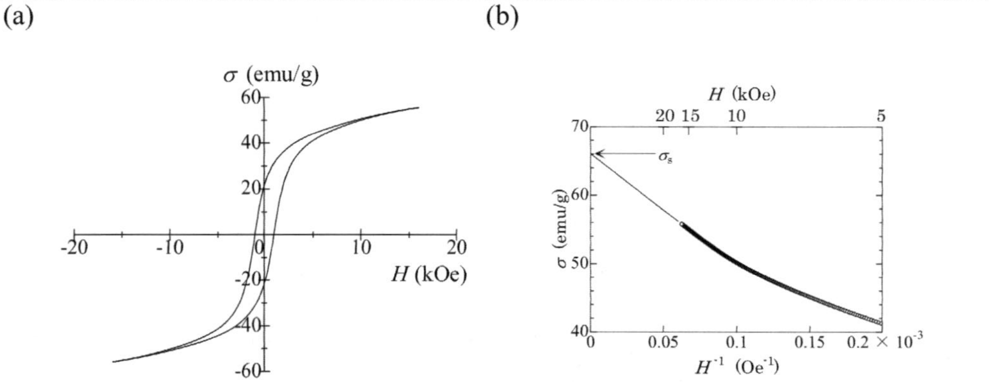

Figure 5. (a) Hysteresis of 55%Mn-bal. Al alloy (x=0) after homogenizing at 1150°C followed by heating at 500°C for 30 min. (b) High-field magnetization curve of 55%Mn-bal. Al alloy as a function of reciprocal field ($1/H$).

4. Conclusions

Supersaturated-solution α-Mn (in the vicinity of 50%Mn-50%Al composition) is obtained after 100 h milling for Mn-Al and Mn-Al-C alloys. The lattice constant of supersaturated-solution α-Mn decreases by Al inclusion. The almost single ε-phase has ferromagnetism having saturation magnetization of 6.5 emu/g. The largest saturation magnetization (σ_s) of 66 emu/g by the homogenization at 1150°C followed by heating at 500°C in this experiment. High H_c of 2.6 kOe is obtained for C-including alloys which have σ_s of 52 emu/g. To obtain τ-phase alloys and consequently higher magnetization, short-time milling (10 h) is better than long-time milling (200 h).

References

[1] Kono H 1958 *J. Phys. Soc. Jpn.* **13** 1444
[2] Massalski T B 1990 *Binary alloy phase diagrams* Vol. 1, 2nd ed. (ASM International) p.172
[3] Koch A J J, Hokkeling P, v. d. Steeg M G and de. Vos K J 1960 *J. Appl. Phys.* **31** 75S
[4] Sugihara M and Tsuboya I, 1962 *J. Appl. Phys.* **33** 1338
[5] Otani T, Kato N, Kojima S, Kojima K, Sakamoto Y, Konno I, Tsukahara M and Kubo T 1977 *IEEE Trans Magn.* **13** 1328
[6] Crew D C, McCormick P G and Street R 1995 *Scr. Metal. Mater.* **32** 315
[7] Kim K J, Sumiyama K and Suzuki K 1995 *J. Alloys Comp.* **217** 48
[8] Saito T 2003 *J. Appl. Phys.* **93** 8686.
[9] Ibrahim S, Shamah A M, Abbas Y M, Hanna F F, Marei L K and Hannora A 2005 *Egypt .J. Solids* **28** 315
[10] Kohmoto O, Yamaguchi N and Mori T 1994 *J. Material Sci.* **29** 3221
[11] Butters, R G and Myers H P 1955 *Phil. Mag.* **46** 895
[12] Zeng Q, Baker I, Cui J B, and Yan Z C 2007 *J. Magn. Magn. Mater.* **308** 214
[13] Kohmoto O 1982 *J. Appl. Phys.* **53** 7486

Journal of Physics: Conference Series **266** (2011) 012017 doi:10.1088/1742-6596/266/1/012017

Investigation of magnetic field effect on $L1_0$ FePt films during two-step tempering process

Y B Li[1], H Lv[1], J M Bai[1,*], F L Wei[1], S Yoshimura[2] and H Saito[2]
[1] Key Laboratory for Magnetism and Magnetic Materials of the Ministry of Education, Lanzhou University, Tianshui Road, No. 222, Lanzhou, 730000, Gansu, P. R. China
[2] Center for Geo-environmental Science, Faculty of Engineering & Resource Science, Akita University, Tegatagakuen-machi 1-1, Akita, 010-8502, Japan

E-mail: jmbai@lzu.edu.cn

Abstract. A two-step tempering process was performed on $L1_0$ FePt films in a magnetic field. FePt (20 nm) films were firstly annealed at 700 °C for 30 minutes and then tempered at lower temperature in a magnetic field of ~ 3500 Oe along the normal direction of films. The influence of magnetic field on (001) texture of FePt films during the two-step tempering process were investigated. It was shown that (001) orientation was improved and, at the same time, distinct perpendicular anisotropy was obtained in FePt films by two-step tempering in a perpendicular magnetic field when the second temperature was around the Curie temperature of the $L1_0$ FePt films.

1. Introduction

$L1_0$ FePt films have been studied extensively as one of promising candidates for ultrahigh density magnetic recording media because of its large magnetocrystalline anisotropy energy ($K_u \sim 7 \times 10^7$ ergs/cm^3) [1]. As well known, as perpendicular magnetic recording media, the magnetic easy axis (c axis) of FePt films should be normal to the films. However, the FePt films deposited directly on amorphous substrates tend to possess (111) texture because (111) plane is the close-packed plane [2,3]. Moreover, the c axis of (111) textured $L1_0$ FePt films deviates ~ 36° away from the films [2,3]. Therefore, one of the key challenges is how to align c axis to normal direction of FePt films. To date, much efforts have been devoted to and several methods have been used to improve the (001) orientation of FePt films: hetero-expitaxial growth on single crystal substrates [4-8] or on textured underlayers/seedlayers [9-20]; alternate deposition of Fe and Pt layers [21-26]; FePt – X (X = B_2O_3, SiO_2, MgO, C, Ag, etc.) nanocomposite fabrication [27-34]; aligning the magnetic particles in a magnetic field [35-38].

As well known, for ferromagnetic materails, the magnetic moments align spontaneously to the same direction inside magnetic domains when the temperature is lower than the T_c. In contrast, when the temperature is above the T_c, the magnetic moments disperse randomly. If FePt films were annealed around its T_c in a perpendicular magnetic field, the preferential direction of c axis should be defined by the external magnetic field energy during the rearrangement process of Fe and Pt atoms. Here, one new method was proposed to improve the (001) orientation of FePt films by two-step tempering. FePt (20 nm) films were annealed firstly at 700 °C for 30 minutes (min) to promote the formation of perfect

* To whom any correspondence should be addressed.

$L1_0$ FePt phase. Then, FePt (20 nm) films were tempered at lower temperature which is around its Curie temperature in a magnetic field of ~ 3500 Oe applied along the normal direction of FePt films. The influence of magnetic field on (001) texture of FePt films during the two-step tempering process were investigated.

2. Experiments

FePt films with structure of Si (substrate)/Fe (2 nm)/FePt (20 nm)/Pt (2 nm) were prepared in a CMS-18 magnetron sputtering system. The background pressure in chamber is 2 × 10^{-6} Pa. The nominal composition of the FePt layer is $Fe_{50}Pt_{50}$. The substrate temperature was raised up to 550 °C and then kept at this temperature for 30 min after the deposition. The as-prepared FePt films were first annealed at 700 °C (T_{A1}) for 30 min and then annealed at different temparature (T_{A2} = 430 – 600 °C) below 700 °C for 60 min either in a magnetic field of ~ 3500 Oe along the normal direction of the films (Route A) or without magnetic field (Route B). The deposition rate of the films was monitored by quartz crystal oscillator. The crystal structure of FePt films was characterized by X-ray diffraction (XRD) with Cu K_α radiation. The magnetic properties were measured at room temperature by superconducting quantum interference device (SQUID) with applied field up to 7 T.

3. Results and discussion

Figure 1 show the XRD patterns of FePt (20 nm) films annealed through Route A and Route B. Only (111) peak was observed in the as-prepared FePt films. From Figure 1(a), we can see that, as T_{A2} increases from 430 °C to 600 °C, the integrated intensity of peaks (001) and (002) of FePt (20 nm) films annealed through Route A increase firstly and then decrease on over trend. When T_{A2} is in a narrow range of 470 - 482 °C, the integrated intensity of peak (001) is much larger than that of peak (111). Moreover, the intensity ratio $I_{(001)}/I_{(111)}$ shows the maximum value of ~ 7 when T_{A2} is 478 °C, which is around the T_c of $L1_0$ FePt films.[1] When the FePt (20 nm) films were annealed through Route B, as shown in Figure 1(b), no significant influence of T_{A2} on the integrated intensity of peaks (001) and (111) can be observed. By comparison, we can see that external magnetic field did improve the (001) orientation of FePt films when they were annealed at the temperature which is around the T_c of $L1_0$ FePt films.

Figure 2 gives the typical hysteresis loops of FePt (20 nm) films: (a) as prepared, (b) annealed through Route B, and (c) annealed through Route A. Comparing with the as prepared state, the in-plane and out-of-plane coercivities of FePt films were enhanced drastically after the thermal treatment, which can be attributed to the formation of perfect $L1_0$ FePt phase. It should be noted that the in-plane remanence and the squareness ratio of as-prepared FePt films were both larger than the out-of-plane ones, while the situation reversed for annealed FePt films. Furthermore, as shown in figure 2(c), the out-of-plane squareness ratio increases significantly when FePt films were annealed at 478 °C (T_{A2}) in a magnetic field applied along the normal direction of FePt films. This indicates that the (001) orientation of FePt films were much improved and distinct perpendicular anisotropy were obtained, which is consistent with the XRD patterns shown in figure 1. Meanwhile, the reason for those kinks around zero field in some hysteresis loops is not clear yet.

4. Conclusion

In summary, FePt (20 nm) films were thermal treated by two-step tempering in a perpendicular magnetic field. The influence of magnetic field on (001) orientation of FePt films during two-step tempering process were investigated. The results indicated that the (001) orientation was improved and distinct perpendicular anisotropy was obtained in FePt films during two-step tempering process. This may be a new method to improve the (001) orientation and the perpendicular anisotropy of FePt films.

Acknowledgements

This work was supported by the National Natural Science Foundation of China (Grant No. 6080 3035).

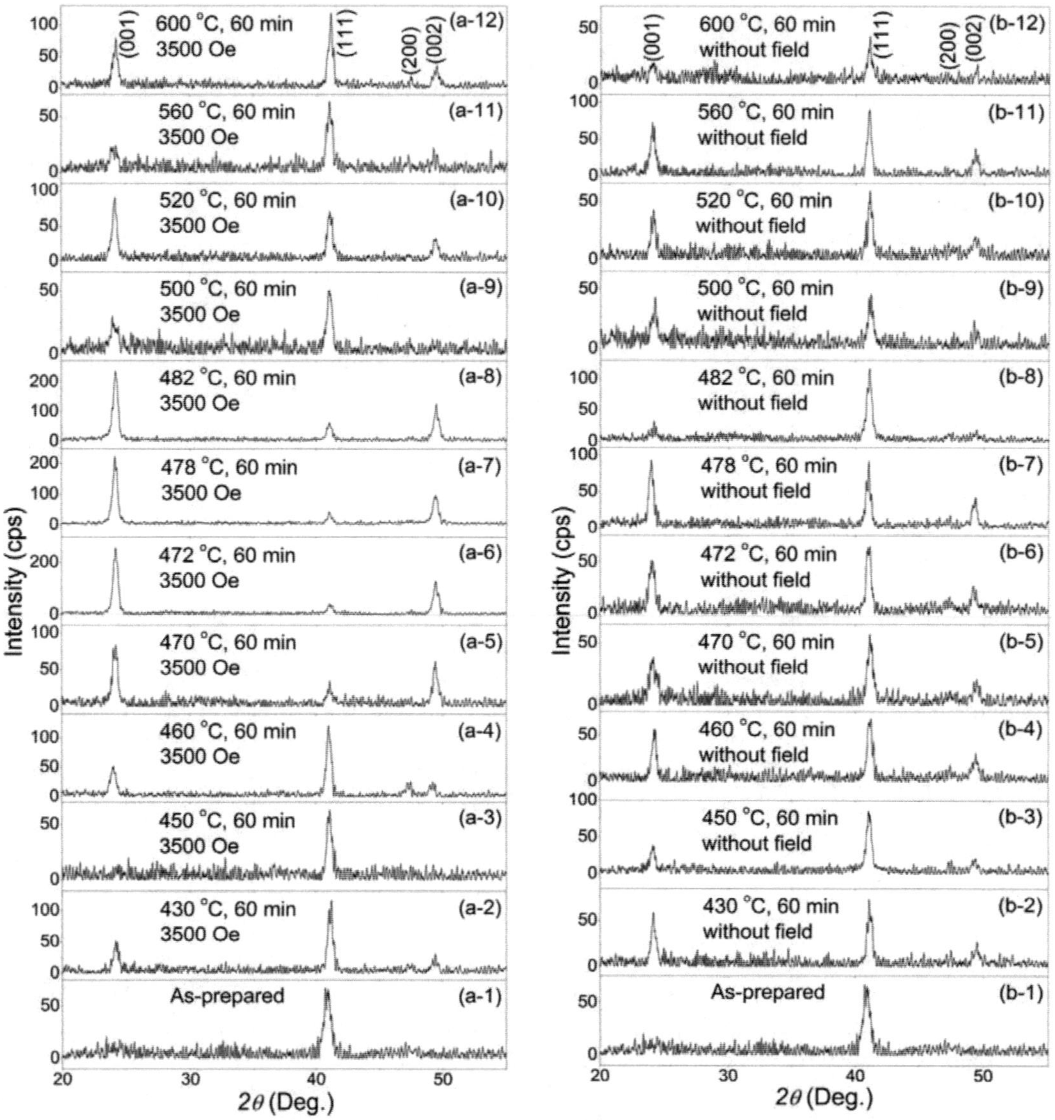

Figure 1. XRD patterns of FePt (20 nm) films annealed through (a) Route A and (b) Route B.

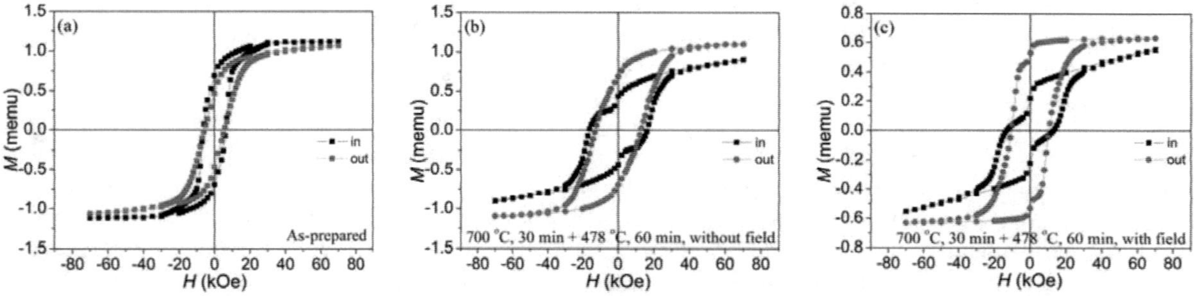

Figure 2. Typical hysteresis loops of FePt (20 nm) films: (a) as prepared, (b) annealed through Route B, and (c) annealed through Route A.

References

[1] Weller D, Moser A, Folks L, Best M E, Lee W, Toney M F, Schwickert M, Thiele J-U and Doerner M F 2000 *IEEE Trans. Magn.* **36** 10
[2] Hsu Y-N, Jeong S, Laughlin D E and Lambeth D N, 2001 *J. Appl. Phys.* **89** 7068
[3] Chen S C, Kuo P C, Chou C Y and Sun A C 2005 *J. Appl. Phys.* **97** 10N107
[4] Shima T, Takanashi K, Takahashi Y K and Hono K 2002 *Appl. Phys. Lett.* **81** 1050
[5] Barmak K, Kim J, Lewis L H, Coffey K R, Toney M F, Kellock A J and Thiele J-U 2005 *J. Appl. Phys.* **98** 033904
[6] Laenens B, Almeida F M, Planckaert N, Temst K, Meersschaut J, Vantomme A, Rentenberger C, Rennhofer M and Sepiol B 2009 *J. Appl. Phys.* **105** 073913
[7] Ding Y F, Chen J S and Liu E 2005 *J. Cryst. Growth* **276** 111
[8] Svedberg E B, Mallett J J, Sayan S, Shapiro A J, Egelhoff, Jr. W F and Moffat T, 2004 *Appl. Phys. Lett.* **85** 1353
[9] Farrow R F C, Weller D, Marks R F, Toney M F, Cebollada A and Harp G R 1996 *J. Appl. Phys.* **79** 5967
[10] Thiele J-U, Folks L, Toney M F and Weller D K 1998 *J. Appl. Phys.* **84** 5686
[11] Ding Y F, Chen J S, Liu E, Sun C J and Chow G M 2005 *J. Appl. Phys.* **97** 10H303
[12] Seki T, Shima T, Takanashi K, Takahashi Y, Matsubara E, Takahashi Y K and Hono K 2004 *J. Appl. Phys.* **96** 1127
[13] Xu Y F, Chen J S and Wang J P 2002 *Appl. Phys. Lett.* **80** 3325
[14] Chen J S, Lim B C and Wang J P 2002 *Appl. Phys. Lett.* **81** 1848
[15] Zhao Z L, Wang J P, Chen J S and Ding J, 2002 *Appl. Phys. Lett.* **81** 3612
[16] Chen J S, Xu Y F and Wang J P 2003 *J. Appl. Phys.* **93** 1661
[17] Chen J S, Lim B C, Hu J F, Lim Y K, Liu B and Chow G M 2007 *Appl. Phys. Lett.* **90** 042508
[18] Sun A-C, Kuo P C, Hsu J-H, Huang H L and Sun J-M 2005 *J. Appl. Phys.* **98** 076109
[19] Trichy G R, Narayan J and Zhou H 2006 *Appl. Phys. Lett.* **89** 132502
[20] Trichy G R, Chakraborti D, Narayan J and Prater J T 2008 *Appl. Phys. Lett.* **92** 102504
[21] Shima T, Moriguchi T, Mitani S and Takanashi K 2002 *Appl. Phys. Lett.* **80** 288
[22] Wu Y-C, Wang L-W and Lai C-H 2007 *Appl. Phys. Lett.* **91** 072502
[23] Wu Y-C, Wang L-W, Chiang C C, Yang C-H and Lai C-H 2009 *JOM* **61** 84
[24] Zeng H, Yan M L, Powers N and Sellmyer D J 2002 *Appl. Phys. Lett.* **80** 2350
[25] Yan M L, Powers N and Sellmyer D J 2003 *J. Appl. Phys.* **93** 8292
[26] Nakagawa S and Kamiki T 2005 *J. Magn. Magn. Mater.* **287** 204
[27] Luo C P, Liou S H, Gao L, Liu Y and Sellmyer D J 2000 *Appl. Phys. Lett.* **77** 2225
[28] Yan M L, Zeng H, Powers N, Sellmyer D J 2002 *J. Appl. Phys.* **91** 8471
[29] Ko H S, Perumal A and Shin S-C 2003 *Appl. Phys. Lett.* **82** 2311
[30] Luo C P and Sellmyer D J 1999 *Appl. Phys. Lett.* **75** 3162
[31] Shao Y, Yan M L, Sellmyer D J 2003 *J. Appl. Phys.* **93** 8152
[32] Sellmyer D J, Yan M L, Xu Y F and Skomski R 2005 *IEEE Trans. Magn.* **41** 560
[33] Yang E and Laughlin D E 2008 *J. Appl. Phys.* **104** 023904
[34] Wu Y-C, Wang L-W and Lai C-H 2008 *Appl. Phys. Lett.* **93** 242501
[35] Wang H Y, Ma X K, He Y J, Mitani S and Motokawa M 2004 *Appl. Phys. Lett.* **85** 2304
[36] Qiu J-M, Bai J M and Wang J-P 2006 *Appl. Phys. Lett.* **89** 222506
[37] Tamada Y, Yamamoto S, Nasu S and Ono T 2008 *Phys. Rev.* B **78** 214428
[38] Tamada Y, Yamamoto S, Nasu S and Ono T 2009 *Appl. Phys. Express* **2** 123001

Journal of Physics: Conference Series **266** (2011) 012018 doi:10.1088/1742-6596/266/1/012018

Transition from weak ferromagnetism to strong paramagnetism in $Zn_{1-x}Cr_xO$ ($0 \leq x \leq 0.026$) thin films

S S Li and Y M Hu

Department of Applied Physics, National University of Kaohsiung, Kaohsiung 81148, Taiwan, R.O.C.

E-mail: ymhu@nuk.edu.tw

Abstract. This paper aims to investigate the effect of Cr doping on the magnetic properties of $Zn_{1-x}Cr_xO$ ($0 \leq x \leq 0.026$) thin films deposited using a magnetron co-sputtering method under $Ar+O_2$ growth atmosphere. All of the films are highly *c*-axis-oriented wurtzite structures. The x-ray photoelectron spectroscopy results revealed that some Cr ions have substituted into Zn sites, showing the Cr^{2+} signal, in all of the Cr-doped ZnO films. The remaining Cr ions are in a 3+ valence state. A distinct paramagnetic response, as well as weak ferromagnetic signal persisting up to RT, has been observed for the pure ZnO film, while only a distinct paramagnetic response is present for the $Zn_{1-x}Cr_xO$ films with x = 0.026. The variation in ferromagnetic signals seems to coincide with that in the surface morphology features for the films. The results suggest that ferromagnetism in the $Zn_{1-x}Cr_xO$ films can be rationalized in terms of surface defect states, and the substituted Cr^{2+} ions would contribute only to paramagnetism or show an antiferromagnetic coupling.

1. Introduction

Zinc oxide (ZnO) doped with various transition metal (TM) elements has been intensively investigated as a promising dilute magnetic semiconductor (DMS), wherein a spontaneous magnetization would persist to above room temperature (RT), for potential spintronic device applications [1]. While ferromagnetism (FM) has been frequently reported for many of ZnO-based DMSs, no consensus has been reached about the coupling mechanism or even the origin of magnetic moments. Coey *et al.* [2] proposed that the TM ions within the hydrogenic orbital of a localized donor electron would be aligned, forming a bound magnetic polaron (BMP): the spin-split impurity band that forms due to the overlap of BMPs enables a ferromagnetic coupling *via* the *s-d* exchange interaction. While this Heisenberg-type model could be applicable to insulating TM-doped ZnO, there is an inconsistency concerning the type of defect that enables the ferromagnetic order [3-5]. There are also indications that the TM ion itself may not be the source of the observed magnetic moment. Some researchers argued that magnetic moments could stem solely from the intrinsic defects, such as oxygen vacancies (V_O) [6] and zinc vacancies (V_{Zn}) [7], residing on surfaces or in grain boundaries of pure ZnO films and nanoparticles. Much evidence has shown that defect-induced ferromagnetism is a universal feature of inorganic nanoparticles [8]. To date, the actual effect of TM dopant on the magnetic properties of TM-doped ZnO is unconfirmed.

Recently, considerable effort has been devoted to Cr-doped ZnO films due to theoretical predictions [9] and experimental evidence [10] of RT-FM. In order to clarify the role of the Cr dopant and to elucidate the possible origin of ferromagnetism, we have prepared $Zn_{1-x}Cr_xO$ ($0 \leq x \leq 0.026$) thin films by using a magnetron co-sputtering method, and examined the structural, optical and magnetic properties by utilizing comprehensive and rigorous characterization techniques.

2. Experimental details

$Zn_{1-x}Cr_xO$ ($0 \leq x \leq 0.026$) thin films with thickness of about 350 nm were deposited on *c*-plane sapphire substrates at 400 °C by sputtering ZnO ceramic (99.99%) and/or Cr metallic (99.99%) targets in an ultra-high vacuum system at a pressure of ~ 20 mTorr. Ultra-pure (99.999%) Ar and O_2 gases were introduced at flow rates of 50 and 20 sccm, respectively, via two independent mass flow controllers. The crystal structure was examined by an x-ray diffractometer (XRD, PANalytical X'Pert PRO) with Cu K_α radiation (λ = 1.5406 Å). The surface morphology was detected using a thermal field-emission scanning electron microscope (TF-SEM, JEOL JSM 6500-F) with an electron beam energy of 15 keV. Room temperature photoluminescence (RT-PL) spectra were recorded, exciting with the 325 nm-line of a He-Cd laser. Chemical bonding states and the compositions of films were analyzed using an x-ray photoelectron spectroscope (XPS, VG Scientific Microlab 310F) with Mg K_α radiation ($h\nu$ = 1253.6 eV); the binding energy was determined with reference to the C *1s* component set. Electrical properties were investigated using the four-point probe method. Magnetic properties were probed by a commercial MPMS / SQUID / VSM dc magnetometer (Quantum Design).

3. Results and Discussion

Figure 1 shows the XRD θ-2θ scans of $Zn_{1-x}Cr_xO$ thin films on a logarithmic scale. All the films were highly *c*-axis-oriented wurtzite structures, evidenced by the prominent ZnO (002) peak. No chromium metal or oxide phases were observed within the detection limit of XRD. The inset of figure 1 depicts the variations in the *c*-axis lattice constant (L_c) with different *x*. The values of L_c were calculated directly from the position of the ZnO (002) peak, using the Bragg law. All L_c values of the films were found to be larger than that of bulk ZnO. This demonstrates that all the films suffer an in-plane compressive stress from the underlying sapphire substrate due to lattice mismatch. In addition, the L_c values of the Cr-doped ZnO films are all smaller than that of the pure ZnO film. The contraction of L_c, corresponding to a relaxation of strain along the *c* axis, in $Zn_{1-x}Cr_xO$ films with low *x* indicates Cr substitution into Zn sites due to the smaller ionic radii of Cr^{2+} compared to Zn^{2+}. A small expansion of L_c for $x = 0.026$ could result from some redundant Cr ions entering into interstitial sites [11].

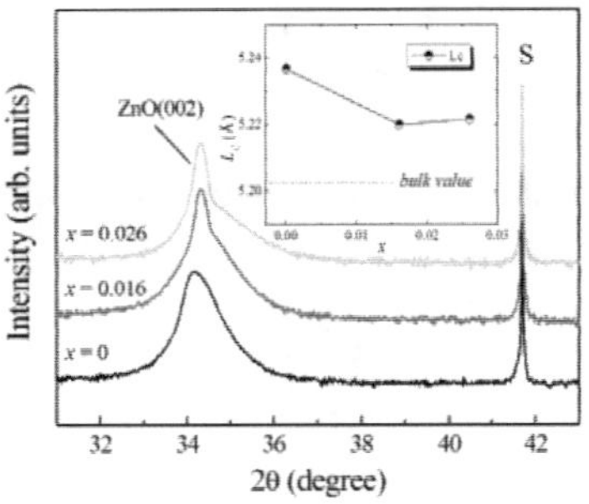

Figure 1. XRD θ-2θ scans of $Zn_{1-x}Cr_xO$ thin films on a logarithmic scale. The "S" designates the peak from the sapphire substrate. The inset illustrates the variation in the calculated *c*-axis lattice constant with differing *x*.

The TF-SEM images of figure 2 illustrate the modifications of the surface morphology due to Cr doping. In figure 2(a), compact spherical grains are distributed over the surface of the ZnO film. Interestingly, the Cr doping reduces the grain size and blurs the grain boundaries, and brings about a sanded-like surface, as can be seen in figures 2(b) and 2(c). These results suggest that a minute amount of Cr dopants could limit the growth and coalescence of ZnO grains along the in-plane direction.

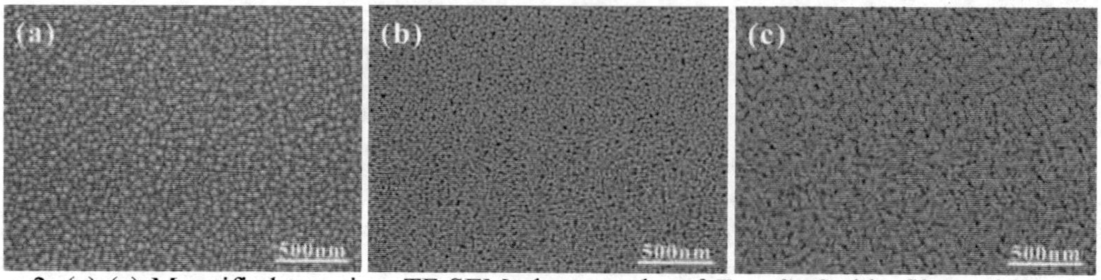

Figure 2. (a)-(c) Magnified top-view TF-SEM photographs of $Zn_{1-x}Cr_xO$ thin films with $x = 0$, 0.016 and 0.026, respectively.

Figure 3(a) displays the RT-PL spectra of the $Zn_{1-x}Cr_xO$ films. The pure ZnO film exhibits a sharp near-band-edge (NBE) emission at 3.25 eV with a FWHM of about 103 meV, suggesting good optical performance and high crystal quality. The absence of the V_O-related emission (~ 2.4 eV) is expected for the films deposited under O_2 atmosphere in this study. However, there is an emission shoulder centered around 3.1 eV. In order to find possible defect-related emission on this shoulder, we performed a multi-peak Gaussian fit for the pure ZnO film and the result is shown in figure 3(b). Clearly, this shoulder is comprised of two peaks D1 (3.11 eV) and D2 (3.02 eV). Theoretical calculations have predicted that the zinc interstitial (Zn_i) level is located 0.22 eV below the conduction band (CB) [12], and that the V_{Zn} level is located 0.3 eV above the valence band (VB) [13]. Therefore D1 peak was assigned to be a transition from the Zn_i level to the VB, and the D2 peak was assigned as a transition from the CB to the V_{Zn} level. A corresponding schematic band diagram for this statement is also shown in figure 3(b). The result suggests that there are few intrinsic Zn_i and V_{Zn} defects in the pure ZnO film: the latter will be demonstrated to be crucial to the observed FM. Although Zn_i has been suggested to be mobile and unstable in ZnO [14], the existence of few Zn_i in grain boundaries is possible to some extent. In contrast, V_{Zn} has been detected by positron annihilation spectroscopy to be one of the main defects in ZnO films [15]. Furthermore, the NBE emission intensity decreases significantly with increasing x for the $Zn_{1-x}Cr_xO$ films, as can be seen in figure 3(a). This rapid quenching phenomenon has been observed for TM-doped ZnO [16], and can be attributed to the formation of localized Cr impurity levels within the ZnO band-gap.

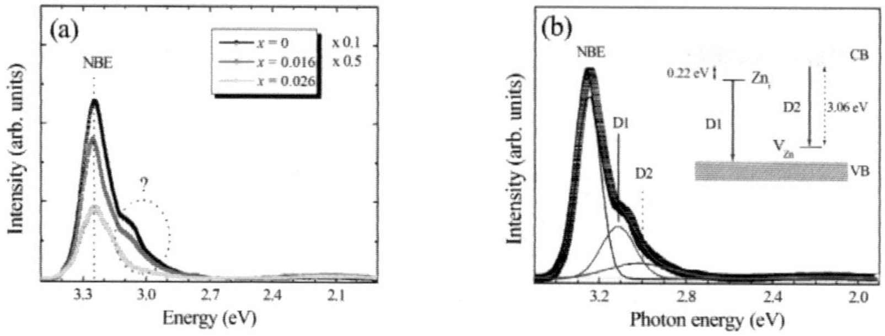

Figure 3. (a) PL spectra of the $Zn_{1-x}Cr_xO$ films at RT. (b) The multi-peak Gaussian fitting result for pure ZnO film. A schematic band diagram explains the D1 and D2 peaks.

Figures 4 (a)-(b) show the XPS spectra of Zn *2p* and Cr *2p* core levels in the $Zn_{1-x}Cr_xO$ films, respectively. For the pure ZnO film, the Zn $2p_{3/2}$ and $2p_{1/2}$ states are observed at 1021.4 eV and 1044.5 eV, respectively, with a spin-orbit splitting of 23.1 eV. The binding energy (BE) values are lower than those of bulk ZnO (1022 eV and 1045 eV), indicating that the pure ZnO film has a zinc deficiency.

Under an oxygen-rich environment, it has been suggested that the electron-compensator V_{Zn} may easily form [17], leading to poor carrier conduction in ZnO. In agreement with this perspective, we found that all of the films possess a high resistivity of about 10^6 Ω cm. It should also be noted that the BE values of the Zn $2p_{3/2}$ state for Cr-doped ZnO films are lower than that for the pure ZnO film, as indicated by the solid line in figure 4(a). The increased screening effect of the valence electrons at Zn sites corresponds to a reduction of Zn^{2+} into Zn^{1+} and/or Zn^0, and could be result of Cr substitution into Zn sites. The multi-peak-fitting result reveals that the asymmetric Cr $2p_{3/2}$ state is comprised of two peaks: Cr^{2+} (~576 eV) and Cr^{3+} (~577.4 eV) for the $Zn_{1-x}Cr_xO$ films, as shown in figure 4(b). Evidently, some Cr ions have substituted into Zn sites, showing the Cr^{2+} signal, in all of the Cr-doped ZnO films. The remaining Cr ions are in a 3+ valence state. From the earlier structural analyses, it is suggested that some Cr^{3+} ions are located at interstitial sites within the wurtzite lattice or at grain boundaries.

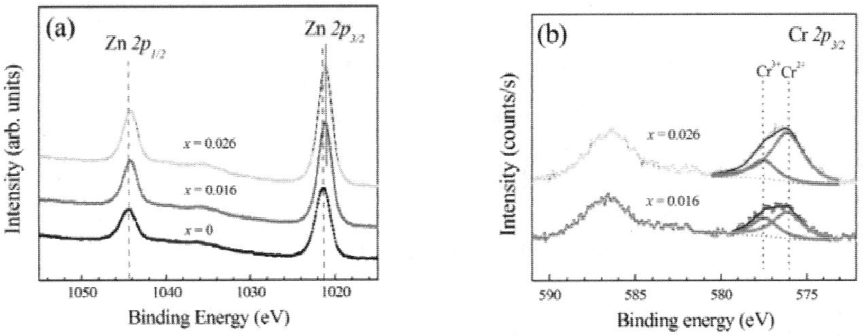

Figure 4. XPS spectra of (a) Zn $2p$ and (b) Cr $2p$ core level regions of the $Zn_{1-x}Cr_xO$ films.

Figure 5 (a) show M-H curves of the $Zn_{1-x}Cr_xO$ films at RT, where the diamagnetic contribution from sapphire substrate has been subtracted directly from the row data. Clearly, all the $Zn_{1-x}Cr_xO$ films show a dominant PM response. After subtracting the dominant PM response, we obtain weak FM signals, as shown in figure 5(b). The saturation magnetization (0.4 emu/cm^3) of the pure ZnO film is comparable with that (0.5 emu/cm^3) of ZnO nanoparticles [18]. As already mentioned, the magnetic moments in pure ZnO could stem from the V_O or V_{Zn}, residing on surfaces or in grain boundaries. Using first-principles calculation, Wang *et al.* [7] have showed that ZnO with V_O is nonmagnetic but with V_{Zn} is magnetic. They also indicated that V_{Zn} prefer to cluster and reside on surfaces, and may introduce spin polarization of the top of valence band. From an energetic point of view, the magnetic moments arising from the unpaired $2p$ electrons at O sites surrounding V_{Zn} prefer to couple ferromagnetically, if neighbored. As evidenced from the XPS results, the pure ZnO film is oxygen abundant, and the formation of V_{Zn} will be more favored. The PL results also suggest the existence of V_{Zn}, not V_O. Accordingly, the FM and PM contributions in the pure ZnO film may stem from the clustered and isolated V_{Zn}, respectively, residing on surfaces or in grain boundaries.

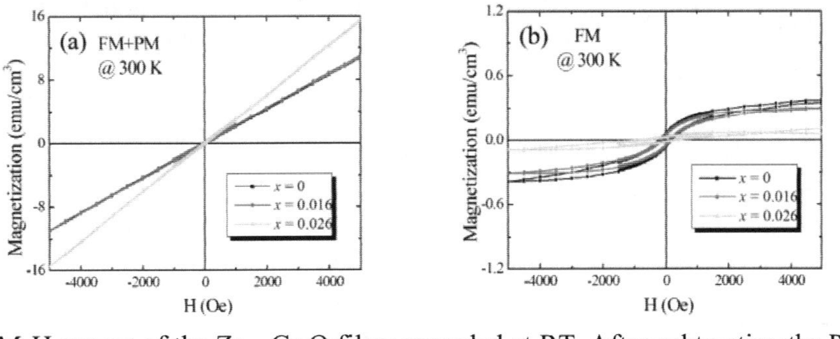

Figure 5. (a) M-H curves of the $Zn_{1-x}Cr_xO$ films recorded at RT. After subtracting the PM contribution, the remaining M-H curves of the films are displayed in (b).

It should be noticed that the FM signal of the $Zn_{1-x}Cr_xO$ film almost disappears at $x = 0.026$, as can be seen in figure 5(b). We also notice that the variation in these FM signals seem to coincide with that in the surface morphology features shown in figure 2. Noting that V_{Zn} prefer to cluster and reside on the surface region, rather than in the interior, of ZnO films, it is suggested that the Cr doping will modify the surface and grain boundaries of films, lessening the quantity of clustered V_{Zn}, and thereby lead to the much weak FM signal for $x = 0.026$. Further, the XPS results have shown that the Cr substitution into Zn sites happens in all Cr-doped ZnO films. Accordingly, we may conclude that the substituted Cr^{2+} ions contribute only to paramagnetism or exhibit an antiferromagnetic coupling.

4. Conclusions

Pure and Cr-doped ZnO films have been examined using comprehensive and rigorous characterization techniques. The PL spectra show that there are zinc interstitials and vacancies in the pure ZnO film, where a zinc deficiency is confirmed by the XPS results. A distinct PM response, as well as weak FM signal persisting up to RT, has been observed for the pure ZnO film, wherein the FM and PM responses could stem from the clustered and isolated V_{Zn}, respectively, residing on surfaces or in grain boundaries. The Cr doping is found to limit the growth and coalescence of ZnO grains along the lateral direction and finally result in a sanded-like surface. The quantity of clustered V_{Zn} would be decreased and consequently the FM signal weakened. The results suggest that ferromagnetism in the $Zn_{1-x}Cr_xO$ films can be rationalized in terms of surface defect states, and the substituted Cr^{2+} ions would contribute only to paramagnetism or show an antiferromagnetic coupling.

Acknowledgments

This study was financially supported by National Nano Device Laboratories (Grant No. NDL-97-C03SG-151) and National Science Council (Grant No. NSC-98-2112-M-390-002).

References

[1] Žutić I, Fabian J and Sarma S D 2004 *Rev. Mod. Phys.* **76** 323
[2] Coey J M D, Venkatesan M and Fitzgerald C B 2005 *Nat. Mater.* **4** 173
[3] Hong N H, Sakai J, Huong N T, Poirot N and Ruyter A 2005 *Phys. Rev. B* **72** 045336
[4] Karmakar D, Mandal S K, Kadam R M, Paulose P L, Rajarajan A K, Nath T K, Das A K, Dasgupta I and Das G P 2007 *Phys. Rev. B* **75** 144404
[5] Liu H, Zhang X, Li L, Wang Y X, Gao K H, Li Z Q, Zheng R K, Ringer S P, Zhang B and Zhang X X 2007 *Appl. Phys. Lett.* **91** 072511
[6] Sundaresan A, Bhargavi R, Rangarajan N, Siddesh U and Rao C N R 2006 *Phys. Rev. B* **74** 161306(R)
[7] Wang Q, Sun Q, Chen G, Kawazoe Y and Jena P 2008 *Phys. Rev. B* **77** 205411
[8] Sundaresan A and Rao C N R 2009 *Nanotoday* **4** 96
[9] Sato K and Yoshida H K 2000 *Jpn. J. Appl. Phys.* Part 2 **39** L555
[10] Bradley B K, Pakhomov A B, Voll P and Krishnan K M 2008 *Appl. Phys. Lett.* **92** 162511
[11] Hu Y M, Hsu C W, Wang C Y, Chiou J W, Han T C, Chen G J, Chou W Y and Chang J 2010 *Thin Solid Films* **518** 2916
[12] Bylander F G 1978 *J. Appl. Phys.* **49** 1188
[13] Lin B, Fu Z and Jia Y 2001 *Appl. Phys. Lett.* **79** 943
[14] Look D C, Farlow G C, Reunchan P, Limpijumnong S, Zhang S B and Nordlund K 2005 *Phys. Rev. Lett.* **95** 225502.
[15] Tuomisto F, Ranki V, Saarinen K and Look D C 2003 *Phys. Rev. Lett.* **91** 205502.
[16] Luo J T, Zhu X Y, Fan B, Zeng F and Pan F 2009 *J. Phys. D: Appl. Phys.* **42** 115109.
[17] Kohan A F, Ceder G, Morgan D and Van de Walle C G 2000 *Phys. Rev. B* **61** 15019.
[18] Straumal B B, Mazilkin A A, Protasova S G, Myatiev A A, Straumal P B, Schütz G, van Aken P A, Goering E and Baretzky B 2009 *Phys. Rev. B* **79** 205206.

Journal of Physics: Conference Series **266** (2011) 012019 doi:10.1088/1742-6596/266/1/012019

Transport and Magnetic Properties of Ferromagnetic Co–Al–O and Fe–Al–O Granular Films

S Nakamura[1,5], **T Iwashita**[1], **T Nojima**[1,5], **A Yoshihara**[2], **S Ohnuma**[3,6], **H Fujimori**[4]

[1] Institute for Materials Research, Tohoku University, Sendai 980-8577, Japan
[2] Department of Basic Sciences, Ishinomaki Senshu University, Ishinomaki 986-8580, Japan
[3] The Research Institute for Electric and Magnetic Materials, Sendai 982-0807, Japan
[4] Professor emeritus, Tohoku University, Sendai 980-8577, Japan
[5] Center for Low Temperature Science, Tohoku University, Sendai 980-8577, Japan
[6] Center for Interdisciplinary Research, Tohoku University, Sendai 980-8578, Japan

E-mail: `nakamura@imr.tohoku.ac.jp`

Abstract. We investigate the transport and magnetic properties of ferromagnetic Fe–Al–O and Co–Al–O granular films. A negative linear magnetoresistivity is found at least up to 27 T for Co–Al–O films. The temperature dependence of the magnetization of Fe–Al–O films can be explained in terms of the $T^{1.5}$-law for three dimensional spin waves theory, but that of Co–Al–O films does not obey the simple power law expected from this theory. The coercive field of a Co–Al–O film is strongly dependent on temperature.

1. Introduction

Granular Co–Al–O and Fe–Al–O films consist of an insulating Al–O matrix and dispersed ferromagnetic nanoparticles. They have attracted considerable interest as materials for electrical devices because of their high resistivity and soft magnetization. Their high resistivity suppresses the eddy currents and the soft magnetization suppresses hysteresis loss. Co–Al–O granular films are also interesting materials from a scientific view-point. It is well known that the superparamagnetic Co–Al–O films exhibit giant magnetoresistance due to spin-dependent tunneling between granules [1]. Their resistivity at zero field increases steeply with decreasing temperature, following a formula $\rho(T) \propto \exp(2C/\sqrt{k_B T})$, where C is related to the tunneling activation energy [2, 3]. On the other hand, the resistivities of ferromagnetic Co–Al–O and Fe–Al–O granular films exhibit a metallic temperature dependence due to percolation [4, 5]. For these films, ferromagnetic spin waves with long wave-lengths (~300 nm) are observed by Brillouin light scattering (BLS), and exchange fields H_E (11-64 mT), which stem from inter-granule exchange coupling are obtained [6]. Their resistivities have a T^2 behavior at low temperatures, which is interpreted in terms of scattering of conduction electrons by the spin waves [7]. The resistivities of ferromagnetic films increase logarithmically with a further reduction in temperature [4, 7]. These films exhibit linear negative magnetoresistivities but their changes are small [7, 8], therefore investigations under higher fields have been required. Similar logarithmic resistivities and negative magnetoresistivities have been found in amorphous alloys [9]. However, the origin of this logarithmic resistivity has not been determined [10, 11]. In the present study we investigate the characteristic linear magnetoresistivity of the ferromagnetic films for an extended magnetic field range. The exchange field H_E obtained in the BLS measurements is found to be proportional to the inverse-square of the electrical resistivity at room temperature [6]. This provide the view that the magnetic properties of the ferromagnetic granular films strongly correlate to the transport properties, but little attention has been given to the magnetic properties of these films at low temperatures. Therefore, we also investigate the magnetic properties of these films at low temperatures, which are expected to reflect the characteristics of thermally excited spin waves.

2. Experimental

Co–Al–O films were prepared on glass and sapphire substrates using radio-frequency magnetron sputtering in $Ar+O_2$ gas [1, 2]. In the present study, only a $Co_{66}Al_{13}O_{21}$ film was deposited on sapphire substrate; the other films were deposited on glass substrates. During film deposition, an external dc field of 15 mT was applied parallel to the substrates to induce uniaxial anisotropy. The films were 1–2 μm thick, which is much greater than granule diameters (3–5 nm). Although the logarithmic resistivity can be expected from two-dimensional Anderson localization, the samples in the present study are regarded as three-dimensional systems [12]. Electrical resistivity and magnetoresistivity measurements were performed by a conventional dc four-terminal method. For the transport measurements, we cut films into rectangles using a diamond cutter and attached lead wires to the samples using silver paste and indium. The temperature dependence of the resistivity was measured in a ^{4}He cryostat with superconducting magnet. To perform magnetoresistivity measurements at high field range we used a cryogen-free hybrid magnet (CHM) with a dilution refrigerator at the High Field Laboratory for Superconducting Materials (HFLSM), Institute for Materials Research, Tohoku University. This CHM consists of a large superconducting magnet cooled by a Gifford-McMahon refrigerator and a water-cooled resistive magnet inserted in the superconducting magnet [13]. To suppress eddy currents arising from vibration of the water-cooled magnet, the mixing chamber of the dilution refrigerator is made of plastic. Samples were set in the mixing chamber and were cooled using ^{4}He exchange gas for measurements at 4.2 K. For measurements below 1 K, the samples were directly cooled by the ^{3}He/^{4}He mixture. In the present study we performed the measurements up to the maximum field of the CHM 27 T. Samples for the magnetization measurements were cut using a diamond cutter into rectangles that were 2-3 mm wide and 3-6 mm long. The magnetization was measured using a SQUID magnetometer in reciprocating sample option mode.

3. Results and Discussion

Figure 1 shows the magnetoresistivity of $Co_{63}Al_{13}O_{24}$. In the present study, we focus on the magnetoresistivity at high fields. Because the residual magnetic field of the superconducting magnet of the CHM is greater than 20 mT and depends on history, we cannot accurately determine the applied field when the field is low. The inset of figure 1 shows the logarithmic dependence of the resistivity on temperature at zero field. The resistivity increases by 7 % when the temperature is reduced from 60 to 5 K. When a field is applied, the resistivity decreases linearly from 2 to 27 T at 4.2 K (figure 1 (a)). The change in the resistivity from 0 to 27 T is about 0.35 %. This linear negative magnetoresistivity is also observed at 35 mK (figure 1(b)). The change in the resistivity is almost the same as that at 4.2 K. To elucidate the sample dependence, we performed the same measurements for $Co_{66}Al_{11}O_{23}$. The inset of figure 2 shows the logarithmic temperature dependence of the resistivity of this sample at zero field. The resistivity increases by 1.4 % when the temperature is reduced from 40 to 5 K, while the resistivity decreases linearly by 0.2 % from 0 to 27 T at 4.2 K (figure 2 (a)). Similar behavior of the magnetoresistivity is also observed at 40 mK (figure 2 (b)). We found that the negative linear magnetoresistivity of Co–Al–O films can be observed at least up to 27 T and that the logarithmic behavior of the resistivity of these films is not strongly affected even by such high fields. However, the resistivity still decreases with increasing field in the present field range, suggesting that further reduction occurs when much higher fields are applied. The thermal fluctuations of the moments of the granules will increase the resistivity via spin-dependent tunneling and will be suppressed by fields. However, the present results at very low temperatures indicate that the negative magnetoresistivity is not due to the suppression of thermal fluctuations by fields. The change in the magnetoresistivity at high fields may point to an unsaturation of magnetization. Further investigation on the magnetization at high fields is required.

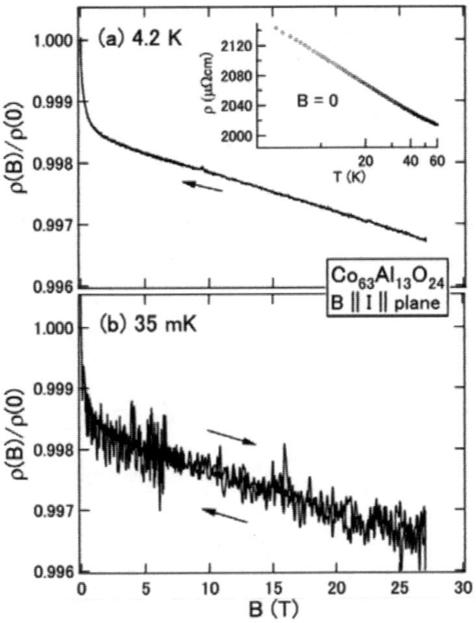

Figure 1. Magnetoresistivity of $Co_{63}Al_{13}O_{24}$ at (a) 4.2 K and (b) 35 mK. The data is normalized by the resistivity at zero field. The field is applied along the current direction and parallel to the film plane. The temperature dependence of the resistivity at zero field is shown in the inset using a log$T$ scale.

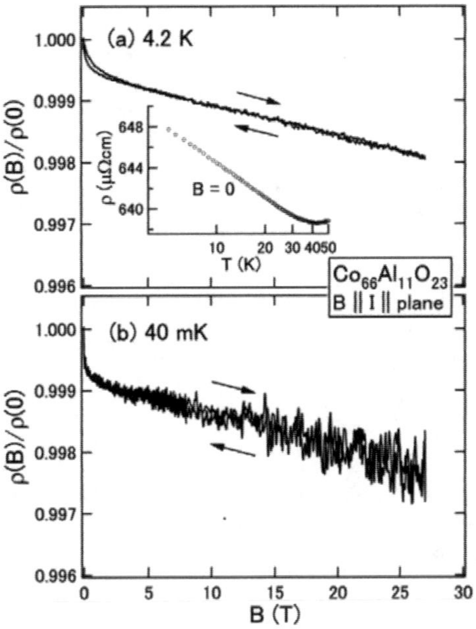

Figure 2. Magnetoresistivity of the $Co_{66}Al_{11}O_{23}$ at (a) 4.2 K and (b) 40 mK. The data is normalized by the resistivity at zero field. The field is applied along the current direction and parallel to the film plane. The temperature dependence of the resistivity at zero field is displayed in the inset using a log$T$ scale.

Figure 3 shows a plot of the magnetization of $Fe_{76}Al_{12}O_{13}$ as a function of $T^{1.5}$. The magnetization is measured during a cooling under a field of 10 mT applied parallel to the film plane. Induced uniaxial anisotropy in the plane was not observed in the Fe–Al–O films examined. As shown in figure 3, the magnetization varies with temperature, exhibiting $T^{1.5}$ behavior. Only a slight deviation is observed below 100 K. According to the three dimensional spin wave theory, thermally excited spin waves reduce the magnetization in proportion to $T^{1.5}$ [14]. The solid line in figure 3 is the fit to the formula $M(T)/M(5\mathrm{K}) = M(0\mathrm{K})/M(5\mathrm{K}) - AT^{1.5}$. The coefficient A is found to be 9.1×10^{-6} K$^{-1.5}$. This is an intermediate value between that for crystalline Fe (3.5×10^{-6} K$^{-1.5}$) and that for amorphous Fe (2.2×10^{-5} K$^{-1.5}$) [15]. In the case of $Fe_{83}Al_9O_8$, the magnetization follows the function $M(T)/M(5\mathrm{K}) = M(0\mathrm{K})/M(5\mathrm{K}) - BT^{1.8}$, where B is a proportionality constant (figure 4). The exponent of 1.8 deviates slightly from 1.5. This deviation may be explained by the interaction between spin waves or the anisotropy in the

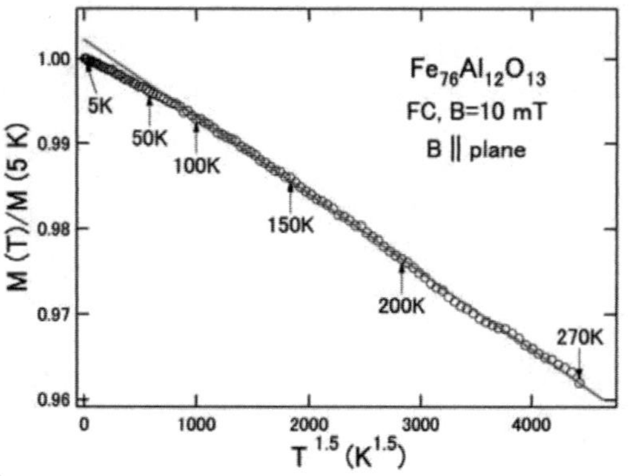

Figure 3. The temperature dependence of the magnetization of $Fe_{76}Al_{12}O_{13}$ as a function of $T^{1.5}$. The data is normalized by the value at 5 K. The measurements were performed with cooling under a field of 10 mT applied parallel to the film plane. The solid red line is a fit for the range $90 < T < 270$ K.

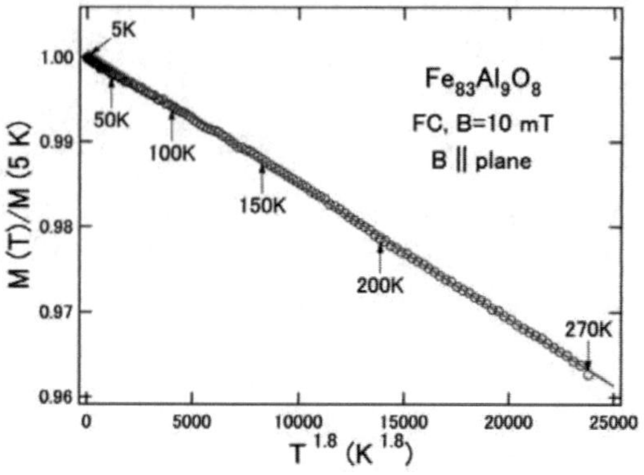

Figure 4. The temperature dependence of the magnetization of $Fe_{83}Al_9O_8$ as a function of $T^{1.8}$. The data is normalized by the value at 5 K. The measurements were performed with cooling under a field of 10 mT applied parallel to the film plane. The solid red line is a fit for the range $100 < T < 270$ K.

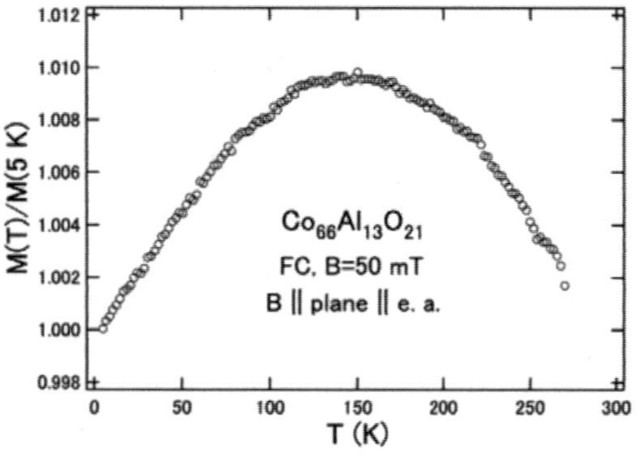

Figure 5. The temperature dependence of the magnetization of $Co_{66}Al_{13}O_{21}$. The data is normalized by the value at 5 K. The measurements were made with cooling in the field of 50 mT applied parallel to the film and the easy axis.

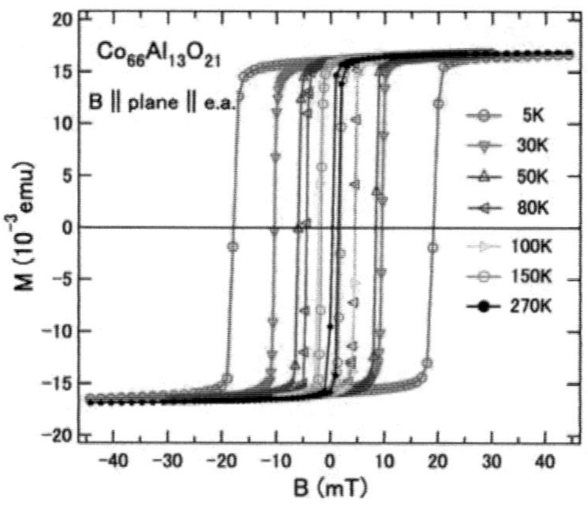

Figure 6. Hysteresis loops of $Co_{66}Al_{13}O_{21}$ at various temperatures. The fields are applied parallel to the film and the easy axis.

exchange interaction neglected in the spin wave theory.

Figure 5 shows the temperature dependence of the magnetization of $Co_{66}Al_{13}O_{21}$. The measurements were performed during cooling under a field of 50 mT directed along the easy axis in the film plane. For $Co_{66}Al_{13}O_{21}$, induced uniaxial anisotropy in the plane was observed (data not shown). As shown in figure 5, the magnetization increases with decreasing temperature, reaching a maximum at about 150 K, and then decreasing about 1 % from 150 K to 5 K. In contrast to Fe–Al–O films, a pronounced deviation from the $T^{1.5}$ law is observed in $Co_{66}Al_{13}O_{21}$. Although the electrical resistivity of Co–Al–O films exhibit a T^2 behavior and is explained by spin waves [7], the magnetization behavior of $Co_{66}Al_{13}O_{21}$ cannot be explained only by the conventional three dimensional spin wave theory.

The temperature dependence of the magnetization of $Co_{66}Al_{13}O_{21}$ suggests that other magnetic properties of this material are temperature sensitive. We measured the magnetization curves of this film at various temperatures. Figure 6 displays the M-B loops of $Co_{66}Al_{13}O_{21}$. Fields are applied parallel to the film plane and along the easy axis. At every temperature a rectangular loop is observed, but it rapidly becomes large with decreasing temperature. The coercive field at 270 K obtained is only 0.75 mT, but it reaches 18.5 mT at 5 K.

4. Conclusion

We have investigated the transport and magnetic properties of ferromagnetic Fe–Al–O and Co–Al–O granular films over a wide range of magnetic field and temperature. We found that the characteristic linear negative magnetoresistivity of $Co_{63}Al_{13}O_{24}$ and $Co_{66}Al_{11}O_{23}$ is observed at least up to 27 T. The negative magnetoresistivity cannot be explained in terms of the field suppression of thermal fluctuation of the moments. The slope of the magnetoresistivity does not become gentle in the present field range. Further investigation under higher field strengths is therefore required. The examination for the unsaturation of the magnetization is also necessary. The temperature dependence of the magnetization of $Fe_{76}Al_{12}O_{13}$ and $Fe_{83}Al_9O_8$ may be explained by the three dimensional spin wave theory. On the other hand, that of $Co_{66}Al_{13}O_{21}$ is not understandable only in terms of the conventional spin wave theory. A dramatic increase in the coercive field is found for a soft magnetic material $Co_{66}Al_{13}O_{21}$ at low temperatures.

Acknowledgments

The authors thank Prof. S. Awaji of HFLSM, Tohoku University for valuable advice on the experiments using CHM. The authors also express their gratitude to S. Tanno, K. Hosokura, A. Ogata and Y. Watanabe for assistance with the low temperature experiments.

References

[1] Fujimori H, Mitani S and Ohnuma S 1995 *Mate. Sci. and Eng.* B **31** 219
[2] Mitani S, Fujimori H and Ohnuma S 1997 *J. Mag. Mag. Mate.* **165** 141
[3] Mitani S, Takahashi S, Takanashi K, Yakushiji K, Maekawa S and Fujimori H 1998 *Phy. Rev. Lett.* **81** 2799
[4] Fujimori H, Mitani S, Ohnuma S, Ikeda T, Shima T and Masumoto T 1994 *Mate. Sci. and Eng.* A **181-182** 897
[5] Ohnuma S and Mitani S 1995 *J. Mag. Soc. Jpn.* **19** 19
[6] Yoshihara A, Ohnuma S, Fujimori H, Nakamura S and Nojima T 2008 *J. Phy. Soc. Jpn.* **77** 094704
[7] Nakamura S, Nojima T, Yoshihara A, Ohnuma S and Fujimori H 2009 *J. Phy. Soc. Jpn.* **78** 074708
[8] Nakamura S, Iwashita T, Nojima T, Yoshihara A, Ohnuma S and Fujimori H 2010 *J. Phy.: Conf. Series* **200** 012141
[9] Hasegawa R and Tsuei C C 1971 *Phy. Rev.* B **3** 214
[10] Grest G S and Nagel S R 1979 *Phy. Rev.* B **19** 3571
[11] Cochrane R W, Harris R, Ström-Olson J O and Zuckermann M J 1975 *Phy. Rev. Lett.* **35** 676
[12] Anderson P W, Abrahams E and Ramakrishnan T V 1979 *Phy. Rev. Lett.* **43** 718
[13] Watanabe K, Nishijima G, Awaji A, Takahashi K, Koyama K, Kobayashi N, Ishizuka M, Itou T, Tsurudome T and Sakuraba J 2006 *IEEE. Trans. Appl. Supercon.* **16** 934
[14] Kittel C 1996 *Introduction to Solid State Physics 7th edition* (New York: John Wirey and Sons) Chapter 15 p. 455
[15] Grinstaff M W, Salamon M B and Suslick K S 1993 *Phy. Rev. B* **48** 269

Journal of Physics: Conference Series **266** (2011) 012020 doi:10.1088/1742-6596/266/1/012020

Effect of surface roughness on magnetism of ultrathin Co films

M Sakamaki[1], **K Amemiya**
Institute of Materials Structure Science, High Energy Accelerator Research Organization, Tsukuba, 305-0801 Ibaraki, Japan

E-mail: masako.sakamaki@kek.jp

Abstract. Interface roughness effect on magnetic anisotropy of Co/Au are studied by using a Au substrate with "sharp" and "rough" surfaces by means of conventional and depth-resolved XMCD techniques. From the Co *L*-edge XMCD and sum rule analysis, the critical Co thickness of SRT from in-plane to perpendicular magnetization are found to be shifted from ~0.8 to ~1.1 nm by Au capping. We also found the sample with "rough" interface favors in-plane magnetization more than the sample with "sharp" interface. Compared to "sharp" sample, "rough" sample has smaller m_l/m_s values, and no significant probing-depth dependence is observed. These differences in the magnetic state at the interface between "sharp" and "rough" might be caused by structural changes. Moreover, m_l/m_s is lager for "sharp" sample especially at the interface. This can be interpreted as a large perpendicular anisotropy at the "sharp" interface, i.e. large K_s, while the magnetic anisotropy of the "rough" sample is determined by K_v more than by K_s.

1. Introduction

It is known that surface/interface roughness of magnetic thin films affects magnetic properties, such as coercivity, domain structure, and magnetic anisotropy [1-4]. These magnetic properties greatly affect the application to e.g. magnetic recording, and data storage devices. We here focus on roughness-induced anisotropy for ultrathin Co films grown on a Au(111) substrate. Several mechanisms have been invoked to explain its origin: (i) edge atom, whose electronic structure is different from atoms inside the island, would feel anisotropic potential and change the anisotropy of whole film via exchange coupling [5], (ii) strain at the interface as a magneto-elastic coupling [6], and (iii) dipolar interaction as a magnetostatic contribution [7]. However, there are few experimental/theoretical

[1] To whom any correspondence should be addressed.

discussions that include all of these contributions. The structural properties strongly depend on the preparation conditions, quality of the substrate and so on. It is therefore important to perform magnetic and structural investigations on the same samples. In this respect, we used wedge-shaped Co thin layers in combination with *in-situ* local probes such as x-ray magnetic circular dichroism (XMCD) and low-energy electron diffraction (LEED) allowing a thickness dependence study of magnetic and structural properties. In addition, we applied depth-resolved XMCD technique [8] to the sample with different degree of roughness, which was controlled by Ar^+ sputtering on the substrate, and evaluated the surface/interface and inner-layer magnetic moments as well as magnetic anisotropy of Co.

2. Experimental

2.1. Sample preparation

A Au(111) single crystal was cleaned by repeated cycles of Ar^+ sputtering (2 keV) and subsequent annealing at ~950 K. We then separately fabricated rough and sharp surfaces on single substrate by sputtering only the half area of the sample. The roughness of the surface by Ar^+ sputtering was confirmed by LEED. Wedge-shaped Co films were deposited on Au(111) at the substrate temperature of 300 K with the electron bombardment evaporation of a Co rod. The deposition rate of the Co films was determined by Auger electron spectroscopy (AES) prior to the sample preparation. The thickness of the Co layer varied from 0 to 2.0 nm. The Au-covered Co films were also prepared by subsequently evaporating Au at the substrate temperature of 300 K with the electron bombardment evaporation of a Au wire supported on a W rod. The Au deposition rate was also calibrated by AES prior to the sample preparation.

2.2. X-ray magnetic circular dichroism measurement

In addition to the total electron yield XMCD, which provides averaged information over the whole sample, the depth-resolved XMCD technique [8] was adopted. The depth-resolved XMCD provides depth-resolved and element specific magnetic information. In this technique, the probing depth of the spectrum is determined by the effective escape depth of the electrons, which depends on the electron emission angle θ_d. The emitted electrons were collected separately at different detection angles by using an imaging type detector consisting of a microchannel plate (MCP), a phosphor screen and a CCD camera. All the spectra with different probing depths were recorded at once. Partial electron

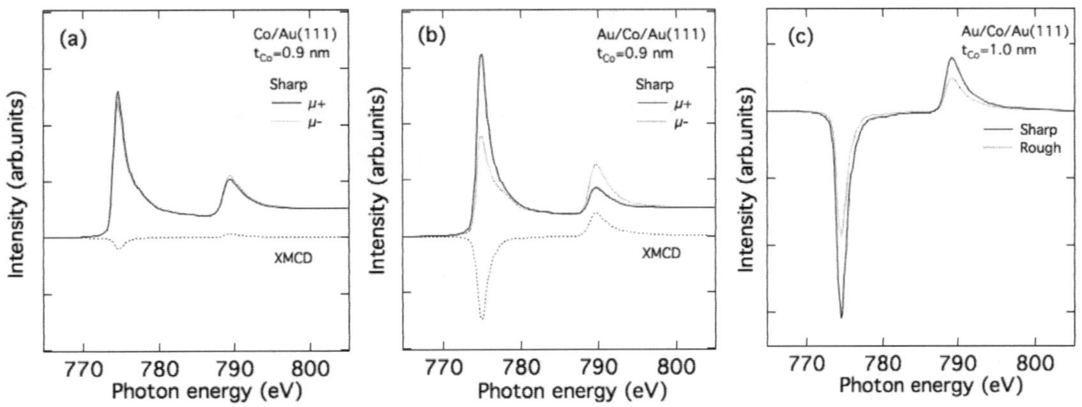

Figure 1. Co *L*-edge XMCD spectra for (a) Co/Au(111) and (b) Au/Co/Au(111) with "sharp" interface at the Co thickness of 0.9 nm. (c) XMCD spectra for Au/Co/Au(111) with "sharp" and "rough" interfaces at the thickness of 1.0 nm. All the spectra were taken at NI.

yield mode with a retarding voltage of 500 V was adopted, so that the Co *LMM* Auger electrons were mainly collected. The effective electron escape depth was experimentally determined at each θ_d from the thickness dependence of the edge-jump intensity of the films. XMCD measurements were performed on the samples remanently magnetized by a pulse magnetic field (~700 Oe) before each measurement. The spectra were recorded at normal (NI) and grazing (GI) x-ray incidences (0° and 60° from normal) to study magnetic anisotropy. All the measurements were performed at room temperature in a vacuum chamber connected to beamline BL-16A at the Photon Factory.

3. Results and discussion

It was reported that Co thin films grown on a Au(111) substrate exhibit the spin reorientation transition (SRT) from in-plane to perpendicular magnetization by capping with Au [9,10]. In the present study, the magnetic structure at the surface and interface has been investigated in order to clarify the origin of the SRT. Figure 1 shows Co *L*-edge XMCD spectra of Co/Au(111) with and without Au capping at the Co thickness of 0.9 nm. Here we show the result of the sample without Ar^+ sputtering (we call this sample as "sharp" interface sample). The spectra were taken at NI, in which only the surface normal component of sample magnetization is detected, in the total electron yield mode. The XMCD intensity drastically increases upon Au capping, indicating that Au capping surely enhances perpendicular magnetic anisotropy. Co showed perpendicular magnetic anisotropy at the thickness of ~0.7-1.1 nm for "sharp" sample after Au capping. In the case of "rough" sample, perpendicular anisotropy was seen at ~0.6-1.0 nm. Figure 1(c) shows Co *L*-edge XMCD spectra of Au/Co/Au(111) with "sharp" and "rough" interfaces at the Co thickness of 1.0 nm. The sample with "rough" interface shows smaller XMCD intensity, which means more in-plane preference compared to the sample with "sharp" interface at the Co thickness of 1.0 nm.

3.1. XMCD sum rule analysis

Figure 2 shows XMCD sum rule [11] analysis for Co/Au(111) with and without Au capping. 3*d* hole number n_{hole}, effective spin magnetic moment m_s, and orbital to spin moment ratio m_l/m_s are plotted as a function of Co thickness. The critical thickness of SRT, which is estimated from the thickness dependence of m_s, is shifted from ~0.8 to ~1.1 nm by Au capping. The sample with "rough" interface favors in-plane magnetization more than the sample with "sharp" interface at thicker region for both Co/Au(111) and Au/Co/Au(111). On the other hand, the reduction in m_s at thinner region (<~0.6 nm) is not attributed to the SRT but to the phase transition to the super-paramagnetic region.

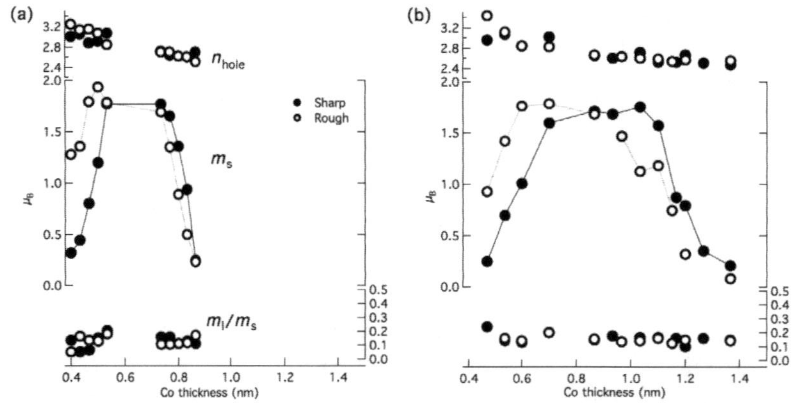

Figure 2. Sum rule analysis of (a) Co/Au(111) and (b) Au/Co/Au(111). 3*d* hole number n_{hole}, effective spin magnetic moment m_s, and orbital to spin moment ratio m_l/m_s are plotted.

Journal of Physics: Conference Series **266** (2011) 012020 doi:10.1088/1742-6596/266/1/012020

We thus conclude magnetostatic contribution, which stabilizes perpendicular magnetization under the presence of surface roughness [7], is negligible in the investigated samples. For m_l/m_s, the sample with "sharp" interface shows slight larger values. The orbital moment corresponds to magnetic anisotropy constant $\boldsymbol{K}$ [12], and it means the "sharp" interface has larger $\boldsymbol{K}$. This is consistent with the result of m_s. In the case of Co/Au(111), m_l/m_s shows slight decrease as the Co thickness increased (>~0.6 nm), and the slope of "sharp" sample is gentler than that of "rough". The change in the slope cannot be due to the difference in K_s, surface component of $\boldsymbol{K}$, but the difference in K_v, volume part of $\boldsymbol{K}$. Generally the structural change, e.g. strain, structural phase transition, affects the magnitude/sign of K_v. From these results, the structural difference between "sharp" and "rough" is expected. In order to separately discuss K_s and K_v, structural/magnetic depth profiling is required.

3.2. Depth-resolved XMCD analysis

In the depth-resolved technique, a set of spectra at different probing depths is obtained. The XMCD spectrum recorded at smaller probing depth contains more contribution from the top Co layer, and is more surface/interface sensitive. According to XMCD sum rule, we obtained effective spin magnetic moment m_s, and orbital to spin moment ratio m_l/m_s at each probing depth. Figure 3 shows probing depth dependence of m_s and m_l/m_s for Au/Co/Au(111) at the Co thickness of 0.9 nm. Both "sharp" and "rough" samples show decrease in m_s at smaller probing depth, indicating that the Co atoms at the interface have smaller effective spin moment compared to the inner layers. On the other hand, m_l/m_s increases at smaller probing depth for "sharp" sample, suggesting that the orbital magnetic moment is enhanced at the interface. These results agree with the theoretical calculation [13], and they can be explained as e.g. structural changes at the interface, lack of magnetic atoms above the film, and hybridization with the non-magnetic atoms. So far, the enhancement of the orbital moment at the interface had not been experimentally observed directly. Compared to "sharp" sample, "rough" sample shows smaller m_l/m_s values, and no significant probing-depth dependence is observed. It indicates the interface has similar orbital moment to that in the inner layers for "rough" sample. We suppose these differences in the magnetic state at the interface between "sharp" and "rough" might be caused by structural changes. Especially the uniform magnetic state observed on "rough" sample might be due to the structural phase transition or structural relaxation.

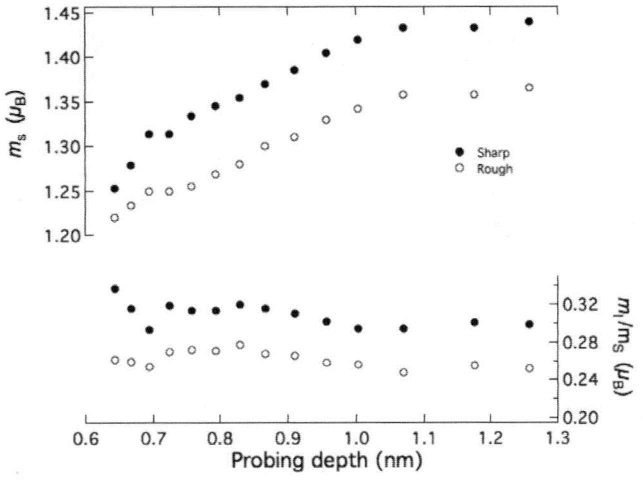

Figure 3. Probing depth dependence of spin magnetic moment m_s, and orbital to spin moment ratio m_l/m_s for Au/Co/Au(111) at the Co thickness of 0.9 nm.

Moreover, the orbital moment is lager for "sharp" sample especially at the interface. This can be interpreted as a large perpendicular anisotropy at the "sharp" interface, i.e. large K_s. On the other hand, the magnetic anisotropy of the "rough" sample might be determined by K_v more than by K_s.

4. Summary

We applied conventional and depth-resolved XMCD techniques to study the interface roughness effect on magnetic anisotropy of Co ultrathin films using a Au substrate with "sharp" and "rough" surfaces. From the Co *L*-edge XMCD and sum rule analysis, we found the critical Co thickness of SRT from in-plane to perpendicular magnetization is shifted from ~0.8 to ~1.1 nm by Au capping. We also found the sample with "rough" interface favors in-plane magnetization more than the sample with "sharp" interface. No significant probing-depth dependence of orbital moment was observed for "rough" sample, indicating that the orbital moment at the interface is similar to that in the inner layers. On the other hand, the orbital moment is lager for "sharp" sample especially at the interface. This might be due to a large perpendicular anisotropy at the "sharp" interface, i.e. large K_s. We suppose these differences in the magnetic state at the interface between "sharp" and "rough" might be caused by structural changes. Especially the uniform magnetic state observed on "rough" sample might be due to the structural phase transition or structural relaxation. A depth-structural study is underway.

5. References

[1] *Ultrathin Magnetic Structures I and II*, ed J A C Bland and B Heinrich (Springer, Berlin, 1994)
[2] Chang C H and Kryder M H 1994 *J. Appl. Phys.* **75** 6864.
[3] Bruno P, Bayureuther G, Beauvillain P, Chappert C, Lugert G, Renard D, Renard J P, and Seiden J 1994 *J. Appl. Phys.* **68** 5759.
[4] Zhang S and Levy P M 1996 *Phys. Rev. Lett.* **77** 916.
[5] Jiang Q, Yang H-N, and Wang G-C 1997 *Surf. Sci.* **373** 181.
[6] Li M, Wang G-C, and Min H-G 1998 *J. Appl. Phys.* **83** 5313.
[7] Bruno P 1988 *J. Appl. Phys.* **64** 15.
[8] Amemiya K, Matsumura D, Abe H, Kitagawa S, Ohta T and Yokoyama T 2004 *Phys. Rev.* B **70** 195405.
[9] Weller D, Stöhr J. Nakajima R, Carl A, Samant M G, Chappert C, Mégy R, Beauvillain P, Veillet P and Held G A 1995 *Phys. Rev. Lett.* **75** 3752.
[10] Andersson C, Sanyal B. Eriksson O, Nordström L, Karis O, Arvanitis D, Konishi T, Holub-Krappe E and Hunter Dunn J 2007 *Phys. Rev. Lett.* **99** 177207.
[11] Thole B T, Carra P, Sette F and Van der Laan G 1992 *Phys. Rev. Lett.* **68** 1943.
[12] Bruno P 1989 *Phys. Rev. B* **39** 865.
[13] Šipr O, Minár J, Vackář J and Ebert H 2007 *Phys. Rev. B* **75** 134422.

Journal of Physics: Conference Series **266** (2011) 012021 doi:10.1088/1742-6596/266/1/012021

Magnetic anisotropy and domain structure in spin-valves based on MTJ

A S Samardak[1], E V Sukovatitsina[1], A V Ognev[1], M V Anisimova[1], E Wahlström[2], L A Chebotkevich[1]

[1]Laboratory of thin film technologies, FENU, Vladivostok, 690950, Russia
[2]Department of Physics, NTNU, N-7491 Trondheim, Norway

E-mail: asamardak@gmail.com

Abstract. We have studied the crystal structure, surface morphology, magnetic anisotropy, domain structure and tunnel magnetoresistance of spin-valves with a single and double MgO barrier layers. We have demonstrated the domain structure for soft and hard magnetic layers and observed significant changes after low temperature annealing. We have carried out magnetoresistance measurements using current-in-plane (CIP) four-probe technique and discovered a substantial difference in the values of TMR ratio for single and double MTJ spin-valves. It is shown that domain structure and magnetization reversal are the same for both systems, but otherwise the behavior of tunnel magnetoresistance is different, because of second MgO barrier effects on the system conductivity. The ability to manipulate the magnetization direction in MTJ systems using temperature annealing is demonstrated. It makes these structures possible for new applications in nanoelectronics as magnetic recording media and high sensitive sensors.

1.Introduction

The multicomponent magnetic systems are new class of spintronic materials having unique combination of magnetic and electrical properties for prospective nanoelectronic applications as a non-volatile magnetoresistance memory (MRAM), spin-valve sensors and microwave spin-current generators [1-3]. The spin-valves based on magnetic tunnel junctions are very interesting objects for investigation, because of they combine all remarkable properties of both devices: high magnetic field sensitivity and high value of magnetoresistance [4,5]. The classical spin-valve is based on two ferromagnets with different coercive forces separated by a thin metal spacer [5]. The relative magnetization orientation of the two ferromagnetic layers strongly modulates the charge flow through the structure. Using this property of the spin-valve, external magnetic field acting on the magnetic moments of the ferromagnetic layers can be sensed electrically. Spin-valves with insulating interlayers, such as Al_2O_3 [6] or MgO [7], are called magnetic tunnel junctions (MTJs) and designed to have two stable parallel and anti-parallel states, can also serve as non-volatile solid-state memory elements [8]. These applications, as read head sensors in hard magnetic disks and memory cells in MRAM are perfect examples of a fast transformation of a fundamental physics discovery [9–11] in to large scale practical devices. This makes spintronic nanostructures not only of fundamental interest,

[1] To whom any correspondence should be addressed.

but also of great technological concernment [3, 12]. Recently the MTJs with double tunnel barrier [13] have began to attract an attention of scientists and engineers due to be an effective means to improve the bias dependence of the tunnel magnetoresistance (TMR) ratio [14] and to reduce the critical current density of spin torque transfer [15]. Moreover, for double barrier MTJs the exciting phenomena such as quantum confinement effect, Coulomb blockade and spin-dependent resonant tunnelling have been predicted and observed [16,17]. The most effective method to improve MTJs' properties is a thermal annealing in vacuum. An annealing can generate better interfaces between magnetic layers and insulating barrier. It is important to find optimal annealing parameters to produce MTJs with clearly defined spin-valve character and high TMR ratio.

The main purpose of this paper is to study temperature dependence of magnetization reversal and an effect of thermal annealing on magnetic and magneto-transport properties of spin-valves with single and double insulating MgO barriers.

2.Experimental details

The structures *Ti(30)/Au(20)/Co(20)/MgO(2)/[Co(0.3)/Pt(0.8)]*$_5$*/MgO(x)/Au(5)*, where x is 0 (single magnetic tunnel junction – SMTJ) or 2 *nm* (double magnetic tunnel junction – DMTJ), were fabricated on natural oxidized single crystal *Si* square substrates (1 × 1 cm^2) with an electron-beam deposition system in vacuum 10^{-7} *Torr*. The layers' thickness is shown in nanometers. The top Au layer is exploited as a cap layer to prevent metal oxidation. The surface morphology and domain structure were studied with atomic force microscope (AFM) and magnetic force microscope (MFM). Transmission electron microscopy (TEM) was exploited to observe a grain size. Hysteresis curves were obtained by longitudinal magneto-optical Kerr effect and with vibrating sample magnetometer (VSM). The magnetic measurements were done at room and low temperatures using SQUID magnetometer. TMR ratio was measured with four-probe station in current-in-plane (CIP) geometry [18] at different configurations: $\mathbf{I} \perp \mathbf{H}$, $\mathbf{I} \perp \mathbf{n}$, $\mathbf{H} \perp \mathbf{n}$ (transverse TMR) and $\mathbf{I} \parallel \mathbf{H}$, $\mathbf{I} \perp \mathbf{n}$, $\mathbf{H} \perp \mathbf{n}$ (longitudinal TMR), where **I** is an applied current (100uA), **n** is a perpendicular to the surface plane, **H** is an applied magnetic field. Temperature annealing was performed in commercial annealing setup in vaccum 10^{-2} *Torr* during 15 minutes at each temperature.

3.Results and Discussion

The TEM images have shown that all as-deposited samples have polycrystalline structure with an average grain size 6 nm. An average amplitude and period of roughness are h =1,5 nm and l = 40 nm, correspondingly. The VSM measurements have indicated that our films possess an in-plane magnetic anisotropy only.

The temperature dependence of the coercive forces of hard magnetic $(Co/Pt)_5$ stack $H_C^{(1)}$ and soft magnetic cobalt layer $H_C^{(2)}$ for SMTJs is represented in Fig.1. For DMTJs we have obtained the same curves. If measurement temperature is dropping from 293 K to 4.2 K, the coercive force $H_C^{(2)}$ changes from 35 Oe to 40 Oe. The enhancement of $H_C^{(2)}$ is due to a contraction of thermal fluctuations of magnetic moments. $H_C^{(1)}$ changes significantly. At room temperature (RT) $H_C^{(1)}$ is 350 Oe and it increases in three times (to 1kOe) at 4.2 K. It is known, that $H_c=2K/M_s$, where M_s is saturation magnetization and K is an anisotropy constant. Therefore, the behavior of $H_C^{(1)}$ can be explained by the change in the values of anisotropy constant and saturation magnetization at different measurement temperatures. According to [19] the anisotropy constant increases in about 6 times in temperature interval from RT to 4.2 K, but M_s in Co/Pt system grows in about 1.3 times only [20].

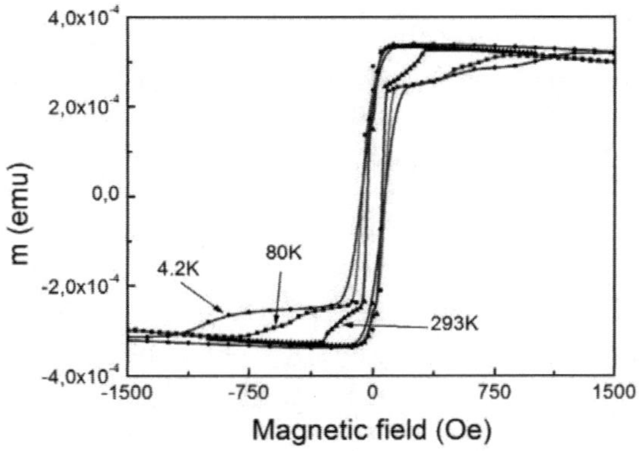

Figure 1. The hysteresis loops of MTJ in dependence on measurement temperature obtained with SQUID magnetometer.

We have studied an angular dependence of coercive force for both magnetic layers at RT. The dependences $H_C = f(\alpha)$ for SMTJs and DMTJs, α is an azimuth angle (this angle was chosen between one side of the square sample and the direction of an applied magnetic field), were measured with the longitudinal magneto-optical Kerr effect, Fig.2. The coercive force of hard magnetic $(Co/Pt)_5$ multilayer stack $H_C^{(1)}$ is 350 Oe and for bottom soft magnetic Co layer $H_C^{(2)}$ is 35 Oe. As shown in Fig.2, the minimum value of $H_C^{(1)}$ repeats itself every 180°. At the same time the easy axes of soft and hard magnetic layers are displaced on 90° relatively to each other. The same behavior has been observed for SMTJs. This behavior of the coercivity can be explained by the following. By the reason of high roughness of the layers (h =1,5 nm for as-deposited films) we suggest, that thin MgO interlayer is structurally nonsolid (i.e. it has fluctuation of thickness). Then, at the places, where MgO thickness less than 1.0 nm, the realization of an indirect exchange coupling between soft and hard layers is possible [21]. According to Slonczewski [22] the fluctuations of interlayer's thickness can lead to an indirect biquadratic exchange coupling between layers. Another possible mechanism of biquadratic coupling is the ferromagnetic pinhole formation through the non-magnetic interlayer [23]. Biquadratic exchange coupling leads to a perpendicular alignment of magnetic moments in adjacent ferromagnetic layers, as we have found in our as-deposited films, Fig.2.

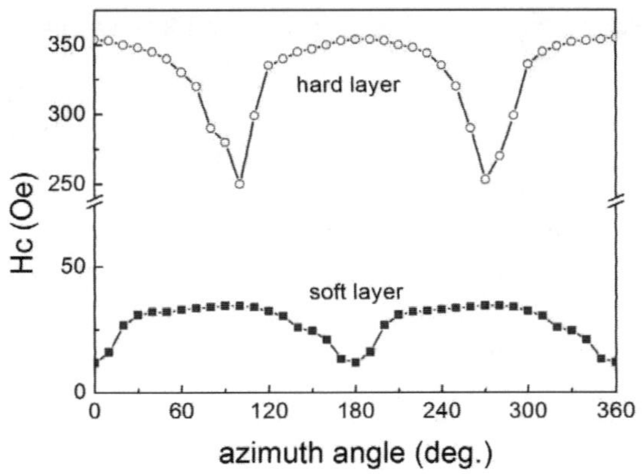

Figure 2. The angular dependence of the coercive force for DMTJs. The empty dots and filled squares are related to the hard and soft magnetic layers, correspondingly.

Journal of Physics: Conference Series **266** (2011) 012021 doi:10.1088/1742-6596/266/1/012021

After annealing of the samples at $T_{ann}=250^oC$ the coercive fields $H_C^{(1)}$ and $H_C^{(2)}$ increase on 80% and 29% correspondingly. The coercive force in polycrystalline films is caused by the domain wall pinning on grain boundaries, by dispersion of magnetization of easy axes and by surface roughness. During annealing the grain size increases. This is the reason for rise of the coercive force component responsible for dispersion of magnetization of easy axes in multilayer film [24]:

$$H_{ck} = 1{,}1\frac{K_i^{4/3}R^{4/3}}{\mu_o M_s \delta \gamma^{1/3}}, \tag{1}$$

where K_i is an induced anisotropy, R is an average grain size, δ is a domain wall width, γ is an energy density of a domain wall. After annealing at $T_{ann}=250^oC$ the average grain size increases in 4 times. This is the result in growth of H_{ck} in 6 times. It explains the increase of the coercive force in both magnetic layers after annealing.

As shown in Fig.3, with increase of the annealing temperature the easy axis of soft magnetic layer is rotated and aligned in parallel with easy axis of hard magnetic layer. The turn of the easy axis of soft layer occurs under influence of the magnetostatic field of the hard layer during annealing. We did mention significant influence of further annealing at higher temperatures on magnetic anisotropy in the films. As shown in [25], temperature annealing in layered magnetic systems can affect on interface roughnesses, hereupon the strength and type (ferromagnetic or antiferromagnetic) of an exchange coupling is changing. As result, biquadratic exchange coupling can decrease and even vanish after annealing. It means that magnetostatic interaction becomes to play major role in alignment of magnetic moments in adjacent soft and hard magnetic layers.

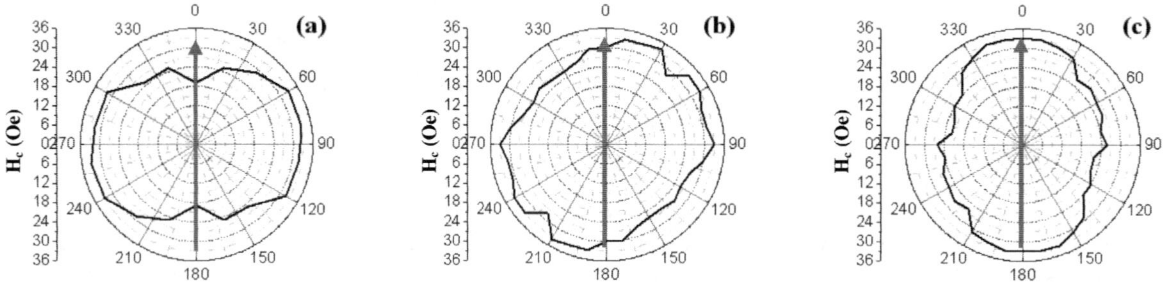

Figure 3. The polar charts represent the dependence of coercive force of soft magnetic layer on an azimuth angle at different annealing temperatures of DMTJs. (a) as-deposited, (b) $T_{ann}=100\ ^oC$, (c) $T_{ann}=150^oC$. The red arrows indicate the direction of easy axis in the hard magnetic layer.

The difference in coercive forces for *[Co/Pt]*$_5$ and *Co* layers allows to observe the domain structure of soft and hard magnetic layers separately by applying different magnetic fields to the samples. Similar investigation is slightly unique, because the partition of domain structure layer-by-layer is a non-trivial task. The analysis of MFM images, received at step-by-step incremented magnetic field in range from -600 to + 600 Oe, has displayed the big difference in the size and the shape of magnetic domains for soft and hard layers, Fig. 4.

At the demagnetized state the size of domains in soft layer is about 30-60 um, Fig.4(a). If to magnetize sample from 0 to the saturation field (600 Oe) and go back to -50 Oe, we have found very complicated domain structure of the hard layer, like in films with perpendicular anisotropy, Fig. 4(b). The domain size is 0.5-1 um. After annealing at $T_{ann}=150\ ^oC$ the domain structure of soft magnetic layer changes appreciably, Fig. 4(c). The domains become smaller. Domain walls are extensive and unilateral. It was an issue to receive MFM image of domain structure of hard magnetic layer, because after annealing the surface roughness of the film increased substantially. This is a reason why we have done domain imaging just for samples after low temperature annealing.

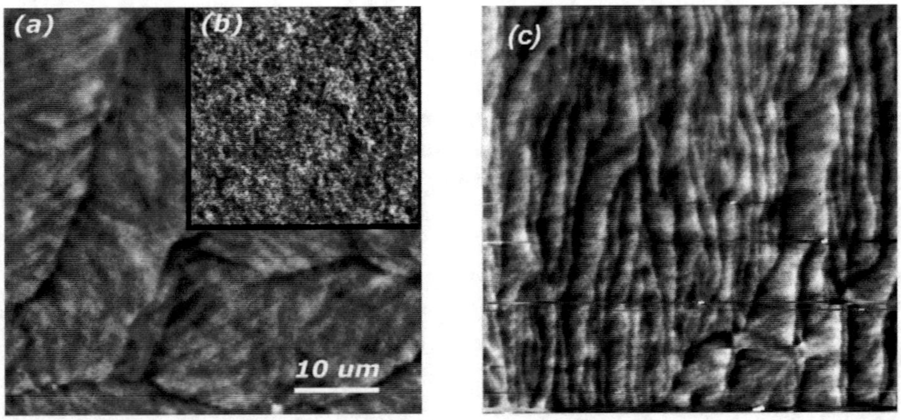

Figure 4. MFM images of domain structure for as-deposited (a, b) and annealed (c) samples. (a) – at magnetic field $H = H_C^{(2)}$, (b) – at *H=-50 Oe*, (c) – at magnetic field $H = H_C^{(2)}$. The scale is the same for all images.

Using the magneto-optical Kerr effect magnetometer, we have measured hysteresis loops at RT for SMTJs and DMTJs. The behavior of magnetization reversal for SMTJs before and after thermal treatment is shown in Fig.5. The characteristic steps on the hysteresis loops prove the spin-valve nature of our samples. After annealing the coercive force of the soft magnetic layer changes slightly, but coercive field and relative magnetization of the hard layer become larger, due to increasing of grain sizes. The magnetization reversal in DMTJs has the same behavior. Due to the different coercivity of soft and hard magnetic layers, the magnetization switching occurs at variant applied magnetic fields. As seen in Fig. 5 there are two mechanisms during switching process: spin-flip in soft layer at low magnetic field (30-35 Oe) and domain wall displacement in the hard layer at higher applied magnetic fields.

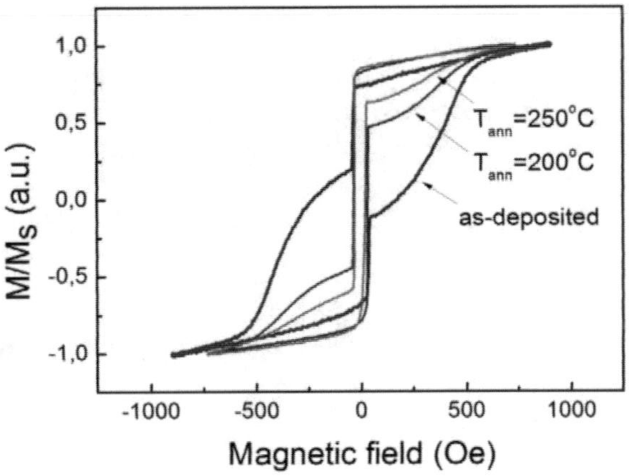

Figure 5. The typical hysteresis loop of as-deposited and annealed spin-valves based on SMTJs.

To study an influence of crystal structure modification on TMR ratio we measured magneto-transport properties of our samples before and after annealing in vacuum. TMR curves, proving spin-valve behavior and spin polarized tunneling of samples, are represented in Fig. 6 and 7. We have measured TMR for as-deposited samples in longitudinal and transverse geometries at RT. TMR ratio has been defined as $(R_{AP} - R_P)/R_P$, where R_{AP} and R_P are the resistances for anti-parallel (AP) and parallel (P) alignment of the magnetic moments, respectively [26]. The difference between the coercive forces of soft and hard layers allows to easily form the parallel and antiparallel magnetization states by changing an external magnetic field. This is the essential property of a spin-valve structure. The scheme of a spin-valve in Fig. 6 demonstrates magnetization switching process at change of applied magnetic field from negative to positive value. If go in backwards, we usually obtain symmetric TMR curves. If we apply negative saturation magnetic field, we get small resistance compare to the high resistance state, when the magnetization in the hard and soft magnetic layers aligned antiferromagneticaly. Due to the spin-dependent tunneling, the TMR curves have a peak, indicating the maximum value of TMR ratio. In SMTJs the longitudinal and transverse TMR ratios are 20% and 17%, correspondingly. We have found bigger TMR ratio for SMTJs, than for DMTJs. In DMTJs these values are 7% and 12.5% accordingly, Fig. 7. Apparently, the second MgO barrier decreases the tunnel conductivity of the DMTJs, which is subject to less TMR ratio [27].

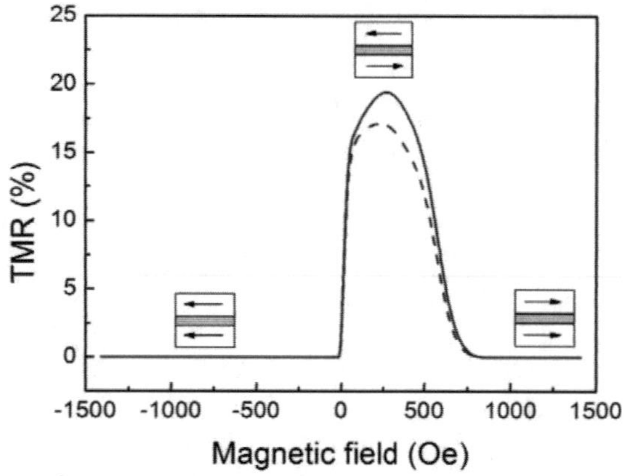

Figure 6. The longitudinal (solid line) and transverse (dash line) TMR in SMTJs.

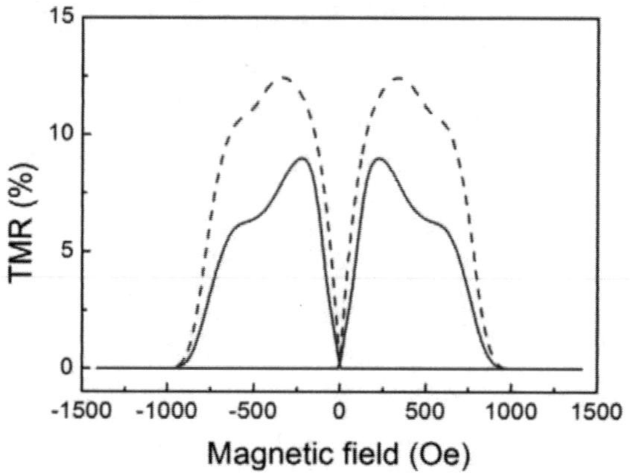

Figure 7. The longitudinal (solid line) and transverse (dash line) TMR in DMTJs.

In Fig. 8 the affect of the thermal treatment on magnetoresistance is shown. In our MTJs TMR ratio is expected to decrease, because the presence of defects like interface roughness, pinholes, interdiffused interfaces, impurities or vacancies, or stacking faults, would provide additional conduction channels. There is no interdiffusion in Co/Pt stack after annealing at low temperatures [28]. Thus, the most likely reason in the decrease of TMR ratio is pinholes formation in MgO layers [21], due to current leakage through them. The TMR ration for DMTJs changes slightly, because of smaller initial tunnel conductance of electrical current, passing through two MgO barriers.

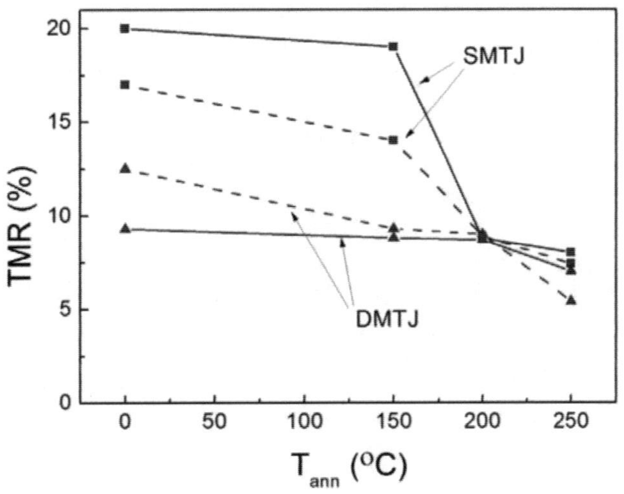

Figure 8. The TMR ratio as a function of annealing temperature. Solid lines show longitudinal TMR, dash lines correspond to transverse TMR.

4. Conclusions

We have presented the recent results on spin-valves with single and double MgO barriers. We have studied magnetization reversal at low temperatures and magnetic and magneto-transport properties after thermal annealing. The investigation is shown, that using temperature annealing we can fabricate a spin-valve structure with predefined TMR ratio and tailored direction of magnetization in soft magnetic layer. The represented results demonstrate the possibility to design novel devices for magnetic memory and programmable logic.

Acknowledgments

This work is supported by the Federal Program "Scientific and research-educational cadres of innovative Russia", Russian Foundation for Basic Research and the Research Council of Norway. The authors thank PhD students Alexey Il'in and Alexander Davidenko for technical support and help in measurements.

[1] Johnson M 2005 *J. Phys. Chem. B* **109** 14278.
[2] Parkin S S P et al. 2003 *Proc. IEEE* **91** 661.
[3] Wolf S A et al. 2001 *Science* **294** 1488.
[4] Miyazaki T, Tezuka N. 1995 *J. Magn. Magn. Mater.* **139** L231.
[5] Dieny B, Speriosu V S, Parkin S S P et al. 1991 *Phys. Rev. B* **43** 1297.
[6] Moodera J S, Nowak J, van de Veerdonk R J M 1998 *Phys. Rev. Lett.* **80** 2941.
[7] Parkin S S P, Kaiser Ch et al. 2004 *Nature materials* **3** 862.
[8] Gallagher W J, Parkin S P P et al. 1997 *J. Appl. Phys.* **81** 3741.
[9] Grunberg P, Schreiber R, Pang Y et al. 1986 *Phys. Rev. Lett.* **57** 2442.
[10] Baibich M N, Broto J M, Fert A et al. 1988 *Phys. Rev. Lett.* **61** 2472.
[11] Binasch G, Grunberg P, Saurenbach F, Zinn W 1989 *Phys. Rev. B* **39** 4828.
[12] Zutic I, Fabian J, Sarma S D 2004 *Rev. Mod. Phys.* **76** 323.
[13] Gan H D, Ikeda S, Shiga W et al. 2010 *Applied Physics Letters* **96** 192507.
[14] Inomata K, Saito Y, Nakajima K, Sagoi M 2000 *J. Appl. Phys.* **87** 6064.
[15] Diao Z, Panchula A, Ding Y et al. 2007 *Appl. Phys. Lett.* **90** 132508.
[16] Zhang X, Li B, Sun G, Pu F 1997 *Phys. Rev. B* **56** 5484.

[17] Jiang L, Naganuma H, Oogane M, Ando Y 2009 *Appl. Phys. Express* **2** 083002.
[18] Worledge D C and Trouilloud P L, 2003 *Appl. Phys. Lett.* **83** 84.
[19] Carr W J, 1958 *Phys. Rev.* **109** 1971.
[20] Zhang H and Wang Y, 1994 *Acta Phys. Sin. (Overseas Edn)* **3** 780.
[21] Katayama T et al. 2006 *Appl. Phys. Lett.* **89** 112503.
[22] Slonczewski J C, 1991 *Phys. Rev. Lett.* **67** 3172.
[23] Bobo J F, et al. 1999 *Phys. Rev. B* **60** 4131.
[24] Malyutin V I, Osukhovskii V E, Lobov I V et al. 1986 *Phys. Stat. Sol. (a)* **98** 585.
[25] Samardak A S et al, 2004 *The Physics of Metals and Metallography* **98** 360.
[26] Moodera J S, Kinder L R, Wong T M et al. 1995 *Phys. Rev. Lett.* **74** 3273.
[27] Iovan A, Andersson S, Naidyuk Yu G, Vedyaev A et al. 2008 *Nano Letters* **8** 805.
[28] Tsay J S and Shern C S, 1996 *J. Appl. Phys.* **80** 3777.

Journal of Physics: Conference Series **266** (2011) 012022 doi:10.1088/1742-6596/266/1/012022

Micromagnetic simulations of magnetization reversal of iron nanowire

H Xiang[1], D M Jiang[1], J C Yao[1], Y P Zheng[1], W Lu[2], G Q Li[1, 4], H Saito[3], S Ishio[3], X W Tan[1] and Y Q Lin[1]

[1] School of Physics, Southwest University, Chongqing, 400715, China

[2] School of Material, Tongji University, Shanghai, 200092, China

[3] Venture Business Laboratory, Akita University, Akita, 010-8502, Japan

E-mail: gqli@swu.edu.cn

Abstract. Spontaneous magnetization and magnetization reversal of individual Fe nanowire with different diameter (d) and length (l) were simulated using a micromagnetic method. Results indicate that the coercivity of a Fe nanowire thinner than 20 nm approximately increases linearly with the increase of l/d ratio in the range of $l/d \leq 3$, instead of $1/d^2$ ratio as reported in a reference, and reaches a maximum at $l/d \geq 10$. At $d = 6$ nm and $l = 200$ nm, the coercivity can be as large as 1294 mT. At $l = 200$ nm, the reversal mechanism significantly depends on the diameter of nanowire, in the subsequence of nucleation, vortex-like nucleation and vortex with the increase of d from 6 nm to 30 nm. At $l/d \geq 10$, the rectangle degree of magnetization curve exceeds 0.95, fitting the prediction for the model of infinite cylinder. The rectangle degree decreases gradually with the decrease of l/d, consistent with the reported experimental behaviors of Fe nanowires deposited on AAO templates.

1. Introduction

Nanostructured materials have at least one relevant physical dimension ranging from one to tens of nanometers (*e. g.*, the thickness of a film, the diameter of a wire, the radius of a particle, etc.). In recent years, regularly aligned arrays composed of one-dimensional soft-magnetic nanowires with high uniaxial shape anisotropy have been researched intensively [1-5], due to their potential application in the fields of magnetic storage and magnetic field sensing. By electrochemical deposition, nanowires with uniform diameter ranging from 4 nm to 200 nm can be fabricated into the packed columnar holes in self-assembled porous anodic aluminium oxide (AAO) templates, known as nanoporous alumite membranes [3, 6]. The magnetization reversal mechanisms of these soft-magnetic nanowires also have been investigated by various methods such as calculation, simulation and experiment to understand the interesting high coercivity and vertical magnetic anisotropy [7-9]. For the cylinders with an infinite length, it has been reported that the reversal mechanisms include curling, buckling, rotation in unison [10, 11], and various nucleations [3, 12] subsequently with the decrease of diameter. In this article, we simulate the spontaneous magnetization and magnetization reversal of individual Fe nanowire with finite length by solving Landau-Lifshitz-Gilbert (LLG) equation to investigate the effects of both length (l) and diameter (d) on the magnetic properties.

[4] To whom any correspondence should be addressed.

2. Micromagnetic model

The micromagnetic simulation was performed by a public domain micromagnetics program developed at the American National Institute of Standards and Technology, known as the OOMMF (Object Oriented MicroMagnetics Framework) [13]. OOMMF is composed of a collection of application programs, each specializing in some task needed as part of the micromagnetic simulation system. The programs work together by providing services to one another. The user controls the launching of programs through the interface provided by a program named as mmLaunch, which is most closely connected to the account service directory. The extensible micromagnetic computation engine, Oxs (OOMMF eXtensible Solver), can solve problems defined on three-dimensional grids of rectangular cells holding three-dimensional spins. By loading a problem definition file written in the extensible format (MIF 2.1), the micromagnetic problem is communicated to the Oxsii (Interactive Interface to Oxs). Oxsii is a client of display/storage application programs for data table and vector field, and is the server of a solver control service for which the only client is mmLaunch. The user can save and display intermediate results, either interactively or via scheduling based on iteration and stage counts. A cylinder-shaped magnet can be specified in conjunction with the scalar field objects and vector field objects via either the Oxs extension object for geometric volume of space, Oxs_ImageAtlas, or a user defined procedure for a dot with diameter of d and length of l.

A finite element method is employed to simulate the spontaneous magnetization and magnetization reversal of individual Fe nanowire (including nanodot) with diameter in the range of d = [6, 50] nm and length in the range of l = [20, 1000] nm (but $l \geq d$). Wires are separated into uniform cubic meshes with a size of $0.2 \times 0.2 \times 0.2$ nm^3, in the magnitude order of an iron atom. The spontaneous magnetization is evolved from an initial configuration of randomly oriented spins, absence of any external field. The hysteresis loops in external fields along the longitude axis of wire (z) are subsequently simulated. The wires are assumed as geometrical symmetrical finite cylinders. The material parameters of Fe for simulation include [14]: $K_1 = 4.\ 8 \times 10^4$ J/m^3, $M_s = 1.7 \times 10^6$ A/m, and $A = 2 \times 10^{-11}$ J/m, where K_1 denotes the cubic magnetocrystalline anisotropy energy constant with the respective anisotropy directions along x, y, and z, M_s is the saturation magnetization, and A is the uniform exchange energy constant.

3. Results and discussion

Figure 1 shows the simulated spontaneous magnetization process of an individual Fe nanowire with l = 200 nm and d = 20 nm. Starting from the initial paramagnetic state (the vectors of spins in meshes are randomly aligned, absence of applied field) as shown in figure 1a, the configuration of spins in meshes firstly evolves into small magnetic clusters to minimize the free energy (see figure 1b), and finally becomes a single-domain structure containing precessions of spins (see figure 1c). In all range of chosen l and d, the structure of spontaneous magnetization is the same single-domain type. The shape anisotropy and cubic magnetocrystalline anisotropy lead to the existence of two stable orientations of the magnetic moment, namely parallel or antiparallel to the wire axis. These orientations correspond to a symmetric potential with two energy minima separated by the barrier of anisotropy.

As the increase of magnetic field applied opposite to the spontaneous magnetization, the energy minima become asymmetric and the barrier decreases. The magnetization will switch irreversibly if the barrier vanishes. Figure 2 gives the x-z cross-section maps of spin vector at the instantaneity when the reversal is triggered by the respective switching field of Fe nanowires with l = 200 nm and d = 6, 10, 20, 30 nm, respectively. The simulation starts at the evolved final state of spontaneous magnetization as shown in figure 1c. At d = 6 and 10 nm (figures 2a and 2b), the reversal mechanism is nucleation. In the region of a nucleation, all the spins in an x-y plane rotate coherently, as the diameter of nanowire is less than the exchange length $\lambda_{ex} = \sqrt{A / \mu_0 M_s^2} \approx 20$ nm. The reversals nucleate in several regions cross the whole x-y section, and then the domain walls propagate quickly. With the increase of d, the regions of nucleation move toward the ends of nanowire. Increasing d to 20 nm, as shown in figure 2c, the magnetization reversal mechanism transforms to vortex-like nucleation which initiates only at two ends of nanowire. Different from the ideal nucleation as shown in figures 2a and 2b, the spins in the edges rotate at first, and in a plane of *x-y* cross section, the rotation angle is different at different position. At d = 30 nm (about a half of the domain wall thickness for Fe) as shown in figure 2d, the magnetization reversal is controlled by the buckling-like motion of larger vortices, similar to the situation in bulk material. Besides at the ends, the reversals also initiate in the middle sections. The vortices always initiate in the edge instead of the centre.

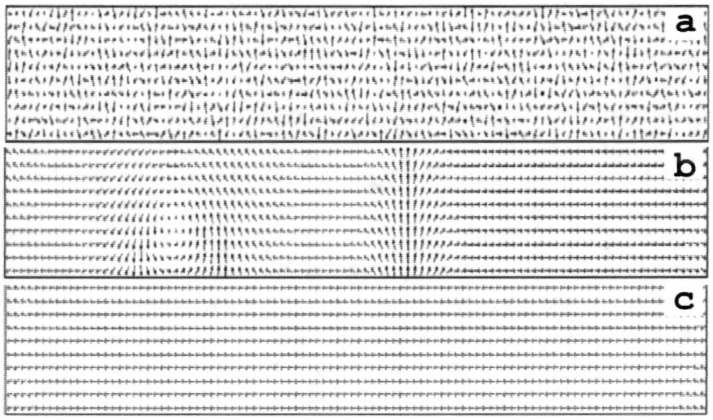

Figure 1. Simulation of spontaneous magnetization of an individual Fe nanowire with l = 200 nm and d = 20 nm: initial state with randomly oriented spins (a), evolving forward ordered configuration (b) and final state (c, single domain).

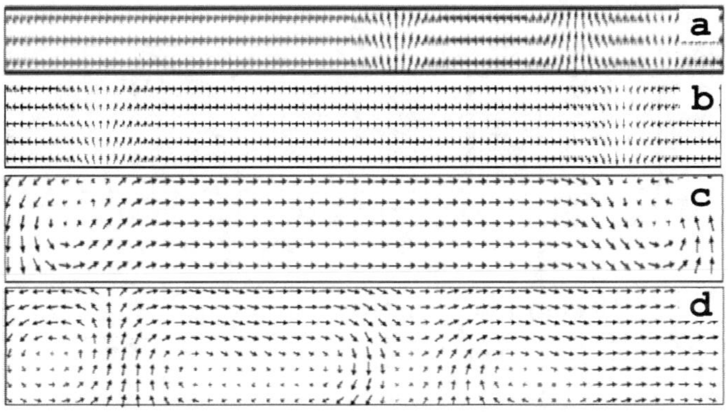

Figure 2. Vector maps of spins at the instantaneity when reversal is triggered in Fe nanowires with l = 200 nm and d = 6 (a), 10 (b), 20 (c), 30(d) nm.

Simulated hysteresis loops of Fe nanowires with typical sizes of l = 200 nm and d = [6, 50] nm are shown in figure 3. The rectangle degree of hysteresis loop is related to the ratio of length to diameter (l/d). At $l/d > 10$, the rectangle degree exceeds 0.95, consistent with the prediction of the model for an infinite cylinder (announcing that the hysteresis loop of a single wire always has a rectangular shape [9, 10]). With the decrease of l/d, the rectangle degree decreases gradually, agreeing with the experimental behaviors of Fe nanowires deposited on AAO templates [3]. At a given l, the coercivity (B_c) decreases in a wide range with the increase of d. The d dependence of B_c at l = 200 nm is shown as figure 4. At d = 6 nm and l = 200 nm, B_c = 1294 mT, greatly higher than the experimental coercivity [3]. This implies that the fabricated nanowires were not long enough, due to the defects. The inset in figure 4 is the replotted $1/d$ dependence of B_c. B_c is approximately proportional to $1/d$, instead of $1/d^2$ as reported in the reference 15.

We should focus on discussing the properties of nanowires thinner than a half of the domain wall thickness (in the diameter range of $d < 30$ nm), as the magnetic behaviours are different from the behaviours of the bulk material. Figure 5 shows the hysteresis loops of Fe nanowires with d = 20 nm and l = [20, 1000] nm. The coercivity increases apparently in the range of l = [20, 200] nm, and reaches a maximum of 464 mT at $l \geq 200$ nm ($l/d \geq 10$). The l dependence of B_c at d = 20 nm is shown in figure 6. B_c is approximately proportional to l in the range of $l \leq 60$ nm ($l/d \leq 3$).

Journal of Physics: Conference Series **266** (2011) 012022 doi:10.1088/1742-6596/266/1/012022

4. Conclusions

The spontaneous magnetization and magnetization reversal of individual Fe nanowire with finite length was simulated to investigate the effects of both length (l) and diameter (d) on the magnetic properties. The reversal mechanism significantly depends on the diameter. For the nanowires with l = 200 nm, the reversal mechanism is typical nucleation at d = 6 and 10 nm with the spins in a plane of x-y cross section rotating coherently, as the diameter of nanowire is less than the exchange length. With the increase of d, the regions of nucleation move toward the ends of nanowire. At d = 20 nm, the magnetization reversal mechanism transforms to vortex-like nucleation, initiating only at the ends of nanowire. At d = 30 nm, the magnetization reversal is vortex similar to the situation in bulk material. The vortices always initiate in the edge instead of the centre. The coercivity of Fe nonowire thinner than 20 nm approximately increases linearly with the increase of l/d ratio in the range of $l/d \leq 3$, and reaches a maximum at $l/d \geq 10$.

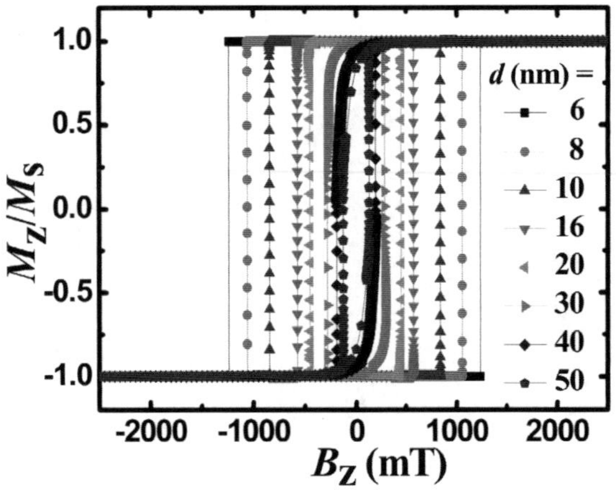

Figure 3. Hysteresis loops of Fe nanowires with l = 200 nm and d = [6, 50] nm.

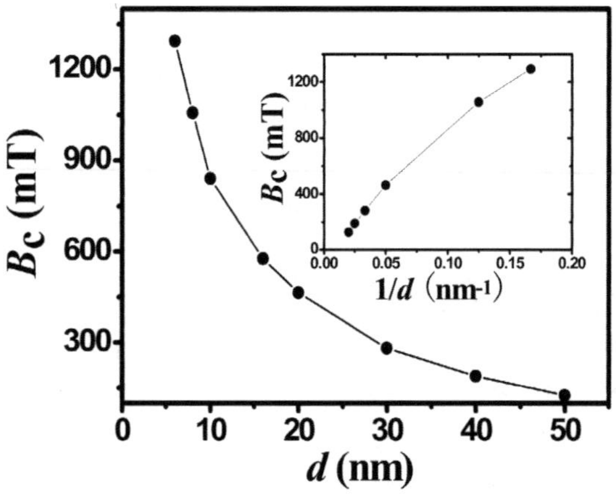

Figure 4. d and $1/d$ (inset) dependences of B_c for Fe nanowires with l = 200 nm and d = [6, 50] nm.

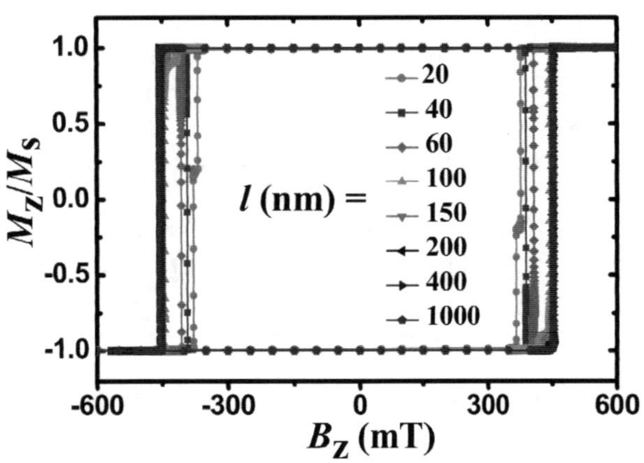

Figure 5. Hysteresis loops of Fe nanowires with d = 20 nm and l = [20, 1000] nm.

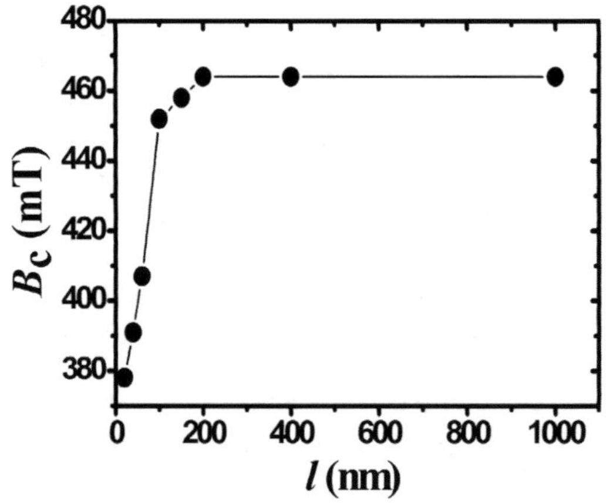

Figure 6. l dependence of B_c for Fe nanowires with d = 20 nm.

Acknowledgement
This work was supported by Chongqing Natural Science Foundation (Grant No. CSTC2009BB8102) and Fundamental Research Funds for the Central Universities (Grant No. XDJK2009C193).

References

[1] Huang Q, Davis D and Podlaha E J 2006 *J. Appl. Electrochem.* **36** 871
[2] Simonds J L 1995 *Phys. Today* **48** 26
[3] Paulus P M, Luis F, Kröll M, Schmid G and Jongh L J 2001 *J. Magn. Magn. Mater.* **224** 180
[4] Wernsdorfer W, Doudin B, Mailly D, Hasselbach K, Benoit A, Meier J, Ansermet J P and Barbara B 1996 *Phys. Rev. Lett.* **77** 1873
[5] García J M, Asenjo A, Velázquez J, García D, Vázquez M, Aranda P and Ruiz-Hitzky E 1999 *J. Appl. Phys.* **85** 5480
[6] Masuda H and Fukuda K 1995 *Science* **268** 1466
[7] Ferré R and Ounadjela K 1997 *Phys. Rev.* B **56** 14066
[8] Lodder J C and Li C Z 1989 *IEEE Trans. Magn.* **25** 4171
[9] Brown G, Novotny M A and Rikvold P A 2000 *J. Appl. Phys.* **87** 4792
[10] Frei E H, Shtrikman S and Treves D 1957 *Phys. Rev.* **106** 446
[11] Aharoni A and Shtrikman S 1958 *Phys. Rev.* **109** 1522
[12] Hertel R and Kirschner J 2004 *Physica* B **343** 206
[13] Donahue M J and Porter D G 1999 OOMMF User's Guide Version 1.0
[14] Dao N, Homer S R and Whittenburg S L 1999 *J. Appl. Phys.* **86** 3262
[15] Xiao J J, Sun C, Xue D S and Li F S 2001 *Acta Phys. Sin.* **50** 1605

Journal of Physics: Conference Series **266** (2011) 012023 doi:10.1088/1742-6596/266/1/012023

Pressure effect on the Curie temperature of $La(Fe_{0.88}Si_{0.12-y}Al_y)_{13}$

H Yako[1], S Fujieda[2], A Fujita[1] and K Fukamichi[2]

[1] Department of Materials Science, Graduate School of Engineering, Tohoku University, Aoba-yama 6-6-02, Sendai 980-8579, Japan
[2] Institute of Multidisciplinary Research for Advanced Materials, Tohoku University, Katahira 2-1-1, Sendai 980-8577, Japan

E-mail: b0td5309@s.tohoku.ac.jp

Abstract. The pressure dependence of the Curie temperature T_C has been investigated up to 1GPa for $La(Fe_{0.88}Si_{0.12-y}Al_y)_{13}$ itinerant-electron metamagnet. The magnetization change at the thermally induced first-order transition at T_C becomes sluggish with increasing y. On the other hand, the magnitude of the pressure coefficient of T_C, $|d\ln T_C/dP|$, is enhanced by the partial substitution of Al for Si. From the relation between the thermal variation of the spin fluctuations and the critical magnetic field of the metamagnetic transition, the change of thermal variation of spin fluctuation is closely related with the enhancement of $|d\ln T_C/dP|$ after partial substitution of Al.

1. Introduction

$La(Fe_xSi_{1-x})_{13}$ compounds with $0.86 \leq x \leq 0.90$ exhibit the thermal induced first-order transition between the paramagnetic and the ferromagnetic states at the Curie temperature T_C [1, 2]. The itinerant-electric metamagnetic (IEM) transition, that is, the first-order transition from the paramagnetic to the ferromagnetic state above T_C is induced by applying magnetic field [1-3]. This compound exhibits the magnetocaloric effects: the isothermal magnetic entropy change ΔS_m and adiabatic temperature change ΔT_{ad} due to IEM transition [4-7]. Enlargement of temperature range for the large ΔS_m is established by the decrease of the thermal variation of IEM critical field, $\mu_0 dHc/dT$, after the partial substitution of Al for Si in $La(Fe_{0.88}Si_{0.12})_{13}$, although the decrease of $\mu_0 dHc/dT$ results in a slight decrease of the maximum of ΔS_m. It has been pointed out that the IEM transition of $La(Fe_xSi_{1-x})_{13}$ becomes sharper on applying pressure P, accompanied by an increase of the magnetization change ΔM at the transition temperature [8, 9]. Since ΔS_m is enhanced by increase of ΔM, an application of pressure is one of the convenient methods to enhance ΔS_m. However, the large pressure-induced change of the IEM transition is originated from the influence of spin fluctuations, and the change of $\mu_0 dHc/dT$ means the change of thermal variation of spin fluctuations after the partial substitution of Al. The pressure effect of T_C is a subtle measure for the relation between spin fluctuations and influence of pressure on the magnetic states. In the present study, therefore, the pressure effect on T_C has been investigated for $La(Fe_{0.88}Si_{0.12-y}Al_y)_{13}$ with y = 0.01, 0.02 and 0.03.

2. Experimental details

$La(Fe_{0.88}Si_{0.12-y}Al_y)_{13}$ were prepared by arc-melting by using 99.9 mass% pure La, Fe and Al and 99.999 mass% pure Si in an argon gas atmosphere. The ingots were re-melted several times, and subsequent heat-treatment for homogenization was carried out in an evacuated quartz tube. The annealing temperature and duration were 1323 K and 10 days, respectively. The crystal structure and the lattice constant were determined by powder x-ray diffraction measurements with Cu Kα radiation at room temperature. The magnetization under hydrostatic pressure was measured with a SQUID magnetometer setting up with a Cu-Ti nonmagnetic pressure clamp cell. The pressure medium was a Daphne 7373 oil. Hydrostatic pressures were calibrated by measuring the shift of the superconducting transition temperature of Pb at low temperatures. By considering the pressure drift due to the solidification and volume contraction of the Daphne oil at temperatures around 100-200 K [10], the pressure calibration has been carried out by referring the available data for compounds such as $Lu(Co_{1-x}Ga_x)_2$ [11].

3. Results and discussion

Shown in Fig. 1 are the thermomagnetization curves for $La(Fe_{0.88}Si_{0.12-y}Al_y)_{13}$ with $y = 0.01$, 0.02 and 0.03 in a magnetic field of 0.3 T under hydrostatic pressures $p = 0.00$, 0.25, 0.50, 0.75 and 1.00 GPa. At ambient pressure, the compounds exhibit the first-order magnetic phase transition from a ferromagnetic to paramagnetic transition at T_C, exhibiting a drastic change of magnetization. With increasing y, the magnetization change at T_C becomes sluggish. The value of T_C for $La(Fe_{0.88}Si_{0.12})_{13}$ has been reported to be 195 K [2], and after partial substitution, T_C becomes 194, 192 and 189 K for $y = 0.01$, 0.02 and 0.03, respectively. Note that the itinerant-electron metamagnetic transition just above T_C is confirmed after partial substitution of Al up to $y = 0.03$. With increasing hydrostatic pressures, a discontinuous change of thermomagnetization curves at T_C becomes clear as seem from Fig.1. In addition, a significant decrease in T_C is observed by applying hydrostatic pressure.

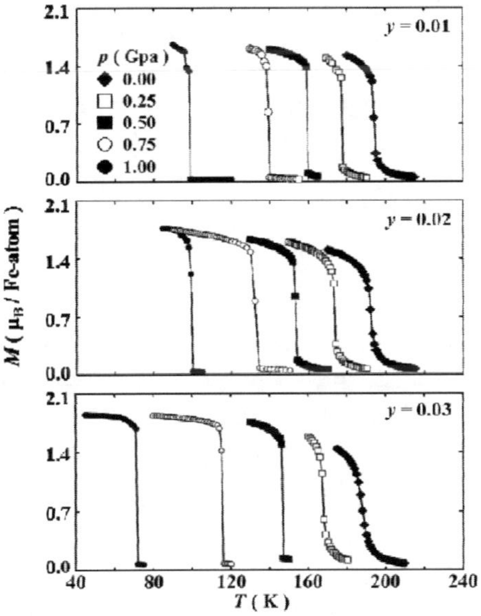

Figure 1. Thermomagnetization curves at various pressures for $La(Fe_{0.88}Si_{0.12-y}Al_y)_{13}$ with $y = 0.01$, 0.02 and 0.03 in a magnetic field of 0.3 T.

Figure 2 displays the pressure P dependence of T_C for $La(Fe_{0.88}Si_{0.12-y}Al_y)_{13}$ with $y = 0.01$, 0.02 and 0.03. For all the specimens, T_C decreases by applying pressure, namely, T_C has a negative pressure coefficient. At low pressure ranges, the value of T_C against P shows almost linear variation. On the other hand, the decrease of T_C gradually deviates from the linear variation with increasing P and the rate of the decrease of T_C becomes larger above 0.5 GPa. For $La(Fe_{0.89}Si_{0.11})_{13}$, the pressure dependence of T_C in wide pressure ranges obeys the following relation [9,12,13],

$$T_C = \beta(P_C - P)^{1/2} \tag{1}$$

where P_C is the critical pressure at which T_C becomes 0 K, and the square-root P dependence in Eq.(1) is explained by the spin fluctuation theory of the IEM transition based on the Landau expansion. According to the theory, free energy $F(M)$ as a function of magnetization M is given as [8, 14]

$$F(M) = \frac{1}{2}A(T)M^2 + \frac{1}{4}B(T)M^4 + \frac{1}{6}C(T)M^6 \tag{2}$$

with

$$\begin{aligned} A(T) &= a + \frac{5}{3}b\xi(T)^2 + \frac{35}{9}c\xi(T)^4 + 2\kappa C_{mv}P, \\ B(T) &= b + \frac{14}{3}c\xi(T)^2,\ C(T) = c \end{aligned} \tag{3}$$

where κC_{mv} and $\xi(T)^2$ are the magnetovolume coupling coefficient and the mean square amplitude of spin fluctuations, respectively. The parameters a, b and c are related to the 3d band structures. Note that the coefficients $A(T)$ and $B(T)$ are modified by the magnetovolume coupling [8], although such contributions are neglected for simplicity. The first-order transition at T_C takes place in the range $5/28 < ac/b^2 < 3/16$ [15, 16]. From the comparison of the phase diagrams obtained by theory and our experiments, the increase of the Fe concentration is associated with the increase of ac/b^2, resulting in a clear first-order transition. For $La(Fe_xSi_{1-x})_{13}$ at higher Fe concentrations, in addition, the clearer the first-order transition, the larger the pressure effect of T_C and the value of $d\ln T_C/dP$ with $x = 0.86$, 0.88 and 0.89 were reported to be -0.31, -0.46 and -0.54 GPa^{-1}, respectively [8, 9]. After partial substitution of Al, the change of thermomagnetization curve at T_C becomes vague with increasing y, namely, the phase transition of $La(Fe_{0.88}Si_{0.12-y}Al_y)_{13}$ comes gradually closer to the second-order. On the other hand, the initial slope in low P ranges and curvature in high P ranges become larger with increasing y as shown in Fig. 2.

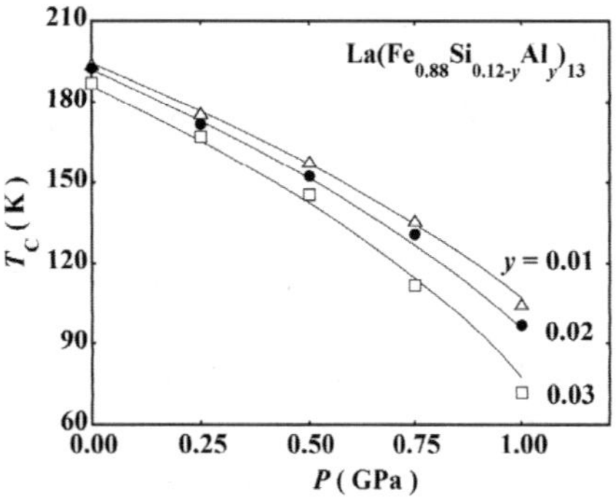

Figure 2. Pressure dependence of the Curie temperature T_C of $La(Fe_{0.88}Si_{0.12-y}Al_y)_{13}$ with $y = 0.01$, 0.02 and 0.03.

To reveal the change of the pressure effect of T_C, the Al concentration dependence of $d\ln T_C/dP$ and β is displayed in Fig. 3. For comparison, the data of $La(Fe_{0.88}Si_{0.12})_{13}$ [8] before partial substitution is also shown. After partial substitution of Al, the value of $|d\ln T_C/dP|$ decreases for the compound with $y = 0.01$. Such a change is advisable in terms of the distinctness of the IEM transition and the magnitude of $d\ln T_C/dP$ in $La(Fe_xSi_{1-x})_{13}$ [3, 9]. However, the concentration dependence of $|d\ln T_C/dP|$ turns into an increasing tendency above $y = 0.01$. The concentration dependence of β is similar to that of $|d\ln T_C/dP|$. This similarity is quite expected because the root function in Eq. (1) becomes asymptotical when a value of P departs from P_C. Accordingly, we focus on the variation of β to explain the influence of Al on the pressure effect of T_C. From Eqs. 1 and 2, β is derived as [9]

$$\beta \propto \{\kappa C_{mv} / (b\delta)\}^{1/2} \quad (4)$$

The parameter δ is the proportional coefficient of the relation $\xi(T)^2 = \delta T^2$ [17]. What has to be noticed is that the value of δ is related to the energy distribution of spin-fluctuation spectrum. Accordingly, it is apparent that β governs the interrelation between the temperature and pressure scales. The value of κC_{mv} of $La(Fe_xSi_{1-x})_{13}$is about 0.5 %/μ_B^2 for $x = 0.86$ and 0.7 %/μ_B^2 for $x = 0.88$ [2]. Since the reported value of κC_{mv} for $La(Fe_{0.87}Al_{0.13})_{13}$ is about 0.5 %/ μ_B^2 [18], the influence of partial substitution of Al up to $y = 0.03$ on κC_{mv} is expected to become small. Direct observation of the thermal variation of $\xi(T)^2$ is difficult, however, the following relation between the temperature dependence of the critical field $\mu_0 H_C$ of the IEM transition and the thermal variation of $\xi(T)^2$ is useful [19].

$$\mu_0 H_C \propto \Delta M\, b\{\xi(T)^2 - \xi(T_C)^2\} = \Delta M\,(b\delta)\{T^2 - T_C^2\} \quad (5)$$

where ΔM is the magnetization change at the IEM transition. Note that the T^2-T_C^2 term is approximated as T around T_C, therefore, $\mu_0 H_C$ is proportional to $\Delta M(b\delta)\,T = \lambda T$. From the magnetization measurements, the value of λ is found to be smaller with increasing y. The magnitude of $\mu_0 dH_C/dT$ becomes smaller with increasing Al concentration and the value for y = 0.00, 0.01, 0.02, 0.03 is evaluated to be about 0.23, 0.21, 0.20 and 0.19 T/K, respectively. Although, ΔM slightly decreases against y, the change is not so drastic because the IEM transition is kept even at $y = 0.03$. Accordingly, the increase of y brings about the decrease of the magnitude of $b\delta$. As a result, the magnitude of β shown by Eq. (4) is enhanced, turning out the larger $|d\ln T_C/dP|$. In our previous results, the change of b is sensitively induced by change of Fe concentration or the partial substitution of Ce for $La(Fe_xSi_{1-x})_{13}$ [19, 20], while the value of λ is almost the same as that of the partial substitution of Ce and also the hydrogenation [21]. Namely, the change of λ is mainly dominated by the change of δ. Consequently, the enhancement in the pressure effect on T_C is closely associated with the decrease of the thermal variation of $\xi(T)^2$ with a small δ after the partial substitution of Al

One may notice that not only β but also P_C influences the variation of T_C at high P ranges. Actually, the decrease of P_C with increasing y is confirmed by extrapolating the T_C-P line by using Eq. (1). This tendency is also opposite to that of $La(Fe_xSi_{1-x})_{13}$, in which a broader first-order transition corresponds to the larger value of P_C. Furthermore, the Landau expansion theory gives no straightforward explanation for decrease of P_C followed by a broadening of the IEM transition. To this issue, further investigations such as direct determination of P_C are necessary.

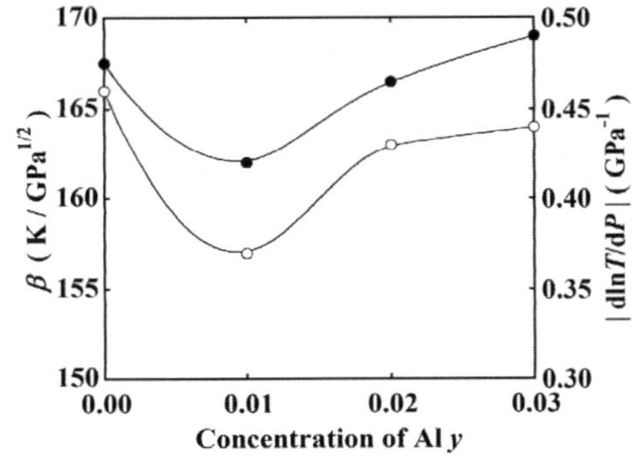

Figure 3. The Al concentration dependence of the coefficient β(●) and the magnitude of $|d\ln T_C/dP|$(○) for $La(Fe_{0.88}Si_{0.12-y}Al_y)_{13}$ with y = 0.01, 0.02 and 0.03.

4. Conclusion

The influence of partial substitution of Al on the pressure effect on the Curie temperature T_C has been investigated for itinerant-electron metamagnetic $La(Fe_{0.88}Si_{0.12-y}Al_y)_{13}$. The pressure P dependence of T_C shows a linear variation at low pressures, and gradually deviates from the linear variation with increasing P, following a line of the square-root dependence, $T_C = \beta(P_c-P)^{1/2}$, where P_C is the critical pressure. Although the phase transition comes close to the second-order type after partial substitution, the value of β becomes larger. In addition, the magnitude of the pressure coefficient $|d\ln T_C/dP|$ is also enhanced with increasing the Al concentration y. From the spin fluctuation theory of itinerant-electron metamagnetic transition, both the value of $d\ln T_C/dP$ and the thermal variation of the critical field of the itinerant-electron metamagnetic transition $\mu_0 dHc/dT$ are highly associated with the change of thermal variation of $\xi(T)^2$. Since the decrease of $\mu_0 dHc/dT$ with increasing y indicates the small thermal variation of $\xi(T)^2$, the enhancement of pressure effect on T_C of $La(Fe_{0.88}Si_{0.12-y}Al_y)_{13}$ is explained in terms of the change of thermal variation of $\xi(T)^2$.

Acknowledgement

This work was partly supported by a Grant-in-Aid for Scientific Research (B), No. 20360289 from JSPS and Global COE Program "Materials Integration (International Center of Education and Research), Tohoku University," MEXT, Japan.

References

[1] Fujita A, Akamatu K and Fukamichi K 1999 *J. Appl. Phys.* **85** 4756
[2] Fujita A, Fujieda S and Fukamichi K 2001 *Phys. Rev. B* **65** 014410
[3] Fukamichi K and Fujita A 2000 *J. Mater. Sci. Technol.* **16** 167
[4] Fujieda S, Fujita A and Fukamichi K 2002 *Appl. Phys. Lett.* **81** 1276
[5] Fujita A, Fujieda S, Hasegawa Y and Fukamichi K 2003 *Phys. Rev. B.* **67** 014410
[6] Hu F. X, Shen B. G, Sun J. R, Cheng Z. H, Rao G. H and Zhang X. X 2001 *Apple. Phys. Lett.* **78** 3675
[7] Fujita A, S. Fujieda, Fukamichi K, Yamasaki Y and Iijima Y 2002 *Mater. Trans.* **43** 1276
[8] Yamada H, Fukamichi K and Goto T 2002 *Phys. Rev. B* **65** 024413
[9] Fujita A, Fukamichi K, Yamada M and Goto T 2006 *Phys. Rev. B* **73** 104420
[10] Murata K, Yoshino H, Yadav H O, Honda Y and Shirakawa N 1997 *Rev. Sci. Instrum.* **68** 2490
[11] Saito H, Yokoyama T, Fukamichi K, Kamishima K and Goto T 1999 *Phys. Rev. B* **59** 8725
[12] Pfleiderer C, McMullan G. J, Julian S. R and Lonzarich G.G 1997 *Phys. Rev. B* **55** 8330
[13] Pfleiderer C, Reznik D, Pintschovius L, Löhneysen H. V, Garst M and Rosch A 2004 *Nature* **427** 227
[14] Yamada H and Terao K 1994 *J. Phys.: Condens. Matter* **6** 10805
[15] Moriya T 1986 J. Phys. Soc. Jpn. **55** 357
[16] Yamada H, Fukamichi K and Goto T 2003 Physica B **327** 148
[17] Yamada H 1993 *Phys. Rev. B* **47** 11211
[18] Irisawa K, Fujita A, Fukamichi K, Yamada M, Mitamura H, Goto T and Koyama K 2004 *Phys. Rev. B* **70** 214405
[19] Fujita A and Fukamichi K 2005 *IEEE Trans. Magn.* **41** 3490
[20] Fujita A, Fujieda S and Fukamichi K 2006 *J. Appl. Phys.* **99** 08K910
[21] Fujita A, Fujieda S and Fukamichi K 2009 The 3rd Int. Conf. on Magnetic Refrigeration at Room *Temperature* ed by Egolf P. W pp 33

Journal of Physics: Conference Series **266** (2011) 012024 doi:10.1088/1742-6596/266/1/012024

Domain wall propagation in single crystalline iron wires

A Yamaguchi[1, 2 †], Y Kasatani[1] and H Miyajima[1]

[1] Department of Physics, Keio University, Hiyoshi, Yokohama, Kanagawa, 223-8522, Japan

[2] PRESTO, JST, Honcho 4-1-8, Kawaguchi, Saitama, 332-0012, Japan

E-mail: yamaguch@phys.keio.ac.jp

Abstract. The characterization of magnetization reversal in a magnetic domain wall (DW) in a single crystalline Fe wire with crystalline anisotropy and shape anisotropy has been investigated by the giant magnetoresistance (GMR) measurement. The DW propagation velocity reached about 1 km/s in 110 Oe field at 77 K. The broad distribution of the propagation velocity versus the switching field in the Fe wire with longitudinal axis parallel to the $\langle 110 \rangle$ direction was found. The magnetic anisotropy energy including the crystalline and shape anisotropies play an important role in the nucleation and propagation of DW.

1.Introduction

General interest in the magnetization reversal in nano- and micron-sized magnets has been studied intensively from both fundamental and applied points of views. Since the majority of magnetization reversal in ferromagnets takes place by the propagation of magnetic domain wall (DW), understanding DW dynamics is a vital part of the study on the magnetizing process [1 - 12]. Much attention has been focused on the studies of DW. One of the recent findings is the experimental observation of Walker breakdown [1, 9, 10]. Just above the Walker breakdown field, the DW velocity against the driving field causes negative differential DW mobility.

The characterization of DW dynamics considering the crystalline anisotropy and pinning force is an important issue, owing to their fundamental physics and implementation of high-speed operation of information devices. In this paper, we present an experimental study of a single DW motion in a single crystalline Fe micro-wire via the giant magnetoresistance (GMR) measurement [7, 11, 12]. We also discuss the magnetization reversal process of the DW from the viewpoint of micromagnetics.

2. Experimental Procedure

A high-quality single crystalline film was grown on a polished MgO(001) substrate using molecular beam epitaxy (MBE). Further details are found in Refs. 11 and 12. We prepared a single crystalline wire Fe(50 nm)/Au(10 nm)/$Fe_{19}Ni_{81}$ (10 nm)/Au(3 nm) by cutting out the film. The crystalline orientation of the longitudinal axis of the wire can be selected as shown in Fig. 1(a). Here, we focus on the magnetization reversal process of Fe(001)$[110]_{bcc}$/Au/$Fe_{19}Ni_{81}$ wire, hereafter called the [110]-wire. The structural characterization was performed by using a reflection high-energy electron diffraction

† Corresponding to AY:yamaguch@phys.keio.ac.jp.

Journal of Physics: Conference Series **266** (2011) 012024 doi:10.1088/1742-6596/266/1/012024

(RHEED) image and X-ray diffraction (XRD). The RHEED image indicates good crystal orientation and flatness [11, 12]. The XRD result is shown in Fig.1(b). Only the (00*n*) diffraction peaks are observed. This indicates that the film is strongly textured in the (001) planes and that it is epitaxially grown on the MgO substrate. The layers of Au and $Fe_{19}Ni_{81}$ were subsequently deposited on 50 nm thick epitaxial Fe layers at room temperature. Then, multi-layer wires of 1 μm width were fabricated by combination of electron beam lithography, Ar ion milling and lift-off methods. One end of the wire is tapered to a sharp point in order to prevent the formation of DW nucleation and to control the DW injection into the wire.

In order to study the magnetization reversal in micron-sized wires, we investigate the giant magnetoresistance (GMR) effect as a probe of the DW position in the wire [7, 11, 12]. The change in the GMR serves to probe the magnetization dynamics of a micron-sized Fe wire. A magnetic field *H* is applied along the longitudinal axis of the wire, and the resistance *R* at 77 K is determined using a standard four-point dc technique. The time evolution of magnetization reversal in the wire was monitored by a differential preamplifier and real-time storage oscilloscope with a sampling rate of 20×10^9 per second and bandwidth of 4 GHz. The magnetoresistance ratio and the time variation of the voltage derived from the time variation of the DW's position in the single-crystalline $Fe(001)[110]_{bcc}/Au/Fe_{19}Ni_{81}$ wire is shown in Fig. 2(a). The linear variation in resistance with time indicates that the propagation velocity *v* of the DW is constant during magnetization reversal in the wire.

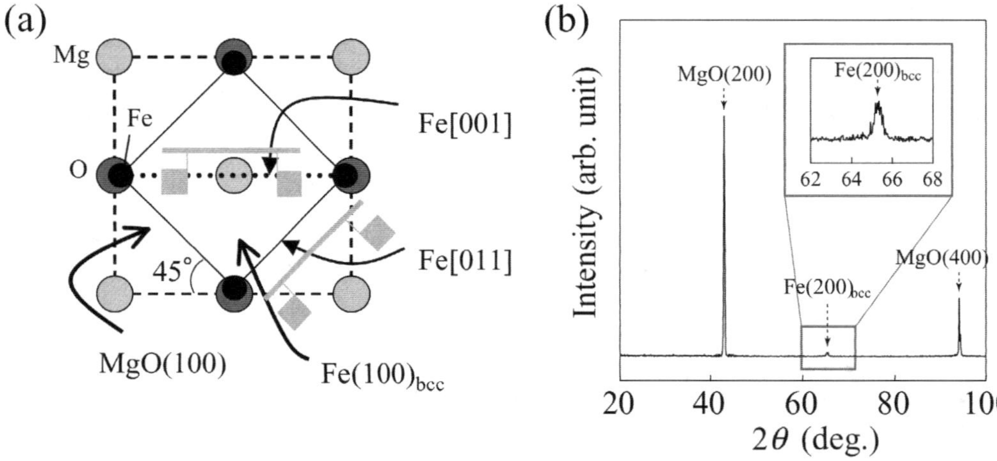

Figure 1. (a) Crystallographic relationship and interface structure of epitaxial bcc Fe(001). Relationship between orientation of longitudinal axis of wire and crystalline direction. (b) XRD pattern of the epitaxial Fe(001) film.

3. Results and Discussion

As the DW velocity *v* measurements were repeated while sweeping the external magnetic field *H*, the switching filed of the [110]-wire ranges from 60 to 110 Oe. The *v*-*H* characteristics in Fig. 2(b) are homogeneously distributed. For the [110]-wire, we can estimate the mobility from $v=\mu(H-H_{SW})$ and evaluate $\mu=13.3\,\mathrm{m}/(\mathrm{Oe}\cdot\mathrm{s})$ and $H_{SW}=60\,\mathrm{Oe}$.

In order to reveal the microscopic magnetization reversal process, a simple model has been developed [4, 5, 11, 12, 13]. Figure 3 summarizes the geometry relation between the anisotropy and applied field. It is assumed that the in-plane uniaxial shape anisotropy along the wire axis is assumed to superimpose the cubic magnetocrystalline anisotropy and that DW propagation as opposed to DW

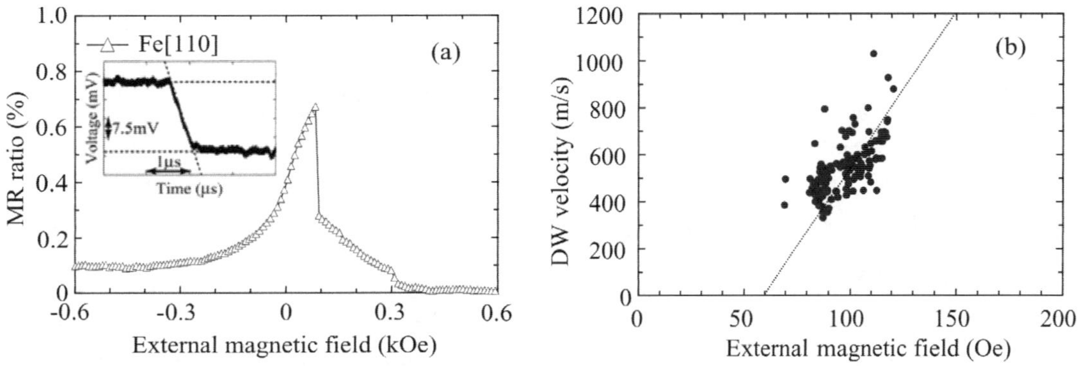

Figure 2. (a) Magnetoresistance ratio as a function of applied field. The inset shows the time variation of the resistance during magnetization reversal. (b) Dependence of DW velocity v on the amplitude of the applied magnetic field H.

nucleation is carried out in the magnetic switching process. We focus the magnetization at the centre magnetization in the DW. The resistance in the absence of magnetic field has already risen as shown in Fig. 2(a). This indicates that the magnetocrystalline anisotropy forced the magnetization to tilt from the longitudinal axis of the wire [9, 10]. Therefore, according to Cowburn *et al.* [4], Ebels *et al.* [5] and Zhan *et al.* [13], we consider the stable single domain spin states aligned along one of the effective magnetization easy axis determined by the competition between the shape anisotropy and the magnetocrystalline anisotropy [6, 11, 12]. Figure 3(a) shows the present coordinate model. The angle γ is the tilt angle of the effective axis from the wire axis in the plane.

The energy of magnetocrystalline anisotropy E_{K} and shape anisotropy E_{D} is described by

$$\begin{aligned} E_{\mathrm{K}} &= \frac{K_1}{64}\left[(3-4\cos 2\theta'+\cos 4\theta')(1-\cos 4\phi')+8(1-\cos 4\theta')\right] \\ &\quad+\frac{K_2}{256}\left[(2-\cos 2\theta'-2\cos 4\theta'+\cos 6\theta')(1-\cos 4\phi')\right] \end{aligned} \tag{1}$$

and

$$E_{\mathrm{D}} = -\frac{M_S^2}{2\mu_0}\left[\begin{array}{l} N_x\left(-\sin\theta'\cos\phi'\sin\gamma+\cos\theta'\cos\gamma\right)^2+ \\ N_y\sin^2\theta'\sin^2\phi'+N_z\left(\sin\theta'\cos\phi'\cos\gamma+\cos\theta'\sin\gamma\right)^2 \end{array}\right] \tag{2}$$

where K_1 and K_2 are the first- and send-order magnetic anisotropy energy, respectively. M_{S} and μ_0 denote the saturation magnetization and vacuum permeability. θ' and ϕ' are the angular components of the magnetization in the spherical coordinate. N_x, N_y and N_z are the demagnetizing coefficient along x-, y-, z-axis, respectively.

Here, we neglect the second-order magnetocrystalline energy K_2 because $K_1=(5.2\pm0.1)\times10^4\ \mathrm{J/m^3}$ is further larger than $K_2=(0\pm0.5)\times10^4\ \mathrm{J/m^3}$ [14].

Figure 3(b) shows the whole anisotropy energy E_{total} given by $E_{\mathrm{K}}+E_{\mathrm{D}}$ as function of θ' and ϕ'. The contour map of Fig. 3(b) is drawn in Fig. 3(c). The calculation indicates that the shape magnetic anisotropy is much larger than the magnetocrystalline anisotropy and basically dominates magnetization reversal. Figure 3(c) reveals that there is the region where anisotropy energy is equivalent to zero around $(\theta',\phi')=(0^\circ,\phi')$ and $(0\le\theta'\le90^\circ,0)$. As is evident from Fig. 3(c), the expected magnetization reversal proceeds as the following. When the increase in the applied field, the magnetization within the DW rotates from the initial direction to the hard axis out of the plane. Next, the magnetization rotates from the previous direction to the nearest in-plane easy axis as the angle θ' does not rotate but ϕ' rotates because of the flat energy level along the direction of ϕ. Thus, the magnetization rotates from the initial direction to the nearest easy axis of the magnetocrystalline anisotropy $\langle 100 \rangle$ in the plane. The *v*-*H* characteristics in Fig. 2(b) are attributed to the random DW nucleation due to the flat energy level.

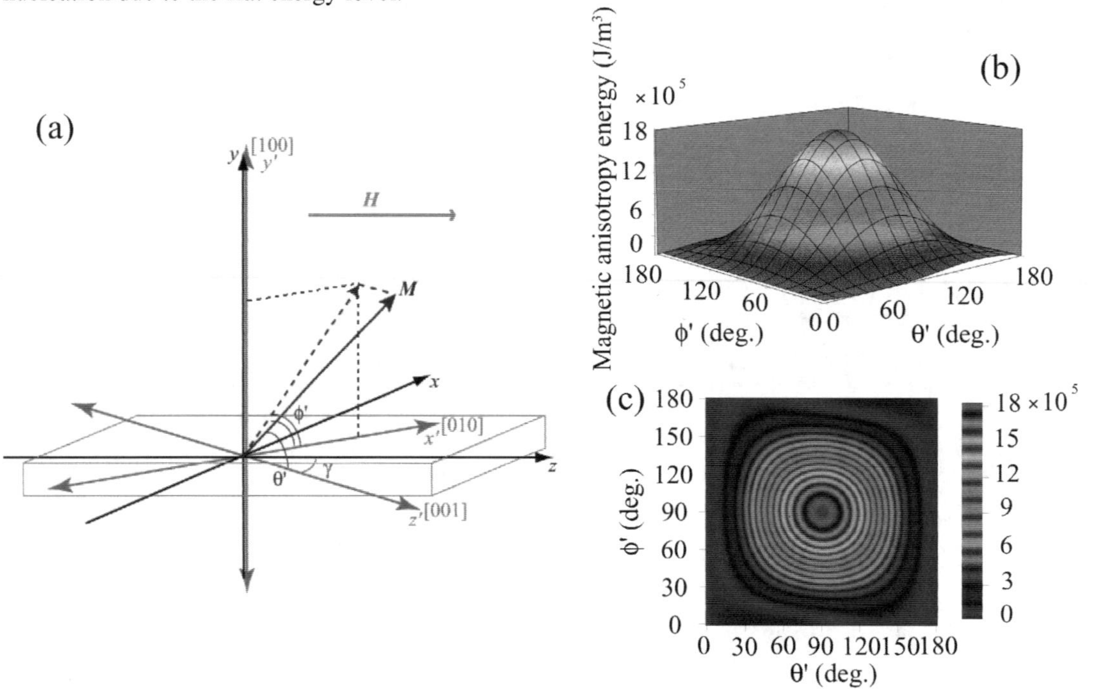

Figure 3. (a) Schematic model geometry and symbol definitions. Image plot and contour plot of magnetic anisotropy energy as a function of polar angle θ' and ϕ'.

4. Conclusions

In summary, the magnetization reversal in a micron-sized epitaxial Fe(001)[110]/Au/$Fe_{19}Ni_{81}$ wire is investigated and it is shown that the switching process is dominated by not only the shape anisotropy but also magnetocrystalline anisotropy. The result is an important step toward a complete microscopic understanding of DW dynamics. It also raises a possibility of the controlling the magnetization reversal by modifying the anisotropy energy barriers.

Acknowledgements

The present study was partly supported by MEXT Grants-in-Aid for Scientific Research in a Priority Area, a JSPS Grant-in-Aid for Scientific Research and the JST PRESTO program.

References

[1] Schryer N L and Walker L R 1974 *J. Appl. Phys.* **45** 5406
[2] Slonczewski J C 1972 *Int. J. Magn.* **2** 85; 1974 *J. Appl. Phys.* **44** 1759; 1974 *J. Appl. Phys.* **45** 2705
[3] Thiaville A, Garcia J M and Miltat J, 2002 *J. Magn. Magn. Mater.* **242-245**, 1061

[4] Cowburn R P, Gray S J, Ferré J, Bland J A C and Miltat J, 1995 *J. Appl. Phys.* **78**, 7210; Cowburn R P, Gray S J and Balnd J. A. C., 1997 *Phys. Rev. Lett.* **79**, 4018

[5] Ebels U, Adeyeye A O, Gester M, Cowburn R P, Daboo C and Bland J A C, 1997 *Phys. Rev.* B **56**, 5443

[6] Hassel C, Römer F M, Meckenstock R, Dumpich G and Lindner J, 2008 *Phys. Rev.* B **77**, 224439.

[7] Ono T, Miyajima H, Shigeto K, Mibu K, Hosoito N and Shinjo T, 1999 *Science* **284**, 468

[8] Atkinson D, Allwood D A, Xiong G, Cooke M D, Faulkner C C and Cowburn R P, 2003 *Nature Mat.* **2**, 85.

[9] Beach G S D, Nistor C, Knutson C, Tsoi M and Erskine J L, 2005 *Nature Mat.* **4**, 741

[10] Kondou K, Ohsima N, Kasai S, Nakatani Y and Ono T, 2008 *App. Phys. Exp.* **1**, 061302

[11] H. Yamamoto, Y. Kasatani, A. Yamaguchi and H. Miyajima, 2010 *J. Phys.: Conf. Series.* **200**, 042028

[12] Kasatani Y, Yamaguchi A and Miyajima H, to be published in 2010 Phys. Rev. B **86**

[13] Zhan Q F, Vandezande S, Temst K and VanHaesendonck C, 2009 *Phys. Rev.* B **80**, 094416

[14] Graham C D, 1958 *Phys. Rev.* **112**, 1117

High Microwave Magnetic Permeability of Composites with Submicron Iron Flakes

M A Abshinova[1], S M Matitsine, L Liu, C R Deng, L B Kong
Temasek Laboratories, National University of Singapore, 117411, Singapore

E-mail: tslma@nus.edu.sg

Abstract. An effective method for the development of magnetic materials with high microwave permeability on the basis of carbonyl iron powders has been proposed. This method consists in the modification of particle morphology from the spherical shape into flaky one with subsequent silica coating of the flakes by sol-gel method. Composites with submicron flaky iron powders (thickness < 1 μm) exhibit remarkable improvement in the microwave magnetic spectra compared to the spherical powder based composites. Even a thin silica overlayer provides a sufficient insulation of the metal particle and decreases the permittivity values.

1. Introduction

The development of magnetic composite materials utilized in various microwave devices has attracted intensive research interest in recent years. Such composites are made by embedding magnetic particles into nonmagnetic polymer resins. Among different magnetic fillers, carbonyl iron powders have received much attention due to the high values of relative complex permeability in GHz range, high saturation magnetization $4\pi M_s$ and high Curie temperature [1–3]. Produced by the thermal decomposition of iron pentacarbonyl, carbonyl iron powders contain more than 96 % of iron, along with carbon, oxygen and nitrogen. However, depending on the manufacturing conditions, carbonyl iron powders of the various types are classified into two groups with different microstructural properties of their particles [4, 5].

Primary carbonyl iron powders are characterized by onion microstructure of their particles and by the presence of the impurities which leads to the formation of iron carbide (Fe_3C), iron nitride (Fe_4N) and iron oxide (Fe_2O_3) phases [4, 6]. In contrast to the primary powders, carbonyl iron powders reduced with hydrogen are extremely pure, mechanically soft, and onion microstructure is replaced by the polycrystalline one [4, 5]. In terms of magnetic properties it leads to the intensification of domain wall motion and higher values of permeability. However, eddy current losses decrease the values of permeability at high frequencies. In its turn, eddy current losses in composite materials with spherical inclusions are given by

$$P_{eddy\ current} = 0.2\pi^2 \frac{d^2\mu_m}{\rho p} 10^{-9} \quad (1)$$

where d is the particle diameter, ρ is the resistivity of the particle, p is the loading factor, and μ_m is the initial permeability of the composite material [6]. Generally, it is better to use particles with the sizes

[1] To whom any correspondence should be addressed.

not exceeding the skin depth, in order to decrease eddy current losses. The skin depth δ is defined as follows

$$\delta = \sqrt{\frac{2}{\omega\mu\sigma}} \quad (2)$$

where ω is the angular frequency, μ is the intrinsic permeability, and σ is the conductivity [7]. For pure iron the skin depth does not strongly depend on the frequency starting from 5 GHz, and is about 1 – 2 μm in the frequency range above 5– 6 GHz. Therefore, the ideal particle size should be within 1 – 2 μm.

Eddy current losses can be further reduced by modifying the shape of the particles from spherical one to flaky [8, 9]. For high frequency applications, the optimum thickness of the flakes should be less than the skin depth (around 1 μm) with aspect ratios from 10 to 100 [10]. However, despite the increasing values of permeability at high frequencies, flaky composites possess high values of permittivity, which is undesirable for some microwave devices. In order to avoid permittivity enhancement, a good insulation of the particles should be provided by coating process.

In this work, we studied the effect of particle morphology, as well as the influence of silica coating of iron particles on the changes in electromagnetic properties of iron composites. For this purpose, the reduced carbonyl iron powder having the smallest available particle size was chosen as a starting material.

2. Experimental results and discussion

Reduced carbonyl iron powder SM (median particle size from 2.8 to 3.5 μm) characterized by spherical particle shape and polycrystalline structure of the particles (BASF, Germany) was used as a starting material. The purity of SM carbonyl iron powder was above 99 % containing carbon < 0.1%, nitrogen < 0.1%, and oxygen < 0.55 %. A commercially available silicone elastomer (SYLGARD 184, Dow Corning, USA) was used as a polymer matrix material. Tetraethyl orthosilicate (TEOS) (95%, Sigma-Aldrich) and absolute ethanol (200 proof, Fisher Scientific Co) were used for silica coating. Ammonium hydroxide solution (NH_4OH, 28–30 % NH_3 basis, Sigma-Aldrich) was used as a catalyst for a sol-gel reaction of TEOS.

SM powder was milled in a PM 400 Retsch planetary ball mill at 100 rpm. The starting powder was loaded into a jar with steel milling balls of 5 mm in diameter. The ball to powder ratio was 10:1. The jar was filled with ethanol to avoid agglomeration. The mechanical milling was performed for 8, 12 and 16 hours. SM flakes were dried at 70°C for 15 hours. As-milled carbonyl iron flakes were named SM-8, SM-12, and SM-16 respectively. The morphology of raw material and powders milled for 8, 12, and 16 hours was studied by JEOL JSM-6340F field emission scanning electron microscope as shown in Figure 1. With increasing ball milling time, spherical particles deform into flaky ones, where after 8 hours of ball milling spherical particles still exist, after 12 hours no spherical particles can be found and the thickness of flakes is about 1μm, after 16 hours the thickness of flakes is below 1μm.

Silica coating of carbonyl iron particles was performed by the hydrolysis of tetraethyl orthosilicate (TEOS) via conventional sol-gel method [11]. The alkoxide solution was obtained by mixing TEOS and ethanol. In the other beaker we mixed ethanol and water with ammonium hydroxide stock solution in order to get the catalyst solution. Then the catalyst solution was poured into the alkoxide solution and stirred. Eventually, we mixed all solutions in glass vessel; the sol-gel process was initiated. The mixture was stirred with 300 rpm at 70°C. After 2 hours of stirring, coated carbonyl iron particles were filtered off, washed with ethanol and dried. Core-shell structure with the thickness of the shell about 10-15 nm was revealed by TEM image. The existence of a SiO_2 nanoshell can be also confirmed indirectly by permittivity measurements.

Composite fabrication procedure involved preparation of the composites with spherical inclusions (SM powder) and flakes (SM-8, SM-12, and SM-16). The volume concentrations of the filler in composite materials were calculated using the densities of the carbonyl iron and silicone elastomer, 7.8 g cm^{-3} and 1.05 g cm^{-3} respectively.

The static magnetic properties of iron powders were measured by a Vibrating Sample Magnetometer (VSM, EV9, ADE Magnetics, USA) with a maximum magnetic field of 20 kOe. The saturation magnetization of iron powders increases with ball milling time (Figure 2, left).

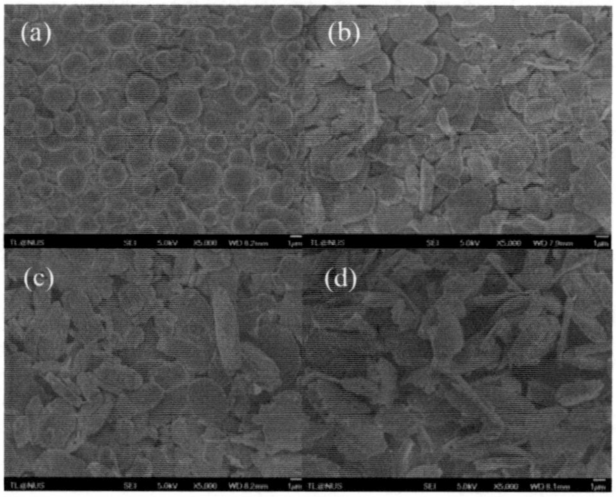

Fig. 1. SEM images of iron powders for different milling times: (a) 0 h, (b) 8 h, (c) 12 h, (d) 16 h.

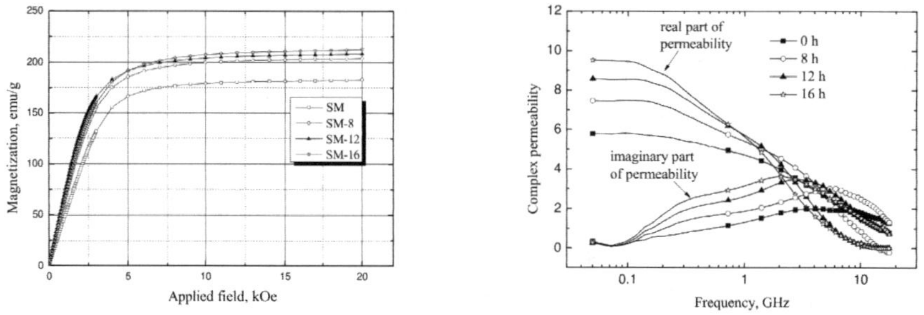

Fig. 2. Magnetization curve of SM, SM-8, SM-12, and SM-16 powders measured on pressed powder samples (left) and complex permeability spectra of composites containing 40 vol. % of powders with various milling time (right).

The relative complex permeability $\mu^* = \mu' - j\mu''$ and permittivity $\varepsilon^* = \varepsilon' - j\varepsilon''$ of the specimens were obtained by 7 mm coaxial line method. Toroidal samples with inner diameter of 3.04 mm and outer diameter of 7 mm were inserted into a coaxial line and the scattering parameters (reflection coefficient S_{11}, transmission coefficient S_{21}) were measured in the frequency range from 50 MHz to 18 GHz by a Vector Network Analyzer (VNA), Agilent N5230A. The complex permeability and permittivity were calculated from the scattering parameters.

Figure 2 (right) shows the real and imaginary parts of complex permeability for the composites containing 40 vol. % powders. The initial values of μ' increase from 5.76 to 7.46, 8.58 and 9.52 SM, SM-8, SM-12 and SM-16 powders respectively. The peak value of the μ'' increases from 1.8 to 3.0, 3.5 and 3.7 with increasing of ball milling time. It is caused by the reduction of eddy current losses and demagnetizing features of the flaky particles. It is known that the permeability values also decreases by the demagnetization field depending on the shape of the magnetic inclusion. The demagnetizing field is determined by the following equation

$$H = H_0 - NM \quad (3)$$

where H_0 is the external applied field, N is the demagnetizing factor and M is the magnetization. For spherical particles, N is equal to 1/3, while for flaky particles N is smaller than 1/3 along the elongated directions. Based on these, higher magnetic permeability can be obtained in flaky composites.

However, a disadvantage of flaky composites consists in the enhancement of permittivity values with ball milling time. It is caused by the enhancement of space-charge polarization between metallic particles with increase of surface area and is undesirable for some microwave applications. In order to avoid permittivity enhancement, a good insulation of the particles should be provided by coating process. In this work we performed silica coating of iron particles by sol-gel method (Figure 3).

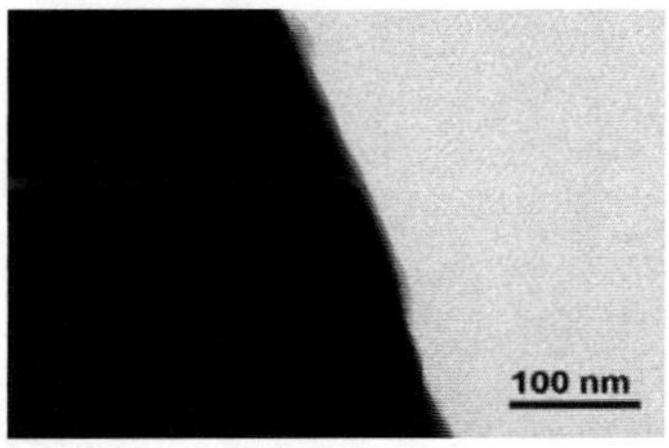

Fig. 3. TEM image of the SiO_2 coated flake.

It results in the slight decrease of permittivity of the composites filled with coated flakes while permeability values remain the same (Figure 4).

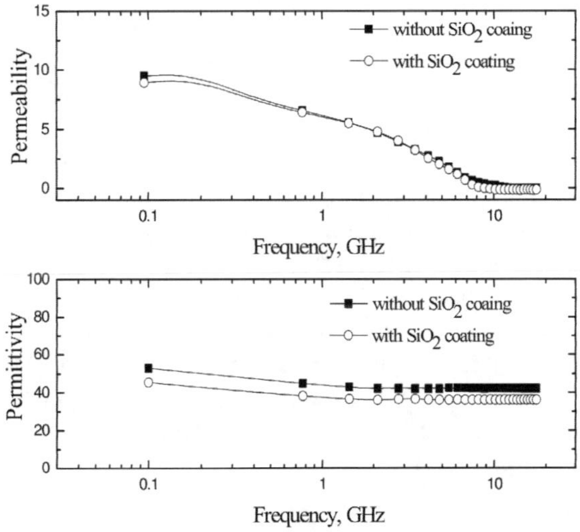

Fig. 4. Permeability and permittivity spectra for composites with 40 vol. % of SM-16 without and with SiO_2 coating.

Microwave losses in magnetic composite materials are attributed to hysteresis, domain wall resonance, natural ferromagnetic resonance and eddy current losses. The domain wall resonance usually occurs in frequency range from 1 to 100 MHz. Hysteresis loss originates from the time lag of magnetization vector behind the external magnetic field vector and is negligible in weak fields. In the GHz frequency range neither hysteresis loss no domain wall resonance contributes to the magnetic losses. The eddy current loss contribution to the imaginary part of permeability can be expressed as follows

$$\mu'' = \frac{2}{3}\pi\mu_0(\mu')^2 d^2 f\sigma \tag{4}$$

where μ_0 is permeability of vacuum, σ – electrical conductivity. This equation can be transformed to

$$\mu''(\mu')^{-2} f^{-1} = \frac{2}{3}\pi\mu_0 d^2\sigma \tag{5}$$

If the magnetic loss originates only from eddy current loss, the values of left-hand side of this equation will remain constant with frequency changes [12]. Figure 5 shows the value of $\mu''(\mu')^{-2}f^{-1}$ of composites filled with spherical and flaky iron particles. As it can be seen, the values for spherical particles are close to each other at frequencies above 1 GHz, while in the case of flaky filled composites they increase with the frequency increasing. Therefore, it means that magnetic losses in composites with spherical inclusions are mainly caused by the eddy current loss over 1–18 GHz, whereas in the flaky composites by natural resonance. Moreover, the natural resonance peak broadens which can be explained by the decreasing of crystallite size with ball-milling time increasing (Figure 2, right).

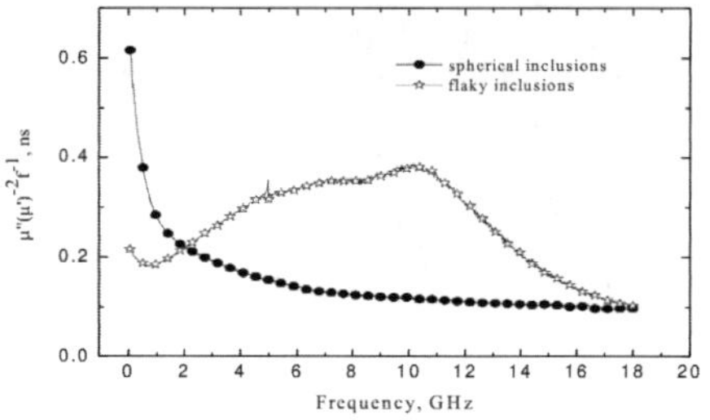

Fig. 5. Values of $\mu''(\mu')^{-2}f^{1}$ for composites with spherical (SM) and flaky (SM-16) iron powders versus frequency.

3. Conclusions

Comparative analysis of the electromagnetic parameters of magnetic composites filled with iron powders differ in their morphology was carried out. It was shown that the changes in the morphology of the particles have the crucial effect on the values and the character of electromagnetic spectra. Iron flakes were prepared by the high-energy planetary ball milling of the starting material with different milling time (8, 12, and 16 hours). The value of initial permeability and the peak value of imaginary part increases with the increasing of ball milling time. In order to avoid permittivity enhancement, silica coating of the particles was performed. It results in slight decrease of permittivity of the composite materials filled with coated flakes while permeability values remain the same. It was established that magnetic losses in composites with spherical inclusions are mainly caused by the eddy current losses over 1-18 GHz, whereas in the flaky composites by natural resonance.

From the practical point of view, composites on the basis of submicron silica coated flakes can be considered as candidates for producing high magnetic performance materials in microwave absorber design, electromagnetic interference (EMI) shielding.

References

[1] Chen P, Wu R X, Zhao T, Yang F and Xiao J Q 2005 *J. Phys. D: Appl. Phys.* **38** 2302
[2] Feng Y B, Qiu T, Shen C Y and Li X Y 2006 *IEEE Trans. Magn.* **42** 363
[3] Kim I, Bae S and Kim J 2008 *Mater. Lett.* **62** 3043
[4] Volkov V S, Syrkin V G and Tolmasskii I S 1969 *Carbonyl Iron*, Moscow Metallurgy
[5] Friederang A W and Schlegel R 1999 *Metal Powder Report* **54** 18
[6] Tolmasskii I S 1968 *High Frequency magnetic materials,* Moscow Energy
[7] Chen L F, Neo C P, Varadan V V, Ong C K, Varadan V K 2004 *Microwave Electronics: Measurement and Materials Characterization,* Wiley
[8] Wen F S, Qiao L, Yi H B, Zhou D and Li F S 2008 *Chinese Phys. Lett.,* **25** 751
[9] Wen F S, Qiao L, Zhou D, Zuo W L, Yi H B and Li F S 2008 *Chinese Phys B* **17** 2263
[10] Walser R M, Win W and Valanju P M 1998 *IEEE Trans. Magn* **34** 1390
[11] Wang G and Harrison A 1999 *J. Colloid Inter Sci* **217** 203
[12] Wu M Z, Zhang Y D, Hui S, Xiao T D, Ge S H, Hines W A, Budnick J I and Taylor G W 2002 *Appl. Phys. Lett.* **80** 4404

Journal of Physics: Conference Series **266** (2011) 012026 doi:10.1088/1742-6596/266/1/012026

Core losses of ring-shaped $(Fe_{0.75}B_{0.20}Si_{0.05})_{96}Nb_4$ bulk metallic glasses

T Bitoh, T Ishikawa, H Okumura

Department of Machine Intelligence and Systems Engineering, Faculty of Systems Science and Technology, Akita Prefectural University, Yurihonjo, 015-0055, Japan

E-mail: teruo_bitoh@akita-pu.ac.jp

Abstract. The soft magnetic properties of ring-shaped $(Fe_{0.75}B_{0.20}Si_{0.05})_{96}Nb_4$ cast bulk metallic glass (BMG) with thickness of 0.3–1.0 mm have been investigated. The BMG specimens exhibit high relative permeability of $(9–29)\times10^3$ at 0.40 A/m and 50 Hz and low coercivity of 4.0 A/m. The core losses of the 0.3 mm thick BMG specimen are lower than those of commercial Fe-6.5 mass% Si steel (6.5Si) with the same thickness, and are comparable to those of the 0.10 mm thick 6.5Si. The low core losses of the BMG originate from the low coercivity and high electrical resistivity.

1. Introduction

The development of new soft materials with lower core losses than that of Si-steels, which are being widely used for power devices, is expected from a viewpoint of energy conservation. It is well known that ordinary Fe-based amorphous alloys exhibit the very low core losses. However, the glass-forming ability (GFA) of the ordinary Fe-based amorphous alloys is low and hence their formation has required extremely high cooling rate, typically $10^5–10^6$ K/s. Therefore the resultant specimen thickness by an ordinary melt-spinning method has been limited to be less than approximately 50 μm [1].

Since 1995, a new class of Fe-based soft magnetic bulk metallic glasses (BMGs) with a large supercooled liquid region (ΔT_x) between the glass transition temperature (T_g) and the crystallization temperature (T_x) has been found [2, 3]. The BMGs have high GFA and, therefore, they can be used to prepare bulk amorphous alloys with a thickness of the order of a few millimeters by casting. Furthermore, the soft magnetic properties of Fe-based BMGs are better than those of ordinary amorphous alloys; therefore, these BMGs have wider applications as soft magnetic materials [4–6].

Typical thickness of commercial Si steels is 0.10–0.35 mm. However, the data of soft magnetic properties of the Fe-based BMGs with a thickness comparable to the Si-steels are hardly available. In this study, the soft magnetic properties including the core losses of the $(Fe_{0.75}B_{0.20}Si_{0.05})_{96}Nb_4$ cast BMG [7] have been investigated.

2. Experimental procedure

The mother alloy with a nominal composition of $(Fe_{0.75}B_{0.20}Si_{0.05})_{96}Nb_4$ was prepared by arc-melting the mixtures of pure Fe (99.99%), B (99.5%), Si (99.999%), and Nb (99.9%) in an argon atmosphere. The ring-shaped specimens with 10 mm in outer diameter, 6 mm in inner diameter and 1.0 mm in thickness were prepared by copper mold casting in an argon atmosphere. The ring-shaped bulk specimens with 0.30 and 0.50 mm in thickness were prepared by mechanically grinding with emery papers. Annealing treatment of the specimens for 600 s at 803 K [8] was carried out with no applied magnetic field in a vacuum by using infrared image furnace. The specimen structures were examined by X-ray diffractometry (XRD) with Cu $K\alpha$ radiation. The relative permeability (μ_r), the hysteresis curves and the core loss (P_{cm}) of the ring-shaped specimens were measured with a vector impedance analyzer, a DC *B-H* loop tracer under a maximum magnetic field (H_m) of 2.0 kA/m and an AC *B-H* analyzer, respectively. The saturation magnetization (J_s) and the electrical resistivity (ρ) of the annealed melt-spun tapes, prepared by a single-roller melt-spinning apparatus using a part of the same mother alloy, with 1.5 mm in width and approximately 20 μm in thickness were also measured with a magnetic balance under H_m of 0.88 MA/m and a DC four-probe method under a constant DC current density of 0.40 A/mm^2, respectively. All the measurements were performed at room temperature.

3. Results and discussion

The XRD profiles of the ring-shaped specimen are shown in figure 1. The XRD profile of the as-cast specimen with thickness (t) of 1.0 mm shows an amorphous halo with some superimposed Bragg lines. On the other hand, the profile of the specimen with t = 0.50 mm (both sides of it were removed 0.25 mm each) exhibit a halo pattern. Therefore, the crystalline phases which formed during casting exist only to the vicinity of the surface of the specimens. The 0.30 mm thick specimen also has the fully amorphous structure in an as-cast state and it is maintained after annealing at 803 K.

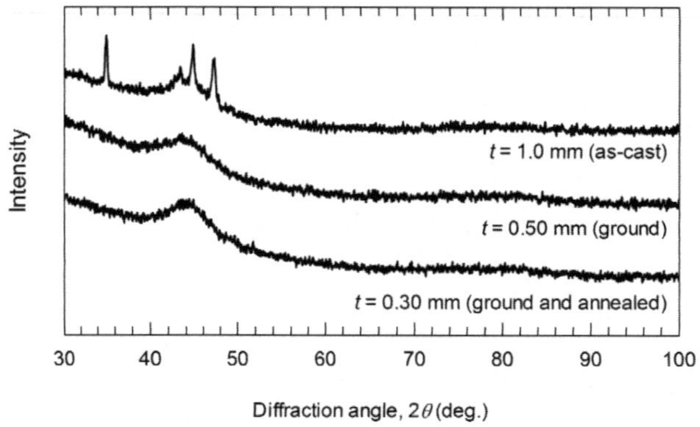

Figure 1. X-ray diffraction profiles of ring-shaped bulk specimens.

Table 1. Fundamental properties of $(Fe_{0.75}B_{0.20}Si_{0.05})_{96}Nb_4$ metallic glass.

Saturation magnetization (T)	1.20
Curie temperature (K)	598[a]
Electrical resistivity (μΩ m)	1.45

[a] Ref. 7.

The fundamental properties of the $(Fe_{0.75}B_{0.20}Si_{0.05})_{96}Nb_4$ metallic glass are summarized in table 1. Figure 2 shows the enlarged hysteresis curves ($-200 \le H \le 200$ A/m) of the specimens with 0.30 and 1.0 mm in thickness after annealing, respectively. Both the specimens exhibit low coercivity (H_c) of 4.0 A/m. The magnetic flux density of the 0.30 mm thick specimen reaches to 1.18 T at 2.0 kA/m. However, the 1.0 mm thick specimen is not easily magnetized (1.06 T at 2.0 kA/m) because of the magnetic anisotropy induced by the surface crystallization. The frequency (f) dependence of the permeability at H_m of 0.40 A/m after annealing is shown in figure 3. Among the three specimens, the 1.0 mm thick one exhibits lowest permeability. This due to pinning of magnetic domain wall displacements and/or magneto-elastic anisotropy caused by the surface crystallized layers. In the low frequency region below 150 Hz, the real part of the relative permeability (μ_r') of the 0.30 mm thick specimen is lower than that of the 0.50 mm thick one because the pinning of magnetic domain wall displacements by surface ruggedness becomes remarkable with decreasing the specimen thickness [9].

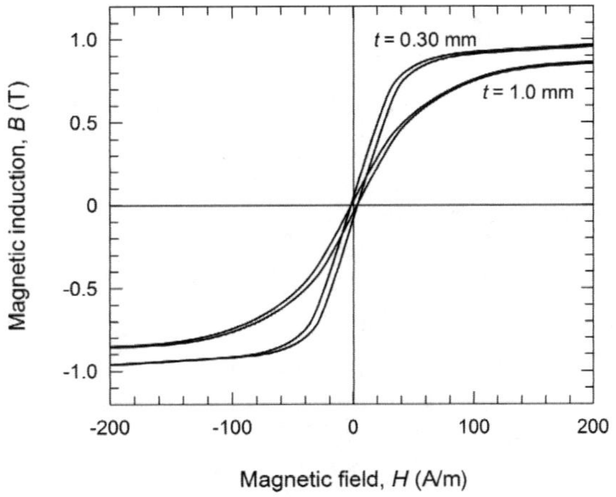

Figure 2. Hysteresis curves of 0.30 and 1.0 mm thick specimens after annealing.

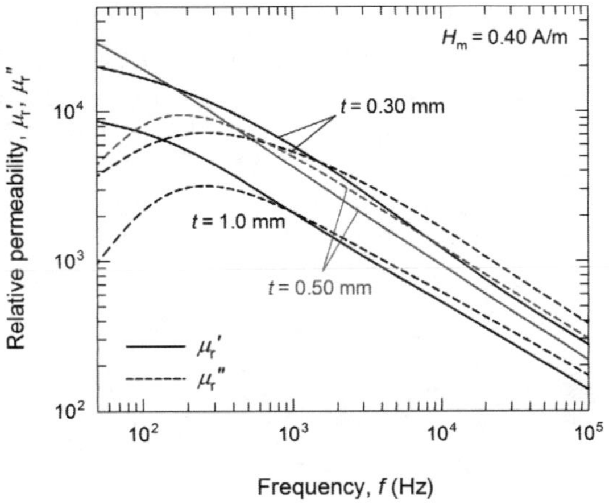

Figure 3. Frequency dependence of relative permeability at 0.40 A/m after annealing.

Figure 4 shows the cycle losses (P_{cm}/f) after annealing as a function of the frequency. The classical eddy current loss (P_e) for sheets at a frequency f and a maximum magnetic induction B_m are given by [10]

$$P_e = \frac{3}{x}\frac{\sinh x - \sin x}{\cosh x - \cos x}\frac{(\pi t f B_m)^2}{6\rho} \tag{1}$$

with $x = 2\sqrt{f/f_w}$, where $f_w = 4\rho/(\pi\mu_0\mu_i t^2)$ is the limiting frequency above which the magnetic field no longer fully penetrates the specimen and μ_i is the initial permeability. The dashed lines in figure 4 represent the estimated P_e. The observed P_{cm} is two times or more as large as P_e due to anomalous eddy current contributions as well as other many soft magnetic materials.

Journal of Physics: Conference Series **266** (2011) 012026 doi:10.1088/1742-6596/266/1/012026

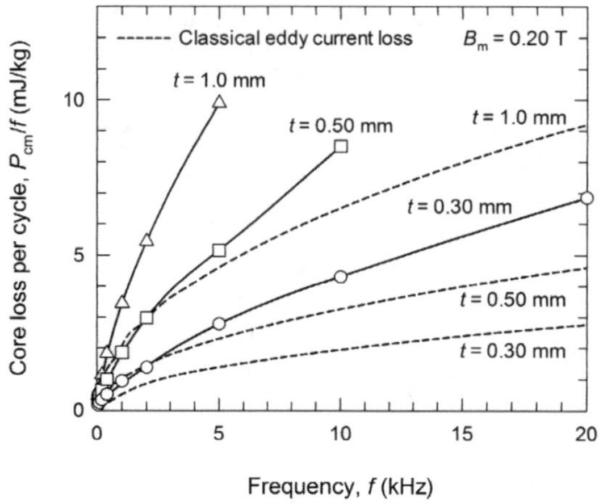

Figure 4. Cycle loss (P_{cm}/f) at 0.20 T as a function of frequency (f). The dashed lines represent estimated classical eddy current losses.

The core losses at 50 Hz–10 kHz of the 0.30 mm thick specimen after annealing as a function of B_m are shown in figure 5. The data for commercial Fe-6.5 mass% Si steel (6.5Si, t = 0.10 and 0.30 mm) are also shown for comparison [11]. In the low frequency region of 50–400 kHz, the BMG with t = 0.30 mm exhibits the lower P_{cm} than that of the 6.5Si steel with t = 0.10 mm. It should be noted that the low P_{cm} of the BMG is maintained to high B_m up to 0.9 T though the saturation magnetization is only 2/3 times as large as that of the 6.5Si steel. The core loss of the 0.3 mm thick BMG increases more rapidly than that of the 6.5Si steel with increasing frequency, and exceeds that of the 6.5Si steel at 10 kHz. However, the BMG maintains the low loss compared with the 6.5Si steel with the same thickness.

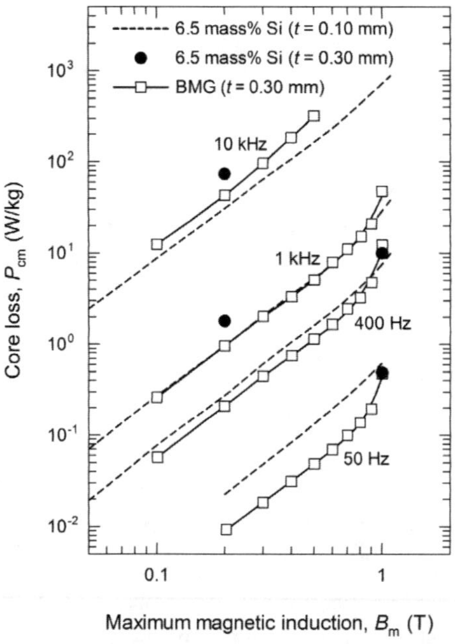

Figure 5. Core loss (P_{cm}) at 50 Hz–10 kHz of 0.30 mm thick specimen as a function of maximum induction (B_m). The data for commercial Fe-6.5 mass% Si steel (t = 0.10 and 0.30 mm) are also shown for comparison [11].

4. Summary

It has been confirmed that the ring-shaped $(Fe_{0.75}B_{0.20}Si_{0.05})_{96}Nb_4$ cast BMG specimens with thickness of 0.30–1.0 mm exhibit the high permeability, the low coercivity and the low core loss. The low core loss of the BMG originates from the low coercivity and the high electrical resistivity. It should be paid attention that the low loss characteristic of the BMG is maintained up to high magnetic induction of approximately 0.9 T that is comparable to the commercial Fe-6.5 mass% Si steels. If thin sheets of Fe-B-Si-Nb BMGs are prepared directly, the saturation magnetization could be improved in exchange for decrease in GFA, e.g., decrease in Nb content. Therefore, it can be concluded that the Fe-B-Si-Nb BMGs have potential for development as a core material for power devices.

Acknowledgement
This work was partly supported by Japan Society for the Promotion of Science (JSPS), Grant-in-Aid for Scientific Research (C) (KAKENHI), Nos. 17560585 and 21560690.

References

[1] Chen H S 1980 *Rep. Prog. Phys.* **43** 353–432
[2] Inoue A and Gook J S 1995 *Mater. Trans., JIM* **36** 1180–3
[3] Inoue A, Takeuchi A and Shen B 2001 *Mater. Trans.* **42** 970–8
[4] Bitoh T, Makino A and Inoue A 2003 *Mater. Trans.* **44** 2020–4
[5] Bitoh T, Makino A and Inoue A 2004 *Mater. Trans.* **45** 1219–27
[6] Bitoh T, Makino A and Inoue A 2006 *J. Appl. Phys.* **99** 08F102
[7] Inoue A, Shen B L and Chang C T 2004 *Acta Mater.* **52**, 4093–9
[8] Bitoh T and Shibata D 2009 *J. Appl. Phys.* **105** 07A312
[9] Kronmüller H and Gröger B 1981 *J. de Phys.* **42** 1285–92
[10] Herzer G 1997 *Handbook of Magnetic Materials* vol. 10 ed K H J Buschow (Amsterdam: Elsevier Science) p 415–62
[11] Takeda Y, Abe M, Masuda S and Inagaki J 1988 *J. Appl. Phys.* **64** 5367–9

Journal of Physics: Conference Series **266** (2011) 012027 doi:10.1088/1742-6596/266/1/012027

Magnetic properties of Nd-Fe-B/ α -Fe multi-layered thick film magnets

H. Fukunaga [1], H. Nakayama[1], T. Kamikawatoko[1], T. Yanai[1], M. Nakano[1], F. Yamashita[2], S. Ohta[3], M. Itakura[3] and M. Nishida[3]

[1] Faculty of Engineering, Nagasaki University, Nagasaki 852-8521, Japan
[2] Rotary Component Tech., Div., Minebea Co., Ltd., Shizuoka 437-1193, Japan
[3] Department of Applied Science for Electronics and Materials, Kyushu University, Kasuga, Fukuoka 816-8580, Japan

E-mail:fukunaga@nagasaki-u.ac.jp

Abstract. Nd-Fe-B/α-Fe multi-layered nanocomposite film-magnets were prepared from a rotating $Nd_{2.6}Fe_{14}B/\alpha$-Fe composite target by the pulse laser deposition method with a Nd-YAG laser (λ=355 nm). The prepared film-magnets were composed of approximately 800 layers with the thickness of about several tens nanometre, and exhibited hard magnetic properties after annealing. The laser power for the ablation of the target was varied between 3 and 7 W, and it was found that the composition, magnetic properties, and the surface roughness of the prepared films depend on the laser power. A reduction in the laser power caused increases in remanence and the maximum energy product $(BH)_{max}$ as well as decreases in Nd content and coercivity of the films. All the samples had Nd-poor compositions compared with that of the $Nd_{2.6}Fe_{14}B$ target and the Nd-contents of the films deposited with laser power of 3 W were less than that of the stoichiometric composition of $Nd_2Fe_{14}B$. The highest $(BH)_{max}$ and remanence values were obtained for a film deposited with laser power of 3 W, and were 90 kJ/m^3 and 1.0 T, respectively. The obtained high remanence value is discussed from the viewpoint of the remanence enhancement as well as the large saturation magnetization of α-Fe .

1. Introduction

Recently, magnets with several tens of microns are needed for applications to small machines such as milli-size motors [1]. As thinning of a bulk magnet has a tendency of degrading its magnetic properties [2], thick film-magnets are hopeful candidates for magnets used in such applications [3]. Although excellent magnetic properties have been reported for anisotropic thick film-magnets [4]-[6], isotropic ones are also attractive, because they are easily magnetized multi-polarly, and the multi-polar magnetization increases output torque of a motor. We have already reported a design example of a small motor consisting isotropic film-magnets which has output torque larger than that of a motor consisting of an anisotropic magnet [7].

Nanocomposite magnets are expected to exhibit isotropic and superior magnetic properties by controlling their nanostructures [8], [9]. One of promising methods of controlling nanostructures is

[1] To whom any correspondence should be addressed.

synthesis of an artificial layered structure, because the layered structure prevents the growth of grains and achieves a fine nanostructure [10], [11].

In order to synthesis multi-layered film-magnets with several tens microns, we need to prepare 10^3-10^4 layers. The PLD method with a composite target is suitable to prepare the layered structure for this purpose, and we reported the results of evaluation of magnetic properties of Nd-Fe-B/Fe_3B multi-layered films prepared by the PLD (Pulsed Laser Deposition) method [11]. In this contribution, we prepared Nd-Fe-B/α-Fe nanocomposite film-magnets using a Nd-YAG laser with different laser powers, and evaluated their magnetic properties and morphology. Subsequently, the obtained magnetic properties were discussed in terms of the remanence enhancement.

2. Experimental Procedure

2.1 Preparation of multi-layered films

A composite target, which were composed of $Nd_{2.6}Fe_{14}B$ and α-Fe segments, were prepared and the area fraction of the α-Fe segment was set to 1/4 in this investigation. Subsequently, the target was ablated by a Nd-YAG laser beam (λ=355 nm) in a vacuum chamber for 1 hour under the vacuum of 5×10^{-7}-2×10^{-6} Torr. As shown schematically in Fig.1, the target was rotated with the speed of 6.5-7.0 rpm during the deposition, which enabled us to deposit 13-14 layers/min. Thus, the deposited films were composed of approximately 800 layers. The repetition frequency of the laser and the distance between a target and a substrate were set to 30 Hz and 10 mm, respectively. The laser power was varied between 3 and 7 W.

We used a Ta substrate and composite films with the thickness of 8.1–28.4 μm were obtained. The Nd-Fe-B layers in an as-deposited film were amorphous after deposition, and the as-deposited films were crystallized and hardened magnetically by pulse-annealing for 1.8-2.0 s with an infrared furnace.

2.2 Evaluation of magnetic properties and microstructures

The prepared films were isotropic magnetically, and in-plane magnetic properties were measured with a vibrating sample magnetometer after magnetization under a pulse field of 6.4 MA/m. The thickness of a film was determined from the in-plane and perpendicular magnetization curves measured for the corresponding film in the as-deposited [12].

Microstructures of film-magnets were analyzed after crystallization with SEM, TEM, and XRD. The composition of a film was determined with an energy dispersive X-ray spectrometer (EDX). The roughness of the surface of a film was evaluated with a surface roughness meter with a contact probe. The average roughness R_a was calculated from the results of 3-5 scans (2.4 mm in length) with the probe for as-deposited films.

3. Results and discussion

Multi-layered films were deposited with the laser powers of 3, 5, and 7 W, and their microstructures and magnetic properties were evaluated after magnetic hardening. Results of X-ray diffraction analysis and TEM observation are shown n Fig.2 for a film deposited with the laser power of 5W. The multi-layer structure, approximately 20 nm in thickness, was clearly seen in Fig.2 (a). This thickness

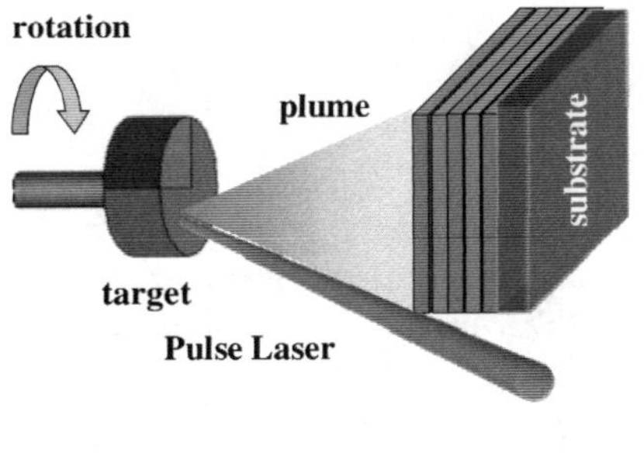

Figure 1. Schematic representation of PLD method for preparing multi-layered nano-composite film.

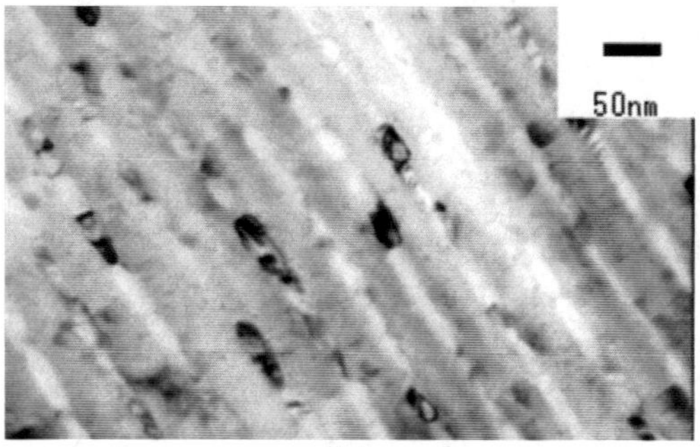

(a)

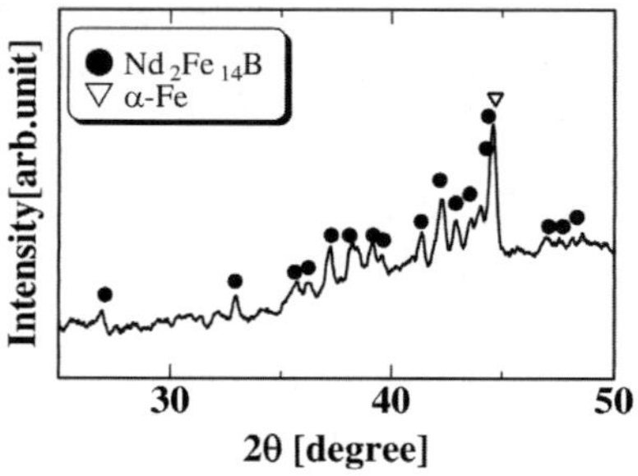

(b)

Figure 2. TEM micrograph (a) and X-ray diffraction pattern (b) of Nd-Fe-B/α-Fe multi-layered film-magnet deposited with laser power of 5W.

is suitable for obtaining a nanocomposite magnet [8], [9]. The X-ray diffraction pattern indicates that the film is composed of $Nd_2Fe_{14}B$ and α-Fe mainly and that the $Nd_2Fe_{14}B$ grains have no preferred orientations. The diffraction pattern obtained by TEM were also supported these results.

The compositions of the prepared films were scattered from a film to a film, and the obtained coercivity H_c, remanence M_r, and maximum energy product $(BH)_{max}$ after crystallization are shown in Fig.3 as a function of the content of Nd, Nd/(Nd+Fe). As the Nd contents of $Nd_{2.6}Fe_{14}B$ and $Nd_2Fe_{14}B$ are 0.156 and 0.125, respectively, all the samples shown in the figure have Nd-poor compositions compared with that of $Nd_{2.6}Fe_{14}B$. It is also seen that a small laser power results in a poor Nd composition, and the samples deposited with the laser power of 3 W were Nd-poor compared with the stoichiometric composition of $Nd_2Fe_{14}B$. Origins of the scattering and shift of Nd-content are not clear at the present stage.

The coercivity H_c decreased with decreasing the Nd content as expected from magnetically soft

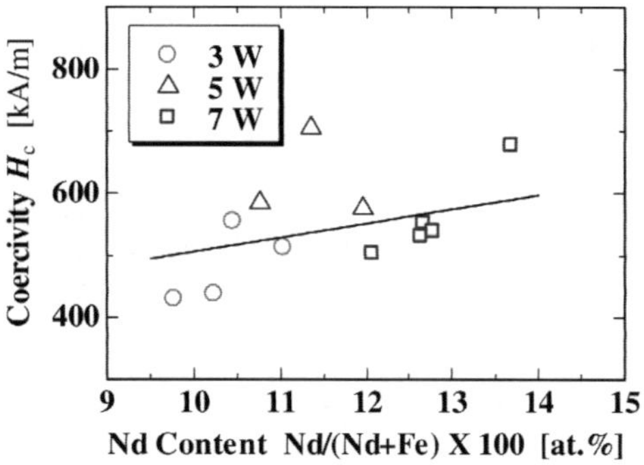

(a) Coercivity H_c

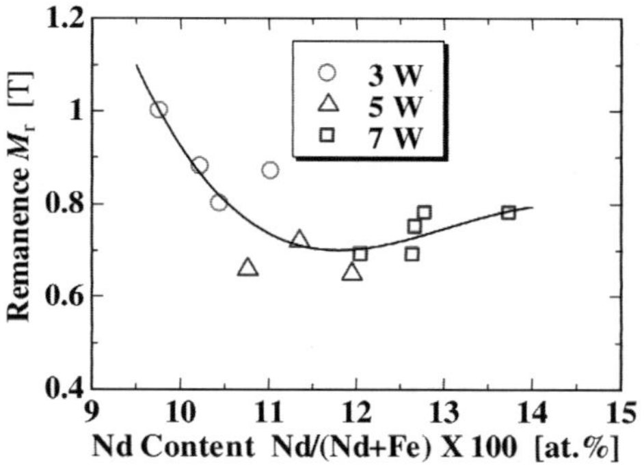

(b) Remanence M_r

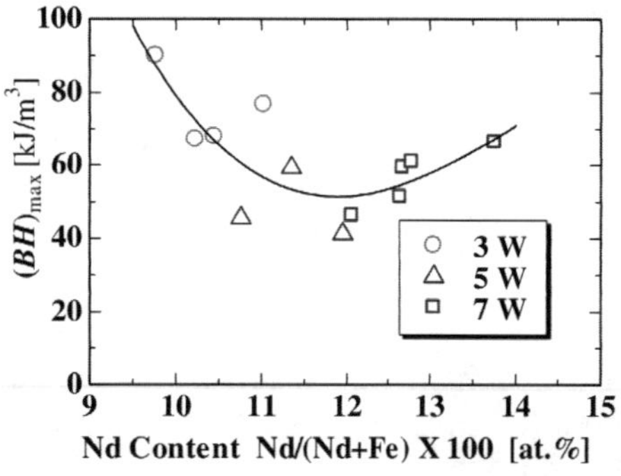

(c) Maximum energy product $(BH)_{max}$

Figure 3. Coercivity H_c, remanence M_r, and maximum energy product $(BH)_{max}$ after crystallization as a function of the content of Nd, Nd/(Nd+Fe). The samples were deposited with the laser powers of 3, 5, and 7 W.

Table 1 Magnetic properties of isotropic thick RE-TM thick film-magnets.

Alloy	Method	Film Thickness (μm)	H_c (kA/m)	M_r (T)	$(BH)_{max}$ (kJ/m^3)	Ref.
Nd-Fe-B	PLD	14.9	430	1.0	90	This Work
Nd-Fe-B	Plasma Spray	500	1380	0.59	61	13
Nd-Fe-B	Tape Casting	300	760	0.45		14
Nd-Fe-B	Screen Printing	10-50	860	0.5		15
Sm-Co	Sputtering	30	1200	0.75	90	16
Sm-Fe-N	Aerosol Deposition	~45	1440	0.55		17

properties of α-Fe. The remanence has a tendency of increasing with a decrease in the Nd content as expected from the small saturation magnetization of $Nd_2Fe_{14}B$, although a broad minimum was seen at around N/(Nd+Fe) = 0.12, which suggests that the squareness of a demagnetization curve is not good at this content. The maximum energy product $(BH)_{max}$ also showed a broad minimum at around N/(Nd+Fe) = 0.12, reflecting the behaviour of the remanence. The largest $(BH)_{max}$ value was obtained for the film with N/(Nd+Fe) = 0.098, and the hysteresis loop of this film is shown in Fig.4. The average layer-thickness calculated from its thickness was 17.7 nm. The obtained $(BH)_{max}$ and M_r, 90 kJ/m^3 and 1.0 T respectively, are larger than those of the Nd-Fe-B/Fe_3B multi-layered film reported previously [11]. Magnetic properties for other RE-TM isotropic film-magnets are shown in Table 1. The film obtained in this study has the largest M_r and $(BH)_{max}$ values in those shown in the table.

In order to study origins of a large M_r value of the films deposited with the laser power of 3 W, we evaluated the spring back ratio *SBR* which is the ratio of the recoiled magnetization from the coercive field to the remance, and found that the *SBR* values (0.46-0.60) of the films deposited with laser power of 3 W were larger than those (0.38-0.46) deposited with the laser power of 5 W. This result suggests strong intergrain exchange interactions in the former films. As a decrease in laser power results in reduction of film-thickness, the average layer-thickness is also reduced with decreasing the laser power. Therefore the intergrain exchange interaction and, resultantly, the remanence would be enhanced [8], [9].

It is well known that small particles called "droplet" are emitted from a target during deposition, and they would break the layered structure of deposited films. From this point of view, the surface roughness was evaluated for films deposited from $Nd_{2.6}Fe_{14}B$ target.. The results are shown in Fig.5 as a function of the film-thickness. It is clearly seen that the reduction in the laser power decreases not

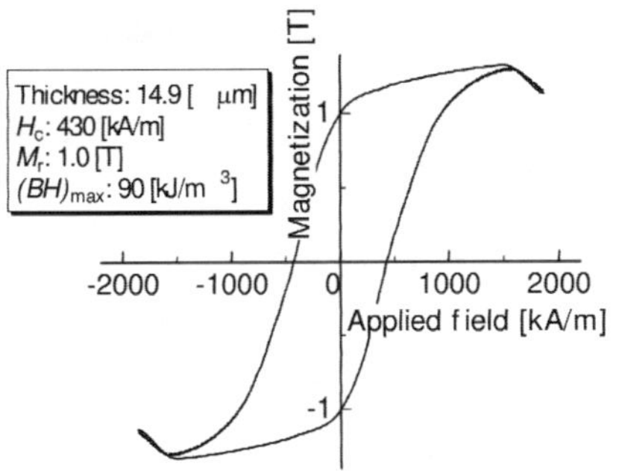

Figure 4. Hysteresis loop of Nd-Fe-B/α-Fe multi-layered film-magnet deposited with laser power of 3W.

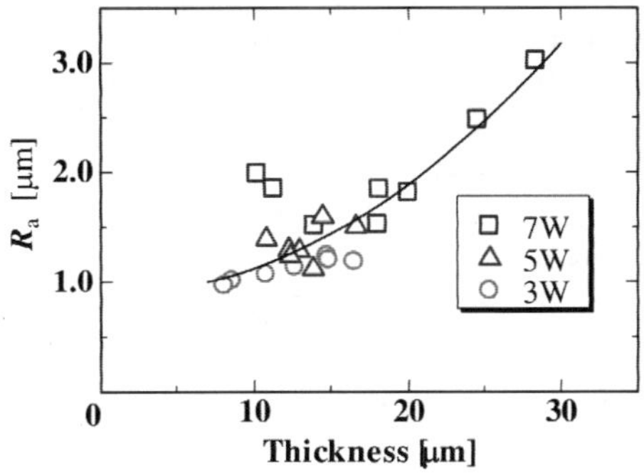

Figure 5. Average surface roughness of films of films deposited from $Nd_{2.6}Fe_{14}B$ target. The samples were deposited with laser powers of 3, 5, and 7 W.

only the film-thickness but also the average roughness R_a, which suggests that the reduction in the laser power decreases the number/or size of droplets and improves the layered structure. Thus, an improvement in the surface roughness seems to enhance the remanence through the intergrain exchange interaction.

4. Conclusions

Nd-Fe-B/α-Fe multi-layered nanocomposite film-magnets were prepared from a rotating $Nd_{2.6}Fe_{14}B$/α-Fe composite target by the pulse laser deposition method with a Nd-YAG laser (λ=355 nm). The laser power for ablation was varied between 3 and 7 W, and the rotating speed of a target was set so as to deposit approximately 800 layers with the thickness of about several tens nanometre. The synthesis of the layered structure was confirmed by TEM observation.

It was found that the composition, magnetic properties, and the surface roughness of the prepared films depend on the laser power. A reduction in the laser power caused increases in remanence and the maximum energy product $(BH)_{max}$ as well as decreases in Nd content and coercivity. All the films had Nd-poor compositions compared with that of the $Nd_{2.6}Fe_{14}B$ target and the Nd-contents of the films deposited with laser power of 3 W were less than that of the stoichiometric composition of $Nd_2Fe_{14}B$. The highest $(BH)_{max}$ and remanence values were obtained for a film deposited with laser power of 3 W, and were 90 kJ/m^3 and 1.0 T, respectively.

The spring back ratio of the films deposited with laser power of 3 W was larger than that with laser power of 5 W, which suggests strong intergrain exchange interaction. As a reduction in the laser power causes the decrease in the layer-thickness and the number/size of droplets, the enhancement of exchange interaction due to the reduction in laser power seems to be reasonable. Therefore, the large remanence and resultant large $(BH)_{max}$ obtained in this investigation can be attributed to the large saturation magnetization of α-Fe and the remanence enhancement due to intergrain exchange interaction.

References

[1] Nakano M, Sato S, Yamashita F, Honda T, Yamasaki J, Ishiyama K, Itakura M, Fidler J, Yanai T and Fukunaga H 2007 *IEEE Trans. Magn.* **43** 2672.
[2] Nakamura H, Hirota K, Shindo M, Minowa T and Masakatu H 2005 *IEEE Trans. Magn.* **41** 3844.
[3] Yamashita S, Yamasaki J, Ikeda M and Iwabuchi N 1991 *J. Appl. Phys.***70** 6627.
[4] Kapitanov B A, Kornilov N V, Linetsky Ya L and Tsvetkov V Yu 1993 *J. Magn. Magn. Mater.* **127** 289.
[5] Uehara M 2004 *J. Magn. Soc. Japan* **28**, pp. 1043-1048. (in Japanese)
[6] Dempsey N M, Walther A, May F, Givord D, Khlopkov K and Gutfleisch O 2007 A*ppl. Phys. Lett.* **90** 092509.
[7] Yamashita F, Nishimura S, Menjo N, Kobayashi O, Nakano M, Fukunaga H and Ishiyama K 2010 *IEEE Trans. Magn.* **46** 2012.
[8] Kneller E F, Hawig R 1991 *IEEE Trans. Magn* **27** 35880.
[9] Fukunaga H, Kuma J and Kanai Y 1999 *IEEE Trans. Magn.* **35**3235.
[10] Shindo M, Ishizone M, Sakuma A, Kato H, Miyazaki T 1997 *J. Appl. Phys.* **81** 4444.
[11] Fukunaga H, Nakano M, Matsuura Y, Takehara H, and Yamashita F 2006 *J. Alloys and Compounds* **408-412** 1355.
[12] Nakano M, Tsutsumi S, Fukunaga H 2002 *IEEE Trans. Magn.* **38** 2913.
[13] Rieger G., Wecker J., Rodewalt W, Sattler W, Bach Fe.-W, Duda T and Unterberg W 2000 *J. Appl. Phys.* **87** 5329.
[14] Pawlowski B, Schwarzer S, Rahmig A and Töpfer J 2003 *J. Nagn. Magn. Mater.* **265** 337.
[15] Schwarzer S, Pawloeski B, Rahmig A and Töpfer J 2004 *J. Mater. Sci.: Mater. Electron.* **15** 165.
[16] Budde T and Gatzen H H 2006 *J. Appl. Phys.* **99** 08N304.
[17] Aketo J and Sugimoto S 2005, *J. Magn. Soc. Jpn.* **29** 20.

Journal of Physics: Conference Series **266** (2011) 012028 doi:10.1088/1742-6596/266/1/012028

Interacting or non-interacting nanoparticles of $(Mn_{1-x}Zn_x)_{0.5}\ Co_{0.5}Fe_2O_4$(x= 0 -1) ferrite synthesized by reverse micelle

A Ghasemi[1,2,*], S E Shirsath[3], A P Júnior[4], X Liu[1] and A Morisako[1]
[1] Spin Device Technology Center, Faculty of Engineering, Shinshu University, Nagano, Japan
[2] Department of Materials Engineering, Malek Ashtar University of technology, Shahin Shahr, Isfahan, Iran
[3] Department of Physics, Dr. Babasaheb Ambedk ar Marathwada University, Aurangabad, India
[4]Centro de Ciências Exatas, Departamento de Física, Universidade Estadual de Maringá, Brazil
*E-mail: ali13912001@yahoo.com

Abstract. $(Mn_{1-x}Zn_x)_{0.5}\ Co_{0.5}Fe_2O_4$ (x= 0 – 1) ferrite nanoparticles were prepared using reverse micelle process. X- ray diffraction patterns demonstrated that single phase spinel ferrites are obtained. Vibrating sample magnetometer (VSM) was employed to probe magnetic properties of samples. The compositional dependence of blocking temperature was measured using superconducting Quantum Interference Device (SQUID). Magnetic dynamics of the samples was studied by the measurement of AC magnetic susceptibility versus temperature at different frequencies. The phenomenological Néel – Brown and Vogel – Fulcher models were employed to distinguish between the interacting or non-interacting system. Results exhibited that there is strong interaction between fine particles.

1. Introduction

Reverse micelles are water-in-oil microemulsions in which molar ratio of constitutive components and pH control the size of water pools and resultant particles. There are several literatures about the synthesizing of Ni-Zn and Mn-Zn ferrite with reverse micelle method [1-5].

In the recent years, we have focused our studies on the preparation and magnetic characteristics of ferrite nanoparticles [6-10]. $(Mn_{1-x}Zn_x)_{0.5}\ Co_{0.5}Fe_2O_4$ has a verity of promising technological applications in high-density recording devices, high frequency devices and microwave-linked applications. The magnetic character of the particles depends crucially on size, shape, purity and magnetic stability of these nanoparticles. The particles should be single domain, of pure phase, suitable coercivity, moderate magnetization and stable blocking effects. From the application point of view, in recording technologies the superparamagnetic blocking temperature of the particles should be well above the room temperature in order to have a stable data recorded in these nanoparticles. In biomedical applications, the nanoparticles are used as drug carriers to the areas of the body where conventional drug delivery systems may not work. For this purpose the nanoparticles used should be magnetically in superparamagnetic unblocked state with relatively low blocking temperature and coercivity. So it is very important to tailor various magnetic properties in these materials according to their requirements. To the best of our knowledge no studies on the preparation of this ferrite via reverse micelle method have been reported.

With this view in mind, current interest is to make $(Mn_{1-x}Zn_x)_{0.5}Co_{0.5}Fe_2O_4$ ferrite nanoparticles using reverse micelle technique at a low temperature.

2. Materials and methods

$(Mn_{1-x}Zn_x)_{0.5}Co_{0.5}Fe_2O_4$ nanoparticles were prepared by a reverse micelle process. In the preparation of ferrites, two microemulsions were synthesized. The first, using a metal solution and the second one, using ammonium hydroxide. Stock solutions of 0.5 M sodium dioctylsulfosuccinate (AOT) were prepared in isooctane. A metal solution was synthesized using $FeCl_2$, $ZnCl_2$, $CoCl_2.6H_2O$ and $MnCl_2$ dissolved in 8 ml of water and 66 ml of AOT- isooctane. The second microemulsion was prepared by mixing 2 ml of ammonia, 2 ml of water and 66 ml of AOT- isooctane. The second microemulsion was then added to the first one and stirred for one hour at room temperature. After rapid magnetic stirring, methanol was used to extract the surfactant and organic solvent. The colloidal precipitate was sedimented by centrifuging and washing with methanol and water. To obtain spinel ferrite in the final product, the pH must be carefully controlled. The pH was adjusted to 10.3 in this experiment. The synthesis temperature was kept 200 ºC.

X-ray powder diffraction was performed on the nanoparticles using a Phillips X-ray diffractometer employing Cu *Kα* radiation from a sealed tube (50 kV, 30 mA) source.

The vibrating sample magnetometer was employed to investigate the variation of magnetization with magnetic field. The temperature dependence of the magnetic susceptibility of nanoparticles was studied under an applied field of 115 Oe at different temperature ranges using Lake Shore magnetic susceptometer model 7000. Magnetization versus temperature variation in the range of 100-300 K was measured with a superconducting Quantum Interference Device (SQUID). In this case, after the ZFC process, a field of 115 Oe was applied.

3. Results and Discussions

3.1. Structural and magnetic analysis

The XRD patterns of samples in figure 1 revealed that in all samples, by substituting Mn with Zn, no extra peak corresponding to any secondary phases were identified. Based on the XRD results, it seems that zinc cations could be re-arranged in the spinel structure of Mn ferrite. It is reasonable to conclude from the patterns that the spinel phases can be formed in all the specimens without the existence of any secondary phases at relatively low temperature.

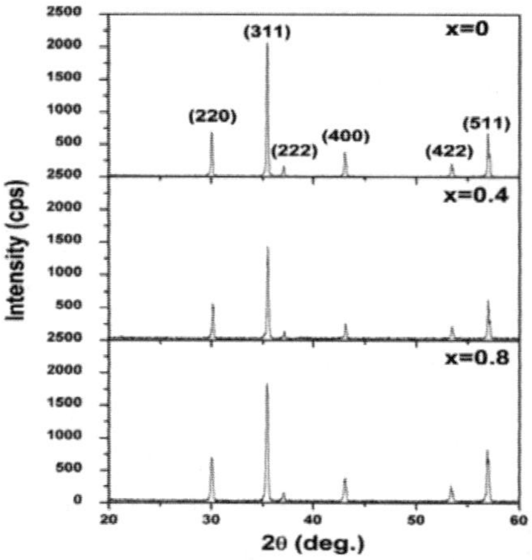

Figure 1. XRD patterns of nanoparticles synthesized by reverse micelle.

Table 1 reflects the values of coercivity and saturation magnetization of samples prepared by reverse micelle. Interestingly, it has been revealed that the coercivity and remanence are almost zero for all samples. The samples were not saturated even at magnetic field of 10 kOe, which reflects the characteristic of superparamagnetism. The value of saturation magnetization obtained by fitting the high field data to the function $M=M_s(1-\alpha/H)$, where M and M_s are magnetization and saturation magnetization respectively, α is the fitting parameter and H is the applied field, appear in Table 1.

Table 1. Magnetic parameters of nanoparticles of spinel $(Mn_{1-x}Zn_x)_{0.5}\,Co_{0.5}Fe_2O_4$ ferrite obtained at room temperature

x	*Hc* (Oe) Reverse Micelle	*Ms* (emu/g) Reverse Micelle
0	0.02	47
0.2	0.05	43
0.4	0.03	38
0.6	0.02	36
0.8	0.06	32
1	0.04	23

3.2. Interacting or non-interacting nanoparticles?

Figure 2 illustrates the temperature dependence of magnetization obtained with an applied field of 115 Oe after zero-field cooling (ZFC) from 300 to 100 K for sample synthesized by reverse micelle method. The ZFC curves indicate that the prepared nanoparticles exhibit crystallite size distribution. There is a maximum in the ZFC curve which indicates the value of blocking temperature. It is observed that addition of zinc gives a decrease in the blocking temperature. This can be attributed to the particle size. Actually, with increasing zinc, the particle size decreases and consequently blocking temperature was decreased. To study the magnetic dynamic behavior of the nanoparticles, we measured the AC magnetic susceptibility of prepared ferrite nanoparticle versus temperature at different frequencies. Figure 3 reflects the temperature dependence of real and imaginary parts of AC magnetic susceptibility of sample with x=1 at different frequencies in the range of 111-1000 Hz and at an AC magnetic field of 115 Oe. In figure 3, χ'' (T) shows a peak near 225 K which is frequency dependent and shifted to a higher temperature with increasing frequency. This characteristic maximum is the signature of blocking/freezing process of the superparamagnetic/spin glass systems [11-15]. Since the nanoparticles and consequently their easy axes are randomly orientated, overall susceptibility is reduced with decreasing temperature as figure 3 shows.

It was found that the blocking temperature increases almost linearly with a decrease in Zn concentration in the reverse micelle synthesized samples. Based on the TEM observation, with an increase in Zn content the size of particles decreases. Consequently, it is reasonable to have the mentioned trend in blocking temperature versus zinc content.

The interacting or non-interacting behavior of the fine powders was evaluated by Neél- Brown (non-interacting model) and Vogel-Fulcher (interacting law). According to Neél- Brown model, the blocking temperature measured at a given working frequency, is related to the considered frequency as

$$\ln 1/2\pi v = \ln \tau_0 + KV/\,kT_B \tag{1}$$

T_B can be obtained by assuming that at the maximum of the out-of phase component of the susceptibility, the relaxation time matches the working frequency. Attempt time is calculated from the curves and it is found that it spreads between $\tau_0 = 10^{-23}$ to 10^{-21} s. These values, however, are unphysical, being several orders of magnitude lower than the typical value found in the literature for a non-interacting assembly of superparamagnetic nanoparticles (10^{-8}-10^{-11} s) [11]. This suggests that the Neél- Brown model is not appropriate to describe the dynamic behavior of our system. In this case, deviations from the model can be ascribed to the presence of strong interactions. The interaction between nanoparticles affected the blocking/freezing temperature by modifying the potential barrier. By increasing the strength of interaction, T_B shifts to higher temperatures. For interacting magnetic nanoparticles, the frequency dependence of T_B is given by the Vogel-Fulcher law [11]

$$\ln 1/2\pi v = \ln\tau_0 + E_a/k(T-T_0) \tag{2}$$

here E_a is the enegy of barrier, T_0 is an effective temperature which reveals the existence of the interaction between nanoparticles and T is the characteristic temperature indicating the onset of the blocking process. We tried to fit the experimental data of susceptibility for our samples, using Eq. (2). It was found that the values of attempt time were spread between 3.8×10^{-9} to 2.3×10^{-10} s. The good agreement between the experimental data and Vogel-Fulcher model confirms the existence of interactions between nanoparticles.

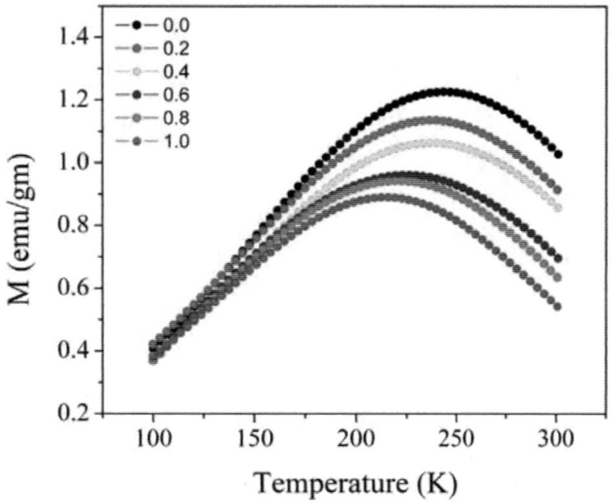

Figure 2. Zero field cooling curves of the samples obtained by SQUID.

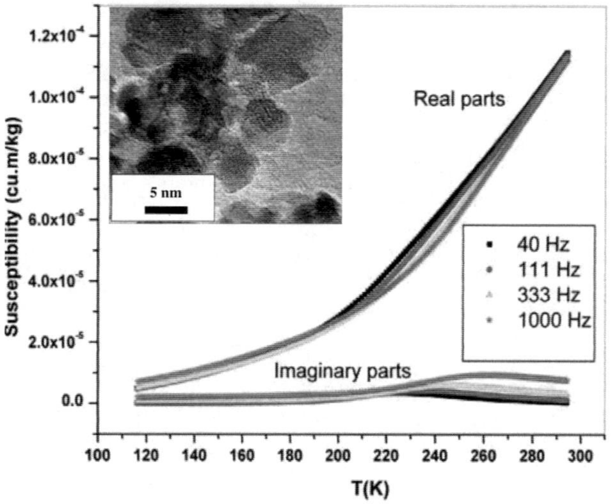

Figure 3. Thermal variation of effective magnetic susceptibility for sample x=1.

4. Conclusions

Fine particles single phase of $(Mn_{1-x}Zn_x)_{0.5}Co_{0.5}Fe_2O_4$ were prepared by reverse micelle method. The value of blocking temperature has straightforward relation with particle size. Temperature dependence of real and imaginary parts of effective magnetic susceptibility was measured. Fitting the experimental data of susceptibility with Neél-Brown model gives unphysical high values for relaxation time and indicates that there are strong interactions between nanoparticles of ferrite. The good agreement between the experimental data and Vogel-Fulcher model confirms the existence of strong interactions between nanoparticles.

Acknowledgements
This work was supported in part by Research (B) from JSPS and Grant-in-Aid for Scientific Research Tokobetstu Kenkyuin Shorei-hi.

References
[1] Thakur S, Latyal S C and Singh M 2009 *J. Magn. Magn. Mater.* **321** 1
[2] Uskokovic V, Drofenik M and Ban I 2004 *J. Magn. Magn. Mater.* **284** 294
[3] Shultz M D, Allsbrook M J and Carpenter E E 2007 *J. Appl. Phys.* **101** 09M518
[4] Morrison S A, Cahill C L, Carpenrer E E, Calvin S, Swaminathan R, McHenry M E and Harris V G 2004 *J. Appl. Phys.* **95**, 6392
[5] Morrison S A, Cahill C L, Carpenrer, E E, Calvin S and Harris, V G 2003 *J. Appl. Phys.* **93** 7489
[6] Ghasemi A, Šepelák V, Liu X X and Morisako A *J. Appl. Phys,* accepted for publication.
[7] Ghasemi A, Šepelák V, Liu X X and Morisako A *J. Appl. Phys.* accepted for publication.
[8] Ghasemi A, Šepelák V, Liu X X and Morisako A 2009 *IEEE Trans. Magn.* **45** 2456
[9] Ghasemi A, Liu X X and Morisako A 2009 *IEEE Trans. Magn.* **45** 4420

[10] Ghasemi A and Morisako A 2008 *J. Magn. Magn. Mater.* **320** 1167
[11] Dormann J L, Fiorani D and Tronc D 1999 *J. Magn. Magn. Mater.* **202** 251
[12] Jönsson P E 2004 *Adv. Chem. Phys.* **128** 191
[13] Suzuki M, Fullem S I and Suzuki I S 2009 *Phys. Rev. B* **79,** 024418
[14] Parker D ,Dupuis V , Ladieu F, Bouchaud J P, Dubois E, Perzynski R and Vincent E 2008 *Phys. Rev. B* **77,** 104428
[15] Vasseur S, Duguet E, Portier J, Goglio G, Mornet S, Hadová E, Knížek K, Maryško M, Veverka P and Pollert E 2006 *J. Magn. Magn. Mater.* **302** 315

Journal of Physics: Conference Series **266** (2011) 012029 doi:10.1088/1742-6596/266/1/012029

Nd-Fe-B sintered magnets fabrication by using atomized powders

R Goto[1], S Sugimoto[1, 2], M Matsuura[2], N Tezuka[2], Y Une[3] and M Sagawa[3]
[1] New Industry Creation Hatchery Center (NICHe), Tohoku University, 6-6-10, Aoba, Aramaki, Aoba-ku, Sendai 980-8579, JAPAN
[2] Department of Materials Science, Graduate School of Engineering, Tohoku University, Aoba-yama 6-6-02, Aoba-ku, Sendai, 980-8579, JAPAN
[3] Interemetallics Co., ltd., Kyodaikatsura #303, 1-36 Goryo Ohara, Nishikyo-Ku, Kyoto-city 615-8245, JAPAN

E-mail: rgoto@material.tohoku.ac.jp

Abstract.

Nd-Fe-B sintered magnets are required to achieve high coercivity for improvement of their thermal stability. Dy is added to increase coercivity, however, this element decrease magnetization and energy products. Therefore, Dy-lean Nd-Fe-B sintered magnets with high coercivity are strongly demanded. To increase coercivity, it is necessary that microstructure of sintered magnets is consisted of both fine main phase particles and homogeneously distributed Nd-rich phases around the main phase. To meet those requirements, Nd-Fe-B atomized powders were applied to the fabrication process of sintered magnets. Comparing with the case of using strip casting (SC) alloys, jet-milled powders from atomized powders show homogeneous distribution of Nd-rich phase. After optimized thermal treatment, coercivities of sintered magnets from atomized powders and SC alloys reach 1050 kA·m-1 and 1220 kA·m-1, respectively. This difference in coercivity was due to initial oxygen concentration of starting materials. Consequently, Nd-rich phases became oxides with high melting points, and did not melt and spread during sintering and annealing.

1. Introduction

Nd-Fe-B sintered magnets, which have been invented by Sagawa *et al.* [1], possesses high energy products ($(BH)_{max}$) and are used for various applications, for example, voice coil motors for hard disk drives and motors of hybrid electric vehicles (HEVs) or electric vehicles (EVs). Recently, HEVs and EVs are greatly paid attention due to a growing concern about environmental issues and the market of HEVs grows. Although Nd-Fe-B sintered magnets have high $(BH)_{max}$, there are several problems for use of motors in HEV. One is their low Curie temperature and the other is intrinsic coercivity so far realized reaches only 12% of the anisotropy field of $Nd_2Fe_{14}B$ compound, which is the main phase of Nd-Fe-B sintered magnets. Therefore, to maintain high magnetic properties at high temperature environment in HEVs, additive elements such as Dy or Tb are used in the Nd-Fe-B sintered magnets because $Dy_2Fe_{14}B$ and $Tb_2Fe_{14}B$ compounds have higher anisotropy field than $Nd_2Fe_{14}B$. However, the magnetic moment of Dy is aligned antiparallel with those of Fe and Nd, and addition of Dy decreases remanence and $(BH)_{max}$. Moreover, the natural abundance of these additive elements is quite

low and their material costs increase very rapidly. If the ideal coercivity of $Nd_2Fe_{14}B$ is obtained, the thermal stability is considered to be enhanced without these additive elements. Therefore, two methods to achieve high coercivity have been proposed. One is obtaining a uniform distribution of the Nd-rich phase that plays important roles in decreasing the nucleation site of reversed magnetic domains. If there are some defects on the surface of the $Nd_2Fe_{14}B$ grains, the reversed magnetic domains appear at low reverse magnetic field. Once a reversed magnetic domain appears in the grain, the domain wall moves and magnetic reversal occurs very easily. Therefore, it is required to decrease defects at the surface of $Nd_2Fe_{14}B$ grains in order to increase coercivity. The Nd-rich phase plays an important role in decreasing nucleation sites of the reversed magnetic domain during annealing process, then, contributes to increase the coercivity [2] –[7]. Another approach is controlling grain sizes below the single magnetic domain size [8]-[10] in sintered magnets. These approaches show that the control of microstructures of Nd-Fe-B magnets is quite important to achieve high coercivity.
The microstructure of the sintered magnets is influenced by each production process such as melting, casting, hydrogen decrepitation (HD) treatment, jet milling, magnetic pressing, sintering, and heat treatments. When the distribution of Nd-rich phase is not homogeneous, the Nd-rich phase segregates, and hence the decreasing the nucleation sites of the reversed magnetic domains occurs partially during the sintering or subsequent annealing process. The degree of increase in coercivity of Nd-Fe-B magnets is small in such a case. When the distribution of Nd-rich phase is homogeneous, which means the Nd-rich phase does not segregate, it is easy for the Nd-rich phase to spread between main phase grains during the sintering or subsequent annealing processes. In this case, the Nd-rich phase spreads along grain boundaries and decreases the nucleation sites. Then high coercivity may be obtained. Therefore, the distribution of Nd-rich phase is an important factor to achieve high coercivity for Nd-Fe-B sintered magnets.
In our previous report, the relationship between the microstructure of strip casting (SC) alloys, which was the conventional starting material, and distribution of the Nd-rich phase in the jet-milled Nd-Fe-B powder particles, have been evaluated [11]. The relationships between the lamella interval of SC alloy and jet-milled Nd-Fe-B powder size were evaluated. The Nd-rich phase is aggregated and the distribution of Nd-rich phases becomes worse when the average diameter of Nd-Fe-B jet-milled powders decreases. It is caused by the fact that the powder size of Nd-Fe-B jet-milled powder is much smaller than the average lamella interval of strip cast alloys.
To achieve small lamella interval, we focused on atomized powders. Sakaguchi *et al.* [12] have been reported microstructure of Nd-Fe-B atomized powders. When the diameter of atomized powder is less than 40 μm, the interval of Nd-rich phase becomes less than 2 μm, which is close to the diameter of Nd-Fe-B jet-milled powders. If Nd-Fe-B atomized powders use as the starting materials, it is expected to improve Nd-rich phase distribution by using atomized powders. However, there are few reports about microstructure during the fabrication processes using Nd-Fe-B atomized powders as a starting material. In this report, Nd-Fe-B atomized powders were applied to the fabrication process of sintered magnets with small grain size.

2. Experimental
Atomized powders were prepared as a starting material to fabricate Nd-Fe-B sintered magnets. The composition was $Nd_{14.49}B_{6.14}Cu_{0.10}Al_{0.66}Fe_{bal.}$ (in at%). SC alloys with same composition that are used as conventional staring materials were also prepared. Fine powders were fabricated by a jet milling after rough pulverization by the conventional HD process. Fine powders were packed into a mold by press-less process (PLP) [13] and aligned by pulse magnetic field. The mold in which powders were packed through the PLP was sintered in furnace. The base pressure of the furnace for sintering was less than 2.4×10^{-4} Pa. After sintering, two step annealing were promoted. The temperature of first annealing was 700 and 800 °C and annealed for 1 hour. The temperature of secondary annealing was varied from 450 to 600 °C. After annealing, samples were quenched in vacuum.
The powder size of the Nd-Fe-B jet-milled powders was evaluated by the laser diffraction method. The lamella interval of atomized powders and SC alloys and the distribution of Nd-rich phase of jet-

milled powders were observed and evaluated by Scanning Electron Microscope (SEM) - Energy Dispersive X-ray Spectroscopy (EDS) and Wave-length Dispersive Spectroscopy (WDS). The distribution of Nd-rich phase was evaluated by the ratio of grains that contained the Nd-rich phase. Magnetic properties of sintered magnets were measured by the pulsed magnetometer.

3. Results and discussion

Microstructures of starting materials were evaluated [14]. The lamellar interval of SC is 3.9 μm. The Nd-rich interval of atomize powder depends on the particle size of atomized powders. The lamellar interval is smaller than 2 μm when the diameter of the atomized powder becomes smaller than 40 μm. Consequently, atomized powders of whose diameter was less than 40 μm were used to fabricate sintered magnets. Both starting materials were pulverized by the jet milling after the HD. The diameter of Nd-Fe-B powders was evaluated. The average diameter of Jet-milled powders from atomized powders and SC alloys were 1.32 μm and 1.42 μm, respectively. For jet-milled powders from SC alloys, the maximum diameter of Nd-rich phase was 1.33 μm and this value was comparable to average diameter of jet-milled powders as we evaluated in previous paper [11]. However, for jet-milled powders from atomized powders, the maximum diameter of Nd-rich phase was only 0.41 μm. This value was about a third compared to the case in SC alloys. Moreover, the ratio of the jet-milled powders containing Nd-rich phases shows 71% by using atomized powder and it is higher than the powders from SC alloys. These Nd-rich phase distribution shows that decreasing the difference between Nd-rich phase interval in starting materials and particle size achieved homogeneous microstructure in jet-milled powders.

Jet-milled powders were sintered after packing through PLP method. Sintering temperatures were fixed at 930 °C and 950 °C for jet-milled powders pulverized from SC alloys and atomized powders, respectively. Each sintering temperature was decided on condition that it should be the lowest temperature that keeps the density of sintered bodies over 7.5 g/cm^3. In spite of smaller average diameter of jet-milled powders from atomized powders, sintering temperature for atomized powders is higher than that for SC alloys. This discrepancy is discussed later. Coercivities of as sintered magnets were 710 kA·m^{-1} and 640 kA·m^{-1} for SC alloys and atomized powders, respectively. Saturation magnetization is almost 1.44 T for both sintered magnets. For sintered magnets fabricated from each starting materials, squareness ratio, that is ratio of remanence to saturation magnetization, is about 0.96. This shows that atomized powders were sufficiently pulverized. If jet-milled powders contains atomized powders that is not pulverized, jet-milled powders do not aligned by the magnetic field because atomized powders are consisted of several Nd-Fe-B grains with randomly aligned.

Coercivity change after secondary annealing is shown in figure 1. Open and closed markers show the coercivity of magnets from SC alloys and atomized powders, respectively. Over 525 °C, magnets which were annealed lower temperature, show higher coercivity. Coercivity increases after secondary annealing. The highest coercivity is obtained after annealed at 550 °C, and coercivity of magnet that is fabricated from atomized powders and SC alloys reach 948 kA·m^{-1} and 1120 kA·m^{-1}, respectively. The demagnetization curves for sintered magnets from atomized powders and SC alloys are shown in figure 2 and 3, respectively. Only the magnet from SC alloys annealed at 800 °C show the highest coercivity after annealing at 575 °C. However, the coercivity is lower than magnets annealed at 700 °C.

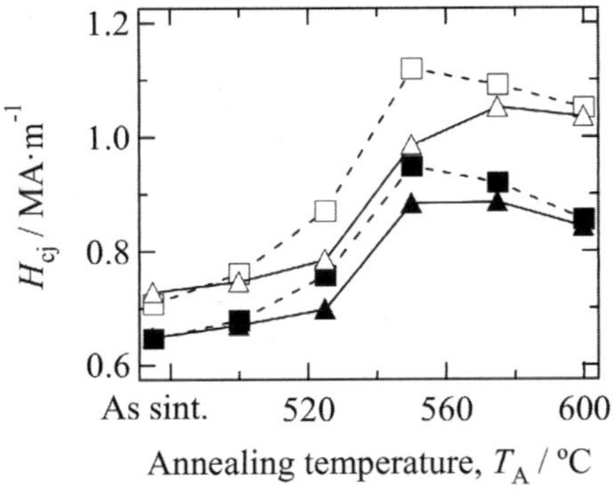

Figure 1. Secondary annealing temperature dependence of coercivity. Open and closed markers show the coercivity of magnets from SC alloys and atomized powders, respectively. Shapes of marker shows intermediate annealing temperature. Squares and triangles show that first annealing temperatures are 700 °C and 800 °C, respectively

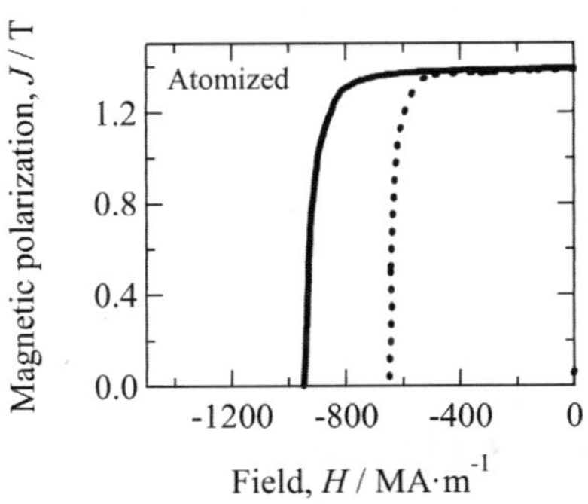

Figure 2. Demagnetization curves for sintered magnets fabricated from atomized powders. Dashed and solid line shows as-sintered and secondary annealed at 550 °C, respectively.

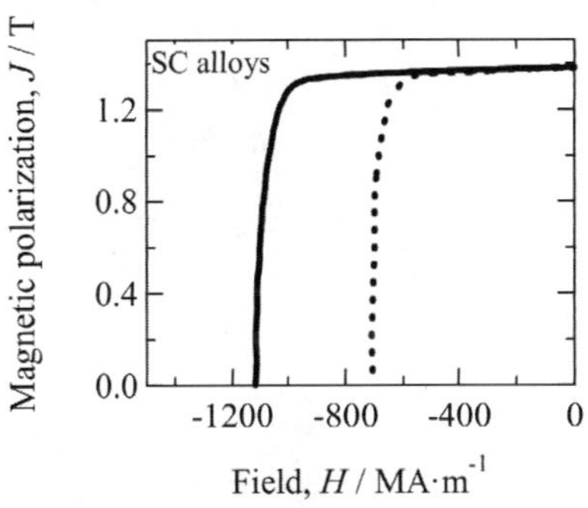

Figure 3. Demagnetization curves for sintered magnets fabricated from SC alloys. Dashed and solid line shows as-sintered and secondary annealed at 550 °C, respectively.

To improve magnetic properties, the cooling rate of thermal treatments was also compared. The first and secondary annealing temperature was fixed at the 700 °C and 550 °C, respectively. Cooling in furnace was promoted as a low speed cooling process and quenching under inert gas atmosphere was promoted as a high speed cooling. For sintered magnets from atomized powder, coercivities become 808 kA·m^{-1} for low speed cooling and 1050 kA·m^{-1} for high speed cooling, respectively. After high speed cooling, high coercivity was obtained for magnets from SC alloys and it becomes 1220 kA·m^{-1}. However, atomized powders have small Nd-rich interval, coercivity of magnets fabricated from atomized powders is lower than magnets from SC alloys. One reason is that the oxygen concentration of sintered magnets fabricated from atomized powders is about 2200 ppm and it was almost 1.5 times those from SC alloys. In starting materials, oxygen concentration of atomized powders was 460 ppm, which was about twice as large as that in the SC alloy. It shows that decreasing the oxygen content in initial atomized powders for reduction of oxygen concentration in sintered bodies. Therefore the fabrication condition of atomized powders, especially control of fabrication atmosphere is required. High oxygen content in Nd-rich phase results in increasing melting point of Nd-rich phase. Consequently, Nd-rich phases did not melt and spread during sintering and subsequently annealing. From this reason, it is considered that the optimized sintering temperature for jet-milled powders from atomized powders was higher than that from SC alloys.

Microstructure observation was promoted to discuss the increase in coercivity. Microstructures of sintered magnets from atomized powders and SC alloys are shown in figure 4. Dark and bright regions show Nd-Fe-B and Nd-rich phases respectively. Although size and distribution of Nd-rich phase are similar in microstructure for both starting materials, it was observed that Nd-rich phase spread into two grain boundaries for sintered magnet from SC alloys. Due to high oxygen concentration, better

Nd-rich phase distribution did not contribute effectively in atomized powders during sintering and annealing. Figure 5 shows Nd and Cu mapping by WDS. For as sintered magnets, Cu is aggregated. After annealing, aggregated Cu regions were disappeared and divided small sizes in Nd-rich phases. After sintering, samples were cooled in furnace. Consequently, equilibrium phases were appeared. Because Cu-Nd binary system has a low melting point, the phase contained Cu is remained as liquid phase at low temperature and aggregated in Nd-rich phase. After annealing and quenching, aggregated Cu was dispersed into Nd-rich phases in the sintered body.

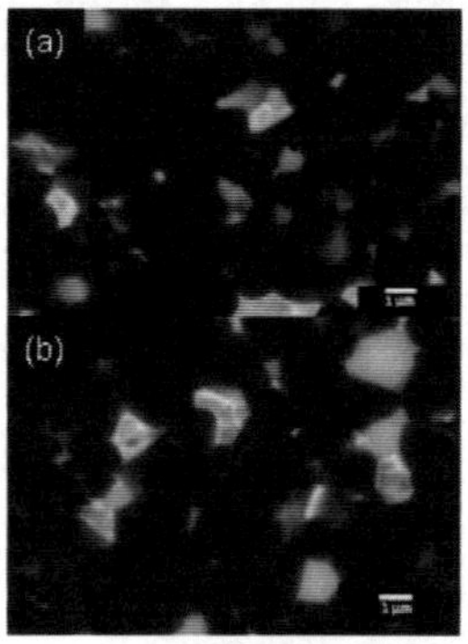

Figure 4. Microstructure of sintered magnets from (a) atomized powders and (b) after secondary annealed at 550 °C.

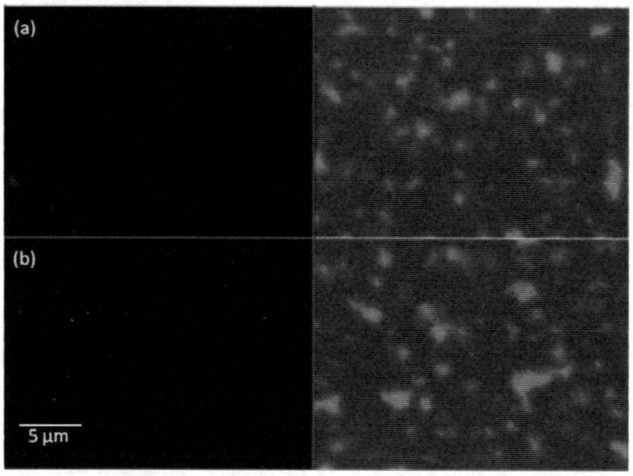

Figure 5. Cu (left) and Nd (right) mapping of sintered magnets from atomized powders for (a) as-sintered and (b) annealed at 550 °C.

4. Summary

Comparing with the case of SC alloys, jet-milled powders from atomized powders show homogeneous distribution of Nd-rich phase. However, after sintering and optimized thermal annealing, coercivities of sintered magnets from atomized powders and SC alloys reach 1050 $kA \cdot m^{-1}$ and 1220 $kA \cdot m^{-1}$, respectively. This difference in coercivity was due to initial oxygen concentration of starting materials. Because the Nd-rich phase became oxide, its melting point raised and Nd-rich phase did not melt and spread during sintering and annealing.

Acknowledgments

This work was partially supported by the METI and NEDO, Japan, Grant-in-Aid for "Development of technology for reducing dysprosium usage in a rare-earth magnet in Rare Metal Substitute Materials

Development Project". The authors also thank to Santoku Corporation for providing SC alloys and atomize powders and to Mr. M. Tsuda for his assistance in microstructure observations.

References

[1] Sagawa M, Fujimura S, Togawa N, Yamamoto H and Matsuura Y 1984 *J. Appl. Phys.* **55** 2083
[2] Strnat K J 1989 *Ferromagnetic Materials vol. 4*, ed. Buschow K H J (Amsterdam: North-Holland Publishing) p 131
[3] Sagawa M, Fujimura S, Yamamoto H, Matsuura Y and Hiraga K 1984 *IEEE Trans. Magn.* **MAG-20** 1584
[4] Eckert D, Hinz D, Handstein A and Schneider J 1987 *Phys. Status. Sol. A* **101** 563
[5] Tsubokawa Y, Hirosawa S and Shimizu R 1990 *Jpn. J. Appl. Phys.* **29** 2737
[6] Vial F, Joly F, Nevalainen E, Sagawa M, Hiraga K and Park K T 2002 *J. Magn. Magn. Mater.* **242-245** 1329
[7] Zhou G F, Fu S Y, Sun X K and Chuang Y C 1990 *Phys. Stat. Sol. A* **121** 257
[8] Livingston J D 1985 *J. Appl. Phys.* **57** 4137
[9] Muller K H, Eckert D, Handstein A, Nothnagel P, and Schneider J. 1991 *J. Magn. Magn. Mater.* **83** 195
[10] Weizhong T, Shouzeng Z, and Bing H 1991 *J. Magn. Magn. Mater.* **94** 67
[11] Hattori T, Ishihara H, Sugimoto S, Goto R, Tezuka N, Une Y and Sagawa M 2009 *Mater. Trans.* **50** 2347
[12] Sakaguchi Y, Harada T and Kuji T 1984 *Mater. Sci. Eng A* **181/182** 1232
[13] Sagawa M 2008 *Proc. REPM in Greece* 103
[14] Goto R, Matsuura M, Sugimoto S, Tezuka N, Une Y and Sagawa M 2010 *Proc. REPM in Slovenia* 193

Effect of DyH_x addition on the magnetic properties and microstructure of $Nd_{14.1}Co_{1.34}Cu_{0.04}Fe_{bal}B_{5.84}$ magnets

S Guo[1], R J Chen, Y Ding, G H Yan, D Lee and A R Yan
Zhejiang Province Key Laboratory of Magnetic Materials and Application Technology; Key Laboratory of Magnetic Materials and Devices, Ningbo Institute of Material Technology and Engineering, Chinese Academy of Sciences, Ningbo, 315201, China

E-mail: gshuai@nimte.ac.cn

Abstract. The sintered $Nd_{14.1}Co_{1.34}Cu_{0.04}Fe_{bal}B_{5.84}$ permanent magnets with DyH_x addition have been prepared by powder-blending technique. The magnetic properties and microstructures of the magnets were investigated. It was found that Dy element is enriched in the intergranular phase, being evenly distributed throughout the grain boundary and improving the orientation of magnets as a lubricant, and 1.2% to 2% DyH_x additions are proved to be very effective in improving the permanent magnetic properties of $Nd_{14.1}Co_{1.34}Cu_{0.04}Fe_{bal}B_{5.84}$ magnets. Further analysis revealed that the increase of coercivity and slight decrease of remanence due to DyH_x addition can be attributed to the reduction of the primary phase volume, formation of the (NdDy)-rich phase and grain refinement of the main phase.

1. Introduction

The production and application of Nd-Fe-B magnets has an incredible growth in recent years. The development of material, process of production, exploration of higher quality of magnets and ultimate upper limits for permanent magnet properties are paid more and more attention to. It was reported that the energy product of Nd-Fe-B reached high up to 59.6MGOe[1]. Nd-Fe-B magnets with excellent properties, especially high-coercivity, are extensively applied to the field of permanent magnet motors. It was well known that the addition of heavy rare-earth element content in Nd-Fe-B system can be very effective to enhance the magnetocrystalline anisotropy field of 2:14:1 phase and thus high coercivity. However, based on the dramatically high cost of magnets, it is very important to develop highly coercive Nd-Fe-B magnets with low cost.

It is well known that the coercivity is strongly dependent on the microstructure of sintered Nd-Fe-B. The distribution of liquid phase as a thin layer separating the grains of the hard magnetic phase is considered a unique microstructure feature that brings significant enhancement in $_iHc$. Besides, the sintering behaviour of sintered Nd-Fe-B magnets can be effectively improved by introducing the liquid phase. Many elements have been tried to modify the composition of grain boundary. However, these additives improve the coercivity at the expense of other magnetic properties such as Br[2-4]. Dy or Tb is known to cause the significant improvement in $_iHc$, which is attributed to the better distribution of the Nd-rich intergranular phase with the solubility of the additive element in it.

[1] Corresponding Author email: gshuai@nimte.ac.cn

However, because Js of $Dy_2Fe_{14}B$ and $Tb_2Fe_{14}B$ compounds only equals to about half of the $Nd_2Fe_{14}B$'s, the addition of Dy or Tb will lead to a sharp decline of Js, thus resulting in the decrease of Br and $(BH)_{max}$. The perfect structure of this way is the heavy rear-earth distributing uniformly in the grain boundaries, and not substituting into the Nd sites of the main phase. Although researchers have attempted to achieve this structure for a long time, no way has been developed in efficiently condensing the heavy rear-earth near the grain boundaries [5, 6].

Our recent results show that the intergranular addition of metal hydrides is helpful in improving the coercivity without significant reduction of other properties (Br, α). In this paper, the $Nd_{14.1}Co_{1.34}Cu_{0.04}Fe_{bal}B_{5.84}$ magnets with different contents of DyH_x additives were prepared by means of the powder blending technique, and their structures, magnetic properties and magnetic orientation were investigated.

2. Experiment

The original alloy with nominal composition of $Nd_{14.1}Co_{1.34}Cu_{0.04}Fe_{bal}B_{5.84}$ (at %) is prepared by the strip casting (SC) technique, using the 99.9 wt% Nd, Co, Cu, Fe, and FeB alloy with 22.74 wt% B content as raw materials. The alloy strips are crushed into coarse powders with particle size about 10~100μm by HD technique. In the nitrogen atmosphere, the hydrogen powders are grinded into powders with an average particle size of 3.38μm through the jet milling. The different proportions of DyH_x (x=1-3) powders with the average size of 1.8μm are mixed, compacted under a magnetic field of 1800kA/m and isostatically pressed at 300MPa. The green compacts are sintered at 1040-1060°C for 2h in vacuum followed by gas quenching and then tempered at 900 and 500°C, respectively, for 2h.

The magnetic properties are measured using a magnetic measurement device NIM-500C. The microstructure investigations of the as-cast and homogenized samples are carried out using Hitachi TM-1000 Scanning Electron Microscope and Hitachi 4800 Field Emission Scanning Electron Microscope (FE-SEM). The grain orientation degree of sintered magnets is studied by Bruker D8 Advance Diffractometer X-ray diffraction (XRD) with Cu Kα radiation.

3. Results and discussion

The sintering function and the gas pressure character in vacuum sinter process are shown in figure 1. It can be seen that large amounts of gas are released in the low temperature stage of 300 - 800°C, attributing to the release of hydrogen from the alloy. The gas pressure peaks have been associated with hydrogen desorbing from the main phase and the intergranular phase with DyH_x addition, same as desorption stage in HD process [7]. Besides, the sintering environment keeps high vacuum in the sintering temperature. And the residual hydrogen content of magnets after the sinter process was less than 60ppm. Thus, this powder blending technique with DyH_x addition can effectively avoid the deterioration in the magnetic properties of the final magnets owing to the residual hydrogen, and ensure the Dy element diffusing uniformly in the grain boundary.

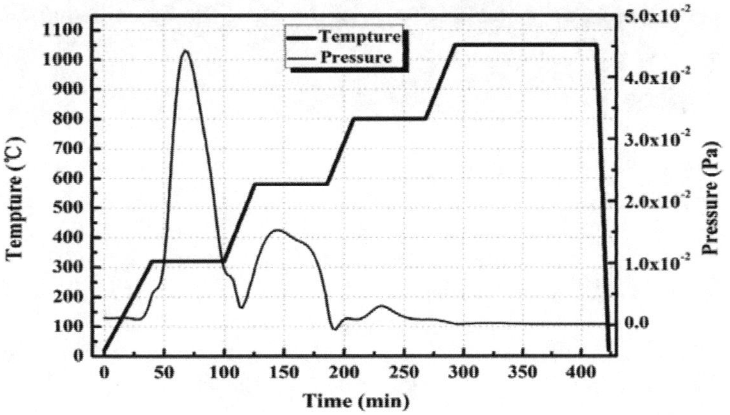

Figure 1. Sintering function and gas pressure character in vacuum sinter process

Journal of Physics: Conference Series **266** (2011) 012030 doi:10.1088/1742-6596/266/1/012030

The magnetic properties of the sintered magnets, with different proportions of DyH_x powder, are shown in figure 2. It can be seen that, with the increase of DyH_x, the coercivity increases and the remanence as well as the energy product decreases monotonously. When the addition of DyH_x is between 1.2-2 wt%, the DyH_x additives result in a considerable increase of $_iHc$, without significant reduction of Br and (BH)max. When the DyH_x additions are added up to more than 2 wt%, the remanence together with the magnetic energy product is dramatically reduced.

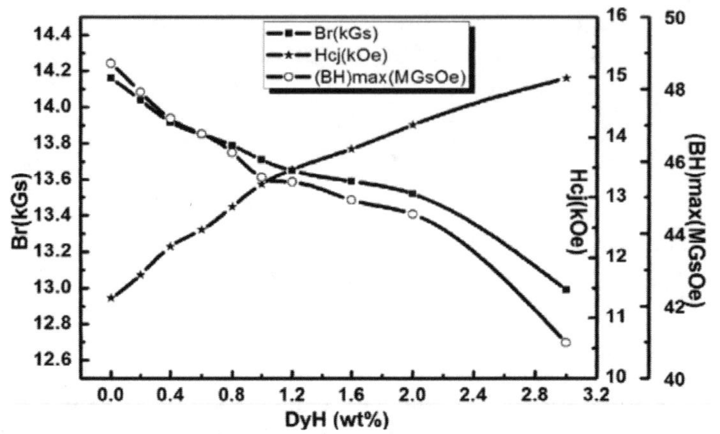

Figure 2. Variation of magnetic properties as a function of DyH_x additions

The SEM images of the magnets with 0.2 wt% and 3 wt% DyH_x additions are shown respectively in figure 3(a) and 3(b). The result indicates that a very uniform Dy distribution along the grain boundary is observed in the magnet with 0.2 wt% DyH_x addition, as presented in figure 3(a). However, there exist some large pieces of agglomerate phase accompanied by some grains of main phase in the magnets with 3 wt% DyH_x additions, in accordance to the observation of X M Li et al. [8] where the excess RE-rich phase reunion in the grain boundary cross-corner and an appropriate amount of RE-rich phase can be better diffuse to the grain boundary precipitates.

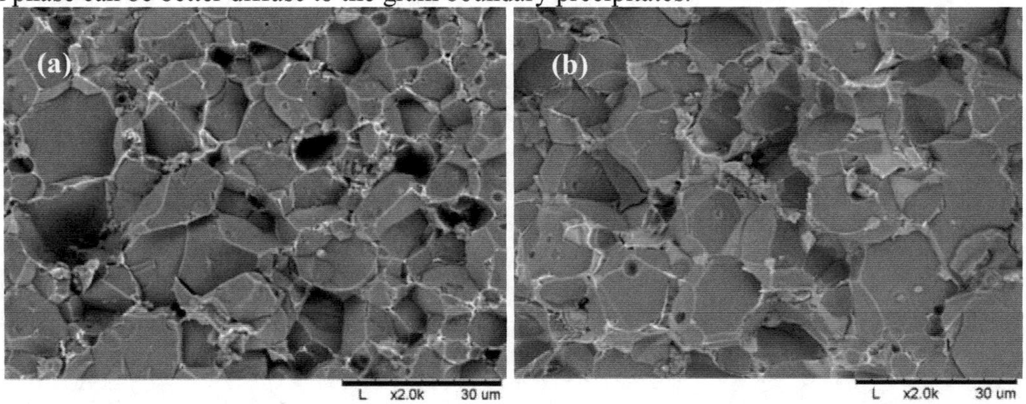

Figure 3. Microstructures of magnets, with 0.2%wt DyHx additions (a) and 3%wt DyH_x additions (b)

Figure 4 shows the EDX results of the areas of grain boundary and grain for magnet with 2 wt% DyH_x additives. It is seen that the additive Dy element presents in the grain boundaries, and no Dy is found in main grains. According to Ref. [9], Dy element can increase the coercivity and decrease both the remanence and the energy product due to a higher magnetic crystalline anisotropy and lower saturation magnetic polarization of $Dy_2Fe_{14}B$. Here, the Dy element is only distributed in grain boundaries and thus the magnetocrystalline anisotropy field of 2:14:1 grain surface is increased. This prevents the nucleation of magnetization reversal domain and effectively improves the coercivity, similar to that observed by F Vial et al. [10]. At the same time, Dy has not entered the main phase grains and the $Nd_2Fe_{14}B$ phase accordingly keeps its high saturation magnetization, so the values of Br and (BH)max should not decrease.

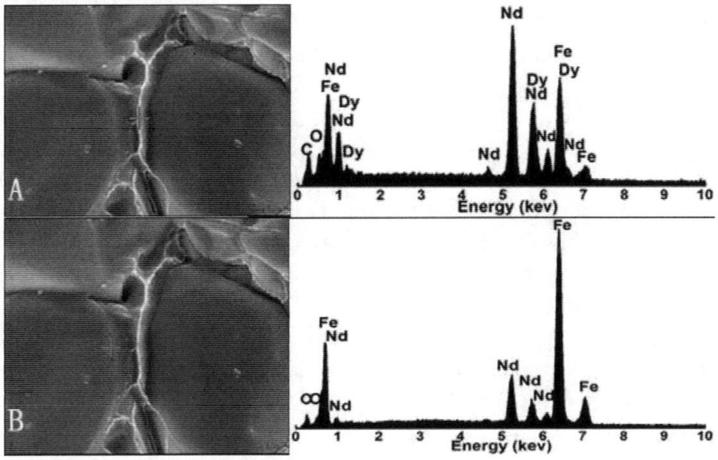

Figure 4. EDX of grain boundary (A) and grain (B) with 2%wt DyH_x addition.

The preferred orientation or crystallographic alignment of final magnets has been studied by the x-ray diffraction. Endoh et al. [11] indicated that the crystal alignment degree of the anisotropic sintered Nd-Fe-B magnet can be qualitative described by (006) and (105) diffraction peak intensity ratio A= I(006)/I(005). The increasing relative intensity of the (006) and (105) peaks reveals the improvement of the degree of crystal alignment. Figure 5 shows the x-ray diffraction patterns of the sintered magnets, with various proportions of the DyH_x additions. The result demonstrates that the orientation of magnets is obviously improved with increasing the content of DyH_x additives, which can be attributed to the addition of DyH_x that plays a role as a grain boundary lubricant at the initial stage of pressing process and the easy c-axis of the powders that is aligned more easily with the increase of intergranular additions.

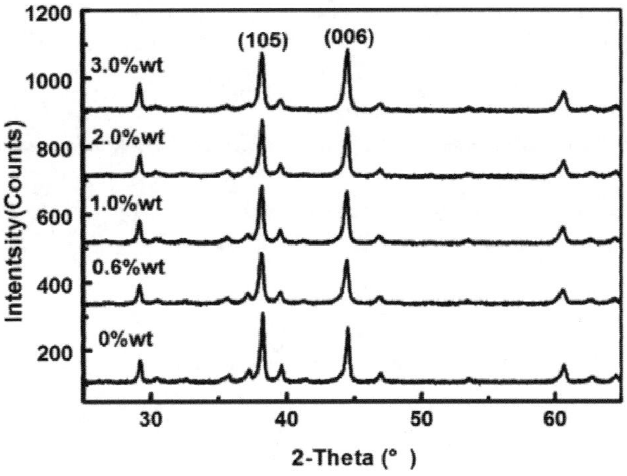

Figure 5. The XRD patterns of magnets with different proportions of DyH_x additions

From the above results, it is clear that the intergranular additions do not enter into the interior of grains, but distribute in the grain boundaries as an intergranular lubricant. According to the viewpoint of Y Matsuura[1], the intrinsic coercivity of Nd-Fe-B sintered magnet decreased with the improvement of grain orientation. When more than 1.2 wt % DyH_x powders are added in the magnets, the uptrend of the coercivity slows down due to the improvement of grain alignment. The hard magnetic phase is diluted by the intergranular additions, leading to the slight decrease of both remanence and maximum energy product.

4. Conclusions

The addition of DyH_x powder is an effective way to enhance the coercivity of Nd-Fe-B type magnets with the slight decrease of remanence. When entering the intergranular phase, the Dy element is evenly distributed throughout the grain boundary, eventually forming (NdDy)-rich phase and refining the grain of main phase. The coercivity of Nd-Fe-B magnets can be enhanced by the addition of DyH_x in the form of the fine powder before the sintering process. The coercivity significantly increases with the increasing of the DyH_x addition. Besides, the intergranular addition as a grain boundary lubricant forming at the initial stage of pressing process induces the easy c-axis of the grains more easily aligned and thus almost keeps the remanence constant.

Acknowledgements

The work is financially supported by Major Scientific and Technological Special of Zhejiang (NO.2007C11046), Key Science and technology project of Ningbo (2008B10024), National Basic Research Program (NO.2010CB934601) of China and Ministry of Science and Technology China and the National Nature Science Foundation of China.

References

[1] Y Matsuura 2006 *J. Magn. Magn. Mater.* **303** 344-47
[2] K H J Buschow 1994 *IEEE Trans. Magn.* **30** 565-70
[3] S Pandian, V Chandrasekaran, G Markandeyulu, K J L Iyer and K V S R Rao 2002 *J. Appl. Phys.* **92** 6082-86
[4] S Li and et al. 2002 *J. Phys. D: Appl. Phys.* **35** 732
[5] S Pandian, V Chandrasekaran, G Markandeyulu, K J L Iyer and K V S R Rao 2002 *J. Appl. Phys.* **92** 6082-86
[6] O Filip, A M El-Aziz, R Hermann, K Mummert and L Schultz 2001 *Mater. Lett.* **51** 213-18
[7] A J Williams, P J McGuiness and I R Harris 1991 *Journal of the Less Common Metals* **171** 149-55
[8] X M Li, T Liu, Z H Guo, M G Zhu and W Li 2008 *Acta Phys. Sinica* **57** 3823-27
[9] G Bai, R W Gao, Y Sun, G B Han and B Wang 2007 *J. Magn. Magn. Mater.* **308** 20-23
[10] F Vial, F Joly, E Nevalainen, M Sagawa, K Hiraga and K T Park 2002 *J. Magn. Magn. Mater.* **242** 1329-34
[11] M Endoh, M Shindo 1994 *Proc. 13th Int. workshop on Rare Earth Permanent Magnets and Their Applications* (Birminghan) Ed. by University of Birminghan, U.K. 397

Magnetic properties and high frequency characteristics of sputtered FeAl and FeAlB

C. C. Hsieh[1], M. Y. Jian[1], H. W. Chang[2], X. G. Zhao[1], W. C. Chang[1]

[1] Department of Physics, National Chung Cheng University, Chiayi, Taiwan 62102 R.O.C.

[2] Department of Physics, Tunghai University, Taichung, Taiwan 40704 R.O.C.

E-mail: cchjeremy@gmail.com, phywcc@ccu.edu.tw

Abstract. The magnetic properties and high frequency characteristics of sputtered $Fe_{81}Al_{19}$ and $Fe_{81}Al_{10}B_9$ thin films have been investigated. The above two thin films were sputtered at different oblique deposition angles in order to induce magnetic anisotropy in the film. The saturation magnetization of the films is easily above 12 kG, while the coercivity and the magnetic anisotropy field of the films are increased with increasing the oblique deposition angle, and decreased for the film with B. The residue stress induced by the oblique deposition is presumed the main factor to influence the magnetic anisotropy field of the thin films, the larger the residue stress the higher the magnetic anisotropy field and ferromagnetic-resonance frequency. The ferromagnetic-resonance frequency of $Fe_{81}Al_{10}B_9$ films can be above 1.3 GHz, while the permeability of the films is as high as 300-750 in the frequency range of 0-1 GHz, depending on the oblique deposition angles.

1. Introduction

Soft magnetic thin films with high saturation magnetization, low coercivity, high permeability and high electrical resistivity are useful for modern inductive devices, such as magnetic recording head and thin film inductors. The ferromagnetic-resonance frequency (f_{FMR}) of such films is proportional to the root of multiplication of saturation magnetization ($4\pi M_s$) and magnetic anisotropy field (H_k), i.e. $f_{FMR} = (\gamma/2\pi)(4\pi M_s \times H_k)^{1/2}$.[1] Therefore, the values of saturation magnetization and magnetic anisotropy field are very crucial for the films to exhibit high ferromagnetic-resonance frequency. Furthermore, the materials with high permeability (μ) can be suitably incorporated into an "air-core" inductor without producing additional loss, and substantially higher inductance and quality factor (Q) would be achievable without increasing the size of the device, or alternatively, for the same inductance value a much smaller substrate area would be needed.[2, 3]

Using an oblique deposition method, FeCoB and FeGaB thin films with high ferromagnetic-resonance frequencies of 1-3.3 GHz have been successfully developed.[4-6] On the other hand, FeAl alloys have been shown to exhibit high magnetostriciton and relatively high saturation magnetization.[7] It has been reported that Fe-19 at% Al alloy ribbons have a high magnetostriction coefficient of around -700 ppm, while it is 350 ppm for Fe-19 at% Ga single crystal and 270 ppm for FeGa polycrystal.[8-10] It is of interest to see the possibility of FeAl series films to play the same role as that of FeCo or FeGa series films.

Therefore, at first, we deposited the $Fe_{81}Al_{19}$ thin films by using oblique deposition method to see if it is suitable for high frequency applications. The impact of oblique angle on the magnetically soft properties was especially investigated. In addition, to reduce the coercivity of the film, part of the Al was replaced with B in preparation of FeAlB thin films because B element is helpful in decreasing the crystallinity in many metallic films.[11] Comparison on the magnetic properties and high frequency characterisics of FeGaB and FeAlB films were reported in this manuscript.

2. Experiment

The $Fe_{81}Al_{19}$ and $Fe_{81}Al_{10}B_9$ thin films, about 110 nm in thickness, were prepared by dc magnetron sputtering onto silicon substrate using oblique deposition method together with an external in-plane magnetic field of 1200 Oe, as show in Figure 1. The base pressure of the sputtering chamber was better than 1×10^{-6} torr, and the working pressure and input power were maintained at 4.5 mtorr and 20 W during the deposition, respectively. The magnetic hysteresis loops along easy axis and hard axis of the films were measured by vibrating sample magnetometer (VSM), respectively. High-frequency permeability characteristics were measured with AC permeance meter (Ryowa: PMF-3000, 0–3 GHz). The crystal structures of the thin films were studied by X-ray diffraction (XRD) with Cu-kα radiation.

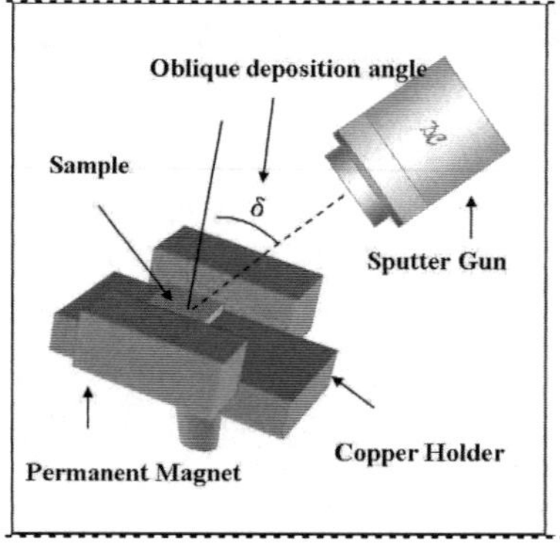

Figure 1. Schematic diagram of oblique deposition method.

3. Results and discussion

Previous studies have shown that using the oblique deposition method can effectively enhance the residue stress of the magnetically soft thin films and, consequently, increase the magnetic anisotropy field and magnetic resonance frequency of the films.[4, 5] Same trial has been made to prepare $Fe_{81}Al_{19}$ film at first in this study. Figure 2 shows the downward part of magnetic hysteresis loops along the hard magnetization axis of the $Fe_{81}Al_{19}$ films prepared at different oblique deposition angles.

Figure 3 shows the change of magnetic properties of the $Fe_{81}Al_{19}$ thin films as a function of oblique deposition angle. It can be found that with increasing the oblique deposition angle the saturation magnetization of the films decreased rapidly, from 12.8 kG for $\delta = 15°$ to 11.3 kG for $\delta = 40°$. Meanwhile, the coercivity measured along easy axis (H_{ce}) and along hard axis (H_{ch}) increased with the increase of oblique deposition angle. The $Fe_{81}Al_{19}$ film deposited at $\delta = 15°$ had H_{ce} of 66.5 Oe and H_{ch} of 15.7 Oe, whereas higher coercivity, $H_{ce} = 143.3$ and $H_{ch} = 51.3$ Oe, appeared for the film deposited at $\delta = 40°$. In addition, the magnetic anisotropy field, obtained from the interception of hysteresis loops along the easy axis and along the hard axis of the film, increased effectively from 68 Oe for $\delta = 15°$ to 191 Oe for $\delta = 40°$. The residue stress induced by the oblique deposition is presumed the main factor to influence the magnetic anisotropy field of the films, the larger the oblique deposition angle the larger the residue stress and induced magnetic anisotropy field. Based on the above magnetic data, the ferromagnetic-resonance frequency can be estimated following the formula of $f_{FMR}=(\gamma/2\pi)(4\pi M_s \times H_k)^{1/2}$, it increases from 2.6 GHz for $\delta = 15°$ to 4.2 GHz for $\delta = 40°$ for $Fe_{81}Al_{19}$ films.

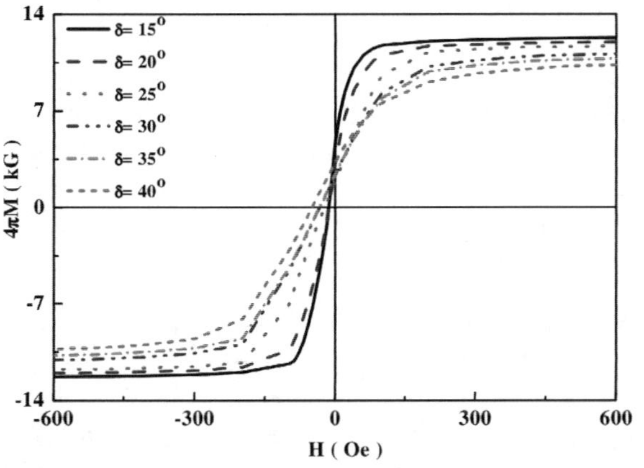

Figure 2. The downward part of magnetic hysteresis loop of $Fe_{81}Al_{19}$ thin films along the hard magnetization axis with different oblique deposition angles.

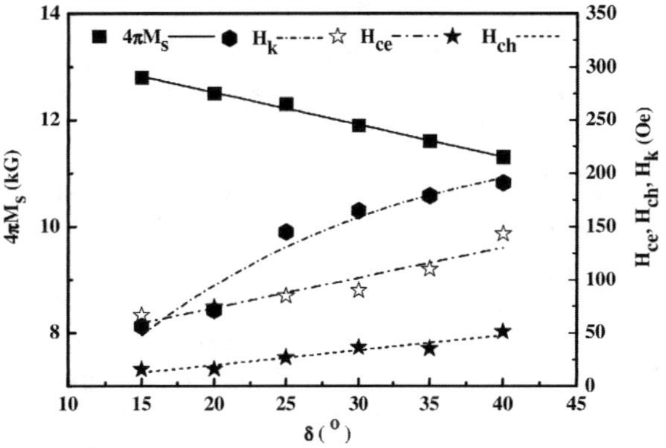

Figure 3. Magnetic properties of the $Fe_{81}Al_{19}$ films prepared at different oblique deposition angles.

For high frequency applications, small coercivity of thin films is requisite in order to reduce the hysteresis loss. To reduce the coercivity and improve the permeability for the studied films, the third element B is adopted to add into FeAl alloy based on Inoue's empirical rules for the alloys with high glass forming ability[12]. Fig.4 displays the XRD patterns of the $Fe_{81}Al_{19}$ and $Fe_{81}Al_{10}B_9$ thin films prepared at different oblique deposition angles. From the change of full width at half maximum (FWHM) of α-Fe (110) peak in $Fe_{81}Al_{19}$ films, it is obvious that FWHM value increased with the increase of oblique angle, whereas the diffraction peak of α-Fe (110) almost vanished in $Fe_{81}Al_{10}B_9$ films. As expected, B substitution for Al did decrease the crystallinity of the films effectively, which is helpful for reducing the coercivity of the films.

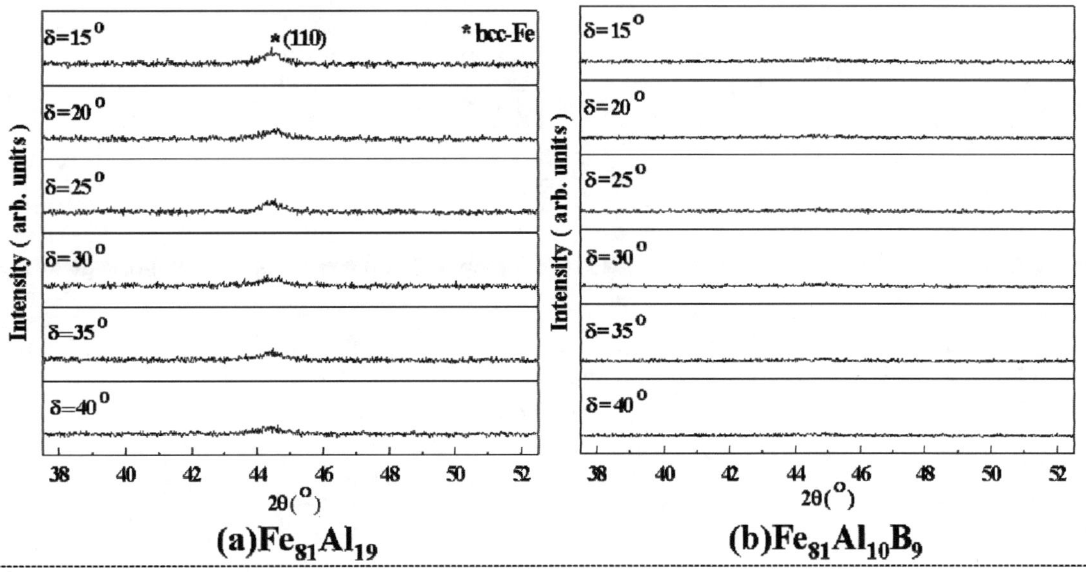

Figure 4. XRD patterns of the (a) $Fe_{81}Al_{19}$ and (b) $Fe_{81}Al_{10}B_9$ films prepared at different oblique deposition angles.

Figure 5 shows the variation of magnetic properties of the $Fe_{81}Al_{10}B_9$ films prepared at different oblique deposition angles. The trend of the variation of magnetic properties for $Fe_{81}Al_{10}B_9$ films prepared at different oblique deposition angles is similar to those of $Fe_{81}Al_{19}$ thin films. With increasing the oblique deposition angle, the saturation magnetization of the thin films decreased from 13.3 kG for δ = 15 ° to 12.7 kG for δ = 40 °, while the anisotropy field was enhanced from 17 Oe for δ = 15 ° to 121 Oe for δ = 40 °. The coercivity of the hysteresis loops increased from H_{ce} = 14.7 and H_{ch} = 5.5 Oe for δ = 15 ° to H_{ce} = 62.3 and H_{ch} = 25.4 Oe for δ = 40 °, respectively. The calculated ferromagnetic-resonance frequency of the films could be improved from 1.3 GHz for δ = 15 ° to 3.5 GHz for δ = 40 ° for $Fe_{81}Al_{10}B_9$ films.

It was evidenced that the coercivity along easy axis and hard axis and the magnetic anisotropy field were all decreased by the substitution of B for Al. The coercivity of the films deposited at δ=20 °decreased from H_{ce} = 74.0 and H_{ch} = 15.7 Oe for $Fe_{81}Al_{19}$, to H_{ce} = 24.8 and H_{ch} = 6.3 Oe for $Fe_{81}Al_{10}B_9$. Besides, the magnetic anisotropy field of the films decreased from 77 Oe for $Fe_{81}Al_{19}$ to 27 Oe for $Fe_{81}Al_{10}B_9$. Meanwhile, the calculated ferromagnetic-resonence frequency of the films was decreased from 2.8 GHz for $Fe_{81}Al_{19}$ film to 1.7 GHz for $Fe_{81}Al_{10}B_9$ film sputtered at δ = 25 °.

Figure 6 shows the frequency dependence of permeability for the $Fe_{81}A_{10}lB_9$ films with an oblique deposition angle of (a) δ = 15 °, (b) δ = 20 °, and (c) δ = 25 °. Table I lists the calculated and experimental ferromagnetic-resonance frequency, and the average maximum permeability at frequencies below 1 GHz of $Fe_{81}A_{10}lB_9$ films. It can be seen that the ferromagnetic-resonance frequency of $Fe_{81}A_{10}lB_9$ films increased from 1.3 GHz for 15° to 2.2 GHz for 25° because of the increase in magnetic anisotropy field of the films, and the average permeability at frequencies below 1 GHz of $Fe_{81}A_{10}lB_9$ films prepared at the above three oblique angles is 780, 590 and 300, respectively. It was also shown that the calculated and experimental ferromagnetic-resonence frequencies are quite closer, reflecting that the former adopted formula is acceptable for estimating the magnetic-resonance frequency in the present alloy films.

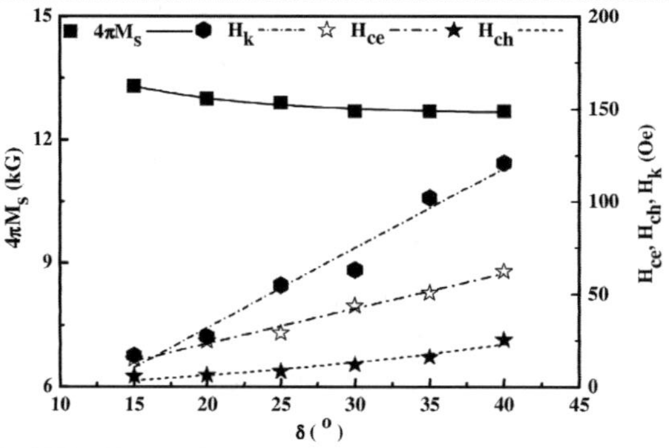

Figure 5. Magnetic properties of the $Fe_{81}Al_{10}B_9$ films prepared at different oblique deposition angles.

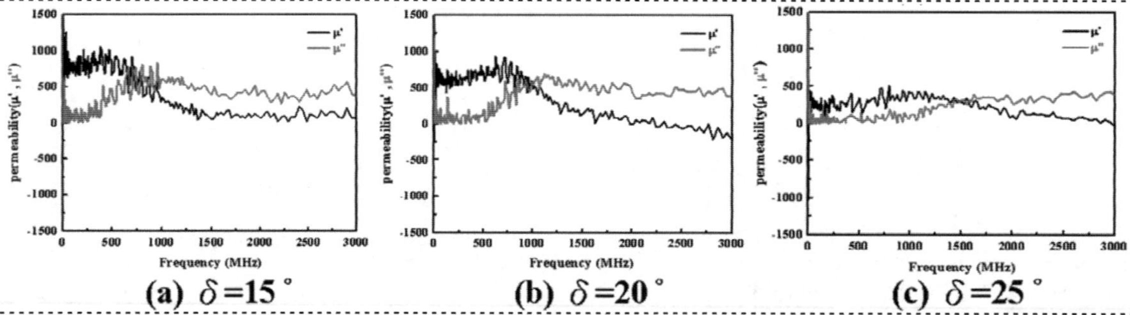

Figure 6. The frequency dependence of permeability for $Fe_{81}Al_{10}B_9$ films prepared at different oblique deposition angles of (a) δ = 15 °, (b) δ = 20 °, and (c) δ = 25 °.

Table 1. The calculated and measured ferromagnetic-resonance frequencies, and the average permeability of the $Fe_{81}Al_{10}B_9$ films prepared at different oblique deposition angles.

δ (°)	$f_{FMR, cal.}$ (GHz)	f_{FMR} (GHz)	μ_{ave} (< 1 GHz)
15	1.3	1.3	780
20	1.7	1.8	590
25	2.4	2.2	300

As shown in the previous results, it is found that the average permeability of $Fe_{81}Al_{10}B_9$ film at the frequencies below 1 GHz is above 500 and the ferromagnetic- resonance frequency is about 1.7 GHz for $\delta = 20$ °, which are identical to the values that obtained in FeGaB films.[5] In short, $Fe_{81}Al_{10}B_9$ thin films prepared at suitable oblique deposition angle may display identical or even better soft magnetic properties than those of FeGaB thin films. Inasmuch as the element Al is much cheaper than the element Ga, FeAlB film has the potential to be an alternative material for high frequency applications.

4. Conclusions

The effect of oblique deposition angles on soft magnetic properties and high frequency characteristics of $Fe_{81}Al_{19}$ and $Fe_{81}Al_{10}B_9$ thin films were examined. The magnetic properties and high frequency characteristics of the $Fe_{81}Al_{19}$ and $Fe_{81}Al_{10}B_9$ thin films can be monitored by different oblique deposition angles. Furthermore, B substitution for Al decreased the crystallinity of the films and magnetic anisotropy field, resulting in the decrease of coercivity. Nevertheless, high permeability of the thin film is obtainable by decreasing the oblique angle. In addition, high ferromagnetic-resonance frequencies between 1-3 GHz can be attained by increasing the oblique deposition angle.

5. Acknowledgements

This paper was supported by the National Science Council, Taiwan under grant no. NSC-98-2112-M-194-005-MY3.

References

[1] A.G. Gurevich, Ferrites at Microwave Frequencies (Consultants Bureau, New York), 42 (1963)
[2] A. Inoue, H. Koshiba, T. Itoi and A. Makino 1997 Appl. Phys. Lett. **73** 744
[3] V. Korenivski and R.B. van Dover 1997 J. Appl. Phys. **82** 5247
[4] I. Kim, J. Kim, K. H. Kim, and M. Yamaguchi 2004 IEEE. Trans. Magn. **40** 2706
[5] T.J. Klemmer, K.A.Ellos, L.H. Chen, B. van Dover, and S. Jin 2000 J. Appl. Phys. **87**, 830
[6] J. Lou, R. E. Insigneares., Z. Cai, K.S. Ziemer, M. Liu and N.X. Sun 2007 Appl. Phys. Lett. **91**, 182504
[7] A. E. Clark, J. B. Restorff, M. Wun-Fogle, T.A. Lograsso, and D. L. Schlagel 2000 IEEE. Tran. Magn. **36**, 3238
[8] J. H. Liu, G. D. Liu, M. Zhang, G.H. Wu, F.B. Meng, H.Y. Liu, L.Q. Yan, J.P. Qu, and Y.X. Li 2004 Appl. Phys. Lett. **85** 1751
[9] A. E. Clark, M. Wun-Fogle, J. B. Restorff, T.A. Lograsso, and J.R. Cullen 2001 IEEE. Tran. Magn. **37** 2678
[10] N. Srisukhumbowormchai, S. Guruswamy 2001 J. Appl. Phys. **90** 5680
[11] C.L. Platt, M.K. Minor, and T.J. Klemmer 2001 IEEE. Tran. Magn. **37**, 2302
[12] A. Inoue 1995 Mater. Trans. JIM **36** 866

Journal of Physics: Conference Series **266** (2011) 012032 doi:10.1088/1742-6596/266/1/012032

Effects of the Hf content on the microstructure and magnetic properties of Co-Hf-Ta thin films

S W Huang, Y T Lai, J G Duh*

Department of Materials Science and Engineering, National Tsing Hua University, Hsinchu, Taiwan

e-mail: jgd@mx.nthu.edu.tw

Abstract. Effects of the Hf content in Co-Hf-Ta thin films on the microstructure and magnetic properties were investigated in this study. It was found that appropriate Hf addition can effectively refine the Co grain size. Co grain sizes sharply decreased from 50 nm down to 2.3 nm with increasing the Hf content from 1.02 at.% to 2.81 at.%, leading to the reduced magneto-crystalline anisotropy. The Co-Hf-Ta thin films with small Co grains reveal low anisotropy field, low coercivity, and high resistivity. By optimizing the Hf content, the film with Hf concentration of 2.81 at.% exhibits excellent soft magnetic properties: high saturation magnetization ($4\pi M_S$ ~ 13.6 kG), and low coercivity (H_C ~ 0.6 Oe). The effective permeability of the film reaches 800 and remains constant up to 1 GHz.

1. Introduction

Recently, lots of attention has been focused on the size-reduction of dc-dc converters for the application in portable electronic products. In these devices, passive components, especially inductors, occupy the largest space. Thus miniaturizing the inductors is strongly desired [1-5]. High permeability ferromagnetic thin films, performing the crucial task of concentrating and shaping magnetic flux, could efficiently enhance the inductance of power inductors in dc-dc converters [6]. With the increasing demand of small inductors operated at several megahertz, the development of high permeability soft magnetic thin films with large saturation magnetization ($4\pi M_S$), low coercivity (H_C) and suitable anisotropy field (H_K) is required. Co-M-Ta (M = nonmagnetic metal) is an attractive class of high permeability materials [7-9]. To obtain superior permeability and optimal magnetic properties, doping a suitable quantity of other elements (M = Hf, Zr, Ce…) is necessary. The amount of added M element is minor in comparison with the Co content. However, it is critical to microstructure control [10]. Appropriate addition of the M element could effectively prevent Co grain growth and improve soft magnetic properties. In this study, soft magnetic Co-Hf-Ta thin films were fabricated for applications as power inductors. The effects of the Hf content on the microstructure and magnetic properties of Co-Hf-Ta thin films would be probed and discussed.

2. Experimental procedures

Soft magnetic films with composition of $Co_{100-x-y}Hf_xTa_y$ ($1.02 \leqq x \leqq 8.22$, $y \sim 5.4$) were fabricated onto n-type Si (100) substrates at ambient temperature by dc reactive magnetron co-sputtering with the base pressure better than 3×10^{-6} torr. The targets Co, Hf and Ta were utilized to deposit these films. The Hf content was varied by adjusting the gun power on the Hf target from 5 W to 30 W. The film thickness was about 750 nm for all samples. Magnetization versus field and the complex permeability were measured by vibrating-sample magnetometer (VSM) and permeameter (Ryowa PMF-3000), respectively. The domain structure in Co-Hf-Ta films was investigated by MFM (Dimension 3100, Veeco). The resistivity was obtained from standard four-point measurement. The microstructure was examined using x-ray diffractometer (XRD) and transmission electron microscopes (TEM, JEOL JEM-2010). The composition of the films was evaluated by field emission electron probe X-ray microanalyzer (FE-EPMA, JEOL JXA-8500F).

3. Results and discussion

Figure 1 reveals the x-ray diffraction (XRD) patterns of Co-Hf-Ta thin films with various Hf contents. A significant change in microstructure is revealed as the Hf content increases from 1.02 at.% to 2.81 at.%. However, further Hf addition from 2.81 at.% to 8.22 at.% has little impact on the microstructure. The sharp (002) peak in the XRD patterns of films with 1.02 at.% Hf shows obviously the preferred orientation in Co grains. When the Hf content increases to 2.81 at.% or more, a broad peak appears in the XRD patterns, implying the nanocrystalline structure with relatively random-oriented Co grains. Besides, according to Scherer's equation, Co grain size sharply decreases from 50 nm (1.02 at.% Hf) down to 2.3 nm (2.81 at.% Hf) and then remains almost constant, around 2 nm for 2.81 ~ 8.22 at.% Hf. Therefore, appropriate Hf addition to the Co-Hf-Ta films can effectively reduce the Co grain size.

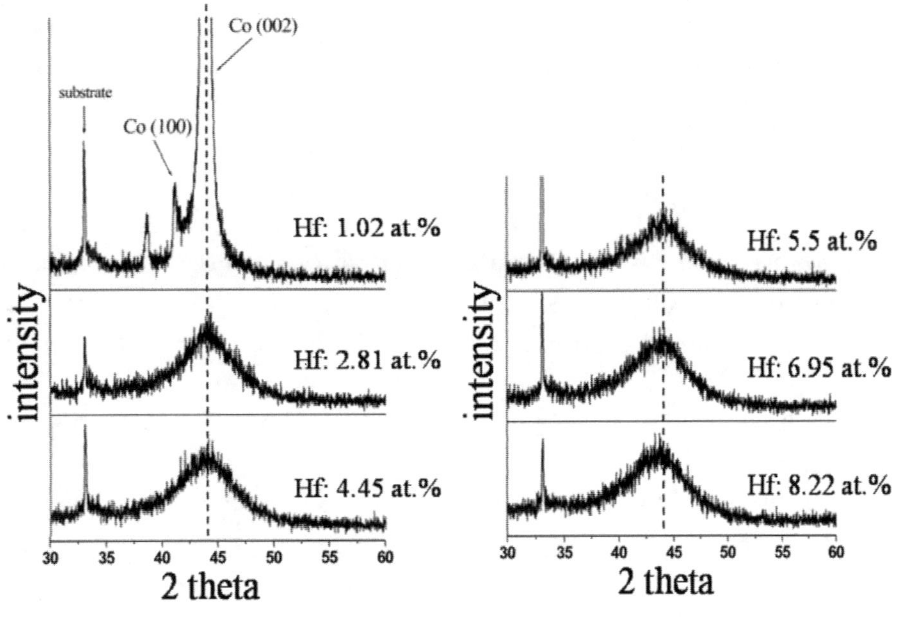

Fig. 1. X-ray diffraction patterns for the Co-Hf-Ta films with a Hf content variation of 1.02-8.22 at.%.

The dependence of anisotropy field (H_K) and coercivity (H_C) of Co-Hf-Ta films on the Hf content is shown in Fig. 2. H_K drops drastically as the Hf content increases from 1.02 at.% to 2.81 at.% and then remains almost constant. H_C shows similar trends with H_K. The change in H_K and H_C for films with different Hf contents is suggested to originate from the microstructure variations.

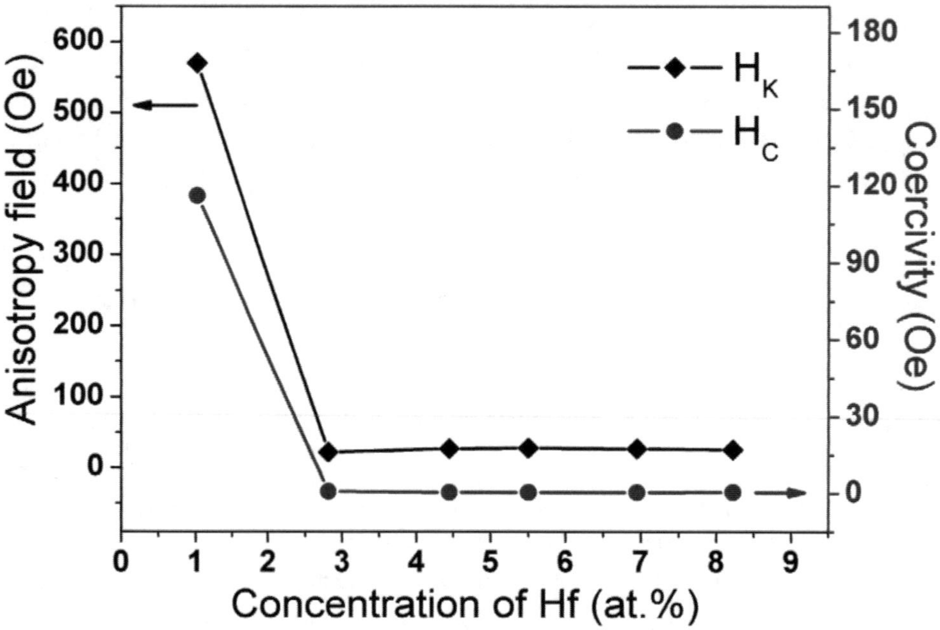

Fig. 2. Anisotropy field (H_K) and coercivity (H_C) for the Co-Hf-Ta films with various Hf contents.

At 1.02 at.% Hf content, the film reveals columnar structure with large Co grains of 50 nm, as shown in Figs.1 and 2. For large grains, magnetization follows the easy magnetic directions in the single grains [11]. Domains can be formed within the grains. Therefore, the magneto-crystalline anisotropy K_1 of the crystallites determines the magnetization process.

When the Hf content increases to 2.81 at.%, the film exhibits nanocrystallin structure with small Co grain size of 2.3 nm. According to the random anisotropy model [11-12], the effective anisotropy $<K>$ affecting the magnetization process results from averaging the $N = (L_{ex}/D)^3$ grains of size D within the volume $V = L_{ex}^3$, where L_{ex} is the exchange length. Thus, $<K>$ is determined by the mean fluctuation amplitude of anisotropy energy in the N grains, i.e. $< K > \approx K_1/\sqrt{N}$. In other words, in the films with small grain sizes, several grains are magnetically coupled to each other by ferromagnetic exchange interaction. As a result, the magneto-crystalline anisotropy is averaged out and thus reduced.

Based on the theory of ferromagnetism, anisotropy field (H_K) is proportional to the effective anisotropy $<K>$, i.e. $H_K \approx 2 < K >/M_S$ [13]. Besides, coercivity (H_C) is also related to $<K>$, i.e. $H_C = p_c < K >/M_S$, where p_c is a material parameter [11,14]. Thus, for the 1.02 at.% Hf film with large grain size of 50 nm, $<K>$ approaches K_1, resulting in high H_K and H_C. For the films with Hf concentration of 2.81 at.% or more, $<K>$ approximates to $K_1/\sqrt{N}$, leading to low H_K and H_C. The

Journal of Physics: Conference Series **266** (2011) 012032 doi:10.1088/1742-6596/266/1/012032

arguments addressed above clarify the significant decrease in H_K and H_C as the Hf content increases from 1.02 at.% to 2.81 at.% in Fig. 3. Furthermore, H_K and H_C for the films with higher Hf content (≧ 2.81 at.%) remain almost constant due to the slight difference in grain sizes.

The dramatic decrease in H_K and H_C as the Hf content increases from 1.02 to 2.81 at.% leads to two entirely different permeability spectra. The corresponding hysteresis loop and MFM image reveal that the film with 1.02 at.% Hf content has characteristic stripe-domain behavior with high H_C and H_K. Stripe domains lead to high coercivity and low permeability, both of which correspond to the frequency dependence of complex permeability spectra. The real part permeability (μ') is very low (~20-30) due to the high H_K. Besides the imaginary part permeability (μ'') rises up at low frequencies because of large hysteresis loss originating from the high H_C. On the other hand, the film with 2.81 at.% Hf content exhibits excellent soft magnetic properties with large $4\pi M_S$ and low H_K. According to Stoner-Wohlfarth theory, permeability (μ') is inversely proportional to H_K ($\mu' \approx 4\pi M_S/H_K$) [15]. Thus, this film is expected to have relatively high μ'. Fig. 3 shows the frequency dependence of complex permeability spectra for the film with 2.81 at.% Hf content. The experimental results fit well with the calculated values. μ' is up to 800 and flattens almost to 1 GHz. Besides, the cut-off frequency ($f_{cut\text{-}off}$), the ferromagnetic resonance frequency (f_{FMR}) and the half width in FMR ($\Delta f_{1/2}$) are 1.1 GHz, 1.6 GHz and 1.02 GHz, respectively.

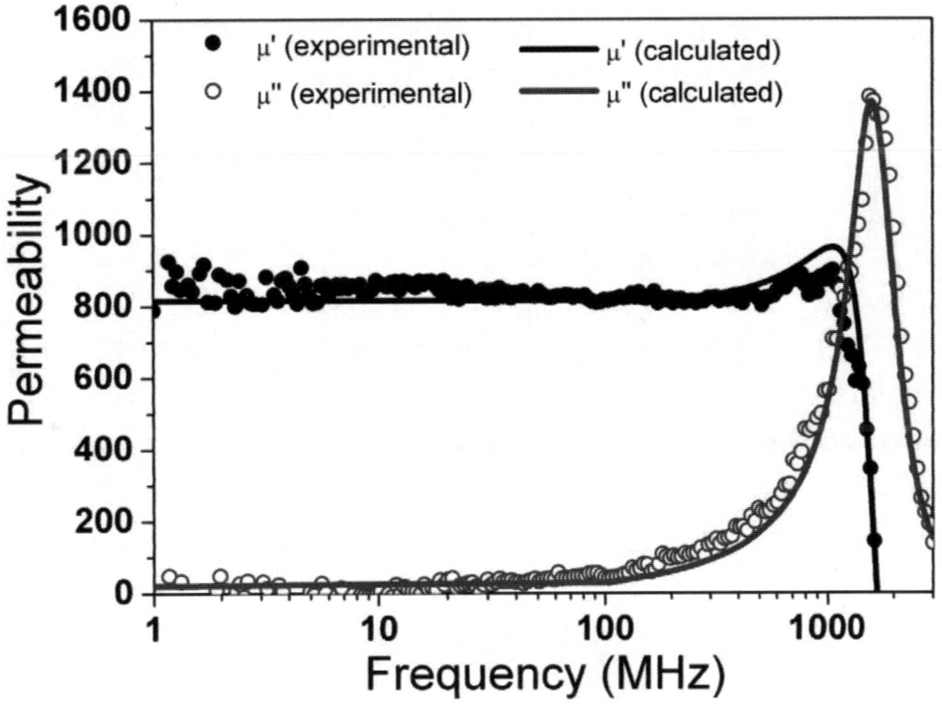

Fig. 3. Frequency dependence of complex permeability spectra for the Co-Hf-Ta films with 2.81 at.% Hf content.

To sum up, appropriate Hf addition to the Co-Hf-Ta magnetic thin films can effectively reduce grain sizes, enhance soft magnetic properties, and achieve a high permeability.

4. Conclusions

In this study, effects of the Hf content on the microstructure and magnetic properties of the Co-Hf-Ta thin film were investigated. Appropriate Hf addition can effectively reduce Co grain sizes and improve magnetic properties. By optimizing the Hf content, the film with Hf concentration of 2.81 at.% exhibits excellent soft magnetic properties: high saturation magnetization ($4\pi M_S$ ~ 13.6 kG), and low coercivity (H_C ~ 0.6 Oe). The effective permeability of the film reaches 800 and remains constant up to 1 GHz. The high value of permeability and the soft magnetic properties reveal that the Co-Hf-Ta thin films possess potential for applications as power inductors.

References

[1] K.H. Kim, J. Kim, H. J. Kim, S. H. Han, H. J. Kim, *IEEE Trans. Magn.* **38**, 3162 (2002).

[2] Y. Zhou, Z. M. Zhou, Y. Cao, X.Y. Gao, W. Ding, *J. Magn. Magn. Mater.* **320**, e963 (2008).

[3] D. H. Shin, C. S. Kim, J. H. Jeong, S.E. Nam, H.J. Kim, *IEEE Trans. Magn.* **35**, 3511 (1999).

[4] T. Sato, E. Komai, K. Yamasawa, *IEEE Trans. Magn.* **33**, 3310 (1997).

[5] M. Mino, T. Yachi, A. Tago, K. Yanagisawa, K. Sakakibara, *IEEE Trans. Magn.* **28**, 1969 (1992).

[6] D. W. Lee, K.P. Hwang, S. X. Wang, *IEEE Trans. Magn.* **44**, 4089 (2008).

[7] J. RE, *Acta Mater.* **46**, 3805 (1998).

[8] Y. Arimoto, K. Ohkubo, T. Fumoto, H. Yamasaki, *IEEE Tra*ns. Magn. **8**, 528 (1993).

[9] K. Ohkubo, Y. Arimoto, T. Fumoto, H. Yamasaki, *IEEE Trans. Magn.* **29**, 2530 (1993).

[10] Y. Hayakawa, A. Makino, *Nanostruct. Mater.* **6**, 989 (1995).

[11] G. Herzer, *IEEE Trans. Magn.* **26**, 1397 (1990).

[12] Y. Liu, Z.W. Liu, C.Y. Tan, C.K. Ong, *J. Appl. Phys.* **100**, 093912 (2006).

[13] B. D. Cullity, C. D. Graham, Second Ed., Wiley-IEEE Press, U.S.A., 2008, pp. 197-238.

[14] G. Herzer, *IEEE Trans. Magn.* **25**, 3327 (1989).

[15] V. Korenivski, *J. Magn. Magn. Mater.* **215**, 800 (2000).

Journal of Physics: Conference Series **266** (2011) 012033 doi:10.1088/1742-6596/266/1/012033

Modified Process for High-Performance Anisotropic $Sm_2Fe_{17}N_3$ Magnet Powder

T Ishikawa[1], K Yokosawa, K Watanabe and K Ohmori[2]
Ichikawa Research Laboratory, Sumitomo Metal Mining Co., Ltd., 3-18-5, Nakakokubun, Ichikawa, Chiba, 272-8588, Japan

E-mail: Takashi_Ishikawa@ni.smm.co.jp

Abstract. The magnetic properties of an anisotropic $Sm_2Fe_{17}N_3$ magnet powder were improved by enhancement of nitrogenation homogeneity. We made the Sm_2Fe_{17} parent material powder by a reduction-diffusion method which contains a reduction-diffusion heat treatment and a subsequent wet process for removal of a CaO by-product. In the present study, as the modified process for the magnet powder, the nitrogenation heat treatment was done between the reduction-diffusion heat treatment and the wet process, though the nitrogenation was normally done after the wet process. The surface of the Sm_2Fe_{17} powder before the wet process is rather clean than the surface after the wet process. It was found that the modified process seemed to promote the nitrogenation and decreased the number of $Sm_2Fe_{17}N_3$ particles with un-nitrogenated phase. The magnetic properties of the magnet powder through this process reached B_r = 1.46 T, H_c = 874 kA/m and $(BH)_{max}$ = 353 kJ/m^3 without demagnetizing field correction and $(BH)_{max}$ = 371 kJ/m^3 with the correction assuming the density 7.67 Mg/m^3.

1. Introduction

An anisotropic $Sm_2Fe_{17}N_3$ magnet powder is a key material for high-performance rare-earth bonded magnets. It is mass-produced by nitrogenation of an Sm_2Fe_{17} parent material powder made by a reduction-diffusion method and its maximum energy product $(BH)_{max}$ was up to 323 kJ/m^3 without a surface treatment for stabilization [1] and 290 kJ/m^3 for the surface-treated stabilized powder [2]. The saturation magnetization J_s of the $Sm_2Fe_{17}N_3$ compound was reported to be 1.54 - 1.57 T [3], [4], and therefore the theoretical $(BH)_{max} = J_s^2/4\mu_0$ is 472 - 490 kJ/m^3. There is much difference between these values.

In the reduction-diffusion process, the mixture of Sm_2O_3 powder and Fe powder is heat-treated with Ca metal granule. The reaction product is a sintered composite which consists of Sm_2Fe_{17} alloy particles and a CaO by-product. In the conventional process shown in figure 1(a), the composite is subjected to water and it turns to be a slurry of the alloy particles and $Ca(OH)_2$. For separation of the Sm_2Fe_{17} alloy powder, the slurry is washed and rinsed several times. After the decantation and drying, the powder is nitrogenated and the $Sm_2Fe_{17}N_3$ coarse powder was obtained. There is a case that the $Sm_2Fe_{17}N_3$ powder contains a un-nitrogenated Sm_2Fe_{17} phase which degrades the saturation magnetization and the squareness ratio of demagnetization curve. In the conventional process, the

[1] Present address: Sumiko Kunitomi Denshi Co., Ltd., 351-1, Kunitomi, Kyowa-cho, Iwanai-gun, Hokkaido, 048-2143, Japan
[2] Present affiliation: The Japan Association of Bonded Magnetic Materials

Sm_2Fe_{17} parent material powder has to be treated very carefully in order to avoid the residue of un-nitrogenated phase. The surface contamination of the Sm_2Fe_{17} alloy powder by water-washing and rinsing operation can be a barrier against homogeneous nitrogenation and it causes particles with the un-nitrogenated phase. The purpose of this study is to develop a fabrication process which reduces these undesirable particles effectively.

2. Experimental procedures

The water-washing and rinsing operation makes the Sm_2Fe_{17} alloy surface oxidized [1]. We thought that the alloy surface just after the reduction-diffusion heat treatment would be rather clean. Therefore it was tried to nitrogenate the sintered composite which consists of Sm_2Fe_{17} alloy particles and a CaO by-product. The modified fabrication process we proposed is shown in figure 1(b) [5]. The particle size of raw materials was under 300 mesh for the Sm_2O_3 powder and under 150 mesh for the Fe powder. The size of Ca metal granule was under 4 mesh. These were mixed together and heated at 1000-1190 ℃ under an Ar atmosphere for the reduction-diffusion reaction. In this experiment, the input composition of raw materials was in the range of $Sm_{2.1}Fe_{17}$ to $Sm_{2.7}Fe_{17}$ and the amount of Ca was 1.2 times the chemical equivalent for reducing the Sm_2O_3. The reacted product was heated at 435-480 ℃ under the NH_3-H_2 mixed gas flow for the nitrogenation. The flow ratio of NH_3/H_2 was in the range of 0.2 to 1.0. After the nitrogenation heat treatment, the product was put into water and washed in the same way as the conventional process to remove $Ca(OH)_2$. The $Sm_2Fe_{17}N_3$ coarse powder was obtained after drying and it was ball milled in ethanol down to 2-3 μm in average particle size.

The nitrogen and samarium content was determined by a thermal conductivity detector and an inductively-coupled plasma atomic emission spectrometry respectively. The phases in the samples were examined by an X-ray diffraction (XRD) apparatus using Cu Kα radiation. The microstructure of the samples was observed with a scanning electron microscope (SEM). The magnetic properties of $Sm_2Fe_{17}N_3$ magnet powder were measured by a vibrating sample magnetometer (VSM) after pulse magnetization of 3.2 MA/m. In order to align the crystal axis, the powder sample was made put in a wax, heated up to 70 ℃ and cooled down to room temperature in a magnetic field of 1.6 MA/m. The magnetization was calibrated by a Ni cylinder with saturation magnetization of 485.6 kA/m [6]. The density of $Sm_2Fe_{17}N_3$ magnet powder was determined as 7.67 Mg/m^3 by the XRD analysis.

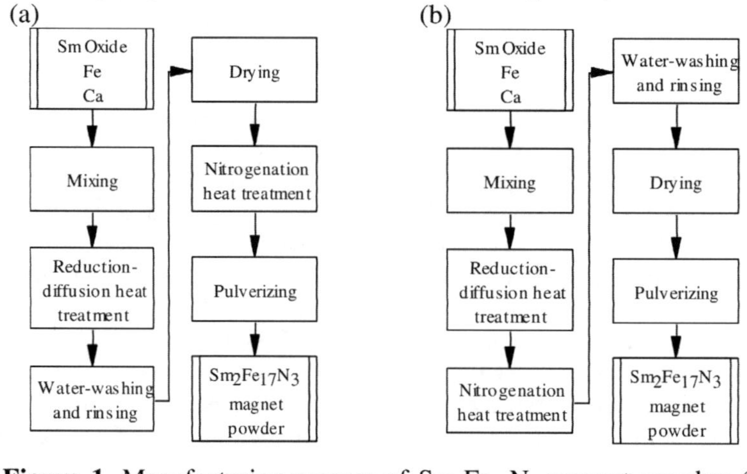

Figure 1. Manufacturing process of $Sm_2Fe_{17}N_3$ magnet powder: (a) conventional, (b) modified.

3. Results and discussion

Figure 2(a) shows an appearance of the reduction-diffusion reaction product. It is a mass of the sintered composite comprised of Sm_2Fe_{17} alloy particles and CaO. The SEM (BEI) photograph of the cross section in the product is shown in figure 2(b). It is observed that the Sm_2Fe_{17} alloy particles of several ten microns are surrounded by CaO particles of a few microns and a lot of voids. Though the reaction product is a bulky material, these voids seem to play an important role in nitrogenation as a diffusion pass of NH_3 and H_2 gasses. Moreover, the bulky product was crushed and turned to be a granule under 1 mm in size during the nitrogenation heat treatment as shown in figure 2(c). Hence we believed that the nitrogenation was possible even if the Sm_2Fe_{17} particles were embedded in the sintered composite.

The dependence of coercive force H_c and nitrogen content of $Sm_2Fe_{17}N_3$ magnet powder on nitrogenation time are shown in figure 3. The nitrogen content increases with increasing the nitrogenation time. As the anisotropy field H_A takes the maximum value at x=3 for $Sm_2Fe_{17}N_x$ [4], the coercive force also indicates a peak at a certain nitrogenation time. The optimum nitrogenation time is in the range of 120 to 180 minutes for the modified process, which is shorter than 180 minutes for the conventional process. Thus the modified process seems to nitrogenate the Sm_2Fe_{17} alloy more easily than the conventional process, which indicates that the voids worked as the diffusion pass of the nitrogenation gasses.

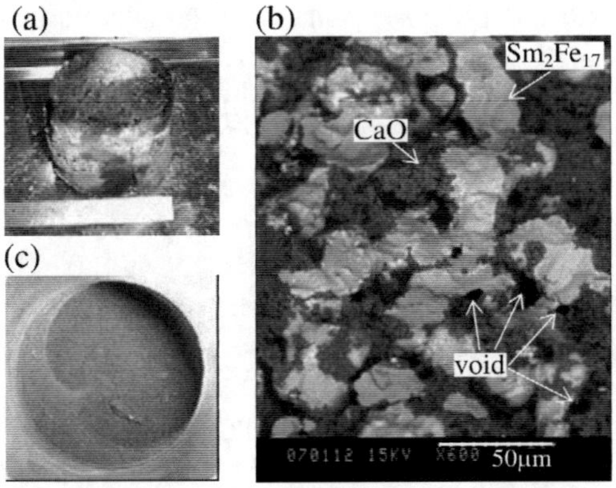

Figure 2. Photographs of the samples: (a) the reduction-diffusion reaction product, (b) SEM back-scattering electron image of fracture cross section in the product (a), (c) the reaction product after the nitrogenation heat treatment.

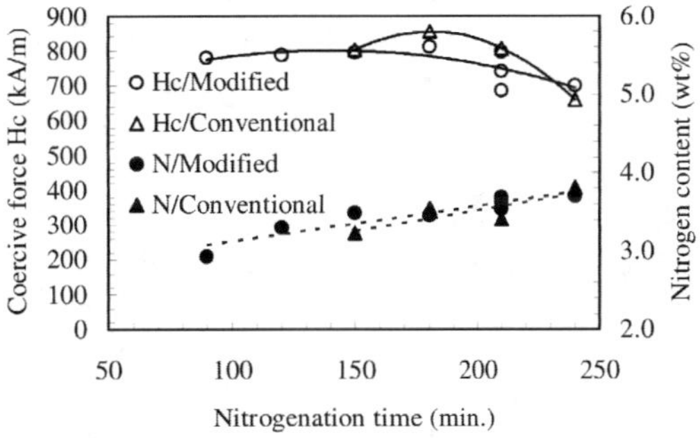

Figure 3. The dependence of coercive force H_c and nitrogen content of $Sm_2Fe_{17}N_3$ magnet powder on nitrogenation time. (input composition $Sm_{2.5}Fe_{17}$, r-d temperature 1190℃, nitrogenation temperature 465℃, flow ratio of NH_3/H_2=1.0, pulverized down to 2.5 μm)

In order to improve the magnetic properties of the magnet powder, the optimum conditions were examined for each operation in the modified process. In consequence, the conditions were set as shown in table 1. In particular, the temperature of the reduction-diffusion heat treatment affected the shape of Sm_2Fe_{17} alloy particles. Figure 4 indicates SEM photographs of the alloy particles reacted at 1050℃ and 1190℃. The powder morphology for 1050℃ was botryoidally-aggregated in contrast with that for 1190℃ and the grain under 10 μm in size was observed. The specific surface area and the number of the grain boundary is obviously large for 1050℃, which appears to enhance the homogeneous nitrogenation. Though a slight α-Fe peak in the XRD profile was detected for the $Sm_2Fe_{17}N_3$ magnet powder made of a 1050℃-reaction product, 1050℃ was the best in the range of 1000-1190℃ in terms of both the residual flux density B_r and the coercive force H_c of the powder.

Table 1. Optimum conditions for the modified process.

Mixing	Reduction-diffusion heat treatment	Nitrogenation heat treatment	
Input composition	Temperature	Temperature	Flow ratio of NH_3/H_2
$Sm_{2.5}Fe_{17}$	1050℃	450℃	0.50

Figure 4. SEM photographs of Sm_2Fe_{17} alloy particles reacted at different reduction-diffusion heat treatment temperatures: (a) 1050℃, (b) 1190℃. (input composition $Sm_{2.5}Fe_{17}$)

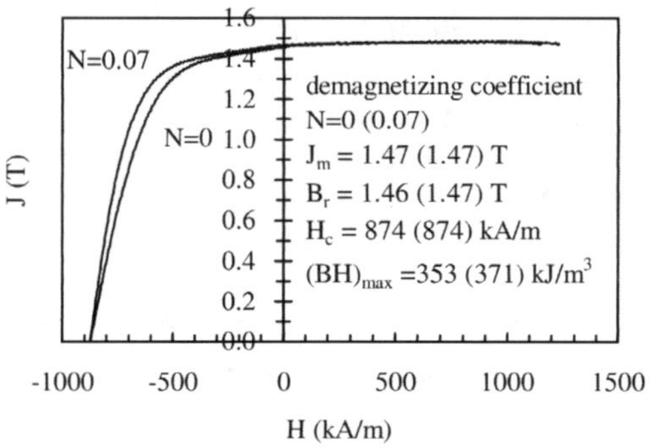

Figure 5. Demagnetization curves of $Sm_2Fe_{17}N_3$ magnet powder fabricated by the modified process with optimum conditions with and without demagnetizing correction.

Figure 5 indicates the demagnetization curve of $Sm_2Fe_{17}N_3$ magnet powder fabricated by the modified process with optimum conditions. The residual flux density B_r and the maximum energy product $(BH)_{max}$ reached 1.46 T and 353 kJ/m^3 respectively without demagnetizing field correction. We evaluated the demagnetizing coefficient N with Ni fine powder (2-3 μm in size) and the value 0.07 was obtained as the coefficient N. After the demagnetizing field correction with N = 0.07, the magnetic properties of the powder were B_r = 1.47 T, H_c = 874 kA/m, $(BH)_{max}$ = 371 kJ/m^3 respectively. The Sm and N contents were determined before pulverizing the $Sm_2Fe_{17}N_3$ coarse powder and they were 23.6 wt% and 3.3 wt% respectively. In order to identify the number of particles with un-nitrogenated phase, cross sections of the $Sm_2Fe_{17}N_3$ coarse powder was observed by SEM. Figure 6(a) is a back-scattering electron image of the cross section for the coarse powder made by the conventional process. The core with the white contrast inside dark-gray particles corresponds to the un-nitrogenated Sm_2Fe_{17} phase. The cross section image for the powder made by the modified process is shown in figure 6(b). It is clearly observed that the number of the particles with the un-nitrogenated phase was reduced drastically for the modified process. The image analysis showed that the ratio of white area to the whole was 0.04 % for the modified process in contrast with 0.73 % for the conventional process. This result shows that the modified process could promote the homogeneous nitrogenation.

Table 2 lists the magnetic properties of $Sm_2Fe_{17}N_3$ magnet powder reported previously. Tajima et al. and Izumi et al. improved the coercive force by adding surfactants to a milling solvent [8], [9]. Sasaki et al. developed high coercivity by mild pulverizing [10]. On the other hand, Kawamoto et al. adopted the reduction-diffusion method for Sm_2Fe_{17} parent material preparation and got high coercive force [1]. In this study, we have succeeded in reducing the particles with un-nitrogenated phase and improving the residual flux density B_r. However, the difference between 1.47 T in B_r and 1.54 T in J_s [3] is 5 % and the difference between 371 kJ/m^3 in $(BH)_{max}$ and 472 kJ/m^3 in $J_s^2/4\mu_0$ is 12 %, which are still large. One of the ways to close these values is to cut off the fine particles in the powder. Figure 7 shows the particle size distribution of the $Sm_2Fe_{17}N_3$ magnet powder made by the modified process. The volume ratio of the particle under 1 μm in size was 12 %. The powder was reported to be surrounded by an amorphous Oxygen-rich layer of about 10 nm in thickness [11]. As the volume ratio of the surface layer ("δr" in thickness) on the spherical particle ("2r" in diameter) is 3δr/r, the volume ratio of the amorphous layer on a spherical 1 μm-$Sm_2Fe_{17}N_3$ magnet powder, for example, is 6 % (δr = 0.01 μm, r = 0.5 μm). Assuming that the amorphous layer is nonmagnetic, the saturation magnetization J_s of 1.54 T drops to 1.45 T. In this manner it was found that the saturation magnetization of the spherical $Sm_2Fe_{17}N_3$ magnet powder was calculated to be 1.49 T based on the particle size distribution shown in figure 7. This value is close to the measured value 1.47 T in figure 5.

To cut off not only the fine particles but also coarse particles is additional way to improve the magnetic properties. Since the particle size affects the coercive force, to sharpen the size distribution improves the squareness ratio of the demagnetization curves.

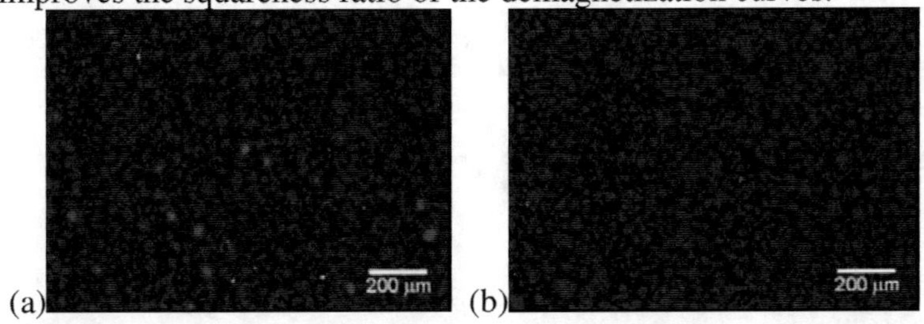

Figure 6. Back-scattering electron images of the cross section for the $Sm_2Fe_{17}N_3$ coarse powder made by (a) the conventional process, (b) the modified process. The white core inside the particles corresponds to the un-nitrogenated Sm_2Fe_{17} phase.

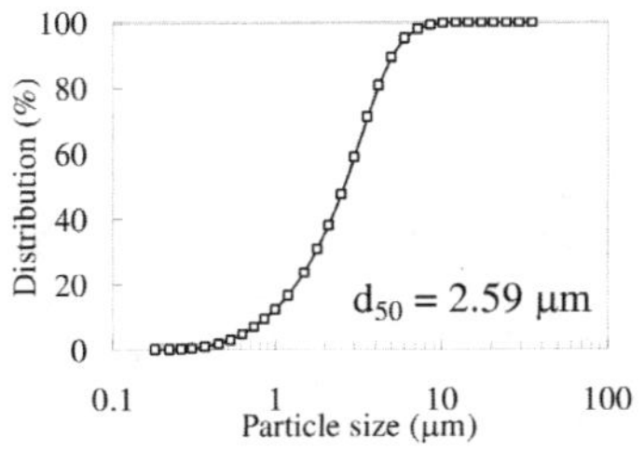

Figure 7. Particle size distribution for the $Sm_2Fe_{17}N_3$ magnet powder made by the modified process.

Table 2. The magnetic properties of $Sm_2Fe_{17}N_3$ magnet powder.

	B_r (T)	H_c (kA/m)	$(BH)_{max}$ (kJ/m^3)	demagnetizing field correction	ref.
Iriyama et al. (1992)	-	-	279	N = 0.13	[7]
Tajima et al. (1995)	1.41	740	271	none	[8]
			318	N = 0.13	
Izumi et al. (1996)	1.37	860	313	?	[9]
Kawamoto et al. (1999)	1.40	899	323	none	[1]
Sasaki et al. (1999)	-	979	329	?	[10]
Ishikawa et al. (2010)	1.46	874	353	none	This work
	1.47	874	371	N = 0.07	

4. Conclusion

We have developed a fabrication process for high-performance $Sm_2Fe_{17}N_3$ magnet powder based on the reduction-diffusion method. In this process, the nitrogenation heat treatment is set between the reduction-diffusion heat treatment and the water-washing and rinsing operation. The undesirable particles with un-nitrogenated phase are readily reduced. The magnetic properties of the magnet powder through this process reached B_r = 1.46 T, H_c = 874 kA/m, $(BH)_{max}$ = 353 kJ/m^3 without demagnetizing field correction and $(BH)_{max}$ = 371 kJ/m^3 with the correction.

5. References

[1] Kawamoto A, Ishikawa T, Yasuda S, Takeya K, Ishizaka K, Iseki T and Ohmori K 1999 *IEEE Trans. Magn.* **35** 3322
[2] Ishikawa T, Kawamoto A and Ohmori K 2003 *J. Jpn. Soc. Powder Powder Metallurgy* **50** 885
[3] Coey J M D and Sun H 1990 *J. Magn. Magn. Mat.* **87** L251
[4] Iriyama T, Kobayashi K, Imaoka N, Fukuda T, Kato H and Nakagawa Y 1992 *IEEE Trans. Magn.* **28** 2326
[5] Hashikawa T 1993 *Japanese Published Patent Application* **1993-148517** 1993-6-15
[6] 1998 *Japanese Industrial Standards* Japanese Industrial Standards Committee **C 2501**
[7] Iriyama T, Kobayashi K, Kato H and Nakagawa Y 1992 *J.Magnetics Society Jpn.* **16** 153
[8] Tajima S, Hattori T and Kato Y 1995 *J.Magnetics Society Jpn.* **19** 221
[9] Izumi H, Machida K, Shiomi A, Iguchi M and Adachi G 1996 *Jpn. J. Appl. Phys.* **35** L894
[10] Sasaki I, Fujii H, Okada H and Suzuki S 1999 *IEEE Trans. Magn.* **35** 3319
[11] Kobayashi K, Iriyama T, Yamaguchi Y, Kato H and Nakagawa Y 1993 *J. Alloys Compounds* **193** 235

Journal of Physics: Conference Series **266** (2011) 012034 doi:10.1088/1742-6596/266/1/012034

Harmonic response of GMR sensor under sinusoidal excitation field

Jen-Tzong Jeng[1,*], Ting-Yu Xu[1], Chih-Cheng Lu[2], and Bo-Ren Cai[1]
[1] Graduate Institute of Mechanical and Precision Engineering, National Kaohsiung University of Applied Sciences, Kaohsiung 80778, Taiwan
[2] Institute of Mechatronic Engineering, National Taipei University of Technology, Taipei 10608, Taiwan.

E-mail: jeng.jt@gmail.com

Abstract. In this work, we report the readout technique of the GMR magnetometer by detecting the odd-harmonic response under an sinusoidal excitation field. The time domain output waveform of the sensor is measured, and an analytical model is used to simulate its behaviour. In the frequency domain, it is found that the major Fourier components of the sensor response are the even harmonics of excitation frequency. The even harmonics are insensitive to the applied field, whereas the odd-harmonic amplitudes change significantly with the increasing applied field. To maximize the field-to-voltage transfer coefficient, the mixing circuit with a square wave reference is used to detect the odd-harmonic components of the output signal. In a small signal mode, the measured field-to-voltage coefficient of the sensor is 138 mV/mT under a 9 V bias voltage, achieving the noise level of 17 nT/Hz$^{1/2}$ at 1 Hz. The noise level can be further reduced by utilizing the multiple-harmonic reference. The proposed device is suitable for measuring the Earth's dc magnetic field.

1. Introduction

In recent years, the researches on miscellaneous magnetic materials have lead to the development of extremely low-noise magnetic field sensors, e.g., giant magnetoresistance (GMR) sensors [1], fluxgates [2,3], and anisotropic magnetoresistance (AMR) sensors [4,5]. Among these devices, the GMR is a relatively low-power detector [6]. As the output characteristics of these magnetic sensors is typically nonlinear and hysteretic, the signal processing circuits are always needed to achieve a linear output in response to the external magnetic field with negligible hysteresis. Owing to hysteresis and nonlinearity, the low frequency noise of GMR sensor is largely limited by the mode of operation [7]. For detecting the dc magnetic field less than 1 μT, the output drift owing to the hysteresis of magnetization in the free layer must be minimized. The field-modulation technique has been used for the fluxgate sensor [2,3], the AMR magnetometer [8,9], and the superconducting quantum interference devices [10] to achieve a linear output in response to the external field signal (B_{sig}). It was also reported that the GMR sensor biased with an ac excitation field (B_{exc}) exhibits lower hysteresis and higher linearity [9]. However, the field noise of 16 nT/Hz$^{1/2}$ at 1 Hz is still larger than expected noise limit [7]. In this work, we investigate the harmonic response of the GMR sensor with an excitation coil wound around its package. The harmonic spectrum is analyzed, and the field-to-voltage transfer

* To whom any correspondence should be addressed.

coefficient of the sensor driven by an ac readout circuit is predicted by using an empirical analytical model. The field noise can be minimized by optimizing the excitation amplitude and the reference waveform.

2. Sensor configuration and voltage-field characteristics

Figure 1(a) shows the schematic diagram of a GMR sensor with four magnetoresistors in a Whitestone-bridge arrangement. The sensor is the AA-002 from the NVE corporation [11]. The magnetoresistors covered by the shielding magnetic film are the reference resistors, while the resistors in-between two shielding/flux-concentrating magnetic films are the sensor resistors. The excitation coil, which has 100 turns within 3 mm, is wound around the package of the sensor. In absence of external magnetic field, the *V-B* curve of the GMR sensor is symmetrical about $B_{exc} = 0$, as shown in figure 1(b). When an external field signal is applied, the *V-B* curve shifts by the amount of the external field signal. The time trace of the output voltage is shown in figure 1(c). The sinusoidal curve is the 1-kHz excitation field. The solid and broken curves with anti-peaks are the sensor responses under the zero and 0.4 mT dc fields, respectively. In the time domain, the time difference between the "V-shaped" anti-peaks varies in proportion to the external field signal. The time difference between anti-peaks can be detected by using the demodulation circuit working at the fundamental frequency of excitation field.

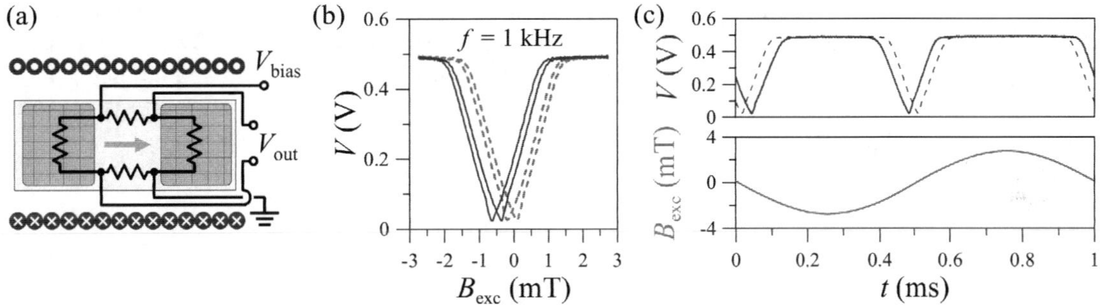

Figure 1. Principle of the ac-driven GMR sensor, (a) schematic layout; (b) *V-B* curves under zero (dashed curve) and non-zero (B_{sig} = 0.4 mT, solid) applied field; (c) Time trace of the output voltage under a sinusoidal excitation field. The bias voltage is 9 V.

3. Harmonic spectrum

Figure 2 shows the harmonic spectrum of the GMR output voltage under the sinusoidal excitation field. The data is the periodic Fourier transform of the measured time trace of the AA-002 GMR sensor [11]. It is found that the odd harmonics, i.e. the 1st, 3rd, 5th, 7th, and 9th, etc., of the excitation frequency responses to the external field signal. Although the 2nd and 4th harmonics have the largest amplitude, they do not change significantly under the external applied field. Among the odd harmonics, the 3rd has the largest amplitude. Since the 1st, 3rd and 5th harmonics dominates the output voltage, one can make use of the three components only to achieve a large field-to-voltage transfer coefficient. It is also found the phases of the 1st, 3rd and 5th harmonics are 178°, 175°, and 171° respectively, which are almost constant. As a square wave with the same the fundamental frequency of the sinusoidal excitation comprises only odd harmonics with constant phases of 0°, it suggests that the odd harmonics of the GMR output can be detected coherently with a square wave reference. By utilizing an analogy mixer to multiply the GMR sensor output with a coherent square wave, one can achieve a high-responsivity linear output in proportion to the applied external field signal in a small signal mode.

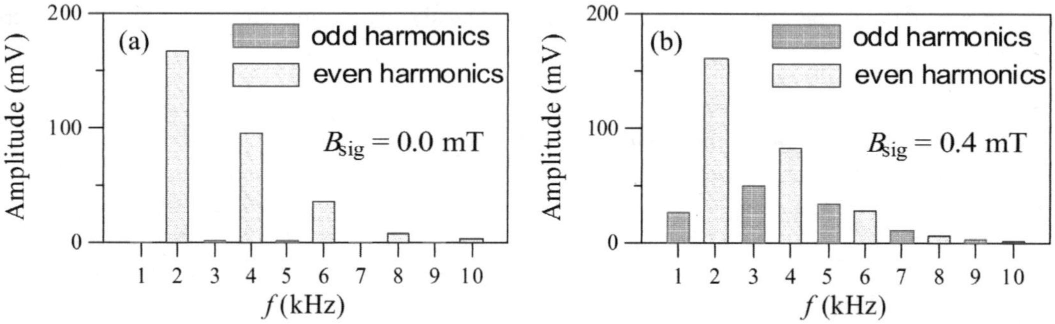

Figure 2. Harmonic spectrum of the GMR sensor under a sinusoidal excitation field at 1 kHz (a) without and (b) with external field.

4. Design of the driving circuit

The schematics of the driving circuit utilizing the odd-harmonic components to maximize the field-to-voltage transfer coefficient of GMR sensor is shown in figure 3(a). The oscillator, phase-shifter and the subsequent power amplifier generates the sinusoidal excitation field in the sensor. The ac response of the sensor is the imbalance voltage of the Whitestone bridge of giant magnetoresistors. The mixer analyzes the synchronous Fourier component of the sensor voltage by multiplying it with the signal from the reference generator, which can be the harmonics of the excitation field. The integrator filters out the ac signals, which extracts the dc level of the mixer voltage, V_m. In a small signal mode, V_m is in proportional to the external applied field. When the reference is a square wave at the excitation frequency, the circuit can be replaced by a lock-in amplifier and a function generator.

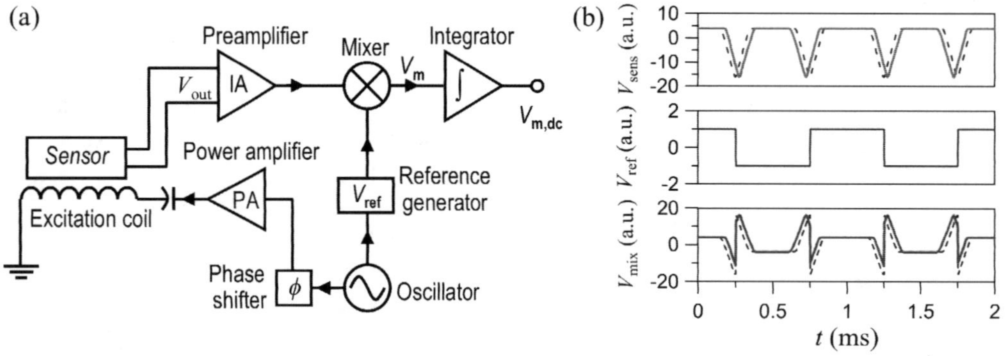

Figure 3. (a) Design of the driving circuit for the ac-excited GMR magnetometer. (b) Simulated mixer output with a square-wave reference.

To analyze the behaviour of the signal mixing process, we model the voltage-field characteristics of the GMR sensor by using the following analytic expression:

$$\begin{aligned} V(B) &= s\cdot\left|B-B_{hys}\right|+C, && \text{when } V<V_{max}, \\ V(B) &= V_{max}, && \text{when } V\geq V_{max}, \end{aligned} \tag{1}$$

in which B is the field in the sensor, s is the maximum slope of the V-B curve, C is the dc offset, V_{max} is the maximum voltage of the sensor output, and B_{hys} is hysteretic shift of the V-B curve when B is increasing. The time trace of the V-B curve under the ac excitation and external applied fields can be simulated by setting $B = B_{sig} + B_{exc}$, in which B_{sig} is the external field signal, and B_{exc} is the sinusoidal excitation field. With proper parameters of s, B_{hys}, C and V_{max}, the above equation can fit quantitatively to the voltage-field characteristics for the AA-002 sensor of NVE [11] at frequencies below 20 kHz. The parameters s, C and V_{max} are independent of the excitation frequency up to 1 kHz. The parameter B_{hys} is constant only for $f < 10$ Hz, and it becomes greater with the increasing frequency. In the time domain, the hysteresis in the V-B curve is proportional to the phase shift between the voltage waveform and the sinusoidal excitation wave. To predict the harmonic spectrum, we utilize equation (1) to generate the time traces of the sensor output under the external applied field. The result is shown in figure 3(b). The predicted wave form is similar to that in figure 1(c), and its harmonic spectrum is almost identical to that of figure 2 with a proper external field signal. By using the analytic model, it is found that the field-to-voltage coefficient of the mixer output with a square-wave reference is maximized when a proper excitation field is applied. To verify this finding, we use a function generator and a lock-in amplifier (SR530, from Stanford Research Systems) to probe the optimal transfer coefficient under the sinusoidal excitation and a square-wave reference. For a 9 V bias voltage at 10 kHz excitation, it is found the transfer coefficient is 47 mV/mT for a excitation current of 210 mA, and it increases to 138 mV/mT for the optimal current of 180 mA. The field noise corresponding to the optimal excitation amplitude is 17 nT/Hz$^{1/2}$ at 1 Hz, which can be further reduced with the circuit in figure 3(a) when the multiple-harmonic reference voltage is applied to enhance the transfer coefficient. Further works are under taken to investigate the noise characteristics with various reference waveforms.

5. Conclusion

By detecting the odd-harmonics of the output voltage, the GMR magnetometer excited with a sinusoidal field shows a high responsivity to applied field. By using a square-wave reference, the measured field-to-voltage transfer coefficient is 138 mV/mT and the field noise is 17 nT/Hz$^{1/2}$ at 1 Hz. The proposed device is promising in miscellaneous applications requiring the accurate measurement of Earth's magnetic field, e.g. the electronic compass.

6. Acknowledgements

This work is supported by the National Science Council of Taiwan under Grant No. NSC98-2112-M-151-002-MY3 and NSC98- 2221-E-027-066.

7. References

[1] Daughton J 2003 *Proc. IEEE* **91** 681
[2] Hinnrichs C, Pels C, and Schilling M, 2000 *J. Appl. Phys.* **87** 7085
[3] Paperno E, Weiss E, and Plotkin A, 2008 *IEEE Trans. Mag.* **44** 4018
[4] Platit A, Kubik J, Vopalensky M, Ripka P, 2003 *Proceedings of IEEE Sensors* **2003** 472.
[5] Ripka P, Vopálenský M, Platil A, Döscher M, Lenssene K-M H, Hauser H, 2003 *J. Magn. Magn. Mater.* **254–255** 639
[6] Rife J C , Miller M M , Sheehan P E, Tamanaha C R, Tondra M , Whitmana L J, 2003 *Sens. Actuators A* **107** 209
[7] Stutzke N A, Russek S E , Pappas D P, 2005 *J. Appl. Phys.* **97** 10Q107
[8] Kaplan B Z and Papemo E, 1994 *IEEE Tram. Magn.* **30** 4614
[9] Ripka P, Tondra M, Stokes J, and Beech R, 1999 *Sens. Actuators A* **76** 225
[10] Kleiner R, Koelle D, Ludwig F, and Clarke J, 2004 *Proc. IEEE* **92** 1534
[11] GMR Sensor Catalog, NVE corporation, the website is http://www.nve.com

Magnetic properties of ion irradiated epitaxial Fe films

Y Kamada[1], H Watanabe[2], S Mitani[3], J Echigoya[1], J N Mohapatra[1], H Kikuchi[1], S Kobayashi[1] and K Takanashi[4]

[1]NDE and Science Research Center, Faculty of Engineering, Iwate University, 4-3-5 Ueda, Morioka, 020-8551, Japan
[2]Research Institute for Applied Mechanics, Kyusyu University, Kasuga, 816-8580, Japan
[3]National Institute for Materials Science, Tsukuba, 305-0047, Japan
[4]Institute for Materials Science, Tohoku University, Sendai, 980-8577, Japan

E-mail: kamada@iwate-u.ac.jp

Abstract. Magnetic properties of a heavy-ion irradiated single crystalline iron film were investigated. A high quality Fe (001) film with a thickness of 250 nm was fabricated on MgO (001) using the molecular beam epitaxy technique. The film was irradiated by 3.2 MeV Ni ions at room temperature using a tandem accelerator. Formation of dislocation loops with nanometer size was observed by TEM observation, and that of sub-nanometer size vacancy clusters was confirmed indirectly from a resistivity increase. However, *M-H* hysteresis curves and magnetic domain structure did not change significantly. These results indicate the formation of irradiation defects of pure iron in nanometer scale range has little influence on the magnetization process of the iron.

1. Introduction

Materials degradation in radiation environment is a crucial issue for safe operation of nuclear fission power plants. Especially irradiation embrittlement of reactor pressure vessel (RPV) steels, iron based alloys, becomes a greater concern for prolonging a lifetime of the present power plants. Therefore, development of nondestructive evaluation (NDE) techniques of the embrittlement is definitely needed and many researchers have been tackling this problem. Magnetic measurement is one of the candidates, which utilizes the changes in magnetic properties caused by the formation of irradiation defects [1-4]. Although the clarification of the underlying mechanism is essential for establishing a reliable NDE technique, there has been little experiment executed from fundamental viewpoints. Various types of lattice defects including vacancy- and interstitial-types are formed in RPV steels and the effect of each defect should be elucidated. High energy ion irradiation has several advantages in simulating the neutron irradiation experiment, such as a production of cascade effect similar to the neutron damage, and an easy handling of non-radioactive specimens [5]. Combination with single crystalline film specimens enables us to overcome the weakness of a short penetrating depth of ions, and exclude complicated polycrystalline boundary effects. Under this concept, magnetic properties of ion-irradiated epitaxial iron film with a thickness of 30 nm were investigated in the previous study [6]. However, in such thin specimen, it was difficult to clarify the effect of interstitial-type defects because large amounts of interstitials move out to the surface due to its high mobility. In this study, a thicker single crystalline iron film was newly fabricated to produce both vacancy- and interstitial-type defects and these effects on magnetic properties were discussed.

2. Experimental procedures

A single crystalline iron film with a thickness of 250 nm was fabricated using the molecular-beam epitaxy technique [6]. The iron was prepared by electron-beam deposition on single crystalline MgO (001) substrate at room temperature, and then annealed at 773 K. The base pressure was 2×10^{-10} Torr and the pressure during deposition was higher than 10^{-8} Torr. The crystalline structure was checked by *in-situ* reflection high-energy electron diffraction (RHEED). After the fabrication, the specimen was cut into small pieces along the cleavage plane of magnesia and the following experiments were executed.

Ion beam irradiation of 3.2 MeV Ni^{3+} was carried out at room temperature, using a tandem accelerator at the Research Institute for Applied Mechanics. In this work, typical irradiation time and total fluence were 160 minutes and 6.4×10^{18} ions/m^2, respectively. Figure 1 shows the damage profile of bulk iron calculated from SRIM (Stopping and Range of Ions in Matter) code with a threshold energy of 40 eV [7]. The displacement damage at a depth of 250 nm was estimated to 0.5 dpa. Almost all injected Ni ions stop at the position well deeper than 250 nm, which means the Ni ions passed through iron film to substrate. As for the specimen for magnetic domain observation, the surface was partially masked by an aluminum foil to make both unirradiated and irradiated regions in one specimen. Irradiation dose was fixed to 0.1 dpa for this specimen.

Resistivity was measured by direct current four points method from 15 K to 295 K in step of 1 K. For each measurement, a constant current of 5 mA was applied reversibly to remove the thermoelectric power effect. Transmission electron microscope (TEM) was used for characterization of irradiation defects. The TEM specimen for plan-view observation was carefully prepared by the conventional techniques of mechanical polishing and Ar ion-milling. In order to clarify the irradiation effect on magnetic properties, *M–H* hysteresis curves and magnetic domain structure were evaluated at room temperature using a vibrating sample magnetometer (VSM) and a polarized light microscope (NEOARK Kerr Microscope), respectively.

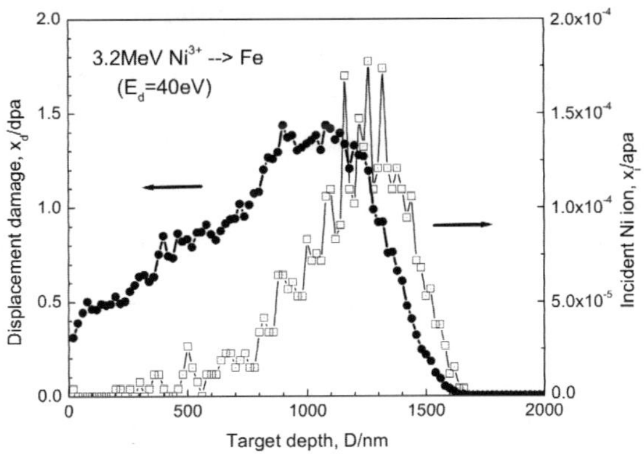

Figure 1. Range of damage and incidented Ni ion distribution calculated by the SRIM code.

3. Results

3.1. Resistivity and microstructure

Fabrication of single crystal iron film with high crystalline quality was confirmed by RHEED patterns. The epitaxial growth relationship was Fe(001)[100]//MgO(001)[110]. Using this film, Ni ion irradiation was carried out. Figure 2 shows the temperature dependence of the resistivity of unirradiated and irradiated iron films. Irradiation caused an increase of residual resistivity, which suggests that irradiation lattice defects, scattering sources of electrons, were formed inside the film. A typical TEM plan-view image of the irradiated specimen is shown in figure 3. Many small black contrasts with a small size (less than 10 nm) can be confirmed with high density (10^{22} m^{-3} range order). Although the detail of defect character has not been analyzed, those defects are expected to be interstitial-type dislocation loops [8, 9].

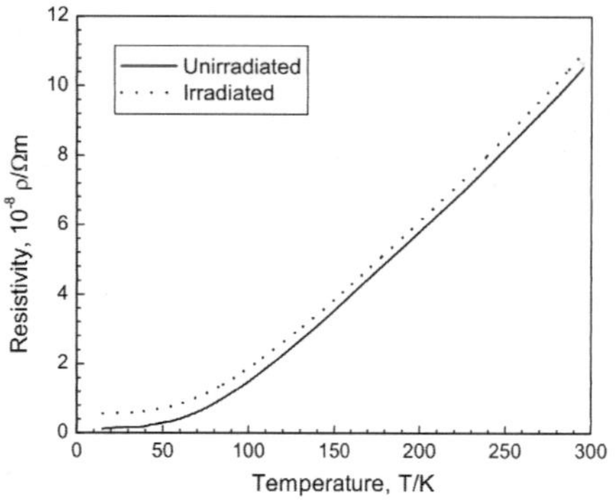

Figure 2. Temperature dependence of the resistivity of unirradiated and irradiated iron films.

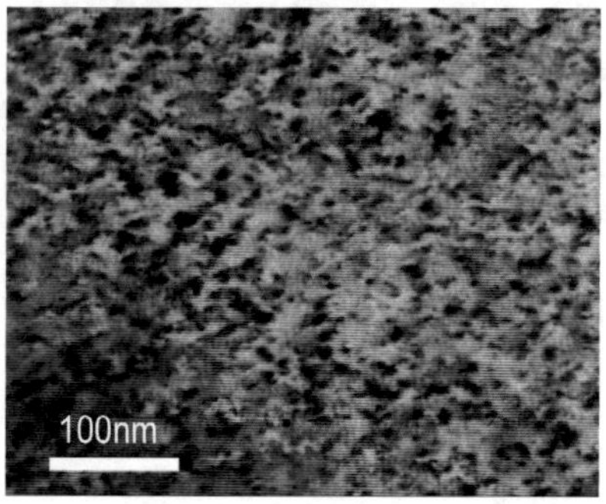

Figure 3. TEM plan-view image of irradiated iron film.

3.2. Magnetic properties

Figure 4 shows *M-H* curves of unirradiated and irradiated iron films. Magnetization was normalized by saturation magnetization. The applied magnetic field was parallel to the $[100]_{Fe}$ direction (an easy magnetization direction in bulk iron) in figure 4 (a) and the $[110]_{Fe}$ direction in figure 4 (b). *M-H* curves in the both directions do not change drastically by irradiation. The coercive forces of unirradiated and irradiated specimens, typical parameters reflecting the impedance of magnetic domain wall motion by lattice defects [10], are almost the same within experimental error. Figure 5 shows the picture of specimen with an aluminium mask and the corresponding image of domain structure under the magnetic field just below the coercive force. The edge of MgO substrate, a cleavage plane of $(100)_{MgO}$, is marked by white solid lines. The regions enclosed by white dotted lines denote the unirradiated region. Large domain patterns are observed in both unirradiated and irradiated regions as shown in figure 5. At the boundary between the two regions, irradiation effects such as abrupt changes of domain structure and/or domain wall pinning are not noticeable, which is consistent with the non-variational behaviour of *M-H* curves by irradiation.

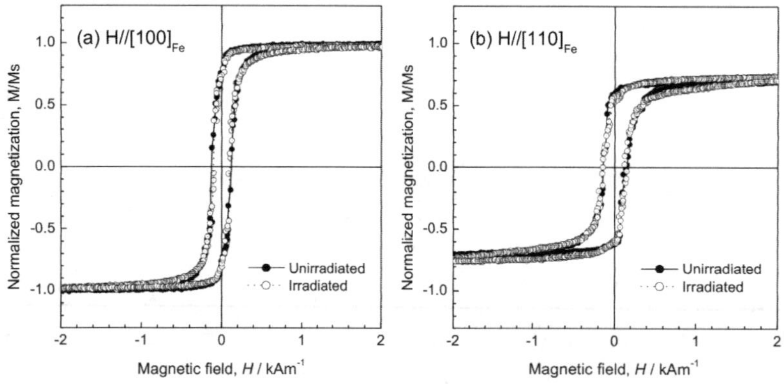

Figure 4. Magnetization curves of unirradiated and irradiated iron films. Applied field is parallel to (a) $[100]_{Fe}$ and (b) $[110]_{Fe}$.

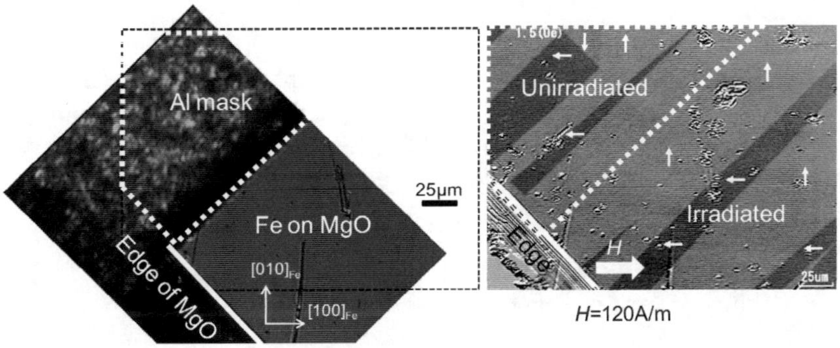

Figure 5. Picture of the specimen with an aluminium mask and the domain structure at the same position under the magnetic field of 120 A/m.

4. Discussion

Irradiation defects of bulk iron have been systematically studied using positron annihilation spectroscopy (PAS), TEM and a conductivity measurement [8, 9]. After the neutron irradiation at 320 K to a dose of 0.2 dpa, sub-nanometer vacancy clusters and small dislocation loops (probably interstitial-type) were confirmed by PAS and TEM, with a density of the order of 10^{24} m^{-3} and 10^{22} m^{-3}, respectively. A decrease in conductivity of about 15 %, corresponding to the increase in resistivity of 2 $\times$ 10^{-8} Ωm, was also reported. The resistivity change reflects the formation of vacancy cluster because its density is much higher than that of dislocation loop. In this study, interstitial dislocation loops were observed by TEM as shown in figure 3, but vacancy-type defects were not confirmed directly by experiment. However, the increase in resistivity of the present iron film (0.4 $\times$ 10^{-8} Ωm from figure 2) was in the same order compared to the above study, which is an indirect evidence of the formation of sub-nanometer vacancy clusters with high density.

Although the present ion-irradiation produced both vacancy- and interstitial-type defects inside the iron film, no noticeable change was found in *M-H* curves and domain structure. The hindrance of magnetic domain wall motion effectively occurs when the size of the defects approaches the same order of magnitude as a domain wall width [11, 12]. The domain wall width in bulk iron is estimated to about 40 nm [13]. Since the size of the defects is much smaller than domain wall width, they did not have much effect on magnetic properties of the iron film.

For neutron irradiation of RPV steel, the coercive force initially increased, then changed to decrease with increasing neutron fluence [2]. Defect nature of RPV steels is slightly different from that of pure iron because RPV steels contain several kinds of elements. During formation of the defects involving such elements, changes of chemical composition in the matrix simultaneously occur. In addition, initial dislocation density of RPV steels is much higher compared to that of well-annealed pure iron. In some cases, irradiation defects form inhomogeneouly near the region of dislocations [14], which probably produce a variation of strain field with the scale well above the domain wall width. Such chemical and inhomogeneous effects are expected to influence the magnetization process. The results of this study indicate the importance to consider also such other factors, and further studies are needed for clarification of irradiation effects for establishing a magnetic NDE technique.

5. Conclusion

An epitaxial single crystalline iron film with a thickness of 250 nm was fabricated and the effects of Ni ion irradiation on magnetic properties were investigated. Formation of dislocation loops with nanometer size was observed by TEM observation, and that of sub-nanometer size vacancy clusters was confirmed indirectly from a resistivity increase. Since such defects have a little effect on domain wall movement due to their small size, the *M-H* curves of the iron film did not change by irradiation.

Acknowledgments
This work was supported in part by a Grant-in-Aid for Scientific Research (B), No.20360420 and 20360416, from the Ministry of Education, Culture, Sports, Science and Technology of Japan. This work was performed under the collaborative research program of the Research Institute for Applied Mechanics, and the Advanced Research Center of Metallic Glasses, Institute for Materials Research.

References

[1] Vandenbossche L, Konstantinović M J and Dupré L 2008 *J. Magn. Magn. Mater.* **320** e562
[2] Takahashi S, Kikuchi H, Ara A, Ebine N, Kamada Y, Kobayashi S and Suzuki M 2006 *J. Appl. Phys.* **100** 023902
[3] Dobmann G, Altpeter I and Kopp M 2006 R*ussian J. Nondestructive Testing* **42** 272
[4] Park D G, Hong J H, Kim I S and Kim H C 1999 *IEEE Trans on Magnetic*s **35** 3898
[5] Abromeit C. 1994 *J. Nucl. Mater.* **216** 78
[6] Kamada Y, Watanabe H, Mitani S, Echigoya J, Kikuchi H, Kobayashi S, Yoshida N, Takanashi K, 2009 *Mater. Trans.* **50** 2234
[7] Ziegler J F 2004 *Nucl. Instrum. Methods in Phys. Res.* B **219-220** 1027, SRIM code is available at http://www.srim.org/
[8] Victoria M, Baluc N, Bailat C, Dai Y, Luppo M I, Schäublin R and Singh B N 2000 *J. Nucl. Mater.* **276** 114
[9] Eldrup M, Singh B N, Zinkle S J, Byun T S and Farrell K 2002 *J. Nucl. Mater.* **307-311** 912
[10] Kronmüller H and Fähnle M 2003 *Micromagnetism and the Microstructure of Ferromagnetic Solids* (Cambridge University Press, UK)
[11] O' Handley R C 2000 *Modern Magnetic Materials: Principles and Applications* (Wiley-Interscience Publication, New York) p 334
[12] Dijkstra L J and Wert C 1950 *Phys. Rev.* **79** 979
[13] Chikazumi S 1997 *Physics of Ferromagnetism* (Clarendon Press, Oxford) p 413
[14] Ghoniem N M, Singh B N, Sun L Z and Diaz de la Rubia T 2000 *J. Nucl. Mater.* **276** 166

Journal of Physics: Conference Series **266** (2011) 012036 doi:10.1088/1742-6596/266/1/012036

Microstructures of Mischmetal–Fe-B Hot-pressed Magnets with Ti Addition

K. Y. Ko[1], S. Yoon[2] and S. W. Ko[3]
[1]Faculty of Mechanical Engineering, Ulsan College University, Ulsan, 680-749, S. Korea
[2]School of Materials Science and Engineering, Ulsan University, Ulsan, 680-749, S. Korea
[3]Division of Natural Sciences, Sungkyunkwan University, Suwon, 440-746, S. Korea

E-mail: kyko@uc.ac.kr

Abstract. Microstructures of MM-Fe-B hot-pressed magnets with and without Ti addition were investigated. Both materials showed hard magnetic behaviours with more or less same (BH)max of ~4MGOe. Microstructures were, however, quite different to each other. Grain size of specimen without Ti were ~500 nm while MM-Fe-B(Ti) showed drastically reduced grain sizes of ~40nm. Increases in coercivity were found in specimens with Ti. DSC results indicated that Ti triggers and speeds the crystallization of $Nd_2Fe_{14}B$ phase when pressure of >1MPa is applied.

1. Introduction

MM-Fe-B alloys from more abundant and inexpensive Mischmetal (MM; mixture of rare earths; La, Ce, Nd, Pr, Fe, Ba etc) and ferroboron (Fe, B, C, Si, Al) have proved to be a possible alternative to expensive rare-earth permanent magnets, although their magnetic properties are far inferior. Nano-composites of $Nd_2Fe_{14}B$ phase and soft magnetic Fe_3B or α -Fe phase permanent magnets were also proved to be a way to get magnets with reasonable performances at lower costs. Microstructures of those materials are crucial for their magnetic properties and many investigations have been made on microstructure control for those materials [1, 2, 3]. No attempts are to control microstructure of MM-FeB permanent magnets to make nano-composite permanent magnets. Ko and Yoon et. al. reported that for MM-FeB-(Al, Ti and Al-Co) alloys better magnetic properties in die-upset magnets were resulted from the alignment of c-axis along the die-upsetting direction [4]. It was shown that magnetic anisotropy of die-upset magnets increases with partial substitution of Al and Al-Co for Fe [5]. Recently Ko reported that crystallization rate in MM-FeB permanent magnet was increased by partial substitution of Ti for Fe [6]. In this paper, microstructures of MM-Fe-B hot-pressed magnets were studied in conjunction with effects of Ti addition.

2. Experimental

MM-Fe-B and MM-Fe-B-Ti alloys were melted in an argon arc furnace from commercial grade Mischmetal, 10 wt.% FeB and 2N5 Ti element. The actual atomic compositions were $MM_{15}Fe_{77}B_8$ and

[1] kyko@uc.ac.kr Tel +82 (0)52 279 3122 Fax +82 (0)52 279 3137

$MM_{15}(Fe_{76}Ti_1)B_8$. Melt-spun ribbons (or flakes) were obtained in an argon atmosphere of ~70 mmHg using a rotating Cu wheel whose surface speed was kept as ~40m/s. Obtained ribbons, 25 - 40 μm thick and 2 - 3 mm wide, were crushed into powder with sizes less than ~150 μm under argon atmosphere using a mortar for fabrication of magnets. Hot-pressed magnets were made in the form of tablet of 4~5 mm thick and 8 mm dia., with applying pressure of 100 MPa and holding time of 2-5 min., at about 700℃ under argon atmosphere. Differential scanning calorimetry (PDSC and DSC-2910, TA instruments) were used to investigate phase formations as a function of temperature and pressure. A vibrating sample magnetometer with a superconducting magnet was employed for magnetic measurements. A high temperature x-ray diffractometer (Rigaku D/Max-2500) was used to analyse the crystal structures in temperature range of R.T - 600 ℃. High resolution transmission electron microscopy (Joel 2100F STEM) and EPMA were used for microstructure and phase analyses.

3. Results and Discussions

Fig 1 shows TEM micrographs of hot-pressed magnets of MM-Fe-B and MM-Fe-B(Ti). It is seen that grain sizes are drastically reduced from ~500nm to ~40nm. EPMA results indicated that main phases for both cases with and without Ti addition were largely 2-14-1 phase and grain boundaries were of somewhat poorer composition of Fe. In the case of MM-Fe-B(Ti), Ti was found at the grain boundaries only as is given in table 1. It revealed that segregation of Ti element at grain boundary led to the grain refinement by slowing interface kinetics and therefore could improve coercivity and so maximum energy product [6].

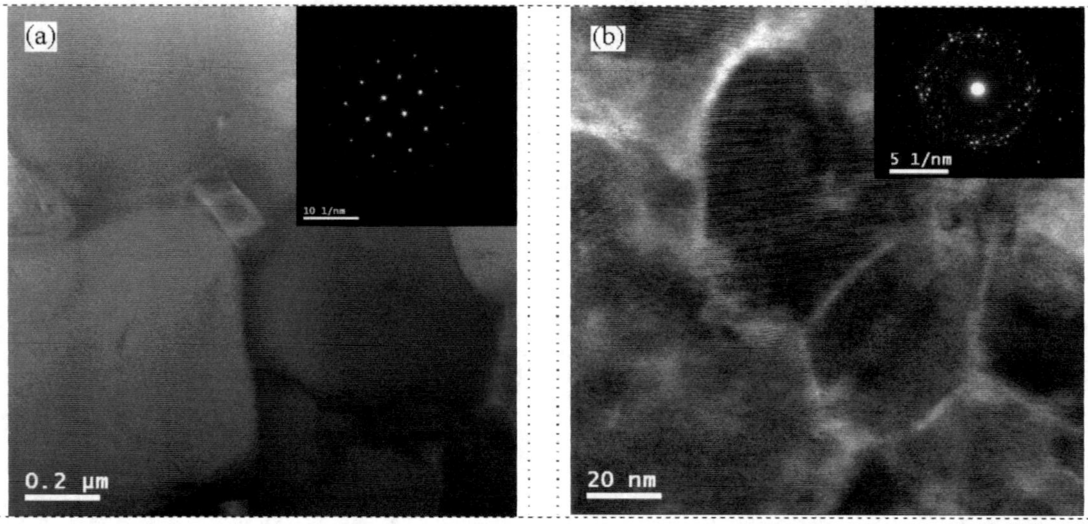

Figure 1. TEM micrographs of hot-pressed magnets: (a) MM-Fe-B and (b) MM-Fe-B(Ti).

Fig 2 shows XRD patterns on the as-melt spun flakes of MM-Fe-B and MM-Fe-B(Ti). For both cases, it is seen that at the melt-spinning stage crystallization is almost completed. Most peaks were identified as for the 2-14-1 phase and some peaks could be indexed as those of Fe_3B phase. As temperature increased over 300℃, peaks became broader and extra peaks disappeared. This is considered due to interfusion and homogenization at elevated temperatures. It is also noticed that interfusion and/or homogenization in the alloy with Ti addition occur somewhat higher temperatures than in the alloy without Ti addition.

In Fig 3, DSC results on the melt-spun specimens of both with and without Ti addition under various pressures are given. It is seen that application of pressure and Ti addition affect on the crystallization and interfusion between existing phases. Comparing DSC curves of the ribbons without and with Ti addition at zero pressure (bottom curves in the figure), crystallization of 2-14-1 phase from amorphous part is suppressed by Ti up to 600℃ while alloys without Ti showed a peak at ~550℃ corresponding to the crystallization of 2-14-1 phase.

Table 1. EPMA results of hot-pressed MM-Fe-B(Ti) magnet.

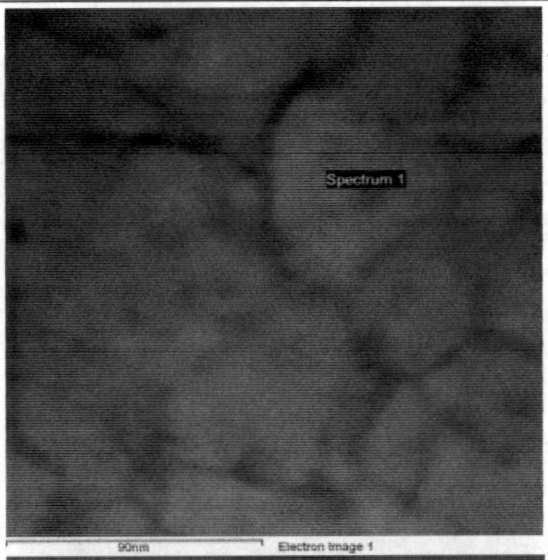

Element	Weight%	Atomic%
C K	1.39	7.34
Al K	0.34	0.79
Si K	0.55	1.24
Fe K	68.21	77.13
Ba L	1.99	0.92
La L	3.43	1.56
Ce L	18.32	8.28
Pr L	0.88	0.40
Nd L	4.89	2.15

Element	Weight%	Atomic%
C K	1.31	7.19
Al K	0.50	1.22
Si K	0.56	1.32
Ti K	1.98	2.73
Fe K	59.68	70.60
Ba L	0.59	0.28
La L	9.87	4.69
Ce L	20.13	9.49
Nd L	5.38	2.47

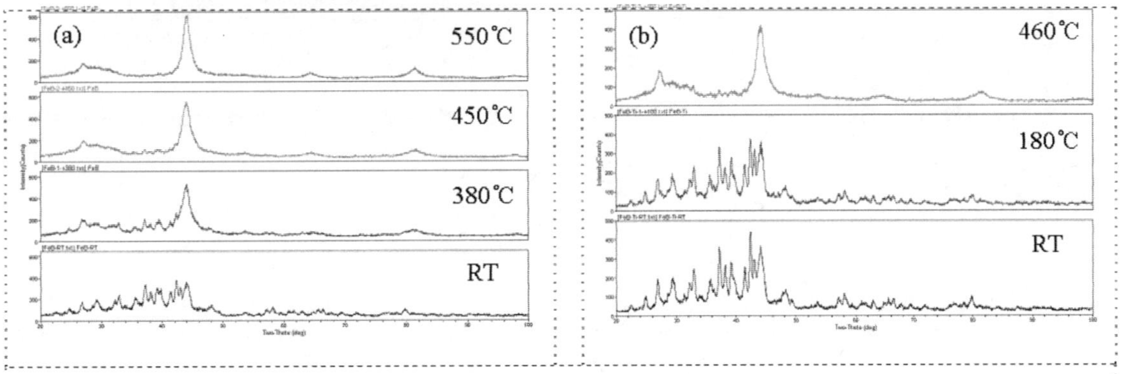

Figure 2. X-ray diffraction data of melt-spun (a) MM-Fe-B and (b) MM-Fe-B(Ti) ribbons at room and elevated temperatures.

Application of pressure seemed to promote both crystallization of 2-14-1 phase and interfusion between existing phases. DSC curves at 1.0 MPa (uppermost curves in the figure) showed rather broad peak at ~350℃ for alloys without Ti and a sharp peak at ~450℃ for alloys with Ti addition. In this case also, Ti seemed to act as pinning site of atomic migration in the crystal and once the internal energy override the pinning, avalanche of crystallization and/or interfusion might occur. This trigger effect of Ti addition might attribute to the refinement of grain sizes down to nano-scale. Ti element recently was reported to increase crystallization velocity in a similar alloy system [7]. This also might be similar to the role of Cu addition for grain size refinement in nano-composite $Fe_3B/Nd_2Fe_{14}B$ permanent magnets [8].

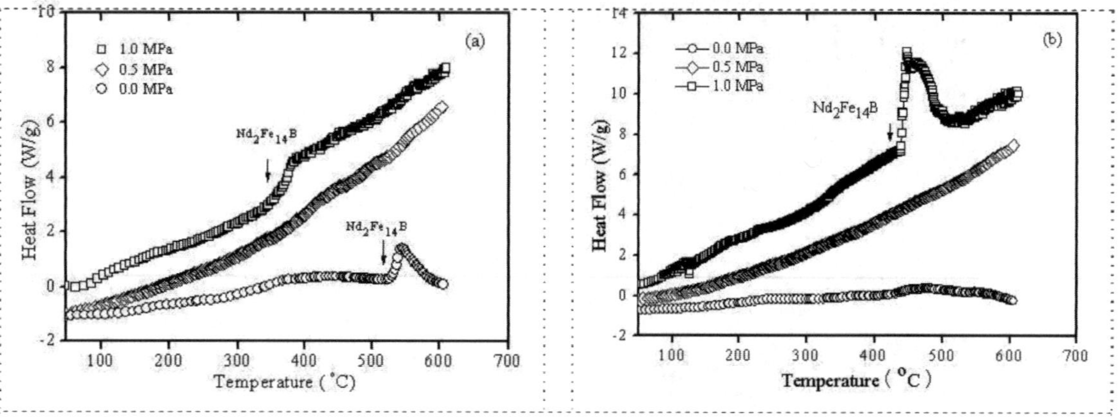

Figure 3. DSC curves of melt-spun (a) MM-Fe-B and (b) MM-Fe-B(Ti) ribbons under various pressures.

Magnetic properties of hot-pressed MM-Fe-B and MM-Fe-B(Ti) magnets are summarized in Table 2. Quite a increase in the coercivity but small changes in remanence with the addition of Ti could be straightforwardly understood from their microstructures. Curie temperatures of magnets without and with Ti were measured as ~575℃ and ~550℃ which are similar to the values for multiphase Nd-Fe-B melt spun ribbons [9].

Table 2. Magnetic properties of hot-pressed MM-Fe-B and MM-Fe-B(Ti) magnets.

Specimens	B_r (kG)	$_iH_c$ (kOe)	$_bH_c$ (kOe)	$(BH)_{max}$ (MGOe)
MM-FeB	5.0	3.7	2.5	4.1
MM-Fe-B(Ti)	5.3	4.8	3.0	4.3

4. Conclusions

Effects of small addition of Ti on microstructures of hot-pressed MM-Fe-B magnets were investigated. HRTEM micrographs showed drastic reduce in grain sizes in Ti added alloys. Grain sizes of hot-pressed MM-Fe-B(Ti) were ~40nm while those for alloys without Ti were found ~500nm. DSC results combined with high temperature XRD patterns, suggest triggering role of Ti additives. Pressure was proved to be a major parameter in phase transformations in melt-spun specimens. Pressures higher than 1MPa seemed to greatly enhance the formation and/or increase of denser 2-14-1 phase. Coercivity($_iH_c$) of the hot-pressed MM-Fe-B magnets was increased to 4.8 kOe when Ti was added from a value of 3.7 kOe without Ti. This seemed mainly due to refinement of grain sizes. Nano-scale grains might provide the way to fabricate $MM\text{-}Fe\text{-}B/Fe_3B$ exchange coupled magnets.

References

[1] Cosovic V, Talijan N, Grujic A, Stajic-Trosic J, Zivkovic D, Lee Z and Radmilovic V 2009 *Association of Metallurgical Engineers of Serbia MJoM* **15**(1) 39.
[2] Talijan N, Cosovic V, Grujic A, Stajic-Trosic J and Zak T 2008 *Acta physica polonica* A **113** 525..
[3] Orru R, Licheri R, Locci A M, Cincotti A and Cao G 2009 *Mat. Sciand Eng.* R **63** 127.
[4] Ko K Y, Yoon S, Booth J G, Al-Kanani H J and Cho S K 2002 *J. Mat. Sci.* **37** 1421.
[5] Ko K Y, Yoon S and Booth J G 1997 J. Mag. Mag. Mat. **176** 313.
[6] Shield J E, Kappes B B, Branagan D J and Bentley J 2002 J. Mag. Mag. Mat. **246** 73.
[7] Ko K Y 2004 *Physica Status Solidi.* A **201** 1834.
[8] Ping D H, Hono K, Kanekiyo H and Hirosawa S 1999 *J. Appl. Phys.* **85** 2448.
[9] Grujic A, Cosovic V, Stajic-Trosic J, Maricic A and Talijan N 2007 *Science of Sintering* **39** 193.

Acknowledgements
The authors are grateful to Prof. H. W. Davies (Sheffield University, UK) and T. Davies (UMIST, UK) for their help in making the hot-pressed magnets, Prof. J. G. Booth (Salford University, UK) for magnetic measurements.

The effect of underlayers on FeCo thin films

Xiaoxi Liu, Hidekazu Kanda, Akimitsu Morisako
Department of Information Engineering, Shinshu University
Wakasato 4-17-1, Nagano 380-8553, Japan

E-mail: liu@cs.shinshu-u.ac.jp

Abstract. Microstructure, mechanical and magnetic properties of FeCo films deposited onto underlayers of Ti, V, Cr, Co, Ni, Cu, Ta, and Pd have been investigated. The microstructure, magnetic, electric properties and internal stress of the FeCo films sensitively depended on the underlayers. The grain size, magnetic softness and electrical resistivity can be tailored in a wide range by the unerlayers. The coercive mechanism of the FeCo films qualitatively follows the random anisotropy model. FeCo films deposited onto underlayers with *bcc* structure (Ti, V, Cr) show relatively low compressive stress. Large tensile stress was found for FeCo films deposited onto underlayers with *fcc* structure (Ni, Cu, Pd). FeCo films with smaller grain size and low internal stress normally preferred to have lower coercivity. The electric resistivity of the films can be tailored in a wide range by choosing different underlayers.

1. Introduction

$Fe_{65}Co_{35}$ alloy has saturation magnetization as high as 23.5 kG. It has been considered as one of the prominent candidate materials for the applications in magnetic write head [1]. However, the alloy has relatively high magnetostriction, which results in higher coercivity [2]. The resistivities of the film also need to be increased for working at a higher frequency. NiFe [3], Co [4], Ru [5], and other underlayers have been reported that can dramatically reduce the coercivities of FeCo films. However, the effect of the underlayers on the magnetic, electric properties and microstructure is not fully understood.

In this experiment, we have systematically prepared FeCo thin films on underlayers from group 4 to group 11 metals. Those underlayers include Ti, V, Cr, Co, Ni, Cu, Ta, and Pd. The magnetic properties, microstructures and crystallographic properties of the films are investigated. Of particular interest is that, although the underlayers is as thin as 3 nm, the resistivity of the 100 nm thick FeCo films increase almost three times from 20 $\mu\Omega$-cm for group 4 metals to about 58 $\mu\Omega$-cm group 11 metals. The coercivity of the samples sensitively depended on the microstructure and stress of the samples.

2. Experimental procedure

Films were prepared by a DC magnetron sputtering system onto glass substrate. The thicknesses of all the underlayers were 3 nm and the thickness of the FeCo layer was 100 nm. Static magnetic properties of the films were measured by a vibrating sample magnetometer (VSM). The surface morphology of the samples was observed by a scanning electron microscope (SEM). The crystallographic properties of the samples were evaluated by an X-ray diffractometer (XRD). The resistivity of the films was measured by four point method. Stress and magnetostriction were measured to understanding their effects on the magnetic properties of the films. Stress was measured from the stain of the cover glass after deposition of the FeCo films.

Journal of Physics: Conference Series **266** (2011) 012037 doi:10.1088/1742-6596/266/1/012037

3. Results and discussion

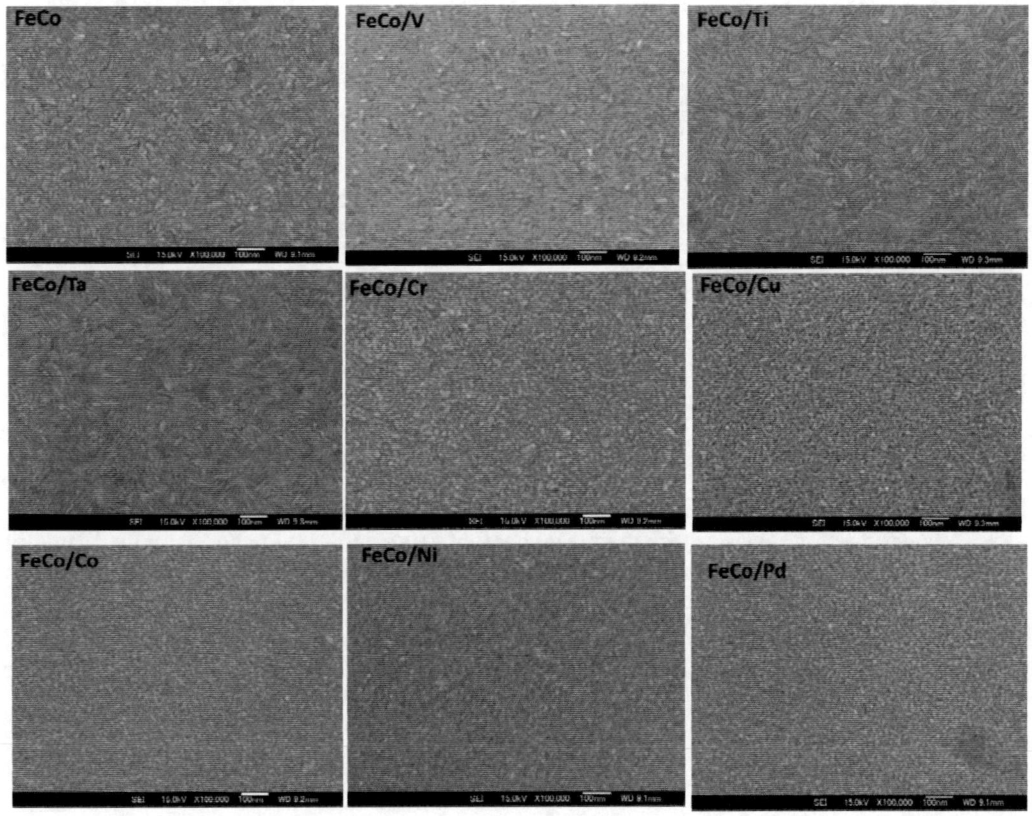

Fig.1. SEM images of FeCo films prepared onto various underlayers.

Fig.1 shows the surface morphology of the FeCo films prepared onto the underlayers. For all the films the thickness of the FeCo layer is 100 nm and the underlayer thickness is 3 nm. As show in Fig.1, the surface morphology of the films sensitively depended on the underlayers. There are typically three kinds of grain structures according to Fig.1. FeCo films deposited onto Ti and Ta underlayer show almost the same microstructure as FeCo films deposited directly onto glass substrate. Those films show needle like grain structures. The typical grain length is about 100 nm and the typical grain width is about 20 nm. Films deposited onto V underlayers show grains as large as 70 nm coexist with fine grains of 15-25 nm. FeCo films deposited onto Cr, Cu, Co, Ni, Pd underlayers show uniform grain structures with the typical grain size as small as 15 nm. It should be noted that cross sectional TEM observations show all the films have columnar structure [4]. This means the grain structure of the films almost not change with the increase of the film thickness.

Fig. 2 shows the X-ray diffraction patterns of FeCo films deposited onto the various underlayers. All the FeCo films show almost similar X-ray diffraction patterns. (110) diffraction line is strongest diffraction line for all the FeCo films. It should be noted that films deposited onto Ni, Cu, and Ta show relatively large shift of the FeCo (110) diffraction line position compared with other underlayers. The shift of the peak position may be caused by the internal stress in the films, which will be discussed later.

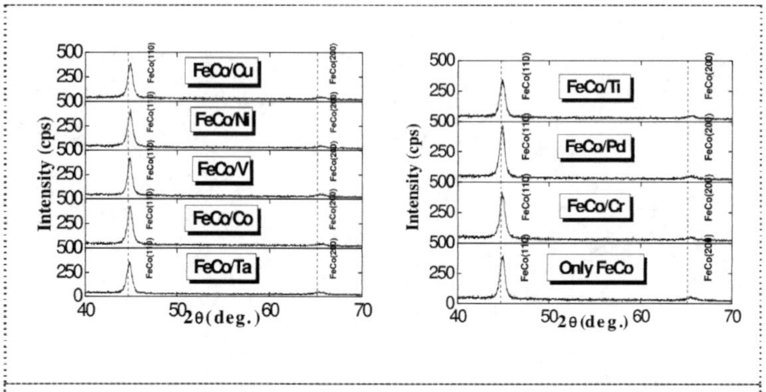

Fig. 2. X-ray diffraction diagrams of FeCo films deposited onto various underlayers.

All the FeCo films show anisotropic in-plane magnetic properties. Fig. 3 shows the dependence of coercivities, both in the hard axis and easy axis direction, on the underlayers. H_{ce} denotes easy axis coercivity and H_{ch} denotes hard axis coercivity. The structures of the underlayers, which are detected by reflection high energy electron diffraction (RHEED), also show in Fig.3. FeCo films deposited onto Ti, Ta underlayers and glass substrate show coercivities around 100 Oe. On the other hand, the hard axis coercivities of FeCo films deposited onto Cr and Co are less than 10 Oe. Those results show that suitable underlayers can dramatically reduce the coercivities in FeCo films.

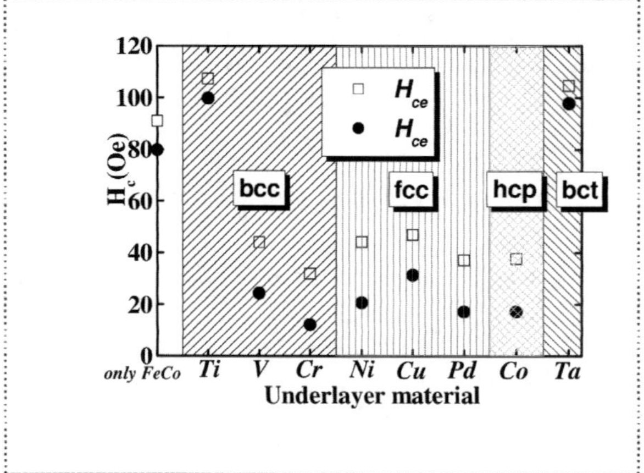

Fig. 3. Hard axis coercivity H_{ch} and easy axis coercivity H_{ce} of FeCo films deposited onto various underlayers.

The coercive mechanism in nanocrysalline thin films has been studied in detail by the random anisotropy model proposed by Herzer [6]. When the grain size in nanocrystalline films is less than the exchange length, the effective anisotropy can be estimated from the equation of $K_{eff} \sim K_1^4 D^6 / A^3$, where K_1 is the magnetocrystalline anisotropy, D is the average grain size, and A is the exchange stiffness. The exchange length ($l_{ex} = (A/K)^{1/2}$) of FeCo is estimated to be about 30 nm by assuming exchange stiffness $A = 1.7\text{x}10^{-11}$ J/m and $K \approx 15$ kJ/m^3 [1]. According to the SEM images shown in Fig.1, FeCo films deposited onto Cr, Cu, Co, Ni, Pd all show grain size smaller than 30 nm. Hence those films are expected to have smaller coercivities following the random anisotropy model. According to the experimental results of Fig.2, FeCo films deposited onto Cr, Cu, Co, Ni, and Pd unerlayers show lower coercivities compared the films deposited onto Ti, or Ta underlayer. Those experimental results qualitatively support the random anisotropy results. However, FeCo films deposited on Cr underlayer obviously have grain size larger than those on Ni unerlayer. But films deposited onto Cr unerlayer have smaller coercivity. Those results indicate that coercivities in nanocrystalline FeCo films could not be fully understood by the random anisotropy model.

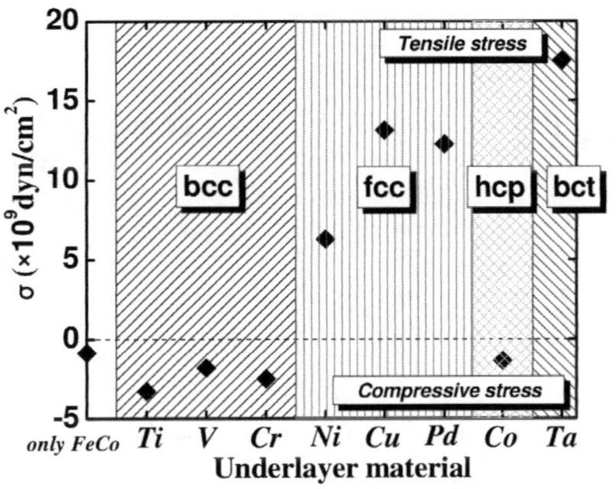

Fig. 4. Internal stress FeCo films deposited onto various underlayers.

Fig. 4 shows the dependence of internal stress on the underlayers. Films deposited onto Ni, Cu, and Ta show large tensile stress. Normally, magnetocrystalline anisotropy and stress are all contribute to the increase of coercivities in soft magnetic thin films. Hence it is easy to understand that FeCo films deposited onto Ni underlayer, although have relatively smaller grain size, show relatively higher coercivity due to the large tensile stress. Of particular interest is that underlayers with *bcc* structure show relatively low internal stresses compared with underlayers with *fcc* structure.

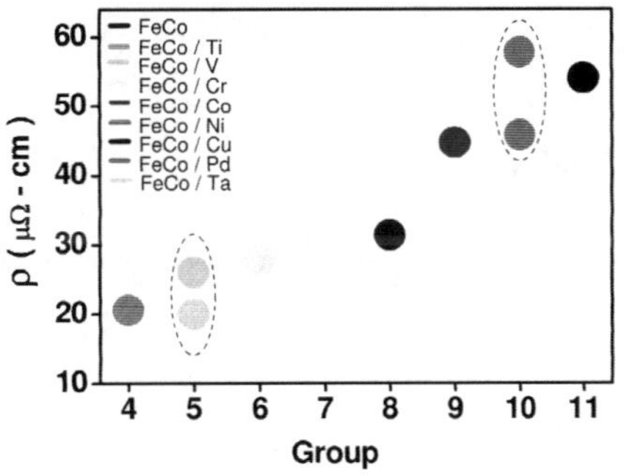

Fig. 5. Electric resistivity of FeCo films deposited onto various underlayers.

Fig. 5 shows the dependence of the electric resistivity of the films on the group number of the underlayers. The total film thickness is about 103 nm and the underlayer thickness is only about 3 nm for all the films. However, the resistivity of the films change from about 20 $\mu\ \Omega$-cm for Ti underlayer to about 58 $\mu\ \Omega$-cm for Ni underlayer. The resistivity of the films increased to about three times by changing the underlayers. It is clearly that the resistivity of the FeCo film increase with the increase of the group number of the underlayer metals. It is well know that impurities and crystal imperfections such as grain boundaries can result in increase of resistivities in metals. FeCo films prepared onto Ni underlayer have small grain sizes and larger amount of grain boundaries; hence the films have higher resistivity. It should be noted that films with higher resistivity is preferred to extend the application for high frequency applications.

4. Results and discussion

In conclusion, it is found that the underlayer plays an important role on the microstructure, and magnetic softness of the FeCo films. Films deposited onto Ti, Ta underlayer and glass substrate show large coercivity. While films deposited onto Co, Cr underlayer show small coercivity. Experimental results show the coercivities of the films not only depended on the grain size of the films, but also depended on the internal stress of the films. Of particular interest is that the electric resistivity can be tailored in a wide range by suitable underlayers.

References

[1] Jung H S, Doyle W D, Matsunuma S 2003 *J. Appl. Phys.* **93** 6462
[2] Mchenry M, Willard M, Laughlin D, *Progress in Materials Science*1999 **44** 291
[3] Wang S X, Sun N X, Yamaguchi M and Yabukami S 2000 *Nature,* **406** 150
[4] Liu X, Morisako A 2008 *J. Appl. Phys.* **103** 07E726
[5] Hashimoto A, Hirata K, Matsuu T, Saito S, Nakagawa S 2008 *IEEE. Trans. Mag.* **44** 3899
[6] Hezer G 1989 *IEEE. Trans. Mag.* **25** 3327

Journal of Physics: Conference Series **266** (2011) 012038 doi:10.1088/1742-6596/266/1/012038

Magnetic properties of FePt thin films with multilayered structure

S Matsumoto, T Shima
Division of Engineering, Graduate School of Tohoku Gakuin University, 1-13-1, Chuo, Tagajo 985-8537, Japan

E-mail: *shima@tjcc.tohoku-gakuin.ac.jp*

Abstract. The structure and magnetic properties of FePt films with multilayered structure have been investigated. Intermediate layer of Cu or Fe/Au was inserted between top and bottom FePt layers. From X-ray diffraction patterns, the (002) fundamental peak and (001) and (003) super lattice peaks of the $L1_0$ phase have been clearly observed for all the samples. The peak from FePt (200) has begun to observe for FePt/Cu/FePt films at the thickness of more than 3 nm. With increasing the intermediate layer thickness (t), coercivity measured in the direction perpendicular to the film plane was increased. However, an uniaxial magnetocrystalline anisotropy (K_u) was decreased for FePt/Cu/FePt films, while K_u maintains a large value of 4×10^7 (erg/cm^3) for FePt/Fe/Au/FePt films. It was confirmed that the top FePt layer was grown epitaxially on the intermediate layer of Fe/Au without deterioration of the large K_u.

1. Introduction

Magnets of nanometer size are of great technological and scientific interest, because nano-scale patterned or particulate magnets may guide to the applications in various magnetic devices such as ultra high density magnetic recording media, magnetic field generator in monolithic microwave integrated circuits and spin electronics devises. For these applications, nanomagnets with a large uniaxial anisotropy is required. Recently, a lot of studies have been made on $L1_0$ ordered [CuAu (I) type] FePt thin films[1]-[11] and nano particles[12],[13] by conventional thin film preparation route and chemical synthesis, since they possess high uniaxial magnetocrystalline anisotropy (K_u=7.0$\times10^7$ erg/cm^3)[14] and high corrosion resistance. For practical use, it is essential to realize highly oriented FePt thin films grown epitaxially on multilayered structure. However, the orientation control and the magnetization behaviour of FePt thin film with multilayer arrangement have not yet been fully elucidated. In this study, in order to see the effect of intermediate layer of Cu or Fe/Au on the film structure and the magnetic properties for FePt films, FePt / (Cu, Fe/Au) /FePt thin films have been investigated.

2. Experimental procedure

All the samples were prepared using an ultrahigh vacuum magnetron sputtering system with separate targets of Fe, Pt, Cr, Cu and Au. The base pressure was under 5.0×10^{-8} Pa and high-purity argon of 0.680 Pa was flown during sputtering. A Cr seed layer of 1nm was deposited on a MgO(100) single crystalline substrate. An epitaxial Au(001) buffer layer of 40 nm was deposited at room temperature, and then annealed at 300°C for 1 hour. Consecutively the bottom FePt layer of 20 nm was deposited at 300°C. An intermediate layer of Cu or Fe/Au layer was deposited at room temperature. The thickness of Cu and Au intermediate layer was varied from 0 to 10 nm. A thin Fe layer of 0.5-nm-thick also plays a role to maintain the (100) orientation of Au layer. Then top FePt layer of 10 nm was deposited at 300°C. Finally, the samples were annealed at 600°C to promote the $L1_0$ ordering of FePt layer. The magnetic properties were measured by a superconducting quantum interference device (SQUID) magnetometer. The structural characterization was performed by X-ray diffraction with Cu Kα radiation.

3. Results and discussion

3.1. FePt/Cu/FePt films

X-ray diffraction (XRD) patterns for FePt (20 nm) /Cu (t_{Cu} nm) / FePt (10 nm) films with various Cu intermediate layer thicknesses (t_{Cu}) are shown in Figure 1. t_{Cu} is varied for 0 nm (a), 1 nm (b), 3 nm (c), 4 nm (d), 5 nm (e), and 10 nm (f). The film with t_{Cu} = 0 nm is identical to be FePt single layer. In addition to fundamental (002) peak, (001) and (003) superlattice peaks of the $L1_0$-FePt phase have been clearly observed for all the samples. The unlabeled peaks are due to the MgO substrate. With increasing t_{Cu}, however, the peak from FePt (200) began to observe for the films with $t_{Cu} \geq 3$ nm. It is thought that the crystal orientation of the top FePt layer tends to lie in the in-plane direction with increasing t_{Cu}. Furthermore, the peak from Au (200) has been shifted to higher angle with increasing t_{Cu}, indicating that the Cu layer was diffused into Au buffer layer after the annealing of the $L1_0$ ordering in FePt layer.

Magnetization curves for FePt/Cu/FePt films with various t_{Cu} are shown in Figure 2. All the measurements were performed at room temperature. The solid and broken lines denote the curves measured in the field perpendicular direction to the film and in the in-plane direction, respectively. The magnetic easy axis is perpendicular to the film plane for all the samples, since the [001] axis of the tetragonal $L1_0$ ordered structure is perpendicular to the film plane as demonstrated in Figure 1. Relatively large uniaxial magnetic anisotropy with perpendicular orientation and low coercivity (H_c) of 0.93 kOe were obtained for the film with t_{Cu} = 0 nm (FePt single layer). Since the magnetization process of FePt film is classified as the nucleation-type and in this case, it was mainly governed by the domain wall displacement with scarce existence of "pinning" site. The magnetization measured in the in-plane direction was hard to be saturated under the maximum applied field of 70 kOe in the SQUID. This result is well consistent with the previous results [6], [7]. With increasing t_{Cu}, remarkable

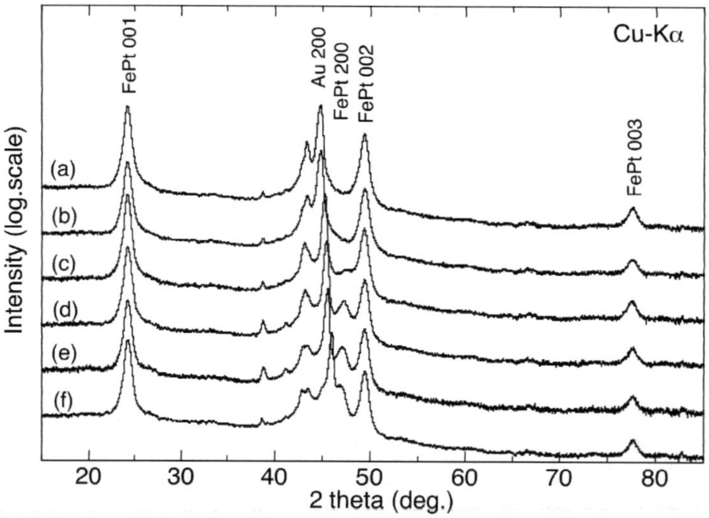

Figure 1. XRD patterns for FePt/Cu/FePt films with various Cu intermediate layer thicknesses (t_{Cu}). t_{Cu} = 0 nm: FePt single layer (a), 1 nm (b), 3 nm (c), 4 nm (d), 5 nm (e), and 10 nm (f).

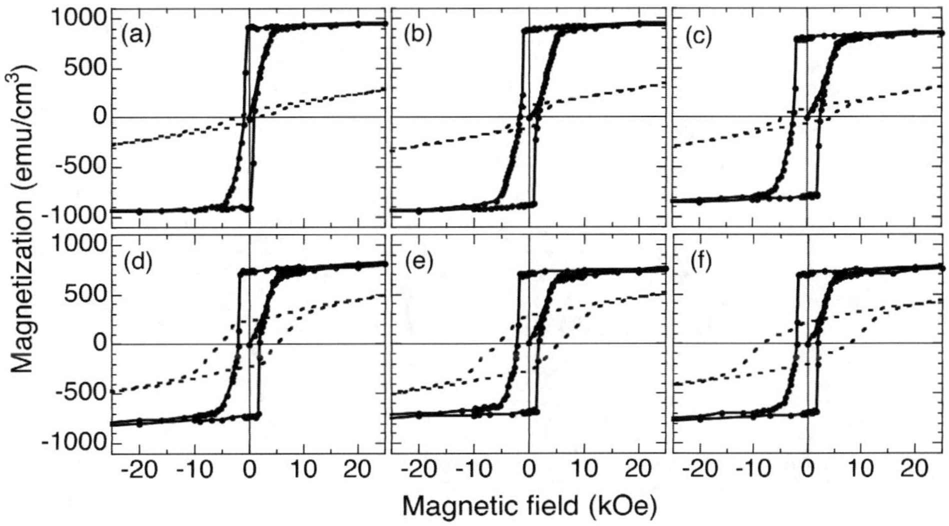

Figure 2. Magnetization curves for FePt/Cu/FePt films with various t_{Cu}. t_{Cu} = 0 nm (a), 1 nm (b), 3 nm (c), 4 nm (d), 5 nm (e) and 10 nm (f). The magnetic field was applied in the perpendicular direction to the film (solid line) and in the in-plane direction (broken line).

expansion of the hysteresis of the magnetization curve in the in-plane direction began to observe for $t_{Cu} \geq 3$ nm, indicating that the magnetic easy axis of top FePt layer was inclined toward the in-plane direction as suggested in Figure 1.

3.2. FePt/Fe/Au/FePt films

XRD patterns for FePt (20 nm) /Fe (0.5 nm) /Au (t_{Au} nm) /FePt (10 nm) films with various Au intermediate layer thicknesses (t_{Au}) are shown in Figure 3. t_{Au} is varied for 0 nm (a), 1 nm (b), 3 nm (c), 5 nm (d), and 10 nm (e). Both fundamental and superlattice peaks of the $L1_0$-FePt phase have been clearly observed for all the samples. While on the contrary to the results with Cu intermediate layer,

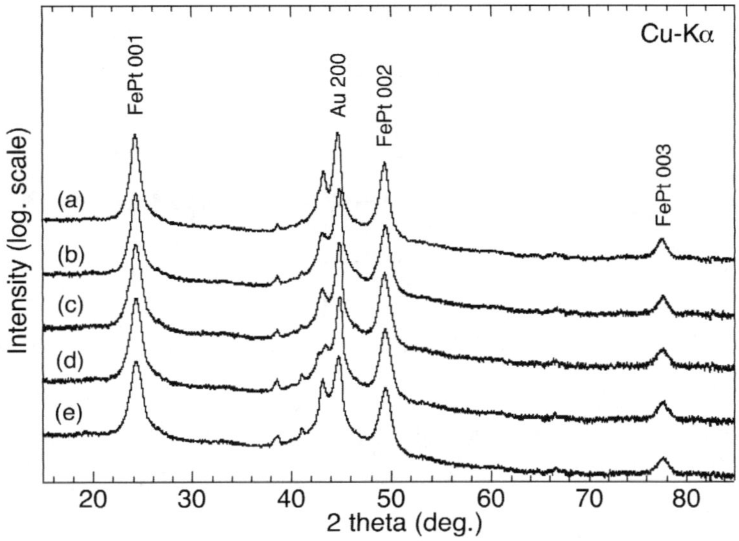

Figure 3. XRD patterns for FePt/Fe/Au/FePt films with various Au intermediate layer thickness (t_{Au}). t_{Au} = 0 nm (a), 1 nm (b), 3 nm (c), 5 nm (d) and 10 nm(e).

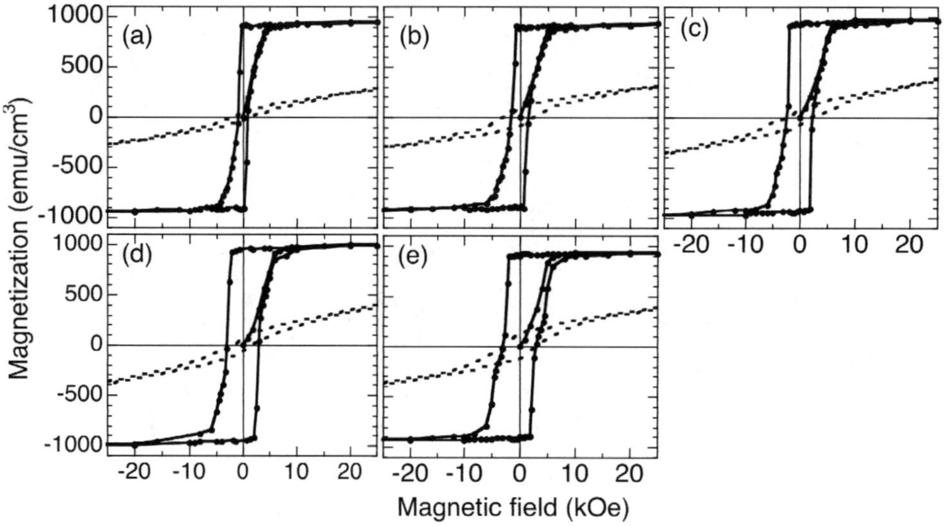

Figure 4. Magnetization curves for FePt/Fe/Au/FePt films with various t_{Au}. t_{Au} = 0 nm (a), 1 nm (b), 3 nm (c), 5 nm (d) and 10 nm (e).

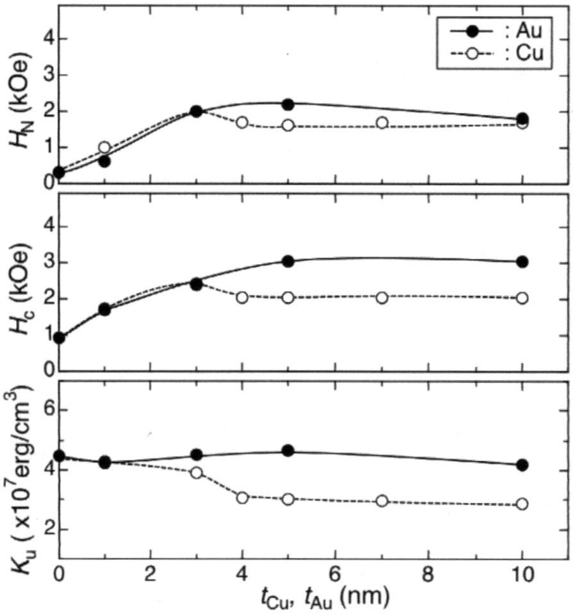

Figure 5. Effect of interlayer thicknesses (t_{Cu}, t_{Au}) on H_N, H_c and K_u for FePt/(Cu, Fe/Au)/FePt thin films.

the peak from FePt (200) could not observe. It is confirmed that c-axis of both bottom and top FePt layers was successfully aligned perpendicular to the film plane for the films with Fe/Au intermediate layer.

Magnetization curves for FePt/Fe/Au/FePt films with various t_{Au} are shown in Figure 4. The magnetic easy axis is perpendicular to the film plane for all the samples. With increasing t_{Au}, H_c was increased. A remarkable step of magnetization curve was observed in the third quadrant of the magnetization curve at t_{Au} = 10nm. This is thought to arise from the different thickness of bottom and top FePt layer, since the nucleation field (H_N), at which the magnetization begins to drop with decreasing the field after saturation of each layer is different and occurred individually.

In Figure 5, the magnetic properties as a function of intermediate layer thickness are summarized. In order to consider the magnetization process, H_N is estimated. The tendency of H_N is almost similar to that of H_c at $t \leq 3$ nm. With further increasing t, both H_N and H_c for the films with Cu intermediate layer became smaller than that with Au layer. The uniaxial magnetic anisotropy (K_u) determined from the area enclosed between the magnetization curves in applied fields parallel and perpendicular to the film plane is also shown in Figure 5(c). With increasing t_{Cu}, K_u decreased to 3×10^7 (erg/cm^3). However, regardless of the intermediate layer thickness, large value of $K_u \approx 4\times10^7$ (erg/cm^3) was obtained for the films with Fe/Au intermediate layer. It was confirmed that top FePt layer was successfully grown epitaxially on Fe/Au intermediate layer.

Summary

The structure and magnetic properties of FePt films with multilayered structure have been investigated. Intermediate layer of Cu or Fe/Au was inserted between top and bottom FePt layers. It was revealed that the FePt layer was successfully grown epitaxially on the intermediate layer of Fe/Au without deterioration of the large perpendicular magnetic anisotropy. However, in order to understand the exchange coupling behavior between top and bottom FePt layer around the critical thickness of $t_{Au} = 3$ nm, further detailed investigation such as magnetic domain observation should be needed.

Acknowledgments

This work was performed at the Hi-tech Research Center of Tohoku Gakuin University.

Reference

[1] B M Lairson, M R Viosokay, R Sinclair, and B M Clemens, 1993 *Appl. Phys. Lett.* **62**, 639
[2] B M Lairson and B M Clemens, 1993 *Appl. Phys. Lett.* **63**, 1438
[3] A Cebollada, D Weller, J Sticht, G R Harp, R F C Farrow, R F Marks, R Savoy, and J C Scott, 1994 *Phys. Rev. B* **50**, 3419
[4] M R Visokay and R Sinclair, 1995 *Appl. Phys. Lett.* **66**, 1692
[5] J -U Tiele, L Folks, M F Toney, and D K Weller, 1998 *J. Appl. Phys.* **84**, 5686
[6] T Shima, T Moriguchi, S Mitani and K Takanashi, 2002 *Appl. Phys. Lett.* **80** 288
[7] T Shima, K Takanashi, Y K Takahashi and K Hono 2002 *Appl. Phys. Lett.* **81** 1050
[8] T Maeda, T Kai, A Kikitsu, T Nagase and J Akiyama 2002 *Appl. Phys. Lett.* **80** 2147
[9] T Seki, T Shima, K Takanashi, Y Takahashi, E Matsubara, and K Hono, 2003 *Appl. Phys. Lett.* **82** 2461
[10] T Shima, K Takanashi, Y K Takahashi and K Hono 2004 *Appl. Phys. Lett.* **85** 13
[11] T Shima, K Takanashi, Y K Takahashi and K Hono 2005 *Appl. Phys. Lett.* **88** 063117
[12] S Sun, C B Murray, D Weller, L Folks, and A Moser, 2000 *Science* **287**, 1989
[13] S Yamamoto, Y Morimoto, T Ono, and M Takano 2005 *Appl. Phys. Lett.* **87**, 032503
[14] O A Ovanov, L V Solina, and V A Demshina 1973 *Phys. Met. Metallogr.* **35** 81

Journal of Physics: Conference Series **266** (2011) 012039 doi:10.1088/1742-6596/266/1/012039

Influence of microstructural change of the interface between $Nd_2Fe_{14}B$ and Nd-O phases on coercivity of Nd-Fe-B films by oxidation and subsequent low-temperature annealing

Masashi Matsuura[1], Ryota Goto[2], Nobuki Tezuka[1] and Satoshi Sugimoto*[1,2]
[1]Department of Materials Science, Graduate School of Engineering, Tohoku University, Aoba-yama 6-6-02, Sendai, Japan
[2]New Industry Creation Hatchery Center (NICHe), Tohoku University, Aoba-yama 6-6-10, Sendai, Japan

E-mail: a9td5312@s.tohoku.ac.jp

Abstract This study provides the influence of microstructural change of the interface between $Nd_2Fe_{14}B$ and Nd-O phases on coercivity of Nd-Fe-B thin films during annealing at low temperature (~350 °C). All films were prepared by using ultra high vacuum (UHV) magnetron sputtering, and the Nd-Fe-B layer was oxidized under Ar gas atmosphere (O_2 content; ~2 Vol.ppm). Then, the films were annealed at 250-350 °C under UHV condition. After oxidation, the coercivity of Nd-Fe-B film decreased to around 40% of the coercivity of as-deposited Nd-Fe-B film. The Nd-rich phase changed from α-Nd to amorphous Nd(-O), and the interface of $Nd_2Fe_{14}B$/Nd(-O) became rough. In the Nd-Fe-B films oxidized and subsequent annealed at 350 °C, the coercivity decreased to around 20%. In the films, polycrystalline hcp Nd_2O_3 phase crystallized in Nd-rich phase, and there were some steps at the surface of $Nd_2Fe_{14}B$ phase contacting with hcp Nd_2O_3 phase. Regardless of crystal orientation of $Nd_2Fe_{14}B$, the microstructural changes of the interface described above were observed.

1. INTRODUCTION

Nd-Fe-B sintered magnets show the highest energy products and are used in various applications. However, the coercivity of the magnets without additional elements is only about 12% of the anisotropy field of $Nd_2Fe_{14}B$ compound. The coercivity of Nd-Fe-B sintered magnets is strongly related to their interfacial microstructure between the main phase ($Nd_2Fe_{14}B$) and the grain boundary phase (Nd-rich). In order to obtain the guiding principles for enhancement of coercivity, many researchers have observed the interfacial microstructure for the Nd-Fe-B sintered magnets[1-7]. They have reported that the Nd-rich phase contains some amount of oxygen, and has various crystal structures. In our previous papers[8-10], we investigated the influence of oxidation on the coercivity and the interfacial microstructure by using Nd-Fe-B thin films. We reported that the oxidation of Nd-Fe-B layer under Ar gas atmosphere decreased the coercivity to ~40% of the coercivity of as-deposited

* Correspondence to; Masashi Matsuura. E-mail; a9td5312@*s.tohoku.ac.jp,* tel/fax; +81-22-795-7332

Nd-Fe-B film. In addition, subsequent annealing at low temperature (250-350 °C) decreased the coercivity more about 20%. In the films, some steps were observed at the surface of the $Nd_2Fe_{14}B(00l)$ plane contacting with the hcp Nd_2O_3 phase[10]. However, the dependence of crystal orientation of the $Nd_2Fe_{14}B$ on interfacial microstructure between $Nd_2Fe_{14}B$ and Nd-O phases is not clear. In this study, the influence of phase change of Nd-O on the microstructure of $Nd_2Fe_{14}B$ phase with various crystal orientations by oxidation and subsequent annealing at low temperature was investigated.

2. EXPERIMENTAL PROCEDURE

All films were prepared by UHV magnetron sputtering system with base pressure of 1×10^{-7} Pa. A Nd-Fe-B layer with the thickness of 100 nm was deposited on Ta buffered MgO(001) substrate, which were annealed at 680 °C. The composition of target used for the preparation of Nd-Fe-B layer was $Nd_{18}Fe_{66}B_{16}$. After deposition of Nd-Fe-B, films were cooled to room temperature and moved to another chamber. The films were kept for 120 min under Ar gas atmosphere to be oxidized. Oxygen content was estimated about 2 vol. ppm in the Ar gas atmosphere. After oxidation, a Nd layer with the thickness of 2 nm was deposited on the oxidized Nd-Fe-B layer, and the films were annealed at 250-350 °C for 60 min in UHV condition. Finally, a Ta layer was deposited on the films for preventing oxidation from air.

Magnetic properties were measured by a vibrating sample magnetometer (VSM) along the direction perpendicular to the film plane after applying a pulsed magnetic field of 6360 kAm^{-1}. Crystalline structure and microstructure were analyzed by X-ray diffraction (XRD) and transmission electron microscopy (TEM), respectively.

3. RESULTS AND DISCUSSION

Figure 1 shows the XRD patterns of Nd-Fe-B films as a function of annealing temperature. A (111) peak of fcc NdO_x (a=0.547 nm) was observed in as-oxidized film, which had coercivity of about 40% of as-deposited film. However, the (111) peak of fcc NdO_x decreased, and a (002) peak of hcp Nd_2O_3 phase increased after annealing at 250 and 350 °C. The coercivity of these films, which were oxidized and subsequent annealed at low temperature, was only about 20% of the value of as-deposited film.

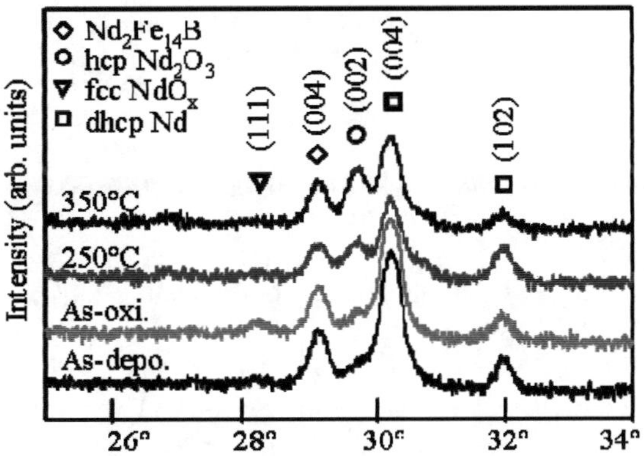

Fig. 1 XRD patterns of Nd-Fe-B films after oxidation, and subsequent annealed at 250 °C and 350 °C in comparison with that of as-deposited film.

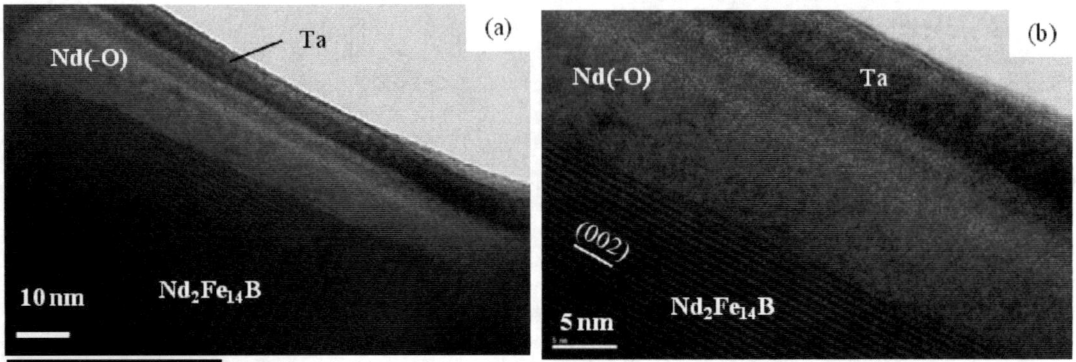

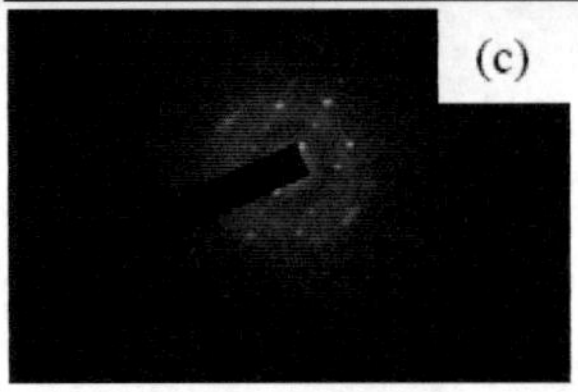

Fig. 2 (a) Bright field and (b) high resolution TEM images of the interface of $Nd_2Fe_{14}B$ (001) / Nd-rich phase in a as-oxidized Nd-Fe-B film. and (c) SAD pattern taken from Nd-rich phase corresponding to (b).

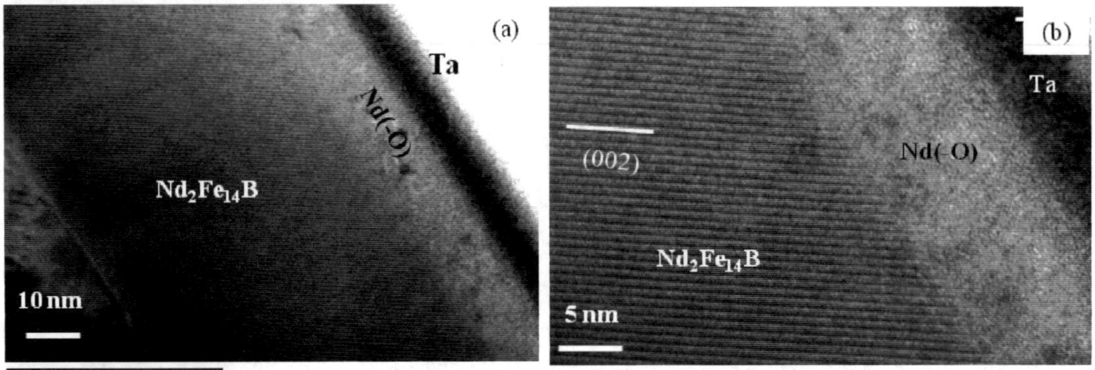

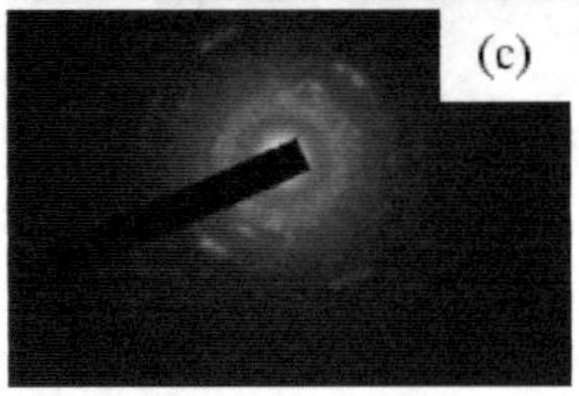

Fig. 3 (a) Bright field and (b) high resolution TEM images of the interface of $Nd_2Fe_{14}B$ (hkl) / Nd-rich phase in a as-oxidized Nd-Fe-B film. and (c) SAD pattern taken from Nd-rich phase corresponding to (b).

From XRD analysis, it is indicated that the phase change from fcc NdO_x into hcp Nd_2O_3 phases related to the decrease of coercivity. Then, the microstructural changes between $Nd_2Fe_{14}B$ and Nd-rich phase were investigated.

Figure 2 and 3 show TEM images of the interface between $Nd_2Fe_{14}B$ and Nd-rich phases in the as-oxidized Nd-Fe-B film. The SAD patterns taken from the Nd-rich phase corresponding to Fig. 2(a) and 3(a) are also shown in Fig. 2(c) and 3(c), respectively. The $Nd_2Fe_{14}B$(001)/Nd-rich and $Nd_2Fe_{14}B$(hkl)/Nd-rich interface (h, k, $l \neq 0$) are shown in HRTEM images of Fig. 2(b) and 3(b), respectively. As reported in our previous papers[8-10], there are $Nd_2Fe_{14}B$ and Nd-rich phases in the Nd-Fe-B layer with quite different contrast because of using Nd rich Nd-Fe-B ternary alloy as a target for sputtering. You can also see that there are $Nd_2Fe_{14}B$ and Nd-rich phases, which have different contrast, in Nd-Fe-B layer, as shown in Fig. 2(a) and 3(a). In SAD patterns, a halo ring was observed in both films (Fig. 2(c) and 3(c)). Since some spots shown in Fig. 2(c) are indexed as $Nd_2Fe_{14}B$ phase, the halo rings originate from Nd-rich phase. The halo rings mean that the Nd-rich phase shown in Fig. 2(b) and 3(b) are amorphous. Considering the fact that the Nd-rich phase in as-deposited Nd-Fe-B film was α-Nd[10], it is indicated that the oxidation of Nd-Fe-B film made α-Nd to amorphous Nd(-O). Furthermore, the interface between $Nd_2Fe_{14}B$ and amorphous Nd(-O) was to be rough, as compared with flat interface of $Nd_2Fe_{14}B$/α-Nd in as-deposited film[10]. Regardless of crystal orientation of $Nd_2Fe_{14}B$ phase, the rough interface was observed, as shown in Fig. 2(b) and 3(b).

Figure 4(a) and 5(a) show HRTEM images of the interface between $Nd_2Fe_{14}B$ and Nd-rich phases for the Nd-Fe-B films, which were oxidized and then annealed at 350 °C. Corresponding SAD patterns taken from the Nd-rich phase are also shown in Fig. 4(b) and 5(b), respectively. There are some spots and ring patterns in both SAD patterns. For both SAD patterns, the spots and rings are indexed as $Nd_2Fe_{14}B$ and hcp Nd_2O_3, respectively. Therefore, annealing after oxidation crystallizes the polycrystalline hcp Nd_2O_3 phase from amorphous Nd(-O). HRTEM image (Fig. 4(a)) shows that there were some steps at the surface of $Nd_2Fe_{14}B$(00l) plane contacting with hcp Nd_2O_3 phase. In addition, some steps at the surface of (hkl) plane of $Nd_2Fe_{14}B$ were also observed at the interface of $Nd_2Fe_{14}B$/hcp Nd_2O_3, as shown in HRTEM image (Fig. 5(a)). These results reveal that there is no dependence of crystal orientation of $Nd_2Fe_{14}B$ on the microstructural changes of the interface between $Nd_2Fe_{14}B$ and Nd-O phases during annealing at low temperature.

From the present study, the decline of coercivity of oxidized Nd-Fe-B films can be explained as being due to the phase changes of Nd-rich phase. The Nd-Fe-B film in as-deposited state has a flat interface of $Nd_2Fe_{14}B$/α-Nd phases[10]. However, oxidation of Nd-Fe-B film makes the Nd-rich phase from α-Nd to amorphous Nd(-O). It is considered that the amorphous Nd(-O) phase contains some amount of oxygen. Therefore, one possibility explanation for decreasing coercivity of Nd-Fe-B film by oxidation is that some nano-ordered hcp Nd_2O_3 crystallize in amorphous Nd(-O) phase and exist at surface of $Nd_2Fe_{14}B$. As a result, the surface of $Nd_2Fe_{14}B$ phase becomes rough, and the coercivity decrease. After oxidation and subsequent annealing at low temperature, the polycrystalline hcp Nd_2O_3 phase crystallizes and induces damages to the surface of $Nd_2Fe_{14}B$ phase regardless of crystal orientation of $Nd_2Fe_{14}B$. Some researchers reported that a few nm thick disordered $Nd_2Fe_{14}B$ phase decreases their coercivity[11-12]. Consequently, the rough surface of $Nd_2Fe_{14}B$, which is observed in this study, can decrease the coercivity of Nd-Fe-B films.

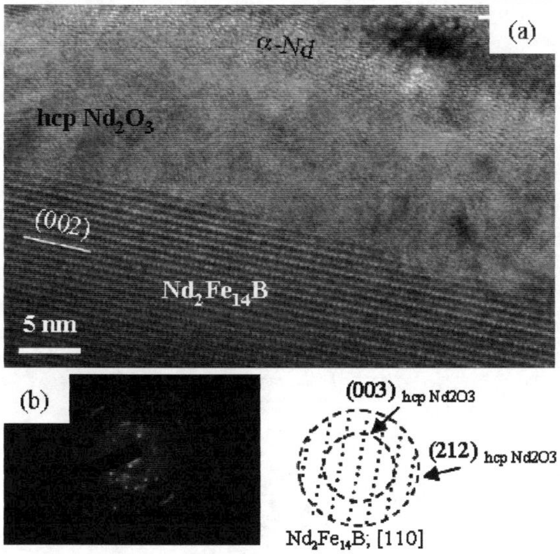

Fig. 4 (a) HRTEM image of the interface of $Nd_2Fe_{14}B$ (00l) / Nd-rich phase in Nd-Fe-B film oxidized and subsequent annealed at 350 °C, and (b) SAD pattern taken from Nd-rich phase corresponding to (a).

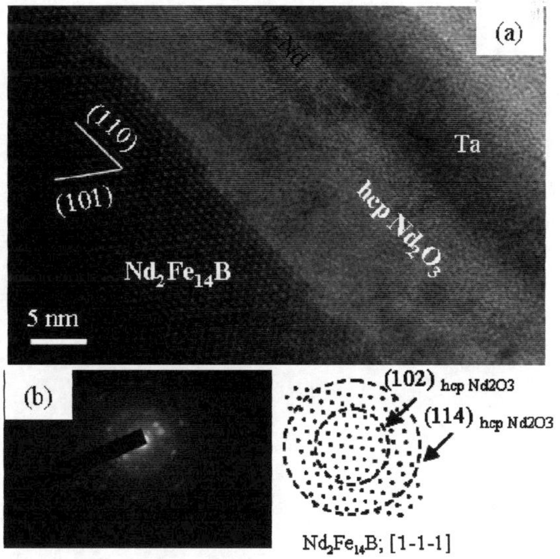

Fig. 5 (a) HRTEM image of the interface of $Nd_2Fe_{14}B$ (hkl) / Nd-rich phase in Nd-Fe-B film oxidized and subsequent annealed at 350 °C, and (b) SAD pattern taken from Nd-rich phase corresponding to (a).

ACKNOWREDGEMENT

The authors would like to thank Dr. K. Kobayashi and Dr. T. Miyazaki of Tohoku University for their support in TEM observation.
This work was partially supported by NEDO, Japan, Grant-in-Aid for "Development of technology for reducing dysprosium usage in a rare-earth magnet in Rare Metal Substitute Materials Development Project", by Global COE Program "Materials Integration (International Center of Education and Research), Tohoku University," MEXT, Japan, and by Grant-in-Aid for JSPS Fellows, Japan Society for the Promotion of Science(JSPS).

REFERENCES

[1] Sagawa M, Fujimura S, Yamamoto H, Matsuura Y and Hiraga K 1984 *IEEE Trans. Magn.* **MAG-20** 1584
[2] Ramesh R, Chen J K and Thomas G 1987 *J. Appl. Phys.* **61** 2993
[3] Vial F, Joly F, Nevalaien E, Sagawa M, Hiraga K and Pard K T 2002 *J. Magn. Magn. Mater.* **242-245** 1329
[4] Makita K and Yamashita O 1999 *Appl. Phys. Lett.* **74** 2056
[5] Fukagawa T and Hirosawa S 2008 *Scr. Mater.* **59** 183
[6] Shinba Y, Konno T J, Ishikawa K and Hiraga K 2005 *J. Appl. Phys.* **97** 053504
[7] Mo W, Zhang L, Liu Q, Shan A, Wu J and Komuro M 2008 *Scripta Mater.* **59** 179
[8] Matsuura M, Sugimoto S, Goto R and Tezuka N 2009 *J. Appl. Phys.* **105** 07A741
[9] Matsuura M, Fukada T, Goto R, Tezuka N and Sugimoto S 2009 *Mater. Trans.* **50** 2139
[10] Matsuura M, Goto R, Tezuka N and Sugimoto S 2010 *Mater. Trans.* **51** 1901
[11] Kronmüller H, Durst K D and Sagawa M 1988 *J. Magn. Magn. Mater.* **74** 291
[12] Fukagawa T, Ohkubo T, Hirosawa S and Hono K 2010 *J. Magn. Magn .Mater.* **322** 3346

Continuity of perpendicular FePt film with a compositional gradient design

X J Mo[1], H Xiang[1], W Lu[2], Y P Zheng[1], G Q Li[1, 4], H Saito[3], S Ishio[3], D M Jiang[1], X W Tan[1] and Y Q Lin[1]
[1] School of Physics, Southwest University, Chongqing, 400715, China
[2] School of Material, Tongji University, Shanghai, 200092, China
[3] Venture Business Laboratory, Akita University, Akita, 010-8502, Japan

E-mail: gqli@swu.edu.cn

Abstract. By magnetron sputtering, $[Fe(0.6\ nm)/Fe_{30.5}Pt_{69.5}(1.9\ nm)]_{10}$ continuous multilayer films with the nominal composition of $Fe_{50}Pt_{50}$ and total thickness of 25 nm were deposited on MgO(001) substrates heated to 400℃, and then subjected to a vacuum annealing at temperatures in the range of T_a = [500, 900]℃. The magnetocrystalline anisotropic energy exceeded 2.5×10^7 erg/cc at $T_a \geq 700$℃, and the morphology kept continuous at $T_a \leq 800$℃. Though the interlayer diffusion during sputtering was fairly sufficient, a weak periodical compositional gradient was left in the films, which is critical for the morphological continuity of film subjected to subsequently annealing. The slight periodical compositional gradient might suppress the segregation of Pt and sufficiently prevent from the Pt enrichment that would result in the discontinuous boundaries. A modified diffusion-recombination model of carrier was adopted to discuss the morphology continuity of annealed FePt films. The continuous film would fit the need for microfabrication of patterned arrays of $L1_0$ FePt.

1. Introduction

The (001) textured $L1_0$ FePt films with a large magnetocrystalline anisotropy (K_u) above 10^7 erg/cc have drawn great attentions due to the potential applications in data storage [1-3]. FePt films synthesized at room temperature normally adopt its disordered $A1$ phase. Post annealing or synthesis at an elevated temperature is necessary to transform to the desired ordered $L1_0$ phase. It is efficient to induce the (001) texture by depositing the FePt films on heated MgO(001) substrates [4,5]. To suppress the dispersion of grain size, researching on the artificially patterned magnetic arrays microfabricated by the methods such as electron beam lithography and focused ion beam lithography has become popular [6,7]. However, thermal treatment always causes sintering of FePt films and leads to destruction of the morphological continuity, conflicting with the need of uniform and continuous film for microfabrication. In order to keep the morphological continuity of film, the circumstantial temperature is generally limited to about 400℃, at which the ordering is insufficient yet and the films have vanishing small K_u [8]. In this article, we report a so-called compositional gradient method to synthesize $[Fe(0.6\ nm)/Fe_{30.5}Pt_{69.5}(1.9\ nm)]_{10}$ continuous multilayer films on MgO(001) substrates, by which the stoichiometric FePt films can undergo an annealing at temperature as high as 800℃ without destruction of morphological continuity.

[4] To whom any correspondence should be addressed.

2. Experiment

By using a magnetron sputtering system with the base pressure better than 8×10^{-6} Pa, multilayer films of [Fe(0.6 nm)/$Fe_{30.5}Pt_{69.5}$(1.9 nm)]$_{10}$ were sputtered on MgO(001) substrates heated to 400℃. The nominal composition of film was $Fe_{50}Pt_{50}$, and the total thickness was 25 nm. The stander with 9 substrate holders was rotating during deposition to ensure the uniformity of composition and thickness. All the samples were prepared at the same time to further ensure the consistence of composition. The elemental purity of target was 99.99 % for Pt and 99.93% for Fe, respectively. The sputtering chamber was pre-baked at ~100℃ for 1 h to remove the water and the gas. The working pressure of argon gas was ~2.8 Pa. The sputtering rate was ~0.1 nm/s for Fe and ~0.3 nm/s for Pt, respectively. As the later discussion on figure 2a, despite the interlayer diffusion was sufficient during sputtering, it is reasonable that a subtle wave of composition (periodical compositional gradient) was left in the films. After sputtering, the samples were encapsulated in the tube of a rapid thermal annealing equipment and annealed for 3 hours at temperatures of T_a = 500, 600, 700, 750, 780, 800, and 900℃, respectively. The crystallographic characterization of the films was performed with *x*-ray diffraction (XRD) using a Cu-K_α radiation. The morphology was observed by a scanning electron microscopy (SEM). The magnetic properties were measured by using a vibrating sample magnetometer (VSM) with a maximum applied field of ±30 kOe.

3. Results and discussions

Figure 1 shows the SEM images of [Fe(0.6 nm)/$Fe_{30.5}Pt_{69.5}$(1.9 nm)]$_{10}$ multilayer films annealed at different temperature [as-deposited (a), T_a = 700℃(b) and 900℃(c)], and the SEM image of FePt alloy film co-deposited at 700℃ (d) for comparison. The thickness of all films was 25 nm. The as-deposited film had a continuous morphology with a number of small holes (the dark areas in figure 1a). At T_a = 700℃ (figure 1b), the small holes disappeared and the surface of film became smoother. Compared to figure 1d, the morphological continuity of this multilayer film was obviously improved. No obvious change of roughness or continuity was observed for the films annealed at T_a in the range of [700, 800]℃. After annealed at 900℃ (figure 1c), the continuity was seriously destroyed.

Figure 2 shows the XRD spectra of [Fe(0.6 nm)/$Fe_{30.5}Pt_{69.5}$(1.9 nm)]$_{10}$ multilayer films annealed at different temperature [as-deposited (a), T_a = 700℃(b) and 900℃(c)]. The strong (002) peak from MgO substrate suggests that the *x*-ray penetrated the FePt film. For the as-deposited film, only the (200) peak attributed to disordered *A*1 FePt was seen. There is no evidence for the presence of Pt and Fe phases, indicating that the interlayer diffusion during deposition was fairly sufficient. FePt (200) peak had a wider full width at half maximum (FWHM), which should be resulted from the difference of lattice constant at different depth of film. The distortion of lattice implies the existence of compositional fluctuation. Therefore, it is reasonable that a subtle wave of composition (periodical compositional gradient along the perpendicular direction) was left in the film which was deposited layer by layer. Annealed at 700℃ and 900℃ (figure 2b and 2c), the onset of (001) and (003) reflections from superlattice appeared, indicating that the film became ordered $L1_0$ phase with significant (001) orientation. The position of FePt (200) fundamental peak also shifted to a bigger angle [labeled as (002)], too, as a result of shortened lattice constant of *c*. In figure 2c (T_a = 900℃), two additional unlabeled peaks respectively split from MgO (002) and FePt (002) appeared. They are likely to be caused by the weaker Cu-K_β radiation. The FePt (002) peak split due to its narrower FWHM, as the periodic composition fluctuation had vanished. However, the split of MgO (002) peak might be the result of the exposure of substrate surface to the *x*-ray radiation directly. This indicates that the morphological continuity of FePt film was seriously destroyed due to the sintering. As a result, the discontinuous FePt film could not completely absorb the Cu-K_β diffraction by MgO substrate any more. The change of XRD pattern agrees with the morphology evolution as seen in figures 1a-1c.

Figure 3 shows the hysteresis loops of [Fe(0.6 nm)/$Fe_{30.5}Pt_{69.5}$(1.9 nm)]$_{10}$ multilayer films. The loops of as-deposited FePt film with disordered *A*1 phase (figure 3a) showed typical soft behaviors with an in-plane easy axis controlled by the shape anisotropy; the in-plane and out-of-plane coercivities were both small. At T_a = 500℃ (figure 3b), the shape anisotropy became weaker and the coercivities increased regardless along in-plane and out-of-plane directions, indicating the occurrence of nucleation of $L1_0$ phase. The film was a mixture of $A1$ and $L1_0$ phases with dominant $A1$ phase. The in-plane and out-of-plane coercivities reached maxima at T_a = 600℃ (loops were not given here). At T_a = 700℃ (figure 3c), the volume of $L1_0$ phase became dominant, and the perpendicular coercivity began to decrease with an out-of-plane magnetic easy axis. In the range of T_a = [700, 780]℃ (figures 3c and 3d), the slope of out-of-plane curve near the coercive force became more abrupt, companied by an increased curve squareness and an decreased perpendicular coercivity. The loops showed obvious hard magnetic behaviors of continuous $L1_0$ films with a significant (001) texture, referring the morphology change as showed in figure 1 and the XRD change as showed in figure 2. The slope of perpendicular curve relaxed contrarily as T_a was slightly elevated from 780℃ to 800℃ (figure 3e). The maximum of positive magnetization was somewhat different from that of negative

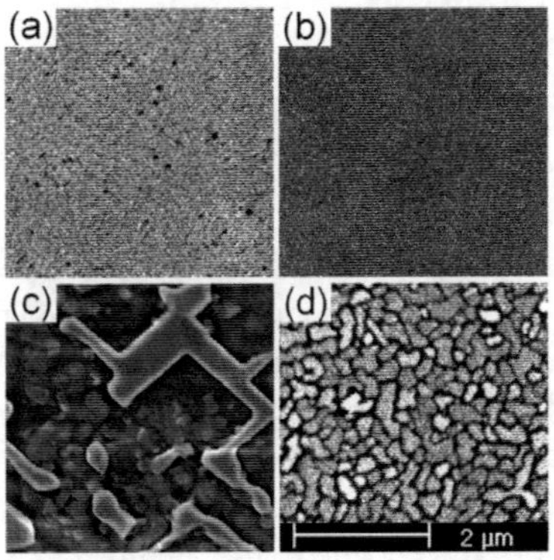

Figure 1. SEM images of [Fe(0.6 nm)/$Fe_{30.5}Pt_{69.5}$(1.9 nm)]$_{10}$ film [(a) as-deposited, (b) T_a = 700℃, (c) T_a = 900℃], and FePt film co-deposited at 700℃ (d).

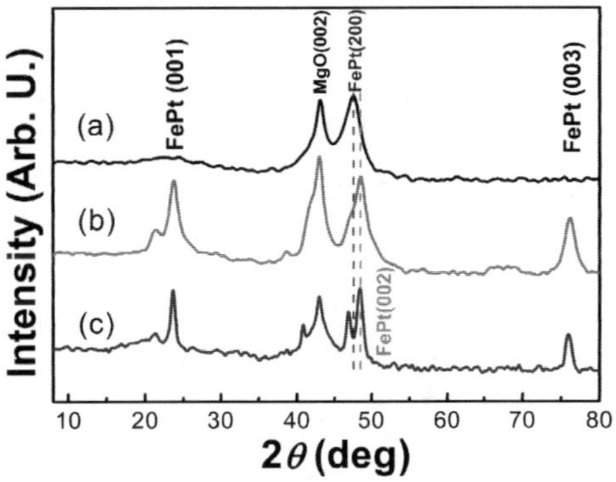

Figure 2. XRD spectra of [Fe(0.6 nm)/$Fe_{30.5}Pt_{69.5}$(1.9 nm)]$_{10}$ film. (a) as-deposited, (b) T_a = 700℃, (c) T_a = 900℃.

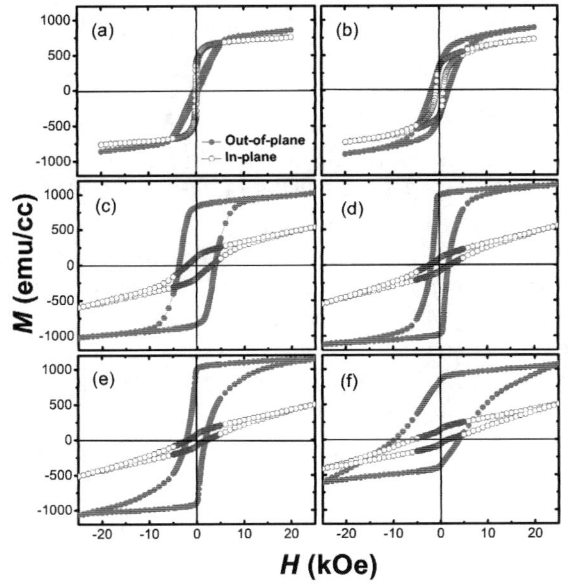

Figure 3. Hysteresis loops of [Fe(0.6 nm)/ $Fe_{30.5}Pt_{69.5}$(1.9 nm)]$_{10}$ film. (a) as-deposited, (b) T_a = 500℃, (c) T_a = 700℃, (d) T_a = 780℃, (e) T_a = 800℃, and (f) T_a = 900℃.

magnetization, suggesting a subtle ruin of the morphological continuity. A small number of separated small $L1_0$ FePt particles with multi-domain magnetic configurations were magnetized to saturation along the direction of *ab initio* applied field, and the maximum of opposite field was not enough to overcome the barrier of large K_u to reverse the magnetization in them. The faint discontinuity should merely have taken place at the edges or locations of defect. At T_a = 900℃, the out-of-plane loop was obviously asymmetric (figure 3f), due to the serious morphological discontinuity. The maximal magnetization of ~1070 emu/cc was less than that of ~1150 emu/cc for the film annealed at 800℃. This could be explained by the presence of fine single-domain particles with large switching field, which would contribute nothing to the magnetization as their magnetic moments aligned randomly upward and downward. Detailed analysis to this phenomenon can be found in references 9-12. Table 1 summarizes the T_a dependence of K_u determined from the enclosed area of the prolonged out-of-plane and in-plane curves.

Table 1. K_u of [Fe(0.6 nm)/$Fe_{30.5}Pt_{69.5}$(1.9 nm)]$_{10}$ film annealed at different T_a.

T_a (℃)	700	750	780	800
K_u($\times 10^7$ erg/cc)	2.5	2.8	3.0	3.5

4. Analysis of the reason for improvement of morphological continuity

The kinetics for the polymorphic phase transition *A*1 to $L1_0$ is rather complex. Recently, the compositional profiles by atom probing indicated a preference of Pt to the grain boundaries in $L1_0$ FePt films, even though the films were Fe-rich [13], and modeling predictions also suggested the same preferential segregation events for changes in crystal symmetry [14-16]. These reports demonstrate that the phase transition provides an additional driving force for Pt migration to induce the segregation and subsequently the ruin of morphological continuity. The critical technique to improve the continuity of FePt film is how to control the segregation of Pt. The ordering in hot FePt film was suggested to initiate at the grain boundaries [17]. The subtle compositional fluctuation at boundaries assists in the atomic rearrangement and therefore is essential for the initiation of phase transition. In this paper, the film was designed as a multilayer structure of [Fe(0.6 nm)/$Fe_{30.5}Pt_{69.5}$(1.9 nm)]$_{10}$ with interlayer diffusion during temperature-elevated sputtering. As a subtle periodic compositional fluctuation was designed into the film along the perpendicular direction (assumed as *z* axis), the nucleation of $L1_0$ phase during annealing would mainly occur in the interior of film, and the periodical compositional gradients of Fe and Pt would suppress the segregation of Pt which can result in grain boundaries. Conveniently, we discuss the diffusion of Fe atoms. The Fe concentration can be roughly expressed as the atom density, n_{Fe},

$$n_{Fe} = n_0 + n_0' \cos(\omega z), \tag{1}$$

where $\omega = 2\pi/2.5$, n_0 represents the averaged n_{Fe}, and n'_0 the fluctuation amplitude of n_{Fe}. In the subsequent ordering by annealing, Fe can suppress the segregation of Pt by preferentially migrating toward Pt-rich directions. Due to lack of grain boundaries, the film can maintain its morphological continuity at a higher temperature, until the depletion of periodical compositional gradient. By referring to the diffusion and recombination of carriers in semiconductor [18], considering the migration of Fe atoms in *z* direction as non-equilibrium diffusion with an activation energy *Q*, and the ordering jumping as recombination with an activation energy *Q*' (*Q*' > *Q*) and with an ordering probability $p = \exp(-Q'/k_BT)$, where k_B is the Boltzmann constant, and *T* is the temperature, the equilibrium in every half of the period of (1) will satisfy the following formula,

$$-\frac{d}{dz}\left(-\frac{1}{6}\frac{c^2}{\tau}e^{-\frac{Q}{k_BT}}\frac{dn}{dz}\right)-\frac{1}{\tau}ne^{-\frac{Q'}{k_BT}}=0. \tag{2}$$

In (2), $n = n_{Fe} - n_0$, *c* is the diffusion length, and $1/\tau$ is the frequency of thermal vibration. The first item represents the accumulation of *n*, caused by the diffusion with a diffusion coefficient of $c^2\exp(-Q/k_BT)/(6\tau)$. The second item represents the loss of *n*, caused by the ordering with a modification of n replacement by np. The 'loss' means that the ordered atoms will no longer contribute to diffusion. z starts at every position of $n = n'_0$ and always toward the Pt-rich direction. The solution of (2) is

$$n = (n_{Fe} - n_0)_0 \exp\left[-\sqrt{6}\left(\frac{z}{c}\right) / e^{\frac{Q'-Q}{2k_BT}} \right]. \tag{3}$$

Therefore, the factor of $\exp[(Q'-Q)/2k_BT]$ decides the new diffusion length. (3) tells us that overlong annealing time is no use to urge the ordering as the initiation time is in the scale of seconds [17]. With the increase of T, the diffusion length becomes shorter, indicating that more volume of the film transits into $L1_0$ phase. $(n_{Fe}-n_0)_0$ should be modified with the increase of T. At $k_BT >> (Q'-Q)/2$, $n \rightarrow 0$, as $(n_{Fe}-n_0)_0 \rightarrow 0$.

5. Conclusions

$[Fe(0.6nm)/Fe_{30.5}Pt_{69.5}(1.9nm)]_{10}$ continuous multilayer films were sputtered on MgO(001) substrates heated to 400℃. By interlayer diffusion during the deposition, a subtle periodical compositional gradient was left in the films along the perpendicular direction. By annealing at $T_a \geq 700$℃, the films became ordered $L1_0$ phase with a significant (001) orientation and a magnetocrystalline anisotropic energy exceeding 2.5×10^7 erg/cc. The morphology of film kept continuous at $T_a < 800$℃. The compositional gradient design is an effectual method to suppress the segregation of Pt in the phase transition and provide a convenience for microfabrication with continuous $L1_0$ FePt films.

Acknowledgement

This work was supported by Chongqing Natural Science Foundation (Grant No. CSTC2009BB8102) and Fundamental Research Funds for the Central Universities (Grant No. XDJK2009C193).

References

[1] Weller D and Moser A 1999 *IEEE. Trans. Magn.* **35** 4423
[2] Sun S 2006 *Adv. Mater.* **18** 393
[3] Rong C B and Liu J P 2007 *Appl. Phys. Lett.* **90** 222504
[4] Shima T, Takanashi K, Takahashi Y K and Hono K 2004 *Appl. Phys. Lett.* **85** 2571
[5] Li G Q, Takahoshi H, Ito H, Saito H, Ishio S, Shima T and Takanashi K 2003 *J. Appl. Phys.* **94** 5672
[6] Lodder J C 2004 *J. Magn. Magn. Mater.* **272–276** 1692
[7] Koike K, Matsuyama H, Hirayama, Tanahashi K, Kanemura T, Kitakami O and Shimada Y 2001 *Appl. Phys. Lett.* **78** 784
[8] Seki T, Shima T, Yakushiji K, Takanashi K, Li G Q and Ishio S 2006 *J. Appl. Phys.* **100** 043915
[9] Shima T, Takanashi K, Takahashi Y K and Hono K 2004 *Appl. Phys. Lett.* **85** 2571
[10] Shima T, Takanashi K, Takahashi Y K, Hono K, Li G Q and Ishio S 2006 *J. Appl. Phys.* **99** 033516
[11] Li G Q, Takahoshi H, Ito H, Washiya T, Saito H, Ishio S, Shima T and Takanashi K 2005 *J. Magn. Magn. Mater.* **287** 219
[12] Li G Q, Saito H, Ishio S, Shima T, Takanashi K and Xiong Z H 2007 *J. Magn. Magn. Mater.* **315** 126
[13] Torres K L and Thompson G B 2009 *Ultramicroscopy* **109** 606
[14] Yang B, Asta M, Mryasov O N, Klemmer T J and Chantrell R W 2005 *Scripta Mater.* **53** 417
[15] Chepulskii R V and Butler W H 2005 *Phys. Rev.* B **72** 134205
[16] Srivastava C, Balasubramanian J, Turner C H, Wiest J M, Bagaria H G and Thompson G B 2007 *J. Appl. Phys.* **102** 104310
[17] Li X H, Liu B T, Li W, Sun H Y, Wu D Q and Zhang X Y 2007 *J. Appl. Phys.* **101** 093911
[18] Huang K and Han R Q 1988 *Solid State Physics* (Beijing: China Higher Education Press) p 350

Journal of Physics: Conference Series **266** (2011) 012041 doi:10.1088/1742-6596/266/1/012041

Study of thermal ageing behaviour of Fe-Cr model alloys by magnetic hysteresis loop technique

J N Mohapatra[1], Y Kamada[1], H Kikuchi[1], S Kobayashi[1], J Echigoya[1], D G Park[2] and Y M Cheong[2]

[1]NDE and Science Research Center, Faculty of Engineering, Iwate University, 4-3-5 Ueda, Morioka, 020-8551, Japan
[2]Korea Atomic Energy Research Institute, Yusung, Taejon 305-600, South Korea

E-mail: jnmohapatra@gmail.com

Abstract. Fe-Cr alloys with different Cr content (5-20 wt. %) were prepared by arc melting technique. The alloys were solution annealed at 1050 °C/ 2h and air quenched and then tempered at 750 °C/ 1h followed by air cooling. Thermal ageing was carried at 475 °C for various lengths of time up to 1000 hours (hrs) to produce thermal embrittlement. The hardness and coercivity of Fe-20% Cr alloys were increased and the remanence was decreased, being due to the precipitation of Cr-rich α′ phase and their subsequent growth. In Fe-15% Cr and Fe-10% Cr alloys, it was found that the coercivity decreased at the initial ageing period and then increased subsequently due to the competing effect of recovery of dislocations and precipitation of Cr rich phase. The change in magnetic properties in Fe-5% Cr alloy was different than other alloys since it was within the solubility limit of Cr in Fe. A linear relationship was found between the coercivity and hardness in Fe-20% Cr alloy, indicating that Magnetic Hysteresis Loop (MHL) technique would be a good tool for the Non-Destructive Evaluatiion (NDE) of embrittlement in Fe-Cr alloys.

1. Introduction

Fe-Cr base steels are widely used in nuclear industries due to superior mechanical properties and good resistance to radiation induced swelling [1]. Such alloys undergo embrittlement in temperature range 450-550 °C called as 475° embrittlement, produces hardening due to α-α′ phase separation [2]. Precipitation of Cr-rich isolated α′-phase and their subsequent linkage to form a highly interconnected region largely affect mechanical properties particularly increase the hardness of the material and decrease the toughness [3-9]. Such behavior of phase separation is enhanced under irradiation and produced even at much lower temperature [10, 11]. A change in microstructure of materials alters both the mechanical as well as physical (electrical, magnetic, ultrasonic etc.) properties. Hence a good correlation of the physical properties with mechanical properties which alter due to microstructural modifications will be helpful for the evaluation of material properties in non-invasive way. An in-depth knowledge on mechanism of change in properties and their correlation are needed for the development of NDE based methodologies for the safe operation of the materials in industrial components. Magnetic NDE techniques have been used to evaluate the ageing creep and fatigue behaviour due to its better sensitivity to the microstructural modifications in steels [12-16]. In the present study, thermal ageing have been carried out in various Fe-Cr alloys to produce the 475°C embrittlement, and magnetic properties and mechanical properties have been found to be correlated to use magnetic technique as a NDE tool for evaluation of embrittlement in nuclear components.

2. Experimental

Fe-Cr model alloys with different Cr (5, 10, 15, 20 wt. %) content were prepared by arc melting technique using Fe with 99.99 % and Cr 99.9 % purity in argon atmosphere. The alloys were solution annealed at 1050 °C/ 2 h followed by air quenching. EDM cutting was used for the sample preparation. Toroid (thick: 2 mm, ID: 10 mm, OD: 15 mm) samples were prepared for the magnetic measurement. All the samples were tempered at 750 °C/ 1h and air-cooled to remove the residual stress generated during air quenching from high temperature. The samples were aged at 475 °C for various lengths of time up to 1000 hrs in high vacuum. Magnetic measurement, hardness measurement and microstructural characterizations were carried out during interruption. The MHL measurements were carried out at a maximum field of 6 kAm^{-1} and frequency of 50 mHz. Transmission Electron Microscopy (TEM) was used to investigate the microstructural modifications during thermal ageing. An automatic Vickers hardness measurement tester was used to measure the micro-hardness of the samples at a load of 100 g.

3. Results

3.1. Change in properties by tempering

Change in hardness of the Fe-Cr alloys against Cr content before and after tempering is shown in figure 1. The hardness of the alloys was increased with increase in Cr content due to the solid solution hardening. With tempering no significant change in hardness was found.

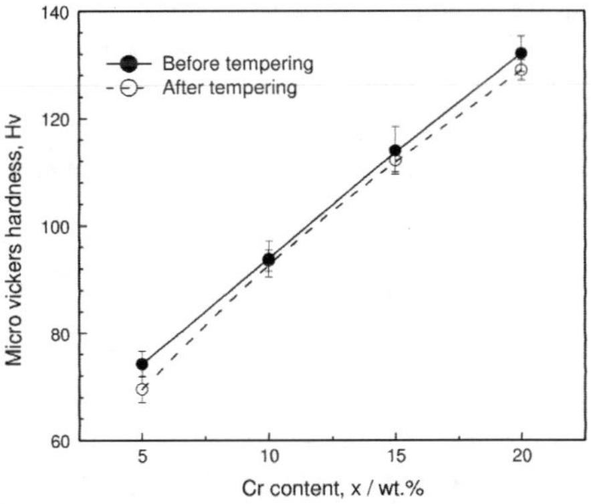

Figure 1. Change in hardness of the Fe-Cr alloys against Cr content before and after tempering.

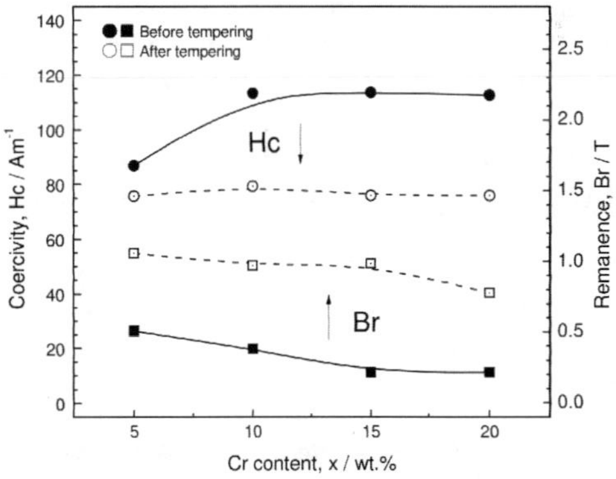

Figure 2. Change in coercivity and remanence of the Fe-Cr alloys against Cr content before and after tempering.

The change in coercivity and remanence of the Fe-Cr alloys before and after tempering is shown in figure 2. Before tempering, coercivity of the alloys was increased and a corresponding decrease in remanence was observed with Cr content. After tempering, the coercivity of all the alloys was dropped significantly whereas remanence of the alloys was increased. The result indicated that the difference in coercivity in the four samples before tempering was due to the difference in their thermal stress imposed during air-quenching. The removal of thermal stress after tempering was resulted in drastic decrease in coercivity and increase in remanence in the alloys.

3.2. Change in properties by thermal ageing

Change in hardness of the alloys with ageing time is shown in figure 3. In Fe-20% Cr alloy, it was found that the hardness increased slowly up to 50 hrs of ageing and then increased rapidly. In Fe-15% Cr alloy after 100 hrs hardness was found to increase. However, hardness of Fe-5% Cr and Fe-10% Cr alloys did not show any change with ageing time.

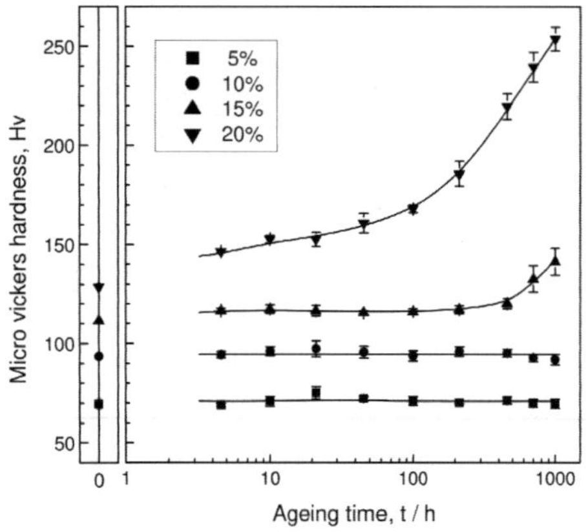

Figure 3. Change in hardness of the Fe-Cr alloys with ageing time.

Coercivity and remanence of the alloys with ageing time is shown in figure 4 and figure 5 respectively. In Fe-20% Cr alloy, the coercivity was increased from the beginning of the ageing and the rate of increase was enhanced after 50 hrs of ageing which was very similar to the change in hardness in the alloy. In Fe-15% Cr and Fe-10% Cr alloy the coercivity was decreased initially up to 50 hrs and after that it was increased. The remanence of the three alloys decreased with ageing time whereas the rate of decrease depends on the Cr content. In Fe-5% Cr alloy, the behavior was completely different than the other three alloys. The coercivity was increased from the beginning of ageing and falls down after 50 hrs. There was no significant change in the remanence found in the alloy containing 5% Cr in the entire period of ageing.

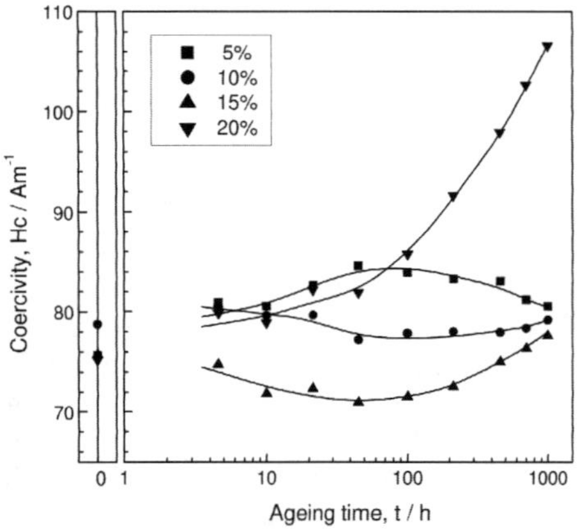

Figure 4. Change in coercivity of the Fe-Cr alloys with ageing time.

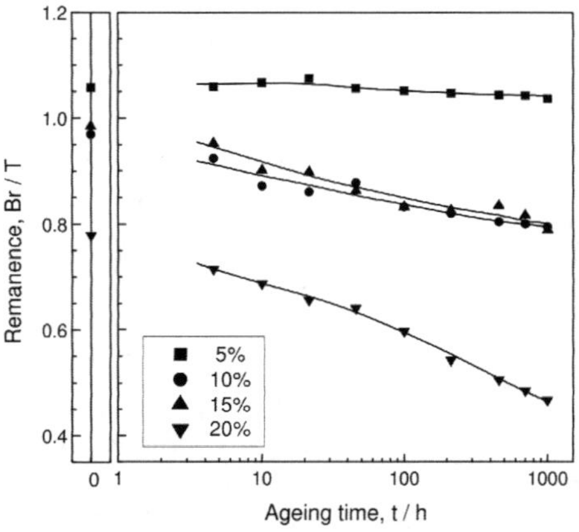

Figure 5. Change in remanence of the Fe-Cr alloys with ageing time.

4. Discussion

The coercivity in Fe-20% Cr alloy was increased substantially and remanece was decreased sharply than the other alloys possibly due to the formation of Cr rich α′ phase precipitations. The TEM study could not able to detect such precipitations in the present study due to similar lattice structure of Fe and Cr. However, results from atom probe field ion microscopy on Fe-Cr alloy shows that the nanometer size precipitations of Cr-rich α′ phase are either isolated or interconnected depending on the Cr content and heat treatment conditions [6, 7]. In alloy containing 24% Cr, formations of interconnected precipitates are seen after 500 hrs aging at 500 °C [7]. Since the ageing conditions and Cr content are almost similar, formation of Cr-rich α′ phase and their growth to form interconnected region are also expected in the present Fe-20% Cr alloy. Formation of the coherent precipitates is responsible for the increase in hardness in the alloy due to their local strain field whereas with the increase in volume fraction of the precipitates and their interconnected network enhanced the hardness furthermore. The local residual stress around the coherent precipitates is quite higher as previously reported in Fe-Cu alloy. The coherent Cu precipitation produces magnetic hardening in the alloys due to hindrance of domain wall motion [17, 18]. In Fe-20% Cr alloy, the increase of volume fraction of Cr rich precipitates and their interconnection lead to the increase of coherent strain field and its enhancement. This is one of the promising mechanisms to explain the behavior of coercivity and remanence of thermally aged Fe-20% Cr alloy. In addition to the strain effect, it would be noticeable to consider the change of chemical composition of the matrix. During the Cr precipitation, reduction of Cr content also occurs in the matrix. Such change in matrix conditions also alters the magnetic hysteresis properties of the alloys through change in intrinsic parameters such as crystalline anisotropy and magnetostriction coefficient.

In Fe-15% Cr and Fe-10% Cr alloys, it was found that coercivity decreased up to 50 hrs. Such behavior of magnetic softness was due to the recovery of dislocation structures in the alloys [19]. Although such effect was also expected to be present in the Fe-20% Cr alloy, the production of more number of precipitates and their increased in volume fraction overwhelmed the effect of recovery of dislocations. There was an increased in the coercivity after 50 hrs in Fe-15% Cr alloy which was more than the Fe-10% Cr alloy. Such behavior can be corroborated with the more volume fraction of Cr-rich precipitates in Fe-15% Cr alloy.

The magnetic properties changed in Fe-5% Cr alloy were quite different than the alloys having higher Cr content. The Fe-5% Cr alloy is below the solubility limit of Fe in Cr, hence there is no possibility of Cr rich α′-phase. In Fe-Cr alloy with Cr concentration ≤ 5%, the short range ordering is strong [20]. Due to such strong short range ordering, the change in magnetic properties in Fe-5% Cr alloy was different from the other alloys.

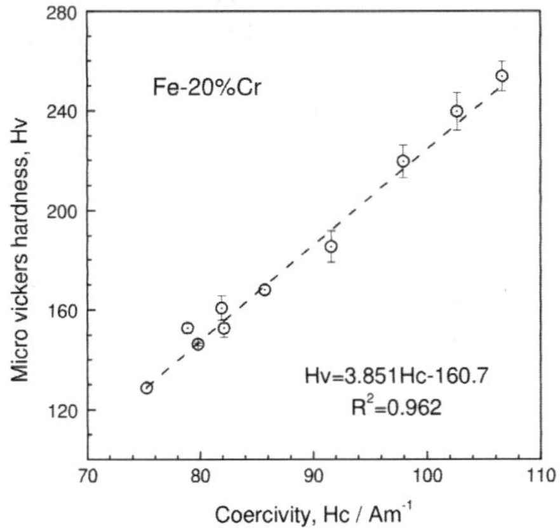

Figure 6. Change in hardness against coercivity in Fe-20% Cr alloy.

The change in hardness with the coercivity in Fe-20% Cr alloy is shown in figue 6. A linear relationship between the hardness and coercivity has been found with the embrittlement. Many reports say that the effect of embrittlement is enhanced by radiation environment and observed at much lower temperature than 475 °C. The change in hardness and hence coercivity in radiation embrittlement materials are expected to be quite more. So the alloys having lower Cr content is expected to be harden more in radiation embrittlement than the present study and can be detected by MHL technique.

5. Conclusion

Hardness of Fe-Cr alloys particularly with higher Cr content was found to increase due to the increase in volume fraction of Cr-rich α′-phase. A linear relationship was found between the hardness and coercivity during thermal ageing in Fe-20% Cr alloy. With increase in volume fraction of the Cr rich α′-phase the remanence was decreased substantially. Due to low volume fraction of such phase the change in properties was not very significant in the alloys having lower Cr content. The results enables that the Magnetic Hysteresis Loop technique would be a good NDE tool for the evaluation of embrittlement in nuclear power plant components made up of Fe-Cr alloys.

Acknowledgements

This work was supported in part by Japan Society for the Promotion of Science and Korea Science and Engineering Foundation under the contract of the Japan-Korea Basic Scientific Cooperation program.

References

[1] Mathon M H, De Carlan Y, Geoffroy G, Averty X, Alamo A, De Novion C H 2003 *J. Nucl. Mater.* **312** 236.
[2] Grobner P J 1973 *Metall. Trans.* **4** 251.
[3] Brenner S S, Miller M K and Sofa W A 1982 *Scr. Metall.* **16** 831.
[4] Y. De Carlan, A. Alamo, M.H. Mathon, G. Geoffroy, A.Castaing 2000 *J. Nucl Mater.* **672** 283.
[5] Soriano-Vargas O, Avila-Davila E O, Lopez-Hirata V M, Cayetano-Castro N, Gonzalez-Velazquez J L 2010 *Mater. Sci. & Eng.* A **527** 2910.
[6] Miller M K, Hyde J M, Cerezo A, Smith G D W 1995 *Appl. Surface Sci.* **87/88** 323.
[7] Miller M K, Hyde J M, Hetherington M G, Cerezo A, Smith G D W and Elliott C M 1995 *Acta Metal. Mater.* **43** 3385.
[8] Yustinovshikov Y, Shirobokova M and Pushkarev B 1996 *Acta Mater.* **44** 5021.
[9] Little E A, Stoter LP 1982 *STM-STP* **782** 207.
[10] Miller M K, R E Stoller, K F Russell 1996 *J. of Nucl. Mater.* **230** 219.
[11] Bonny G, Terentyev D, Malerba L, 2008 *Scripta Mater.* **59** 1193.
[12] Mohapatra J N, Panda A K, Mitra A, 2009 *J.Phys.D: Appl. Phys.***42** 095006.
[13] Mitra A, Mohapatra J N, Swaminathan J, Ghosh M, Panda A K, Ghosh R N 2007 *Scripta Mater* **57** 813.
[14] Mohapatra J N, Ray A K, Swaminathan J, Mitra A 2008 *J.Magn. and Magn.Mater.* **320** 2284.
[15] Moorthy V, Choudhary B K, Vaidyanathan S, Jayakumar T, Rao K B S and Raj B 1999 *Int. J. of Fatigue* **21** 263.
[16] Lo C C H, Tang F, Jiles D C, Biner S B 1999 *Magnetics IEEE Trans.* **35** 3977.
[17] Aleshin N P, Altpeter I, Dobmann G, Kroening M, Reddy A 2002 *National Seminar of ISNT* **5-7** 12.
[18] Kamada Y, Takahashi S, Kikuchi H, Kobayashi S, Ara K, Echigoya J, Tozawa Y, Watanabe K, 2009 *J. Mater. Sci.* **44** 949.
[19] Martinez-de-Guerenu A, Arizti F, Diaz-Fuentes M, Gutierrez I *Acta Mater.* 2004 **52** 3657.
[20] Erhart P, Caro A, Caro M S D, Sadigh B 2008 *Phys. Rev. B* **77** 134206.

Journal of Physics: Conference Series **266** (2011) 012042 doi:10.1088/1742-6596/266/1/012042

Effect of structural transition on the temperature-dependent magnetic properties of epitaxial FePd alloy nanoparticles

Hiroshi Naganuma[1,2], Kazuhisa Sato[1,3], and Yoshihiko Hirotsu[1]

[1] *The Institute of Scientific and Industrial Research, Osaka University, 8-1 Mihogaoka, Ibaraki, Osaka 567-0047, Japan*
[2] Present address: *Department of Applied Physics, Graduate School of Engineering, Tohoku University, 6-6-05 Aoba Aramaki, Aoba-ku, Sendai, Miyagi 980-8579, Japan*
[3] Present address: *Institute for Materials Research, Tohoku University, 2-1-1 Katahira, Aoba-ku, Sendai, Miyagi 980-8577, Japan*

Electronic mail: naganuma@ mlab.apph.tohoku.ac.jp

Abstract. We have studied atomic ordering and magnetic properties of epitaxial FePd nanoparticles 3-14 nm in diameter. Thermal fluctuation of magnetic moment of $L1_0$-FePd nanoparticles becomes prominent when particle size is smaller than about 8 nm. Atomic images of high-resolution transmission electron microscopy revealed the formation of the ordered phase in FePd nanoparticles as small as 4 nm in diameter, while the superlattice reflections were quite weak compared to those of the larger sized particles (>8nm). The experimental results indicate that the degradation of magnetic anisotropy is induced by the size-dependent decay of long-range atomic order.

1. Introduction

The demand for higher density magnetic recording media is one of the fundamental motivations for the recent attentions to hard magnetic nanoparticles that are characterized by high magnetocrystalline anisotropy energy (MAE) [1]. One of the candidates for ultrahigh density magnetic storage media is FePd alloy nanoparticles with the $L1_0$-type ordered structure. The hard magnetic properties of the $L1_0$-FePd alloy nanoparticles originate from the tetragonal ordered structure with a high MAE [2], and thus the atomic ordering and the stability of the ordered phase are key issues. According to our previous work [3], the degree of order depends on particle size and the decrease of the long-range order (LRO) parameter was found for FePd nanoparticles below 8 nm in diameter. The latest atomic resolution imaging based on the aberration-corrected high-resolution transmission electron microscopy (HRTEM) also revealed a loss of clear-cut LRO for particles 2-5 nm in size [4]. Such a depression of LRO, namely, structural transition towards disordered phase, is a size-dependent event frequently observed in nanoparticles, and will lead to a rapid decrease in MAE. The effect of particle size including such a size-dependent structural transition on magnetic properties can be detected by measuring temperature dependence of magnetization. Understanding of the aforementioned size-dependent phenomenon may also contribute to the recent high-density magnetoresistive random access memory (MRAM) using a thin $L1_0$–ordered layer as well as a magnetic storage media.

This study aims at clarifying particle size dependence of magnetic properties of epitaxial FePd nanoparticles at the onset of the structural transition. We show that superparamagnetic behavior becomes prominent when a clear-cut LRO is lost. Thermal relaxation of remanent magnetization is discussed based on the particle size distribution.

2. Experimental Procedure

FePd nanoparticles were fabricated by sequential electron-beam deposition of Pd and Fe onto NaCl (001) single crystal substrates at substrate temperature T_s of 476-673 K (base pressure of 1×10^{-6} Pa) [5]. After the deposition of Fe, a 10-nm-thick amorphous film of Al_2O_3 was deposited at a substrate temperature of 523 K to protect the particles. Postdeposition annealing was performed in a high vacuum quartz-tube furnace (2×10^{-5} Pa) for 3.6ks at 873 K. Particle morphology and atomic structures were observed by using 200 kV (JEOL JEM-2010) and 300 kV (JEM-3000F) electron microscopes. Compositional analysis was carried out by using an energy dispersive x-ray spectrometer (EDS) attached to the 300 kV-TEM. Magnetic properties, including hysteresis loops, temperature dependence of magnetization, and remanent magnetization decay, were measured using a superconducting quantum interference device (SQUID) magnetometer (Quantum Design MPMS-XL) in the temperature range between 10 and 300 K with a magnetic field up to 50 kOe. The magnetic field was applied perpendicular to the film plane. The measuring time was set to be 10 sec in this study.

3. Results and discussion

3.1 *Particle morphology and atomic structure*

Figure 1 shows the bright-field (BF) TEM images and the corresponding selected area electron diffraction (SAED) patterns of FePd nanoparticles after annealing for 3.6 ks at 873 K. Substrate temperature (T_s) and the average particle size (d) are as follows: (a) T_s = 573 K, d = 8.2 nm, (b) T_s = 473 K, d = 4.3 nm, respectively. Histograms of particle size distribution of the FePd nanoparticles followed a log-normal-type distribution function. Note that size distribution becomes narrower as the average size reduces. In fact, as can be seen in the histogram of Fig. 1(b), a quite sharp distribution has been realized for nanoparticles 4.3 nm in average size. In the [001] incidence of the $L1_0$ structure, four-fold 110 superlattice reflections can be seen on an SAED pattern. In fact, the SAED pattern in Fig.1(a) clearly indicates a formation of the $L1_0$ structure with preferential c-axis orientation normal to the film plane. Noticeable feature here is the intensity of the superlattice reflections: the intensity is

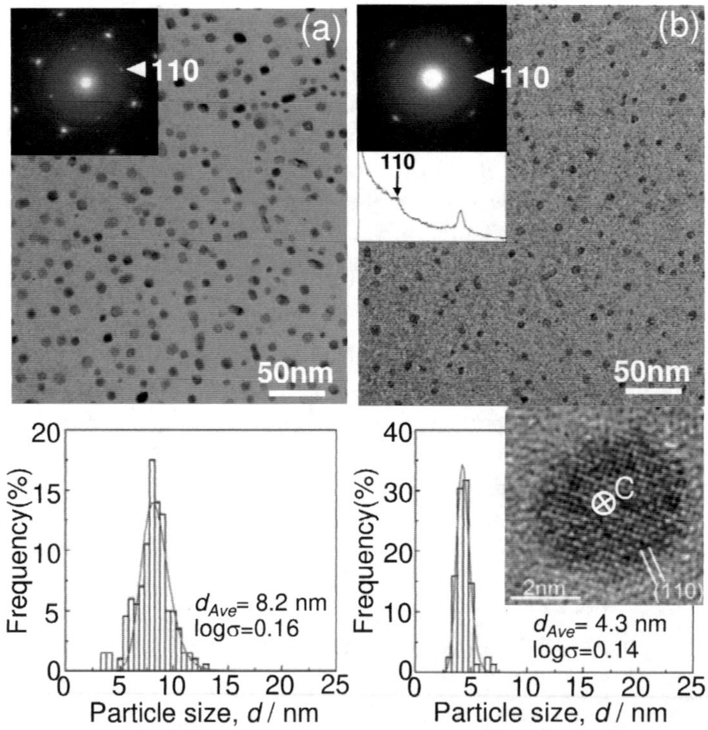

Figure 1 TEM images and the corresponding SAED patterns for (a) 8-nm-sized and (b) 4-nm-sized FePd nanoparticles after annealing. Histograms of particle size distribution are also shown in the inset. In Fig.1(b), an example of the HRTEM image of a locally ordered particle is also shown.

quite weak for nanoparticles shown in Fig.1(b). This result clearly indicates the size-dependent degradation of the LRO in FePd nanoparticles as confirmed by our previous work using nanobeam electron diffraction [3]. We confirmed atomic structures of these 3-4-nm-sized nanoparticles by HRTEM observation. As a result, faint {110} lattice fringes, due to alternate stacking of Fe and Pd in the [001] direction, were still observed, in accordance with the latest HRTEM study using an aberration corrector [4]. An example of HRTEM image of a 4-nm-sized particle is shown in the inset of Fig.1(b).

3.2 *Size and temperature-dependent magnetic properties*

Figure 2(a) shows the average particle size dependence of the coercivity measured at 300 and 10 K with the external magnetic field perpendicular to the film plane. Coercivity as high as 3.5 kOe was obtained for larger sized particles (>11nm), while the coercivity abruptly dropped in the particle size range between 8 and 10 nm. The coercivity reached quite low values (<200 Oe) for particles smaller than 6 nm. Such an apparent decay of coercivity with size reduction below 10 nm can be attributed to superparamagnetic behaviour, namely appearance of thermal fluctuation of magnetization. Noticeable feature here is the size dependence of coercivity at 10 K. As shown in Fig.2(a), the coercivity at 10 K also drastically decreased similar to those at 300 K, indicating the degradation of MAE with size reduction. Note that there are two kind of characteristic particle size regions: one is the region between 6 and 8 nm that is characterized by the decrease in LRO [3]. The other is the region between 2 and 5 nm, where local ordering, indicated by faint superlattice reflections and loss of clear {110} fringes, appeared as shown in Fig.1(b). The rapid decrease in coercivity was observed at the former size region, indicating the loss of magnetocrystalline anisotropy triggered by the decay of LRO. Figure 2(b) shows average particle size dependence of the coercivity ratio, $H_c(10\mathrm{K})/H_c(300\mathrm{K})$. There observed a jump in the region at around 8 nm in diameter, clearly indicating the fact that the effect of thermal fluctuation on coercivity becomes prominent below 8 nm.

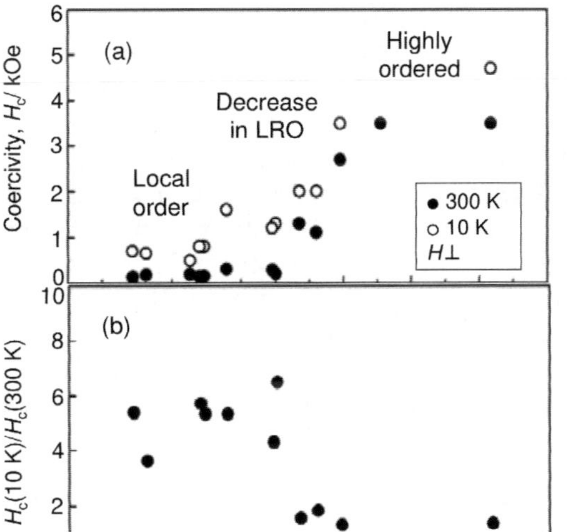

Figure 2 (a) Average particle size dependence of coercivity for FePd nanoparticles after annealing at 3.6ks for 873K. Coercivity abruptly dropped in the particle size range between 8 and 10nm. (b) Size dependence of the coercivity ratio, $H_c(10\mathrm{K})/H_c(300\mathrm{K})$.

To further confirm the effect of thermal fluctuation, we measured field-cooling (FC) and zero-field-cooling (ZFC) magnetization curves in the temperature range between 10 and 400 K. Figures 3(a) and 3(b) show the FC and the ZFC curves for FePd nanoparticles with average sizes of 8.2 nm and 4.3 nm, respectively. The magnetic field of 100 Oe was applied perpendicular to the film plane. In Fig. 3(a), the ZFC curve increases monotonously as the temperature increases, and showed a very broad, faint peak approximately at 350 K. Relatively high blocking temperature observed for the 8-nm-sized particles can be attributed to the coexistence of superparamagentic and ferromagnetic components, depending on the particle size. By decreasing the average particle size to 4.3 nm, blocking temperature shifts towards the lower temperature and a clear peak appeared at 100 K in the ZFC curve as shown in Fig. 3(b). The observed sharp peak can be attributed to the narrower particle size distribution (Fig.1(b)). The FC curve almost overlapped with the ZFC curve for nanoparticles 14 nm in average size in the whole temperature range between 10 K and 400 K. Thus it has been revealed that the thermal fluctuation of magnetic moment becomes prominent when the particle size is smaller than about 8 nm in diameter, as confirmed by the abrupt reduction of coercivity, jump in coercivity ratio, and the appearance of blocking temperature.

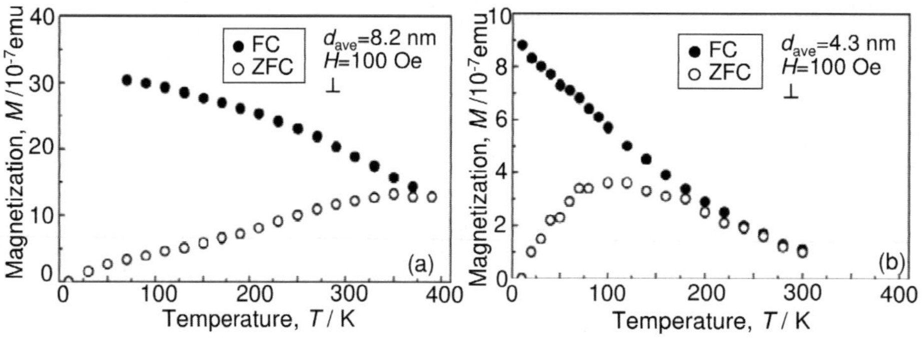

Figure 3 ZFC and FC curves for (a) 8-nm-sized and (b) 4-nm-sized FePd nanoparticles after annealing at 3.6ks for 873 K. The applied field was 100 Oe in the film normal direction.

3.3 *Size-dependent thermoremanent magnetization decay*

Figure 4(a) shows remanent magnetization decay measured at 300 K for FePd nanoparticles with different average sizes. Magnetic field of 50 kOe was first applied perpendicular to the film plane, and then, immediately removed the field, followed by the measurement of the thermoremanent magnetization (TRM) decay. All the TRM curves decrease logarithmically irrespective of the average particle size, but with different slopes, due to thermal fluctuation of magnetic moment. The gradient of the decay curves increase as the average particle size becomes smaller, which can be attributed to the enhancement of thermal fluctuation in smaller sized particles due to the degradation of MAE triggered by the size-dependent reduction of LRO. It should be noted that particle areal density also plays a crucial role on the relaxation behaviour of nanoparticles, since the relaxation time can be affected by interparticle magnetostatic interaction, which is determined by geometrical arrangement of magnetic

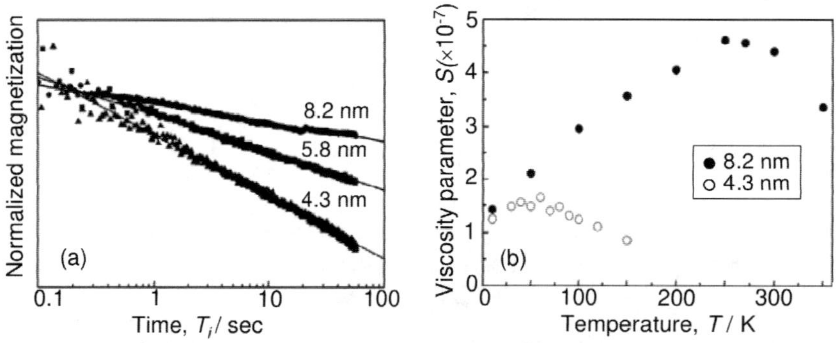

Figure 4 (a) Remanent magnetization decay curves for different sized FePd nanoparticles measured at 300K. All the specimens were annealed for 3.6ks at 873K. (b) Temperature dependence of the viscosity parameter (S) for different sized FePd nanoparticles. A peak position shifts towards lower temperatures as the particle size reduces.

particles and distribution of easy axis orientation. In the present experiment, the particle density decreased from 3.2×10^{11} cm^{-2} to 1.9×10^{11} cm^{-2} as the average particle size reduced from 8.2 nm to 4.3 nm in diameter. Such a reduction of the particle density will lead to weaken the magnetostatic interaction among the nanoparticles, while the details still remain open question.

To examine the temperature dependence of the decay of TRM, we measured the decay curves at different temperatures between 10 and 350 K. The gradient of the decay curve, so called the magnetic viscosity parameter (*S*) [6], is expressed by the following logarithmic form [6,7],

$$M_r(t) = M_0 - S\ln t \qquad (1)$$

where $M_r(t)$, M_0, and t denote the time-dependent remanent magnetization, the initial remanent magnetization, and the time, respectively. Figure 4(b) summarizes the parameter S for two kinds of specimens, 8.2 nm and 4.3 nm in average sizes. For the larger particles, the viscosity parameter shows the maximum at 250 K, while that of the smaller one at 50 K. The origin of such maximum peaks can be attributed to the distribution of relaxation time due to particle size distribution. Note that the amount of relaxation is small at very low temperatures, and small as well at higher temperatures, since most of the small particles with short relaxation times will reach thermal equilibrium within the experimental measurement time at higher temperatures. Thus, a peak appears in the temperature dependence of the viscosity parameter. The smaller particles have shorter relaxation times; hence, the peak temperature shifts towards lower temperatures as the average size reduces. The present study hence demonstrates the degradation of MAE triggered by the size-dependent reduction of LRO (<8nm), and it is further enhanced by the decay of clear-cut LRO (<5nm). To overcome the effect of thermal fluctuation, proper heat treatment for realizing sufficiently high degree of order will be effective, as indicated by recent literatures [8-10].

4. Conclusion

In conclusion, we demonstrated that the hard magnetic properties of epitaxial $L1_0$-FePd nanoparticles are largely affected by the particle size and temperature. Thermal fluctuation of magnetic moment of $L1_0$-FePd nanoparticles becomes prominent when particle size is smaller than about 8 nm, due to the size-dependent reduction of the LRO parameter. The coercivity reached quite low values (<200 Oe) for particles smaller than 6 nm, in accordance with the loss of clear-cut LRO as observed by HRTEM (<5nm). The present study clearly showed the degradation of MAE triggered by the instability of the $L1_0$ ordered phase in FePd nanoparticles 3-14 nm in diameter.

Acknowledgments

This study was carried out at the Institute of Scientific and Industrial Research, Osaka University, under the financial supports by the Grant-in-Aid for Scientific Research (S) (No. 16106008) from the Ministry of Education, Culture, Sports, Science, and Technology, Japan. One of the authors (HN) wishes to thank Professor Y. Ando for his encouragement.

References

[1] Weller D and Doerner M F 2000 *Annu. Rev. Mater. Sci.* **30**, 611.
[2] Shima H, Oikawa K, Fujita A, Fukamichi K, Ishida K, and Sakuma A 2004 *Phys. Rev. **B*** 70, 224408.
[3] Sato K, Hirotsu Y, Mori H, Wang Z, and Hirayama T 2005 *J. Appl. Phys.* **98**, 024308.
[4] Sato K, Konno T. J, and Hirotsu Y 2009 *J. Appl. Phys.* **105**, 034308.
[5] Sato K and Hirotsu Y 2003 *J. Appl. Phys* **93**, 6291.
[6] Street R and Brown S. D 1994 *J. Appl. Phys.* **76**, 6386.
[7] Park Y, Adenwalla S, Felcher G. P. and Bader S. D. 1995 *Phys. Rev. **B*** **52**, 12779.
[8] Chepulskii R. V and Butler W. H 2005 *Phys. Rev. **B*** **72**, 134205.
[9] Alloyeau D, Ricolleau C, Mottet C, Oikawa T, Langlois C, Le Bouar Y, Braidy N and Loiseau A 2009 *Nature Mater.* **8**, 940.
[10] Sato K 2009, *Nature Mater.* **8**, 924.

Journal of Physics: Conference Series **266** (2011) 012043 doi:10.1088/1742-6596/266/1/012043

High performance Sm-Fe-N magnets prepared by compression shearing method

T Saito[1]
Department of Mechanical Science and Engineering, Chiba Institute of Technology, 2-17-1 Tsudanuma, Narashino, Chiba, Japan

E-mail: tetsuji.saito@it-chiba.ac.jp

Abstract. Sm-Fe-N anisotropic magnets were produced from magnetically aligned green compact at room temperature in ambient atmosphere by the compression shearing method. X-ray diffraction studies revealed that the magnets retained their original $Sm_2Fe_{17}N_3$ phase structure without any appreciable decomposition of the phase and that the phase had a pronounced crystallographic alignment. The magnetic properties of the Sm-Fe-N anisotropic magnets were dependent on the magnetic alignment of the Sm-Fe-N powder before implementation of the compression shearing method. A Sm-Fe-N anisotropic magnet with the magnetic alignment produced under an applied field of 1.6 MA/m exhibited a maximum energy product of 161 kJ/m^3 with a high coercivity of 0.88 MA/m.

1. Introduction

With the advent of high-performance Nd-Fe-B permanent magnets, research and development of new permanent magnets has been largely concentrated on rare-earth-based alloys [1-3]. As a result of intensive research, several new rare-earth-based alloys for permanent magnets have been found [4-7]. Among them, Sm-Fe-N alloy is the most promising candidate for high-performance permanent magnets [4]. The superiority of Sm-Fe-N magnets arises from $Sm_2Fe_{17}N_3$ intermetallic compound, which exhibits high saturation magnetization with a large anisotropy field and a high Curie temperature [4]. $Sm_2Fe_{17}N_3$ intermetallic compound has been prepared by the production of Sm_2Fe_{17} alloy powder and subsequent nitrogenation of the powder by a gas-solid reaction. The resultant $Sm_2Fe_{17}N_3$ intermetallic compound has thus been produced in powder form.

In the production of high-performance Nd-Fe-B permanent magnets, Nd-Fe-B alloy powder has been consolidated into green compact and then sintered at a high temperature [2]. However, this sintering technique cannot be applied to the production of Sm-Fe-N bulk magnets because $Sm_2Fe_{17}N_3$ intermetallic compound is not stable at high temperatures and decomposes into α-Fe and SmN above 500-600°C [4]. Up to now, Sm-Fe-N bulk magnets have been produced by shock compaction such as by explosion or a gun method [8,9]. Sm-Fe-N bulk magnets with porosities of 2-10% have been realized by optimizing the shock compression conditions [9]. There has been no practical application of this technique, however, due to the difficulties involved.

A recently developed consolidation technique, called compression shearing, can consolidate powders into bulk form at room temperature by the application of shear stress to them [10]. This makes the compression shearing method advantageous for the production of bulk materials from Sm-

[1] To whom any correspondence should be addressed.

Fe-N powders compared with the conventional consolidation techniques. We have recently produced Sm-Fe-N bulk magnets by the compression shearing method [11,12]. The reported magnetic properties, however, were not entirely satisfactory. The magnetic properties of Sm-Fe-N magnets produced by the compression shearing method depend on several processing conditions, particularly the magnetic alignment of the Sm-Fe-N powder. In this study, the effect of the applied field on the magnetic alignment of the Sm-Fe-N powder before implementation of the compression shearing method was investigated. The magnetic properties of Sm-Fe-N magnets produced by the compression shearing method are discussed here.

2. Experiment

Commercially available $Sm_2Fe_{17}N_3$ powder was used in the experiment. The average particle size of the powder was about 3 μm. The powder was placed on a hardened steel plate and then pressed into green compact under an applied magnetic field of up to 1.6 MA/m before implementation of the compression shearing method. Shear stress was applied to the magnetically aligned green compact by the compression shearing method. Details of the compression shearing method are described elsewhere [11]. The phases of the specimens were examined by X-ray diffraction (XRD) using Cu K_α radiation. The magnetic properties of the specimens were measured using a vibrating sample magnetometer (VSM) with a maximum applied field of 2.0 MA/m. The specimens had been magnetized in a pulsed field of 4.0 MA/m prior to the magnetic measurements. The magnetic measurements were made in the direction parallel to the direction of magnetic alignment. No demagnetization correction was made in the curve. In the calculations of magnetization, the value obtained by Archimedes' method was used for the density of the specimen.

3. Results and discussion

Figure 1 shows a view of the Sm-Fe-N magnet produced from magnetically aligned green compact under an applied magnetic field of 1.6 MA/m at room temperature in ambient atmosphere by the compression shearing method. Virtually the same specimens were obtained from green compact without magnetic alignment and from magnetically aligned green compact produced under applied magnetic fields of 0.4-1.2 MA/m. The Sm-Fe-N powder was successfully consolidated into bulk magnets at room temperature in ambient atmosphere by the compression shearing method regardless of the applied magnetic field. The resultant magnets were rectangular in shape with dimensions of 18 x 10 mm and a thickness of 0.2 mm. No significant voids or cracks were observed in the magnets, which had a smooth metallic surface.

Archimedes' method was used to confirm the density of the Sm-Fe-N magnets produced by the compression shearing method. Figure 2 shows the dependence of the density of the Sm-Fe-N magnets on the applied magnetic field before implementation of the compression shearing method. The density of the Sm-Fe-N magnets increased as the applied magnetic field increased, from 7.02 Mg/m^3 (91.5% of the X-ray density of the $Sm_2Fe_{17}N_3$ phase) for the specimen produced without an applied magnetic field to 7.30 Mg/m^3 (95.1% of the X-ray density of the same phase) for the specimen produced from magnetically aligned green compact under an applied magnetic field of 1.6 MA/m. It was found that

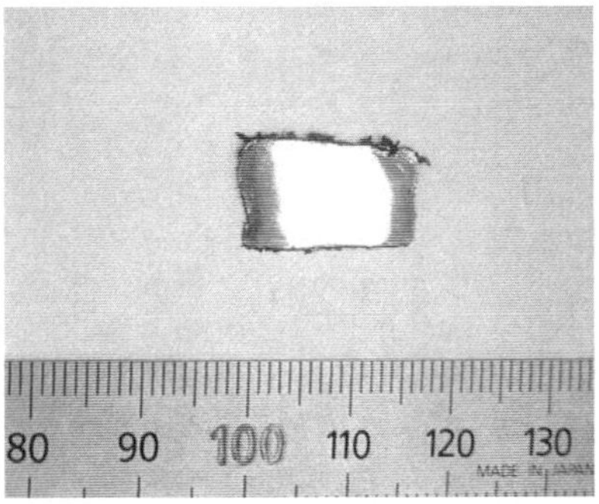

Figure 1. External appearance of Sm-Fe-N magnet produced from magnetically aligned green compact under an applied magnetic field of 1.6 MA/m at room temperature in ambient atmosphere by the compression shearing method.

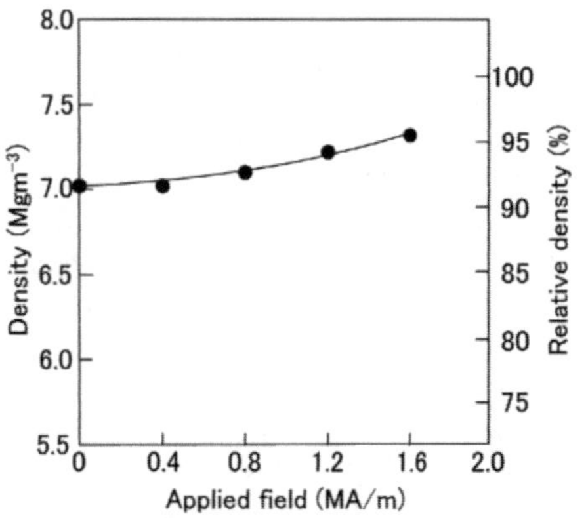

Figure 2. Dependence of the density of the Sm-Fe-N magnets on the applied magnetic field before implementation of the compression shearing method.

the application of a magnetic field before implementation of the compression shearing method is effective in increasing the density of bulk magnets produced by the compression shearing method.

Figure 3 shows the XRD patterns of the Sm-Fe-N magnets produced from magnetically aligned green compact before implementation of the compression shearing method. The specimens were measured parallel to the direction of magnetic alignment. The XRD pattern of the Sm-Fe-N magnet produced without magnetic alignment exhibits the typical powder pattern of the $Sm_2Fe_{17}N_3$ phase. In the compression shearing method, the shear stress applied to the powder during consolidation not only moves the particles so that the cavities are filled, but also causes friction between the particles while closing the gaps between them. The heat generated by the closing of these gaps may affect the $Sm_2Fe_{17}N_3$ compound, which is unstable at high temperatures and decomposes into α-Fe and SmN above 873 K. However, neither the α-Fe phase nor the SmN phase were found in the XRD pattern. This indicates that the Sm-Fe-N magnet was produced by the compression shearing method without decomposition of the $Sm_2Fe_{17}N_3$ phase. On the other hand, the XRD patterns of the Sm-Fe-N magnets produced from magnetically aligned green compact under applied magnetic fields show a pronounced *c*-axis alignment of the $Sm_2Fe_{17}N_3$ phase, which is characterized by the prominent (006) peak. This indicates that the resultant Sm-Fe-N magnets retain the magnetic alignment of the $Sm_2Fe_{17}N_3$ phase in the compression shearing method and thus exhibit the crystallographic alignment of the $Sm_2Fe_{17}N_3$ phase. It was found that the degree of crystallographic alignment of the $Sm_2Fe_{17}N_3$ phase—in this case, the degree of crystallographic alignment of the Sm-Fe-N powder—increased as the applied magnetic field increased.

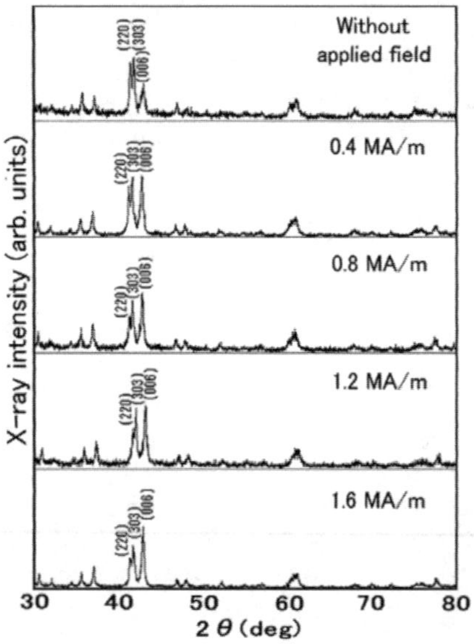

Figure 3. X-ray diffraction patterns of the Sm-Fe-N magnets produced from magnetically aligned green compact before implementation of the compression shearing method. The hysteresis loop of a Sm-Fe-N magnet produced without magnetic alignment is also shown.

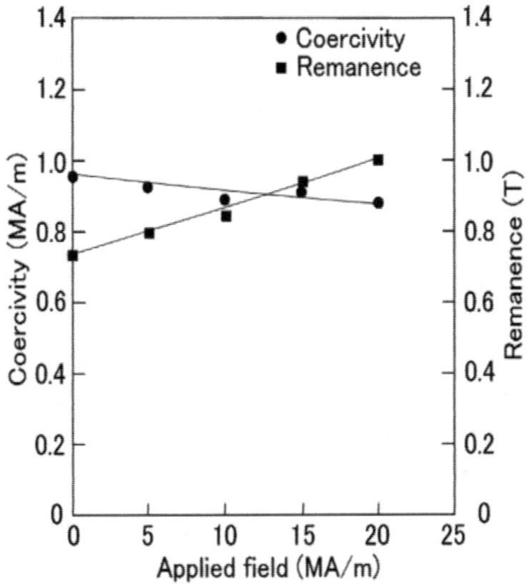

Figure 4. Dependence of the remanence and coercivity of Sm-Fe-N magnets on the applied magnetic field before implementation of the compression shearing method.

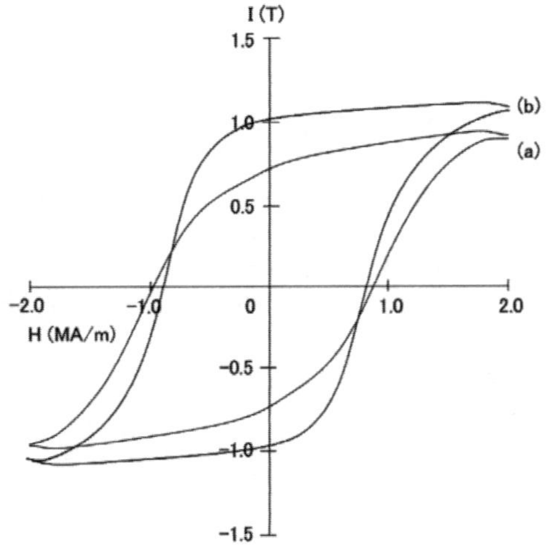

Figure 5. Hysteresis loop of Sm-Fe-N magnet produced from magnetically aligned green compact under an applied magnetic field of 1.6 MA/m by the compression shearing method (b). For comparison, the hysteresis loop of a Sm-Fe-N magnet produced without magnetic alignment is also shown (a).

Figure 4 shows the dependence of the remanence and coercivity of the Sm-Fe-N magnets on the applied magnetic field before implementation of the compression shearing method. The remanence and coercivity of the isotropic Sm-Fe-N magnet produced without an applied magnetic field are 0.72 T and 0.96 MA/m, respectively. Although the coercivity of the Sm-Fe-N magnets slightly decreased as the applied magnetic field increased, their remanence increased. The higher remanence of the Sm-Fe-N magnets produced from the magnetically aligned green compact compared with that of the magnet produced without magnetic alignment confirms that the Sm-Fe-N magnets produced from the magnetically aligned green compact are magnetically anisotropic. The crystallographic alignment of the *c*-axis of the $Sm_2Fe_{17}N_3$ phase in the Sm-Fe-N magnets produced from the magnetically aligned green compact gives rise to the observed higher remanence.

Figure 5 shows the hysteresis loop of the Sm-Fe-N magnet produced from magnetically aligned green compact under an applied magnetic field of 1.6 MA/m by the compression shearing method. For comparison, the hysteresis loop of a Sm-Fe-N magnet produced without an applied magnetic field is also shown. Although the coercivity of the Sm-Fe-N magnet produced from magnetically aligned green compact was slightly smaller than that of the magnet produced without magnetic alignment, the remanence of the former was much higher than that of the latter. The anisotropic Sm-Fe-N magnets produced from magnetically aligned green compact by the compression shearing method exhibited a maximum energy product of 161 kJ/m^3 with a high coercivity of 0.88 MA/m. The maximum energy product of the anisotropic Sm-Fe-N magnets is much higher than that of the isotropic Sm-Fe-N magnet produced without magnetic alignment, which is typically around 64 kJ/m^3.

4. Conclusion

Sm-Fe-N powder was successfully consolidated into bulk materials at room temperature by the compression shearing method. The resultant bulk magnets retained the original $Sm_2Fe_{17}N_3$ phase structure without any appreciable decomposition of the $Sm_2Fe_{17}N_3$ phase. The degree of crystallographic alignment of the $Sm_2Fe_{17}N_3$ phase and the remanence of the Sm-Fe-N magnets increased as the magnetic field applied before implementation of the compression shearing method increased. An anisotropic Sm-Fe-N bulk magnet with a maximum energy product of 161 kJ/m^3 was produced from magnetically aligned green compact under an applied magnetic field of 1.6 MA/m by the compression shearing method.

Acknowledgments
This work was made possible by grants for Rare Metal Substitute Materials Development Project from New Energy and Industrial Technology Development Organization (NEDO).The author would like to thank Mr. T. Ishikawa of Sumitomo Metal Mining Co., Ltd. for providing the Sm-Fe-N powder.

References

[1] Croat J J, Herbst J F, Lee R W, and Pinkerton F E 1984 *J. Appl. Phys.* **55** 2078
[2] Sagawa M, Fujimura S, Togawa N, Yamamoto H, and Matsuura Y 1984 *J. Appl. Phys.* **55** 2083
[3] Coehoorn R, de Mooij D B, Duchateau J P W B, and Buschow K H J 1988 *J. de Phys.* **49** 669.
[4] Coey J M D and Sun H 1990 *J. Magn. Magn. Mater.* **87** L251
[5] Buschow K H J 1988 *J. Appl. Phys.* **63** 3130
[6] Cadgan J M, Li H S, Margarian A, Dunlop J B, Ryan R H, Collocott S J, and Davis R L 1994 *J. Appl. Phys.* **76** 6138
[7] Sakurada S, Tsutai A, Hirai T, Yanagida Y, and Sahashi M 1996 *J. Appl. Phys.* **79** 4611
[8] Leonowicz M, Kaszuwara W, Jezierska E, Januszewski D, Mendoza G, Davies H A, and Paszula J 1998 *J. Appl. Phys.* **83** 6634
[9] Mashimo T, Huang X, Hirosawa S, Makita K, Kato Y, Mitsudo S, and Motokawa M 2000 *J. Magn. Magn. Mater.* **210** 109
[10] Takeishi H, Nakayama N, and Miki H 2005*J. Soc. Mater. Sci. Jap.* **545** 233
[11] Saito T, Fukui M, and Takeishi H 2005 *Scripta Mater.* **53** 1117
[12] Saito T, Sato H, Takeishi H, and Nakayama N 2006 *Appl. Phys. Lett.* **89** 162511

Journal of Physics: Conference Series **266** (2011) 012044 doi:10.1088/1742-6596/266/1/012044

Preparation and magnetic properties of amorphous Co-Fe-B nanoparticles

K Shimba, N Tezuka, S Sugimoto[1]

Department of Materials Science, Graduate School of Engineering, Tohoku University, 6-6-02, Aramaki Aza Aoba, Aoba-ku, Sendai, 980-8579, Japan

E-mail: b0td9506@s.tohoku.ac.jp.

Abstract. Amorphous Co-Fe-B nanoparticles coated with PVP (polyvinyl pyrrolidone) were prepared by reduction of metal ions in aqueous solution by use of $NaBH_4$ at room temperature. The size of nanoparticles was in the range of 10-20 nm and from the XRD patterns, these nanoparticles were amorphous. FE-SEM-WDX analyses showed the B content to be in the range of 38-45 at%. The saturation magnetization and coercivity for the samples was in the range of 45-50 emu·g^{-1} and 31-110 Oe, respectively. The composites of nanoparticles with the minimum coercivity and PVP as the binder (particle weight percent = 75 wt.%) showed permeability of $\mu''_{r\,max}$ = 1.9 at 0.7 GHz and a good microwave absorption property (reflection loss: RL < -20 dB).

1. Introduction

Powder material has been studied extensively in the last years because of their novel properties and technological applications in optical, magnetic and electronic fields. Specifically, Fe, Co based amorphous powders have a high potential as soft magnetic materials for high frequency applications [1-5]. Amorphous powders are usually synthesized by various methods such as atomizing and rapid quenching. However, the size of these particles is a few micrometers, and the frequency range of their applications is limited due to eddy current loss. The particle size must be less than about 1 μm to avoid eddy current loss when used in the gigahertz range, where a wide variety of new applications of high permeability particles such as planer inductors, antennas, and EM noise suppressors, are expected. Therefore, the size of powders must be decreased in nanometer size with monodisperse form to avoid eddy current loss. On the other hand, chemical reduction methods give a possibility to obtain the alloys in the form of nanoparticles with a wide range of compositions [6-10]. The chemical reduction method involves the addition of a reductor (usually $NaBH_4$ or KBH_4) into metallic salt aqueous solution, resulting in the precipitation of nanometer scale metallic particles.

In this study, we paid attention to Co-Fe-B that has zero magnetostriction and tried to prepare Co-Fe-B amorphous nanoparticles with soft magnetism by chemical reduction method. However, it is difficult to obtain the nanoparticles with soft magnetism and good dispersibility because the rapidly response set off extraction with aggregated form, generally. To avoid

[1]Corresponding author. E-mail addresses: sugimots@material.tohoku.ac.jp (S. Sugimoto). Tel. & fax: +81-22-795-3785.

aggregation, PVP was used as a dispersing agent. We report on properties of PVP coated amorphous Co-Fe-B nanoparticles and their magnetic properties in a gigahertz range.

2. Experimental

Co-Fe-B nanoparticles were prepared by chemical reduction of Co(II), Fe(II) ions in aqueous solution by $NaBH_4$. PVP (polyvinyl pyrrolidone) was used as a dispersing agent. First, PVP (11.1 g) and $CoCl_2 \cdot 6H_2O$ (2.4 g), $FeSO_4 \cdot 7H_2O$ (0.28 g) was dissolved in deionized water (125-250 mL). The concentration of metal ions, [Co + Fe] is 40-100 mM. Then, aqueous solutions (25-50 mL) of $NaBH_4$ (0.38 g) and NaOH (adjust solution's pH to 13) were added to the above solutions under vigorous stirring with the drop rate of 30 drops/min. The solution changed to black immediately, the mixtures were stirred for another 1 hr to finish the reaction completely. And the entire operation was performed under a nitrogen atmosphere. Then, Precipitated nanoparticles were collected by centrifugation at 5,000 rpm for 5 min and rinsed in distilled water and ethanol. Then, the nanoparticles were dried by vacuum pomp. After drying at room temperature in vacuum box, the gel was ground into a powder for the measurements. For measurements of high frequency magnetic properties, the nanoparticles were mixed with PVP at ratio of about 75 wt.% (particle weight percent) and then pressed into toroidally shaped samples. (ϕ_{out} = 7.00 mm, ϕ_{in} = 3.04 mm, t = 0.5–1.5 mm).

The phases present in samples were characterized by X-ray powder diffraction (XRD) and composition of samples was determined by FE-SEM-WDX. Microstructure was checked with transmission electron microscopy (TEM). Magnetic properties were measured using a vibrating sample magnetometer (VSM). The relative complex permeability (μ_r) and permittivity (ε_r) were calculated by vector network analyzer. Microwave absorption properties were evaluated by the reflection loss (*R.L.*) which was derived from the following formulae (1), (2)

$RL = 20 \log | (Z_{in} - Z_0) / (Z_{in} + Z_0) |$, (1)

$Z_{in} = Z_0 (\mu_r / \varepsilon_r)^{-1/2} \tanh\{ 2 jfd / c (\mu_r \varepsilon_r)^{-1/2} \}$, (2)

Where Z_{in} is the input impedance of absorber and Z_0 is the impedance of air, c is the velocity of light, f is frequency, d is the absorber thickness. When RL was less than -20 dB (matching situation), the sample shows adequate microwave absorption.

3. Result and Discussion

Co-Fe-B nanoparticles were prepared by chemical reduction method with PVP. The reduced particles were found to consist of amorphous particles for all concentrations of metal ions from XRD patterns. Figure 1 shows the relationship between concentration of metal ions and (a) magnetic properties, (b) composition of Co-Fe-B nanoparticles. The saturation magnetization σ_S (emu g^{-1}) were measured with applying a magnetic field of 10 kOe. The saturation magnetization changed slightly regardless of concentration of metal ions. The coercivity was decreasing with approaching the ratio of Co to Fe in the particles to 0.90, that corresponded to composition of zero magnetostriction, and showed minimum value 42 Oe for sample of [Co + Fe] = 0.06 M. Although the B contents increased 38 ~ 43 at.% in proportion to concentration of metal ion, the correlation with the magnetic properties was not seen in Fig. 1.

Next, the concentration of metal ions was fixed to 0.06 M where the ratio of Co to Fe was 0.90, and the relationship between the concentration of reductor, [$NaBH_4$] and particle characteristics was investigated (Figure 2). The saturation magnetization changed slightly same with Fig. 1 except for sample of [$NaBH_4$] = 0.1 M. The reason of this is seemed that the reaction has not completed because of low concentration of reductor solution. The ratios of Co to Fe in the samples of [$NaBH_4$] = 0.2 ~ 0.4 M were 0.90. And in same Co to Fe ratio, the coercivity lowered value with decreasing B content and showed minimum value 31 Oe for sample of [$NaBH_4$] = 0.3 M. The decrease of B content is due to increasing the frequency of particles formation with increasing of [$NaBH_4$]. This value is considerably lower than other amorphous nanoparticle (about 200 Oe [7, 8, 10]). This may be because of isolated amorphous nanoparticle form coated with PVP, from the XRD pattern and TEM images of this sample (shown in Figure 3). Fig. 3 indicates that these nanoparticles were consists of amorphous with particle size of 10 - 50 nm. However, coercivity of this sample is larger than the particles prepared by atomizing and rapid quenching. From larger amount of B (40 at%) than zero magnetostriction composition (about 26 at.%) and thin ring pattern (inset Fig. 3 (c)), it seems that coercivity increased by the precipitation of Co-B nanocrystalline phase. Therefore, Co-Fe-B amorphous nanoparticles with lower coercivity can be expected by decreasing the amount of B content.

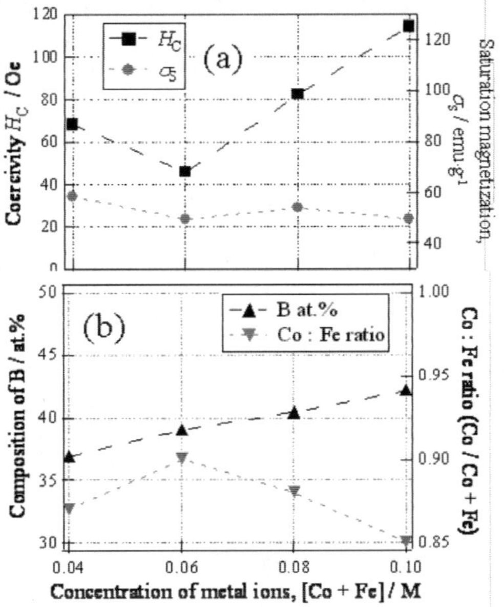

Fig. 1. The relationship between concentration of metal ions, [Co + Fe] and (a) magnetic properties, (b) composition of Co-Fe-B nanoparticles ([$NaBH_4$] = 0.2 M).

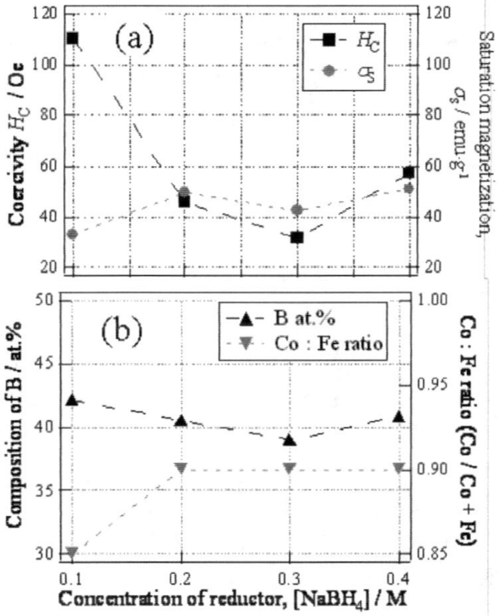

Fig. 2. The relationship between concentration of reductor, [$NaBH_4$] and (a) magnetic properties, (b) composition of Co-Fe-B nanoparticles ([Co + Fe] = 0.06 M).

Last, the high frequency magnetic properties of Co-Fe-B nanoparticles which showed the minimum coercivity were investigated and as an example of application of high frequency device, the microwave absorption properties were evaluated. Figure 4 shows the frequency dependences of (a) the relative permeability, (b) reflection loss, RL of the composites of Co-Fe-B nanoparticles and PVP (75 wt.%, particles). When RL was less than -20 dB (matching situation), the sample shows adequate microwave absorption. The PVP composite showed relative permeability of μ'_r = 3.9 at 0.1 GHz and $\mu''_{r\,max}$ = 1.9 at 0.7 GHz, permittivity of $\varepsilon_{r\,avg.}$ = 50 – 3.1j. The resonance profile in Fig. 4 (a) is broad because of the influence of local demagnetizing fields and local anisotropies. This is because the possible factors that determine resonance profile in particles are demagnetizing fields imposed on individual particles and local defects giving rise to dispersed local anisotropy. The PVP composite RL values less than -20 dB in the 0.2–0.35 GHz frequency range. In particular, a minimum RL value of -32 dB is obtained at 0.3 GHz with thickness, d of 17.0 mm. This suggested that these PVP-coated Co-Fe-B nanoparticles prepared can be used as microwave absorber.

To conclude, Co-Fe-B amorphous nanoparticles were obtained by chemical reduction using PVP as dispersing agent. The nanoparticles exhibited lower coercivity than other amorphous nanoparticles. They are small enough to suppress eddy current loss at high frequencies and the PVP composite had good microwave absorption properties.

Journal of Physics: Conference Series **266** (2011) 012044 doi:10.1088/1742-6596/266/1/012044

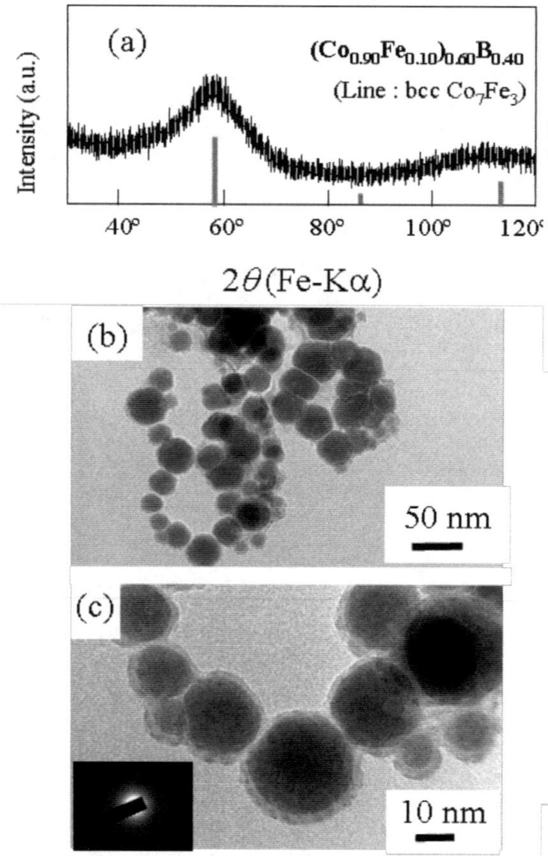

Fig. 3. (a) XRD pattern and (b,c) TEM images of Co-Fe-B nanoparticles (H_C = 31 Oe).

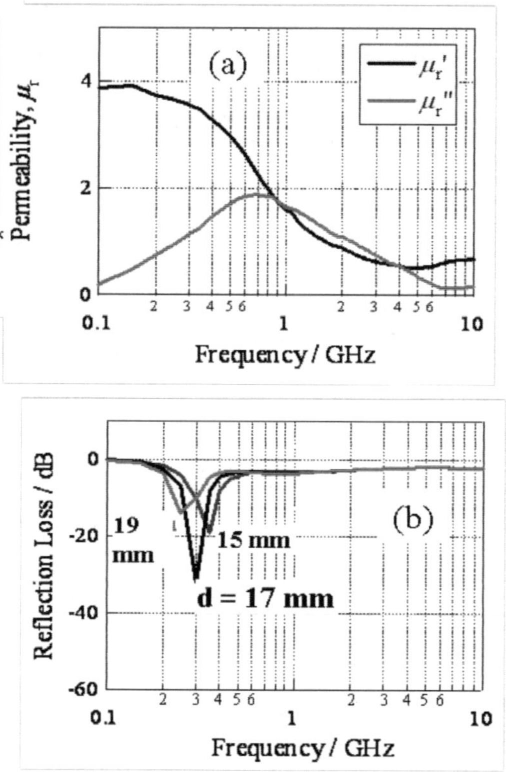

Fig. 4. The frequency dependencies of (a) the relative permeability, (b) reflection loss, RL of PVP composite (75 wt.% particle) of Co-Fe-B nanoparticles (H_C = 31 Oe).

References

[1] Luborsky F E 1983 *Amorphous Metallic Alloys* (London: Butterworths)

[2] I Endo, I Otsuka, R Okuno, A Shintani, M Yoshino, M Yagi 1999 *IEEE Trans. Magn.* **35** 3385

[3] M Yagi, I Endo, I Otsuka, H Yamamoto, R Okuno, H Koshimoto, A Shintani 2000 *J. Magn. Magn. Mater.* **55** 146

[4] H Matsumoto, A Urata, Y Yamada, A Makino 2009 *J. appl. Phys*. **105** a7A317

[5] Y Shimada, Y Endo, M Yamaguchi, S Okamoto, O Kitakami, Y Imano, H Matsumoto, S Yoshida 2009 *IEEE Trans. Magn.* **45** 4298

[6] J van Wonterghem, S Morup, C J W Koch, S W Charles, S Wells 1986 *Nature* **322** 622

[7] L Yiping, G C Hadjipanayis, C M Sorensen, K J Klabunde 1989 *J. Magn. Magn. Mater.* **79** 321

[8] S Wells, S W Charles, S Morup, S Linderoth, J van Wonterghem, J Larsen, M B Madsen 1989 *J. Phys. Condens. Matter* **1** 8199

[9] A Yedra, L Fernandez Barquın, R Garcıa Calderon, Q A Pankhurst, J C Gomez Sal 2001 *J. Non-Cryst. Solids* **287** 20

[10] L Fernandez Barquın, A Yedra, S N Kaul, M L Fdez-Gubieda, J F W Mosselmans, Q A Pankhurst 2007 *Journal of Non-Crystalline Solids* **353** 733

2nd Int. Symp. on Advanced Magnetic Materials and Applications (ISAMMA 2010) IOP Publishing
Journal of Physics: Conference Series **266** (2011) 012045 doi:10.1088/1742-6596/266/1/012045

First-principles calculation of crystal field parameters of Dy ions substituted for Nd in Nd-Fe-B magnets

S Tanaka[1], **H Moriya**[1,2], **H Tsuchiura**[1], **A Sakuma**[1], **M Diviš**[3] **and P Novák**[4]

[1] Department of Applied Physics, Tohoku University, Sendai 980-8579, Japan
[2] Mechanical Engineering Research Laboratory, Hitachi Ltd., Ibaraki 312-0034, Japan
[3] Department of Condensed Matter, Charles University, FMF, Prague, Czech Republic
[4] Institute of Physics of ASCR, Cukrovarnická 10, 162 53 Prague 6, Czech Republic

E-mail: tanaka@olive.apph.tohoku.ac.jp

Abstract. We study the electronic structures of crystalline $Nd_2Fe_{14}B$, $Dy_2Fe_{14}B$ and Dy-doped Nd-Fe-B, and estimate the crystal field parameter $A_2^0\langle r^2\rangle$ of the rare earth ions of these systems based on the first principles calculations. We find that the crystal field of the Dy ions is appreciably insensitive to its crystallographic location than that of Nd ions.

1. Introduction

Nd-Fe-B sinterd magnets have had few high-temperature applications because of their relatively low Curie temperature despite their excellent room temperature properties. Addition of Dy to the Nd-Fe-B magnets is known to increase the coercivity of the $Nd_2Fe_{14}B$ hard phase. In particular, the Dy substitution has a significant beneficial effect on the thermal stability of the magnetic properties. It has been believed, from a phenomenological viewpoint, that this is due to the higher value of the anisotropy field H_a of $Dy_2Fe_{14}B$ than that of $Nd_2Fe_{14}B$. However, the microscopic study for this mechanism has not been reported. Moreover, nowadays, technology that can increase the coercivity of Dy-free Nd-Fe-B sintered magnets is strongly desired. Thus it is important to reveal the role of the substituted Dy ions, and to give a guiding principle for improving the coercivity of Dy-free Nd-Fe-B sintered magnets.

Thus in this paper, as a first step, we theoretically study the electronic structures, the magnetic moments and the crystal field (CF) parameters of Nd and Dy ions in Dy-doped crystalline $Nd_2Fe_{14}B$, by using a first principles method based on density functional theory. Needless to say, in general, there would exist much complicated factors in the mechanisms which dominates the coercivity of rare-earth (RE) permanent magnets. However, we believe that it is useful to estimate the local magnetic anisotropy of the substituted Dy ions.

Figure 1 shows the crystal structure of $Nd_2Fe_{14}B$ (space group $P4_2/mnm$). There are two crystallographically inequivalent Nd sites, denoted as f and g, in this structure as shown in Fig. 1. Here, for simplicity, we focus on the following two extreme cases: (i) Dy is substituted for *all* of the Nd at the f-sites of the system, which is denoted here by $DyNdFe_{14}B$, and (ii) for all of the Nd at g-sites, denoted by $NdDyFe_{14}B$.

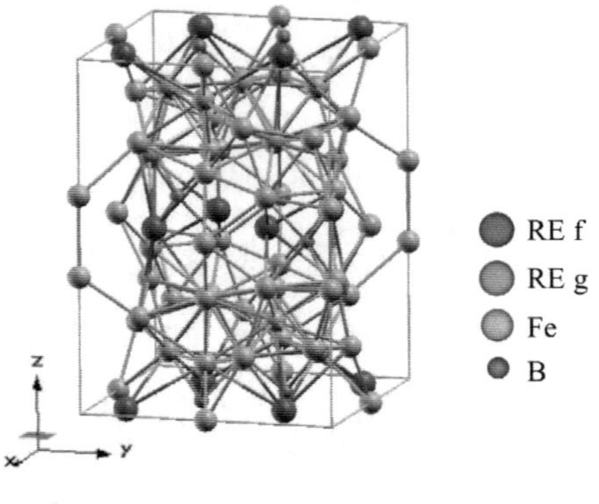

Figure 1. The unit cell of $(RE)_2Fe_{14}B$. Nd and Dy ions are located at RE f- or g-sites.

2. Computational details

The magnetic anisotropy energy of RE ions is described by the Hamiltonian consisting of spin-orbit interaction, magnetic exchange, and crystal field interactions. The crystal field interaction energy depends on the orientation of $4f$ magnetic moment, because a rotation of magnetic moment leads to a rotation of the $4f$ aspherical charge cloud due to strong spin-orbit interaction.Thus the $4f$ anisotropic magnetic energy depends on the crystal field energy described by the standard model Hamiltonian [1]

$$\mathcal{H} = \sum_{L,M} \Theta_L A_L^M \langle r^L \rangle \hat{O}_L^M, \tag{1}$$

where $\hat{O}_L^M$ are the Stevens *operator equivalents*, and Θ_L are the reduced matrix elements [2].

Following Diviš *et al.* [3, 4], the CF parameters $A_L^M\langle r^L\rangle$ originating from the aspherical part of the total single particle DFT potential in the crystal can be obtained from

$$A_L^M\langle r^L\rangle = a_{LM}\left(\int_0^{R_{MT}} drr^2|R_{4\mathrm{f}}(r)|^2 V_L^M + \int_{R_{MT}}^{\infty} drr^2|R_{4\mathrm{f}}(r)|^2 W_L^M\right), \tag{2}$$

with

$$\langle r^l\rangle = \int_0^{R_{MT}} drr^L|R_{4\mathrm{f}}(r)|^2, \tag{3}$$

where $V_L^M(r)$ and $W_L^M(r)$ are the components of the total (Coulomb and exchange-correlation) potential inside the atomic sphere of the rare-earth ions with the muffin-tin sphere radius R_{MT}, and its continuation in the interstitial region, respectively. The $R_{4f}(r)$ is the radial part of the $4f$-orbital wave function. The conversion factors a_{LM} show the relation between the symmetrized spherical harmonics and the real tesseral harmonics which transfers in the same way as the tensor operators O_L^M. In particular, $a_{20} = \frac{1}{2}\sqrt{\frac{5}{2\pi}}$.

To obtain $V_L^M(r)$ and $W_L^M(r)$, we use the full-potential linearized augmented plane wave plus local orbitals method (APW+lo) implemented in the WIEN2k code [5]. The Kohn-Sham equations are solved within the generalized-gradient approximation (GGA). At this stage, however, we neglect the term $W_L^M(r)$ because the radial wave function $R_{4f}(r)$ quickly decays around $r \geq R_{MT}$, as described below. For the Brillouin zone integration, a modified tetrahedron method with 12 k points in the irreducible wedge was used in constructing the charge density in each self-consistency step. The number of the basis functions is 7183, including 848 local

Table 1. The magnetic moments in μ_B for the crystalline $Nd_2Fe_{14}B$, $Dy_2Fe_{14}B$, and Dy-doped systems. The results obtained by using the LDA+U[12] and FP-LMTO methods[13] are also shown for comparison.

	$Nd_2Fe_{14}B$ (this work)	(LDA+U)	(FP-LMTO)	$Dy_2Fe_{14}B$	$DyNdFe_{14}B$	$NdDyFe_{14}B$
Fe(16k$_1$)	2.29	2.27	2.22	2.27	2.30	2.29
Fe(16k$_2$)	2.38	2.38	2.28	2.36	2.38	2.38
Fe(8j$_1$)	2.31	2.34	2.67	2.30	2.32	2.33
Fe(8j$_2$)	2.74	2.73	2.16	2.73	2.75	2.75
Fe(4e)	2.16	2.16	1.96	2.15	2.17	2.18
Fe(4c)	2.47	2.47	2.43	2.43	2.48	2.48
B(4g)	-0.17	-0.18	-0.13	-0.16	-0.17	-0.17
interstitial	-5.42	-4.92	-	-5.00	-5.42	-5.35

orbitals. Throughout the present calculation, we take $R_{MT} = 2.8, 2.02, 1.8$ a.u. for RE, Fe, B ions, respectively.

To simulate localized $4f$ states in the $(RE)_2Fe_{14}B$ systems, we switch off the hybridization between $4f$ and valence states, and treat the RE $4f$ states in the spherical part of the potential as atomic-like core states (the so-called open-core treatment). Here, $R_{4f}(r)$ is obtained by performing separate atomic calculations of the electronic structure of an isolated RE atom. In these calculations the correction for the self-interaction (SIC) was included, which leads to better approximation for the single electron densities. This approach [6] was found to provide a $4f$ charge density being very close to that obtained from more rigorous SIC-DFT band calculations [1]. We obtain $\langle r^2 \rangle = 1.02a_0$ for the Nd $4f$-density and $\langle r^2 \rangle = 0.727a_0$ for Dy, where a_0 is the Bohr radius.

We should note here that the lattice constants of $Nd_2Fe_{14}B$ and $Dy_2Fe_{14}B$ are set to the experimental values of $a = b = 8.80$Å and $c = 12.19$Å [7], and $a = b = 8.76$Å and $c = 11.99$Å [8], respectively. We also use these values for the $DyNdFe_{14}B$ and the $NdDyFe_{14}B$ systems.

3. Results and discussion

We show the magnetic moments obtained inside each of the atomic spheres in Table 1, in comparison with the previous works for $Nd_2Fe_{14}B$ using the LDA+U method [12] and by the full-potential linear-muffin-tin-orbital (FP-LMTO) method [13]. We can see that, for $Nd_2Fe_{14}B$, the present results obtained by the open-core treatment are in good agreement with both the LDA+U and the FP-LMTO results, especially with the former. We also find that the systems including Dy, that is, $Dy_2Fe_{14}B$, $DyNdFe_{14}B$ and $NdDyFe_{14}B$ have quite similar but slightly smaller magnetic moments to those of $Nd_2Fe_{14}B$. As noted above, these results for the Dy-doped systems are obtained using the lattice constants for the crystalline $Nd_2Fe_{14}B$. We find that the magnetic moment of each Fe ion in $DyNdFe_{14}B$ and $NdDyFe_{14}B$ shrinks down about 1 percent if we use the lattice constants for the crystalline $Dy_2Fe_{14}B$, which are smaller than those for $Nd_2Fe_{14}B$. This can be understood from the well-known property of ferromagnets, that is, ferromagnetism favors narrower band structure. We should also note here that, in the present calculations, all magnetic moments are aligned parallel to the crystallographic c-axis, while the magnetic moments undergo the well-known spin-reorientation into non-collinear structure at very low temperatures in real $Nd_2Fe_{14}B$ compounds. Thus the present results should be interpreted as extrapolated ones from the situation above the reorientation temperature to $T = 0$.

Next, we show the results for the CF parameter $A_2^0\langle r^2 \rangle$ of Nd and Dy ions in these systems. Here we show the results in Table 2 in the same manner as in Table 1. It should be noted

Table 2. The crystal field parameters $A_2^0\langle r^2\rangle$ obtained by the present calculations, LDA+U[12], and FP-LMTO methods[13].

rare-earth site	$Nd_2Fe_{14}B$ (this work)	$Nd_2Fe_{14}B$ (LDA+U)	$Nd_2Fe_{14}B$ (FP-LMTO)	$Dy_2Fe_{14}B$	$DyNdFe_{14}B$	$NdDyFe_{14}B$
f	552	427	540	537	556 (Dy)	507 (Nd)
g	790	464	323	602	736 (Nd)	604 (Dy)

here again that the Dy-doped systems are obtained using the lattice constants for $Nd_2Fe_{14}B$. For $Nd_2Fe_{14}B$, we find that our result is in good agreement with the previous studies for the f-site, but not for the g-site. The possible reasons of such significant discrepancy are mainly the difference of the adopted theoretical schemes, such as the open-core method used here, the LDA+U method and the LMTO method. In particular, it is known that the self-interaction of $4f$-electrons can be only partially corrected within the LDA+U method.

Here let us look at the results for the systems containing Dy. We find that, in $Dy_2Fe_{14}B$, the difference between the value of $A_2^0\langle r^2\rangle$ for the f- and g-site is rather smaller than in $Nd_2Fe_{14}B$. This implies that the CF of Dy is less affected by crystallographic location than that of Nd. The variation of $A_2^0\langle r^2\rangle$ from the pure systems, such as $Nd_2Fe_{14}B$ and $Dy_2Fe_{14}B$, is 6.8 percent less for Nd(g) in $DyNdFe_{14}B$ and 8.2 percent less for Nd(f) in $NdDyFe_{14}B$, while 0.3 percent more for Dy(g) in $NdDyFe_{14}B$ and 3.5 percent more for Dy(f) in $DyNdFe_{14}B$. Moreover, this tendency is more marked when we use the lattice constants for the crystalline $Dy_2Fe_{14}B$. Indeed, we obtain the 16.8 percent less for Nd(f) and 5.0 percent more for Dy(g) in $NdDyFe_{14}B$ with the smaller lattice constants. Thus, indeed, the crystal field of Dy is appreciably more insensitive to its surroundings than that of Nd. We speculate that this is due to the more localized $4f$-electronic density $|R_{4f}(r)|^2$ of Dy than that of Nd. Recently, three of the present authors have pointed out that Nd ions located around a (001) surface of $Nd_2Fe_{14}B$ exhibits $A_2^0\langle r^2\rangle < 0$, which implies local, in-plane magnetic anisotropy [12]. Also, we have confirmed quite recently that such Nd ions with negative $A_2^0\langle r^2\rangle$ can be nucleation sites of magnetization reversal of the whole system by using micromagnetic simulation technique [14]. Thus it is interesting to study the CF parameter of Dy substituted for the surface Nd, and to reveal whether it can be negative or not.

4. Summary

In summary, we have estimated the crystal field parameter $A_2^0\langle r^2\rangle$ of Nd and Dy ions in crystalline $Nd_2Fe_{14}B$, $Dy_2Fe_{14}B$, and Dy-doped Nd-Fe-B systems. We found that the crystal field of the Dy ions is appreciably more insensitive to their surrounding than that of Nd ions.

Acknowledgments

A part of numerical computations in this work was carried out at the Yukawa Institute Computer Facility, and also at Cyberscience Center, Tohoku University. Some parts of this work are the results of the research supported by Rare Metal Substitute Materials Development of technology for reducing dysprosium usage in a rare-earth magnet, from METI, Japan. The work of M. D. is a part of the research program MSM 0021620834 financed by the Ministry of Education of the Czech Republic and was also supported by the Grant Agency of Czech Republic (Project 202/09/1027). Also this work is partly supported by Grant-in-Aid for Scientific Research (C) and also by Next Generation Supercomputing Project, Nanoscience Program, from the Ministry of Education, Culture, Sports, Science and Technology of Japan.

Journal of Physics: Conference Series **266** (2011) 012045 doi:10.1088/1742-6596/266/1/012045

References

[1] M. Richter, J. Phys. D: Appl. Phys. **31**, 1017 (1998).
[2] M. T. Hutchings, Solid State Phys. **16**, 227 (1964).
[3] M. Diviš, K. Schwarz, P. Blaha, G. Hilscher, H. Michor and S. Khmelevskyi, Phys. Rev. B **62**, 6774 (2000).
[4] M. Diviš, J. Rusz, H. Michor, G. Hilscher, P. Blaha and K. Schwarz, J. Alloys. Compd. **403**, 29 (2005).
[5] P. Blaha, K. Schwarz, G. Madsen, D. Kvasnicka, and J. Luitz, WIEN2k, an Augmented Plane Wave + Local Orbitals Program for Calculating Crystal Properties, Karlheinz Schwarz, TU Wien, Austria, 2001, ISBN 3-9501031-1-2.
[6] P. Novák, phys. stat. sol. (b) **198**, 729 (1996).
[7] J. F. Herbst, J. J. Croat, F. E. Pinkerton and W. B. Yelon, Phys. Rev. B **29**, 4176 (1984).
[8] S. Hirosawa, Y. Matsuura, H. Yamamoto, S. Fujita and M. Sagawa, J. Appl. Phys. bf 59, 873 (1986).
[9] M. Yamada, H. Kato, H. Yamamoto and Y. Nakagawa, Phys. Rev. B **38**, 620 (1988).
[10] M. Sagawa, S. Fujimura, H. Yamamoto, Y. Matsuura and K. Hiraga, IEEE Trans. Magn. MAG-20, 1584 (1984).
[11] T. Fukagawa and S. Hirosawa, Scripta Materialia **59**, 183 (2008).
[12] H. Moriya, H. Tsuchiura and A. Sakuma, J. Appl. Phys. **105**, 07A740 (2009).
[13] K. Hummler and M. Fähnle, Phys. Rev. B **53**, 3290 (1996).
[14] Y. Toga, H. Moriya, H. Tsuchiura and A. Sakuma, unpublished.

Journal of Physics: Conference Series **266** (2011) 012046 doi:10.1088/1742-6596/266/1/012046

First principles study on interfacial electronic structures in exchange-spring magnets

Yuta Toga[1], **Hiroshi Moriya**[2,1], **Hiroki Tsuchiura**[1], **Akimasa Sakuma**[1]

[1]Department of Applied Physics, Tohoku University, 6-6-05 Aoba, Aoba-ku, Sendai 980-8579, Japan
[2]Mechanical Engineering Research Laboratory, Hitachi Ltd., 832-2 Horiguchi, Hitachinaka, Ibaraki 312-0034, Japan

E-mail: toga@solid.apph.tohoku.ac.jp

Abstract. We study the magnetic properties of $Nd_2Fe_{14}B/\alpha$-Fe multilayer systems, especially the exchange coupling between the two layers and the crystal field parameters $A_2^0\langle r^2\rangle$ for the Nd ions located around the interfaces, based on first principles calculations. We find that the exchange coupling depends on the Miller indices of the crystallographic planes at the interface, and can be negative which favors anti-parallel alignment of the magnetizations of the hard- and soft-magnet layers.

Exchange-spring magnets are nanocomposite materials that are composed of hard- and soft-magnetic phases that interact by magnetic exchange coupling[1, 2, 3, 4]. These are promising systems for advanced permanent magnetic applications because they have been expected, based on micromagnetic simulations, to have a large energy product compared to conventional single-phase materials.

However, in practice, it is well known that such high values of the energy product have been hard to achieve in real materials. This would be due to the difficulty in controlling the nanostructure such as grain sizes of hard and soft crystallites, and crystallographic alignment of the hard grains. Moreover, there might exist much complicated factors in the mechanisms which dominates the coercivity and energy product of exchange-spring magnets, which have inevitably been neglected in micromagnetic simulations and should be described microscopically. Thus it is important to have detailed experimental and theoretical information on the electronic states around well-fabricated interfaces in exchange-spring magnets.

Exchange-spring multilayer thin films will have such well-fabricated interfacial structures, and thus will serve as a desirable model system for studying fundamental physical properties of exchange-spring magnets. Actually, Zhang *et al.* have recently succeeded to fabricate the $SmCo_5$/Fe multilayer film that approach the ideal exchange-spring magnets with aligned hard- and soft-layers [5]. The largest $(BH)_{max}$ of the film has been reported as 32MGOe, which is larger than the theoretical limit for a $SmCo_5$ single phase bulk magnet, 28.8MGOe. This was the first example that exchange-spring magnets can exhibit larger $(BH)_{max}$ than the theoretical limit. We can naturally expect that, experimental realization of exchange-spring thin film magnets based on $Nd_2Fe_{14}B$ is only a matter in time. Therefore, it is urgent to have detailed information on the electronic states around the interfaces in such multilayers based on $Nd_2Fe_{14}B$.

Motivated by these backgrounds, we theoretically study the magnetic properties of $Nd_2Fe_{14}B/\alpha$-Fe multilayer systems, especially the crystal field parameters (CFPs) of Nd ions and the exchange coupling between hard- and soft-layers. Very recently, three of the present authors have shown that the CFPs $A_2^0\langle r^2\rangle$ of Nd ions located around a (001) *surface* of crystalline $Nd_2Fe_{14}B$ are negative, based on the

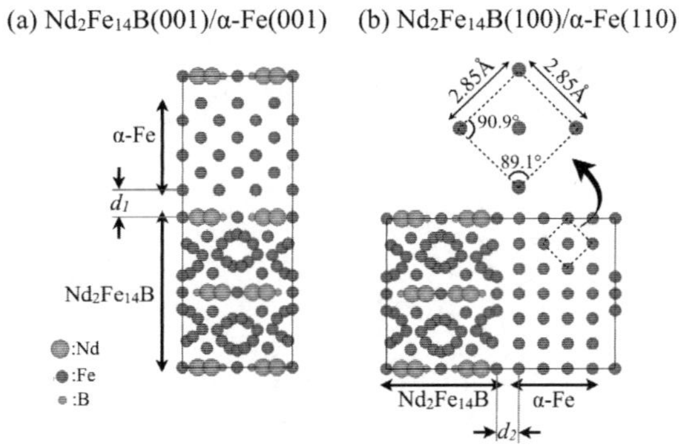

Figure 1. Models of composite system of (a) $Nd_2Fe_{14}B(100)/\alpha$-Fe(110) and (b) $Nd_2Fe_{14}B(001)/\alpha$-Fe(001) . The large spheres indicate rare-earth ions. This figure is plotted by using VESTA. [7]

first principles calculations [6]. This implies that those Nd ions would be nucleation center of reversed magnetic domains, and that the coercivity of $Nd_2Fe_{14}B$-based exchange-spring magnets may also be affected by such Nd ions around interfaces. Thus in this paper, as a first step, we focus on the magnetic properties of the (a) $Nd_2Fe_{14}B(001)/\alpha$-Fe(001) and (b) $Nd_2Fe_{14}B(100)/\alpha$-Fe(110) multilayered systems, shown in Fig. 1. We note that the crystallographic planes at the interfaces of α-Fe in the cases (a) and (b) are chosen so as to realize commensurate interfacial structures with $Nd_2Fe_{14}B$ (001) and (100) planes. Also we should note here that the lattice constants of $Nd_2Fe_{14}B$ are set to experimental values of $a = b = 8.8$Å, $c = 12.19$Å, while those for α-Fe are slightly modified for commensurability of the interfacial structures: $a = b = 2.93$Å, $c = 2.83$Å for the $Nd_2Fe_{14}B(001)/\alpha$-Fe(001) system, and $a = b = 2.85$Å, $c = 2.93$Å, $\alpha = \beta = 90°$, $\gamma = 89.1°$ for the $Nd_2Fe_{14}B(100)/\alpha$-Fe(110) system. We confirmed that α-Fe with lattice distortion assumed here also exhibit ferromagnetism even in the bulk system. Furthermore, the lattice distortion does not affect significantly the magnetic properties of α-Fe such as magnetic moments of each Fe.

In this work, the electronic structures of the multilayer models are calculated with first-principles density functional calculations using a plane-wave basis set. We use the Vienna Ab-initio Simulation Package (VASP) [8, 9]. The ionic potentials are described by projector-augmented-wave (PAW) method and the exchange-correlation energy of the electrons is described within a generalized gradient approximation (GGA). We use the exchange-correlation functional determined by Ceperly and Alder and parameterized by Perdew and Zunger. The energy cutoff of 318.6 eV is used to generate a plane wave basis set. For the k-point integration, we use a $3 \times 3 \times 3$ mesh for the unit cell of a bulk $Nd_2Fe_{14}B$ model and equivalent densities of mesh points for the multilayer models. The total energy convergence is less than 0.1meV per unit cell.

The CF parameter $A_2^0\langle r^2\rangle$ of a Nd ion in the systems can be given as [10]

$$A_2^0\langle r^2\rangle \;=\; \frac{1}{4}\int_{|\boldsymbol{R}|<R_{MT}} d\boldsymbol{R}\left(3\cos^2\theta - 1\right)\rho(\boldsymbol{R})\int_{r<R_{MT}} dr\left(\frac{r_<^2}{r_>^3}\right) r^2|R_{4f}(r)|^2,$$

where $r_< \equiv \min(r, |\boldsymbol{R}|)$, $r_> \equiv \max(r, |\boldsymbol{R}|)$, and $\rho(\boldsymbol{R})$ is the valence charge density within the atomic sphere ($r < r_{AS}$) of the Nd ions of the system. The $R_{4f}(r)$ describes the radial shape of the localized 4f charge density of the Nd ion. In the present study we used a self-interaction corrected (SIC) wave function from atomic calculations. This approach [11] was found to provide a 4f charge density being very close to that obtained from more rigorous SIC-DFT band-calculations [12].

Table 1. The magnetic moments in μ_B for the crystalline $Nd_2Fe_{14}B$ obtained by the present calculations, APW+lo[6], and FP-LMTO methods[13].

	this work	APW+lo	FP-LMTO
Fe(16k$_1$)	2.18	2.27	2.22
Fe(16k$_2$)	2.28	2.38	2.28
Fe(8j$_1$)	2.15	2.34	2.67
Fe(8j$_2$)	2.65	2.73	2.16
Fe(4e)	2.11	2.16	1.96
Fe(4c)	2.41	2.47	2.43
B(4g)	-0.15	-0.18	-0.13
interstitial	-	-4.92	-

Table 2. The CF parameters $A_2^0\langle r^2\rangle$[K] of Nd ions obtained for the crystalline $Nd_2Fe_{14}B$.

	this work	APW+lo	FP-LMTO
Nd(4f)	547	427	540
Nd(4g)	320	464	323

First, let us briefly look at the results for the crystalline $Nd_2Fe_{14}B$. We show the magnetic moments obtained inside each of the atomic spheres in Table 1, and also show the crystal field parameters $A_2^0\langle r^2\rangle$ for Nd ions in Table 2, in comparison with the previous works [6, 13] using the full potential linearized augmented plane wave plus local orbitals method (APW+lo) using WIEN2k [14] and by the full-potential linear-muffin-tin-orbital (FP-LMTO) method. The present results are in good agreement with both the APW+lo and the FP-LMTO results, especially with the latter. In the present calculations, all magnetic moments are aligned parallel to the crystallographic c-axis, while the magnetic moments undergo the well-known spin-reorientation into non-collinear structure at very low temperatures in real $Nd_2Fe_{14}B$ compounds. Thus we also note here that the present results should be interpreted as extrapolated ones from the situation above the reorientation temperature to $T = 0$.

Next, we show the results for the $Nd_2Fe_{14}B$ (001)/α-Fe (001) interface model, shown in Fig. 1 (a). Figure 2 shows the total electronic energy of the system for parallel and anti-parallel magnetization alignments between $Nd_2Fe_{14}B$ and α-Fe layers, as functions of the interlayer distance d_1 shown in Fig. 1 (a). We can easily see that, in this case, the parallel alignment of the magnetizations is always favored in $2.0 \leq d_1 \leq 2.6$[Å], and that d_1 is optimized at $d_1 \simeq 2.2$Å for the parallel and $d_1 \simeq 2.3$Å for the anti-parallel cases. The exchange coupling per unit area between the magnetizations of the two layers is estimated as $J \simeq 0.06$ J/m^2 at $d_1 = 2.2$Å. The CFPs obtained for the Nd ions are shown in table 3(a). Since the $P4_2/mnm$ symmetry of the crystalline $Nd_2Fe_{14}B$ is broken in this multilayer model, we have four Nd ions crystallographically inequivalent. We label them as Nd(f)-a and Nd(g)-a, located at the (001) interface, and Nd(f)-b and Nd(f)-b, inside. In contrast to the $Nd_2Fe_{14}B$ (001) *surface* model[6] mentioned above, the CFPs $A_2^0\langle r^2\rangle$ are all positive as shown in Table 3(b), that is, they are restored as in the bulk system. In the present multilayer system, there exist Fe ions in the α-Fe layer lie directly on the interfacial Nd ions. The valence electrons of the Nd ions couple with the $3d$ electrons of those Fe ions and thus slightly extend outward from the interfaces, then they form the crystal field which gives the uniaxial magnetic anisotropy perpendicular to the (001) interface of the $Nd_2Fe_{14}B$ layer. We speculate that this is the reason for the restored CFPs of the Nd ions.

Here we see how these results would be changed in the $Nd_2Fe_{14}B$ (100)/α-Fe (110) model, shown in Fig. 1 (b). Figure 3 shows the total electronic energies for parallel and anti-parallel magnetization

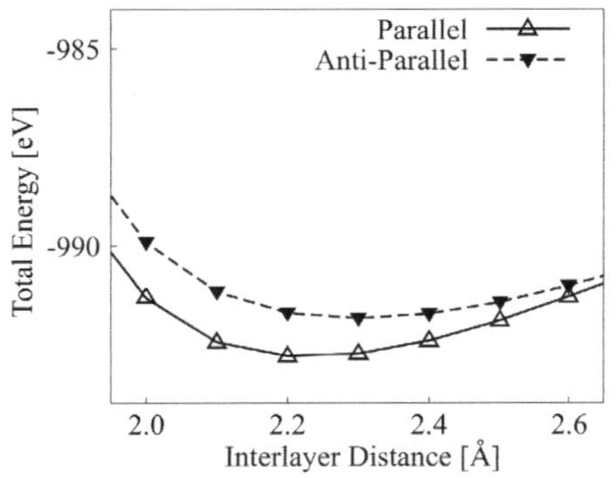

Figure 2. The total electronic energy of the system for parallel (△) and anti-parallel (▼) magnetization alignments between $Nd_2Fe_{14}B$ and α-Fe layers.

Table 3. The CF parameters $A_2^0\langle r^2\rangle$[K] of Nd ions obtained for the (a) $Nd_2Fe_{14}B$ (001)/α-Fe (001) and (b) $Nd_2Fe_{14}B$ (100)/α-Fe (110) models.

(a) $Nd_2Fe_{14}B$ (001)/α-Fe (001)

	$A_2^0\langle r^2\rangle$ [K]
Nd(4f)-a	491 - 976
Nd(4g)-a	309 - 1403
Nd(4f)-b	538
Nd(4g)-b	253

(b) $Nd_2Fe_{14}B$ (100)/α-Fe (110)

	$A_2^0\langle r^2\rangle$ [K]
Nd(4f)-c	639 - 695
Nd(4g)-c	31 - 98
Nd(4f)-d	423 - 464
Nd(4g)-d	221 - 224

alignments as functions of the interlayer distance d_2, in the same manner as in Fig. 2. At this time, the *anti-parallel* alignment is stabilized around the minimum of the total energy, $1.6 \leq d_2 \leq 2.0$[Å]. The exchange coupling per unit area between the magnetizations of the two layers is estimated as $J \simeq 0.013$ J/m^2 at $d_2 = 1.8$Å. The CFPs obtained for the Nd ions are shown in table 3(b). Here we label the Nd ions in this model as Nd(f)-c and Nd(g)-c, located around the (100) interface, and Nd(f)-d and Nd(f)-d, inside. In this case, the CFPs are all positive. We would note here that, we have confirmed very recently [15] that they are also all positive even if the Nd ions are located around the (100) *surfaces* of the crystalline $Nd_2Fe_{14}B$ using the WIEN2k code.

An existence of such anti-parallel coupling between the hard- and soft-layers, which might have been realized partially in random dispersed geometry such as melt-spun ribbons of exchange-spring magnets, would have given rise to a negative contribution to the magnetic properties, especially to the maximum energy product. We believe that the present results, that is, the exchange coupling between the hard- and soft-phases depends on the Miller indices of the interfacial planes, can be checked by experimental study on well-fabricated $Nd_2Fe_{14}B$ (100)/Fe multilayer systems

In our calculation, so far, the structural relaxation around the interfaces has not been taken into account. We have actually tried such relaxation calculations for our models and have seen that they support the present results. Needless to say, there would exist much complicated structures around the grain boundaries of real samples of exchange-spring magnets, or even around the interfaces of exchange-spring multilayer thin films, while the number of atoms nowadays available in first principles calculations are far from sufficient to reproduce such realistic structures. In fact, even if we fully optimize the local structures of the present models, they would still be simple mimics of the real systems. However, the first aim of this paper is to point out that the anti-parallel coupling between hard- and soft-layers in exchange-spring magnets can be realized, and thus we believe that the modes should not necessarily be

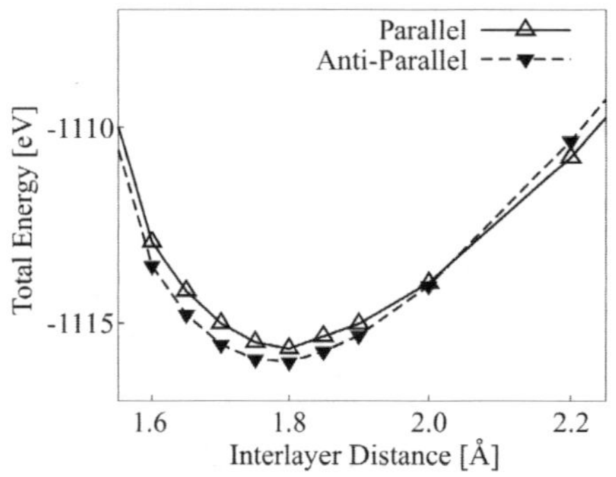

Figure 3. The total electronic energy of the system for parallel (△) and anti-parallel (▼) magnetization alignments between $Nd_2Fe_{14}B$ and α-Fe layers.

well-realistic for this aim.

In summary, we have theoretically investigated the magnetic properties of the $Nd_2Fe_{14}B/\alpha$-Fe multilayer models which can be fundamental systems to reveal the important factors in the mechanisms which dominates the coercivity and energy product of exchange-spring magnets. On the basis of first principles calculations, we predict that the exchange coupling between the two layers of the models depends on the Miller indices of the interfacial planes, and can be negative.

Acknowledgments
One of the authors (H. T.) is grateful for helpful discussion with P. Novák and M. Diviš. A part of numerical computations in this work was carried out at the Yukawa Institute Computer Facility, and also at Cyberscience Center, Tohoku University. Some parts of this work are the results of the research supported by Rare Metal Substitute Materials Development of technology for reducing dysprosium usage in a rare-earth magnet, from METI, Japan. Also this work is partly supported by Grant-in-Aid for Scientific Research (C) and also by Next Generation Supercomputing Project, Nanoscience Program, from the Ministry of Education, Culture, Sports, Science and Technology of Japan.

References

[1] Coehoorn R, de Mooij D B and de Waard C 1989 *J.Mag.Mag.Mater.* **80** 101
[2] Kneller E F and Hawig R 1991 *IEEE Trans. Magn.* **27** 3588
[3] Skomski R and Coey J M D 1993 *Phys. Rev. B* **48** 812
[4] Sabiryanov R F and Jaswal S S 1998 *Phys. Rev. B* **58** 12071
[5] Zhang J, Takahashi Y K, Gopalan R and Hono K 2005 *Appl. Phys. Lett.* **86** 122509
[6] Moriya H, Tsuchiura H and Sakuma A 2009 *J. Appl. Phys.* **105** 07A740
[7] Momma K, Izumi F 2008 *J. Appl. Cryst.* **41** 653
[8] Kresse G and Furthmuller J 1996 *Phys. Rev. B* **54** 11169; 1996 *Comput. Mater. Sci.* 6 15
[9] Kresse G and Hafner J 1993 *Phys. Rev. B* **47** R558
[10] Fähnle M, Hummler K, Liebs M and Beuerle T 1993 *Appl. Phys. A* **57** 67
[11] Novák P 1996 *Phys. Stat. Sol.* **198** 729
[12] Richter M 1998 *J. Phys. D: Appl. Phys.* **31** 1017
[13] Hummler K and Fähnle M 1996 *Phys. Rev. B* **53** 3290
[14] Blaha P, Schwarz K, Madsen G, Kvasnicka D, and Luitz J 2001 *WIEN2k, an Augmented Plane Wave + Local Orbitals Program for Calculating Crystal Properties* Karlheinz Schwarz, TU Wien, Austria, ISBN 3-9501031-1-2
[15] Moriya H, Toga Y, Tsuchiura H and Sakuma A, unpublished.

Journal of Physics: Conference Series **266** (2011) 012047 doi:10.1088/1742-6596/266/1/012047

Significant changes in crystallization kinetics of $Nd_2Fe_{14}B/\alpha$-Fe nanocomposites induced by Nb addition

C Wang[1], W C Chang[2]
[1] College of Materials Science and Engineering, Fuzhou University, Fuzhou 350108, China
[2] Department of Physics, National Chung Cheng University, Chia-Yi 621, Taiwan

E-mail: msewang@fzu.edu.cn

Abstract. The effects of Nb addition on the crystallization kinetics, microstructure and magnetic properties of melt-spun $Nd_2Fe_{14}B/\alpha$-Fe nanocomposite magnets have been studied. The analysis of crystallization kinetics for the alloys indicates that the growth of crystallites in the Nb-doped alloy needs more activation energy than that in the Nb-free alloy. This crystallization behaviour contributes to the formation of a finer and more homogeneous microstructure in the Nb-doped alloy than in the Nb-free alloy. And the Nb addition makes the grains more equiaxed shape. The intrinsic coercivity and the maximum energy product increase from 600 kA/m, 105 kJ/m^3 for $Nd_{10}Fe_{84}B_6$ alloy to 750 kA/m, 120 kJ/m^3 for $Nd_{10}Fe_{83}Nb_1B_6$ alloy, respectively.

1. Introduction

Nanocomposite magnets have attracted considerable attention due to their high remanence ratio, great energy product and low rare-earth content. The high performance is attributed to the intergrain exchange interactions between the hard-magnetic grains and the soft-magnetic grains. For the effective exchange coupling, a homogeneous and fine-grained microstructure is required.

Microalloying is an effective way to improve the magnetic properties by optimizing the microstructure and improving the intrinsic magnetic properties of magnetic phases [1-3]. The addition of Nb was reported to be advantageous in refining the grain size and improving the energy product of nanocomposite magnets [4-6]. On the other hand, Jin et al. [7] reported that some large grain clusters were observed in Nb-doped alloy, thus degrading the squareness of hysteresis loops. Ping et al. [8] reported that the Nb-doped ribbons prepared by annealing amorphous precursors exhibit coarser microstructures than the Nb-free ribbons prepared by melt spinning at a wheel speed of 20 m/s. Therefore, the effects of Nb in nanocomposite magnets need to be further studied.

One of the practical ways of preparing nanocomposite magnets is through the crystallization of amorphous phase into a mixture of hard and soft phases. Thus, the crystallization behaviour of the alloy plays a key role in the microstructure development. Heretofore, the influences of Nb addition on the crystallization behaviour of the $Nd_2Fe_{14}B/\alpha$-Fe magnets have not been understood very well. The investigation on the crystallization kinetics [9,10] can provide an insight into the nucleation and growth mechanisms for the crystalline phases. However, work dealing with the crystallization kinetics of $Nd_2Fe_{14}B/\alpha$-Fe magnets, especially the Nb-doped alloys, is limited. Our previous study [11] showed that the crystallization behaviour of the amorphous Nd-Fe-B alloy changes from a two-step process to a single-step process with the addition of Zr, leading to the difference in the activation energy of crystallization between Zr-free and Zr-doped alloys, thus refining the microstructure and improving the magnetic properties of magnets. Nb is adjacent to Zr in the periodic table of elements. It raises an interesting question: whether or not the Nb addition has similar effect in the $Nd_2Fe_{14}B/\alpha$-Fe magnets? In this paper, the effects of Nb addition on the crystallization kinetics, microstructure and magnetic properties of $Nd_2Fe_{14}B/\alpha$-Fe magnets are investigated.

2. Experiment

Alloy ingots with nominal compositions of $Nd_{10}Fe_{84}B_6$ and $Nd_{10}Fe_{83}Nb_1B_6$ were prepared by induction melting under purified argon. Alloy ribbons were obtained by melt spinning in an Ar atmosphere at a wheel speed of 35 m/s, and then annealed between 650 and 750 ℃ for 10 minutes under vacuum to develop the nanocrystalline microstructure. The magnetic properties were measured using a LakeShore 7410 vibrating sample magnetometer (VSM) with a maximum magnetic field of 2.5 T. X-ray diffraction (XRD, Rigaku D/max 2550Pc) with Cu K_α radiation was used to identify the phase composition of samples. The microstructures of samples were observed by a Philips CM200 transmission electron microscope (TEM). The crystallization behaviours of as-spun ribbons were traced using a SDT Q600 differential scanning calorimeter (DSC) at heating rates of 5, 10, 15, and 20 ℃/min, respectively.

3. Results and discussion

Table 1 lists the magnetic properties (B_r, ${}_iH_c$ and $(BH)_{max}$) and the corresponding annealing parameters for $Nd_{10}Fe_{84}B_6$ and $Nd_{10}Fe_{83}Nb_1B_6$. It can be seen that, for both alloys, the magnetic properties first increase and then decrease with increasing annealing temperature. The optimal annealing temperature at which the samples exhibit the optimized magnetic properties is 715 ℃ for both samples. At the optimal annealing temperature, the intrinsic coercivity ${}_iH_c$ increases from 600 kA/m for the Nb-free sample to 750 kA/m for the Nb-doped sample, and the maximum energy product $(BH)_{max}$ increases from 105 kJ/m^3 to 120 kJ/m^3 correspondingly. The non-magnetic Nb element leads to the dilution of the overall magnetization, which may be the reason for the little decrease in the remanence for the Nb-doped alloy.

Table 1. Magnetic properties and corresponding annealing conditions of $Nd_{10}Fe_{84}B_6$ and $Nd_{10}Fe_{83}Nb_1B_6$ ribbons.

Compositions	Annealing condition	B_r (T)	${}_iH_c$ (kA/m)	$(BH)_{max}$ (kJ/m^3)
	700℃, 10min	0.91	592	101
$Nd_{10}Fe_{84}B_6$	715℃, 10min	0.92	600	105
	730℃, 10min	0.89	590	100
	700℃, 10min	0.88	735	115
$Nd_{10}Fe_{83}Nb_1B_6$	715℃, 10min	0.90	750	120
	730℃, 10min	0.88	730	113

The XRD studies revealed that the as-spun Nb-free and Nb-doped ribbons contain fully amorphous phase. Figure 1 illustrates the bright-field TEM images with selected area electron diffraction (SAED) patterns of both the optimally annealed samples. The SAED patterns show that the microstructures of both alloys are crystallographically isotropic, and the rings in SAED patterns can be indexed with those arising from the $Nd_2Fe_{14}B$ and α-Fe phases. It means that, for both alloys, the phase composition transforms from an as-spun amorphous state to a mixture of $Nd_2Fe_{14}B$ and α-Fe phases after optimal annealing. Figure 1a indicates that the $Nd_{10}Fe_{84}B_6$ has a heterogeneous microstructure consisting of large $Nd_2Fe_{14}B$ and α-Fe grains, and abnormally coarse grains with grain sizes of ~70 nm can been observed. A significant reduction in grain size with an improvement in microstructural homogeneity is observed in the Nb-doped alloy, as shown in Figure 1b, although this alloy still contains a few large grains ~50 nm. So the exchange coupling between the hard and soft magnetic phases is strengthened and the magnetic properties are improved. At the same time, it is shown that the Nb addition optimizes the shape of grains, making the grains more equiaxed shape. It was reported [12] that a large stray field originating from the sharp edges of grains can reduce the coercivity. Thus, a microstructure with more equiaxed shaped grains is helpful to improve the coercivity in the Nb-doped alloy.

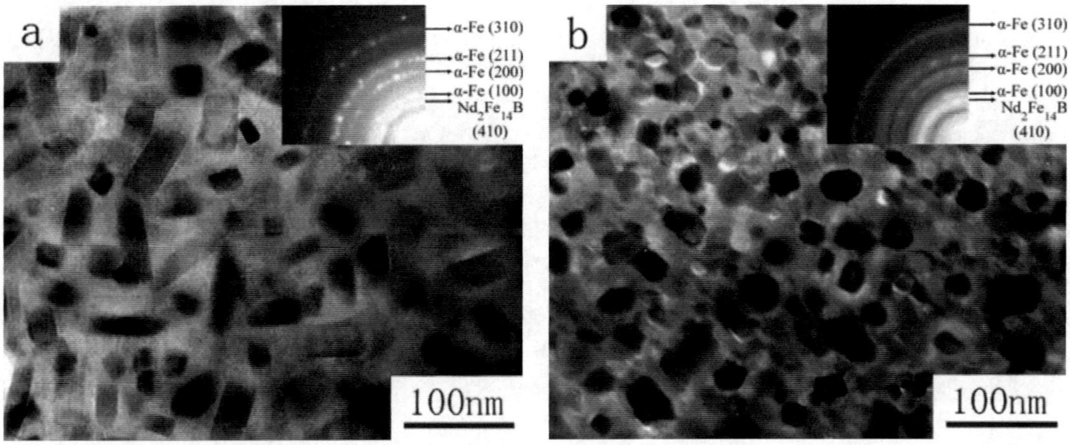

Figure 1. TEM bright-field images and selected area diffraction patterns of optimally annealed (a) $Nd_{10}Fe_{84}B_6$ and (b) $Nd_{10}Fe_{83}Nb_1B_6$ ribbons.

Figures 2 and 3 show the DSC curves for the crystallization behaviours of as-spun $Nd_{10}Fe_{84}B_6$ and $Nd_{10}Fe_{83}Nb_1B_6$ alloys at different heating rates, respectively. There are two exothermic peaks in the crystallization curves, which mean the crystallization behaviour is a two-step process for both alloys. According to our previous study [11] and other literature [13], the phase transformations of the as-spun $Nd_{10}Fe_{84}B_6$ and $Nd_{10}Fe_{83}Nb_1B_6$ ribbons during annealing take place with the same following sequences: Amorphous $\rightarrow$ $Nd_3Fe_{62}B_{14}$+α-Fe+$Nd_2Fe_{14}B$ $\rightarrow$ α-Fe+$Nd_2Fe_{14}B$. For the both alloys, the first exothermic peak can be attributed to the formation of α-Fe phase, $Nd_2Fe_{14}B$ phase and $Nd_3Fe_{62}B_{14}$ metastable phase from the amorphous phase, and the second one is related to the transformation from the $Nd_3Fe_{62}B_{14}$ metastable phase to the $Nd_2Fe_{14}B$ and α-Fe phases.

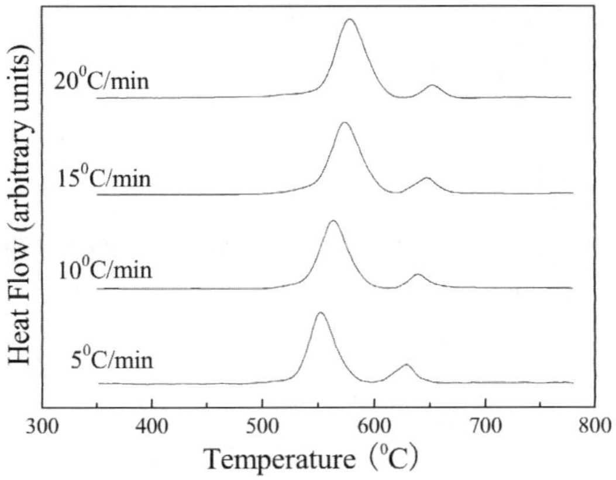

Figure 2. DSC traces for as-spun $Nd_{10}Fe_{84}B_6$ alloy at different heating rates.

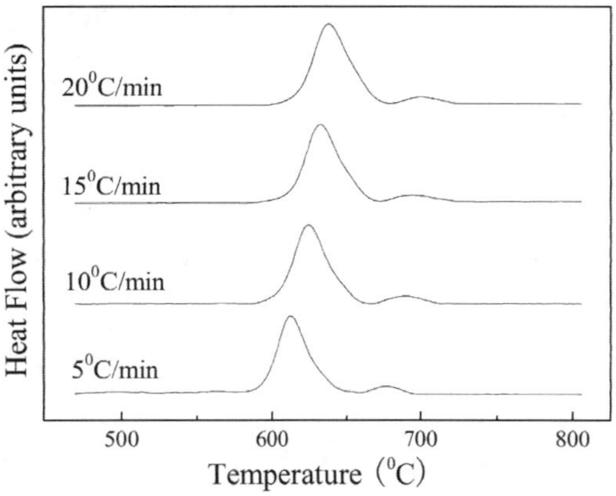

Figure 3. DSC traces for as-spun $Nd_{10}Fe_{83}Nb_1B_6$ alloy at different heating rates.

By measuring the peak positions of the crystallization exotherms shown in figures 2 and 3 at different heating rates, the average activation energies of crystallization (E_c) for the two amorphous alloys are calculated using the Kissinger method [14]. For the $Nd_{10}Fe_{84}B_6$ alloy, the average activation energy of crystallization is 281±5 kJ/mol for the first exothermic peak and 365±5 kJ/mol for the second exothermic peak. For the $Nd_{10}Fe_{83}Nb_1B_6$ alloy, the average activation energy of crystallization is 351±5 kJ/mol for the first exothermic peak and 398±5 kJ/mol for the second exothermic peak.

To better understand the growth behaviour of crystalline phases in the amorphous alloy, the activation energy of crystallization at a given crystallized fraction x, $E_c(x)$, can be calculated by the Doyle equation [15] using the DSC experimental data. The details of the calculation have already been published elsewhere [9-11,15]. Figure 4 shows the dependence of activation energy of crystallization on their crystallized fraction in Nb-free and Nb-doped alloys calculated from the DSC curves shown in figures 2 and 3. As illustrated in figure 4a, for both alloys, the activation energy of the first exothermic peak decreases with increasing the crystallized fraction. Obviously, the decrease of the activation energy in Nb-doped alloy is only from 390 to 340 kJ/mol, which is much smaller than that in the Nb-free alloy (from 390 to 270 kJ/mol). The activation energy of the second exothermic peak, as shown in Figure 4b, presents quite different behaviours in the two alloys. Significantly, in a wide range of crystallization fraction, the activation energy in the Nb-doped alloy is higher than that in the Nb-free alloy. These results imply that the growth of crystallites in Nb-doped alloy needs more energy than that in the Nb-free alloy. Thus, the growth of crystallites in Nb-doped alloy would be relatively slower than that in the Nb-free alloy.

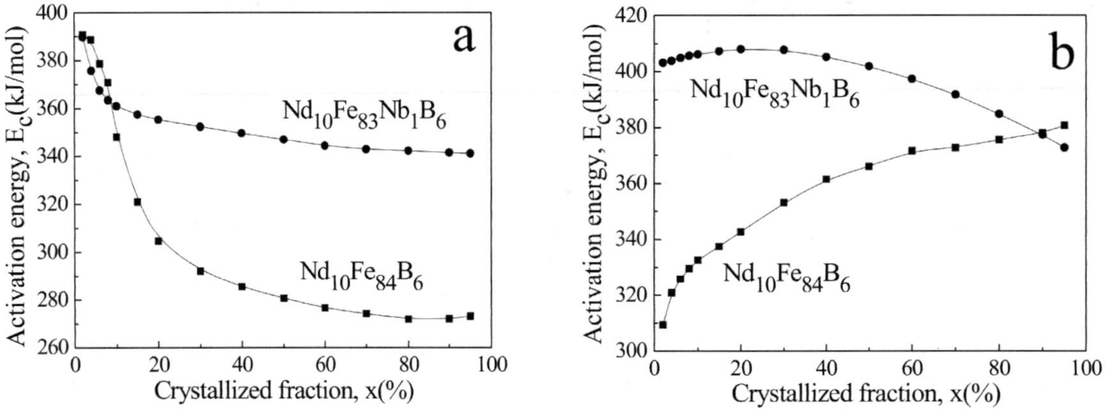

Figure 4. Dependence of activation energy on crystallized fraction in as-spun $Nd_{10}Fe_{84}B_6$ and $Nd_{10}Fe_{83}Nb_1B_6$ alloys: (a) the first crystallization exothermic peak; (b) the second crystallization exothermic peak.

The growth of crystallites in an amorphous alloy is strongly affected by the atomic diffusion. Due to the large atomic diameter, the diffusivity of Nb atoms is low. Moreover, after Nb addition there are more elements and atom radius (of which, $R_{Nd} > R_{Nb} > R_{Fe} > R_B$) variety in the as-spun ribbons. Thus the packing density of atoms would be enhanced, which also inhibits the atomic diffusion.

Pervious studies [6, 16] reported that Nb has very low solubility in both α-Fe and $Nd_2Fe_{14}B$ phases, the Nb atoms tend to enrich at the interface between the α-Fe and $Nd_2Fe_{14}B$ grains detected by three-dimensional atom probe (3DAP) and nano-probe energy dispersive spectroscopy. The segregation of Nb to the interfaces, in combination with the low atomic diffusion, could reduce the mobility of the interfaces. Thus the growth of crystallites in $Nd_{10}Fe_{83}Nb_1B_6$ alloy needs more energy than that in the $Nd_{10}Fe_{84}B_6$ alloy, leading to a finer and more equiaxed grain structure in the Nb-doped alloy than in the Nb-free counterpart.

However, the addition of Nb does not change the two-step crystallization process of amorphous phase. In the Nb-doped alloy, some α-Fe and $Nd_2Fe_{14}B$ grains precipitated earlier from the amorphous matrix in the first-step crystallization process still tend to grow larger during the decomposition of metastable phase. It is one of the reasons why the Nb-doped alloy still contains some large grains in the present work and in the previous works [7, 8]. In addition, some Nb atoms may enter the metastable phase, leading to the difference in the activation energy of the second exothermic peak between the Nb-free and Nb-doped alloys. Further investigations are needed.

4. Conclusions

For both $Nd_{10}Fe_{84}B_6$ and $Nd_{10}Fe_{83}Nb_1B_6$ alloys, the crystallization behaviour is a two-step process. The analysis of crystallization kinetics for the two alloys indicates that the growth of crystallites in the Nb-doped alloy needs more activation energy than that in the Nb-free alloy. These crystallization behaviours result in a finer and more homogeneous microstructure in the Nb-doped alloy than in the Nb-free alloy. And the Nb addition makes the grains more equiaxed shape. The $Nd_{10}Fe_{83}Nb_1B_6$ alloy annealed at 715 ℃ for 10min exhibits the improved magnetic properties, B_r=0.90T, $_iH_c$=750kA/m, $(BH)_{max}$=120kJ/m^3. The intrinsic coercivity and the maximum energy product increase by 25% and 14%, respectively, compared with those of $Nd_{10}Fe_{84}B_6$ alloy.

Acknowledgements

This work is financially supported by the National Natural Science Foundation of China (Grant No. 51001027), the Natural Science Foundation of Fujian Province of China (Grant No.E0710005), the Science and Technology Foundation of Fuzhou University (Grant No.2007-XY-01) and the Elitist Foundation of Fuzhou University (Grant No. 0626).

References

[1] Yang B, Shen B G, Zhao T Y and Sun J R 2008 *J. Magn. Magn. Mater.* **320** 2406–10
[2] Tanaka K, Fukuzaki T, Nishio K and Tamura R 2009 *J. Phys.: Conf. Ser.* **191** 012025-1–5
[3] Chang H W, Cheng Y T, Chang C W, Hsieh C C, Guo Z H, Chang W C and Sun A C 2009 *J. Appl. Phys.* **105** 07A742-1–3
[4] Tamura R, Kobayashi S, Fukuzaki T, Isobe M and Ueda Y 2009 *J. Phys.: Conf. Ser.* **144** 012068-1–4
[5] Betancourt I and Davies H A 2003 *J. Magn. Magn. Mater.* **261** 328–36
[6] Chen Z M, Wu Y Q, Kramerb M J, Smith B R, Ma B M and Huang M Q 2004 *J. Magn. Magn. Mater.* **268** 105–13
[7] Jin Z Q, Okumura H, Zhang Y, Wang H L, Muñoz J S and Hadjipanayis G C 2002 *J. Magn. Magn. Mater.* **248** 216–22
[8] Ping D H, Wu Y Q and Hono K 2002 *J. Magn. Magn. Mater.* **239** 437–40
[9] Wang Z C and Davies H A 2003 *J. Appl. Phys.* **93** 8149–51
[10] Zhang X Y, Zhang J W and Wang W K 2001 *Acta mater.* **49** 3889–97
[11] Wang C, Yan M and Li Q 2007 *J. Phys.* D **40**:3551–6
[12] Fischer R, Schrefl T, Kronmüller H H and Fidler J 1996 *J. Magn. Magn. Mater.* **153** 35–49
[13] Hadjipanayis G C, Withanawasam L and Krause R F 1995 *IEEE Trans. Magn.* **31** 3596–601
[14] Kissinger H E 1957 *Anal. Chem.* **29** 1702–6
[15] Doyle C D 1961 *J. Appl. Polym. Sci.* **5** 285–92
[16] Wu Y Q, Ping D H, Hono K, Hamano M and Inoue A 2000 *J. Appl. Phys.* **87** 8658–65

Journal of Physics: Conference Series **266** (2011) 012048 doi:10.1088/1742-6596/266/1/012048

Magnetic properties of iron films on anodized aluminum underlayer

H Xiang[1], Y Zhang[1], J C Yao[1], X J Mo[1], G Q Li[1, 3] and H Saito[2]
[1] School of Physics, Southwest University, Chongqing, 400715, China
[2] Venture Business Laboratory, Akita University, Akita, 010-8502, Japan

E-mail: gqli@swu.edu.cn

Abstract. A one-step anodization process was used to prepare the anodic alumina (AA) film on glass. Using the AA as an inserted underlayer, iron films with thickness of t_N in the range of 10 ~ 35 nm were deposited by argon ion sputtering. The iron film deposited on AA underlayer exhibited different magnetic behaviors from the iron film deposited on glass or on aluminum underlayer. The perpendicular coercivity of film deposited on AA underlayer reached a maximal value of about 1 kOe at t_N = 30 nm. We believe that the improvement of magnetic properties came from the modulation of the morphology of Fe film by the porous structure of AA underlayer.

1. Introduction

Anodic alumina (AA) generated by anodizing aluminum in acid solution is favorable for the fabrication of ferromagnetic nanostructures such as patterned magnetic recording media, as it is known to have perpendicular holes normal to the surface with diameter ranging from 4 to 200 nm [1-3]. Considerable interest has been attracted to the magnetic nanocylinders filled into the holes of AA, including Fe, Co, Ni, alloys, and ferromagnetic or ferromagnetic-nonmagnetic multilayers, in purpose of enhancing the coercivity and shape magnetic anisotropy [4-12]. AA was obtained by anodizing the bulk or sheet of aluminum in most of the previous work [13-15]. In many occasions of application, using a thin layer of AA is expected to be able to improve the properties of magnetic film. In this article, the aluminum films deposited on glass were anodized and the magnetic properties of Fe film grown on the AA underlayer were investigated.

2. Experiment

Aluminum films with thickness of 400 nm were deposited on glass at a rate of 0.17 nm/s by magnetron sputtering at room temperature. The base pressure before introducing argon gas was better than 2.5×10^{-5} Pa. The pressure of argon was 1 Pa during sputtering. A one-step anodic oxidation process was used to prepare the AA film by dipping the Al film into a 0.3 mol/l oxalic acid solution cooled in a mixture of ice and water. The Al-coated glass substrate acted as the anode, and a copper plate was used as the cathode. The anodizing voltage was 40 V. Using argon ion sputtering, Fe films were deposited at room temperature on glass, Al and AA, respectively. The nominal thickness of Fe film was in the range of t_N = [10, 35] nm, by controlling the sputtering power and time. The base pressure was better than 6×10^{-5} Pa and working pressure was about 0.03 Pa. Deposition rate of iron

[3] To whom any correspondence should be addressed.

was measured to be 0.9 nm/min. The surface of Fe film was *in situ* coated with a protecting layer of Pt with thickness of 2 nm. The purity of Al, Fe and Pt targets was higher than 99.99%. The morphology of film was observed using a scanning electron microscope (SEM). The magnetization curves were measured using a vibrating sample magnetometer (VSM) with the maximum applied field of ±27 kOe.

3. Results and discussion

Figure 1 shows the SEM images of the as-deposited aluminum film (figure 1a) with thickness of 400 nm on glass substrate and the anodized aluminum film (figure 1b) in oxalic acid solution at 40 V for 30 min. The surface of the oxidized film was much rougher than that of the unoxidized one. There are a number of small shallow pin holes in figure 1a, which might be caused by the imperfect of the glass substrate or the local temperature fluctuation during deposition. These shallow pin holes would act as seeds for pore nucleation [16]. The dark areas in figure 1b represent the deep holes (pores) generated in the oxidation process. The average diameter of these holes is about 70 nm.

Figure 1. SEM images of as-deposited Al film (a) and anodized Al film in oxalic acid solution at 40 V for 30 min (b).

Figure 2 shows the magnetization curves of Fe films at t_N = 10 nm deposited on glass, Al/glass and AA/glass, respectively. Both the in-plane loops (figure 2a) and the out-of-plane loops (figure 2b) were measured at room temperature. The insert of figure 2b shows the out-of-plane loop of Fe film deposited on glass in a large range of applied field. The saturation magnetic field exceeded 15 kOe, whereas it was only about 50 Oe measured in parallel direction. This strongly implies that the Fe/glass sample had a continuous morphology. Its easy magnetization axis was obviously parallel to the film plane due to the shape anisotropy. The in-plane coercivity of Fe/glass, Fe/Al/glass and Fe/AA/glass was 11 Oe, 62 Oe and 160 Oe, respectively. The change of in-plane coercivity should be induced by the ruin of morphological continuity oriented from the Al and AA underlayer. Obviously the in-plane coercivity of Fe/AA/glass sample was much larger than that of Fe/glass or Fe/Al/glass sample. Comparing with figure 1, it is clear that the vertical pores in AA underlayer denaturalized the magnetic properties of the overlaid Fe films. According to the magnetic hysteresis loops shown in figure 2b, it can be seen that both the Al underlayer and the AA underlayer induced a drastic decrease of the saturation magnetic field when the sample was magnetized in perpendicular direction, which is greatly different with the in-plane case. The perpendicular coercivity of Fe/glass, Fe/Al/glass and Fe/AA/glass was 0, 184 Oe, and 657 Oe, respectively. The easy magnetization axis of Fe/AA/glass sample switched to perpendicular direction, most probably influenced by the vertical pores in AA layer. Therefore, it is likely that the Fe layer in Fe/AA/glass sample was composed of cylinder-like grains in AA holes with higher aspect ratios (vertical height to horizontal size of a grain) aligned in vertical direction and network on the top of walls of AA hole with 'anti-dots' aligned in horizontal plane. The cylinder-like Fe grains in AA holes should mainly contribute the perpendicular magnetization behavior. The small shallow pin holes in deposited Al underlayer (as show in figure 1a) also resulted in a small amount of discontinuous structure of Fe film, which are able to modulate the magnetic properties of overlaid Fe layer. Due to the defects in Fe layer generated at these pin holes, Fe/Al/glass sample showed higher coercivity than Fe/glass sample did. The in-plane and out-of-plane saturation field (H_s) and coercivity (H_c) of Fe film with thickness of 10 nm are all listed in Table 1.

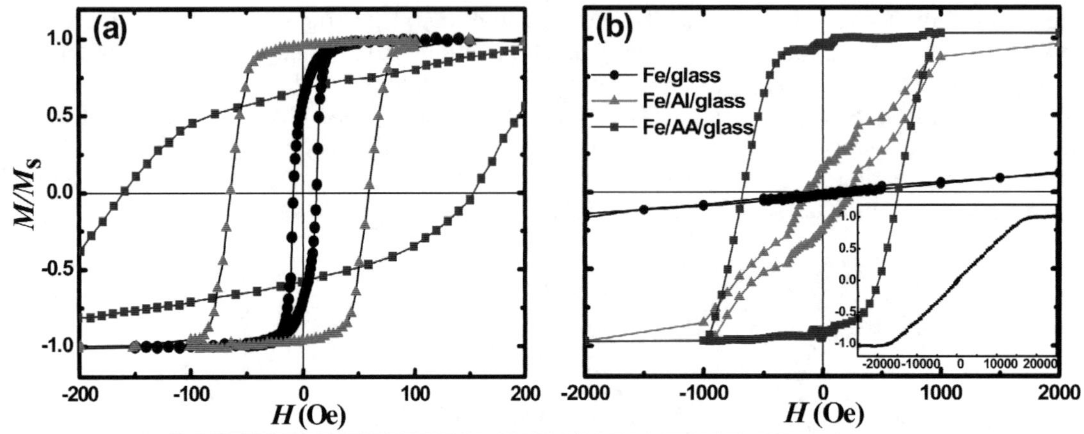

Figure 2. Hysteresis loops of Fe films at $t_N = 10$ nm deposited on glass, Al and AA, with the magnetic fields applied in plane (a) and perpendicular to plane (b). The insert in b shows the out-of-plane loop of Fe film deposited on glass in a large range of applied field.

Table 1. Underlayer dependence of saturation field (H_s) and coercivity (H_c) of Fe film at $t_N = 10$ nm.

Underlayer	H_s (Oe)		H_c (Oe)	
	In-plane	Out-of-plane	In-plane	Out-of-plane
glass	50	15000	11	0
Al/glass	100	5000	62	184
AA/glass	500	1000	160	657

The aspect ratios of Fe grain will surely influence the magnetization process. The aspect ratio should be mainly determined by the thickness of overlaid Fe layer. Figure 3 shows the perpendicular hysteresis loops of Fe films with different thickness on AA underlayer. The coercivity increased with t_N and reached a maximal value of about 1040 Oe at $t_N = 30$ nm. This should be oriented from the aspect ratio increase of Fe in the holes. At $t_N > 30$ nm, the coercivity began to decrease, as the holes were filled out with Fe. With the further increase of t_N, the morphological modulation of Fe film by AA underlayer switched off eventually, consistent with the report in reference 17. It is expected that an even larger coercivity contributed by the Fe grains in holes can be obtained if the Fe outside holes is managed to remove.

4. Conclusions

A one-step anodic oxidation process was used to prepare the anodic alumina film, and the magnetic properties of Fe films deposited on glass, Al/glass, and AA/glass were analyzed. The Fe film deposited on AA underlayer exhibited different magnetic behaviors from the iron film deposited on glass or on aluminum underlayer. The anodized aluminum underlayer obviously influenced the properties of overlaid Fe film by modulating the morphology. The perpendicular coercivity contributed by the iron filled into the holes in the AA underlayer reached a maximal value of about 1040 Oe at the nominal thickness of $t_N = 30$ nm.

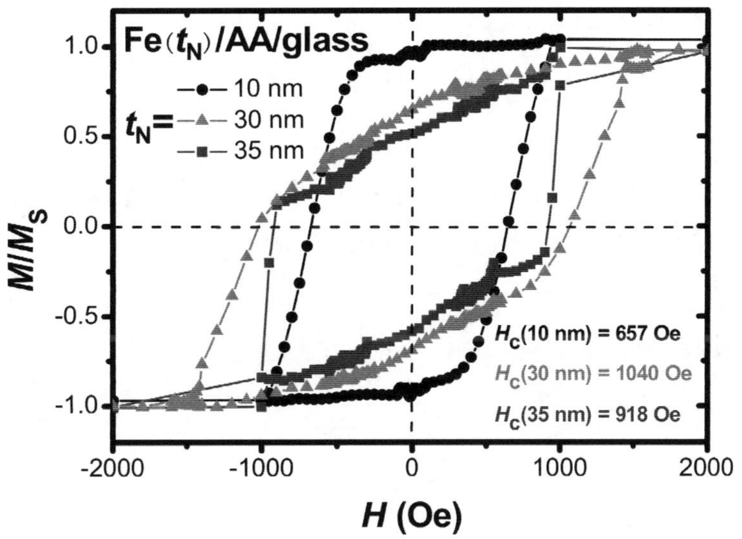

Figure 3. Hysteresis loops of Fe(t_N nm)/AA/glass with magnetic fields applied in perpendicular direction.

Acknowledgement
This work was supported by National Natural Science Foundation of China (Grant No. 51071132), Chongqing Natural Science Foundation (Grant No. CSTC2009BB8102), and Fundamental Research Funds for the Central Universities (Grant No. XDJK2009C193).

References

[1] Masuda H and Fukuda K 1995 *Science* **268** 1466
[2] Foss C A, Jr, Lhornyak G, Stockert J A and Martin C R 1994 *J. Phys. Chem.* **98** 2963
[3] Manalis S, Babcock K, Massie J, Elings V and Dugas M 1995 *Appl. Phys. Lett.* **66** 2585
[4] Ferre R, Ounadjela K, George J M, Piraux L and Dubois S 1997 *Phys. Rev.* B **56** 14066
[5] Zeng H, Zheng M, Skomski R, Sellmyer D J, Liu Y, Menon L and Bandyopadhyay S 2000 *J. Appl. Phys.* **87** 4718
[6] Nielsch K, Wehrspohn R B, Barthel J, Kirschner J, Gösele U, Fischer S F and Kronmüller H 2001 *Appl. Phys. Lett.* **79** 1360
[7] Zhu H, Yang S G, Ni G, Yu D L and Du Y W 2001 *Scripta Mater.* **44** 2291
[8] Dubois S, Marchal C, Beuken J M, Piraux L, Duvail J L, Fert A, George J M and Maurice J L 1997 *Appl. Phys. Lett.* **70** 396
[9] Su Y K, Qin D H, Zhang H L, Li H and Li H L 2004 *Chem. Phys. Lett.* **388** 406
[10] Ji G B, Tang S L, Gu B X and Du Y W 2004 *J. Phys. Chem.* B, **108** 8862
[11] Gao D Q, Fu J L, Xu Y and Xue D S 2008 *Mater. Lett.* **62** 3070
[10] Sun Z G and Akinaga H 2005 *Appl. Phys. Lett.* **86** 181904
[13] Peng Y, Zhang H L, Pan S L and Li H L 2000 *J. Appl. Phys.* **87** 7405
[14] Zhang X Y, Wen G H, Chan Y F, Zheng R K, Zhang X X and Wang N 2003 *Appl. Phys. Lett.* **83** 3341
[15] Xu X J, Fei G T, Zhu L Q and Wang X W 2006 *Mater. Lett.* **60** 2331
[16] Li F, Zhang L and Metzger R M 1998 *Chem. Mater.* **10** 2470
[17] Barnard J A, Fujiwara H, Inturi V R, Jarratt J D, Scharf T W and Weston J L 1996 *Appl. Phys. Lett.* **69** 2758

Journal of Physics: Conference Series **266** (2011) 012049 doi:10.1088/1742-6596/266/1/012049

Preparation and characterization of Co epitaxial thin films on Al_2O_3(0001) single-crystal substrates

Osamu Yabuhara, Mitsuru Ohtake, Yuri Nukaga, and Masaaki Futamoto

Faculty of Science and Engineering, Chuo University, 1-13-27 Kasuga, Bunkyo-ku, Tokyo 112-8551, Japan

E-mail: yabuhara@futamoto.elect.chuo-u.ac.jp

Abstract. Co epitaxial thin films were prepared on Al_2O_3(0001) single-crystal substrates in a substrate temperature range between 50 and 500 °C by ultra high vacuum molecular beam epitaxy. Effects of substrate temperature on the structure and the magnetic properties of the films were investigated. The films grown at temperatures lower than 150 °C consist of fcc–Co(111) crystal. With increasing the substrate temperature, hcp–Co(0001) crystal coexists with the fcc crystal and the volume ratio of hcp to fcc crystal increases. The films prepared at temperatures higher than 250 °C consist primarily of hcp crystal. The film growth seems to follow island-growth mode. The films consisting primarily of hcp crystal show perpendicular magnetic anisotropy. The domain structure and the magnetization properties are influenced by the magnetocrystalline anisotropy and the shape anisotropy caused by the film surface roughness.

1. Introduction

Co and Co-based alloy thin films are widely used as magnetic recording media, magnetic heads, etc. The crystal structure of Co film relies on a stacking sequence of close-packed plane, and it varies between hcp ($K_u=5\times10^6$ erg/cm^3) and fcc ($K_1=6\times10^5$ erg/cm^3) phases depending on the material selected for the substrate and the film growth condition. hcp–Co films have been prepared employing underlayers such as Ru, Ag, and Au [1–3], whereas fcc–Co films have been obtained on Cu underlayers [3]. However, stacking faults are generally recognizable in these films. The crystallographic quality gives strong influence on the magnetic properties [4].

Evaluation of stacking sequence has been performed for Co films by transmission electron microscopy [5,6], perturbed angular correlation [7], in-plane X-ray diffraction (XRD) [4], etc. In order to investigate the basic structural and magnetic properties, it is useful to employ well-defined epitaxial films, since the film uniformity and the magnetic anisotropy are well controlled. In the present study, Co films were prepared on Al_2O_3(0001) single-crystal substrates. The effects of substrate temperature on the structure and the magnetic properties were investigated.

2. Experimental procedure

Co thin films of 40 nm thickness were prepared on polished α–Al_2O_3(0001) single-crystal substrates at temperatures ranging between 50 and 500 °C by using a molecular beam epitaxy chamber under base pressures lower than 3×10^{-8} Pa. Substrates were heated at 500 °C for 1 hour in the UHV chamber to obtain clean surfaces. The surface structure was checked by reflection high energy electron diffraction (RHEED). The RHEED pattern observed for the substrate exhibited *Kikuchi* patterns, indicating that

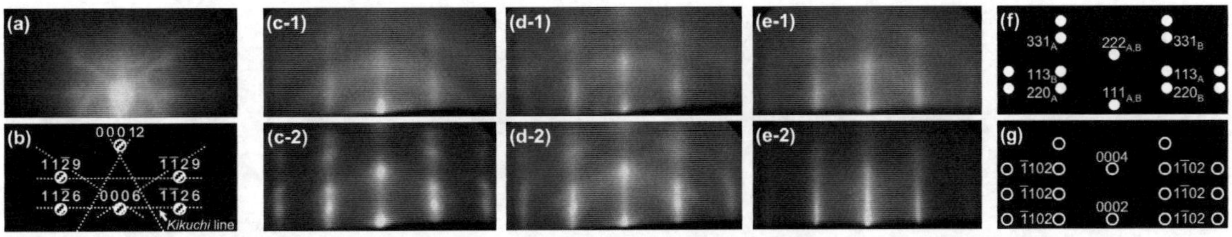

Figure 1. RHEED patterns observed for (a) an $Al_2O_3(0001)$ substrate and (c)–(e) Co films grown on $Al_2O_3(0001)$ substrates at (c) 50 °C, (d) 200 °C, and (e) 250 °C. The film thicknesses are (c–1,d–1, e–1) 2 nm and (c–2,d–2,e–2) 40 nm. RHEED spot maps of (b) α–corundum(0001), (b) fcc(111), and (c) hcp(0001). The incident electron beam is parallel to the $Al_2O_3[1\bar{1}00]$ direction.

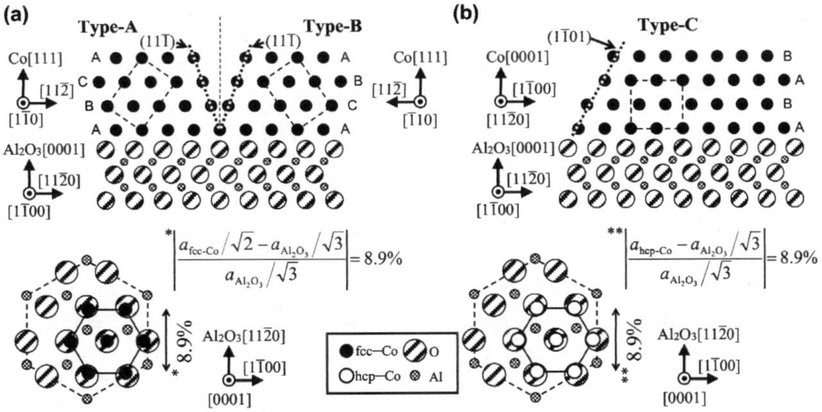

Figure 2. Epitaxial orientation relationships of (a) $Co(111)_{fcc} \parallel Al_2O_3(0001)$ and (b) $Co(0001)_{hcp} \parallel Al_2O_3(0001)$. The lattice mismatches are calculated from the lattice constants of the bulk Al_2O_3, fcc–Co, and hcp–Co crystals.

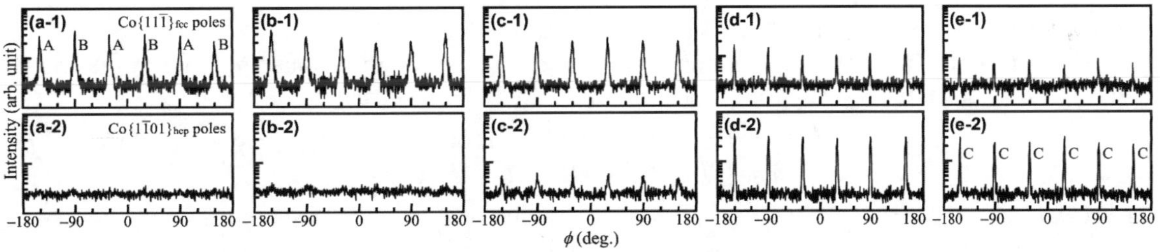

Figure 3. ϕ–scan XRD spectra of (a–1,b–1,c–1,d–1,e–1) $Co\{11\bar{1}\}_{fcc}$ and (a–2,b–2,c–2,d–2,e–2) $Co\{1\bar{1}01\}_{hcp}$ poles measured for Co films grown on $Al_2O_3(0001)$ substrates at (a) 50 °C, (b) 150 °C, (c) 200 °C, (d) 250 °C, and (e) 500 °C. The intensity is shown in a logarithmic scale. The symbols, A, B, and C, respectively correspond to the epitaxial orientation relationships of Type A, B, and C explained in the text.

the surface was clean and smooth. The surface roughness (R_a) value examined by atomic force microscopy (AFM) was 0.12 nm. Pure Co (99.99%) was evaporated by electron beam heating. The deposition rate was kept constant at 0.01 nm/s.

The surface structure during Co deposition was studied by *in-situ* RHEED. The film structure was investigated by out-of-plane ($2\theta/\omega$–scan), in-plane ($2\theta\chi/\phi$–scan), and ϕ–scan XRD with Cu–$K\alpha$ radiation (λ=0.15418 nm). The surface morphology and the magnetic domain structure were observed by AFM and magnetic force microscopy (MFM), respectively. The magnetization curves were measured by using a vibrating sample magnetometer (VSM).

3. Results and discussion

Co epitaxial thin films with close-packed planes parallel to the substrate surfaces were obtained on

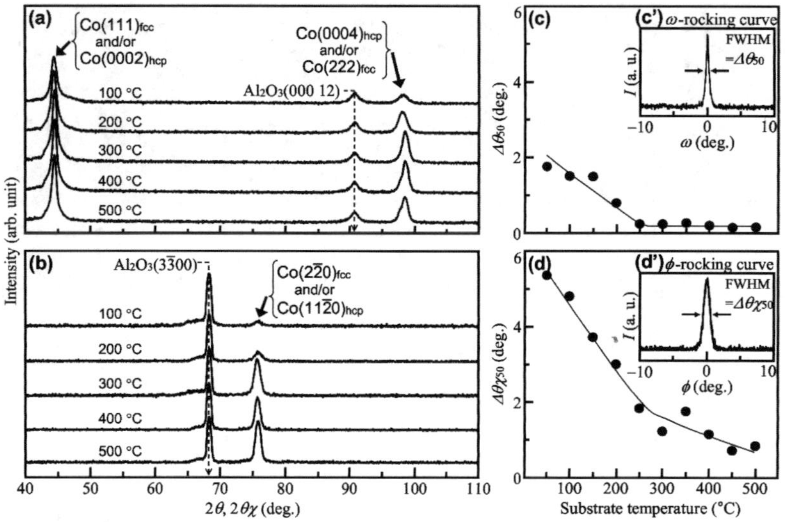

Figure 4. (a) Out-of-plane and (b) in-plane XRD spectra of Co films grown on $Al_2O_3(0001)$ substrates. Rocking curves of (c') $Co(222)_{fcc}$ and/or $Co(0004)_{hcp}$ and (d') $Co(220)_{fcc}$ and/or $Co(11\bar{2}0)_{hcp}$ XRD peaks of the films grown at 500 °C. The intensity is shown in (a,b) a logarithmic or (c',d') a linear scale. Substrates temperature dependences on (c) the $\Delta\theta_{50}$ and (d) the $\Delta\theta\chi_{50}$ values.

$Al_2O_3(0001)$ substrates for the investigated temperatures. Figure 1 shows the RHEED patterns observed during Co deposition on $Al_2O_3(0001)$ substrates heated at different temperatures. Clear diffraction patterns are observed from the beginning of Co deposition and they remain unchanged during film formation for all samples, as shown, for example, in figures 1(c)–(e). With increasing the substrate temperature, the diffraction spots and streaks become sharper, suggesting that the strain in the film decreases by employing high substrate temperatures. For the films prepared at temperatures ranging between 50 and 200 °C (figures 1(c) and (d)), the RHEED pattern consists of spots corresponding to fcc(111) texture and streaks along the fcc[111] direction, as shown in the RHEED spot map of figure 1(f). The streaks indicate that the film has atomically flat terraces and/or involves stacking faults along the fcc[111] direction. The epitaxial orientation relationship between the fcc–Co crystal and the substrate (figure 2(a)) is determined by the RHEED observation as follows:

$$Co(111)[1\bar{1}0]_{fcc} \parallel Al_2O_3(0001)[1\bar{1}00] \text{ (Type A)},$$
$$Co(111)[\bar{1}10]_{fcc} \parallel Al_2O_3(0001)[1\bar{1}00] \text{ (Type B)}.$$

The fcc–Co crystal consists of two variants whose orientations are rotated around the film normal by 180° each other, similar to the cases of fcc–Co film growth on MgO(111) [8] and $SrTiO_3(111)$ [9]. For the films prepared at temperatures higher than 250 °C (for example, figure 1(e)), the RHEED spots are not recognized and only the streaks along the fcc[111] and/or the hcp[0001] direction are observed, as shown in the spot maps of figures 1(f) and (g). By comparing the patterns observed for the film and the substrate, the hcp–Co crystal is considered to be formed in the following crystallographic relationship (figure 2(b)),

$$Co(0001)[11\bar{2}0]_{hcp} \parallel Al_2O_3(0001)[1\bar{1}00] \text{ (Type C)}.$$

When the crystal structure of Al_2O_3 is considered, there are two possibilities of Co crystal nucleation on $Al_2O_3(0001)$. The crystal structure of Al_2O_3 is α–corundum and the (000 l_{odd}) and the (000 l_{even}), respectively, consist of Al and O atoms, as shown in figure 2. The atomic arrangements of Al and O layers are rotated around the film normal by 30° each other. One possibility is that Co atoms of fcc(111) or hcp(0001) fit with the Al atoms of Al_2O_3(000 l_{odd}), where the lattice mismatch is calculated to be 5.6%. Another possibility is that Co atoms fit with the O atoms of Al_2O_3(000 l_{even}), where the mismatch increase up to 8.9%. The experimental result shows that the arrangement of Co atoms is similar to that of O atoms, as shown in figure 2. Theoretical simulation has shown that 3*d* ferromagnetic transition metal film growth on Al_2O_3 substrate is strongly influenced by the O atoms located on the substrate surface [10]. The Co film growth on $Al_2O_3(0001)$ is also considered to be influenced by the atomic arrangement of O atoms of Al_2O_3.

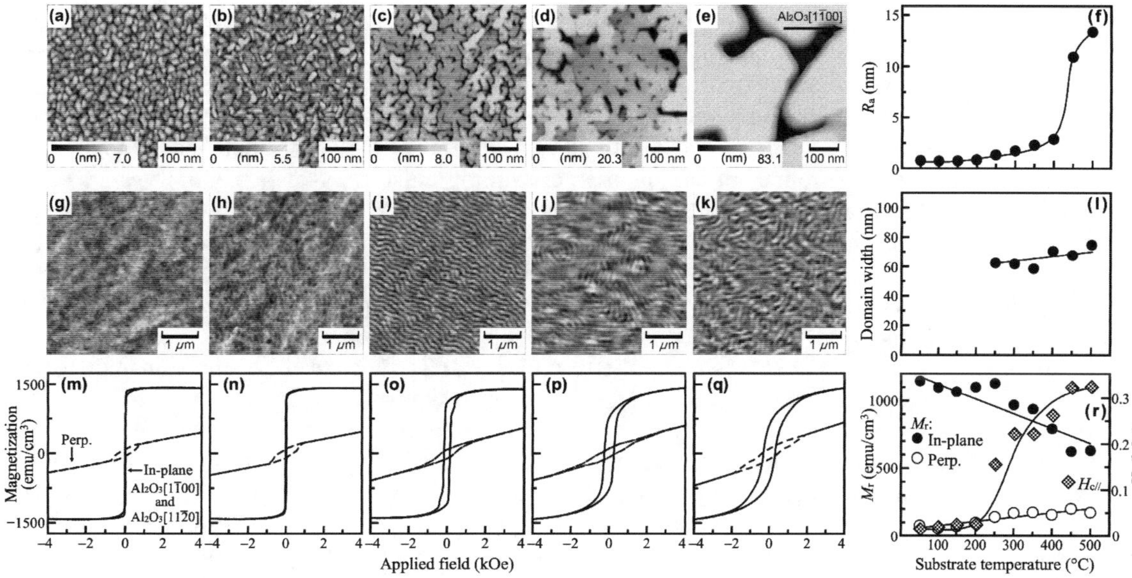

Figure 5. (a)–(e) AFM images, (g)–(k) MFM images, and (m)–(p) magnetization curves of Co films grown on $Al_2O_3(0001)$ substrates at (a,g,m) 50 °C, (b,h,n) 200 °C, (c,i,o) 250 °C, (d,j,p) 400 °C, and (e,k,q) 500 °C. Substrate temperature dependences on (f) the R_a, (l) the magnetic domain width, and (r) the M_r and the H_c of the films.

In order to investigate the stacking sequence of close-packed plane for the Co epitaxial films, ϕ–scan XRD measurements were performed. Figure 3 shows the ϕ–scan XRD spectra of fcc–Co$\{11\bar{1}\}$ and hcp–Co$\{1\bar{1}01\}$ poles, where the intensity is shown in a logarithmic scale. For the films prepared in a temperature range between 50 and 150 °C, six diffraction peaks, which originate from the Type A and the Type B fcc–Co(111) variants, are observed with 60° separation for the spectra of fcc$\{11\bar{1}\}$ poles, as shown in figures 3(a) and (b). This result indicates that the films consist of fcc crystals. When the substrate temperature exceeds 200 °C, six-fold symmetrical diffraction peaks appear in the spectra of hcp$\{1\bar{1}01\}$ poles, as shown in figure 3(c–1). The film prepared at 200 °C consists primarily of fcc crystal and includes very small amount of stacking faults. For the films prepared at temperatures between 250 and 500 °C, strong hcp$\{1\bar{1}01\}$ and weak fcc$\{11\bar{1}\}$ reflections are observed, as shown in figures 3(d) and (e). The films consist primarily of hcp crystal coexisting with small amount of fcc crystal. With increasing the substrates temperature, the intensity of fcc$\{11\bar{1}\}$ reflection slightly decreases, whereas that of hcp$\{1\bar{1}01\}$ reflection increases. The volume ratio of hcp to fcc crystal apparently increases with increasing the substrate temperature.

Figures 4(a) and (b) show the out-of-plane and the in-plane XRD spectra of Co films grown on $Al_2O_3(0001)$ substrates, respectively. The $Co(111)_{fcc}$ and/or $Co(0002)_{hcp}$ out-of-plane and $Co(2\bar{2}0)_{fcc}$ and/or $Co(11\bar{2}0)_{hcp}$ in-plane XRD reflections are clearly observed for all samples. The rocking curves were measured for the out-of-plane and the in-plane XRD peaks, as shown, for example, in the insets of figures 4(c) and (d). Figures 4(c) and (d) show the substrate temperature dependence on the values of full width at half maximum of rocking curves observed for the out-of-plane and the in-plane peaks, $\Delta\theta_{50}$ and $\Delta\theta\chi_{50}$. With increasing the substrate temperature, $\Delta\theta_{50}$ and $\Delta\theta\chi_{50}$ decrease, indicating that the film strain decreases by employing higher substrate temperatures. The respective in-plane lattice strains are slightly larger than the out-of-plane lattice strains, possibly due to the accommodation of lattice mismatch between the film and the substrate.

Figures 5(a)–(e) show the AFM images. The film growth seems to follow island-growth mode. fcc(111) and/or hcp(0001) islands with hexagonal shape are recognized for the Co epitaxial films prepared in a temperature range between 300 and 500 °C, as shown in figures 5(c)–(e). Figure 5(f) shows the substrate temperature dependence on the surface roughness (R_a). The R_a does not increase much up to the substrate temperature of 200 °C. With increasing the substrate temperature beyond 250 °C, the R_a starts to increase. Figures 5(g)–(k) show the MFM images. Stripe magnetic domain structure typical for films with perpendicular anisotropy is observed for the films prepared in a temperature range between 250 and 500 °C, as shown in figures 5(j)–(k). Such magnetic domain structure is considered to be reflecting the magnetocrystalline anisotropy of hcp–Co crystal with *c*–axis normal to the substrate surface. Figure 5(l) shows the substrate temperature dependence on magnetic domain width. The domain width slightly increases with increasing the substrate temperature. Figures 5(m)–(q) show the magnetization curves. The films are easily magnetized when the magnetic field is applied parallel to the in-plane direction, and isotropic in-plane magnetization curves are observed. The perpendicular direction is the hard axis due to demagnetizing field, although the crystallographic easy axis of hcp–Co is perpendicular to the substrate surface for the films consisting of hcp crystal. The saturation magnetization (M_s) is similar to that of bulk fcc–Co (M_s=1422 erg/cm^3 [11]) or hcp–Co (M_s=1479 erg/cm^3 [12]) crystal for all samples. Figure 5(r) shows the substrate temperature dependence on the remanent magnetization (M_r) and the coercivity (H_c) of the films. With increasing the substrate temperature, the out-of-plane $M_{r\perp}$ increases, whereas the in-plane $M_{r//}$ decreases. This result is considered to be partially due to the magnetocrystalline anisotropy of hcp–Co crystal and partially due to the shape anisotropy of large islands which form the Co films prepared at higher substrate temperatures. Furthermore, with increasing the substrate temperature beyond 250 °C, the in-plane $H_{c//}$ starts to increase, possibly due to suppression of domain wall motion by the crevasses in the film existing between Co islands, as shown in figures 5(c)–(e). The crystal structure and the surface roughness influence the magnetic properties of Co epitaxial thin films.

4. Conclusion

Co epitaxial thin films were prepared on Al_2O_3(0001) single-crystal substrates. The effects of substrate temperature on the structure and the magnetic properties were investigated. The films grown at temperatures lower than 150 °C consist of Co$(111)_{fcc}$ crystal. With increasing the substrate temperature, Co$(0001)_{hcp}$ crystal coexists with fcc crystal and the volume ratio of hcp to fcc crystal increases. The film growth seems to follow island-growth mode. The film strain decreases by employing higher substrate temperatures. Perpendicular magnetic anisotropy is observed for the films consisting primarily of hcp phase. The domain structure and the magnetization properties are influenced by the crystal structure and the surface roughness of the films.

Acknowledgment

A part of this work was supported by NEDO–Japan, MEXT–Japan and STFJ (Science and Technology Foundation of Japan). M. Ohtake is a Research Fellow of the Japan Society for the Promotion of Science.

References

[1] Krishnan K M, Takeuchi T, Hirayama Y and Futamoto M 1994 *IEEE Trans. Magn.* **30** 5115
[2] Gong H, Rao M, Laughlin D E and Lambeth D N 1999 *J. Appl. Phys.* **85** 4699
[3] Ohtake M, Futamoto M, Kirino F, Fujita N and Inaba N 2008 *J. Appl. Phys.* **103** 07B522
[4] Hinata S, Yanagisawa R, Saito S and Takahashi M 2009 *J. Appl. Phys.* **105** 07B718
[5] Wong B Y, Shen Y and Laughlin D E 1993 *J. Appl. Phys.* **73** 07C303
[6] Nukaga Y, Ohtake M, Kirino F and Futamoto M 2010 *J. Phys.: Conf. Ser.* **200** 072071
[7] Barradas N P, Wolters H, Melo A A, Soares J C, da Silva M F, Leal J L, Melo L V and Freitas 1994 *J. Appl. Phys.* **76** 6537
[8] Nukaga Y, Ohtake M, Yabuhara O, Kirino F and Futamoto M 2010 *J. Magn. Soc. Jpn.* **34**
[9] Yabuhara O, Nukaga Y, Ohtake M, Kirino F and Futamoto M 2010 *J. Magn. Soc. Jpn.* **34** 78
[10] Zhang W, Smith J R and Evans A G 2002 *Acta Mater.* **50** 3803
[11] Kachkachi H and Dimian M 2002 *Phys. Rev.* B **66** 174419
[12] Bozorth R M 1993 *Ferromagnetism* (New York: Wiley–IEEE press) p 264

Journal of Physics: Conference Series **266** (2011) 012050 doi:10.1088/1742-6596/266/1/012050

Direct Iron Coating onto Nd-Fe-B Powder by Thermal Decomposition of Iron Pentacarbonyl

S Yamamuro[1#], M Okano[1], T Tanaka[1], K Sumiyama[2], N Nozawa[3], T Nishiuchi[3], S Hirosawa[3], T Ohkubo[4]

[1] Department of Materials Science and Engineering, Ehime University, Matsuyama 790-8577, Japan
[2] Nagoya Institute of Technology, Nagoya 466-8555, Japan
[3] Magnetic Materials Research Laboratory, NEOMAX Company, Hitachi Metals, Ltd., Osaka 618-0013, Japan
[4] Magnetic Materials Center, National Institute for Materials Science, Tsukuba 305-0047, Japan

E-mail: yamamuro@eng.ehime-u.ac.jp

Abstract. Iron-coated Nd-Fe-B composite powder was prepared by thermal decomposition of iron pentacarbonyl in an inert organic solvent in the presence of alkylamine. Though this method is based on a modified solution-phase process to synthesize highly size-controlled iron nanoparticles, it is in turn featured by a suppressed formation of iron nanoparticles to achieve an efficient iron coating solely onto the surfaces of rare-earth magnet powder. The Nd-Fe-B magnetic powder was successfully coated by iron shells whose thicknesses were of the order of submicrometer to micrometer, being tuneable by the amount of initially loaded iron pentacarbonyl in a reaction flask. The amount of the coated iron reached to more than 10 wt.% of the initial Nd-Fe-B magnetic powder, which is practically sufficient to fabricate Nd-Fe-B/α-Fe nanocomposite permanent magnets.

1. Introduction

Nanocomposite permanent magnets currently gain a growing attention because of their potential superiorities of enhanced remanence and high energy product in the range of reduced rare-earth content [1-3]. These magnets are comprised of a fine mixture of a magnetically hard phase with a large magnetic anisotropy and a magnetically soft phase with a high magnetization. The two phases are preferred to have intimate contacts to be exchange-coupled via their interfaces. Since such interaction is effective at the exchange length scale, the two phases are preferably mixed in the nanoscale order [1]. Since the strict nanostructural control of such a bulk nanocomposite is hardly achieved by the metallurgical processes, the fabrication of soft/hard magnetic composite powder is an alternative approach that is followed by their subsequent consolidation and sintering. Some attempts have been made to prepare composite magnetic powders consisting of rare-earth magnetic powder (either Sm-Co or Nd-Fe-B system) mixed/coated by metallic α-Fe using mechanical [4-6], sonochemical [7] and colloidal approaches [8]. A major difficulty ascribes to the highly reactive nature of both rare-earth

[#] To whom any correspondence should be addressed.

magnet and α-Fe phases, which restricts the available processes to make composite powders. A sufficiently low content of impurities such as oxygen is also strongly required.

Here, we present a new solution-phase coating process of soft magnetic iron onto the surfaces of rare-earth magnet powders. Since the coating reaction is carried out under the inert conditions using hydrocarbon solvent and argon atmosphere, severe oxidation and undesired reactions between rare-earth magnet powder and reagents are largely suppressed. The process is also characterized by the use of highly anisotropic Nd-Fe-B powder that is prepared by the hydrogenation-disproportionation-desorption-recombination (HDDR) process. Such HDDR-processed Nd-Fe-B powder is anticipated to ensure a high magnetic coecivity even in the reduced size range of powder particles and lead to a magnetically anisotropic bulk nanocomposite [3].

2. Experimental

The present method of iron coating is based on a modified wet process that is used for high-quality iron nanoparticles synthesis via thermal decomposition of iron pentacarbonyl ($Fe(CO)_5$) [9,10]. However, the process used here is featured by the suppressed formation of iron nanoparticles to achieve an efficient iron coating onto the surfaces of rare-earth magnet powder. This carbonyl complex is used as an iron source that generates neutral iron atoms when it is thermally decomposed in an inert organic solvent in the presence of surfactant. These iron atoms give rise to a direct coating without any reducing treatment of metal ions. The chemical stability of organic reagents at elevated temperature is also important to avoid undesirable reactions with highly reactive Nd-Fe-B magnet powders. After a series of examinations, hydrocarbons and alkylamines were found to be suitable for solvent and surfactant, respectively. Even after a mixture of these reagents (either dodecane or dioctylamine) and HDDR powder was heated at 200 °C for 2 h in an argon gas atmosphere, a severe damage was not detected in the x-ray diffraction profiles and demagnetiztion curves of the reacted HDDR powder, though a slight decrease in remanence and coecivity was observed (see Fig. 1(a) and 1(b)). In this reaction, the surfactant plays a crucial role to preferentially coat onto the rare-earth magnet powder.

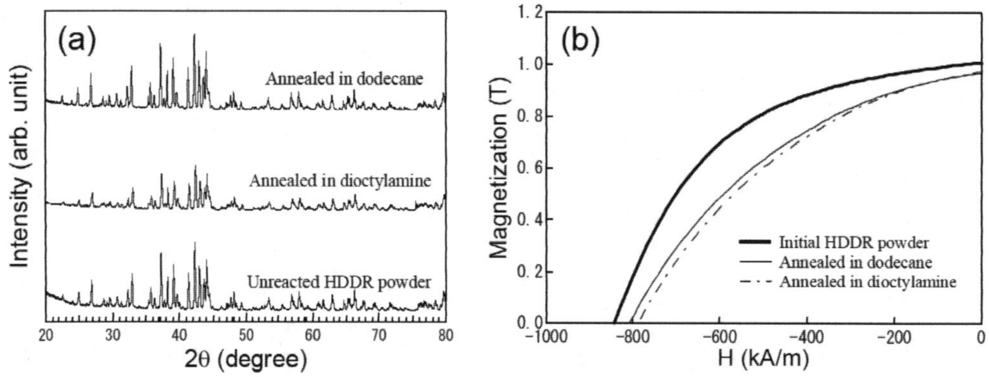

Figure 1. The chemical reactivity between HDDR magnetic powder and chemical reagents (dodecane and dioctylamine) after being heated at 200 °C for 2 h: (a) x-ray diffraction patterns and (b) demagnetization curves.

In a typical reaction, 4 g of Nd-Fe-B powder (50~100 μm) was mixed in 30 mL of dodecane with 0.2 mL of dioctylamin in a flask. The solution mixture was mechanically stirred for 1 hour under an Ar gas flow, and then rapidly heated to 100 °C. 1~5 mL of $Fe(CO)_5$ was then introduced into the flask via a syringe and the temperature was hold in a couple of minutes. Temperature was then increased to 180 °C and was hold for 4 h. On the reaction, the yellow color originating from $Fe(CO)_5$ was gradually diminished in the flask. After the reaction was finished, the solution was air-cooled to room temperature. The powder is then collected via filtration and washed by hexane under nitrogen gas atmosphere. The product was examined with a Carl Zeiss Crossbeam 1540 energy selective backscattered scanning electron microscope (SEM). The compositional analysis was carried out by the inductively coupled plasma (ICP) method.

3. Results and Discussion

The appearance of Fe-coated HDDR magnetic powder exhibited an increased gray color. When these powder samples were handled in air, they were rapidly oxidized with a smoke due to the existence of highly reactive Fe surface layers. Since the coated iron formed by this process has an amorphous-like crystal structure [10], the x-ray diffraction measurement does not provide a meaningful result. The SEM observation was instead performed for Fe-coated HDDR powders. As shown in Fig. 2(a), the surface of each powder particle is attached by a number of small granules that are presumably the fragments of the coated iron layers. One of these particles is etched with a focused ion beam (FIB) of Ga ions and the cross section is observed with a backscattered electron (BSE) SEM. As shown in Fig. 2(b), the surface of the particle is covered by an iron shell that has a bumpy morphology and dose not uniformly coat over the entire surface of the particles (see the upper dark region in Fig. 2(b)). However, it can be seen that the iron surface layer fills well into the morphologically uneven surface of the HDDR particle. The thickness of the Fe shell is estimated to be about submicrometer-to-micrometer order (The arrows in Fig. 2(b) show the cross-sectional thickness of coated iron layer). Since such a thick layer of soft magnetic iron is apparently too large for exchange coupling in comparison to the theoretically estimated dimension [1], the fabrication of finer HDDR powders and thinner iron layers is strongly desirable.

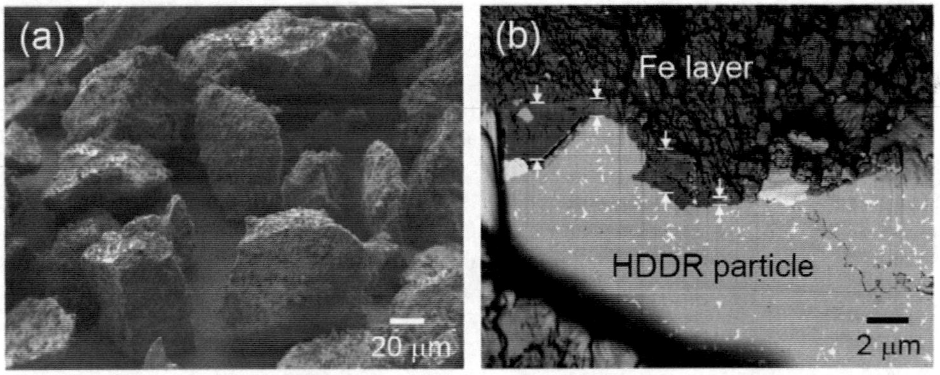

Figure 2. (a) Low magnification SEM image of Fe-coated HDDR magnetic powder and (b) cross-sectional BSE image of a Fe-coated HDDR particle etched by a focused Ga ions beam. The images were taken by tilting the specimen by an angle of 35 °. 1 mL of $Fe(CO)_5$ was used in the Fe coating reaction.

To examine the tuneability of coated Fe amount, a straightforward way is varying the amount of iron source during the reaction. As shown in Fig. 3, the mass of coated Fe layers is increased by increasing the amount of initially loaded $Fe(CO)_5$ into the reaction flask. In the present experiments, the maximum value reached to more than 10 wt.% of the initial Nd-Fe-B magnetic powder when 5 mL of $Fe(CO)_5$ was served for the coating reaction. Though this value is comparable to the one required for the fabrication of Nd-Fe-B/α-Fe nanocomposite permanent magnets, the coating efficiency is not satisfactory yet; our simple calculation showed that only 1/4~1/3 of loaded $Fe(CO)_5$ was used for Fe coating onto HDDR powder. The major cause of such a low efficiency is ascribed to the undesirable coating onto the inner wall of flask. Another issue is the incorporation of impurities such as oxygen and carbon elements into the Fe-coated HDDR powder though inert environment was employed during the reaction. This largely deteriorates the magnetic properties. The EPMA mapping analysis showed that these elements are preferentially accompanied by the coated Fe phase. The minimization of these impurities is currently underway.

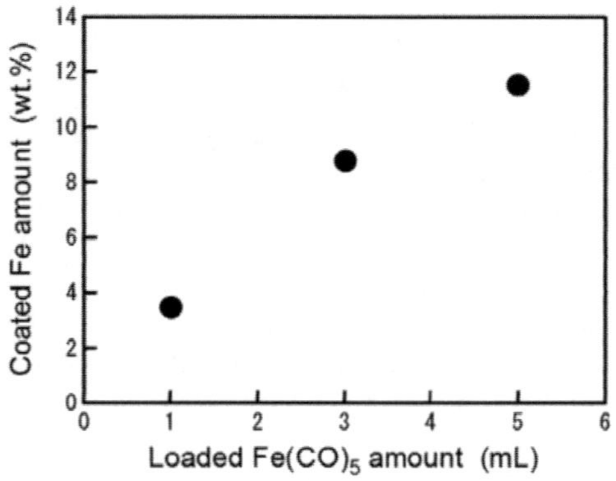

Figure 3. The relation between the amounts of loaded $Fe(CO)_5$ and coated Fe.

4. Conclusion

A new solution-phase technique of iron coating onto Nd-Fe-B magnetic powder was reported. This is based on a modified process to synthesize high-quality iron nanoparticles by thermal decomposition of iron pentacarbonyl in an inert organic solvent in the presence of alkylamine. The Nd-Fe-B magnetic powder was successfully coated by iron shells that are bumpy and uneven in thickness. The typical thickness of the coated iron shell was of the order of submicrometer to micrometer. The amount of coated Fe was tuneable by the amount of initially loaded iron pentacarbonyl in a reaction flask. The maximum amount of the coated iron reached to more than 10 wt.% of the initial Nd-Fe-B magnetic powder, which is practically sufficient to fabricate Nd-Fe-B/α-Fe nanocomposite permanent magnets.

Acknowledgments

This work was supported by a grant for the Elements Science and Technology Project, 'High Performance Anisotropic Nanocomposite Permanent Magnets with Low Rare-Earth Content,' given by the Ministry of Education, Science, Culture, and Sports, Japan.

References

[1] Kneller E F and Hawig R 1991 *IEEE Trans. Magn.* **27** 3588
[2] Lee D, Hilton J S, Chen C H, Huang M Q, Zhang Y, Hadjipanayis G C and Liu S 2004 *IEEE Trans. Magn.* **40** 2904
[3] Hirosawa S, Nishiuchi T, Nozawa N, Ohkubo T, Hono K, Seperhi-Amin H, Takezawa M, Yamasaki J, Yamamuro S, Tanaka T, Okano M and Sumiyama K 2010 *Proc. of 21st International Workshop on Rare-Earth Permanent Magnets and their Applications* (submittted)
[4] McCormick P G, Miao W F, Smith P A I, Ding J and Street R 1998 *J. Appl. Phys.* **83** 6256
[5] Cha H G, Kim Y H, Kim C W, Kwon H W and Kang Y S 2007 *Curr. Appl. Phys.* **7** 400
[6] Rama Rao N V, Gopalan R, Manivel Raja M, Chandrasekaran V, Chakravarty D, Sundaresan R, Ranganathan R and Hono K 2007 *J. Magn. Magn. Mater.* **312** 252
[7] Niu P, Yue M, Li Y, Huang W and Zhang J X 2007 *Phys. Status Solidi* **204** 4009
[8] Hou Y, Sun S, Rong C and Liu J P 2007 *Appl. Phys. Lett.* **91** 153117
[9] Farrell D, Majetich S A and Wilcoxon J P 2003 *J. Phys. Chem.* B **107** 11022
[10] Peng S, Wang C, Xie J and Sun S 2006 *J. Am. Chem. Soc.* **128** 10676

Multi-polarly micro rotor prepared from isotropic nano-crystalline films with self-bonding layer

F Yamashita[1], N Menjo[1], S Nishimura[1], O Kobayashi[1], M Itoh[2], K Terada[3], M Nakano[3], H Fukunaga[3] and K Ishiyama[4]

[1]Rotary Component Tech., Div., Minebea Co., Ltd., Shizuoka 437-1193, Japan
[2]Center for Advanced Science and Innovation, Osaka Univ., Osaka, 565-0871, Japan
[3]Faculty of Engineering, Nagasaki Univ., Nagasaki, 852-8521, Japan
[4]RIEC, Tohoku Univ., Sendai, 980-8577, Japan

E-mail: fmtsymst@minebea.co.jp

Abstract. A magnetized micro rotor such as 1.6 mm in outer diameter was prepared from a laminated isotropic nanocrystalline film with an optimized self-bonding layer. A relative density(RD) higher than 85 % in perpendicular direction of the above film could be obtained, and an estimated permeance coefficient, B/μoH, exceeded 20 by 3-D finite element method. Furthermore, an enhancement of magnetic torque and relative torque constant, dT/dH-gradient, of the above-mentioned sample with the single pole pair could be achieved. Consequently, it was found that use of laminated isotropic film with optimized self-bonding layer is effective in obtaining a multi-polarly magnetized micro rotor with highly dense torque compared with those of a bonded magnet prepared from consolidated flake with the same materials.

1. Introduction

The film magnets with the thickness range between 10 - 300 microns are indispensable for advancing a size reduction in millisize-motors and micro-actuators [1]. In particular, further improvement in the rotating torque of the above motor and actuator is required for practical applications. We have already investigated laminated films such as $Fe_3B/Nd_2Fe_{14}B$ nanocomposite material [2] with double pole-pairs for highly dense torque compared with that of conventional anisotropic bulk magnet with single pole-pair [3].

In this contribution, an enhancement in the torque for a multi-polarly micro cylindrical magnet rotor was achieved by using isotropic nanocrystalline films with an optimized self-bonding layer. An approximately 45 μm-thick and 2 mm-width isotropic nanocrystalline film with self-bonding layer was used as a starting material. A relative density, RD, in perpendicular direction of the above film exceeded 85 %, and the estimated value of permeance coefficient, B/μoH, showed higher than 20 by using a 3-D finite element. Namely, the magnetic torque and relative torque constant, dT/dH-gradient, of the obtained sample could be improved compared with that of a conventional bonded magnet rotor. The superior magnetic torque of the obtained rotor is attributed to a use of the shape-magnetic-anisotropy in the plane together with the improvement of relative density into perpendicular direction. Consequently, the usage of isotropic nanocrystalline films with optimized self-bonding layer is expected to be suitable to multi-polarly magnetized micro rotor for highly dense torque millisize-motors and micro-actuators.

Journal of Physics: Conference Series **266** (2011) 012051 doi:10.1088/1742-6596/266/1/012051

2. Experimental procedure

2.1. Preparation processes

In the section, the detailed preparation processes of a nanocrystalline film, a laminated film and a bonded magnet, respectively, were described (see Fig.1).

$Nd_{4.5}Fe_{bal}Co_5B_{18.5}Cr_2$ molten alloy was melt spun into an amorphous ribbon with approximately 45 μm in thickness and 2 mm in average width using a Cu wheel in Ar atmosphere at the pressure of 10 kPa. In the melt-spinning process, the ejection pressure of 50 kPa by Ar gas, the orifice of 0.7 mm in the quartz crucible, and the wheel speed of 30 m/s, respectively, were used. The amorphous ribbons were cut to 50 mm or less in length and were crystallized at the elevated temperature of 680 ºC for 10 min with heating rate of 200 ºC/min under the vacuum pressure of 10^{-2} Pa. After cooling them to room temperature, the Cr-Co-doped $Fe_3B/Nd_2Fe_{14}B$ nanocrystalline films were coated with approximately 5 μm-thick or less self-bonding layers including amino active hydrogen in macromolecule chain and blocked-isocyanate as seen in Fig.2.

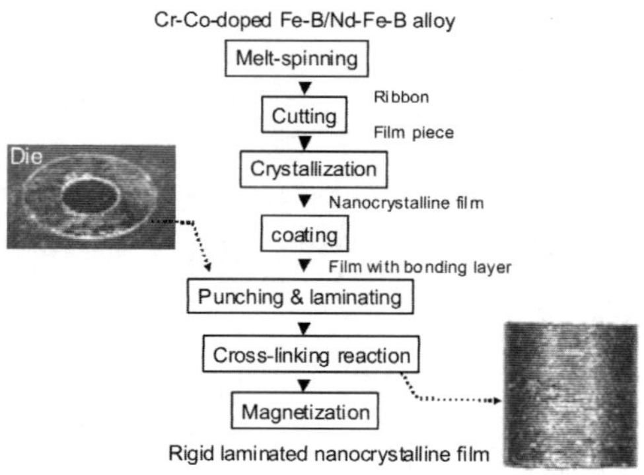

Figure 1 Process for producing a multi-polarly micro rotor prepared from isotropic nanocrystalline films with self-bonding layer. The photographs show external views of a hollow disc shaped film and an obtained rotor magnet, respectively.

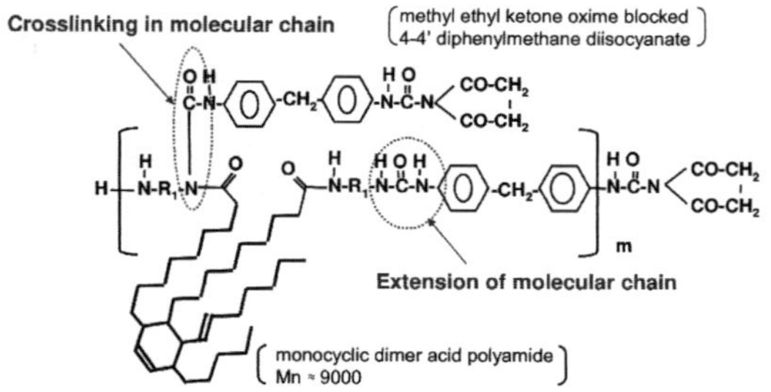

Figure 2 Example of molecular design in self-bonding layer for nanocrystalline film lamination.

Figure 1, further, shows a laminated film fabricated from the above-mentioned films. Each film was pierced into a hollow disc shape under a pressure up to 50 kN. The intended dimensions are 1.6 mm in outer diameter and 0.7 mm in inner diameter. In addition, the laminated film was prepared into several specific lengths such as approximately 1.0 - 1.5 in L/D ratio (L: length, D: diameter) in the punching. Namely, the laminated film was solidified under a pressure of 50 MPa at the temperature of 180 ºC for 5 min. In the cross-linking to obtain a rigid laminated film, the blocked-isocyanate dissociated at elevated temperature. Subsequently, the generated isolation isocyanate reacted to amino active hydrogen in the macromolecule chain including in a self-bonding layer (See Fig.2). The laminated film was magnetized to an in-plane direction under the pulsed magnetic field of 4.8 MA/m.

We also prepared a bonded magnet by using the above-mentioned nanocrystalline film. The particle size of the compound was adjusted to 350 μm or less by crushing and sieving-classification. The obtained compound was compacted into specific shape with 1.6 mm in outer diameter under a pressure of 1000 MPa at 180 °C. Through the compaction, a rigid bonded magnet could be obtained by cross-linking reaction of self-bonding material.

2.2. Evaluations

A magnetization reversal rate and magnetic torque in a different rotating field were measured with a VSM and a magnetic torque meter, respectively, and a permeance coefficient, B/μoH, was estimated by a 3-D finite element model (EFM).

3. Results and Discussion

3.1. Improvement in permeance coefficient, B/μoH

The typical magnetic properties of obtained film are 1.1 T in remanence, 330 kA/m in coercivity, 95 kJ/m^3 in $(BH)_{max}$, respectively.

Figure 3 shows the permeance coefficient, B/μoH, of an above-mentioned laminated film with specific shape such as L/D=1.5. The maximum value of B/μoH could be obtained as the relative density into perpendicular direction became approximately 90 %. From a outer side view of B/μoH distribution, B/μoH little by little decreasing with the relative density of the laminated film can be observed (see Fig. 4). The decrease is attributed to the thickness of the self-bonding layer into perpendicular direction of a film seen in Fig.3. These results indicate that the use of a shape-magnetic-anisotropy together with a high relative density into perpendicular direction of approximately 90 % is effective to obtain an isotropic multi-polarly miniaturized rotor with approximately 1.6 mm in outer diameter. Consequently, a relative density into perpendicular direction by using optimized self-bonding layer could exceed 85 % (see Fig.3).

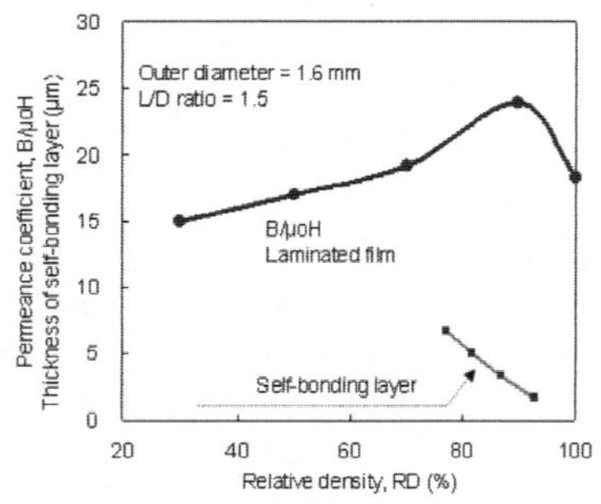

Figure 3 Relationship between relative density into perpendicular direction and permeance coefficient, B/μoH, together with thickness of a self-bonding layer before cross-linking.

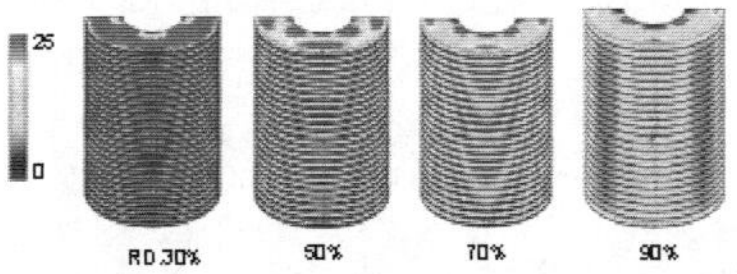

Figure 4 B/μoH distribution of a laminated film magnet including self-bonding layer. The number of pole pair was 2 and estimated by 3-D FEM. The elements of a fine mesh for the model were 1, 277, 316. The dimensions of a model are 1.6 mm in outer diameter, 0.7 mm in inner diameter and 1.5 in L/D ratio, respectively.

3.2. Magnetic torque

A magnetic torque of an isotropic laminated film for a miniaturized rotor was evaluated.

A relational expression of motor torque is denoted by $T = P_n \times \Phi_a \times I_q + P_n \times (L_d \pm L_q) \times I_d$. In the right side of the equation, the clauses 1 and 2 show magnetic and reluctance torques, respectively. Each symbol of P_n, Φ_a, I, and L denote the number of pole pair, inter-linkage flux, current, and inductance, respectively. In addition, the subscripts q and d show the axis of different direction of magnetic circuit design of a motor. In the generated torque by the interaction between each magnetic field and the obtained magnetized specimens, the reluctance torque does not occur.

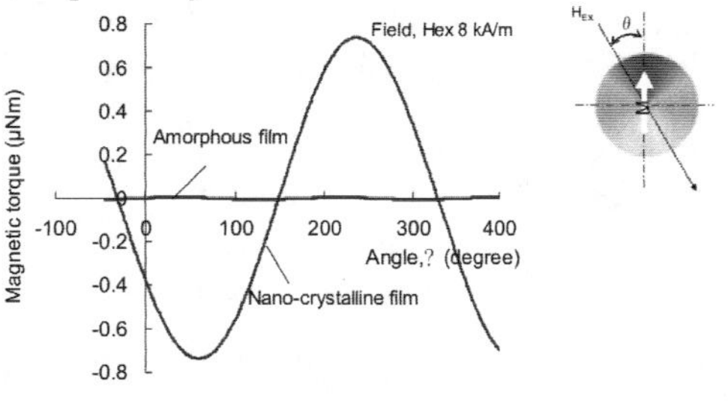

Figure 5 Magnetic torque curve of a magnetized nanocrystalline film magnet together with that for amorphous film with the same alloy composition. The intended dimensions of samples are 1.6 mm in outer-diameter and 45 μm in average thickness, respectively.

The magnetic torque of a magnetized laminated film after magnetization was examined with a magnetic torque meter because the measured magnetic torque was applicable compared with general S-T equipment in the micro rotor with a single pole pair.

Figure 5 shows a magnetic torque curve of a disc shaped film with single pole pair compared with that of an amorphous film with the same alloy composition. The maximum torques of a nanocrystalline and an amorphous film with 1.6 mm in outer diameter were 0.74 and 1.6×10^{-4} μNm, respectively.

As the rotating field range became between 0 and 80 kA/m, good correlation coefficient of a magnetic torque as well as magnetization reversal rate could be observed as shown in Fig.6. These results suggest that the small magnetic torque in the small rotating field for a miniaturized rotor can be evaluated clearly.

Table 1 and Figure 7 show the magnetic torques and dT/dH-gradients, respectively, in laminated films and consolidated flake as a function of L/D ratio. In the laminated films with the single pole pair magnetization to in plane direction, an enhancement in magnetic torque occurs compared with that of a bonded magnet preparing consolidated flake with the same materials and dimensions. It is clarified that a usage of the shape-magnetic-anisotropy to the in-plane and the relative density higher than 85 % into perpendicular direction are effective.

From the above-mentioned results, the values of dT/dH-gradient for a laminated film magnet (L/D=1) could be improved by 121 % compared with that of a bonded magnet. We also considered that the above dT/dH-gradient is corresponding to a torque constant of a micro DC brushless motor. In particular, as their L/D ratio increased, the superior magnetic torque and the dT/dH-gradient of the laminated film due to a use of the shape-magnetic-anisotropy could be achieved.

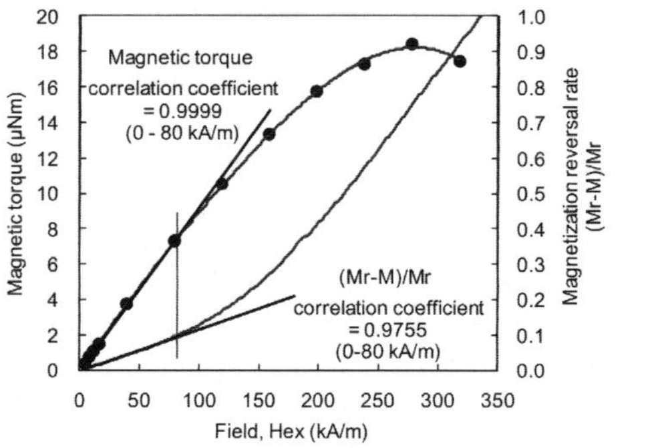

Figure 6 Magnetic torque and magnetization reversal rate of a magnetized film as a function of field strength. The magnetization reversal rate of a film was measured by VSM.

Table 1 Magnetic torque of a laminated film magnet together with those of a bonded magnet for comparison.

Sample	RD (%)	Magnetic torque (μNm) L/D=1	L/D=1.5
Laminated film	85.25	17.5	27.8
Flake (bonded megnet)	78.11	14.5	20.0
Ratio	1.09	1.21	1.39

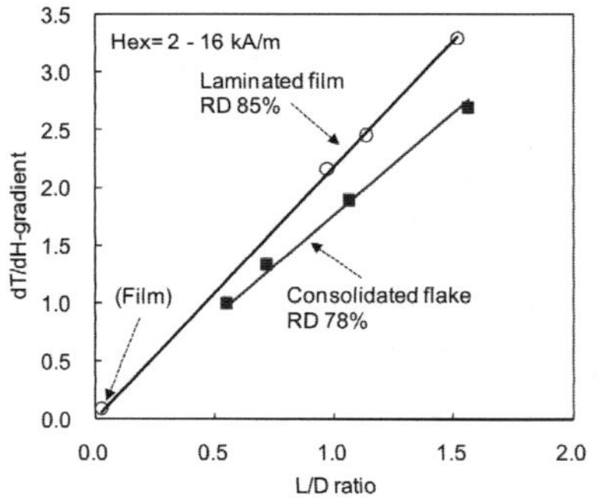

Figure 7 Relative torque constant as a function of L/D ratio in laminated films magnet together with those of conventional bonded magnets.

4. Conclusion

In a miniaturized rotor comprising an isotropic nanocrystalline film with the optimized self-bonding layer, highly dense torque and relative torque constant, dT/dH-gradient, could be achieved by taking advantage of a shape-magnetic-anisotropy and an improved relative density into perpendicular direction.

Acknowledgments

This work was supported from New Energy and Industrial Technology Development Organization (NEDO, Project P03033 and P07026) of Japan.

References

[1]K. Makita, O. Yamashita, and A. Nakanishi, *J. Magn. Soc. Jpn.* 22 (1998) 365 (in Japanese).
[2]S. Hirosawa, H. Kanekiyo, Y. Shigemoto, *Proc. Mat. Res. Soc. Symp.*, 577, (1999) 141.
[3]F. Yamashita, S. Nishimura, N. Menjo, O. Kobayashi, M. Nakano, H. Fukunaga, K. Ishiyama, *IEEE Trans. Magn.* 46, 6 (2010) 2012 - 2015.

Journal of Physics: Conference Series **266** (2011) 012052 doi:10.1088/1742-6596/266/1/012052

The preparation of sintered NdFeB magnet with high-coercivity and high temperature-stability

G H Yan, R J Chen, Y Ding, S Guo, Don Lee and A R Yan
Zhejiang province Key Laboratory of Magnetic Materials and Application Technology; Key Laboratory of Magnetic Materials and Devices, Ningbo Institute of Material Technology & Engineering, Chinese Academy of Sciences, Ningbo, 315201, P. R. China

E-mail: yangh@nimte.ac.cn

Abstract. The NdFeB magnets with high intrinsic coercivity have been produced by using the conventional powder metallurgy method (including SC, HD and JM) of sintered NdFeB magnets. The effects of grain boundary phases on the microstructure and magnetic properties of as-sintered and annealed magnets have been tried to investigate. Also the Curie temperature of the magnets was studied. By adopting suitable component ratio of some heavy rare-earth atoms and some micro-quantity additives, we have prepared high-coercivity sintered NdFeB magnets with magnetic properties of ${}_{j}H_c$=36.3kOe, B_r=11.7kGs and $(BH)_{max}$=34.0MGOe. The temperature coefficient of residual magnetic flux of the magnets (between 20 and 200°C) is -0.113%/°C, while the temperature coefficient of intrinsic coercivity -0.355%/°C. The Curie temperature of the magnets is about 342°C.

1. Introduction

NdFeB magnets have been employed into various applications because of their excellent magnetic properties, since they were discovered by Sagawa [1]. In recent years, there is a trend of permanent magnet motors replacing traditional motors, because they have high efficiency and can run on lower energy. The NdFeB magnets with excellent properties have extended the applications in the field of motors such as Hybrid Electric Vehicles and Electric Vehicles. Therefore, the sintered NdFeB magnets with high coercivity have much appeal for the scientists and engineers. In order to attain high coercivity of sintered NdFeB magnets, various methods have been employed. The intrinsic coercivity can be obviously increased by adding heavy rare-earth elements [2-3]. The elements Al and Cu with a low melting point mainly affect the wetting behavior of the RE-rich phase, and can modify the boundaries phase. Subsequently, they can increase the intrinsic coercivity [4-5]. At the same time, permanent magnet motors often work at high temperature, even to 200°C. So there is a requirement of high temperature-stability for the sintered NdFeB magnets to ensure high efficiency of the motors. By adding Co atoms, the high Curie temperature and high temperature-stability of the magnets can be obtained [6-7]. In this paper, we adopted suitable component ratio of heavy rare-earth elements and some micro-quantity additives, and adjusted optimized processing parameters. The sintered NdFeB magnets with high-coercivity and relatively high temperature-stability have been produced eventually.

2. Experimental

The alloy with nominal composition of $Nd_{11.13}Fe_{bal}Tb_{2.67}Dy_{0.61}Co_{1.12}Zr_{0.14}Al_{0.49}Cu_{0.24}B_{6.07}$ (at.%) was produced by the strip casting (SC) technique from industrial grade raw materials with purity above 99.5%. The strips were employed in hydrogen decrepitation (HD) into coarse fragile particles. These coarse particles were then sent into jet milling (JM) procedure under a N_2 gas flow. The dry powder was compacted of 14 mm diameter and 15 mm height with the alignment made with the maximum 2 T magnetic field. This was followed by a cold isostatic pressing under a pressure of 300 MPa. The compacts were sintered in the vacuum of 10^{-3} Pa at the temperature of 1060-1090°C for 2-4 hours. The magnets were then rapidly quenched from the sintering temperature to room temperature under an over pressure of argon gas. Two step heat treatments were employed on the sintered magnets under the same vacuum of sintering procedure. A post heat treatment was made at 800-900°C for 2 h, including rapidly quench to the room temperature under the gas atmosphere of argon. And the low temperature annealing was made at 480-550°C for 2 h, followed by the same quenching techniques of the post heat treatment.

The microstructure of SC strips was observed by scanning electron microscope (TM-1000). And the size distribution of powder attained by JM was measured by laser diffraction particle size analyzer. The magnetic properties of the magnets at room temperature were measured with physical properties measurement system (PPMS) (Quantum Design Model-9). And the demagnetization curves at high temperature were measured after full magnetization under a pulsed magnetic field of 3 T, using the B-H measuring instrument.

In order to correlate the intrinsic coercivity with annealing processing and the microstructure of the boundary phase, the high-resolution field emission gun-scanning electron microscopy (FEG-SEM) was applied. All the test samples were prepared using as-sintered and two steps heat treatment. And the microstructure was investigated on cross sections perpendicular to the alignment direction.

The curves of magnetization depended on temperature for the sintered magnets were measured by VSM (Lakeshores 7410) with the magnetizing field of 1000Oe with temperature zone from room temperature (298K) to 1273 K.

3. Results and discussion

Figure 1(a) and 1(b) show the microstructure of the strips made at 500X and 10kX magnification, respectively.

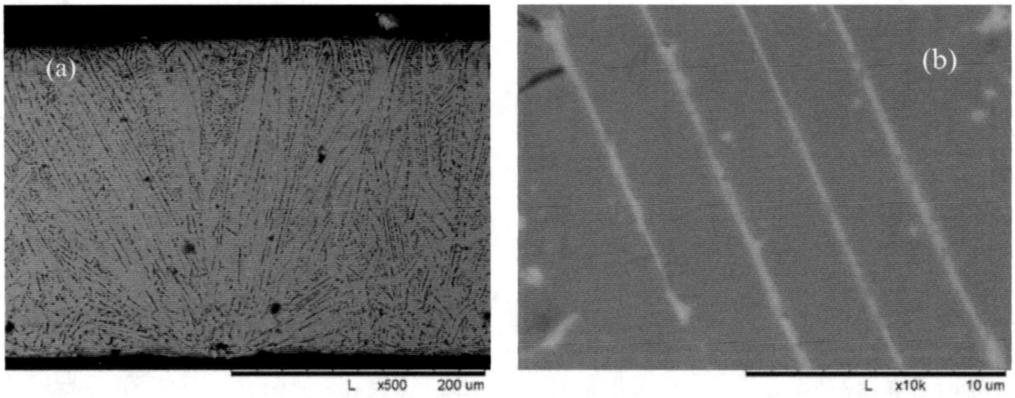

Figure 1. The microscopy pictures of the SC strip with v(wheel speed)=1.7m/s.

Figure 1(a) indicates that a high uniform columnar grain of matrix phase has formed through the strips with the thickness between 0.2 and 0.3mm. From Figure 1(b), it can be seen that the matrix phases ($RE_2Fe_{14}B$) with the thickness between 4 and 6μm were obviously separated by the grain boundaries of RE-rich phases with the thickness of 0.3-0.5μm. No alpha iron phase can be observed from the microstructure of SC strips.

After procedures of strip casting and hydrogen decrepitation, the attained coarse particles were then sent into jet milling procedure. The average size of the powder produced by JM is about 3.60μm.

In order to verify whether the alpha iron is absent in the sintered magnets or not, we employed VSM to measure the magnetization of sintered NdFeB magnets dependent on temperature at the magnetic field H=1000Oe and under increasing temperature from 25-980°C applied on them. Also the Curie temperature of the magnet was easily obtained.

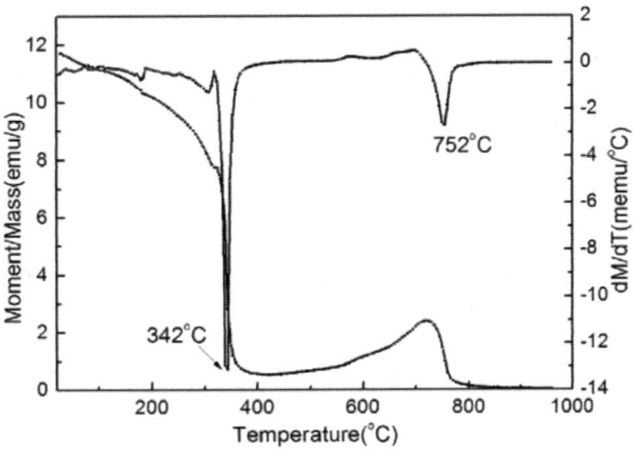

Figure 2. Temperature dependence of magnetization and dM/dT of the sintered magnets measured at H=1000Oe and above 25°C.

Figure 2 shows the curves of magnetization and dM/dT depended on temperature. The Curie temperature of the sintered magnet reaches a higher value of about 342°C than that of $Nd_2Fe_{14}B$ (312°C). The additive cobalt atoms can enter the matrix phase to substitute for iron, and formed additional $RE_2(Fe, Co)_{14}B$ during sintering process. The Curie temperature of cobalt (1403K) is higher than that of iron (1043K). Thereby, the Curie temperature can increase by about 11°C per at% cobalt, reported by R. S. Mottram et al. [6]. And additive Tb (or Dy) atoms are able to partially substitute into Nd sites in the matrix phase and form $(Nd, Tb)_2Fe_{14}B$ (or $(Nd, Dy)_2Fe_{14}B$), which can also increase the Curie temperature of the magnets. The Curie temperature of the magnets can increase by about 30°C under the effect of the additive atoms substituting.

One can find a peak in M-T curve from 600-800°C, as shown in Figure 2. It indicates that some soft magnetic phase (α-Fe) may exist in the sintered NdFeB magnets. That may attribute to alpha iron dendrites in SC strips which attained under not optimized conditions. The soft magnetic phase plays a negative role on the intrinsic coercivity as well as the magnetic energy product [8]. Therefore, the sintered NdFeB magnets have some potential in both intrinsic coercivity and energy product when the strip casting technique has been improved.

The density of sintered NdFeB magnets is roughly 7.57 g/cm^3. The magnetic properties were measured to investigate the influence of the annealing conditions on the coercivity of the demagnetizing loops. Also the demagnetization curves measured at difference temperature (room temperature, 100°C, 200°C) were conducted to investigate the temperature coefficient of residual magnetic flux $\alpha(B_r)$ and intrinsic coercivity $\beta({}_jH_c)$.

Figure 3 shows the demagnetizing loops of as-sintered state and as-annealed state. We employed the vacuum sintering parameter of 2h at 1085°C, post heat treatment of 2h at 875°C, and dual heat treatment of 2h at 875°C and 2h at 515°C. The intrinsic coercivity increases from 28.2kOe to 33.1kOe after post annealing. And it exceeds 36kOe after dual heat treatment. The high coercivity of the as-sintered magnets is attributed to the effects of Dy and Tb addition on the anisotropy field of the matrix phase of magnets. At the same time, the heavy rare-earth elements can be able to enter the grain boundaries, which may increase the nucleation field of sintered magnet and then make the intrinsic coercivity increase.

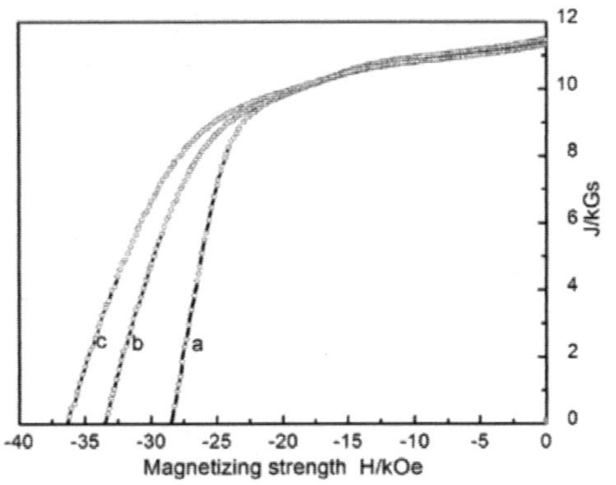

Figure 3. The demagnetization curves of the sintered magnets. a) as-sintered; b) post heat treatment; c) dual heat treatment.

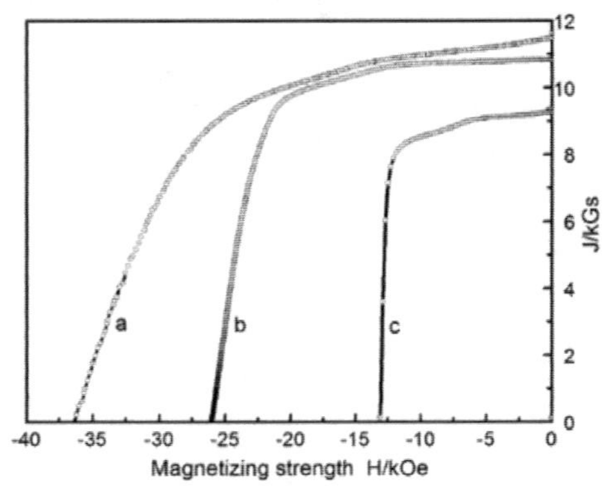

Figure 4. The demagnetization curves of the magnets with dual heat treatment. a) room temperature; b) 100°C; c) 200°C.

Figure 4 shows the demagnetizing curves measured at difference temperature for the magnet annealed at both high and low temperature. Both the residual magnetic flux B_r and intrinsic coercivity ${}_jH_c$ decrease with the temperature of circumstance increasing. According to Figure 4, the temperature coefficient of residual magnetic flux $\alpha(B_r)$ (20-200°C) is calculated as -0.113%/°C while the one of coercivity $\beta({}_jH_c)$ -0.355%/°C.

The low temperature coefficients display a high temperature-stability of sintered magnets. And this may be due to the effect of temperature compensation for residual magnetic flux and magnetocrystalline anisotropy field for coercivity. The low temperature coefficient of B_r may due to the effect of temperature compensation of heavy rare-earth elements as Tb and Dy. The compounds containing heavy rare-earth (Tb and Dy) have the positive temperature coefficient zone of B_r, in comparison to the compounds containing light rare-earth elements, such as Nd and Pr, with negative temperature coefficient zone [9]. Thereby, Tb (or Dy) atoms partially substituting into Nd sites could be able to compensate the negative coefficient of $Nd_2Fe_{14}B$, which result the lower temperature coefficient of residual magnetic flux. By adding the heavy rare-earth elements, the magnetocrystalline anisotropy field can be increased, which may result a slight decrease in coercivity with the temperature outside increasing in a certain extent of temperature. Because of all the reasons above, the temperature coefficient of coercivity may become to get lower. With the low temperature coefficient of B_r and ${}_jH_c$, the temperature-stability of the sintered NdFeB magnets can be enhanced.

Trying to study the effect of annealing processing on the enhancement of magnetic properties, the microstructure of the sintered magnets was analyzed by FEG-SEM, which is a possible technique to detect the morphology of grain boundaries affected by heat treatment.

Figure 5 shows FEG-SEM images of sintered magnet made at 30kX magnification. Figure 5(a) corresponds to the as-sintered sample, in which the interfaces between grains seem not well defined and relatively discontinuous. Figure 5(b) corresponds to the sample after dual heat treatment at same magnification. The grain boundary phase of annealed sample looks more defined, continuous and clean compared to the as-sintered state. The annealing processing may modify the grain boundaries of the magnets into well defined, continuous and clean interfaces, which may result the increasing of intrinsic coercivity. Besides the change of morphology of grain boundaries, the diffusion and distribution of the elements affected by heat treatment may be another factor on coercivity.

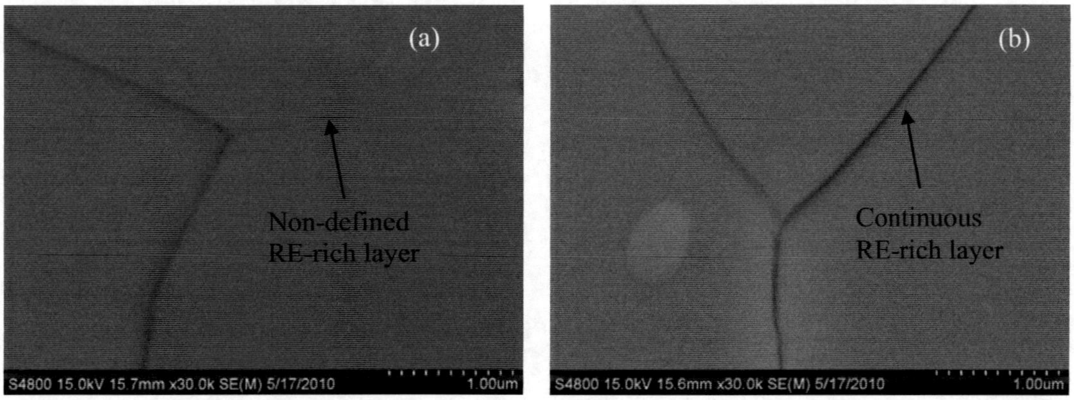

Figure 5. FEG-SEM pictures. (a) Sample as-sintered. (b) Sample with dual heat treatment.

There are some visible white inclusions in the $RE_2Fe_{14}B$ phase, which is similar to the results reported by F. Vial et al [10]. The EDX shows that the visible spherical inclusions in the $RE_2Fe_{14}B$ phase are RE-rich inclusions. These probably correspond to liquid of RE-rich caught in matrix phase during the re-crystallization processing.

4. Conclusions

Through appropriate processing conditions of traditional powder metallurgy of sintered NdFeB magnets, the magnet with high intrinsic coercivity was produced. After optimized heat treatment, the magnet with the nominal composition of $Nd_{11.13}Fe_{bal}Tb_{2.67}Dy_{0.61}Co_{1.12}Zr_{0.14}Al_{0.49}Cu_{0.24}B_{6.07}$ (at.%) reached high coercivity with the value of 36.3kOe, which the energy product was 34MGOe. The temperature coefficients of residual magnetic flux and intrinsic coercivity (20-200°C) were about -0.113%/°C and -0.355%/°C, respectively.

The relation between the magnetic properties, especially the intrinsic coercivity, and heat treatment applied on the magnets of as-sintered state had been investigated.

The analysis of microstructure of the magnet indicates that the intrinsic coercivity may be closely related to the morphology of grain boundary phases. The FEG-SEM technique may provide a method to detect the change of morphology of grain boundaries affected by heat treatment.

Acknowledgements

The work is financially supported by Major Scientific and Technological Special of Zhejiang (NO.2007C11046), Key Science and technology project of Ningbo (2008B10024), National Basic Research Program (NO.2010CB934601) of China and Ministry of Science and Technology China and the National Nature Science Foundation of China.

References

[1] M Sagawa, S Fujimura, N Togawa, H Yamamoto and Y Matsuura 1984 *J. Appl. Phys.* **55** 2083
[2] F Pourarian, S G Sankar and W E Wallance 1988 *J. Magn. Magn. Mater.* **74** 177-180
[3] G Bai, R W Gao, Y Sun, G B Han and B Wang 2007 *J. Magn. Magn. Mater.* **308** 20-23
[4] B M Ma, W L Liu, Y L Liang, D W Scott and C O Bounds 1994 *J. Appl. Phys.* **75** 6628
[5] K G Knoch, B Grieb, E T Henig, H Kronmüller and G Petzow 1990 *IEEE Trans. Magn.* **26** 951
[6] R S Mottram, A J Williams and I R Harris 2000 *J. Magn. Magn. Mater.* **217** 27-34
[7] R S Mottram, A J Williams and I R Harris 2001 *J. Magn. Magn. Mater.* **234** 80-89
[8] A Kianwash and I R Harris 1998 *J. Alloys Comp.* **279** 245-251
[9] Dong Li, Erda Xu, Jailin Liu and Yuxian Du 1980 *IEEE Trans. Magn.* **16** 988-990
[10] F Vial, F Joly, E Nevalainen, M Sagawa, K Hiraga and K T Park 2002 *J. Magn. Magn. Mater.* **42-245** 1329-1334

Journal of Physics: Conference Series **266** (2011) 012053 doi:10.1088/1742-6596/266/1/012053

The analysis of adhesion failure between Ni-coating and sintered NdFeB substrate

Heng xiu Y[1]**, Yong D, Zhen lun S**
Ningbo Institute of Materials Technology and Engineering, Chinese Academy of Sciences, 315201 China

E-mail: yanghengxiu@nimte.ac.cn.

Abstract. Ni-coating was widely used to protect the sintered NdFeB magnet from corrosion by Watt electro-deposition solution. However, the protection failure always occurs due to poor adhesion strength between Ni-coating and NdFeB substrate. In present work, the adhesion strength of the Ni-coating on NdFeB substrate was measured by vertical tensile method to strip Ni-coating from NdFeB substrate. The results revealed that the adhesion failure was occurred in the side of the NdFeB substrate due to a weak zone sometimes shown cracks located inside of NdFeB substrate, rather than in the interface between Ni-coating and NdFeB substrate. Comparing with cross section morphology of NdFeB magnet after pretreatment, it is concluded that the crack could be formed during the electro-deposition process. The effect of the pH value of bath on adhesion strength indicated that the crack could be induced due to electrochemical hydrogenation of NdFeB substrate during electro-deposition.

1. Introduction

NdFeB permanent magnets have been widely used for their excellent magnetic performance [1]. However, the corrosion resistance of the NdFeB permanent magnets is poor due to its multiphase microstructure consisting of matrix phase $Nd_2Fe_{14}B$, Nd-rich phase, and B-rich phase [2-6]. Protective coatings had to be deposited on the NdFeB substrate surface to isolate the substrate from the environment [7, 8]. Ni-coating was widely used in industry by Watt electro-deposition solution. Adhesion strength of the Ni-coating on the NdFeB substrate is one of the most important crucial factors to protect the NdFeB magnet form corrosion. However, the failure of protection always occurs due to poor adhesion strength between the Ni coating and the NdFeB substrate. The mechanism of the adherence evaluation between electro-deposited Ni-coating and NdFeB permanent magnet was reported by a thermal shock method in previous paper [9]. In present report, the adhesion strength of Ni-coating on NdFeB substrate was evaluated by a vertical tensile method, the effect of pH on adhesion strength was researched and the adhesion failure mechanism between electro-deposition Ni-coating and sintered NdFeB substrate was discussed.

2. Experiment

The specimens of the sintered NdFeB magnets were shaped in dimension of 25×20×3mm. The NdFeB magnet is composed of matrix phase, Nd-rich phase, and B-rich phase. The matrix phase, a tetragonal

[1] Corresponding author: yanghengxiu@nimte.ac.cn.

compound of $Nd_2Fe_{14}B$, is the most significant of the ferromagnetic phase. The Nd-rich phase was located at the grain boundary surrounding the matrix phase. B-rich phase disperses randomly inside the Nd-rich phase.

Before electro-deposition, pretreatment was performed as follows: degreasing the surfaces of the specimens in an alkaline solution for 2 minutes, pickling in a 3% nitric acid solution for 3 minutes, activated in a 3% sodium citrate solution for 2 minutes, and finally, rinsing with de-ionized water. The Ni-coating was deposited on NdFeB substrate by electro-deposition in watt solution, and the composition of bath as follows: $NiSO_4$ 1.8mol/L, $NiCl_2$ 0.35mol/L, HBO_3 45g/L.The density of the deposition current was 0.3A/dm^2, with deposition time 100 minutes and bath temperature 323K. The Ni-coating was deposited in different pH value of bath, and the pH value as following: 4.75, 4.5, 4.16, 3.8.

The adhesion strength of electro-deposition Ni-coating on NdFeB substrate was tested by vertical tensile method, during test, the coating was stripped out from substrate by tensile load. The schematic diagram of test mount was shown in fig.1. A stainless steel rod with 18 mm diameter was glued onto the Ni-coating surface and dried at room temperature for 72 hours before tests. During adhesion strength test, with increasing tensile load, the load-displacement curve was recorded until the Ni-coating was stripped out from NdFeB substrate. After adhesion strength test, the NdFeB specimen was stripped into two parts: the stripped layer and the remained NdFeB substrate as its model was shown in fig.2. The cross section morphologies were observed by scanning electron microscopy (SEM).

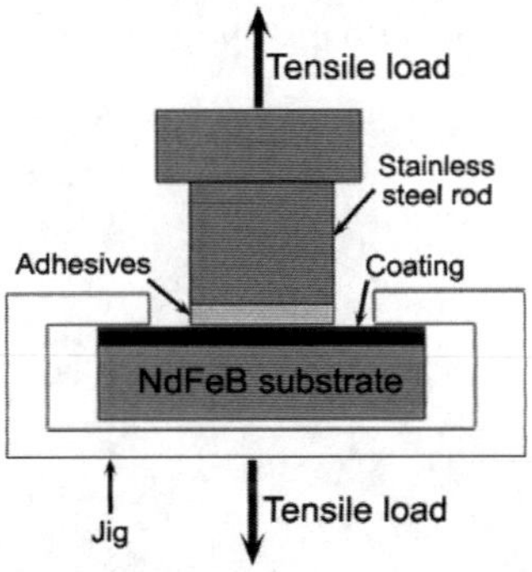

Figure 1. Schematic diagram of test mount

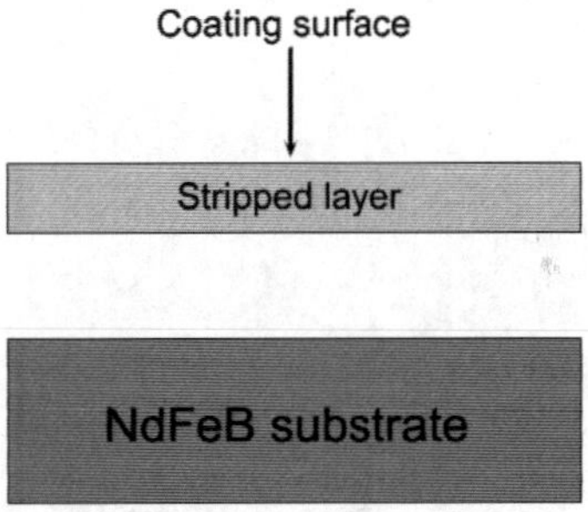

Figure 2. A model of the NdFeB specimen was stripped into two parts

3. Results and Discussion

For the tensile load applied on the coating, the NdFeB specimen was stripped into two parts: the stripped layer and the remained NdFeB substrate. Fig.3 shows the cross section morphology of the stripped layer, and its Ni-coating was deposited with the bath pH value of 4.16. Both the Ni-coating and NdFeB substrate can be seen from fig.3. It can be found that the fracture did not happen at the interface between the Ni coating and the substrate. The stripped plane was located inside of NdFeB substrate close to the interface between the Ni-coating and the NdFeB substrate. It indicated that the adhesion failure was happen at inside of NdFeB substrate.

Fig.4a shows the cross section morphology of the Ni-coated NdFeB substrate, and the Ni-coating was deposited with the bath pH value of 4.16. There is an excellent interface bonding between the Ni-coating and NdFeB substrate. However, a crack was found in the NdFeB substrate close to the interface between the Ni-coating and NdFeB substrate. As a weak zone, the crack of the substrate could lead to material fracture while a tensile load was applied to the as-deposited NdFeB specimens. The existence of the crack could induce the stripped plane occurred in the NdFeB substrate close to the interface between the Ni-coating and NdFeB substrate, and the NdFeB substrate was divided into surface layer and inner layer due to the crack. The adhesion strength was measured by vertical tensile

Journal of Physics: Conference Series **266** (2011) 012053 doi:10.1088/1742-6596/266/1/012053

method reflected only the adhesion property between the surface layer and inner layer of the NdFeB substrate. The adhesion strength could depend intensively on the crack shape, and the wider or longer crack can lead to the lower adhesion strength. Therefore, the formation reason of the crack is a crucial factor to influence the adhesion strength. It is obvious that the crack can be formed during the whole Ni-coating preparation process including pretreatment and electro-deposition process. However, no crack was found in the cross section morphology of NdFeB substrate after pretreatment in Fig4b and this indicated that the crack should be formed in the course of electro-deposition.

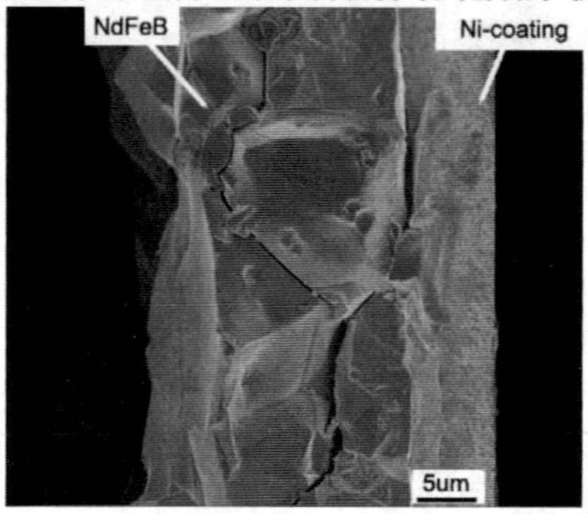

Figure 3. The cross-section morphology of the stripped layer.

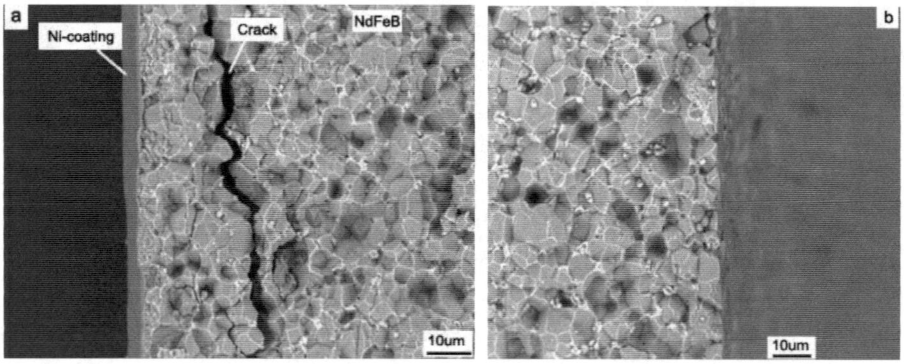

Figure 4. a: The cross section morphology of Ni-coated NdFeB substrate; b: The cross section morphology of NdFeB substrate after pretreatment

Fig.5 shows adhesion strength of the NdFeB specimens deposited with the Ni-coating at different pH value of bath. The adhesion strength was increased gradually with the increase of pH value of bath. During electro-deposition in Watt solution, a small amount of hydrogen ions were reduced to atomic hydrogen on the electrode surface is inevitable. Generally, the decrease of pH value of bath can make more hydrogen atoms reduce on the substrate surface during electro-deposition, thereby the more hydrogen enter into the coating. This leads to the increase in the brittleness and decrease in the plasticity of the coating, and it could deteriorate the interface bonding and decrease the adhesion strength between the coating and substrate. However, for the Ni-coated NdFeB, the adhesion failure did not happen in the interface between the Ni-coating and NdFeB substrate, and the adhesion strength could depend intensively on the width and length of crack. The change of adhesion strength indicated that the pH value of bath had a strong influence on the crack and the reduced hydrogen atoms could diffuse into the NdFeB substrate and induced the formation of crack. W. Bloch had reported that hydrogen embrittlement could happen in the NdFeB substrate during electro-deposition of the Ni-coating [10]. It was well known that the hydrogen can be absorbed forcefully into the sintered NdFeB magnet [11, 12]. The affinity of hydrogen with the Nd-rich phase is stronger than that of the $Nd_2Fe_{14}B$ matrix phase [13], and in the case of electrochemical hydrogen charging, the hydrogen absorption process is rather slow and hydrogen is mainly located in intergranular regions surrounding the grains from all sides [14]. Therefore, for the electro-deposition of Ni-coating on NdFeB substrate, most of reduced hydrogen atoms could be absorbed into the Nd-rich phase and diffused into inside of the NdFeB substrate. Those hydrogen atoms could finally concentrate in the defects (pore, cavity etc) of the grain boundary. Subsequently hydrogen can react with Nd-rich grain boundary phase to produce Nd hydride, which can cause tremendous crystal lattices expansion of the Nd-rich phase and generate local stresses. The mechanical effect of the expansion is the disadhesion of the $Nd_2Fe_{14}B$ matrix grains, and finally resulted in the crack formation. When tensile load was applied to the NdFeB specimen, the adhesion failure should be occurred in the weakest location, thus, the NdFeB specimen could be stripped along the crack. Even without load was applied to the NdFeB specimen, the bubbling and peeling of the Ni-coating could happen due to the existence of the inner crack

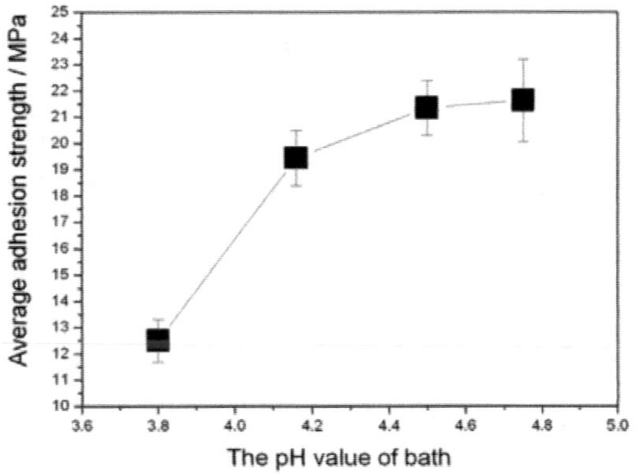

Figure 5. Adhesion strength of NdFeB specimens deposited with Ni-coating at different pH value of bath.

4. Conclusion

Adhesion failure of the Ni-coated NdFeB substrate was occurred in the NdFeB substrate close to the interface between the Ni-coating and NdFeB substrate due to the existence of the crack. The crack formation was attributed to the electrochemical hydrogenation of the Nd-rich grain boundary phase during electro-deposition.

References

[1] J F Herbst, and J J Croat 2004 *Journal of Magnetism and Magnetic Materials* **279** 353
[2] Y W Song, H Zhang, H X Yang, and Z. L. Song 2008 *Materials and Corrosion* **59** 794
[3] S M Tamborim, Takeuchi, D S Azambuja, A M Saliba-Silva, and I Costa 2006 *Surf Coat Technol* **200** 6826
[4] H Bala, S Szymura, and J J Wyslocki 1990 *IEEE Tran Magn* **26(5)** 571
[5] A S Kim, and F E Camp 1991 *J Mater Eng* **131** 175
[6] A Saliba-Silva, R N Faria, M A Baker, and I Costa 2004 *Surf Coat Technol* **185** 321
[7] H H Man, H C Man, and L K Leung 1996 *Journal of Magnetism and Magnetic Materials* **152** 40

[8] H H Man, H C Man, and L K Leung 1996 *Journal of Magnetism and Magnetic Materials* **152** 47
[9] O B G ASSIS, and M FERRANTE 1997 *Journal of Materials science Letters* **16** 1902
[10] W Bloch, K Grendel, and H staubach 1990 *Proceedings 11th intl. Workshop on RE Magnets and Their Application, Pittsburgh* 108
[11] E Claude, S. Ram, I Gimenez, P chaudouet, D Boursier, J Joubert 1993 *IEEE Trans Magn* **29** 2767
[12] K E Chang, Garry, and W Warren 1995 *IEEE Trans Magn* **31** 3671
[13] K Turek, P Liszkowski, and H Figiel 2000 *J Alloy Comp* **309** 239
[14] M Rada, and A Gebert 2006 *Journal of Alloys and Compounds* **415** 111

Journal of Physics: Conference Series **266** (2011) 012054 doi:10.1088/1742-6596/266/1/012054

Preparation and Properties of Anisotropic Nano-crystalline NdFeB Powders Made by Hydrogen Decrepitation of Die Upsetting Magnets

P P Yi[1], D Lee, A R Yan
Ningbo Institute of Material Technology & Engineering,
Chinese Academy of Sciences, Ningbo 315201, China

E-mail: ypp@nimte.ac.cn

Abstract. Anisotropic nanocrystalline NdFeB powders were prepared by hydrogen decrepitation (HD) of die upsetting magnets. The effects of varying temperatures of HD on the microstructure and magnetic properties of the anisotropic NdFeB particles were studied. It shows that the powders which obtained by HD process at higher temperature were larger than that at lower temperature, and the HD powders show a well anisotropy at 723 K, the remanence (B_r) was more than 12.46 kG, the maximum energy product ($(BH)_{max}$) was 19.06 MGOe, and the coercivity (H_{cj}) was 7.2 kOe. The microstructure of the anisotropic powders revealed that with a reasonable HD temperature, the platelet grains were not destroyed. They were nearly 150-300 nm long and 30-50 nm wide. The results indicate that HD process was an effective way to prepare the anisotropic NdFeB powders.

1. Introduction

Hot pressing and hot deformation have been used for producing fully dense anisotropic magnets in commercial quantity for many years. The combination of these two techniques has been successfully shown to transform the isotropic grains of melt-spun ribbons into fully oriented grains in fully dense magnets [1, 2]. HD process has been developed to a well-established method for producing sintered NdFeB magnets [3, 4]. The effects of hydrogenation on the structural, chemical, physical and mechanical properties have been well established [5, 6]. In $R_2Fe_{14}BH_x$ lattice (R = Nd, Pr, Dy, Tb), hydrogen insertion concludes four types of R-rich tetrahedral interstitial sites in the lattice of tetragonal phase, among which two sites are mostly active for hydrogen absorption: site (8j) (r_i = 0.045 nm, 3R-1Fe) and site (16k) (r_i = 0.04 nm, 2R-2Fe) [7]. The "r_i" means the radius of the insertion site in the NdFeBH lattice. The numbers before R and Fe denote those of ligand atoms. So, r_i = 0.045nm, 3R-1Fe means that: 1) The radius of the insertion site 8j is 0.045nm; 2) The surrounding environment of the site 8j has 3 atoms of rare earth element and 1 atom of iron. For light R elements site (16k) is favoured. The relatively large expansion in lattice parameters induces a HD process and very fine magnetic particles can be obtained easily [8]. It was reported that the anisotropic NdFeB powders could be obtained from hot rolled ingots by HD process [9]. However, few works have been reported to adopt the approach that combine the die upset process with the HD process to get the anisotropic

[1] Postal address: No.519 Zhuangshi Road, Ningbo 315201, China. Tel: +86-574-86685145, Fax: +86-574-87911392.

NdFeB nano-crystalline powders. It should be a useful tool for the production of anisotropic NdFeB powders from dies upset material.

In this paper, the authors report the method and results of the anisotropic NdFeB nano-crystalline powders prepared by hydrogen decrepitation of die upsetting magnets.

2. Experimental

Anisotropic oriented NdFeB magnets were prepared by die upsetting with nominal compositions of $Nd_{29.6}Pr_{0.1}Co_{5.9}Ga_{0.44}B_{0.91}Fe_{bal}$ in an argon atmosphere at 1133 K. The above die upset magnets were decrepitated in hydrogen at temperature range from 473 K to 883 K for 2h with a partial pressure of 70 kPa, followed by a slow desorption in high vacuum at 653 K for 2h. The heating rate of hydrogen absorption and hydrogen desorption as shown in Fig. 1 and Fig.2 were 5 K/min. The two steps treatments were performed in a special reactor equipped with a fast electric heating device and a powerful vacuum pumping system. Subsequently, the above powders were pestle and sieved to get homogeneous anisotropic powders. The orientation of the anisotropic powders which were cemented with AB glues was detected by X-ray Diffraction (XRD). Magnetic properties and phase transitions were measured with a vibrating- sample magnetometer. The sweeping rate as shown in Fig. 4 was 224.39 Oe/s. The heating rate for the thermo-magnetic curve measurements were 0.5 K/min with the applied field of 200 Oe. Microstructure observations of the anisotropic powders were carried out using scanning electron microscopy (SEM).

3. Results and Discussion

Fig. 1 shows the curves of the hydrogen absorption as a function of temperature. It can be seen that there are two peaks in the Hydrogen absorption curves, one is between 473 K and 723 K and the other is between 723 K and 973K. There is a little inflection point at 473 K. During hydrogen absorption procedure, the pressure in the reactor was in a non-dynamic equilibrium, concluding the positive pressure caused by heat expansion and negative pressure caused by hydrogen absorption. The inflection point at 473 K indicates that pressure transformation from dynamic equilibrium to non-dynamic equilibrium.

There are also two peaks in the hydrogen desorption curves, see in Fig. 2. The first peak between 450 K and 653 K, which can be attributed to the hydrogen desorption associated with both the partial dihydride of Nd hydride from Nd_2H_5 to NdH_2 as interpreted by Book et al. [10] and the dehydrogenation of hydrogenated nano-crystalline grain boundaries which are at least hundreds of times more than that in the conventionally disproportionate alloy. The second peak between 800 K and 900 K indicates the full de-hydriding of the Nd hydride from NdH_2 to Nd.

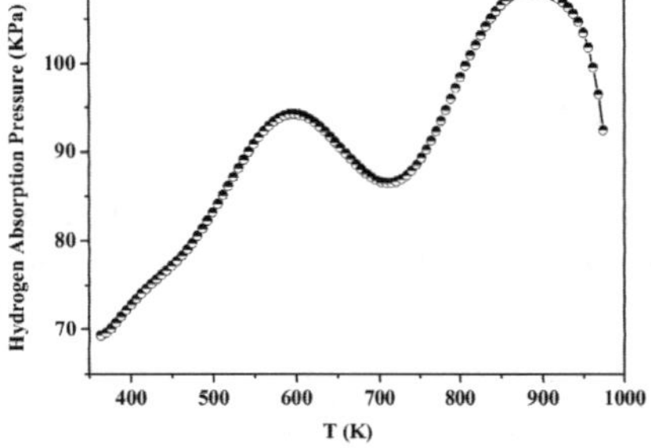

Figure 1. Hydrogen absorption as a function of temperature.

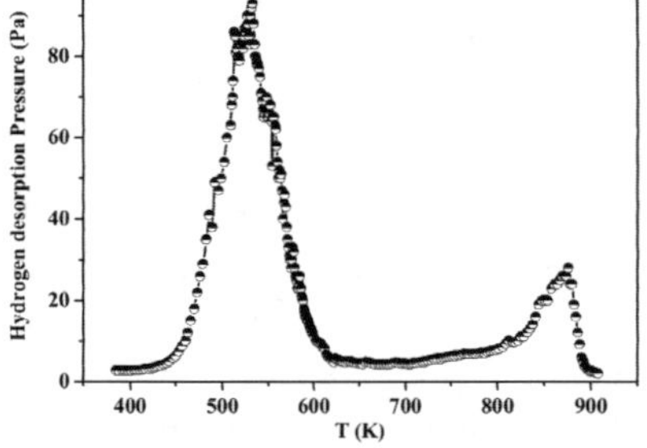

Figure 2. Hydrogen desorption as a function of temperature.

According to Fig.1 and Fig. 2, hydrogen absorption of samples was performed at 473 K, 603 K, 723 K, and 883 K, respectively. Hydrogen desorption of all the samples was carried out at 653 K for 2h.

Fig. 3 shows the XRD patterns of the oriented anisotropic NdFeB powders which were cemented with AB glue (A: B=1:1) after hydrogen absorption, without desorption. It can be seen that the (004), (105), (006) and (008) peaks are very obvious, and the intensity of (006) peak is stronger than the (105) peak. The (004) peak is increased with increasing the temperature of hydrogen absorption. However, the (221) peak is enhanced first and than dropped down as the temperature of hydrogen absorption increased. The peaks move about 0.4° to the left compared with the NdFeB standard patterns (PDF: 36-1296, Cu kα1), which the (004), (105) and (006) peaks are located at 29.320°, 38.305° and 44.581° respectively, indicating the lattice expansion after hydrogen absorption.

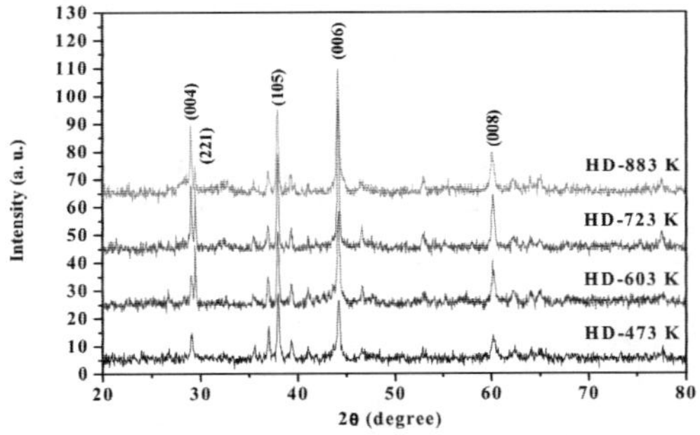

Figure 3. XRD patterns of the oriented anisotropic NdFeB powders, without hydrogen desorption.

Fig. 4 (a) shows the hysteresis loops of the oriented NdFeB powders after hydrogen decrepitation. The powders are not dehydrogenated. It can be clearly seen that without hydrogen desorption, the magnetic properties of powders are drastically worse, even no anisotropy can be seen from the curves for the powders. Fig. 4 (b) gives magnetic hysteresis loops of the oriented NdFeB powders after hydrogen decrepitating, the powders are dehydrogenated. It can be seen that the magnetic properties of the powders are enhanced greatly as compared with the powders without hydrogen desorption. The H_{cj} is increased first and than declined markedly with increasing the temperature of hydrogen absorption. With a reasonable temperature of hydrogen absorption, it shows a well anisotropy. The B_r is more than 12.46 kG; the $(BH)_{max}$ is 19.06 MGOe, but the H_{cj} is very small, only 7.2 kOe. Interestedly, the B_r and H_{cj} are quite lower with the temperature of hydrogen absorption of 883 K. It may ascribe to the damaged of the grain boundary phase, especially the Nd-rich phase. It is well known that [11, 12] the Nd-rich phase in the grain boundary plays an important role on improving the H_{cj}. However, the hydrogen decrepitation reaction is major in the Nd-rich at the grain boundary. As a result, the Nd-rich phase in the grain boundary is destroyed badly, thus the H_{cj} is lower.

Table 1 gives the hydrogen content in the decrepitation NdFeB powders which are not dehydrogenated. The hydrogen content rose sharply with increasing the temperature of hydrogen absorption. This is also a major reason why the H_{cj} is very lower. Moreover, the size of the powders with a higher hydrogenation temperature is bigger than that with a lower hydrogenation temperature. It can be also verified by SEM microstructures.

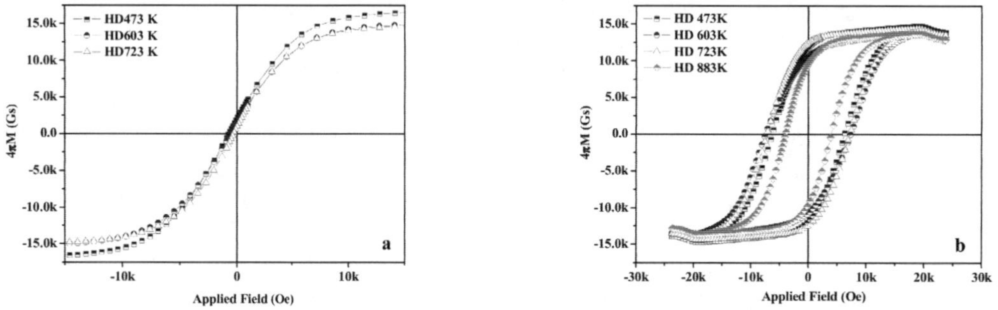

Figure 4. Hysteresis loops of the oriented NdFeB powders obtained by hydrogen decrepitation, (a) without dehydrogenation, and (b) after dehydrogenation.

Table 1. Hydrogen content in the hydrogen decrepitation NdFeB powders as a function of temperature, the powders were not dehydrogenated.

Hydrogenation temperature(K)	473	603	723	883
Hydrogen content (ppm)	2408	2939	3142	4008

The microstructure of the hydrogen decrepitation NdFeB powders is highly anisotropic with a reasonable temperature of hydrogen decrepitation, as shown in cross-section view (Figs. 5 (a)).The materials mainly consist of small platelet-shaped grains of the $Nd_2Fe_{14}B$ phase. Most small grains are stacked tighter, such that their flat faces (a, b planes) are aligned predominantly perpendicular to the pressure direction. The elongated grains are about 150-300 nm long and 30-50 nm wide. Such c-axis aligned platelet grains contribute to the high remanence of the hydrogen decrepitation samples. However, the textures in Fig. 5 (b) are damaged considerably. The grains growth largely and link up into stretches.

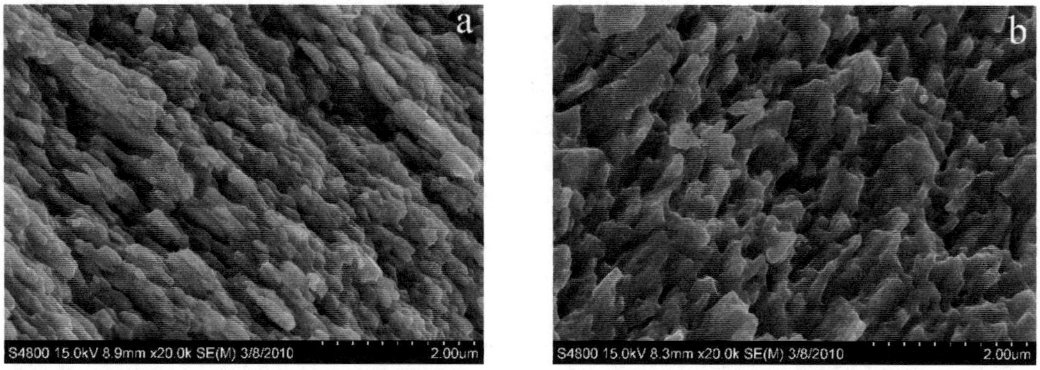

Figure 5. SEM images of the hydrogen decrepitation NdFeB powders, (a) at 723 K, and (b) at 883K. The powders were dehydrogenated.

Fig. 6 shows the thermo-magnetic curves of the hydrogen decrepitation powders without hydrogen desorption. It can be seen that there is no α-Fe phase in the high-temperature zone. It indicates that the disproportionation not occurred during hydrogen decrepitation procedure. So the decline of the H_{cj} is not due to the disproportionation reaction, but mainly because of the grain growth.

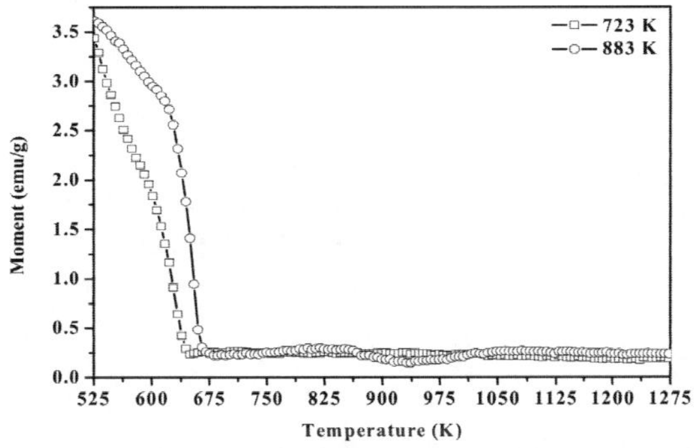

Figure 6. Thermo-magnetic curves of the hydrogen decrepitation NdFeB powders, without dehydrogenation.

4. Conclusions

HD powders prepared from die upset magnets show a well anisotropy at a reasonable hydrogenation temperature. With proper hydrogenation temperature, the textures in the decrepitation powders are not destroyed. It was found that the hydrogen dicrepitation powders shown a well anisotropy at 723 K, the B_r was more than 12.46 kG, the $(BH)_{max}$ was 19.06 MGOe, and the H_{cj} was 7.2 kOe. It is an effective tool for the production of anisotropic NdFeB powders from dies upset material.

5. References

[1] Lee R W 1985 *Appl. Phys. Lett.* **46** 790
[2] Mishra R K 1987 *J. Appl. Phys.* **62** 967
[3] Liszkowski K T, Figiel H 2000 *J. Alloy Compd.* **307** 297
[4] Hinz D, Kirchner A, Brown D N, Ma B M and Gutfleisch O 2003 *J. Mater. Proc. Tech.* **137** 358
[5] McGuiness P J, Harries I R, Rozendaal E, Ormerod J and Ward M 1986 *J. Mater. Sci.* **21** 4107
[6] Nakayama R, Takeshita T, Itakura M, Kuwano N, Oki K 1991 *J. Appl. Phys.* **70** 3770
[7] Fruchart D, Miraglia S, De Rango P and Wolfers P 2004 *J. Alloy Compd.* **383** 17
[8] Williams A J, McGuiness P J and Harries I R 1991 *J. Less-Common Met.* **171** 149
[9] Hinz D, Handstein A 1994 *IEEE Trans. Magn.* **30** 601
[10] Book D, Harries I R 1995 *J. Alloy Compd.* **221** 187
[11] Volkov V V and Zhu Y 1999 *J. Appl. Phys.* **85** 3254
[12] Krentz J E, Lee R W and Waldo R A 1991 *J. Appl. Phys.* **70** 6474

Acknowledgments

The project is supported by National Basic Research Program of China (Grant No. 2010CB934601) and the Major Scientific and Technological Special of Zhejiang (Grant No. 2007C11046).

Magnetoacoustic Emission Characteristics on Cold Rolled Low Carbon Steel

H Kikuchi[1], K. Matsumura, T. Fujiwara, K. Ara, Y. Kamada, S. Kobayashi
Iwate University, *Faculty of Engineering*
4-3-5 Ueda, Morioka, Iwate, Japan

E-mail: hkiku@iwate-u.ac.jp

Abstract. This paper describes the behaviours of magnetoacoustic emission (MAE) of cold rolled low carbon steel specimens. The MAE rms voltage and the moving average analysis were used for evaluating the profiles of MAE signals. The MAE rms voltage and the peak value of averaged rms voltage rise rapidly with increasing cold rolling below 10 % reduction ratio, and then they decreases at higher reduction ratio, whereas the magnetizing current when the averaged rms voltage takes a peak decreases up to 10 % reduction ratio, then becomes constant value. As a result of the observation of microstructures by transmission electron microscopy, the dislocations increases homogeneously up to 10 % reduction ratio, and then forms cell structures without a significant increase in dislocation density. The behaviours of MAE are attributed to the combined effects of cell texture and dislocation density.

1. Introduction

Magnetic Barkhausen noise (MBN) occurs when mainly 180° domain walls are unpinned from microstructural obstructions such as grain boundaries, inclusions and dislocations in a ferromagnetic material. Therefore MBN has interactions with microstructures of the materials. There are many investigations concerning MBN changes under various conditions including fatigue [1], irradiation [2], residual stress [3], creep [4], grain size [5, 6], etc. Since mechanical properties of a material also strongly depend on microstructures, those properties have good correlation with magnetic properties [7]. Consequently, MBN has been widely used as a tool for a characterization or a nondestructive evaluation (NDE) of ferromagnetic structural steels. On the other hand, magnetoacoustic emission (MAE) includes acoustic pulses generated by discontinuous motions of non-180° domain walls and localized strains. Because discontinuous domain wall motions arise from an interaction between domain walls and microstructures involving dislocations, grain boundary, etc., MAE is also a useful candidate for NDE technique for a degradation of mechanical properties and a characterization of microstructures. In addition, since MAE propagates through whole specimen unimpeded by eddy currents, MAE has an advantage over MBN technique. However, there are few researches that MAE characteristics are explained related to the microstructure changes observed experimentally [8, 9]. In this work, MAE of low carbon steel, typical structural steel, deformed by cold rolling was evaluated and its behaviour was discussed based on the observation of microstructure by transmission electron microscope (TEM).

[1] H. Kikuchi, E-mail: hkiku@iwate-u.ac.jp.

2. Experimental procedure

Low carbon steels (S15C) were prepared and annealed at 1173 K for one hour, followed by air-cooling. Then, those steels were cold rolled with different reduction ratios of 5, 10, 20 and 40 %. The chemical composition of the steels is 0.16 wt.% C, 0.2 wt.% Si, 0.44 wt.% Mn and Fe in balance. For measuring MAE, five kinds of plates (undeformed and cold rolled with 5 – 40 % reduction ratio) were ground into a dimension of $40 \times 60 \times 10$ mm^3.

The plate specimens were magnetized parallel to the rolling direction by a triangle current of 1 Hz and 1A amplitude using a single-yoke. The yoke was composed of a U-shaped Fe-Si yoke and a magnetizing coil as shown in figure 1. The dimension of the yoke is also shown in figure 1. The number of the magnetizing coil is 1000 turns. The MAE signals were detected by an acoustic emission (AE) sensor, *i.e.*, a PZT (lead zirconate titanate) transducer (NF 903N) with a diameter of 3 mm, attached on a surface of specimen with a medium for good contact between the PZT sensor and the specimen. The resonance frequency of the PZT sensor is 350 kHz. The measurement system for MAE signals is illustrated in figure 2. The original MAE signals induced at the PZT sensor were amplified (60 dB), band pass filtered (300 – 400 kHz), and finally acquired with a sampling rate of 100 kHz. The MAE signals of each specimen were measured several times and averaged. The dependence of rms (root mean square) voltage on reduction ratio was studied. The rms voltage RMS_{MAE} is defined as equation (1).

$$RMS_{MAE} = \sqrt{\frac{1}{T}\int_0^T V_{MAE}^2 dt} \tag{1}$$

where V_{MAE} is MAE signals acquired and T is a half period of excitation field.

In addition, rms voltage calculated with a moving average method [10] was adopted as an evaluated parameter in order to acquire much information. The data from each run consisted of a set of points (t_n, I_n, V_n), where t_n is the time at point n, I_n is the magnetizing current at point n, V_n is the value of V_{MAE} at point n. Since sampling frequency is 100 kHz in this study, the index n runs from 1 to 1×10^5 during 1 cycle. The data were divided into n/m sample sets in each of which V_n data of m points are stored consequently. Here, m is 50. For each sample set, rms voltage and averaged magnetizing current were computed as follows.

$$V_{rms} = \sqrt{\frac{1}{m}\sum_{i=1}^{m} V_i^2} \tag{2}$$

$$I_a = \frac{1}{m}\sum_{i=1}^{m} I_i \tag{3}$$

According to the calculation by equations (2) and (3), new data set (t_j, I_j, V_j) was obtained and the

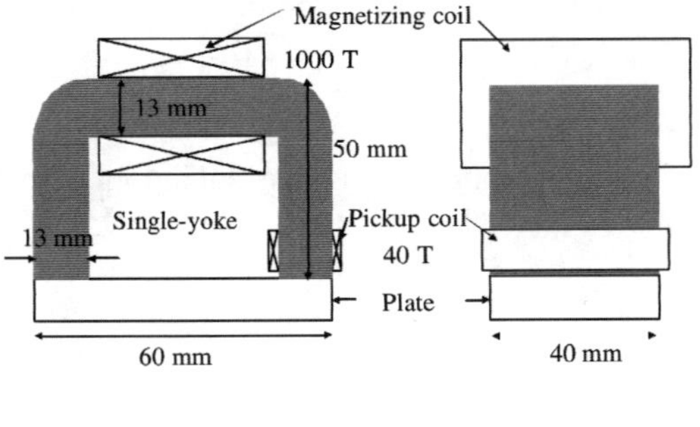

Figure 1. Dimension of single-yoke.

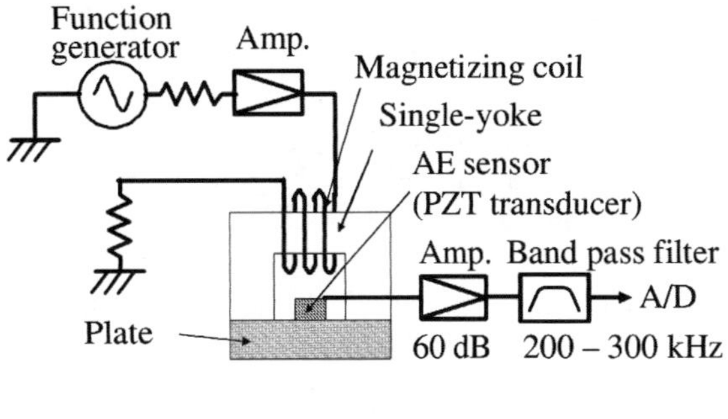

Figure 2. Measurement system for MAE.

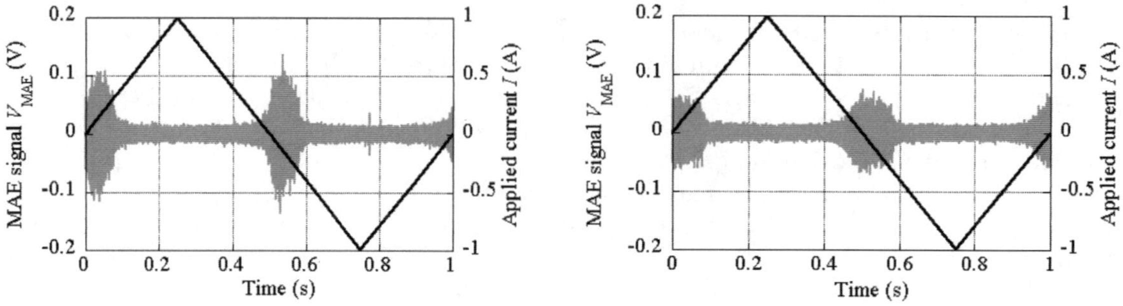

(a) undeformed specimen (0 %) (b) deformed with 40 % reduction ratio

Figure 3. Time dependent of MAE signals and excitation current applied to yoke on specimens with 0 % and 40 % reduction ratio.

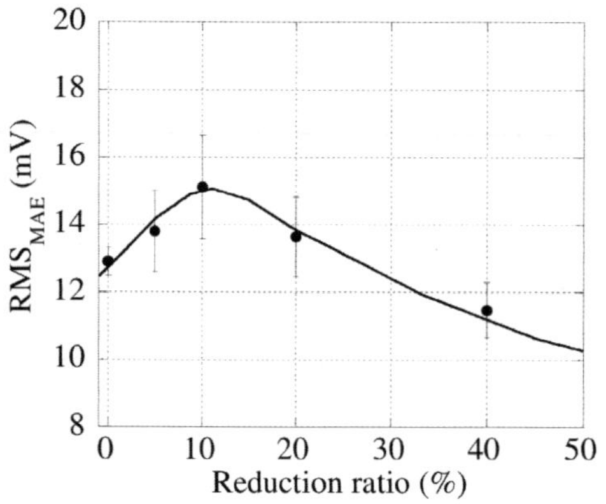

Figure 4. Dependence of rms voltage RMS_{MAE} on reduction ratio.

index j runs from 1 to n/m (= 2000). For the data set (t_j, I_j, V_j), moving average of rms voltage for k sample sets was calculated using the following equation.

$$ {}^{j}V_{rms\_ave} = \frac{1}{k}\sum_{q=1}^{k} V_{j+q} \tag{4} $$

When the index is j, the data sets (*i.e.*, V_j, V_{j+1}, ..., V_{j+k}) were used for calculation of jth value of ${}^{j}V_{rms\_ave}$. Finally the data sets of (I_a, ${}^{j}V_{rms\_ave}$) was obtained for each of n/m points. Here, k is 20.

To investigate the microstructures of the specimens, disk specimens of cold rolled steel were prepared and observed by TEM (Philips-tecnai 30).

3. Experimental results

Figure 3 shows a typical wave form of MAE signals and the magnetizing current applied to the magnetizing coil of the yoke against time for specimens undeformed (0 %) and deformed with 40 % reduction ratio. One cycle of MAE signal wave appears in a half cycle of the magnetizing current. The peak intensity of the MAE signal V_{MAE} for 0 % is greater than that for 40 %. The rms voltage RMS_{MAE} was calculated using equation (1) for each specimen based on the profiles shown in figure 3. Figure 4 shows the dependence of RMS_{MAE} on the reduction ratio. The value of RMS_{MAE} increases firstly and takes a maximum at 10 % followed by its reduction.

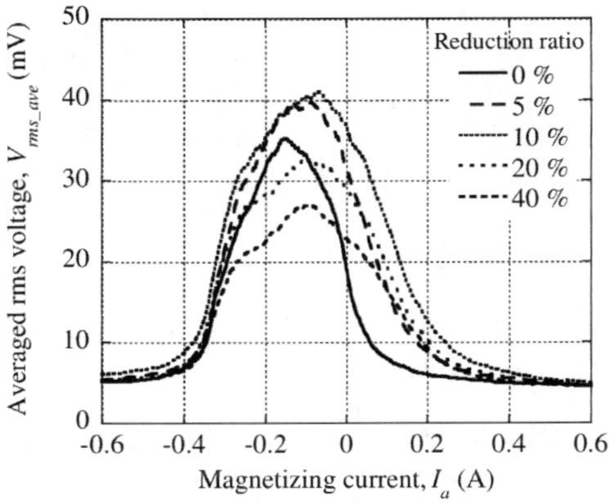

Figure 5. Dependence of moving averaged rms voltage on averaged magnetizing current.

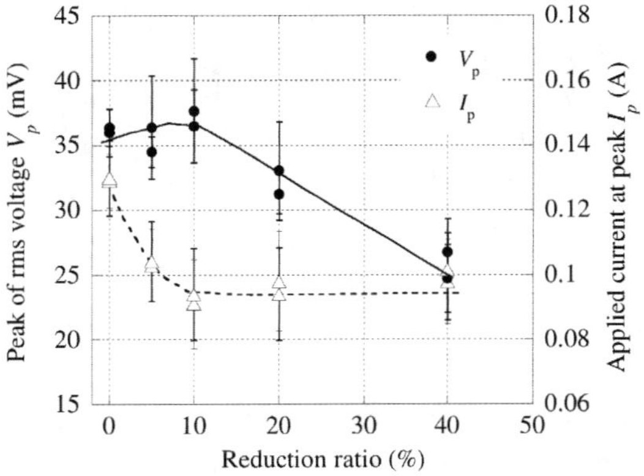

Figure 6. Dependence of peak in averaged rms voltage V_p and magnetizing current at peak I_p on reduction ratio.

Figure 5 shows the changes of the moving averaged rms voltage V_{rms_ave} against the averaged magnetizing current I_a for a half cycle of magnetizing current. One can see an averaged rms voltage in each specimen has one peak and an intensity of magnetizing current at a peak for the undeformed specimen (0 %) is larger than that of deformed specimen. Figure 6 shows the dependence of the peak intensity in averaged rms voltage V_p and the magnetizing current I_p at which averaged rms voltage reaches to the peak on the reduction ratio. The peak V_p increases at first up to 10 % reduction ratio, and then decreases, while the magnetizing current at peak I_p decreases rapidly up to 10 %, and becomes nearly constant above 10 %.

Figure 7 shows the photographs of TEM observation for S15C steels [10]. For undeformed specimen, dislocations distributed homogeneously and a large amount of dislocations are generated with the increase in reduction ratio up to 10 %. Above 10 %, dislocations form tangles and then cell structures. The dislocation density estimated by the Ham method [11] is 5×10^9 cm^{-2}, 2×10^{10} cm^{-2}, and 6–10 × 10^{10} cm^{-2} for 0, 5 and 10 % reduction ratio, respectively; for higher reduction, is not shown because of the formation of cell structures, which makes it difficult to estimate a reliable value of dislocation density.

4. Discussion

TEM observation shows that dislocations increase homogeneously firstly, and then tend to aggregate at the boundary of cell structure during cold rolling, and the dislocation density inside the cell texture is relatively low. These results show that dislocation density increases sharply during the reduction develops to 10 %, and then a higher reduction ratio does not induce a significant increase in dislocation density but induce only morphological changes in dislocations. As dislocations act as pinning sites for domain wall (DW) motion, the number of pinning sites for DW increases with the increasing dislocation density. When the number of pinning site increases, the event of MAE activity increases. Since the rms voltage RMS_{MAE} depends on the total event of MAE activity during one cycle of magnetization, RMS_{MAE} increases due to the increase in dislocation density up to 10 %. A large number of smaller cell structures may induce the reduction of the number of effective pinning sites for non-180° DW above 10 %; this causes the RMS_{MAE} decreases at higher reduction ratio.

The tendency in the peak of averaged rms voltage shows the same behaviour as the RMS_{MAE} values. However, below 10 % the parameter increases not remarkable as compared with RMS_{MAE} values, while the reduction above 10 % shows rapid decreases. These results may indicate that the number of pinning sites with weaker pinning forces increases with increasing dislocations up to 10 % reduction ratio. The magnetizing field at peak of the averaged rms voltage means a large number of MAE activity occurs at this magnetic field strength; many non-180° DWs were unpinned at these fields.

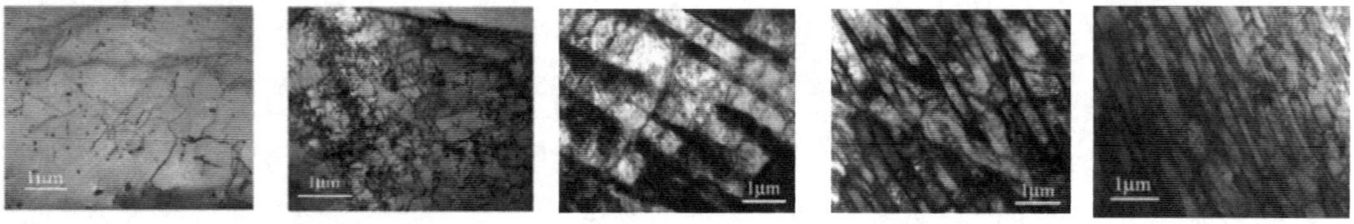

(a) 0 % (b) 5 % (c) 10 % (d) 20 % (e) 40 %

Figure 7. Photographs of microstructure in cold rolled S15C steel observed by TEM [10]. Reduction ratios are (a) 0 % (b) 5 %, (c) 10 % (d) 20 % and (e) 40 %, respectively.

Since pinning sites with weaker pinning forces increases, the magnetizing field at peak decreases below 10% reduction. Contrary, above 10 %, the forming of cell texture decreases the effective pinning sites, however, the pinning force does not change notably. Thus, only the peak of the averaged rms voltage decreases, and there is no change in the magnetizing field at peak.

Thus, MAE parameter changes can be understood on the basis of microstructure changes: the combined effects of cell texture and dislocations. However, further investigations are required to understand interactions between non-180° DWs and pinning sites in detail. Especially, DW observation is important and direct observation of interactions between DWs and pinning site is desirable.

5. Conclusion

This paper investigated the behaviours of MAE signals for low carbon steel with cold rolling related to their microstructure changes and the following results are clarified.

The rms voltage and the peak in averaged rms voltage reflect the number of pinning sites for non-180° domain wall motion. They rise sharply in an initial stage of cold rolling due to the increases in dislocation density. Then they decrease at higher reduction ratio since cell structures of dislocation may decrease the effective pinning sites. On the contrary, the magnetizing current at peak in the averaged rms voltage decreases monotonically due to the increase in dislocation with weaker interactions between pinning sites and DWs below 10 % reduction ratio. Since the development of cell structures of dislocation only decreases the number of pinning sites, the magnetizing current at peak is constant above 10 % reduction ratio.

References

[1] Palit Sagar S, Parida N, Das S, Dobmann G and Bhattacharya D. K., 2005 *Int. J. Fatigue*, **27**, 317.
[2] Park D-G, Hong J-H, Ok C-I, Kim J-W and Kim H.C, 1998 *IEEE Trans. Magn.*, **34**, 2036.
[3] Lo C. C. H, Paulsen J and Jiles D. C, 2004 *IEEE Trans. Magn.*, **40**, 2173.
[4] Mitra A, Chen Z J., Laabs F and Jiles D. C, 1997 *Philos. Mag.*, **A 75**, 847.
[5] Ranjan R, Jiles D. C, Buck O, and Thompson R. B, 1987 *J. Appl. Phys.*, **61**, 3199.
[6] Anglada-Riveraa J, Padoveseb L. R, Capó-Sáncheza J, 2001 *J. Magn. Magn. Mater.*, **231**, 299.
[7] H. Kronmüller, 1972 *Int. J. Nondestruct. Testing*, **3**, 315.
[8] Kusanagi H, Kimura H and Sasaki S, 1976 *J. Appl. Phys.*, **50**, 2985.
[9] Ono K and Shibata M, 1980 *Mater. Eval.*, **38**, 55.
[10] Kikuchi H, Ara K, Kamada Y and Kobayashi S, 2009 *IEEE Trans. Magn.*, **45**, 2744.
[11] Ham R, 1961 *Philos. Mag.*, **6**, 1183.

Analysis of Magnetic Minor Hysteresis Loops in Thermally Aged and Cold-rolled Fe-Cu Alloys

F. Takahashi, S. Kobayashi, T. Murakami, S. Takahashi, Y. Kamada, and H. Kikuchi
NDE&Science Research Center, Iwate University, Morioka 020-8551, Japan

E-mail: t2209018@iwate-u.ac.jp

Abstract. Neutron irradiation causes the formation of Cu precipitate in reactor pressure vessel steel and makes the steel susceptible to rupture. In the present study, we have examined magnetic minor hysteresis loops of Fe-1wt%Cu alloy after thermally ageing at 753 K and subsequent cold rolling to elucidate the effects of Cu precipitation on magnetic properties. Minor-loop coefficients, obtained from scaling power laws between field-dependent parameters of minor hysteresis loops, decrease with ageing time and show a local maximum around 200 min, reflecting the growth of Cu precipitates with ageing. For the alloy cold-rolled after ageing, the minor-loop properties linearly increase with reduction and show a good relationship with mechanical properties such as DBTT and hardness. These observations indicate that the analysis method using magnetic minor loops can be an useful technique of nondestructive evaluation of irradiation embrittlement and subsequent deformation hardening in reactor pressure vessel steels.

1. Introduction

Formation of Cu precipitate causes embrittlement of iron-based materials such as reactor pressure vessel (RPV) [1]. Neutron irradiation enhances the nucleation and formation of nano-size defects such as Cu-rich precipitates (CRPs) and increases in matrix damage. Both CRPs and the matrix damage impede the dislocation motion, resulting in undesired changes in mechanical properties such as a decrease in toughness and ductility, and an increase in ductile-to-brittle transition temperature, hardness. Hence, the embrittlement of RPV steels is a main concern for the nuclear industry for a long-term operation of nuclear power plants.

Magnetic method using minor hysteresis loops is one of useful one to get information on lattice defects [2,3]. It was revealed that there exist some universal scaling power laws between the field-dependent parameters and the coefficients are sensitive to defect density compared with coercivity. In addition, the coefficients were obtained with very low measurement field, being advantageous for the on-site integrity assessment.

Traditionally, thermally aged Fe-Cu alloys with Cu precipitates have been extensively studied to elucidate the detailed mechanism of Cu precipitation because of great cost in time and expense for preparation of neutron irradiated RPV steel samples as well as a difficulty of the handling of the irradiated samples. In this study, we have developed the magnetic method to investigate the effects of formation of Cu precipitates and subsequent cold rolling in Fe-1wt.% Cu alloys where nanometer Cu precipitates are induced by thermal ageing.

2. Experimental

2.1 Materials

Fe-1wt%Cu alloy was annealed at 1123 K for 5 hours, followed by a water quench. The alloy was then cold rolled (CR) up to 10% reduction in order to introduce dislocations. The alloy was thermally aged at 753 K up to a desired time (0, 20, 200, 1800 min) and subsequently cold rolled up to 20%

reduction. The thermal ageing was performed to induce Cu precipitation and simulate irradiation embrittlement, whereas the subsequent cold rolling was performed to simulate the effect of plastic deformation such as due to earthquake on irradiated RPV steels during operation.

Charpy impact test and hardness measurements were performed to investigate mechanical property changes due to ageing and subsequent cold rolling. Charpy impact test was performed with a pendulum of 26.6 kG and life angle of 138.5 ° in the temperature range of -100~200 °C. Ductile-brittle transition temperature (DBTT) was determined from a midpoint between low toughness brittle and high toughness ductile fracture regimes in the absorption energy curve. Vickers hardness was measured with the standard indentation technique. The applied load was 300 gf and 10 indents were taken for each sample.

2.2 Magnetic measurements

We prepared samples with different sample geometries; toroid (external and internal diameters of 20 and 18 mm, respectively, and thickness of 3 mm), Charpy (10×10×55 mm), disk (ϕ100×10 mm). For the toroidal sample, 110-turn exciting and 114-turn detecting coils were wound around the sample, whereas for the Charpy and disk samples a magnetic yoke was used to form a magnetically closed circuit. After demagnetizing the sample, a set of magnetic minor hysteresis loops was measured with step-by-step increasing field amplitude H_a. The major hysteresis loop was obtained with H_a = 5 kA/m.

2.3 Analysis method of minor loops

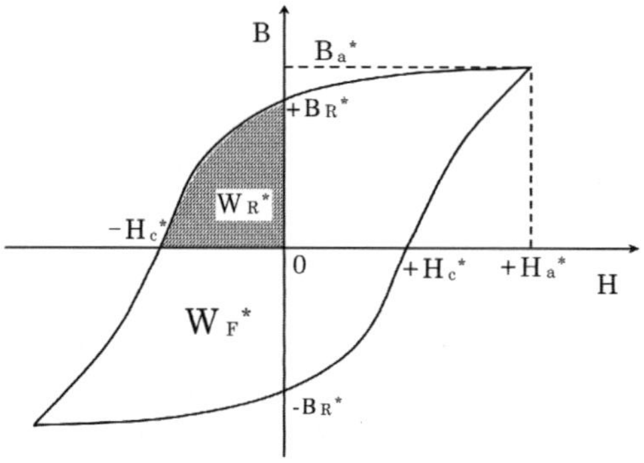

Figure 1. Parameters of a minor loop; B_a^*, B_R^*, H_c^*, W_F^*, and W_R^*, are magnetization, remanence, coercive force, hysteresis loss, remanence work of a minor loop, respectively.

Our analysis method of minor loops is based on our previous experimental results for Fe single crystals [2,3]. As shown in Fig. 1, several minor loop properties are introduced; minor-loop magnetization B_a^*, minor-loop coercive force H_c^*, minor-loop remanence B_R^*, minor-loop hysteresis loss W_F^*, minor-loop remanence work W_R^*.

The magnetization process before saturation can be divided conveniently into three stages. With increasing a field amplitude H_a after demagnetizing the sample, M_a^* shows a gentle increase in the first stage and then steeply increases in the second stage, followed by a gradual increase of M_a^* toward the saturation in the third stage. The most part of magnetization proceeds in the second stage through the movement of Bloch walls. We found that there exist several scaling laws between a pair of minor-loop parameters in the first or second stage as follows.

$$W_F^* = W_F^0 \left(\frac{B_a^*}{B_s} \right)^{n_F} \qquad (1)$$

Journal of Physics: Conference Series **266** (2011) 012056 doi:10.1088/1742-6596/266/1/012056

$$W_R^{\ *} = W_R^{\ 0}\left(\frac{B_R^{\ *}}{B_R}\right)^{nR} \qquad (2)$$

$$H_C^{\ *} = H_C^{\ 0}\left(\frac{B_R^{\ *}}{B_R}\right)^{nC} \qquad (3)$$

The exponents n_F, n_R, and n_C are independent of materials; $n_F \sim n_R \sim 1.5$ and $n_C \sim 0.45$. Here, Eq. (1) is well known as the Steinmetz [4]. $W_F^{\ 0}$, $W_R^{\ 0}$, and $H_c^{\ 0}$ are minor-loop coefficients and sensitive indicators of internal stress. In this study, the behaviour of the coefficients after ageing and cold rolling was investigated in detail.

3. Experimental result and discussion

Figure 2 shows a set of minor loops before and after thermal ageing and subsequent cold rolling. Loop width of each minor loop becomes narrower after thermal ageing, whereas it largely becomes wider after cold rolling due to the increase of density of dislocations which disturb the Bloch wall movement. Figures 3(a) and 3(b) show double logarithmic plots of W_F*-B_a* and W_R*-B_R* curves, respectively, before and after thermal ageing and subsequent cold rolling. As is seen in the linear relationship on a double logarithmic scale, there exist a power-law relation in these curves. Least-squares fits of W_F*-B_a* and W_R*-B_R* curves to Eqs. (1) and (2), respectively, yielded minor-loop coefficients $W_F^{\ 0}$ and $W_R^{\ 0}$.

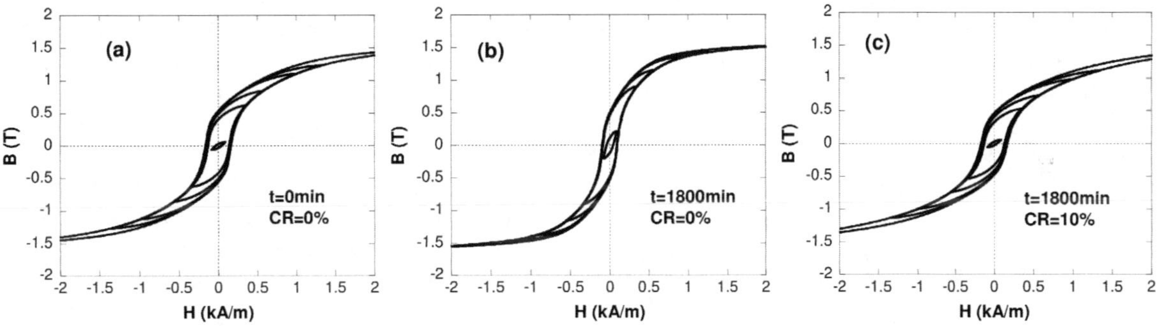

Figure 2. A set of minor loops for (a) t = 0 min and CR=0%, (b) t = 1800 min and CR=0% and (c) t = 1800 min and CR=10%. (toroid samples)

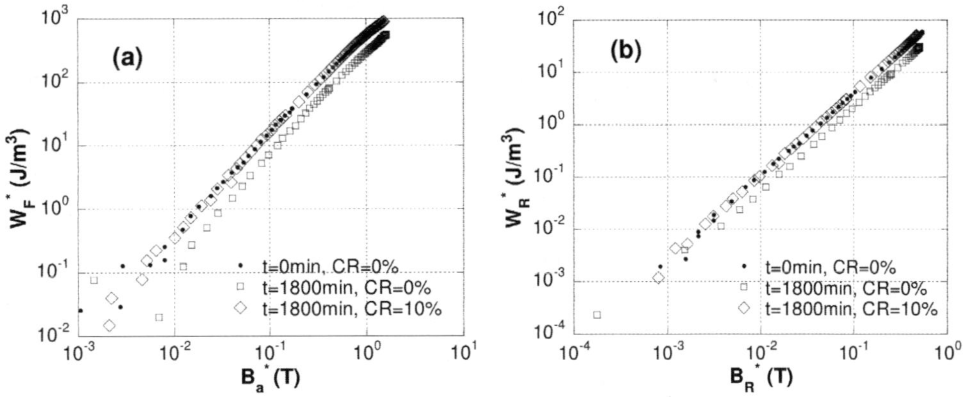

Figure 3. Double logarithmic plot of (a) $W_F^{\ *}$-$B_a^{\ *}$ curves and double logarithmic plot of (b) $W_R^{\ *}$-$B_R^{\ *}$ curves. (toroid samples)

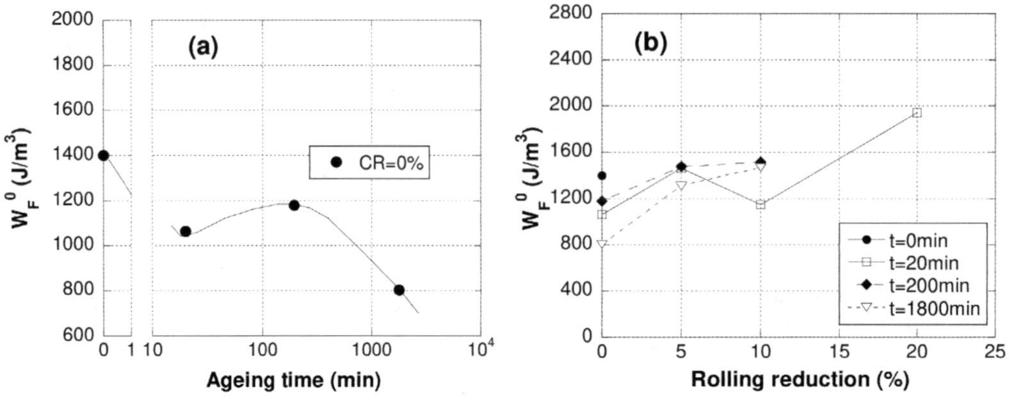

Figure 4. (a) Dependence of W_F^0 on aging time for CR=0%, (b) Dependence of W_F^0 on rolling reduction, taken at various ageing time. (toroid samples)

Note that minor loops with B_a* in the range of 0.1-1.0 T were used for the fits. Since both the coefficients show a similar dependence to ageing time and rolling reduction, we will only show the behaviour of W_F^0.

First, we show the dependence of W_F^0 on aging time before cold rolling in Fig. 4(a). W_F^0 steeply decreases at an early stage of ageing, followed by a local maximum around 200 min. The initial decrease might be due to the relaxation of residual stress, while the local maxima reflects the domain-wall pinning due to Cu-rich precipitates [5]. With increasing rolling reduction after thermal ageing, W_F^0 monotonically increase for each ageing time. These behaviors were commonly observed for all the sample geometries (toroid, Charpy, and disk). These observations reflect the sensitivity of minor-loop properties to lattice defects i.e. nano-size Cu precipitates and dislocations.

Figure 5 shows the relation between W_F^0 and DBTT for each ageing time. W_F^0 is almost in a linear proportion to DBTT for all the ageing time and the relation shifts toward a higher DBTT with increasing ageing time. This shift reflects the formation and growth of Cu precipitates due to thermal ageing. Such a good relationship was also observed for the relation between W_F^0 and hardness.

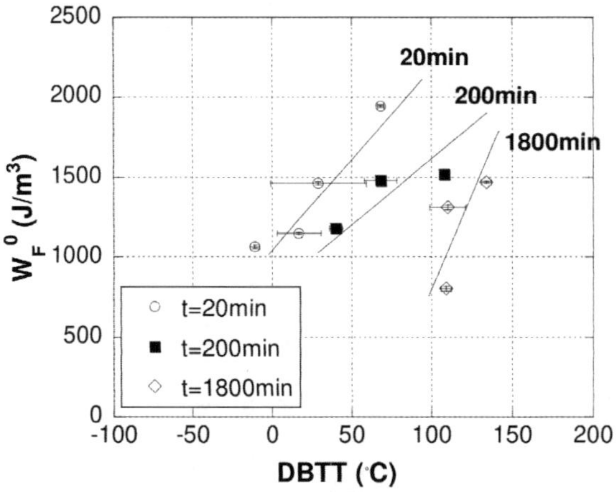

Figure 5. Relation between W_F^0 and DBTT for each ageing time. (toroid samples)

4. Conclusion

We analyzed a set of minor hysteresis loops for thermally aged and cold rolled Fe-1wt%Cu alloy in order to investigate the effect of irradiation embrittlement and subsequent plastic deformation due to such as earthquake on magnetic property changes. Minor-loop properties change depending on ageing time as well as rolling reduction, and show a good relationship with DBTT and hardness, which are a sensitive indicator of irradiation embrittlement in RPV steels. These findings indicate that the magnetic method using minor hysteresis loops can be an useful technique for nondestructive evaluation of degradation in nuclear RPV.

5. References

[1] Odette G R and Lucas G E 1998 *Radiat. Eff. Def. Solids* **144** 189
[2] Takahashi S, Ueda T and Zhang L 2004 *J. Phys. Soc Jpn.* **73** 239
[3] Takahashi S and Zhang L 2004 *J. Phys. Soc Jpn.* **73** 1567
[4] Steinmetz C P 1892 *Trans. Am. Inst. Electr. Eng.* **9** 3
[5] Vandenbossche L P, Konstantinovic M J, Almazouzi A and Dupre L R 2007 *J. Phys. D* **40** 4114

Journal of Physics: Conference Series **266** (2011) 012057 doi:10.1088/1742-6596/266/1/012057

Magnetic Isolation for Hf–capped CoPtCr–SiO_2 Perpendicular Media by Postannealing

S Hinata*, S Saito and M Takahashi

Department of Electronic Engineering, Graduate School of Engineering, Tohoku University, 6–6–05, Aoba, Aramaki, Aoba–ku, Sendai 980–8579, Japan.

E–mail: s_hinata@ecei.tohoku.ac.jp,

Effect of post–annealing on magnetic properties for Hf–capped CoPtCr–SiO_2 perpendicular granular recording media was investigated. It was found that Hf diffusion improved magnetic isolation among magnetic grains of the post–annealed media. Detailed analysis revealed that; non–magnetic layer was formed after post–annealing by interdiffusion of Hf and magnetic grains, retaining saturation magnetization and uniaxial magnetic anisotropy of a columnar part. The results suggest that capped Hf reacted with ferromagnetic Co into non–magnetic CoHf compound at the boundaries of the columns by diffusing through Si–O rich grain boundary.

1. Introduction

Co–based alloy–oxide granular layer was adopted for perpendicular magnetic recording media to obtain a good signal–to–noise (S/N) ratio [1]. For further improvement of writing properties, suppression of intergranular exchange coupling in the initial growth state of Co–based grains is indispensable [2]. Post–annealing is known to be one of the effective methods for enhancing the magnetic isolation for Cr–segregation type media [3–7], especially through diffusion of a non–ferromagnetic cap element to amorphous grain boundaries (boundary diffusion) [4–7]. Thus intergranular exchange coupling is considered to be suppressed by diffusion of non–magnetic elements into the bottom of the magnetic grains through amorphous SiO_2 boundary. However, few study reported the degree of diffusion of non–magnetic element into oxide grain boundary. In the present study, post–annealing process was applied to the practical CoPtCr–SiO_2 granular media in order to diffuse non–magnetic element from cap material into grain boundaries.

2. Experimental method

All samples in the present study were fabricated by dc magnetron sputtering onto glass substrates at room temperature. Hf was selected as the diffusion element due to its high cohesion with Co [7]. The media were in–situ post–annealed up to 350 °C by an infrared lamp running at constant power. The film stacking structure was C (7 nm)/ Hf (d_{Hf} nm)/ $Co_{74}Pt_{16}Cr_{10}$–8mol (SiO_2) (16 nm)/ Ru (20 nm)/ Pt (6 nm)/ Ta (5 nm)/ substrate. Magnetic hysteresis loops were measured by the vibrating sample magnetometer (VSM) and polar Kerr equipment. Saturation magnetization M_s was determined from loops measured by the VSM. Uniaxial magnetic anisotropy K_u was evaluated using the torque

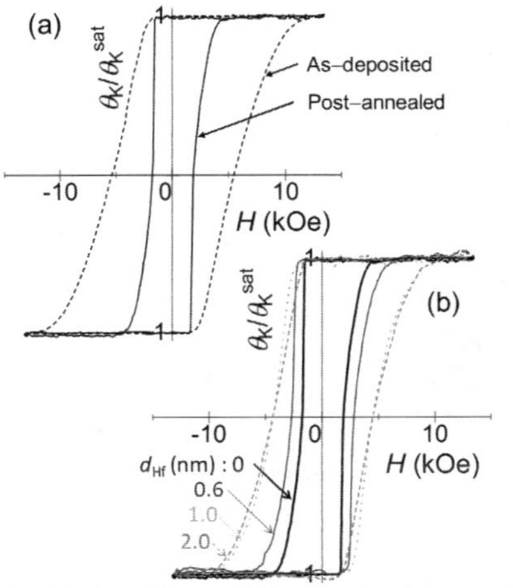

Figure 1. Polar Kerr hysteresis loops for (a) as–deposited and post–annealed media without Hf cap layer and (b) the post–annealed media with various thicknes of Hf cap layer.

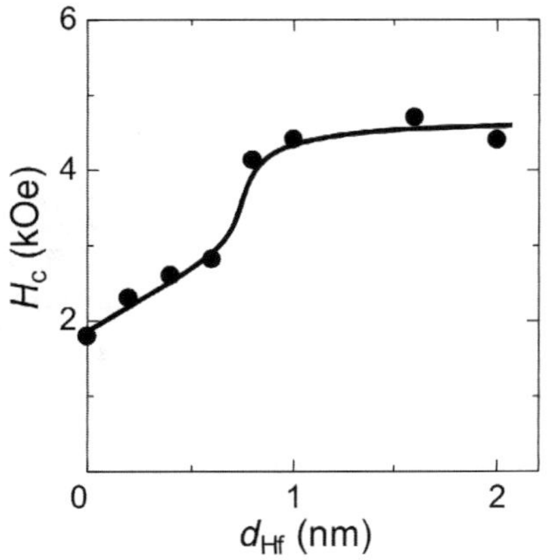

Figure 2. Dependence of coercivity H_c on d_{Hf} for post–annealed media.

magnetometer. Taking into account of the self–energy caused by the demagnetizing field, K_u was determined by

$$K_u = L_{2\theta}^{sat} + 2\pi M_s^2 \; . \quad (1)$$

where, $L_{2\theta}^{sat}$ is the saturated torque coefficient of twofold–component. Anisotropy field is $H_k = 2K_u/M_s$. The microstructure of the media was examined by cross–sectional transmission electron microscopy (TEM). The composition analysis was performed by energy dispersive X–ray spectroscopy attached to TEM (TEM–EDX).

3. Result and discussion

3.1. Magnetic properties for post–annealed media.

Figure 1 (a) shows polar Kerr loops for as–deposited (dashed) and post–annealed (solid) media without Hf cap layer. Coercivity H_c drastically decreased from 5.4 kOe to 1.8 kOe and loop slope increased by postannealing. This indicates that the intergranular exchange of the media was increased by postanneling. Figure 1 (b) shows polar Kerr loops for post–annealed media with various d_{Hf}–cap layer. H_c increased from 1.8 kOe to 4.2 kOe with increasing d_{Hf} for the media with $d_{Hf} \leq 1$ nm and

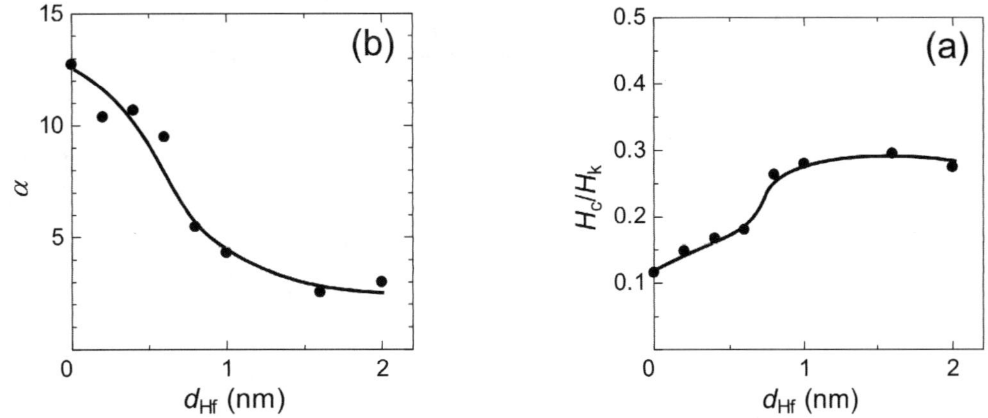

Figure 3. Dependence of (a) loop slope parameter $\alpha = 4\pi|dM/dH|$ around H_c and (b) normalized coersivity H_c/H_k for post–annealed media.

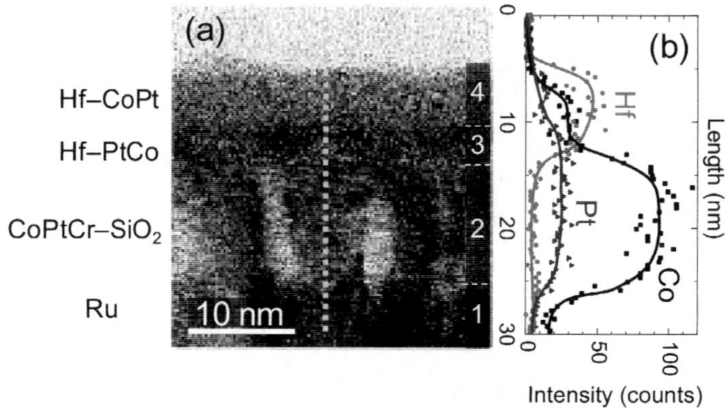

Figure 4. (a) Bright–field image obtained by cross–sectional TEM for the post–annealed medium with the Hf cap and (b) TEM– EDX profile on a green dotted line in (a).

saturated for $d_{Hf} \geq 1$ nm. Degree of the isolation of post–annealed media with Hf cap were also estimated from the loops. Figure 3 shows (a) loop slope $\alpha = 4\pi|dM/dH|$ around H_c and (b) normalized coercivity H_c/H_k for post–annealed media with Hf cap. α decreased from 13 to 2.5 and H_c/H_k increased from 0.1 to 0.3. And the values of these quantities tend to be saturated for $d_{Hf} \geq 1$ nm as well as the H_c value. These results suggest that the increasing of H_c was originated from increase of the magnetic isolation among magnetic grains.

3.2. Structure of the post–annealed media

Figure 4 shows (a) Bright–field image obtained by cross–sectional TEM for post–annealed medium with Hf cap with the layer structures of Hf (3 nm)/ $Co_{74}Pt_{16}Cr_{10}$–8mol (SiO_2) (16 nm)/ Ru (20 nm)/ Pt (6 nm)/ Ta (5 nm)/ glass sub, and (b) Energy Dispersive X–ray spectroscopy (EDX) profile on a green dotted line in the figure 4 (c). Contrast of the bright–field image is divided into four regions of (1) gray continuous–, (2) black continuous– (3) stripe– (4) black continuous– layers. Composition distributions of these layers were evaluated by the TEM–EDX. Co and Pt atoms were detected at the region 2–4. Co–atom intensity was high around the region 2. Most of Hf atoms existed in region 3–4, and also were a few detected in the region 2. From the stripe contrast of cross–sectional image and existence of large amount of the Co and Pt atoms, region 3 can be concluded to be CoPtCr–SiO_2 layer. Moreover, there was interdiffusion at the interface of CoPtCr–SiO_2 and Hf–rich region 3, 4 and it is suggested that they formed B2–type Hf (CoPt) or Hf (PtCo) judging from the composition. This suggestion is also supported by in–plane XRD profiles (not shown). Therefore, stacking structure of the media was thought to be B2–type Hf (CoPt) / CoPtCr–SiO_2/ Ru, and small amount of Hf existsed in CoPtCr–SiO_2 layer.

3.3. Evaluation of Hf diffusion

We tried to evaluate degrees of inter and boundary diffusion through analysis of dependence of magnetic properties on the magnetic layer thickness d_{mag}. Taking into account of the formation of non–ferromagnetic reacted layer between cap and granular layer, we can simply describe a total saturation magnetization M_s^{exp} as

$$M_s^{exp} = \frac{M_s^{react} \times d_{mag}^{react} + M_s^{ave} \times \left(d_{mag} - d_{mag}^{react}\right)}{d_{mag}}, \tag{2}$$

where M_s^{react} and d_{mag}^{react} are the saturation magnetization and thickness of the reacted layer, M_s^{ave} is averaged saturation magnetization of magnetic layer. Eq. (2) can be simply rewritten into the following form assuming that M_s^{react} of B2–type CoHf equals to 0:

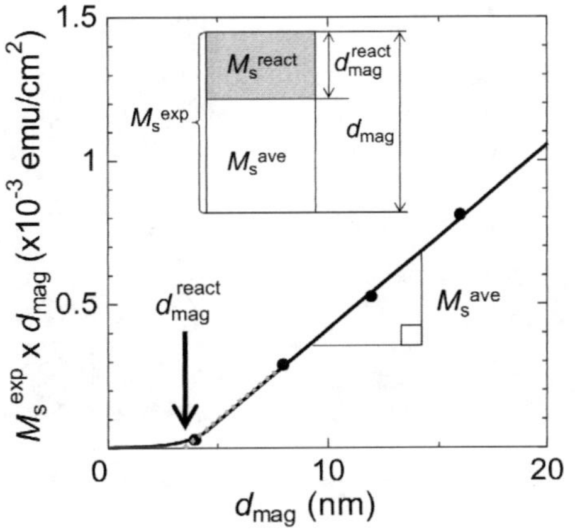

Figure 5. $M_s^{exp}\times d_{mag}$ as a function of $M_s^{exp}\times d_{mag}$ for the post–annealed media with 3 nm–Hf cap. A gray dotted line indicates extended approximation line of the data.

$$M_s^{exp} \times d_{mag} = M_s^{ave} \times \left(d_{mag} - d_{mag}^{react}\right). \qquad (3)$$

By a linear portion of the $M_s^{exp}\times d_{mag}$ vs. d_{mag} plot, d_{mag}^{react} and M_s^{ave} can be determined from the intersection of an extended line with the d_{mag} axis and the slope of the $M_s^{exp} \times d_{mag}$ line. Here, the value of d_{mag}^{react} and M_s^{ave} correspond to degree of the inter– and boundary– diffusion. Figure 5 shows typical dependence of $M_s^{exp} \times d_{mag}$ on d_{mag} for the post–annealed media with 3 nm–Hf cap. The gray dotted line indicates extended approximation line of the data. $M_s^{exp} \times d_{mag}$ shows linear correlation with d_{mag} in the region of $d_{mag} > 4$ nm. d_{mag}^{react} was evaluated to be 3.5 nm. In the same way, averaged uniaxial magnetic anisotropy K_u^{ave} can be determined by the slope of a $K_u^{exp} \times d_{mag}$ vs. d_{mag} plot, where K_u^{exp} is the total uniaxial magnetic anisotropy of the media [8].

Figure 6 shows the d_{Hf} dependence of (a) d_{mag}^{react}, (b) M_s^{ave} and (c) K_u^{ave} for the post–annealed media. Values for the as–deposited medium were indicated by white circles as a reference. d_{mag}^{react} increases from 0.5 nm to 3.5 nm with increasing d_{Hf} from 0 nm to 3 nm, and it tends to be saturated for $d_{Hf} > 1$ nm. On the other hand, M_s^{ave} and K_u^{ave} were almost unvaried from 650 emu/cm^3 and 5.5×10^6 erg/cm^3 for different d_{Hf}. These results suggest that interdiffusion hardly grew in the region of $d_{Hf} > 1$ nm and boundary–diffusion region was relatively narrow.

3.4. Mechanism of the Hf diffusion into post–annealed CoPtCr–SiO$_2$ media

Based on the results described in section 3.1–3.3, mechanism of Hf diffusion into the CoPtCr–SiO_2 layer was discussed. Figure 7 shows schematic views of (a) as–deposited and (b) post–annealed media with Hf–cap layer. In the case of the as–deposited media, CoPtCr columns grew on the rough–surface

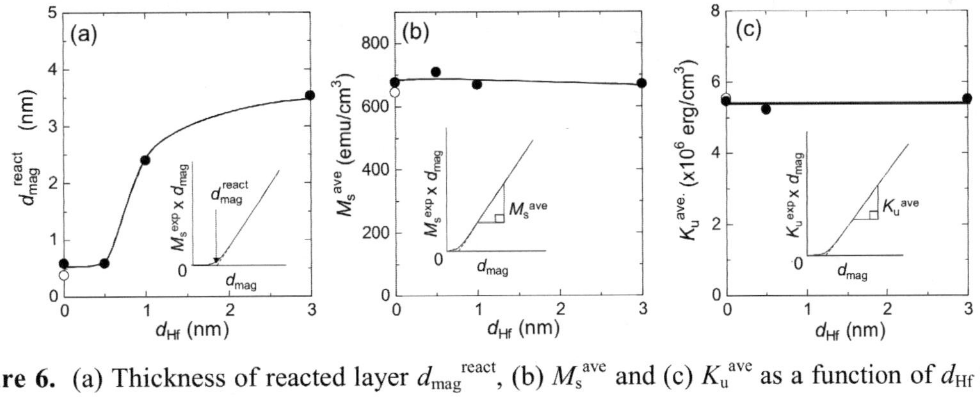

Figure 6. (a) Thickness of reacted layer d_{mag}^{react}, (b) M_s^{ave} and (c) K_u^{ave} as a function of d_{Hf} for the post–annealed media. White circles indicate values for as-deposited medium as a reference.

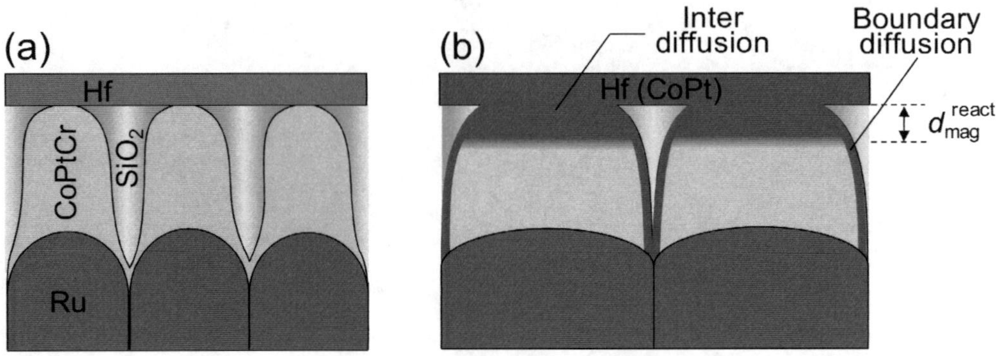

Figure 7. Schematic view of diffusion phenomena for (a) as–deposited and (b) post–annealed medium with Hf–cap layer.

Ru grains. Contact of each column occurred around dip of the Ru (figure 7 (a)) [2]. After post–annealing, the diameters of columns were expanded. Co and Hf atoms mainly interdiffused around the Hf/ CoCrPt interface and B2–type Hf (CoPt)–reacted layer was formed. Hf atoms also diffused into grain boundaries and reacted with a couple of Co or Pt layer (figure 7 (a)). As a result, exchange coupling of each CoPtCr columns was reduced by formation of non–ferromagnetic Hf (CoPt).

4. References

[1] T. Oikawa, M. Nakamura, H. Uwazumi, T. Shimatsu, H. Muraoka, and Y Nakamura, 2002, *IEEE Trans. Magn.*, **38**, 1976
[2] I. Takekuma, R. Araki, M. Igarashi, H. Nemoto, I. Tamai, Y. Hirayama, and Y. Hosoe, 2006, *J. Appl. Phys.* **99**, 08E713
[2] Mukai R, Yamanaka K, and Oshiki M, 1997 *J. Appl. Phys*., **81**, 3931
[3] Roy A G, Nuhfer N T, and Laughlin D E, 2000 J. *Appl. Phys*, **99**, 8179
[4] Saito S, Itagaki N and Takahashi M, 2004, *J. Magn. Soc. Jpn*, **28**, 295
[5] Saito S, Itagaki N and Takahashi M, 2004, *IEEE Trans. Magn.*, **40**, 2467
[6] Itagaki N, Saito S and Takahashi M, 2005, *IEEE Trans. Magn.*, **41**, 3187
[7] Saito S, Hasegawa D, Hoshi F, Djayaprawira D D and Takahashi M, 2002, *Appl. Phys. Lett.*, **40**, 811

Journal of Physics: Conference Series **266** (2011) 012058 doi:10.1088/1742-6596/266/1/012058

Fabrication of constricted compositionally-modulated Ni_xFe_{1-x} nanowires

H F Liew, S C Low and W S Lew*
School of Physical & Mathematical Sciences,
Nanyang Technological University, 21 Nanyang Link, Singapore 637371

*E-mail: wensiang@ntu.edu.sg

Abstract We demonstrate the fabrication of compositionally-modulated Ni_xFe_{1-x} nanowires using templated growth electrodeposition techniques. The composition of the Ni and Fe in a grown magnetic layer was modulated by varying the deposition potential during growth. Periodic constrictions were created along the Ni_xFe_{1-x} nanowires via diluted nitric acid etching. Conceptually, the constrictions created along nanowires may induce a low energy well which has an advantage as a potential barrier to trap domain walls in magnetic storage application.

1. Introduction

Magnetic nanowires have received strong interest recently because it has promising potential for application in magnetic solid state memory devices[1]. For instance, IBM has proposed the magnetic domain-wall based "racetrack memory" that uses planar and cylindrical magnetic nanowires as the storage medium for horizontal and vertical memory design, respectively[2]. The vertical "racetrack memory" design has the advantage of higher density but the implementation is extremely challenging. Generally, a magnetic domain-wall based memory device works by controlling the movement of the domain wall. When a nanosecond pulsed current is applied, the domain wall propagates along the nanowire until it meets an obstacle where it requires higher energy to overcome the barrier. The movement of domain wall enables binary data stored in the ferromagnetic nanowire be retrieved or written by a magnetic recording head. To precisely control the movement of the domain wall intentional defects[3-10] is usually created along the nanowires. Such defect acts as a low energy well that is readily to trap the moving domain wall. Systematic creation of such intentional defects so that domain wall can be effectively trapped at this region for planar nanowires is considered established; however, the technique required for making similar defects in cylindrical nanowire is still to be explored[11].

A significant number of dry and wet chemical deposition methods of producing magnetic nanowires have been reported in literature. Template-assisted electrodeposition is one widely-used technique because of its simplicity and flexibility, specifically in making multilayered structures. Here, we demonstrate the fabrication of constricted cylindrical Ni_xFe_{1-x} nanowires, which consist of alternate layers of Ni_xFe_{1-x} with two different of x values. The constrictions are created along the nanowires by etching the as-deposited nanowires in diluted HNO_3.

2. Experimental

Compositionally-modulated Ni_xFe_{1-x} nanowires were fabricated by using templated growth pulse electrodeposition technique. The deposition cell consists of 0.5 M nickel sulfate, 0.01 M iron sulfate and 0.5 M boric acid. Figure 1 shows the step-by-step fabrication process of the constricted Ni_xFe_{1-x} nanowires. A 200-nm-thick aluminium layer was first deposited on one side of an anodized aluminum oxide (AAO) template (Anodisc 13, Whatman). The Ni_xFe_{1-x} nanowires were alternately deposited between –1.0 V and –1.4 V vs SCE for a period of 50 sec using similar pulse potential, as shown in Figure 2. The pH and temperature during the growth were maintained at 4 and 40 °C, respectively. The length of the alternating segments is determined by the electrodeposition time. Upon the completion of growth, the nanowires were released from the AAO template by dissolving them into 3M NaOH solution for 30 minute. Then, the nanowires were thoroughly cleaned from the AAO template by using deionized water, with the assistance of ultrasonic agitation. The Ni_xFe_{1-x} nanowires segment that grown at potential –1.0 V (higher Fe composition) were then selectively etched by 1% HNO_3 for 40 sec. The segment deposited at a reduction potential of –1.0 V has a higher etching rate compared to that of the segment deposited at potential –1.4 V. Thus, the depth of the constriction can be controlled by altering the concentration of the HNO_3 and the etching time.

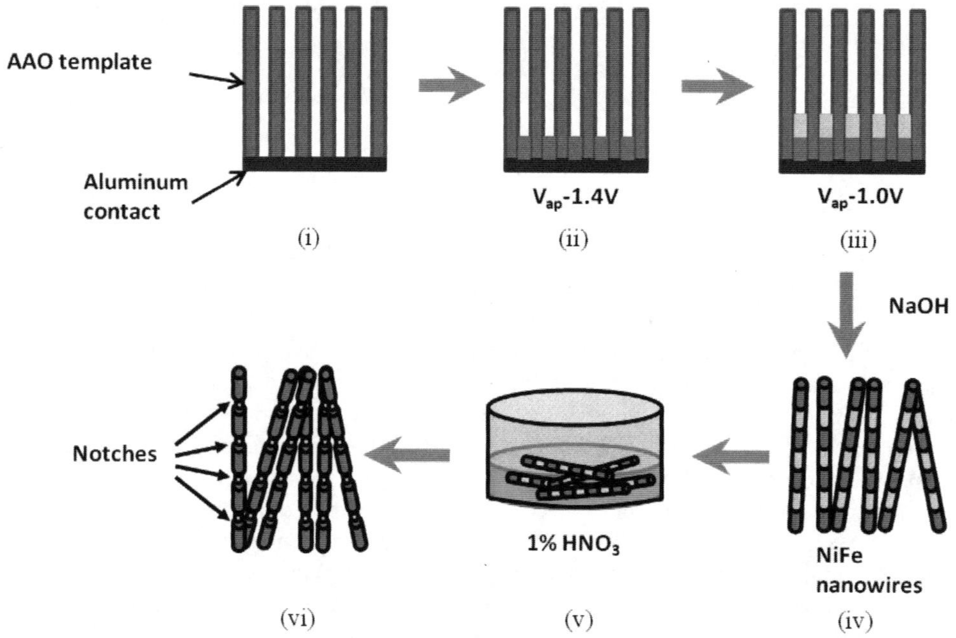

Figure 1. Schematic diagram illustrating the fabrication process of notched Ni_xFe_{1-x} nanowires

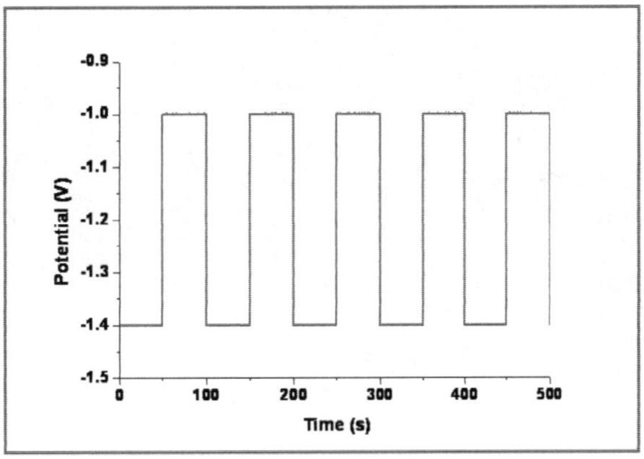

Figure 2. The pulse potential applied during the growth of compositionally-modulated Ni_xFe_{1-x} nanowires.

3. Results and Discussion

Field emission scanning electron microscopy (FE-SEM) imaging was carried out by using a JEOL JSM-6700F operated at 10kV. Figure 3 shows the bamboo like feature of the Ni_xFe_{1-x} cylindrical nanowires with alternate constrictions along the nanowires. The constrictions obtained at potential – 1.0V are well defined and uniform in shape and size. The constrictions are clearly distinguishable and orderly formed along the nanowires. In addition, the bath contains Ni^+ and Fe^+ ions in a single bath which produced the multilayered nanowires in an ordered fashion when compared with separate bath deposition. A close-up view of the image (Figure 4) revealed that the segments grown at potential– 1.4V have a length of 1.5μm and diameter of 350 nm, while segments growth at potential of –1.0V have a length of 100 nm and a diameter of 150 nm. The above multilayered nanowires with optimized diameter of 150 nm which produces the constrictions easily and mechanically stable. Figure 5 shows transmission electron microscope (TEM) image of a released Ni_xFe_{1-x} nanowire with constrictions. The Ni and Fe composition along the two different potential segments were determined by elemental analysis by using energy dispersive X-Ray spectroscopy (EDX). EDX result shown in Figure 6 indicates that higher atomic percentage of Fe (13%) was obtained in the constriction and less trace of Fe (5%) was analysed in the non-constriction region. The variation in the measured Ni and Fe compositions confirms the modulation of the composition in the Ni_xFe_{1-x} nanowires.

Figure 3. SEM image of compositionally-modulated Ni_xFe_{1-x} nanowires with notches along the nanowires.

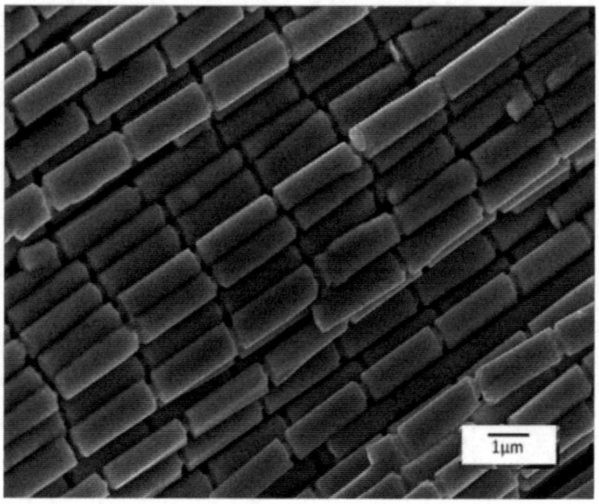

Figure 4. A close-up view of well defined notches structure along the Ni_xFe_{1-x} nanowires.

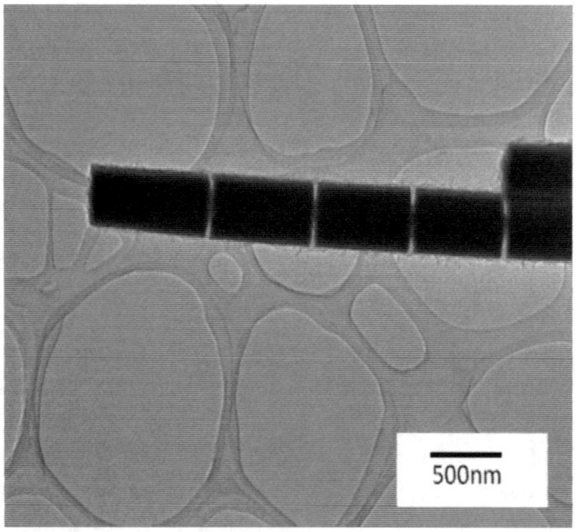

Figure 5.TEM image of single notched Ni_xFe_{1-x} nanowires.

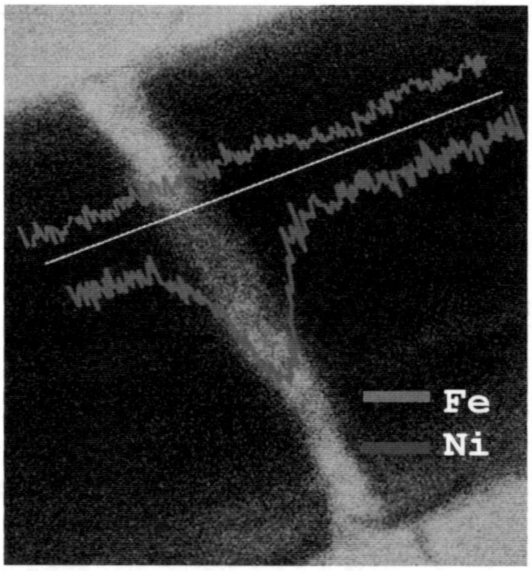

Figure 6. TEM elemental line scanning of Ni and Fe elements (Ni red, Fe blue) along the compositionally-modulated Ni_xFe_{1-x} nanowires.

The magnetic properties of the notched Ni_xFe_{1-x} nanowires with alternately different composition were measured using an alternating gradient force magnetometer (AGFM). The hysteresis loop of the released Ni_xFe_{1-x} nanowires sample is shown in Figure 7, where $H^{//}$ and $H^{\perp}$ represent the applied field, H being parallel and perpendicular to the nanowire axis, respectively. The coercivity $H_c(//)$ and ratio of the remnance to the saturation magnetization $(M_r/M_s)(//)$ are 77 Oe and 0.27, respectively compared to $H_c(\perp)$ at 118 Oe and $(M_r/M_s)(\perp)$ at 0.10. The squareness ratio in the parallel direction is higher than the perpendicular direction which indicates that the easy axis is parallel to the orientation of nanowire direction. The hysteresis measurement indicates the fabricated Ni_xFe_{1-x} nanowires are of soft ferromagnetic material due to its high magnetic saturation and low coercivity. The magnetization contributions from the contricted segments could have been negligible as the length of the contrictions is insignificant compared to the other segment.

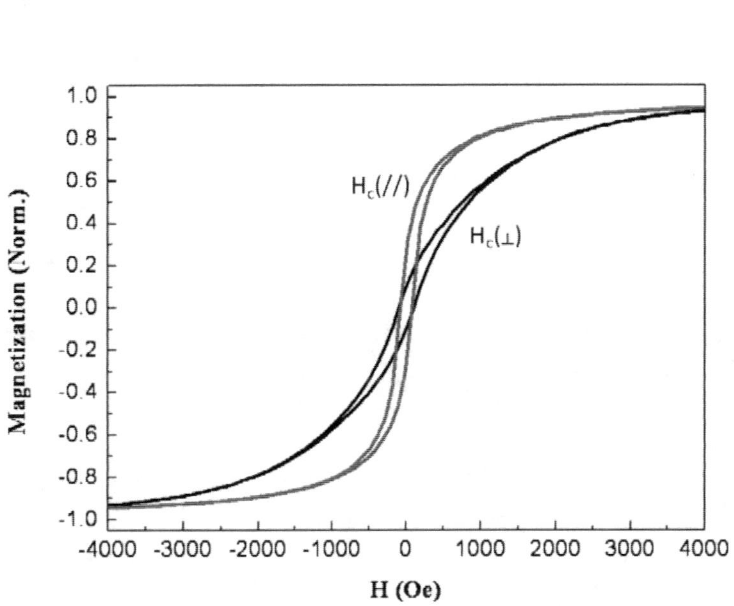

Figure 7. Hysteresis loop for the compositionally-modulated Ni_xFe_{1-x} nanowires with notches at room temperature.

4. Conclusion

We have successfully fabricated constricted Ni_xFe_{1-x} nanowires with different composition of Ni and Fe by using pulse electrodeposition techniques. By selectively chemical etching, the diameter of the Ni_xFe_{1-x} nanowires at constriction region reduces from 350 nm to 150 nm. EDX measurement confirms the variation percentage of Fe along the compositionally-modulated Ni_xFe_{1-x} nanowires. The constricted Ni_xFe_{1-x} nanowires have an easy magnetization axis parallel to the nanowire axis. This constricted nanowire has the potential to be used as the storage medium for magnetic domain-wall memory device.

References

[1] Parkin S S P, Hayashi M, and Thomas L 2008 *Science* **320** 190
[2] Moriya R, Parkin S S P, Thomas L 2010 *U.S.Patent* 0046268A1
[3] Bogart L K, Atkinson D, O'Shea K, McGrouther D and McVitie S 2009 *Phys. Rev. B* **79** 054414
[4] Im M Y, Bocklage L, Fischer P, and Meier G 2009 *Phys. Rev.Letts.* **102** 147204
[5] Kunz A 2009 *Appl. Phys. Letts.***94** 132502
[6] O'Shea K J, McVitie S, Chapman J N, and Weaver J.M.R. 2008 *Appl. Phys. Letts* **93** 202505
[7] McGrouther D, McVitie S and Chapman J N 2007 *Appl. Phys. Letts.* **91** 022506
[8] Hirohata A, Xu Y B, Yao C C, Leung H T, Lee W Y, Gardiner S M, Hasko D G and Bland J A C 2000 *J. Appl. Phys.* **87** 4727
[9] Petit D, Jausovec A, Read D and Cowburn R P 2008 *J. Appl. Phys.* **103** 114307
[10] Li S P, Lew W S, Bland J A C, Lopez-Diaz L, Natali M, Vaz C A F and Chen Y 2002 *Nature* **415** 600
[11] Rheem Y W, Yoo B Y, Koo B K and Myung N V 2007 *Phys. Stat. Sol. (a)* **204** 4021

Journal of Physics: Conference Series **266** (2011) 012059 doi:10.1088/1742-6596/266/1/012059

Thermal fluctuations in perpendicular recording media: New methodology for estimation of activation moment

H Mamiya[1], H Hagiya[2,3], Y Oba[1], M Ohnuma[1], T Oku[2], J Suzuki[2], M Yokoyama[3], T Katayama[4], Y Nishihara[3], H Awano[5], and T Koda[6]

[1] National Institute for Materials Science, Tsukuba 305-0047, Ibaraki, JAPAN
[2] Japan Atomic Energy Agency, Tokai 319-1184, Ibaraki, JAPAN
[3] Ibaraki University, Mito 310-8512, Ibaraki, JAPAN
[4] Research and Development Association for Future Electron Devices, Minato-ku 105-0001, Tokyo, JAPAN
[5] Toyota Technological Institute, Tenpaku 468-8511, Nagoya, JAPAN
[6] Tohoku University, Sendai 980-8577, Miyagi, JAPAN

E-mail: MAMIYA.Hiroaki@nims.go.jp

Abstract. In nanoparticulate films with perpendicular magnetic anisotropy, a large demagnetizing field almost compensates for the experimentally designed change in the magnetic field applied parallel to the film normal. We propose a new method based on field-cycling to eliminate the uncertainty in the demagnetizing response from the analysis of the activation volume of the reversal or magnetic activation moment in such films. In this method, the applied field induced variation in the magnetic relaxation rate is measured before the effect of the demagnetizing field becomes dominant. We also discuss an analogical thermal-cycling method to clarify the temperature dependence of the barrier height for magnetization reversal in a magnetic field. We apply these methods to a $Co_{74}Pt_{16}Cr_{10}$-SiO_2 nanoparticulate film as an example. The results demonstrate that these methods are useful for studying thermal fluctuations in perpendicular recording media.

1. Introduction

Recently, thermal fluctuations in nanoparticulate films have been intensively studied, since thermally assisted switching of magnetization vectors in a desired region and the thermal stability in the direction of the vectors for the other recorded regions are central topics of current research on ultrahigh-density magnetic recording media. Therefore, a number of methods have been developed to clarify the effect of the thermal fluctuations [1-8]. For example, the typical energy barrier height for switching can be derived from the dependence of the *M-H* curve on the field sweep rate dH/dt [3]. Another example is the waiting time method [5,6], wherein the activation volume of the switching or magnetic activation moment μ_A is determined by the analysis of two relaxation curves at two different applied fields H_{appl}. However, such analyses are complicated for perpendicular recording media, because a large demagnetizing field H_d perpendicular to the film plane compensates for the experimentally designed change in the applied field. In the former case, the variation in the actual field inside the film $H = H_{appl} - H_d$ abruptly slows down as the switching starts partially. Similarly, the latter method may be confusing if it is applied to perpendicular recording media, because at any measuring time during the observation, each nanoparticle is exposed to almost the same H between the two relaxations with significantly different H_{appl}, as shown later. Thus, a new methodology is urgently required for the in-depth analysis of the thermal fluctuations in perpendicular recording media.

We propose a field-cycling method to simply clarify μ_A in perpendicular recording media, where H_{appl} is temporarily changed to $H_{appl} + \Delta H_{appl}$ halfway through the relaxation. The new method was applied to a $Co_{74}Pt_{16}Cr_{10}$-SiO_2 nanoparticulate film provided by Hitachi Maxell, Ltd. [9]. The results demonstrate the proposed method's simplicity and convenience.

2. Conventional methodology and its results

In nanoparticulate films, it is not easy to determine the magnitude of the magnetic moment $\boldsymbol{\mu}$ for each nanoparticle. For this reason, we consider the fictitious magnetic moment μ_A given by $-\partial E_A/\partial H$ because the height of the potential energy barrier E_A separating the direction of $\boldsymbol{\mu}$ parallel to $\boldsymbol{H}$ from the anti-parallel direction is lowered by the Zeeman energy $-\boldsymbol{\mu}\cdot\boldsymbol{H}$ in $\boldsymbol{H}$. In general, variation in the total magnetization in H is given by

$$\Delta M = (\partial M/\partial \ln t)|_H \cdot \Delta \ln t + (\partial M/\partial H)|_{\ln t} \cdot \Delta H = S \cdot \Delta \ln t + \chi \cdot \Delta H, \quad (1)$$

where S is the magnetic viscosity and χ is the susceptibility. For perpendicular recording media, rotation of $\boldsymbol{\mu}$ is almost negligible when $\boldsymbol{H}$ is applied parallel to the film normal, because $\boldsymbol{H}$ is parallel to the easy axis of the large perpendicular magnetic anisotropy. In other words, the reversal of $\boldsymbol{\mu}$ due to thermally fluctuations is predominant. In this case, the relaxation time τ is expressed as $f_0^{-1}\exp(E_A(H)/k_BT)$, where f_0 is the attempt frequency and k_B is the Boltzmann constant. If $E_A(H)$ is common in every nanoparticle at any H, magnetic relaxations may be scaled by the ratio t/τ: $\Delta M = F(t/\tau)$, where $F(x)$ is a scaling function. This assumption leads to the relationship:

$$\mu_A = -\partial E_A/\partial H = k_BT \cdot \chi/S = k_BT \cdot (\Delta \ln t/\Delta H)|_M. \quad (2)$$

Therefore, μ_A or the activation volume of reversal μ_A/M_s are estimated by using the ratio χ/S (Street-Woolly method [1]), where M_s is the saturation magnetization. If S significantly varies with ln t in the experimental time window, the waiting time method is also employed, where the times (t_1, t_2) required for the magnetization to decay to a constant level at slightly differing fields (H_1, H_2) are measured. μ_A is then estimated at $k_BT\ln(t_1/t_2)/(H_2 - H_1)$ [5,6].

Figure 1(a) shows a $Co_{74}Pt_{16}Cr_{10}$-SiO_2 nanoparticulate film with a thickness of 16 nm as an example; the film was completely saturated in a high magnetic field of –20 kOe at 300 K. The field was then isothermally reversed to H_{appl} = 4.1 kOe or $H_{appl} + \Delta H_{appl}$ = 4.15 kOe at t = 0, and the magnetic relaxation of the film was measured at each field. M was easily found to reach zero at t_1 = 2 ks in H_{appl} and t_2 = 0.6 ks in $H_{appl} + \Delta H_{appl}$. Thus, $\ln(t_1/t_2)$ is 1.2. In contrast, consideration of H_d is required to estimate the actual fields H_1 and H_2 at the applied fields. Generally, the self-demagnetizing field H_{ds} of each nanoparticle contributes to the shape anisotropy. Therefore, H_d is expressed as $4\pi M - H_{ds}$. If we assume that the nanoparticles are disks with a diameter of roughly 50 nm [9], H_{ds} is approximately $3\pi M$. Figure 1(b) shows the actual fields: in this case, $H_1 = H_{appl} - H_d$ and $H_2 = H_{appl} + \Delta H_{appl} - H_d$. Note that $H_2(t)$ is almost identical to $H_1(t)$ at any measuring time t, despite ΔH_{appl} = 0.05 kOe. On the other hand, the difference under concern $(H_2(t_2) - H_1(t_1))$ is certainly equal to ΔH_{appl}. As a result, μ_A was calculated to be 1.1×10^5 μ_B. This non-intuitive estimation is confusing although the formalism is valid. The question arises as to why the two relaxation curves in Figure 1(a) are different from one another in the same actual field. One interpretation is that the difference is generated by a field-induced-reversal of $\boldsymbol{\mu}$ in the very early stage of the relaxations, where $H_1(t) \neq H_2(t)$. However, the details are unclear since it occurs out of the experimental time window. Although the accuracy of the estimated H_d is debatable, this issue is crucial as long as H_d is large enough.

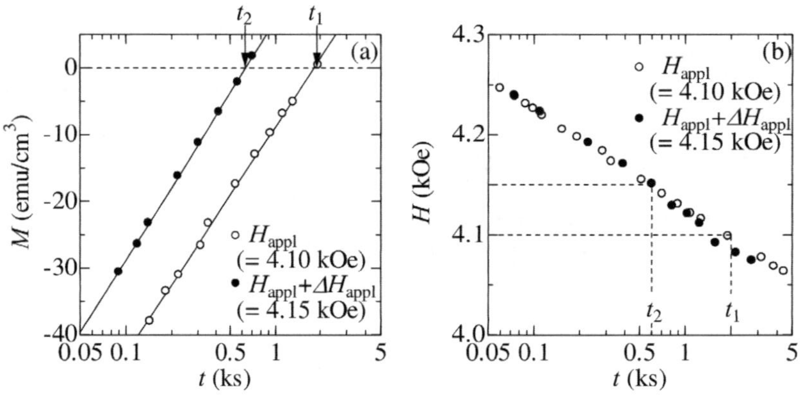

Figure 1. (a) Magnetic relaxation curves of $Co_{74}Pt_{16}Cr_{10}$-SiO_2 nanoparticulate film at applied fields of H_{appl} = 4.1 kOe and of $H_{appl} + \Delta H_{appl}$ = 4.15 kOe. The film was completely saturated in a field of −20 kOe, applied parallel to the film normal, at 300 K. The field was then isothermally reversed at t = 0. (b) Estimated variation in actual field H inside film during magnetic relaxation.

3. New methodology and its results

Now, we consider that the magnetic field is changed to $H_{appl} + \Delta H_{appl}$ at t_0 after measurement starts at H_{appl} at t = 0. In this case, all of the states just before the changes are identical each other. Hence, H_d is uniform at t_0. After the changes, the relaxation of total magnetization M accelerates or slows down according to the variation in the barrier height $E_A(H)$. Consequently, H_d starts to branch off in response to the divergence of M due to the acceleration/deceleration. Point is that, in the initial stage of the period, the difference of H_d is still small in comparison with ΔH_{appl}, i.e. $\Delta H \sim \Delta H_{appl}$. If the field returns to H_{appl} at short time Δt later, we can estimate the extent of the acceleration or deceleration since contribution from field-reversible components such as the canting of μ is almost the same between the periods before and after field-cycling. For this reason, the field-cycling method allows us to directly compare the two relaxations that start from the same initial state and proceed in different actual fields.

Figure 2(a) shows the same $Co_{74}Pt_{16}Cr_{10}$-SiO_2 nanoparticulate film being used as an example. First, the film was completely saturated in H_{appl} = −20 kOe at 300 K. The field was then isothermally reversed to H_{appl} = 4.1 kOe at t = 0, and the magnetic relaxation of the film was measured. Halfway through the relaxation at t_0 = 1.3 ks, the applied field was temporarily weakened to $H_{appl} + \Delta H_{appl}$ = 4.05 kOe. At that time, M slightly dropped without delay. One of the causes is the decrease in the canting angle of μ. M then started to increase again, but the relaxation rate in $H_{appl} + \Delta H_{appl}$ was apparently slower than that in H_{appl}. The inset shows the variation in the actual field H that was estimated in the same way as above. As expected, H decreased by 0.05 kOe when H_{appl} was weakened by the same magnitude. In contrast, M instantly rose when the applied field returned to H_{appl} at $t_0 + \Delta t$ = 10 ks. This can be attributed to the recovery of the canting angle. Subsequently, a gradual increase in M continued in H_{appl}. When the relaxation curve obtained after the field-cycling is shifted leftward by 4.5 ks, it can be superimposed on the relaxation without the field-cycling (represented by open circles on the solid curve); that is, the variation ΔM in the period of Δt = 8.7 ks in $H_{appl} + \Delta H_{appl}$ is almost equivalent to ΔM in the period of Δt_{eff} = 8.7 − 4.5 = 4.2 ks in H_{appl}. Note that the instant variations of M due to the changes in the canting angle at the beginning and the end of the field-cycling cancel each other out in this analysis. Figure 2(b) shows the ratio of Δt_{eff} to Δt obtained for various $\Delta H_{appl} \sim \Delta H$. The obtained exponential relationship $\Delta t_{eff}/\Delta t \propto \exp(\Delta H_{appl})$ was consistent with equation (2). More simply, this relationship is explained by the ratio $\tau(H+\Delta H)/\tau(H) = \exp[(E_A(H+\Delta H) - E_A(H))/k_BT] \sim \exp[\Delta H\cdot(\partial E_A/\partial H)/k_BT]$. Therefore, $\mu_A \sim 0.7 \times 10^5$ μ_B can be derived from the gradient of the relationship. This value is comparable to μ_A of 1.1×10^5 μ_B estimated by the waiting time method (see above), and μ_A of 1.0×10^5 μ_B by the Street-Woolly method, where $S = 13$ emu·cm^{-3} and $\chi_{irr} = 0.3$ emu·Oe^{-1}·cm^{-3} obtained from dc demagnetization curve. As discussed here, the field-cycling method enabled us to achieve a reasonable μ_A in the straightforward framework.

As stated in the introduction, thermally assisted switching of μ in nanoparticulate films is a topical issue for high-density magnetic recording. Thus, we applied the cycling method to the temperature dependence of $E_A(H, T)$, i.e,

$$\partial E_A(H, T)/\partial T - E_A/T = k_BT\cdot(\Delta \ln t\ /\Delta T)|_{M,H}. \tag{3}$$

Figure 3(a) shows the magnetic relaxation curves of the same nanoparticulate film. After the film was completely saturated by −20 kOe at $T = 300$ K, the field was isothermally reversed to $H_{appl} = 4.1$ kOe at $t = 0$, and the measurement was started. Halfway through the relaxation, the temperature was temporarily lowered to $T + \Delta T = 297$ K, as shown in the inset. The relaxation was readily shown to slow down after cooling at $t_0 = 1.3$ ks. The relaxation rate then recovered the original extent after reheating at $t_0 + \Delta t = 10$ ks. When the relaxation curve obtained after thermal-cycling is shifted leftward by 5.2 ks, it can be superimposed on the isothermal relaxation (represented by open circles on the solid curve). In other words, the variation ΔM in the period $\Delta t = 8.7$ ks at $T + \Delta T$ is almost equivalent to ΔM in the period $\Delta t_{eff} = 3.5$ ks at T. Figure 3(b) shows the ratio of Δt_{eff} to Δt obtained for various ΔT. We are certain the relationship $\ln(\Delta t_{eff}/\Delta t) \propto \Delta T/T$ for the weak thermal perturbations ΔT. On the basis of the equation (3), the gradient of the relationship indicates that $E_A(H, T)^{-1}\cdot\partial E_A(H, T)/\partial T \sim -0.005$ (K^{-1}) at $H \sim 4.1$ kOe, where $E_A(H, T)/k_BT$ was assumed to be 30 in our experimental time window. This temperature dependence was significantly large in comparison with that of the anisotropy constant reported previously [10]. Although further information on $E_A(H, T)$ cannot be derived from the results obtained here, this example demonstrates the possible applicability of the thermal-cycling method.

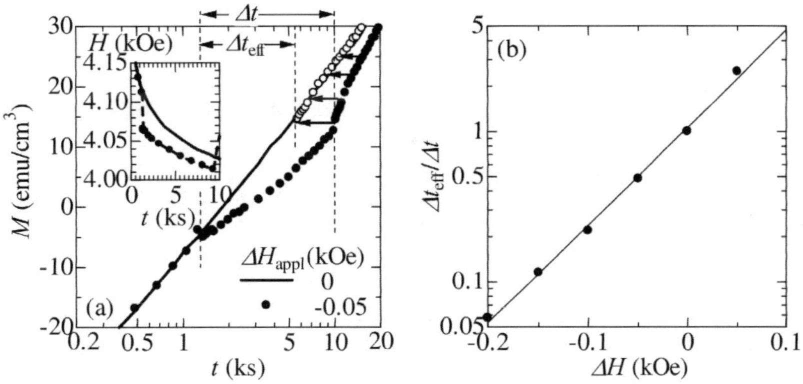

Figure 2. (a) Magnetic relaxation curves of $Co_{74}Pt_{16}Cr_{10}$-SiO_2 nanoparticulate film with a field-cycling. The film was completely saturated in a field of −20 kOe, applied parallel to the film normal, at 300 K. The field was then isothermally reversed to $H_{appl} = 4.1$ kOe at $t = 0$. Halfway through the relaxation, the actual field H was temporarily changed, as shown in the inset. The open circles indicate time-shifted curves. (b) Ratio of Δt_{eff} to Δt as a function of amplitude of the field-cycling ΔH.

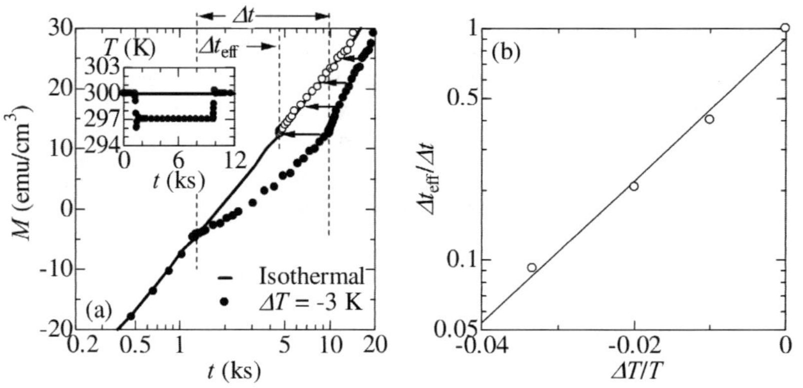

Figure 3. (a) Magnetic relaxation curves of the same film with a thermal-cycling. After complete saturation in $H_{appl} = -20$ kOe, the field was isothermally reversed to $H_{appl} = 4.1$ kOe at $t = 0$. Halfway through the relaxation, the sample was temporarily cooled, as shown in the inset. The open circles indicate time-shifted curves. (b) Ratio of Δt_{eff} to Δt as a function of amplitude of the thermal-cycling.

4. Summary

In nanoparticulate films with perpendicular magnetic anisotropy, a large demagnetizing field, whose amplitude is often unknown, compensates for the designed change in the applied field. Therefore, experimental results obtained by changing the applied field are always controversial as long as uncertainty remains in the demagnetizing response. Therefore, new methodology based on field cycling is proposed to eliminate the influence of the large demagnetizing field from the analysis of the activation volume of the reversal or magnetic activation moment in the nanoparticulate films. Additionally, an analogous argument shows that the relaxometry with thermal-cycling is also helpful in discussing the temperature dependence of the barrier height for magnetization reversal in a magnetic field. Thus, thermal/field cycling is found to be a promising methodology for investigating thermal fluctuations in perpendicular recording media.

References

[1] Street R and Woolley J C 1949 *Proc. Phys. Soc.* **A62** 562
[2] Wohlfarth E P 1984 *J. Phys. F: Met. Phys.* **14** L155
[3] Sharrock M P 1994 *J. Appl. Phys.* **76** 6413
[4] Lyberatos A, Chantrell R W 1997 *J. Phys.: Condens. Matter* **9** 2623
[5] O'Grady K, Dova P and Laidler H 1998 *Mater. Res. Soc. Symp. Proc.* **517** 231
[6] El-Hilo M, O'Grady K and Chantrell R W 2002 *J. Magn. Magn. Mater.* **248** 360
[7] Lauhoff G and Suzuki T 2000 *J. Appl. Phys.* **87** 5702
[8] Dutson J D, Litvinov D, Gibbs M R J, Inaba Y, Muraoka H and O'Grady K 2006 *J. Magn. Magn. Mater.* **304** 51
[9] Hagiya H, 2010 *Master Thesis*, Ibaraki Univ.
[10] Inaba N, Uesaka Y and Futamoto M 2000 *IEEE Trans. Magn.* **36** 54

Acknowledgments

This study was partly supported by a Grant-in-Aid for Scientific Research (#21241023, #21681014), the Ibaraki Society for Promotion of Use of Neutrons, and the NIMS-RIKEN-JAEA Cooperative Research Program on Quantum Beam Science and Technology.

Journal of Physics: Conference Series **266** (2011) 012060 doi:10.1088/1742-6596/266/1/012060

High-data-transfer-rate read heads composed of spin-torque oscillators

K Mizushima[1], **K Kudo, T Nagasawa and R Sato**
Corporate Research & Development Center, Toshiba Corporation, 1 Komukai-Toshiba-cho, Saiwai-ku, Kawasaki-shi, 212-8582, Japan

E-mail: koichi.mizushima@toshiba.co.jp

Abstract. The signal-to-noise ratios (SNRs) of the high-data-transfer-rate read heads beyond 3 Gbits/s composed of spin-torque oscillators (STOs) are calculated under the thermal magnetization fluctuations by using the recent nonlinear theories. The STO head senses the media field as a modulation in the oscillation frequency, enabling high signal transfer rates beyond the limit of ferromagnetic relaxation. The output (digital) signal is obtained by FM (frequency modulation) detection, which is commonly used in communication technologies. As the problem of rapid phase diffusion in STOs caused by the thermal fluctuations is overcome by employing a delay detection method, the sufficiently large SNRs are obtained even in nonlinear STOs less than 30 x 30 nm^2 in size.

1. Introduction

The areal density of commercial hard disk drives is expected to reach 2 Tbit/in^2 around 2015 [1] and the signal to noise ratio (SNR) of the magnetoresistive (MR) reader faces the challenge from the thermal magnetic noise [2]. The noise originates from the magnetization fluctuations in the thermal equilibrium and generally exists in the passive MR devices. Here, we consider an application of active devices, that is, spin-torque oscillators (STOs) to read heads [3][4], and calculate the SNRs of the output signals by using the recent nonlinear theories of STOs [5][6]. The STO heads are compatible with the high data-transfer rate beyond 3 Gbps (Gbits per second), which are important in large-scale magnetic storage. In conventional read heads, the magnetization vector **M** in the free layer changes directions, responding to the media fields $\pm h_s$ applied perpendicular to the bias field $\mathbf{H}_{\text{bias}}$, as shown in figure 1 (a). The rate of the change is, however, limited by the energy relaxation rate, *i. e.*, the linewidth of ferromagnetic resonance $\Gamma_G = \alpha\gamma(4\pi M)/2 \leq 1\ \text{ns}^{-1}$, where $\alpha \sim 0.01$ and $\gamma = 17.6 \times 10^6$ rad/s are the Gilbert damping parameter and the gyromagnetic ratio, respectively, and $4\pi M = 4\pi|\mathbf{M}| \leq 10^4$ emu/cm^3. It is, therefore, difficult for the conventional read heads to comply with the high-speed signal processing far beyond 1 Gbps. On the contrary, the precession frequency of the magnetization in STO is expected to change promptly irrespective of the relaxation rate in response to $\pm h_s$ applied along $\mathbf{H}_{\text{bias}}$ as shown in figure 1 (b), followed by a small change of the precession angle θ. The prompt frequency changes are also predicted by micromagnetic simulations [7]. An FM (frequency modulation) signal from the STO head is schematically shown in figure 2, in which the small change in amplitude caused by that of the angle θ is suppressed by a power limiter (PWL in figure 3). T is the 1 bit period (1 time slot) and 1/T is the data-transfer rate. The FM signal is converted to a train of

[1] To whom any correspondence should be addressed.

voltage pulses (a digital signal) by FM detection commonly used in communication technologies. It is, however, noted that frequency (phase) stability of STOs is very poor compared with that of conventional electronic oscillators. Although the thermal magnetization fluctuations in STOs are much different from those in the passive MR devices [5], the fluctuation cause rapid phase diffusion (phase noise) in the STO signal. The phase correlation time, *i. e.*, the characteristic time of phase stability is less than 0.1 μs at room temperature for most devices, making the synchronous detection of the signal difficult. A delay detection method is instead employed for the FM detection, in which phase correlation only in the 1 bit period T < 1 ns is important as explained in the next section. The SNRs of the output (digital) signals are calculated for 3, 6 and 9 Gbps heads of various sizes under the thermal magnetization fluctuations. The sufficiently large SNRs are obtained even in nonlinear STOs less than 30 x 30 nm^2 in size, showing that the STO read heads beyond 3 Gbps are promising. Electrical noises such as Johnson and shot noises are also considered, which cause phase fluctuations in the FM signal depending on the amplitude (power) of the signal.

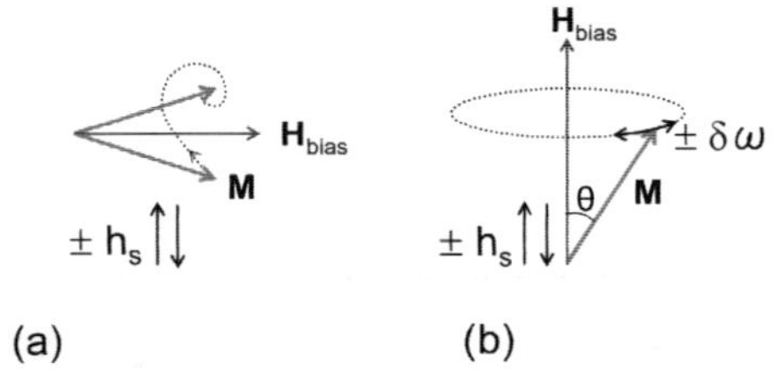

Figure 1. (a) Magnetization vector **M** in the free layer changes directions, responding to the media field $\pm h_s$ applied perpendicular to the bias field $\mathbf{H}_{bias}$. (b) Precession frequency of **M** in STO changes in response to $\pm h_s$ applied along $\mathbf{H}_{bias}$. The frequency changes are $\pm \delta\omega$.

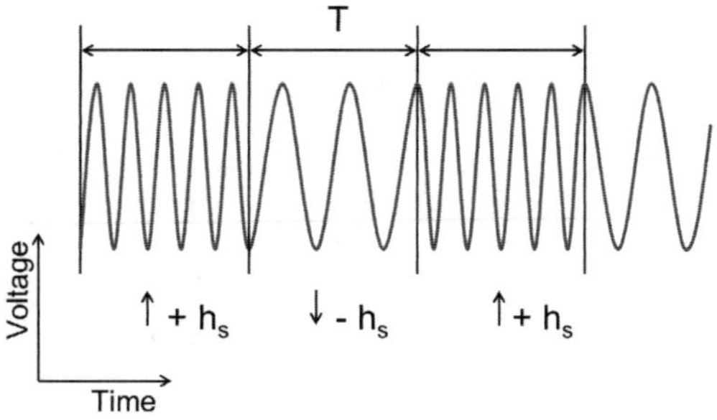

Figure 2. FM signal from the STO head in the media field $\pm h_s$. T is the 1 bit period and 1/T is the data-transfer rate.

2. Delay Detection of FM Signals from STO and SNRs of Output Signals

A typical circuit for delay detection is shown in figure 3. Signal $s_0(t)$ from STO is divided into signals $s(t)$ and $s(t - \mathrm{T})$ of equal amplitude, the latter of which is delayed 1 bit period T. Amplitude (power) fluctuations in $s_0(t)$ are suppressed by a power limiter (PWL) placed before the divider. The output voltage $v(t)$ is obtained by multiplying $s(t)$ and $s(t - \mathrm{T})$ and passing through a low-pass filter (LPF). $s(t)$ and $s(t - \mathrm{T})$ are expressed, respectively, as

$$s(t) = A\cos[\omega_0 t + \varphi(t)], \tag{1}$$

$$s(t-\mathrm{T}) = A\cos[\omega_0 (t-\mathrm{T}) + \varphi(t-\mathrm{T})], \tag{2}$$

where ω_0 is the STO frequency in the absence of the media field. Setting the cutoff frequency of LPF appropriately, the output voltage $v(t)$ is given by

$$\begin{aligned} v(t) = Cs(t)\cdot s(t-\mathrm{T})\,|_{\mathrm{LPF}} &= C(A^2/2)\cos[\omega_0\mathrm{T} + \varphi(t) - \varphi(t-\mathrm{T})] \\ &= (A^2/2)\cos[\omega_0\mathrm{T} + \varphi(t) - \varphi(t-\mathrm{T})] \end{aligned}, \tag{3}$$

where the proportionality constant C of the dimension $(\text{voltage})^{-1}$ is made unity for simplicity. Under the quadrature condition of $\omega_0\mathrm{T} = 2n\pi - \pi/2$ (n: integer), equation (3) reduces to

$$v(t) = (A^2/2)\sin[\varphi(t) - \varphi(t-\mathrm{T})]. \tag{4}$$

The oscillation frequency shifts $+\ \delta\omega$ ($-\ \delta\omega$) in the media field $+\ h_s$ ($-\ h_s$) and the phase $\varphi\ (t)$ increases (decreases) linearly with time, *i. e.*, $d\varphi = \pm\ \delta\omega dt$. For example, the phases $\varphi\ (t)$ and $\varphi\ (t - \mathrm{T})$ and the phase difference $\varphi\ (t) - \varphi\ (t - \mathrm{T})$ vary as curves (a), (b) and (c), respectively, in figure 4, where bit boundaries are at $t = m\mathrm{T}$ (m: integer). Phase diffusion caused by the thermal magnetization fluctuations is neglected for the moment. In the MSK (minimum shift keying) method, which is widely used in FM communications, the condition of $\delta\omega\mathrm{T} = \pi/2$ is imposed on $\delta\omega$ and $\varphi\ (t) - \varphi\ (t - \mathrm{T})$ is $\pm\ \pi/2$ at bit boundaries. Detecting $v\ (t)$ at the boundaries, the output voltages of $\pm\ A^2/2$ are obtained corresponding to the media fields $\pm\ h_s$, respectively, in the preceding bit. Considering STO composed of a free layer of in-plane magnetization with H_{bias} applied in the plane, the oscillation frequency $f_0 = \omega_0/2\pi$, the frequency shift $\delta f = \delta\omega/2\pi$, the in-plane field $H = H_{\mathrm{bias}} + H_{\mathrm{K}}$ and the media field h_s, which satisfy both the quadrature and the MSK conditions are summarized in table 1 in the cases of $1/\mathrm{T} = 3$ ns^{-1} (3 Gbps), 6 ns^{-1} and 9 ns^{-1}, where H_{K} is the in-plane anisotropy field in the free layer. H and h_s are calculated from ω_0 and $\delta\omega$, respectively, by using the Kittel equation $\omega_0 = \gamma(4\pi MH)^{1/2}$ and the differential $\delta\omega = h_s(\gamma/2)(4\pi M/H)^{1/2}$ with a practical value of $4\pi M = 10^4$ emu/cm^3 ($M = 796$ emu/cm^3).

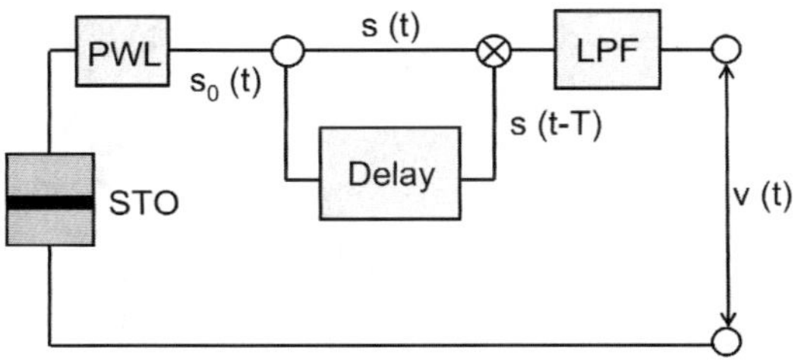

Figure 3. Typical circuit for delay detection. Signal $s_0(t)$ from STO is divided into signals $s(t)$ and $s(t\text{-T})$ of equal amplitude, the latter of which is delayed 1 bit period T. Power (amplitude) fluctuations in $s_0(t)$ are suppressed by a power limiter (PWL) placed before the divider.

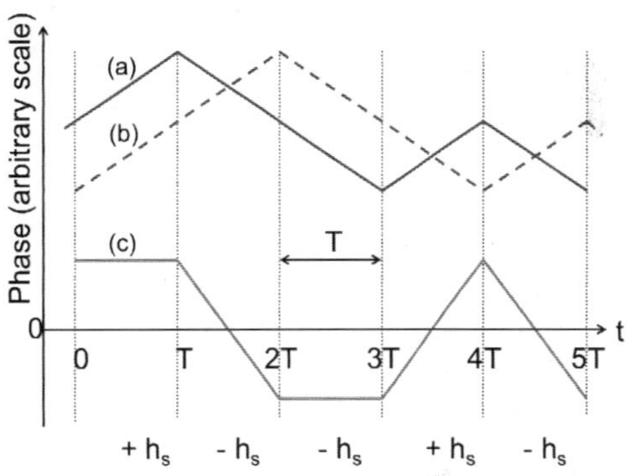

Figure 4. Curves (a), (b) and (c) show the time-dependence of $\varphi\ (t)$, $\varphi\ (t - \mathrm{T})$ and $\varphi\ (t) - \varphi\ (t - \mathrm{T})$, respectively.

Table 1. The oscillation frequency f_0, the frequency shift δf, the in-plane field $H = H_{\mathrm{bias}} + H_{\mathrm{K}}$ and the media field h_s, which satisfy both the quadrature $2\pi f_0\mathrm{T} = 2n\pi - \pi/2$ and the MSK $2\pi\delta f\mathrm{T} = \pi/2$ conditions in the cases of $1/\mathrm{T} = 3$ ns^{-1} (3 Gbps), 6 ns^{-1}, and 9 ns^{-1}.

1/T (ns^{-1})	n	f_0 (GHz)	δf (GHz)	H (Oe)	h_s (Oe)
3	2	5.25	0.75	352	100
3	3	8.25	0.75	868	158
3	4	11.25	0.75	1614	215
6	1	4.5	1.5	258	172
6	2	10.5	1.5	1406	402
9	1	6.75	2.25	581	387

Taking into account the phase diffusion $\varphi_{\mathrm{D}}\ (t)$ caused by thermal magnetization fluctuations, equations (1) and (2) are replaced by equations (5) and (6), respectively.

$$s(t) = A\cos[\omega_0 t + \varphi(t) + \varphi_{\mathrm{D}}(t)] \quad (5)$$

$$s(t - \mathrm{T}) = A\cos[\omega_0 (t - \mathrm{T}) + \varphi(t - \mathrm{T}) + \varphi_{\mathrm{D}}(t - \mathrm{T})] \quad (6)$$

φ (t) is a phase in the absence of the thermal fluctuations and ω_0 is a peak frequency in the power spectrum in the absence of the media fields $\pm h_s$. Assuming $d\varphi_D(t)/dt \ll \omega_0$, the output voltage $v(t)$ is given by equation (7) corresponding to equation (3).

$$v(t) = s(t)\cdot s(t-\mathrm{T})\,|_{\mathrm{LPF}} = (A^2/2)\cos[\omega_0\mathrm{T} + \varphi(t) - \varphi(t-\mathrm{T}) + \varphi_\mathrm{D}(t) - \varphi_\mathrm{D}(t-\mathrm{T})] \qquad (7)$$

Under both the quadrature and the MSK conditions mentioned above, the output voltage at the bit boundary $t = t_\mathrm{m} = m\mathrm{T}$ is

$$v = v(t_\mathrm{m}) = \pm(A^2/2)\cos[\varphi_\mathrm{D}(t_\mathrm{m}) - \varphi_\mathrm{D}(t_\mathrm{m} - \mathrm{T})]\,. \qquad (8)$$

The ensemble average of the output voltage $<v>$ is expressed as

$$< v >= \pm(A^2/2) < \cos[\varphi_\mathrm{D}(t_\mathrm{m}) - \varphi_\mathrm{D}(t_m - \mathrm{T})] >. \qquad (9)$$

According to the recent theories [5], the ensemble average $K(\tau) = <\cos[\varphi_\mathrm{D}(t) - \varphi_\mathrm{D}(t-\tau)]>$ in nonlinear STOs sufficiently above the threshold is given by

$$K(\tau) =< \cos[\varphi_\mathrm{D}(t) - \varphi_\mathrm{D}(t-\mathrm{T})] >= \exp[-\Delta\varphi_\mathrm{D}{}^2(\tau)/2]\,, \qquad (10)$$

where the phase variance $\Delta\varphi_\mathrm{D}{}^2(\tau)$ is

$$\Delta\varphi_\mathrm{D}{}^2(\tau) =< \varphi_\mathrm{D}{}^2(\tau) > - < \varphi_\mathrm{D}(\tau) >^2 = \Delta\omega_0[(1+\nu^2)\tau - \nu^2(1-e^{-2\Gamma_\mathrm{p}\tau})/(2\Gamma_\mathrm{p})]\,. \qquad (11)$$

ν and Γ_p are the nonlinear frequency shift coefficient and the effective damping rate, respectively. $\Delta\omega_0$ is the full linewidth of the corresponding linear STO ($\nu = 0$) [4] given by

$$\Delta\omega_0 = \frac{2H + 4\pi M}{2H}\frac{\gamma\alpha k_\mathrm{B}T}{MV}\frac{1}{p} \cong \frac{4\pi M}{2H}\frac{\gamma\alpha k_\mathrm{B}T}{MV}\frac{1}{p}\,, \qquad (12)$$

where $p \le 1$ is the normalized power of oscillation. $k_\mathrm{B}T$ and V are the thermal energy and the volume of the free layer, respectively. Defining the noise voltage v_n and the SNR by equations (13) and (14), respectively,

$$v_\mathrm{n} = \sqrt{< v^2 > - < v >^2} \qquad (13)$$

$$\mathrm{SNR} =< v > / v_\mathrm{n}\,, \qquad (14)$$

the SNR of the output signal is given by

$$\mathrm{SNR} = \sqrt{2}\exp[-\Delta\varphi_\mathrm{D}{}^2(\mathrm{T})/2] / \sqrt{1+\exp[-2\Delta\varphi_\mathrm{D}{}^2(\mathrm{T})] - 2\exp[-\Delta\varphi_\mathrm{D}{}^2(\mathrm{T})]}\,. \qquad (15)$$

Considering the free layer of the size $L \times L$ nm^2 and the thickness d nm ($V = L \times L \times d$ nm^3), dependence on L of the SNR is calculated [4]. The results are shown (in decibels) in figure 5 in the case of $1/\mathrm{T} = 3$ ns^{-1} (3 Gbps). Solid and dashed curves in the figure are under the fields of $H = 352$ Oe and 868 Oe (see table 1), respectively, where $M = 796$ emu/ cm^3 ($4\pi M = 10^4$ emu/ cm^3), $\Gamma_\mathrm{p} = 0.15$ ns^{-1} [8], $\alpha = 0.01$, $d = 3$ nm, $T = 300$ K and $p = 0.5$. The nonlinear coefficient ν is varied as 1, 3, and 5, and the top and the bottom solid/dashed curves correspond to $v = 1$ and $v = 5$, respectively. Figure 6 shows similar results in the cases of 6 Gbps (solid curves) and 9 Gbps (dashed curves) under the fields of 258 Oe and 581 Oe, respectively, where the same values are used for M, Γ_p, α, d, T and p as those in figure 5. The top and the bottom solid/dashed curves correspond to $v = 1$ and $v = 5$, respectively. Figures 5 and 6 show that the SNRs larger than 30 dB are attainable even for $L \leq 30$ nm by reducing v and/or increasing H appropriately.

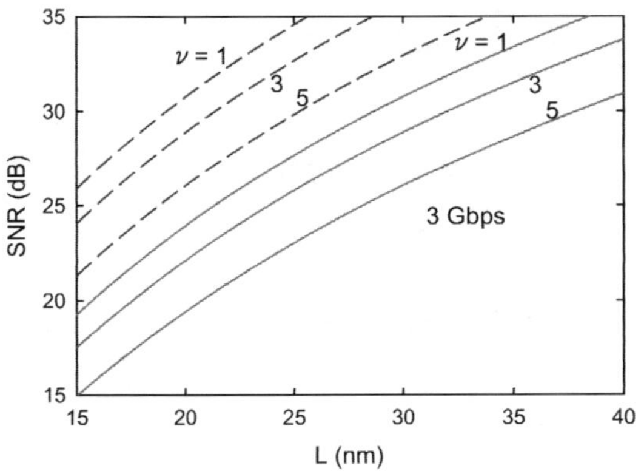

Figure 5. Dependences of SNR in decibels on L in the case of 3 Gbps. Solid and dashed curves in the figure are under the fields of $H = 352$ Oe and $H = 868$ Oe, respectively. The nonlinear coefficient v is varied as 1, 3, and 5.

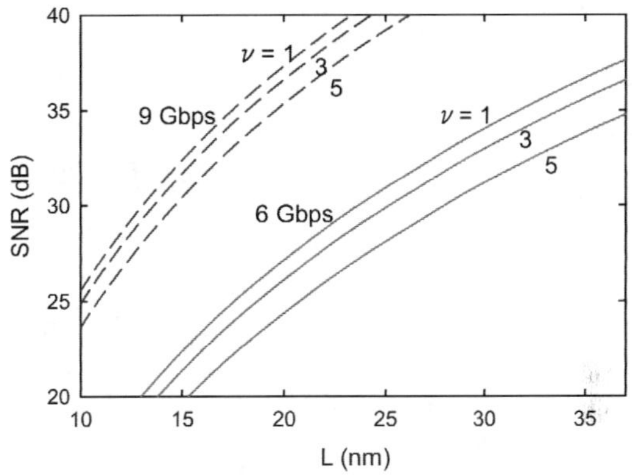

Figure 6. Dependences of SNR in decibels on L in the cases of 6 Gbps (solid curves) and 9 Gbps (dashed curves) under the fields of 258 Oe and 581 Oe, respectively. The nonlinear coefficient v is varied as 1, 3, and 5.

The decrease of the SNR by electrical noises such as Johnson and shot noises is calculated by adding the electrical noise $n(t) = n_c(t)\cos\omega_0 t + n_s(t)\sin\omega_0 t$ to $s_0(t)$ [4]. In the case of the total electrical noise of 100 pW, the output power of STO larger than 10 nW is sufficient to avoid the further degradation of the SNRs by the electrical noises.

3. Summary

Application of active devices, i. e., STOs to high-signal-transfer-rate read heads beyond 3 Gbps is considered and the SNRs under the thermal magnetization fluctuations are calculated by using the recent nonlinear theories. The STO head senses the media field as a modulation of the oscillation frequency, enabling the high signal transfer rates beyond the limit of ferromagnetic relaxation. The problem of rapid phase diffusion caused by the thermal fluctuations is overcome by employing a delay detection method, which is commonly used in communication technologies, and the sufficiently large SNRs are obtained in STOs less than 30 x 30 nm^2 in size.

References

[1] Chen Y et al. 2009 *TMRC Digest* A1
[2] Smith N and Arnett 2001 *Appl. Phys. Lett.* **78** 1448-50
[3] Braganca P M et al. 2010 *Proc. 11 th Joint MMM-Intermag Conf.* EA-06
[4] Mizushima K et al. 2010 *J. Appl. Phys.* **107** 063904
[5] Slavin A and Tiberkevich V 2009 *IEEE Trans. Magn.* **45** 1875-1918
[6] Kudo K et al. 2009 *J. Appl. Phys.* **105** 07D105
[7] Kudo K et al. 2010 *Appl. Phys. Express* **3** 043002
[8] Boone C et al. 2009 *Phys. Rev. B* **79** 140404(R)

Journal of Physics: Conference Series **266** (2011) 012061 doi:10.1088/1742-6596/266/1/012061

Damping Constant Influence on Spin Dynamics in Field Generating Layer of STO for MAMR Writing Head

SeungMo Noh, Daisuke Monma, Kousaku Miyake
Masaaki Doi and Masashi Sahashi

Department of Electronic Engineering, Tohoku University, 980-8579 Sendai, Japan

E-mail: nohsm@ecei.tohoku.ac.jp

Abstract. For Microwave Assisted Magnetic Recording (MAMR), large amplitude of ac stray field and coherent motion of magnetization and also low driving power are necessary as a performance of spin torque oscillator for writing head. The damping constant is characterized by both the dynamical motion of magnetization and critical current for the precession-motion. The damping constants for Fe_xCo_{1-x} dot were investigated by a network analyzer-ferromagnetic resonance measurement. From these results, suitable material for the oscillator and the effect of finite magnetic element size are discussed.

1. INTRODUCTION

As the recording bit density of hard-disk drive becomes higher and higher, the volume of a bit becomes still smaller with larger media anisotropy energy to keep thermal stability. As a result, a required magnetic field is approaching a physical limit of the magnetic field generated by writer. Microwave Assisted Magnetic Recording (MAMR)[1] is a promising candidate for solving a writability problem in future hard-disk drives. It is an important issue to realize a writing head generating a high frequency ac magnetic field superimposed on dc field. In STO (Spin Torque Oscillator), magnetic material for Field Generating Layer (FGL) is needed to have a high saturation magnetization (M_s) and a low damping constant (α) for a large ac field and a low driving power. What's more, the lower α value should be needed in a nano-sized dot shape to generate a coherent ac field in STO element. Therefore, understanding the mechanism responsible for magnetic damping in dot-materials has become an important issue.

Among the high frequency techniques, ferromagnetic resonance (FMR) measurement, which provides important parameters for magnetic dynamics, such as Gilbert damping constant and magnetic anisotropy, is especially crucial. Network analyzer-ferromagnetic resonance (NA-FMR) techniques [2-4] have been developed. In standard FMR the used frequency is fixed by the employed microwave cavity. NA-FMR measurements on a coplanar waveguide (CPW) can help resolve this: The impedance of the CPW can be chosen such that the usable frequency range extends to several tens of gigahertz. In contrast to classical FMR, frequency and magnetic field can both be varied continuously.

In this paper, the suitable material of FGL for MAMR was searched from the view point of α and Ms. Furthermore the size effect on α i.e. coherent motion of FGL was also investigated.

2. EXPERIMENTAL PROCEDURE

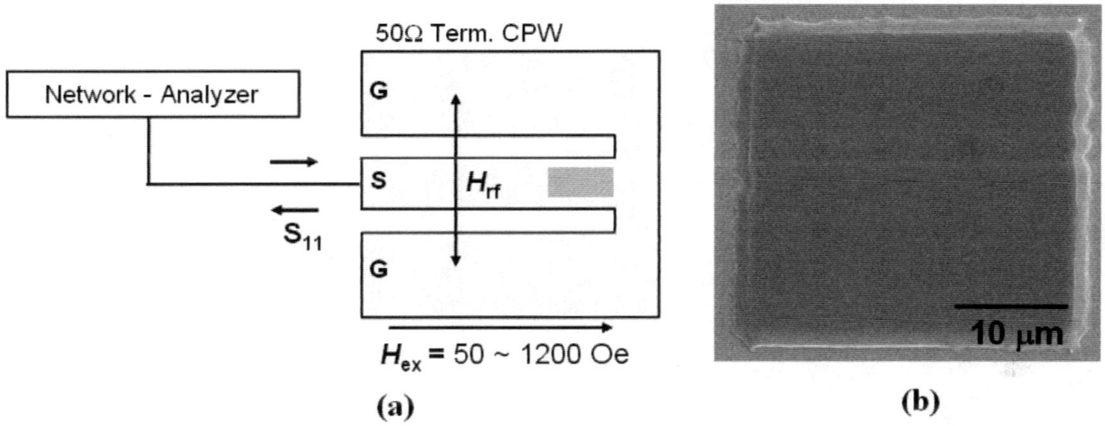

Fig 1.Schematic diagram of the network analyzer-ferromagnetic resonance (NA-FMR) measurement. The sample is deposited on the coplanar waveguide. The mutually right-angled external field (H_{ex}) and microwave ac field (H_{rf}) are in plane of a dot pattern. (b) SEM image of dot of 30×30 [μm^2]

The NA-FMR technique allows for operating over a wide frequency band and yields FMR parameters from standard microwave *S*-parameter measurements as function of frequency under various field. Figure 1 shows a diagram of the measurement system. The microwave drive was provided by CPW excitation structure with the thin film sample deposited across the signal line. The mutually right-angled external field (H_{ex}) and the microwave field (H_{rf}) were in plane of dot patterned sample. The signal analysis was carried out with a standard network analyzer. The electrically shorted coplanar waveguide transmission line consisting of Sub./Ta(5 nm)/Cu(300 nm)/ Ta(25 nm) and dot patterns of Fe_xCo_{1-x} are fabricated on quartz substrate by a photolithography, an Ar^+ ion milling and a resist removal processes. The CPW and dot pattern were insulated by a 60 nm thick SiO_2 film. The widths of signal and ground line of the CPW were 100μm, 250 μm respectively. The spacing distance between the ground lines and the signal line was 10 μm. The characteristic impedance of the CPW evaluated from the relative permittivity of 3.8 for the quartz substrate and the geometrical parameters of CPW was approximately 50 Ω.

Each magnetic dot with a thickness of 20 nm and various lateral sizes, i.e. 30 × 30, 30 × 60, 30 × 90, 30× 120 and 30×150 μm^2, were fabricated. The magnetization curves and FMR spectra of the sputtered ferromagnetic films, i.e. Fe_xCo_{1-x}, were measured by using vibrating sample magnetometor and NA-FMR spectroscopy, respectively. The NA-FMR was measured with an Agilent N5230A network analyzer and 40A-GSG-250 EDP of high frequency probe in the frequency range of 0.01–20 GHz. Two ground-signal-ground (GSG) pin type wafer probes were in mechanical contact with both ends of the CPW transmission line. Therefore, the ac magnetic field caused by applying microwaves to the CPW and the input power of -15 dBm. The microwave magnetic field (H_{rf}) was generated in plane by the ac current through signal line around CPW. H_{rf} was right-angled to the external field (H_{ex}). H_{ex} from 0 Oe to 1200 Oe was applied parallel to the signal line during measurement.

Journal of Physics: Conference Series **266** (2011) 012061 doi:10.1088/1742-6596/266/1/012061

3. RESULT

The dynamics of magnetic dots were deduced from the frequencies and linewidths of each FMR spectrum. For a magnetic thin film with an in-plane shape anisotropy and with magnetization direction along with applied H_{ex}, the resonance frequency (f_{res}) is written as the Kittel formula,

$$f_{res} = \frac{|\gamma|}{2\pi}\sqrt{\left(H_{k,//}+H_{ex}\right)\left(H_{k,\perp}+H_{ex}\right)} \qquad (1)$$

, where $H_{k,//}$ and $H_{k\perp}$ are contribution due to the crystalline and shape anisotropy of in-plane and out-plane respectively. $\gamma$is the gyromagnetic factor of Fe_xCo_{1-x}.[5]

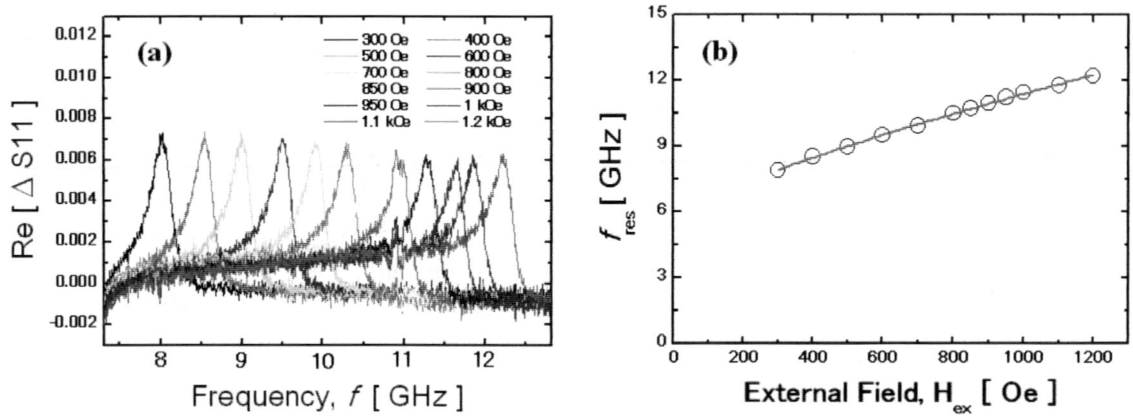

Fig 2. (a) FMR spectra of Re[ΔS_{11}]resonance in the 30×150 μm² size of 20 nm thickness $Fe_{70}Co_{30}$ at the various H_{ex} between 300-1200 Oe. (b) Resonance frequency (f_{res}) in the 30×150 um² size of 20 nm thickness $Fe_{70}Co_{30}$ film as a function of H_{ex}. Black open circle indicates the experimental result and red solid line indicates the fitting line by the Kittel formula Eq. (1).

Figure 2(a) shows FMR spectra of Re[ΔS_{11}] and f_{res} in the 30×150 μm² size of 20 nm thickness $Fe_{70}Co_{30.}$ The external field of 300-1200 Oe was applied parallel to the signal line i.e. the easy axis of patterned dot because the external field below 250 Oe are thought to be too weak to saturate the magnetization of specimens. The spectrum measured at 0 Oe was taken as the background signal to renormalize the data. f_{res} shifted to the right side with increasing H_{ex}. Figure 2(b) shows the dependence of f_{res} on H_{ex}. In Figure (b), black open circle and red solid line show the experimental result and the fitting line by Kittel formula, respectively. The plot posses good linearity and indicates the approximation of Eq. (1) works well. The anisotropies ($H_{k,//}$ and $H_{k,\perp}$) were obtained from the fitting. The obtained parameters for the samples are summarized in Table 1.

Table 1. In-plane and out-plane anisotropy fields ($H_{k,//}$ and $H_{k,\perp}$) of various size $Fe_{70}Co_{30}$ dots.

	30×30 [μm²]	30 60 [μm²]	30 90 [μm²]	30×120 [μm²]	30×150 [μm²]
$H_{k,//}$ [Oe]	276	246	250	254	247
$H_{k,\perp}$ [Oe]	12049	13175	13378	13384	13410

In order to study quantitatively, the damping constant (α) was obtained from Full-Width-Half-Maximum (FWHM: (Δf) of FMR spectra and anisotropy fields (H_k : $H_{k,//}+H_{k,\perp}$) of Eq. (1).

$$\alpha = -\frac{2\pi \Delta f}{\gamma\left(2H_{ex} + H_{k,//} + H_{k,\perp}\right)} \tag{2}$$

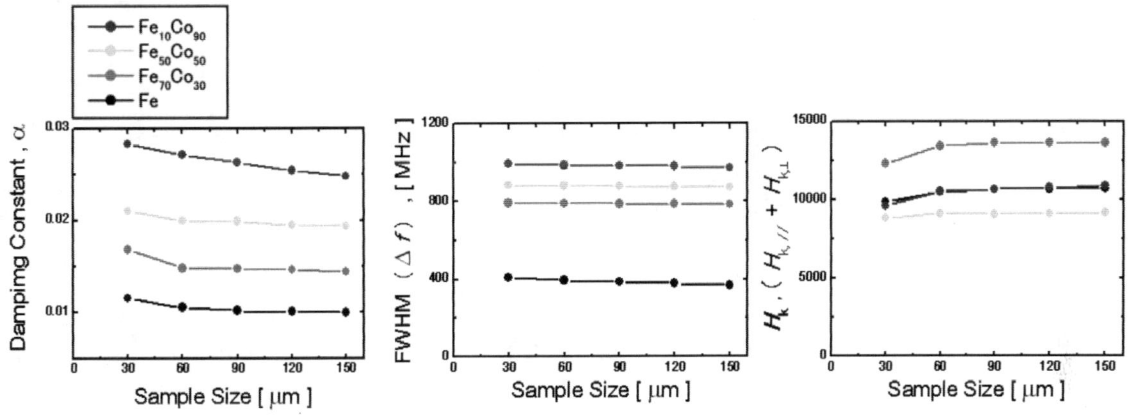

Fig 3. Sample-size dependence on the damping constant (α), FWHM (Δf), and H_k : $H_{k,//}+H_{k,\perp}$ for various composition of Fe_xCo_{1-x} film-dot with 20nm in thickness.

The results are summarized in Fig. 3. The damping constant increases with the sample size decrease. Δf shows slightly increase with decrease sample size and large decrease of H_k is observed. This variation is regarded to be an influence of dispersion of anisotropy or stray field effect at edge part of the sample. In Table 1, The in-plane anisotropy field ($H_{k,//}$) slightly change almost constant. In contrast, the out of plane anisotropy field ($H_{k,\perp}$) largely decrease with smaller sample size.
The magnetic property of single dot depends on the balance of the spins alignment; the in-plane magnetic anisotropy which favours alignment of spins along a particular direction, and the out of plane demagnetizing field is created by the thickness of the dots. If the shape of the dots is not a rectangular at the edge part, the demagnetizing field is non-uniform. H_k or stray field influenced by the curvature distortion of the dot outline of lithographically defined structures as observed with the size variations (insets SEM image of Fig. 1(b)).

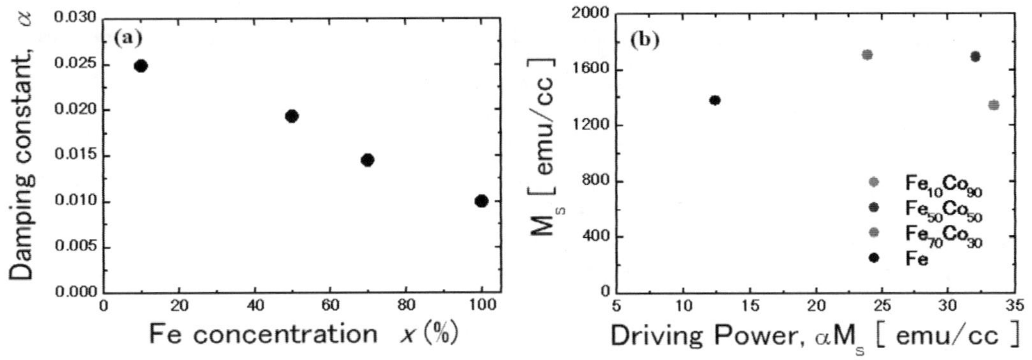

Fig 4.(a) Fe concentration dependence on damping constant and (b) relations between M_s ans αM_s for Fe_xCo_{1-x} 30 × 150 μm^2 dots.

Figure 4(a) shows the Fe concentration dependence on α for the Fe_xCo_{1-x} film-dots with 30×150 μm^2 size. The damping constant almost linearly decreases with Fe concentration increasing. This behavior of damping constant for Fe_xCo_{1-x} film-dots are consistent with published conventional FMR data[7]. But the α value is fairly larger than the reported conventional FMR data which might be attributed to the fact that NA-FMR spectrum also contains the FWHM due to a non-uniformity originating from the rf field.

4. DISCUSSION AND CONCLUSION

In the macrospin approximation[8], the critical current for spin-torque magnetization reversal of a FGL is given by

$$J_c = \frac{e\alpha}{\hbar g}(H_k + 2\pi M_s) \qquad (3)$$

$$J_c \propto \alpha M_s \qquad (4)$$

, where g is the spin polarization of the current. J_c, which related to the driving power of STO, is proportional to αM_s. Figure 4(b) shows the relation between α and M_s. Within the assumptions of Eq. (4), low critical currents can be achieved by reducing the αM_s. On the other hand, High M_s material generate large amplitude of H_{ac}. Therefore, the material parameters should be locate at upper left side in the Fig. 4(b). So, for Fe_xCo_{1-x}, Fe is the most suitable material for FGL. Note that the above evaluation does not account g parameter effect of Eq. (3). The g parameter should be accounted from the other experiments of MR effects.

For MAMR, STO is necessary to be nano-dot size for magnetization motion of finite size element. The α increased with smaller dot-size as shown in the Fig. 3, which is regarded to be anisotropy dispersion due to the curvature distortion of the dot outline. As a result, Δf increases. Although it is extremely difficult to quantify, this anisotropy dispersion could explain why only size variations alone are not sufficient to explain our experimental data.

In conclusion, both dynamics of processional motion and performance of STO can be evaluated by NA-FMR. Fe rich material is much better than Co rich material in the aspect of M_s and αM_s for FGL generating high H_{ac}.

This research was in part supported by the Storage Research Consortium (SRC) and Center of Education and Research for Information Electronics systems of Global COE Program, Japan.

[1] Jian-Gang.Zhu, Xiaochun Zhu and yuhui Tang 2008 *IEEE.Trans. Magn.* **44** 125
[2] T. J. Silva, C. S. Lee, T. M. Crawford and C. T. Rogers 1999 *J. Appl. Phys.* **85** 7849
[3] S. S. Kalarickal, Pavol Krivosik, Mingzhong Wu and Carl E. Patton, Michael L. Scheider, Pavel Kabos and T. J. Silva, John P. Nibager 2006 *J. Appl. Phys.* **99** 093909
[4] N. Toda, K. Saito, H. Maekawa, M. Mizuguchi, M. Shiraishi and Y. Suzuki 2007 *J. Magn. Soc.* **31** 435
[5] Richard M. Bozorth 2003 John Wiley and Sons INC. New Jersey
[6] Justin M. Shaw and T. J. Silva, Michael L. Schneider, Robert D. McMichael 2009 *Phys. Rev.* B. **79** 184404
[7] F. Schreiber, J. Pflaum, Z. Frait, Th. Muhge, and J. Pelzl 1995 *Solid State Commun,* **93** 965
[8] S. Mangin, Y. Henry, D. Ravelosona, J. A. Katine and Eric E. Fullerton 2009 *Appl. Phys. Lett.* **94** 012502

Journal of Physics: Conference Series **266** (2011) 012062 doi:10.1088/1742-6596/266/1/012062

Effect of Ar-gas-pressure on oxygen content of Co–Si–O granular film sputter-deposited by using sintered targets

S Sasaki [1], **S Saito** [2], **and M Takahashi** [2]
[1]Department of Intelligent Systems Engineering, Ichinoseki National College of Technology, Ichinoseki, Iwate 021-8511, Japan
[2] Department of Electronic Engineering, Graduate School of Engineering, Tohoku University, Sendai 980-8579, Japan

Abstract. A systematic study of the effect of Ar-gas-pressure (P_{Ar}) in Co–Si–O granular film by sputter deposition using sintered targets is carried out for applications in CoPtCr–SiO_2 perpendicular media. Detailed analyses of composition, phase formation and microstructure of granular films revealed that: (1) Co–Si–O granular films with similar compositions could be prepared by different sintered targets, (Co)+(SiO_2), (Co)+(CoSi)+(CoO), (Co)+(Si)+(CoO), and (Co)+(Si)+(Co_3O_4), (2) oxygen content in film increases with increasing P_{Ar}, (3) oxygen content in film could be adjustable by the target composition (e.g. (CoO)/(CoSi)). Taking these results into account, we tried low P_{Ar} deposition of Co–Si–O granular film. Under P_{Ar} = 0.6 Pa, oxygen-poor Co–Si–O granular film was obtained using a conventional (Co)+(SiO_2) target, while by using a new type of sintered (Co)+(CoSi)+enriched(CoO) target, the oxygen composition in the granular film could be adjusted to be O/Si= 2 with grain isolated structure.

1. Introduction

In a perpendicular recording medium, a CoPtCr–SiO_2 granular film as a magnetic layer has been deposited by dc sputtering with a (CoPtCr)–(SiO_2) composite target under high Ar-gas-pressure (P_{Ar}) from 4.0 to 8.0 Pa [1, 2]. We have previously reported that low P_{Ar} deposition of a Co–Si–O granular film with stoichiometric SiO_2 phase is impossible by using conventional composite target using SiO_2 powder, because oxygen-poor Co–Si–O granular film was formed [3, 4]. For low P_{Ar} deposition, reactive sputtering in (Ar+O_2) gas mixture was reported [5-7]. However, the film deposited by reactive sputtering has the problems of compositional and structural inhomogeneities. In this work, we selected a simple system of Co–Si–O as a granular layer material and investigated the Co–Si–O granular film deposited by various sintered targets without SiO_2 and examined the effect of P_{Ar} on the magnetic properties, composition, forming phase, and microstructure.

2. Experimental procedure

Co–Si–O ternary system was investigated in order to simplify the phase formation in the film. CoSi (melting point 1460 °C), CoO (1935 °C) and Co_3O_4 (895 °C) compounds were selected as raw materials for the target from the viewpoint of stability during the sintering process and independent control of Si and O content in the film. Mixed powders of 92.3(Co)+7.7(SiO_2) (mol%) (target A), 74.8(Co)+8.4(CoSi)+16.8(CoO) (target B), 76.9(Co)+7.7(Si)+15.4(CoO) (target C), and

[1] Corresponding author: E-mail: shingo@ichinoseki.ac.jp, Phone: +81-191-24-4755, Fax: +81-191-24-4755

87.4(Co)+8.4(Si)+4.2(Co_3O_4) (target D) were prepared as sintered targets. The atomic composition of all targets A, B, C, and D was $Co_{79.9}Si_{6.7}O_{13.4}$ (at%). Next, (Co)+(CoSi)+enriched(CoO) target (E) was tried to adjust the O/Si ratio to 2 in a granular film under a low P_{Ar} process. Whereas target E had a higher oxygen content than target A, but retained a relative atomic ratio of Co and Si. The details of the target E are described in the latter section. The grain sizes of the (Co), (SiO_2), (CoSi), (Si), (CoO), and (Co_3O_4) raw materials were 0.5–2.0 μm, 0.5–1.0 μm, 0.5–3.0 μm, 1.0–7.0 μm, 1.0–5.0 μm, and 0.5–1.0 μm, respectively.

Granular films were prepared by dc magnetron sputtering on glass substrates at room temperature. The base pressure in the deposition chamber was 7×10^{-6} Pa. The sputtering conditions, i.e., discharge power (W) of 0.4–4.0 W/cm^2 and P_{Ar} of 0.6–8.0 Pa, were varied in order to fabricate films with a composition of $Co_{79.9}Si_{6.7}O_{13.4}$ (at%) (the objective composition). The layered structure of the film was Ta (5 nm)/Pt (6 nm)/Ru (20 nm)/granular film (16 nm)/capping layer. The capping layer was deposited in order to avoid oxidation after fabrication [3]. The film thickness varied by the sputtering time, and the sputtering rate was determined by measuring the thickness of the standard sample with a scanning step height analyzer. The magnetic properties were measured using a vibrating sample magnetometer (VSM). The total amount of each atomic element in the granular film was evaluated using X-ray fluorescence analysis (XRF) [3, 4]. The crystalline structure was examined by X-ray diffraction (XRD) using Cu K_α radiation and transmission electron microscopy (TEM).

3. Results and discussion

In a simple system of Co–Si–O as a granular layer material, we tried sputter depositions of a Co–Si–O granular film made by using sintered targets, A (Co)+(SiO_2), B (Co)+(CoSi)+(CoO), C (Co)+(Si)+(CoO), and D (Co)+(Si)+(Co_3O_4). As shown Fig. 1, O/Si ratio of target composition cannot be changed with the conventional target A, but that can be changed by ratios of (CoO)/(CoSi), (CoO)/(CoSi), (CoO)/(Si), (Co_3O_4)/(Si) with targets B, C, and D , respectively. Figure 2 show in-plane and out-of-plane hysteresis loops of Co–Si–O films deposited by targets A, B, C, and D under P_{Ar} of 4.0 Pa and W of 2.0 W/cm^2 on a Ta/Pt/Ru underlayer. All films indicate a similar perpendicular anisotropy. Figure 3 shows (1) out-of-plane and (2) in-plane XRD profiles for granular films deposited using targets A, B, C and D. For all samples, the diffraction peaks from Co grains appeared only around 44.6° in (1) and 41.6° in (2). These diffractions were identified as those from hcp (00.2) and

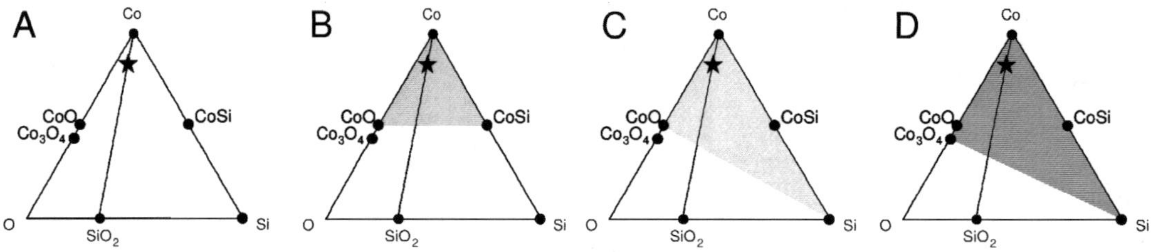

Figure 1. Objective composition (★: $Co_{79.9}Si_{6.7}O_{13.4}$ (at.%)) and adjustable range of sintered targets, A (Co)+(SiO_2), B (Co)+(CoSi)+(CoO), C (Co)+(Si)+(CoO), and D (Co)+(Si)+(Co_3O_4) plotted on ternary map, respectively.

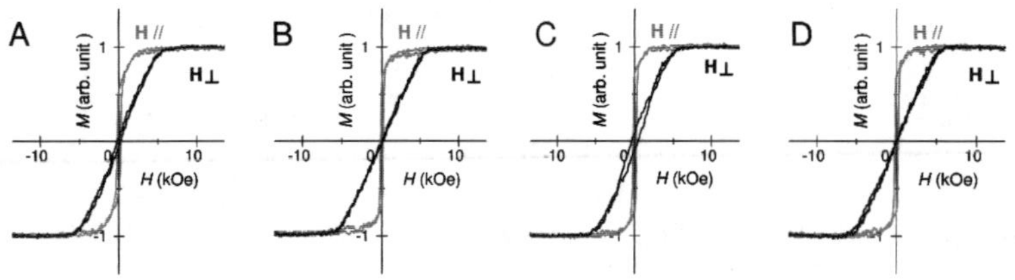

Figure 2. Hysteresis loops of media deposited by sintered targets, A, B, C, and D under high P_{Ar} of 4.0 Pa and W of 2.0 W/cm^2, respectively.

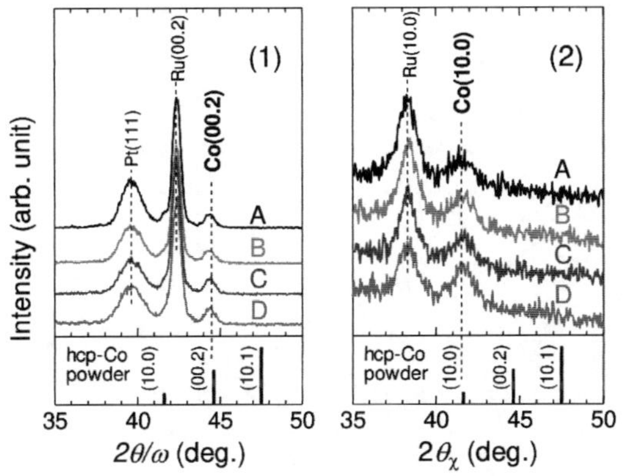

Figure 3. XRD profiles for sputtered Co–Si–O films using targets A, B, C, and D under P_{Ar} = 4.0 Pa and W of 2.0 W/cm^2 on a Ta/Pt/Ru underlayer: (1) out-of-plane, (2) in-plane.

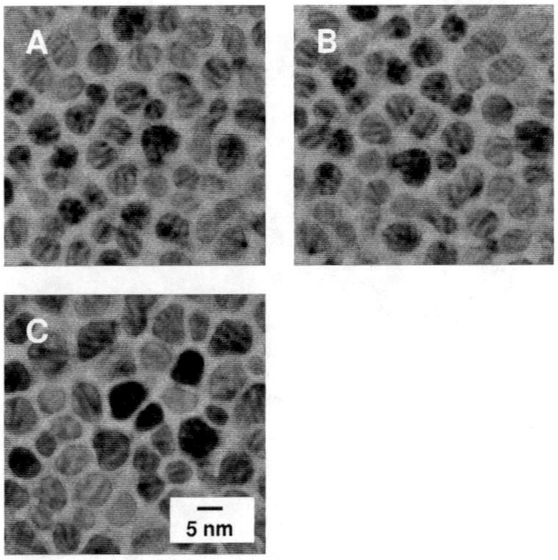

Figure 4. In-plane TEM images of the films deposited using targets for A, B, and C under P_{Ar} = 4.0 Pa and W of 2.0 W/cm^2 on a Ta/Pt/Ru underlayer, respectively.

(10.0) planes, respectively, which indicates that the sheet texture of Co grains in the granular films is the *c*-plane. Fig. 4 shows in-plane TEM images of the granular films deposited under P_{Ar} = 4.0 Pa using targets A, B, and C. In all granular films, gray or black grains with 3–8 nm in diameter surrounded by white grain boundaries can be clearly observed in the image. Therefore, Co-based magnetic grains were isolated by SiO_2-rich grain boundaries. These results suggested that the same Co–Si–O granular film is prepared by any target with the same atomic composition in sintered targets, (Co)+(SiO_2), (Co)+(CoSi)+(CoO), (Co)+(Si)+(CoO), and (Co)+(Si)+(Co_3O_4).

Figure 5 (1) shows film composition plotted on Co–Si–O enlarged ternary map for granular films prepared under various sputtering conditions (P_{Ar} = 0.6–8.0 Pa and W = 0.4–4.0 W/cm2) using targets A, B, C, and D. The area of the film composition distribution for all targets is almost the same as that for conventional target A. Fig.5 (1) indicates that the films deposited under low P_{Ar} are oxygen poor and oxygen content of films made by any target increases with increasing P_{Ar}. Figure 5 (2) shows the relationship between M_s^{VSM} evaluated by VSM, and M_s^{XRF} that was calculated using Co volume density, for Co–Si–O films under various sputtering conditions. For clarity, the condition of M_s^{XRF} = M_s^{VSM} is shown by a dotted line. M_s^{VSM} almost equals M_s^{XRF}, except for those samples deposited under high P_{Ar}. Fig. 6 (1) and (2) show compositional ratios in the films, as Co–$n$$SiO_x$ film, n mol% is the ratio of Co to Si, and x denotes O/Si ratio. In Fig. 6 (1) and (2), dashed lines show the nominal target composition of each target. Sputtering conditions around P_{Ar} = 3.0–4.0 Pa and W = 1.0–2.0 W/cm^2 enabled granular films with the same composition as the targets. Fig. 6 (3) shows a contour map of deposition rate R relative to P_{Ar} and W. The tendency of changes in R approximately corresponds to that of n in Fig. 6 (1). This indicates that film thickness depends strongly on the SiO_x phase formation.

Figure 6 (2) show that oxygen content of film increases with increasing P_{Ar} using any sintered target. To reveal the origin of this phenomenon, we examined the effect of high P_{Ar} process in Co film made by Co target. Fig. 7 shows the atomic density of (1) Co and (2) O with respect to various sputtering conditions for Co films prepared using melted Co target. Co atomic density decreases with increasing P_{Ar} and decreasing W. However, oxygen atomic density increases with increasing P_{Ar} and decreasing W. These results indicate that oxygen content increases with increasing P_{Ar}. Whereas, H_2O impurity in ultraclean Ar process gas is measured within 1 ppb. Therefore, this suggests that oxygen gas is supplied from high P_{Ar} process.

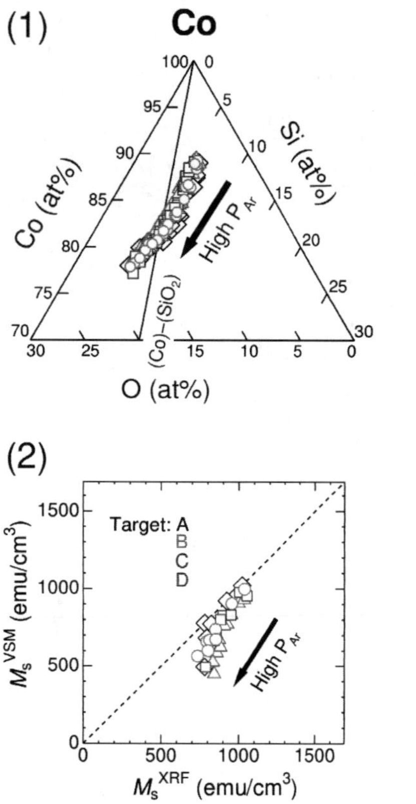

Figure 5. (1) Changes in film composition under various sputtering conditions on enlarged ternary map, and (2) relationship between saturation magnetizations, M_s^{VSM} evaluated by VSM and M_s^{XRF} that was calculated using Co volume density, of the Co–Si–O sputtered films deposited by using targets, respectively.

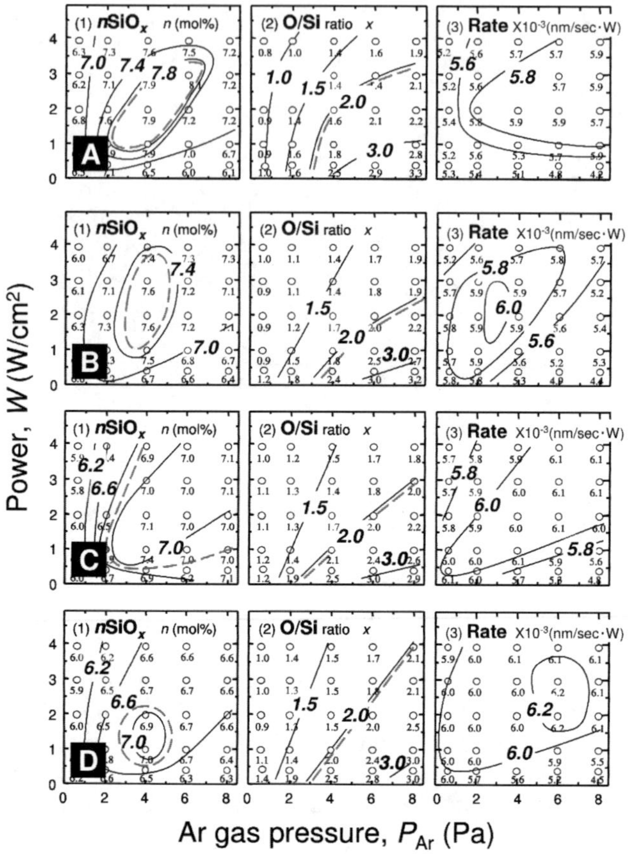

Figure 6. Changes in (1) ratio of Co to Si, *n* mol%, (2) O/Si ratio, *x*, and (3) deposition rate, relative to the sputtering conditions of discharge power, *W*, and Ar gas pressure, P_{Ar}, for granular films made by sintered targets of A (Co)+(SiO_2), B (Co)+(CoSi)+(CoO), C (Co)+(Si)+(CoO), and D (Co)+(Si)+(Co_3O_4), respectively.

O/Si ratio in films deposited under $P_{Ar} = 0.6$ Pa is found to be proportional to the one in targets. Consequently, target E (Co)+(CoSi)+enriched(CoO) is prepared for the deposition of Co–SiO_2 granular film under P_{Ar} = 0.6 Pa. The composition of target E was 53.4(Co)+8.4(CoSi)+38.2(CoO) (mol%). Whereas, target E had higher oxygen content than conventional target A, but retained a relative atomic ratio of Co and Si. Figure 8 (1) shows the film compositions of Co–Si–O films under

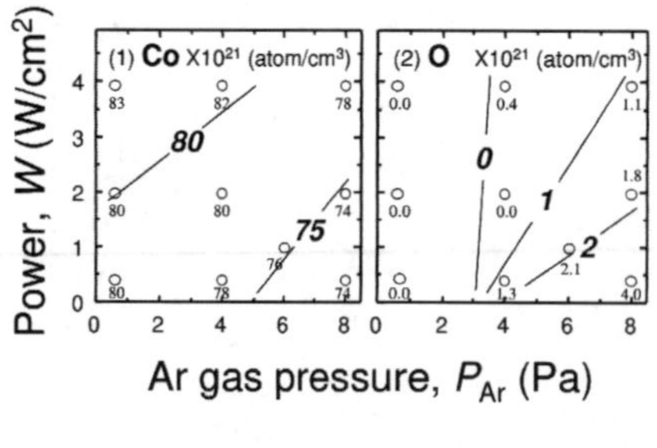

Figure 7. Changes in (1) Co, (2) O density, relative to the sputtering conditions of discharge power, *W*, and Ar gas pressure, P_{Ar}, for Co films sputtered by Co target.

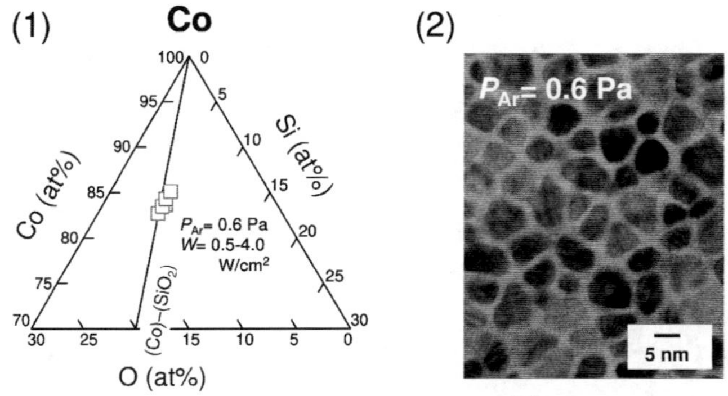

Figure 8. (1) Film composition plotted on enlarged ternary map, and (2) in-plane TEM image of the Co–Si–O film sputtered under P_{Ar}= 0.6 Pa by sintered target E (Co)+(CoSi)+enriched(CoO) on a Ta/Pt/Ru underlayer, respectively.

the sputtering condition of P_{Ar} = 0.6 Pa and W =0.4–4.0 W/cm^2. According to the compositions on the (Co)–(SiO_2) line, the granular film with O/Si ratio = 2 was realized. Fig.8 (2) shows the in-plane TEM image of the Co–Si–O granular film. Gray or black grains with 5-10 nm in diameter surrounded by white grain boundaries can be clearly observed in the image. Therefore, Co-rich grains are well isolated by SiO_2-rich grain boundaries.

4. Conclusion

A systematic study of the effect of Ar-gas-pressure (P_{Ar}) in Co–Si–O granular film by sputter deposition using sintered targets is carried out for applications in CoPtCr–SiO_2 perpendicular media. Detailed analyses of composition, phase formation and microstructure of granular films revealed that: (1) Co–Si–O granular films with similar compositions could be prepared by different sintered targets, (Co)+(SiO_2), (Co)+(CoSi)+(CoO), (Co)+(Si)+(CoO), and (Co)+(Si)+(Co_3O_4), (2) oxygen content in film increases with increasing P_{Ar}, (3) oxygen content in film could be adjustable by the target composition (e.g. (CoO)/(CoSi)). Taking these results into account, we tried low P_{Ar} deposition of Co–Si–O granular film. Under P_{Ar} = 0.6 Pa, oxygen-poor Co–Si–O was obtained using a conventional (Co)+(SiO_2) target, while by using a new type of sintered (Co)+(CoSi)+enriched(CoO) target, the oxygen composition in the granular film could be adjusted to O/Si= 2 with grain isolated structure under P_{Ar} = 0.6 Pa.

Acknowledgments

The authors would like to thank Y. Nakamura, Y. Ikeda and A. Satoh from JX Nippon Mining & Metals Corporation for their cooperation with the fabrication of targets used in these experiments. This work was supported in part by the Industrial Technology Research Grant Program in 2006 from the New Energy and Industrial Technology Development Organization (NEDO) of Japan and the Storage Research Consortium (SRC) of Japan.

References

[1] Oikawa T, Nakamura M, Uwazumi H, Shimatsu T, Muraoka H and Nakamura Y 2002 *IEEE Trans. Magn.* **38** 1976
[2] Uwazumi H, Enomoto K, Sakai Y, Takenoiri S, Oikawa T and Watanabe S 2003 *IEEE Trans. Magn.* **39** 1914
[3] Sasaki S, Saito S and Takahashi M 2008 *J. Magn. Magn. Mat.* **320** 3023
[4] Sasaki S, Saito S and Takahashi M 2009 *J. Magn. Soc. Jpn.* **33** 362 [in Japanese]
[5] Zheng M, Acharya B, Choe G, Zhou J, Yang Z, Abarra E and Johnson K 2004 *IEEE Trans. Magn.* **40**, 2498
[6] Choe G, Roy A, Yang Z, Acharya B and Abarra E 2006 *IEEE Trans. Magn.* **42** 2327
[7] Jung H, Kwon U, Kuo M, Velu E, Malhotra S, Jiang W and Bertero G 2007 *IEEE Trans. Magn.* **43** 615

Journal of Physics: Conference Series **266** (2011) 012063 doi:10.1088/1742-6596/266/1/012063

Evaluation of interlayer ferromagnetic coupling for stacked media by adding reference layer

K K Tham[1, 3], S Saito[1], D Hasegawa[2], N Itagaki[1], S Hinata[1], and M Takahashi[1]

[1] Department of Electronic Engineering, Graduate School of Engineering, Tohoku University, 6-6-05, Aoba, Aramaki, Aoba-ku, Sendai 980-8579, Japan
[2] Waseda Institute for Advanced Study, Waseda University, 1-6-1 Nishi Waseda, Shinjuku-ku, Tokyo 169-8050, Japan

E-mail: thamaya38@r8.dion.ne.jp

Abstract. The trial for quantitative evaluation of interlayer ferromagnetic coupling between granular and cap layer in stacked media is reported. The evaluation is realized by analyzing *M-H* loop of stacked media with another reference layer added on the cap layer. The reference layer is antiferromagnetically coupled with the cap layer through non-magnetic spacer layer. In this experiment, Rh which leads to antiferromagnetic coupling constant along film normal direction of around 2 erg/cm^2 was used as non-magnetic spacer layer. According to the evaluation result done by this method, when thickness of the spacer Pd layer between granular layer and cap layer is increased to 1.1 nm, ferromagnetic coupling constant is weakened to 7.2 erg/cm^2 which results in reduction of saturation field.

1. Introduction

The stacked media has been widely used for high density perpendicular recording media [1], [2]. The media consists of a-CoCrPt-oxide granular layer and a cap layer with continuous structure. In this media, the control of intergranular coupling through the cap layer is very important to reduce the saturation field which results in improvement of the writing ability [3]. Excessive intergranular coupling will increase the magnetic cluster size which limits the recording density.

Exchange spring media with spacer layer inserted in between granular layer and cap layer has been widely studied in order to further control the intergranular coupling [4]. In such as media, the weakened interlayer ferromagnetic coupling between granular layer and cap layer can suppress the increase of intergranular coupling in granular layer without sacrificing writing ability which results in smaller magnetic clusters. Therefore quantification of interlayer ferromagnetic coupling between granular layer and cap layer is indispensable. However, currently there is no detail report regarding the qualitative evaluation of the interlayer ferromagnetic coupling constant.

In this paper, we report the trial for quantitative evaluation of interlayer ferromagnetic coupling between granular and cap layer in stacked media. We also show the correlation between interlayer ferromagnetic coupling and saturation field.

[3] To whom any correspondence should be addressed.

Journal of Physics: Conference Series **266** (2011) 012063 doi:10.1088/1742-6596/266/1/012063

2. Experimental procedure

All layers were fabricated on glass substrate at room temperature with DC magnetron sputtering system. The stacking structure of the samples for interlayer ferromagnetic coupling evaluation: Glass/Ta(5 nm)/Pt(6 nm)/Ru(20 nm)/Granular layer/Pd spacer layer/Cap layer/ Rh/ Reference layer. $Co_{74}Cr_{10}Pt_{16}$-SiO_2 (8 mol%) (16 nm) with M_s value of around 640 emu/cm^3 and grain size of around 10 nm was used as granular layer. [Co(0.6 nm)/Pd(1.1 nm)]$_n$ multilayer with n=3 and n=5 were used for cap layer and reference layer. M_s value of the cap layer and reference layer were around 580 emu/cm^3. For evaluation of saturation field (H_s), stack media without reference layer were also prepared. The thickness of Pd spacer layer (s-Pd) is varied from 0 to 3.1 nm to change the strength of ferromagnetic coupling (F-coupling) between granular layer and cap layer. Reference layer was deposited above Rhodium to generate antiferromagnetic coupling (AF-coupling) between cap layer and reference layer. The samples for evaluation of interlayer ferromagnetic coupling were prepared with Rhodium thickness of 0.8, 0.9 and 1.1 nm. During the experiment for finding non-magnetic spacer layer which leads to larger AF-coupling, s-Pd thickness was fixed at 3.1 nm to avoid the effect of F-coupling. The exchange coupling field (H_{ex}) of AF-coupling was determined from minor *M-H* loop measured by VSM. H_{ex} value is defined as the distance from centre of the minor *M-H* loop to *H*=0 axis.

3. Result and discussion

3.1. Evaluation of ferromagnetic coupling with reference layer

3.1.1. Evaluation method

Figure 1 shows the schematic *M-H* loop for stacked media with and without reference layer. (a) *M-H* loop for stacked media without reference layer. (b) *M-H* loop for stacked media with reference layer when F-coupling is larger than AF-coupling. (c) *M-H* loop for stacked media with reference layer when F-coupling is smaller than AF-coupling. The arrows at *M-H* loop of figure 1(a) show the magnetization of cap layer and granular layer from the top. The arrows at *M-H* loop of figure 1(b) and 1(c) show the magnetization of reference layer, cap layer and granular layer respectively from the top. The interlayer F-coupling constant (J_F) between granular layer and cap layer was quantified by observing whether the cap layer is coupled with granular or reference layers during magnetization reversal. When F-coupling is larger than AF-coupling, magnetization reversal of cap layer is mainly affected by F-coupling. On the contrary, when F-coupling is smaller than AF-coupling, both AF-coupling and F-coupling influenced magnetization reversal of cap layer.

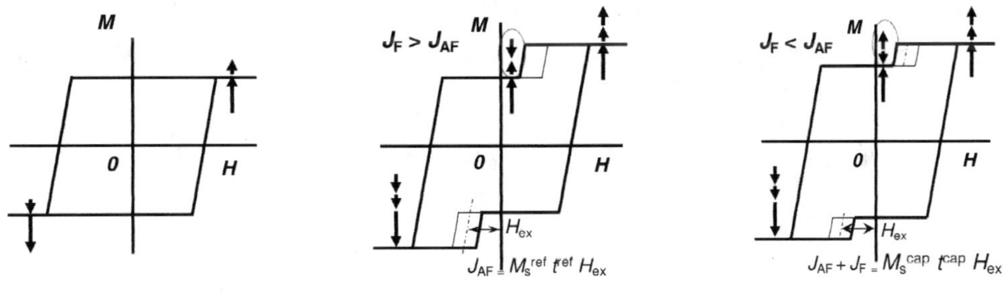

(a) (b) (c)

Figure 1. The schematic *M-H* loop for stacked media with and without reference layer. (a) *M-H* loop for stacked media without reference layer. (b) *M-H* loop for stacked media with reference layer when F-coupling is larger than AF-coupling. (c) *M-H* loop for stacked media with reference layer when F-coupling is smaller than AF-coupling.

3.1.2. Finding spacer layer material for larger anti ferromagnetic coupling

In order to quantify the large F-coupling, we search for the non-magnetic spacer layer material which has larger AF-coupling. We tried Rh as the spacer layer which was reported to have very large AF-coupling along in-plane direction between Co layers [5].

Figure 2 shows the dependence of AF-coupling constant (J_{AF}) along film normal direction on Rh and Ru thickness. We found that J_{AF} shows the peak at Rh thickness of 0.9 nm with value of 1.95 erg/cm^2 which is around twice as compared to Ru. In this experiment, we prepared samples with Rh thickness of 0.8, 0.9 and 1.1 nm to vary the strength of AF-coupling.

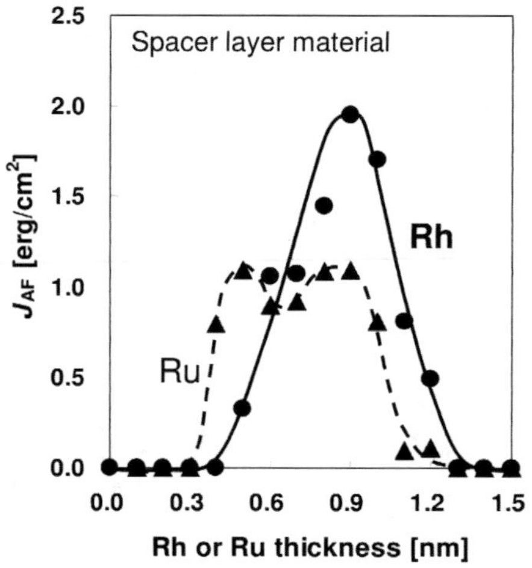

Figure 2. The dependence of AF-coupling constant (J_{AF}) along film normal direction on Rh and Ru thickness.

3.1.3. Evaluation of ferromagnetic coupling

Figure 3 shows the dependence of J_F and J_{AF} on s-Pd layer thickness at Rh thickness of around 0.9 nm. The thickness of s-Pd was changed from 0 to 3.1 nm to vary the F-coupling. Arrows at the *M-H* loop shows the magnetization at reference layer, cap layer and granular layer from the top. For s-Pd thickness less than 2.1 nm J_F is larger than J_{AF} that magnetization at cap layer is fixed by F-coupling and magnetization at reference layer is reversed by AF-coupling. J_{AF} can be determined with following equation.

$$J_{AF} = M_s^{ref} t^{ref} H_{ex} \tag{1}$$

M_s^{ref}: Saturation magnetization of reference layer, t^{ref}: thickness of reference layer.

When s-Pd thickness is thicker than 2.2 nm, J_{AF} becomes large enough that magnetization reversal at cap layer is influenced by AF-coupling and F-coupling. Therefore the exchange coupling constant becomes the combination of J_{AF} and J_F which can be determined with following equation.

$$J_{AF} + J_F = M_s^{cap} t^{cap} H_{ex} \tag{2}$$

M_s^{cap}: Saturation magnetization of cap layer, t^{cap}: thickness of cap layer

From this result J_F for s-Pd thicker than 2.2 nm can be calculated by equation (1) and (2). The same experiment was conducted also for Rh thickness 0.8 nm and 1.1 nm.

Figure 4 shows the dependence of J_F on s-Pd thickness at various Rh thicknesses. When s-Pd thickness is 0 nm, J_F=100 erg/cm^2 is plotted [6]. Concerning exponential decay of J_F against s-Pd thickness [7], J_F can be estimated as the magnitude of around 7.2 erg/cm^2 for cap-media with typical s-Pd thickness of 1.1 nm.

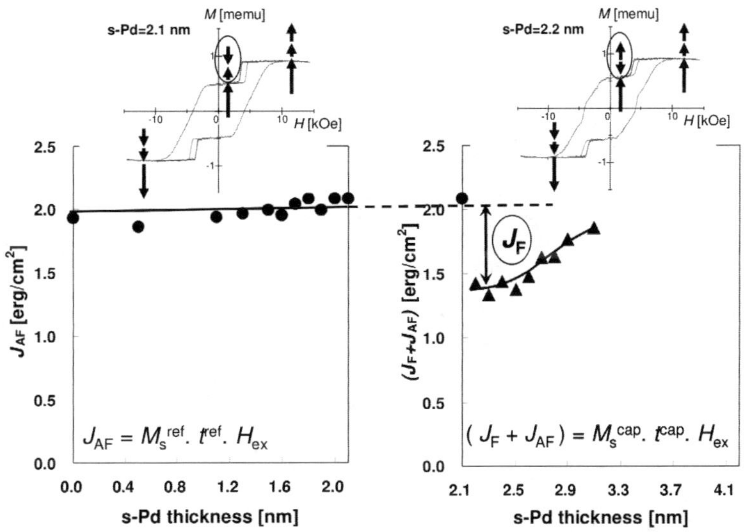

Figure 3. The dependence of J_F and J_{AF} on s-Pd layer thickness at Rh thickness of 0.9 nm.

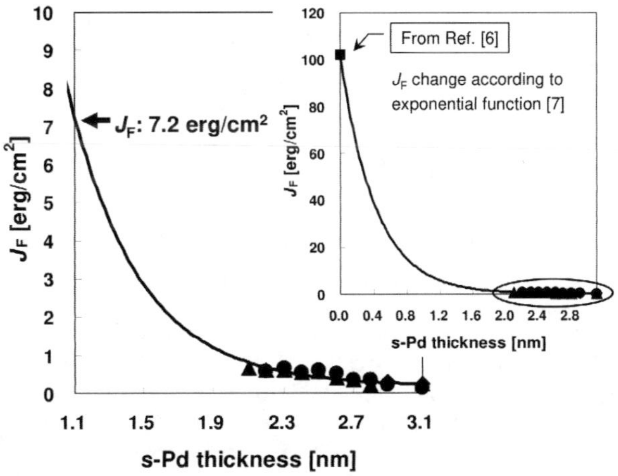

Figure 4. The dependence of J_F on s-Pd thickness at various Rh thicknesses.

3.2. Magnetic properties of stacked media

Figure 5 shows *M-H* loop of stacked media with various s-Pd thickness. When the thickness of s-Pd is increased to 1.1 nm, slightly decrease of H_s can be observed from the *M-H* loop. Further increase of the s-Pd thickness will increase the H_s. At s-Pd thickness of around 2.3 nm, a kink can be observed at around zero field which means magnetization reversal for granular layer and cap layer starts to happen separately.

Figure 6 shows the dependence of H_s on J_F and s-Pd thickness. When 1.1 nm of s-Pd is inserted J_F decreases from 100 erg/cm^2 to 7.2 erg/cm^2 which results in the reduction of H_s from 7.2 kOe to 6.8 kOe. The tendency of H_s reduction by inserting spacer layer agrees with previous report [8]. Further decrease of J_F will increase H_s of the media. From this result, by inserting s-Pd thickness 1.1 nm, interlayer ferromagnetic coupling is reduced that decreases H_s.

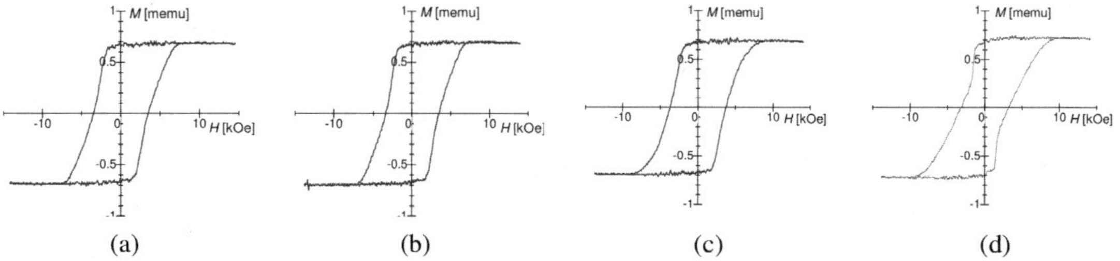

Figure 5. *M-H* loop of stacked media with various s-Pd thicknesses. (a) s-Pd: 0 nm, (b) s-Pd: 1.1 nm, (c) s-Pd: 1.9 nm, (d) s-Pd: 2.3 nm.

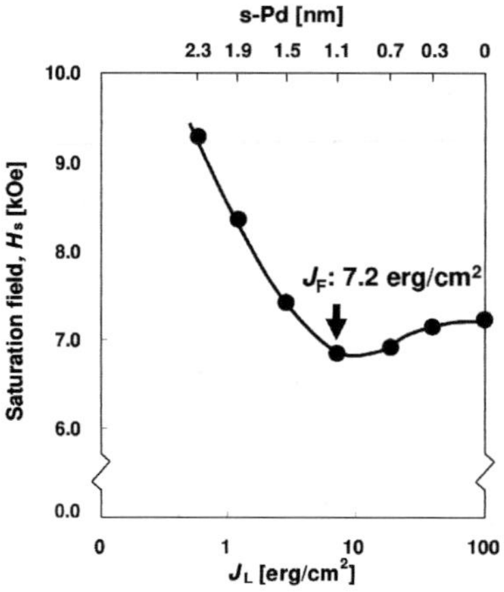

Figure 6. H_s dependence on J_F and s-Pd thickness.

4. Conclusions

The qualitative evaluation of interlayer exchange coupling between granular and cap layer by adding another reference layer was shown practicable. According to this method J_F between granular layer and cap layer can be evaluated by analyzing *M-H* loop. At s-Pd layer thickness of around 1.1 nm J_F is decreased from 100 erg/cm^2 to 7.2 erg/cm^2 which results in H_s reduction of around 0.4 kOe.

5. References

[1] Tham K K, Sonobe Y and Wago K 2007 *IEEE Trans. Magn.*, **43**, 671

[2] Tanahashi K, Nakagawa H, Araki R, Kashiwase H and Nemoto H 2009 *IEEE Trans. Magn.*, **45**, 799

[3] Choe G, Zheng M, Acharya B R, Abarra E N and Zhou J N 2005 *IEEE Trans. Magn.*, **41**, 3172

[4] Supper N F, Berger A, Ikeda Y and Fullerton 2007 presented at the 52nd MMM Conf.,

[5] Zoll S, Dinia A, Stoeffler D, Gester M, Van Den Berg H A M and Ounadjela K 1997 *Europhys. Lett.,* **39** **(3)**, 323

[6] Ounadjela K, Muller D, Dinia A, Arbaoui A, Panissod P and G Suran 1992 *Phys. Rev. B,* **45**, 7768

[7] Moritz J, Garcia F, Toussaint J C, Dieny B and Nozieres J P 2004 *Europhys. Lett.,* **65**, 123

[8] Tang K, Takano K, Choe G, Wang G, Zhang J, Bian X, Mirzamaani M 2008 *IEEE Trans. Magn.*, **44**, 3507

Journal of Physics: Conference Series **266** (2011) 012064 doi:10.1088/1742-6596/266/1/012064

Effect of Rh spacer on Synthetic-Antiferromagnetic Coupling in FeCoB/Rh/FeCoB Films

W X Xia, K Inoue, S Saito and M Takahashi
Department of Electronic Engineering, Graduate School of Engineering, Tohoku University, 6–6–05, Aoba, Aramaki, Aoba–ku, Sendai 980–8579, Japan

E-mail: ssaito@ecei.tohoku.ac.jp

Abstract. Effect of Rh spacer on synthetic antiferromagnetic (SAF) coupling was investigated in subnano-crystalline $(Fe_{65}Co_{35})_{88}B_{12}$ (30 nm)/ Rh/ $(Fe_{65}Co_{35})_{88}B_{12}$ (30 nm) films. The flopping field (H_f) showed oscillatory behavior with respect to Rh thickness, d_{Rh}. The 1st peak of H_f appeared at $d_{Rh} = 0.9$ nm, and the 2nd at $d_{Rh} = 1.7$ nm. These results are analyzed in terms of interlayer coupling effect including the bilinear (J_1) and biquadratic (J_2) coupling energy, and found to be 0.65 erg/cm^2 for J_1 and 0.12 erg/cm^2 for J_2 at the 1st peak. Compared to SAF coupling with Ru spacer, the 1st peak thickness with Rh spacer is thicker. This is because the effective spacer layer thickness decreases by the polarization of Rh neighboring FeCo based material.

1. Introduction

For high density perpendicular recording media, a thick soft magnetic underlayer (SUL) is required for the magnetic flux path for writing. The thick SUL introduces spike noise and wide adjacent track erasure (WATE). To avoid these problems, the SUL with synthetic antiferromagnetic (SAF) coupled structure have been widely used [1-3] as following reasons [4]; 1) Néel walls are formed in the top and bottom soft magnetic (SM) layers in a pair [5], which is effective for suppression of spike noise, and 2) Low susceptibility are realized when the angle of applied field is slightly tilted from the film normal, which can efficiently restrain WATE. We have already reported that large flopping field (H_f) appeared for $(Fe_{65}Co_{35})_{88}B_{12}$ (30nm)/ Ru/ $(Fe_{65}Co_{35})_{88}B_{12}$ (30 nm) stacked SUL at the 1st peak with Ru spacer thickness (d_{Ru}) of 0.3 nm due to the extremely large unidirectional interlayer coupling energy J_1 of 1.9 erg/cm^2 as a result of composition dependence of SM materials [6-7]. In this paper, we investigate interlayer coupling by using Rh as a spacer material to obtain the material guide of spacer for further enhancement of H_f.

2. Experimental Procedure

All samples were fabricated by the dc magnetron sputtering method at room temperature under an Ar gas pressure of 0.6 Pa. The stacking structure of the samples was glass sub./ $(Fe_{65}Co_{35})_{88}B_{12}$ (30nm)/ Rh (d_{Rh} nm)/ $(Fe_{65}Co_{35})_{88}B_{12}$ (30 nm). Ru was also chosen as a spacer material for reference. The $(Fe_{65}Co_{35})_{88}B_{12}$, Rh and Ru layers were sputtered with deposition rate of 2.38, 0.26 and 0.15 nm/s, respectively. For confirmation of polarization of Rh and Ru, $Co_{100-x}Rh_x$ and $Co_{100-x}Ru_x$ films were also deposited by cosputtering with Co and Rh or Ru by varying the discharge power for each target. Magnetic properties of the samples were evaluated by vibrating sample magnetometer.

Journal of Physics: Conference Series **266** (2011) 012064 doi:10.1088/1742-6596/266/1/012064

3. Results and Discussion

Figure 1 show *M–H* loops along the easy magnetization direction for FeCoB/ Rh (d_{Rh} nm)/ FeCoB films where d_{Rh} is (a) 0.9 nm, and (b) 1.7 nm. As seen in Figures 1(a) and (b), both samples show a small hysteresis in their *M-H* loop, which suggests that the FeCoB in these samples become soft magnetic. In the magnetization process for a stacked film with SAF interlayer coupling, the magnetization is almost cancelled out near zero field, as shown for both samples, i. e. the magnetic moments of the top and bottom SM layers are antiparallelly aligned. As the external field increases, the magnetization flops at 25 Oe for (b) and 60 Oe for (a). Increasing the external field further brings the magnetic moments to saturation through a rotation-magnetized process. Figure 2 shows a change in H_f for samples using Rh spacer (solid circle) as a function of spacer thickness. For comparison, change in H_f using Ru spacer (open triangle) is also shown in the figure. With decreasing d_{Rh}, H_f oscillates and takes the local maximum at d_{Rh} = 2.2, 1.7, and 0.9 nm. Each peak is indicated as the 3rd, 2nd, and 1st peak with d_{Rh}. At the 1st peak, H_f shows 60 Oe for the maximum value. In case of Ru spacer, H_f increases, attaining the local maxima at d_{Ru} = 3.3, 1.9, 0.9, and 0.3 nm. Each peak is indexed as the 4th, 3rd, 2nd, and 1st peak with d_{Ru} as well as the case of Rh spacer. It should be noted that SAF coupling does not appear when d_{Rh} < 0.9 nm even using the same SM layer material as Ru spacer.

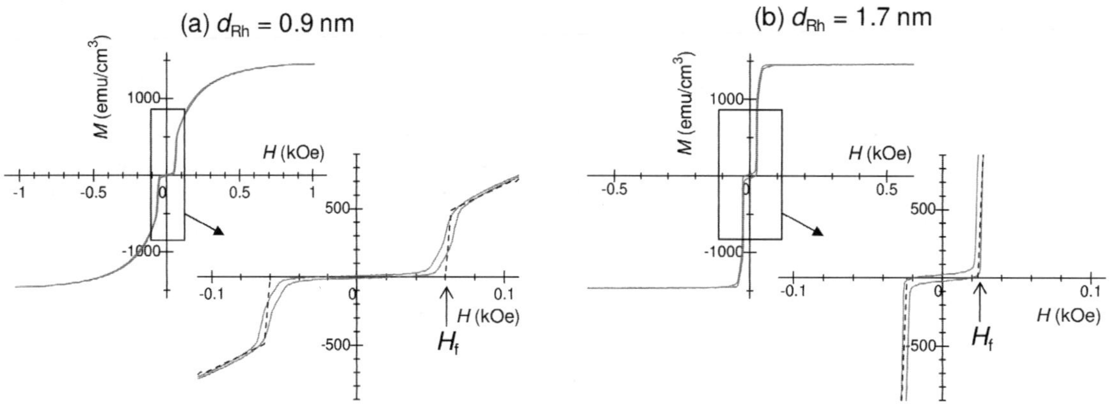

Figure 1. *M–H* loops along the easy magnetization direction for a FeCoB (30 nm)/ Rh (d_{Rh} nm)/ FeCoB (30 nm) samples where d_{Rh} is (a) 0.9 nm, (b) 1.7 nm. The figures at the left show a wide *H* range; the ones at the right are expansions around the flop field. Solid and dashed lines are experimental and calculated data, respectively.

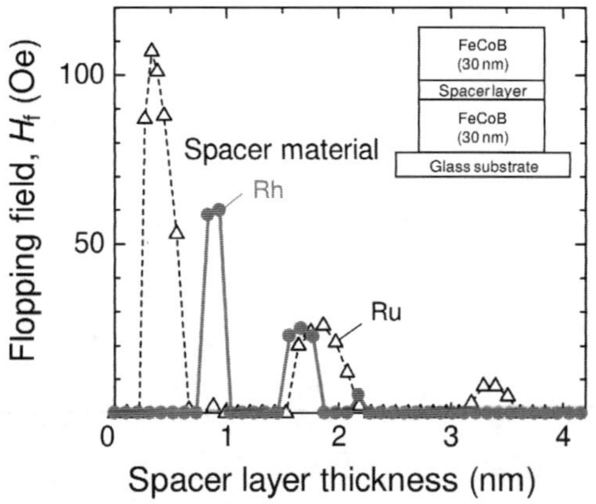

Figure 2. Change in H_f for samples with SAF structure using (solid circle) Rh and (open triangle) Ru spacers as a function of spacer layer thickness.

The mechanism of enhancement of the H_f at the 1st peak in FeCoB/ spacer/ FeCoB films is phenomenologically analyzed in terms of interlayer coupling effect. The total magnetic energy per unit area (E_T) including bilinear coupling energy, $J_1\cos(\phi_1-\phi_2)$, and biquadratic coupling energy, $J_2\cos^2(\phi_1-\phi_2)$, in SAF structure is expressed as follows [8];

$$E_T = K_u d_{SM}(\sin^2\phi_1 + \sin^2\phi_2) + J_1\cos(\phi_1-\phi_2) + J_2\cos^2(\phi_1-\phi_2) - M_s H d_{SM}(\cos\phi_1 + \cos\phi_2). \quad (1)$$

The d_{SM} and K_u denote the thickness and uniaxial magnetic anisotropy energy of SM layer, respectively. And ϕ_1 and ϕ_2 are the angles between easy magnetization direction and magnetic moment of top and bottom SM layers. Note that in this equation, domain structure and dispersion of magnetic anisotropy are not considered. The net magnetization of the two SM layers is calculated by the equilibrium angle of ϕ_1 and ϕ_2, which were determined from the minimum energy condition of equation (1). Then M–H loop is derived using following equation;

$$M = M_s(\cos\phi_1 + \cos\phi_2). \quad (2)$$

The values of J_1 and J_2 for FeCoB/ Rh/ FeCoB films were evaluated from the best fitting of measured M–H loop to that calculated by equation (1) [5]. In fitting, correspondence of M–H loops until a half of rotation reversal region was considered because higher-order terms were ignored (see figure 3). The values of K_u and M_s used in the calculation were experimentally determined on 1.5×10^4 erg/cm^3, 1450 emu/cm^3, respectively.

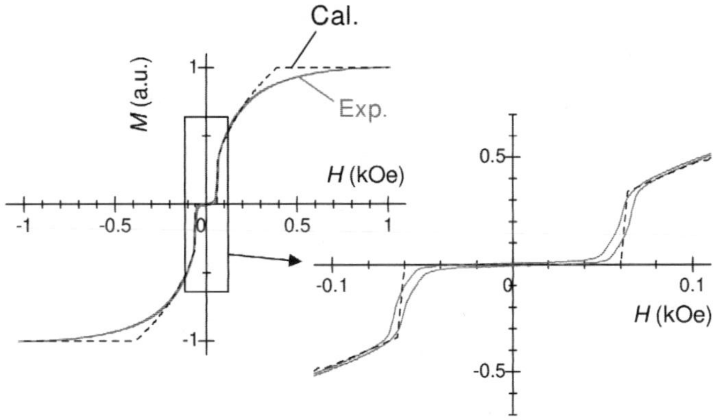

Figure 3. A typical result of fitting for *M-H* loop. Solid and broken lines correspond to experimental and calculated results, respectively, for the sample with d_{Rh} = 0.9 (nm).

Figures 4(a) and (b) show evaluated J_1 and J_2 against spacer thickness, respectively. J_1 and J_2 take the local maximum on the same d_{Rh} where H_f takes peaks.

Now, discussion is focused on the 1st peak. At the 1st peak of H_f, concerning that J_1 and J_2 are one or two figures larger than $K_u d_{SM}$ and that J_1 is larger than $2J_2$ (see figure 4(a) and (b)), H_f can be expressed as follows;

$$\begin{aligned} H_f &= (2/M_s d_{SM})\sqrt{K_u d_{SM}(J_1 - 2J_2 - K_u d_{SM})} \\ &\approx (2/M_s d_{SM})\sqrt{K_u d_{SM}(J_1 - 2J_2)} \end{aligned} . \quad (3)$$

According to equation (3), H_f has linear relationship with $\sqrt{J_1 - 2J_2}$. As shown in figure 4(a) and 4(c), in the present films, J_1–$2J_2$ takes the maximum at the same d_{Rh} as J_1. Therefore, the enhancement of H_f of the 1st peak is mainly originated from the enhancement of J_1. This result is similar to the stacked SUL with Ru spacer on our previous report [6]. FeCoB/ Rh / FeCoB film exhibits larger 1st peak thickness of Rh and less interlayer coupling strength in comparison with the film with Ru spacer; this result is in good agreement with Co/Rh and Co/Ru multilayers [9]. In [9], it is shown that in the case of multilayer the indirect exchange-coupling strength increases from the *5d* to *4d* to *3d* transition metals and increases exponentially with the number of *d* electrons along each period; it is also pointed out that the intermixing or the local interaction between the magnetic and spacer-layer atoms is one possible reason for the different spacer-layer thickness corresponding to the 1st peak. Here, the polarization of Rh neighboring ferromagnetic atoms is considered to be the most possible reason for the thicker Rh than Ru at the 1st peak and is evaluated as follows.

In order to consider the large thickness of the 1st peak of Rh spacer, polarization of Rh neighboring ferromagnetic atoms were evaluated by using glass Sub./NiFeCr (20 nm)/ Ru (20 nm)/ CoRh (20 nm)/ C (7 nm) films were fabricated. Figure 5 shows the magnetization per 1 Co atom m_{Co} for $Co_{100-x}Rh_x$ alloy films as a function of Rh content, *x*. m_{Co} for $Co_{100-x}Ru_x$ alloy films with the same underlayer is also shown. m_{Co} is evaluated by assuming that the whole magnetization of the film comes from Co atoms in the film. A Co film has m_{Co} of 1.5–1.6 μ_B/atom, which is smaller than reported value of 1.7 μ_B/atom [10]. This may be because slight carbonization and/or oxidization after deposition [11]. m_{Co} monotonically decreases with increasing Ru content for CoRu films, whereas increment of m_{Co} was observed with Rh content up to 20 at.%. Here, 20 at.% corresponds to the existence of at least 1 Rh atom in neighbors of Co. Applying this result to the case of FeCoB/ Rh/ FeCoB films with SAF coupling, the effective Rh layer thickness decreases, since the monolayers of top and bottom surface of Rh are polarized (see figure 6).

In summary, the intrinsic strong J_1 was realized in FeCoB/ Rh/ FeCoB system similar to that using Ru spacer. On the other hand, the 1st peak thickness of H_f using Rh is 0.9 nm, which is thicker than that for sample with Ru spacer. The decrease of effective spacer thickness due to polarization is the most possible reason.

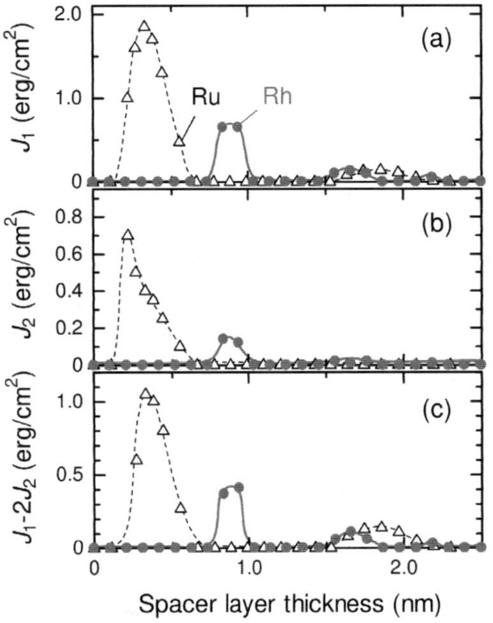

Figure 4. The variation of (a) J_1, (b) J_2, and (c) J_1–2J_2 for samples with SAF structure using (solid circle) Rh and (open triangle) Ru spacers as a function of spacer layer thickness.

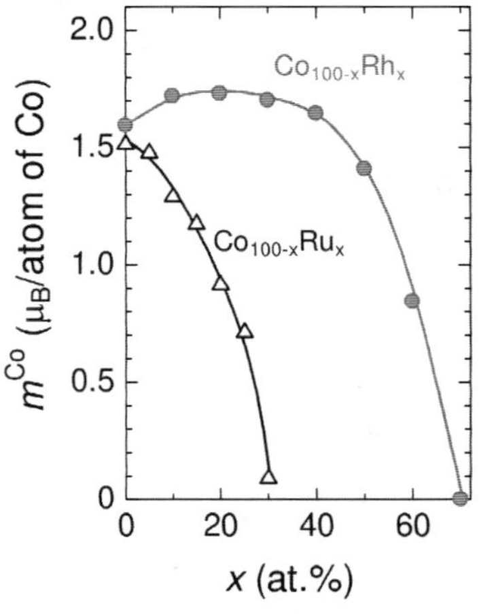

Figure 5. Dependence of moment per Co atom m^{Co} for $Co_{100-x}Rh_x$ and $Co_{100-x}Ru_x$ films on Rh or Ru content *x*.

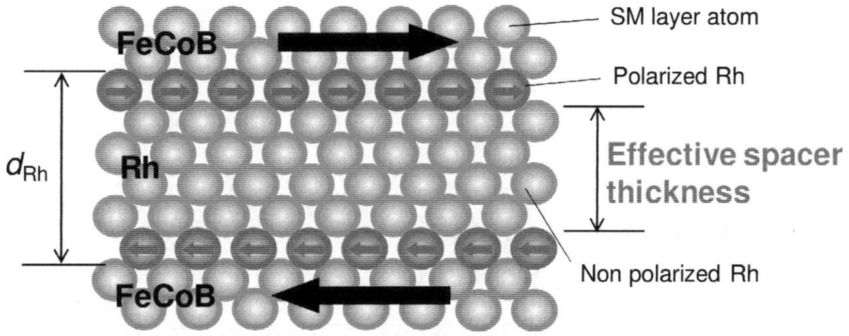

Figure 6. Schematic view of interlayer between soft magnetic (SM) and Rh spacer layers.

4. References

[1] Saito S, Hirai K, Hashimoto A, Tsunoda M and Takahashi M 2003 *J. Magn. Soc. Jpn.* **27** 224
[2] Acharya B R, Zhou J N, Zheng M, Choe G, Abarra E N and Johnson K E 2004 *IEEE Trans. Magn.* **40** 2383
[3] Byeon S C, Misra A and Doyle W D 2004 *IEEE Trans. Magn.* **40** 2386
[4] Hashimoto A, Saito S and Takahashi M 2006 *J. Appl. Phys.* **99** 08Q907
[5] Iida S, Iwasaki S, Iwama Y, Kobayashi H, Sakurai Y, Nagashima T and Watanabe S 1977 *Engineering of Magnetic Thin Films (in Japanese)* (Maruzen) p. 231
[6] Hashimoto A, Saito S, Kim D Y, Takashima H, Ueno T and Takahashi M 2006 *IEEE Trans. Magn.* **42** 2342
[7] Hashimoto A, Saito S, Omori K, Takashima H, Ueno T and Takahashi M 2006 *Appl. Phys. Lett.* **89** 032511
[8] Bobo J F, Kikuchi H, Redon O, Snoeck E, Piecuch M and White R L 1999 *Phys. Rev. B* **60** 4131
[9] Parkin S S P 1999 *Phys. Rev. Lett.* **67** 3598
[10] Meyer A J P and Taglang P 1950 *Comptes Rendus Hebdomadaires des Seances de l Academie des Sciences. R.* **231** 612
[11] Sasaki S, Saito S and Takahashi M 2008 *J. Magn. Magn. Mat.* **320** 3023

MFM analysis of magnetic inhomogeneity in recorded area for perpendicular magnetic recording media by simultaneous imaging of perpendicular and in-plane magnetic field gradient

Satoru Yoshimura[1], Genta Egawa[1], Tasuku Miyazawa[1], Zhenghua Li[1], Hitoshi Saito[1], Jianmin Bai[2], Guoqing Li[3]
[1] Center for Geo-environmental Science, Faculty of Engineering & Resource Science, Akita University,
Tegatagakuen-machi 1-1, Akita, 010-8502, JAPAN
[2] Key Laboratory for Magnetism and Magnetic Materials of the Ministry of Education, Lanzhou University,
TianShui Road, S 222, LanZhou, GanSu, 730000, P.R. CHINA
[3] School of Physical Science and Technology, South Western University,
No.2 Tiansheng Road, BeiBei District, Chongqing, 400715, P.R. CHINA

E-mail:syoshi@gipc.akita-u.ac.jp

Abstract. A new MFM method which can measure the perpendicular and in-plane magnetic field gradient simultaneously was developed. The measuring direction of in-plane magnetic field can be easily changed only by selecting the scanning direction of MFM tip with respect to the sample. This method was used to successfully estimate the magnetic domain structure of recorded area on CoCrPt-SiO_2 perpendicular magnetic recording media. When the scanning direction is along the down-track direction, the component of strong magnetic field gradient along the down-track direction can be observed with high spatial resolution at the bit transition. On the other hand, when the scanning direction is along the corss-track direction, the new MFM method is effective to mask the component of magnetic field from homogeneous recorded bits, and consequently the components of magnetic field from bended recorded bits at the track edge can be observed. The simultaneous measurement of perpendicular and in-plane magnetic field is useful to evaluate the magnetic inhomogeneity of recorded area such as bit transition and track edge.

1. Introduction

In order to analyse the microscopic magnetic domain structure of perpendicular magnetic recording media, magnetic force microscopy (MFM) vector imaging of perpendicular and in-plane magnetic fields with high spatial resolution is quite useful. However, the conventional MFM can only detect the gradient of a vector component of the magnetic field determined by the direction of the tip magnetization [1]. J. G. Zhu et al. proposed the imaging method for perpendicular and in-plane magnetic field gradient [2]. This method uses the change of MFM tip magnetization direction. Here, we showed that MFM tip coated by L10-FePt film, whose magnetized direction is strongly perpendicular to the sample surface due to its very large coerctivity, is very effective to obtain high spatial resolution image for very high density recording media. Therefore, the change of MFM tip magnetization direction is not effective to obtain high spatial resolution. We developed a new MFM imaging method which can measure the perpendicular and in-plane magnetic field gradient simultaneously during the same scan by using the transient oscillation of a MFM tip coated by $L1_0$-FePt film [3]. It was showed that frequency change of MFM tip, which is occurred by the transient oscillation of the MFM tip, corresponds to the in-plane magnetic field gradient [4]. The measuring direction of the in-plane magnetic field gradient, which is the same as the scanning direction of the MFM tip, can be easily selected only by changing the scanning direction of the MFM chip.

In this study, the perpendicular magnetic recording media were observed by using simultaneous imaging of the perpendicular and in-plane magnetic field gradient to analyze the magnetic inhomogeneity of the recorded area.

2. Experimental procedure

A phase locked loop (PLL: easyPLL, Nanosurf) circuit was used to measure the frequency shift of the MFM tip resonance in a conventional phase detection MFM. Here, the MFM tip was driven at a constant frequency near the resonant frequency of the MFM tip. The frequency shift of the MFM tip was only detected when the MFM tip has moved. The phase shift was measured after frequency shift disappeared. Therefore, the phase and frequency of the oscillation of MFM tip were measured at the same scan, though the phase and frequency shift were measured at the respective time. The MFM tip-sample distance was kept constant (15 nm) after the topographic scan by using the so-called "lift mode" measurement. We used a high-coercivity MFM tip with a coating of 20 nm $L1_0$-FePt film whose coercivity is about 8 kOe, which is made by Nitto Optical Co., Ltd. The magnetized direction of the MFM tip was perpendicular to the sample surface by applying a magnetic field of 18 kOe in advance. The MFM tip with high coercivity is effective to suppress the oscillation of its magnetization, and consequently to obtain microscopic magnetic domain images with high spatial resolution.

The CoCrPt-SiO_2 perpendicular magnetic recording media were observed, which was prepared by using an in-line-type magnetron sputtering system. Transmission electron microscopy showed CoCrPt nanograins with an averaged diameter of 5.9 nm and an average intergranular distance of 1.6 nm. Recording signals (50 or 500 or 1000 kfci) were written by using a perpendicular single-pole inductive head. The MFM measurement was done in vacuum atmosphere. The value of Q was in the range from 6000 to 8000. The resonant frequency of MFM tip was 280-320 kHz.

3. Results and Discussions

Figure 1 (a) shows the MFM images of a CoCrPt-SiO_2 perpendicular magnetic recording media with the recorded linear density of 50 kfci measured by the conventional phase detection method, and (b) shows the MFM image measured by the presented frequency detection method. The images of Figs. 1 (a) and (b) were measured at the same scan position from left to right in horizontal direction. Figs. 1 (c) and (d) show the signal profiles of the lines in image (a) and image (b). The phase detection image (a) and its corresponding signal profile (c) are well characterized as those of the perpendicular magnetic field gradient. The areas of up and down magnetizations with respect to the sample plane show the bright and dark contrasts. The boundaries between up and down magnetization areas show the neighboring bright-dark contrast. On the

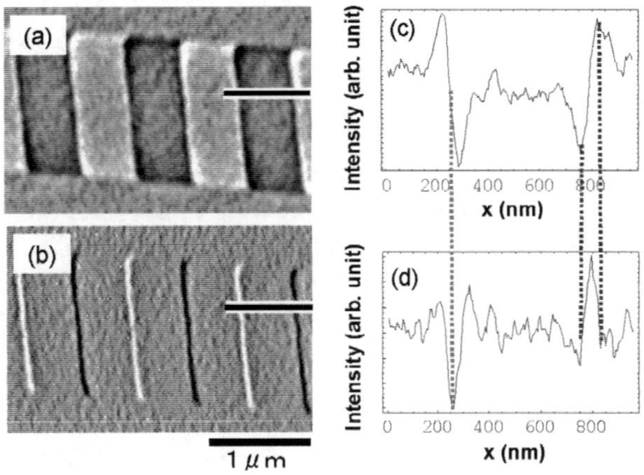

Figure 1. Phase detection image (a) and frequency detection image (b) for a CoCrPt-SiO_2 perpendicular magnetic recording media. Signal profiles (c) and (d) on the line in image (a) and (b) of the perpendicular magnetic recording media.

other hand, the frequency detection image (b) and its corresponding signal profile (d) are well characterized as those of the in-plane magnetic field gradient, and show the neighboring bright-dark-bright and dark-bright-dark contrasts at the boundaries. In the case of perpendicular magnetic recording media, the maximum and minimum intensities of the in-plane field gradient are obtained at the boundaries of neighboring recorded bits where the intensities of the perpendicular field gradient is zero. This indicates that frequency detection method is effective to observe the magnetic fluctuation at recorded bit transition.

Figure 2 (A) shows (a) phase detection image and (b) frequency detection image of the CoCrPt-SiO_2 perpendicular magnetic recording media with the recorded linear density of 500 kfci and their spatial resolution results. The spatial resolution is defined by the value of k_x at which MFM signal intensity becomes same level as noise intensity, as shown by arrows. The spatial resolution of the in-plane magnetic field gradient measured by the new MFM method was 9.1 nm, nevertheless the case of the perpendicular magnetic field gradient measured by same scan with conventional phase detection method was 9.5 nm. This indicates that frequency detection method is effective to observe the magnetic domain structure with high resolution.

Figure 2 (B) shows the simulation results of MFM resolution for (a) phase detection image and (b) frequency detection image. A quadrilateral pyramid is used as the MFM tip shape in this simulation, which was observed by using SEM. The spatial resolution of the in-plane magnetic field gradient measured by the new MFM method was 10.0 nm, nevertheless the case of the perpendicular magnetic field gradient measured by the conventional phase detection method was 10.5 nm. The simulation results agree with experimental data.

From these results, it can be said that the frequency detection method is helpful to analyze the magnetic inhomogeneity of recorded bits with high resolution.

Figures 3 (A) (a) - (c) and (d) - (f) show the images of perpendicular and in-plane magnetic field gradient for the CoCrPt-SiO_2 perpendicular magnetic recording media with the recorded linear density of 500 kfci, with the scanning angle θ of 0 ((a), (d)), 45 ((b), (e)), and 90 ((c), (f)) degree. Here the θ is assumed angle between the scanning direction (from the left side to the right side in the images) and the down-track direction. The clear contrast of perpendicular magnetic field gradient at the recorded bits is observed among each θ. However, the contrast of in-plane magnetic field gradient at the recorded bits decreases with the increasing θ.

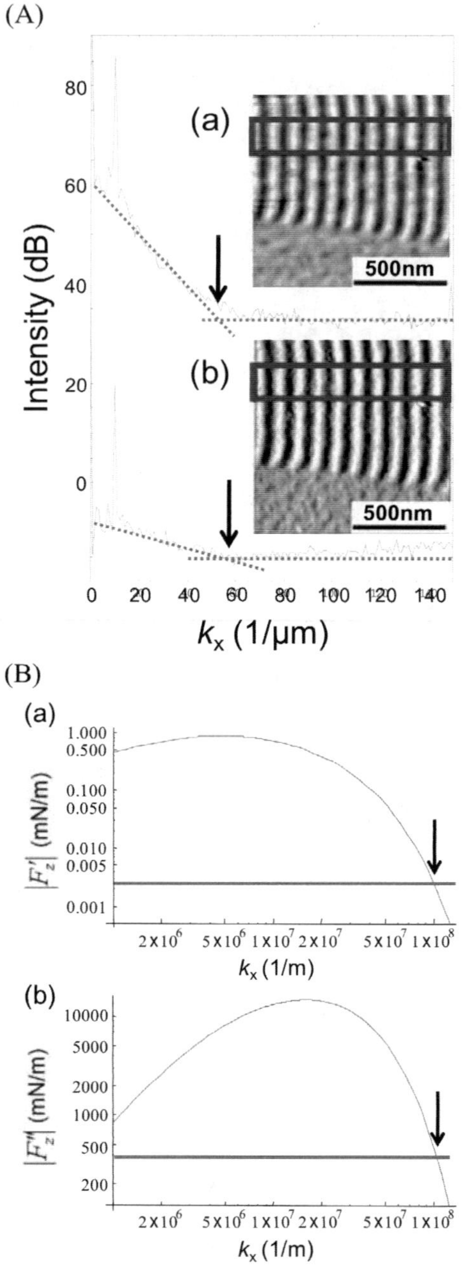

Figure 2. (A) MFM images and its spatial resolution results of (a) phase detection and (b) frequency detection method. (B) Simulation results of MFM resolution for the (a) phase detection and (b) frequency detection image. The arrows indicate the spatial resolution which is defined by the value of k_x at which MFM signal intensity becomes the same as noise level.

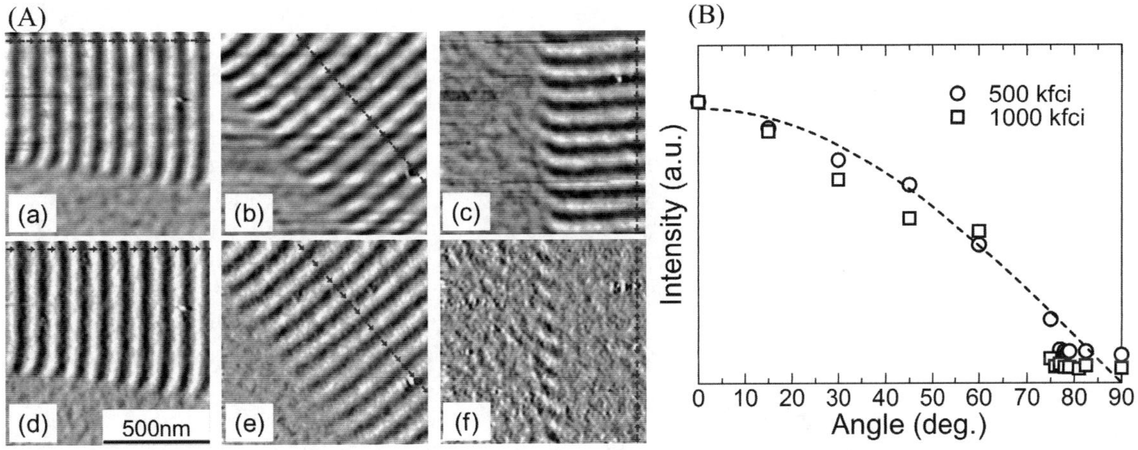

Figure 3. (A) Phase detection images (a), (b), and (c), and frequency detection images (d), (e), and (f) for a CoCrPt-SiO_2 perpendicular magnetic recording media, with the scanning angle θ of 0, 45, 90 degree. The recorded linear density is 500 kfci. (B) Dependence of the average intensity of in-plane magnetic field on the angle between the scanning direction and the down-track direction for CoCrPt-SiO_2 perpendicular magnetic recording media with the recorded linear density of 500 kfci. The dotted line indicates the cosine curve.

Figure 3 (B) shows dependence of the averaged signal intensity of in-plane field at the recorded bits. In the case of $\theta = 0$ degree (scanning direction is parallel to the down-track direction), largest signal intensity is obtained. With the increasing θ, the signal intensity decreases following the cos(θ) curve (dotted line). In the case of θ of more than 75 degree, the signal intensity becomes constant. These results indicate that signal output corresponds to the in-plane magnetic field gradient along with the scanning direction, and signal output with θ of around 90 degree is due to media noise. It can be said that the in-plane magnetic field gradient with the scanning angle of $\theta = 90$ degree is effective to mask the ideally recorded signal.

Figure 4 shows (a) the perpendicular magnetic field gradient image and (b) in-plane magnetic field gradient image obtained by the same scan for CoCrPt-SiO_2 perpendicular magnetic recording media with the recorded linear density of 500 kfci. The scanning angle θ is 90 degree (scanning direction is along the cross-track direction.).

The recorded bits can be clearly observed by the conventional imaging as shown in Fig. 4(a), while, they cannot be seen from Fig. 4(b). However, the in-plane magnetic field imaging detects the bended recorded bits around the track edges. Moreover, the magnetic contrast is observed in the recorded area. If the recorded patterns have no media noise, no magnetically contrast can be observed in the recorded area. Therefore, the imaging method seems to be effective to mask homogeneous magnetic field from recorded bits and to observe magnetic inhomogeneity in recorded area as can be seen in Fig. 4(b).

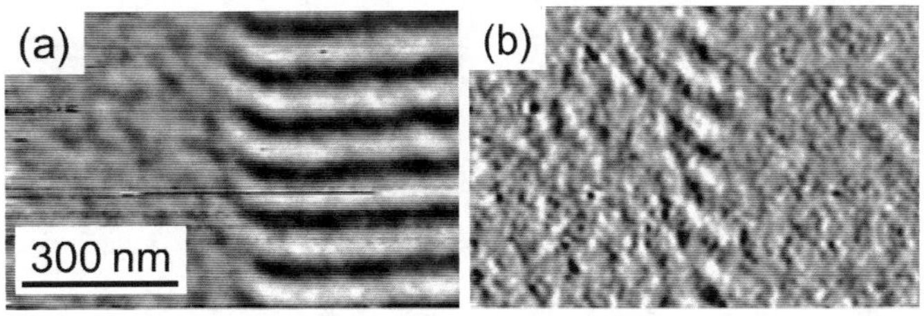

Figure 4. (a) Perpendicular and (b) in-plane magnetic field gradient image for CoCrPt-SiO_2 perpendicular magnetic recording media, with the scanning angle θ of 90 degree. The recording density is 500 kfci.

From these results, the present simultaneous imaging of perpendicular and in-plane magnetic field gradient with high spatial resolution is useful to evaluate the magnetic inhomogeneity of recorded area for high-density perpendicular magnetic recording media.

4. Conclusions

The simultaneous measurement of the perpendicular and in-plane magnetic field gradient for perpendicular magnetic recording media was examined by using a new MFM method which detects the transient oscillation of a tip. The obtained results can be summarized as follows:

(1) The in-plane magnetic field gradient detection method is effective to observe the recorded bit transition when the scanning direction is parallel to the down-track direction.

(2) The spatial resolution of the in-plane magnetic field gradient measured by the new MFM method is higher than the case of the perpendicular magnetic field gradient measured by the same scan with conventional MFM method.

(3) The in-plane magnetic field gradient detection method is effective to mask the component of homogeneous magnetic field from recorded bits, and to observe the bended magnetic field from recorded bits at the track edges when the scanning direction is along the cross-track direction.

(4) The simultaneous measurement of perpendicular and in-plane magnetic field is effective to evaluate the magnetic inhomogeneity of recorded area.

5. References

[1] Rugar D, Mamin H J, Guethner P, Lambert S E, Stern J E, McFadyen I and Yogi T 1990 *J. Appl. Phys.* vol 68 p 1169

[2] Zhu J G, Lin X and Shi R C 1998 *J. Appl. Phys.* vol 83 p 6223

[3] Saito H, Egawa G, Ishio S and Li G Q 2008 *J. Appl. Phys.* vol 103 p 07D921

[4] Egawa G, Yoshimura S, Miyazawa T, Li G Q and Saito H 2009 *Proceedings of International Conference of Materials for Resources (ICMR 2009)* p 345

Fabrication and application of a wireless inductance-capacitance coupling microsensor with electroplated high permeability material NiFe

YH Chen[1], HC Chang[2], CC Lai[2], IN Chang[1]
[1] Ph. D. Program in Electrical and Communications Engineering, Feng-Chia University, Taichung 40724, Taiwan, ROC.

[2] Dept. of Automatic Control Engineering, Feng-Chia University, Taichung 40724, Taiwan, ROC.

E-mail: hcchang@fcu.edu.tw

Abstract. A fully integrated wireless inductance-capacitance (LC) coupling microsensor was designed and fabricated by MEMS technology. The sensing loop was formed by connecting a deformable parallel-plated capacitor and a planar spiral inductor with a Ni(80)Fe(20) core. Polyimide and PMMA were used to isolate and package the devices. Typical dimension of the sensors was 5×5 mm^2 $\times$ 0.77 mm. Different electroplated inductive coils (30, 40, and 60 turns) were fabricated to connect with a 4×4 mm^2 plate capacitor in series. The LC sensing module for measuring liquid-level induced frequency responses was setup. Experimental results show that frequency response decreased as liquid level increased and sensitivity is about 7.01 kHz/cm with deviation less than 2%. Developed planar spiral inductor with high permeability magnetic core can provide a wide range of frequency variation in LC sensing applications.

1. Introduction

The monitoring and control of liquid level is of critical importance in a variety of industrial processes. Many measurement methods were in wide use exclusively based on hydrostatic pressure, visual inspections, mechanical floats, ultrasonics, fiber optics or capacitive sensor. The power delivery, data acquisition, deep depth and far distance measurement were always limited while traditional methods used. The demand for more potable devices continues to increase rapidly. A wireless fluid-level measurement technique using magnetic response and capacitive sensors for various substance levels were presented [1]. The integrated inductors with high permeability core to have large inductance density and low coil resistance show a great potential in various application such as filter, power conversion, and electromagnetic interference noise reduction [2-5]. The performance of a variety of magnetic thin films (nickel-iron and cobalt-iron-copper alloys) for magnetic components has been characterized [6]. Capacitive pressure sensors developed to meet critical required for reliable, low cost, high performance pressure measurements [7-9]. In this paper, an LC-coupling wireless microsensor was developed with electroplating NiFe core by MEMS-based compatible technology. A planar inductor shows high inductance enhancement over the air core equivalent and small area consumption.

[2] hcchang@fcu.edu.tw

An application of the device was achieved by signal processing technique using transmitting and receiving antennas, a signal function synthesizer, and a spectrum analyze. The various detected signal in the sensor arises from a change in effective electrode area. The LC-coupling wireless microsensor can be reliably fabricated for various applications to realize RF integrated electronics.

2. Principle of operation

The LC coupling microsensor consists of a variable capacitive element and a reliable inductive element to form a wireless resonance circuit. Its operation principle is by detecting the resonance frequency variation due to the deformation of capacitive electrode. Based on inductance and capacitance variation (ΔL and ΔC), the resonance frequency variation Δf of the LC coupling microsensor is given by equation (1), and f_0, L_0 and C_0 are the resonance frequency, inductance and capacitance in equilibrium, respectively.

$$\Delta f = f_0 \mp f = \frac{1}{2\pi\sqrt{L_0 \cdot C_0}} - \frac{1}{2\pi\sqrt{(L_0 \pm \Delta L)\cdot(C_0 \pm \Delta C)}} \tag{1}$$

In general with permittivity ε, electrode area A and electrode separation d can change if the parallel-plated type capacitor is pressured so that the change ratio in capacitance $\Delta C/C$ is given by equation (2). For a capacitor with a constant permittivity, the pressure-induced plate variable area will dominate the capacitance variation ratio. For a capacitive sensor, the capacitance variation due to membrane deflection is given by equation (3), $w(x, y)$ is the deflection at the point (x, y), and the product of ε_0 and ε_r is permittivity ε between plates [7]. Equation (4) is given to calculate the inductance L with a planar spiral rectangular coil [8], where N is the number of turns, μ is permeability, L_{eff} and W_{eff} are effective length and width respectively, h and w are coil thickness and line-width respectively, k_1 and k_2 are constant.

$$\frac{\Delta C}{C} = \frac{\Delta\varepsilon}{\varepsilon} + \frac{\Delta A}{A} - \frac{\Delta d}{d} \tag{2}$$

$$\Delta C = C - C_0 = \frac{\varepsilon_0 \cdot \varepsilon_r}{d} \iint \frac{dx \cdot dy}{\left[1 - \frac{w(x,y)}{d}\right]} - \frac{\varepsilon_0 \cdot \varepsilon_r \cdot A}{d} \tag{3}$$

$$L = 9.21 \times 10^{-7} N^2 \cdot \mu \cdot \left[\left(L_{eff} + W_{eff}\right) \cdot \log\left(\frac{8L_{eff} \cdot W_{eff}}{N(h+w)}\right) - k_1 - k_2 \right] \tag{4}$$

3. Design and simulation

The microstructure is designed as an LC resonated circuit which one electrode of the capacitor is a fixed metal film to connect the planar spiral inductor with high permeability core in series, the other is flexible flat rectangular membrane clamped around its circumference, dielectric are polyimide and air. To achieve magnetic measurement with high inductance, some electrodeposited magnetic materials, FeTaN (μ_r = 200 ~ 500), NiFe (μ_r = 2000), CoTaZr (μ_r = 600) and CoFeCu (μ_r = 300), were used as inductor core [5-7]. Compared with these thin film materials, the NiFe for magnetic core was chosen and fabricated to develop device with high permeability, high saturation flux, high resistivity and low coercivity. The core resistance is proportional to both the frequency and the permeability ratio, but do not influence by resistivity of the core. The resistance of the inductor with a NiFe core sources from the materials of coil and core, and it keeps a constant in the frequency range of 8 - 12 MHz [5]. By developing compatible fabrication process, the inductor has a stable inductance 24.1μH that shows large inductance and Q-factor to enhance sensor quality for various capacitance operations with high frequency response. The Q-factor can be increased by reducing coil resistance and by increasing coil thickness. The sensing membrane is an elastic element which is bent into a curve by the applied pressure. A deformation of the electrode plate means that the average separation of the plates is reduced, the resulting increases in capacitance. The sensing capacitance was determined by Equation

(3). For the application of liquid-level measurement, the liquid-depth related resonant frequency change is proportioned to fluidic pressure. The LC coupling microsensor has completed using (100)-orientation silicon wafers. The designed sensor has two parts, the upper and the lower structures. The upper structure is designed as a sensing part contains a flexible sensing membrane with a metal electrode and an isolated dielectric polyimide film. The lower structure consists of a fixed electroplated electrode and a planar spiral inductor that connect together in series as a passive resonated circuit. The integrated microsensor keeps a deformable sensing capacitance and a robust inductor with high permeability NiFe core that can reduce structure area effectively. To protect the inductor and the magnetic core, thick polyimide and PMMA are used to package the device for preventing device from leakages and damages. Simulation of the micro structures was analyzed using software ANSYS 10.0, and simulation of the liquid-level measurement with the effective capacitor electrode of 4×4 mm^2, 5 μm thick polyimide and 30 μm air gap. The simulation results indicated the thicker sensing membrane with the electrode area of 4×4 mm^2, and air gap of 30 μm, can sense the larger range of pressure, but sensitivity is reduced as shown in Figure 1. Where, the sensitivity is defined as the ratio of resonate frequency variation (kHz) to liquid level (cm). For the robust inductor designed, the higher the capacitance variation is, the higher the resonant frequency variation is, and then the higher the sensitivity has. Higher sensitivity can be achieved by increasing metal thickness of the sensing electrode. With a constant sensing gap between two electrodes of the capacitor, a result shows the sensitivity increased by increasing polyimide thickness because of its high permittivity. As the space increases between capacitor electrodes, the sensitivity decreases. The capacitance change is proportional to the electrode area of capacitor. Two different sizes, 5×5 mm^2 and 4×4 mm^2, were compared shown in Figure 2. All results show that the sensing capacitance is proportional to liquid-level pressure. The inductance was affected evidently by coil turns and permeability of magnetic core, but the influence from coil thickness can be neglected. The planar spiral inductors were fabricated and inductance at the line-width and space of 20, 30, 40, and 50 μm, were calculated by equation (4) [8].

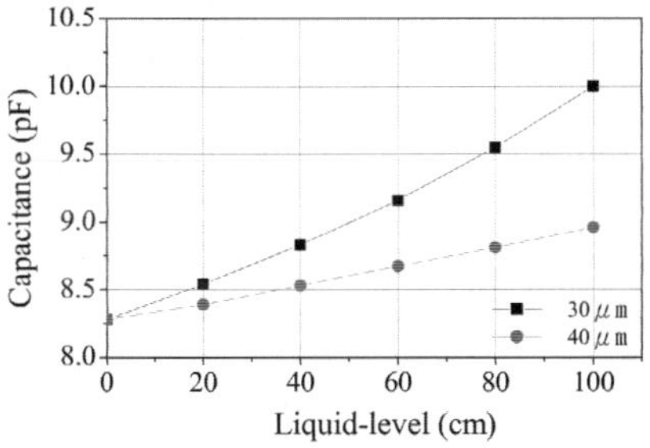

Figure 1. Capacitance variation related liquid-level on different separations of dielectric space.

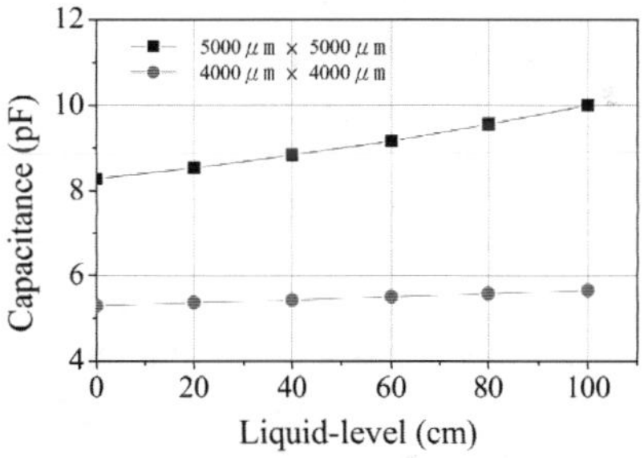

Figure 2. The liquid-level dependent capacitance variation in various sensing electrode area.

4. Fabrication

A six-mask process has been developed to fabricate the LC coupling microsensor with major process flow shown in Figure 3. The fabrication processes have two parts, the upper and the lower structure processes. The lower substrate was fabricated on 4 inch (100), 250 μm thick, p-type Si wafer with thermal-growth silicon dioxide thickness of 1.2 μm shown in Figure 3(a). The backside SiO_2 was patterned by photolithography and wet etching. The Si substrate was etched anisotropically using 20% KOH to define conductive through holes shown in Figure 3(b). The SiO_2 was patterned again to define a location for the planar inductor shown in Figure 3(c). The topside SiO_2 was patterned and etching to define a location for the parallel-plated capacitor, then the Si substrate was double-sided etching away to connect the conductive holes shown in Figure 3(d). The chromium-nickel-based seed layer was deposited to increasing adhesion of electroplate nickel film, and photo-lithographic defined planar inductor and then electroplating shown in Figure 3(e-g). The Ni-Fe alloy as the magnetic core was electroplated as Figure 3(h). The electroplated NiFe has high permeability (≈ 2000) used as magnetic core to keep inductor stable [4]. The upper substrate was fabricated on 4 inch (100), 525 μm thick, p-type Si wafer with thermal-growth silicon dioxide thickness of 1.2 μm shown in Figure 3(i). The topside SiO_2 was patterned by photolithography and wet etching to open Si etching windows, then Si was etched anisotropically using KOH to define deformable membrane shown in Figure 3(j). A Ni electrode was deposited and a 5μm thick polyimide was spun and cured as an insulator and then was patterned to open electrical connection windows shown in Figure 3(k-l). Each device of the lower and the upper structure was aligned to each other and bonded together to assemble a sensing LC chamber that was a crucial step. The packaging was completed using PMMA as protection material and the sensing area was exposed for satisfying environmental measurements. The whole schematic diagram of a fully-integrated wireless LC coupling microsensor was shown in Figure 3(m).

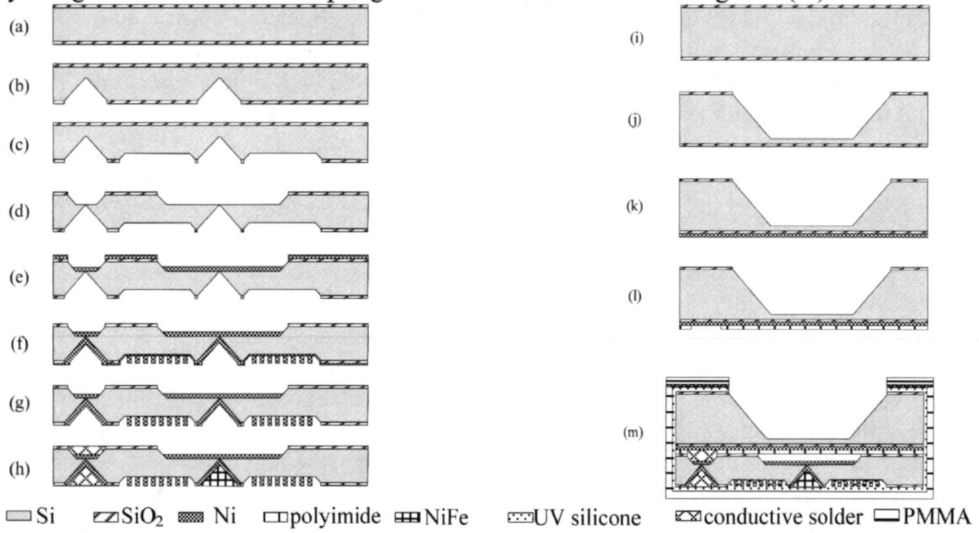

Figure 3. Fabrication processes of the LC coupling microsensor: (a)-(h) lower structure processes, (i)-(l) upper structure processes, and (m) assembly with packaging.

5. Experimental result and discussion

Preliminary experiments were performed to evaluate the performance of the fabricated LC coupling microsensor for the liquid-level measurements. An experimental apparatus consists of a signal function synthesizer, amplifiers, antennas, a spectrum analyzer, a computer and the microsensor. The waveforms generated by an arbitrary waveform generator were used as a signal which was amplified. Amplified signal was transmitted by a transmit antenna at a fixed position exciting the LC sensor to generate sensing resonant waveform. The inductor was characterized by RLC meter. The fluid-level related resonant frequency was received by an antenna for analyzing by a spectrum analyzer shown as Figure 4. In this work, the 5 × 5 mm^2 capacitor consists of two parallel 5 μm thick electrodes with a 30 μm separation and a 5 μm thick polyimide isolation layer. For the fixed device dimension, design parameters of a planar spiral inductor are composed of magnetic core material, coil line width (*W*) and separation (*d*), and coil turns (*N*). The *W* and *d* were designed and tested in 30, 40, and 50 μm, respectively. The different *N*s were tested in 60, 40, and 25 turns. The experimental results by measuring resonant frequency response with different *W*, *d* and *N* in the liquid-level range from 0 to 100 cm were shown in Figure 5. To electroplate NiFe core must be cared because the seed layer may not attach well

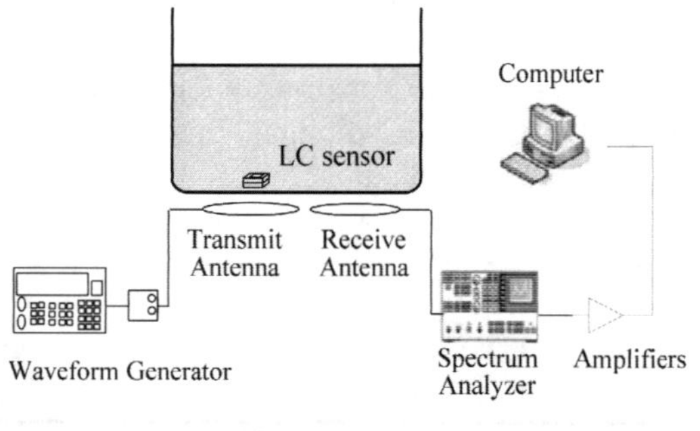

Figure 4. An experimental setup for measuring various liquid levels.

to deeply etched substrate due to more than 30 μm roughness on silicon surface. And the capacitance was also affected from the wet etching that induced lower measured resonant frequency than the simulation value. The sensitivity of 7.01 kHz/cm for liquid-level measurement is shown in Figure 6.

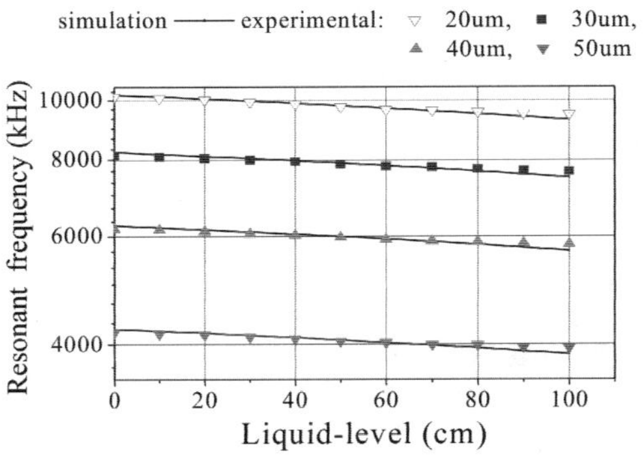

Figure 5. Frequency response versus liquid-level, the inductor has 60 × 20 μm^2 winding area, 20 μm line spacing, and widths as 50, 40, 30 and 20 μm.

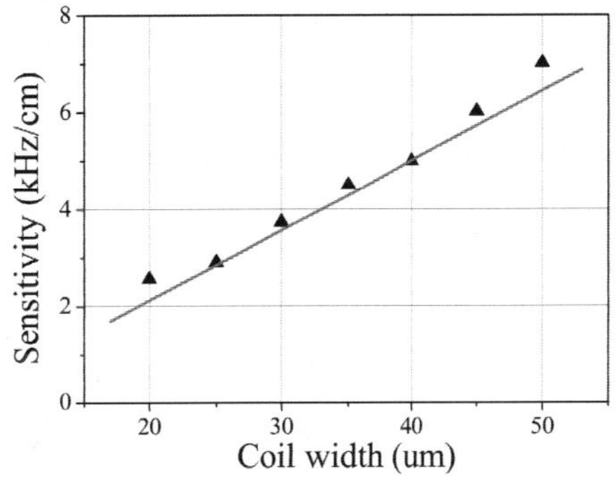

Figure 6. The sensitivity versus coil width with an equal separation between adjacent coils.

6. Conclusion

A high performance LC-coupling microsensor integrated wireless transmitting module has been developed using MEMS technology. A polyimide film is coated between deformable electrodes as an insulator to prevent short-circuited problem. The silicon-based microsensor (5 × 5 mm^2 × 0.77 mm) formed by a planar inductor with electroplated Ni-Fe core connected an electrode-deformable capacitor in series. The inductance 24.1 μH with coil resistance less than 1 Ω and capacitance 6 pF are obtained, corresponding to the inductance enhancement of 18 times over the air core equivalent. A wireless sensing and communication module achieves room temperature operations over a telemetry distance of 2 meters with a power consumption of 160 μW. The liquid-level measurements have been executed by measuring variation of resonant frequency with achieved sensitivity 7.01 kHz/cm and deviation less than 2% in range of 0 - 100 cm. The proposed LC coupling microsensor shows a great potential for various application in pressure-related measurements. Further work into a real-time automatic measurement system is to be performed using developed microdevice to integrate a human machine interface.

Acknowledgment

This work was partly supported by the National Science Council of Taiwan (NSC 98-2221-E-035-040-MY2).

7. References

[1] Woodard S E and Taylor B D 2007 *Sens. Actuat. A* **137** 268
[2] Shin D H, Kim C S and Jeong J H 1999 *IEEE Trans. Magn.* **35(5)** 3511
[3] Graford A M, Hwang D K P and Wang S X 2008 *IEEE Trans. Magn.* **44(11)** 4089
[4] Flynn D, Sudan N S and Toon A 2006 *Microsyst. Technol.* **12** 923
[5] Lee D W, Hwang K P and Wang S X 2008 *IEEE Trans. Magn.* **44(11)** 4089
[6] Flynn D, Toon A and Allen L, Dhariwal R and Desmulliez M P Y 2007 *IEEE Trans. Magn.* **43** 3171
[7] Lee K W and Wise K D 1982 *IEEE Trans. Electron Devices* **29** 34
[8] Akar O, Akin T and Najafi K 2001 *Sens. and Actuat. A* **95** 29
[9] Luo J K, Lin M, Fu Y Q, Wang L, Flewitt A J, Spearing S M, Fleck N A and Milne W I 2006 *Sens. Actuat. A* **132** 139

Journal of Physics: Conference Series **266** (2011) 012067 doi:10.1088/1742-6596/266/1/012067

Development of a Switched Reluctance Motor made of Permendur

Y Hasegawa[1], K Nakamura[1] and O Ichinokura[1]
[1]Graduate School of Engineering, Tohoku University,
6-6-05 Aoba Aramaki Aoba-ku, Sendai, Japan

E-mail: power15@ec.ecei.tohoku.ac.jp

Abstract. A switched reluctance (SR) motor consists of stator and rotor cores, and windings. Both the stator and rotor have salient poles. The stator has concentrated windings on each salient pole. On the other hand, the rotor has no windings and no permanent magnets. Therefore, the SR motor is a low cost, extremely robust, and wide-range variable-speed motor. The performance of the SR motor greatly depends on magnetic properties of core material since it consists of only iron cores and windings. This paper presents the development of a novel SR motor made of permendur which has extremely high saturation flux density and very low core loss. Two types of SR motors, one is made of conventional non-oriented Si steel, the other is made of permendur, are compared by simulation and experiment. It is demonstrated that the torque of the SR motor made of permendur is greater than that of the conventional Si steel by 20%.

1. Introduction

The demand for electric motors continues to increase in various fields such as home appliances, industries, and transportations. Hence, the improvement of size and weight, torque, and efficiency of the motors is strongly required.

A switched reluctance (SR) motor has a doubly salient pole structure. Concentrated windings are individually coiled around each stator pole, while the rotor is only made of iron core. Therefore, the SR motor is a low cost, extremely robust, wide-range variable-speed motor [1]. The performance of the SR motor depends on magnetic properties of core material since it consists of only iron cores and windings. In previous papers, we have investigated SR motors with divided-core [2] and with C-shaped-core [3], respectively, in order to utilize grain-oriented Si steel. These SR motors exhibited the high power and efficiency, but the productivity is low due to a complicated structure.

This paper presents the development of a novel SR motor made of permendur which has high saturation flux density and low core loss [4]. The permendur is an alloy created by mixing 49% cobalt with iron. The performance of the SR motor made of permendur is investigated by finite element method (FEM) and experiment.

2. Basic configuration and rotating principle of the SR motor

Figure 1 shows a basic configuration of an SR motor drive system. The drive circuit is called asymmetric half bridge converter. Each phase winding is connected to a couple of transistors and free-wheeling diodes. In the motoring mode, electric power from DC battery is supplied to the windings through the transistors. Contrary to this, in the regenerating mode, stored energy of the SR motor is back to DC buttery through the free-wheeling diodes.

Figure 2 illustrates the rotating principle of the SR motor. If the B-phase winding is excited when the A-phase stator pole and the rotor pole of • are aligned as shown in the figure, the rotor pole of □ is attracted to the B-phase stator pole by the electromagnetic force, and the rotor rotates to the position where the B-phase stator pole and the rotor pole of □ are aligned. After that, by switching the excitation phase to the C-phase, the rotor rotates continuously to the position where the C-phase stator pole and the rotor pole of • are aligned. A continuous rotation can be obtained by switching the excitation phase with the appropriate order and position described above.

Now, let the rotor position angle be θ, which is 0 deg. when a rotor pole is aligned with the A-phase stator pole as shown in Figure 3(a), while $\theta = \pm 45$ deg at the unaligned position. Figure 3(b) illustrates the A-phase inductance L_A, exciting voltage v_A, and current i_A, respectively. In the figure, the excitation beginning angle and the excitation width are θ_b and θ_w, respectively.

If the magnetic nonlinearity is ignored, the A-phase torque τ_A of the SR motor is given by,

$$\tau_A = \frac{1}{2} i_A^2 \frac{dL_A(\theta)}{d\theta} \tag{1}$$

It is understood from (1) that the positive torque can be obtained when the A-phase winding is excited in *Region I* of Figure 3(b). On the other hand, the negative torque is generated when the A-phase winding is excited in *Region II*. Therefore, the SR motor requires the position detection by a rotary encoder or a hole sensor.

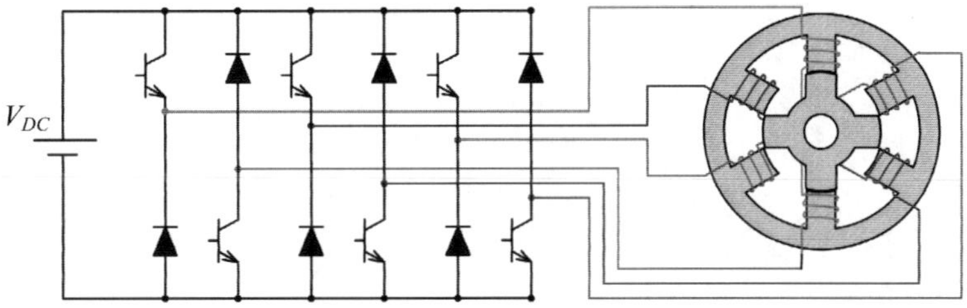

Figure 1. Basic configuration of the SR motor drive system.

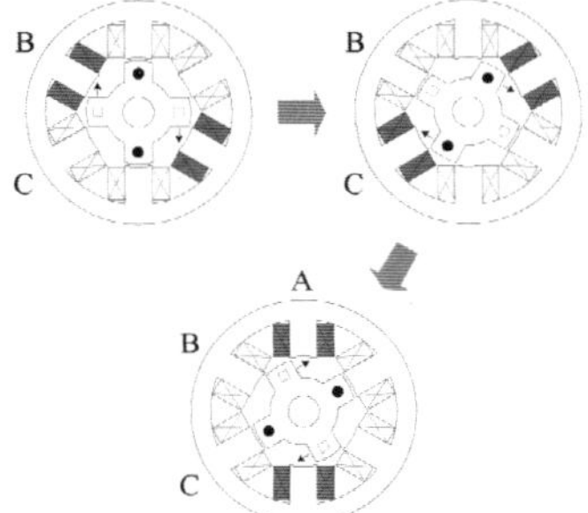

Figure 2. Rotating principle of the SR motor.

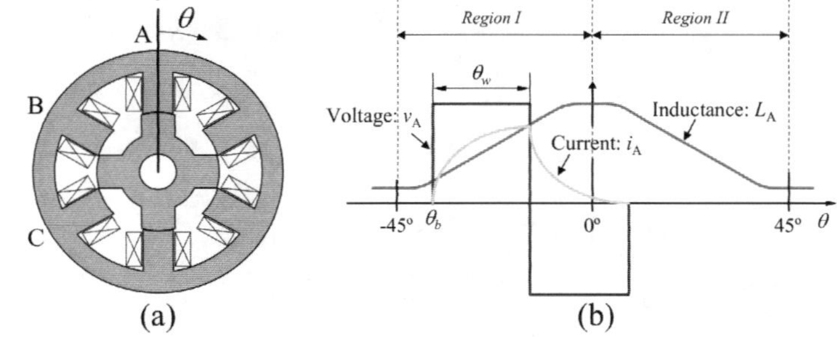

Figure 3. (a) Definition of the rotor position angle $\theta = 0°$, and (b) schematic waveforms of A-phase inductance L_A, exciting voltage v_A, and current i_A.

3. Trial machines and experimental and analytical conditions

Figure 4 shows the specifications of an SR motor used in the consideration. It has 12 stator and 8 rotor poles, respectively. Three-phase concentrated windings are arranged individually around each stator pole. The number of winding turns per phase is 140 and the winding resistance is 0.37 Ω (the average value of three-phases).

Figure 5 shows the appearance of the trial SR motors, which have the same shape and dimensions shown in Figure 4 but have different core materials. One is made of conventional non-oriented Si steel with a thickness of 0.35 mm and the other is made of permendur with a thickness of 0.2 mm. Figure 6 shows the *B-H* curves of core material. The figure indicates that the saturation flux density of permendur is larger than that of conventional non-oriented Si steel by 40% or more.

Figure 7(a) shows a three-dimensional FEM model of the SR motor. Since the SR motor used in the study has a relatively flat body that the motor diameter is about three times as large as the stack length, the influence of the leakage flux in the axial direction cannot be neglected. Hence the three-dimensional electromagnetic field analysis is necessary. Figure 7(b) indicates an electric circuit model of the drive circuit, which is coupled with the FEM model. The transistors are approximately expressed as the ideal switches as shown in the figure.

In the experiment and FEM simulation, the DC battery voltage of the SR motor made of non-oriented Si steel is V_{DC} = 60 V. On the other hand, V_{DC} = 84 V in the case of permendur since it has 1.4 times high saturation flux density as shown in Figure 6. The excitation beginning angle and the excitation width are θ_b = -22.5 deg. and θ_w = 15 deg., respectively.

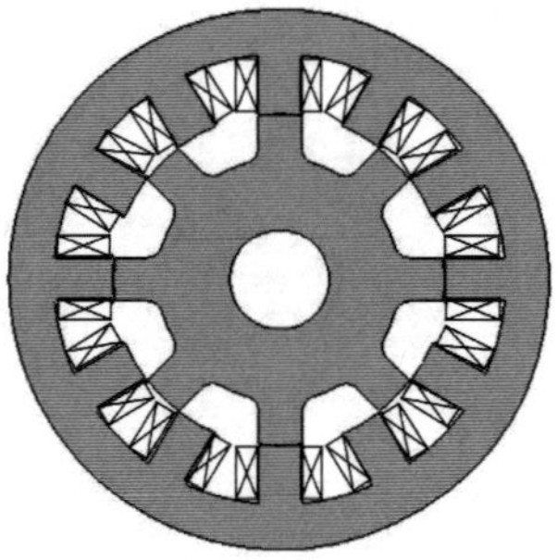

Stack length:	40 mm
Stator diameters :	136 mm
Rotor diameters :	83.2 mm
Gap :	0.2 mm
Number of windings/phase:	140 turns
Winding resistance/phase :	0.37 Ω

Figure 4. Specifications of the 12/8-pole SR motor.

Figure 5. Appearance of the trial SR motors.

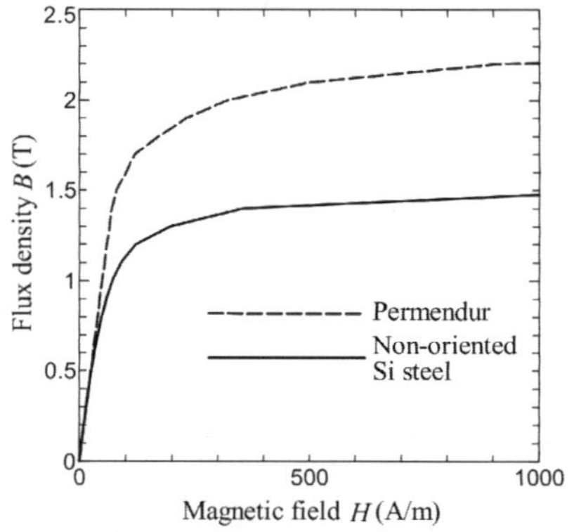

Figure 6. Comparison of the *B-H* curves of core material.

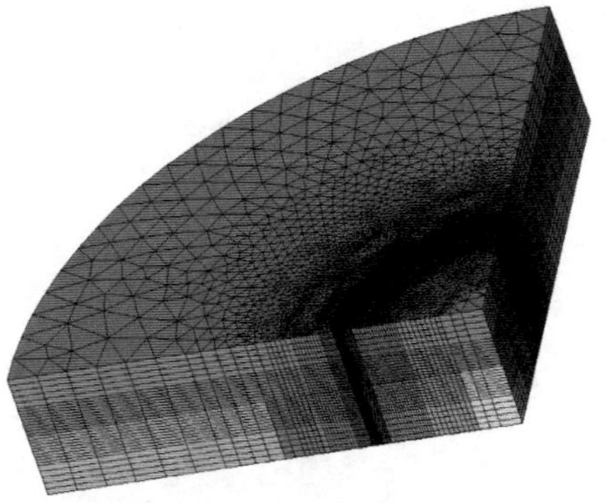

(a)

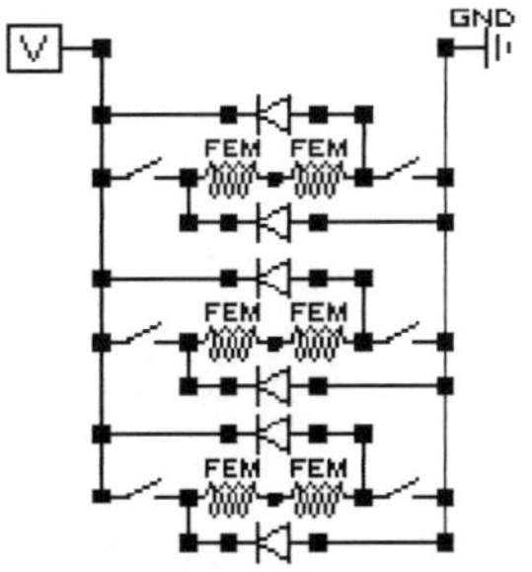

(b)

Figure 7. Specifications of the 12/8-pole SR motor. (a) Three-dimensional FEM model of the 12/8-pole SR motor. (b) Drive circuit model.

4. Calculation and experimental Results

Figure 8 (a) and (b) indicate the observed and calculated voltage and current waveforms of the SR motors made of non-oriented Si steel and permendur. The measured rotational speed is conformed to calculate ones. These figures clearly indicate that both waveforms are in good agreement.

Figure 9 shows the comparison of the current density versus torque characteristics. It is understood that the torque of permendur at a current density of 10 A/mm^2 is 1.2 times larger than that of non-oriented Si steel, and that the measured and calculated values are in good agreement.

Figure 10 indicates the rotational speed versus torque characteristics. The SR motor made of permendur has wider speed and torque range.

Figure 11 shows the efficiency characteristics of the both motors. The efficiency of the SR motor made of permendur is improved from the conventional SR motor, especially under heavy load region.

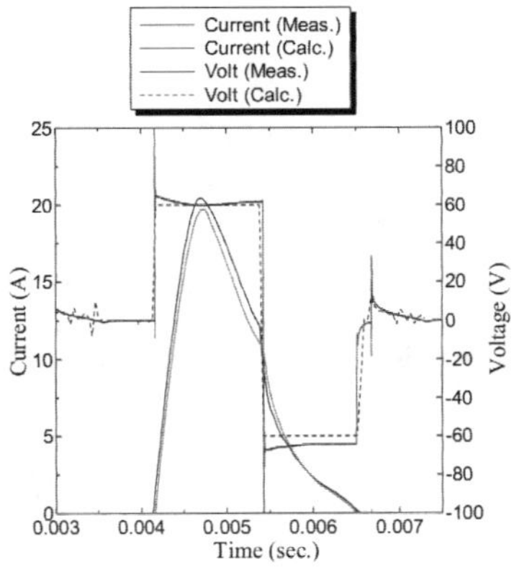

(a)

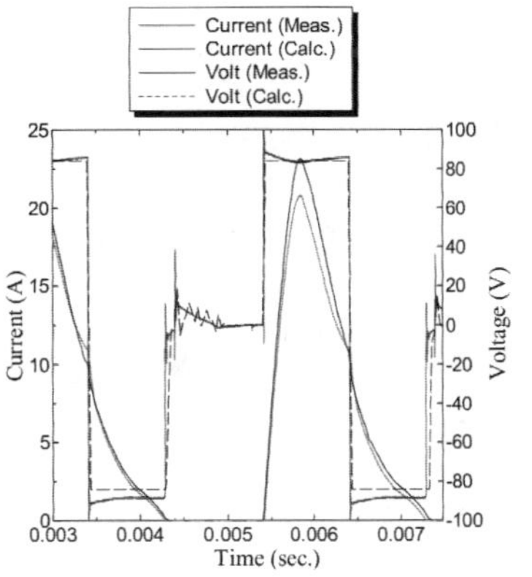

(b)

Figure 8. Current and volt waveforms. (a) Non-oriented Si steel at a speed of 2000 rpm. (b) Permendur at a speed of 2500 rpm.

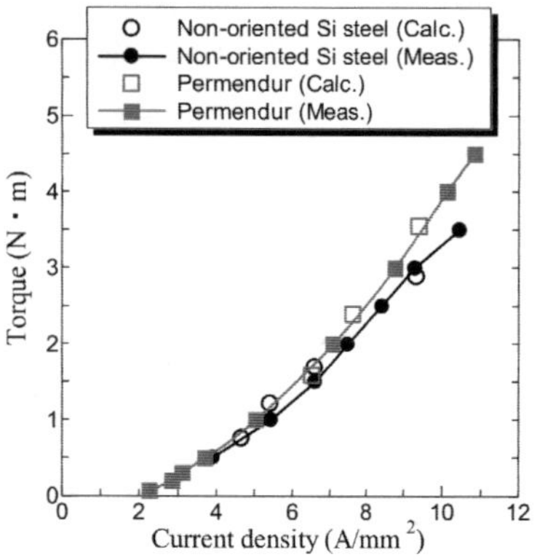

Figure 9. Current density versus torque characteristics.

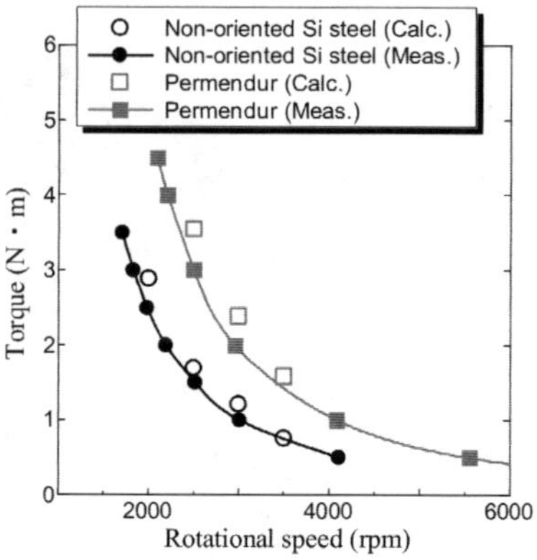

Figure 10. Rotational speed versus torque characteristics.

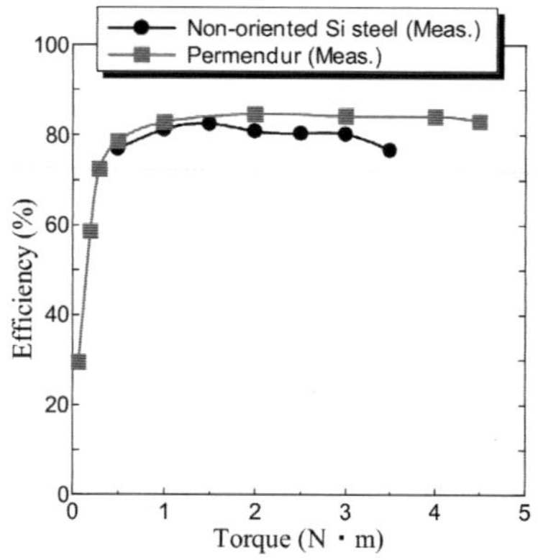

Figure 11. Efficiency characteristics.

5. Conclusions

This paper presented the development of a novel SR motor made of permendur which has extremely high saturation flux density and very low core loss. The performance of the new developed SR motor was compared with the conventional SR motor by the FEM simulation and experiment. As a result, it was demonstrated that the torque of the SR motor made of permendur is greater than that of non-oriented Si steel by 20%, and that the speed range is wider and the efficiency is also improved.

6. References

[1] Becerra C R, Ehsani M, and Miller E J T 1993 Commutation of SR motors *IEEE Trans. Power Electron.* **8** 257–263

[2] Nakamura K, Ono T, Goto H, Watanabe T, and Ichinokura O 2005 A Novel Switched Reluctance Motor With Wound-Cores Put on Stator and Rotor Poles *IEEE Trans. Magn.* **41** 919-3921

[3] Murota K, Nakamura K, and Ichinokura O 2006 Calculation of Characteristics of an SR Motor with Divided Stator Core made of Grain-Oriented Silicon Steel *17th International Conference on Electrical Machines* (ICEM 2006) OTA2-2

[4] Ish-Shalom J 1995 Composite Magnetic Structure for Planar Motors *IEEE Trans. Magn.* **31** 4077-4079

Journal of Physics: Conference Series **266** (2011) 012068 doi:10.1088/1742-6596/266/1/012068

Design of rotor shape to reduce torque ripple in IPM motors

CC Hwang[1], CM Chang[2], PL Li[2] and CT Liu[3]
[1]Dept. of EE, Feng Chia University, Taichung 407, Taiwan, ROC
[2]Ph.D. Program in ECE, Feng Chia University, Taichung 407, Taiwan, ROC
[3]Dept. of EE, National Sun Yat-sen University, Kaohsiung 804, Taiwan, ROC

E-mail: cchwang@fcu.edu.tw

Abstract. This paper presents a shape design sensitivity analysis of the effect of geometrical parameters on torque pulsations of an interior permanent magnet (IPM) synchronous motor. The analysis was conducted in two dimensions using the finite element method (FEM). Calculation results have been compared with experimental results.

1. Introduction

An IPM synchronous motor is a highly efficient motor and can operate in a wide speed range. It is widely used in many industrial applications and electric household appliances. However, the presence of torque pulsations (i.e. cogging torque and torque ripple) in IPM motors is a major problem in some applications where speed and position control accuracy is of great importance. Several techniques for reducing the torque pulsations in IPM motors have been proposed in the literature [1-7]. Most of the techniques used for reduction of pulsating torque based either on an adequate motor design [1-4, 6-7] or on control techniques [5, 6]. Although control-based techniques for reduction of pulsating torque can be quite effective, this paper concentrates on the design-based approaches for sinusoidal drives.

Based on the design-based approaches, a decrease of the torque pulsations can be expected by drilling three small circular holes [3] and varying the geometry shape of the saliencies in the magnetic island of the rotor of the IPM motor. Therefore, this paper presents a shape sensitivity analysis of torque pulsations with respect to the holes and rotor saliency of an IPM motor using the FEM [8-11]. Experimental results to support the simulation findings are included in this paper.

2. Model of IPM motor

The initial design of the motor is a three-phase, 6.9-kW, 3900-rpm, four-pole, twelve-slot, Y-connected IPM motor as shown in Figure 1. It uses four V-shaped sintered NdFeB magnets with a remanence $B_r = 1.1$ T and coercivity $H_c = 840$ kA/m. The stator and rotor core use the nonoriented silicon steel with saturated flux density $B_s = 1.8$ T.

3. Geometry optimization using gradient method

The geometry optimization problem arises in finding the maximum performance, e.g. minimum torque pulsations of an IPM motor subject to constraints on magnetic (available magnetic saturation level) requirement and manufacturing and geometry limits. Such a constrained problem involving finite

[1] cchwang@fcu.edu.tw

elements may be expressed as

$$\begin{aligned} &\text{minimize} \quad f(\mathbf{x}) \\ &\text{subject to} \quad g_i(\mathbf{x}) \leqq 0 \quad i = 1, 2, \ldots\ldots, m \\ &\text{and} \quad h_j(\mathbf{x}) = 0 \quad j = 1, 2, \ldots\ldots, \ell . \end{aligned} \tag{1}$$

where $\mathbf{x} = (x_1, x_2, \ldots, x_n)$ is a column vector of n real-valued design variables that control the shape or topology of the motor. The finite element-based optimization problem in (1) is solved using gradient methods. Consider an implicit function q which can be represented an objective or constraint function in (1), as

$$q = q\,(\mathbf{x}) \tag{2}$$

Let $\mathbf{x}^0$ be the current design. Then the gradient of q evaluated at $\mathbf{x}^0$ is

$$\nabla q = \frac{dq}{dx} = \left[\frac{dq}{dx_1}, \frac{dq}{dx_2}, \ldots\ldots, \frac{dq}{dx_n}\right]^T \tag{3}$$

It can be computed by the direct method or the adjoint method. In this analysis, since the number of design variables is less than that of constraints, the direct method is adopted because it is more efficient than the adjoint method [11].

4. Geometry optimization of the IPM motor

To reduce the torque pulsations, one of the effective methods is to modify the distribution of flux in the rotor. In this study, three small circular holes have been drilled in the magnetic island and the saliencies of the rotor have been also created. Figure 2 shows the definition of parameters R, r and θ for the three circular holes and parameters d and r_d for the variation of saliencies of a rotor core. Where R and $R+r$ are the radius of the holes, θ is the angle between the center hole and side holes, and d is the distance from the center of shaft to the center for the radius r_d used to cut the saliencies of the rotor. The air gap length and the bridge width are fixed to 1 mm and 1 mm to allow for clearance and tolerances. Figure 3 Summarizes the steps involved in the complete shape design sensitivity analysis of the motor. To have a more realistic design, the constraints of parameters R, r, θ and d are also shown in Figure 3.

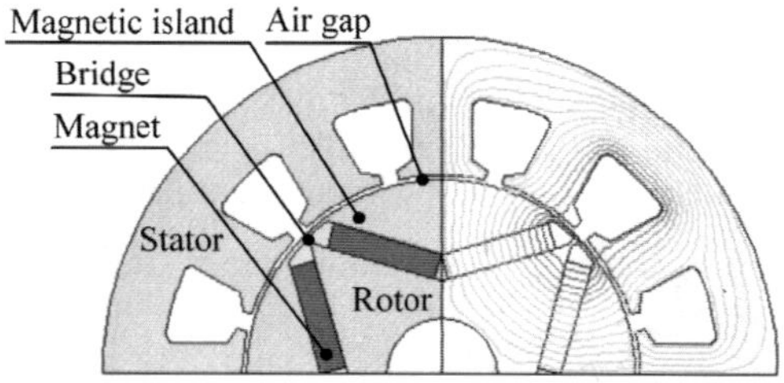

Figure 1. Initial design of IPM motor.

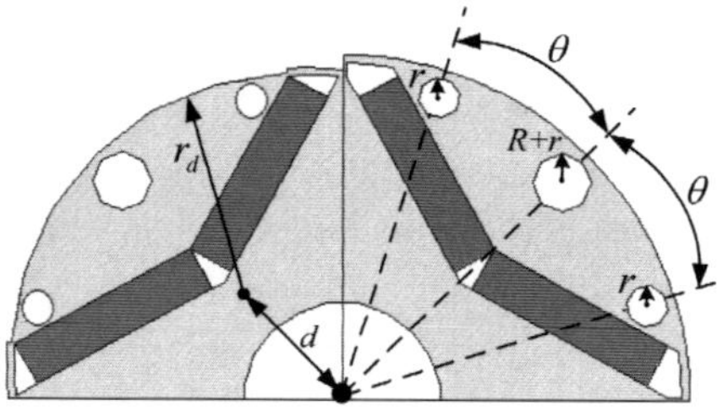

Figure 2. Definition of parameters R, r, θ, d and r_d.

5. Results and discussions

5.1. Cogging torque

5.1.1. Influence of angle θ

In this part of analysis, the only variable is θ. The other parameters R and r are held constant (R = 0 mm and r = 2 mm). The results are shown in Figure 4. As noted in the figure, as the angle θ increases, the peak-to-peak value of cogging torque (T_{cog}) decreases. It is shown that T_{cog} = 3.8 N.m with $\theta = 27.4^0$.

5.1.2. Influence of radius r

In this part of analysis, the only variable is r. The other parameters R and θ are held constant (R = 0 mm and $\theta = 27.4^0$). The effects of length d on cogging torque are illustrated in Figure 5. As the radius r increases, the peak-to-peak value of cogging torque (T_{cog}) decreases. It is shown that T_{cog} = 2.6 N.m with r = 2.5 mm.

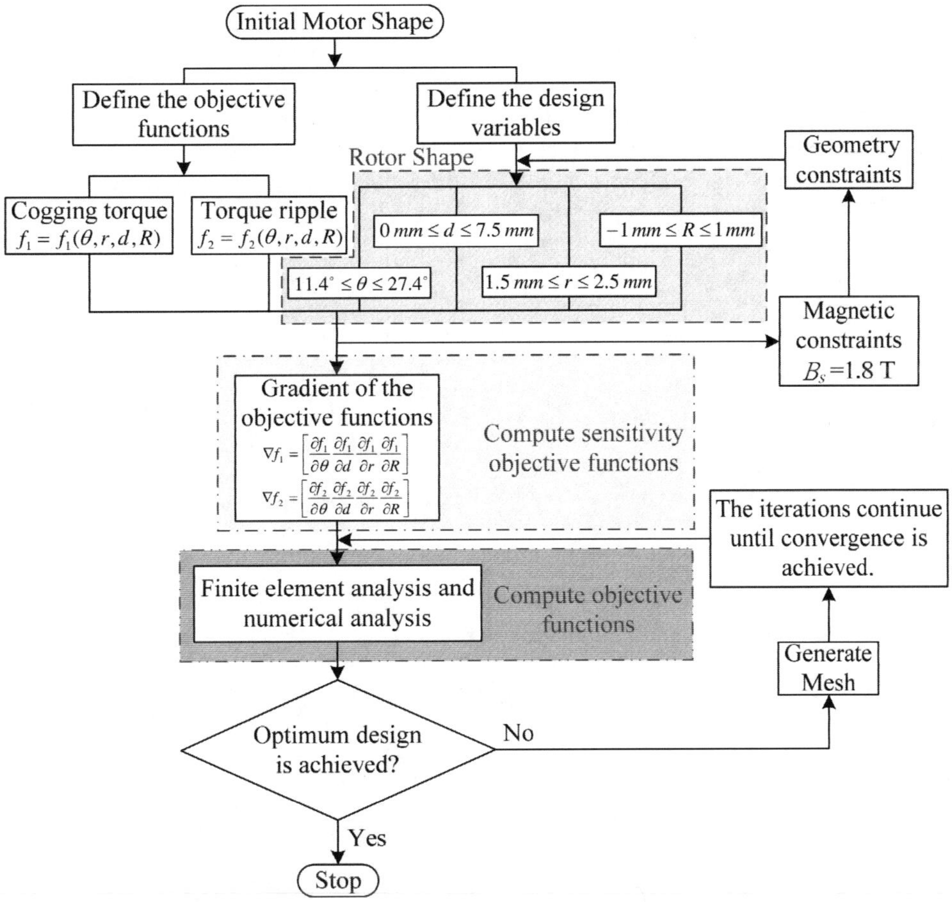

Figure 3. Main steps of shape design sensitivity analysis.

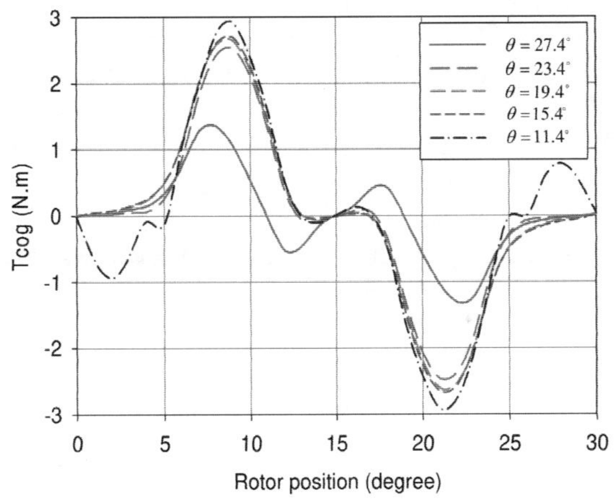

Figure 4. Sensitivity of θ.

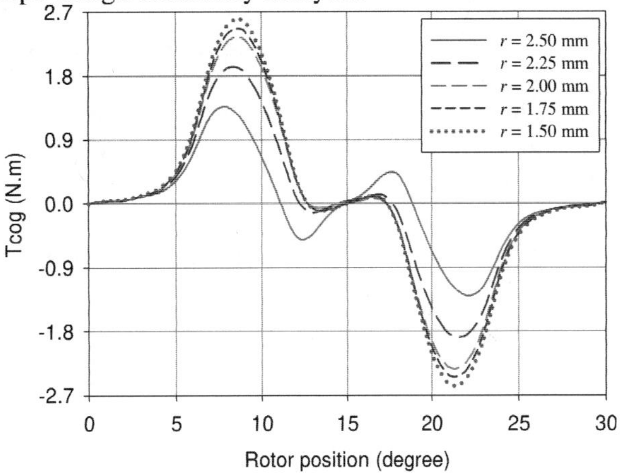

Figure 5. Sensitivity of r.

5.1.3. Influence of length d

In this part of analysis, the only variable is d. The other parameters R, r and θ are held constant (R = 1 mm, r = 2.5 mm and $\theta = 27.4^0$). The results are shown in Figure 6. As noted in the figure, the minimum value of cogging torque is 0.8 N.m with d = 5.5 mm. From the above analysis, the best parameter set for lower level of cogging torque is R = 1 mm, r = 2.5 mm, $\theta = 27.4^0$ and d = 5.5 mm.

5.2. Torque ripple

Torque ripple is defined as a ratio of the diffenence between the values of maximum and minimum torque to the average torque (T_{avg}) multiplied by 100%. In this part of analysis, the only variable is *d*. The other parameters *R*, *r* and θ are held constant (R = 1 mm, r = 2.5 mm and θ = 27.4^0). Sinusoidally excited currents are applied to the motor. The sensitivity of *d* on the torque ripple (T_{rip}) is plotted in Figure 7. As the figure shows, increasing the value of *d*, decreasing the torque ripple. It is found that the minimum value of torque ripple is 6.48% with *d* = 7.5 mm. Based on the above analysis, varying the saliencies of the rotor core is more effective in reducing the magnitude of both cogging torque and torque ripple.

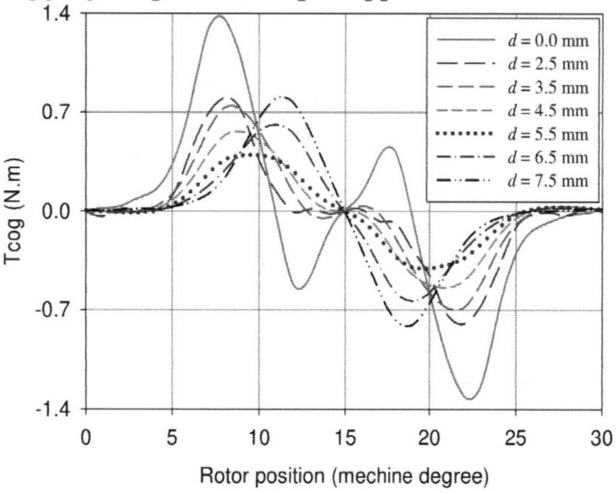

Figure 6. Sensitivity of *d*.

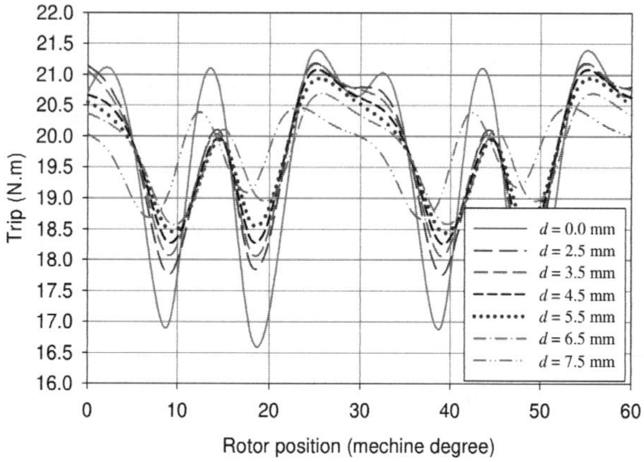

Figure 7. Sensitivity of *d*.

5.3. Summary of results

The effect of drilling three small circular holes and varying the saliencies of the rotor on the torque pulsations of the IPM motors has been studied. The simulation results for the optimal design motor with θ = 27.4^0, r = 2.5 mm, R = 1 mm, and d = 5.5 mm are investigated. Figure 8(a) and (b) compare the cogging torques and average torques. It is seen that the values of predicted and measured cogging torque of the optimized machine are good agreement with each other. Also, it can be seen that the peak to peak values of cogging torque and torque ripple reduce from the initial design of 5.16 N.m and 51.7 % to the optimal design of 0.8 N.m and 6.48 %, respectively, and the average torque increases from the initial of 19.45 N.m to the optimal design of 20.14N.m. In particular, the reduction of torque pulsations does not result in any reduction of the average torque. Furthermore, the core loss reduces from the initial design of 226.303 W to the optimal design of 197.860 W, results in the enhancement of efficiency from 95.05% to 95.5%.

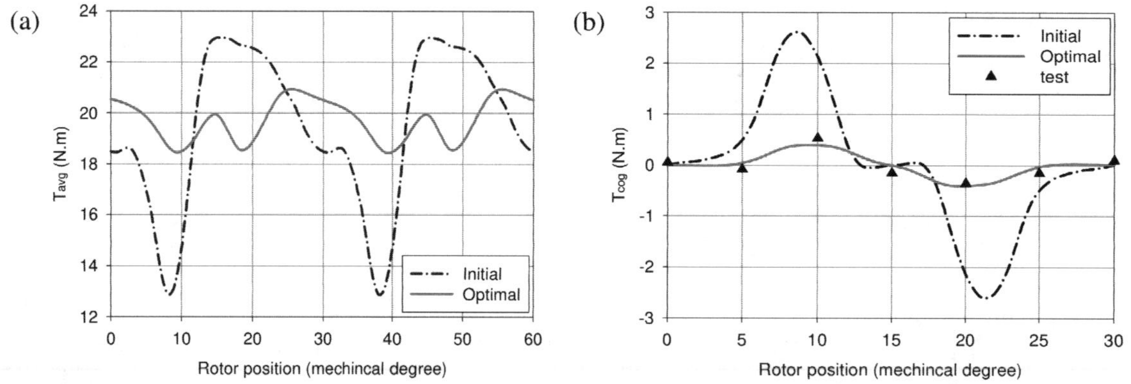

Figure 8. Comparison of simulation and test results between initial and optimal design of (a) cogging torques and (b) average torques.

To drill three circular holes in the magnetic island and shape the saliency of the rotor may affect the inductances of d-axis (L_d) and q-axis (L_q). Figure 9 compares the values of L_d and L_q between initial and optimal design. As noted in the figure, initially, the value of L_q drops rapidly with increasing current, but the value of L_d reduces slightly, and then they almost do not change. The holes and saliency of the rotor do affect the flux path of q-axis then d-axis around rated operation condition. The saliency ratio (e.g. L_q/L_d) reduces from 2.586 of initial design to 2.185 of optimal design at rated current of 68 A. Figure 10 shows the optimized cross section of the IPM motor.

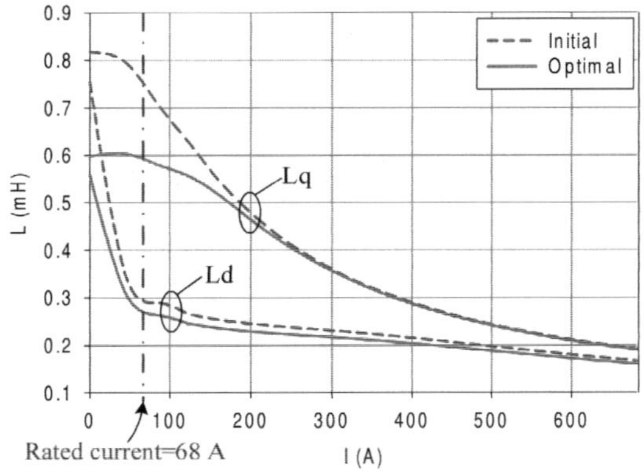

Figure 9. Compares the values of L_d and L_q.

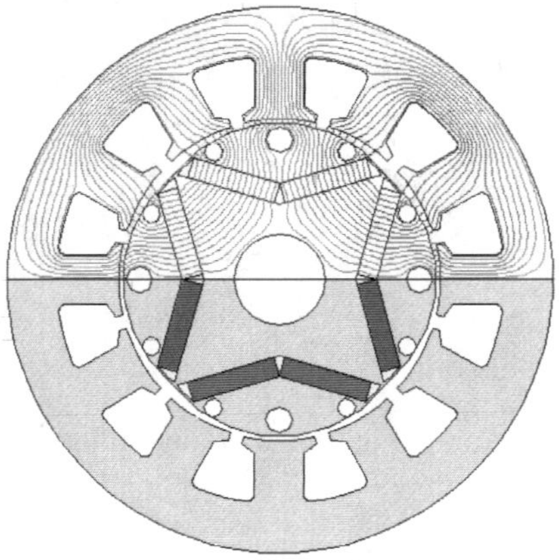

Figure 10. Optimized cross section of the IPM motor.

6. Conclusion

In this paper, the influence of drilling three small circular holes in the magnetic island and varying the geometry shape of the saliencies of the rotor on the torque pulsations of three-phase IPM motors has been studied. Experimental results confirm a good correlation between torque pulsations and the proposed finite element sensitivity analysis. It was shown that the technique presented in this paper is effective for minimizing the production of pulsating torque in IPM motors.

Acknowledgment

This work was supported by the National Science Council of Taiwan through grant numbers NSC 97-2221-E-035-095-MY2.

7. References

[1] Kim D H, Park I, Lee J H and Kim C E 2003 *IEEE Trans. Magn.* **3** 1456
[2] Zhu Z Q, Ruangsinchaiwanich S, Schofield N and Howe D 2003 *IEEE Trans. Magn.* **5** 3238
[3] Kioumarsi A, Moallem M and Fahimi B 2006 *IEEE Trans. Magn.* **11** 3706
[4] Kang G H, Son Y D and Kim G T 2007 *Proc. IEEE IAS Annu. Meeting* (New Orleans) 558-565
[5] Lee G H, Kim S I, Hong J P and Bahn J H 2008 *IEEE Trans. Magn.* **6** 1582
[6] Parasa L and Hao L 2008 *IEEE Trans. Ind. Electroics* **2** 602
[7] Han S H, Jahns T M and Zhu Z Q 2010 *IEEE Trans. Ind. Appl.* **1** 187
[8] Biedinger J M and Lemoine D 1997 *IEEE Trans. Magn.* **3** 2309
[9] Wang S and Kang J 2000 *IEEE Trans. Magn.* **4** 1119
[10] Lee J H, Kim D H and Park I H 2003 *IEEE Trans. Magn.* **3** 1269
[11] Belegundu A D and Chandrupatla T R 1999 *Optimazation Concepts and Applications in Engineering* (New Jersey: Prentice Hall, Inc.)

2nd Int. Symp. on Advanced Magnetic Materials and Applications (ISAMMA 2010) IOP Publishing
Journal of Physics: Conference Series **266** (2011) 012069 doi:10.1088/1742-6596/266/1/012069

Optimal design of an IPM motor using Taguchi and Rosenbrock's methods

C C Hwang[1], P L Li[2], CM Chang[2] and C T Liu[3]
[1]Dept. of EE, Feng Chia University, Taichung 407, Taiwan, ROC
[2]Ph.D. Program in ECE, Feng Chia University, Taichung 407, Taiwan, ROC
[3]Dept. of EE, National Sun Yat-sen University, Kaohsiung 804, Taiwan, ROC

E-mail: cchwang@fcu.edu.tw

Abstract. Techniques for the design optimization for cogging torque minimization and average torque maximization of a high-speed 2-pole interior permanent magnet (IPM) synchronous motor are presented. It is shown by the finite element method (FEM) and measurement, that combined the Taguchi and Rosenbrock's methods is a very efficient and effective approach in robust design a high performance motor.

1. Introduction

High performance IPM synchronous motors have been used extensively in recent years for industrial applications [1]-[3]. The maximum speed of such motors is about 6000 rpm with four magnet poles in the rotor. But in some applications, such as the fiber industry or hand dryers, a higher motor speed is required. To keep the same inverter, because of cost, the use of a two-pole motor is necessary [4]. Unfortunately, machines equipped with two magnet poles exhibit a high level of cogging torque. It is well known that cogging torque produces vibration and noise which may be detrimental to the performance of position and speed control systems.

In the literature, several methods for reducing the cogging torque in brushless permanent magnet motors have been proposed [5-9]. However, these techniques increase the complexity of the motor construction and reduce the average output torque. In this paper, technique is proposed in order to minimize the level of cogging torque without any sacrifice of average output torque. These goals can be accomplished using the optimization methods.

Stochastic methods of optimization have gained popularity over the last few years for optimization of engineering problems due to their fast convergence on a global optimum. Among the developed of stochastic methods, the Rosenbrock's method is robot and flexible [10-12]. However, the major concern of this technique is that it requires a large computation time of numerical filed computation. It can not consider for reducing the cogging torque by means of shaping the magnets either [7]. On the other hand, the Taguchi method is not required to use sophisticated algorithms and additional programming aside from the FEM of electromagnetic field analysis [13, 14]. It also allows many settings of as many as necessary design parameters in optimization computation simultaneously. Hence, effects of several factors on cogging torque and average torque can be investigated.

[1] cchwang@fcu.edu.tw

In this paper, the Taguchi method is used for the determination of the initial optimal values of the design variables for minimum cogging torque and maximum average torque [13, 14], and the Rosenbrock's method is applied to determine the optimal settings of the design parameters [10-12].

2. Model of Motor

Figure 1 shows a three-phase, two-pole, six-slot IPM motor. The initial design parameters are stator outside diameter = 90 mm, inner diameter = 53 mm, stack length = 65 mm, and rotor outside diameter = 52 mm. The core is made of nonoriented silicon steel, and two sintered NdFeB magnets with a remanence of 1.2 T and a relative recoil permeability of 1.05 are inserted in the rotor. The rotating speed is 12000 rpm.

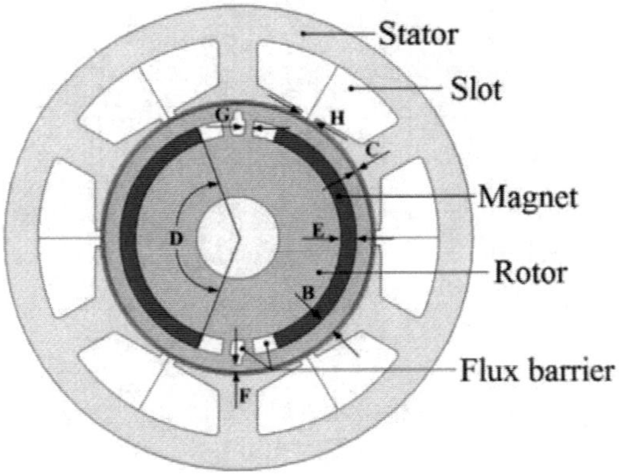

Figure 1. Model of IPM motor.

3. Taguchi's Methodology

3.1. Establish orthogonal array

Taguchi's method provides a systematic and efficient approach for conducting the numerical experimentation to determine near optimum settings of design parameters. An orthogonal array representing factors and their levels allowable in the design is used to study the parameter space. Considering feasible refinements, Table 1 shows the selected eight factors, A, B, C, D, E, F, G and H affecting the values of cogging torque and average torque. Each factor corresponding to one design variable as shown in Figure 1, and each at three levels except factor A that has two levels. The values with the underline denote the initial design.

Table 1. Levels of design variable.

Factor	Definition	Level 1	Level 2	Level 3
A	magnetization of magnets	parallel	radial	--
B (mm)	magnetic island width	2	2.5	3
C (mm)	air gap length	0.5	0.6	0.7
D	ratio of magnet pitch to pole pitch	0.778	0.806	0.833
E (mm)	thickness of magnet	3	3.5	4
F (mm)	magnetic bridge width	1	1.2	1.4
G (mm)	web width	1	1.2	1.4
H (mm)	slot opening	2.5	3	3.5

3.2. Conduct the experiment

There are 18 experiments required for us to determine the optimum combination of the levels of these factors as shown in Table 2. To obtain the values of cogging and average torques for each case, 2D FEM analysis is conducted. Table 2 tabulates the motor performance indexes. Where T_c is the peak to peak value of cogging torque, and T_{ave} is the value of average torque.

Table 2. Motor performance assessments from Taguchi's experiments.

Experiment	Factor level combination	T_c (Nm)	T_{ave} (Nm)
1	A1-B1-C1-D1-E1-F1-G1-H1	0.663	2.137
2	A1-B1-C2-D2-E2-F2-G2-H2	0.619	2.215
3	A1-B1-C3-D3-E3-F3-G3-H3	0.525	2.327
4	A1-B2-C1-D1-E2-F2-G3-H3	0.503	2.161
5	A1-B2-C2-D3-E3-F3-G1-H1	0.455	2.231
6	A1-B2-C3-D3-E1-F1-G2-H2	0.199	2.028
7	A1-B3-C1-D2-E1-F3-G2-H3	0.183	2.004
8	A1-B3-C2-D3-E2-F1-G3-H1	0.270	2.013
9	A1-B3-C3-D1-E3-F2-G1-H2	0.300	2.105
10	A2-B1-C1-D3-E3-F2-G2-H1	0.658	2.603
11	A2-B1-C2-D1-E1-F3-G3-H2	0.165	2.351
12	A2-B1-C3-D2-E2-F1-G1-H3	0.791	2.454
13	A2-B2-C1-D2-E3-F1-G3-H2	0.852	2.440
14	A2-B2-C2-D3-E1-F2-G1-H3	1.063	2.371
15	A2-B2-C3-D1-E2-F3-G2-H1	0.526	2.319
16	A2-B3-C1-D3-E2-F3-G1-H2	1.151	2.367
17	A2-B3-C2-D1-E3-F1-G2-H3	1.138	2.362
18	A2-B3-C3-D2-E1-F2-G3-H1	0.850	2.236

3.3. Analysis of results

The means of all results can be calculated as

$$m = \frac{1}{18}\sum_{i=1}^{18} T_i \tag{1}$$

The means are $T_c = 0.6062$ Nm and $T_{avg} = 2.2624$ Nm.

3.4. Calculate average effect

The peak to peak value of cogging torque of setting factor C at level 3 are calculated as

$$m_{C_3}(T_c) = \frac{1}{6}\left(T_c(3)+T_c(6)+T_c(9)+T_c(12)+T_c(15)+T_c(18)\right) \tag{2}$$

where the factor C is set to level 3 only in experiments 3, 6, 9, 12, 15 and 18. The peak to peak values of cogging torque for all levels of factors can be obtained in a similar way. Figures 2 and 3 illustrate the main factor effects on cogging torque and average torque, respectively.

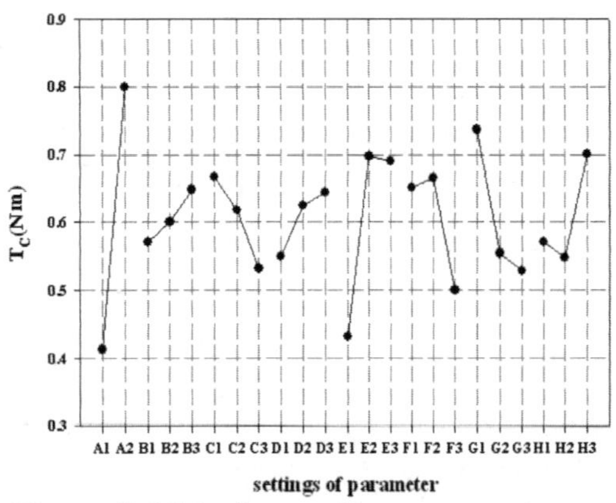

Figure 2. Main factor effects on cogging torque.

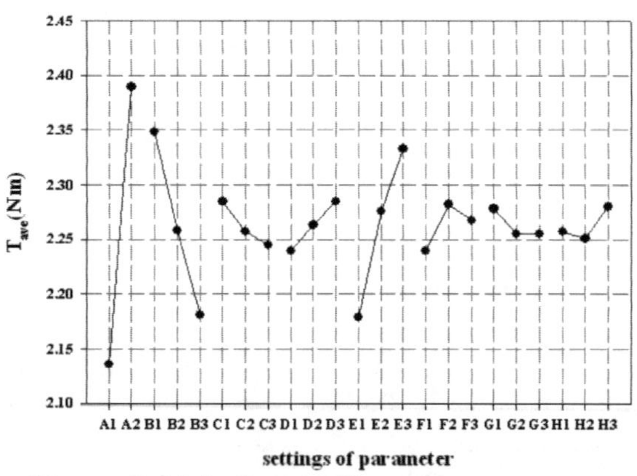

Figure 3. Main factor effects on average torque.

3.5 Analysis of variance

Analysis of variance (ANOVA) provides a measure of confidence. To conduct ANOVA, the sum of squares (SS) is calculated first, and it can be calculated as,

$$SSF_A = n\sum_{i=1}^{n}\left(m_i - m\right)^2 \qquad (3)$$

where n denotes the number of levels for each factor. The SSF_B ~ SSF_H can be obtained in the same way. These results are summarized in Table 3.

Table 3. Effects of various factors on motor performance indexes.

Factor		A	B	C	D	E	F	G	H	Total
T_c	SSF	1.326	0.009	0.029	0.015	0.138	0.050	0.078	0.041	1.686
	Factor effects(%)	78.619	0.559	1.696	0.900	8.191	2.976	4.640	2.412	100
T_{avg}	SSF	15.452	0.042	0.003	0.003	0.036	0.003	0.001	0.001	15.541
	Factor effects(%)	99.427	0.269	0.017	0.020	0.234	0.018	0.007	0.009	100

3.6 Design Optimization from Taguchi method

It is noted in Figure 2 that the factor-level combination (A1, B1, C3, D1, E1, F3, G3, H2) contributes to minimization of cogging torque, and it can be seen in Figure 3 that the factor-level combination (A2, B1, C1, D3, E3, F2, G1, H3) contributes to maximization of average torque. Therefore, the element B1 is immediately selected to constitute the elements of the optimum design for minimum cogging torque and maximum average torque. It is shown in Table 3 that factor A has large effect on average torque (99.427%) to cogging torque (78.619%), and factors from C to H have large effect on cogging torque to average torque. Therefore, the best combination of design parameters for minimum cogging torque and maximum average torque is determined to be (A2, B1, C3, D1, E1, F3, G3, H2). Based on this combination of design parameters, 2-D FEM analysis was used to determine the performance of the optimized motor. Results show that cogging torque reduces from the initial design of 1.343 Nm to Taguchi method of 0.241 Nm, and average torque increases from the initial design of 2.307 Nm to Taguchi method of 2.384 Nm.

4. Rosenbrock's method and results

The Rosenbrock's method of rotating coordinate is used to search the optimal value. The coordinate system is rotated at any stage for finding the minimum. The details of the steps to reach the solutions are described in [11, 12]. In this section, the optimization using Rosenbrock's method is carried out so that the cogging torque becomes minimal under the constraint that the average torque is not less than 2.4 Nm. Using the initial design variables from the Taguchi method, the constraints of design variables B, C, D, E, F, G and H are defined as follows:

$$\begin{gathered} 2.0 \le B \le 2.5;\ 0.6 \le C \le 0.7 \\ 0.778 \le D \le 0.8;\ 3 \le E \le 3.5 \\ 1.4 \le F \le 1.5;\ 1.4 \le G \le 1.5 \\ 3 \le H \le 3.3 \end{gathered} \qquad (4)$$

The performance of the motor was obtained using 2-D FEM analysis. Table 4 compares the data of the motor between the initial design, Taguchi, Rosenbrock's methods and measurement. Figures 4 and 5 compare the results of cogging torque and average torque. It is seen that the values of predicted and measured of cogging torque and average torque are good agreement with each other. Also, it can be seen that cogging torque reduces from the initial design of 1.343 Nm and Taguchi method of 0.241 Nm to Rosenbrock's method of 0.206 Nm, and average torque increases from the initial design of 2.307 Nm and Taguchi method of 2.384 Nm to Rosenbrock's method of 2.432 Nm.

Table 4. Comparison results.

	A	B(mm)	C(mm)	D	E(mm)	F(mm)	G(mm)	H(mm)	T_c (Nm)	T_{ave} (Nm)
Initial	radial	3	0.5	0.833	3	1	1	3	1.343	2.307
Taguchi	radial	2	0.7	0.778	3	1.4	1.4	3	0.241	2.384
Rosenbrock	radial	2.212	0.652	0.793	3.152	1.412	1.423	3.123	0.206	2.432
Test									0.210	2.410

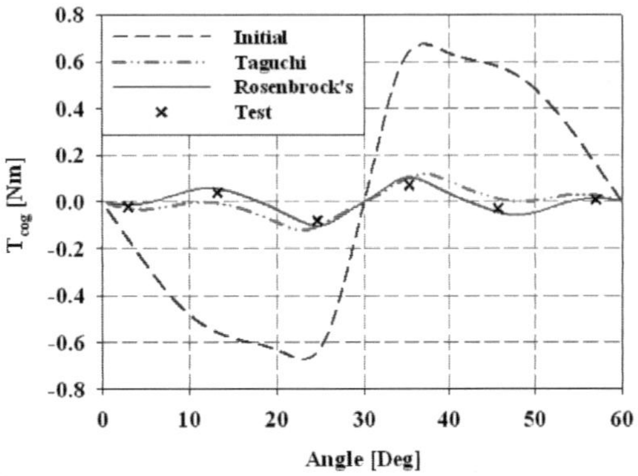

Figure 4. Comparison of cogging torques.

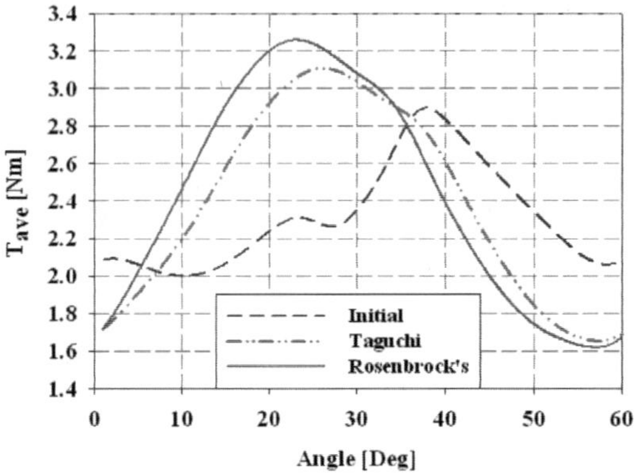

Figure 5. Comparison of average torques.

5. Conclusion

This paper applied the combination of the Taguchi and Rosenbrock's methods to the design optimization for cogging torque minimization and average torque maximization of a high-speed 2-pole IPM synchronous motor. It was shown that the technique presented in this paper is effective for obtaining the design parameters having low levels of cogging torque. The technique would link the existing FEM packages to complete an iterative design loop.

Acknowledgment

This work was supported by the National Science Council of Taiwan through grant numbers NSC 97-2221-E-035-095-MY2.

6. References

[1] Jahns T M, Kliman G B, and Neumann T W 1986 *IEEE Trans. Ind. Appl.* **IA 22** 738-47
[2] Murakami H, Honda Y, and Higaki T 1996 *Proc. 13th Int. Conf. Electric Vehicle Symposium* **1** 135-41
[3] Bianchi N, Pre M D, and Bolognani S 2006 *IEEE Trans. Vehicular Technology* **55** 1102-11
[4] Miyashita K, Yamashita S, Tanabe S, Shimozu T, and Sento H 1980 *IEEE Trans.* **PAS-99** 2175-81
[5] Hwang S and Lieu D K Design 1994 *IEEE Trans. Magn.* **30** 4287-89
[6] Deodhar R P, Staton D A, Jahns T M, and Miller T J E 1996 P *IEEE Trans. Ind. Appl.* **IA-32** 569- 76
[7] Hamler A and Hribernik B 1996 *IEEE Trans. Magn.* **32** 1545-48
[8] Hwang C C, John S B, and Wu S S 1998 *IEEE Trans. Magn.* **34** 468-70
[9] Kioumarsi A, Moallem M, and Fahimi B 2006 *IEEE Trans. Magn.* **42** 3706-11
[10] Ohnishi T and Takahashi N 2000 *IEEE Trans. Magn.* **36** 3537-39
[11] Takahashi N 2001 *Optimal Design Using Finite Element Method for Magnetic Field Analysis* (Tokyo: Morikita Publishing Company)
[12] Rao S S 1978 *Optimazation,Theory and Applications* (Wiley: Eastern Limited)
[13] Hwang C C, Lyu L Y, Liu C T, and Li P L 2008 *IEEE Trans. Magn.* **44** 4325-28
[14] Roy R K 1990 *A Primer on the Taguchi Method* (New York: Van Nostrand Reinhold)

Journal of Physics: Conference Series **266** (2011) 012070 doi:10.1088/1742-6596/266/1/012070

Repeatable Shape Memory Effect and Mechanical Resonance of TiNiCu-Coated Magnetic Ribbons

O Ishii[1], Y Miyahara[1], S Kambe[1], N Kutsuzawa[1] and A Ishida[2]
[1]Yamagata University, Yonezawa, Yamagata 992-8510, Japan
[2]National Institute for Materials Science, Tsukuba, Ibaraki 305-0047, Japan

E-mail: oishii@yz.yamagata-u.ac.jp

Abstract. This paper describes a new bimorph-type actuator composed of a shape-memory-alloy coated magnetic ribbon. A high magnetostrictive amorphous ribbon (Metglas 2605SC, 37 mm×6 mm×0.025 mm) coated with a 1.0 μm thick sputter-deposited $Ti_{48.5}Ni_{33.5}Cu_{18}$ film exhibited a repeatable shape memory effect in the temperature range from 10 °C to 70 °C; reverse martensitic transformation upon heating bent the ribbon and martensitic transformation upon cooling flattened it. Simultaneous application of AC and DC magnetic fields excited the longitudinal mechanical vibration which can be monitored wirelessly with a pickup coil. The resonance frequency was proportional to the displacement of the ribbon within an accuracy of 0.76 %. Consequently, it is confirmed that a TiNiCu-coated magnetic ribbon actuator with high positioning resolution can be realized by monitoring its resonance frequency and feeding back it to the heating and/or cooling power control algorithms.

1. Introduction

Recently, MEMS (Microelectromechanical System) technology is studied intensively. For an actuator in MEMS, large force and displacement are required. Shape-memory-alloy (SMA) is considered as a promising actuator material for MEMS because of its large force and displacement. The SMA films have been applied to various actuators such as micropump[1,2], microgripper[1,3] and positioner[4]. Ishida *et al.* reported that substrate temperature during sputtering process can be reduced by adding Cu in TiNi-alloy and that a TiNiCu-coated polyimide sheet shows repeatable shape-memory effect[5,6]. Since the repeatable shape-memory effect is based on the phase transformations, the displacement has large temperature hysteresis which results in large positioning error.

On the other hand, Shibata *et al.* have developed a wireless hydrogen sensor that uses the mechanical resonance of a PdNi-coated magnetic ribbon[7]. By applying AC and DC magnetic fields to a highly magnetostrictive ribbon, the ribbon vibrates mechanically. The behavior of the vibration can be detected wirelessly by monitoring the induced voltage on a pickup coil placed away from the ribbon. As hydrogen concentration is increased, the volume of the PdNi-film increases and the ribbon is bent remarkably. Because the curvature of the ribbon greatly affects the mechanical resonance frequency, the hydrogen concentration can be detected by monitoring the resonance frequency.

In this context, it is expected that a TiNiCu-coated magnetic ribbon also shows the repeatable shape-memory effect and that its mechanical resonance frequency changes with temperature. In order to achieve heat-driven SMA actuator with high positioning resolution, the relationship between displacement and mechanical resonance frequency was investigated for a TiNiCu-coated magnetic ribbon.

2. Experiments

Magnetic ribbons (Metglas 2605SC) which have high permeability (15000) and high magnetostriction (30×10^{-6}) were shaped into rectangular with length, width and thickness of 37 mm, 6 mm and 0.025 mm, respectively. In order to relieve the internal stress and introduce a transverse magnetic anisotropy, the ribbons were annealed at 410 °C for 10 min in N_2 flow under a transverse magnetic field of 100 Oe. Films of $Ti_{48.5}Ni_{33.5}Cu_{18}$ alloy with thickness of 1.0 μm were deposited on the ribbons with a carousel-type magnetron sputtering apparatus. Three targets of pure Ti (100 mm in dia.), Ni (100 mm in dia.) and Cu (75 mm in dia.) were used to obtain a desired composition. Sputtering conditions were as follows; Ar pressure was 0.2 Pa. Substrate temperature was 310 °C. The sputtering powers of Ti, Ni and Cu targets were 1000, 224 and 85 W, respectively. Sputtering duration was 18 min.

The phase transformation temperatures of the film peeled from the substrate were determined by differential scanning calorimetry (DSC) with a sweep rate of 1.0 °Cmin^{-1}. The peak temperatures of martensitic and reverse martensitic transformations were determined to be 43.8 °C and 54.9 °C, respectively.

Figure 1 shows a resonance frequency measurement setup. DC bias magnetic field (H_{dc}) was applied on a magnetic ribbon with Helmholtz coil. A pick-up coil was connected to port 1 of a network analyzer to measure reflection coefficient (S_{11}). The mechanical resonance frequency (f_r) was defined as the dip point of S_{11}. The temperature of the sample was swept in a range of 0~100 °C by blowing a hot or cold air on the quartz tube in which the ribbon was placed. The temperature sweep rate was 2.5 °Cmin^{-1}.

Figure 2 shows a definition of displacement (h) of the magnetic ribbon. The h value was calculated by counting the number of the pixels in a digitized picture. The measurement accuracy of h was 0.02 mm.

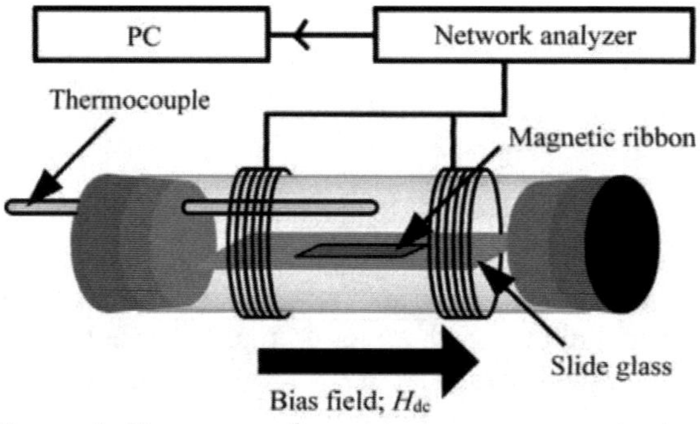

Figure 1. Resonance frequency measurement setup.

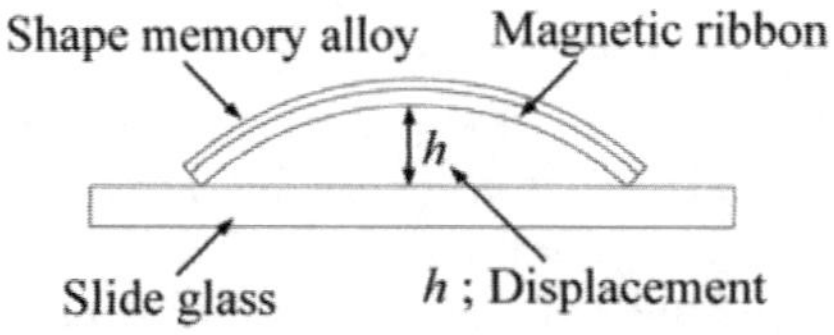

Figure 2. Definition of displacement (h) of TiNiCu-coated magnetic ribbon.

3. Results and discussions

Figure 3 shows temperature dependencies of h upon heating (open circles) and cooling (open squares). Reverse martensitic transformation upon heating bent the ribbon and h increased from 1.0 mm at 0 °C to 2.2 mm at 70 °C. In contrast, martensitic transformation upon cooling flattened the ribbon and h decreased from 2.2 mm at 70 °C to 1.0 mm at 0 °C. The fact that h=2.2 mm at 70 °C means that internal stress introduced to the TiNiCu-film during sputtering process remains after deposition. The maximum temperature difference between the martensitic and reverse martensitic transformations was determined to be 7.3 °C by the displacement measurement. These features are consistent with the DSC measurements. As a result, it is proved that a TiNiCu-coated magnetic ribbon actuator can be driven by heating and/or cooling and that its displacement has large temperature hysteresis which results in large positioning error.

Figure 4 shows temperature dependencies of f_r upon heating (open circles) and cooling (open squares). The bias magnetic field was 3.5 Oe at which the largest ΔE effect was attained. The temperature dependencies of f_r are quit similar to those of h as shown in figure 3; f_r increased upon heating and decreased upon cooling. Moreover, f_r had the maximum temperature hysteresis of 7.6 °C which is almost equal to that of h.

Figure 5 shows the relation between h and f_r upon heating (open circles) and cooling (open squares). The f_r values are proportional to the h. Furthermore, there is no hysteresis during thermal cycling within an accuracy of 0.76 %. Consequently, it is confirmed that precise position control of a TiNiCu-coated magnetic ribbon actuator can be realized by feeding back the monitored f_r values to the heating and/or cooling power control algorithms.

The mechanism of the change in f_r with temperature is considered to be as follows; TiNiCu alloy film on magnetic ribbon expands on heating and bends the ribbon. Figure 6 shows the H_{dc} dependencies of f_r at (a) 25 °C and (b) 60 °C. The f_r shows deep and shallow dips at 25 °C and 60 °C, respectively. Figure 7 shows the schematic drawings of the magnetic anisotropies for (a) flat and (b) bent magnetic ribbons. When the magnetic ribbon is flat and has single axis anisotropy in the width direction, large ΔE effect brings about the considerable decrease in f_r at $H_{dc} = H_k$ as shown in figure 6(a). In contrast, as temperature is increased, the ribbon bends and the tension is introduced on the TiNiCu-coated side of the ribbon. This tension causes the longitudinal magnetic anisotropy due to the reverse magnetostriction effect. Accordingly, the average magnetic anisotropy becomes isotropic as shown in figure 7(b), which weakens ΔE effect and increases the f_r at $H_{dc} = H_k$. This interpretation is consistent with the report on the PdNi-coated magnetic ribbon which was investigated as a hydrogen sensor[7].

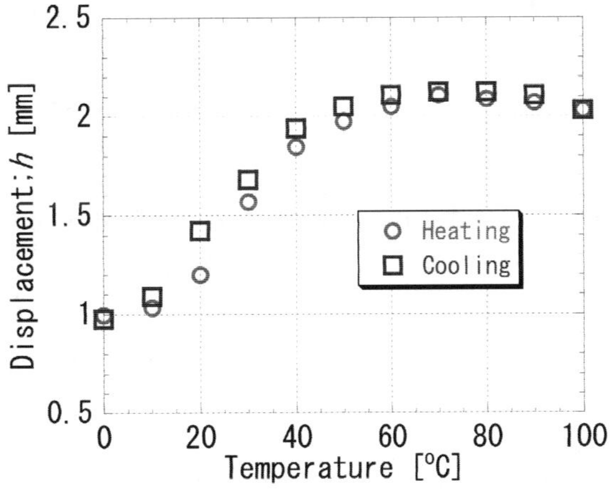

Figure 3. Temperature dependencies of displacement (h) upon heating (open circles) and cooling (open squares).

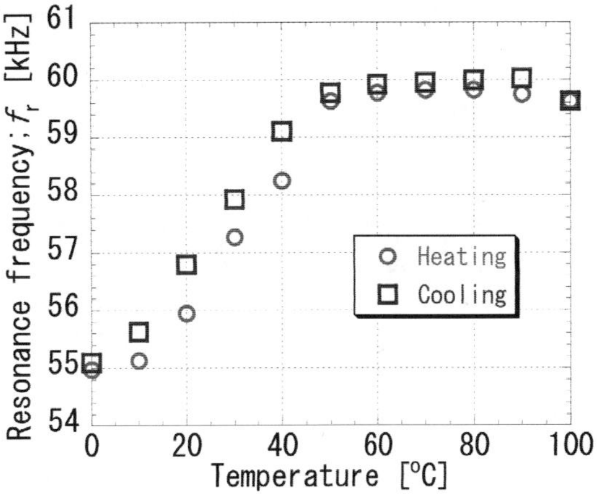

Figure 4. Temperature dependencies of resonant frequency (f_r) upon heating (open circles) and cooling (open squares).

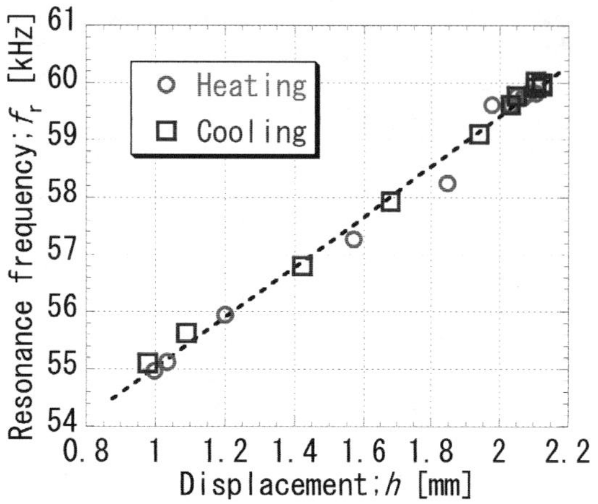

Figure 5. Relation between displacement (h) and resonance frequency (f_r) upon heating (open circles) and cooling (open squares).

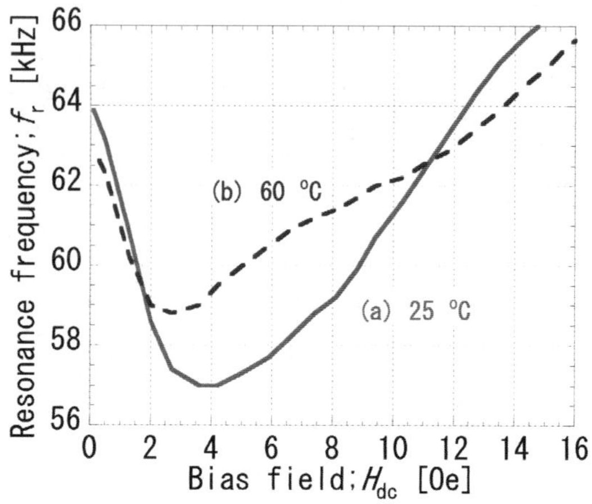

Figure 6. Dependencies of resonance frequency (f_r) on bias magnetic field (H_{dc}) at (a) 25 °C and (b) 60 °C.

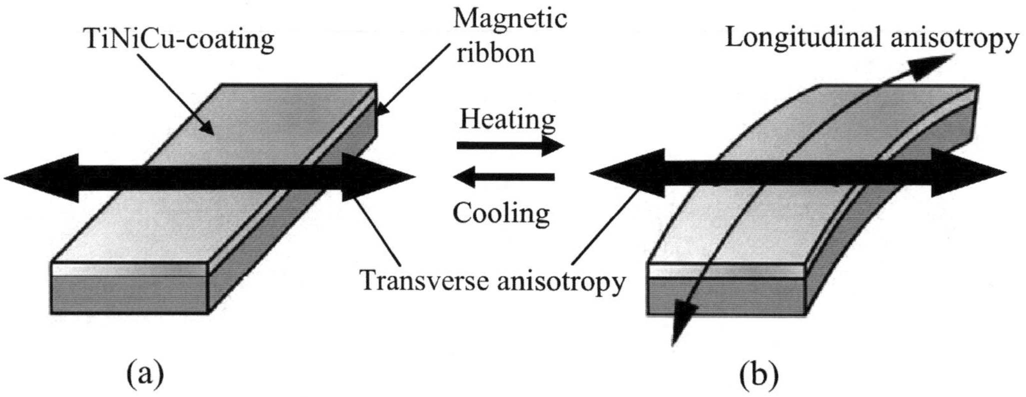

Figure 7. Schematic drawings of the magnetic anisotropies for (a) flat and (b) bent magnetic ribbons.

4. Coclusion

A high magnetostrictive amorphous ribbon (Metglas 2605SC, 37 mm×6 mm×0.025 mm) coated with a 1.0 μm thick sputter-deposited $Ti_{48.5}Ni_{33.5}Cu_{18}$ film exhibited a repeatable shape memory effect in the temperature range from 0 °C to 70 °C; reverse martensitic transformation upon heating bent the ribbon and martensitic transformation upon cooling flattened it. The temperature difference between the martensitic and reverse martensitic transformations was 7.3 °C. The resonance frequency of the longitudinal mechanical vibration excited by simultaneous application of AC and DC magnetic fields indicated the similar temperature dependence to the displacement. As a result, the mechanical resonance frequency was proportional to the displacement within an accuracy of 0.76 % and the relationship was not affected by the thermal hysteresis. Consequently, it was confirmed that a TiNiCu-coated magnetic ribbon actuator with high positioning resolution can be realized by monitoring its resonance frequency and feeding back it to the heating and/or cooling power control algorithms.

5. References

[1] Fu Y, Du H, Huang W, Zhang S and Hu M 2004 *Sensors and Actuators* **A112** 395

Journal of Physics: Conference Series **266** (2011) 012070 doi:10.1088/1742-6596/266/1/012070

[2] Makino E, Shibata T and Kato K 1999 *Sensors and Actuators* **78** 163
[3] Kohl M, Just E, Pfleging W and Miyazaki S 2000 *Sensors and Actuators* **83** 208
[4] Asua E, Garcia-Arribas A and Etxebarria V 2008 *Eur. Phys. J. Special Topics* **158** 231
[5] Ishida A and Sato M 2008 *Thin Solid Films* **516** 7836
[6] Ishida A, Sato M, Yoshikawa W and Tabata O 2002 *Mater. Sci. Forum* **394-395** 487
[7] Shibata T, Kutsuzawa N, Yoshizawa S, Okuyama S, Kambe S and Ishii O 2009 *J. Magn. Soc. Jpn.* **33** 50

Acknowledgments
We thank Dr. Masaki Nakamura at Hitachi Metals, Ltd. (former Yamagata University) for his useful discussion on this study.

Journal of Physics: Conference Series **266** (2011) 012071 doi:10.1088/1742-6596/266/1/012071

Structure and magnetic properties of γ-Fe_2O_3-multi walled carbon nanotubes nanocomposites prepared by sol-gel method

Taichi Kanazawa, Xiaoxi Liu, Akimitsu Morisako

Department of Information Engineering, Shinshu University,
4-17-1 Wakasato, Nagano, Japan.

E-mail: liu@cs.shinshu-u.ac.jp

Abstract. In this paper, the magnetic and crystallographic properties of nano-composite composing multi walled carbon nanotubes (MWCNTs) and maghemite (γ-Fe_2O_3) were investigated. The nano-composite was prepared by sol-gel method. The nano-composite prepared by sol-gel method shows unique structure with nano-sized maghemite particles decrease the surface of MWCNTs. The grain size of maghemite increase with the increase of the MWCNT composition. Coercivities of the nano-composite decrease from about 265 Oe to 216 Oe with the increase of the MWCNT/ Fe_2O_3 composition from 0 wt% to 23.0 wt%.

Key words sol-gel, Ferrite, carbon nanotube, maghemite, nano-composites

1. Introduction

Carbon nanotube has novel applications due to unique mechanical, electrical, thermal and optical properties such as small diameter, high aspect ratio, superior mechanical strength, and large flexibility. As one of the characteristics, MWCNT has potential applications for microwave absorption applications [1] [2] [3].

Pure ferrites particles with high anisotropy have been proved as one of the most excellent microwave absorber. However, the poor mechanical properties of ferrite limited the application of ferrite as

microwave absorber. In this paper, we show our effort to prepare nano-composite of MWCNTs and ferrite as microwave absorber. The nano-composite is believed to have good mechanical properties for microwave absorber. Most importantly, it is expected that the nano-composite show wide band of absorption properties that can cover the absorption band of both MWCNT and ferrite [4].
In this paper, the samples were prepared by sol-gel method. Sol-gel method is a well-known method to prepare ferrite nano-particles. The composition and grain size of ferrite can be easily manipulated by this method. The crystallization temperature of ferrite can be reduced dramatically by sol-gel method. Magnetic properties and crystallographic properties of obtained nano-composite were characterized. We found that uniform nano-composite can be formed by sol-gel method.

2. Experimental method

Maghemite was fabricated by sol-gel method. The starting materials were $Fe(NO_3)_3{\cdot}9H_2O$(99 %), citric acid (99 %), ethylene glycol(99 %) and distilled water. Firstly, aqueous solutions of citric acid and ethylene glycol were obtained by dissolving the complex in distilled water. Secondly the solution was prepared by mixing the complex and ferric salt as starting material. The ratio of $Fe(NO_3)_3{\cdot}9H_2O$: citric acid: ethylene glycol: molar ratio distilled water is 1:1:1:200. Then the fabricated solution was added to MWCNT/Fe_2O_3 from 0 wt% to 23.0 wt%. The average diameter of MWCNT is 100 nm and the typical length is 1~5 um. The pH of the solution was adjusted to 6.0 with ammonia. The precursor was heated to 210 °C at the rate of 210 °C/h. The final heat treatment was carried out at relatively at 210 °C for 1h. The differential thermogravimetry (DTG), thermogravimetry (TG) and differential thermal analyses (DTA) were used to probe the heating characteristic of the gelation after heating solvent at 150 °C for 1h. Vibrating sample magnetometer (VSM) and X-ray diffractometer (XRD) were employed to characterize structural and magnetic properties of samples at room temperature. The microstructure of the samples was observed by a scanning electron microscopy (SEM).

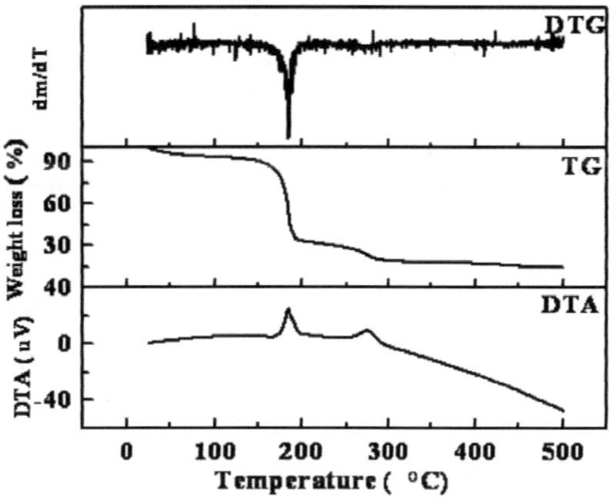

Fig.1 TG, DTG and DTA curve for the gelation of citric acid:ethylene glycol 1:1

3. Result and discussion

A typical TG, DTG and DTA curve for the gelation is shown in Fig.1. The TG curve exhibits three distinct weight loss steps and the DTA curve shows exothermic peaks. The first weight loss step was the loss of residual water in the gel. The second loss in the range of 150-220 °C corresponds to the decomposition occurs with elimination of NO_2 and NO gas products. Leading to a strong exothermic effect on the DTA curve is presumed to be associated with the crystallization of spinel maghemite phase. The thread loss in the range of 270-280 °C corresponds to be associated with the crystallization of spinel hematite phase. The mass of these oxides disappeared completely at 210 °C.

Figure 2 shows the XRD patterns of the sample after annealing at 210 °C. The MWCNT/ Fe_2O_3 composition is from 0 to 16 wt%. We can see only diffraction lines from the spinel structure except the diffraction line from MWCNT. This means the nano-composite composing of MWCNT and single phase of spinel structure of maghemite. It is clearly that the diffraction density of spinel phase decrease with the increase of the composition of MWCNT.

In the spherical particle model, the critical diameter d_{cr} (single-domain particle) can be calculated with the following formula [5]

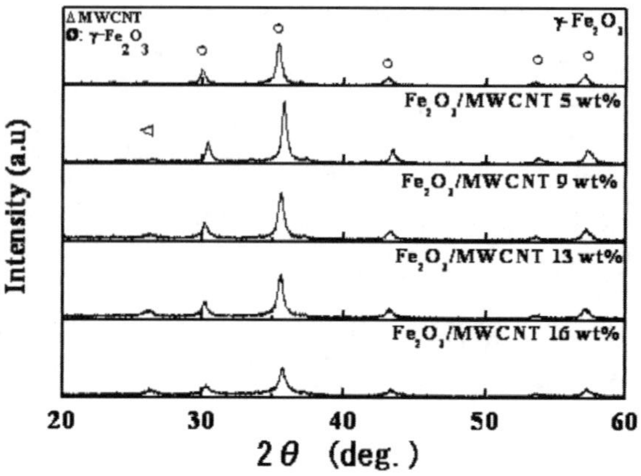

Fig.2 XRD pattern of nanocomposites prepared by sol-gel method.

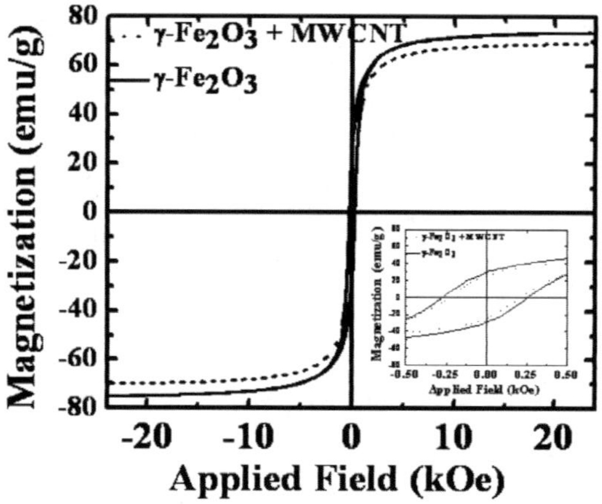

Fig.3 M-H loop for the maghemite and Fe_2O_3/MWCNT 5 wt% nanocomposite

$$d_{cr} = \frac{9\varepsilon_p}{2\pi M^2_s}$$

wheree ε_p is the surface energy of the domain wall and M_s is the saturation magnetization. ε_p can be calculation the T_C, curie temperature, K_1 the magetocrystalline anisotropy constant, K_B the Boltzmann constant and a the lattice constant.

$$\varepsilon_p = \left(\frac{2K_B T_C K_1}{a}\right)^{1/2}$$

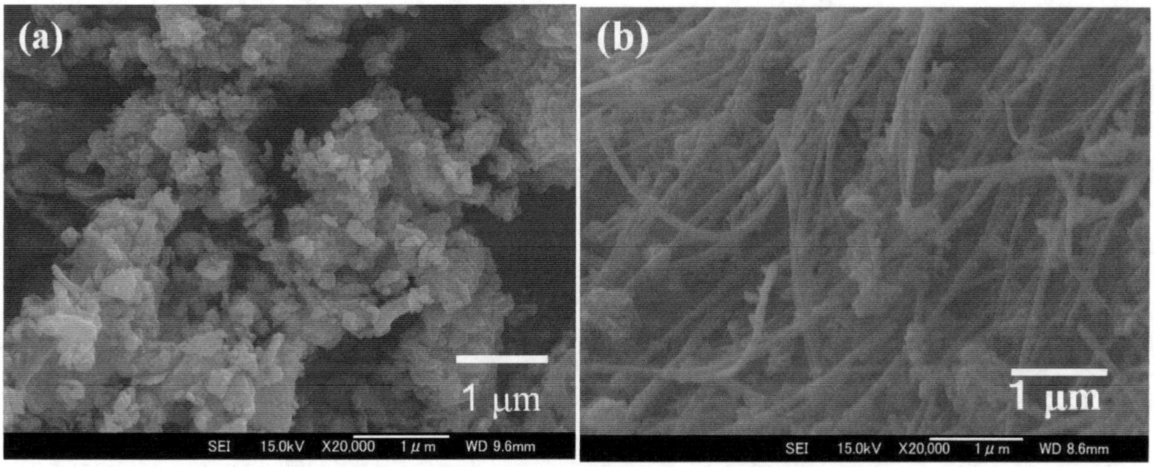

Fig.4. The SEM images of pure maghemite (a) and maghemite MWCNT nanocomposite after filtering solution and MWCNT, take out MWCNT, heat to 210 °C (b)

In the case of magheimite, K_1 is 4.7×10^4 erg/cm^3 ,T_c is 918 K ,M_s is 420 G and a is 8.34×10^{8} cm. The calculated ε_p is 0.378 erg/cm^2 and d_{cr} is 30.6 nm. In our sample, the grain size of maghemite can be calculated according to Scherrer formula. The grain size of maghemite for all the samples is less than 30 nm. This is much smaller that the single domain particle size of maghemite. This means that the maghemite particles behaved as single domain magnetic particles.

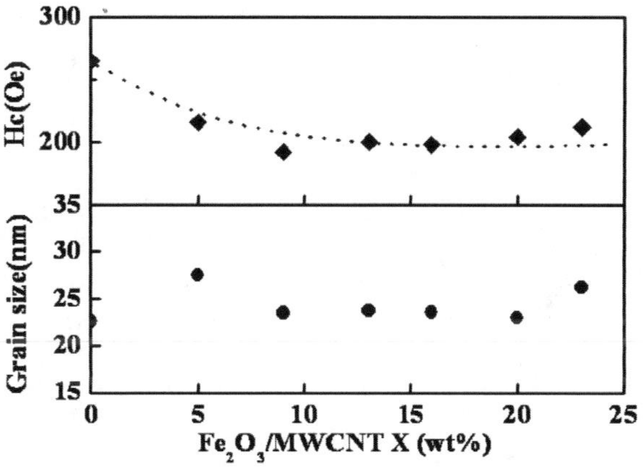

Fig.5 The coercivity and particle size of maghemite and MWCNT nano-composite.

Figure 3 shows M-H loops at room temperature for the typical spinel maghemite and nano-composite composing MWCNT of 5.0 wt%. Pure maghemite shows the saturation magnetization (M_s) of 73.6 emu/g, and coercivity (H_c) of 265 Oe. It should be noted that the value of saturation magnetization and coercivity are almost in bulk value [6]. The typical M_s is 68.9 emu/g, H_c is 216 Oe for nano-composite composing Fe_2O_3/MWCNT of 5.0 wt%. Due to the doping of MWCNT, it is easily to understand the decrease of saturation magnetization in the nano-composite. However, the decrease of coercivity in the nano-composite can be explained by increasing maghemite particle size with the introducing of the MWCNT.

Figure 4 shows SEM image of pure maghemite sample and the nano-composite composing Fe_2O_3/MWCNT. The pure maghemite samples grain size is between nano-size to micro-size and nano-composite is less than pure maghemite sample. The nano-composite was filtered using by isopropanpanol. The grain size decreases with increasing MWCNT. The result shows the strong attachment of maghemite nano-particles on the surface of MWCNT. Maghemite particles and MWCNT are uniformly dispersed with each other. This kind of microstructure is expected to have good microwave absorption properties.

Figure 5 shows the coercivity and grain size of maghemite and MWCNT nano-composite. The grain size was determined with the Scherrer formula:

$$D = \frac{0.9\lambda}{\beta_{1/2}\cos\theta}$$

where $\beta_{1/2}$ is the full width at half-maximum (in 2θ), λ is the x-ray wavelength and θ is the Bragg angle corresponding to the diffraction peak. The grain size was calculated for maghemite main peak (311) in the spectrum.The sample particle size was less than d_{cr} 30 nm. The coercivity decreases slowly with increasing MWCNT addition.

4. Conclusion

We have successfully prepared nano-composite of MWCNT and maghemite by sol-gel method at temperature as low as 210 °C. The average grain size of maghemite is smaller than the single domain size. Maghemite particles and MWCNT were found uniformly dispersed with each other. Our experimental results show that sol-gel method is a good method to prepare nano-composite of MWCNT and ferrite at relatively low temperature.

Acknowledgements
The authors gratefully acknowledge helpful discussions with JSPS fellow Ali Ghasemi.

Reference
[1] P.C.P. Watts, D.R. Ponnampalam, W.K. Hsu, A. Barnes, B. Chambers. *Che Phys.* 378 (2003) 609–614
[2] Paul C. P. wattas, W. K. Hsu, A. Barnes B. chambers *Adv. Mat.* 15 (2003) 7-8
[3] J. Wu, L. Kong *App. Phys.* 84 (2004) 24
[4] P. Xu et al, *Mat. Che. and Phys.* 114 (2009) 556-560
[5] C.Caizer. *Mat Sci and Eng.* 100 (2003)63-68
[6] T. Tuutijarv, J. Lu, M. Sillanpaa, G. Chen . *Haz Mat.* 166 (2009) 1415-1420

Journal of Physics: Conference Series **266** (2011) 012072 doi:10.1088/1742-6596/266/1/012072

Centrifugal Force Based Magnetic Micro-Pump Driven by Rotating Magnetic Fields

S H Kim[1], S Hashi[1] and K Ishiyama[1]
[1] Research Institute of Electrical Communication, Tohoku University, 2-1-1 Katahira, Aoba-ku, Sendai 980-8577, Japan

E-mail: kshoon@riec.tohoku.ac.jp

Abstract. This paper presents a centrifugal force based magnetic micro-pump for the pumping of blood. Most blood pumps are driven by an electrical motor with wired control. To develop a wireless and battery-free blood pump, the proposed pump is controlled by external rotating magnetic fields with a synchronized impeller. Synchronization occurs because the rotor is divided into multi-stage impeller parts and NdFeB permanent magnet. Finally, liquid is discharged by the centrifugal force of multi-stage impeller. The proposed pump length is 30 mm long and19 mm in diameter which much smaller than currently pumps; however, its pumping ability satisfies the requirement for a blood pump. The maximum pressure is 120 mmHg and the maximum flow rate is 5000ml/min at 100 Hz. The advantage of the proposed pump is that the general mechanical problems of a normal blood pump are eliminated by the proposed driving mechanism.

1. Introduction

Numerous studies have been conducted on micro-pumps for various applications. Moreover, the developments in materials and micro/nano fabrication techniques have led to the development of various types of micro-pumps. However, most of the micro-pumps are controlled by wires or batteries. Recently, magnetic micro-pumps with external magnetic field controls have been introduced in order to overcome these problems [1]. Micro-pumps have been receiving attention because of their important role in drug delivery and uTAS [2-3]. But, macro size pumps have been used as blood pumps. Pumps for medical applications can be classified into three types according to their actuating mechanisms: centrifugal pumps, axial pumps, and vibrating flow pumps. Moreover, blood pumps can be classified into two types: pulsating flow pumps and continuous flow pumps (rotary type). In general, the pulsating flow pumps are valve type, expensive, very heavy, low efficiency, complex controls, high power consumption, and low productivity. To solve these problems, continuous flow pumps have been developed for blood pumps [4-6]. Their features are very simple structure, small size, easy control, low power consumption and high productivity. In this paper, we focus on blood pumps used in the medical field. The main features of proposed pump system are that it is wireless, battery-free, and valve-less. It generates no heat and delivers continuous flow by centrifugal force and rotating magnetic fields. General centrifugal pumps consist of a shaft, a bearing, and an electrical motor. However, the proposed pump system does not require these mechanical components.

2. Principle of operating mechanism

The magnetic micro-pump is driven by a rotating magnetic field. The original energy source is magnetic torque. An NdFeB permanent magnet on the rotor is synchronized with a rotating magnetic field. The rotating speed is dependent on the operating frequency. Figure 1 illustrates the principles of the rotating magnetic field and synchronized state within the rotating magnetic field. To generate uniform rotating magnetic field, the angle of cross-point between coil_1 and coil_2 is 90° and the difference of phase of input current signal is 90°. In this state, the rotating magnetic field is produced as a sum of the vector as shown in Fig. 1 (c) and (d). Also, the magnetic torque between the rotating magnetic field and the magnetic moment on the NdFeB permanent magnet can be express as

$$T = MH \sin\theta \ \text{[Nm]} \tag{1}$$

where M is the magnetic moment of the magnet, H is the rotating magnetic field, and θ is the angle between m and H.

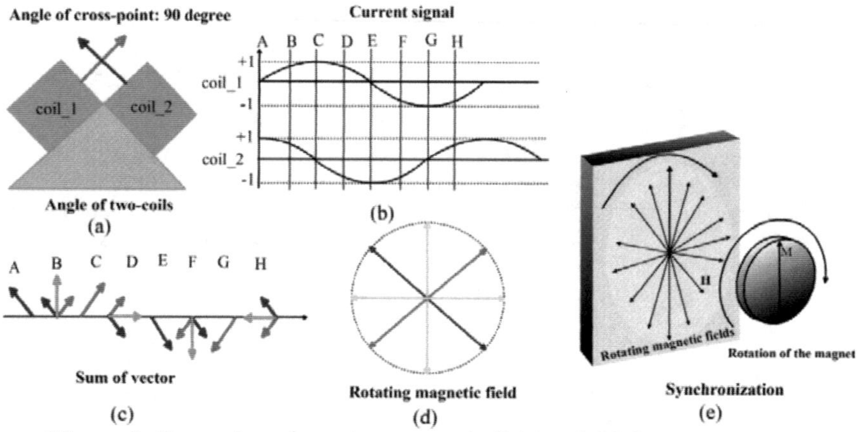

Figure 1. Generation of rotating magnetic field and driving method

3. Basic theory of centrifugal pump and impeller

A centrifugal pump is analyzed by the angular momentum theory and the principle of moment of momentum. The centrifugal pump's main idea is the energy conversion of kinetic energy to pressure energy. The amount of energy given to the liquid is proportional to the velocity at the edge or blade tip on the impeller. Figure 2 illustrates velocity at the edge or blade on the single impeller. Characteristics of the impeller or pump are dependent on the blade shape which is decided by blade-angle β_2. According to angular momentum theory, torque and power can be express as

$$T_{torque} = \rho Q(r_2 v_2 \cos\alpha_2 - r_1 v_1 \cos\alpha_1) \tag{2}$$

$$\begin{aligned} P_{power} &= T_{torque} \times \omega = \rho g Q H_p \\ &= \rho Q(u_2 v_2 \cos\alpha_2 - u_1 v_1 \cos\alpha_1) \end{aligned} \tag{3}$$

Where ρ is fluid density, Q is volume of flowing liquid, H_p is pump head, g is gravity, and ω is angular velocity. To obtain real pump head, we assume the angle ($\alpha_1 = 90^{\circ}$). In this case, the pump head (H_p) can be express as

$$H_p = \frac{1}{g} u_2 v_2 \cos\alpha_2 \tag{4}$$

To analyze influence of the blade-angle (β_2), equation 4 is converted in term of β_2, then, α_2 is converted by v_{2u}, as shown in Figure 2. The pump head (H_p) can be rewritten as

$$H_p = \frac{1}{g}(u_2^2 - u_2 v_{n2} \cot \beta_2) \tag{5}$$

The pump head is dependent on the angle (β_2) according to three kinds of conditions (at constant rpm).

1. $\beta_2 > 90^\circ$: $\cot \beta_2 < 0$ and $v_{n2} \cot \beta_2 < 0$, Head is increased according to decrease of flow rate.
2. $\beta_2 = 90^\circ$: $\cot \beta_2 = 0$ and $v_{n2} \cot \beta_2 = 0$, Head is a constant value without relation to flow rate.
3. $\beta_2 < 90^\circ$: $\cot \beta_2 > 0$ and $v_{n2} \cot \beta_2 > 0$, Head is decreased according to increase of flow rate.

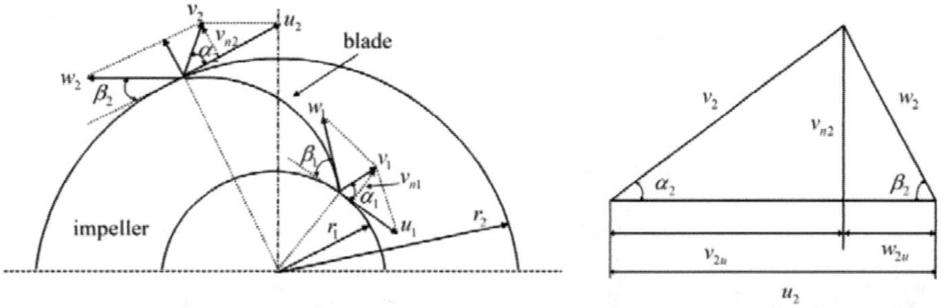

Figure 2. Velocity diagram between input and output of impeller: w is the relative velocity of fluid particle, v is the absolute velocity of fluid particle, u is the peripheral velocity, r is the radius, α is the angle between u and v, and β is the angle of the blade.

4. Pump design and basic property

To realize a wireless pump and to overcome various problems, the proposed pump consists of a multi-stage impeller and an NdFeB permanent magnet (diameter: 18.8 mm and thickness: 4 mm). The pump does not require a shaft and bearing because its impeller is of a floating type. Therefore, general mechanical problems are overcome. The magnetic pump has various advantages for medical application. It is wireless, battery-free, of simple structure with no mechanical problems, and it generates no heat. The basic properties of the pump are dependent on the magnetic field and operating frequency. The cross-point of two coils and the magnetic field density decide the distance between the pump and the driving coil. Then, the frequency can control the discharge pressure. In the case of a vibrating flow pump, it is dependent on the resonant frequency. However, a centrifugal pump is proportional to the operating frequency up to saturating point. The proposed pump is possibly bi-directional in rotation according to rotating direction (counter-clockwise or clockwise) of the rotating magnetic field. In this case, flow rate and pressure is decided by equation (5). Fabricated blade-angle (β_2) on the impeller is less than 90° when the rotating direction is counter-clockwise. But when the rotating direction is clockwise, blade-angel (β_2) is more than 90°. Figure 3 shows a 3D model, fabricated impeller, and pump case. Each measurement of the single impeller is 1mm and the gap between the rotor and inner wall of the pump case is 0.2 mm. This size is an important factor in order to decide the starting torque because this rotor is of a floating type. The output port diameter decides flow rate and dynamic pressure. For example, a small diameter can increase the dynamic pressure but flow rate is decreased at a constant rpm. Figure 4 illustrates changes of pressure by discharge part on the pump case. Through the flow simulation, we can expect optimized output port size and design.

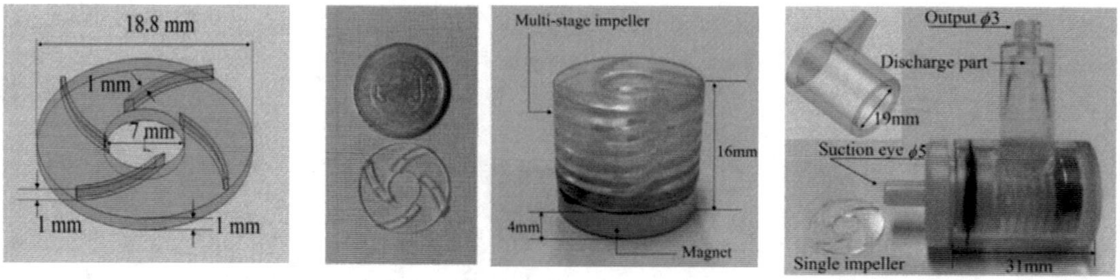

(a) (b) (c)

Figure 3. Design and fabrication of magnetic micro-pump: (a) 3D model of impeller and blades: it produces pressure energy and (b) configuration of the rotor (multi-stage impeller with disc type NdFeB permanent magnet): rotation of the rotor brings about the centrifugal force in the pump, and (c) fully implemented magnetic micro-pump

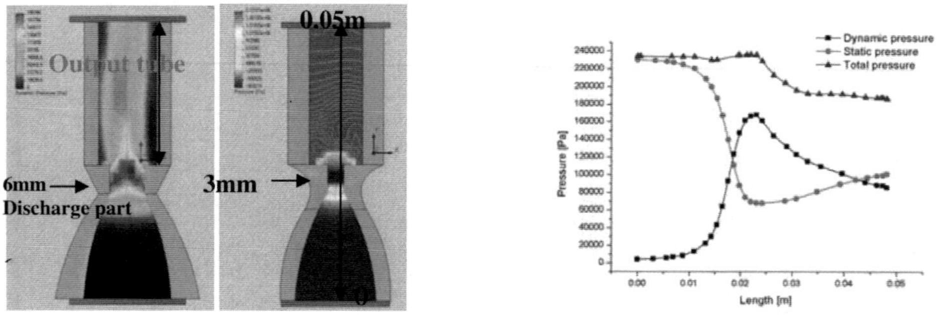

(a) (b)

Figure 4. Flow dynamics simulation of output port: (a) pressure distribution of diameter of discharge part (6 mm and 3 mm) and (b) results of pressures (dynamic, static, and total: discharge part 6mm)

5. Experimental results

We constructed a magnetic driving centrifugal pump system. Its configuration consists of a pump, driving coil pair, and power supply. As mentioned before, the distance between the pump and the driving coil is decided by the angle of cross-point. In this paper, all experiments are carried out at 90°. Also, the difference of phase of two current signals is fixed at 90°. To drive the magnetic rotor, the operating frequency is 10 Hz up to 100 Hz (rpm: up to 6000 rpm). In this experiment, we used two kinds of pump case (output diameters of ϕ 3 mm and ϕ 6 mm) and tubes each of ϕ 6 mm, ϕ 8 mm, and ϕ 10 mm. The magnet properties (size, magnetic moment) on the rotor are an important factor within the rotating magnetic field because it produces torque. Figure 5 illustrates these properties. First, the relation between flow rate and pressure is in inverse proportion. Second, the flow rate and pressure is proportional to the operating frequency (rpm). However, impedance on a coil is dependent on the frequency. After all, increasing the operating frequency causes a decrease in the driving current. Finally, output port size decides flow rate and pressure with the operating frequency. We compared discharge parts (ϕ 3 mm and ϕ 6 mm) for flow rate and pressure. In this case, the difference in pressure is 200 Pa and flow rate is 500 ml/min at 70 Hz when the diameter of the output tube is ϕ 6 mm and ϕ 3mm as shown in figure 5 (b). However, a discharge part of ϕ 6 mm and an output tube of ϕ 10 mm produces a high flow rate (3200 ml/min at 70 Hz and 4800 ml/min at 100 Hz), but variation of the pressure is a maximum of 400 Pa at 70 Hz as shown in figure 5 (b) and (c).

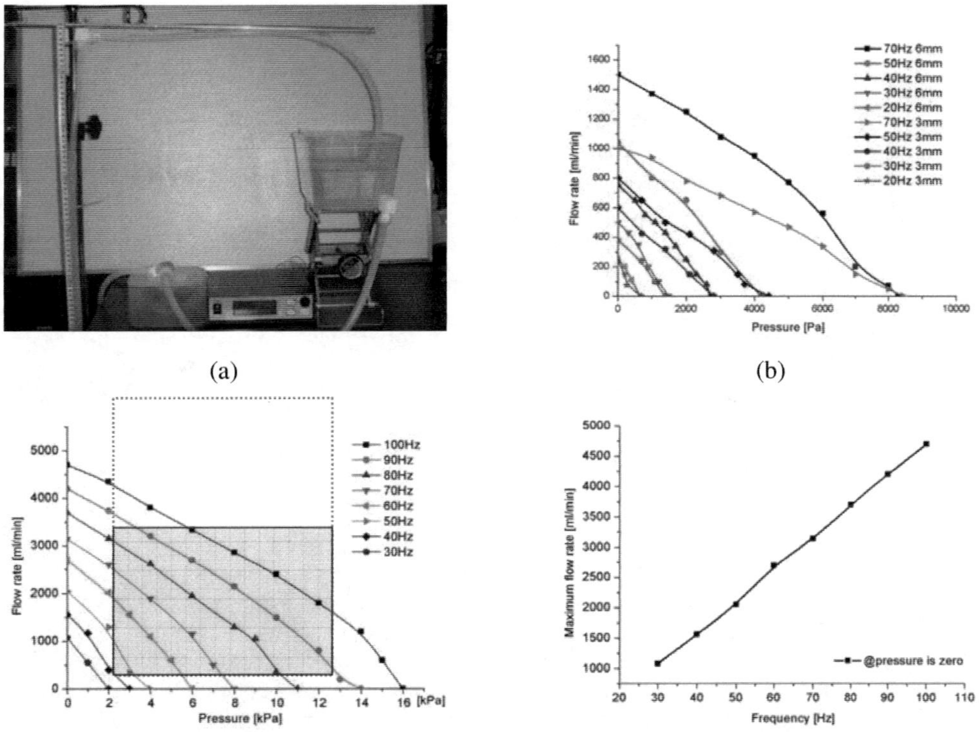

(a) (b)

(c) (d)

Figure 5. (a) Test-bed for circulation (b) the comparison of discharge parts ϕ 3mm and ϕ 6 mm, (c) the pump performance of discharge part ϕ 6 mm and output tube ϕ 10 mm: this result satisfies with the general blood pump: the red box is operating zone of the general artificial heart pump and the blue box is operating zone of the proposed pump, and (d) the change of flow rate at increasing frequency.

7. Conclusion

This proposed pump system was developed by mathematical analysis, flow simulation and experimentation. The developed pump is very small size but it produced flow rate and pressure sufficient to apply to a blood pump. Moreover, it provides a wireless and battery-free function which has important advantage for medical application. However, an innovated coil system is required to generate constant magnetic field at the fixed condition because heating problem on the coil decrease the magnetic field.

References

[1] A. Yamazaki, M. Sendoh, K. Ishiymam, K.I Arai, and T. Hayase (2003), *IEEE trans on magnetics* **39** 5
[2] D. R Reyes, D. Issofidis, P.A Auroux, A. Manz (2002), *Anal. Chem* **74** 2623-2636
[3] Jen, C. P and Y. C Lin (2002) *J. Micromech. Microeng* **12** 115-121
[4] T. Yamane (2002) *J Artif Organs* **5** 149-155
[5] Jarvik RK (1995) *Artif Organs* **19** 565-570
[6] Masuzawa T, Kita T, Okada Y (2001), *Artif Organs* **25** 395-399

Journal of Physics: Conference Series **266** (2011) 012073 doi:10.1088/1742-6596/266/1/012073

Applicability of radioactive $^{99m}Tc\text{-}O_4^-$ magnetic fluid to nuclear medicine

Jong-Hee Kim[1], Seong-Min Kim[2], Keun-Ho Kim[2] and Chong-Oh Kim[1]

[1]Research Center for Advanced Magnetic Materials, Chungnam Nat'l University, Daejeon 305-764, Korea
[2]Dept. of Nuclear Medicine, Chungnam Nat'l University Hospital, Daejeon 301-721, Korea

E-mail: sjh@cnu.ac.kr

Abstract Magnetite nanoparticles were synthesized with solution of ferrous and ferric chlorides and ammonia water by sonochemical method. The hydrophilically radioactive magnetic fluids were prepared by labeling technetium pertechnetate ($^{99m}Tc\text{-}O_4^-$) and then adsorbing alginic acid on the magnetite particles. In order to measure some properties of the dispersed particles, the magnetic fluids were freezed down to -70 °C, and were dried in vacuum. The total size of the particles was about 15 nm with the core diameter of 12 nm and their superparamagnetic saturation magnetization was 63 emu/g for the core-shell of Fe_3O_4/Algin and 52 emu/g for that of $Fe_3O_4/^{99m}Tc\text{-}O_4^-$/Algin. The labeling of radioactive $^{99m}Tc\text{-}O_4^-$ to the magnetite particles was efficient to about 70 %. The fluid of magnetic particles on which the radioisotopic substance is labeled with such an efficiency level may be applied as a tracer for diagnosis in nuclear medicine.

1. Introduction

Functional and biocompatible particles can be prepared by conjugating organic matters to an inorganic core. In general, such a core-shell combination is important in nano-bio technology: e.g., superparamagnetic magnetite nanoparticles [1] are moved in an external magnetic or electromagnetic field and adsorbing chemically drugs onto a colloid carrier [2] influences favorably their properties by decreasing the toxicity because a side effect due to direct administration can be avoided [3,4].

It is well-known that in order to transport the drugs to organs and tissues via blood flow, the intravenous injection is very effective. From this point of view, the permeability of colloid particles through the blood vessel is important in delivering the drugs to the targeted area for diagnosis and treatment. Since the capillary endothelial lining cells are finely connected in nervous systems of back bone, the very small particles can pass through the vessel wall and diffuse toward the outside. Therefore, a limit size of the particles is only about 20 nm for the kupffer cell in liver and the reticuloendothelial system of macrophage in spleen, medulla and lymph. The particles above the size of 50 nm are practically difficult to diffuse freely to the internal tissue because the tissues of muscle, skin and lung possess a continuous capillary vessel [5]. In order to be delivered to application parts by passing the endoderm of capillary and to be discharged out of the living body, it is desired that the magnetic particles are spherical with 10~15 nm diameter. Accordingly, the magnetic fluids need to be superparamagnetic with high saturation magnetization and also to satisfy simultaneously the

Journal of Physics: Conference Series **266** (2011) 012073 doi:10.1088/1742-6596/266/1/012073

controlled particle size, the biocompatibility and the colloidal dispersibility for medical applications [6,7].

The substitution of Fe elements in the magnetite with a nuclide of copper is very effective for tumor therapy due to its β-emission and also γ-radiation close to the energy of Technetium-99m(^{99m}Tc) [8]. The ^{99m}Tc is widely used as a radioisotope-tracer for diagnosis in nuclear medicine because it radiates only a single gamma ray and has a short physical half life of 6.02 h. In this work, superparamagnetic magnetite nanoparticles were synthesized by sonochemical method, whose colloid was obtained by adsorbing alginic acid ($C_6H_8O_6$), a biocompatible surfactant, on the magnetic particles. Technetium pertechnetate ($^{99m}Tc\text{-}O_4^-$) was labeled as a diagnosis agent on the magnetic particles. The applicability of the radioactive magnetic fluids was investigated by estimating the medical effect with calculation of the labeling efficiency.

2. Experimental

$FeCl_2$• $4H_2O$(1.72 g) and $FeCl_3$• $6H_2O$(4.70 g) with a stoichiometric ratio of 8.65 mmol Fe^{2+}/17.30 mmol Fe^{3+} = 0.5 were solved in 80 ml of distilled water. The black precipitate of magnetic oxide was obtained by adding up to pH 10 ammonia water (28~30 % pure) under ultrasonication of 20 kHz at 140 W for 1 h to the mixed iron solution heated previously to 80 °C under Ar atmosphere [9]. The black gel was washed ten times by magnetic decantation until the pH value reached around 8. 50 ml of distilled water was added to the washed gel, which was heated to 80 °C under Ar atmosphere again. After 0.4~0.5 ml of technetium pertechnetate corresponding to triple of the practical dosage was diluted with 5 ml of saline, the radioactive solution was poured at a time into the gel solution in the ultrasonication of 30 min. The suspension was washed five times by magnetic decantation to remove the $^{99m}Tc\text{-}O_4^-$ substance unlabeled on the particles. 60 ml of saline was added to the labeled magnetic particles, which was reheated to 80 °C under Ar atmosphere. The stable colloid was prepared by adding 25 ml of alginic acid 0.2 g to the solution of labeled particles in the ultrasonication of 1 h. The colloid volume was about 80 ml when cooled to room temperature. A process for preparing the $^{99m}Tc\text{-}O_4^-$ labeled magnetic fluids by ultrasonication is shown in figure 1.

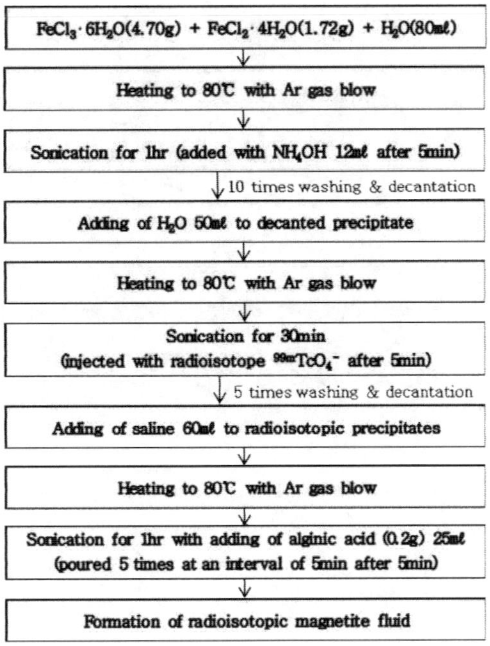

Figure 1. Preparation process of technetium pertechnetate-labeled magnetic fluids by sonochemical method.

Microstructure of the fluid particles were observed by transmission electron microscopy (TEM) where the crystallinity was analyzed by selected-area electron diffraction (SAED) and crystal phase of the particles were analyzed by X-ray diffraction (XRD). Their magnetic properties were measured in the field range of ±10 kOe using vibrating sample magnetometer.

3. Results and Discussion

In general, when a cation surfactant are adsorbed on the particles of iron oxide, the polar radicals of internal chemical adsorption layer are attached directly on the surface of core particles inwards and those of external physical adsorption layer face towards the outside. Since the weakly bonded external layer can be removed by washing with acid solution, the chemically bonded internal layer finally remains as the shell. The addition of anion surfactants to the colloid with this mono-molecular layered particles leads to the formation of the dual adsorption layer on the core particles by which the hydrophilic group is located outwards. Alginic acid, a kind of polysaccharide, possesses excellent chemical affinity due to the typical hydrophilic group. The water-based magnetic fluids can be prepared even with the monolayer of alginic acid on the magnetic particles which is formed by the adsorption process of only one step.

In order to obtain the layered particles from the magnetic fluids, the fluids were freezed down to -70 °C and then dried in vacuum using freezing drier. In result, the ratio of weight of layered particles to fluid volume, i.e. colloid concentration was 27 mg/ml. Since the ratio of weight of core particles to fluid volume is 25 mg/ml from the formation amount of magnetite compound of 0.00865 mol × 231.54 g/mol = 2.0 g, the amount of shell is 2 mg/ml and totally becomes 160 mg. This shell weight corresponds mostly to that of the surfactant layer. Therefore, the addition of alginic acid 0.2 g seems to be sufficient to prepare the water-based magnetic fluid in this experimental condition.

Figure 2 shows that both of the magnetic particles with and without $^{99m}Tc\text{-}O_4^-$ are spherical with the mean core diameter of 12 nm and the layer thicknesses of 2 nm to 4 nm, in which the SAED results exhibit high crystallinity.

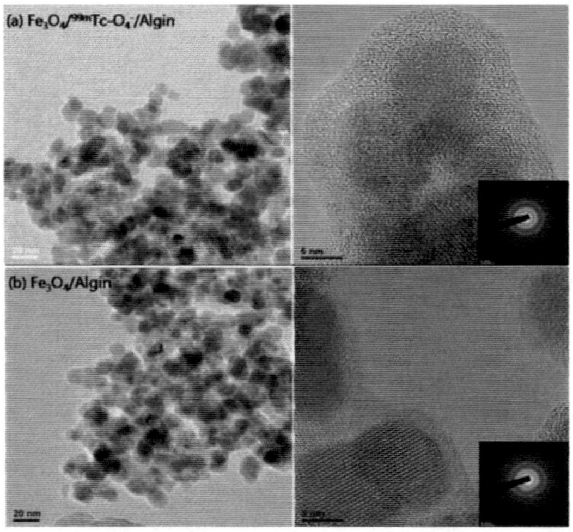

Figure 2. TEM images of magnetic particle fluids algin-adsorbed (a) with and (b) without $^{99m}Tc\text{-}O_4^-$ labeling. The inserted SAED patterns represent high crystallinity.

In figure 3, the XRD analysis revealed that the prepared fluid particles were highly crystallized as in the SAED with a Fe_3O_4(magnetite) phase. Although the Fe_3O_4 may coexist with a γ-Fe_2O_3(maghemite) phase because their diffraction pattern is much the same [10], the iron oxide seems to consist mainly of Fe_3O_4 in the process environment of Ar atmosphere. Also, the color of the synthesized powder was very close to the black of magnetite rather than the brown of maghemite. Lattice constants of the magnetite particles algin-adsorbed with $^{99m}Tc\text{-}O_4^-$ label and with no label were measured as 0.83736 nm and 0.83727 nm, respectively. Such almost the same values in the lattice parameter suggest that labeling of a very small amount of technetium constituent has no effect on the crystalline state of magnetite particles.

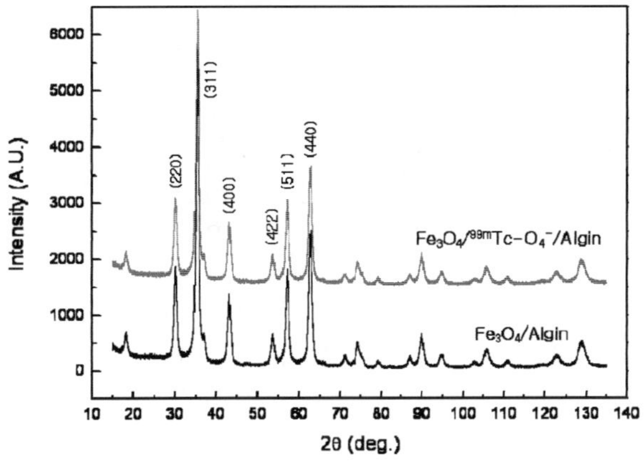

Figure 3. XRD patterns of magnetic nanaoparticles algin-adsorbed with and without $^{99m}Tc\text{-}O_4^-$ labeling. The peak indices indicate crystal planes of magnetite.

Figure 4 shows that the synthesized particles are superparamagnetic with no remanent magnetization. Also, saturation magnetization for the core-shell systems of Fe_3O_4/Algin and $Fe_3O_4/^{99m}Tc\text{-}O_4^-$/Algin was 63 emu/g and 52 emu/g, respectively, while that of the only core particles of Fe_3O_4 was 66 emu/g. This decrease in the magnetization value is caused practically by the increse of weight ratio of nonmagnetic alginic acid and technetium pertechnetate layers.

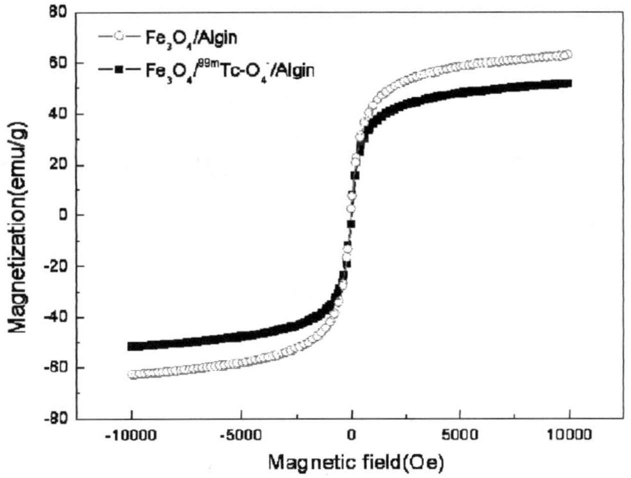

Figure 4. Magnetization curves of magnetite nanoparticles algin-adsorbed with and without $^{99m}Tc\text{-}O_4^-$ labeling.

In general, the decay activity A_d of radioisotope is calculated as in equation (1).

$$A_d = A_i\,(1/2)^{T/T'} \qquad (1)$$

where A_i is the initial activity measured from a radioactive substance in radiation counter before labeling, T is the time elapsed until the last measurement after labeling and T' is the physical half life of radioisotope. The percentage of labeling efficiency L_{eff} is evaluated with A_d as in equation (2).

$$L_{eff} = A_t / A_d \times 100 \qquad (2)$$

where A_t is the terminal activity measured from a radioisotope-labeled material in radiation counter. The addition of technetium pertechnetate to the magnetite fluids was carried out three times repeatedly for reliablity. Table 1 represents the numerical values of main factors observed in the labeling experiment.

Table 1. Data obtained in labeling of $^{99m}Tc\text{-}O_4^-$ (phys. half time T': 361.2 min) on magnetite nanoparticles.

Experi. No.	Initial activity A_i (mCi)	Elapsed time T (min)	Decay activity A_d (mCi)	Terminal activity A_t (mCi)	Labeling efficiency L_{eff} (%)
1	35.6	120	28.3	20.4	72.1
2	33.0	160	24.3	16.6	68.3
3	22.7	250	14.1	9.6	68.1

The labeling efficiency of technetium pertechnetate was satisfactory at the level of about 70 % even though the improved adsorption of the radioisotope would be more effective in observing the internal organs. It is suggested that the $^{99m}Tc\text{-}O_4^-$ is chemically well bonded on the surface of the magnetite particles. It has been also reported that the adsorption rate of 5-aminolevulinic acid for photodynamic therapy onto the magnetite nanoparticles was only 12.5 % [11]. A distribution experiment on animals of such radioactive fluids will be hereafter carried out to apply to nuclear medicine.

4. Conclusions

Magnetite nanoparticles were chemically synthesized with the addition of ammonia water to a solution of ferrous and ferric chlorides in ultrasonication. Radioactive magnetic fluids were prepared by labeling technetium pertechnetate ($^{99m}Tc\text{-}O_4^-$) and then adsorbing alginic acid on the particles. The saturation magnetization of superparamagnetic magnetite particles was 63 emu/g for the core-shell system of Fe_3O_4/Algin and 52 emu/g for that of Fe_3O_4/$^{99m}Tc\text{-}O_4^-$/Algin. The size of the fluid particles was 12 nm for the diameter of magnetite core and 2~4 nm for the thickness of $^{99m}Tc\text{-}O_4^-$ and algin. The labeling efficiency of $^{99m}Tc\text{-}O_4^-$ on the magnetite core particles was a high level of 70 %. Therefore, the $^{99m}Tc\text{-}O_4^-$ labeled magnetite fluids could be potentially used as a tracer for diagnosis in nuclear medicine.

Acknowledgments
This work was supported from Research Center for Advanced Magnetic Materials of Chungnam National University and by Basic Research Program of Agency for Defense Development.

References

[1] Xiaoxiao He, Kemin Wang, Weihong Tan, Dan Xiao, Xiaohai Yang and Jun Li 2001 *Proc.*

SPIE **4414** 394
[2] Zhengrong Cui and Russell J. Mumper 2002 *Bioconjugate Chem.* **13**(6) 1319
[3] Dunkan R, Dimitoijevic S and Evagorou EG 1996 *STP Pharma Sci.* **6** 237
[4] Couvreur P, Kante B, Grislain L, Roland M and Speiser P 1982 *J. Pharm. Sci.* **71**(7) 790
[5] Jeong Y K 2002 *J. Korean Powder Metallurgy Institute* **9**(3) 141
[6] Lee S W, Woo K J and Kim C S 2004 *J. Magnetics* **9**(3) 83
[7] Lacava Z G M, Azevedo R B, Martins E V, Lacava L M, Freitas M L L, Garcia V A P, Rébula C A, Lemos A P C, Sousa M H, Tourinho F A, Da Silva M F and Morais P C 1999 *J. Magn. Magn. Mater.* **201** 431
[8] Chong-Oh Kim, Jong-Hee Kim, Yuqiang Huang and Sang-Im Park 2008 US Patent 7407673
[9] Park S I, Lim J H, Hwang Y H, Kim J H, Kim C G and Kim C O 2007 *phys. stat. sol. (a)* **204**(12) 3913
[10] Jongnam Park, Kwangjin An, Yosun Hwang, Je-Geun Park, Han-Jin Noh, Jae-Young Kim, Jae-Hoon Park, Nong-Moon Hwang and Taeghwan Hyeon 2004 nature materials **3** 891
[11] Park S I, Lim J H, Kim J H, Yun H I and Kim C O 2005 *IEEE Transactions on Magnetics* **41**(10) 4111

Journal of Physics: Conference Series **266** (2011) 012074 doi:10.1088/1742-6596/266/1/012074

Effect of nitrogen contents on the high-frequency characteristics and magnetic properties of FeHfN films

Yu-Ming Kuo and Jenq-Gong Duh*

Department of Material Science and Engineering, National Tsing Hua University, Hsinchu 30013, Taiwan
e-mail: jgd@mx.nthu.edu.tw

Abstract. High permeability magnetic films can enhance the inductance of thin-film inductors in DC-DC converters. In order to obtain high permeability, uniaxial anisotropy and coercivity should be as low as possible. This study employed dc reactive magnetron sputtering to fabricate the nanocrystalline FeHfN thin films. The influence of nitrogen flows on composition, microstructure, permeability characteristics, as well as magnetic properties was investigated. Increasing nitrogen contents can alter the FeHfN films from amorphous-like to crystalline phases, and the magnetic properties and permeability depend on the variations of microstructures. With this optimum condition of the nitrogen flows at 1.2 sccm, low anisotropy (H_K = 20 Oe), low coercivity (H_C = 1.1 Oe) and permeability (μ' > 600 at 50 MHz) were obtained for the nanocrystalline FeHfN film with a thickness around 700 nm. The as-fabricated FeHfN films with permeability over 600 at 50 MHz should be promising for high-permeability ferromagnetic material applications.

1. Introduction

The magnetic devices and thin-film inductors have been confronted with a strong demand for further miniaturization in DC-DC converters. High-permeability magnetic films can reduce the dimensions of the magnetic devices and enhance the inductance of thin-film inductors [1-3]. According to the Stoner-Wohlfarth theory [1], the permeability is defined as

$$\mu_r = \frac{4\pi M_S}{H_K} \tag{1}$$

where μ_r is relative permeability, M_S is saturation magnetization, H_K is in-plane uniaxial anisotropy, and H_K should be as low as possible to increase the permeability for inductors applications.

FeCo-based soft magnetic thin films featuring high saturation magnetization, low coercivity and high electrical resistivity have attracted much attention and are extensively incorporated in many applications, such as micro-magnetic devices and magnetic sensors [4–7]. However, the FeCo alloys have relatively high magnetocrystalline anisotropy constant (K_1) up to 10 kJ/m^3 [8]. Thus, the uniaxial anisotropy of the FeCo-based soft magnetic films is unacceptable for high-permeability applications. Recently, Fe-based soft magnetic thin films have been employed in thin-film inductors, exhibiting high saturation magnetization and low uniaxial anisotropy [9]. The FeHfN system is selected for further investigation in this study.

In this work, to adjust the nitrogen contents is exploited to improve the soft magnetic properties in FeHfN, aiming to obtain high permeability for thin-film inductors applications. The effects of the variations of nitrogen contents on composition, microstructure, anisotropy and coercivity will be investigated. In addition, the correlations of the above-mentioned properties with the permeability characteristics of FeHfN films are discussed.

2. Experimental

The FeHfN films were deposited on Si (100) substrates by dc reactive magnetron co-sputtering with Ar + N_2 gases in a working pressure 1.3×10^{-3} Torr. The power of the Fe and Hf targets was kept 200 and 40 W, respectively. The FeHfN films were prepared with various nitrogen flows from 1.0 to 2.0 sccm. The film thickness was measured by a surface profiler (α-step), and all the films had a thickness about 700 nm. Magnetic properties were measured with a vibrating sample magnetometer (VSM), and the permeability was evaluated using a PMF-3000 permeameter with a maximum frequency of 3 GHz. The composition of FeHfN films was quantitatively analyzed with a field emission electron probe microanalyzer (FE-EPMA, JEOL JXA-8500F). The microstructure of films was further revealed by transmission electron microscopes (TEM, JEOL JEM-2010) and the X-ray diffractometer (XRD).

3. Results and discussion

Figure 1 shows the quantitative analysis of the FeHfN films with various nitrogen flows. When the nitrogen flows increase from 1.0 to 2.0 sccm, the Fe contents are nearly fixed at 81 at%. With the increases of nitrogen flows, the N contents increase almost linearly from 7.17 to 10.97 at% and the Hf contents decrease from 10.59 to 8.59 at%. It is revealed that the N contents in the film are proportional to the N_2 flows.

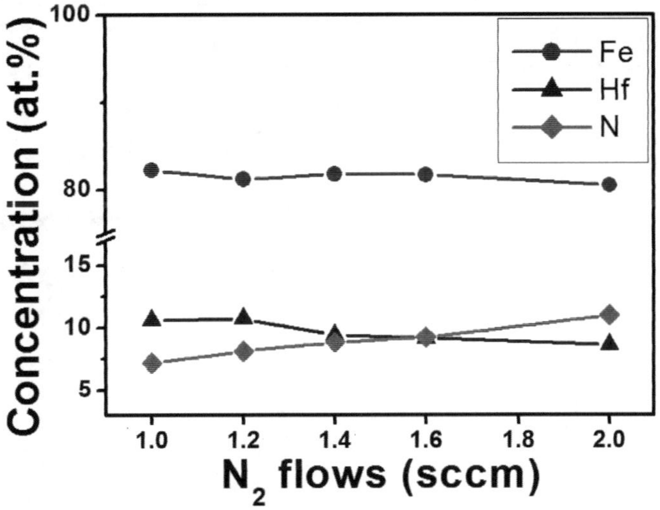

Fig. 1. The compositions of FeHfN films deposited at various nitrogen flows.

To realize the structure of the film, the XRD measurements were performed. Fig. 2 shows the XRD patterns of FeHfN films at various N_2 flows from 1.0 to 2.0 sccm. Only the diffraction peak of the Fe (110) is found in the diffraction patterns, indicating that the (110) texture forms in the thin films. The (110) diffraction peak becomes sharper with increasing N_2 flows, implying an increase in grain size with increasing N contents. It appears that increasing N contents enhance the crystallization of the FeHfN films from amorphous-like (N_2: 1.0 sccm) to crystalline (N_2: 2.0 sccm) phases.

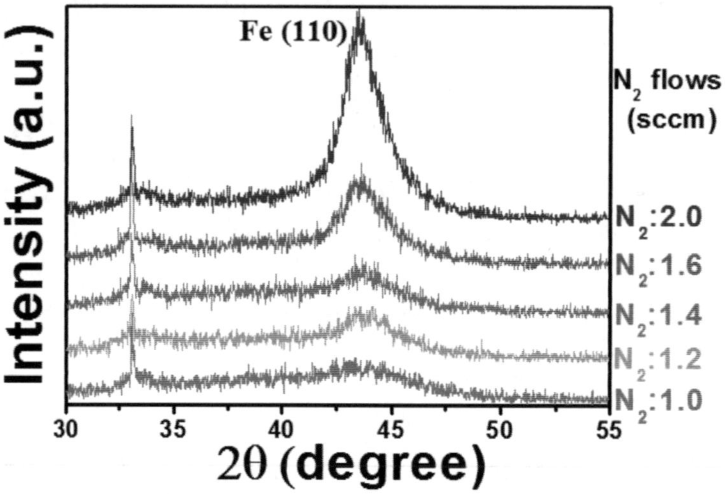

Fig. 2. XRD patterns of FeHfN films at various nitrogen flows from 1.0 to 2.0 sccm.

The variations of anisotropy (H_K) and coercivity (H_C) with the N_2 flows in FeHfN films are illustrated in Fig. 3. At N_2 flows of 1.0 sccm, the H_C is relatively large. For an ultrafine grain (amorphous-like), the distance within one grain and from adjacent grains is too long. Hence, the ferromagnetic exchange force within the grains is not strong enough to hold up the neighbouring magnetic moments, bringing out a large H_C. As compared with the sample with 1.2 sccm N_2 flows, the H_C is small. With the increases of N contents, the grain size increases to 3-4 nm. Increasing grain size makes the distance within the grains shortened. The ferromagnetic exchange force becomes strong enough to rotate the neighbouring magnetic moments, exhibiting a small H_C.

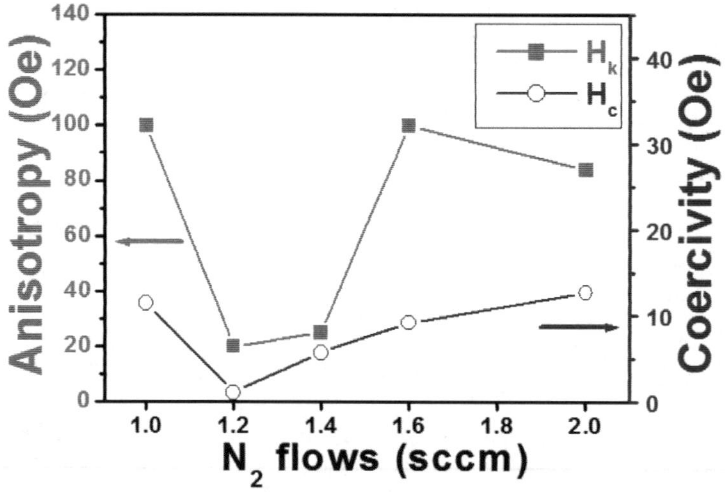

Fig. 3. Anisotropy and coercivity of the as-deposited FeHfN films as a function of nitrogen flows.

At N_2 flows > 1.2 sccm, the H_C increases further as the N contents increase. Based on the random anisotropy model [10-11], the H_C is defined as

$$H_C = p_c \frac{\langle K \rangle}{M_S} \sim \frac{p_c K_1 D^6}{M_S A^3} \qquad (2)$$

where $<K>$ is average anisotropy, M_S is the saturation magnetization, K_1 is the magnetocrystalline anisotropy constant, D is grain size, A is the exchange constant and p_c is a dimensionless quantity related to the crystal structure. The H_C is proportional to the D^6. At N_2 flows > 1.2 sccm, the grain size is raised with the increases of N contents, as shown in Figs. 2. Therefore, the increases of N contents make the grain size grow and a large H_C is achieved. In addition, the H_K and H_C exhibit the similar trends, as shown in Fig. 4. In fact, Eq. (2) presents the proportional relationship between H_C and anisotropy ($<K>$). It is argued that both H_K and H_C might be affected by grain size.

With respect to the permeability in this series of FeHfN films, lower H_K is observed in the sample with the N_2 flows of 1.2 sccm, as shown in Fig. 3. According to the Stoner-Wohlfarth theory [1], permeability is inversely proportional to the anisotropy. Therefore, the permeability of the sample with 1.2 sccm N_2 flows is expected to be relatively higher than that of other samples. Fig. 4 shows the permeability spectra of the films deposited at N_2 flows of 1.2 sccm. The permeability (μ') can maintain above 600 for frequencies ranging from 10 to 100 MHz. Besides, the cut-off frequency ($f_{cut\ off}$) is 466 MHz, and the ferromagnetic resonance frequency (f_{FMR}) is around 640 MHz.

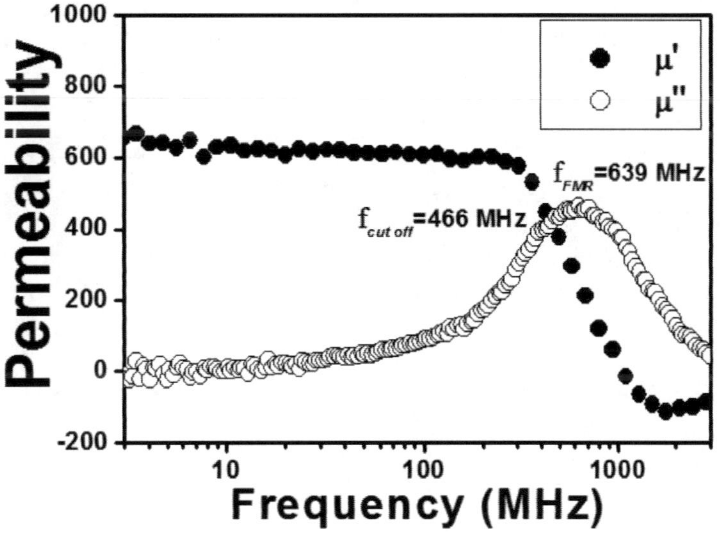

Fig. 4. Frequency dependence of complex permeability spectra for the FeHfN thin films deposited at nitrogen flows of 1.2 sccm.

In summary, the FeHfN film prepared with 1.2 sccm N_2 flows exhibits excellent soft magnetic and permeability properties. The anisotropy and coercivity are only 20 and 1.1 Oe, respectively, and the permeability was up to 600 at 50 MHz.

4. Conclusions

FeHfN soft magnetic thin films with various N contents were fabricated by varying the N_2 flows. This study demonstrated that FeHfN films with favorable high permeability characteristics were successfully deposited by dc sputtering. The N contents may play an important role in affecting magnetic and permeability properties. The microstructures of FeHfN films changed from amorphous-like (1-2 nm) to crystalline (4-6 nm) phases by increasing the N contents. The variations of grain size would influence the H_K and H_C. The minimum value of H_K and H_C was obtained when the N_2 flows reached 1.2 sccm. As a result, the permeability at 1.2 sccm N_2 flows was over 600 at 50 MHz. It is revealed that the FeHfN films possess potential applications in high-permeability electromagnetic devices.

Acknowledgement
This work was financially supported by Ministry of Economics, Taiwan, under the contract No. 98-EC-17-A-08-S1-003.

References
[1] V. Korenivski, *J. Magn. Magn. Mater.*, **215**, 800 (2000).
[2] M. Yamaguchi, K. Suezawa, K. I. Arai, Y. Takahashi, S. Kikuchi, Y. Shimada, W. D. Li, S. Tanabe, K. Ito, *J. Appl. Phys.*, **85**, 7919 (1999).
[3] I. Fergen, K. Seemann, A. V. D. Weth, A. Schűppen, *J. Magn. Magn. Mater.*, **242**, 146 (2002).
[4] N. X. Sun, S. X. Wang, *J. Appl. Phys.*, **92**, 1477 (2002).
[5] S. Ohnuma, N. Kobayashi, T. Masumoto, *J. Appl. Phys.*, **85**, 4574 (1999).
[6] Y. M. Kuo, C. C. Lee, J. G. Duh, *Mater. Lett.*, **64**, 759 (2010).
[7] Y. M. Kuo, C. C. Lee, J. G. Duh, *Appl. Surf. Sci.*, **256**, 6437 (2010).
[8] N. X. Sun, Q. F. Xiao, B. York, *J. Appl. Phys.*, **97**, 10F906 (2005).
[9] Y. Hayakawa, A. Makino, H. Fujimori, A. Inoue, *J. Appl. Phys.*, **81**, 3747 (1997).
[10] G. Herzer, *IEEE Trans. Magn.*, **26**, 1397 (1990).
[11] S. Thomas, S. H. Al-Harthi, D. Sakthikumar, I. A. Al-Omari, R. V. Ramanujan, Yasuhiko Yoshida, M. R. Anantharaman, *J. Phys. D: Appl. Phys.*, **41**, 155009 (2008).

Optimal Design of a Direct Driven Slotless Tubular Linear Generator for Renewable Energy Extraction

Cheng-Tsung Liu[1], H N Lin[1], H C Yeh[1] and C C Hwang[2]

[1]Dept. of Electrical Eng., National Sun Yat-Sen University, Kaohsiung 80424, Taiwan
[2]Dept. of Electrical Eng., Feng Chia University, Taichung 40724, Taiwan

E-mail: ctliu@ieee.org

Abstract. This paper is aimed to present the systematic design scheme of a direct driven slotless tubular linear generator for renewable energy retraction. To reduce operational cogging and increase energy conversion efficiency, the generator stator is equipped with slotless concentrated annular multiphase windings and its mover is encircled by quasi-Halbach arranged permanent magnets. By systematic comparisons and classifications, the preliminary design and operational specifications of supplying maximum flux that can link the stator windings are fulfilled by setting appropriate factor level combinations from the Taguchi's method. Along with the stator winding selections, potential energy generations from the proposed slotless tubular linear generator can then be estimated. Verified by three-dimensional finite element analyses, the constructed machine prototype can show its adequacy for the operational and design requirements.

1. Introduction

With the demands for generating electricity from various carbon-free renewable energy sources, many efforts have been developed on the designs of appropriate mechanical and electrical conversion systems that can fulfill these requirements. A tubular linear generator (TLG) [1], with slotless stator windings and quasi-Halbach mover permanent magnet (PM) arrangement [2], [3] that can operate in reciprocating manner to incorporate with those reciprocating energy extraction mechanisms, will be reported in this paper. In addition, with specific operational requirements and constraints, a systematic scheme that can assist to design the most appropriate slotless tubular linear generator (STLG) will also be illustrated. Aside from empirical basis, the idea of properly selecting sets of machine parameters that can supply the optimal combinations with desired performance as commonly adopted in the quality engineering, namely as Taguchi's method [4], will be introduced to perform the relative machine parameter selections and performance assessments.

2. The Slotless Tubular Linear Generator

Figure 1 shows the conceptual structure of a STLG. The main flux direction in the mover is determined by the magnetization of permanent magnets and the generated voltage level is dominated by the interactions between stator and mover. For those STLGs with traditional axial and radial magnetizations, ferromagnetic materials are generally required to guide the desired flux paths. Inevitably, the operational temperatures and the induced eddy currents in these materials will produce

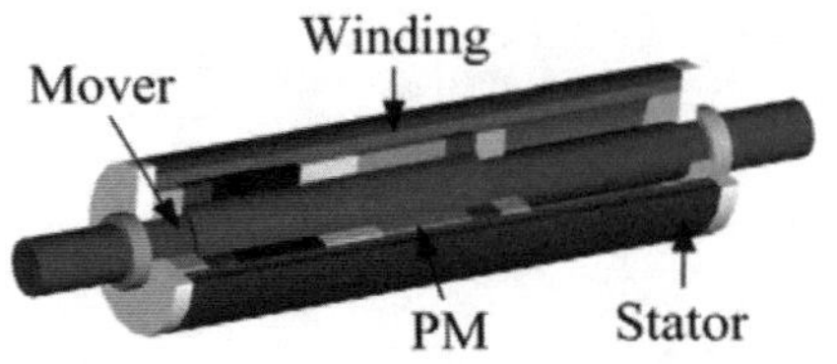

Figure 1. Conceptual structure of a slotless tubular linear generator with quasi-Halbach arranged permanent magnets on the mover.

extra system power losses and also degrade the PMs. For construction simplicity and cost reduction, as well as provide flexibility in the machine axial rod material selections, the quasi-Halbach PM arrays have been selected to fulfill the STLG mover requirements [5], [6]. Due to the asymmetrical stator winding arrangements and mechanical end effects, the induced voltages will be unbalanced in the stator three-phase windings. Besides, with position-dependent operational flux linkages, harmonics on the voltages will also affect the operations of such direct driven STLG. Proper performance indices on the voltages at both the generator AC side and the regulated DC side must be incorporated in the determinations of adequate machine parameters.

3. Optimal Design of the Generator

3.1 Problem Definitions

Figure 2 shows the related design parameters of the STLG. The objective is to design a laboratory-class generator prototype capable of supplying 1/2 hp of power and a minimum regulated DC-side voltage of 60 V. By selecting the key optimization factors and assigning every factor with three feasible levels, the classified factor-level combination for design are illustrated in Table 1.

To find the optimal parameters, the operational objective along with certain constraints are:

$$\begin{cases} Maximize: \ \phi(r_\phi, z, d_m) = \int_S \vec{B}(r, \theta, z, d_m) \cdot d\vec{s} \\ Subject\ to: \ 0 \le z \le l_s, \ 0 \le d_m \le l_s - l_m, \ r_a \le r_\phi < r_c \end{cases}, \tag{1}$$

where ϕ is the axial-directional flux passing through the stator winding and $\vec{B}$ is the flux density enclosed by the corresponding winding coil.

A systematic and efficient search of the optimal factor-level combinations based on the Taguchi's method is developed. There are only total 9 experiments are designed for performing the desired performance index evaluations. The axial-directional magnetic fluxes passing the stator windings of the STLG, can then be calculated by using three-dimensional (3-D) finite element analyses (FEA) [7] with one of the simulated results shown in Figure 3. The exhibited magnetic fluxes and the total harmonic distortion (THD) can be expressed in the forms:

$$\phi(r_\phi, z, d_m) = F_0 + \sum_{n=1}^{\infty} F_n \cos(2\pi n \frac{d_m}{l_m} - \theta_n) \ , \ \text{and THD} = \left(\frac{1}{F_1} (\sum_{n=2}^{\infty} F_n) \right)^{1/2} . \tag{2}$$

where F_i is the magnitude of the stator winding flux at the corresponding i^{th} space harmonic.

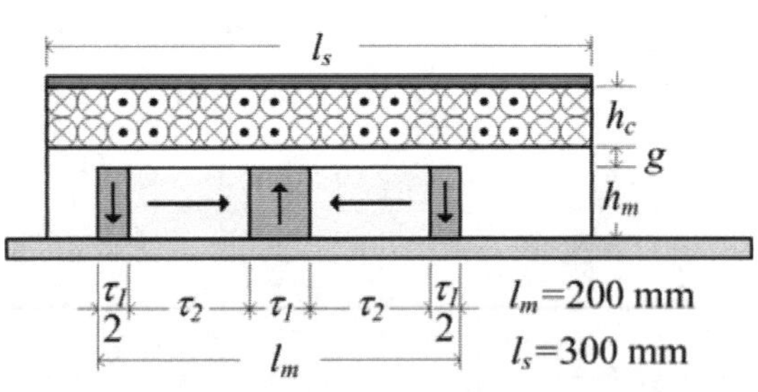

Figure 2. Physical parameters of the STLG.

Table 1. Design factor and level classifications of the slotless tubular linear generator

Design factor	Level 1	Level 2	Level 3
A: τ_1	42 mm	33.4 mm	25 mm
B: h_m	10 mm	11 mm	12 mm
C: h_c+g	9 mm	10 mm	11 mm
D: h_c/g	1.57	1.25	1.0

3.2 Optimization Specifications

By taking either the fundamental component or the THD as the target, different optimization designs can be obtained with the signal-to-noise ratio (SNR) at two distinct objectives as:

$$\textit{Smaller the better}: SNR_{STB} = -10\log(\frac{1}{n}\sum_{i=1}^{n} y_i^2),\ \text{or}\ \textit{Larger the better}: SNR_{LTB} = -10\log(\frac{1}{n}\sum_{i=1}^{n}\frac{1}{y_i^2}). \quad (3)$$

where n is the number of experiments and y_i is the ith experimental result. It is apparent that SNR_{LTB} must be picked for the flux fundamental component F_1, and SNR_{STB} must be picked for the flux THD. By taking the averaged effects contributed from individual factor-level, the resultant optimal combination for providing maximized F_1 is A3-B3-C1-D2, and for exhibiting minimized THD is A2-B3-C3-D1. Thus, to provide optimal system design guidance, compromised selections of the possible factor-level combinations are required. Since there are three levels for every of the design factor, a systematic analysis of variance (ANOVA) scheme that introduces the sum of squares is defined as:

$$SS_i = 3\sum_{j=1}^{3}(m_{ij} - m_{i,all}), \quad (4)$$

where i is the corresponding design factor, m_{ij} is the averaged *SNR* of factor i at the jth level, and $m_{i,all}$ is the averaged *SNR* of factor i at all three levels. For evaluating the impacts on the two optimization objectives, contributions from those design factors are summarized in Table 2. By selecting the dominant combinations, it can be seen that the corresponding level for the individual factor should be:

Factor A: THD → A2, **Factor B**: F_1 → B3, **Factor C**: F_1 → C1, and **Factor D**: THD → D1.

From this systematic ANOVA, the individual dominant factor-level that will affect the machine design can then be determined. Thus the combination that can provide the optimized solution about the STLG structure, with its axial-directional flux as the design objective, will be A2-B3-C1-D1.

3.3 Stator Winding Selections

Since the generated 3-phase AC voltages from STLG will be rectified to DC for further utilization, the generator stator windings are thus designed based on the desired DC-side voltage level. At an input speed of 3 m/s, if the desired DC voltage is 80 V with an adjustable range of 10%, a preliminary investigation shows both of the above devised combinations, A3-B3-C1-D2 and A2-B3-C3-D1, can achieve this voltage level by using the AWG18 or thinner wires. However, with their different physical dimensions, the compromised one with combination of A2-B3-C1-D1 requires less winding turns and thus smaller resistances, and better electric characteristics can then be expected.

From the above investigations, clearly that the selections of winding sizes will greatly affect the performance of designed STLG based on different specifications. By assembling the generator stator windings using different wires, further detailed comparisons have also been performed based on the

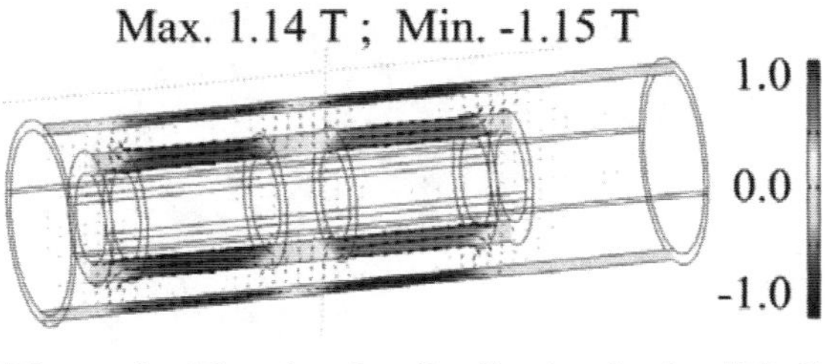

Figure 3. Flux density distribution in the STLG calculated by 3-D FEA.

Table 2. Impacts of design factors on the optimization objectives.

Design factor	F_1 SS_i	F_1 Contribution	THD SS_i	THD Contribution
A	1.4417	24.88%	0.1560	27.04%
B	3.7606	64.90%	0.2662	46.14%
C	0.5551	9.58%	0.0057	1.00%
D	0.0368	0.64%	0.1490	25.82%
Total	5.7942	100.00%	0.5769	100.00%

same structural design (A2-B3-C1-D1). Figure 4 shows the output no-load regulated DC-side voltages of the STLG at different mechanical input speeds. With the same physical space allowances, it is certain that more turns can be wound by using thinner wires and consequently higher voltages will be induced. In addition to the no-load performance, the output power generation capabilities with different stator windings have also been investigated as shown in Figures 5 and 6. From these illustrations, possibility of selecting the AWG18 wire can be eliminated due to the relatively poor power output characteristics. While from the two intersection points shown on the power-speed and power-voltage characteristics, selections of the other two types of wires for the STLG stator winding constructions are clearly determined by the objective operational speed and voltage levels.

4. Design Confirmations

The systematic design scheme as provided in the preceding section showed that the A2-B3-C1-D1 structure combined with AWG19 stator windings, which has a maximum current allowance of 2.57 A, of the STLG can meet the desired design objective, and the laboratory STLG prototype based on such design is shown in Figure 7 for demonstration. By using the same wires for stator winding constructions, further comparison investigations on the other structural parameter combinations will be conducted to validate the optimization scheme. These structures are the empirical design (A3-B1-C3-D2), the maximized fundamental flux design (A3-B3-C1-D2), the minimized total harmonic distortion design (A2-B3-C3-D1), and the optimized design (A2-B3-C1-D1) through the proposed systematic

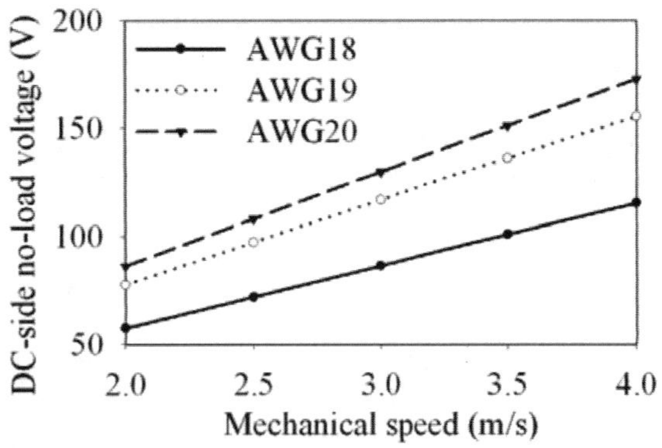

Figure 4. No-load DC-side voltages of the STLG with various driven speeds and stator windings.

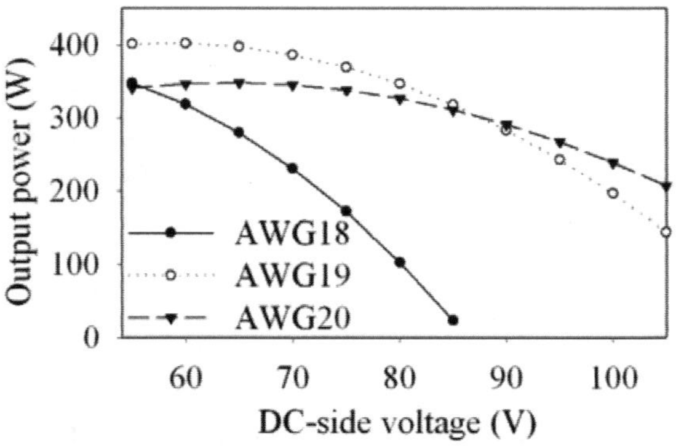

Figure 5. Output power capabilities of the STLG at a fixed speed of 3 m/s with various DC-side voltages and stator windings.

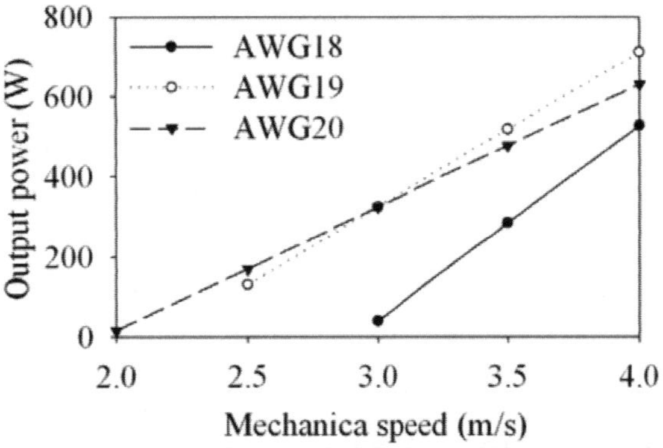

Figure 6. Output power capabilities of the STLG at a fixed 84 V DC-side voltage with various driven speeds and stator windings.

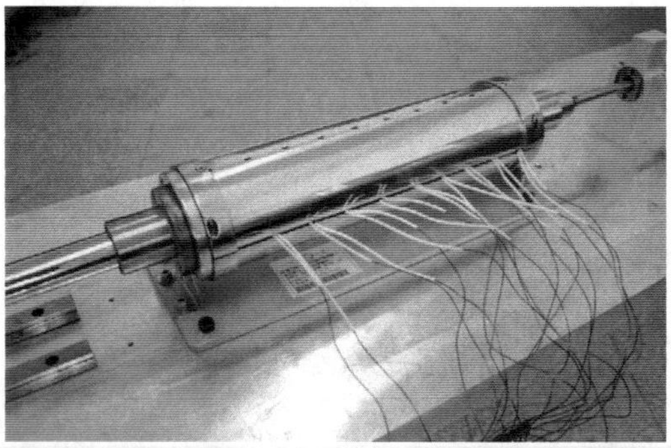

Figure 7. Constructed laboratory STLG prototype based on the optimal design scheme.

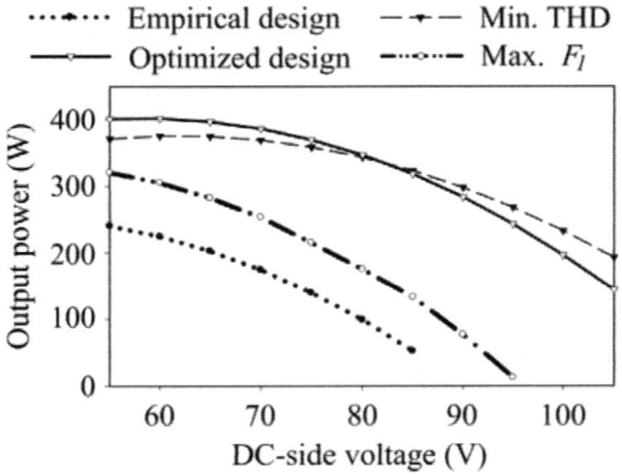

Figure 8. Output power capabilities of the STLG at a fixed speed of 3 m/s with various DC-side voltages and optimization objectives.

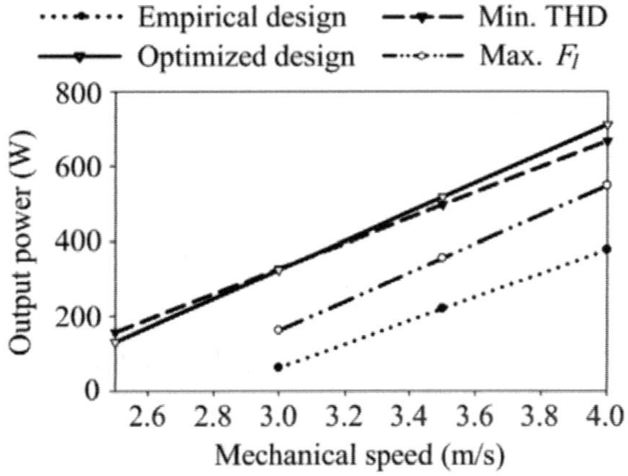

Figure 9. Output power capabilities of the STLG at a fixed 84 V DC-side voltage with various speeds and optimization objectives.

scheme. As illustrated in Figures 8 and 9, the STLG output powers at various operational specifications are provided for comparisons. Clearly, the one based on either empirical or maximized fundamental flux design will not exhibit competitive performance. Though the one based on minimization of THD will supply larger power output at certain operational ranges (higher output voltages or lower input speeds), its relatively larger power losses in these conditions also downgrade the possibility of selecting such structure. At a rated input mechanical speed of 3 m/s, the STLG with optimized design will supply electric power outputs larger than 340 W if the regulated DC-side voltages are smaller than 84 V. Or if this terminal voltage is fixed, the output powers will be larger than 340 W with higher mechanical speed inputs. The design objectives based on the optimized structure can thus certainly be confirmed.

5. Conclusions

This paper presents a systematic design scheme of a direct driven slotless tubular linear generator for direct renewable energy retraction. Based on Taguchi's method, comparisons and classifications were performed to identify the optimal machine structure for the design objectives. Appropriate stator windings, by thoroughly estimating the feasible STLG power generations, can then be selected. From the developed scheme and the machine prototype, adequate guidance for designing the related renewable energy retraction mechanisms has successfully demonstrated.

6. Acknowledgment

The authors want to express their appreciations to Mr. C.-H. Tu for his assistance on the STLG prototype construction. Financial supports from the National Science Council, Taiwan, R.O.C. under contracts NSC 96-2221-E-110-099-MY3 and NSC 98-3114-E-110-004 are also to be acknowledged.

7. References

[1] Cancelliere P, Marignetti F, Delli Colli V, Di Stefano R and Scarano M 2005 *Int. Conf. Elec. Mach. and Drives, 2005 IEEE* p 1473.
[2] Jang S, Choi J, Cho H and Lee S 2005 *IEEE Trans. Magn.* **41** 2028.
[3] Wang J, Jewell G and Howe D 1999 *IEEE Trans. Magn.* **35** 1986.
[4] Ross P 1996 *Taguchi Techniques for Quality Engineering: Loss Function, Orthogonal Experiments, Parameter and Tolerance Design* (New York: McGraw-Hill).
[5] Wang J and Howe D 2005 *IEEE Trans. Magn.* **41** 2470.
[6] Wang J and Howe D 2005 *IEEE Trans. Magn.* **41** 2479.
[7] COMSOL AB 2007 *COMSOL Multiphysics User's Guide* Version 3.4, (Stockholm).

Journal of Physics: Conference Series **266** (2011) 012076 doi:10.1088/1742-6596/266/1/012076

Surface Analysis of Ion Irradiated Amorphous Ribbon by GIXD and NEXAFS

D. G. Park[1], C. S. Angani[1,2] , G. D. Kim[1], C. G. Kim[2] and Y. M. Cheong[1]

[1] Nuclear Material Research Division, Korea Atomic Energy Research Institute, Daejeon, South Korea
[2] Dept, of material Science Engineering, Chungnam National University, Daejeon, South Korea

E-mail: dgpark@kaeri.re.kr

Abstract. Recently a great increase of giant magneto impedance (GMI) ratio due to ion irradiation was found in the amorphous ribbon and intensive research into this mechanism will result in a technological importance in the field of sensor applications. The mechanism of the abrupt increase of GMI ratio by ion irradiation was not identified yet. The samples used in the present study were commercial amorphous ribbons of $Co_{66}Fe_4B_{15}Si_{15}$, in the size of 2 mm x 40 mm x 20 μ m. Various kinds of ions such as N, Ar, and Xe have been irradiated with energy of 70 keV by an ion implanter and an ion dosage was set to 1.0×10^{16} ion/cm^2 at a beam flux of 3.7 μ A/cm^2. The grazing incident X-ray diffraction (GIXD) and near edge X-ray absorption fine structure (NEXAFS) experiments have been conducted in order to know the origin of abrupt increase of GMI ratio by ion irradiation. The GMI ratio was increased by about 6 times by an ion irradiation for the 1 MHz measuring frequency, and the change of the GMI ratio by an ion irradiation decreased with an increasing measuring frequency. The creation of crystalline phase and the new phase change were not observed from GIXD and NEXAFS experiments in case of Ar ion irradiation, but the drastic changes of the spectrum in the GIXD and NEXAFS experiment in case of Xe and N show that the crystalline phase and new phase change may affect the enhancement of GMI properties.

1. Introduction

The great increase of giant magneto-impedance (GMI) ratio has been recently observed in the ion irradiated amorphous ribbon [1] and intensive research into this mechanism will result in a technological importance in the field of sensor applications. It has been shown experimentally and theoretically [2], [3] that magneto-impedance value depends on the properties of the layer, in particular, magnetic anisotropy because anisotropy determines the magnetic permeability. The GMI phenomenon is due to the change in skin depth arising from the change in the magnetic permeability of materials with applied magnetic fields. The GMI is understood as the effect of magnetic field on transverse permeability which is associated with domain-wall motion and rotational magnetization. The GMI properties of an amorphous ribbon can be modified by the heat treatment or ion irradiation, which induces a structural relaxation and/or creates the appropriate anisotropy. In the case of Co-

[1] To whom any correspondence should be addressed.

based alloys, significant improvement of soft magnetic behaviour has been reached with surface crystalline layer [4]. The usual method of getting these crystallites is a suitable annealing. On the other hand, the crystallization mostly starts at the irregularities of the surface. Even a crystallized surface creates itself a perpendicular easy direction (through magnetization) and consequently preferred domain rotation against wall motion [5]. Ion irradiation techniques have the advantage that the crystallization in a controllable manner can be restricted to a desired depth. Ion irradiation at the same time has annealing and disordering effects. When the energetic ions penetrate into a thin film, the collision cascade causes the displacement of the target atoms, which produces a new mixed phase through the intermixing of cobalt, boron, and silicon substrate, and results in a metastable one as a result of rapid quenching. It is well known that a sudden rearrangement of domain-wall structure induces a relaxation behaviour of the low-field magnetic permeability [6]. According to the prior study [1], the effectiveness of GMI response on the amorphous ribbon shows quite different behavior with respect to the irradiated ion species and the applied driving frequencies range. In this study, N, Ar, and Xe ion have been irradiated in the $Co_{66}Fe_4B_{15}Si_{15}$ amorphous ribbon, and the effects have been investigated according to the irradiated ion species.

2. Experiment

The samples used in this study were commercial amorphous ribbons of $Co_{66}Fe_4B_{15}Si_{15}$, in the size of 2 mm x 40 mm x 20 μ m. The N, Ar, and Xe ions have been irradiated by an ion implanter with ion dose of 1.0 x 10^{16} ion/cm^2 and ion energy of 70 keV at a beam flux of 3.7 μ A/cm^2. The acceleration energies of the ions were designed to match their calculated mean projectile ranges to the corresponding thickness of the sample. The damage depth profile of deposited energy as a function of penetration depth calculated by computer simulation can be found elsewhere [1]. The GMI ratio was measured as a function of applied direct current (dc) magnetic field (H_{max} = 35 Oe) at different driving frequencies up to 10 MHz with 5-mA amplitude to the sample, which is irradiated with ions having different energies. The GMI ratio profile was obtained by plotting $\Delta Z / Z(\%) = [(Z(H) - Z_{sat}) / Z_{sat}] \times 100$ % for the applied cyclic dc field; here Z_{sat} is the impedance at saturation field. The impedance Z was measured using an HP4192A impedance analyzer with four terminal contacts. X-ray diffraction and absorption experiment were done by the grazing incident diffraction (GID) method and near edge X-ray absorption fine structure (NEXAFS) method using the POSCO beam line at Pohang light source (PLS) with the wavelength 1.5402 A.

3. Results and discussion

Fig. 1(a) shows the change in the GMI ratio at 1 MHz driving frequencies of the sample, which is irradiated by N, Ar, and Xe ions having the energy of 70 keV. Fig. 1(b) shows the frequency dependence of the maximum value of GMI ratio for the unirradiated and irradiated samples. The GMI ratio increased by factors from 2 to 8 with respect to frequencies for the ion irradiated samples, as shown in Fig. 1. It is worth noting that the slope obtained from the ion irradiated sample is steep in comparison to the unirradiated sample. Ion irradiation on an amorphous ribbon results in quite different behaviors at low and high frequencies. It is instructive to observe the gradual appearance of a double peak structure with an ion irradiation, which reflects the increasing anisotropy field. Furthermore, the depth of a ridge of the GMI ratio increases with an increase in frequency of the driving current. GMI ratio significantly increased in the low magnetizing frequency region and showed a strong dependency on the driving frequencies irrespective of the irradiated ion species.

The GMI phenomena can be explained by the skin depth model of classical electrodynamics. The skin depth in a magnetic medium with the transverse permeability μ_T and conductivity of σ is expressed as $\delta_m = (c)/(\sqrt{2\pi\omega\sigma\mu_T})$, where c is the speed of light and $f = \omega / 2\pi$ is the frequency of the alternating current (ac), which is used to drive the sample. In this context, the GMI results can be interpreted by considering the change in δ_m in relation to the change of μ_T caused by the application of external magnetic field according to the skin depth equation [7]. At frequencies below 1 MHz, the maximum value of GMI was relatively low due to the contribution of the induced magneto-inductive voltage to magneto-impedance. The shape of GMI curves of a sample is composed as a single peak at zero field for the lower driving frequencies but it is divided as double peaks at higher frequencies. A double peak is good indication of a magnetization rotation process. The GMI curves of irradiated samples exhibit a double peak which indicates anisotropy associated with domain rotation. The field dependence and split of double peak in the ion irradiated amorphous ribbon can be found elsewhere [1]. The process of a single peak evolving into a double peak indicates that the susceptibility due to domain-wall movement is being reduced and the susceptibility due to rotational effects is important. When the energetic ions with energy of several tens of keV penetrate into a thin film, the collision cascade causes displacement of the target atom that induces a uniaxial anisotropy in the sample [8], [9]. To obtain a significant MI effect, it has been generally found by many authors that materials should possess a domain structure in which the domain walls are perpendicular to the current direction. The anisotropy induced in the sample is sufficiently large that it prevents the formation of closure domains, which would require the domain magnetization to point along the hard axis.

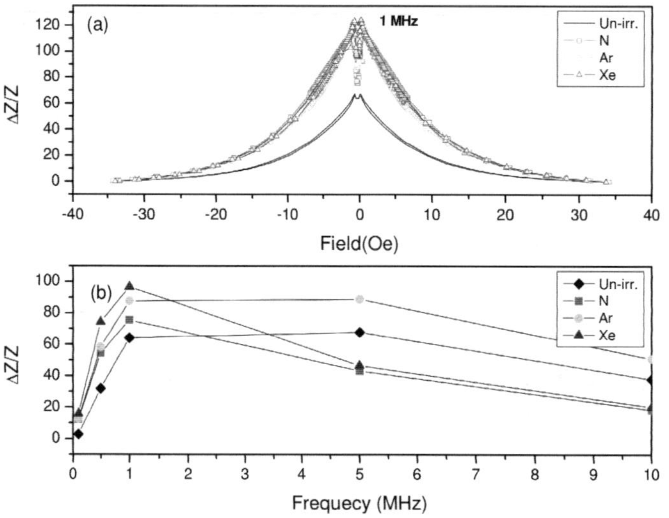

Figure 1. GMI curves obtained at 1 MHz frequency of the samples which are irradiated at 70 keV (b) Maximum GMI response as a function of frequency for the un-irradiated and N, Ar, and Xe ion irradiated amorphous ribbon

In order to check how the amorphous and crystalline phases are distributed in depth of the ribbon, X-ray diffraction experiment was done by GID. The incident beam angle of X-ray was changed from 1° to 5°. As shown in Fig. 2, the drastic changes of the spectrum were observed in the N irradiated sample and the sharp crystalline peaks of surface have been dominated by great amorphous fraction at 3°. In a common heat treatment the first step of crystallization (after precrystallization) of Fe-X-B-type metallic glasses is the formation of eutectic α -(Fe, X) and tetragonal $(Fe, X)_3B$ [10], [11]. The amorphous $Co_{66}Fe_4B_{15}Si_{15}$ contains a large amount of crystalline phase, and $(Fe, X)_3B$ is very easy to be amorphized by irradiation [12]. So it is assumed that the irradiation blocked the way of crystallization via tetragonal Fe_3B, because it was continuously amorphized. It is known that, during N ion irradiation, formation of amorphous BN-rich regions may increase the Fe concentration in the residue [11]. It should be stressed that the concentration of implanted ions and the deposited energy is almost three times larger in the case of Ar ion in the implanted layer, and even this was ineffective in crystallizing the surface up to 1.0 x 10^{16} ion/cm^2. Figure 3 shows the Co L-edge NEXAFS spectra of the various ion irradiated sample. There is no changes of peak at around 780 eV of L_3 edge for ion irradiation, but small variation of peak at around 794 eV of L_2 edge for N and Xe ion irradiation, suggesting that electrical state of N and Xe ion irradiated samples are changed by irradiation.

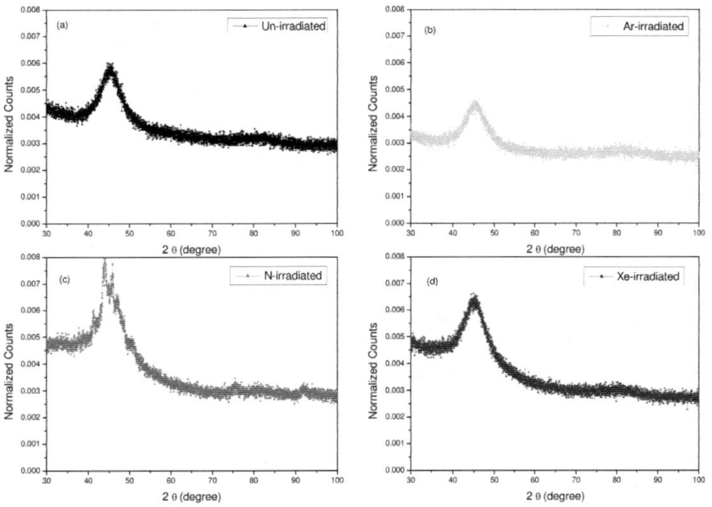

Figure 2. GID spectrum of (a) unirradiated, (b) Ar ion irradiated, (c) N ion irradiated, and (d) Xe ion irradiated samples.

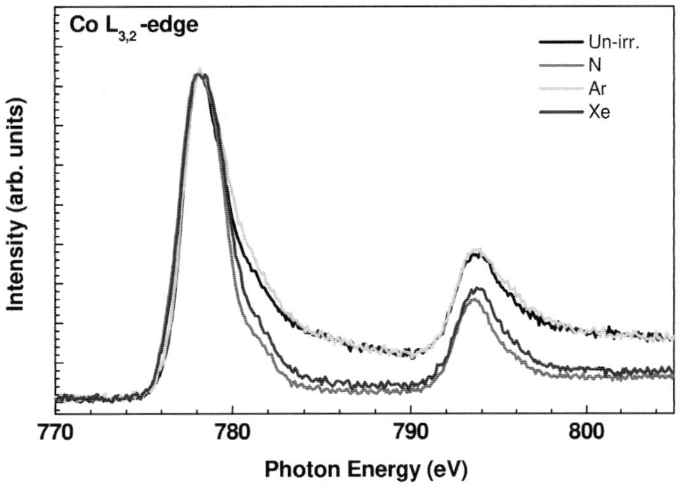

Figure 3. NEXAFS spectrum of unirradiated, N, Ar, and Xe ion

irradiated samples at $L_{3,2}$ edge.

4. Conclusion

Various kinds of ions such as N, Ar, and Xe have been irradiated to find the mechanism of the abrupt increase of GMI ratio by ion irradiation. The GMI ratio was increased by about 2 times by an ion irradiation for the 1 MHz measuring frequency, and the change of the GMI ratio by an ion irradiation decreased with an increasing measuring frequency. The creation of crystalline phase and the new phase change were not observed from GIXD and NEXAFS experiments in case of Ar ion irradiation, but the drastic changes of the spectrum in the GIXD and NEXAFS experiment in case of Xe and N show that the crystalline phase and new phase change may affect the GMI properties.

5. References

[1] D. G. Park, C. G. Kim, J. H. Lee, W. W. Kim, and J. H. Hong, J. Appl. Phys., vol. 101, no. 09N109, 2007.
[2] N. A. Usov, A. S. Antonov, and A. N. Lagar'kov, J. Magn. Magn. Mater., vol. 185, p. 159, 1998.
[3] A. Gromov and V. J. Korenivski, Phys. D, Appl. Phys., vol. 33, p. 773, 2000.
[4] K. J. Jang, C. G. Km, S. S. Yoon, and S.-C. Yu, J. Magn. Magn. Mater., pp. 488–491, 2000.
[5] G. Herzer and H. R. Hilzinger, J. Magn. Magn. Mater., vol. 62, p. 143, 1986.
[6] P. Allia, C. Beatrice, and F. Vinai, Physica Scripta, vol. 39, pp. 489–491, 1989.
[7] K. S. Byon, S. C. Yu, J. S. Kim, and C. G. Kim, IEEE Trans. Magn., vol. 36, p. 3439, 2000.
[8] K. H. Chae, Y. S. Lee, S. M. Jung, Y. Jeon, M. Croft, and C. N. Whang, N. Instrum. Methods B, vol. 106, p. 60, 1995.
[9] D. G. Park, H. Song, C. Y. Park, C. S. Angani, C. G. Kim, and Y. M. Cheong, IEEE Trans. Magn., 2009, accepted for publication.
[10] L. Battezati, C. Antonione, and G. Riontino, J. Non-Cryst. Solids, p.114, 1987.
[11] Cs. S. Daroczi, G. Serfozo, and P. Kollar, J. Magn. Magn. Mater., vol. 112, pp. 367–368, 1992.
[12] T. Kemeny, I. Vince, and B. Fogarassy, Phys. Rev. B, Condens. Matter, vol. 20, p. 401, 1979.

Journal of Physics: Conference Series **266** (2011) 012077 doi:10.1088/1742-6596/266/1/012077

Magnetocaloric effect in (Tb,Dy,*R*)(Co,Fe)$_2$ (*R* = Ho, Er) multicomponent compounds

I S Tereshina[1], G A Politova[1], E A Tereshina[2,3], G S Burkhanov[1], O D Chistyakov[1] and S A Nikitin[3]

[1] Baikov Institute of Metallurgy and Material Sciences RAS, Moscow 119991, Russia
[2] Institute of Physics ASCR, Prague 18221, Czech Republic
[3] Lomonosov Moscow State University, Moscow 119991, Russia

E-mail: teresh@imet.ac.ru

Abstract. The magnetic, magnetocaloric, and magnetoelastic properties of $Tb_xDy_yR_z(Co,Fe)_2$ (R = Ho, Er; $x + y + z = 1$) single-phase alloys prepared by arc melting were investigated. The high-purity samples with various compositions were used for the study, and a comprehensive investigation of the properties was performed. The measurements of the magnetocaloric effect (MCE) were carried out by a direct method; and the magnetostriction was determined using the strain-gauge technique. In the vicinity of the magnetic phase transition, an abrupt change of the magnetization, a large MCE, and the maxima of volume magnetostriction were observed for the compounds studied. Obtained results of MCE in (Tb,Dy)Co_2 agree well with available theoretical calculations.

1. Introduction

Owing to the relatively simple crystallographic and magnetic structures, the Laves phase intermetallic compounds RCo_2 (R is the rare earth element) remain an attractive subject for testing various physical models in the field of the Solid State Physics [1,2]. RCo_2 crystallize in the $MgCu_2$-type structure (Laves C15 phase, space group $Fd\bar{3}m$). The compounds with non-magnetic R, YCo_2 and $LuCo_2$, are Pauli paramagnets. The magnetic state of the Co-sublattice can be stabilized by applying the magnetic field higher than ~70 T or by the molecular field arising from the magnetic R-sublattice. The RCo_2 with R = Gd – Er are ferrimagnets. As the temperature is varied, $GdCo_2$ and $TbCo_2$ pass through the second-order magnetic phase transition from ferri- to the paramagnetic state while the first-order magnetic phase transition takes place for $DyCo_2$, $HoCo_2$ and $ErCo_2$ [3]. It is known that the highest values of the magnetocaloric effect (MCE) are usually observed in the vicinity of the magnetic phase transitions [4], particularly of the first-order, where the magnetization and other characteristics change drastically upon the temperature or magnetic field variation, suggesting the potential application of the compounds as magnetic refrigerants.

In the present work, the search for new materials with a large MCE was carried out among the multicomponent alloys of the RCo_2 type. Though during the past decade numerous investigations of MCE performed on RCo_2 with different substitutions either in the rare-earth (RR')Co_2 [5-8] or in the

[1] I S Tereshina

Co sublattices $R(Co,T)_2$ [9-12] (T is another transition metal) were carried out, we are not familiar with the works on MCE of the multicomponent compounds of the $(RR'R'')(Co,T)_2$ type.

Among the $(RR')Co_2$ alloys, $Tb_xDy_{1-x}Co_2$ system deserves a special attention. $TbCo_2$ and $DyCo_2$ with the neighboring R and different disorder-order transition types have similar lattice parameters. According to Refs. [13,14], the change of the transition type from the first to the second-order in $Tb_xDy_{1-x}Co_2$ occurred at $0.1 < x < 0.3$. The compounds were shown to have the compensated magnetic anisotropy (MA) within the rare-earth sublattice due to the opposite sign of the single-ion MA constants of Terbium and Dysprosium [15]. (It should be mentioned that the Fe-containing analogue of $(Tb,Dy)Co_2$, $Tb_{0.23}Dy_{0.73}Fe_2$, also known as terfenol-D, possesses giant values of magnetostriction and is widely used in practice.) In Ref. [16], the magnetocaloric properties of $Tb_xDy_{1-x}Co_2$ were discussed in terms of the theoretical model using the treatment of the 4*f* spin-spin interaction with respect to the rare earth ion type.

The aim of the present work was to investigate the magnetic, magnetothermal and magnetoelastic properties of the multicomponent compounds $(Tb,Dy,R)(Co,T)_2$ (R = Ho, Er; T = Fe) with the compensated magnetic anisotropy, selected on the account of theoretical predictions based on the single-ion model of magnetic anisotropy, as well as to compare the experimental data on MCE with known theoretical values. Partial substitution of Fe atoms for Co was performed in order to bring the Curie temperatures (T_C) of the compounds under study towards the room temperature [17].

2. Experimental details

The rare-earth metals purified by a double vacuum distillation (Tb) and sublimation (Dy, Ho, Er) with the use of original furnace designed at Baikov Institute of Metallurgy and Materials Science RAS [18] were used as starting materials. The purified metals (99.956-99.983 wt. %) were characterized by low content of metallic and interstitial elements. An Armco iron and Co with the purity of 99.9% were also used. The following alloys have been prepared: $Tb_{0.27}Dy_{0.73}Co_2$, $Tb_{0.3}Dy_{0.7}Co_2$, $(Tb_{0.45}Dy_{0.55})_{1-x}Ho_xCo_2$ (x = 0.1, 0.2, 0.3 and 0.5), $Tb_{0.23}Dy_{0.27}Ho_{0.5}Co_{1.8}Fe_{0.2}$ and $(Tb_{0.45}Dy_{0.55})_{1-x}Er_xCo_2$ (x = 0.1, 0.2, 0.3). Details of the sample preparation and quality control have been described earlier [19].

The MCE measurement was conducted by a direct method. The specific-heat measurements in 0 T were carried out by a standard relaxation method on PPMS installation (Quantum Design). The field and temperature dependencies of magnetization were measured by the standard SQUID magnetometer and the magnetostriction and thermal expansion was determined using the strain-gauge technique.

3. Results and discussion

The magnetic anisotropy constants are known to have non-zero values at the Curie point. In general, the MA constant of the rare-earth sublattice is an additive superposition of the product of single-ion anisotropy constants of the R ions multiplied by their concentrations [20]. The total lth-order MA constant can be calculated from the relation [11]

$$K_l = \sum X_n K_{nl}^0 \hat{I}_{l+(1/2)} \{L_m^{-1}\}, \quad (1)$$

where K_{nl}^0 is the lth-order single-ion anisotropy constant at 0 K, $\hat{I}$ is the reduced hyperbolic Bessel function, $L_m^{-1}(m)$ is the inverse Langevin function of reduced magnetization $m = I_s(T)/I_s(0)$ ($I_s(0)$ and $I_s(T)$ are the spontaneous magnetizations at 0 K and at a temperature T, respectively). In order to attain the compensation of MA within the rare-earth sublattice, the compositions with the R ions having different signs of the single-ion MA constants (of the first and higher orders) were chosen. So that we defined and prepared the multicomponent compounds with general formulas $Tb_xDy_yR_zCo_2$ (R = Ho or Er, $x+y+z = 1$). From the analysis of X-ray diffraction patterns, it was concluded that the samples were single phase with the cubic $MgCu_2$ structure. The prepared alloys were also examined by the X-ray fluorescent microanalysis in order to determine the exact compositions.

The magnetization curves for the compounds under study were measured (see, for example Fig.1(1)). A weak magnetic hysteresis (not more that 10% of magnetization value) was observed in fields below

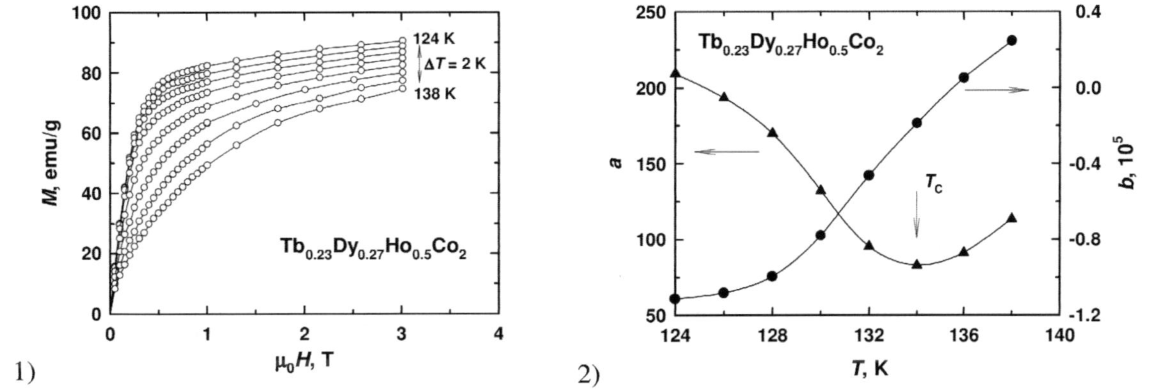

Figure 1. Magnetization isotherms (*1*) and the temperature dependence of Landau coefficients *a* and *b* (*2*) for $Tb_{0.23}Dy_{0.27}Ho_{0.5}Co_2$. The units for $a(T)$ and $b(T)$ are T^2kg/J and T^4kg/J, respectively.

0.05 T, which disappeared at higher applied fields (such a small hysteresis can be explained by the compensated MA constants near T_C). The magnetic phase transitions were examined in terms of Landau theory by analyzing the temperature dependencies of Landau coefficients in the magnetic free energy (*F*) equation as a function of total magnetization (in case of neglecting the higher order terms):

$$F = (1/2)a(T)M^2 + (1/4)b(T)M^4 + (1/6)c(T)M^6 - \mu_0 MH . \qquad (2)$$

The description of magnetization near T_C follows from the equilibrium condition $(\partial F / \partial M) = 0$. The $a(T)$ dependence was used to determine the Curie temperatures of the compounds. A group of alloys demonstrating the first-order transition defined in accordance with the sign of $b(T_C)$ (negative at T_C while the parameters $a(T_C)$ and $c(T_C)$ are positive) was revealed [8,21]. The lattice parameters, the refined alloys compositions, the Curie temperatures and the type of order of magnetic phase transition are listed in Table 1.

Table 1. Lattice parameters, Curie temperatures and the types of magnetic phase transition in (Tb,Dy,*R*)$(Co,Fe)_2$ (*R* = Ho, Er).

Compounds	*a*, Å	*V*, $Å^3$	T_C, K	Transition type
$Tb_{0.27}Dy_{0.73}Co_2$	7.1930	372.11	168	First-order
$Tb_{0.3}Dy_{0.7}Co_2$	7.1937	372.27	170	First-order
$Tb_{0.4}Dy_{0.5}Er_{0.1}Co_2$	7.1917	371.96	168	First-order
$Tb_{0.35}Dy_{0.45}Er_{0.2}Co_2$	7.1878	371.36	162	First-order
$Tb_{0.23}Dy_{0.27}Ho_{0.5}Co_2$	7.1820	370.46	134	First-order
$Tb_{0.23}Dy_{0.27}Ho_{0.5}Co_{1.8}Fe_{0.2}$	7.2126	375.21	290	Second-order

Figure 2 shows the temperature dependence of MCE of selected (Tb,Dy,*R*)$(Co,Fe)_2$ compounds. For $Tb_{0.27}Dy_{0.73}Co_2$, $Tb_{0.35}Dy_{0.45}Er_{0.2}Co_2$ and $Tb_{0.23}Dy_{0.27}Ho_{0.5}Co_2$, MCE is observed in the 130 – 170 K temperature range and consists of 1.6 – 1.8 K adiabatic temperature change under the variation of magnetic field from 0 up to 1.2 T. The partial substitution of Fe for Co has allowed us to move the transition temperature towards the room temperature, however, resulting in a change of the transition type from the first to the second order in $Tb_{0.23}Dy_{0.27}Ho_{0.5}Co_{1.8}Fe_{0.2}$. Such substitution has also led to a threefold decrease in the MCE value (curve 4 in Fig. 2). As it is seen from the field dependence of MCE presented in the inset in Fig. 2, and in contrast with $Tb_{0.27}Dy_{0.73}Co_2$, the $Tb_{0.35}Dy_{0.45}Er_{0.2}Co_2$ and $Tb_{0.23}Dy_{0.27}Ho_{0.5}Co_2$ compounds demonstrate an almost linear behavior and higher MCE values in a

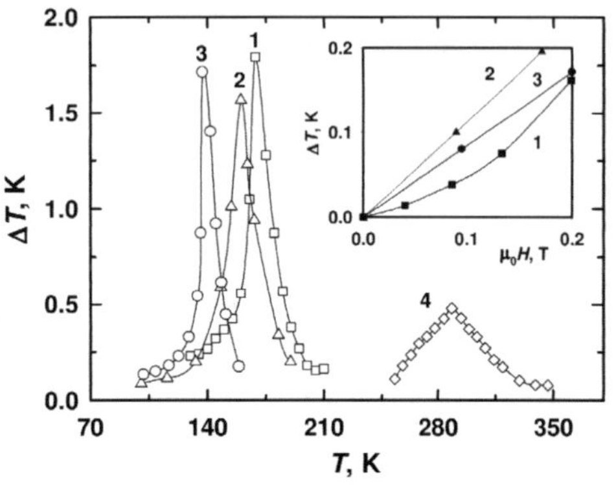

Figure 2. Adiabatic temperature change ΔT vs. absolute temperature for $Tb_{0.27}Dy_{0.73}Co_2$ (*1*), $Tb_{0.35}Dy_{0.45}Er_{0.2}Co_2$ (*2*), $Tb_{0.23}Dy_{0.27}Ho_{0.5}Co_2$ (*3*) and $Tb_{0.23}Dy_{0.27}Ho_{0.5}Co_{1.8}Fe_{0.2}$ (*4*) near T_C upon the $\mu_0\Delta H$ = 1.2 T field change. The inset: field dependence of ΔT for the same samples.

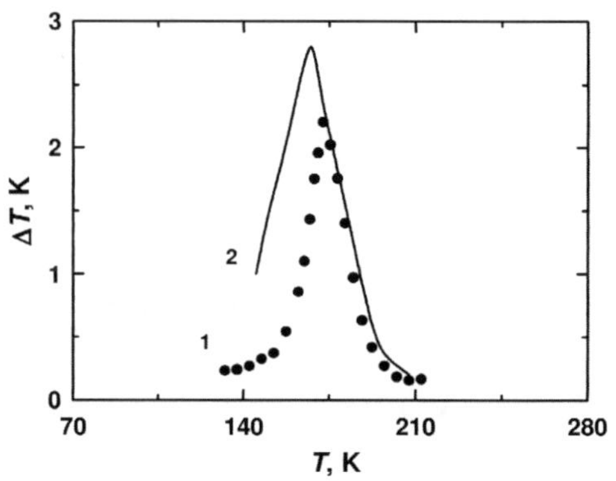

Figure 3. Adiabatic temperature change ΔT vs. absolute temperature in $Tb_{0.3}Dy_{0.7}Co_2$ as measured upon the magnetic field variation from 0 to 1.5 T (*1*). Solid line (*2*) shows theoretical predictions adopted from Ref. [17].

rather weak (< 0.2 T) magnetic field. On our point of view, the latter fact is connected with a total compensation of magnetic anisotropy in the studied compounds.

The theoretical predictions on the magnetocaloric potential of $Tb_xDy_{1-x}Co_2$ were given in Ref. [17] with the $H = H_d + H_f$ model Hamiltonian used for the MCE calculation (H_d and H_f describe the subsystems of itinerant 3*d*-electrons of Co and the localized 4*f*-electrons of the *R* ions, respectively (see the corresponding formulas (1) and (2) in Ref. [17])). The exchange, magnetoelastic, anisotropic and magnetic energy contributing to the observed effects are also taken into account. Experimental adiabatic temperature change upon the magnetic field variation from 0 to 1.5 T in $Tb_{0.3}Dy_{0.7}Co_2$ compound is shown in Fig. 3 in comparison with the theoretical data adopted from Ref. [17]. The obtained results agree rather well with calculations for $T > T_C$, whereas a small deviation from the calculated curve at lowered temperatures may be connected with the domain structure formation.

The zero-field specific heat data for $Tb_{0.3}Dy_{0.7}Co_2$ and $Tb_{0.23}Dy_{0.27}Ho_{0.5}Co_2$ is shown in Fig. 4. The sharp peak in the temperature dependence of C_p is observed at T = 134 K for $Tb_{0.23}Dy_{0.27}Ho_{0.5}Co_2$,

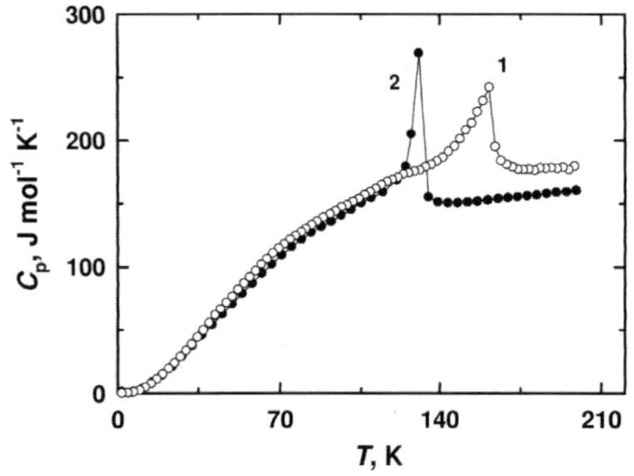

Figure 4. Temperature dependence of specific heat of $Tb_{0.3}Dy_{0.7}Co_2$ (*1*) and $Tb_{0.23}Dy_{0.27}Ho_{0.5}Co_2$ (*2*) compounds in zero magnetic field.

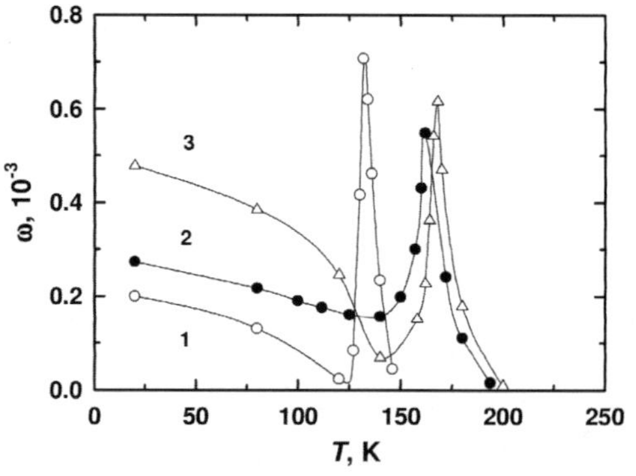

Figure 5. Volume magnetostriction vs. temperature for $Tb_{0.27}Dy_{0.73}Co_2$ (*1*), $Tb_{0.35}Dy_{0.45}Er_{0.2}Co_2$ (*2*) and $Tb_{0.23}Dy_{0.27}Ho_{0.5}Co_2$ (*3*) in field of 1.2 T.

whereas for $Tb_{0.3}Dy_{0.7}Co_2$ it is somewhat reduced and broadened at T = 170 K. Nevertheless, the obtained values of $b(T_C)$ for $Tb_{0.3}Dy_{0.7}Co_2$ were found to be negative, confirming an occurrence of the first order magnetic phase transition, though with the magnitude rather close to zero.

In addition to aforesaid, the investigation of the magnetostriction of (Tb,Dy,R)Co_2 alloys with R = Ho and Er was carried out in the present work in the temperature range from 5 to 200 K and in magnetic fields up to 1.2 T. Figure 5 presents the temperature dependencies of the volume magnetostriction ω for $Tb_{0.27}Dy_{0.73}Co_2$, $Tb_{0.35}Dy_{0.45}Er_{0.2}Co_2$ and $Tb_{0.23}Dy_{0.27}Ho_{0.5}Co_2$ compounds. In the vicinity of the magnetic phase transition, the observed volume magnetostriction reaches the range of "giant" values of $(0.55–0.7)\times 10^{-3}$. As a result of the conducted comprehensive research on magnetic, magnetoelastic and magnetothermal properties, the estimation of various energy contributions (namely, the exchange, magnetoelastic, anisotropic and magnetic energy) to the thermodynamic potential [22,23] of the (Tb,Dy,R)Co_2-type compounds with the first order magnetic phase transitions was carried out (will be published elsewhere). At the Curie temperature, the ratio between the values of exchange and the magnetoelastic energies accounts for 3:1. The energy of magnetic anisotropy is found to affect insignificantly the magnetic phase transition.

The work is supported by RFBR, pr. No. 10-03-00848 and No. 10-02-00721.

References

[1] Duc N C , Brommer P E Formation of 3d-moments and spin fluctuations in some rare-earth-cobalt compounds , in: *Handbook of Magnetis Materials* K.H.J. Buschow (Ed.) 1999 vol. 12 (Amsterdam, Elsevier Science)
[2] Gratz E and Markosyan A S 2001 *J. Phys.:Condens. Matter* **13** R385
[3] Khmelevski S and Mohn P 2000 *J. Phys.:Condens. Matter* **12** 9453
[4] Tishin A M Magnetocaloric effect in the vicinity of phase transitions in: *Handbook of Magnetis Materials* K.H.J. Buschow (Ed.), 1999 vol. 12 (Amsterdam, Elsevier Science).
[5] Liu X B and Altounian Z 2005 *J. Magn. Magn. Matter.* **292** 83
[6] Gomes A M , Reis M S, Oliveira I S , Guimaraes A P , Takeuchi A Y 2002 *J. Magn. Magn. Matter.* **242-245** 870
[7] Ouyang Z W , Rao G H , Yang H F, Liu W F , Liu G Y , Feng X M , Liang J K 2004 *J. Alloys Compd.* **372** 76
[8] Balli M, Fruchart D, Gignoux D 2007 *J. Magn. Magn. Matter.* **314** 16
[9] Liu H, Wang D, Tang S, Cao Q, Tang T, Gu B, Du Y 2002 *J. Alloys Compd.* **346** 314
[10] Prokleska J, Vejpravova J, Vasylyev D, Danis S, Sechovsky V 2005 *J. Magn. Magn. Matter.* **290-291** 676
[11] Chen X, Zhuang Y, Yan J, Zhou K, Li K. 2008 *J. Rare metals* **27** 350
[12] Wada H, Tanabe Y, Shiga M, Sugawara H, Sato H 2001 *J. Alloys Compd.* **316** 245
[13] Gu K, Li J, Ao W, Jian Y, Tang J 2007 *J. Alloys Compd.* **441** 39
[14] Zhuang Y, Chen X, Zhou K, Li K 2008 *J. Rare metals* **26** 749
[15] Klimker H, Dariel M P, Rosen M 1980 J.Phys.Chem.Solids **41** 215
[16] Han Z, Hua Z, Wang D, Zhang C, Gu B, Du Y 2006 *J. Magn. Magn. Matter.* **302** 109
[17] de Oliveira N A 2008 *J. Magn. Magn. Matter.* **320** e150
[18] Chistyakov O D, Burkhanov G S, Kol'chugina N B 1994 *Vysokochistye veshchestva* **3** 57 (in Russian)
[19] Tereshina I, Politova G, Tereshina E, Nikitin S, Burkhanov G, Chistyakov O, Karpenkov A 2010 *J. Phys.: Conf. Series* **200** 092012
[20] Koon N C, Williams C M, Das B N 1991 *J. Magn. Magn. Matter.* **100** 173
[21] Yamada H, Goto T 2003 *Phys. Rev. B* **68** 184417
[22] Nikitin S A 1984 *Sov. Phys. JETP* **59(5)** 1010
[23] Spichkin Y I and Tishin A M 2005 *J. Magn. Magn. Matter.* **290-291** 700

Journal of Physics: Conference Series **266** (2011) 012078 doi:10.1088/1742-6596/266/1/012078

Thickness dependence of the magnetic properties on FeHfN thin film by DC sputtering

Z S Wang, Y T Lai and J G Duh *

Department of Material Science and Engineering, National Tsing Hua University, Hsinchu 30013, Taiwan

E-mail: jgd@mx.nthu.edu.tw

Abstract. As-deposited nanocrystalline Fe–Hf–N thin films were investigated to improve soft magnetic properties by controlling thickness of the films. On the XRD analysis, the strain of α-Fe was due to the incorporation of interstitial nitrogen atoms, which might induce the soft magnetic properties. However, the poorer magnetic properties of 1.1-μm FeHfN film was attributed to the stripe domain structure, for the higher residual stress of 1.1-μm FeHN film produced stress-induced anisotropy. Nevertheless, the residual stress decreased with increasing thickness of the films. Thus, the striped domain structure disappeared in both 3.3-μm and 5.5-μm thick FeHfN. The 5.5-μm thick nanocrystalline FeHfN film exhibited high magnetization ($4\pi M_S \sim 16.9$ kG) and lower coercivity ($H_C < 1$ Oe). The superior soft magnetic properties show the potential application of this films for power inductor.

1. Introduction

In recent years, miniaturization has been a crucial issue in portable electronic devices, such as mobile communication equipment and consumer electronic products. To overcome the miniaturization problem, the small size passive components is required, especially for inductor.

Integration of micro-inductor with soft ferromagnetic thin films not only enhances the inductance of magnetic inductors but also reduces the device dimensions. Studies of soft magnetic film for inductor had been reported, such as FeCoAlO [1], NiFe and CoZrTa [2]. The properties of magnetic thin film for inductor applications require good soft magnetic characteristics, such as high saturation magnetization (M_S) and low coercivity (H_K) [3].

FeXN nanocrystalline films with the third elements X (X = Ti, Hf, Ta...) have shown excellent soft magnetic properties and thermal stability [4]-[6]. In addition, the modulation of film thickness could improve the magnetic properties in other nanocrystalline films [7, 8]. However, studies seldom discussed the optimum thickness for the nanocrystalline thick films.

This study investigated the dependences of film thickness on structural and soft magnetic properties of FeHfN films. Furthermore, the optimum thickness of FeHfN film was further employed by the combination with NiZnCu substrate. The synthetic magnetic property of FeHfN/NiZnCu ferrite was discussed.

2. Experiment

The FeHfN film with various thickness (t = 1.1, 3.3, and 5.5 μm) were co-deposited onto (100) silicon and NiZnCu ferrite substrates by dc magnetron sputtering. The power of Fe and Hf targets was 200W and 40W, respectively. The base pressure was 4.0×10^{-6} Torr and the working pressure was 1.3×10^{-3} Torr in a mixed Ar + N_2 atmosphere. The compositions of FeHfN thin films were quantitatively determined by a field emission electron probe microanalyzer (FE-EPMA JXA-8500F, JEOL). In addition, the phase identification and grain size estimation were analyzed by X-ray diffraction (XRD). The magnetic properties were characterized by using a vibrating sample magnetometer (VSM). The magnetic domain structure was observed by magnetic force microscope (MFM). Besides, the residual stress of films was determined by the curvature measurement.

3. Results and discussion

The compositions of FeHfN films with various thickness are presented in Fig.1. Under the identical sputtering condition, the content of Fe, Hf, and N was nearly fixed at 88, 8, and 4 at. %, respectively. As a result, there was no appreciable variation in the compositions of FeHfN films with different thickness.

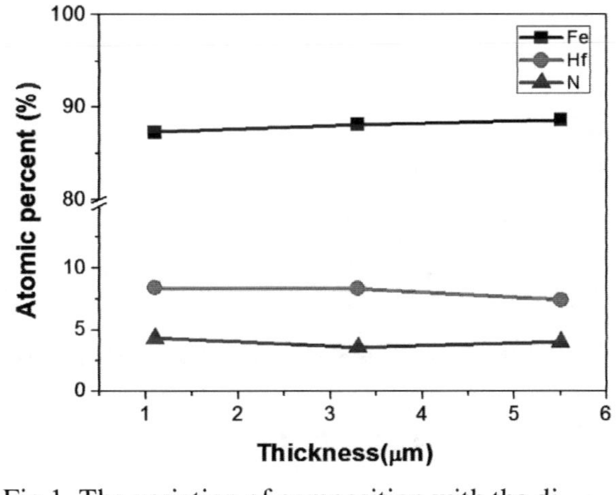

Fig.1. The variation of composition with the di fferent thickness for the FeHfN films.

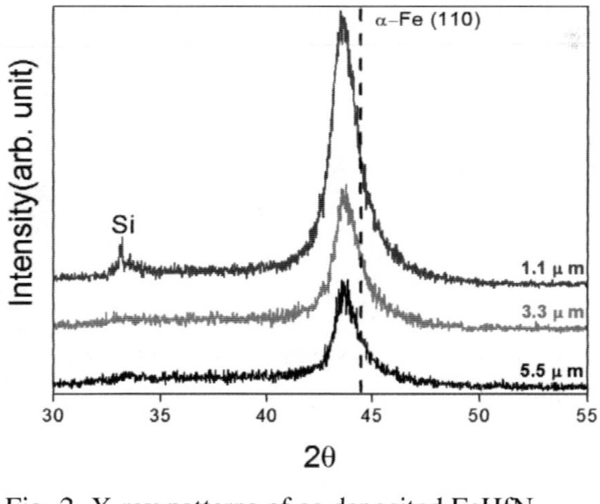

Fig. 2. X-ray patterns of as-deposited FeHfN films with various thickness.

X-ray diffraction of FeHfN as a function of thickness was conducted for phase identification in Fig. 2. The XRD patterns showed a strong texture of α-Fe (110), which proved the crystallization of these films. As the thickness of FeHfN films increased, the (110) peak intensity of α-Fe decreased.

According to Scherrer formulas, the average grain size (*D*) of FeHfN films were estimated around 5.5 - 6.5 nm. Thus, the grain size of FeHfN films corresponded to that of nanocrytalline [9].

The positions of α-Fe peaks shifted to low angle side with increasing film thickness, because the nitrogen atoms dissolved in the interstitial sites of the α-Fe phase, leading to lattice expansion. The position of α-Fe peak for 1.1-μm, 3.3-μm, and 5.5-μm FeHfN films was 43.65°, 43.76°, and 43.75°, respectively. A significant shift of 2θ angle was revealed between 1.1-μm and 3.3-μm thick film. It was argued that the changes of 2θ for α-Fe position might be caused by the internal stress in FeHfN films.

Figure 3 demonstrates the relationship among compressive stress (σ), the inter-planar spacing of the (110) plane (d_{110}) and thickness of FeHfN films. The residual stresses in the FeHfN films with the variation of thicknesses were calculated by Stoney's equation [10]. The compressive stress in the FeHfN films dropped from 1.1 GPa to 0.8 GPa with increasing thickness. In addition, as thickness increased, the spacing d_{110} of FeHfN films was close to a bulk value for α-Fe of 2.027Å. Thus, the residual stress dominated the lattice expansion of α-Fe. Under the identical sputtering conditions, the mechanism for the variations of residual stress might be resulted from the defect distribution at the interface between the FeHfN film and silicon substrate [11].Therefore, a thinner film implied a higher residual stress in FeHfN films.

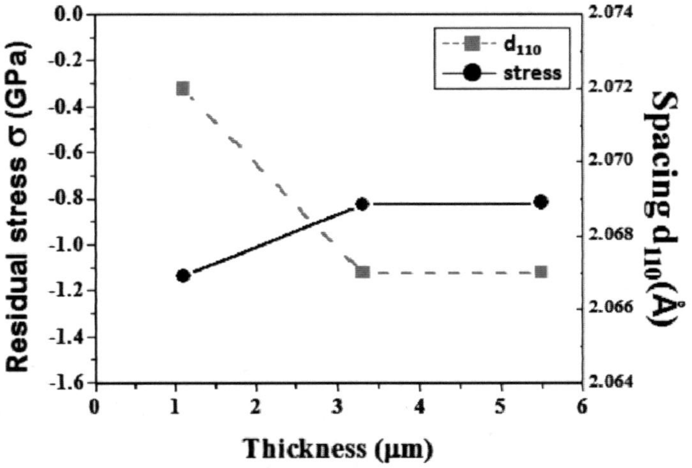

Fig.3. Residual stress and inter-planar spacing of (110) planes d_{110} of as-deposited FeHfN films as a function of thickness

In-plane hysteresis loops of FeHfN films with varying thickness are shown in Fig. 4. The in-plane hysteresis loop of 1.1-μm FeHfN film exhibited the relatively large coercivity (H_C), as shown in Fig. 4(a), which was characteristic of typical stripe domain structure [12]. Moreover, MFM images of 1.1-μm FeHfN film revealed a distinguished feature of bright-to-dark contrast, indicating that the presence of the strip domain was from perpendicular anisotropy. The sources of the perpendicular anisotropy might be affected by magnetoelastic anisotropy and magnetocrystalline anisotropy. According to the random anisotropy model [9], a smaller average grain size less than the ferromagnetic exchange coupling length in films implied a stronger exchange coupling between α-Fe grains. The magnetocrystalline anisotropy would be averaged over several grains and finally reduced in magnitude. Therefore, the effects of magnetocrystalline anisotropy could be ignored.

On the other hand, the magnetoelastic anisotropy energy (K) is calculated by using the formula:

$$K = (-3/2)\,\lambda_s\sigma \tag{1}$$

where σ is the residual stress of films. The saturation magnetostriction λ_s is almost invariable for identical preparation condition. In addition, the magnitude of magnetostriction of FeHfN films was positive and estimated to be in the order of 10^{-6} [5]. From this equation, the magnitude of the stress-induced magnetoelastic anisotropy was proportional to compressive stress in the films. Thus, the 1.1-μm FeHfN had the largest magnetoelastic anisotropy energy than 3.3-μm and 5.5-μm thick FeHfN films. The magnetization direction of 1.1-μm FeHfN was toward out-planes and perpendicular anisotropy formed. Finally, the stripe domain structure was revealed in 1.1-μm FeHfN film. To sum up, the largest compressive stress in the 1.1-μm FeHfN film with positive magnetostriction producesd largest perpendicular anisotropy among various thickness of FeHfN films. Thus, 1.1-μm FeHfN film had out-plane orientation of magnetization.

Moreover, several studies [13-15] have confirmed that the relationship between lattice strain, $\Delta d_{(110)}/d_{(110)}$,and magnetic anisotropy energy, $U_{(110)}$. In the case of shear deformation of Fe, $U_{(110)}$ enhances with increasing strain $\Delta d_{(110)}/d_{(110)}$.The relatively large value of magnetic anisotropy energy of Fe degrades soft magnetic properties of Fe-based alloy films. In this study, 1.1-μm FeHfN showed the largest lattice strain than the others, and thus had the relatively large magnetic anisotropy energy $U_{(110)}$.Therefore, the 1.1-μm FeHfN showed poor soft magnetic properties.

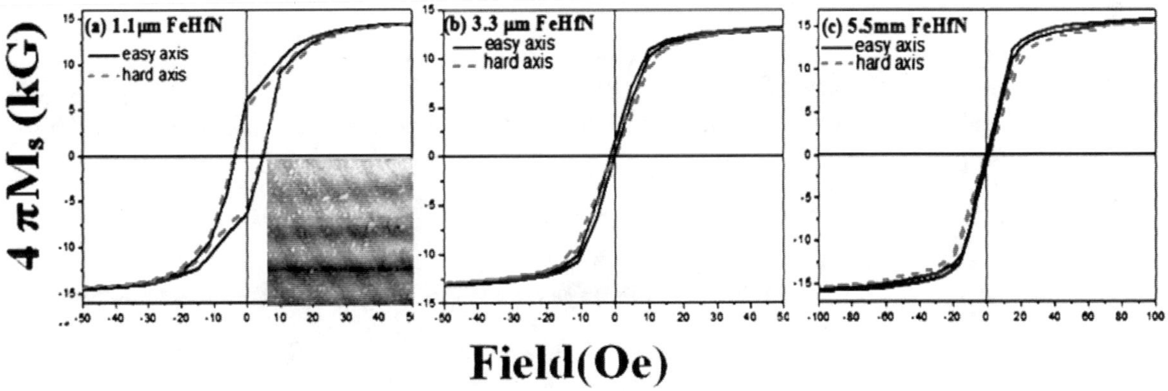

Fig. 4. In-plane hysteresis loops for as-deposited FeHfN films with thickness of (a) 1.1μm (b) 3.3μm (c) 5.5μm.

The magnetic properties, grain size, spacing d_{110} and residual stress as a function of thickness are summarized in Table I. The 5.5-μm FeHfN film had a nanocrystalline structure and good soft magnetic properties, such as Hc less than 1 Oe and $4\pi Ms$ = 16.9 kG.

Table I. Compositions, magnetic properties, grain size, spacing d_{110} and stress of FeHfN films with different thickness

Thickness (μm)	Composition (at.%)	4πMs (kG)	Hc (Oe)	Grain size(nm)	d_{110}(Å)	σ(GPa)
1.1	$Fe_{87.3}Hf_{8.4}N_{4.3}$	15.1	4.77	5.78	2.072	-1.137
3.3	$Fe_{88.1}Hf_{8.3}N_{3.6}$	14.1	0.62	5.85	2.067	-0.827
5.5	$Fe_{88.6}Hf_{7.4}N_{4.0}$	16.9	0.95	6.25	2.067	-0.817

4. Conclusions

This study revealed the magnetic properties of nanocrystalline FeHfN films with varying thicknesses as deposited by dc sputtering. The soft magnetic properties of the FeHfN films strongly depended on film thickness. The relatively large coercivity (Hc) and the stripe domain structure were revealed in 1.1-μm FeHfN film. The absence of stripe domain was resulted from the stress-induced magnetoelastic anisotropy. However, the FeHfN films thicker than 1.1μm did not exhibit stripe domain structures and had small coercivity (< 1 Oe). The 5.5-μm FeHfN films with a nanocrystalline structure showed good soft magnetic properties. The superior magnetic properties of the films exhibit the potential application for power inductor.

Acknowledgements

This work was financially supported by Ministry of Economics, Taiwan, under the contract No. 98-EC-17-A-08-S1-003.

References

[1] S. Li, J.G. Duh, C. N. Liao, *Thin Solid Films* **516** (2008)

[2] D.S. Gardner, G. Schrom, F. Paillet, B. Jamieson, and S. Borkar, *IEEE Trans. Magn.*, **45**, (2009)

[3] V. Korenivski, *J. Magn. Magn. Mater.* 215 (2000)

[4] H. Y. Wang, E. Y. Jiang, H. L. Bai, Y. Wang and F. F. Gong, *J. Phys.: Condens. Matter.* **9** (1997)

[5] A. Makino, and Y. Hayakawa, *IEEE Trans. Magn.*, **31** (1995)

[6] B. Viala, M. K. Minor, and J. A. Barnard, *J. Appl. Phys.* **80** (1996)

[7] H. J. Jeon, I. Kim, K. H. Kim, M. Yamaguchi,*J. Magn. Magn. Mater.* **272-276** (2004)

[8] Y. M. Kim, S. H. Han, H. J. Kim, D. Choi, and J. Kim, *J. Appl. Phys.* **91** (2002)

[9] G. Herzer, *IEEE Trans. Magn.* **26**, No. 5 (1990)

[10] G. G. Stoney, Proc. R. Soc. London, Ser. A. 82, 172 (1909)

[11] J. McCord, S. Dieter, and K. Seemann, *J. Magn. Magn. Mater.* **271**, Issue 1 (2004)

[12] N. Saito, H. Fujiwara and Y. Sugita, *J. Phys. Soc. Jpn.* **19** (1964)

[13] M. Takahashi, and T. Shimatsu , *IEEE Trans. Magn.* **26**, No. 5(1990)

[14] M. Takahashi, H. Shoji, T. Shimatsu, H. Komaba, and T. Wakiyama , *IEEE Trans. Magn.* **26**, No. 5 (1990)

[15] M. Takahashi, and T. Shimatsu , *J. Magn. Magn. Mater.*, 101 (1991).

Journal of Physics: Conference Series **266** (2011) 012079 doi:10.1088/1742-6596/266/1/012079

Dependence of pinning on domain wall spin structure and notch geometry

S Goolaup, S C Low, M Chandra Sekhar and W S Lew
Physics and Applied Physics Division, School of Physics and Mathematical Sciences, Nanyang Technological University, 21 Nanyang Link, Singapore 637371

E-mail: wensiang@ntu.edu.sg

Abstract. In this work, we present a systematic investigation of the domain wall spin structure and notch geometries on the pinning field strength. We observed that for transverse domain wall, pinning is strongly dependent on the transversely varying energy profile of the wall. Domain walls are pinned at notches only when the notch provides a barrier to the higher energy component of the domain wall. For all notch shapes investigated, we observed that when the notch height/nanowire width > 0.3, the depinning field reaches a maximum and remains constant. We also note that the pinning of a domain wall at a notch is markedly sensitive to the angle of the notch with respect to the domain wall.

1. Introduction

Magnetic domain walls in ferromagnetic nanowires are the focus of intense interest both from a fundamental perspective and due to their potential applications. Magnetic logic [1] and memory devices [2] based on domain wall (DW) motion and manipulation have been recently proposed and experimentally demonstrated. A key issue in the application of DW devices is the ability to manipulate the DW configurations using magnetic field [3,4] or spin polarized current [5,6]. As such, a detailed knowledge of the DW nucleation and propagation is needed. The control and manipulation of magnetic domain walls are the focus of intense research. Deliberately fabricated defects in ferromagnetic nanowires allow for the control of the position of DW as they create changes in the energy landscape which increase the propagation field. Artificial nanowire (NW) defect of different geometries acting as trapping sites have been investigated using both numerical simulation and experimental observations [7-10]. The strength of the pinning is directly correlated with the wall spin structure so that only the determination of its nanoscale spin structure will enable an in-depth understanding of the energetics governing the pinning strength. However, a detailed description of the pinning effect at the notch is still lacking.

In this work we present a systematic study on the propagation and pinning of transverse domain wall on the wall spin structure and notch geometry in nanowires. We observed that domain wall pinning is strongly dependent on the transversely varying energy profile of the wall. The pinning effect is highly effective when the notch provides a potential well to the higher energy component of the domain wall. For notch height/wire width > 0.3, the depinning field for the domain wall remains constant irrespective of the notch shape. We also observed that the angle of the notch with respect to the domain wall greatly determines the pinning strength of the notch.

2. Micromagnetic Simulation

The magnetization configuration and depinning processes were simulated using the Object Oriented Micromagnetic Framework (OOMMF) simulation software [11]. Domain wall pinning was simulated for a planar nanowire with various notch geometries and configurations. The structure simulated comprised of a large nucleation pad followed by a nanowire with a pointed end as seen in the schematic shown in Fig 1. The nanowire was 2um in length and 200 nm wide. The structure also comprises of a wire along the y-direction. This wire acts as a chirality filter to set the orientation of the transverse component of the domain wall generated. The position of the notch was set at 1um from the filter. The film thickness was maintained at 10nm. For the simulation, the parameters for Permalloy were used; saturation magnetization, Ms = 860 x 10^3 A/m, exchange constant, A = 1.3 x 10^{-11} J/m, with a cell size of 5 nm. The magnetocrystalline anisotropy was set to zero.

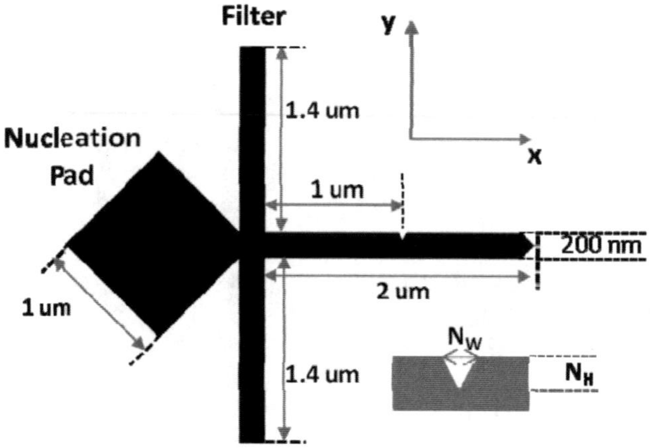

Figure 1: Simulated geometry consisting of a nucleation pad, a chirality filter and a nanowire of width 200nm. The notch is located at 1um from the filter. The notations used to define the notch geometry is also shown

For all the simulated geometries and domain wall type, a field of 1 kOe was applied along the positive(negative) x-direction to saturate the nanowire and then reduced to zero. This was followed by the application of small field of (-)200 Oe along the y-direction to set the orientation of the filter. The field was then increased incrementally in the x-direction to generate a domain wall in the wire. To obtain an accurate depinning field, the simulation parameters for the while were set, so that each step corresponds to a change of 0.5 Oe.

3. Results and Discussion

3.1. Domain Wall Chirality

In this section, we investigate the influence of domain wall spin structure on the pinning strength at notches along the nanowire. Four type of transverse domain walls (TDW) were generated during the simulation; Head-to-Head (HH) or Tail-to-Tail (TT) configuration with an "Up" or "Down" chirality as shown in Fig 2(a). The triangle encompassing the y-component of the wall, Fig 2(a), depicts how the transverse spin components are aligned within the TDW.

For this simulation, a triangular notch with notch width, N_W and height, N_H set at 66nm (~1/3 of nanowire width) was chosen. Three different types of notch configuration were investigated; with the notch positioned at the upper edge or/and lower edge of the nanowire as seen in Fig 2(b). The depinning fields, as extracted from the simulation, for the different types of TDW and notch configurations are shown in Fig 2(b). In our simulation, the field needed to inject a domain wall into the nanowire was (-)100 Oe. At this field, some DW simply passed through the notch without any evidence of pinning. For these DWs, we have set the depinning field to be equal to the nucleation field of (-)100 Oe.

(a)

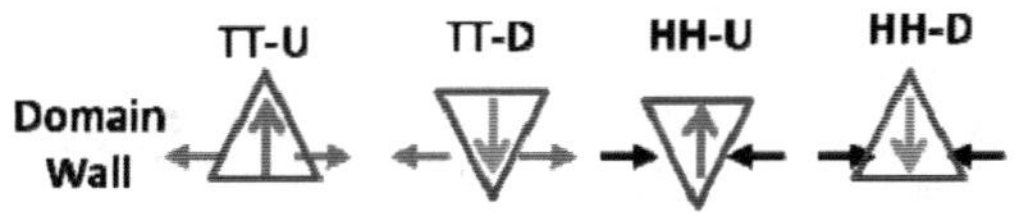

(b)

	Notch	Depinning Field (Oe)			
		TT-U	TT-D	HH-U	HH-D
1		-100	-257.5	257.5	100
2		-257.5	-100	100	257.5
3		-262.5	-262.5	262.5	262.5

Figure 2: (a) Four types of transverse domain walls (b) Depinning field strength for different notch configurations and domain walls. Domain walls that passed through the notch were assigned a value of (-)100 Oe which equals to the nucleation field of the domain wall.

When the notch is on the upper edge of the nanowire, Fig 2(b)-1, we note that the TDW with Tail-to-Tail with "Down" chirality (TT-D) and Head-to-Head with "Up" chirality (HH-U) are pinned. For both these wall, TT-D and HH-U, the field needed to depin from the notch has the same magnitude of 257.5 Oe. On the otherhand, the TT-U and HH-D simply passed through the notch.

For the notch on the lower edge of the nanowire, Fig 2(b)-2, we observed a reversal in the pinning in the TDW. In this case, the TT-U and HH-D are pinned at the notch, while requiring the same exact field of 257.5 Oe to depin from the notch. The HH-D and TT-U walls passed through the notch. For a double notch configuration, Fig 2(b)-3, we observed that all the TDWs are pinned and required a field of 262.5 Oe to depin. Interestingly, the field needed to depin the wall is marginally larger as compared to the field of single notch configuration by 5 Oe, an increase of less than 2%.

As the notch acts as a potential well, since all the TDWs have the same depinning field for a double notch geometry, this implies that all the TDWs configuration have the same total energy. The energy profile of the TDWs though may be different, as evidenced from the pinning of domain wall pairs at the notches on the lower and upper edge of the nanowire.

To better understand this result, we look into the energies of the domain wall. From the domain wall theory, the energy of the domain wall, E_{DW} is given by [13]:

$$E_{DW} \square \frac{\pi^2 A}{\partial} + K\partial \qquad (1)$$

where A is the exchange stiffness constant, K is the anisotropy energy density and δ is the wall thickness. The domain wall thickness, δ, is a function of the number of planes (N) over which the magnetization rotates and the lattice constant (a). As such, for the transverse domain wall, δ vary as function of the width of the transverse component of magnetization, which has a triangular shape as seen in Fig 2(a). Thus, all the TDWs have the same energy but their transverse energy profile varies as a function on their specific spin structure. Due to the triangular shape of the transverse component of the TDW, the energy of the wall varies along the y-direction, with the base of the wall (triangle) having the highest energy. From Fig 2(b), we note that pinning is effective only when the notch provides a potential barier to the highest energy component of the domain wall.

Journal of Physics: Conference Series **266** (2011) 012079 doi:10.1088/1742-6596/266/1/012079

3.2. Geometry Dependence

To gain a better understanding of the pinning mechanism in the nanowire structures, we have investigated of the notch geometry on the pinning field of the nanowires. For comparison, the notch depth was set to 66nm, while the notch width on the upper edge of the nanowire was kept at 133nm. Shown in Fig 3 is the depinning field for a Tail-to-Tail "Down" chirality (TT-D) transverse domain wall for all the different geometries. We observed that the depinning field is highly dependent on the geometry of the notch. The depinning field ranged from 212.5 Oe for a trapezoidal shaped notch to a maximum of 228 Oe for a triangular notch.

Geometry	Trapezium 	Rectangle 	Semicircle 	Triangle 
Depinning Field (Oe)	212.5	220	227.5	228

Figure 3: **Depinning field strength for a Tail-to-Tail "Down" chirality for different notch geometries. The notch width was set to 133nm and notch height was 66nm.**

3.3. Height and Width Variation

The depinning field has been predicted to increase with decreasing constriction width, implying that walls in narrower constriction are more strongly pinned [14,15]. In this section, we investigate the effect of different notch heights on the depinning field of DWs that are trapped at a notch. For this set of simulations, we have investigated the depinning field of a Tail-to-Tail TDW with "Down" chirality, TT-D at a notch on the upper edge of the nanowire. Four notch geometries as discussed in section 3.2 were investigated. In Fig 4(a), we present the plot of the depinning field as a function of the notch height,N_H to nanowire width, NW_W ratio, N_H/NW_W. For all the geometries investigated, we observed that the depinning field increases with the notch height, until a N_H/NW_W ratio of 0.3. Beyond this ratio, the depinning field remains constant, irrespective of the notch height and geometry. This is in contradiction of the assumption that as notch height increases, the depinning strength should increase. Our result, can be explained by looking at the energy profile of the TDW. As the transverse spin structure adopts the shape of an isoceles triangle, a height of 0.3 from the base of the wall, correspond to an area containing almost 50% of the wall energy. From section 3.1, in the double notch configuration, we have shown that once the higher energy component of the wall is impeded, the pinning of the lower energy component results in marginal increase in the depinning field.

We have further investigated the effect of the notch width on the depinning field of the domain wall. In Fig 4(b), we present the plot of the depinning field of a transverse down Tail-to-Tail (TT-D) DW for different notch geometries, as a function of the notch width to wire width, while the notch height is kept fixed. We observed a non-monotonic variation of the depinning field, with a maximum field 277Oe occurring at a N_W/NW_W ratio of 0.65. Further increase in the notch width leads to a gradual decrease of the depinning field until a N_W/NW_W of 1.5. Beyond this N_W/NW_W, the wall just passes through the notch with a depinning field of 100 Oe. This can be explained by the fact that the potential landscape as induced by the notch changes as the notch width increases. The notch can be seen as acting as an abrupt energy barrier to the domain wall. An increase in the notch width while maintaining fixed notch height, leads a gradual change in the width of the wire at the notch, as seen in the inset of Fig 4(b). Thus the angle of the notch with respect to the domain wall decreases, leading to a smoother change in the energy barrier. For the rectangular notch, an almost constant depinning field of 210 Oe is obtained for $N_W/NW_W > 0.5$. This can be explained by the fact that, for the rectangular geometry, the angle of the notch remains fixed irrespective of the notch width.

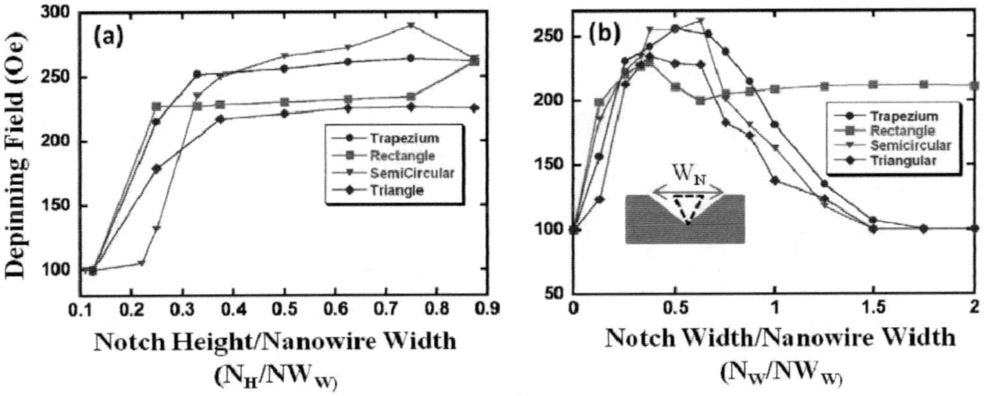

Figure 4: Depinning field strength for a Tail-to-Tail "Down" chirality for different notch geometries as a function of; (a) notch height, N_H while the notch width, N_W is kept fixed (b) notch width, N_W while the notch height, N_H is kept fixed

4. Conclusion

In summary, we have investigated the depinning of transverse domain wall at notches of different geometry and shapes. We note that the pinning of a domain wall at a notch is highly dependent on the energy profile of the domain wall. The pinning of a domain wall by a notch is most effective when the notch acts as a potential barrier to the higher energy component of the domain wall. We observed that depinning field strength does not scale with the notch height. Beyond a height to wire width ratio of 0.3, the depinning field remains constant. As regards to the width of the notch, we observed that the angle of the notch with respect to the domain wall determines the strength of the depinning field. The smaller the angle, the gradual is the change in the potential landscape, resulting in a decrease in the depinning field.

References

[1] Allwood, D.A., et al. 2005 *Science* **309** 5741
[2] Parkin, S.S.P., M. Hayashi, and L. Thomas 2008 *Science* **320** 5873
[3] R. P. Cowburn, M. E. Welland 2000 *Science* **287** 1466
[4] D. A. Allwood, N. Vernier, Gang Xiong, M. D. Cooke, D. Atkinson, C. C. Faulkner, and R. P. Cowburn 2002 *Appl. Phys. Lett.* **81** 4005
[5] L. Berger 1978 *J. Appl. Phys.* **49** 2156
[6] J. Slonczewski 1996 *J. Magn. Magn. Mater* **159** L1
[7] Wilhelm, E.S et al. 2009 *Applied Physics Letters* **95**(25) 252501
[8] Uhlig, W.C. et al. 2009 *Journal of Applied Physics* **105**(10) 103902
[9] Lepadatu, S. et al. 2009 *Physical Review Letters* **102** 127203.
[10] Kunz, A. 2009 *Applied Physics Letters* **94**(13) 132502.
[11] M. J. Donahue and D. G. Porter, OOMMF User's Guide Version 1.0, http://math.nist.gov/oommf
[12] Tchernyshyov, O. and G.-W.Chern 2005 *Physical Review Letters* **95** 197204.
[13] Marrows, C.H. 2005 *Advances in Physics* **54**(8) 585

[14] C. C. Faulkner, M. D. Cooke, D. A. Allwood, D. Petit, D. Atkinson and R. P. Cowburn 2004 J. Appl. Phys. **95** 6717
[15] M. Klaui et al. 2005 *App. Phys. Lett.* **87** 102509

Journal of Physics: Conference Series **266** (2011) 012080 doi:10.1088/1742-6596/266/1/012080

Detection of vortex-core dynamics using current-induced self-bistable rectifying effect

M Goto[1], H Hata[1], A Yamaguchi[1, 2], H Miyajima[1], Y Nakatani[3], T Yamaoka[4] and Y Nozaki[1, 5]
[1]Department of Physics, Keio University, 3-14-1 Hiyoshi, Yokohama, Kanagawa 223-8522, Japan
[2]PRESTO, JST, Honcho, 4-1-8, Kawaguchi, Saitama 332-0012, Japan
[3]University of Electro-Communications, 1-5-1 Chofugaoka, Chofu, Tokyo 182-8585, Japan
[4]SII Nanotechnology Inc., RBM Tsukiji Bldg., Shintomi, Chuo, Tokyo 104-0041, Japan
[5]CRESTO, JST, 5, Sanbancho, Chiyoda-ku, Tokyo, 102-0075, Japan

E-mail to M.G.:mgoto@phys.keio.ac.jp
to A.Y. :yamaguch@phys.keio.ac.jp

Abstract. A magnetic vortex core confined in a micron-scale magnetic disk is resonantly excited by spin-polarized radio-frequency (rf) current and rf field. We show that rectifying voltage spectra caused by the vortex core resonance is dependent on the core polarity. Rectifying voltage spectra are given by the superposition of the polarity-dependent term and the polarity-independent term. The sign of the polarity-dependent rectifying voltage reverses when the sign of polarity *P* or external field *H* is reversed. This experimental result can be explained by the anisotropic magnetoresistance effect caused by the vortex core motion.

I. Introduction

The magnetic vortex state in micron-scale soft magnetic disk often exhibits a characteristic equilibrium configuration in which the magnetization curls within the plane of the disk around a singular point with out-of-plane magnetization [1, 2]. The vortex core is a small object, with a

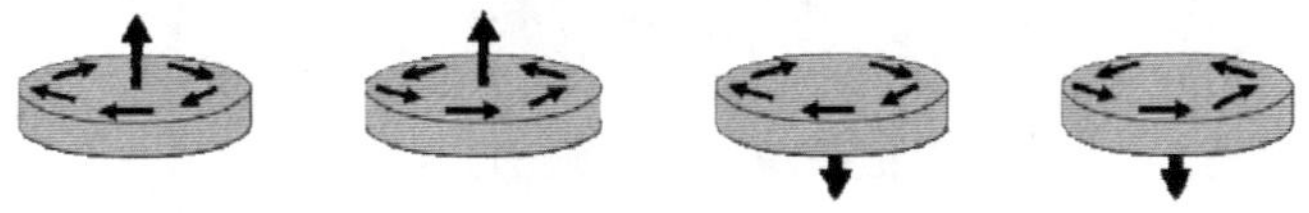

Figure 1. Schematic diagram of magnetic vortices. Black arrows indicate the direction of magnetization. Four magnetic vortex states, polarity (up or down) and chirality (clockwise or counterclockwise), are shown. *P* and *C* denote polarity and chirality, respectively.

diameter of only ~10 nm in $Fe_{19}Ni_{81}$ disk. In particular, when a vortex confined in the disk is driven out of equilibrium by either a magnetic field or a spin-polarized current, the vortex core gyrates around its equilibrium position.

A magnetic vortex is characterized by two Boolean parameters, the polarity and the chirality as shown in figure 1. The polarity is defined by the core magnetization direction pointing out of the plane of the vortex, either up or down ($P = \pm 1$). The chirality is defined by the direction of rotation of the magnetization around the vortex core, either clockwise or counterclockwise ($C = \pm 1$). These four states are degenerated; it is difficult to identify the vortex state [3]. Here, we present a detection of vortex core dynamics using current-induced rectifying effect.

Ⅱ. Experimental Setup

$Fe_{19}Ni_{81}$ disk with diameter of 3 μm and thickness of 30 nm were patterned directly onto a polished MgO substrate by means of electron-beam lithography and lift-off techniques. We had designed the circular disk with two tags on both sides; this shape enabled us to control the vortex polarity and chirality as shown in figures 2a and b. The polarity was controlled by the magnetic field of 5 kOe normal to the plane [4], and the chirality was controlled by the in-plane saturation magnetic field along magnetization reversal process [5]. We controlled the polarity and chirality simultaneously to apply the oblique magnetic field from sample plane. We applied the magnetic field of 5 kOe to the

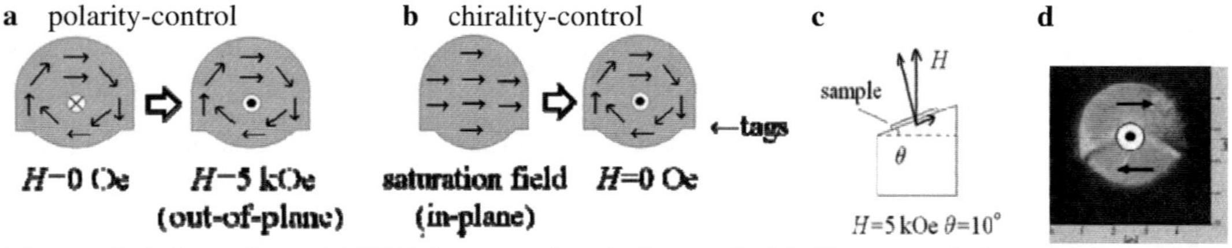

Figure 2. Schematic and MFM images of polarity- and chirality-controlled magnetic disks. **a,** polarity-control method for applying magnetic field of 5 kOe normal to the plane. **b,** chirality-control method for applying in-plane saturation magnetic field. Two tags of circular disk nucleate a chirality-controlled vortex core. **c,** The method for applying in-plane and out-of-plane magnetic field simultaneously. We apply the magnetic field of 5 kOe to the sample on the brazen oblique foundation, and the tilt angle θ is 10°. **d**, MFM image of the designed disk with $(P,C)=(+1,+1)$.

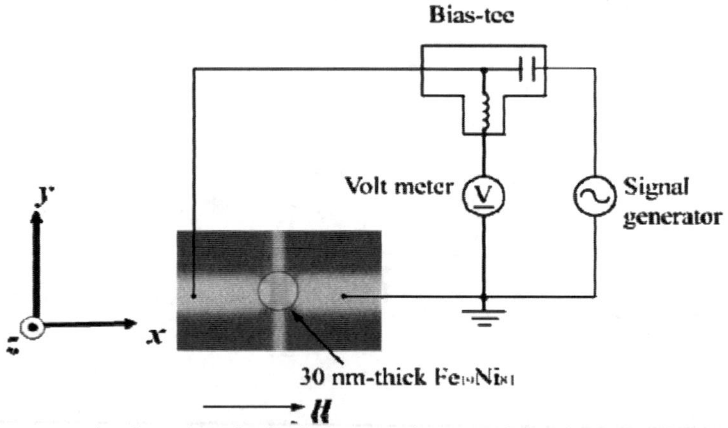

Figure 3. Measurement circuit and optical microscope image of the $Fe_{19}Ni_{81}$ disk in the absence of external magnetic field. The + direction of external magnetic field H is defined as shown in this figure.

sample on the brazen oblique foundation, with a tilt angle $\theta = 10°$ as shown in figure 2c. In order to reveal the relation between the vortex state and rectifying spectra, the vortex state is confirmed by the direct magnetic force microscope (MFM) observation before measurement of the rectifying spectra. Figure 2d shows the MFM image of the circular disk with two tags. This direct observation enabled us to confirm the up polarity and clockwise circulation.

The coplanar waveguide structure made from Au (80 nm) / Cr (5 nm) was connected to the disk, and the center conductive strip line was placed on the disk as shown in figure 3 [6]. A sinusoidal continuous wave rf current with a power of -13 dBm was subsequently injected into the disk by a signal generator in the frequency range between 10 MHz and 200 MHz. A dc voltage along the rf current direction was measured via the bias-tee, which separates the dc and rf components of the current. The measurement circuit and an optical microscope image of the disk are shown in figure 3.

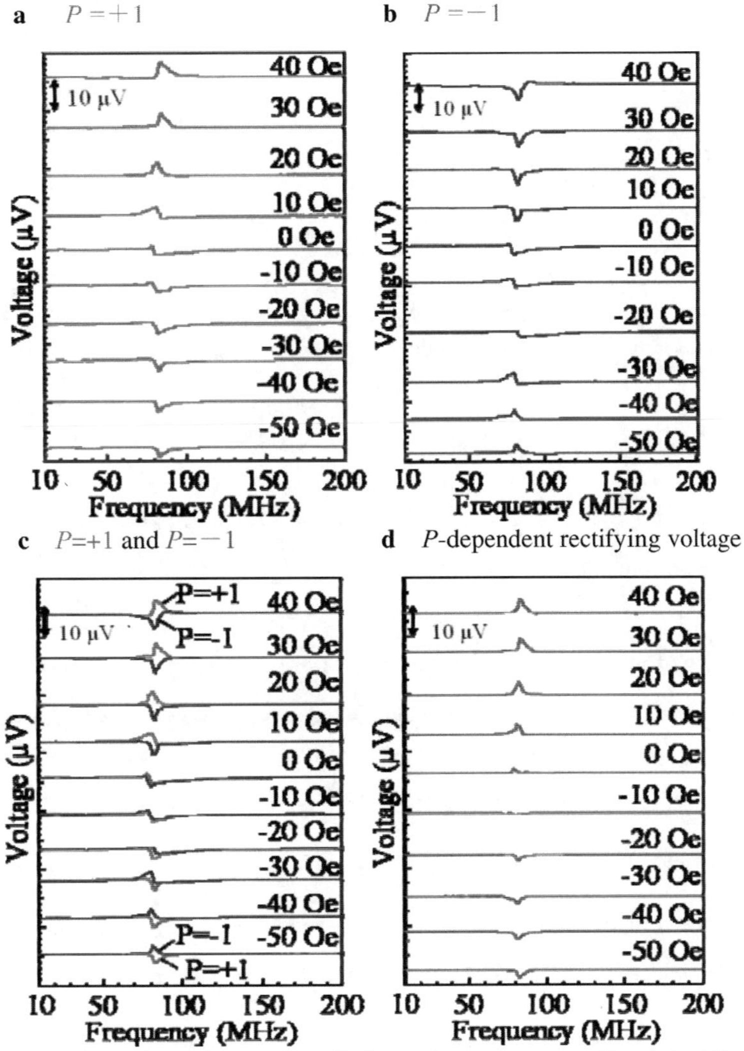

Figure 4 a, b. External field dependence of rectifying voltage spectra at P=+1 (red lines) and P=−1 (blue lines) **c,** polarity and external field dependence of rectifying voltage spectra. **d,** The polarity-dependent rectifying voltage $V_{P\text{-}dep}(P, C, H, \omega)$.

The external field parallel to the rf current was applied in the range of -50 Oe to 40 Oe. Here, the Hall electrode was not used. All measurements were performed at room temperature. The + direction of the external field is defined as shown in figure 3.

Ⅲ. Experimental Result

Figures 4a and b show the external field dependences of the rectifying voltage spectra at $P=+1$ and $P=-1$, respectively. The chirality was kept counterclockwise ($C=-1$) in this study. Above $H=+10$ Oe, the spectra for $P=+1$ are convex upwards, while the spectra for $P=-1$ are convex downwards. In contrast, below $H=-10$ Oe, the spectra for $P=+1$ are convex downwards, while the spectra for $P=-1$ are convex upwards. This result demonstrates that the sign of the rectifying voltage reverses when the sign of polarity P or external field H is reversed. These spectra at variable external field are in the same graph as shown in figure 4c to compare the dependence of rectifying voltage spectra on polarity. This figure in the vicinity of $H = 0$ Oe demonstrate that the polarity dependence of rectifying voltage is vanished, while the rectifying voltage is not vanished. These experimental results demonstrate that rectifying voltage is given by the superposition of polarity-dependent term and polarity-independent term. The experimental rectifying voltage spectra can be explained by the following equations,

$$V(P,C,H,\omega)=V_{P-dep}(P,C,H,\omega)+V_{P-indep}(C,H,\omega) \tag{1}$$

$$V_{P-dep}(-P,C,H,\omega)=-V_{P-dep}(P,C,H,\omega) \tag{2}$$

$$V_{P-dep}(P,C,-H,\omega)=-V_{P-dep}(P,C,H,\omega) \tag{3}$$

Here V (P, C, H, ω), $V_{\text{P-dep}}$ (P, C, H, ω) and $V_{P\text{-indep}}$ (C, H, ω) are total rectifying voltage, polarity-dependent and polarity-independent rectifying voltage, respectively. The rectifying voltage V (P, C, H, ω) is given by the superposition of polarity-dependent term $V_{P\text{-dep}}$ (P, C, H, ω) and polarity-independent term $V_{P\text{-indep}}$ (C, H, ω). P, C, H and ω are polarity, chirality, external magnetic field and frequency of rf current, respectively. $V_{P\text{-dep}}$ (P, C, H, ω) is derived from equation (1) and equation (2),

$$V_{P-dep}(P,C,H,\omega)=\frac{1}{2}\{V(P,C,H,\omega)-V(-P,C,H,\omega)\} \tag{4}$$

The spectrum represented by equation (4) is shown in figure 4d.

Ⅳ. Analysis

Our model is twofold; first we derive the trajectory of the vortex core position $\boldsymbol{r}$ under the rf current density $\boldsymbol{J}$, and second, we calculate the time dependence of the anisotropy magneto-resistance (AMR) based on the trajectory. The (x, y, z) coordinate system is defined as schematically shown in figure 3. In general, the motion of a vortex core can be described by spin torque included Thiele equation [3, 7],

$$\boldsymbol{G}(P)\times(\boldsymbol{u}-\dot{\boldsymbol{r}})=-\frac{\delta U}{\delta \boldsymbol{r}}-\alpha D\dot{\boldsymbol{r}}+\beta D\boldsymbol{u} \tag{5}$$

Here $\boldsymbol{G}(P)=-P\frac{2\pi LM_S}{\gamma}\boldsymbol{e}_z$ is the polarity dependent gyrovector. L and M_S are the thickness and saturation magnetization of the disk, respectively, and γ is the gyromagnetic ratio. $\boldsymbol{u}$ is the spin torque interaction term of the current and the magnetization, which is given by $\boldsymbol{u}=\frac{\mu_B pJ}{eM_S}\boldsymbol{e}_x$ and is proportional to the current density J, with Bohr magneton μ_B, electron charge e and spin polarization ratio p. α is the Gilbert damping, β is the non-adiabatic contribution to the spin-transfer torque and D is the diagonal element of damping tensor. U denotes the sum of exchange energy, magnetostatic energy and Zeeman energy in magnetic disk. We approximate the parabolic potential well; $U(\boldsymbol{r}) = \frac{1}{2}\kappa(\boldsymbol{r} - \boldsymbol{r}_0)^2$, where $\boldsymbol{r}_0$ is the equilibrium position of the vortex core, and κ is the effective stiffness coefficients.

The resistance in magnetic disk depends on the position of vortex core arising from the AMR effect in $Fe_{19}Ni_{81}$ [8]. The resistance decreases with moving the vortex core in the direction of the current ($\pm x$ direction), and increases with moving the vortex core in the direction perpendicular to the current ($\pm y$ direction), respectively. Then, we approximate the disk resistance near the center of the disk by the following equation,

$$R = R_0 - a_x x^2 + a_y y^2 \tag{6}$$

Here x and y are the vortex core position. Proportionality constant are defined a_x and a_y. R_0 is the resistance for x=y=0 (the vortex core position is the center of the disk).

The solution of equation (5) can be written in the form,

$$\begin{aligned} x &= x_0 + \mathrm{Re}\{X(\omega)\exp(i\omega t)\}, \\ y &= y_0 + \mathrm{Re}\{PY(\omega)\exp(i\omega t)\} \end{aligned} \tag{7}$$

where x_0 and y_0 are the equilibrium position of the vortex core. $X(\omega) = X'(\omega) + iX''(\omega)$ and $Y(\omega) = Y'(\omega) + iY''(\omega)$ are the complex oscillation amplitudes [3]. As can be seen from equation (7), the polarity P only exists in y. This is because the direction of gyroforce $\boldsymbol{G}(P) \times \boldsymbol{u}$ in equation (5) depends on the polarity. Then, we substitute x and y into equation (6). As a result, the rectifying voltage V_{dc} is derived from resistance R in equation (6) and rf current $I = I_0 \exp(i\omega t)$, where I_0 is the amplitude of current. Then, we obtain the rectifying voltage,

$$V_{dc} = -a_x I_0 x_0 X'(\omega) + a_y I_0 P y_0 Y'(\omega) \tag{8}$$

By considering the fact that y_0 is proportional to the external field when the vortex core position is near the center of the disk, the sign of the second term spectrum $a_y I_0 P y_0 Y'(\omega)$ in equation (8) reverses when the sign of polarity P or external field H is reversed. Therefore, the experimental rectifying voltage spectrum can be explained by the following equation,

$$V_{P-dep}(P, C, H, \omega) = a_y I_0 P y_0 Y'(\omega) \tag{9}$$

The experimental result as shown in figure 4 can be explained on the basis of the equation (9). The details of the calculation have been given by supplementary information in reference [3].

V. Conclusion

We have demonstrated the dependence of rectifying voltage spectra excited by the rf current on vortex polarity in the micron-scale magnetic disk. These experimental spectra are good agreement with our analytical model. Our experimental result provides a way to detect the vortex core state by using current-induced rectifying effect.

[1] Wachowiak A, et al. 2002 *Science* vol 298, p 577
[2] Shinjo T, Okuno T, Hassdorf R, Shigeto K and Ono T 2000 *Science* vol 298, p 930
[3] Moriya R, Thomas L, Hayashi M, Bazaliy B Y, Pettner C and Parkin P S, 2008 *Nature*. vol 4, p 368
[4] Okuno T, Shigeto K, Ono T, Mibu K and Shinjo T, 2002 *J. Magn. Magn. Mater* vol 1, p 240
[5] Taniuchi T, Oshima M, Akinaga H and Ono K, 2005 *J. Appl. Phys*. vol 97, p 10J904
[6] Yamaguchi A, Motoi K, Miyajima H, Hirohata A, Yamaoka T, Uchiyama T and Utsumi Y, 2009 *Appl. Phys. Lett.* vol 95, p 122506
[7] Thiele A A, 1973 *Phys. Rev. Lett.* vol 30, p 230
[8] Kasai S, Nakatani Y, Kobayashi K, Kohno H and Ono T, 2006 *Phys. Rev. Lett.* vol 97, p 107204

Journal of Physics: Conference Series **266** (2011) 012081 doi:10.1088/1742-6596/266/1/012081

Charge and spin currents in magnetic multi layers in the presence of both electric field and spin dynamics

Daisuke Miura and Akimasa Sakuma

Department of Applied Physics, Tohoku University, 6-6-05 Aoba, Aoba-ku, Sendai 980-8579, Japan

E-mail: d_miura@solid.apph.tohoku.ac.jp

Abstract. The number and the spin angular momentum of electrons on each site, and the charge and the spin currents between adjacent sites are theoretically studied. Considering one dimensional electronic system including impurity atoms with some density, the static electric field (represented by an electric potential) and the s-d interaction with spin dynamics are treated perturbatively. We show that the effect of the applied electric field can be understood from the ideas of the drift by the field and the band shift due to the electric potential.

1. Introduction

In most spintronic devices consisting of magnetic multi layers, an interplay between spin and charge currents is believed to play an essential role for its peculiar phenomena. For that reason, the establishment of methods generating and controlling the currents is the central issue in the study of spintronics, and elucidating the mechanisms of the creation and the control of these currents is strongly desired.

It is confirmed experimentally that spin dynamics induce spin currents [1][2]. In particular, the phenomenon in metallic magnetic multi layers is called the spin-pumping. The spin-pumping, the conversion from spin dynamics to spin currents, have been studied based on phenomenological theories [3][4][5]. Takeuchi *et al.*[6] have recently discussed a spin current in the dynamical spin system from the microscopic viewpoints. They have shown that a spin current is pumped into non-magnetic layers from magnetic layers with spin dynamics such as a precession due to a magnetic field. In this paper, we consider the case that an external static homogeneous electric field is applied to magnetic multi layers with spin dynamics, and calculate microscopically the charge current, the spin current, the number and the spin angular momentum of electrons.

2. Model

We assume one dimensional electronic system described by the tight-binding model. Our model Hamiltonian is $H(t) := H_0 + H_{\rm imp} + H_{\rm ex}(t) + H_{\rm bias}(t)$ where $H_0 := \sum_k E_k c_k^\dagger c_k$ describes itinerant electrons with $E_k := \varepsilon_0 - 2T\cos ka$ (ε_0 is an on-site energy, T is a transfer integral between nearest-neighbor sites and a is a lattice constant), and $H_{\rm imp} := \frac{u}{N}\sum_k \sum_{q\neq 0}\sum_{j=1}^{n_{\rm i}} {\rm e}^{-{\rm i}qr_j a} c_{k+q}^\dagger c_k$ describes the elastic electron scattering by impurity atoms with some density $n_{\rm i}/N$ (u is a strength of the impurity scattering, $n_{\rm i}$ is the number of impurities, N is the number of sites), which gives rise to a lifetime of an electron in the Born approximation $\tau := \left(\frac{2\pi\rho n_{\rm i} u^2}{\hbar N}\right)^{-1}$ where ρ is the density of states per a site and a spin at the Fermi energy μ, and $H_{\rm ex}(t) := -J\sum_{k,q}\boldsymbol{S}_q(t)\cdot c^\dagger_{k+q}\hat{\boldsymbol{\sigma}}c_k$ describes the exchange interaction with time-dependent local spins $\boldsymbol{S}_q(t)$ (J is the exchange coupling constant, $\hat{\boldsymbol{\sigma}}$ is a vector of Pauli matrices and a hat '$\hat{}$' denotes a 2×2 matrix in spin-space), and $H_{\rm bias}(t) := -e\sum_{k,q}\phi_q(t)c^\dagger_{k+q}c_k$ ($e>0$ is the elementary charge) describes at once the time-dependent electric potential $\phi_q(t)$ and the electric field $E_q(t) := -{\rm i}q\phi_q(t)$. $-e\phi_q(t)$ means the energy difference from the μ.

3. Method

The number of electrons at the l site is defined as $n_l(t) := \langle c^\dagger_l(t)c_l(t)\rangle_H$, the spin angular momentum at the l site as $\boldsymbol{\sigma}_l(t) := \langle c^\dagger_l(t)\hat{\boldsymbol{\sigma}}c_l(t)\rangle_H$, the charge current flowing from the l site to the $l+1$ site as $I^{\rm C}_l(t) := \frac{\rm i}{\hbar}T\left[\langle c^\dagger_{l+1}(t)c_l(t)\rangle_H - {\rm c.c.}\right]$ and the spin current flowing from the l site to the $l+1$ site as $\boldsymbol{I}^{\rm S}_l(t) := \frac{\rm i}{\hbar}T\left[\langle c^\dagger_{l+1}(t)\hat{\boldsymbol{\sigma}}c_l(t)\rangle_H - {\rm c.c.}\right]$, where $\langle\cdots\rangle_H$ is the expectation value estimated by $H(t)$ and we calculate it within the Keldysh-Green function framework. We can express these quantities using a lesser Green function as follows

$$n_l(t) = \sum_q\int\frac{{\rm d}\omega}{2\pi\hbar}{\rm e}^{{\rm i}(qla-\omega t/\hbar)}{\rm Tr}_\sigma\left[\hat{\Pi}^0_q(\omega)\right],$$

$$\boldsymbol{\sigma}_l(t) = \sum_q\int\frac{{\rm d}\omega}{2\pi\hbar}{\rm e}^{{\rm i}(qla-\omega t/\hbar)}{\rm Tr}_\sigma\left[\hat{\boldsymbol{\sigma}}\hat{\Pi}^0_q(\omega)\right],$$

$$I^{\rm C}_l(t) = \sum_q\int\frac{{\rm d}\omega}{2\pi\hbar}{\rm e}^{{\rm i}(qla-\omega t/\hbar)}{\rm Tr}_\sigma\left[\hat{\Pi}^1_q(\omega)\right],$$

$$\boldsymbol{I}^{\rm S}_l(t) = \sum_q\int\frac{{\rm d}\omega}{2\pi\hbar}{\rm e}^{{\rm i}(qla-\omega t/\hbar)}{\rm Tr}_\sigma\left[\hat{\boldsymbol{\sigma}}\hat{\Pi}^1_q(\omega)\right],$$

where ${\rm Tr}_\sigma[\cdots]$ denotes taking the trace in spin-space, and

$$\hat{\Pi}^\nu_q(\omega) := \frac{1}{N}\int\frac{{\rm d}E}{2\pi{\rm i}}\sum_k j^\nu_k\hat{G}^<_{k_+,k_-}(E_+,E_-),$$

$$j^\nu_k := \left(1, \frac{{\rm e}^{{\rm i}qa/2}}{a}v_k\right).$$

$v_k := \frac{1}{\hbar}\frac{\partial E_k}{\partial k} = \frac{2Ta}{\hbar}\sin ka$ is a group velocity of an electron having an energy E_k, $k_\pm := k\pm q/2$ and $E_\pm := E\pm\omega/2$. $\hat{G}^<_{k,k'}(E,E') := \int{\rm d}t{\rm d}t'{\rm e}^{{\rm i}Et/\hbar}{\rm e}^{-{\rm i}E't'/\hbar}\hat{G}^<_{k,k'}(t,t')$, and $\hat{G}^<_{k,k'}(t,t')$ is the matrix of the lesser Green function with components $\hat{G}^<_{k\sigma,k'\sigma'}(t,t') := \frac{\rm i}{\hbar}\langle c^\dagger_{k'\sigma'}(t')c_{k\sigma}(t)\rangle_H$, ($\sigma=\uparrow,\downarrow$).

4. Result

We discuss the four quantities in perturbative regimes that are showed in figure 1, and take certain extreme limits: the slowly varying limit $\Omega\tau \ll 1$ where Ω is a frequency of a local spin, the electric potential and the electric field (note that for the electric field we must take the limit after the uniform limit $E_q(\Omega)\to\delta_{q,0}E_q(\Omega)$), the metallic limit $\frac{\hbar}{\mu\tau}\ll 1$, and the low temperature limit. However, to focus on structural dependence of magnetic multi layers, we make an effort to reflect wavenumber dependence of spins in the calculation as much as possible. In particular, we calculate equilibrium components using approximation that $E_k \simeq \varepsilon_0 - 2T(1-\frac{k^2a^2}{2})$ and the scope of k is extended from the first Brillouin zone $(-\frac{\pi}{a},\frac{\pi}{a}]$ to $(-\infty,\infty)$ instead of assuming a spatially smooth spin structure. Unfortunately, only Fourier components of small

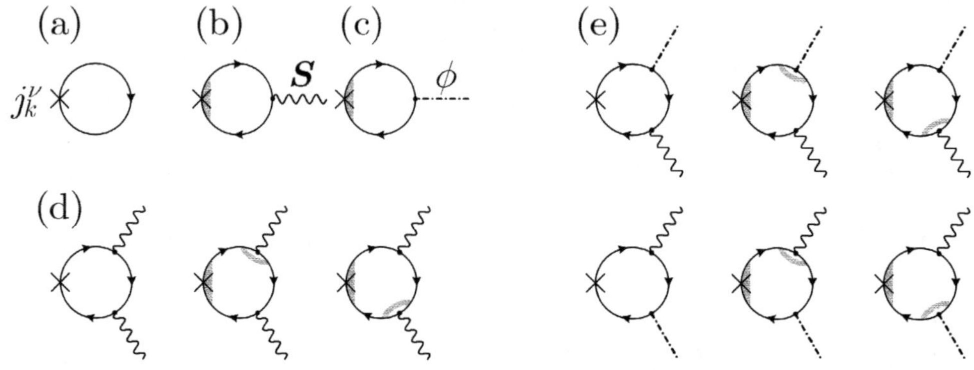

Figure 1. Diagrammatic representation of j_k^ν. Wavy lines and chain lines denote the exchange interaction and the electric potential, respectively. Shaded areas represent the diffusion ladder. Labels distinguish the order of perturbation: (a) denotes the unperturbed system, (b) and (c) the first-order exchange interaction and the first-order electric potential respectively, (d) the second-order exchange interaction, and (e) the processes for both them. The vertex corrections are considered for dynamical components only and the diagrams including a maximal number of diffusion propagators are considered.

wavenumber Q ($v_\mathrm{F}\tau Q \ll 1$, $v_\mathrm{F} := v_{k_\mathrm{F}}$, k_F is the Fermi wavenumber) of spins are considered for dynamical components, and the diagrams including a maximal number of diffusion propagators are considered.

After straightforward calculations, we obtain

$$
\begin{aligned}
n_l(t) =& 2n_0 + \frac{e}{2T}\sum_{l'} F^{(1)}_{2k_\mathrm{F}a(l-l')}\phi_{l'}(t) - 2\rho J P\,(k_\mathrm{F}a)^2 \sum_{l',l''} F^{(2)}_{2k_\mathrm{F}a(l-l'),2k_\mathrm{F}a(l'-l'')}\boldsymbol{S}_{l'}(t)\cdot\boldsymbol{S}_{l''}(t) \\
& - 2\rho J\tau P\,\langle \boldsymbol{S}_l(t)\cdot\dot{\boldsymbol{S}}_l(t)\rangle , \\
\boldsymbol{\sigma}_l(t) =& \frac{J}{2T}\sum_{l'} F^{(1)}_{2k_\mathrm{F}a(l-l')}\boldsymbol{S}_{l'}(t) - 4\rho e P\,(k_\mathrm{F}a)^2 \sum_{l',l''} F^{(2)}_{2k_\mathrm{F}a(l-l'),2k_\mathrm{F}a(l'-l'')}\phi_{l'}(t)\boldsymbol{S}_{l''}(t) \\
& - 2\rho J\tau\,\langle \dot{\boldsymbol{S}}_l(t)\rangle + \frac{4\rho J^2\tau^2}{\hbar}\langle \boldsymbol{S}_l(t)\times\dot{\boldsymbol{S}}_l(t)\rangle - 2e\rho\tau P\,\langle \phi_l(t)\dot{\boldsymbol{S}}_l(t)\rangle , \\
I_l^\mathrm{C}(t) =& -\frac{\sigma}{ea}E_l(t) + \frac{\sigma\tau J}{e^2 a}P\partial_l\,\langle \boldsymbol{S}_l(t)\cdot\dot{\boldsymbol{S}}_l(t)\rangle , \\
\boldsymbol{I}_l^\mathrm{S}(t) =& -\frac{J^2}{2\hbar T}\sum_{l',l''}\mathrm{Si}\left[\pi(l-l')\right] F^{(1)}_{2k_\mathrm{F}a(l'-l'')}\boldsymbol{S}_{l'}(t)\times\boldsymbol{S}_{l''}(t) + \frac{\sigma J\tau}{e^2 a}\partial_l\,\langle \dot{\boldsymbol{S}}_l(t)\rangle \\
& - \frac{2\sigma J^2\tau^2}{e^2 a\hbar}\partial_l\,\langle \boldsymbol{S}_l(t)\times\dot{\boldsymbol{S}}_l(t)\rangle - \frac{\sigma E_l(t)}{ea}P\,\langle \boldsymbol{S}_l(t)\rangle + \frac{\sigma\tau}{ea}P\partial_l\,\langle \phi_l(t)\dot{\boldsymbol{S}}_l(t)\rangle ,
\end{aligned}
$$

where we have introduced quantities, the unperturbative number of electrons $n_0 := \int^{\mu}\mathrm{d}E\rho(E)$, the spin polarization $P := \frac{\rho'}{\rho}J$, the electric conductivity $\sigma := 2e^2\rho D$, the diffusion constant $D := v_\mathrm{F}^2\tau$, the sine integral function $\mathrm{Si}(x) := \frac{2}{\pi}\int_0^x \mathrm{d}t\frac{\sin t}{t}$, $F^{(1)}_\alpha := 1 - \mathrm{Si}(|\alpha|)$, $F^{(2)}_{\alpha,\beta} := \frac{1}{2}\left[f(\alpha+\beta) + f(\beta) + \frac{\mathrm{i}\,\mathrm{sgn}(\alpha)}{2\pi}\int \mathrm{d}y\frac{\mathrm{e}^{\mathrm{i}\beta y}}{y}\int_{-1}^{1}\mathrm{d}x\frac{\mathrm{e}^{\mathrm{i}\alpha x}}{x}\left(\frac{\mathrm{e}^{\mathrm{i}\alpha y}}{x+y} - \frac{1}{x-y}\right)\right]$, $f(x) := \cos x - \frac{\pi}{2}\left(|x| - \mathrm{Si}|x|\right)$, and a spatial derivative operator is interpreted by the expression $\partial_l A_l := \frac{A_{l+1}-A_l}{a}$ where A_l is an arbitrary function of l. $\langle\cdots\rangle$ means diffusion

$$\langle A_l(t)\rangle := \sum_{l'}\int^t \frac{\mathrm{d}t'}{\tau} D_{l-l'}(t-t')A_{l'}(t'),$$
$$\langle A_l(t)B_l(t)\rangle := \sum_{l',l''}\int \frac{\mathrm{d}t'}{\tau}\int \frac{\mathrm{d}t''}{\tau} D_{l-l'}(t-t')A_{l'}(t')D_{l'-l''}(t'-t'')B_{l''}(t''),$$

where, $A_l(t)$ and $B_l(t)$ are arbitrary functions of both l and t. $D_l(t)$ is the diffusion propagator satisfying $\left[\frac{\partial}{\partial t} - D\partial_l^2\right] D_l(t) = \delta_{l,0}\delta(t)$. The results do not break the gauge invariance because $-e\phi_l(t)$ is the energy difference from the μ.

5. Discussion and Summary

First, we focus on the equilibrium components

$$n_l^{(\mathrm{eq})}(t) = 2n_0 + \frac{e}{2T}\sum_{l'} F^{(1)}_{2k_{\mathrm F}a(l-l')}\phi_{l'}(t) - 2\rho JP\,(k_{\mathrm F}a)^2 \sum_{l',l''} F^{(2)}_{2k_{\mathrm F}a(l-l'),2k_{\mathrm F}a(l'-l'')}\boldsymbol{S}_{l'}(t)\cdot\boldsymbol{S}_{l''}(t),$$
$$\boldsymbol{\sigma}_l^{(\mathrm{eq})}(t) = \frac{J}{2T}\sum_{l'} F^{(1)}_{2k_{\mathrm F}a(l-l')}\boldsymbol{S}_{l'}(t) - 4\rho eP\,(k_{\mathrm F}a)^2 \sum_{l',l''} F^{(2)}_{2k_{\mathrm F}a(l-l'),2k_{\mathrm F}a(l'-l'')}\phi_{l'}(t)\boldsymbol{S}_{l''}(t),$$
$$I_l^{\mathrm{C}\ (\mathrm{eq})}(t) = 0,$$
$$\boldsymbol{I}_l^{\mathrm{S}\ (\mathrm{eq})}(t) = -\frac{J^2}{2\hbar T}\sum_{l',l''} \mathrm{Si}\left[\pi(l-l')\right] F^{(1)}_{2k_{\mathrm F}a(l'-l'')}\boldsymbol{S}_{l'}(t)\times\boldsymbol{S}_{l''}(t).$$

Terms containing $F^{(1)}_{2k_{\mathrm F}al}$ in $n_l^{(\mathrm{eq})}(t)$ and $\boldsymbol{\sigma}_l^{(\mathrm{eq})}(t)$ express the Friedel oscillation and the Ruderman-Kittel-Yosida oscillation which decay inversely proportional to al, oscillating with the half Fermi wavelength period [8]. The equilibrium spin current $\boldsymbol{I}_l^{\mathrm{S}\ (\mathrm{eq})}(t)$ describes the exchange interaction between local spins through the Ruderman-Kittel-Kasuya-Yosida interaction. The factor $\mathrm{Si}\left[\pi(l-l')\right]$ reflects a situation that the spin current is induced only in areas between non-parallel spins. The second term of $\boldsymbol{\sigma}_l^{(\mathrm{eq})}(t)$ represents the effect of shift of the local density of states at a site due to the electric potential. Also we can paraphrase that the presence of electric field is essential for this effect because the term exists only in the case that the electric potential possess spatial dependence. The last term of $n_l^{(\mathrm{eq})}(t)$ represents the interference effect between a local spin and a spin of spin-polarized itinerant electrons. The numerical calculation for $F^{(2)}_{\alpha,\beta}$ reveals that the number of electrons in areas between local spins is controllable by the relative angle between the spins. These terms including $F^{(2)}_{\alpha,\beta}$ will be discussed in detail in a separate paper. The equilibrium charge current vanishes identically.

Next, we see non-equilibrium components

$$n_l^{(\mathrm{ne})}(t) = -\,2\rho J\tau P\left\langle \boldsymbol{S}_l(t)\cdot\dot{\boldsymbol{S}}_l(t)\right\rangle,$$
$$\boldsymbol{\sigma}_l^{(\mathrm{ne})}(t) = -\,2\rho J\tau\left\langle \dot{\boldsymbol{S}}_l(t)\right\rangle + \frac{4\rho J^2\tau^2}{\hbar}\left\langle \boldsymbol{S}_l(t)\times\dot{\boldsymbol{S}}_l(t)\right\rangle - 2e\rho\tau P\left\langle \phi_l(t)\dot{\boldsymbol{S}}_l(t)\right\rangle,$$
$$I_l^{\mathrm{C}\ (\mathrm{ne})}(t) = -\,\frac{\sigma}{ea}E_l(t) + \frac{\sigma\tau J}{e^2a}P\partial_l\left\langle \boldsymbol{S}_l(t)\cdot\dot{\boldsymbol{S}}_l(t)\right\rangle,$$
$$\boldsymbol{I}_l^{\mathrm{S}\ (\mathrm{ne})}(t) = \frac{\sigma J\tau}{e^2a}\partial_l\left\langle \dot{\boldsymbol{S}}_l(t)\right\rangle - \frac{2\sigma J^2\tau^2}{e^2a\hbar}\partial_l\left\langle \boldsymbol{S}_l(t)\times\dot{\boldsymbol{S}}_l(t)\right\rangle - \frac{\sigma E_l(t)}{ea}P\left\langle \boldsymbol{S}_l(t)\right\rangle + \frac{\sigma\tau}{ea}P\partial_l\left\langle \phi_l(t)\dot{\boldsymbol{S}}_l(t)\right\rangle.$$

Journal of Physics: Conference Series **266** (2011) 012081 doi:10.1088/1742-6596/266/1/012081

The first term of $I_l^{\mathrm{C}\,(\mathrm{ne})}(t)$ is the drift electric current according to the Ohm's law. This part appears also in the third term of $\boldsymbol{I}_l^{\mathrm{S}\,(\mathrm{ne})}(t)$, thus the terms describe the spin-polarized electric current induced by the electric field. The last term of $\boldsymbol{I}_l^{\mathrm{S}\,(\mathrm{ne})}(t)$ represents correction for the first term from the band shift due to the electric potential. Much the same is true on the last term of $\boldsymbol{\sigma}_l^{(\mathrm{ne})}(t)$. $n_l^{(\mathrm{ne})}(t)$ is the dynamical version of the interference effect, and the second term of $I_l^{\mathrm{C}\,(\mathrm{ne})}(t)$ is the diffusion current of $n_l^{(\mathrm{ne})}(t)$ because these quantities satisfy the Fick's first law of diffusion $\frac{\sigma\tau J}{e^2 a} P \partial_l \langle \boldsymbol{S}_l(t) \cdot \dot{\boldsymbol{S}}_l(t) \rangle = -\frac{D}{a} \partial_l n_l^{(\mathrm{ne})}(t)$. Other parts of contributions from the exchange interaction are compatible with Takeuchi *et al.*[6][7] and Ohe *et al.*[9].

In summary, we have derived the microscopic expressions for the charge current, the spin current, the number and the spin angular momentum of electrons in the magnetic multi layers in the presence of both electric field and spin dynamics. In particular, the analytic expressions have been derived for the equilibrium components, and the effects that the static homogeneous electric field gives to magnetic layers have been discussed.

References

[1] Mizukami S, Ando Y and Miyazaki T 2002 *Phys. Rev.* B **66** 104413
[2] Saitoh E, Ueda M, Miyajima H and Tatara G 2006 *Appl. Phys. Lett.* **88** 182509
[3] Tserkovnyak Y, Brataas A and Bauer G E W 2002 *Phys. Rev. Lett.* **88** 117601
[4] Tserkovnyak Y, Brataas A and Bauer G E W 2003 *Phys. Rev.* B **67** 140404
[5] Taniguchi T and Imamura H 2008 *Mod. Phys. Lett.* B **22** 2909
[6] Takeuchi A, Hosono K and Tatara G 2010 *Phys. Rev.* B **81** 144405
[7] Takeuchi A and Tatara G 2008 *J. Phys. Soc. Jpn.* **77** 074701
[8] Rikitake Y and Imamura H 2005 *Phys. Rev.* B **72** 033308
[9] Ohe J, Takeuchi A and Tatara G 2007 *Phys. Rev. Lett.* **99** 266603

Journal of Physics: Conference Series **266** (2011) 012082 doi:10.1088/1742-6596/266/1/012082

Current-Induced Domain Wall Motion in TbFeCo micro wire

Hajime Nakamura, Songtain Li, Xiaoxi Liu, Akimitsu Morisako

Department of Information Engineering, Shinshu-University Wakasato 4-17-1, Nagano, Japan

E-mail: liu@shinshu-u.ac.jp

Abstract. Current-induced domain wall motion in TbFeCo micro wire with perpendicular anisotropy has been observed by Kerr microscope. The critical current density in ferrimagnetic TbFeCo was found to be as low as 5.4 $\times 10^{10}$ A/m^2, which is at least 1 order smaller than in other metallic materials reported so far. The results show that the critical current density can be reduced by controlling pinning sites in micro wire.

1. Introduction

Recently current-induced domain wall (DW) motion [1] has attracted attention because of it's scientific importance and promising application in data storage [2]. To date, current-induced domain wall motion has mainly studied in Permalloy (NiFe) [3], which has an in-plane magnetization. In Permalloy, a high current density of approximately 10^{12} A/m^2 is needed to drive the DW. From the practical application point of view, the critical current density has to be decreased furthermore. In both experiment and simulation results, the possibility and superiority by using materials have perpendicular magnetic anisotropy for current induced domain wall motion has been reported [4][5]. In this article, we report on the current-induced domain wall motion in TbFeCo micro wire with perpendicular magnetic anisotropy. TbFeCo has very sharp DWs. The calculated domain wall width $\delta = \pi\sqrt{A/K_u}$ (A is the exchange stiffness and K_u is the anisotropy constant) is less than 10 nm. Due to the large perpendicular anisotropy and low saturation magnetization ($K_u = 3\times 10^6$ erg/cm^3, $M_s = 200$ emu/cm^3 for the TbFeCo films used in this experiment), high spin torque efficiency can be expected [5]. Furthermore, magnetic properties such as saturation of magnetization and anisotropy field of TbFeCo film can be manipulated in a wide range by controlling the film composition. Such properties make TbFeCo a prominent material both for application in spintronic devices and for fundamental research of current-induced domain wall motion.

2. Experiments

$Tb_{30}Fe_{58}Co_{12}$ layer with thickness 40 nm was lithographically patterned into Hall elements with 40 μm in width between a set of Au/Ti electrodes on thermally oxidized Si substrate by magnetron sputtering method. Electrodes were deposited for both pulse current injection and resistance measurement. Si_3N_4 layer with thickness 3 nm was deposited on TbFeCo film to protect the wire from oxidation. Fabricated sample is shown in Fig. 1(a). Then micro wire with 10 μm in width and 150 μm in length was structured by processing the sample in a focused ion beam (FIB) system as shown in Fig. 1(b). DWs were first nucleated in the pad regions and then propagated from the two ends to the centre through the micro wire when applying a negative saturating magnetic field followed by a positive field. DW propagation from the right end of the micro wire was successfully suppressed by means of etching the right end by ion beam (etching depth is 2 nm) so only one DW could be introduced into micro wire. After introducing a DW in to the wire, pulse currents were injected via electrode using a pulse generator under zero magnetic field. The pulse currents used in this experiment have width of 500 μs and frequency of 1 Hz. The amplitude of the pulse currents was set to increase step by step in sweep mode. The volt value when DW was moves was recorded as critical voltage and the critical current density was calculated from that value. The DW position before and after application of current pulse was imaged directly by MOKE microscope.

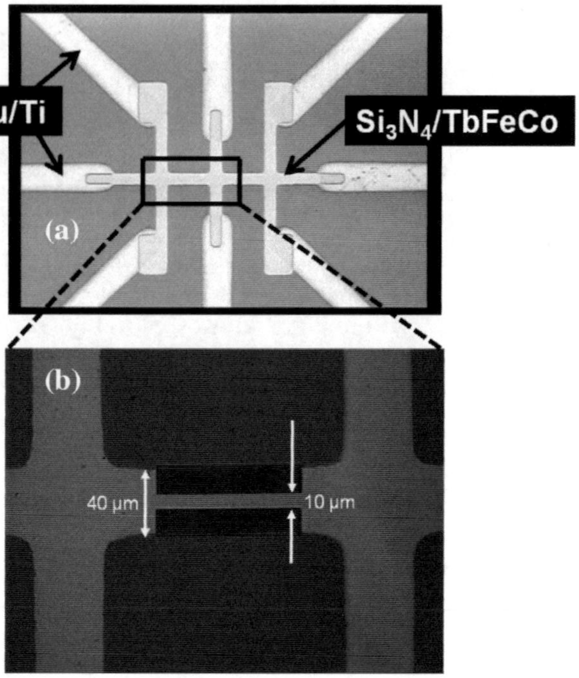

Fig. 1 Process of sample fabrication

After that, to research the current density's dependence on pinning sites, TbFeCo Hall elements with various thick (30 nm to 60 nm) were fabricated to controlled pinning sites and measured the critical current density by same way. The amount of pinning sites was measured from initial curve of M-H loop. To inspect the experimental result, micro magnetic simulation using OOMMF was performed.

3. Results and Discussion

MOKE microscope results show that DW was moved by the injection of pulse current and it's direction depends on pulse direction. Fig.2 (a) shows initial DW position. One DW was introduced to the wire by external field. When the pulse current flows left side to right, DW was moved to left side as shown in Fig.2 (b). On the other hand, when the current flowing from right side to left, it can drive DW to right direction from initial position (Fig.2(c)). Moving direction is always same and depends on the flowing

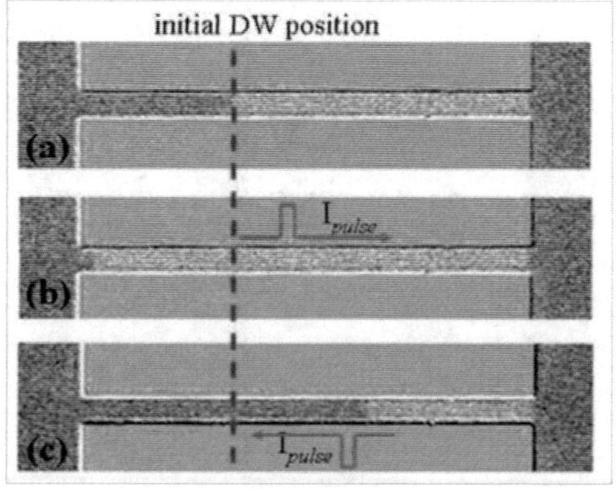

Fig.2 (a) Initial position of DW. (b) , (c) DW position after pulse.

direction of current, so this result is not due to thermal effect. The critical current density is 5.4×10^{10} (A/m^2).

Fig.3 (a) shows the dependence of critical current density on pinning sites. We use the applied field on the rising point of initial curve of M-H loop as pinning field (Fig.3 (b)). The critical current density decreases with the decrease of pinning field. Theoretically, It is known that critical current density is given by $J_c \propto V_o / \alpha$ in material with strong pinning [6] and TbFeCo has strong pinning as

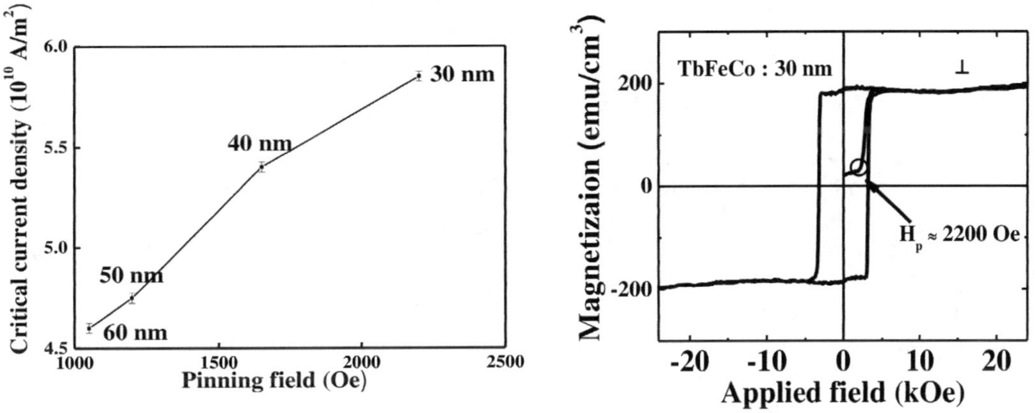

Fig.3 The relation of the critical current density and de-pinning field is shown in (a). The numbers (30~60 nm) is the thickness of TbFeCo. Depinning field H_p was shown in (b). The rising point of initial curve of M-H loop equals to pinning potential.

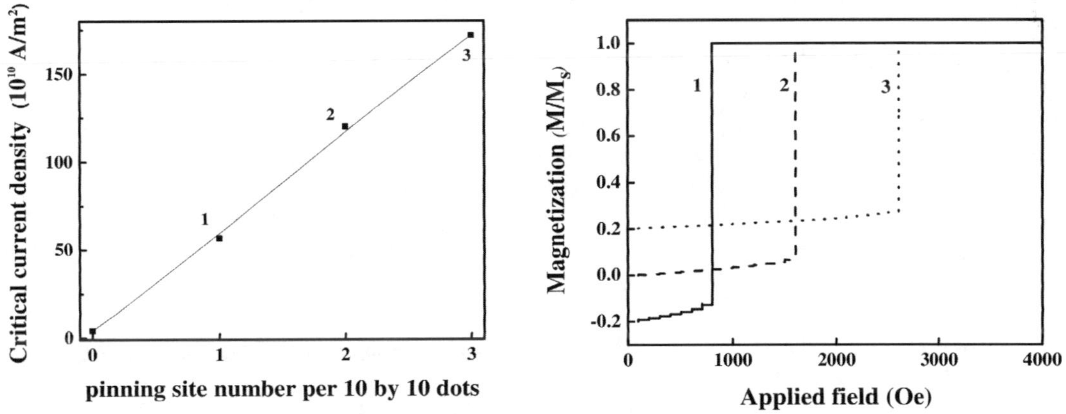

(c)

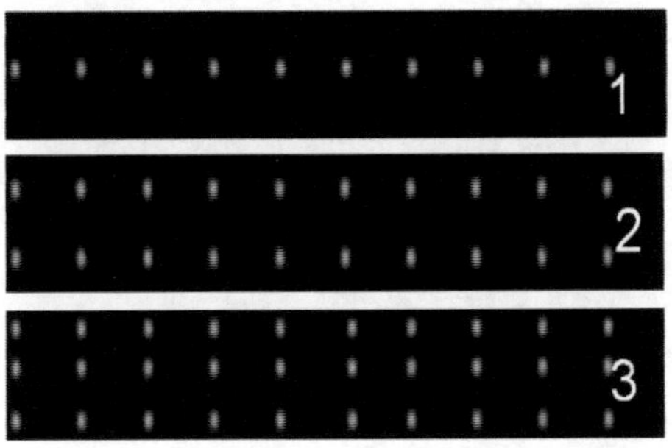

Fig.4 (a) The critical current density and number of pinning sites in OOMMF simulation. Numeral numbers is the mask used in simulation shown in (c). White dots were non-magnetic cell used as pinning site. (b) shows initial curve in simulation per the number of the pinning sites. De-pinning field is in promotion with number of the pinning sites.

shown in initial curve of M-H loop in Fig.3 (b). However, from the results, the decrease width of critical current density is not in proportion to a decrease width of pinning field. Those results suggest that depinning field is different qualitatively from pinning sites and current-induced domain wall motion with strong pinning is depends on hard-axis anisotropy or sample thickness [7, 8].

Fig.4 (a) shows the result of same experiment in micro magnetic simulation. In this case, opposite to the experimental result, pinning field and the critical current density were in direct proportion. This is the same result as Tatara's theory with strong pinning [7]. From those results, it is hopeful that a reduced critical current density can be achieved by fabrication of smooth wire with few impurities.

4. Conclusion

DW was successfully driven by pulse current in TbFeCo micro wire. The current density needed to drive DW was found to 5.4×10^{10} A/m^2, which is at least 1 order smaller than in Permalloy. The critical current density can be further reduced to 4.7×10^{10} A/m^2 by controlling thickness of TbFeCo wires. Films with low pining field show lower critical current density. Micromagnetic simulation results show that the critical current densities are in direct proportion to pinning field. Those results are in agreement with the theory of Tatara. The low critical current density found in TbFeCo films indicated the TbFeCo a potential candidate for domain wall memories.

References

[1] L Berger: *J. Appl. Phys.*, 55, 1954(1984).
[2] S S P Parkin, M. Hayashi, and L. Thoms: *Science,* 320,190(2008)
[3] A Yamaguchi, T Ono, S Nasu, K Miyake, K Mibu, and T Shinjo: *Phys. Rev. Lett.*, 92 077205(2004)
[4] H Tanigawa, K Kondou, T Koyama, K Nakano, S Kasai, N Ohshima, S Fukami, N Ishiwata, and T Ono: *Appl. Phys. Express*, 1 011301
[5] S W Jung, W Kim, T D Lee, and H W Lee: *Appl. Phys. Lett.*, 92, 202508(2008)
[6] G. Tatara et al.,: J. Phys. Soc. Jpn. 75, 64708 (2006)
[7] Gen Tatara, Hiroshi Kohno: *Phys. Rev. Lett.*, 92, 086601(2004)
[8] A Yamaguchi, U Tanigawa, T Ono, S Nasu, K Miyake, K Mibu and T Shinjo: *Journal of the Magnetics Society of Japan*, 28, 343-346(2004)

Journal of Physics: Conference Series **266** (2011) 012083 doi:10.1088/1742-6596/266/1/012083

Magnetization Reversal of Single Co/Pd Multilayer Nanodot by Nanoseconds Pulse Field

Y Suyama, Y Murayama, N Kikuchi, S Okamoto, and O Kitakami

Institute of Multidisciplinary Research for Advanced Materials, Tohoku University, Sendai 980-8577, Japan

E-mail: suyama.y@mail.tagen.tohoku.ac.jp

Abstract. Magnetization switching experiments on nanosized [Co(δ_{Co} = 0.3 – 0.8 nm)/Pd (2 nm)]$_3$ multilayer dots with the diameter of D = 240 - 500 nm are carried out using pulse fields with the duration of 1.6 – 10 ns. The switching fields in the nanoseconds regime are significantly enhanced with respect to the values measured by conventional quasistatic magnetometry. The energy barriers are evaluated from the time-dependence of the switching field based on the Néel-Arrhenius law and are found to be proportional to the Co layer thickness.

1. Introduction

Magnetic recording technology has shown continuous progresses in both density and operation speed for last decades. For development of further advanced devices using nanomagnets as memory elements, such as bit patterned media (BPM) and magnetoresistive random access memory (MRAM), understanding of the magnetization behavior of nanoscale structures becomes more important. Dynamic magnetization behavior is also a key issue since their operation frequency is supposed to be of the order of GHz being comparable with the magnetization precession frequency. A number of studies on magnetization dynamics [1] with the time resolution of femtoseconds have been reported [2]. However, few studies have been reported on irreversible switching of hard magnetic materials [3], due to difficulty in large pulse field generation. In the previous work, we have reported the pulse field generator with the maximum field of 3 kOe and investigated the magnetization switching behavior of a single Co/Pt multilayer dot of 300 nm in diameter [4, 5]. In this study, we present experimental results on magnetization switching of nanosized Co/Pd multilayer dots with the Co layer thickness of δ_{Co} = 0.3 – 0.8 nm by nanoseconds pulse fields.

2. Experimental

A series of Co/Pd multilayers were deposited on quartz substrates by dc magnetron sputtering at room temperature. The base pressure before deposition was 5.0×10^{-5} Pa and Ar gas pressure during deposition was 2.0 Pa. The film structure was Ta (2 nm)/Pd (2 nm)/[Co (δ_{Co} nm)/Pd (2 nm)]$_3$. The Co layer thickness δ_{Co} was varied from 0.3 to 0.8 nm. The saturation magnetization M_s was measured by vibrating sample magnetometry (VSM). The effective uniaxial magnetic anisotropy constant K_u^{eff} and the effective anisotropy field H_k^{eff}, which includes demagnetization terms, were evaluated from anomalous Hall effect (AHE) curves using the generalized Sucksmith-Thompson (GST) method [6]. K_u^{eff} per Co layer was varied from 2.3×10^7 to 2.1×10^6 erg/cc by controlling δ_{Co} from 0.3 to 0.8 nm. The multilayer films were patterned into circular dots with the diameter D = 240 – 500 nm by using electron beam lithography and Ar ion etching. Subsequently, a 9 nm thick Pt layer with Ta adhesion layers was deposited onto the dots and then patterned into cross shape electrodes for AHE measurements. Then, an insulating layer of 200 nm in thickness was formed by hardening hydrogen silsequioxane (HSQ) with e-beam. Finally, a Ag coil with the inner diameter of 6 μm and width of 5 μm for pulse field generation was fabricated by the lift-off technique. Figure 1 shows the SEM image for the device without HSQ insulating layer for SEM observation. Magnetic pulse fields were generated by using an electric pulse generator equipped with a coaxial cable as a capacitor [4, 5, 7]. The field amplitude and duration were varied by adjusting charging voltage V_0 and the cable length L, respectively. Figure 2 shows a series of typical pulse profiles for $V_0 = 400$ V and $L = 0.1$ – 1.0 m. All the pulses show well rectangular shape with the rise time of less than 0.4 ns. The pulse duration increases from 1.6 to 10.0 ns with increasing L from 0.1 to 1.0 m. Ideally, the peak voltage in Fig. 2 of the pulse should be $V_0 / 2 = 200$ V, but the experimentally obtained value is 80 % of the ideal one. This amplitude decay is probably due to electric leakage and/or partial reflection of the pulse due to impedance mismatching. Finite element method calculations revealed that the maximum pulse field reaches 4 kOe at the coil center for $V_0 = 400$ V.

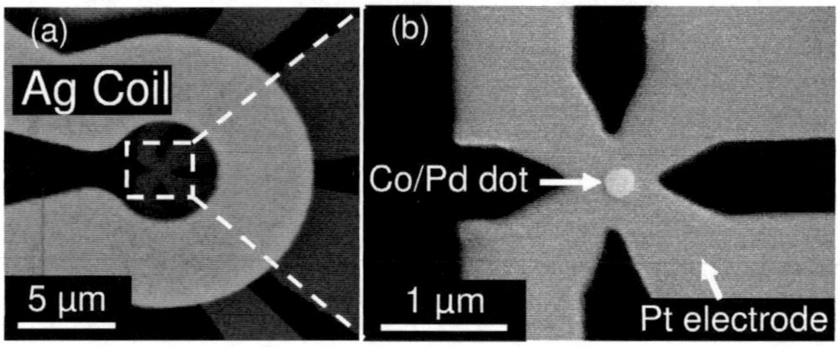

Figure 1. (a) SEM image for a Co/Pd dot with a cross electrode and a Ag coil. The inner diameter of the coil is 6 μm. (b) Magnified image of the dot.

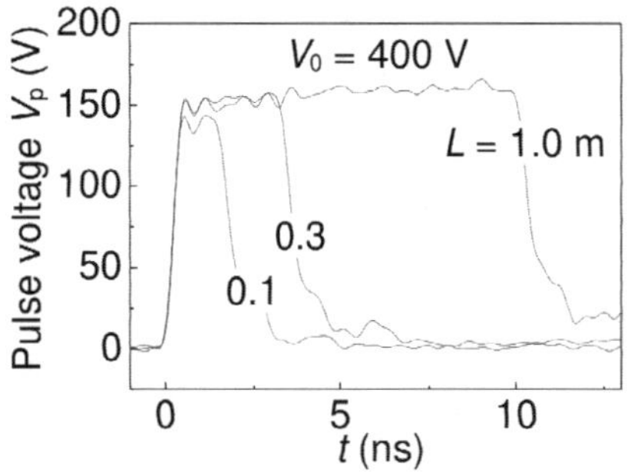

Figure 2. Pulse waveforms for charging voltage V_0 = 400 V and the coaxial cable length L = 0.1 – 1.0 m.

3. Results and Discussion

Pulse field switching experiments were carried out in the presence of the dc bias field H_{dc}. The dot magnetization was negatively saturated by the dc field of -5 kOe before application of each pulse field. The switching probability p for the dot with $D = 400$ nm and $\delta_{Co} = 0.6$ nm is plotted in Fig. 3 (a) as a function of H_{dc}, here p is obtained as the average of 10 times measurements. The pulse field duration and charging voltage were $\tau_p = 10.0$ ns and $V_0 = 0 - 300$ V, respectively. It was confirmed that the dot was always single-domained both before and after the pulse field application. Note that the switching probability curves shifts toward smaller H_{dc} with increasing V_0. In Fig. 3 (b) the mean dc field required for switching H_{dc}^{sw} which gives $p = 0.5$ in Fig. 3 (a) is plotted as a function of V_0. The linear decrease in H_{dc}^{sw} for $V_0 >$ 150 V indicates that the magnetization switching occurs when the condition $H^{sw} \leq H_p + H_{dc} = \xi V_0 + H_{dc}$ is satisfied, where ξ is the proportional coefficient given by H_p / V_0. For $V_0 <$ 150 V, however, H_{dc}^{sw} remains constant and equals to $H_c = 0.94$ kOe measured by the quasistatic magnetometry. The reason for this result is explained elsewhere [4]. The dotted line in Fig. 3 (b) is the best fitted linear function for $V_0 >$ 150 V. The intersection with the vertical axis ($V_0 = 0$ V) gives the switching field on the time scale of pulse duration $\tau_p = 10$ ns.

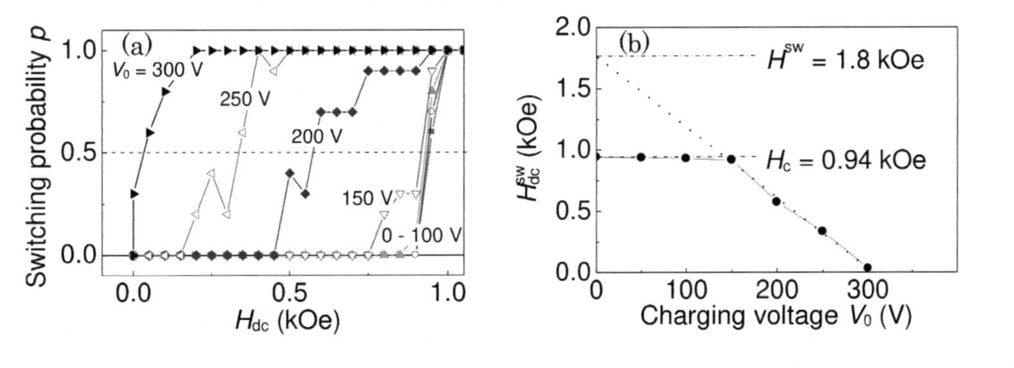

Figure 3. (a) Switching probability p of a [Co (0.6 nm)/Pd (2 nm)]$_3$ dot as a function of H_{dc} for V_0 = 0 - 300 V (D = 400 nm). (b) The charging voltage V_0 dependence of the mean dc field required for switching H_{dc}^{sw}. The dotted line indicates the best fitted linear function for V_0 > 150 V. The intersection at V_0= 0, H^{sw} = 1.8 kOe, corresponds to the switching field on the time scale of pulse duration τ_p = 10 ns.

In Fig. 4, the H^{sw} for Co/Pd dots with (a) δ_{Co} = 0.3 nm, D = 240 nm, (b) 0.6 nm, D = 360 nm, and (c) 0.8 nm D = 240 nm are plotted as a function of the pulse field duration τ_p. The dc coercivity H_c measured by the quasistatic magnetometry is also plotted at τ_p = 1 s. Assuming the single energy barrier for the same magnetization switching over the whole time range examined in the present study, the median switching field $H^{sw}(\tau_p)$ can be expressed by the following Sharrock equation based on the classical Néel-Arrhenius law [8],

$$H^{sw}(\tau_p) = H_0^{sw}\left\{1 - \left[k_B T/E_0 \cdot \ln(f_0\tau_p/0.693)\right]^{1/\alpha}\right\} \quad , \qquad (1)$$

where f_0 is the attempt frequency of the order of 10^{10} Hz, E_0 is the energy barrier height, H_0^{sw} is the switching field without thermal agitation, k_B is the Boltzmann constant, T is temperature, α is the constant depending on the magnetization reversal mode, respectively. Here we tentatively assumed α = 2 as in the previous work [4]. The best fitted results using eq. (1) are shown as the solid lines in Figs. 4 (a) – (c). For all the δ_{Co}, the obtained H_0^{sw} was significantly smaller than the anisotropy field $2K_u^{eff}/M_s$. This discrepancy might be related with the intrinsic and extrinsic inhomogeneity in the film, and further study to reveal the origin of the nucleation site would be required for more detailed discussion. In Fig. 5, thus obtained energy barrier E_0 for all Co/Pd dots is plotted as a function of K_u^{eff}. Here, the effect of the dot diameter on K_u^{eff} is negligibly small since the diameter is considerably larger than the film thickness. Although the data are widely scattered, E_0 tends to decrease with K_u^{eff} and increase with δ_{Co}. The median values E_0 for δ_{Co} = 0.3, 0.6, and 0.8 nm are, respectively, 60, 150, and 180 k_BT, indicating that E_0 is roughly proportional to δ_{Co}. Now, let us discuss this result in the following. As has been revealed in the previous papers [4, 5, 9], the magnetization reversal in a nanosized magnet proceeds via nucleation of a reversed embryo followed by its rapid expansion. Therefore the nucleation process determines the value of E_0. It is considered that a reversed embryo with minimum energy has the dimension of the domain wall width l_w. Therefore its energy is roughly estimated to be $\sigma_w l_w^2$, where σ_w is the wall density, as suggested by Givord [10]. In the present study, however, since δ_{Co} is much smaller than l_w, the energy may be reduced to $\sigma_w l_w \delta_{Co}$. Thus, the E_0 may be proportional to δ_{Co} as,

$$E_0 \approx \sigma_w \cdot l_w \cdot \delta_{Co} \sim \sqrt{AK_u} \cdot \sqrt{A/K_u} \cdot \delta_{Co} \propto \delta_{Co} \quad , \qquad (2)$$

where A denotes exchange stiffness. The above crude discussion roughly explains the δ_{Co} dependence of E_0 for Co/Pd dots but not for Co/Pt data [9], which exhibit the monotonic increase in E_0 with K_u. To solve the contradictory results between Co/Pd and Co/Pt, further intensive studies, such as detailed, magnetic and structural analyses, are obviously required.

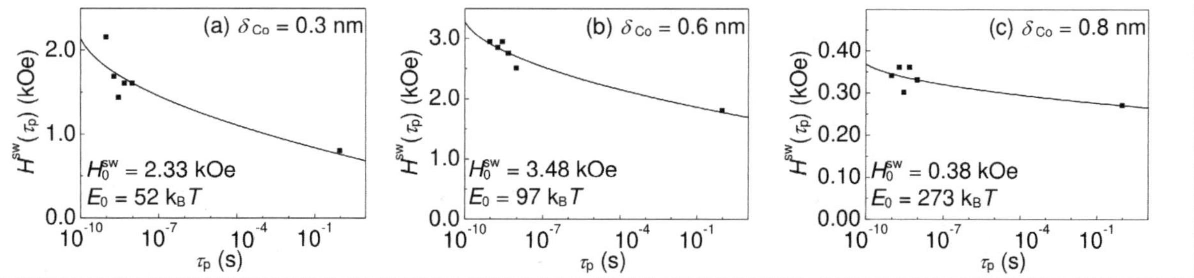

Figure 4. Switching field H^{sw} of [Co (δ_{Co} nm)/Pd (2 nm)]$_3$ dots with (a) δ_{Co} = 0.3 nm, (b) 0.6 nm, and (c) 0.8 nm as a function of τ_p. The points plotted at τ_p = 1 s indicate the coercivity measured by quasistatic magnetometry. The solid lines show the best fitted results based on the Néel-Arrhenius law.

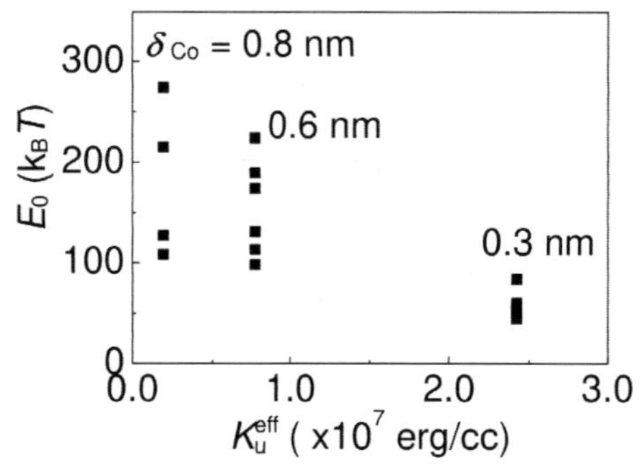

Figure 5. Energy barrier E_0 for [Co (δ_{Co} nm)/Pd (2 nm)]$_3$ dots (D = 240 – 500 nm) as a function of the effective perpendicular anisotropy K_u^{eff}.

4. Summary

Magnetization switching experiments are carried out on nanosized Co/Pd multilayer dots with the Co layer thickness of δ_{Co} = 0.3 – 0.8 nm by applying pulse fields with nanoseconds duration. Significant increase of the switching field in the nanoseconds regime is found for all Co/Pd dots prepared in this study. The analysis of the time dependence of the switching field based on the Néel-Arrhenius law has revealed that the energy barrier for magnetization switching monotonically increases with δ_{Co}. This result is roughly explained by assuming that the nucleation of a reversed embryo has the literal size of the domain wall width and the thickness of the Co layers. However, this crude model fails in explains the experiments of Co/Pt dots of while energy barrier are exclusively dominated by the magnetic anisotropy. This contradictory result between Co/Pd and Co/Pt data obviously needs further intensive studies.

Acknowledgments

This work was partially supported by the "Research and Development for Next-Generation Information Technology" of the Ministry of Education, Culture, Sports, Science and Technology (MEXT) and Grant-in-Aid for Scientific Research from MEXT, and the Storage Research Consortium in Japan.

5. Reference

[1] Hiebert W K, Stankiewicz A, and Freeman M R 1997 *Phys. Rev. Lett.* **79** 1134
[2] Beaurepaire E, Merle J C, Daunois A, and Bigot J Y, 1996 *Phys. Rev. Lett.* **76**, 4250
[3] Weisheit M, Bonfim M, Grechishkin R, Barthem V, Fähler S, and Givord D 2006 *IEEE Trans. Magn.* **42** 3072
[4] Ito A, Kikuchi N, Okamoto S and Kitakami O 2008 *IEEE Trans. Mag.* **44** 3446
[5] Kikuchi N, Okamoto S and Kitakami O 2009 *J. Appl. Phys* **105** 07D506

[6] Okamoto S, Kikuchi N, Kitakami O, Miyazaki T, Shimada Y, Fukamichi K 2002 *Phys. Rev. B* **66** 024413
[7] He L, Doyle W D, Varga L, Fujiwara H, and Flanders P J 1996 *J. Magn. Magn. Mater.* **155** 6; Doyle W D, and He L, 1993 *IEEE Trans. Magn.* **29** 3634
[8] Sharrock M P and Mckinney J T 1981 *IEEE Trans. Mag.* **17** 3020
[9] Okamoto S, Kato T, Kikuchi N, Kitakami O, Tezuka N and Sugimoto S 2008 *J. Appl. Phys.* **103** 07C501
[10] Givord D, Tenaud P, Viadieu T 1988 *IEEE Trans. Magn.* **24** 1921

Microscopic theory on the Gilbert damping due to spin pumping effects in the magnetic multi-layer system

Nobuyuki Umetsu, Daisuke Miura and Akimasa Sakuma

Department of Applied Physics, Tohoku University, Sendai, Japan

E-mail: `umetsu@olive.apph.tohoku.ac.jp`

Abstract. We have calculated the Gilbert damping coefficient α in the magnetic multi-layer structures consisted of ferromagnetic (FM) and non-magnetic (NM) layers which are FM/NM and FM/NM/FM from microscopic theory. In this study, our model Hamiltonian doesn't include the spin-orbit interaction term, but the kinetic electron hopping term. For obtaining α in theory, we used Kubo formula and Projection method. We found α is dependent on structure of systems and both thickness of the FM and NM layer. Spin pumping effect which is predicted by experimental results is confirmed by this calculation in theory. We conclude that transferring conduction electrons carrying angular momentum to adjacent layers are the origin of the Gilbert damping for multi-layer systems.

1. Introduction

In the field of spintronics, there is rapidly growing interest in controlling the damping behavior of dynamical spins in the magnetic devices such as magnetic random access memory (MRAM) and so on, and then the microscopic foundation on the origin of the Gilbert damping coefficient (α) is strongly desired to understand completely. Phenomenologically, the low-temperature magnetization $\boldsymbol{M}$ dynamics in ferromagnets is well described by the following Landau-Lifshitz-GIlbert (LLG) equation [1, 2, 3].

$$\frac{\mathrm{d}\boldsymbol{M}}{\mathrm{d}t} = -\gamma \boldsymbol{M} \times \boldsymbol{H}_{\text{eff}} + \frac{\alpha}{M_s} \boldsymbol{M} \times \frac{\mathrm{d}\boldsymbol{M}}{\mathrm{d}t}. \tag{1}$$

Here, $\boldsymbol{H}_{\text{eff}}$ is the "effective" magnetic field (including external, exchange and anisotropy fields), $\gamma(> 0)$ is the gyromagnetic ratio, M_s is the saturation magnetization, and α is the Gilbert damping coefficient, often measured in terms of the Landau-Lifshitz damping rate [4] $\lambda = \alpha\gamma M_s$ (in cgs units). As regards MRAM, α has a close relation to speed of switching a bit of information encoded by the magnetization direction of a ferromagnetic grain and energy demands of switching [5, 6]. A detailed understanding of the Gilbert damping mechanisms in metallic ferromagnets and precise calculation of damping constant would greatly facilitate the design of new materials appropriate for a variety of applications.

Kamberský [7] and Gilmore *et al.* [8] have recently proposed a convenient way to calculate the Gilbert damping coefficient for the ferromagnetic bulks by using the first principles technique for the electronic structure calculations. This approach is based upon the theory called torque correlation model, which gives the Gilbert damping coefficient as

$$\alpha = \lim_{\omega \to 0} \frac{1}{\hbar\omega} \mathrm{Im} \frac{1}{\langle M^- \mid M^- \rangle} \left\langle \mathcal{L}M^- \left| \frac{1}{\hbar\omega + \mathcal{L} + \mathrm{i}\delta} \right| \mathcal{L}M^- \right\rangle, \tag{2}$$

where $\langle\mathcal{A}|\mathcal{B}|C\rangle \equiv \left\langle\left[\mathcal{B}C,\mathcal{A}^\dagger\right]\right\rangle$ for any operator $\mathcal{A},\mathcal{B},C$, $\delta(>0)$ is the infinitesimal quantity and $\mathcal{L}\mathcal{A} \equiv [\mathcal{H},\mathcal{A}]$ is Liouville superoperator that commutate with the Hamiltonian $\mathcal{H}$ for any operator $\mathcal{A}$. Kamberský and Gilmore *et al* chose the spin-orbit interaction $\mathcal{H}_{so}$ for the $\mathcal{H}$ and obtained quantitative results for α of transition metals such as Fe, Co and Ni metals.

On the other hand, the Gilbert damping mechanisms in the magnetic inhomogeneous systems are different from that in the magnetic homogeneous bulk structures. Mizukami [9, 10] have experimentally measured the ferromagnetic resonance (FMR) linewidth for NM (non-magnetic)/FM (ferromagnetic) /NM structures and found that α is strongly related to thickness of the NM and FM layers. This phenomenon is due to spin current pumped into NM layer by the localized spins in FM layer, which is so-called spin pumping effect. In theory, we consider that not only spin-orbit interaction but also the kinetic (electron hopping) term $\mathcal{H}_t$ contribute to the origin of α in magnetic inhomogeneous systems.

From these viewpoints, we have calculated α for the multi-layer structures consisted of FM and NM layers from microscopic theory using Hamiltonian which includes $\mathcal{H}_t$.

2. Model

Our model is one dimension and described by the total of the tight-binding model Hamiltonian (electron hooping term) $\mathcal{H}_t$ and *s*-*d* model Hamiltonian $\mathcal{H}_{sd}$:

$$\mathcal{H} = \mathcal{H}_t + \mathcal{H}_{sd} = -\sum_{i,j,\sigma} t_{ij} c^\dagger_{i\sigma} c_{j\sigma} - J_{sd} \sum_i^{FM} \boldsymbol{S}_i \cdot \hat{\boldsymbol{\sigma}}_i , \tag{3}$$

where $c^\dagger_{i\sigma}(c_{i\sigma})$ create (annihilate) a electron of σ spin on i site, $t_{i,j}$ is the hopping rates, $\boldsymbol{S}$ denotes localized d spins, $\hat{\boldsymbol{\sigma}}\,(\equiv c^\dagger \boldsymbol{\sigma} c$, $\boldsymbol{\sigma}$ is a vector of Pauli spin matrices) denotes conducting s electrons, $J_{sd}(>0)$ is *s*-*d* exchanging coupling constant, $\sum_{i,j}$ means summing when i and j sites are adjacent and $\sum_i^{FM}$ means summing when i sites are in FM layer.

3. Theory

Linear response to a harmonic field $(h_x - \mathrm{i}h_y)\exp(-\mathrm{i}\omega t)$ is described by a single susceptibility function $\chi_m(\omega)$. One can obtain the macroscopic susceptibility:

$$\chi_m(\omega) = \frac{\gamma M_s}{\gamma H_{\text{eff}} - \omega - \mathrm{i}\alpha\omega} \tag{4}$$

from LLG equation and $\boldsymbol{M} \equiv \chi_m \boldsymbol{H}_{\text{eff}}$. On the other hand, in microscopic theory, the transverse spin susceptibility can be written as

$$\chi_{-+}(\omega) = -\frac{1}{2\mathrm{i}\hbar}\int_0^\infty dt\, e^{\mathrm{i}(\omega+\mathrm{i}\delta)t}\left\langle [M^-(t), M^+]\right\rangle, \tag{5}$$

from formal first-order response theory (Kubo formula) [11, 12]. Here, $\langle\quad\rangle$ denotes the statistics average in adiabatic approximation and $M^-(t)$ denotes the Heisenberg representation of $M^\pm = \sum_i -(2S_i^\pm + \sigma_i^\pm)\mu_B$ (μ_B is Boltzmann constant), which may be written as $M^-(t) = \exp(i\mathcal{L}t/\hbar)M^-$. The projection formalism [13] allows us to rewrite Eq.(5)

$$\chi_{-+}(\omega) = -\frac{1}{2}\frac{\langle M^- \mid M^-\rangle}{\hbar\omega - \Omega + \Gamma(\hbar\omega + \mathrm{i}\delta)}, \tag{6}$$

$$\Omega = -\frac{1}{\langle M^- \mid M^-\rangle}\langle M^- | \mathcal{L} | M^- \rangle, \tag{7}$$

$$\Gamma(\hbar\omega + \mathrm{i}\delta) = -\frac{1}{\langle M^- \mid M^-\rangle}\left\langle \mathcal{L}M^- \left| \frac{1}{\hbar\omega + \mathcal{L} + \mathrm{i}\delta} \right| \mathcal{L}M^- \right\rangle \tag{8}$$

and we obtain Eq.(2) from comparing Eqs.(4) with (6). Substituting Eq.(3) for Eq.(2) gives the average of α in FM layer :

$$\alpha = -\lim_{\omega\to 0}\frac{1}{\hbar\omega}\mathrm{Im}\,\Gamma(\hbar\omega + i\delta)$$
$$= \frac{\pi}{4\mu_B\sum_i\langle M_i\rangle}\sum_{m,n}\left|\left\langle m,\downarrow\left|\left[\mathcal{H},M^-\right]\right|n,\uparrow\right\rangle\right|^2\delta(\epsilon_F-\epsilon_{m\downarrow})\,\delta(\epsilon_F-\epsilon_{n\uparrow}) \qquad (9)$$

where $\epsilon_{m\downarrow(n\uparrow)}$ is a one-electron eigenvalue, ϵ_F is the Fermi energy and $|\,m(n),\downarrow(\uparrow)\rangle$ denotes an eigenstate. We introduce a life time of conduction electrons and use the Lorentzian instead of delta-function.

4. Results and Discussion

Figure 1 shows NM layer thickness dependence of α in FM layer for FM(100 sites)/NM. We found that α increases with increasing of the thickness of NM layer. The oscillating behavior is an interference effect appearing in the finite size system. We consider this result is due to transferring of conduction electrons having a spin. Concretely, current of conduction electrons carrying spin angular momentum from FM layer to NM layer increase because of increasing the thickness of NM layer, and then more angular momentums in FM layer go through the interface to NM layer. It can be considered that this phenomenon is occurred by spin-current induced by the magnetizations rotation in FM layer, which is spin pumping effect.

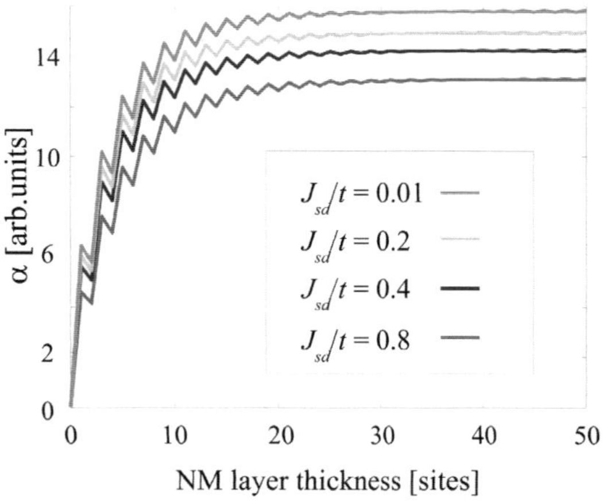

Figure 1. NM layer thickness dependence of α for FM(100 sites)/NM. (t is the hopping rate and J_{sd} is the s-d exchanging coupling constant)

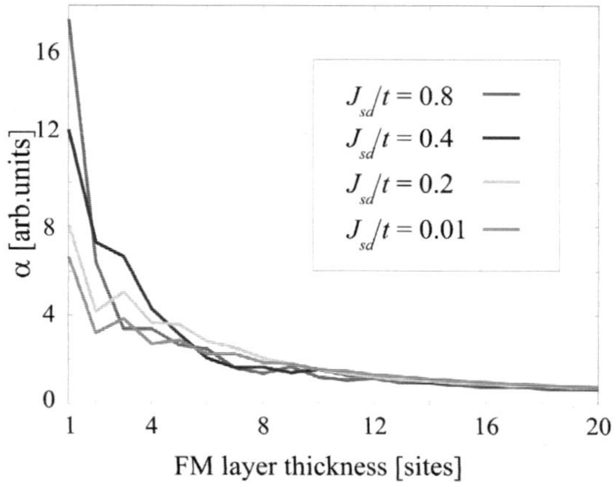

Figure 2. FM layer thickness dependence of α for FM /NM(100 sites).

Figure 2 shows FM layer thickness dependence of α in the FM layer for FM /NM(100 sites). We confirmed that α tends to decrease by increasing the FM layer. This may reflect the situation that the fraction of interface layer relative to the FM layer region decreases with increasing FM layer, leading to depress of the spin pumping efficiency.

After obtaining α for FM/NM systems, we calculated α for three layers system of FM1(J_{sd} = 0.2, 100 sites)/NM(10 sites)/FM2 (Figure 3). In the system, we suppose that magnetizations in FM1 layer are rotating while magnetizations in FM2 layer are motionless. Figure 4 shows FM2 layer thickness dependence of α in FM1 layer for (red line). (NM layer thickness dependence of α in FM1 layer for FM1/NM (blue line) is shown for comparison with red line on Figure 3.) Comparing red and blue lines, we found that α becomes larger by substituting FM2 layer for a part of NM layer in FM1/NM. This suggests a possibility that conduction electrons carrying angular momentum from FM1 layer to NM layer go to FM2 layer to rotate the magnetizations in FM2 layer. We conclude that since magnetizations in FM2 layer need more spin torques to rotate itself, spin-current is larger and that causes α to increase.

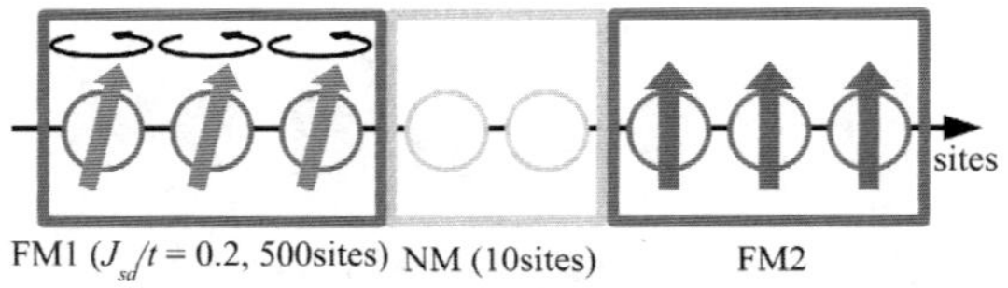

Figure 3. FM1(J_{sd} = 0.2, 100 sites)/NM(10 sites)/FM2. While magnetizations in FM1 layer (J_{sd}/t = 0.2) are rotating, magnetizations in FM2 layer are motionless.

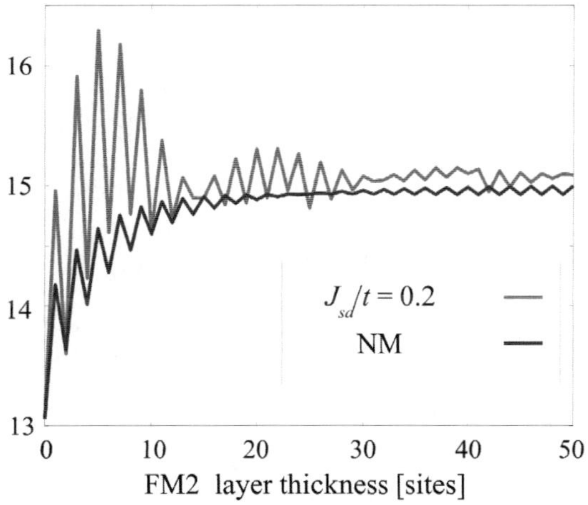

Figure 4. FM2 layer thickness dependence of α for FM1(J_{sd} = 0.2, 100 sites)/NM(10 sites)/FM2 (red line). The blue line shows α, if one substitutes NM layer for FM2 layer.

5. Summary

We have calculated Gilbert damping coefficient α in the magnetic multi-layer systems consisted of ferromagnetic (FM) and non-magnetic (NM) layers from microscopic theory using Hamiltonian which includes the kinetic term $\mathcal{H}_t$. We found α is dependent on structure of systems and both thickness of the FM and NM layer. We conclude that transferring conduction electrons carrying angular momentum to adjacent layers are the origin of the Gilbert damping for multi-layer systems. In other words, we could confirm that spin pumping effect in theory.

[1] E. M. Lifshitz and L. P. Pitaevskii 1980 *Statistical Physics, Part2,3rded., Course of Theoretical Physics* Vol.9
[2] T. L. Gilbert 1955 *Phys. Rev.* **100** 1243 ; see also T. L. GIlbert 2004 *IEEE Trans.Magn.* **40** 3443
[3] L. Landau and E. M. Lifshitz 1935 *Phys. Z. Sowjet union* **8** 153
[4] S. M. Bhagat and P. Lubitz 1974 *Phys. Rev. B* **10** 179
[5] W. Chen, M. J. Rooks, N. Ruiz, J. Z. Sun and A. D. Kent 2006 *Phys. Rev. B* **74** 144408
[6] R. H. Koch, J. A. Katine and J. Z. Sun 2004 *Phys. Rev. Lett* **92** 088302
[7] V. Kamberský 2007 *Phys. Rev. B* **76** 134416
[8] K. Gilmore, Y. U. Idzerda, and M. D. Stiles 2007 *Phys. Rev. Lett* **99** 027204
[9] S. Mizukami, Y. Ando, and T. Miyazaki 2001 *Jpn. J. Appl. Phys.* **40** 580
[10] S. Mizukami, Y. Ando, and T. Miyazaki 2002 *Phys. Rev. B* **66** 104413
[11] R.Kubo 1955 *Rep. Prog. Phys.* **29** 255
[12] D.N.Zubarev 1960 *Sov. Phys. Usp.* **3** 320
[13] H. Mori 1965 *Prog. Theor. Phys.* **33** 423

Journal of Physics: Conference Series **266** (2011) 012085 doi:10.1088/1742-6596/266/1/012085

Field effect on domain wall trapping and domain wall motion in notch free NiFe asymmetric nanorings

X.H. Wang, I. Purnama,S. Goolaup and W.S. Lew
Physics and Applied Physics Division, School of Physical and Mathematical Sciences, Nanyang Technological University, 21 Nanyang Link, Singapore 637371

E-mail: wensiang@ntu.edu.sg

Abstract. We report on the simulation of domain wall trapping and field hystery effect on domain wall in notch free asymmetric ring structures. Our results reveal that a domain wall is trapped at the narrow arm of the asymmetric ring when a vortex state is formed. The chirality of the vortex is found to be dependent on the field history of the asymmetric nanoring. The domain wall behavior under the externally applied magnetic field is found to be related with the chirality of the vortex of the asymmetric ring.

1. Introduction

Domain walls and domain wall(DW) motions have attracted intense research interest due to their potential significance in spintronic appliacations, for example, data storage[1] and magnetic logic devices[2]. For optimum performance of these devices, understanding the details of DW formation and motion plays an essential role. A lot of research efforts have been dedicated to manipulating the DWs using magnetic fields[3] and/or spin-polarized current[4] originating from the spin-transfer torque[5]. In the conventional field switching devices, the reduction of device size down to nanoscale causes significant increase[6] in the switching field, thus causing the disadvantage of high energy consumption. As a result, finding a way to trap a domain wall and depin it with a small field is of significant importance. In most experiments examining the domain walls, a DW is trapped at an artificially fabricated constriction[7] in nanowires or nanostripes. The artificially fabricated geometric defects such as notches traps the domain wall at the expense[8] of large DW depin field or current. To investigate magnetic switching mechanisms, a configuration in which different magnetic states can easily be realized is required[9]. A possible configuration that fulfills these criteria is the ring structure. Elongated ring structure and asymmetric ring[10] which is a ring structure with the inner circle off-centered, were also used to investigate the domain wall behavior. Previously we have reported simulation work on the domain wall trapping behaviour in the asymmetric ring structure[10].

In this paper, we present a systematic investigation of the influence of the field history on the domain wall trapping behavior in an asymmetric ring structure as shown in Fig. 1. The dimensions are D_o= 1200 nm, D_i=600 nm. l is the distance between the centres of the two circle. The motion of the trapped domain wall under an externally applied field is also studied via micromagnetic simulations.

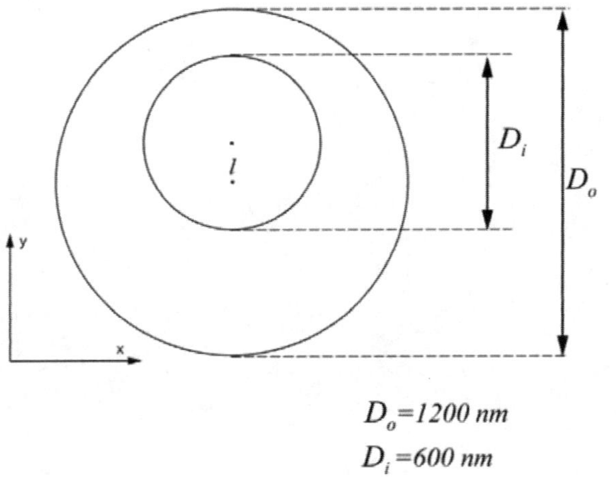

Fig.1 Schematic of the geometric model of the asymmetric ring.

2. Simulations

The magnetization configuration and switching properties were simulated using the Object Oriented MicroMagnetic Framework (OOMMF) simulation program[11]. In this simulation, the micromagnetic equilibrium equations were solved in a two dimensional square mesh with a cell size of 4 nm and the spins are free to rotate in three dimensions. The field was ramped from -1000 Oe to +1000 Oe in 200 steps. Other parameters used for the permalloy were crystalline anisotropy constant $K_1 = 0$, saturation magnetization $M_s = 8.6 \times 10^5$ A·m^{-1}, exchange stiffness $A = 1.3 \times 10^{-11}$ J·m^{-1} and the damping coefficient α= 0.5. We assumed that the intrinsic uniaxial anisotropy of the bulk NiFe film is negligible when compared with the shape-induced anisotropy of the rings.

3. Results and discussions

3.1. Effect of field history on chirality control

In order to study the effect of the field history of the asymmetric magnetic ring on the switching processes and mechanisms, the hysteresis behavior of a magnetic asymmetric ring was simulated with the magnetic field applied both in x axis and y axis. Repeated simulations using the same parameters have been carried to confirm the reproducibility of the results Fig.2 shows the simulated switching process of the 8-nm-thick Ni80Fe20 asymmetric nanorings with the field applied in y axis. When a negative field of 1000 Oe is applied, the magnetization of the asymmetric ring adopts a negative saturation state. Reducing the field, an onion with two transverse domain walls nucleated at the narrow arm and wide arm of the asymmetric ring, forming an onion state with magnetization aligns along the -y direction. When the field is reversed, the domain wall at the wide arm begins to move towards the narrow arm, while the domain wall at the narrow arm stays as shown in process I in Fig.2.

Journal of Physics: Conference Series **266** (2011) 012085 doi:10.1088/1742-6596/266/1/012085

However, the route is selected randomly. The domain wall either moves via the left arm or the right arm of the asymmetric ring. These two kinds of switching both result a domain wall trapping vortex state. However, the chirality of the domain wall trapping vortex states is not determined. Shown in Fig.2 are the two kinds of switching process. The captured remanence states of the two kinds of switching processes are anticlockwise and clockwise domain wall trapping vortex.

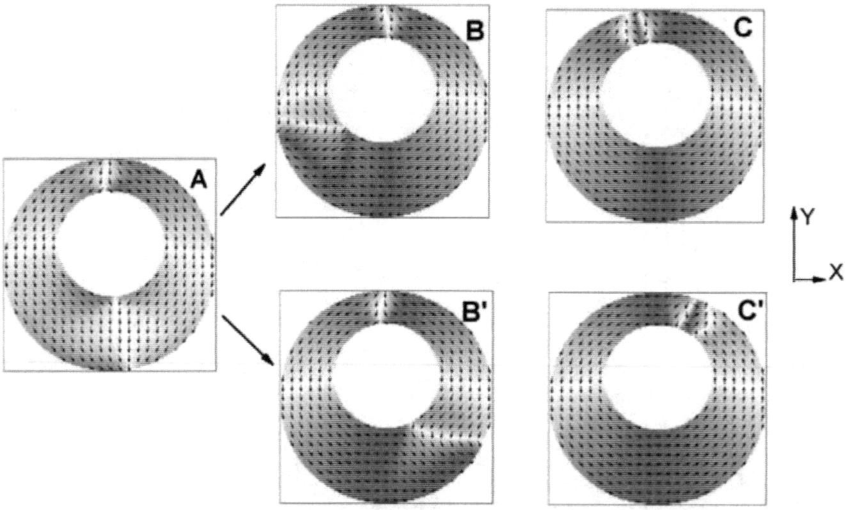

Fig.2 Two different switching process after saturating the magnetization along negative y axis: A-B-C, forming a clockwise domain wall trapping vortex; A-B'-C', forming an anticlockwise domain wall trapping vortex. State A is the negative onion after saturation; state B, B' are domain wall movements via left and right arm of the asymmetric ring; state C, C' are clockwise and anticlockwise domain wall trapping vortex.

When the negative field was applied along the x axis of the asymmetric ring and then reduces, the magnetization of the asymmetric ring forms a negative onion with two transverse domain walls nucleated at the asymmetric axis. Reversing the field, the two domain walls both start to move towards the narrower arm of the asymmetric ring and form a vortex state with clockwise direction with a 360 degree domain wall trapped at the narrow arm. Shown in Fig.3 is the corresponding switching process. The chirality of the domain wall trapping vortex is thus determined by the field history. When the asymmetric ring is at the right beginning saturated in negative direction, the vortex is clockwise and vise verse.

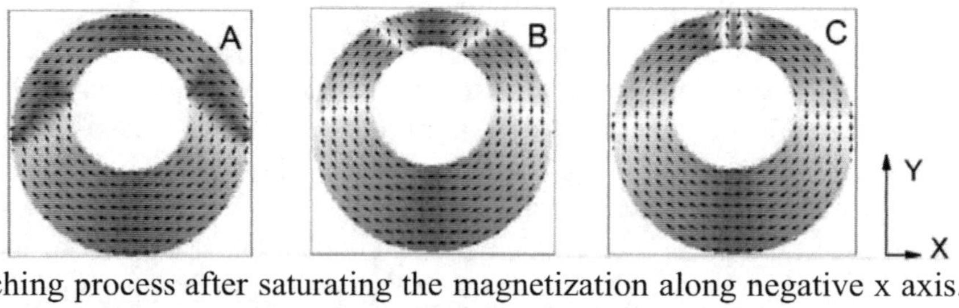

Fig. 3 Switching process after saturating the magnetization along negative x axis. State A is the onion state after saturation along x axis; state B is the domain wall movement via narrow arm

of the asymmetric ring; state C is the clockwise domain wall trapping vortex.

3.2. Chirality dependence of the field driven domain wall motion

After the domain wall trapping vortex state is obtained, we further study the field driven domain wall motion. By saturating the asymmetric ring along the –y axis and then relaxed it, the domain wall trapping vortex states are formed. The chirality is by chance clockwise or anticlockwise. The field is to apply both in –x and +x direction. Considering the external fields direction and the two kinds of the chirality, there are four cases, which are: (i), clockwise vortex state with field applied in the -x direction; (ii), clockwise vortex state with field applied in the +x direction; (iii), anticlockwise vortex state with field applied in the -x direction, and (iv) anticlockwise vortex state with field applied in the +x direction. However, case (i) is same as case (iv), so as (ii) and (iii). As a result, only the domain wall motion with fields applied in –x direction were studied, which are case (i) and (iii).

The simulation shows that the domain wall behaviors are markedly dependent on the chirality of the vortex. Starting from the domain wall trapping vortex as shown in Fig.4.(a), when the chirality is anti-clockwise, applying a magnetic field to (-x) direction does not annihilate the trapped domain wall. Rather, another domain wall is nucleated from the wider part of the ring as can be seen as a switch at -160 Oe as shown in Fig. 4.(b). As the field is increased, the new wall moves to the left edge of the ring and stayed there. The trapped domain wall then annihilates at -620 Oe as shown in Fig.4.(c). The new domain wall does not annihilate even when the applied field is increased to a large value of 1000 Oe. Interestingly, when the direction of the vortex state is anti-clockwise, the field influences the trapped domain wall more significantly. A –x field of 50 Oe breaks the trapped 360 degree domain wall in the anti-clockwise vortex (Fig.4(d)) into two transverse walls. The separated domain walls move apart quickly and annihilate, forming the a saturation magnetic state with the magnetization aligns to –x which is shown in Fig.4(e).

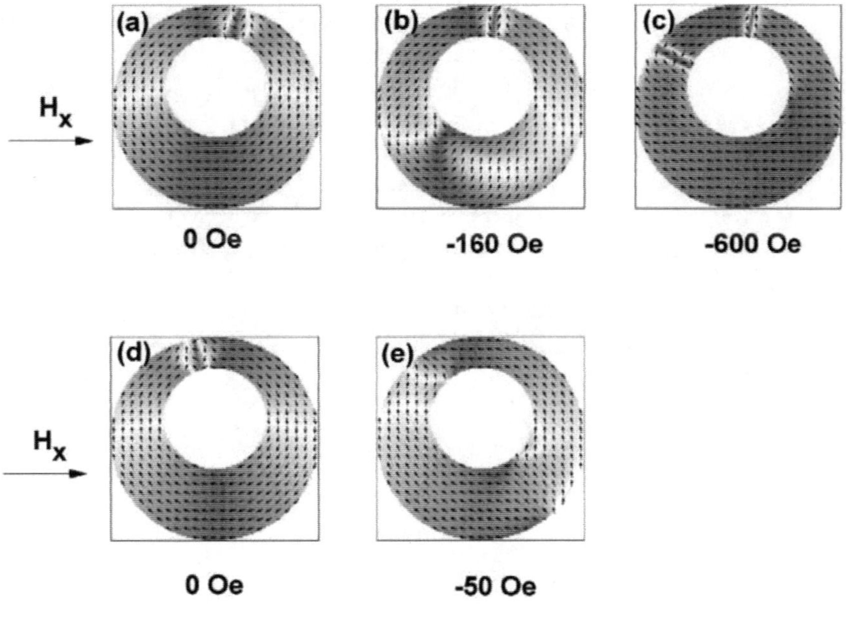

Fig.4 Simulated domain wall motion with different chirality in a 10-nm-thick NiFe asymmetric ring.(a) anticlockwise vortex with a 360 degree domain wall trapped at the

narrow arm; (b) nucleation of a second domain wall at the wide arm with a field of -160 Oe applied in the x axis; (c) two domain walls under a large field of -600 Oe; (d) clockwise vortex with a 360 degree domain wall trapped at narrow arm; (e) negative 'onion' in a small field of -50 Oe in the x axis.

4. conclusions

We have studied the magnetic field orientation effects on the domain walls behavior in asymmetric nanoring using micromagnetic simulation. At certain thickness a 360 degree domain wall is trapped at the narrow arm of the vortex state at remanence. The chirality of the vortex is found to be dependent on the field history of the asymmetric nanoring. The magnetic field orientation effects on the domain walls behavior are also studied. The simulation results show that the domain wall behavior under the externally applied magnetic field is found to be related with the chirality of the vortex of the asymmetric ring.

5. acknowledgement

This work was supported in part by the ASTAR SERC grant (082 101 0015) and the NRF-CRP program (Multifunctional Spintronic Materials and Devices).

6. References

[1] Parkin S. S. P., Hayashi M., Thomas L., 2008, *Science*, **320** 190-194.
[2] Allwood D. A., Xiong G., Faulkner C. C., Atkinson D., Petit D., Cowburn R. P. 2005, *Science*, **309** 1688, .
[3] Beach G. S. D., Nistor C., Knutson C., Tsoi M. & Erskine J. L., 2005, *Nature Mater,* **4** 741-744.
[4] Krause S., Berbil-Bautista L., Herzog G., Bode M, Wiesendanger R., 2007 *Science*, **317** 1538-1540.
[5] San Emeterio Alvarez L., Wang K.-Y., Lepadatu S., Landi S., Bending S. J., Marrows C. H. 2010, *Phys. Rev. Lett.*, **104** 137205.
[6] García N., Saveliev I. G., Zhao Y. -W. and Zlatkine A.,2000, *J. Magn. Magn. Mater.*, **214** 7-12.
[7] Lepadatu S., Wessely O., Vanhaverbeke A., Allenspach R., Potenza A., Marchetto H., Charlton T. R., Langridge S., Dhesi S. S., Marrows C. H., Phys. Rev. B., **81** 060402.
[8] McMillen M., McQuaid R. G. P., Haire S. C., McLaughlin C. D., Chang L. W., Schilling A., Gregg J. M., 2010, *Appl. Phys. Lett.*, **96** 042904.
[9] Singh D. K., Krotkov R., and Tuominen M.T., 2009, *Phys. Rev. B*, **79** 184409.
[10] Wang X. H., Peng W. K., Lew W.S, 2009, *J. Appl. Phys.*, **106** 043905-4.
[11] "Availabe at http://math.nist.gov/oommf."

Fabrication of MgO-based magnetic tunnel junctions for subnanosecond spin transfer switching

Tatsuya Aoki[1], Yasuo Ando, Mikihiko Oogane and Hiroshi Naganuma
Department of Applied Physics, Graduate School of Engineering, Tohoku University
Aoba-yama 6-6-05, Sendai, Japan

E-mail: aoki@mlab.apph.tohoku.ac.jp

Abstract. In the development of spin-transfer torque (STT)-based magnetic random access memories (spin-RAMs), characterizing the physics in sub-ns STT switching is a crucial issue for designing operation schemes. This paper describes the procedure we used to fabricate MgO-based magnetic tunnel junctions for measuring sub-ns STT switching. We also discuss the sub-ns STT switching properties based on the quasi-static properties.

1. Introduction

Spin-transfer torque (STT)-based magnetic random access memories (spin-RAMs) have been widely developed as next generation non-volatile memories[1]. Since one of the features of these memories is a fast write cycle, characterization of the physics in fast STT switching is a crucial issue for the designing of the fast spin-RAMs. Since MgO-based magnetic tunnel junctions (MTJs), which consist of CoFeB ferromagnetic electrodes[2] are promising candidates for the storage element in spin-RAMs, STT switching properties have been widely investigated[3]. Recently, demonstration and the detailed examinations about sub-ns STT switching for MgO-based MTJ were acheived[4]. The higher resistance of MTJ devices than that of metallic spin valves[5] enhances electrical leakage loss via the parasitic capacitance of devices. Therefore, in this paper, we mainly show the details of how to fabricate MgO-based MTJs for sub-ns STT switching measurements.

2. Sample fabrication procedures

The stacking structure of the MTJs used in this study is buffer layers/PtMn/$Co_{90}Fe_{10}$/Ru/$Co_{40}Fe_{40}B_{20}$/Mg/MgO/free layer/Ta/Ru on sapphire wafers fabricated by using a UHV magnetron sputtering system. The synthetic anti-ferromagnet (SyF) free layer consists of $Co_{40}Fe_{40}B_{20}$ (2 nm)/Ru (0.9 nm)/$Co_{40}Fe_{40}B_{20}$ (2 nm). The MTJs, annealed at 275°C for 1 hour in a 10-kOe magnetic field, were patterned using electron beam lithography and Ar ion milling into a 0.1×0.2 μm^2 rounded rectangle shape. The devices used in this study had a resistance area product in parallel state of 5.5 $\Omega\mu m^2$ and an MR ratio of 46 % at room temperature. Figure 1(a) shows a microscopic image of a fabricated device.

---

[1] To whom correspondence should be addressed.

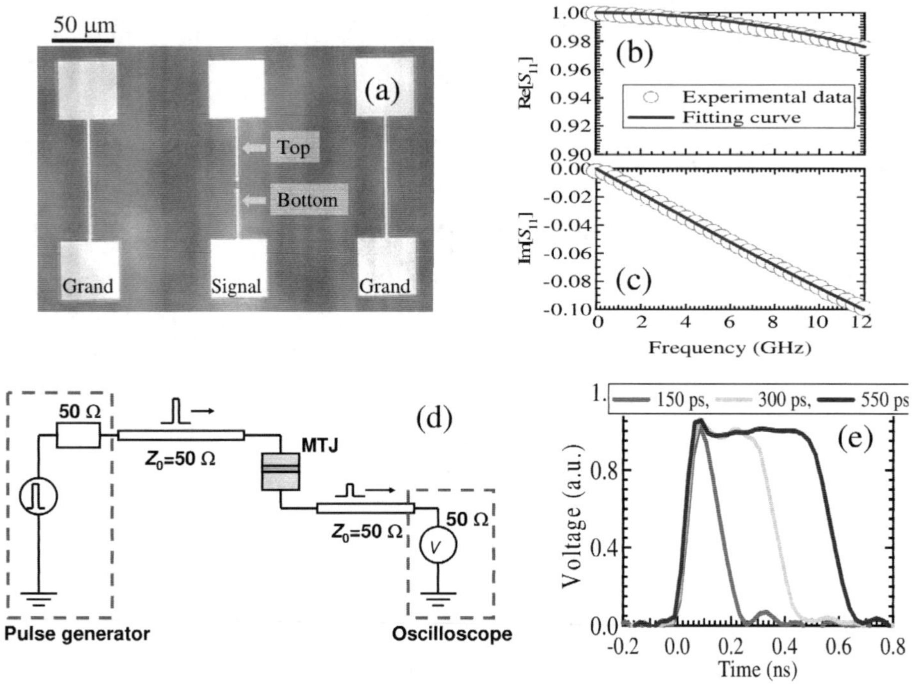

Figure 1 (a) Microscope picture of a device. Real (b) and imaginary (c) parts of S_{11} for a device without any junctions. Data for the equivalent circuit model (solid blue lines) well matched the experimental data (rounded red marks). (d) Schematic picture of the system for measuring voltage transmit through a MTJ. (e) Observed waveforms of the sub-ns width pulse signals transmitted through a junction by the oscilloscope illustrated in (d).

The electrode was patterned into a grand-signal-grand (G-S-G) transmit type[5] micro-stripe structure. Here, 40 nm thick SiO_2 was deposited as insulation material for the 12 μm^2 cross sectional area of the top and bottom electrodes. Since the parasitic capacitance of MTJ devices is considered as the main microwave loss factor[6], we estimated it from the analysis of reflectivity S_{11}[7] as shown in Fig. 1(b) and (c). The obtained value of parasitic capacitance was 14 fF, an extremely low value because the barrier capacitance in sub nanometer scaled MTJs is only a few fF. To confirm the pulsed current properly transmitted through the MTJs, without incurring significant losses, we used the circuit diagram shown in Fig. 1(d) to evaluate the actual pulse waveforms through the devices. Figure 1(e) shows the pulse voltage waveforms we observed by the oscilloscope illustrated in (d). Sub-ns pulse signals were clearly observed without significant distortions or losses. In Ref.6, transmit pulse waveforms in the waveguide devices were shown, however, they are not passed through the MTJs.

3. Spin-transfer properties

To show the basic spin transfer properties, we first experimentally analysed the current (I) – field (H) diagram. The methodology is described below. DC two probe magneto-resistance measurements were performed by applying the fixed sweep rate R (=12 Oe) quasi-static magnetic field. These measurements were performed five times for each sensing DC current value (I_{dc}) which was systematically changed from -2 to 2 mA in 0.1 mA steps. We converted R to the effective relaxation time τ_P shown in Table I following the procedure described in Ref. 8 by assuming the parameters of saturation magnetization M_S, volume of free layer V, temperature T, and attempt time τ_0 shown in Table I. In this way, we performed a fitting analysis based on a thermal activation model on the plots of the coercive force H_C versus I_{dc} obtained through above measurement following the literature in

Table I. Parameters for obtaining analysis results for the I - H diagram. τ_P, M_S, V, T, and τ_0 are assumed parameters. H_{C0}, I_{C0}, β, and C are estimated parameters from the analysis.

Assumption					Estimation			
τ_P (s)	M_S (T)	V (m^3)	T (K)	τ_0 (s)	H_{C0} (Oe)	I_{C0} (mA)	β (K^2/mA^2)	C (Oe/mA)
0.09	1.2	8×10^{-23}	300	1×10^{-9}	50	5	$0.6\text{-}1.1 \times 10^5$	0

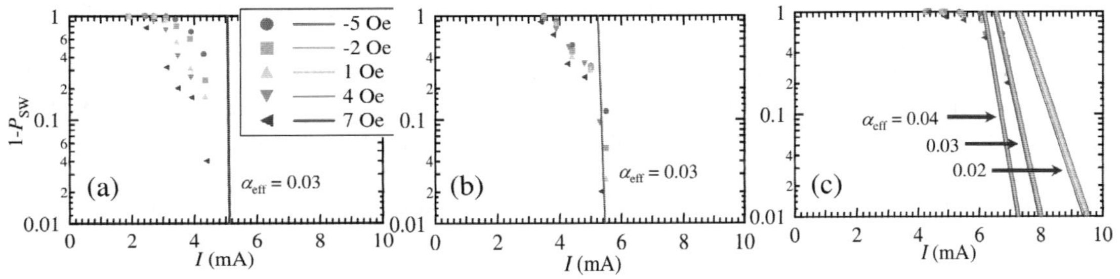

Figure 2 1-P_{SW} depending on I for pulse widths of (a) 10 ns, (b) 2.5 ns, and (c) 0.4 ns. Data (colored marks) were measured under various effective easy axis fields. Solid lines in (a)-(c) show fitting or calculation results based on Eq.(1).

Ref. 9. Here, we analysed data independently for each quadrant defined by the polarity of I_{dc} (positive or negative) and magnetization configuration (parallel (P) or anti-parallel (AP)). Consequently, we obtained estimation values of the zero temperature switching current I_{C0}, the zero temperature coercive force H_{C0}, the hating coefficient β and the coefficient of the effective field C which was assumed to be linearly depending on I_{dc}. Table I shows the positive I_{dc} and P to AP switching values for our MTJs. Similar analysis results have been reported for MgO-based MTJs[10]. In the next paragraph, we discuss the spin transfer switching properties induced by sub-ns current pulses based on the above analytical results.

To show the fast STT switching properties, we measured STT switching probabilities (P_{SW}) in the following manner. First we applied a hard-axis external field to cancel the hard axis offset field estimated from an asteroid curve and fixed it. Then, to reset the magnetization configuration into P state, an easy-axis external field (500 Oe) was applied. After that, we approached the field to desirable strength remaining P state and fixed it at that level. We then injected a current pulse into the MTJs, just after we had detected the resistance in order to determine whether or not the magnetization had switched into an AP configuration or not. P_{SW} was obtained through 150 time repeats of the cycle of above easy axis field reset, current pulse injection, and resistance detection. Here, we used current pulses whose rise time and width were 55 ps and 100 - 10000 ps, respectively. The signal line length L_{line} and its width are 150 μm and 3 μm, respectively. The L_{line} was much smaller than the maximum voltage wavelength (= 3.3×10^4 μm) corresponding to the pulse rise time of 55 ps used in this study. Therefore, the electrode structure is regarded as a lumped constant circuit. The measurement circuit was constructed by radio frequency (RF) assemblies, each consisting 26 GHz bandwidth coaxial lines, and DC assemblies consisting of lock-in amplifiers in order to measure quasi-static resistance measurement.

Figure 2 shows 1-P_{SW} depending on current pulse strength I. Pulse widths (= τ_P) of (a) 10 ns, (b) 2.5ns, and (c) 0.4 ns were used for the measurements. Fitting analyses were performed using the following switching probability formula based on an adiabatic model derived from the thermal distribution of the initial magnetization direction[11].

$$1 - P_{SW} = \exp\left\{ -2\alpha_{eff}\gamma \cdot (H_{C0} + 2\pi M_S) \cdot \tau_P \cdot \frac{I - I_{C0}}{I_{C0}} + \ln\left[\frac{\pi^2}{8} \cdot \frac{H_{C0} M_S V}{k_B V} \cdot \left(1 - \frac{H_{easy}}{H_{C0}} \right)^2 \right] \right\}, \quad (1)$$

Here, α_{eff}, γ, and H_{easy} are effective Gilbert damping constants of the free layer, the gyro-magnetic momentum ratio, and the effective easy axis external field, respectively. We assumed that H_{C0} and I_{C0} were 50 Oe and 5 mA, respectively. α_{eff} was treated as a fitting parameter. The solid lines in Fig. 2(a)-(c) show the fitting results based on Eq.(1). As shown in Fig. 2(b) and (c), the model well reproduced the experimental data, which means the switching mechanisms in the sub-ns regime are precession dominated and that the thermal activation effect is negligible except for the initial condition. In addition, here, α_{eff} was estimated to be 0.03. Hence, the calculation results using Eq.(1) were also shown in Fig. 2(c), in which the assumed α_{eff} values were 0.02 and 0.04 (α_{eff} in I_{C0} was also taken into account), clearly disagree with the data. This indicates that the spin transfer property in the sub-ns regime is strongly affected by α_{eff}. The obtained α_{eff} value is larger than intrinsic damping constant of a 2 nm thick $Co_{40}Fe_{40}B_{20}$ film (~0.015)[12]. The reason for this, however, still unclear. We found that α_{eff} can be enhanced by the roughness of a free layer[13], by the dynamic interaction between magnetic layers in SyF via interlayer exchange coupling[14], or by angular momentum dissipation due to multi-mode spin wave excitation caused by larger current injection[15].

4. Conclusions

The device fabrication of MgO-based MTJs for sub-ns STT switching measurement were described. The spin transfer switching properties induced by sub-ns current pulses were strongly affected by α_{eff}.

5. Acknowledgements

This work was supported by the Research and Development for Next-Generation Information Technology by MEXT. The first author received support from a JSPS Research fellowship. The MgO-based MTJ stacks used in this work were provided by ULVAC Inc.

References

[1] Ikeda S *et al.* 2007 *IEEE Trans. Ele.* **54** 991.
[2] Djayaprawira D D *et al.* 2005 *Appl. Phys. Lett.* **86** 092502.
[3] Kubota H *et al.* 2005 *Jpn. J. Appl. Phys. Part 2* **44** L1237; Hayakawa J *et al.* 2005 *Jpn. J. Appl. Phys. Part 2* **44** L1267.
[4] Aoki T, Ando Y, Oogane M, and Naganuma H 2010 *Appl. Phys.Lett.* **96** 142502.
[5] Devolder T *et al.* 2006 *Appl. Phys. Lett.* **88** 152502.
[6] Aoki T *et al.* 2007 *J. Magn. Soc. Jpn.* **31** 94.
[7] Deac A M *et al.* 2008 *Nat. Phys.* **4** 803. (in Supplementary Information)
[8] Sharrock M P 1990 *IEEE Trans. Magn.* **26** 193; Kitakami O *et al.* 2004 *Jpn. J. Appl. Phys.* **43** L115.
[9] Fuchs G D *et al.* 2005 *Appl. Phys. Lett.* **86** 152509.
[10] Yoshikawa M *et al.* 2007 *J. Appl. Phys.* **101** 09A511.
[11] He J, Sun J Z, and Zhang S 2007 *J. Appl. Phys.* **101** 09A501.
[12] Oogane M *et al.* 2007 *J. Magn. Soc. Jpn.* **31** 209.
[13] Mizukami S, Ando Y, and Miyazaki T 2002 *Phys. Rev. B* **66** 104413.
[14] Ando Y unpublished.
[15] Lee K J *et al.* 2004 *Nat. Mat.* **3** 877; Fuchs G D *et al.* 2007 *Appl. Phys. Lett.* **91** 062507.

Journal of Physics: Conference Series **266** (2011) 012087 doi:10.1088/1742-6596/266/1/012087

Magnetoresistance and electronic structure of granular films with MgO or MgF_2 matrices

Y. Fujiwara[1], **H. Matsuda**[1], **K. Sato**[1], **M. Jimbo**[2], **T. Kobayashi**[1]
[1]Graduate school of Engineering, Mie University: Mie 514-8507, Japan
[2]Faculty of Engineering, Daido University: Nagoya 457-8530, Japan

E-mail: fujiwara@phen.mie-u.ac.jp

Abstract. The magneto-transport properties and electronic structures of FeCo-MgO, FeCoB-MgO, FeCo-MgF_2 and FeCoB-MgF_2 granular films were investigated. From the X-ray photoelectron spectroscopy (XPS) measurements, it was recognized that the oxide layer formed on the surface of granules in FeCo-MgO. Especially Fe oxides were preferentially formed. By using $(Fe_{50}Co_{50})_{80}B_{20}$, the formation of Fe oxides on the granule surface was suppressed since B was oxidized instead of Fe. The MR ratio of FeCoB-MgO was approximately 2%. By using MgF_2 matrix, the surface states of granules were drastically improved. XPS measurements indicated that an amount of metallic Fe in FeCoB-MgF_2 increased compared with FeCoB-MgO. Reflecting the improvement of surface states of granules, a large MR ratio of approximately 6 % was obtained in FeCoB-MgF_2.

1. Introduction

Magnetic granular films, consisting of magnetic granules and insulating matrix, exhibit a large tunnel magnetoresistance (TMR) effect. The granular films have been attracted attention as one of promising candidates for magnetic sensor elements because of their technical advantages. The granular films can be generally prepared in one process, and various deposition techniques for preparation are developed, such as co-sputtering, reactive sputtering or tandem deposition. A high break down voltage is expected, since the granular films are regarded as a network of a large number of tunnel junctions. Furthermore, in the granular films, the TMR effects have been observed in wide resistivity range [1]-[6] from 10^4 – 10^{10} $\mu\Omega$cm. The granular films are therefore expected to be integrated into various magnetic sensor devices.

The granular films employing various insulating matrices such as Al_2O_3 [1], SiO_2 [2], MgF_2 [3], MgO [4] [5] and ZrO_2 [6] have been investigated so far, and it has been known that the TMR effects are influenced by the kind of insulating materials. Furthermore, since the TMR ratio of granular materials depends on the spin polarization of granules [7], the choice of magnetic materials is also an important factor to obtain a large TMR ratio. Recently, it has been reported that tunneling spin polarization (TSP) of FeCo increased by an addition of B, and $(Fe_{50}Co_{50})_{80}B_{20}$ showed a large TSP of 49 % [8].

In this work, we made an attempt to investigate the influence of insulating and magnetic materials on TMR effects of granular films. MgO and MgF_2 were employed for insulating matrix, and $Fe_{50}Co_{50}$ and $(Fe_{50}Co_{50})_{80}B_{20}$ alloys were employed for the magnetic materials. The granular films were prepared by a tandem deposition method. The magneto-transport properties and electronic structure of granules were investigated. In this paper, our experimental results are presented.

2. Experimental

The tandem deposition method was employed to prepare the granular films by dc and rf magnetron sputtering. The base pressure was less than 4 × 10^{-4} Pa. $Fe_{50}Co_{50}$ and $(Fe_{50}Co_{50})_{80}B_{20}$ sputtering targets were utilized for the deposition of magnetic layer. Reactive sputtering in a gas mixture of Ar + O_2 was utilized for the deposition of the MgO under layer. Depositions of other layers were carried out in an Ar atmosphere. The sputtering pressure was 0.8 Pa. Four kinds of granular films were prepared. The granular films with MgO/$Fe_{50}Co_{50}$/MgO structure are called FeCo-MgO. MgO/$(Fe_{50}Co_{50})_{80}B_{20}$/MgO, MgF_2/$Fe_{50}Co_{50}$/MgF_2 and MgF_2/$(Fe_{50}Co_{50})_{80}B_{20}$/$MgF_2$ are called FeCoB-MgO, FeCo-MgF_2 and FeCoB-MgF_2, respectively. The sample structure is shown in Fig. 1. The thickness of magnetic layer d_{mag} was varied from 0.7 to 1.4 nm. The microstructures of samples were investigated by transmission electron microscopy (TEM). Depth profiling of X-ray photoelectron spectroscopy (XPS) was made in order to investigate the electronic structure of granules. Resistivity measurements were performed with the two-terminal method. The resistivity was calculated using nominal value of d_{mag}. MR measurements were carried out under the sweep of a magnetic field of ±15 kOe. The magnetic field was applied parallel to both the film plane and current direction. All measurements were carried out at room temperature.

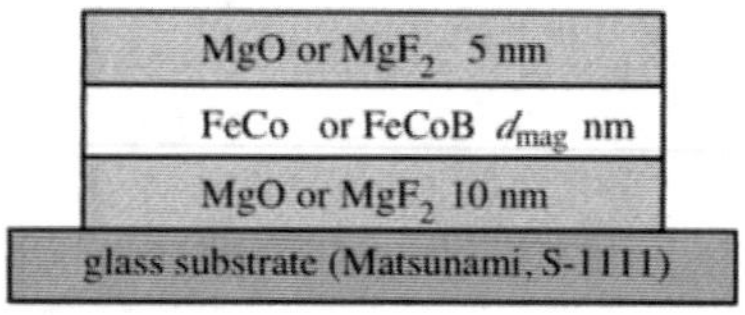

Figure 1. A schematic of sample structure.

3. Results and discussion

Fig. 2 shows comparisons of d_{mag} dependence of resistivity. In Fig. 2 (a), it was recognized that the resistivity of FeCo-MgO increased with decreasing d_{mag}. An abrupt increase in resistivity around d_{mag} = 1.1 nm was observed between d_{mag} = 1.1 and 1.2 nm. This increase indicates the structural change of magnetic layer from continuous film to granular structure. When d_{mag} < 1.1 nm, apparent nonlinear dI-dV curves were obtained, indicating the tunnel conduction of the samples. FeCoB-MgO shows a gradual change of resistivity, and their resistivity is larger than that of FeCo-MgO. The tunnel conduction of FeCoB-MgO was confirmed in dI-dV measurements when d_{mag} < 1.2 nm. Fig. 2 (b) shows the comparison between FeCo-MgF_2 and FeCoB-MgF_2. The feature, that the granular films without B shows a sharp increase in resistivity around d_{mag} = 0.9 nm, is similar to that of FeCo(B)-MgO. The tunnel conductions were also confirmed at d_{mag} < 0.9 nm for FeCo-MgF_2 and d_{mag} < 1.1 nm for FeCoB-MgF_2. Comparing the resistivity at the same d_{mag}, the resistivity of FeCoB-MgF_2 was smaller than that of FeCoB-MgO.

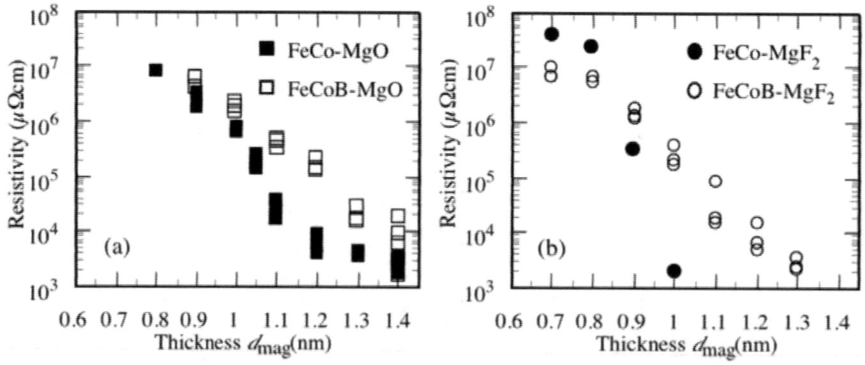

Figure 2. Resistivity as a function of d_{mag}, (a) FeCo-MgO and FeCoB-MgO, (b) FeCoB-MgO and FeCoB-MgF_2.

In order to discuss the structure and resistivity of granular films, TEM and XPS depth profile measurements were carried out. Fig. 3 shows a typical TEM micrograph of FeCoB-MgF_2. It is clearly seen that the granules are separated by MgF_2 matrix. The average diameter of granules was estimated to be 3.0 – 5.0 nm.

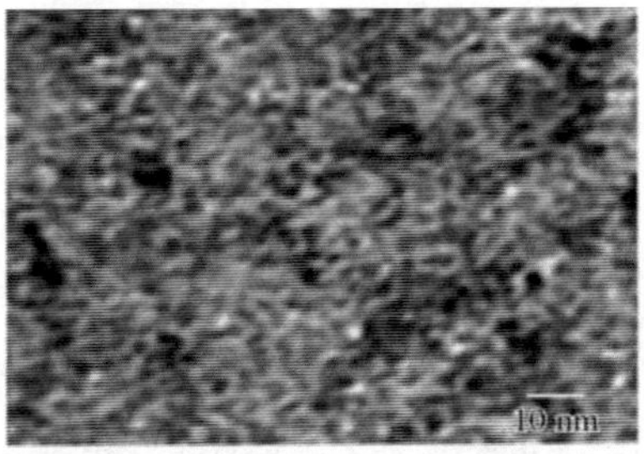

Figure 3. A TEM micrograph of FeCoB-MgF_2, where d_{mag} = 0.9 nm.

Figs. 4, 5, 6 show the depth profile spectra of FeCo-MgO, FeCoB-MgO and FeCoB-MgF_2, respectively. Each spectrum was measured after Ar ion etching every 30 seconds. The total etching time (t_{etch}) is indicated in the figures. In Fig. 4, the left and right panels correspond to the Co 2p and Fe 2p XPS spectra. The position and profile of the Co 2p peaks did not change by t_{etch} variations. On the other hand, varying the t_{etch} changed the profile of Fe 2p spectra. Two peaks can be seen at the Fe 2p peaks, especially at the $2p_{3/2}$ peak. Peak positions are estimated to be 707-708 eV and 711-712 eV, respectively. Although the peak positions were slightly shifted from original positions by charge up effects, it is thought that these peaks correspond to metallic Fe and Fe_2O_3. The high peak intensity of Fe_2O_3 indicates oxidation of granules. As the t_{etch} increased, the Fe_2O_3 peak decreased and then increased again, whereas the Fe peak increased and then decreased. These results indicate that the surface of granules is oxidized and the oxide layer is mainly composed of Fe_2O_3. In this work, Fe was selectively oxidized. The selective oxidation of Fe can be explained by the difference in heat of formation [9][10][11]. In our case, Fe is easily oxidized owing to a large heat of formation of Fe oxide. XPS depth profile spectra of FeCoB-MgO are shown in Fig. 5. The left and right panels correspond to the Fe 2p and B 1s XPS spectra, respectively. The Co 2p XPS spectra were not shown here, since they showed almost the same feature as that of FeCo-MgO. Two peaks originated from Fe and Fe_2O_3 are seen in Fe 2p spectra. It is also confirmed that the almost the same t_{etch} dependence of peak intensity as

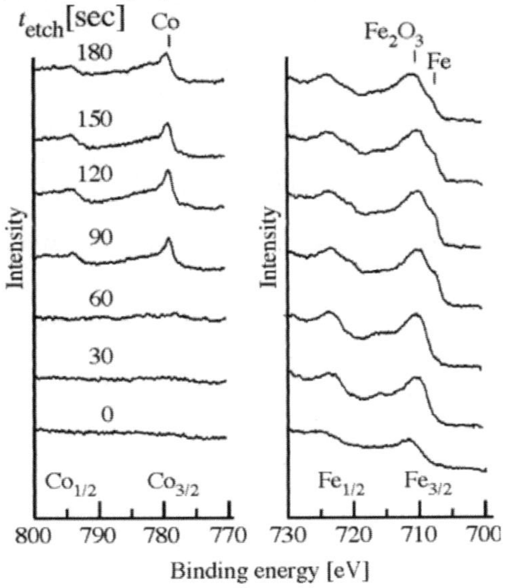

Figure 4. XPS depth profile spectra of FeCo-MgO, where d_{mag} = 1.1 nm.

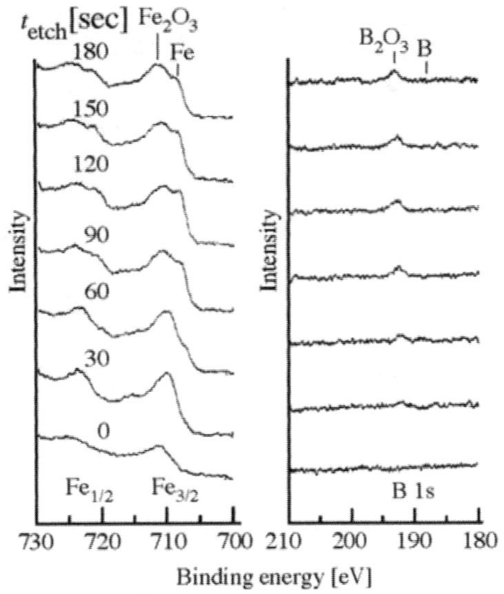

Figure 5. XPS depth profile spectra of FeCoB-MgO, where d_{mag} = 1.1 nm.

that of FeCo-MgO. The difference in spectra is the peak intensity ratio of Fe / Fe_2O_3. These differences are clearly seen in the spectra with t_{etch} > 90 sec. The peak intensity ratios of FeCoB-MgO become larger compared with that of FeCo-MgO, indicating that metallic Fe increased owing to the addition of B. In B 1s XPS depth profile spectra, the peaks originated from B_2O_3 appeared, whereas the metallic B peak could not confirmed, suggesting a preferential oxidization of B. Thus, it is thought that the addition of B prevents Fe from oxidation. Although the surface of granules was oxidized in both FeCo-MgO and FeCoB-MgO, the composition of surface oxidized layer on granules seems to be different. It is thought that this difference had an influence on resistivity shown in Fig. 2 (a).

The XPS depth profile of FeCoB-MgF_2 is shown in the left panel of Fig. 6. Only Fe 2p spectra are presented. In the right panel, Fe 2p spectra FeCoB-MgO are presented for comparison. Two peaks can be seen at 707-708 eV and 712-714 eV, which corresponded to metallic Fe and Fe fluorides, respectively. Although two peaks can also be confirmed, the shape of spectra is quite different from that of FeCoB-MgO. In all spectra, Fe peaks were larger than fluorides peaks excluding the film surface (t_{etch} = 0), although the fluoride peaks become large with increasing t_{etch}. A large peak intensity ratio of Fe / FeF_x suggest less influences of MgF_2 matrix on granule surfaces owing to the large heat of formation of MgF_2. Although the surface fluoride layer exists on granules in FeCoB-MgF_2, its thickness is expected to be much thinner than the surface oxide layer thickness of granules in FeCoB-MgO. In FeCoB-MgO, the electrons must tunnel through both thick surface oxide layer on granules and MgO matrix. Thus, FeCoB-MgO showed high resistivity compared with FeCoB-MgF_2 at the same d_{mag}.

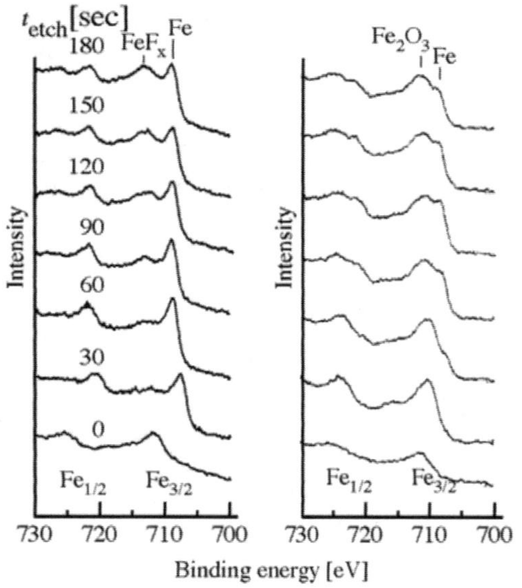

Figure 6. XPS depth profile spectra of FeCoB-MgF_2, where d_{mag} = 1.0 nm (left panel). XPS spectra of FeCoB-MgO are also shown in right panel for comparison.

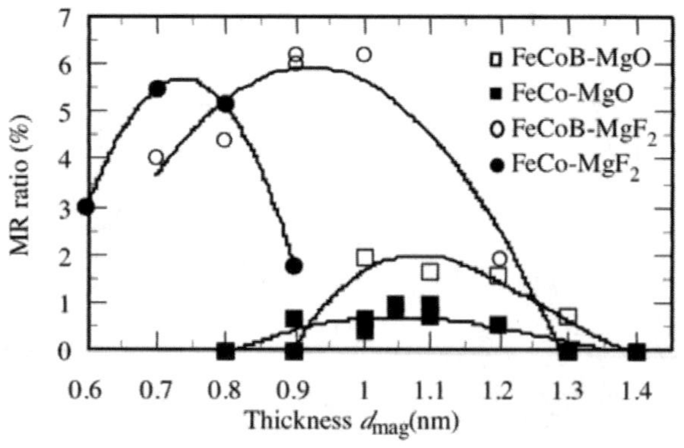

Figure 7. MR ratio of granular films as a function of d_{mag}.

Fig. 7 shows MR ratio of granular films as a function of d_{mag}. For all kinds of granular films, MR ratio shows a maximum at a certain thickness. The maximum MR ratio of approximately 1 % was obtained in FeCo-MgO, whereas approximately 2 % was obtained in FeCoB-MgO. We infer that this increase in MR ratio results from a suppression of formation of Fe oxides on granules owing to the addition of B. Both in FeCo-MgF_2 and FeCoB-MgF_2, large MR ratios were obtained. The maximum MR ratio of FeCoB-MgF_2 reached of approximately 6 %. This drastic increase by using MgF_2 is due to an improvement of the surface state of granules, which confirmed from XPS depth profile measurements. The maximum MR ratio of in FeCo-MgF_2 is approximately 5.5 % which is slightly small compared with the maximum MR ratio in FeCoB-MgF_2. This difference may result from the difference in TSP between $Fe_{50}Co_{50}$ and $(Fe_{50}Co_{50})_{80}B_{20}$.

4. Summary

We investigated the magneto-transport properties of FeCo-MgO, FeCoB-MgO, FeCo-MgF_2 and FeCoB-MgF_2 granular films. The electronic structure of granules was also investigated by the depth profiles of XPS.

From the XPS depth profiles, it was recognized that the oxide layer existed on the surface of granules in FeCo-MgO. On the surface of granules, Fe oxides were preferentially formed owing to its relatively high heat of formation. The maximum MR ratio of approximately 1 % was obtained. In FeCoB-MgO, surface oxidation of Fe was suppressed by the addition of B. Thus, the maximum MR ratio increased to 2 %. High intensity of metallic Fe peak was found in XPS spectra of FeCoB-MgF_2, indicating less influence of MgF_2 matrix on granules. The difference in surface states of granules in between FeCo-MgO and FeCoB-MgF_2 seemed to be due to the difference of heat of formations of insulating materials. A large MR ratio of approximately 6 % was obtained in FeCoB-MgF_2, reflecting the improvement of surface states of granules.

5. References

[1] Mitani S, Fujimori H and Ohnuma S 1997 *J. Magn. Magn. Mater.* **165** 141
[2] Honda S, Hirata M and Ishimaru M 2005 *J. Magn. Magn. Mater.* **290-291** 1053
[3] Kobayashi N, Ohnuma S, Masumoto T and Fujimori H 2001 *J. Appl. Phys.* **90** 4159
[4] Mitani S, Shintani Y, Ohnum S and Fujimori H 1997 *J. Soc. Magn. Jpn.* **21** 465 in Japanese
[5] Arita M, Wakasugi K, Ohta K, Hamada K, Takahashi Y and Choi J B 2008 *Microelec.* **85** 2445
[6] Hattink B J, Carcia del Muro M, Konstantinovic Z, Batlle X, Labarta A and Varela 2006 *Phys. Rev* B **73** 045418
[7] Inoue J and Maekawa S 1996 *Phys. Rev.* B **53** R11927
[8] Kubota T, Daibou T, Oogane M, Ando Y and Miyazaki T 2007 *J. Appl. Phys.* **46** L250
[9] Keavney D J, Park S, Falco C M and Slaughter J M 2001 *Appl. Phys. Lett.* **78** 234
[10] Yoon C S, Lee J H, Jeong H D, Kim C K, Yuh J H and Haasch R 2002 *Appl. Phys. Lett.* **80** 3976
[11] Hong J, Lee Y, Lee M K, Song H. J Shin H J, Yoo Y and Suh J 2003 *Appl. Phys. Lett.* **83** 4803

Journal of Physics: Conference Series **266** (2011) 012088 doi:10.1088/1742-6596/266/1/012088

Room temperature magnetoresistance in $Fe_3Si/CaF_2/Fe_3Si$ MTJ epitaxially grown on Si(111)

K Harada[1], K S Makabe[1], H Akinaga[2] and T Suemasu[1]

[1] Institute of Applied Physics, University of Tsukuba, Ibaraki 305-8573, Japan

[2] Nanodevice Innovation Research Center, AIST, Tsukuba, Ibaraki 305-8569, Japan

E-mail: suemasu@bk.tsukuba.ac.jp

Abstract. A tunnel magnetoresistance has been investigated for $Fe_3Si/CaF_2/Fe_3Si$ magnetic tunnel junctions (MTJs). We fabricated Fe_3Si(20 nm)/CaF_2(2 nm)/Fe_3Si(15 nm) heterostructures on a CaF_2(2 nm)/Si(111) substrate by molecular beam epitaxy (MBE) and processed them into 300×15 μm^2-area MTJs using selective wet chemical etching and conventional photolithography. The current-voltage (*J-V*) characteristics for the MTJs measured at room temperature (RT) were well fitted to Simmons' equation. The fitting yields the barrier height $\varphi = 2.5$ eV and the barrier thickness $d = 1.26$ nm. The magnetoresistance curve for the MTJs reaches a maximum in the range of 100-200 Oe and the magnetoresistance ratio is approximately 0.28% under a bias voltage of 20 mV at RT.

1. Introduction

A magnetic tunnel junction (MTJ) formed by ferromagnet-insulator-ferromagnet has shown a large tunnel magnetoresistance (TMR) effect, and the application of MTJs has been a key issue in the development of spin-based electronics (spintronics). The MTJs have been receiving increasing attention due to their possible technological applications, such as a memory cell in magnetic random access memories (MRAMs) [1-3]. Ferromagnetic silicide Fe_3Si is an attractive material for Si-based spintronics applications because the lattice parameter of Fe_3Si is 0.564 nm, which nearly lattice-matches Si. Furthermore, Fe_3Si has a relatively high Curie temperature of approximately 570 °C [4]. Electrical injection and detection of spin-polarized electrons in silicon using epitaxially grown Fe_3Si/Si Schottky-tunnel-barrier contacts has gained interest in this material [5]. We have chosen CaF_2 as a barrier layer in MTJs because it also lattice-matches Si. We have developed a technique for epitaxial growth of Fe_3Si/CaF_2 heterostructures on Si(111) substrates by molecular beam epitaxy (MBE) [6-8], and obtained the large barrier height for electrons in Fe_3Si to tunnel through the CaF_2 barrier layer of 2.5 eV [9]. Furthermore, we have very recently succeeded in observing a clear negative differential resistance in the current–voltage (*J–V*) characteristics of $CaF_2/Fe_3Si/CaF_2$ ferromagnetic resonant tunneling diodes at room temperature (RT) [10, 11]. In this paper, we report on the epitaxial growth of Fe_3Si(20 nm)/CaF_2(2 nm)/Fe_3Si(15 nm) MTJs on Si(111) substrates by MBE, and successfully demonstrated TMR effect at RT for the first time using ferromagnetic Fe_3Si.

2. Experiments

The Fe_3Si(20 nm)/CaF_2(2 nm)/Fe_3Si(15 nm) heterostructures were fabricated epitaxially by MBE onto a CaF_2(3 nm)/Si(111) substrate. The epitaxial growth of the MTJs was carried out in an MBE chamber with electron beam evaporation sources for Fe and Si, and a high-temperature Knudsen cell for CaF_2. First, a Si buffer layer (10 nm) was deposited on the Si(111) and was annealed at 1000 °C for 15 min to enhance crystallization and flatten the surface of the Si(111) substrates. Next, Fe_3Si(20 nm)/CaF_2(2 nm)/Fe_3Si(15 nm)/CaF_2(3 nm) heterostructures were grown on a Si(111) substrate. The growth temperatures were at 140 and 280 °C, for Fe_3Si and CaF_2, respectively, and each layer was annealed at 300 °C for 20 min. Finally, an Fe capping layer (5 nm) was deposited on the heterostructures at around 150 °C to change a coercive field, H_c, between the top Fe/Fe_3Si and the bottom Fe_3Si layers. The Fe capping layer was not annealed. The grown layers were processed into mesa structures with dimensions of 600×20, 450×15, 300×10, 300×15 and 200×10 μm^2 using a conventional photolithography, selective wet etching, and lift-off processes. Fe_3Si was etched using HF: HNO_3: H_2O = 1: 2: 400 at 0 °C, and CaF_2 was etched using H_2SO_4: H_2O = 1: 20 at 0 °C. Ohmic contacts were formed on top of the MTJs using Au/Cr. Figure 1 shows the schematic cross section of the fabricated MTJs. The crystalline quality of grown layers was investigated by reflection high-energy electron diffraction (RHEED) and X-ray diffraction (XRD) equipment. The *J-V* characteristics and magnetic-field dependence of the resistance for the MTJs were measured by a standard two-probe method at RT.

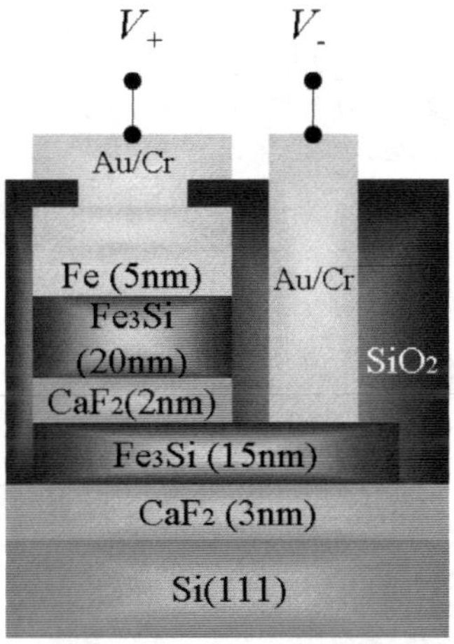

Figure 1. Schematic cross section of fabricated MTJs.

3. Results and Discussion

Figure 2 shows RHEED patterns taken after each growth stage along the $[1\bar{1}0]$ azimuth of Si. The RHEED clearly displays sharp and fine streaky patterns, showing that the Fe_3Si/CaF_2/Fe_3Si MTJ structures were epitaxially grown and the films with good crystalline quality together with a smooth surface were obtained from the 10-nm-thick Si buffer layer through the 20-nm-thick-Fe_3Si upper layer. A typical XRD pattern of the multilayer measured by a θ-2θ scan is shown in figure 3. The multilayer shows the intense diffraction peak due to Fe_3Si(220). The multilayer clearly displays sharp and fine streaky RHEED patterns as shown in figure 2, meaning that the (110)-oriented Fe_3Si layers were epitaxially grown on the Si(111) substrate. Similar result has been already reported for Fe_3Si layers grown by rf-magnetron sputtering on a Si(111) substrate with yttria-stabilized zirconia buffer layers, where the epitaxial relationship between the Fe_3Si film and the Si(111) substrate was also discussed [12]. The (110) orientation of a Fe_3Si film on Si(111) was different from the (111)-oriented one we have obtained in our previous report [7]. We speculate that this difference comes from the decrease of growth temperature of Fe_3Si from 400 °C to 140 °C. We adopted the growth temperature of 140 °C in this work because Fe_3Si films grown at this temperature had better crystalline quality and a smoother surface. Note that the diffraction peak of $Fe_3Si(111)$ is originated by DO_3-type Fe_3Si, but that of $Fe_3Si(222)$ is given by not only DO_3-type Fe_3Si but also B_2-type Fe_3Si. The diffraction peaks of $Fe_3Si(111)$ and $Fe_3Si(222)$ were not observed in figure3. Thus, we don't have enough information to understand which type Fe_3Si was formed [13].

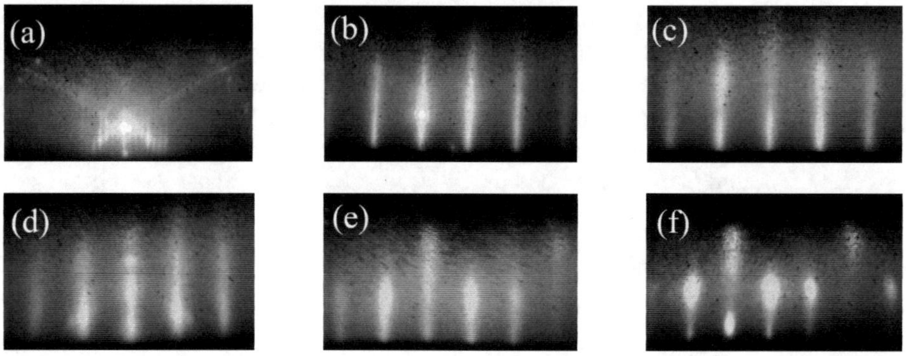

Figure 2. RHEED patterns taken after each growth stage along $[1\bar{1}0]$ azimuth of Si: (a) Si buffer layer, (b) CaF_2 buffer layer, (c) Fe_3Si bottom ferromagnetic layer (d) CaF_2 barrier layer (e) Fe_3Si upper ferromagnetic layer, and (f) Fe capping layer.

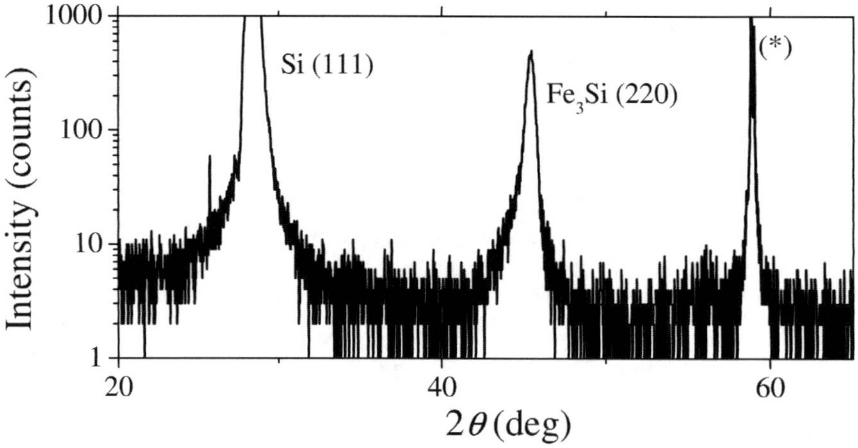

Figure 3. *θ-2θ* x-ray diffraction pattern of the MTJ.

The measured *J-V* characteristics of the MTJ without external magnetic field are shown in figure 4. The voltages were applied between the top and the bottom Fe_3Si layers in the MTJ as shown in figure 1. The positive bias voltage, V_+, is defined as that applied to the top Fe_3Si (20 nm) layer with respect to the bottom Fe_3Si (15 nm) layer as shown in figure.1. The data were well fitted to Simmons' equation as shown by the solid line [14]. The fitting yields the barrier height φ = 2.5 eV for electrons in the Fe_3Si layers to tunnel the CaF_2 barrier layers, which is the same as that previously reported [9], and the barrier thickness d = 1.26 nm. This result shows that the thickness of the CaF_2 barrier layer became substantially thinner than that designed, probably due to rough interface between the CaF_2 barrier layer and Fe_3Si ferromagnetic layers.

Figure 5 shows the magnetic-field dependence of the resistance for the 300×15 μm^2-area MTJs under a bias voltage of 20 mV measured at RT. The magnetic field H was applied parallel to the sample surface along the long axis of the MTJs. Magnetoresistance curve increases gradually at 0 Oe and reaches a maximum in the range of H = ±100-200 Oe. The increase of resistance is attributed to the antiparallel magnetic orientation of both the top and bottom Fe_3Si ferromagnetic layers. The corresponding magnetoresistance ratio ($MR = (R_{AP} - R_P)/R_P$, P: parallel, AP: antiparallel) is approximately 0.28 % at RT. Dependence of the magnetoresistance ratio on bias voltage is now under investigation.

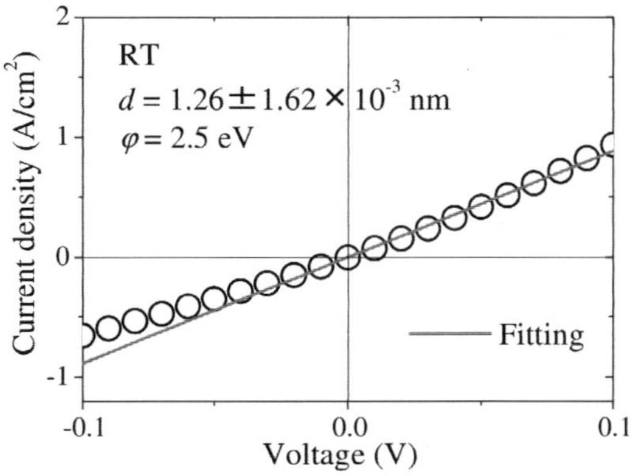

Figure 4. J– V characteristics measured at RT for the MTJ

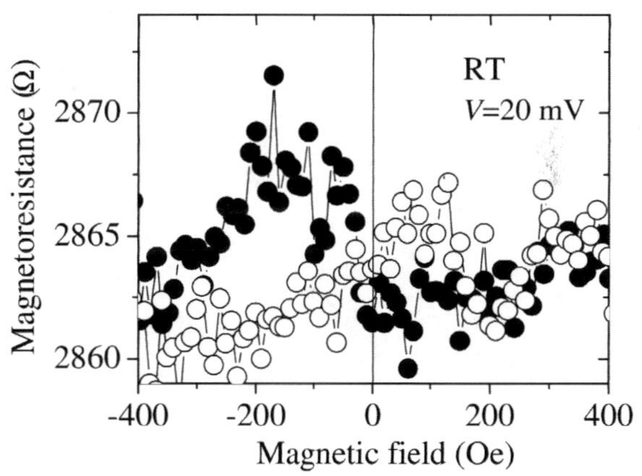

Figure 5. Magnetoresistance characteristics measured at RT for the MTJ

4. Conclusion

We fabricated the Fe_3Si (20 nm)/CaF_2 (2 nm)/Fe_3Si (15 nm) MTJ on Si(111) by MBE, and measured the J-V and magnetoresistance characteristics at RT. The J-V characteristics were well fitted to the Simmons' equation, and the fitting yielded the barrier height φ = 2.5 eV and the barrier thickness d = 1.26 nm. The MTJs showed magnetoresistance ratio of 0.28% at RT.

Acknowledgements

This work was supported in part by a Grant-in-Aid for Scientific Research in the Priority Area of "Creation and Control of Spin Current" (No. 19048029, MEXT). The authors would also like thank Prof. M. Isshiki and Dr. M. Uchikoshi of the Tohoku University for supplying us high-purity 5N Fe sources and Prof. E. Kita and Prof. H. Yanagihara, the University of Tsukuba, for help in J-V and magnetoresistance measurements.

References

[1] Miyazaki T and Tezuka N 1995 *J. Magn. Magn. Mater.* **139** L231
[2] Moodera J S, Kinder L R, Wong T M and Meservey R 1995 *Phys.Rev. Lett.* **74** 3273
[3] Nozaki T, Tezuka N and Inomata K 2006 *Phys. Rev. Lett.* **96** 027208
[4] Waliszewski J, Dobrzynski L, Malinowski A, Satula D, Szmanski K, Prandl W, Brückel T and Schärpf O 1994 *J. Magn. Magn. Mater.* **132** 349
[5] Ando Y, Hamaya K, Kasahara K, Kishi Y, Ueda K, Sawano K, Sadoh T and Miyao M 2009 *Appl. Phys. Lett.* **94** 182105

[6] Sunohara T, Kobayashi K, Umada M, Yanagihara H, Kita E, Akinaga H and Suemasu T 2005 *Jpn. J. Appl. Phys.* **44** L715
[7] Kobayashi K, Sunohara T, Umada M, Yanagihara H, Kita E and Suemasu T. 2006 *Thin Solid Films* **508** 78
[8] Harianto T, Kobayashi K, Suemasu T and Akinaga H 2007 *Jpn. J. Appl.Phys.* **46** L904
[9] Harianto T, Sadakuni K, Akinaga H and Suemasu T 2008 *Jpn. J. Appl. Phys.* **47** 6310
[10] Sadakuni K, Harianto T, Akinaga H and Suemasu T 2009 *Appl. Phys. Express* **2** 063006
[11] Makabe K S, Suzuno M, Harada K, Akinaga H and Suemasu T 2010 *Jpn. J. Appl. Phys.* **49** 060212
[12] Akiyama K, Kaneko S, Kadowaki T, Hirabayashi Y and Funakubo H 2008 *Jpn. J. Appl. Phys.* **47** 577
[13] Yoshitake T, Nakagauchi D, Ogawa T. Itakura M, Kuwano N and Tomokiyo Y 2005 *Appl. Phys. Lett.* **86** 262505
[14] Simmons J G 1963 *J. Appl. Phys.* **34** 1793

Journal of Physics: Conference Series **266** (2011) 012089 doi:10.1088/1742-6596/266/1/012089

Spin current study of spin glass AgMn using spin pumping effect

R Iguchi[1], K Ando[2], E Saitoh[2,3,4], T Sato[1]

[1] Department of Applied Physics and Physico-Informatics, Faculty of Science and Technology, Keio University, Yokohama 223-8522, Japan

[2] Institute for Materials Research, Tohoku University, Sendai 980-8577, Japan

[3] The Advanced Science Research Center, Japan Atomic Energy Agency, Tokai 319-1195, Japan

[4] PRESTO, Japan Science and Technology Agency, Sanbancho, Tokyo 102-0075, Japan

E-mail: `iguchi@az.appi.keio.ac.jp`

Abstract. The role of spin glass as a spin current absorber was investigated by spin pumping effect in $Ni_{81}Fe_{19}$/Cu/spin glass $Ag_{90}Mn_{10}$ trilayer structure. The spin current was pumped to the spin glass layer via the Cu layer by means of the ferromagnetic resonance of the $Ni_{81}Fe_{19}$ layer. The damping constant of the ferromagnetic layer reflecting the intensity of the spin current absorption was evaluated by tting the spectra with Lorentzian. At low temperatures, the spin current absorption in the spin glass layer arises sharply.

1. Introduction

Spin glass (SG) has been investigated as a typical model of random systems. In SG, spins freeze randomly with competing interactions and the cusp in the magnetic susceptibility is observed at a low temperature [1]. Because the strength of the competing interactions is random, these spins show interesting memory phenomena [2][3] related to the multivalued memory, the associated memory and the neural network [4]. Recently SG is starting to be applied to the spintronics devices. The exchange bias e ect, which is one of the topics in the spintronics, is employed to ferromagnetism (FM)/SG system. Curious temperature dependence of the exchange bias eld, which reflects the properties of SG, was reported in the FM/SG systems [5][6]. In recent spintronics devices, understanding of the physics of spin current, which is a flow of electron spins [7], is important. Because spin current can interact with the spins in SG, spin current is expected to modify the spin con gurations in SG. In addition, spin current has a potential to obtain the information related to the spin con gurations which electric current cannot obtain. Hence, it is important to study how SG acts to spin current. We studied SG as a spin current absorber using spin pumping e ect. In FM/non-magnetic metal (NM) system, spin current was injected to the NM layer from the FM layer by the spin pumping e ect through ferromagnetic resonance (FMR)[8][9]. When the spin pumping occurs, the spin dynamics changes in the FM layer. The damping constant, which is evaluated as the line width of the spectra, is enhanced as the spin current is absorbed in NM layer [8]. The contribution of exchange bias e ect in the FM/SG system should be removed to clarify the contribution of the SG layer to the spin current. Because the exchange bias occurs by the short range coupling between the spins, the Cu spacer layer between the FM layer and the SG layer was inserted. On account of long spin di usion length of Cu, in the previous report, the spin current was clearly injected via the Cu layer to the NM layer in FM/Cu/NM system [10]. We fabricated the FM/Cu/SG and FM/Cu samples. The damping constant of the magnetization of the FM layer was analyzed based on the line width by tting the FMR spectra. The external eld angle dependence of line width at room temperature and the temperature dependence of line width at low temperatures were studied.

2. Experimental Procedure

FM (10 nm)/Cu (10 nm)/SG (30 nm) lms, which are described in Figure 1, were prepared. The SG layer of AgMn (30 nm) was fabricated with Ar ion beam sputtering on the oxidized Si substrate. Before the deposition of Cu and FM layers, the surface of the AgMn layer was etched to remove the oxidized surface. Then the spacer layer of Cu and the FM layer of $Ni_{81}Fe_{19}$ (Permalloy (Py)) were deposited with electron beam deposition. In addition, the capping layer of Cu 20 nm was deposited to prevent the oxidization of the Py layer. Also, the FM/Cu samples were prepared to investigate the e ect of the AgMn layer. The Mn concentration of the sputterd AgMn layer was deduced as 10.5 at.% by an electron probe micro analyzer. The magnetic properties of sputtered AgMn were measured using a superconducting quantum interference device (SQUID) magnetometer. The FMR studies were carried out using a X-band

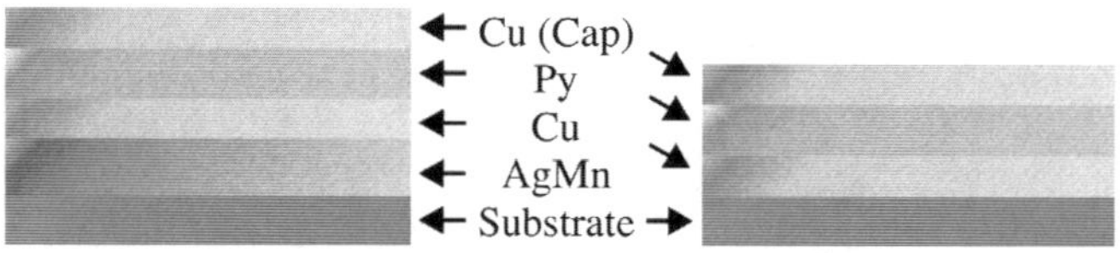

Figure 1. Schematic image of the samples with and without AgMn layer.

spectrometer. In the external eld angle dependence measurement, the microwave frequency f was 9.79 GHz and the out-of-plane angle of the external magnetic eld was varied by goniometer. In the temperature dependent measurement, the microwave frequency f was 9.39 GHz and the external magnetic eld was applied parallel to the lm plane.

3. Result and Discussion

SG shows the transition to the SG phase from the paramagnetic phase at a low temperature T_g when the temperature decreases. The temperature dependence of magnetic moment of the thick AgMn lm is shown in Figure 2. The temperature, at which the magnetizations under zero eld cooling (ZFC) and eld cooling (FC) separate, is the spin glass transition temperature T_g. In Figure 2, the thick AgMn lm shows the transition to the spin glass phase at $T_g = 25$ K. The measured FMR spectrum of the Py/Cu sample at 5 K is shown in Figure 3. The FMR spectra originate from the resonance in Py layer. The tted curve in Figure 3 is obtained using di erentiated Lorentzian with following parameters: $\Delta H = 69.24$ Oe and $H_r = 1075.6$ Oe, where ΔH and H_r represent the line width as the full width at half maximum (FWHM) of the spectrum and the resonance eld, respectively. The temperature dependences of ΔH and the di erence of ΔH between the Py/Cu/AgMn and Py/Cu samples are shown in Figure 4. The value of ΔH of the Py/Cu/AgMn sample is enhanced compared with the Py/Cu sample for the whole of temperature range because of the existence of additional SG layer. Generally, the origins of the enhancement of ΔH are the sample inhomogeneity in the FM layer and the enhanced damping by spin current absorption in the SG layer. To evaluate the sample inhomogeneity, the external eld angle θ_H dependence of ΔH and H_r were measured and the result is shown in Figure 5. θ_H is the angle between the external magnetic eld and the normal direction to the lm plane,

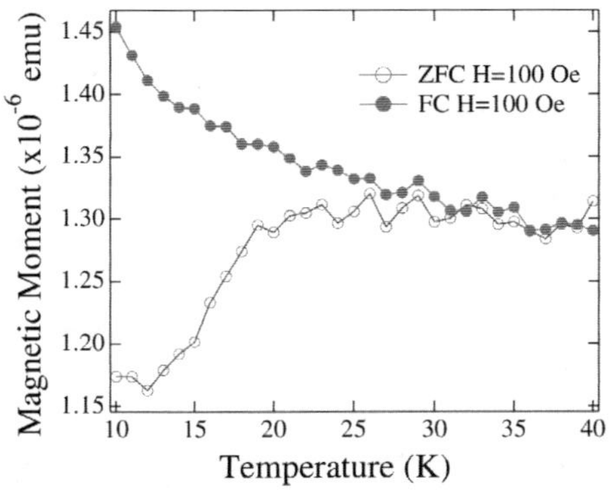

Figure 2. Temperature dependence of magnetic moment of thick AgMn lm.

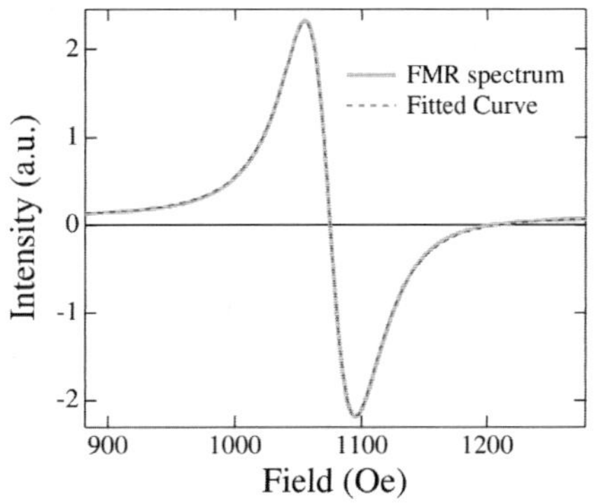

Figure 3. FMR spectrum of Py/Cu sample at 5 K and the tted curve.

which is also illustrated in Figure 5. θ_H dependence of ΔH consists of the terms of the sample inhomogeneity and the damping [11]. The experimental data were tted by Eq. 8 and 9 in Ref. [11] and the tted curve in Figure 5 is reproduced using e ective magnetization 4 M_{ef} = 8.1 kOe, damping constant = 0.011, g factor g = 2.09 and no sample inhomogeneity term. For comparison, we show an overview of sample inhomogeneity term calculated based on Eq. 9 in Ref. [11] using $\Delta 4$ M_{ef} = 40 Oe and $\Delta\theta_H$ = 0 deg, which are respectively the fluctuations of 4 M_{ef} and of θ_H. It is clear that the terms proportional to the sample inhomogeneity were signi cantly small in our samples. In addition, the di erence of ΔH in the individual samples was estimated as 0.5 Oe. This is negligibly small compared with the di erence of ΔH between the Py/Cu/AgMn and Py/Cu samples. Therefore, the di erence of ΔH between the Py/Cu/AgMn and Py/Cu samples is the enhanced part of ΔH that is attributed to the spin current absorption in the SG layer. We note that the amount of the spin current absorption,

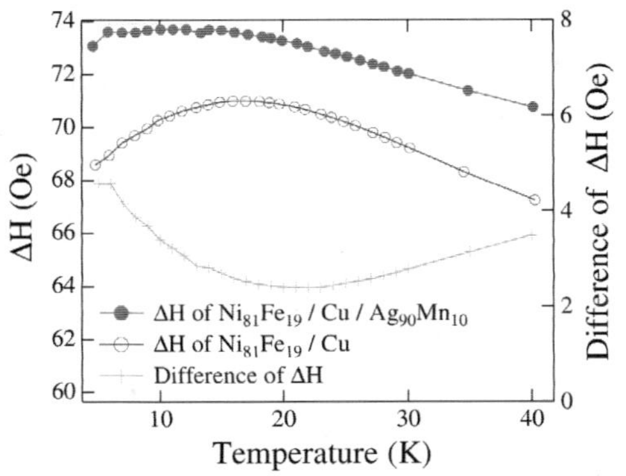

Figure 4. Temperature dependences of FWHM line width ΔH of Py/Cu and Py/Cu/AgMn samples (left) and the di erence of ΔH between Py/Cu and Py/Cu/AgMn samples (right).

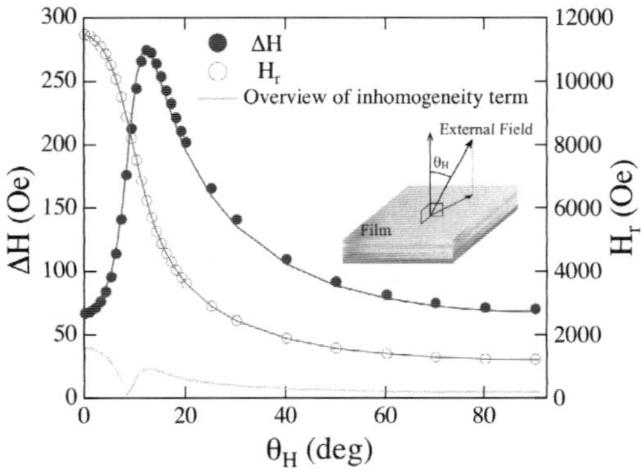

Figure 5. Out-of-plane angle θ_H dependences of ΔH and H_r of Py/Cu/AgMn sample. The solid curves are tted to the experimental data. The dotted curve is an overview of inhomogeneity term.

which is revealed as the di erence of ΔH between the Py/Cu/AgMn and Py/Cu samples in Figure 4, shows a minimum around T_m = 22 K. The amount of the spin current absorption is determined by the mixing conductance and the spin di usion length. When the spin current is injected from the FM layer into the NM layer by the spin pumping e ect using FMR, the damping constant, which is proportional to the line width of spectra, becomes enhanced. Mizukami *et al.* reported the enhanced damping constant of the Py/various NM systems [8] and Tserkovnak *et al.* reported the detailed theoretical analysis [12]. The additional damping constant (which is the di erence between the enhanced damping constant and that of bulk samples) depends on the mixing conductance and the ratio of the layer thickness to the spin di usion length of the NM layer. The mixing conductance determines the amount of the injected spin current and the spin di usion length determines how long the injected spin current can survives. The amount of the spin current absorption is determined considering the back flow of the spin current to the FM layer. In our system, the temperature dependence of the additional line width, caused by the additional damping, is explained by the temperature dependence of the mixing conductance and the spin di usion length of AgMn, because the Cu layer plays no role to the spin current absorption. In the previous report [7], the mixing conductance was constant and governed by FM/NM interface band matching. However this is not clear in the FM/SG interface. The spin di usion length of AgMn with about 9 at.% Mn is reported as 7 nm at 4.2 K but the detailed temperature dependence has not be reported [13]. In our system, the spin di usion length may not be shorter than the SG layer thickness at all temperature range. Generally, the spin di usion length increases as temperature decreases in NM because the phonon scattering is suppressed [14]. If the spin di usion length is longer than the AgMn layer thickness, the injected spin current is not absorbed perfectly, so that the enhancement of ΔH is small. This situation is satis ed in our samples above T_m. Below T_m, spin di usion length decreases as the temperature decreases, so that the enhancement of ΔH becomes larger. The decrease in the spin di usion length below T_m is caused by the other mechanism. Since the SG layer shows SG transition below $T_g = 25$ K, the spin di usion length can be governed by the interactions from the freezing spins in the SG layer below T_m.

4. Summary

We have studied the spin current absorption in spin glass using the spin pumping e ect in the Py/Cu/AgMn system. The amount of the spin current absorption increased sharply at low temperatures. This indicates that either the spin di usion length or the mixing conductance changes when the system enters into the spin glass phase. Therefore, the use of spin current can provide an alternative method to investigate spin glass properties.

Acknowledgments

The authors thank to H. Kura and T. Ogawa for their help with the measurement.

References

[1] Cannella V and Mydosh J 1972 *Phys. Rev. B* **6** 4220–37
[2] Lundgren L, Svedlindh P, Nordblad P and Beckman O 1983 *Phys. Rev. Lett.* **51** 911–4
[3] Jonason K, Vincent E, Hammann J, Bouchaud J and Nordblad P 1998 *Phys. Rev. Lett.* **81** 3243–6
[4] Amit D, Gutfreund H and Sompolinsky H 1985 *Phys. Rev. A* **32** 1007–18
[5] Ali M, Adie P, Marrows C, Greig D, Hickey B and Stamps R 2006 *Nat. Mater.* **6** 70–5
[6] Lin J, Yao Y and Lee S 2010 *Appl. Phys. Lett.* **96** 162502
[7] Brataas A, Nazarov Y V and Bauer G E W 2000 *Phys. Rev. Lett.* **84** 2481–4
[8] Mizukami S, Ando Y and Miyazaki T 2001 *Jpn. J. Appl. Phys.* **40** 580–5
[9] Saitoh E, Ueda M, Miyajima H and Tatara G 2006 *Appl. Phys. Lett.* **88** 182509
[10] Mizukami S, Ando Y and Miyazaki T 2002 *Phys. Rev. B* **66** 104413
[11] Chappert C, Dang K L, Beauvillain P, Hurdequint H and Renard D 1986 *Phys. Rev. B* **34** 3192–7
[12] Tserkovnyak Y, Brataas A and Bauer G 2002 *Phys. Rev. B* **66** 224403
[13] Yang Q, Holody P, Lee S F, Henry L L, Loloee R, Schroeder P A, Pratt W P and Bass J 1994 *Phys. Rev. Lett.* **72** 3274–7
[14] Yataka S, Ando Y, Miyazaki T and Mizukami S 2006 *Jpn. J. Appl. Phys.* **45** 3892–5

2nd Int. Symp. on Advanced Magnetic Materials and Applications (ISAMMA 2010) IOP Publishing
Journal of Physics: Conference Series **266** (2011) 012090 doi:10.1088/1742-6596/266/1/012090

Spin accumulation and mistracking e ects on the magnetoresistance of a ferromagnetic nano-contact

Hiroshi Imamura[1†] **and Jun Sato**[2]

[1] Nanosystem Research Institute, AIST, Tsukuba 305-8568, Japan
[2] Department of Physics, Graduate School of Humanities and Sciences, Ochanomizu University, 2-1-1 Ohtsuka, Bunkyo-ku, Tokyo 112-8610, Japan

E-mail: † `h-imamura@aist.go.jp`

Abstract. We theoretically studied spin accumulation and mistracking e ects on the current-perpendicular-to-plane magnetoresistance of a domain wall confined in a nano-contact which is experimentally fabricated as current-confined-path structure made in a nano-oxide layer between ferromagnetic metal layers. We calculated the magnetoresistance (MR) ratio by evaluating the voltage drop due to the spin accumulation and additional resistivity due to mistracking. We showed that both the spin accumulation and mistracking give the same order of magnitude to the MR ratio for the nano-contact with 2 nm diameter.

1. Introduction

Current-perpendicular-to-plane giant magnetoresistance (CPP-GMR) has attracted much attention for its potential application as a read sensor for high-density magnetic recording. In order to realize high-density magnetic recording, we need MR devices with high MR ratio and low resistance area product (RA). Although the RA value of a CPP-GMR system is much smaller than that of a tunneling magnetoresistance (TMR) system, the MR ratio of a conventional CPP-GMR system still remains as a small value of a few %. Much effort has been devoted to increase the MR ratio of the CPP-GMR system. One of the candidates for such a device is a spin-valve with nano-oxide-layer (NOL) [1]. The NOL is an insulator layer sandwiched between two ferromagnetic metal layers, and has a lot of metallic nano-contacts.

Recently, Fuke et al. obtained the MR ratio about 10% for the CoFe ferromagnetic nano-contacts in the NOL[2]. For CoFe nano-contacts with anti-parallel magnetic alignment the domain wall of nano-meter size is created in the nano-contact, and both the spin accumulation [3, 4, 5] and the mistracking can be the origin of the MR. Therefore, it is important to analyze both the effects of spin accumulation and mistracking on MR of a ferromagnetic nano-contact. The enhancement of the MR ratio due to nano contact was discussed in Refs. [6, 7, 8] from the view point of parasitic resistance of electrode and spin accumulation around contacts.

2. Model and Method

The system we consider is schematically shown in Fig. 1 (a). We assume that when the magnetization vectors of top and bottom ferromagnetic layers are aligned to be anti-parallel to each other the Bloch-type domain wall is created in the contact. The unit vector of the local magnetization of the Bloch-type domain wall is given by $\mathbf{M} = (\cos(\pi z/h), \sin(\pi z/h), 0)$,

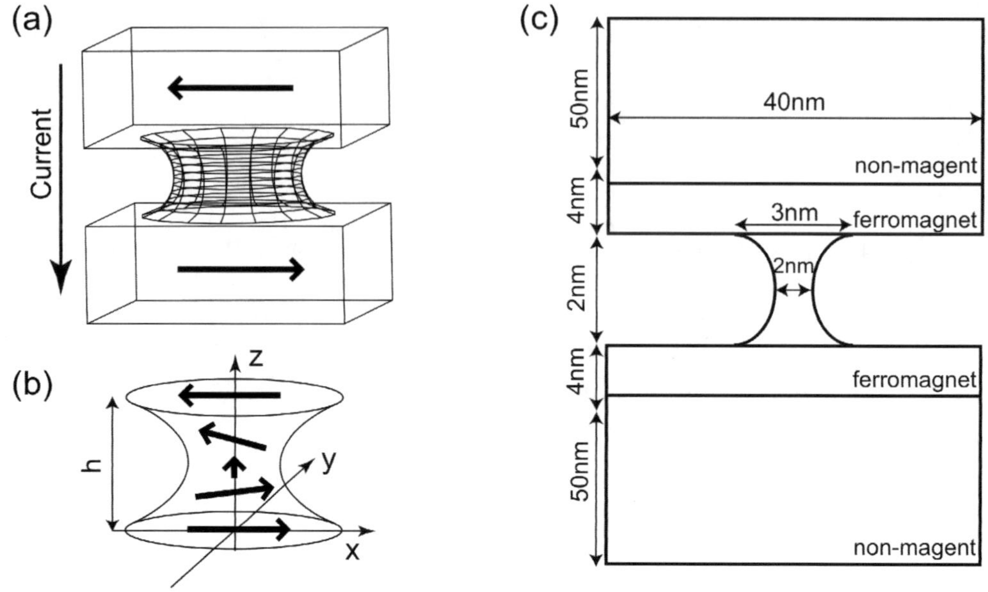

Figure 1. (a) A current-confined-path (CCP) spin-valve with a domain wall. The current is owing perpendicular to the plane. (b) The spin configuration in the contact region. (c) The cross-sectional view of the CCP spin-valve. The curve of the contact region is modeled by the half ellipse, which is tangent to the ferromagnetic layers.

where h is the thickness of the nano-contact (=2nm) and z is the height from the bottom of the nano-contact as shown in Fig. 1 (b). The cross-sectional view of the system is shown in Fig. 1 (c). The boundary of the contact region is modeled by the half ellipse, which is tangent to the ferromagnetic layers.

The MR of a domain wall due to mistracking was calculated by P. M. Levy and S. Zhang in Ref. [9]. Within the perturbation formalism, they obtained the wave function of conduction electron whose spin can not follow the direction of local magnetization in a domain wall. Then they calculate the scattering rate within the Fermi Golden rule with the Born approximation. The spin-up(+) and spin-down(-) components of conductivity is obtained as

$$\sigma^{\pm} = \sigma^{\pm(0)} \left[1 \mp 2\xi^2 \left(\frac{3\beta}{5(1 \mp \beta)} - \frac{2\beta}{3\sqrt{1-\beta^2}} \right) \right]. \tag{1}$$

Where the quantization axis is taken to be along the direction of the rotating local magnetization, $\xi = \pi\hbar^2 k_F / 4mhJ$ is the perturbation parameter which is implicitly assumed to be less than 1, k_F is the Fermi wave number, m is electron mass, J represents the exchange coupling constant between conduction electron and local magnetization. $\sigma^{\pm(0)}$ represents the spin-up(+) and spin-down(-) conductivity of bulk material and the spin polarization β is defined as

$$\beta = \frac{\rho^{+(0)} - \rho^{-(0)}}{\rho^{+(0)} + \rho^{-(0)}}, \tag{2}$$

where $\rho^{\pm(0)} = 1/\sigma^{\pm(0)}$. From Eq. (1) one can easily see that conductivity is suppressed by mistracking, i.e., the resistance increases due to mistracking.

The spin accumulation is another origin of the MR of a ferromagnetic nano-contact. The spin accumulation is obtained by solving the spin diffusion equation. According to Ref. [4], the electric current $\vec{j}_e$ and the spin current $\vec{\mathbf{j}}_m$ are given by

$$\vec{j}_e = 2C_0\vec{E} - 2\mathbf{D}\cdot\vec{\nabla}\mathbf{m}, \tag{3}$$

$$\vec{\mathbf{j}}_m = 2\mathbf{C}\vec{E} - 2D_0\vec{\nabla}\mathbf{m}, \tag{4}$$

where we choose the unit such as the charge of an electron and the Bohr magneton are set to be 1, i.e., $e = \mu_B = 1$. The vectors denoted by the arrow $\vec{v}$ and the bold font $\mathbf{v}$ represent those in the real space and in the spin space, respectively. $\vec{E}$ is the electric field and $\mathbf{m}$ is the spin accumulation. The conductivity $\hat{C}$ and the diffusion constant $\hat{D}$ are written in the spinor form as $\hat{C} = C_0 + \hat{\boldsymbol{\sigma}}\cdot\mathbf{C}$, $\hat{D} = D_0 + \hat{\boldsymbol{\sigma}}\cdot\mathbf{D}$, where the $\hat{\boldsymbol{\sigma}}$ is the vector of the Pauli matrices. $\mathbf{C}$ and $\mathbf{D}$ are proportional to M with the spin polarization parameters β and β' as $\mathbf{C} = \beta C_0\mathbf{M}$ and $\mathbf{D} = \beta' D_0\mathbf{M}$. Hereafter we assume $\beta' = \beta$.

Eliminating the electric field $\vec{E}$ from the Eqs. (3) and (4), and introducing the matrix

$$\hat{\mathbf{A}} = -2D_0\left[\begin{pmatrix}1 & & \\ & 1 & \\ & & 1\end{pmatrix} - \beta^2\begin{pmatrix}M_x^2 & M_xM_y & M_xM_z\\ M_yM_x & M_y^2 & M_yM_z\\ M_zM_x & M_zM_y & M_z^2\end{pmatrix}\right], \tag{5}$$

where $\boldsymbol{M} = (M_x, M_y, M_z)$, the spin current is expressed as

$$\vec{\mathbf{j}}_m = \beta\mathbf{M}\vec{j}_e + \hat{\mathbf{A}}\vec{\nabla}\mathbf{m}. \tag{6}$$

In the stationary state, the continuity equation for the spin accumulation takes the form

$$\vec{\nabla}\cdot\vec{\mathbf{j}}_m + (J/\hbar)\mathbf{m}\times\mathbf{M} + \frac{\mathbf{m}}{\tau_{\rm sf}} = 0, \tag{7}$$

where $\tau_{\rm sf}$ is the spin- ip relaxation time. Introducing the matrix,

$$\hat{\mathbf{B}} = \frac{1}{\tau_{sf}}\begin{pmatrix}1 & & \\ & 1 & \\ & & 1\end{pmatrix} + (J/\hbar)\begin{pmatrix}0 & M_z & -M_y\\ -M_z & 0 & M_x\\ M_y & -M_x & 0\end{pmatrix} \tag{8}$$

we obtain the following diffusion equation for the spin accumulation $\mathbf{m}$ to be solved

$$\vec{\nabla}\cdot\left(\beta\mathbf{M}\vec{j}_e + \hat{\mathbf{A}}\vec{\nabla}\mathbf{m}\right) + \hat{\mathbf{B}}\mathbf{m} = 0. \tag{9}$$

The determination of the MR ratio was made as follows. First we determine the charge current density $\vec{j}_e$ from the continuity equation $\vec{\nabla}\cdot\vec{j}_e = 0$. Next we solve the Eq. (9) for a given local magnetization $\mathbf{M}$ in the domain wall and determine the spin accumulation $\mathbf{m}$. Finally we determine the voltage drop of the system by integrating the electric field $\vec{E} = (\vec{j}_e + 2\mathbf{D}\cdot\vec{\nabla}\mathbf{m})/(2C_0)$. Since we apply the constant electric current density, the voltage drop is proportional to the total resistance, from which we obtain the MR ratio of the system.

We use material parameters for the conventional CPP-GMR spin-valve system; $\rho_{\rm F} = 150\Omega\,{\rm nm}$, $\lambda_{\rm F} = 12$ nm, $\lambda_{\rm N} = 14$ nm, $J = 0.5$ eV, where subscripts F(N) indicates the ferromagnetic (non-magnetic) material and $\lambda_{\rm F(N)}$ denotes the spin diffusion length. We set the resistivity of non-magnetic capping layers to be $\rho_{\rm N} = 1\ \Omega\mu{\rm m}$ so that the total RA without CCP structure becomes about 0.1 $\Omega\mu{\rm m}^2$, which is the value reported in the experiments [1, 2]. The diffusion constant D_0 and the relaxation time τ_{sf} are determined by the relation $C_0 = N_F D_0$ and $\lambda = \sqrt{2\tau_{sf}D_0(1-\beta^2)}$, where we set the density of states at the Fermi level as $N_F = 7.5\ {\rm nm}^{-3}{\rm eV}^{-1}$.

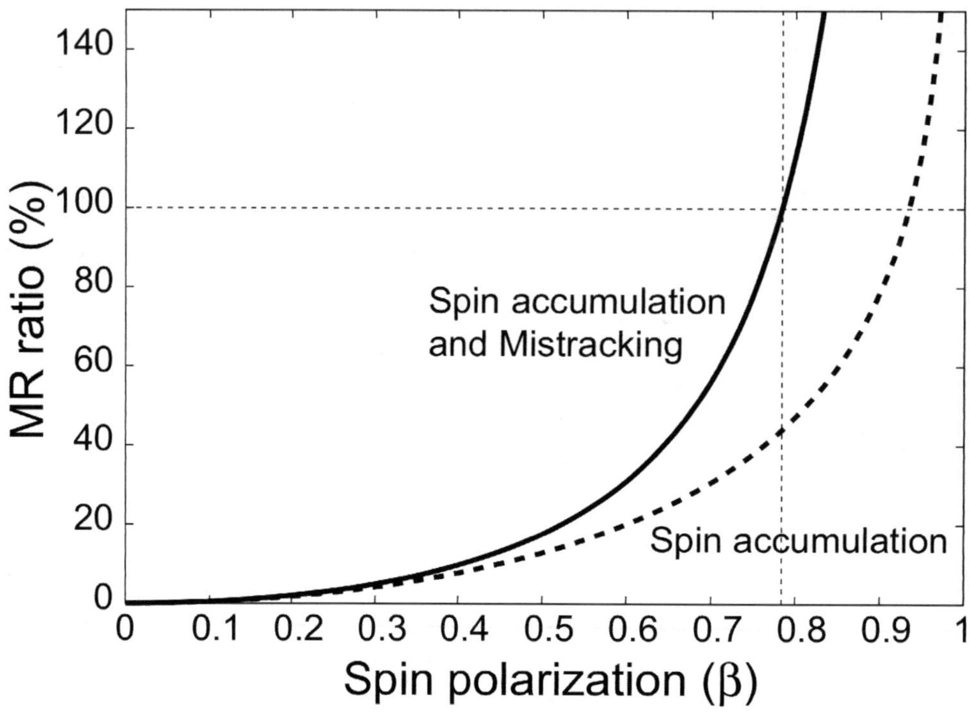

Figure 2. MR ratio of the ferromagnetic nano-contact is plotted against the spin polarization β. Solid line indicates the MR ratio obtained by considering both spin accumulation and mistracking. Dashed lines indicate that obtained by neglecting the contribution from mistracking. Thin dotted lines are guide to eyes for MR=100%.

3. Results

We numerically solve the spin diffusion equation of Eq.(9) by use of the finite element method. We take into account the effect of mistracking by substituting Eq.(1) into $\hat{C}$ of Eq.(4) as

$$\hat{C} = \left(\frac{\sigma^+ + \sigma^-}{2}\right)\hat{I} + \left(\frac{\sigma^+ - \sigma^-}{2}\right)\hat{\boldsymbol{\sigma}} \cdot \boldsymbol{M}. \tag{10}$$

In Fig. 2 we plot the MR ratio obtained by considering both the spin accumulation and mistracking effects by the solid line. The dotted line indicates the MR ratio obtained by neglecting the contribution from mistracking. For $\beta = 0.65$ which is value of the spin polarization of CoFe, the calculated MR ratio (solid line) is 41 %. One can see that both spin accumulation and mistracking give the same order of magnitude to the MR for the nano-contact made of conventional ferromagnetic metal with β around 0.5. The MR ratio rapidly increases with increasing the spin polarization β and it reaches 100% for $\beta = 0.79$.

The MR ratio increases with decreasing the diameter of the contact. If we reduce the diameter of the contact from 2 nm to 1 nm, the MR ratio exceeds 100 % for β >0.71.

4. Conclusion

In conclusion, we theoretically studied spin accumulation and mistracking effects on MR of a ferromagnetic nano-contact. We showed that both the spin accumulation and mistracking give the same order of magnitude to the MR ratio for the nano-contact made of conventional ferromagnetic metals. We also showed that the MR ratio of 100% is available for materials with $\beta = 0.79$.

Acknowledgements
The authors would like to thank K. Matsushita, M. Sahashi, M. Doi, K. Miyake, H. Iwasaki, H. Fuke, and M. Takagishi for valuable discussions they had with. The work has been supported by The New Energy and Industrial Technology Development Organization (NEDO).

References
[1] H. Fukuzawa, H. Yuasa, S. Hashimoto, K. Koi, H. Iwasaki, M. Takagishi, Y. Tanaka and M. Sahashi 2004 IEEE Trans. Magn. Mater. **40** 2236
[2] H. N. Fuke and S. Hashimoto, M. Takagishi, H. Iwasaki, S. Kawasaki, K. Miyake and M. Sahashi 2007 IEEE Trans. Magn. **43** 2848
[3] T. Valet and A. Fert 1993 Phys. Rev. B **48** 7099
[4] S. Zhang, P.M.Levy and A. Fert 2002 Phys. Rev. Lett. **88** 236601
[5] H. Imamura 2007 Physica status solidi B-basic solid state physics **244** 4394
[6] J. Sato, K. Matsushita, and H. Imamura 2008 IEEE Trans. Magn. **44** 2608
[7] J. Sato, K. Matsushita, and H. Imamura 2009 J. Appl. Phys. **105** 07D101
[8] J. Sato, K. Matsushita, and H. Imamura 2010 J. Phys.:Conf. Ser. **200** 062023
[9] P. M. Levy and S. Zhang 1997 Phys. Rev. Lett. **79** 5110

Epitaxial growth of ferromagnetic Fe_4N thin films on $SrTiO_3(001)$ substrates by molecular beam epitaxy

K Ito, G H Lee and T Suemasu
Institute of Applied Physics, University of Tsukuba, 1-1-1 Tennohdai, Tsukuba Ibaraki 305-8573, Japan

E-mail: suemasu@bk.tsukuba.ac.jp

Abstract. We have grown γ'-Fe_4N thin films on $SrTiO_3$(STO)(100) and MgO(100) substrates by molecular beam epitaxy. From the ω scan X-ray diffraction measurement for the γ'-Fe_4N(200) plane, the a-axis of γ'-Fe_4N grown on STO(100) was found to be perpendicular to the substrate. In contrast, the a-axis was inclined to the surface normal for γ'-Fe_4N formed on MgO(100). The saturation magnetization of γ'-Fe_4N/STO(100) deduced from the magnetization versus magnetic field curve was approximately 1900 emu/cm^3, which is higher than that of bulk γ'-Fe_4N value of 1400 emu/cm^3.

1. Introduction

Spintronics, a new research field of electronics based on the combination of magnetic and semiconducting materials, has attracted much attention. Techniques of spin injection, control and detection are required to achieve spintronics devices. Highly spin-polarized ferromagnetic materials are of great importance as spin sources. Thus, numerous different types of half metals and hetero junctions have been studied extensively [1-4]. Iron nitrides have attracted significant attention as spintronic materials. These materials have unique features including high-hardness, chemical inertness, and high electrical conductivity. Among them, ferromagnetic α"-$Fe_{16}N_2$, γ'-Fe_4N and ε-$Fe_{2-3}N$ have been extensively studied for applications in high-density magnetic recording heads and media [5-9]. We have paid special attention to γ'-Fe_4N since highly spin-polarized transport has been theoretically predicted [10]. Recently, we have confirmed a larger spin polarization in γ'-Fe_4N than that in α-Fe by Andreev reflection measurements [11]. There have been a few reports thus far on the magnetic tunnel junctions (MTJ)s using γ'-Fe_4N such as $Fe_4N/Cu_3N/Fe_4N$, Fe_3N/AlN/Fe_4N, Fe_4N/MgO/Fe, and CoFeB/MgO/Fe_4N MTJs [12-15]. In the CoFeB/MgO/Fe_4N MTJ, an inverse tunnel magnetoresistance (TMR) effect was reported [15]. A negative anisotropic magnetoresistance (AMR) effect was also observed in γ'-Fe_4N films [16]. These results were explained by dominant minority-spin electrons at the Fermi level of γ'-Fe_4N. Therefore, we think that γ'-Fe_4N has great possibilities of application for the spintronics devices. However, there have been a small number of reports on the epitaxial growth of γ'-Fe_4N films. In this paper, we demonstrate the epitaxial growth of γ'-Fe_4N thin films on $SrTiO_3$(STO)(100) substrates by molecular beam epitaxy (MBE) for the first time. The lattice mismatch between the two materials is only 2.9% [17]. We also prepared γ'-Fe_4N thin films on MgO(100) and α-Fe/MgO(100) substrates, in which lattice mismatches are 10% and 6.6%, respectively. The saturation magnetization of γ'-Fe_4N thin films was deduced from magnetization versus magnetic field (*M-H*) curves, and was compared with those grown on different substrates.

2. Experiment

γ'-Fe_4N thin films were grown by MBE using $Fe(CO)_5$ or solid Fe as an Fe source, and a nitrogen or an ammonia radical beam produced by ECR plasma as a N source. We prepared four different γ'-Fe_4N thin films (samples A-D). Prior to the growth of γ'-Fe_4N, MgO(100) substrates were heated at 900 °C for 60 min. STO(100) substrates were dipped into BHF solution, and heated at 900 °C for 30 min under an oxygen partial pressure of 5×10^{-6} Torr in order to get an atomically flat surface [18]. All the samples were grown on MgO(100) except that sample D was formed on STO(100). For sample A, the solid Fe and ECR-NH_3 were supplied simultaneously on MgO(100) at 450 °C for 30 min and a 10-nm-thick γ'-Fe_4N film was grown. For sample B, *a*-axis-oriented α-Fe layers were first grown on MgO(100) at room temperature (RT) using solid Fe. After annealing at 450 °C for 10 min, ECR-NH_3 was supplied at 450 °C on the α-Fe layers. During the growth, the α-Fe layers were nitrided to form γ'-Fe_4N and a 20-nm-thick γ'-Fe_4N film was grown in total. For sample C, *a*-axis-oriented α-Fe layers were first epitaxially grown on MgO(100) at 450 °C using $Fe(CO)_5$. Then, $Fe(CO)_5$ and ECR-N_2 were supplied simultaneously at 450 °C on the α-Fe layer. During the growth, the α-Fe layers were nitrided to form γ'-Fe_4N and a 100-nm-thick γ'-Fe_4N film was grown in total. For sample D, the solid Fe and ECR-NH_3 were supplied simultaneously on STO(100) at 450 °C for 120 min and a 20-nm-thick γ'-Fe_4N film was grown. The crystallinity of the grown films was characterized by θ-2θ X-ray diffraction (XRD), ω scan XRD and reflection high energy electron diffraction (RHEED). All the RHEED patterns were taken from the [100] azimuth of γ'-Fe_4N. A superconducting quantum interference device (SQUID) magnetometer was used to measure the magnetic properties of the samples at RT.

3. Results and discussions

Figure 1 shows θ-2θ XRD patterns of samples A-D. Intense diffraction peaks of γ'-Fe_4N{100} are not

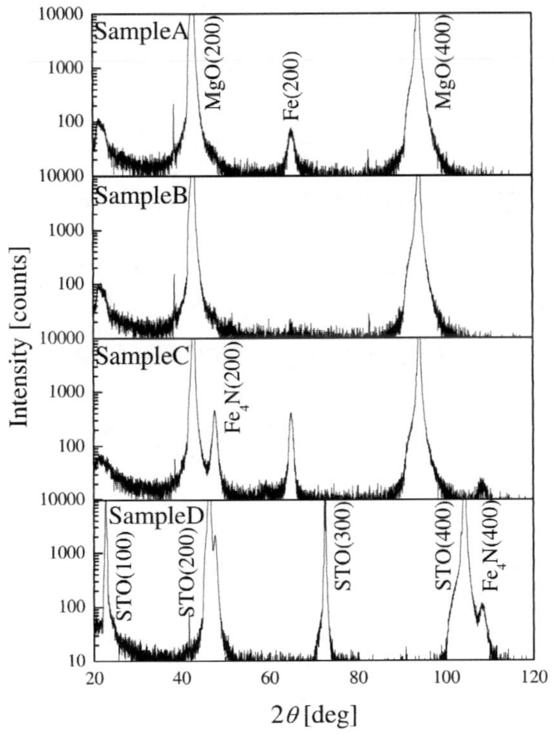

Figure 1. θ-2θ XRD patterns of the samples A-D.

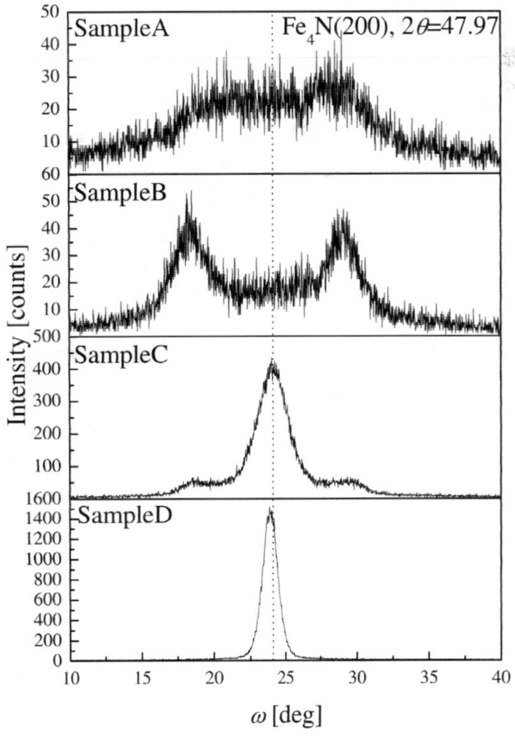

Figure 2. ω scan XRD patterns of the samples A-D for γ'-Fe_4N(200) plane.

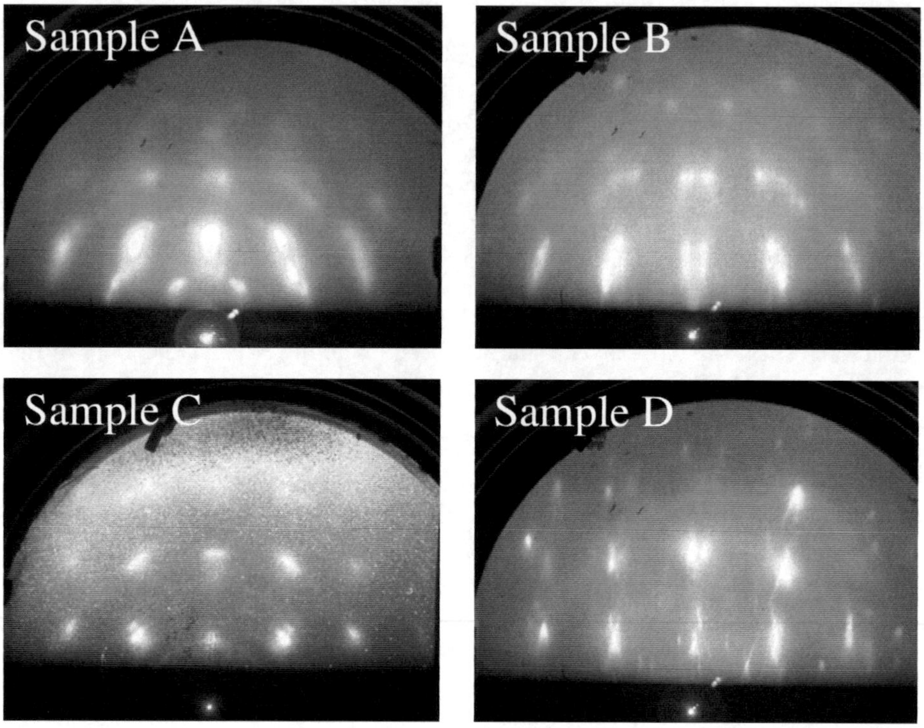

Figures 3. RHEED patterns of samples A-D taken from the [100] azimuth of γ'-Fe_4N.

observed in samples A-C, grown on the MgO(100) substrates. In contrast, intense peaks of γ'-Fe_4N{100} are observed in sample D, grown on the STO(100) substrate. Figure 2 shows ω scan XRD patterns of samples A-D for the γ'-Fe_4N(200) plane. Broad diffraction patterns were observed in samples A-C. We can see two side peaks approximately 5° away from ω=24°, meaning that there exist γ'-Fe_4N domains with their *a*-axes inclined by approximately 5° to the surface normal. In contrast, a sharp diffraction peak at the Bragg angle of γ'-Fe_4N(200) is dominant and the side peaks are not observed in sample D. Thus, we can say that the *a*-axis of the γ'-Fe_4N thin film on STO(100) was perpendicular to the substrate. Figure 3 shows RHEED patterns of samples A-D. Clear spotty RHEED pattern was observed only for sample D. On the basis of these results, it can be stated that we have succeeded the epitaxial growth of γ'-Fe_4N films on a STO(100) substrate, due probably to the smaller lattice mismatch between γ'-Fe_4N and STO than that between γ'-Fe_4N and MgO.

Figure 4 shows the *M-H* curves measured at RT using SQUID magnetometer for an 80-nm-thick single phase γ'-Fe_4N film on MgO(100), grown by the same procedure for sample C, [14] and sample D. The external magnetic field *H* was applied parallel to in-plane [100] azimuth of γ'-Fe_4N. The saturation magnetization of the sample grown on the MgO(100) substrate deduced from the *M-H* curve was approximately 1470 emu/cm^3, which is in good agreement with that of bulk γ'-Fe_4N. In contrast, that of sample D grown on the STO(100) substrate was approximately 1900 emu/cm^3. The increase in saturation magnetization for γ'-Fe_4N thin films on STO(100) was also reported by Atiq *et al.* [19]. They reported that the saturation magnetization in γ'-Fe_4N increases with decreasing a lattice mismatch between γ'-Fe_4N and a substrate used. Our results agree with that reported. We think that the inclusion of α''-$Fe_{16}N_2$ in the grown layers, which has the largest magnetic moment among Fe-N compounds, is not likely to occur, because sample D was prepared at 450 °C, and the decomposition of α''-$Fe_{16}N_2$ into α-Fe and γ'-Fe_4N was reported to happen at temperatures higher than 400 °C [20].

We'd rather expect the change of the interatomic distance between Fe and Fe atoms or Fe and N atoms in γ'-Fe_4N, giving rise to the enhancement in saturation magnetization. However, we don't have enough data to discuss this issue. Thus, further experimental analyses such as X-ray magnetic circular dichroism measurement and theoretical analyses are required to understand the origin of the enhancement in saturation magnetization.

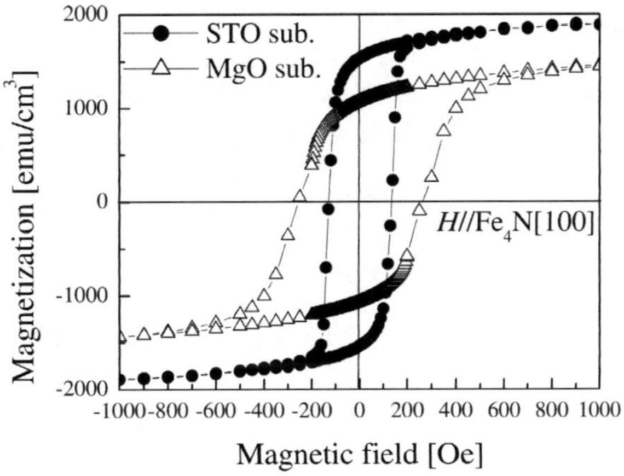

Figure 4. *M-H* curves measured at RT using SQUID magnetometer for an 80-nm-thick single phase γ'-Fe_4N film on MgO(100), grown by the same procedure for sample C, and sample D.

4. Conclusion

We have succeeded the epitaxial growth of γ'-Fe_4N films on a STO(100) substrate by MBE. The *a*-axis of γ'-Fe_4N films grown on MgO(100) was inclined to the surface normal by approximately 5°. The saturation magnetization of the γ'-Fe_4N film grown on the STO(100) substrate deduced from the *M-H* curve was approximately 1900 emu/cm^3, which is larger than that of bulk γ'-Fe_4N value.

Acknowledgements

This work was supported in part by a Grant-in-Aid for Scientific Research on Priority Areas of "Creation and Control of Spin Current" (No. 19048029) from the Ministry of Education, Culture, Sports, Science and Technology of Japan (MEXT) and by the NanoProcessing Partnership Platform (NPPP) at AIST, Tsukuba. Measurements of the magnetic properties were performed with the help of Dr. H. Ikeda and Dr. T. Koyano of University of Tsukuba. High-purity 5N Fe was supplied by Prof. M. Isshiki and Dr. M. Uchikoshi of Tohoku University.

References

[1] Bowen M, Bibes M, Barthe´le´my A, Contour J P, Anane A, Lemaıˆtre Y and Fert A 2003 *Appl. Phys. Lett.* **82** 233
[2] Coey J M D and Venkatesan M 2002 *J. Appl. Phys.* **91** 8345
[3] Ishikawa T, Hakamata S, Matsuda K, Uemura T and Yamamoto M 2008 *J. Appl. Phys.* **103** 07A919
[4] Yuasa S, Nagahama T, Fukushima A, Suzuki Y and Ando K 2004 *Nat. Mater.* **3** 868
[5] Kim T K and Takahashi M 1972 *Appl. Phys. Lett.* **20** 492
[6] Tagawa K, Kita E and Tasaki A 1982 *Jpn. J. Appl. Phys.* **21** 1596
[7] Borsa D M, Grachev S, Boerma D O and Kerssemakers J W J 2001 *Appl. Phys. Lett.* **79** 994

[8] Takahashi T, Takahashi N, Tamura N, Nakamura T, Yoshioka M, Inami W and Kawata Y 2001 *J. Mater. Chemistry* **11** 3154
[9] Yamaguchi K, Yui T, Yamaki K, Kakeya I, Kadowaki K and Suemasu T 2007 *J. Cryst. Growth* **301** 597
[10] Kokado S, Fujima N, Harigawa K, Shimizu H and Sakuma A 2006 *Phys. Rev. B* **73** 172410
[11] Narahara A, Ito K, Suemasu T, Takahashi Y K, Rajanikanth A and Hono K 2009 *Appl. Phys. Lett.* **94** 202502
[12] Borsa D M, Grachev S, Presura C and Boerma D O 2002 *Appl. Phys. Lett.* **80** 1823
[13] Narahara A and Suemasu T 2007 *Jpn. J. Appl. Phys.* **46** L892
[14] Narahara A, Ito K and Suemasu T 2009 *J. Cryst. Growth* **311** 1616
[15] Komasaki Y, Tsunoda M, Isogami S and Takahashi M 2009 *J. Appl. Phys.* **105** 07C928
[16] Tsunoda M, Komasaki Y, Kokado S, Isogami S, Chen C C and Takahashi M 2009 *Appl. Phys. Express* **2** 083001
[17] Nikolaeva K R, Krivorotov I N, Dahlberg E D, Vas'ko V A, Urazhdin S, Loloee R and Pratt Jr W P 2003 *Appl. Phys. Lett.* **82** 4534
[18] Kawasaki M, Takahashi K, Maeda T, Tsuchiya R, Shinohara M, Ishiyama O, Yonezawa T, Yoshimoto M and Koinuma H 1994 *Science* **266** 1540
[19] Atiq S, Ko H S, Siddiqi S A and Shin S C 2008 *Appl. Phys. Lett.* **92** 222507
[20] Sugita Y, Mitsuoka K, Komuro M, Hoshiya H, Kozono Y and Hanazono M 1991 *J. Appl. Phys.* **70** 5077

Journal of Physics: Conference Series **266** (2011) 012092 doi:10.1088/1742-6596/266/1/012092

Magnetic Anisotropy Modulation in Ta/ CoFeB/ MgO Structure by Electric Fields

S Kanai[1], M Endo[1], S Ikeda[1,2], F Matsukura[1,2] and H Ohno[1,2]
[1]Laboratory for Nanoelectronics and Spintronics, Research Institute of Electrical Communication, Tohoku University, Katahira 2-1-1, Aoba-ku, Sendai 980-8577, Japan
[2]Center for Spintronics Integrated Systems, Tohoku University, Katahira 2-1-1, Aoba-ku, Sendai 980-8577, Japan

E-mail: ohno@riec.tohoku.ac.jp

Abstract. We investigate the magnetic anisotropy in as-deposited and annealed Ta/ CoFeB/ MgO samples prepared by sputtering and its CoFeB thickness dependence. The magnetic easy axis changes from in-plane to perpendicular with decreasing CoFeB thickness. The thickness, at which magnetic easy axis direction changes, is increased by annealing. It is also shown that the magnetic anisotropy can be modulated by electric field and its modulation ratio is larger for the annealed samples.

1. Introduction

Magnetic materials are utilized for information storage devices in which data are maintained by magnetization direction controlled conventionally by current generated magnetic fields. The requirement for the reduction of a writing current for high-density memories triggered the extensive studies on spin transfer torque magnetization switching [1,2]. In order to reduce drastically the power consumption for writing, one needs to find a different way of switching. The switching by the application of an electric field E_G is one of the candidates, which is attractive because of not only extremely low power consumption for switching but also a high degree of compatibility with the present technology developed for integrated circuits. For ferromagnetic semiconductors, *e.g.* (In,Mn)As [3] and (Ga,Mn)As [4], it has been shown that E_G can modulate their magnetic properties such as the Curie temperature [3], coercivity [5] and magnetocrystalline anisotropy which results in the change of a magnetization direction [6]. Stimulated by these demonstrations, more recently, studies on E_G modulation of magnetic anisotropy is now progressing in material systems other than ferromagnetic semiconductors; these include ferromagnetic metals [7-10] and multiferroics [11]. Among them, ferromagnetic metals have attracted great attention from the application point of view, because relatively large E_G control of anisotropy can be achieved at room temperature (RT) for the material system used for magnetic tunnel junctions with a high tunnel magnetoresistance ratio [12].
Most of the previous reports on ferromagnetic metals investigated electric-field effects in Fe or Fe alloys combined with a noble metal grown by molecular beam epitaxy [7-9]. In this work, we have prepared Ta/ CoFeB/ MgO stack by rf-sputtering, and investigate the CoFeB thickness dependence of perpendicular magnetic anisotropy and the electric-field effect on it for as-deposited and annealed samples [10].

2. Method

2.1. Sample preparation

A wedged $Co_{40}Fe_{40}B_{20}$ with nominal thickness *t* of 0-2 nm is deposited on underlying layers of Ta(5 nm)/ Ru(10 nm)/ Ta(5 nm) onto 3-inch thermally oxidized Si substrate by rf magnetron sputtering. The CoFeB layer is capped by a 2-nm thick sputtered MgO layer. The sample is cleaved into two parts; one is annealed at 400°C in vacuum under perpendicular magnetic field of 0.4 T. It is known that as-deposited CoFeB is amorphous, whereas annealed one has highly-oriented bcc (001) texture due to the presence of highly oriented MgO [13]. Both layers are processed into a number of Hall bar geometries with different *t* by photolithography and ion milling. After forming Hall bars, the photoresist remaining on the MgO surface is removed by organic solvents and plasma ashing. Then, a 45 nm ZrO_2 insulator (dielectric constant $\kappa \sim 23$) is formed by using atomic layer deposition at 150°C. The deposition takes about 2 hours. Finally, 5 nm Cr/ 50 nm Au gate electrode is evaporated and lifted off (figure 1(a)).

2.2. Magneto-transport measurements

The magnetic properties of the resultant CoFeB films are measured by magneto-transport measurement using lock-in amplifier technique. Schematic of measurement configuration is shown in figure 1(b). We measure the Hall voltage using four terminal method by sweeping a perpendicular external magnetic field *H*. The Hall resistance R_{Hall} of ferromagnetic materials is given by

$$R_{Hall} = \frac{V_{Hall}}{I} = \frac{R_0}{t}\mu_0 H + \frac{R_S}{t} M_\perp, \qquad (1)$$

where V_{Hall} is a Hall voltage, R_0 the ordinary Hall coefficient, R_S the anomalous Hall coefficient, μ_0 permeability in free space and $M_\perp$ the perpendicular component of magnetization. The magnitude of an ac current *I* is determined as applied ac voltage V_i divided by reference resistance R_r of 11 kΩ, which is much larger than sample resistance of ~500 Ω.

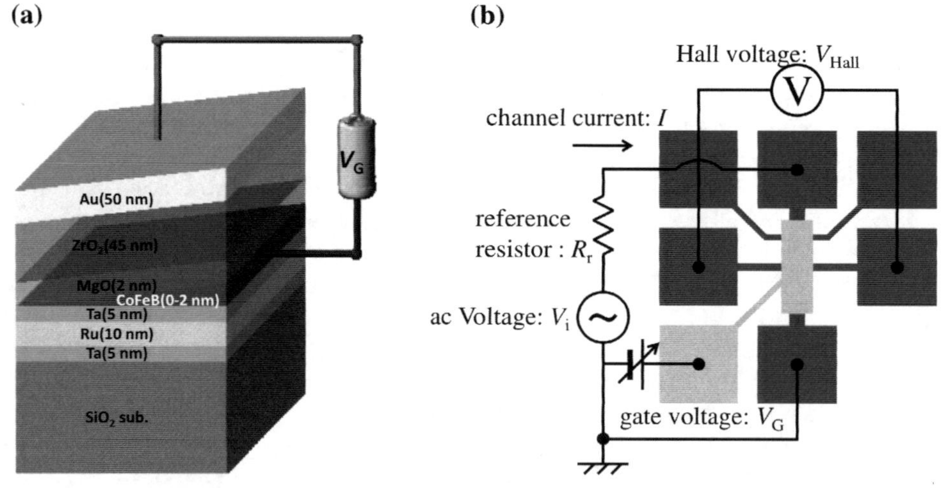

Figure 1. The schematics of (a) sample structure and (b) measurement configuration.

3. Results and discussions

3.1. Thickness dependence of magnetic anisotropy

We investigate t dependence of perpendicular magnetic anisotropy. Figure 2 shows the R_{Hall}-H curves for (a) as-deposited and (b) annealed sample as a function of t at $E_G = 0$ and RT. At low magnetic fields, hysteresis and nonlinear slope are observed, which indicate that one can probe the magnetic properties by magneto-transport measurement through the anomalous Hall effect. In present samples, the sign of R_S is found to be negative. At a high magnetic field region above 300 mT, R_{Hall} increases almost in proportion to H because of the dominant contribution of the ordinary Hall effect. Sheet carrier concentration calculated from the slope is of the order of 10^{17} cm^{-2}. The magnitude of R_{Hall} decreases as t decreases, which is due to the existence of parallel conduction in Ta/ Ru/ Ta underlying layer [10].

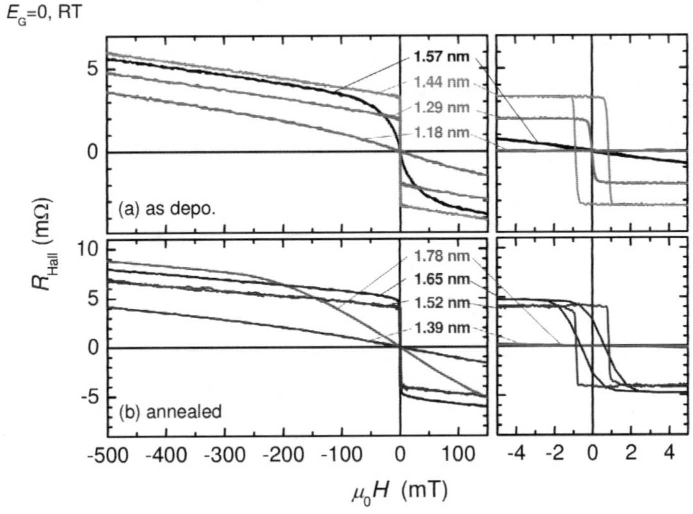

Figure 2. R_{Hall}-H curves under at $E_G = 0$ for (a) as-deposited and (b) annealed samples.

We first focus on the results of as-deposited samples in figure 2(a). A rounded R_{Hall}-H curve without hysteresis is observed at t = 1.57 nm, which indicates that the layer has in-plane magnetic easy axis. As t decreases, the slope of R_{Hall} in the vicinity of zero magnetic field increases rapidly. At t = 1.44 and 1.29 nm, hysteresis loops are observed; this indicates that the layers now have perpendicular magnetic easy axis. The perpendicular easy axis in the present system is considered to be originated from the superexchange coupling between 3d electrons of Co and/or Fe and 2p electrons of O at the interface of CoFeB and MgO [14,15]. Further reduction of t below 1.2 nm results in almost linear curve without hysteresis; this is due to the superparamagnetic behaviour with blocking temperature below RT. The temperature dependence of magnetization shows the blocking temperature is about 100 K for the sample with t = 1.18 nm (not shown).

Next, we focus on the annealing effect on R_{Hall} curves. As noticed from the comparison of figure 2(b) and (a), the annealed sample shows similar thickness dependence of magnetic anisotropy to that of as-deposited sample, but the change of magnetic easy direction occurs at a different thickness. The perpendicular easy axis starts to appear for the samples with t between 1.5 and 1.7 nm. The result indicates that the enhancement of perpendicular magnetic anisotropy energy in the annealed samples compared with the as-deposited ones at the same t. This is probably related to the oxidation of Fe/ Co at the CoFeB/ MgO interface [15].

3.2. Magnetic anisotropy modulation by electric fields

We investigate the electric-field effect on magnetic anisotropy for both as-deposited and annealed samples. E_G denotes the electric field in MgO determined by

$$E_G = \frac{\kappa_{ZrO_2} V_G}{\kappa_{MgO} d_{ZrO_2} + \kappa_{ZrO_2} d_{MgO}}, \tag{2}$$

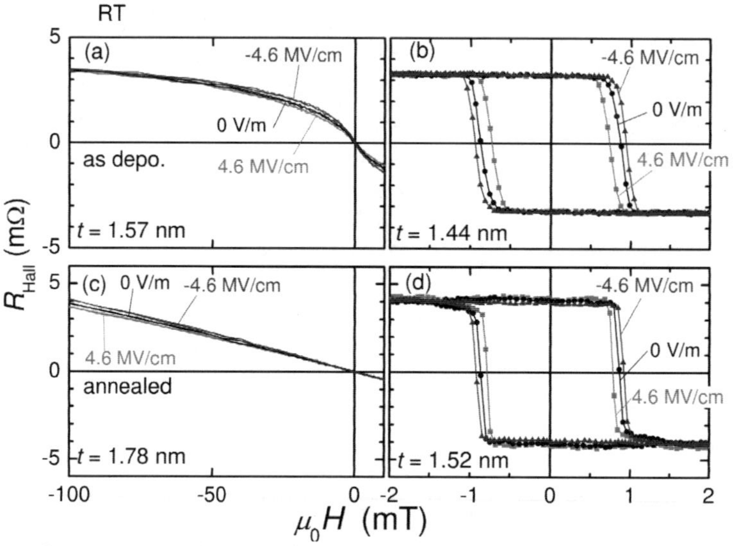

Figure 3. Magnetic-field H dependences of Hall resistance R_{Hall} for as-deposited samples with (a) t = 1.57 nm and (b) 1.44 nm, and for annealed samples with (c) 1.78 nm and (d) 1.52 nm under three different electric fields; E_G = 4.6 MV/cm , 0 V/cm and -4.6 MV/cm.

where κ_{ZrO_2} = 23 and d_{ZrO_2} = 45 nm are the dielectric constant and thickness of ZrO_2 and κ_{MgO} = 10 and d_{MgO} = 2 nm are those of MgO. The application of E_G = ± 4.6 MV/cm used in this study is expected to result in the change of sheet carrier concentration of about ± 2.6 × 10^{13} cm^{-2}. Figure 3 shows the typical result of R_{Hall}-H curves under E_G for the as-deposited sample with (a) t = 1.57 nm (in-plane easy axis) and (b) t = 1.44 nm (perpendicular easy axis) as well as the annealed sample with (c) t = 1.78 nm (in-plane easy axis) and (d) t = 1.52 nm (perpendicular easy axis). The samples with in-plane easy axis shows steeper slope under negative E_G. This tendency is the same as the previous report in Au/ Fe(FeCo)/ MgO [8, 9] and CoFeB/ MgO [10] system. We evaluate the magnitude of magnetic anisotropy modulation per sheet area ΔKt by E_G using equation (2) in [10]. Figure 4 shows the obtained ΔKt-E_G curves, which show almost linear dependence. From the slopes, we obtain the values of $\Delta Kt/E_G$ = 12 and 32 μJ/m^2 per 1 V/nm for the as-deposited sample with t = 1.55 nm and the annealed sample with t = 1.78 nm, respectively. These values are comparable with a theoretical value of 100 μJ/m^2 calculated for Fe/ MgO system [16]. The increase of ΔKt by annealing may be related to the degree of crystallization of CoFeB, because the occupation of d states of the CoFeB at the interface is expected to depend on the crystal orientation. For the samples with perpendicular magnetic easy axis, a coercivity is increased (decreased) by applying negative (positive) E_G, which also reflects the change of ΔKt by E_G. The larger modulation ratio of coercivity for annealed sample than that for as-deposited samples shows that one can expect larger modulation of anisotropy in annealed samples.

Journal of Physics: Conference Series **266** (2011) 012092 doi:10.1088/1742-6596/266/1/012092

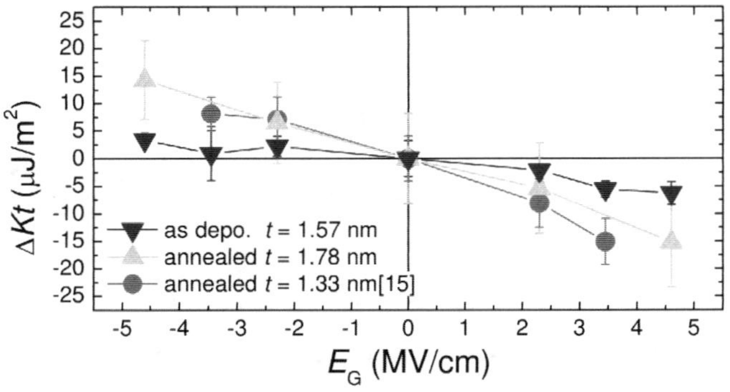

Figure 4. The electric field E_G dependence of the modulation of perpendicular magnetic anisotropy energy per sheet area ΔKt for as-deposited sample (inverted triangles) and annealed sample (triangles). Circles show the previous results in [10].

Summary

We investigate the CoFeB thickness dependence of perpendicular magnetic anisotropy and annealing effect on it in Ta/ CoFeB/ MgO structure deposited by rf magnetron sputtering. Magnetization easy axis is changed from in-plane to perpendicular as CoFeB thickness decreases due to the interfacial magnetic anisotropy. The interfacial magnetic anisotropy is found to be enhanced by annealing. We observed the modulation of the interfacial anisotropy by the application of electric fields and the larger modulation ratio for the annealed sample than as-deposited sample.

References

[1] Hosomi M *et al.* 2005 *IEEE International Electron Devices Meeting* USA, Dec. 5-7, pp 473-476
[2] Kawahara T *et al.* 2007 *IEEE International Solid-State Circuits Conference* USA, Feb. 12-14, pp 480-481
[3] Ohno H, Chiba D, Matsukura F, Omiya T, Abe E, Dietl T, Ohno Y and Ohtani K 2000 *Nature* (London) **408** 944
[4] Chiba D, Yu K M, Walukiewicz W, Nishitani Y, Matsukura F and Ohno H 2008 *J. Appl. Phys.*, **103** 07D136
[5] Chiba D, Yamanouchi M, Matsukura F and Ohno H 2003 *Science* **301** 943
[6] Chiba D, Matsukura F and Ohno H 2008 *Nature* (London) **455** 515
[7] Weisheit M, Fähler S, Marty A, Souche Y, Poinsignon C and Givord D 2007 *Science* **315** 349
[8] Maruyama T *et al.* 2009 *Nature Nanotech.* **4** 158
[9] Shiota Y, Maruyama T, Nozaki T, Shinjo T, Shiraishi M and Suzuki Y 2009 *Appl. Phys. Express* **2** 063001
[10] Endo M, Kanai S, Ikeda S, Matsukura F and Ohno H 2010 *Appl. Phys. Lett.* **96** 212503
[11] Cheong S W and Mostovoy M 2007 *Nature Materials* **6** 13
[12] Ikeda S, Hayakawa J, Ashizawa Y, Lee Y M, Miura K, Hasegawa K, Tsunoda M, Matsukura F and Ohno H 2008 *Appl. Phys. Lett.* **93** 082508
[13] Hayakawa J, Ikeda S, Matsukura F, Takahashi H and Ohno H 2005 *Jpn. J. Appl. Phys.* **44** L587
[14] Manchon A *et al.* 2008 *J. Appl. Phys* **104** 043914
[15] Nistor L E, Rodmacq B, Auffret S, Dieny B 2009 *Appl. Phys. Lett.* **94** 012512
[16] Niranjan M K, Duan C G, Jaswal S S and Tsymbal E Y 2010 *Appl. Phys. Lett.* **96** 222504

Acknowledgement

The authors thank I. Morita and T. Hirata for discussion. This work was supported in part by Grant-in-Aids from MEXT/JSPS, the GCOE Program at Tohoku University, the Research and Development for Next-Generation Information Technology Program from MEXT, and JSPS through its FIRST Program.

Journal of Physics: Conference Series **266** (2011) 012093 doi:10.1088/1742-6596/266/1/012093

Surface morphology and transport properties of Cr nanoparticles in single electron tunneling regime

Tetsunori Koda*[1], Seiji Mitani[2], Masaki Mizuguchi[1] and Koki Takanashi[1]
1 Institute for Materials Research, Tohoku University, Sendai, 980-8577, Japan
2 National Institute for Materials Science, Tsukuba, 305-0047, Japan

E-mail: koda@imr.tohoku.ac.jp

Abstract. We have carried out an investigation of surface structures of Cr nanoparticles grown on a MgO layer for studying single electron tunneling (SET). Reflection high-energy electron diffraction pattern changes from spotty to streak with decreasing Cr nominal thickness, while Cr forms nano size particles with the size of around 3.5 nm in diameter for all the samples. The double tunnel junctions for the Cr nominal thickness of 0.1 nm show clear Coulomb blockade with the threshold voltage of 50 mV, indicating that this Cr nanoparticle satisfies the condition for the appearance of SET. The relatively large volume of the Cr nanoparticles may be related with the fact that tunnel magnetoresistance due to spin accumulation in Cr nanoparticles is small.

1. Introduction

Single electron tunneling (SET) appears in tunnel junctions containing nano size particles due to the increase in the charging energy when an electron tunnels onto the particle in the range that the charging energy is much higher than the thermal energy [1]. Especially, double tunnel junctions containing particles as a center electrode have an ideal structure for obtaining SET.

When top and bottom electrodes consist of ferromagnetic materials in double tunnel junctions, "spin-dependent" SET appears due to spin accumulation in particles. This phenomenon is observed even in double tunnel junctions containing a non-ferromagnetic particle when spin relaxation time in a non-ferromagnetic particle is longer than the time interval of successive tunneling events between top and bottom electrodes. We have reported that the spin-dependent SET exhibits using Au [2], Cr [3], Co [4] and Fe [5] nanoparticles. Interestingly, the estimated spin relaxation time in nanoparticles were much longer than those in bulk.

It is of great importance to control the microstructure of the center electrode for studying spin-dependent SET because the threshold voltage for Coulomb blockade and spin accumulation in nanoparticles is affected by the size of particle [6]. We have reported structure of Fe nanoparticles systematically [7]. In this paper, we report on a investigation of surface morphology of Cr nanoparticles for SET.

2. Experimental

All samples were prepared using molecular beam epitaxy (MBE) on polished MgO (100) single crystal substrates. The film structure is schematically shown in Figure 1. Each layer was deposited at room temperature. Fe was used for top and bottom ferromagnetic electrodes. Because of small lattice mismatch between each layer, all layers grew epitaxially with the relationship of

Fe(001)[110]/MgO(001)[100]/Cr(001)[110]/MgO(001)[100]. Post annealing was carried out to improve the crystallinity and flatness after the depositions of the MgO buffer and the Fe buffer layers at 600°C for 30 minutes and 450°C for 15 minutes, respectively. Epitaxial growth was confirmed by *in situ* reflection high-energy electron diffraction (RHEED). *Ex situ* atomic force microscopy (AFM) observations under ultra high vacuum condition were performed for structural characterizations. A 200 nm × 500 nm size pillar was fabricated using electron beam lithography and Ar ion etching techniques for the characterizations of electronic transport properties. Electronic transport was measured with the two-fold method at 7 K.

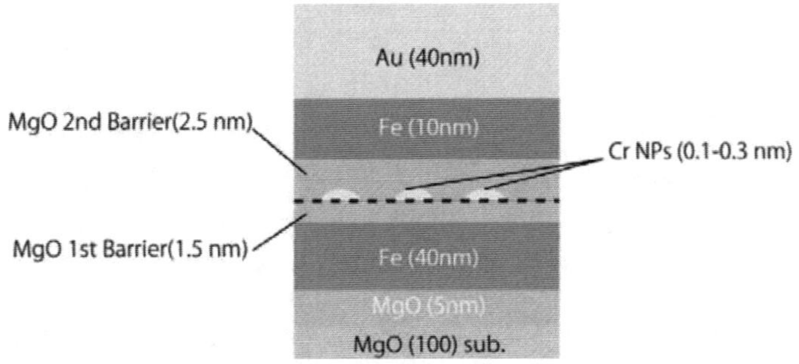

Figure 1. Film structure of double tunnel junctions containing Cr nanoparticles as a center electrode.

3. Results and Discussions

Figure 2 shows AFM images and RHEED patterns for surfaces of Cr layer with several nominal thicknesses. To compare those images and patterns, an AFM image and a RHEED pattern of the MgO 1st barrier are also shown. After the deposition of the MgO 1st barrier, the RHEED showed a streak pattern, suggesting that the MgO layer had a smooth surface. The relatively smooth surface was confirmed from the AFM image, the Ra was estimated to be 0.09 nm. On the other hand, the RHEED patterns for Cr layers showed spotty; the intensity of spots decreased with decreasing the nominal thickness of Cr layers. For the Cr nominal thickness of 0.1nm, the spotty pattern was unclear and similar to the pattern of the MgO layer, while the AFM image showed that Cr aggregated and formed nano size particles onto the MgO 1st barrier. The size of the nanoparticle was around 3.5 nm in diameter. Because of this nanosize particle formation, the surface roughness Ra of Cr increases from 0.10 nm (Cr: 0.1 nm) to 0.16 nm (Cr: 0.3 nm), which is consistent with the RHEED pattern behavior. Thus, the RHEED pattern for the Cr nominal thickness of 0.1nm was similar to that for MgO layer owing to this relatively small Ra. This formation of Cr particles is considered to be due to the large surface energy difference between Cr (2400 erg/cm^2) [8] and MgO (1160 erg/cm^2) [9], which promotes the Volmer-Weber growth mode. The particle density was almost constant at 4.0×10^{12} cm^{-2} in the Cr thickness range, indicating that the volume of Cr nanoparticles increases with increasing nominal thickness of Cr layer. The nucleation sight of Cr might be governed by the defects of the 1st MgO barrier surface [10], which influences this constant particle density with various film thicknesses.

A I-V curve measured at 7K with the Cr nominal thickness of 0.1 nm is plotted in Figure 3. Clear Coulomb blockade with the threshold voltage of 50 mV was confirmed. Clear Coulomb staircases appeared with the steps at ±210 and ±380 mV as indicated by arrows. From this step period, the capacitance of the Cr particle was estimated to be 2.0 aF using the following equation $V_{th}=(2n+1/2)e/C_0$, where V_{th}, n and C_0 are the threshold voltage for the appearance of Coulomb blockade, the number of electrons in a center particle and the capacitance of junctions, respectively. Note that most of the samples showed Coulomb blockade without appearance of Coulomb staircases, implying that most of the samples had several electron current paths and the Coulomb staircase becomes smeared out because of the size distribution of the Cr nanoparticles. Thus, the appearances of

Coulomb blockade and Coulomb staircase in Fig.3 suggest that the pillar has only one electron current paths even though there are several hundreds of nanoparticles in the pillar.

The reason why we obtained Coulomb staircases from the sample that has hundreds of nanoparticles in one pillar is not clear yet. However, the thickness of the 2nd MgO barrier may have fluctuation because the double tunnel junctions have roughness due to the formation of Cr nanoparticles on the 1st MgO barrier. The thinner area could be an electric current path. Thus, the number of current paths is considered to be much less than that of nanoparticles. For this reason, it is possible that a pillar has only one electric current path.

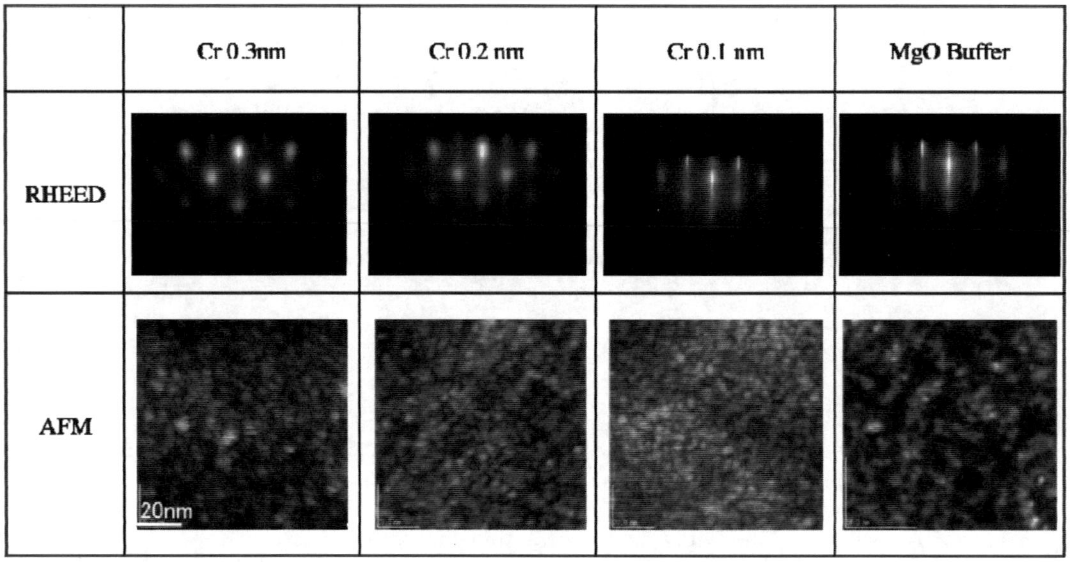

Figure 2. RHEED patterns and AFM images for Cr layer and MgO 1st barrier.

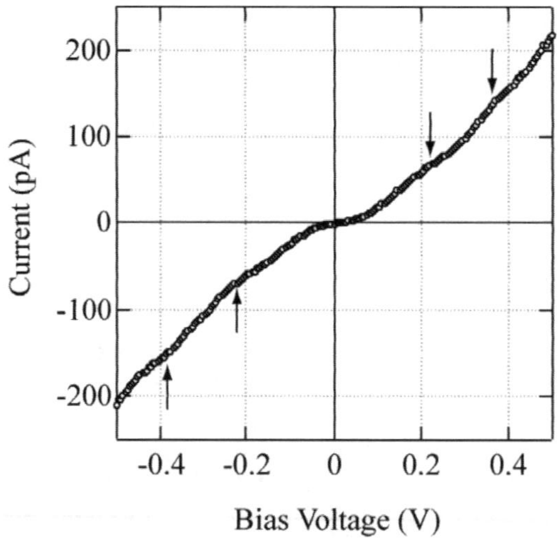

Figure 3. I-V curve for the Cr nominal thickness of 0.1 nm measured at 7 K.

From the point of view of study on "spin-dependent " single electron tunneling using a non-ferromagnetic particle, spin accumulation induced by Fermi level shift in a nanoparticle is essential. This Fermi level shift is in inverse proportional to the volume of a particle [4]. The estimated volume for Cr nanoparticles (2.5×10^{-21} cm^3) [3] is larger than that for Au nanoparticles (2.0×10^{-22} cm^3) [11]. We observed that the TMR for Cr nanoparticles showed small MR ratio (4 %) [3] in comparison with that for Au nanoparticles. (12 %) [2], suggesting that this relatively large volume may affect the appearance of TMR in single electron tunneling regime.

4. Summary

We carried out an investigation of surface structures of Cr nanoparticles grown on a MgO barrier for studying single electron tunneling. The RHEED patterns changed from spotty to streak with decreasing the Cr nominal thickness. The threshold voltage for Coulomb blockade decreased with increasing the Cr nominal thickness due to the increase in the size of the Cr nanoparticles. The double tunnel junctions with the Cr nominal thickness of 0.1 nm showed clear Coulomb blockade with the threshold voltage of 50 mV and clear Coulomb staircases, indicating that this structure satisfies the appearance of SET. The relatively large volume of the Cr nanoparticle may affect the appearance of TMR induced by spin accumulation in the Cr nanoparticles.

5. Acknowledgement

This work was supported in part by a Grant-in-Aid for Scientific Research in Priority Area "Creation and control of spin current" and by a grant from the for Global COE program "Materials Integration International Center of Education Research" from the Ministry of Education, Culture, Sports, Science and Technology, Japan.

References

[1] Likharev K K 1999 *Proc. IEEE* **87** 606
[2] Mitani S, Nogi Y, Wang H, Yakushiji K, Ernult F, Takanashi K 2008 *Appl. Phys. Lett.* **92** 112509
[3] Koda T, Mitani S, Takanashi K 2010 *IEEE Trans. Magn.* **46** 2060
[4] Yakushiji K, Mitani S, Takanashi K, Fujimoti H 2005 *Nature Mater.* **4** 57
[5] Ernult F, Yamane K, Mitani S, Yakushiji K, Takanashi K, Takahashi Y. K, Hono K 2004 *Appl Phys Lett.* **84** 3106
[6] Ernult F, Yakushiji K, Mitani S, Takanashi K 2007 *J. Phys. Cond. Matt.* **19** 165214
[7] Ernult F, Mitani S, Takanashi K, Takanashi K, Takahashi Y. K, Hono K 2006 *Phase Trans.* **79** 717
[8] Renies F, Delplancke M. P, Asskali A, Rooryck V, Van Sinay O 1996 *Appl. Surf. Sci.* **92** 35
[9] Vassent J. L, Dynna M, Marty A, Gilles B, Patrat G 1996 *J. Appl. Phys.* **80** 5727
[10] Haas G, Menck A, Brune H, Barth J. V, Venables J. A., Kern K 2000 *Phys. Rev. B* **61** 11105
[11] Nogi Y, Wang H, Ernult F, Yakushiji K, Mitani S, Takanashi K 2007 *J. Phys. D: Appl. Phys.* **40** 1242

Journal of Physics: Conference Series **266** (2011) 012094 doi:10.1088/1742-6596/266/1/012094

First-principles study for electronic structure and physical property of Co-based Heusler alloys

Yohei Kota and Akimasa Sakuma

Department of Applied Physics, Tohoku University, Sendai, Japan

E-mail: kota@solid.apph.tohoku.ac.jp

Abstract.
The statistical investigation of the physical property of Co-based Heusler alloys with the atomic composition or configuration dependence are carried out in first-principles approach. In particular, the electronic structure, magnetic moment and electrical resistivity due to disorder alloy effects of $Co_2MnAl_{1-z}Si_z$ ($0.0 < z < 1.0$) and $Co_2Mn_{1-y}Fe_ySi$ ($0.0 < y < 1.0$) with the $L2_1$, B2 and A2 structure are calculated by using the tight-binding linear muffin-tin orbital method combined with the coherent potential approximation based on the local spin-density functional approximation and by using the Kubo-Greenwood formula, respectively. The obtained results indicate that these properties of the Co-based Heusler alloys significantly depend on the atomic configuration and composition, especially, they are sensitive to the existence of the half-metallic property.

1. Introduction

Co-based Heusler alloys are one of the strong candidates for a high spin-polarized material after the predictions of their half-metallic property from the results of some electronic structure calculations. Thus they have attracted much attention because of their potential in the application to Spintronics. In these years, there are a lot of experimental reports that large magnetoresistance ratios have been obtained from the tunnel magnet resistance (TMR) [1, 2] and current perpendicular to plane giant magnetoresistance (CPP-GMR) [3, 4] devices by using Heusler alloys as a ferromagnetic electrode. These experimental results show an evidence of the high degree of the spin-polarization of Heusler alloys due to half-metallic property which had been previously expected in the theoretical viewpoints.

The half-metallic property of Heusler alloys, however, is considered to be quite sensitive to their atomic configuration and composition [5, 6]. It is pointed out that if there is some lattice defections in crystal, the half-metallic property is possibly broken. Recently, artificial Fermi level controlling of half-metallic Heusler compounds Co_2MnSi is attempted by Al doping, and the observed data is good agreement with the results obtained by the first-principles calculation [7]. These facts indicate that the electronic structure of Heusler alloys are strongly influenced by both atomic configuration and composition, therefore, their physical properties based on the electronic structure are also influenced.

Furthermore, it is a serious problem that the fundamental properties, in particular transport property, of Heusler alloys are not investigated enough experimentally and theoretically. Motivated by these background, we calculate some fundamental properties of Co-based Heusler

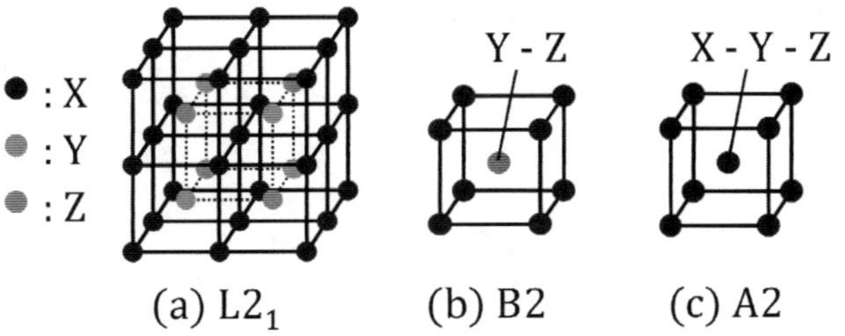

Figure 1. The crystal structure of X_2YZ full-Heusler alloys with (a) $L2_1$, (b) B2 and (c) A2 structure.

alloys such as the electronic structure, magnetic moment and electrical resistivity by first-principles approach in the present study. The statistical investigation of these properties depending on the atomic configuration and composition is carried out to clarify the effect of the atomic defection and substitution in crystal.

2. Methodology

For the present calculation, we adopt the three types of the typical crystal structure of Co_2YZ full-Heusler alloys, which is so called $L2_1$ (full-ordered), B2 (disordered between Y and Z atom) and A2 (full-disordered) structure as shown in Fig. 1. The compounds of $Co_2MnAl_{1-z}Si_z$ (Y=Mn, Z=Al-Si, $0.0 < z < 1.0$) and $Co_2Mn_{1-y}Fe_ySi$ (Y=Mn-Fe, Z=Si, $0.0 < y < 1.0$) alloys are dealt in the bulk crystals with $L2_1$, B2 and A2 structure. At first, we carry out the electronic structure calculation in disordered alloys system by using the tight-binding linear muffin-tin orbital (TB-LMTO) method in conjunction with the coherent potential approximation (CPA) to deal with the substitution-type alloy effects [8]. The local spin-density functional approximation (LSDA) is adopted for the self-consistent electronic structure calculation. Next, the electrical conductivity based on the obtained electronic structure is calculated by using the Kubo-Greenwood formula assuming zero temperature, which is formulated by Turek *et.al.* in the TB-LMTO-CPA scheme [9]

$$\sigma_\mu = \lim_{\delta\to+0} \frac{e^2}{4\pi\hbar V} \mathrm{Tr}\big\langle \big(g^\beta(\epsilon_F^+) - g^\beta(\epsilon_F^-)\big)\big[X_\mu, S^\beta\big]\big(g^\beta(\epsilon_F^+) - g^\beta(\epsilon_F^-)\big)\big[X_\mu, S^\beta\big]\big\rangle_{\mathrm{conf.}} \tag{1}$$

where X_μ, V, ϵ_F are the Cartesian coordinate operator ($\mu = x, y, z$), the volume of the unit cell, the Fermi level ($\epsilon_F^\pm = \epsilon_F \pm i\delta$), respectively, and $\langle\cdots\rangle_{\mathrm{conf.}}$ means the configuration average of random system. S^β, $g^\beta(\epsilon_F^\pm)$ are the structure constant and the retarded and advanced anxiously Green's function in the most localized β representation of the TB-LMTO-CPA method [8]. Note that the vertex correction is evaluated by means of the ladder approximation to satisfy the charge conservation law [10]. The lattice constants are set to a experimental data, $a =$ 5.755 Å for Co_2MnAl, $a = 5.655$ Å for Co_2MnSi [11] and $a = 5.640$ Å for Co_2FeSi [12]. In the intermediate region of these compounds, the lattice constants are estimated by the Vegard's law, *i.e.* $a = 5.755 \times (1-z) + 5.655 \times z$ Å for $Co_2MnAl_{1-z}Si_z$ and $a = 5.655 \times (1-y) + 5.640 \times y$ Å for $Co_2Mn_{1-y}Fe_ySi$.

3. Results and Discussions

First, we show the density of the states (DOS) of Co_2MnSi with the $L2_1$, B2 and A2 structure divided into the majority and minority spin states in Fig. 2. In the $L2_1$ and B2 structure, it is confirmed that their DOS of the minority spin states have an energy gap at the Fermi level and become a half-metal. The energy gap around the Fermi level of Co-based full-Heusler alloys

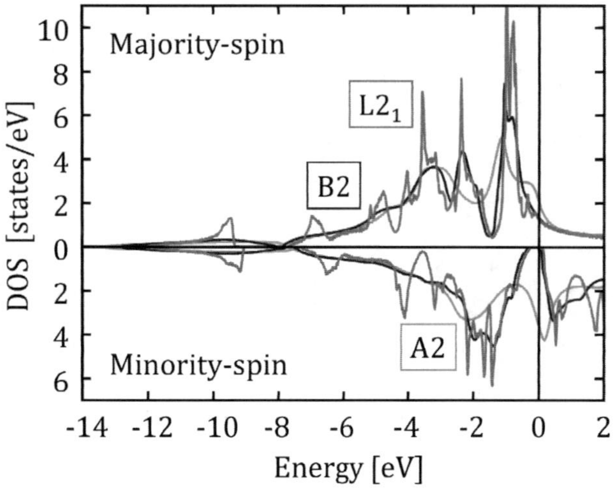

Figure 2. Spin-resolved density of the states in Co_2MnSi with $L2_1$ (red), B2 (blue) and A2 (orange). Zero axis of energy indicates the Fermi level.

in the ordered $L2_1$ structure is mainly constructed by the states based on the second nearest Co-Co hybridization [13]. So the half-metallic property of the B2 structure is still preserved due to the ordering of the Co-sub-lattice as well as Co_2CrAl [6]. The magnetic moment in the $L2_1$ and B2 structure become close to 5.0 μ_B, which reflects the feature of half-metals having integer magnetic moment. On the other hand, in the A2 structure, the energy gap around the Fermi level in the minority spin states is completely disappeared, therefore, the magnetic moment deviates from integer to 4.7 μ_B.

Next, we investigate the physical property such as the magnetic moment and residual resistivity due to substitution type alloy effect in $Co_2MnAl_{1-z}Si_z$ ($0.0 < z < 1.0$) and $Co_2Mn_{1-y}Fe_ySi$ ($0.0 < y < 1.0$) with the $L2_1$, B2 and A2 structure, respectively. The atomic composition dependence of the magnetic moment is shown in Fig. 3 as a function of the number of the valence electron (N_v). Note that the number of the valence electron is corresponding to $N_v = 28$ for Co_2MnAl, $N_v = 29$ for Co_2MnSi and $N_v = 30$ for Co_2FeSi. The left half ($28 < N_v < 29$) of Fig. 3 belongs to the $Co_2MnAl_{1-z}Si_z$ region, while the right half ($29 < N_v < 30$) to the $Co_2Mn_{1-y}Fe_ySi$ region.

When we focus on the $Co_2MnAl_{1-z}Si_z$ region in Fig. 3, there is a linear increase of the magnetic moments in $L2_1$ and B2 structure, which is proportional to the valence electrons number. This tendency is one of the features of half-metals that is so called the Slater-Pauling behavior, and the magnetic moment of half-metallic full-Heusler alloys is expressed in $N_v - 24$ μ_B [13]. In the region of $Co_2Mn_{1-y}Fe_ySi$, however, the Slater-Pauling behavior is not obtained, and the magnetic moments have a peak about 5.2 μ_B around $y = 0.4$. These results are corresponding to the result of a similar calculation in Refs. [14] for $Co_2Mn_{1-y}Fe_ySi$ ($y = 0.05, 0.10, 0.20$) with the $L2_1$ structure. If Fe concentration is increased, there is two significant changes in the electronic structure. One is that the partial DOS of Fe lie in overall lower energy side than that of Mn, so in the minority spin states the edge of states of Fe laps over the energy gap constructed by Co-Co hyblidization. The other is the increase of the valence electrons number, therefore, the Fermi level shifts to higher energy side. Consequently, the position of the Fermi level of $Co_2Mn_{1-y}Fe_ySi$ goes into the conduction band in the minority spin states, so the behavior of the magnetic moment is deviated from $N_v - 24$ μ_B. In the A2 structure, since the half-metallic property is broken in the whole region, the composition dependence of the magnetic moment is different behavior from those of the $L2_1$ and B2 structure.

The calculated resistivities in the bulk system as a function of the valence electrons number are shown in Fig. 4. Some of the experimental measurements of the electrical resistivity of Co-based Heusler alloys are reported by several workers, 7 $\mu\Omega cm$ [15] for the single crystal, 106

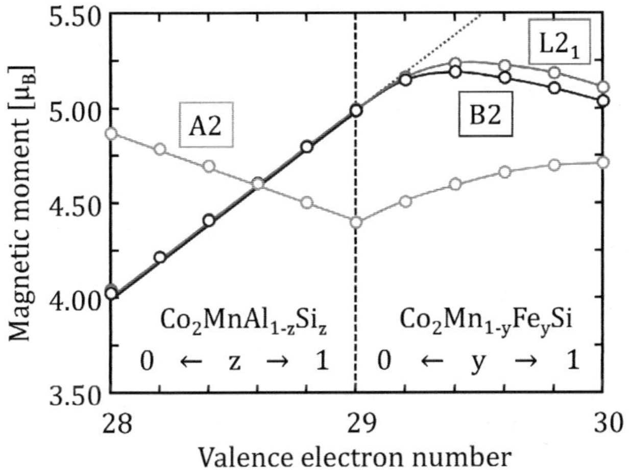

Figure 3. Magnetic moments as a function of valence electrons number. The red, blue and orange line indicates the results in $L2_1$, B2 and A2 structure, respectively.

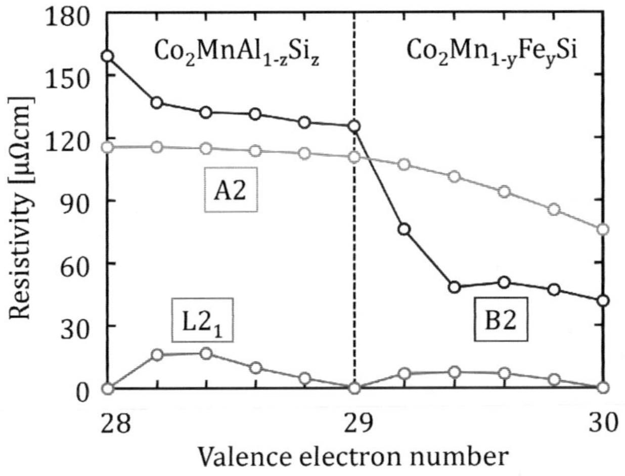

Figure 4. Electrical resistivities as a function of valence electrons number.

$\mu\Omega cm$ [16] for the thin film of Co_2MnSi, and 0.2 $\mu\Omega cm$ [17] for the single crystal, 41 $\mu\Omega cm$ [18] for the thin film of Co_2FeSi, respectively. These data seem to be good agreement with the calculated results, although there is no guarantee that the degree of the order, *i.e.* order parameter, of these samples is corresponding to the situation of the present calculation. Thus we have to be more careful to compare them.

When we pay attention to the calculation results, the electrical resistivities are originated only in the alloy effects, therefore, the electrical resistivity of Co_2MnAl, Co_2MnSi and Co_2FeSi with the full-ordered $L2_1$ structure vanish. The measured resistivities of them are, of course, finite due to the scattering by impurities and lattice defects in samples. In the intermediate region between these compositions in the $L2_1$ structure, the electrical resistivities become finite values about 20 $\mu\Omega cm$ for $Co_2MnAl_{1-z}Si_z$ and 10 $\mu\Omega cm$ for $Co_2Mn_{1-y}Fe_ySi$, which are close to most 3*d* transition metal alloys. They are strictly smaller than those of the B2 and A2 structure in the whole region, because the scattering probability due to random alloys effects, *i.e.* the inverse of electrons relaxation-time, in the ordered structure is clearly decreased comparing with other disordered structures. In the $Co_2MnAl_{1-z}Si_z$ region, the electrical resistivities in the B2 and A2 structure are more than 100 $\mu\Omega cm$. Worth mentioning is that the resistivities of the A2 structure are smaller than those of the B2 structure in spite of the small relaxation-time. Within the phenomenological theory the electrical conductivity of metals is proportional to the electrons velocity, relaxation-time and DOS around the Fermi level, respectively. Total DOS at the Fermi level in the A2 structure is larger than that of the B2 structure due to the breakdown of the half-metallic property. This may lead to the small resistivities in the A2 structure. By the similar reason, the electrical resistivities in $Co_2Mn_{1-y}Fe_ySi$ with the $L2_1$ and B2 structure are relatively smaller than $Co_2MnAl_{1-z}Si_z$, since their Fermi level goes into the conduction band of the minority-spin states with increasing the valence electrons number.

Lastly, these results imply that it is possibly to cause the abrupt change of the electrical resistivities in a series of half-metallic Heusler compounds. In other word, the transport property of Heusler alloys is quite sensitive to whether it is half-metal or not. More detailed theoretical analysis of the transport property in Heusler alloys would be performed soon.

4. Summary

In the present study, we carried out the statistical investigation of the several properties such as the magnetic moment and electrical resistivity in Heusler compounds $Co_2MnAl_{1-z}Si_z$ ($0.0 < z < 1.0$) and $Co_2Mn_{1-y}Fe_ySi$ ($0.0 < y < 1.0$) with the $L2_1$, B2 and A2 structures by using first-principles calculations. As a theoretical result, these properties strongly depend on the atomic configuration and composition that change the electronic structure of these alloys. In particular Co-based Heusler alloys, the existence of the half-metallic property significantly influences their physical properties.

Acknowledgments

This work is supported by Grant-in-Aids from the Ministry of Education, Culture, Sports, Science and Technology (MEXT), Japan. Y. K. is also supported by the Grant-in-Aid for Japan Society for the Promotion of Science (JSPS) Fellows.

References

[1] Sakuraba Y, Hattori M, Oogane M, Ando Y, Kato H, Sakuma A, Miyazaki T and Kubota H 2006 *Appl. Phys. Lett.* **88** 192508
[2] Tsunegi S, Sakuraba Y, Oogane M, Takanashi K and Ando Y 2008 *Appl. Phys. Lett.* **93** 112506
[3] Sakuraba Y, Iwase T, Saito K, Mitani S and Takanashi K 2009 *Appl. Phys. Lett.* **94** 012511
[4] Iwase T, Sakuraba Y, Bosu S, Saito K, Mitani S and Takanashi K 2009 *Appl. Phys. Exp.* **2** 063003
[5] Picozzi S, Continenza A and Freeman A J 2004 *Phys. Rev.* B **69** 094423
[6] Miura Y, Nagao K and Shirai M 2004 *Phys. Rev.* B **69** 144413
[7] Sakuraba Y, Kota Y, Kubota T, Oogane M, Sakuma A, Ando Y and Takanashi K 2010 *Phys. Rev.* B **81** 144422
[8] Turek I, Drchal V, Kudrnovský J, Sŏb M and Weinberger P 1997 *Electronic structure of disordered alloys, surface and interfaces* (Kluwer, Boston, 1997)
[9] Turek I, Kudrnovský J, Drchal V, Szunyogh L and Weinberger P 2002 *Phys. Rev.* B **65** 125101
[10] Carva K, Turek I, Kudrnovský J and Bongone 2006 *Phys. Rev.* B **73** 144421
[11] Webster P J 1971 *J. Phys. Chem. Solids* **32** 1221
[12] Wurmehl A, Fecher G H, Kandpal H C, Ksenofontov V, Felser C and Lin H-J 2006 *Appl. Phys. Lett.* **88** 032503
[13] Galanakis I, Dederichs P H and Papanikolaou N 2002 *Phy. Rev.* B **66** 174429
[14] Galanakis I, Özdoğan K, Aktaş B and Şaşioğlu E 2006 *Appl. Phys. Lett.* **89** 042502
[15] Geiersbach U, Bergmann A and Westerholt K 2002 *J. Magn. Magn. Matter.* **240** 546
[16] Raphael M P, Ravel B, Huang Q, Willard M A, Cheng S F, Das B N, Stroud R M, Bussmann K M, Claassen J H and Harris V G 2002 *Phys. Rev.* B **66** 104429
[17] Schneider H, Jakob G, Kallmayer M, Elmers H J, Cinchetti M, Balke B, Wurmehl S, Felser C, Aeschlimann M and H. Adrian M 2006 *Phys. Rev.* B **74** 174426
[18] Bulm C G F, Jenkins C A, Barth J, Felser C, Wurmehl W, Friemel G, Hess C, Behr G ,Büchner B, Reller A, Riegg S, Ebbinghaus S G, Ellis T, Jacobs P J, Kohlhepp J T and Swagten H J M 2009 *Appl. Phys. Lett.* **95** 161903

Journal of Physics: Conference Series **266** (2011) 012095 doi:10.1088/1742-6596/266/1/012095

Determination of local magnetic moment in $L1_0$-FeNi using photoelectron emission microscopy (PEEM)

Masato Kotsugi[1], Masaki Mizuguchi[2], Shigeki Sekiya[2], Takuo Ohkouchi[1], Takayuki Kojima[2], Koki Takanashi[2] and Yoshio Watanabe[1]
[1]SPring-8/JASRI, 1-1-1 Koto, Sayo, Hyogo 679-5198, Japan
[2]Tohoku University, 2-1-1 Katahira, Aoba-ku, Sendai 980-8577, Japan

E-mail: kotsugi@spring8.or.jp

Abstract. Magnetic domain structure of $L1_0$-type FeNi is observed by angular dependent photoelectron emission microscopy. It reveals the perpendicular magnetic component in magnetic domain structure with Bloch wall. The formation of Bloch wall is ascribed to the large magnetic anisotropy energy of $L1_0$-FeNi.

1. Introduction

$L1_0$-type ferromagnets, such as CoPt or FePt, have attracted much attention for use in next generation electro-magnetic devices because of its large magnetic anisotropy. Understanding the local magnetic moment associating with the macroscopic magnetic anisotropy is one of the most important items to improve areal density in magnetic recording medium. We, here, address the microscopic magnetic behavior in rare-metal-free $L1_0$-FeNi ferromagnetic thin film.

$L1_0$-FeNi — originally found in meteorites — shows entirely large magnetic anisotropy of 3.2×10^5 J/m^3, which is more than one order of magnitude larger than that of common disordered FeNi phase (-2.5×10^3 J/m^3) being close to the required value for perpendicular magnetic recording medium [1-8]. Such hard magnetism in $L1_0$-FeNi phase have brought great benefits to planetary scientists to study the history of solar system using the remanent magnetization in meteorites [2, 3]; meanwhile, its unique magnetic property in $L1_0$-FeNi could be also attractive target for material scientists and related applications (Fig.1) [7]. Owing to the considerable crisis of Pt resource depletion in these days, use of abundant and inexpensive resource of Fe and Ni is expected to offer great benefits in both ecology and economy. T. Shima et al. and M. Mizuguchi et al. recently have succeeded the synthetic fabrication of single-crystalline $L1_0$-FeNi thin film by alternate mono-atomic deposition using molecular beam epitaxy (MBE) technique [4-6].

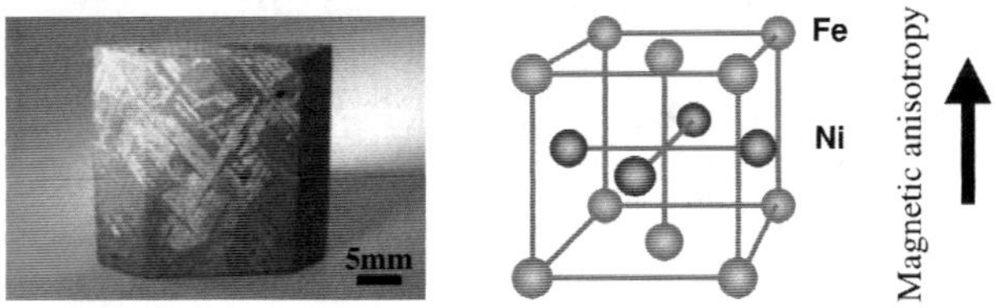

Figure 1. (a) Surface of iron meteorite. (b) Lattice structure of $L1_0$-FeNi alloy.

Here, we discuss how the microscopic magnetic domain structure associates with the macroscopic magnetic property in $L1_0$-FeNi thin film. We measure the specimen described in the reference 5, in which the magnetic anisotropy is reported as 4.8×10^5 J/m^3 [5]. Magnetic domain structure is observed using photoelectron emission microscope (PEEM) with magnetic circular dichroism (MCD) technique [9-11]. Circularly polarized synchrotron radiation (SR) provides the direction of the atomic magnetic moment in ferromagnetic material in connection to the polarization vector of the synchrotron radiation.

2. Sample preparation

$L1_0$-type FeNi is described as a chemically ordered FeNi alloy with the face-centered-tetragonal (fct) superstructure. Néel has first described the synthetic production of $L1_0$-type ordered FeNi phase by irradiating neutron under magnetic field and high temperature [1]. Single crystalline $L1_0$-FeNi film was prepared by alternating deposition of Fe and Ni using MBE technique, which is one of the most realistic methods for synthetic fabrication of $L1_0$-FeNi phase [5]. MBE has been of strong availability to prepare atomically flat thin-film with controlling the thickness accurately, and it should be important technique for further application [4, 5].

$L1_0$-type ordered FeNi superstructure was prepared on MgO substrate using UHV deposition system with independent multiple evaporators [5]. We constructed the multilayer of $(Fe/Ni)_{50}$/Cu/Au/Fe/MgO using ultra high vacuum deposition system with independent multiple evaporators. 1 nm of Fe seed layer was deposited on a MgO(001) substrate at 100 ℃ for first, and 20 nm of Au is deposited, then 50 nm of Cu is deposited at 496.9 ℃. These temperatures were optimised temperature for the preparation of atomically flat substrate. Next, mono-atomic-layers of Fe and Ni were alternately deposited on the top. The alternation of Fe and Ni deposition was repeated for 50 times at 187.3 ℃. The prepared FeNi surface is (001) [5]. The thickness was controlled based on the values observed by a quartz oscillator. The calibration of the oscillator was carefully achieved by observing the oscillation in reflection high energy electron diffraction (RHEED) intensity before the preparation of multilayer films. The surface structure was monitored by RHEED observation. The magnetic anisotropy energy evaluated by the area of magnetic hysteresis curve was 4.8×10^5 J/m^3. Unfortunately, this value was not enough large to obtain spontaneous perpendicular magnetization. To emphasize perpendicular magnetic domain structure, the specimen was magnetized perpendicularly to the film by the magnetic field of 1.1 T.

3. Experiment

Magnetic domain observation was performed by spectroscopic low energy electron microscope (SPELEEM) installed at BL17SU of SPring-8 [12]. PEEM has been of powerful technique to observe the microscopic magnetic information with several tens nanometer in lateral resolution. Magnetic domain images were obtained by the helicity reversal of circularly polarized light. Since, the magnetic contrast in the magnetic domain structure is proportional to the cosine of the angle between atomic magnetic moment and light incident, hence angular dependent magnetic domain observation offers 3 dimensional components (x, y, z) of the sample. For example, a constant angular dependence in magnetic contrast derives out-of-plane magnetization. The sample azimuthal angle ϕ was varied from 0 to 90 degree, where ϕ is the angle between light incidence to [100] direction of the film. Photon energy was set to L_3 absorption edge of Fe. Field of view was set to 30 μm, and exposure time for single image was several minutes. Lateral resolution of apparatus reaches 22 nm.

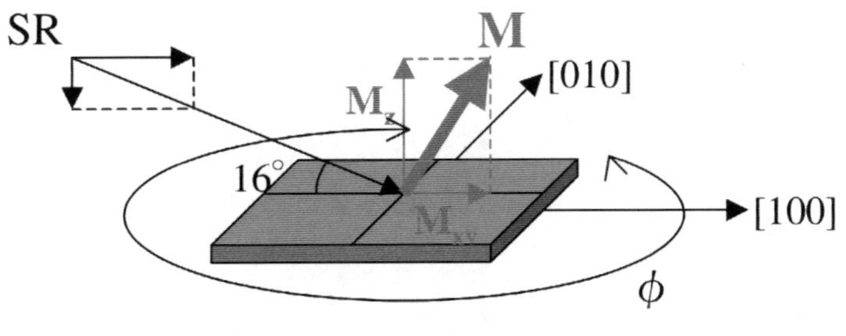

Figure 2. Experimental geometry for angle dependent magnetic domain imaging in photoelectron emission microscopy with synchrotron radiation.

4. Results and discussion

Figures 3 (a-d) show the magnetic domain structures obtained at Fe L_3 absorption edge for various incident angle of synchrotron radiation. The color scale of MCD signal is fixed in all image displayed here. It reveals that the wide bright stripes marked as (A) are separated by the narrow dark stripes marked as (B). The brightness and contrast in wide striped magnetic domain (A) changes gradually with varying SR light incidence; thus (A) is immediately determined as in-plane magnetization. On the other hand, the narrow magnetic domain (B) does not change a lot respect to the SR light incidence.

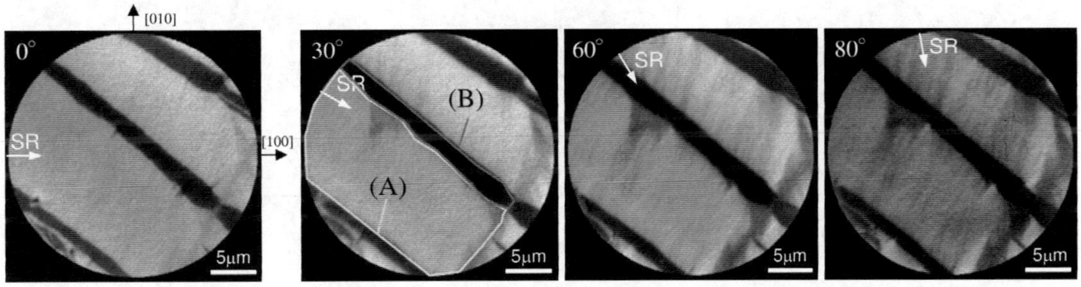

Figure 3. Observed magnetic domain structure of $L1_0$-FeNi for various incidence of synchrotron radiation. The dark narrow area shows identical brightness respect to light incidence.

To understand the detail about the perpendicular magnetization in the narrow magnetic domains (B), we carried out quantitative analysis with pixel-by-pixel manner. The MCD contrast of each pixel was fitted by cosine function as a function of incident angle to decompose quantitatively in-plane and out-of-plane component using following formula.

$$I_{MCD}(\phi) \propto m_{\perp} \cos 16 + m_{//} \sin 16 \cos(\phi + \delta) \qquad (1)$$

where I_{MCD} is the observed contrast in magnetic domain image, 16 degree is incident angle of SR respect to the surface normal, constant factors of cos16 and sin16 are projection of SR to in-plane and out-of-plane direction of the sample, respectively. δ is the angle between in-plane magnetic component and [100] direction of the specimen. Figures 4 (a, b, c) show the decomposed in-plane and out-of-plane components in the magnetic domain structure. Fig. 4(a) displays the vector map of in-plane component derived from fitted δ . The in-plane magnetization vector aligns to [-110] or [1-10], and it looks forming 180 degree domain wall. On the other hand, absolute value of in-plane and out-of-plane magnetic component derived from fitted $m_{//}$ and $m_{\perp}$ are shown in Fig. 4 (b, c). The absolute in-plane component shows clearly small value in the narrow magnetic domain (B) rather than the surrounded wide domains (A). Hence, the narrow domain is not simply in-plane and decrease in in-plane component may move to the other out-of-plane component.

Figure 4(d, e) shows the line profile of in-plane and out-of-plane component. Value is averaged over 21 pixels parallel to the stripe for better statistics and avoiding uncertainty. The in-plane component obviously decreases in the narrow magnetic domain (B) as shown in Fig. 4 (d), meanwhile, we note that the slight increase in out-of-plane component is confirmed in the central narrow domain (B) in Fig. 4 (e). It suggests that the perpendicular magnetic component arises in the narrow magnetic domain. Although the statistic of Fig. 4 (e) is not still enough, the magnetic moment in domain (B) cants perpendicularly with the elevation angle of about 40 degree, which is analytically derived from the ratio between in-plane and out-of-plane component.

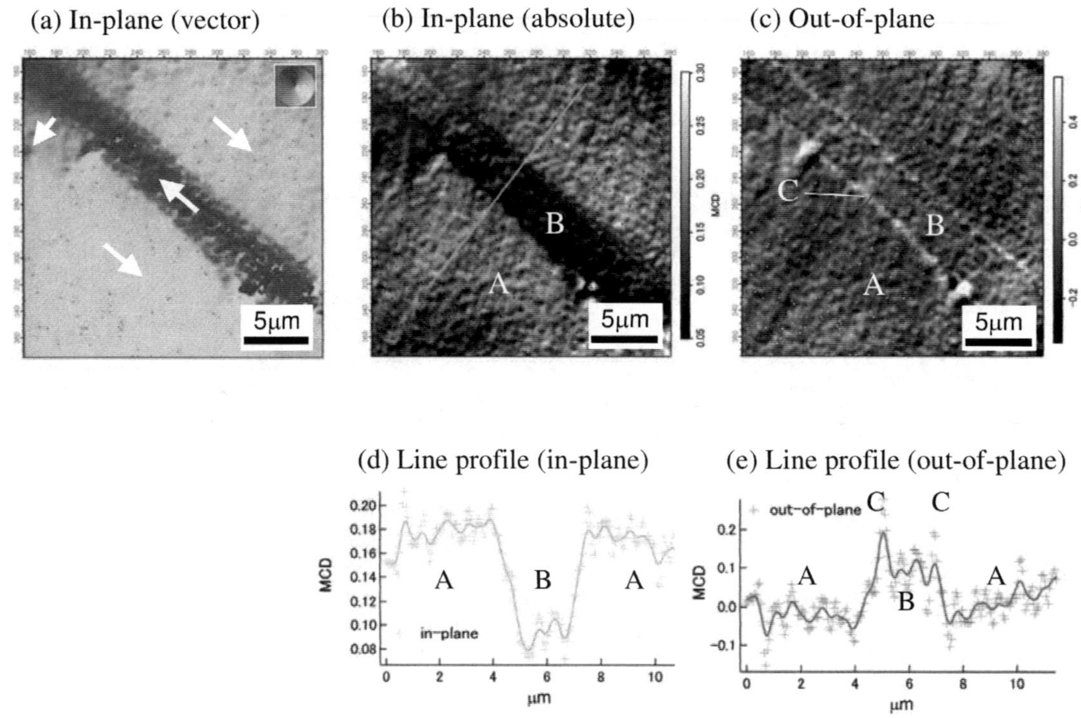

Figure 4. Decomposed 3 dimensional magnetization maps in $L1_0$-FeNi. a) In-plane component basically constructing 180 domain wall. b) Obvious decrease is confirmed at the central magnetic domain structure (B) c) Out-of-plane component increases at the central magnetic domain (B), and bright rim is recognized at the domain wall. d) Line profile of in-plane component in Fig. 4 b). e) Line profile of out-of-plane component in Fig. 4c). Slight increase is recognized at the centre, and rapid increase is also confirmed at the domain wall, indicating the formation of Bloch wall.

Another remark is also found on the domain wall, a bright rim at the domain wall is recognized in the out-of-plane component map as marked (C) in Fig. 4 (c). It is identically found as sharp peaks at the line scan in Fig 4 (e). Finding the out-of-plane magnetization at the domain wall suggests that Bloch wall is constructed here. In common thin film system with in-plane easy-axis, the construction of Néel wall is favourable to decrease the cost of demagnetization energy at the surface that shows in-plane spin rotation over the domain wall. Meanwhile, the formation of Bloch wall requires large crystallographic magnetic anisotropy energy overcoming the cost of demagnetization energy at the surface. Therefore, the formation of Bloch wall is ascribed to the large crystallographic magnetic anisotropy in $L1_0$-FeNi phase.
Consequently, we can conclude that the out-of-plane magnetic component is arisen in a part of magnetic domain structure, especially the formation of Bloch wall at the domain wall is related to the large magnetic anisotropy in $L1_0$-FeNi. Suppressing the shape anisotropy will become important to generate the spontaneous perpendicular magnetization, and the fabrication to micro or nano dot is expected in the future.

5. Summary

We argue the local magnetic moment in connection with macroscopic magnetic property in $L1_0$-FeNi thin film. $L1_0$–FeNi single crystal fabricated by MBE was measured by angular dependent MCD-PEEM. Decomposed magnetization vector map shows perpendicular component in a part of magnetic domain structure. The formation of Bloch domain wall ascribed to the large magnetic anisotropy in $L1_0$-FeNi.

6. Acknowledgement

Part of this study was supported by grants-in-aid from the Ministry of Education, Culture, Sports, Science and Technology (Young scientist (B) 17740198, 19740210). PEEM experiment was performed at the RIKEN beam line (BL17SU) of SPring-8. We thank Drs. Masaki Oura, Haruhiko Ohashi and Yasunori Senba of RIKEN. We also thank Dr. Minoru Funaki of the National Institute of Polar Research for useful discussions of meteorites.

7. References

[1] L. Néel, J. Pauleve, R. Pauthenet, J. Laugier, and D. Dautreppe: J. Appl. Phys. **35** (1964) 873.
[2] P. Wasilewski: Phys. Earth Planet. Int. **52** (1988) 150.
[3] M. Funaki, M. Koshita, and H. Nagai: Antarct. Meteorite Res. **16** (2003) 220.
[4] T. Shima, M. Okamura, S. Mitani, and K. Takanashi: J. Magn. Magn. Mater. **310** (2007) 2213.
[5] M. Mizuguchi, T. Sekiya, K. Takanashi and M. Shirai, in preparation.
[6] M. Mizuguchi, S. Sekiya, and K. Takanashi, J. Appl. Phys. **107**, 09A716 (2010).
[7] M. Kotsugi, C. Mitsumata, H. Maruyama, T. Wakita, T. Taniuchi, K. Ono, M. Suzuki, N. Kawamura, N. Ishimatsu, M. Oshima, Y. Watanabe, and M. Taniguchi Appl. Phys. Express **3** (2010) 013001
[8] H. P. J. Wijn, Magnetic Properties of Metals (Springer-Verlag, Berlin, 1991), pp. 36-38.
[9] W. Kuch, L. I. Cheralu, F. Offi, J. Wang, M. Kotsugi, and J. Kirschner: Nat. Mater. **5** (2006) 128.
[10] M. Kotsugi, W. Kuch, F. Offi, L. I. Chelaru, and J. Kirschner: Rev. Sci. Instrum. **74** (2003) 2754.
[11] J. Stöhr, H. A. Padmore, S. Anders, T. Stammler, and M. R. Scheinfein: Surf. Rev. Lett. **5** (1998) 1297.

[12] M. Oura T. Nakamura T. Takeuchi, Y. Senba, H. Ohashi, K. Shirasawa, T. Tanaka, M. Takeuchi, Y. Furukawa, T. Hirono, T. Ohata, H. Kitamura and S. Shin, J. of Syn. Rad. **14**, 483, (2007).

Journal of Physics: Conference Series **266** (2011) 012096 doi:10.1088/1742-6596/266/1/012096

Magnetoresistance Effect in Co_2MnSi/semimetallic-Fe_2VAl/CoFe Junctions

T Kubota [1, *], M Oogane [2], S Mizukami [1], H Naganuma [1], Y Ando [2] and T Miyazaki [1]
[1] WPI-Advanced Institute for Materials Research, Tohoku University, Katahira 2-1-1, Sendai 980-8577, Japan
[2] Department of Applied Physics, Graduate School of Engineering, Tohoku University, Aoba 6-6-05, Sendai 980-8579, Japan

E-mail: takahide@wpi-aimr.tohoku.ac.jp

Abstract. Epitaxially grown semimetallic-Fe_2VAl Heusler thin films were fabricated on single crystalline MgO (100) substrate. Crystalline structure, magnetic property, and electrical transport property of the films were investigated. Non-magnetic and bulk-like resistivity was achieved in a *B*2-ordred Fe_2VAl film annealed at 700°C. Curret-perpendicular-to-plane (CPP-) Magnetoresistance effect in Co_2MnSi/Fe_2VAl/CoFe junctions was also investigated. Maximum value of MR ratio was 1.1% at room temperature, which is an evidence of spin-dependent transport through Fe_2VAl Heusler alloy.

1. Introduction

Half-metallic full-Heusler alloys [1] are ideal spin-polarizer for giant magnetoresistance (GMR) junctions, which are one of promising candidates of reading head in hard disc drives. In recent works, exceedingly large GMR effects were reported in $Co_2MnSi/Ag/Co_2MnSi$ [2], and $Co_2FeAl_{0.5}Si_{0.5}/Ag/Co_2FeAl_{0.5}Si_{0.5}$ [3] structures, which are experimental demonstrations of applicability of half-metallic Co-based full Heusler alloys. However it is important issue to investigate spin polarizing layer, it is also important to investigate spacer materials in GMR devices. Considering a combination with such half-metallic full-Heusler compounds mentioned above, non-magnetic material with similar crystalline system is desirable. In this study, we are focusing on a *semimetallic* Heusler compound of Fe_2VAl as a spacer of GMR junctions. Fe_2VAl is a non-magnetic Heusler compound with a number of valence electron of 24 [1]. In an earlier work, GMR effect with non-magnetic Heusler spacer has been reported in $Co_2MnGe/Rh_2CuSn/Co_2MnGe$ structures [4]. Rh_2CuSn is considered to have an advantage of good band-matching to half-metallic Co-based full-Heusler alloys as discussed in ref. [4]. On the contrary, the use of Fe_2VAl in this study is motivated to achieve relatively larger resistance compared to metallic compounds, which enables larger change of resistance of a junction compared to conventional metallic junctions with same MR ratio and smaller junction resistance. Bulk [5], poly-crystalline thin films [6], and theoretical investigations [7] of Fe_2VAl have been reported, although there have been no report on fabrication of epitaxial film and investigation of spin-dependent transport. Therefore, in the current study, magnetic and electrical transport properties of epitaxial Fe_2VAl thin films, and magnetoresistance effect in the junctions using Fe_2VAl spacing layer were investigated.

* To whom any correspondence should be addressed.

Journal of Physics: Conference Series **266** (2011) 012096 doi:10.1088/1742-6596/266/1/012096

2. Experimental Procedures

2.1. Fabrication of epitaxial Fe_2VAl thin films
30-nm-thick Fe_2VAl films were deposited directly onto single crystalline-MgO (100) substrate by ultra high vacuum magnetron sputtering technique at room temperature (base pressure of the chamber is below 4×10^{-7} Pa). After the deposition, annealing by infra-red heater was performed in the chamber without breaking the vacuum. Annealing temperature (T_{annl}) was varied from 400°C to 700°C. Finally, 2-nm-thick Ta was deposited on the top as a protection layer. Crystalline structure and magnetic properties were examined by x-ray diffractometer (XRD) using Cu $K\alpha$ radiation, and superconducting quantum interface device (SQUID), respectively.

2.2. Investigations of magnetoresistance effect in $Co_2MnSi/Fe_2VAl/CoFe$ multi-layers
Multi-layers with a stacking structure of MgO (100) substrate/Cr (100 nm)/Co_2MnSi (20 nm)/Fe_2VAl (5 nm)/CoFe (4 nm)/Ru (10 nm) were prepared by magnetron sputtering technique. All the films were deposited at room temperature, and subsequent annealing processes were performed for Cr (at 700°C), Co_2MnSi (at 500°C), and Fe_2VAl (at T_{annl} = 300°C or 500°C) layers. Multi-layers were patterned into pillars designed to elliptical shapes from 70 × 140 to 200 × 400 nm^2 by electron-beam lithography and Ar ion plasma etching techniques. The actual pillar size was determined by scanning electron micro scope. Electrical transport measurements were done by direct current four-probe technique at room temperature. In-plane XRD measurements ($2\theta\chi/\phi$ scans, using Cu $K\alpha$ radiation with a grazing angle of 0.50°) were also performed to confirm structural property of 5-nm-Fe_2VAl layer on Co_2MnSi. Stacking structures for the in-plane XRD investigations were MgO(100) sub./Cr (40 nm), T_{annl} = 500°C /Co_2MnSi (20 nm), T_{annl} = 500°C /Fe_2VAl (5 nm), T_{annl} = 300°C or 500°C /Ta (5 nm).

3. Results and Discussion

3.1. Structural, Magnetic and Transport Properties of Fe_2VAl Thin Films
XRD profiles of MgO(100) / Fe_2VAl (30 nm) / Ta (2 nm) films with different T_{annl} are shown in figures 1. Epitaxial relationships of MgO(001)[100] / Fe_2VAl(001)[110] was confirmed for all the samples using ϕ-scans (not shown here). Supperlattice diffraction of (002) peak from $B2$ ordered structure is confirmed in the samples with T_{annl} = 500°C, 600°C, and 700°C by 2θ scans (fig. 1 (a)). $L2_1$-ordered structure was not confirmed in the current investigations. Long-range order parameter for $B2$ structure (S_{B2}) was calculated by the Webster's model [8], which is plotted in fig. 1 (b). The S_{B2} increased with increasing T_{annl}, and achieved maximum values over 0.8 for the samples with T_{annl} = 600°C and 700°C. Lattice constants of a and c axes are plotted in fig. 1 (c). Both axes are relaxed towards the balk value [5] with increasing T_{annl}. All the samples were rather strained to in-plane direction due to a slightly larger lattice constant of MgO substrate. The dependences of saturation magnetization (M_s) on measurement temperature are shown in figure 2 (a). The M_s was reduced with increasing T_{annl}. In contrast, resistivity was increased with increasing T_{annl}, as shown in fig. 2 (b). All samples exhibit semiconductor-like temperature dependence of resistivity and its slope tends to be larger in the samples with higher T_{annl}. Those tendencies indicate a manifestation of semimetallic pseudo gap in Fe_2VAl, which accompanied by the atomic ordering. Such non-magnetic nature and the order of resistivity are comparable to those reported in bulk samples [5], especially regarding on the sample with T_{annl} = 700°C.

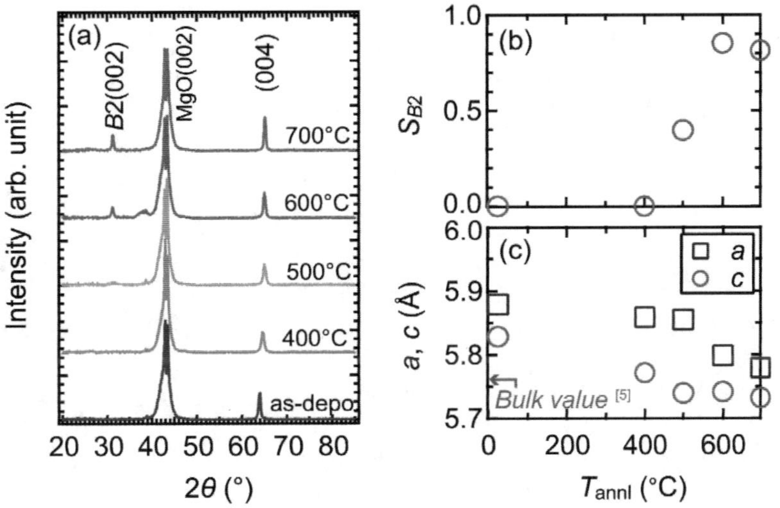

Figure 1. XRD profiles of MgO sub./Fe_2VAl (30 nm)/Ta (5 nm) structures. (a) 2θ scans. Temperatures noted in the figure are post-annealing temperature (T_{annl}) for the Fe_2VAl layer. (b) Long-range order parameter for *B*2 ordered structure in Fe_2VAl as a function of T_{annl}. (c) The dependence of Lattice constants (*a,* and *c*) of Fe_2VAl films on T_{annl}.

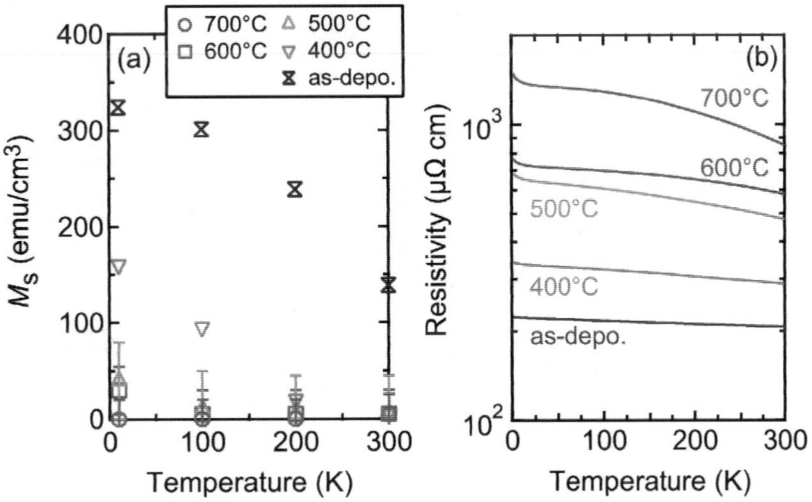

Figure 2. The dependence of (a) saturation magnetization (M_s), and (b) resistivity on measurements temperature.

3.2. Crystalline Structure of Fe_2VAl on Co_2MnSi

XRD profiles obtained by in-plane measurements ($2\theta\chi/\phi$ scans) are shown in figures 3, temperatures annotated in the figures are T_{annl} for the Fe_2VAl layer. In fig. 3 (a), peaks (and shoulders) corresponding to diffractions from (400) plane of Fe_2VAl, Cr, and Co_2MnSi were confirmed around $2\theta\chi/\phi$ = 64°, as pointed with the arrows, on both samples with T_{annl} = 300°C and 500°C. In addition, (200) diffraction from *B*2-ordered structure of Fe_2VAl was confirmed on the sample with T_a = 500°C (see, fig. 3 (b)). *B*2 diffractions from Fe_2VAl and Co_2MnSi are overlapped each other, although $2\theta\chi/\phi$ values can be obtained by use of fitting with Lorenz functions. Obtained values are 31.02° and 31.48°, and those are well matched to those expected from (400) peaks, which are 31.02° and 31.53 for Fe_2VAl and Co_2MnSi, respectively. For the case on the sample with T_{annl} = 300°C, appeared peak is rather broad, although it may be only from *B*2-Co_2MnSi, because reasonable fitting was not obtained for the (200) peak as expected from (400) peaks.

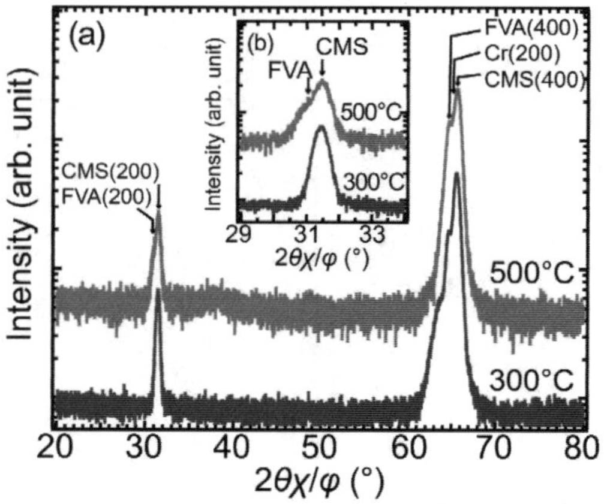

Figure 3. (a) In-plane XRD profiles ($2\theta\chi/\phi$ scans) of MgO sub./Cr (40)/Co_2MnSi (CMS) (30)/Fe_2VAl (FVA) (5)/Ta(5) (unit: nm) structures. Inset (b) is an expanded view around (200) peaks. Annotated temperatures in the figures are post-annealing temperature for Fe_2VAl layer.

3.3. Magnetoresistance Effect in $Co_2MnSi/Fe_2VAl/CoFe$ Junctions

Figure 4 (a) shows the dependence of MR ratio on post annealing temperature for Fe_2VAl layer at room temperature (measured at RT). The inset, fig. 4 (b), is a typical MR curve for the sample with T_{annl} = 500°C. Resistance and area products ($R \times A$) were raging from 60 to 480 mΩ μm^2, which were slightly smaller than that expected from the resistance of single layer film described in the section 3.1. One possible explanation of the small $R \times A$ is proximity effect by contacting the metallic layers and/or atomic disorder of Fe_2VAl layer. MR ratio was increased with the increasing of annealing temperature for Fe_2VAl layer. Such a tendency of increment of MR ratio may be attributed to the appearance of *B*2 ordering in Fe_2VAl as discussed in the section 3.2. Maximum value of MR ratio was 1.1% for the sample with T_{annl} = 500°C measured at RT. Absolute value of the MR ratio was still small, although these results are an evidence of that spin-dependent transport through semimetallic-Fe_2VAl is possible.

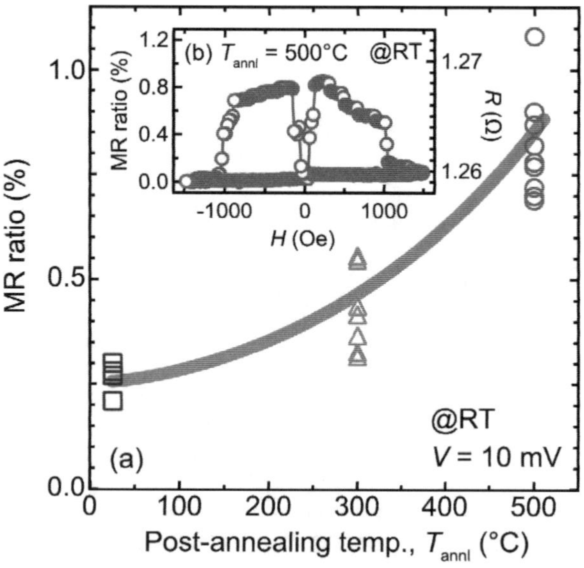

Figure 4. (a) The dependence of MR ratio on the post-annealing temperature (T_{annl}) for Fe_2VAl. Inset (b) is a typical MR curve of a junction with T_{annl} = 500°C.

4. Summary

We have fabricated epitaxially grown semimetallic-Fe_2VAl thin films on single crystalline MgO (100) substrate by magnetron sputtering technique, and clarified those magnetic and electric properties. Magnetoresistance effect in $Co_2MnSi/Fe_2VAl/CoFe$ junctions was observed, and the maximum value was 1.1% for the junction with T_{annl} = 500°C for Fe_2VAl.

5. References

[1] Galanakis I, Dederichs P H and Papanikolaou N 2002 *Phys. Rev. B* **66** 174429
[2] Iwase T, Sakuraba Y, Bosu S, Saito K, Mitani S and Takanashi K 2009 *Appl. Phys Express.* **2** 063003
[3] Nakatani T M, Furubayashi T, Kasai S, Sukegawa H, Takahashi Y K, Mitani S and Hono K 2010 *Appl. Phys. Lett* **96** 212501
[4] Nikolaev K, Kolbo P, Pokhil T, Peng X, Chen Y, Ambrose T, and Mryasov O 2009 *Appl. Phys. Lett.* **94** 222501
[5] Nishino Y, Kato M, Asano S, Soda K, Hayasaki M, and Mizutani U 1997 *Phys. Rev. Lett.* **79** 1909
[6] Mikami M, Kamiya T, Kobayashi K 2010 *Thin Solid Films* **518** 2796
[7] Singh D J, Mazin I I 1998 *Phys. Rev. B* **57** 14352; Weht R, Pickett W E 1998 *Phys. Rev. B* **58** 6855
[8] Webster P J 1971 *J. Phys. Chem. Solids* **32** 1221

6. Acknowledgement

This work was partly supported by the SCOPE grant from the Ministry of Internal Affairs and Communications, Japan.

Journal of Physics: Conference Series **266** (2011) 012097 doi:10.1088/1742-6596/266/1/012097

Structural, optical, and magnetic properties of Ho-implanted GaN thin films

F.-Y. Lo[(a)], J.-Y. Guo[(b)], V. Ney[(c)], A. Ney[(c)], M.-Y. Chern[(d)], A. Melnikov[(e)], S. Pezzagna[(e)], D. Reuter[(e)], A. D. Wieck[(e)], and J. Massies[(f)]

[(a)]Department of Physics, National Taiwan Normal University, 88, Sec. 4, Ting-Chou Rd., Taipei City 11677, Taiwan
[(b)]Department of Physics, National Dong Hwa University, Sec. 2, Da-Hsueh Rd., Hualien 97401, Taiwan
[(c)]Experimentalphysik, Universität Duisburg-Essen, Lotharstr. 1, 47048 Duisburg, Germany
[(d)]Department of Physics, National Taiwan University, 1, Sec. 4, Roosevelt Rd., Taipei City 10617, Taiwan
[(e)]Lehrstuhl für Angewandte Festkörperphysik, Ruhr-Universität Bochum, Universitätsstr. 150, 44780 Bochum, Germany
[(f)]Centre de Recherche sur l'Hétéroépitaxie et ses Application, UPR-CNRS 10, rue Bernard Gregory, 06560 Valbonne, France

e-mail: fangyuhlo@ntnu.edu.tw

Abstract. Ho ions were implanted into highly-resistive molecular-beam-epitaxy grown GaN thin films with a 100kV focused-ion-beam implanter at room temperature (RT). The implantation doses of Ho ions ranges from 10^{14} to 10^{16} cm^{-2}. Without thermal annealing, the structural, optical, and magnetic properties of the Ho-implanted thin films were investigated. Structural properties studied by x-ray diffraction revealed Ho incorporation into GaN matrix without secondary phase. The overall photoluminescence of any implanted sample is weaker than that of the non-implanted one. The spectra show neutral-donor-bound exciton emission and defect-related blue luminescence. Blocked superparamagnetic behavior was identified from Ho-implanted samples at temperatures below RT by measurements with a superconducting quantum interference device. The highest ordering temperature is 100 K.

1. Introduction

Rare earth doped III-V nitride semiconductor materials have received considerable attention primarily for potential use in optoelectronic applications in visible displays or in white light systems that employ color-combining techniques. Recently, rare earth doped GaN and AlN thin films have also shown potential impact for application in spintronics. Among them, Gd-doped GaN attracted great interest because it was reported to be ferromagnetic above room temperature (RT) and have large colossal magnetic moment [1-3]. Holmium (Ho) has the atomic magnetic moment of 10.4 μ_B, which is the second largest value among rare earth elements, and therefore, it would be interesting to incorporate Ho into GaN and compare the properties between Ho- and Gd-doped GaN.

In this paper, we investigate the structural, optical, and magnetic properties of GaN thin films implanted with Ho by focused-ion-beam (FIB) implantation at different doses. Our results show well-incorporation of Ho into GaN, and the impurity-related emission in photoluminescence spectra. The magnetic ordering occurs at temperatures lower than 130 K.

2. Experimental

Wurtzite GaN thin films grown by ammonia-source molecular beam epitaxy on a *c*-oriented sapphire substrate were implanted with Ho ions at three different doses (1×10^{14}, 1×10^{15}, and 1×10^{16} cm^{-2}, and labeled as sample A, B, and C, respectively) at RT by using a FIB implanter (EIKO-100). For samples A and B, the ion energy of Ho is 300keV, and the projected range simulated by SRIM code [4] is 52.6nm. For sample C, the ion energy and the simulated projected range are 200keV and 38.2nm, respectively.

The structural, optical, and magnetic properties of the samples were investigated after implantation without performing thermal annealing. The structural properties of the samples were investigated by x-ray diffraction (XRD) at RT. The optical properties were studied by photoluminescence (PL) spectroscopy with an excitation wavelength of 325 nm. PL spectra in the range of 2.0 to 3.6 eV were recorded with standard lock-in technique at two temperatures, 10 and 300 K, respectively. The magnetic properties were measured with a Quantum Design superconducting quantum interference device (SQUID) magnetometer in the temperature range from 2 to 300 K. The magnetization loops (m-H curves) of each sample were taken at T = 5 and 300 K in the magnetic field between ±50 kOe. The temperature dependence of the magnetization between 2 and 300 K was studied by recording field-cooled (FC) and zero-field-cooled (ZFC) curves at a magnetic field of 100 Oe. The magnetic field was applied parallel to the surface of the samples for all SQUID measurements, and all the data presented here are corrected for the artifacts and the diamagnetic background.

3. Results and discussions

3.1. Structural properties

Wide range ω-2θ scan of the Ho-implanted samples were carried out with a Phillips X'Pert MRD diffractometer. No diffraction peaks other than sapphire(0006) and GaN(0002) were detected from the implanted samples, which shows that no secondary phase were formed during FIB implantation. Figure 1 compares the high-resolution x-ray diffraction rocking curves of the Ho-implanted and non-implanted thin films obtained by using Bede D^3 diffractometer, and the intensity is normalized to the peak intensity of GaN(0002). The GaN(0002) peak of sample A has no shift in peak position but with a slight broadening at the lower-angle side. GaN(0002) peak of sample B broadens on both sides, where lower-angle side has stronger broadening than the higher-angle side. However, the peak position is not shifted. For sample C, the GaN(0002) peak shifts to the lower-angle side and with pronounced broadening on both sides.

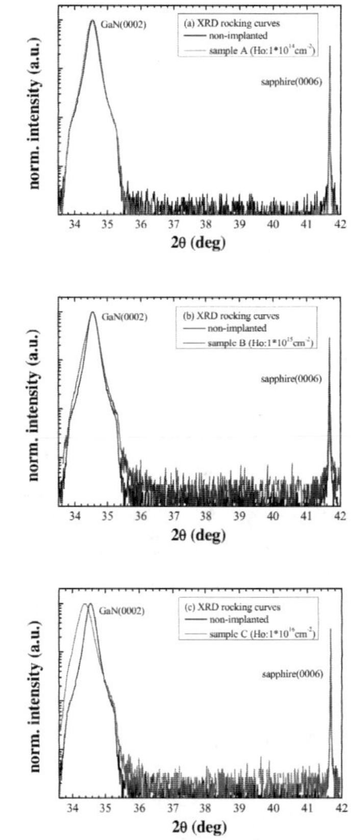

Figure 1 Comparison of rocking curves of Ho-implanted GaN thin films to a non-implanted one. The intensity is normalized to the peak intensity of GaN(0002). (a) Sample A, (b) sample B, and (c) sample C.

The slight broadening on the lower-angle side of GaN(0002) peak of samples A and B and the shift of peak position of sample C are the evidence of incorporation of Ho into GaN matrix, and the reasons are the following: (i) Ho-doped GaN would have larger lattice constant because Ho has larger ionic radius than Ga; and (ii) the GaN(0002) peak is the superposition of the diffraction peaks of the thin Ho-doped GaN and the underneath thick undoped GaN layers. Therefore, the GaN(0002) broadens slight on the lower-angle side for low Ho doses and shifts to lower angle for high Ho dose. The broadening of on the higher-angle side of GaN(0002) peak of sample B and C is attributed to implantation-induced local amorphization of the thin films, which becomes pronounced with doses above 1×10^{15} cm^{-2}.

3.2. Optical properties

Before the PL spectra are discussed, it is worthy to note that the overall integrated emission intensity of any of the four samples is weak – without focusing the laser beam, there was no observable spectrum even for the non-implanted sample. Figure 2 compares the PL spectra of the non-implanted sample to the Ho-implanted ones at T =10 K. For non-implanted sample (black curves), PL spectra show only the near-band-edge emissions, but the typical yellow luminescence was not observed at both temperatures. The near-band-edge emissions consist of neutral-acceptor-bound exciton (A^0X) and its longitudinal phonon replica (A^0X-LO), and donor-acceptor-pair (DAP) emissions. After implantation, near-band-edge emission was observed only for sample A, and it has only one contribution, which is the neutral-acceptor-bound exciton emission [5] – indicating the majority of the optically active defects is acceptor-like, possibly due to either the magnesium impurity or gallium vacancy. For samples B and C, the near-band-edge emission disappeared completely, and the reason can be implantation-induced local defect formation and amorphization. Local amorphous material and the defects modify the band structure and create absorption centers, which render the near-band-edge emission invisible.

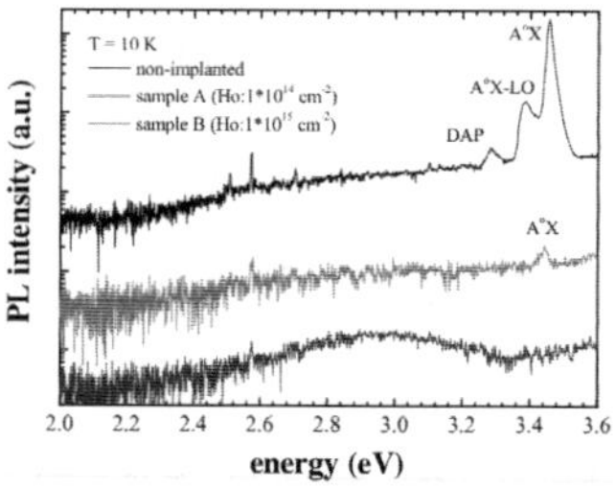

Figure 2 Photolumiescence spectra at T = 10 K of the non-implanted sample and samples A and B.

Furthermore, a broad but weak blue luminescence (BL) emission, whose peak is around 2.9 eV, is barely visible as background for the non-implanted sample and becomes clearer for the Ho-implanted samples. According to Ref. 6, this BL can be related to stacking-fault type of defects, and this indicates good crystal quality of our as-grown GaN thin films and the increase in density of such defects to Ho implantation.

3.3. Magnetic Properties

At T = 5 K, *m-H* curves, shown in Fig. 3, of the all Ho-implanted samples show only paramagnetic-like behavior without saturation at H = 50 kOe after all the artifacts and the diamagnetic background are corrected. Surprisingly, the m-H curves of samples A and B overlap completely. In small magnetic fields, $|H| \leq 3$ kOe, no hysteresis loop was observed from all Ho-implanted samples, and the remenant magnetic moment of samples A and B are at the resolution limit of our SQUID. The non-saturation feature indicates that the magnetic moment of all samples has two contributions, one from the (Ga,Ho)N thin films and the other from defects in the thin films, whether intrinsic or

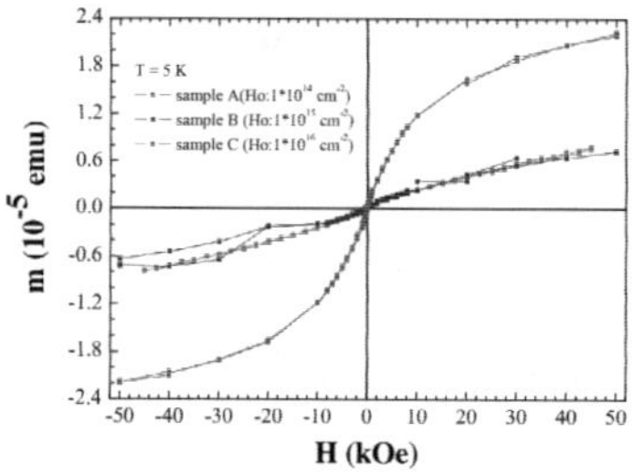

Figure 3 magnetization loops (m-H curves) of all Ho-implanted samples after diamagnetic background and artifacts are subtracted.

implantation-induced. The latter becomes dominant at high magnetic fields, and the defect density can be quite large. The magnetization loops at T = 300 K do not show any hysteretic behavior for all the implanted samples and are not depicted.

Figure 4 shows the temperature dependence of FC and ZFC magnetic moment curves under a magnetic field of 100 Oe for all three Ho-implanted samples. The FC and ZFC curves of sample B coincide each other – representing no long-range magnetic ordering. Separation between FC and ZFC curves were observed only for samples A and C, and this indicates long-range magnetic ordering in these two samples. The long-range ordering of the samples looks like blocking type because the FC curves show the temperature dependence of a paramagnet, and the ordering temperatures are around 100 K and 80 K for samples A and C, respectively. Because both Ho and HoN have the Curie temperature at around 20 K [7,8], the long-range magnetic ordering comes from (Ga,Ho)N.

The surprising magnetic properties of sample B (Ho-dose of 1×10^{15} cm^{-2}) – (i) having the same *m-H* curve as sample A (Ho-dose of 1×10^{14} cm^{-2}), and (ii) showing no long-range magnetic ordering while the other two do – prompted us to postulate that there is possible change in ordering mechanism for GaN implanted with a Ho-dose at around 1×10^{15} cm^{-2}. For lower Ho doses, where the separation between the Ho ions is large, the magnetic ordering is by blocking. For higher Ho doses, where the Ho density is in the percentage range, the magnetic properties of (Ga,Ho)N tend toward HoN, and the magnetic ordering mechanism is ferromagnetic – in our case, it could be impurity-mediated mechanism [9] or exchange mechanism. However, the FC curve shows no (clear) step-wise behavior, and it requires further investigations to verify our postulate.

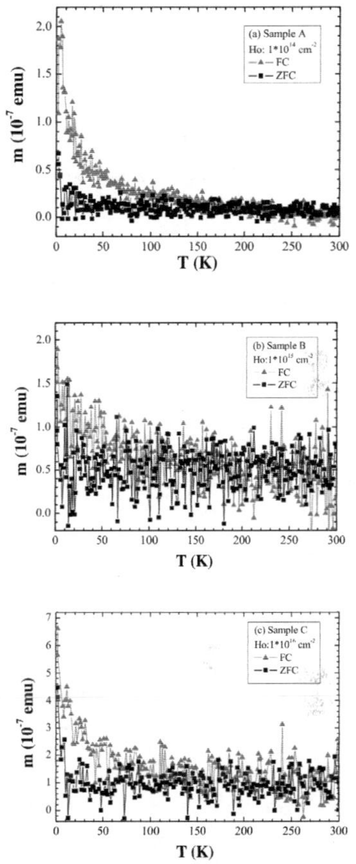

Figure 4 Temperature dependence of the field-cooled (FC) and zero-field-cooled (ZFC) curves of (a) sample A, (b) sample B, and (c) sample C. The measurement was in a magnetic field of 100 Oe.

4. Conclusions

Ho ions were implanted into wurtzite GaN thin films at room temperature with a dose ranging from 10^{14} to 10^{16} cm^{-2}. Without any thermal treatment on the samples, X-ray diffraction showed Ho-incorporation into the GaN matrix without any secondary phase. PL spectra revealed the presence of acceptor-like defects in Ho-implanted GaN with the dominant neutral-acceptor-bound-exciton emission. Ho-implanted GaN shows superparamagnetic behavior at low temperatures for implantation doses of 1×10^{14} cm^{-2} and 1×10^{16} cm^{-2}. The ordering temperature of the samples is around 100 K and 80 K, respectively. The sample implanted with a dose of 1×10^{15} cm^{-2} shows paramagnetic behavior with no magnetic ordering.

This work was partly supported by the National Science Council of Taiwan by the Grant NSC 98-2112-M-003-007-MY3, the European Union under the Marie-Curie Excellence Grant No. MEXT-CT-2004-01495 of the sixth Framework Programme, the Université Franco-Allemande (DFH/UFA) within the CDFA-05-06, and the Deutsche Forschungsgemeinschaft in the framework of the Sonderforschungsbereich 491 and the Priority Program SPP 1285.

References

[1] S. Dhar, O. Brandt, M. Ramsteiner, V. F. Sapega, and K. H. Ploog, Phys. Rev. Lett. **94**, 037205 (2005).
[2] S. Dhar, T. Kammermeier, A. Ney, L. Pérez, K. H. Ploog, A. Melnikov and A. D. Wieck, Appl. Phys. Lett. **89**, 062503 (2006).
[3] F.-Y. Lo, A. Melnikov, D. Reuter, A. D. Wieck, V. Ney, T. Kammermeier, A. Ney, J. Schörmann, S. Potthast, D. J. As, and K. Lischka, Appl. Phys. Lett. **90**, 262505 (2007).
[4] J. F. Ziegler and J. P. Biersack, http://www.srim.org/.
[5] A. K. Viswanath, J. I. Lee, S. Y., D. Kim, Y. Choi, and C.-H. Hong, J. Appl. Phys. **84**, 3848 (1998).
[6] R. Liu, A. Bell, F. A. Poncea, C. Q. Chen, J. W. Yang, and M. A. Khan, Appl. Phys. Lett. **86**, 021908 (2005).
[7] T. A. Yamamoto, T. Nakagawa, K. Sako, T. Arakawa, H. Nitani, J. Alloys Compd. **376**, 17 (2004).
[8] E. Schreier, M. Ekström, O. Hartmann, R. Wäppling, G.M. Kalvius, F.J. Burghart, S. Henneberger, A. Marelius, and A. Kratzer, Physica B **289-290**, 240 (2000).
[9] A. Svane, N. E. Christensen, L. Petit, Z. Szotek, and W. M. Temmerman, Phys. Rev. B **74**, 165204 (2006).

Journal of Physics: Conference Series **266** (2011) 012098 doi:10.1088/1742-6596/266/1/012098

Spin-torque induced rf oscillation in magnetic tunnel junctions with an Fe-rich CoFeB free layer

Y. Masugata[1], S. Ishibashi[1], H. Tomita[1], T. Seki[1], T. Nozaki[1], Y. Suzuki[1,2], H. Kubota[2], A. Fukushima[2], S. Yuasa[2]
[1] *Graduate School of Engineering Science, Osaka Univ., 560-8531, Toyonaka, Japan*
[2] *Nano-Spintronics Center, AIST, 305-8568 Tsukuba Japan*

E-mail: masugata@spin.mp.es.osaka-u.ac.jp

Abstract. A steady magnetization precession induced by spin-torque was studied in magnetic tunnel junctions with an Fe-rich CoFeB free layer under a perpendicular magnetic field. The Fe-rich CoFeB free layer showed a low critical current (I_c) for spin-torque induced switching. The low value of I_c for the Fe-rich CoFeB free layer enabled us to observe the spin-torque induced rf oscillation even at a low dc bias current (I_{dc}). The oscillation frequency increased with increasing I_{dc}, indicating the out-of-plane precessional mode. Owing to the perpendicular magnetic anisotropy in the Fe-rich CoFeB free layer, the large relative angle between the free and pinned layers was achieved under the perpendicular magnetic field. The out-of-plane precession of the magnetization and the large relative angle resulted in the large rf output power of 175 nW.

1. Introduction

Versatile phenomena induced by spin-torque, which originate from the spin angular momentum transfer from conduction electron spins to local spins, [1, 2] have attracted much attention from the fundamental and practical points of view. One of the important phenomena is a steady magnetization precession induced by spin-torque [3-5] because it provides us with potential applications such as a nanometer-sized rf oscillator. The spin-torque oscillator has great advantages: frequency tunability by electric current injection and/or magnetic field, and the narrow generation linewidth of oscillation. These advantages have fascinated us and spin-torque oscillators have extensively been studied using current-perpendicular-to-plane giant magnetoresistance devices, [4, 6-9] point-contact devices, [5, 10-12] and magnetic tunnel junctions (MTJs). [13-19] From the viewpoint of the output power, MTJs with an MgO tunnel barrier are a candidate for the spin-torque oscillator since a huge tunnel magnetoresistance (TMR) ratio of an MgO-MTJ leads to the large rf output power.

Deac *et al.* [13] investigated the spin-torque induced rf oscillation in the CoFeB / MgO / CoFeB MTJs under the in-plane magnetic field. Although the large rf output power of 140 nW was achieved, the multiple peak structures of the rf spectra implied the existence of several precessional modes. Wada *et al.* [19] have reported that the application of the magnetic field perpendicular to the plane of the MTJ is an effective to suppress the additional precessional modes since the perpendicular magnetic field leads to a uniform demagnetization field in the nanosized-pillar of the MTJ in comparison with the in-plane magnetic field. However, the output power under the perpendicular magnetic field was reduced to be 26 nW because the the free and pinned layer magnetizations were almost aligned in the perpendicular direction and the precessional mode was out-of-plane, resulting in a small resistance change. A recent paper [20] has reported that the Fe-rich CoFeB free layer possesses a perpendicular magnetic anisotropy. This suggests that an Fe-rich CoFeB free layer may be suitable to increase the relative angle between the free and pinned layer magnetizations under the perpendicular magnetic field, and may be useful to obtain large output power.

In this study, we examined the spin-torque induced rf oscillation in MgO-MTJs with an Fe-rich CoFeB free layer. The magnetic field (H_{ext}) and the dc bias current (I_{dc}) dependences of the rf oscillation were investigated by applying the perpendicular magnetic field. The critical current (I_c) for the spin-torque switching was also evaluated using the current pulse measurement.

2. Experimental procedure

MTJ films were prepared using sputter-deposition. The stacked structure of the film is SiO_2 substrate / buffer layer / PtMn (15 nm) / CoFe (2.5 nm) / Ru (0.85 nm) / CoFeB (3 nm) / MgO (1.2 nm) / CoFeB (2 nm) /capping layer, where the 2-nm-thick CoFeB layer is the free layer and its composition is $Co_{16}Fe_{64}B_{20}$. This composition is Fe-rich compared with the $Co_{60}Fe_{20}B_{20}$ free layer that was used in a previous study. [19] The microfabrication was carried out through the use of electron beam lithography, Ar ion milling, and lift-off technique. The designed size of the junction is 50 nm × 150 nm. The magnetoresistance (MR) measurement was performed at room temperature using a conventional two-terminal technique. The frequency-domain measurement was carried out using a spectrum analyzer. The MTJ device was connected to the circuit with a two-terminal rf probe, and I_{dc} was applied from a dc power source through a bias-Tee. The rf oscillation generates a high frequency voltage output signal due to the time variation of resistance originating from the TMR effect. Output signals amplified by a preamplifier were fed to a spectrum analyzer. All the rf measurements were performed at room temperature. The positive I_{dc} is defined as the direction of the electron flow from the free layer to the pinned one. The background signal at I_{dc} = 0 mA was subtracted from the measured rf spectra.

3. Results and discussion

Figures 1 (a) and (b) show MR curves measured at the in-plane and perpendicular magnetic fields, respectively. The MR curve measured in the in-plane magnetic field shows that the hysteresis appears in the negative H_{ext} region and the parallel magnetization alignment is stable at H_{ext} = 0 Oe because of the orange-peel coupling between the free and pinned layers. The resistance at the parallel state is 139 Ω and the MR ratio is about 70 %. The coercivity of the Fe-rich CoFeB free layer is obtained to be 110 Oe from the in-plane MR curve. Due to the ferromagnetic coupling between the free and pinned

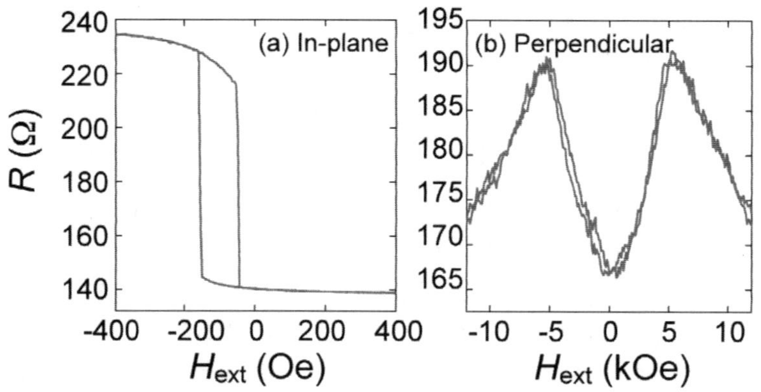

Figure 1. MR curves for MTJs with a Fe-rich CoFeB free layer. The external magnetic fields were applied in (a) the in-plane and (b) the perpendicular direction to the plane of the MTJs.

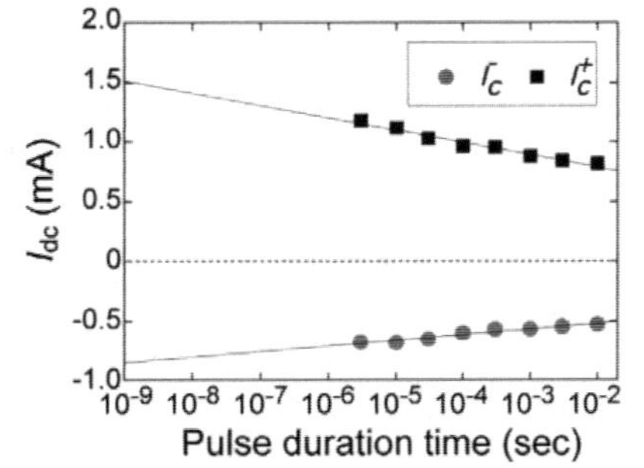

Figure 2. Critical current for spin-torque switching (I_c) as a function of the pulse duration time. H_{ext} = -100 Oe was applied in the in-plane direction. Solid squares denote the results for the switching from the parallel to the antiparallel state (I_c^+), and solid circles denote those from the antiparallel to the parallel state (I_c^-).

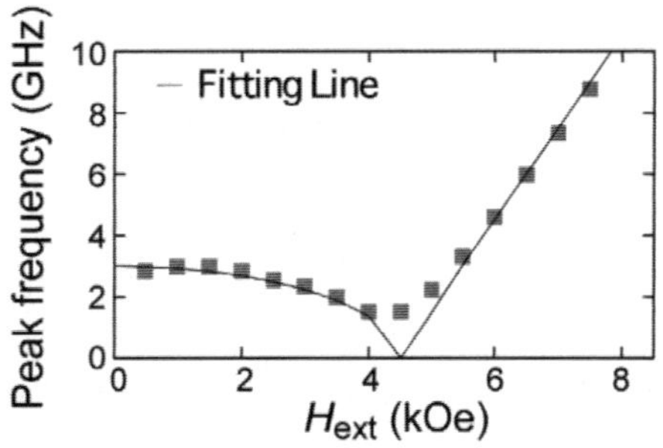

Figure 3. Peak frequency of the rf spectrum as a function of H_{ext}. The magnetic field was applied perpendicular to the plane of the MTJ. I_{dc} was set to 0.3 mA. The solid line is the fitting results.

layers, the resistance of the sample gradually increases as the perpendicular magnetic field increases to 5 kOe. This increment of the resistance is mainly due to the rotation of the free layer magnetization from the in-plane direction to the perpendicular direction. Further increase of the perpendicular magnetic field leads to the decrease of the resistance, which results from the rotation of the pinned layer magnetization.

In order to evaluate the value of I_c for the Fe-rich CoFeB free layer, the spin-torque induced switching measurement was performed using the current pulse sequence. Figure 2 shows the values of I_c as a function of the pulse duration time, where H_{ext} = -100 Oe was applied in the in-plane direction. Solid squares denote the results for the switching from the parallel to the antiparallel state (I_c^+), and solid circles denote those for the switching from the antiparallel to the parallel state (I_c^-). From the linear fitting to the data, I_{c0}^+ and I_{c0}^- were evaluated to be 1.51 mA and -0.85 mA, respectively. The average critical current (I_{c0}) was obtained to be 1.18mA, which corresponds to the critical current density (J_{c0}) of 9.1×10^6 A/cm^2. This J_{c0} is lower than $J_{c0} = 16.1 \times 10^6$ A/cm^2 for the $Co_{60}Fe_{20}B_{20}$ free layer used in a previous study. [19]

Figure 3 shows the peak frequency of the rf spectrum as a function of H_{ext}. The magnetic field was applied perpendicular to the plane of the MTJ. I_{dc} was set to 0.3 mA, which is much lower than I_c. Since I_{dc} = 0.3 mA is small, the peaks in the rf spectra are due to the thermally-excited ferromagnetic resonance mode. The peak frequency takes minimum between 4 kOe < H_{ext} < 5 kOe. From the fitting result (a solid line), which was performed based on the modified Kittel's equation, [19, 20] the demagnetizing field (H_d) of the Fe-rich CoFeB free layers is obtained to be 4.5 kOe. This value of H_d is low compared with the case of the $Co_{60}Fe_{20}B_{20}$ free layer showing H_d = 11 kOe, [19] indicating the existence of the perpendicular magnetic anisotropy in the Fe-rich CoFeB free layers. The value of saturation magnetization (M_s) is not measured in this study. According to a previous study [20], M_s for the present Fe-rich CoFeB free layer is considered to be about 1200 emu/cm^3.

The rf spectra under the application of various I_{dc} are shown in Fig. 4. H_{ext} = 4.5 kOe was applied in the perpendicular direction to the plane of the MTJ. A small peak was observed at I_{dc} = 0.9 mA. With increasing I_{dc}, the peak intensity gradually increases together with the broadening of the peak shape. The frequency jump also occurs at I_{dc} = 1.3 mA, and another peak appears around 4 GHz. The peak intensity around 4 GHz shows the remarkable increase for I_{dc} = 1.5 mA. Figure 5 summarizes (a) the linewidth of the peak, (b) the peak frequency, and (c) the output power as a function of I_{dc} at H_{ext} = 4.5 kOe. The dashed line represents I_c evaluated from the spin-torque induced

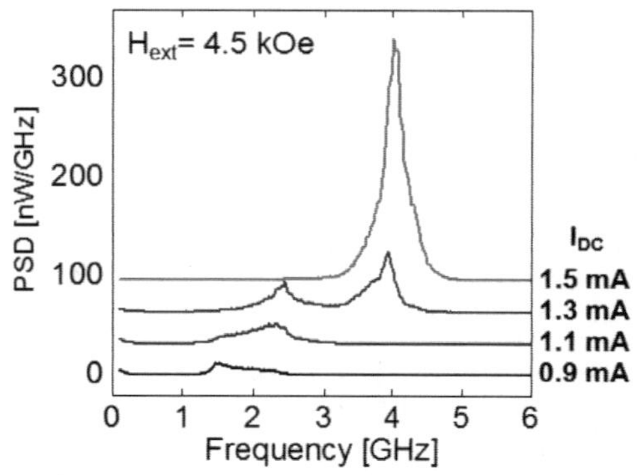

Figure 4. rf spectra for the various I_{dc} under the perpendicular magnetic field. H_{ext} was fixed at 4.5 kOe.

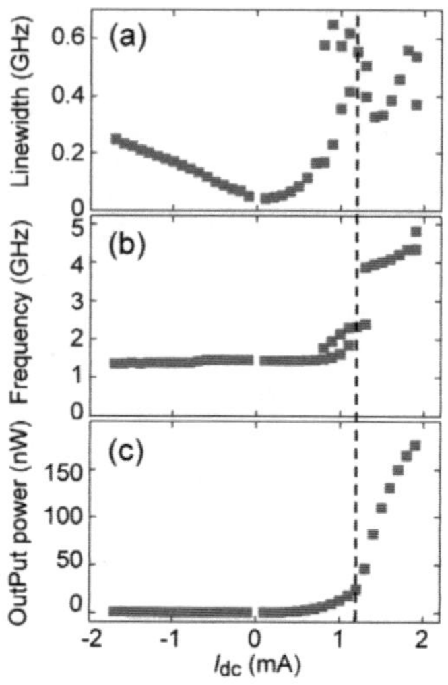

Figure 5. I_{dc} dependence of (a) the linewidth, (b) the peak frequency, and (c) the output power under the perpendicular magnetic field. H_{ext} was fixed at 4.5 kOe.

switching. At I_{dc} around the value of I_c, the linewidth drastically changes, and the peak broadening occurs. This behavior is similar to the previous results. [19] According to a theory for the nonlinear auto-oscillator, [21] the line shape distortion occurs around I_c in the case that the oscillator exhibits the strong nonlinearity. The observed peak broadening indicates the onset of the spin-torque induced rf oscillation with strong nonlinearity. The peak frequency shows discontinuous jumps and the blueshift at $I_{dc} > I_c$, and the output power significantly increases. The blueshift of the peak frequency means that the magnetization precession is the out-of-plane mode. An important point is that the output power reaches up to 175 nW at I_{dc} = 1.9 mA, which is much larger than the previous output power in the $Co_{60}Fe_{20}B_{20}$ free layer under the perpendicular magnetic field. [19] After the calibration of the rf transmission properties of the MTJ and the power loss in the measurement circuit, the output power from the MTJ is evaluated to be 568 nW.

One of the reasons for the enhancement of the output power is the reduction of I_c for the onset of the rf oscillation. The low I_c enables us to apply the large I_{dc} after the onset of the rf oscillation. Another reason is the increase of the relative angle between the free and pinned layer magnetizations. From the MR curve, the relative angle is estimated to be 47.4° at H_{ext} = 4.5 kOe. Considering that the free and pinned layer magnetizations are almost aligned in the parallel configuration in a previous study using the $Co_{60}Fe_{20}B_{20}$ free layer, [19] the relative angle of the present case increases owing to the perpendicular magnetic anisotropy of the Fe-rich CoFeB free layer. Consequently, the increase of the relative angle and the out-of-plane precessional motion of the magnetization lead to the enhancement of the output power. Although the perpendicular magnetic field is useful to achieve the single peak structure in the spectra, the generation linewidths of the spectra are still large (~400MHz). The reason is not clear at present. However, we achieved both the large output power and the spectral shape with a single peak structure by using the perpendicular magnetic field and the MTJ with the Fe-rich CoFeB free layer, which has not been achieved in previous studies [13, 19]. Therefore, the perpendicular magnetic anisotropy of the free layer and the perpendicular magnetic field are keys to develop high power spin-torque oscillators.

4. Summary

We obtained the large rf output power in the MTJ with the Fe-rich CoFeB free layer under the perpendicular magnetic field. The output power reached up to 175 nW, which is mainly attributable to the increase of the relative angle and the out-of-plane precessional mode. The present results clearly indicate that the Fe-rich CoFeB layer with the perpendicular magnetic anisotropy is suitable for the free layer of the spin-torque induced rf oscillation in order to achieve the large output power.

Acknowledgments

The authors thank Canon ANELVA Corporation for their cooperation in the device preparation. This work was partly supported by SCOPE (051407005) from MIC and a Grant-in-Aid for Scientific Research in priority area from MEXT.

References

[1] Slonczewski J C 1996 *J. Magn. Magn. Mater.* **159** L1
[2] Berger L 1996 *Phys. Rev. B* **54** 9353
[3] Tsoi M, Jansen A G M, Bass J, Chiang W-C, Seck M, Tsoi V and Wyder P 1998 *Phys. Rev. Lett.* **80** 4281
[4] Kiselev S I, Sankey J C, Krivorotov I N, Emley N C, Schoelkopf R J, Buhrman R A, and Ralph D C 2003 *Nature* **425** 380
[5] Rippard W H, Pufall M R, Kaka S, Russek S E and Silva T J 2004 *Phys. Rev. Lett.* **92** 027201
[6] Kiselev S I, Sankey J C, Krivorotov I N, Emley N C, Rinkoski M, Perez C, Buhrman R A and Ralph D C 2004 *Phys. Rev. Lett.* **93** 036601
[7] Mistral Q, Kim J V, Devolder T, Crozat P and Chappert C 2006 *Appl. Phys. Lett.* **88** 192507
[8] Boulle O, Cros V, Grollier J, Pereira L G, Deranlot C, Petroff F, Faini G, Barnaś J and Fert A 2007 *Nature Phys.* **3** 492
[9] Seki T, Tomita H, Shiraishi M, Shinjo T and Suzuki Y 2010 *Appl. Phys. Exp.* **3** 033001
[10] Rippard W H, Pufall M R, Kaka S, Silva T J and Russek S E 2004 *Phys. Rev. B* **70** 100406
[11] Kaka S, Rippard W H, Pufall M R, Russek S E, Silva T J and Katine J A 2005 *Nature* **437** 389
[12] Mancoff F B, Rizzo N D, Engel B N and Tehrani S 2005 *Nature* **437** 393
[13] Deac A M, Fukushima A, Kubota H, Maehara H, Suzuki Y, Yuasa S, Nagamine Y, Tsunekawa K, Djayaprawira D D and Watanabe N 2008 *Nature Phys.* **4** 803
[14] A. V. Nazarov, K. Nikolaev, Z. Gao, H. Cho, and D. Song, *J. Appl. Phys.* **103**, 07A503 (2008).
[15] Houssameddine D, Florez S H, Katine J A, Michel J-P, Ebels U, Mauri D, Ozatay O, Delaet B, Viala B, Folks L, Terris B D and Cyrille M-C 2008 *Appl. Phys. Lett.* **93** 022505
[16] Kudo K, Nagasawa T, Sato R and Mizushima K 2009 *Appl. Phys. Lett.* **95** 022507
[17] Cornelissen S, Bianchini L, Hrkac G, Op de Beeck M, Lagae L, Kim J V, Devolder T, Crozat P, Chappert C and Schrefl T 2009 *Euro. Phys. Lett.* **87** 57001
[18] Matsumoto R, Fukushima A, Yakushiji K, Yakata S, Nagahama T, Kubota H, Katayama T, Suzuki Y, Ando K, Yuasa S, Georges B, Cros V, Grollier J and Fert A 2009 *Phys. Rev. B* **80** 174405
[19] Wada T, Yamane T, Seki T, Nozaki T, Suzuki Y, Kubota H, Fukushima A, Yuasa S, Maehara H, Nagamine Y, Tsunekawa K, Djayaprawira D D and Watanabe N 2010 *Phys. Rev. B* **81** 104410
[20] Yakata S, Kubota H, Suzuki Y, Yakushiji K, Fukushima A, Yuasa S and Ando K 2009 *J. Appl. Phys.* **105** 07D131
[21] Kim J V, Mistral Q, Chappert C, Tiberkevich V and Slavin A 2008 *Phys. Rev. Lett.* **100** 167201

Evaluation of Magnetization Interaction at the Interface between FeCo and TbFeCo for p-MTJ

N. Miyamoto, K. Mamiya and S. Nakagawa
Department of Physical Electronics, Tokyo Institute of Technology,
2-12-1 O-okayama, Meguro, Tokyo, 152-8552, Japan

E-mail: nakagawa@pe.titech.ac.jp

Abstract. Magnetic interaction at the interface between FeCo and TbFeCo were investigated to improve perpendicular magnetic tunnel junction. Magnetic properties of TbFeCo/thin and thick FeCo were observed by anomalous Hall effect (AHE). A perpendicular component of magnetization of thin FeCo layer adjacent to TbFeCo layer was larger than that of thick FeCo one. Kerr hysteresis observation clarified the magnetization behavior of the region of the TbFeCo layer adjacent to FeCo layer using the characteristic of a penetration depth of the probe light. The TbFeCo layer attached with thinner FeCo layer was saturated at lower applied field than that with thicker one. This suggested that the magnetization of FeCo layer affected the TbFeCo magnetization and the thicker FeCo layer tilts the TbFeCo magnetization more greatly than thinner one.

1. Introduction

Perpendicular magnetic tunnel junctions (p-MTJs) are promising candidates for magnetoresistive random access memory (MRAM) with high integration and low power consumption [1][2]. There are various perpendicular magnetization films such as $L1_0$-FePt(001) ordering thin films and Co/Pt multilayers [3][4][5], however we have studied p-MTJs with Rare Earth-Transition Metal (RE-TM) alloy films, and obtained 64% tunnel magnetoresistance (TMR) ratio at room temperature [6]. In this p-MTJ structure, highly magnetically polarized layer, such as FeCo layer, should be inserted at the interface of the tunnel barrier in MTJ structure in order to obtain high TMR ratio. In addition, the FeCo polarized layer should be magnetized perpendicularly to the interface plane by exchange coupling with RE-TM alloy films for practical application of this p-MTJ. Therefore, it is important to evaluate coupling of the magnetizations at the interface between FeCo and RE-TM alloy films. In this study, such magnetic interactions between FeCo and RE-TM alloy films were not only clarified by anomalous Hall effect (AHE) and Kerr effect, but also the useful method of measuring the magnetization state by Kerr effect was reported.

2. Experimental

All specimen films were prepared on glass substrates at ambient temperature by ion beam sputtering. $Fe_{70}Co_{30}$, $Tb_{26}(Fe_{90}Co_{10})_{74}$ and Ta were used as targets in this study. Ta layers were prepared to avoid oxidation of specimens and adjust intrusion length of Kerr probe light as discussed later. Magnetic properties were observed by anomalous Hall effect (AHE) and polar Kerr effect. The AHE output voltage V_{AHE} which is proportional to the perpendicular magnetization component of a film is given by,

Journal of Physics: Conference Series **266** (2011) 012099 doi:10.1088/1742-6596/266/1/012099

$$V_{AHE} = \left(\frac{\mu_0 R_S I}{t}\right) M \cos\theta \qquad (1)$$

In the above equation, μ_0, R_s, t and I are permeability of vaccuum, Hall resistivity, film thickness and current, respectively. M and θ are the magnetization and its angle from the normal of the film plane, respectively [7].

3. Results and discussions

3.1. Magnetization states observed by Hall effect

Figure 1 shows the magnetic properties of TbFeCo(50 nm)/FeCo(X nm)/Ta(15 nm) film, where the magnetic field was applied perpendicularly to the film plane with different FeCo layer thickness X. The specimen with 0.6 nm thickness FeCo is saturated at lower applied field than that with 2 nm thickness one. This indicates that a tilt of magnetization in the thinner FeCo layer is less than that of magnetization in the thicker FeCo one as shown in Figure 2. Since an exchange coupling originates just at the interface between FeCo and TbFeCo layers, the upper part of the thick FeCo layer, where the region is far from the interface, is regarded that the magnetization in the region is more tilted than that in the bottom part of the region near the interface between the two layers as shown in Figure 2.

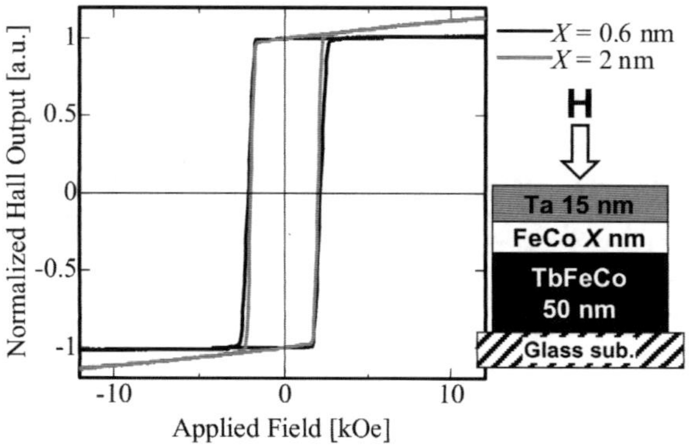

Figure 1. AHE hysteresis loops of TbFeCo(50 nm)/FeCo(X nm)Ta(15 nm) by AHE.

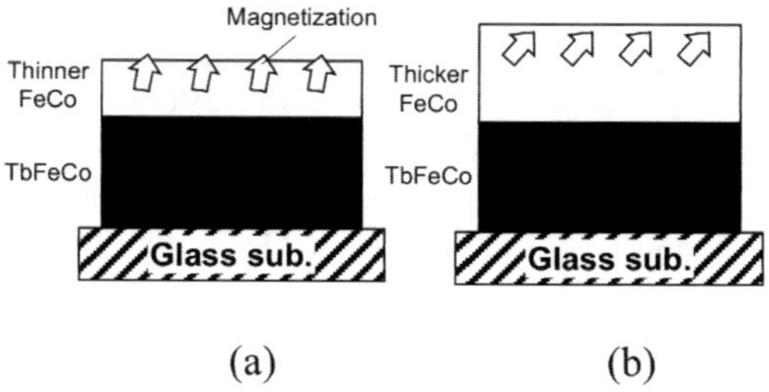

Figure 2. Schematic images of magnetization state of the specimen with (a)thinner FeCo layer (b) thicker FeCo one.

3.2. Determinaton of the penetration length of Kerr probe light

The magnetization state of the interface between FeCo and TbFeCo cannot be fully understood by the observation of AHE loops, thus Kerr hysteresis analysis was chosen to clarify the magnetization of TbFeCo layer at the region near the FeCo layers. Magnetic properties where probe light reaches can be observed by Kerr effect, therefore penetration length of the probe light should be estimated. Figure 3 shows magnetic properties of TbFeCo(50 nm)/Ta(X nm) film with various layer thickness X of Ta by Kerr observation. In the case of X = 40 and 42 nm, typical hysteresis loops of perpendicularly magnetized films can be obtained from the TbFeCo layer beneath the Ta layers. However, these hysteresis loops ware not shown and just a loop derived from paramagnetic was observed when the thickness of Ta was 45 nm. This indicates that Kerr probe light can pass through 42 nm in Ta layer and reach TbFeCo layer. This means that the probe light cannot pass through more than 42 nm of Ta layer. Therefore intrusion length of Kerr probe light is estimated as 42 nm in this study.

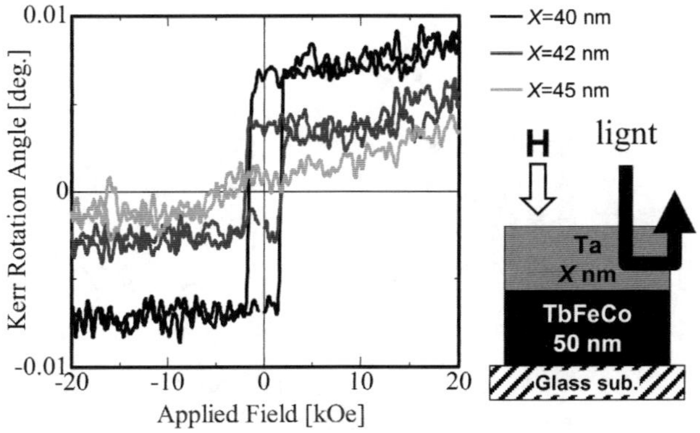

Figure 3. Kerr hysteresis loops of TbFeCo(50 nm)/Ta(*X* nm) films with different Ta layer thickness.

3.3. Observation of magnetization states at the interface between FeCo and TbFeCo

In order to evaluate magnetization states of the interface between FeCo and TbFeCo by Kerr hysteresis loops, FeCo(*X* nm)/TbFeCo(30 nm)/Ta(15 nm) films with various FeCo film thickness were prepared. Here the deposition order of TbFeCo and FeCo was changed in order to observe mainly TbFeCo magnetic characteristics. Supposing that the intrusion length of the Kerr probe light is 42 nm, Figure 4 shows magnetic properties of TbFeCo around the interface between FeCo and TbFeCo of FeCo(X nm)/TbFeCo(30 nm)Ta(15 nm) films. The difference of magnetization characteristics of the two hysteresis loops is regarded as the difference of the magnetization behaviour of the region of the TbFeCo layer near the interface between FeCo and TbFeCo, since the penetration of the probe light detects only the region of the TbFeCo layer. Outputs of each loop increase as increasing applied field, since magnetization of TbFeCo are tilted. However the specimen with thinner FeCo layer was saturated at lower applied field of 10 kOe than that with thicker one. Figure 5 shows the change of $d\theta_K/dH$ as a function of applied field *H*. The decrease of $d\theta_K/dH$ can be recognized around 10 kOe in the specimen with 0.6 nm thickness FeCo layer. This indicates that the magnetization of FeCo layer affects the magnetization of TbFeCo at the interface between FeCo and TbFeCo as shown in Figure 6. Thicker FeCo layer tilts the magnetization of TbFeCo more greatly than the thinner one.

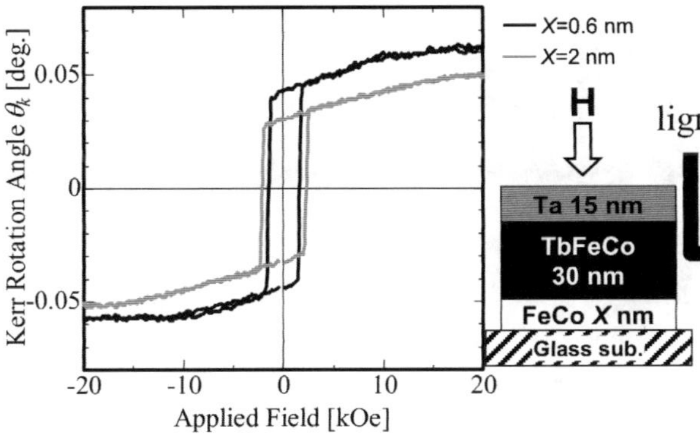

Figure 4. Kerr hysteresis loops of FeCo(X nm)/TbFeCo(30 nm)Ta(15 nm) films with different FeCo layer thickness.

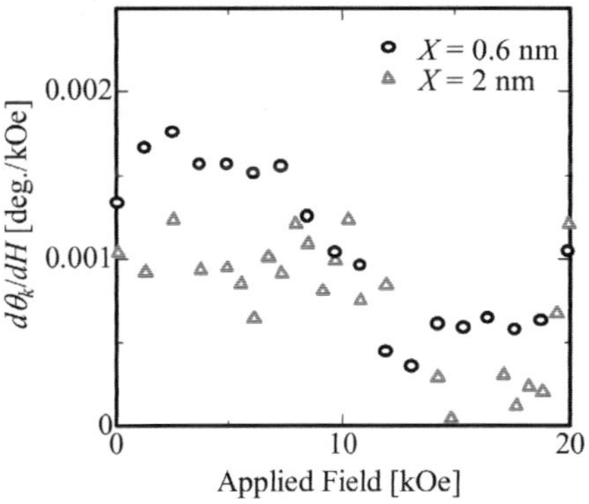

Figure 5. $d\theta_K/dH$ of FeCo(X nm)/TbFeCo(30 nm)Ta(15 nm) films observed by Kerr effect.

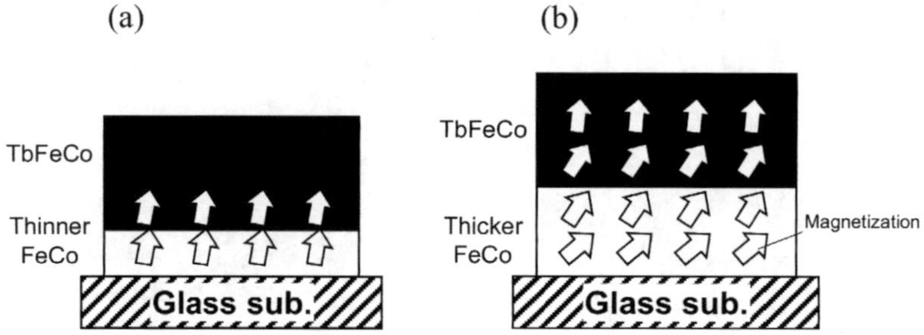

Figure 6. Schematic images of magnetization states of the interface of the specimen with (a) thinner FeCo layer (b) thicker FeCo one.

4. Conclusion

Magnetization states of the region around the interface between FeCo and TbFeCo layers in the FeCo/TbFeCo bilayers were investigated. Magnetic properties of TbFeCo/thinner and thicker FeCo/Ta were observed by AHE. It was found that a tilt of magnetization of thinner FeCo layer is less than that of magnetization of thicker FeCo one. Kerr hysteresis observation clarified that the specimen with thinner FeCo layer can be saturated at lower applied field than that with thicker one. This suggested that FeCo layer affected the TbFeCo magnetization at the interface between FeCo and TbFeCo and thicker FeCo layer tilted the TbFeCo magnetization more greatly than thinner one.

5. References

[1] N. Nishimura, T. Hirai, A. Koganei, T. Ikeda, K. Okano, Y. Sekiguchi and Y. Osada, *J. Appl. Phys.* **91**(8), 5246 (2002)
[2] S. Mangin, D. Ravelosona, J. A. Katine, M. J. Carey, B. D. Terris and E. E. Fullerton, *Nature Mater.* **5**, 210-215 (2006)
[3] M. Yoshikawa, E. Kitagawa, T. Nagase, T. Daibou, M. Nagamine, K. Nishiyama, T. Kishi and H. Yoda, *IEEE Trans. Magn.* **44**(11), 2573 (2008)
[4] C. Ducruet, B. Carvello, B. Rodmacq, S. Auffret, G. Gaudin and B. Dieny, *J. Appl. Phys.* **103**, 07A918 (2008)
[5] J.-H. Park, C. Park, T. Jeong, M. T. Moneck, N. T. Nufer and J.-G. Zhu, *J. Appl. Phys.* **103**, 07A917 (2008)
[6] H. Ohmori, T. Hatori, and S. Nakagawa, *J. Appl. Phys.* **103**, 07A911 (2008)
[7] E. M. Pugh, N. Rostoker, *Rev. of modern phys.* **25**, 151, (1953)

Journal of Physics: Conference Series **266** (2011) 012100 doi:10.1088/1742-6596/266/1/012100

Inverse spin-Hall effect induced by spin pumping in different size $Ni_{81}Fe_{19}$/Pt films

H Nakayama[1], **K Ando**[1], **K Harii**[1], **Y Fujikawa**[1], **Y Kajiwara**[1], **T Yoshino**[1] **and E Saitoh**[1,2,3]

[1] Institute for Materials Research, Tohoku University, Sendai 980-8577, Japan
[2] The Advanced Science Research Center, Japan Atomic Energy Agency, Tokai, Ibaraki 319-1195, Japan
[3] PRESTO, Japan Science and Technology Agency, Sanbancho, Tokyo 102-0075, Japan

E-mail: nakayama@imr.tohoku.ac.jp

Abstract. The inverse spin-Hall effect, conversion of spin currents into electric voltage, has been investigated in $Ni_{81}Fe_{19}$/Pt bilayer films with different length and width. We measured the electric voltage generated by the inverse spin-Hall effect induced by the spin pumping at room temperature. The experimental results indicate that the electric voltage induced by the inverse spin-Hall effect is proportional to the length of $Ni_{81}Fe_{19}$/Pt bilayer films. In contrast, the electric voltage exhibits almost no variation by changing the width of the $Ni_{81}Fe_{19}$/Pt films. These results are consistent with the inverse spin-Hall effect model.

1. Introduction

There has been rapidly growing interest in the field of spintronics, which aims to extend the scope of conventional electronics using the spin degree of freedom of electrons in solid-state systems [1-3]. In this field, generation and detection of a spin current, a flow of electron spins in a solid, are the essential technique that realizes efficient spin-based magnetic memories and computing devices. In this stream, intense theoretical and experimental interests have been focused on the direct and inverse spin-Hall effects (SHE and ISHE). The SHE and the ISHE are responsible for the reciprocal conversion between a spin current and a charge current via the spin-orbit interaction, and thus will be essential in combining spin-current technology with conventional electronics technology based on a charge current.

The ISHE, which converts a spin current into electric voltage, was observed using the non-local method [4-7] and the spin pumping [8-20]. The spin pumping refers to the generation of spin currents from magnetization dynamics; a precessing magnetization in a ferromagnetic metal induces a spin current in a paramagnetic metal attached to the ferromagnetic metal. The spin pumping method requires only a simple ferromagnetic/paramagnetic bilayer system, making it a key technique for generating and detecting spin currents in a wide range of sample system. Thus, the ISHE induced by the spin pumping should be systematized for the development of spin-based devices. In this article, we report the film size dependence of the ISHE induced by the spin pumping in $Ni_{81}Fe_{19}$/Pt bilayer films.

2. Experimental procedure

Figure 1 (a) shows a schematic illustration of the sample used in the present study. The samples are $Ni_{81}Fe_{19}$/Pt films comprising a 10 nm-thick ferromagnetic $Ni_{81}Fe_{19}$ layer and a 10 nm-thick paramagnetic Pt layer. The paramagnetic Pt layer was sputtered on a thermally oxidized Si substrate and then the ferromagnetic $Ni_{81}Fe_{19}$ layer was evaporated in a high vacuum on the Pt layer. The size of the $Ni_{81}Fe_{19}$/Pt bilayer films are $w = 0.6$ with changing the length $l = 0.6$, 0.9, 1.2, 1.6 mm, and $l = 1.2$ mm with changing the width $w = 0.2$, 0.4, 0.6, 0.8 mm, where the length of the Pt layer is $l_{Pt} = 3.2$ mm for all the samples [see figure 1 (a)]. Two electrodes are attached to the Pt layer.

For the measurement, the sample system is placed near the center of the TE_{011} microwave cavity at which the magnetic-field component of the mode is maximized, while the electric-field component is minimized. During the measurement, the TE_{011} microwave mode with frequency $f = 9.43$ GHz is excited in the cavity, and an external magnetic field $\boldsymbol{H}$ along the film plane is applied perpendicular to the direction across the electrodes. Since the magnetocrystalline anisotropy in $Ni_{81}Fe_{19}$ is negligibly small, the magnetization in the $Ni_{81}Fe_{19}$ layer is uniformly aligned along the external magnetic field direction $\boldsymbol{H}$. When $\boldsymbol{H}$ and f fulfill the ferromagnetic resonance (FMR) condition, a dc pure spin current with the spin polarization $\boldsymbol{\sigma}$, which is parallel to the external-field direction $\boldsymbol{H}$, is resonantly injected into the Pt layer by the spin pumping. This injected spin current $\boldsymbol{J}_S$ is converted into the electric field $\boldsymbol{E}_{ISHE}$ by the ISHE, which is expressed in a form as $\boldsymbol{E}_{ISHE} \parallel \boldsymbol{J}_S \times \boldsymbol{\sigma}$, giving rise to an electric potential difference between the edges of the Pt layer [8-16]. We measured the FMR signal and electric voltage between the electrodes attached to the Pt layer. All the measurements were performed at room temperature.

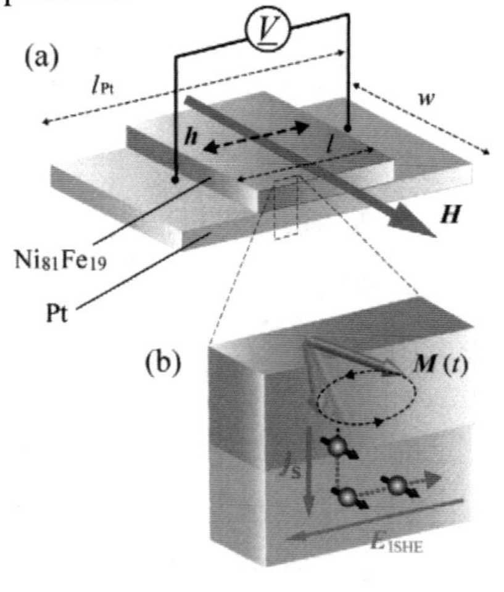

Figure 1. (a) A schematic illustration of the metallic bilayer system used in the present study. $\boldsymbol{H}$, $\boldsymbol{h}$, l_{Pt}, l and w denote the external magnetic field, microwave magnetic field, length of the Pt film, length and width of the $Ni_{81}Fe_{19}$/Pt bilayer film, respectively. (b) A schematic illustration of the spin pumping and the inverse spin-Hall effect (ISHE) in the present system. $\boldsymbol{M}(t)$, $\boldsymbol{j}_S$, and $\boldsymbol{E}_{ISHE}$ denote the precessing magnetization in the ferromagnetic layer, the spin current generated by the spin pumping, the electric field induced by the ISHE, respectively. The red dotted arrow in the paramagnetic layer describes electron motion bent by the spin-orbit interaction in the paramagnetic layer, a motion responsible for ISHE.

3. Results and discussion

Figure 2 (a) shows the FMR spectrum measured for the $Ni_{81}Fe_{19}$/Pt (0.6 mm × 1.6 mm) bilayer film under the 200 mW microwave excitation. Here, I and H_{FMR} denote the microwave absorption intensity and the resonance field, respectively. The FMR spectrum indicates that the magnetization in the $Ni_{81}Fe_{19}$ layer resonates with the applied microwave at $H_{FMR} = 115.8$ mT.

Figure 2 (b) shows the measured voltage V for the $Ni_{81}Fe_{19}$/Pt (0.6 mm × 1.6 mm) bilayer film under the 200 mW microwave excitation. In the V spectrum, an unconventional signal appears around H_{FMR}; evidently, the resonance signal in the V spectrum originates from FMR in the $Ni_{81}Fe_{19}$ layer. This electric voltage is observed for all the samples. Notable is that the shape of the V spectrum is well reproduced using a simple Lorentz function, indicating that the electric voltage observed here is dominated mainly by ISHE induced by the spin pumping.

Figure 3 shows the microwave power dependence of the electric voltage V_{ISHE}, where V_{ISHE} is defined as the height of the V spectrum [see figure 2 (b)], measured for the $Ni_{81}Fe_{19}$/Pt (0.6 mm × 1.6 mm) bilayer films. V_{ISHE} is proportional to the microwave power P_{MW} for all the samples, being consistent with the prediction of ISHE induced by the dc spin pumping [8-15].

In figure 4 (a), we show the $Ni_{81}Fe_{19}$/Pt bilayer film length dependence of the electric voltage induced by the ISHE V_{ISHE}. The magnitude of V_{ISHE} is proportional to the length of the $Ni_{81}Fe_{19}$/Pt bilayer film, which is expected from the series arrangement of ISHE voltage sources. Figure 4 (b) shows the $Ni_{81}Fe_{19}$/Pt bilayer film width dependence of V_{ISHE} measured for the $Ni_{81}Fe_{19}$/Pt bilayer film under the 200 mW microwave excitation. The magnitude of V_{ISHE} exhibits almost no variation by changing the width of the $Ni_{81}Fe_{19}$/Pt bilayer film because ISHE voltage sources are arranged in parallel. These results are consistent with the equivalent circuit model [8] using the relation $I_C = \lambda w i_C$, indicating that the applied microwave is uniform. Here, I_C and λ denote the magnitude of electric currents induced by the ISHE and the spin diffusion length of Pt, respectively.

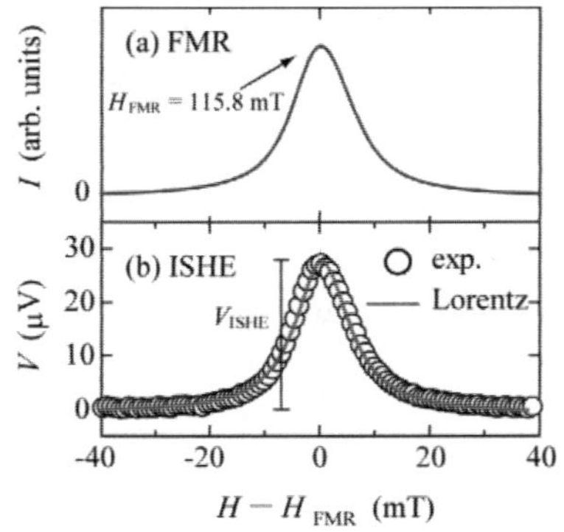

Figure 2. (a) External magnetic field H dependence of the FMR signal measured for the $Ni_{81}Fe_{19}$/Pt bilayer film (0.6 mm × 1.6 mm). Here, I and H_{FMR} denote the microwave absorption intensity and the resonance magnetic field, respectively. (b) H dependence of the eledctric voltage V between the electrodes on the Pt films measured for the $Ni_{81}Fe_{19}$/Pt bilayer film (0.6 mm × 1.6 mm) under the 200 mW microwave excitation. The shape of V is well reproduced by a Lorentz function. The electric voltage induced by the inverse spin-Hall effect V_{ISHE} is estimated as the height of the resonance shape in the V spectrum.

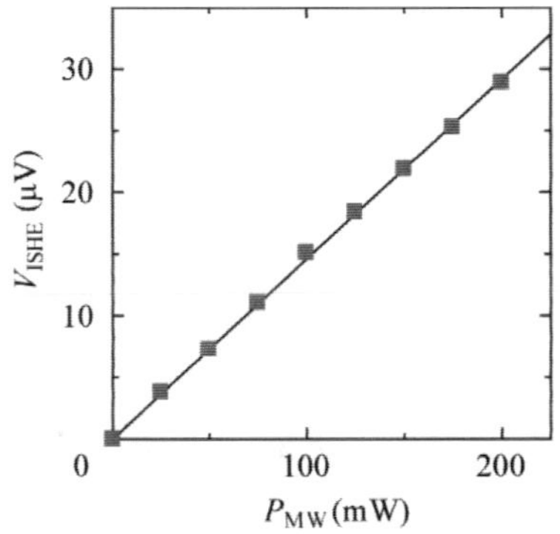

Figure 3. Microwave power P_{MW} dependence of the electric voltage induced by the inverse spin-Hall effect V_{ISHE} measured for the $Ni_{81}Fe_{19}$/Pt bilayer thin film (0.6 mm × 1.6 mm). The solid line shows linear fits to the data: $V_{ISHE} \propto P_{MW}$.

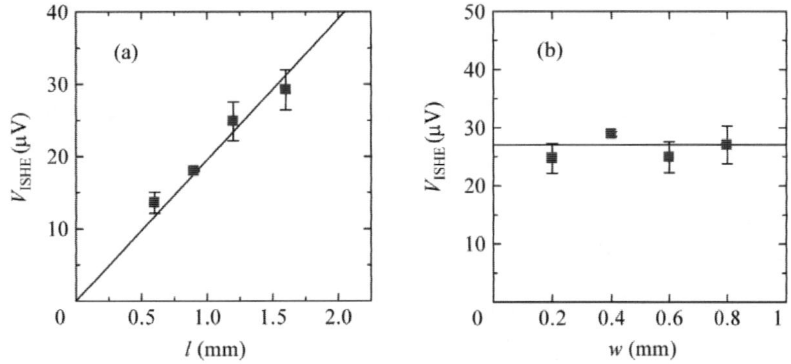

Figure 4. (a) $Ni_{81}Fe_{19}$/Pt bilayer film length dependence of the inverse spin-Hall voltage V_{ISHE} measured for the $Ni_{81}Fe_{19}$/Pt bilayer films (w = 0.6 mm) under the 200 mW microwave excitation. The solid line shows a linear fit to the data. (b) $Ni_{81}Fe_{19}$/Pt bilayer film width dependence of V_{ISHE} measured for the $Ni_{81}Fe_{19}$/Pt bilayer films (l = 1.2 mm) under the 200 mW microwave excitation. The solid line shows $V_{ISHE}^{w=0.8mm}$.

4. Summary

We investigated the inverse spin-Hall effect (ISHE) induced by the spin pumping in $Ni_{81}Fe_{19}$/Pt bilayer films with different length and width. The experimental results indicate that the ISHE voltage V_{ISHE} is proportional to the length of $Ni_{81}Fe_{19}$/Pt bilayer film. In contrast, V_{ISHE} exhibits almost no variation by changing the width of $Ni_{81}Fe_{19}$/Pt bilayer film. These results are consistent with the ISHE model, indicating that the applied microwave is uniform.

Acknowledgments

The authors thank K. M. Itoh and M. Matoba for valuable discussions. This work was supported by a Grant-in-Aid for Scientific Research in Priority Area "Creation and control of spin current" (19048028) from MEXT, Japan, a Grant-in-Aid for Scientific Research (A) from MEXT, Japan, Global COE Program "Materials Integration International Center of Education and Research" from MEXT, Japan, a Grant for Industrial Technology Research from NEDO, Japan, and Fundamental Research Grant from TRF, Japan.

References

[1] Maekawa S 2006 *Concepts in Spin Electronics* (Oxford Univ. Press, Oxford)
[2] Zutic I, Fabian J and Sarma S D 2004 *Rev. Mod. Phys.* **76** 323-410
[3] Wolf S A, Awschalom D D, Buhrman R A, Daughton J M, von Molnar S, Roukes M L, Chtchelkanova A Y and Treger D M 2001 *Science* **294** 1488
[4] Seki T, Hasegawa Y, Mitani S, Takahashi S, Imamura H, Maekawa S, Nitta J and Takanashi K 2008 *Nat. Mater.* **7** 125
[5] Kimura T, Otani Y, Sato T, Takahashi S and Maekawa S 2007 *Phys. Rev. Lett.* **98** 156601
[6] Valenzuela S O and Tinkham M 2006 *Nature* **442** 176
[7] Jedema F J, Filip A T and van Wees B J 2001 *Nature* **410** 345
[8] Nakayama H, Ando K, Harii K, Kajiwara Y, Yoshino T, Uchida K and Saitoh E 2010 *IEEE Trans. Magn.* **46** 2202
[9] Nakayama H, Ando K, Harii K, Kajiwara Y, Yoshino T, Uchida K, Ota T and Saitoh E 2010 *J. Phys.: Conf. Ser.* **200** 062014
[10] Yoshino T, Kajiwara Y, Ando K, Nakayama H, Ota T, Uchida K and Saitoh E 2010 *J. Phys.: Conf. Ser.* **200** 062038
[11] Nakayama H, Ando K, Harii K, Kajiwara Y, Yoshino T and Saitoh E 2010 *J. Magn. Soc. Jpn.* **34** 337-341
[12] Ando K, Yoshino T and Saitoh E 2009 *Appl. Phys. Lett.* **94** 152509
[13] Harii K, Ando K, Inoue H Y, Sasage K, Saitoh E 2008 *J. Appl. Phys.* **103** 07F311
[14] Ando K, Kajiwara Y, Takahashi S, Maekawa S, Takemoto K, Takatsu M and Saitoh E 2008 *Phys. Rev.* B **78** 014413
[15] Inoue H Y, Harii K, Ando K, Sasage K and Saitoh E 2007 *J. Appl. Phys.* **102** 083915
[16] Saitoh E, Ueda M, Miyajima H and Tatara G 2006 *Appl. Phys. Lett.* **88** 182509
[17] Mosendz O, Pearson J E, Fradin F Y, Bauer G E W, Bader S D and Hoffmann A 2010 *Phys. Rev. Lett.* **104** 046601
[18] Tserkovnyak Y, Brataas A and Bauer G E W 2002 *Phys. Rev. Lett.* **88** 117601
[19] Mizukami S, Ando Y and Miyazaki T 2002 *Phys. Rev.* B **66** 104413
[20] Silsbee R H, Janossy A and Monod P 1979 *Phys. Rev.* B **19** 4382

Journal of Physics: Conference Series **266** (2011) 012101 doi:10.1088/1742-6596/266/1/012101

Spin pumping and magnetization-precession trajectory in thin film systems

T. Nonaka[1], K Ando[1], T. Yoshino[1] and E Saitoh[1,2,3]
[1]Institute for Materials Research, Tohoku University, Sendai 980-8577, Japan
[2]The Advanced Science Research Center, Japan Atomic Energy Agency, Tokai 319-1195, Japan
[3]PRESTO, Japan Science and Technology Agency, Sanbancho, Tokyo 102-0075, Japan
E-mail: ando@imr.tohoku.ac.jp

Abstract. The spin pumping, generation of spin currents from magnetization precession, has been investigated in terms of the trajectory of magnetization precession in thin film systems. By using the Landau-Lifshitz-Gilbert equation combined with the model of the spin pumping, we found that the magnitude of the spin current generated by the spin pumping is determined by the elliptical orbit area of magnetization precession, which is maximized when the external magnetic field is applied oblique to the film plane.

1. Introduction

The generation, manipulation, and detection of a spin current, a ow of electron spins, are the main challenges in the field of spintronics, which involves the study of active control and manipulation of the spin degree of freedom in solid-state systems [1, 2, 3]. A spin current interacts with magnetization by exchanging the spin-angular momentum, enabling the direct manipulation of magnetization without using magnetic fields [4, 5].

The interaction between spin currents and magnetization provides also a method for spin-current generation from magnetization precession, which is the spin pumping [6, 7]. The spin pumping is a method for inducing spin currents in a paramagnetic metal attached to a ferromagnetic metal; the precessing magnetization in the ferromagnetic metal pumps a spin current into the paramagnetic metal. The spin pumping was observed electrically in a simple $Ni_{81}Fe_{19}$/Pt film using the strong inverse spin-Hall effect (ISHE) in the Pt layer; in the $Ni_{81}Fe_{19}$/Pt film, when the external magnetic field and the frequency of the applied microwave fulfill the ferromagnetic resonance (FMR) condition, a spin current is injected into the Pt layer by the spin pumping, giving rise to an electric voltage via the ISHE in the Pt layer [8, 9, 10, 11]. In thin film systems, when the external magnetic field is applied oblique to the film plane, the magnetization-precession trajectory, the trajectory of the point of a magnetization vector is distorted due to a demagnetization field. Since the spin pumping generates spin currents from magnetization precession, the precession trajectory is expected to strongly affect the amplitude of the generated spin currents. In this paper, we investigate the amplitude of the spin current generated by the spin pumping in terms of a magnetization-precession trajectory based on the Landau-Lifshitz-Gilbert (LLG) equation combined with the model of the spin pumping.

2. Phenomenological model

The dynamics of magnetization $\mathbf{M}(t)$ under an effective magnetic field $\mathbf{H}_{\text{eff}}$ is described by the LLG equation,

$$\frac{\mathrm{d}\mathbf{M}(t)}{\mathrm{d}t} = -\gamma \mathbf{M}(t) \times \mathbf{H}_{\text{eff}} + \frac{\alpha}{M_s}\mathbf{M}(t) \times \frac{\mathrm{d}\mathbf{M}(t)}{\mathrm{d}t}. \tag{1}$$

Here, γ, α, and M_s are the gyromagnetic ratio, the Gilbert damping constant, and the saturation magnetization, respectively. We consider the magnetization $\mathbf{M}(t)$ precession around the z axis in a ferromagnetic film attached to a paramagnetic film as shown in figure 1, where $\mathbf{M}(t) = \mathbf{M} + \mathbf{m}(t)$. $\mathbf{M}$ and $\mathbf{m}(t)$ are the static and the dynamic components of the magnetization $\mathbf{M}(t)$, respectively. Here, we consider a soft ferromagnetic thin film, e.g. $\mathrm{Ni_{81}Fe_{19}}$, and we neglect the magnetocrystalline anisotropy. Taking into account the external magnetic field $\mathbf{H}$, the static demagnetizing field $\mathbf{H}_{\mathbf{M}}$ induced by $\mathbf{M}$, the dynamic demagnetization field $\mathbf{H}_{\mathbf{m}}(t)$ induced by $\mathbf{m}(t)$, and the external ac field $\mathbf{h}(t)$ as the effective magnetic field $\mathbf{H}_{\text{eff}}$: $\mathbf{H}_{\text{eff}} = \mathbf{H} + \mathbf{H}_{\mathbf{M}} + \mathbf{H}_{\mathbf{m}}(t) + \mathbf{h}(t)$, and using the ferromagnetic resonance condition [9]:

$$\left(\frac{\omega}{\gamma}\right)^2 = \left[H_{\text{FMR}}\cos({}_H - {}_M) - 4\pi M_s \cos 2\,{}_M\right]\left[H_{\text{FMR}}\cos({}_H - {}_M) - 4\pi M_s \cos^2{}_M\right], \tag{2}$$

we obtain the dynamic components of the magnetization $\mathbf{m}(t)$ at ferromagnetic resonance (FMR) condition as

$$m_x(t) = \frac{4\pi M_s h\gamma\left[2\alpha\omega\cos\omega t + \left(4\pi M_s\gamma\sin^2{}_M + \sqrt{(4\pi M_s)^2\gamma^2\sin^4{}_M + 4\omega^2}\right)\sin\omega t\right]}{8\pi\alpha\omega\sqrt{(4\pi M_s)^2\gamma^2\sin^4{}_M + 4\omega^2}}, \tag{3}$$

$$m_y(t) = -\frac{4\pi M_s h\gamma\cos\omega t}{4\pi\alpha\sqrt{(4\pi M_s)^2\gamma^2\sin^4{}_M + 4\omega^2}}. \tag{4}$$

where ${}_M$ and ${}_H$ are the magnetization and the external magnetic field angle to the normal vector of the film plane, respectively [see figure 1]. Here, $\omega = 2\pi f$, f is the microwave frequency. H_{FMR} is the resonance field.

In the model of the spin pumping [6], the dc component of a spin current density j_s is expressed as

$$j_s = \frac{\omega}{2\pi}\int_0^{2\pi/\omega}\frac{\hbar}{4\pi}g_r^{\uparrow\downarrow}\frac{1}{M_s^2}\left[\mathbf{M}(t)\times\frac{\mathrm{d}\mathbf{M}(t)}{\mathrm{d}t}\right]_z \mathrm{d}t, \tag{5}$$

where $\hbar$ and $g_r^{\uparrow\downarrow}$ are the Dirac constant and the real part of the mixing conductance, respectively. Here, $[\mathbf{M}(t)\times\mathrm{d}\mathbf{M}(t)/\mathrm{d}t]_z$ is the z component of $\mathbf{M}(t)\times\mathrm{d}\mathbf{M}(t)/\mathrm{d}t$. z axis is directed along the magnetization-precession axis [see figure 1]. Using equations (3) and (4), we find the phenomenological expression of j_s as

$$j_s = \frac{g_r^{\uparrow\downarrow}\gamma^2 h^2\hbar\left(4\pi M_s\gamma\sin^2{}_M + \sqrt{(4\pi M_s)^2\gamma^2\sin^4{}_M + 4\omega^2}\right)}{8\pi\alpha^2\left((4\pi M_s)^2\gamma^2\sin^4{}_M + 4\omega^2\right)}. \tag{6}$$

3. Discussion

In order to characterize a magnetization-precession trajectory, we define the ellipticity A of a magnetization-precession trajectory as $A \equiv |m_y|/|m_x|$, where $|m_x|$ and $|m_y|$ are the major

Journal of Physics: Conference Series **266** (2011) 012101 doi:10.1088/1742-6596/266/1/012101

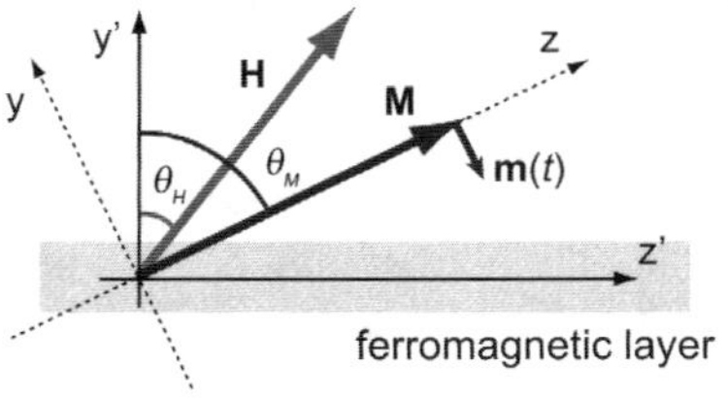

Figure 1. A schematic illustration of the coordinate system used for describing a ferromagnetic film. $\mathbf{M}$ and $\mathbf{m}(t)$ are the static and the dynamic components of the magnetization $\mathbf{M}(t)$. $\mathbf{H}$ is the external magnetic field.

and minor radiuses of the trajectory, respectively. Using equations (3) and (4), we obtain the ellipticity of a magnetization-precession trajectory as

$$A = \frac{2\omega}{4\pi M_s\gamma \sin^2 {}_{M} + \sqrt{(4\pi M_s)^2\gamma^2 \sin^4 {}_{M} + 4\omega^2}}. \tag{7}$$

As described in equation (7), in thin film systems, the magnetization precesses in an elliptical orbit due to the demagnetization field except at ${}_H = {}_M = 0$. When ${}_H = {}_M = 0$, the magnetization precesses in a circular orbit ($A = 1$).

We define the normalized spin current density $\tilde{j}_s$ as $\tilde{j}_s \equiv j_s/j_s^{A=1}$, where $j_s^{A=1}$ is the spin-current density when the magnetization precesses in a circular orbit (the external magnetic field is applied perpendicular film plane: ${}_H = {}_M = 0$). Using equation (6), we find $\tilde{j}_s$ as

$$\tilde{j}_s = \frac{2\omega\left(4\pi M_s\gamma \sin^2 {}_{M} + \sqrt{(4\pi M_s)^2\gamma^2 \sin^4 {}_{M} + 4\omega^2}\right)}{(4\pi M_s)^2\gamma^2 \sin^4 {}_{M} + 4\omega^2}. \tag{8}$$

The relation between the spin current density $\tilde{j}_s$ and the ellipticity A is readily obtained by combining equations (7) and (8), which is expressed as

$$\tilde{J}_s = \frac{4A}{(1+A^2)^2}. \tag{9}$$

This simple expression indicates that the spin current density is maximized when the precession trajectory is distorted: $A = 1/\sqrt{3}$. Here, the magnetization angle ${}_M$ at which $\tilde{j}_s$ is maximized is given as

$$\sin {}_{M} = 3^{-1/4}\sqrt{\frac{2\omega}{4\pi M_s\gamma}}, \tag{10}$$

showing that the spin current density is maximized when the external magnetic field is applied oblique to the film plane.

A magnetization-precession trajectory is characterized also by the elliptical area of a magnetization-precession trajectory, which is defined as $S \equiv \pi|m_x||m_y|$. We define the normalized area of a magnetization-precession trajectory as $\tilde{S} \equiv S/S^{A=1}$, where $S^{A=1}$ is the area of the magnetization-precession trajectory when the magnetization precesses in a circular orbit. Using equations (3) and (4), we obtain $\tilde{S}$ as

$$\tilde{S} = \frac{2\omega\left(4\pi M_s\gamma \sin^2 {}_{M} + \sqrt{(4\pi M_s)^2\gamma^2 \sin^4 {}_{M} + 4\omega^2}\right)}{(4\pi M_s)^2\gamma^2 \sin^4 {}_{M} + 4\omega^2}. \tag{11}$$

This expression of $\tilde{S}$ is exactly the same as that of the spin current density $\tilde{j}_s$ in equation (8): $\tilde{j}_s = \tilde{S}$. This shows that the spin current density $\tilde{j}_s$ is determined by the elliptical area of a

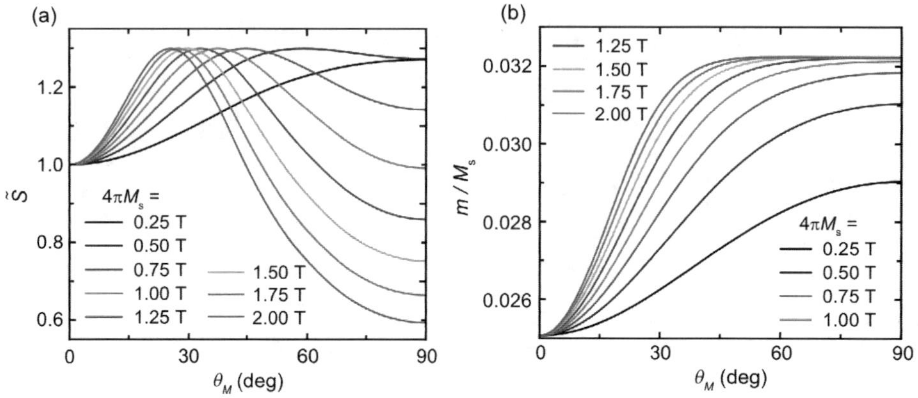

Figure 2. (a) The magnetization angle $_M$ dependence of the normalized area of a magnetization-precession trajectory $\tilde{S}$ for ferromagnetic films with the saturation magnetization $4\pi M_s$. Here, $\tilde{S} \equiv S/S^{A=1}$ and $S \equiv \pi|m_x||m_y|$. (b) The magnetization angle $_M$ dependence of the dimensionless precession amplitude m/M_s. Here, $\omega = 5.93 \times 10^{10} s^{-1}$,$\gamma = 1.86 \times 10^{11} T^{-1} s^{-1}$.

magnetization-precession trajectory $\tilde{S}$, which is maximized when the magnetization precession axis is oblique to the film plane as shown in figure 2(a). In figure 2(b), we show the averaged precession amplitude m, which is defined as $m = (\omega/2\pi) \int_0^{2\pi/\omega} |\mathbf{m}(t)| \mathrm{d}t$. The variation of m is different from that of $\tilde{S}$, showing that the spin pumping is dominated by the elliptical area of a magnetization-precession trajectory not by the precession amplitude.

4. Summary

In summary, we calculated a spin current density generated by the spin pumping in terms of a magnetization-precession trajectory. The spin current density was found to be determined by the elliptical orbit area of the magnetization precession. Since the spin pumping enables the spin-current injection into a wide range of systems, this result will be useful for developing spin-current technologies.

5. Acknowledgments

This work was supported by a Grant-in-Aid for Scientific Research in Priority Area "Creation and control of spin current" (19048028) from MEXT, Japan, a Grant-in-Aid for Scientific Research (A) from MEXT, Japan, a Grant for Industrial Technology Research from NEDO, Japan, and Fundamental Research Grant from TRF, Japan.

6. Reference

[1] Zutic I, Fabian J and Sarma S D 2004 *Rev. Mod. Phys.* **76** 323–410
[2] Wolf S A, Awschalom D D, Buhrman R A, Daughton J M, von Molnar S, Roukes M L, Chtchelkanova A Y and Treger D M 2001 *Science* **294** 1488–95
[3] Maekawa S (ed) 2006 *Concepts in Spin Electronics* (Oxford: Oxford University Press)
[4] Grollier J, Cros V, Hamzic A, George J M, Ja rès H, Fert A, Faini G, Youssef J B and Legall H 2001 *Appl. Phys. Lett.* **78** 3663–3665
[5] Ando K, Takahashi S, Harii K, Sasage K, Ieda J, Maekawa S and Saitoh E 2008 *Phys. Rev. Lett.* **101** 036601
[6] Tserkovnyak Y, Brataas A and Bauer G E W 2002 *Phys. Rev. Lett.* **88** 117601
[7] Mizukami S, Ando Y and Miyazaki T 2002 *Phys. Rev. B* **66** 104413
[8] Saitoh E, Ueda M, Miyajima H and Tatara G 2006 *Appl. Phys. Lett.* **88** 182509

[9] Ando K, Kajiwara Y, Takahashi S, Maekawa S, Takemoto K, Takatsu M and Saitoh E 2008 *Phys. Rev. B* **78** 014413
[10] Ando K, Yoshino T and Saitoh E 2009 *Appl. Phys. Lett.* **94** 152509
[11] Ando K, Ieda J, Sasage K, Takahashi S, Maekawa S and Saitoh E 2009 *Appl. Phys. Lett.* **94** 262505

2nd Int. Symp. on Advanced Magnetic Materials and Applications (ISAMMA 2010) IOP Publishing
Journal of Physics: Conference Series **266** (2011) 012102 doi:10.1088/1742-6596/266/1/012102

Theoretical study of the spin polarization measurement using non-contact Andreev reflection

Hiroyuki Ohtori[1,2†] **and Hiroshi Imamura**[2*]

[1] Institute of Applied Physics, University of Tsukuba, Tsukuba 305-8573, Japan
[2] Nanosystem Research Institute, AIST, Tsukuba 305-8568, Japan

E-mail: † hiroyuki-ohtori@aist.go.jp
* h-imamura@aist.go.jp

Abstract. We investigated the non-contact Andreev reflection of the normal-conductor / superconductor junction as a tool for spin polarization measurement of ferromagnetic metals. The conductance due to the Andreev reflection is obtained by numerically solving the corresponding one-dimensional Bogoliubov-de Gennes equation with use of the recursion-transfer-matrix method. We showed that the conductance suppression due to the exchange splitting of conduction band in ferromagnets. We also derived the analytical expression of the zero-bias conductance due to the Andreev reflection through the square barrier which agrees well with the exact numerical values.

1. Introduction

Andreev reflection (AR) at the interface between the ferromagnetic metal (FM) and super conductor (SC) has been attracted much attention because it provides a powerful tool for measuring the spin polarization of conduction electrons of the ferromagnetic metal. In 1995 de Jong et al. predicted that the AR of a FM/SC point contact is suppressed due to the exchange splitting of conduction band in the FM[1]. In the contact region the number of spin-up transmitting channels ($N_\uparrow$) is larger than that of spin-down transmitting channels ($N_\downarrow$), i.e., $N_\uparrow \geq N_\downarrow$. They suppose that there is no partially-transmitting channels and neglect mixing of channels for simplicity. When the SC is in the normal conducting state, all scattering channels (transverse modes in the point contact at the Fermi level) are fully transmitted, yielding the conductance

$$G_{FN} = \frac{e^2}{\hbar}\left(N_\uparrow + N_\downarrow\right), \tag{1}$$

where subscript FN stands for the FM/normal-conductor(NC) junction. When the SC is in the superconducting state, all the spin-down incident electrons in $N_\downarrow$ channels are Andreev reflected and give a double contribution to the conductance, $2(e^2/\hbar)N_\downarrow$. However, spin-up incident electrons in some channels cannot be Andreev reflected since the density of states for spin-down electrons is smaller than that for spin-up electrons. Then only a fraction $(N_\downarrow/N_\uparrow)$ of the $N_\uparrow$ channels can be Andreev reflected and the resulting conductance is $2(e^2/\hbar)(N_\downarrow/N_\uparrow)N_\uparrow$. The total conductance at zero bias voltage ($V = 0$) is given by the sum of these two contributions:

$$G_{FS} = \frac{e^2}{\hbar}2\left(N_\downarrow + \frac{N_\downarrow}{N_\uparrow}N_\uparrow\right) = 4N_\downarrow. \tag{2}$$

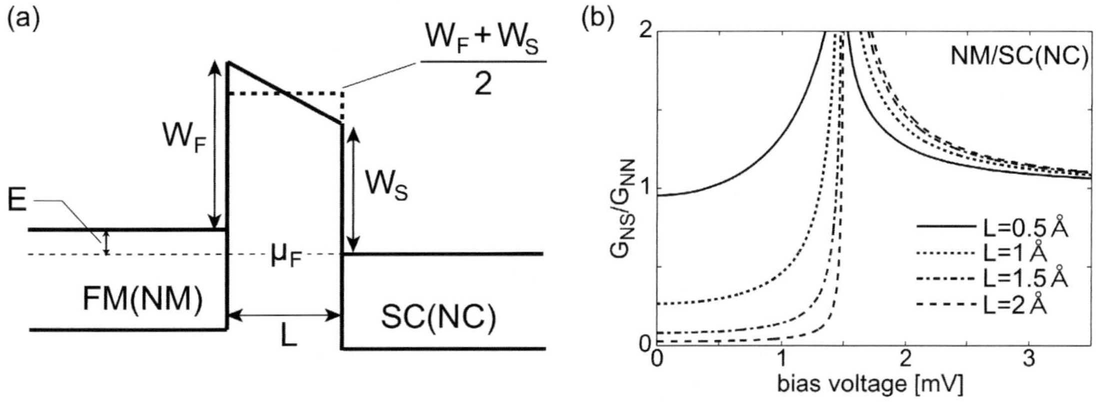

Figure 1. (a) The vacuum barrier we considered. The solid line shows the triangular barrier and the dashed line shows the square barrier. (b) Energy diagram of the FM(NM) / SC(NC) junction is schematically shown. The distance between the ferromagnetic metal and the superconductor is taken to be $L = 0.5, 1, 1.5$, and 2 Å.

Taking the ratio of Eq.(2) to Eq.(1), we obtain the normalized conductance

$$\frac{G_{FS}}{G_{FN}} = 2(1-P), \tag{3}$$

where P is the spin-polarization of transmitting channels defined as $P = (N_\uparrow - N_\downarrow)/(N_\uparrow + N_\downarrow)$ For non-magnetic metal (NM) the spin-polarization is zero, i.e., $P = 0$. Eq.(3) shows that the normalized conductance is a monotonic decreasing function of P and it vanishes if the FM is half metallic ($P = 1$).

In 1998, Soulen *et al* [2, 3] and independently by Upadyay *et al* [4] proposed the measurement technique called Point contact Andreev reflection (PCAR), which is based on the theory of Ref.[1]. The PCAR technique is much easier to put into practice than the tunneling technique. With no restrictions on the sample geometry, one can avoid the complex fabrication steps. They put a superconducting tip on a ferromagnet and measure the bias voltage dependence of the normalized conductance. By analyzing the observed normalized conductance curves they can estimate the spin polarization of FM [5, 6].

However, in the PCAR measurement, the surface of the FM is damaged by the superconducting tip. The surface damage will cause the interfacial scattering as well as the suppression of the spin polarization at surface. The above difficulty can be overcome by considering the non-contact AR through the vacuum, like the STM. Here we propose the non-contact Andreev reflection of the FM/SC junction as an effective tool for spin polarization measurement of ferromagnetic metals.

2. Model and Method

The system we considered is a one-dimensional tunnel junction with vacuum barrier shown in Fig. 1 (a). The work functions for FM(NM) and SC(NC) electrodes are denoted by W_F and W_S, respectively. Two electrodes is separated by vacuum tunnel barrier of thickness L.

We employ the Stoner model for the conduction electrons in FM electrode, where the spin asymmetry is given by the exchange field h. We neglect the proximity effect for simplicity. Then the system is described by the following Bogoliubov-de Gennes (BdG) equation [7]:

$$\begin{pmatrix} H_0(x)-h(x)\sigma & \Delta(x) \\ \Delta^*(x) & -H_0(x)-h(x)\sigma \end{pmatrix} \begin{pmatrix} f_\sigma(\vec{r}) \\ g_\sigma(\vec{r}) \end{pmatrix} = E \begin{pmatrix} f_\sigma(\vec{r}) \\ g_\sigma(\vec{r}) \end{pmatrix}, \tag{4}$$

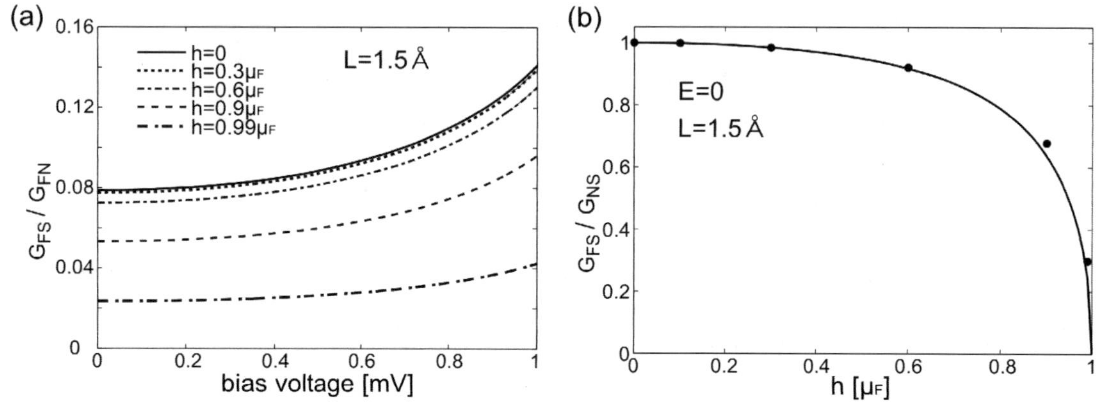

Figure 2. (a) Normalized conductances of the FM/SC junction through the triangular barrier for $h = 0, 0.3, 0.6, 0.9$, and $0.99\ \mu_F$. The distance between the FM and the SC electrodes is set to be 1.5 Å. (b) The zero-bias conductance of the FM/SC junction with L =1.5 Å is plotted against the exchange field h. The conductance is normalized by that for $h = 0$. The filled circles indicate the results obtained by numerically solving the BdG equation of Eq.(4) with use of the recursion-transfer-matrix method. The line shows the system with the square barrier and the plot is the one with the triangular barrier.

where $H_0(x) \equiv -(\hbar^2/2m)\nabla^2 - \mu_F + V - Fx$ is the single particle Hamiltonian, E is the quasiparticle (QP) energy measured from the Fermi energy μ_F, and $\sigma = +(-)$ represents the up-(down-)spin band. The vacuum potential barrier V is given by $V = W_F + \mu_F$ where W_F is the work function of the FM(NM) surface. The electric field F is given by $F = (W_F - W_S)/L$ where W_S is the work function of the SC(NC) tip. The exchange field function $h(x)$ is given by $h(x) = h_0[1 - \Theta(x)]$ where h_0 represents the exchange field in the ferromagnetic metal and $\Theta(x)$ is the Heaviside step function. The superconducting gap function is expressed as $\Delta(x) = \Delta_0\Theta(x - d)$, where Δ_0 represents the superconducting gap in the superconductor.

The probabilities of the AR, $A_\sigma(E)$, and normal reflection, $B_\sigma(E)$ is obtained by solving the scattering problem with appropriate boundary conditions. Using the probabilities $A_\sigma(E)$ and $B_\sigma(E)$ the conductance is expressed as

$$G = \frac{e}{h}\sum_\sigma [1 + A_\sigma(eV) - B_\sigma(eV)] \,. \tag{5}$$

The material parameters we used are the following. The superconducting gap is $\Delta_0 =$ 1.5meV, Fermi energy $\mu_F = 3.8$ eV, the work function of FM(NM) electrode is $W_F = 5.2$eV and that of SC(NC) electrode is $W_S = 4.0$eV.

3. Results

In Fig. 1 (b), we plot the normalized conductance of NM/SC junctions with $L = 0.5$, 1, 1.5, and 2Å. The conductance due to the AR below the superconducting gap $\Delta_0 = 1.5$meV is suppressed by the vacuum tunnel barrier and decreases with increasing L. In Fig. 2 (a) we plot the normalized conductance due to the AR for various values of exchange field h. The distance between two electrodes is set to be 1.5 Å. For $h = 0$, the normalized conductance takes the value of 0.08 and it decreases with increasing the exchange field h. The obtained results are quite similar to those obtained in the PCAR system and imply that the non-contact AR can be used as a tool for spin polarization measurement of ferromagnetic metals.

In Fig. 2 (b) we plot the zero-bias conductance shown in Fig. 2 (a) against the exchange field h by filled circles. The zero-bias conductance is normalized by that for $h = 0$. We also calculate the zero-bias conductance for the junction with rectangular potential indicated by dotted line in Fig.1 (a). The height of the potential is assumed to be $(W_F + W_S)/2$. Under the Andreev approximation we obtain the analytic from of the normalized zero-bias conductance,

$$G_{FS}/G_{NS} = \frac{4\sqrt{1-\eta^2}(V_0-1)^2}{\xi}, \tag{6}$$

where

$$\xi = \left\{\left(\sqrt{1+\eta} - \sqrt{1-\eta}\right) V_0\sqrt{V_0-1}\cosh(L\rho)\sinh(L\rho)\right\}^2 + \left\{\left(\sqrt{1-\eta^2}+1\right)(V_0-1) + \left(\sqrt{1-\eta^2}+V_0-1\right)V_0\sinh^2(L\rho)\right\}^2. \tag{7}$$

Here we introduce the parameters $\eta = h_0/\mu_{\mathrm{F}}$ and $V_0 = V/\mu_{\mathrm{F}}$. In Fig. 2 (b) we plot Eq. (6) by the solid line. One can see that the approximated results of Eq.(6) agree with the exact ones indicated by filled circles very well. Equation (6) is the basis of the non-contact measurement of spin polarization we propose.

4. Conclusion

We theoretically studied the non-contact AR of the FM / SC junction as a tool for spin polarization measurement of FMs. We show that the conductance due to the AR decreases with increasing the exchange field. We derived the analytical expression of the zero-bias conductance due to the AR through the square barrier and showed that the approximated results agree with the exact ones obtained by using the recursion-transfer-matrix method. We found that the conductance with the triangular barrier is similar to the one with square barrier in the range of L we considered. Our results support that the non-contact AR will be a powerful tool for measuring the spin polarization of FMs.

References

[1] de Jong M and Beenakker C 1995 *Phys. Rev. Lett.* **74** 1657
[2] Jr R S, Byers J, Osofsky M, Nadgorny B, Ambrose T, Cheng S, Broussard P, Tanaka C, Nowak J, Moodera J, Barry A and Coey J 1998 *Science* **282** 85
[3] Jr R S, Osofsky M, Nadgorny B, Ambrose T, Broussard P, Cheng S, Byers J, Tanaka C, Nowack J, Moodera J, Laprade G, Barry A and Coey M 1999 *J. Appl. Phys.* **85** 4589
[4] Upadhyay S, Palanisami A, Louie R and Buhrman R 1998 *Phys. Rev. Lett.* **81** 3247
[5] Ji Y, Strijkers G, Yang F, Chien C, Byers J, Anguelouch A, Xiao G and Gupta A 2001 *Phys. Rev. Lett.* **86** 5585
[6] Mazin I I, Golubov A A and Nadgorny B 2001 *J. Appl. Phys.* **89** 7576
[7] de Gennes P G 1966 *Superconductivity of Metals and Alloys* (New York: W. A. Benjamin) chap 5

Journal of Physics: Conference Series **266** (2011) 012103 doi:10.1088/1742-6596/266/1/012103

Finite size effects on the quantum spin Hall state

Yuichi Ohyama, Hiroki Tsuchiura and Akimasa Sakuma

Department of Applied Physics, Tohoku University, 6-6-05 Aoba, Aoba-ku, Sendai 980-8579, Japan

E-mail: yuichigm@solid.apph.tohoku.ac.jp

Abstract. We theoretically study the helical edge states in the quantum spin Hall (QSH) effect based on an effective 4-band tight-binding model for HgTe/CdTe quantum wells. We microscopically describe the spatial dependence of the helical edge modes in the QSH state. It is found that the helical edge states are gapped when the width of sample narrows. This is because the edge states on the opposite boundary of the sample overlap each other, and then the backward scattering channel opens up.

1. Introduction

Recently, Kane and Mele proposed the quantum spin Hall (QSH) effect in a graphene layer with spin-orbit coupling [1,2]. The QSH state has a bulk energy gap, but has gapless edge states with different spins moving in opposite directions. No net charge current but spin current is carried by this edge state. Since the spin current is protected by the time reversal symmetry, it is quite robust against nonmagnetic disorder. Thus the QSH effect is expected to play an important role in spintronics.

Soon after the proposal of the QSH effect in a graphene, Bernevig *et al.* proposed the QSH effect in HgTe/CdTe quantum wells [3]. They predicted a quantum phase transition in HgTe/CdTe quantum wells as a function of a thickness d of the quantum wells. The quantum well system is a conventional insulator for $d < d_c$ and the QSH state for $d > d_c$, which has a single pair of helical edge state. This QSH effect in HgTe/CdTe quantum wells was observed by König *et.al.* [4]. More recently, it is known that a certain class of three-dimensional topological insulator material can have protected surface states which is a higher-dimensional analogue of one-dimensional helical edge states in the QSH effect [5,6,7,8,9].

In order to utilize the QSH effect in microscopically fabricated devices, it is important to reveal the system size effects on the QSH state. Recently, Zhou *et al.* presented an analytical study of the continuum effective 4-band model for a strip of finite width for the HgTe/CdTe quantum wells [10]. They showed that edge states on the two sides can couple together to produce a gap in the spectrum, destroying the QSH state.

In this paper, we carry out the similar analysis on the finite size effect on the QSH state, but based on the tight-binding scheme. Because the tight-binding model can easily describe atomic scale roughness around the edge of the sample and simulate various device geometry.

2. Method

The effective 4-band model for HgTe quantum wells proposed by Bernevig *et al.* is given as

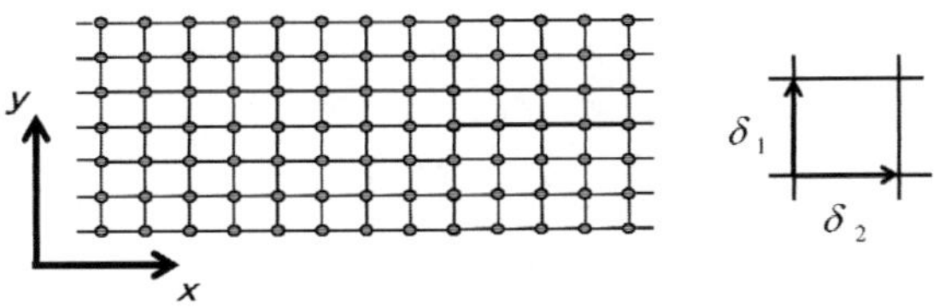

Figure 1. Finite strip geometry. Each site has two orbits between which an electron can transfer.

$$H_{\rm eff}(k_x,k_y)=\begin{matrix} H(\boldsymbol{k}) & 0 \\ 0 & H^{\dagger}(-\boldsymbol{k}) \end{matrix},$$

$$H(\boldsymbol{k})=\epsilon(\boldsymbol{k})\boldsymbol{I}_{2\times2}+\sum_{\alpha=1,2,3} d_{\alpha}(\boldsymbol{k})\sigma^{\alpha}, \tag{1}$$

where k_x and k_y are momenta in the plane of the two-dimensional electron gas, $\boldsymbol{I}_{2\times2}$ is a 2×2 unit matrix, σ_α are the Pauli matrices, $\epsilon(\boldsymbol{k})=C-D(k_x^2+k_y^2)$, $d_1+{\rm i}d_2(\boldsymbol{k})=A(k_x+{\rm i}k_y)$, and $d_3=M-B(k_x^2+k_y^2)$. A, B, C, D, and M are material parameters that depend on the quantum well geometry. The most important property of this model is that the gap parameter M can be continuously tuned from a positive value for thin quantum well with thickness $d<d_{\rm c}$ to a negative value for thick quantum well with $d>d_{\rm c}$. A quantum phase transition between a insulating state and the QSH state occurs at a critical thickness $d_{\rm c}$.

Because the Hamiltonian is block diagonal and the lower block $H^{\dagger}(-\boldsymbol{k})$ is time reversal of the upper block $H(\boldsymbol{k})$, it is sufficient to solve the upper block $H(\boldsymbol{k})$ on the finite strip geometry. Here, we derive a corresponding Hamiltonian based on the square lattice tight-biding scheme given as,

$$\begin{aligned}\mathcal{H}=&\ e_1\sum_{i,l=1,2}c^{\dagger}_{il}c_{il}+e_2\sum_i c^{\dagger}_{i,1}c_{i,1}-e_2\sum_i c^{\dagger}_{i,2}c_{i,2}+e_3\sum_{<i,j>}c^{\dagger}_{i,1}c_{j,1}+e_4\sum_{<i,j>}c^{\dagger}_{i,2}c_{j,2}\\ &+{\rm i}e_5\sum_i(c^{\dagger}_{i,1}c_{i+\boldsymbol{\delta}_2,2}-c^{\dagger}_{i.1}c_{i-\boldsymbol{\delta}_2,2}+c^{\dagger}_{i,2}c_{i+\boldsymbol{\delta}_2,1}-c^{\dagger}_{i,2}c_{i-\boldsymbol{\delta}_2,1}+{\rm h.c.})\\ &+e_5\sum_i(-c^{\dagger}_{i,1}c_{i-\boldsymbol{\delta}_1,2}+c^{\dagger}_{i,1}c_{i+\boldsymbol{\delta}_1,2}+c^{\dagger}_{i,2}c_{i-\boldsymbol{\delta}_1,1}-c^{\dagger}_{i,2}c_{i+\boldsymbol{\delta}_1,1}+{\rm h.c.}),\end{aligned} \tag{2}$$

where i,j are the square lattice sites, $l=1,2$ is the orbital degree of freedom, $e_1=2(C-4D)$, $e_2=-4B(2-\frac{M}{2B})$, $e_3=2(D+B)$, $e_4=2(D-B)$ and $e_5=\frac{A}{2}$. $<i,j>$ is the sum runs over all nearest neighbor sites, $\boldsymbol{\delta}_i(i=1,2)$ are the nearest neighbor vectors. We obtain the energy spectrum and dispersion by numerical diagonalization.

3. Result

To study the system-size dependence of the QSH state, we choose model parameters so as to describe the QSH state regime, for instance, $M=-3.0$ [meV], $A=1.0$ [meV nm], $B=-0.5$ [meV nm^2], $C=0.0$ [meV], $D=0.3$ [meV nm^2]. Figure 2 (a) and (b) show the energy dispersion relation for $L=40$ sites. We can clearly see that gapless helical edge modes exist inside the bulk energy gap of the system. We note here that, since we are treating $H(\boldsymbol{k})$ only, the helical edge mode with positive and negative group velocity are located around opposite side of the sample,

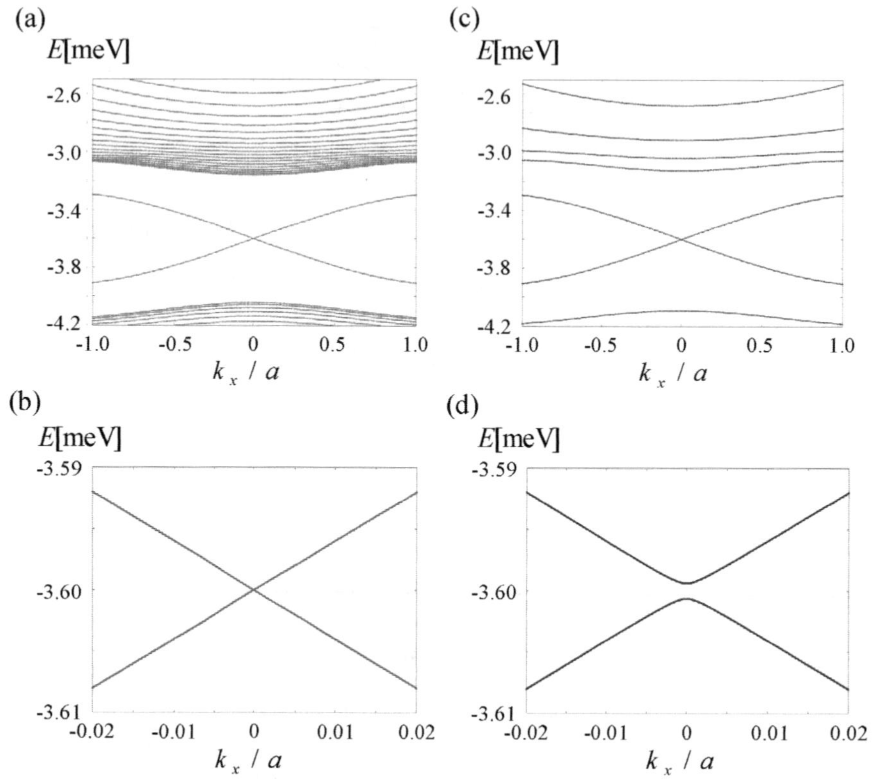

Figure 2. Energy spectrum for the strip with (a) $L = 40$ sites and (c) $L = 10$ sites. Energy spectrum of the edge states for the strip with (b) $L = 40$ sites and (d) $L = 10$ sites.

respectively.

We also show the results for $L = 10$ sites in Figure 2 (c) and (d). Contrary to the $L = 40$ sites case, we can see that the energy spectrum for the edge states is parabolic near $k_x = 0$ and energy gap opens. Thus the system is insulating. This mechanism can be understood as follows. As the strip narrows, the edge states on each boundary start to couple together. Thus the backward scattering channels open up and then the time reversal protection of the edge states has been broken.

To confirm this coupling of the edge states, we plot the density distribution of the edge states for the strip with $L = 40$ sites and $L = 10$ sites in Figure 3. For the strip with $L = 40$ sites, the density distribution is located at each side of the strip and decays exponentially toward the center. Because there are no overlap of the edge states near the center, the edge states can survive in the strip with $L = 40$ sites. On the other hand, for the strip with $L = 10$ sites, it is found that the density distribution for the edge states on opposite boundary overlap each other near the center and then the backscattering channel opens up. Because of this the phase transition from QSH states to the band insulator occurs.

These behaviors are quite similar to those observed by Zhou *et al.* based on the continuum model [10]. However, we also find non-monotonic atomic scale spatial dependences of the edge states as shown in Figure 3.

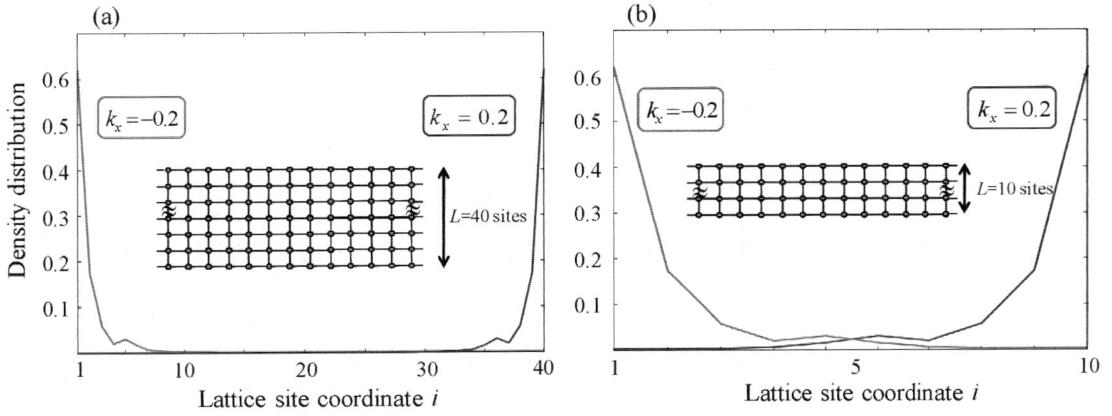

Figure 3. The density distribution of the edge states (a) for $L = 40$ sites and (b) for $L = 10$ sites. The red solid line corresponds to $|\ \Psi(k_x = -0.2, y)\ |^2$, and the blue solid line corresponds to $|\ \Psi(k_x = 0.2, y)\ |^2$.

4. Conclusion and Discussion

We have theoretically studied the system-size effect on the helical edge state in two-dimensional strip geometry based on the effective 4-band tight-binding model for the HgTe/CdTe quantum well. It has been found that the QSH state can be broken when the width of the system is narrows enough for the edge states around each edge to overlap to open the backward scattering channels.

The present model can describe various impurity distributions, or atomic scale roughness of the sample edge. Relating studies including such disorder are being carried out.

Acknowledgments

A part of numerical computations in this work was carried out at the Yukawa Institute Computer Facility, and also at Cyberscience Center, Tohoku University. This work is partly supported by Grant-in-Aid for Scientific Research (C) and also by Next Generation Supercomputing Project, Nanoscience Program, from the Ministry of Education, Culture, Sports, Science and Technology of Japan.

[1] Kane C L and Mele E J 2005 *Phy. Rev.Lett.* **95** 146802
[2] Kane C L and Mele E J 2005 *Phys.Rev.Lett.* **95** 226801
[3] Bernevig B A, Hughes T L and Zhang S C 2006 *Science* **314** 1757
[4] König M, Wiedmann S, Brüne C, Roth A, Buhmann H, Molenkamp L W, Qi X L and Zhang S C 2007 *Science* **318** 766
[5] Fu L, Kane C L and Mele E J 2007 *Phys.Rev.Lett.* **98** 106803
[6] Fu L and Kane C L 2007 *Phys.Rev.* B **76** 045302
[7] Xia Y, Qian D, Hsieh D, Wray L, Pal A, Lin H, Bansil A, Grauer D, Hor Y S, Cava R J and Hasan M Z 2009 *Nat.Phys.* **5** 398
[8] Zhang H, Liu C X, Qi X L, Dai X, Fang Z and Zhang S C 2009 *Nat.Phys.* **5** 438
[9] Moore J 2009 *Nat.Phys.* **5** 378
[10] Zhou B, Lu H Z, Chu R L, Shen S Q and Niu Q 2008 *Phys.Rev.Lett.* **101** 246807

Journal of Physics: Conference Series **266** (2011) 012104 doi:10.1088/1742-6596/266/1/012104

The effect of inserting thin Co_2MnAl layer into the Co_2MnSi/MgO interface on tunnel magnetoresistance effect

E. Ozawa, S. Tsunegi, M. Oogane, H. Naganuma, and Y. Ando
Graduate School of Engineering, Tohoku University, Aoba-yama 6-6-05, Aramaki, Aoba-ku, Sendai 980-8579, Japan

E-mail: ozawa@mlab.apph.tohoku.ac.jp

Abstract. We fabricated an epitaxially grown *B*2-ordered Co_2MnAl Heusler alloy film by optimizing fabrication conditions, such as composition of the films and annealing temperature. Tunnel magnetoresistance (TMR) effect was investigated in Co_2MnSi/Co_2MnAl(0–1.0 nm)/MgO/CoFe magnetic tunnel junctions (MTJs). TMR ratio was enhanced by inserting a thin Co_2MnAl layer at the Co_2MnSi/MgO interface. The MTJ with Co_2MnAl thickness of 0.5 nm exhibited the highest TMR ratio at 310 K and 10 K.

1. Introduction

Development of ferromagnetic materials with high spin polarization is the most important subject in spintronics field. Ideal high spin polarization materials are half-metallic ferromagnets with a band gap at the Fermi energy level for one spin band [1]. Experimentally, we have observed a large tunnel magnetoresistance (TMR) ratio of up to 159% at 2 K in magnetic tunnel junctions (MTJs) with Co_2MnSi/Al–O/CoFe structure, thus proving the halfmetallic property of Co_2MnSi [2].

Similarly, MTJs using a crystalline MgO tunnel barrier producing coherent tunneling through the Δ_1 band of ferromagnetic electrodes have attracting much interest for some spintronics devices, such as magnetic random access memory (MRAM) and read heads for hard disk drives (HDDs). Ikeda *et al.* showed a huge TMR ratio of up to 600% at RT in a CoFeB/MgO/CoFeB MTJ [3]. However, the TMR ratio using a CoFeB electrode and an MgO barrier is coming close to the theoretical limit [4]. Miura *et al.* showed theoretically that coherent tunneling through the Δ_1 band of Co_2MnSi Heusler alloy could also occur in a $Co_2MnSi/MgO/Co_2MnSi$ MTJ [5]. We have demonstrated a giant TMR ratio of 753% at 2 K in a MTJ with Co_2MnSi/MgO/CoFe structure, in which the MgO barrier was deposited by electron beam (EB) evaporation system. This result indicates that Heusler alloys as electrodes of MgO-based MTJs have great potential to achieve giant TMR ratio by both half-metallicity and coherent tunnelling [6]. However, the TMR ratio decreased rapidly with increasing of temperature in the Co_2MnSi/MgO/CoFe-MTJ. We considered that the large temperature dependence of the TMR ratio was originated from the inelastic tunneling at the antiparallel magnetization configration. Y. Miura *et al.* suggested by their theory that Co_2MnSi/MgO interfacial states were created in the half metallic gap due to the non-bonding states [7]. Such interfacial states enhanced the inelastic tunneling [8]. They also suggested that a MTJ having an ultrathin Co_2MnAl layer between a Co_2MnSi electrode and a MgO barrier does not have the interfacial state and show the halfmetallic character at the MgO barrier interface [7]. In this study, we fabricated high-quality Co_2MnSi/Co_2MnAl/MgO/CoFe-MTJs and investigated the TMR effect.

2. Fundamental properties of Co_2MnAl thin films

The Co_2MnAl thin films were prepared by means of magnetron sputtering in order to characterize their crystal structure, surface morphology and magnetic properties. The MgO-buffer(20 nm)/Co_2MnAl(30 nm) films were deposited on single crystalline MgO (001) substrates at ambient temperature and subsequently annealed at 400-600 °C in order to reduce site-disorder. In order to obtain the stoichiometric film composition, we used a composition-adjusted Co-Mn-Al alloy sputtering target (Co: 43.7%, Mn: 27.95%, Al: 28.35%). Crystal structure of the prepared Co_2MnAl thin films were verified by x-ray diffraction (XRD) with out-of plane geometries and surface roughness was verified by atomic force microscopy (AFM). We measured magnetization curves by means of a vibrating sample magnetometer (VSM). Film compositions were examined by means of inductively coupled plasma (ICP) analysis.

The XRD profiles for Co_2MnAl thin film with T_a = as depo.-600°C are shown in Figure 1(a). In addition to peaks from MgO substrates, only the (002) and (004) Co_2MnAl peaks were detected, indicating perfect (001)-preferred orientation. The full width at half maximum (FWHM) of the rocking curve for the (004) Co_2MnAl peak is very small (under 0.5°). Therefore, the structural quality of the film in the direction perpendicular to the plane was quite good. In addition, ideal epitaxial growth was confirmed with the epitaxial relation [110]Co_2MnAl|| [100]MgO|| [100]MgO from the ϕ-scan result shown in the Figure 1(b). Clear (220) Co_2MnAl peaks are observed in the ϕ-scan profile for all the prepared films, indicating that a highly *B*2-order structure can be obtained. For all films, the (111) peak of $L2_1$ structure is undetectable from the (111) pole-figure measurements. These results suggest that prepared epitaxial Co_2MnAl films are identified as having not $L2_1$ structure but *B*2 structure. In addition, the composition of Co_2MnAl film analyzed by ICP spectroscopy was nearly stoichiometric (Co: 49.3%, Mn: 24.2%, Al: 26.5%).

Figure 2 shows the average roughness (R_a) of the prepared Co_2MnAl thin films as function of T_a. Roughness of the films decreases with increasing T_a above T_a = 300 °C. Inset in Figure 2

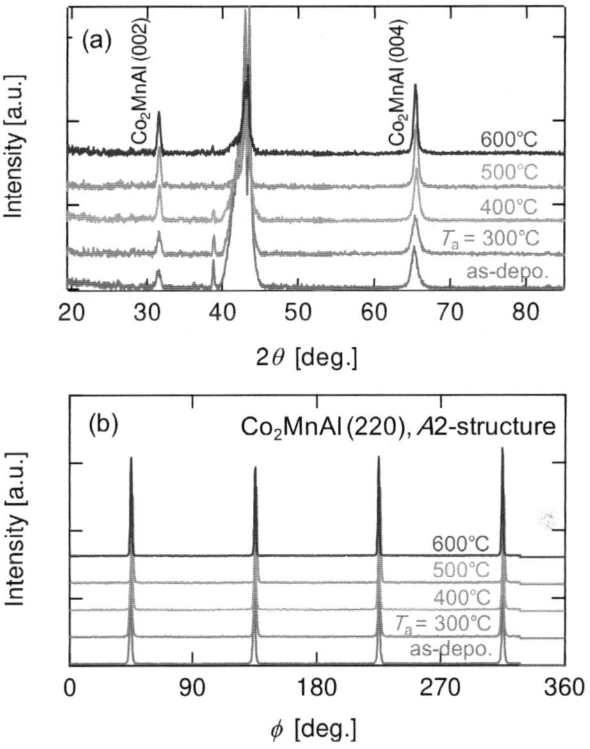

Figure 1. XRD patterns of (*a*) 2θ-scan and (b) ϕ-scan for the MgO(001)-sub./MgO-buffer(20 nm)/Co_2MnAl(30 nm) (T_a = as.depo.-600 °C).

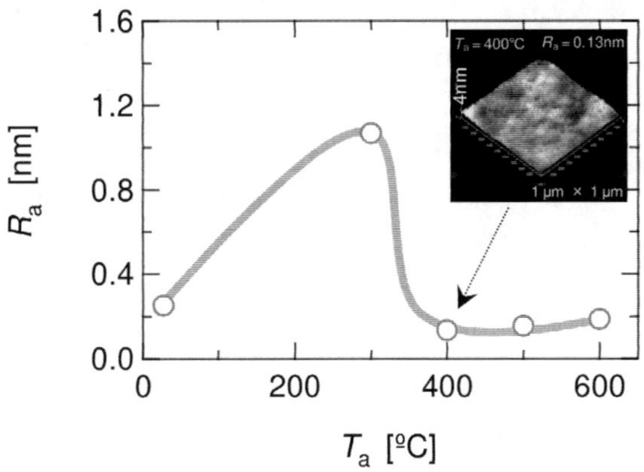

Figure 2. Average roughness (R_a) of the prepared Co_2MnAl films as function of T_a. Inset shows the AFM image for the MgO(001)-sub./ MgO-buffer(20 nm)/ Co_2MnAl(30 nm) (T_a = 400 °C) film.

shows the AFM image for the films with T_a = 400 °C, where R_a is about 0.2 nm. The R_a of 0.2 nm was small enough to be formed a thin tunnel barrier.

Saturation magnetization M_s estimated from *M–H* curves is depicted in Figures 3 as functions of T_a. The maximum value of M_s is about 20% smaller than the bulk M_s.

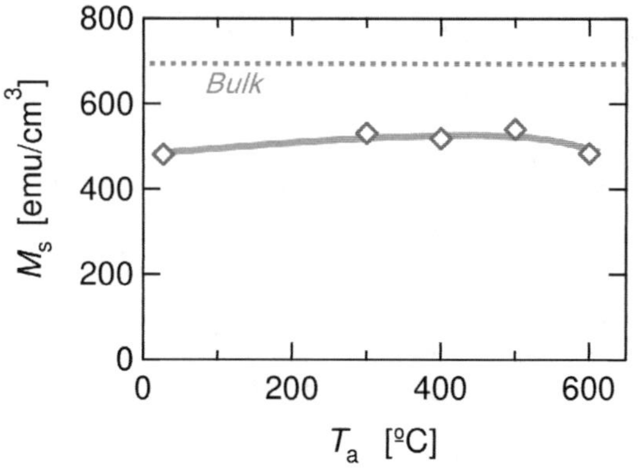

Figure 3. Saturation magnetization of the prepared Co_2MnAl films as function of T_a

3. TMR in MTJs using Heusler alloy thin films

We prepared MTJs with two different structures of bottom electrodes. The stacking structure of MTJs is bottom-electrode/MgO(2.3 nm)/ $Co_{50}Fe_{50}$(5 nm)/ IrMn (10 nm)/Ta(5 nm)/cap layer. The structures of bottom electrodes are as follows.

Sample A: MgO(001)-sub./MgO-buffer(20 nm)/ epitaxial Co_2MnSi(50 nm) (T_a = 600 °C)/ Co_2MnAl(0.0-1.0 nm) (T_a = 400-600 °C).

Sample B: MgO(001)-sub./MgO-buffer(20 nm)/ epitaxial Co_2MnAl(30 nm) (T_a = 400-600 °C).

All the MTJs were patterned into four-terminal structure by using photo-lithography and Ar ion milling. The TMR effect was measured at 310 and 10 K. To improve the crystallinity of the MgO barrier, patterned MTJs were annealed at the 250-525 °C for 1 hour in high vacuum by applying a magnetic field of 1 T.

TMR ratio increased with increasing annealing temperature because of crystallization of the MgO barrier, and shows a maximum at 450-475 °C. TMR ratio showed maximum for the Co_2MnSi/Co_2MnAl(T_a = 400°C)/MgO/$Co_{50}Fe_{50}$ MTJs and for the Co_2MnAl(T_a = 400°C)/MgO/ $Co_{50}Fe_{50}$ MTJs. Figure 4(a) shows maximum TMR ratio measured at 310 K and 10 K in each inserted Co_2MnAl thickness. The MTJ with Co_2MnAl thickness of 0.5 nm shows the highest TMR ratio of 180% at 310 K and 600% at 10 K. These TMR ratios are larger than those of both Co_2MnSi/ MgO/CoFe-MTJ and Co_2MnAl/MgO/CoFe-MTJ. Figure 4(b) shows the ratio of the TMR ratios measured at 310 K and 10 K (TMR_{310K}/TMR_{10K}) for each inserted Co_2MnAl thickness. TMR_{310K}/TMR_{10K} ratio tends to saturate beyond a thickness of 1.0 nm. This result indicates that the temperature dependence of TMR ratio is slightly improved by insertion of Co_2MnAl.

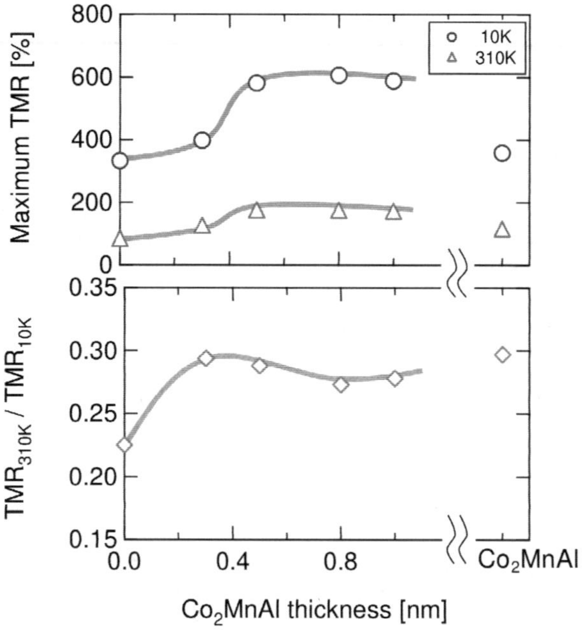

Figure 4. (a) The Co_2MnAl thickness dependence of TMR ratio measured at 310 K and 10 K. (b) The ratio of the TMR ratios measured at 310 K and 10 K (TMR_{310K}/ TMR_{10K}) for each inserted Co_2MnAl thickness.

Some origins of the large temperature dependence of TMR in the MTJs with Heusler alloy electrodes are considered both experimentally and theoretically: e.g., small energy separation between the conduction band and the Fermi level [9], non-quasiparticle state [10], and the weak exchange interaction energy of the Co layer at the Co_2MnSi/MgO interface [11] and interfacial states created in the half-metallic gap as mentioned above [7]. It is not clear which origins of the large temperature dependence were improved by insertion of the thin Co_2MnAl layer. In order to clarify the role of Co_2MnAl insertion layer, detailed investigations on the tunneling conductance properties are necessary.

4. Conclusion

We have fabricated MTJs with *B*2-ordered (001)-epitaxial Co_2MnAl bottom electrodes. Co_2MnAl thin films had good surface flatness (about 0.2 nm) and showed a highly-ordered *B*2 structure for $T_a \geq$ 400 °C. We observed very high TMR ratios of 180% at 310 K and 600% at 10 K and improved the temperature dependence of TMR ratio by insertion of very thin Co_2MnAl films into the Co_2MnSi/MgO interface.

Acknowledgement

This research was supported by Research and Development for Next-Generation Information Technology by MEXT and Funding Program for World-Leading Innovative R&D on Science and Technology (FIRST Program) by JSPS.

References

[1] S. Ishida, S. Fujii, S. Kashiwagi, and S. Asano, J. Phys. Soc. Jpn. **64**, 2152 (1995).
[2] Y. Sakuraba, J. Nakata, M. Oogane, H. Kubota, Y. Ando, A. Sakuma, and T. Miyazaki, Jpn. J. Appl. Phys., Part 1 **44**, L1100 (2005).
[3] S. Ikeda, J. Hayakawa, Y. Ashizawa, Y. M. Lee, K. Miura, H. Hasegawa, M. Tsunoda, F. Matsukura, and H. Ohno, Appl. Phys. Lett. **93**, 082508 (2008).
[4] J. Mathon and A. Umerski, Phys. Rev. B **63**, 220403 (2001).
[5] Y. Miura, H. Uchida, Y. Oba, K. Nagao, and M. Shirai, J. Phys.: Condens. Matter **19**, 365228 (2007).
[6] S. Tsunegi, Y. Sakuraba, M. Oogane, K. Takanashi, and Y. Ando, Appl. Phys. Lett. **93**, 112506 (2008).
[7] Y. Miura, K. Abe, M. Shirai, Journal of Physics : Conference Series **200,** 052016 (2010).
[8] P. Mavropoulos, M. Ležaić, and S. Blügel, Physical Review B **72**, 174428 (2005).
[9] Y. Sakuraba, T. Miyakoshi, M. Oogane, Y. Ando, A. Sakuma, T. Miyazaki, and H. Kubota, Appl. Phys. Lett., **89,** 052508 (2006).
[10] L. Chioncel, Y. Sakuraba, E. Arrigoni, M. I. Katsnelson, M. Oogane, Y. Ando, T. Miyazaki, E. Burzo, and A. I. Lichtenstein, Phys. Rev. Lett., **100,** 086402 (2008).
[11] A. Sakuma, Y. Toga, and H. Tsuchiura, J. Appl. Phys. **105,** 07C910 (2009).

Journal of Physics: Conference Series **266** (2011) 012105 doi:10.1088/1742-6596/266/1/012105

Substrate bias effect on AlO_x based magnetic tunnel junctions

Ajeesh M Sahadevan[1,4], Jae S Son[2], Hyunsoo Yang[1], Aaron J Danner[2] and Charanjit S Bhatia[1,3]

[1] Information Storage Materials Laboratory, Department of Electrical and Computer Engineering, NUS, Singapore

[2] Centre for Optoelectronics, Department of Electrical and Computer Engineering, NUS, Singapore

[3] Institute of Materials Research and Engineering, 3 Research Link, Singapore

E-mail: g0800464@nus.edu.sg

Abstract. The effect of substrate bias during deposition on the properties of magnetron sputtered aluminium oxide-based magnetic tunnel junctions (MTJ) is investigated. Ge buffer layers have been used for MTJs for the first time. Transmission electron microscope (TEM) images clearly indicate that application of substrate bias during the deposition of MTJ layers [IrMn/Co/AlO_x/NiFe/Cu] provides extremely smooth layers and flat interfaces (without annealing) in comparison to the layers deposited without bias. But hysteresis loops (magnetization v/s magnetic field M-H) obtained from alternating gradient force magnetometer (AGFM) make it clear that IrMn deposited with a substrate bias is not effective in providing exchange bias. The reason is the change in composition of IrMn with bias as shown by Rutherford backscattering (RBS) experiments. RBS for AlO_x shows a clear Ar peak with the application of substrate bias.

1. Introduction

Magnetic Tunnel Junctions (MTJs) will play a key role in the future of storage devices such as the read heads of hard disc drives (HDD) and memory elements in magnetic random access memory (MRAM) [1, 2]. Though MTJs have been commercialized, the mechanisms involved as well as the exact structural conditions required for a high tunnel magnetoresistance (TMR) value are not completely clear. Recently TMR close to 1000% has been reported using MgO tunnel barriers [3]. Crystalline barriers like MgO enable huge TMR through coherent tunneling of electrons even using conventional ferromagnetic (FM) elements and alloys. Interface properties such as stoichiometry as well as the crystallinity play a very important role and can be tuned to change the TMR significantly [4]. A recent paper shows the effect of substrate bias during MgO deposition on CoFeB/MgO based MTJs [5]. They show an increase in TMR and a reduction in resistance area (RA) product with the application of bias to the substrate and explain the reason to be the enhanced MgO (001) growth, required for high TMR. Here we study the effect of substrate bias during the deposition on the properties of Aluminium oxide-based MTJs by magnetron sputtering.

[4] To whom any correspondence should be addressed.

Journal of Physics: Conference Series **266** (2011) 012105 doi:10.1088/1742-6596/266/1/012105

2. Experiments

The films are deposited in an ultra-high vacuum (10^{-9} Torr) sputter chamber. Atomic force microscopy is used to study the effect of substrate bias on film roughness and deposition rate. Germanium is deposited in a high vacuum (10^{-7} Torr) electron beam evaporator on Si/SiO_2 (thermally oxidized) substrates. Two sets of the same structure are then deposited [Ge (40nm)/IrMn (25nm)/Co (˜3.5nm)/AlO_x (˜3nm)/Co (0.5 nm)/NiFe (˜5nm)/Cu (2nm)/Ge (5nm)] - one with a 20W RF bias to the substrate during deposition for all the layers and no bias for any of the layers in the other. The aim of this study is simply to see the effect of bias deposition on the properties of an MTJ (mainly the structure) and 20W RF is the maximum bias power possible. A 200kV Transmission electron microscope (TEM) is used to get the cross-sectional images of the two structures. The switching characteristics are obtained form an atomic force gradient magnetometer (AGFM). The effect of substrate bias on the composition of the antiferromagnetic IrMn and its crystal orientation is studied using Rutherford Backscattering (RBS) and X-ray diffraction (XRD) experiments, respectively. RBS is used to study the AlO_x composition as well.

3. Results and discussions

As an initial study the roughness and deposition rates for all the relevant layers in the structure are obtained. For most of the materials, both the parameters are reduced with the application of bias during deposition. E-beam evaporated Ge on Si/SiO_2 substrate is found to be extremely smooth. The root-mean-square value (RMS) roughness is measured to be less than 3 Å. Cross-sectional TEM images for the two structures (without and with bias) are shown in figure 1a) and b).

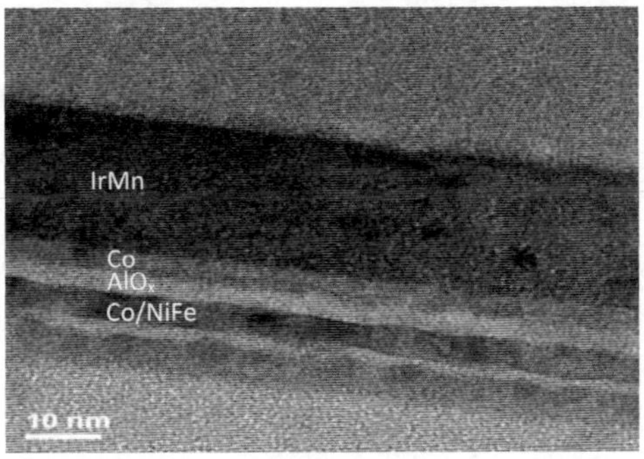

Figure 1 a). Cross-sectional TEM image for MTJ [Ge (buffer) /IrMn/Co/AlOx/Co/NiFe/Cu/Ge] without substrate bias

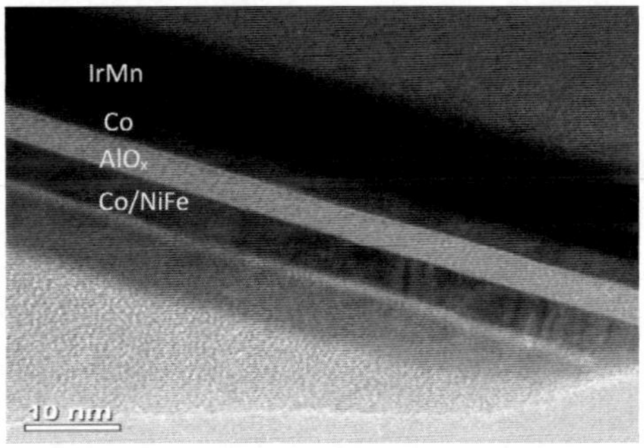

Figure 1 b). Cross-sectional TEM image for MTJ with substrate bias

The effect of bias to improve the uniformity of the layers and sharpness of the interfaces is clear. But alternating gradient force magnetometer (AGFM) measurements for hysteresis loops (magnetization v/s magnetic field M-H) show no switching for the layers deposited with a substrate bias of 20W, while the layers deposited without any bias show clear switching between the two magnetic layers. The M-H loops have been shown in figure 2a) and b). It is clear from these loops that in layers deposited with a bias there is no significant coercivity (H_c) difference between the FM layers to observe independent switching.

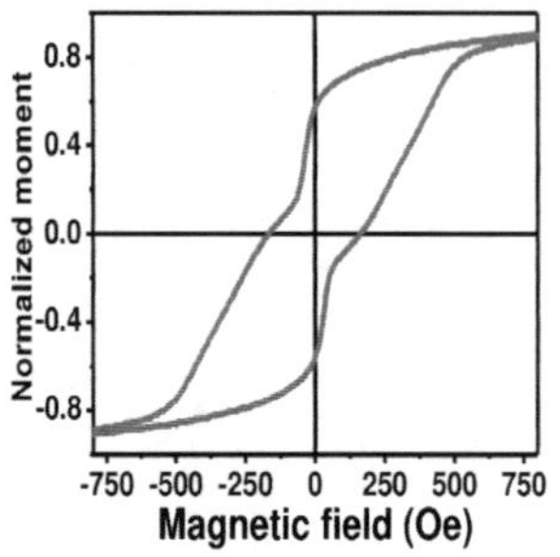

Figure 2 a). AGFM M-H loops for structure without substrate bias for any layer.

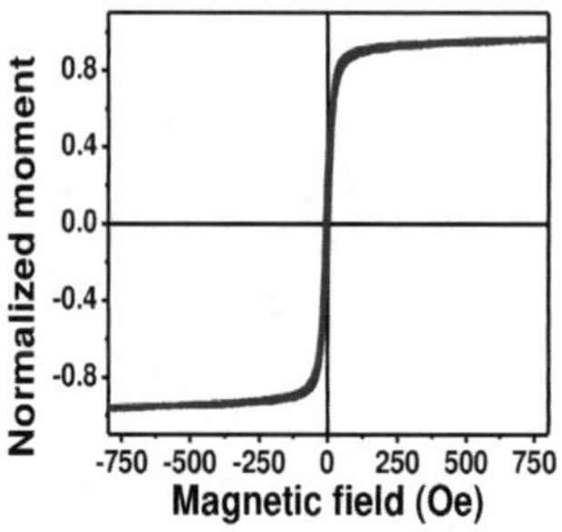

Figure 2 b). AGFM M-H loops for structure with substrate bias for all layers.

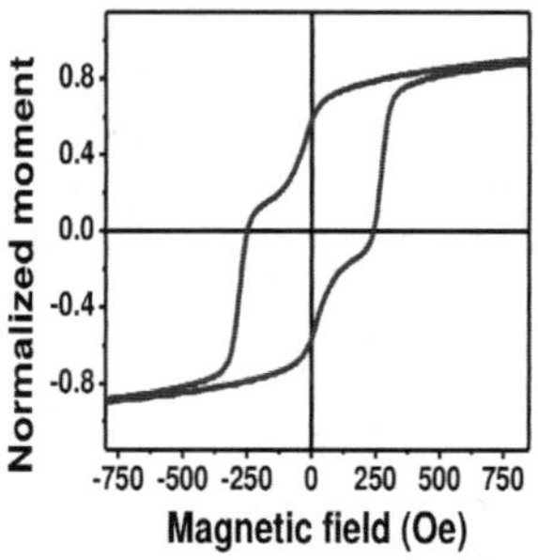

Figure 2 c). AGFM M-H loops for structure with substrate bias for all layers except IrMn.

The reason for above observation is found to be the change in composition of the IrMn antiferromagnet layer. The compositional variation of a sputtered alloy with substrate bias has been studied elsewhere using both experiments and modelling [6]. Substrate bias as such has minimal effect on coercivity of a Co free layer. A 5nm Co layer has a coercivity of about 10 Oe, both with and without substrate bias during deposition. No unidirectional anisotropy is observed in figure 2a) since the deposition was done without an in-situ magnetic field.

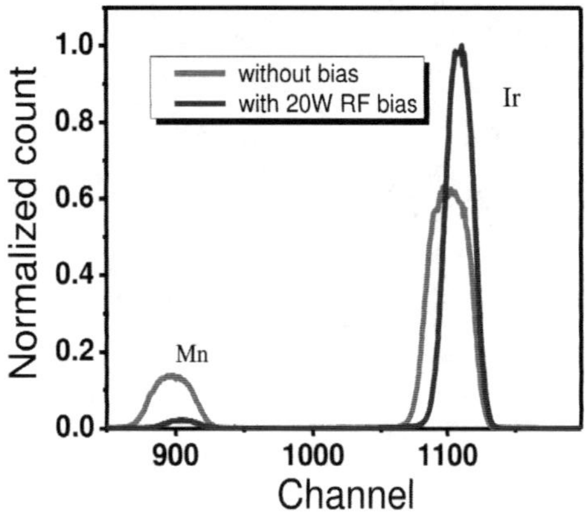

Figure 3. RBS data for IrMn with and without substrate bias

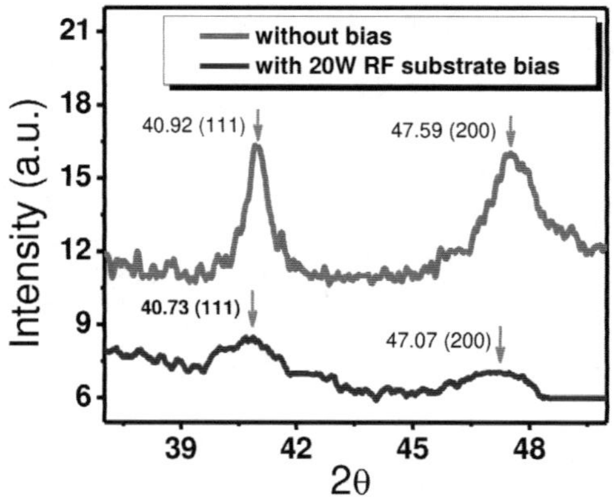

Figure 4. XRD signal for IrMn with and without substrate bias. Peaks broaden with bias.

Although substrate bias improves the surface RMS roughness of IrMn by over 100% (from 9Å to 4Å) as observed from the AFM data, it turns out that its composition no longer makes it an antiferromagnet to exchange bias the fixed layer in the MTJ. RBS experiments indicate that IrMn composition changes from 34:66 to about 82:18 with substrate bias indicating a preferential re-sputtering of Mn from the substrate (due to its lower atomic mass). The results are shown in figure 3. XRD also shows a shift in the (111) peak of IrMn from 40.92 ° to 40.73 °, closer to the elemental Ir peak (40.044 °). At the same time the IrMn (111) peak, which is critical to get a good exchange bias, is broadened with substrate bias as shown in figure 4. A 5 nm film of Co, deposited (with an in-situ field) over IrMn without substrate bias shows an exchange bias of about 150 Oe whereas Co on IrMn with substrate bias, shows no exchange bias as shown in figure 5a) and b). In order to observe the exchange bias effect, the composition of the IrMn is very critical [7]. The M-H loop for a structure with an applied bias for all layers except IrMn shows switching as shown in figure 2 c).

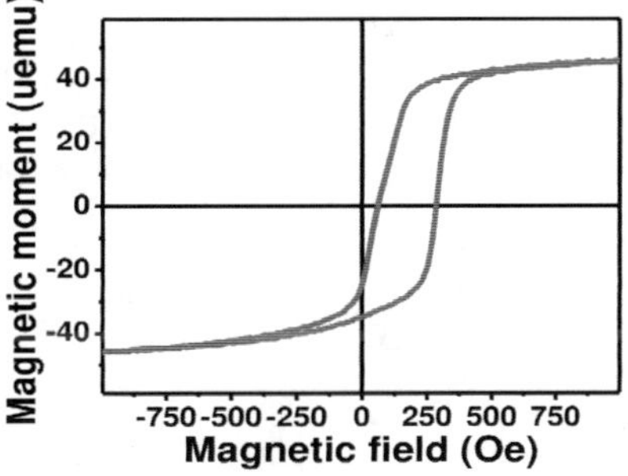

Figure 4 a). AGFM signal showing a clear exchange bias (150 Oe) for IrMn/Co films when IrMn deposited without bias.

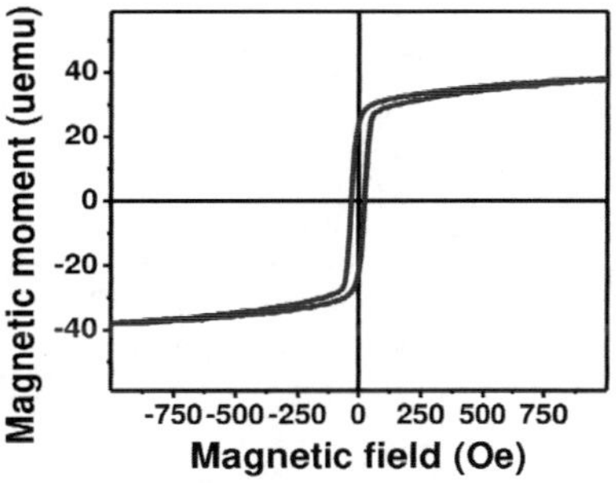

Figure 4 b). AGFM signal showing loss of exchange bias in IrMn/Co structure. IrMn deposited with substrate bias.

Antiferromagnetic materials like FeMn might be used with substrate bias without any significant changes in the composition as Fe and Mn masses are similar, unlike Ir and Mn in IrMn.

The RBS signal for AlO_x did not show any significant differences in the Al:O ratio (being 2:3) except a relatively higher Ar concentration with bias as shown in figures 6 a) and b). Even without a substrate bias presence of Ar is possible and substrate bias provides an efficient way of tuning it. The presence of Ar and its role in ultrathin barrier layers in MTJ structures will be of special interest.

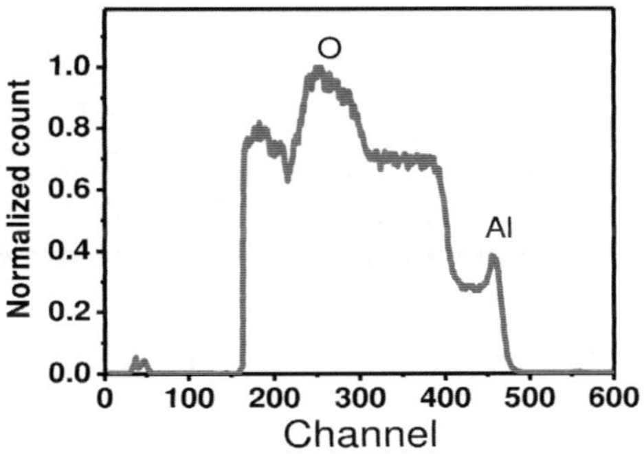

Figure 6 a). RBS signal for AlOx without substrate bias

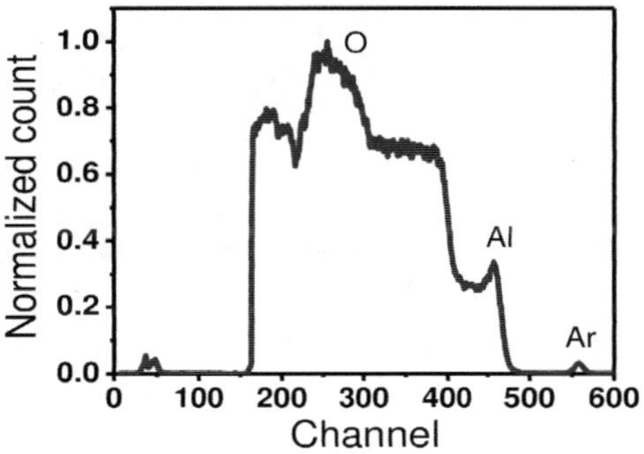

Figure 6 b). RBS signal for AlOx with substrate bias. No difference except higher Ar concentration.

4. Conclusion

The importance of process parameters on the physical structure of thin films and interfaces is demonstrated. Applying a substrate bias during deposition of multilayers is a very effective way to get smooth layers and flat interfaces even without annealing. Post deposition annealing may further improve the uniformity and crystallinity of the layers. But unwanted consequences with substrate bias deposition like change in composition and deterioration of crystallinity need to be taken care of. A Ge buffer layer also provides a smooth surface for the layers to be grown on. Studying the effect of substrate bias on crystallinity and composition of layers other than IrMn and barrier in the MTJ structure is also very important along with studying the effect on the transport properties. Such a study will help to develop a better understanding of MTJs.

5. Acknowledgments

The present work is supported by the Singapore Ministry of Education Academic Research Fund Tier 2 (MOE2008-T2-1-105). The author acknowledges experimental support from Dr. Kalon Gopinadhan, Mallikarjuna Rao Motapothula, Dr. Zhang Jixuan and Kwon Jae Hyun.

References

1. Yuasa, S. and D.D. Djayaprawira, *Giant tunnel magnetoresistance in magnetic tunnel junctions with a crystalline MgO(001) barrier. Journal of Physics D-Applied Physics*, 2007. **40**(21): p. R337-R354.
2. Parkin, S., et al., *Magnetically engineered spintronic sensors and memory. Proceedings of the IEEE*, 2003. **91**(5): p. 661-680.
3. Ikeda, S., et al., *Tunnel magnetoresistance of 604% at 300 K by suppression of Ta diffusion in CoFeB/MgO/CoFeB pseudo-spin-valves annealed at high temperature. Applied Physics Letters*, 2008. **93**(8).
4. Velev, J.P., et al., *Interface effects in spin-polarized metal/insulator layered structures. Surface Science Reports*, 2008. **63**(9): p. 400-425.
5. Choi, G.M., et al., *Substrate Biasing Effect during MgO Deposition in CoFeB/MgO/CoFeB MTJs. IEEE Transactions on Magnetics*, 2009. **45**(6): p. 2371-2373.
6. Sinder, M., G. Sade, and J. Pelleg. *Modeling of substrate bias effect on the compositional variations in sputter-deposited TiB2+x diffusion barrier thin films*. in *Computational and Mathematical Models of Microstructural Evolution. Symposium, 13-17 April 1998*. 1998. Warrendale, PA, USA: Mater. Res. Soc.
7. Devasahayam, A.J., P.J. Sides, and M.H. Kryder, *Magnetic, temperature, and corrosion properties of the NiFe/IrMn exchange couple. Journal of Applied Physics*, 1998. **83**(11): p. 7216-7218.

Journal of Physics: Conference Series **266** (2011) 012106 doi:10.1088/1742-6596/266/1/012106

Magnetic and electrical transport properties on $(La_{0.7}Sr_{0.3}MnO_3)_{1-x}/(CuFe_2O_4)_x$ composites

Y J Seo[1], G W Kim[1], C H Sung[1], C G Lee[1] and B H Koo[1*]

[1]School of Nano & Advanced Materials Engineering, Changwon National University, Changwon, 641-773, Korea

*E-mail: bhkoo@changwon.ac.kr

Abstract. The magnetic and electrical properties of $(La_{0.7}Sr_{0.3}MnO_3)_{1-x}/(CuFe_2O_4)_x$ have been investigated. The $(La_{0.7}Sr_{0.3}MnO_3)_{1-x}/(CuFe_2O_4)_x$ composites with x = 0, 0.01, 0.03, 0.06 and 0.09 wt.% were prepared by conventional solid state reaction method, respectively. The structures, morphology, magnetic and electrical properties of the samples have been studied by X-ray diffraction (XRD), scanning electron microscopy (SEM) and physical property measurement system (PPMS). The resistance measured as a function **of** temperature demonstrates that the pure LSMO samples display metal to semiconductor transition. On the other hand, the others indicated semiconductor/insulator behaviors. The MR was measured in an applied magnetic field of 0.5T. The results clearly show that the magnetoresistance (MR) effect is enhanced at a wide temperature range with $CuFe_2O_4$ composition. It may be helpful to improve the low-field magnetoresistance (LFMR) under the influence of the spin-polarized tunneling and the spin-dependent scattering.

1. Introduction

Perovskite manganites $Re_{1-x}A_xMnO_3$ (Re: rare earth, A= Ca, Sr, Ba) have attracted a lots of interest due to the colossal magnetoresistance(CMR) [1-4]. These manganites have found two CMR effects: one is intrinsic CMR [5]; the other is extrinsic CMR. The intrinsic CMR is caused by the Zener's double exchange (DE) mechanism [6] and the cooperative Jahn-Teller effect [7]. However, the intrinsic CMR effect in the perovskite manganites is normally restricted to a magnetic field of several teslas and a narrow range of the metal-insulator (M-I) transition temperature (T_p). So, this is far from the practical applications. The extrinsic CMR can be explained by spin-polarized tunneling or spin-dependent scattering at grain boundary of charge carriers [8-9]. The extrinsic CMR can be found at low fields, this behavior is the low-field magnetoresistance (LFMR) effect. The existence of an insulating phase such as Al_2O_3 [10], NiO [11], SiO_2 [12], CeO_2 [13] and ZnO [14] in grain boundary improves low field magnetoresistance (LFMR) in a wide temperature range. Many researchers have attempted to synthesize CMR-insulators composites in order to reduce the required magnetic fields of the CMR effects at room temperature.

In this present work, we attempt to introduce the insulating $CuFe_2O_4$ into the manganites and investigate the structure, magnetic and electrical transport properties of ferromagnetic/insulator (1-

* Bon Heun Koo, Corresponding author.

x)$La_{0.7}Sr_{0.3}MnO_3$/$x$$CuFe_2O_4$ composites (referred as LSMO/CFO). We choose CFO as second phase because it is good insulator and nonmagnetic material, thus it can enhance the LFMR effects.

2. Experimental

The (1-x)$La_{0.7}Sr_{0.3}MnO_3$/$x$$CuFe_2O_4$samples ($x$ = 0 ~ 0.09) were prepared as follows: First, the pure polycrystalline LSMO was prepared by the traditional ceramic reaction method. The stoichiometric amount of La_2O_3, Sr_2O_3 and Mn_2O_3, (the high purity 99.9%, Aldrich) were ground completely to achieve the homogeneity of the mixed powder for 12h by using a ball milling technology and then pre-calcinated in air at 1000°C for 12h. The pre-calcinated materials were again ground and pre-sintered at 1200°C for 24h. Secondly the prepared LSMO powder was mixed with CFO of high purity and then the mixed LSMO/CFO powders were re-processed, according to the above experimental system to get the expected composites powder. Finally, the LSMO/CFO composites were ground to fine powder and palletized at 5 tons. Finally the samples were sintered at 1300°C for 24h followed by slow cooling, as is an essential and important process to favor the required oxygen content in the material. The structural properties of the LSMO/CFO composites were analyzed by X-ray diffraction (Cu Kα_1, λ = 0.154 nm) and the scans were performed with 0.02°θ step size in the 2θ range of 20 - 80°. The surface morphologies of the LSMO/CFO composites were observed by a JSM 5610 scanning electron microscope (SEM). The M-T curves and resistance (Ω) was measured by using a Quantum Design Physical Properties Measurement System (PPMS) over a temperature range of 100 – 400K.

3. Results and discussion

The XRD patterns of a series of (1-x)LSMO + xCFO (x = 0, 0.03, 0.06, 0.09 and 1) samples are shown in Fig. 1. For the composite samples, the XRD patterns can be indexed to two set of diffraction peaks corresponding to LSMO and CFO phases. As it can be seen, pure LSMO has a rhombohedral structure, and CFO has a single inverse spinel structure and no new phase appeared in the composites. Compared with the XRD pattern of pure LSMO, CFO with low doping level has no effects on the structure of LSMO. In the composite with x=0.09, small diffraction peak at 2θ of 35.39° related to CFO appears.

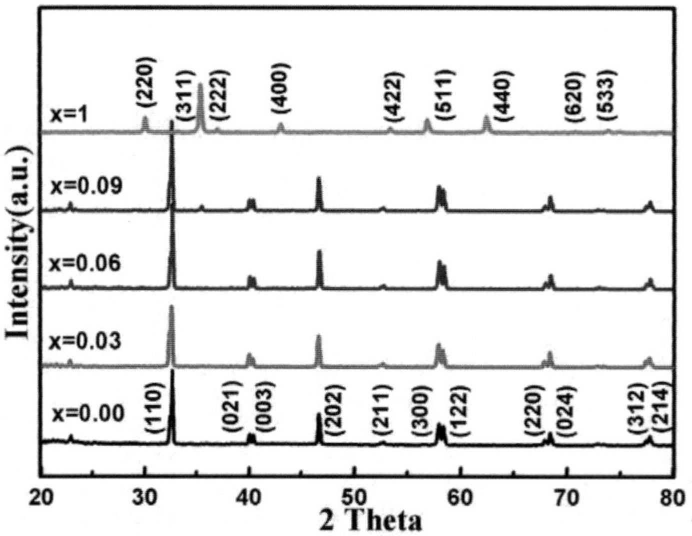

Figure 1. Room temperature X-ray diffraction patterns of (1-x)$La_{0.7}Sr_{0.3}MnO_3$/$x$$CuFe_2O_4$ composite with x=0, 0.03, 0.06, 0.09 and 1, respectively.

To confirm the coexistence of both phases in the composites, all samples were examined by SEM.

Fig. 2 shows the back scattering SEM image of pure LSMO and the CFO doped composites of x= 0.03, 0.06 and 0.09, respectively. From the image of Fig. 2, grain boundaries in pure LSMO(x=0) can be clearly observed. While for the compositions, as shown in Fig. 2(x=0.03, 0.06 and 0.09), two phases (LSMO and CFO) are clearly distinguishable from each other. Fig. 2(a, b) show that the main composition at the bright regions (a) are LSMO and the dark regions (b) are CFO with iron. It is obvious that cubic type structure CFO grains are exists between LSMO grains and CFO grain size increases with increasing the CFO contents. The shape of LSMO grains transform from a hexagon to a quadrangle as composition change from x=0.03 to x=0.09. It is expected that the cause of transition behavior is due to the inter-substitution between manganese in LSMO grains and iron in CFO grains which can transform the shape and volume of LSMO grains.

Figure 2. SEM micrographs of (1-x) $La_{0.7}Sr_{0.3}MnO_3$/$x$$CuFe_2O_4$ composites with x=0 ~ 0.09.

The temperature dependence of magnetization measured at 0.5T is shown in Fig. 3. All samples have a similar behavior of the magnetization and there is the paramagnetic (PM) to ferromagnetic (FM) transition around Curie temperature (T_c). The values of magnetization (M) at 60K are 16.1, 12.8, 11.4 and 4.8 emu/g for x=0, 0.03, 0.06 and 0.09, respectively. This decrease in M with increasing CFO concentrations is due to the dilution effect of CFO on the magnetism of the composites and result in additional magnetic disorder. The transition temperature (T_C) is decreased from 372K (x=0) to 147K (x=0.09) and T_C is defined as the temperature where dM/dT reaches minimum. The cause of T_C decrease can be seen that the DE interaction of the compounds will be destructed. We can suggest that the CFO do not react with LSMO, and only located in the interface or surface of LSMO grains.

The temperature dependent of resistivity in magnetic field of 0.5T and with the temperature from 90K to 400K for the LSMO/CFO composites is shown in Fig. 4. We can see that the samples (x=0, 0.03 and 0.06) exhibit an insulator to metal (I-M) transition behavior at the transition temperature (T_p) and with increasing LSMO content x, T_p decreases from 345K to 190K and resistivity increases. While only sample with x=0.09 show fully insulator behavior. The transition behavior can be explained by the DE mechanism in LSMO grains [15-17] and the disappearance of I-M transition and increase of resistivity are due to the enhanced grain boundaries. Massive interfaces and boundaries between LSMO and CFO grains may play the role of additional barriers, as results in increase in the carrier scattering causing enhancement in the resistivity.

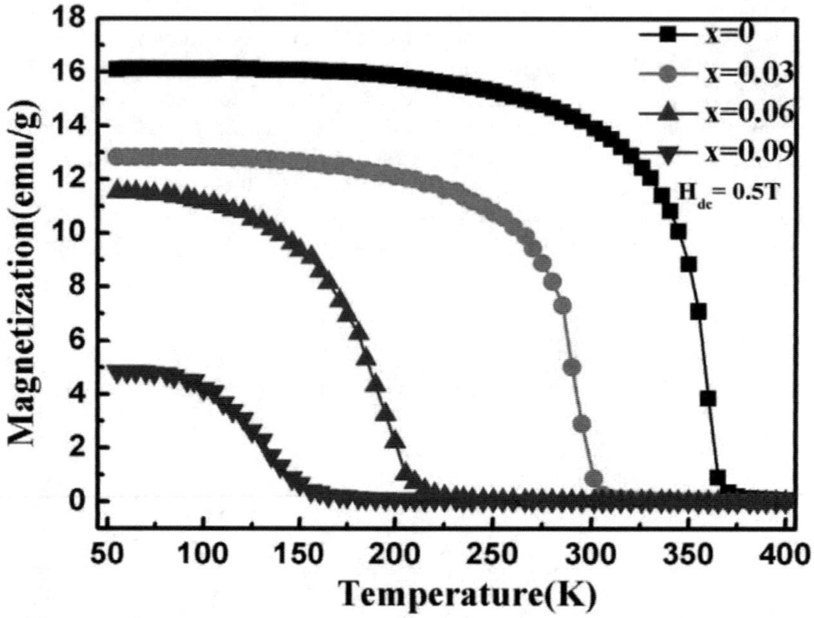

Figure 3. Temperature dependence of magnetization at 0.5T for (1-*x*) $La_{0.7}Sr_{0.3}MnO_3$/$x$$CuFe_2O_4$ combined samples.

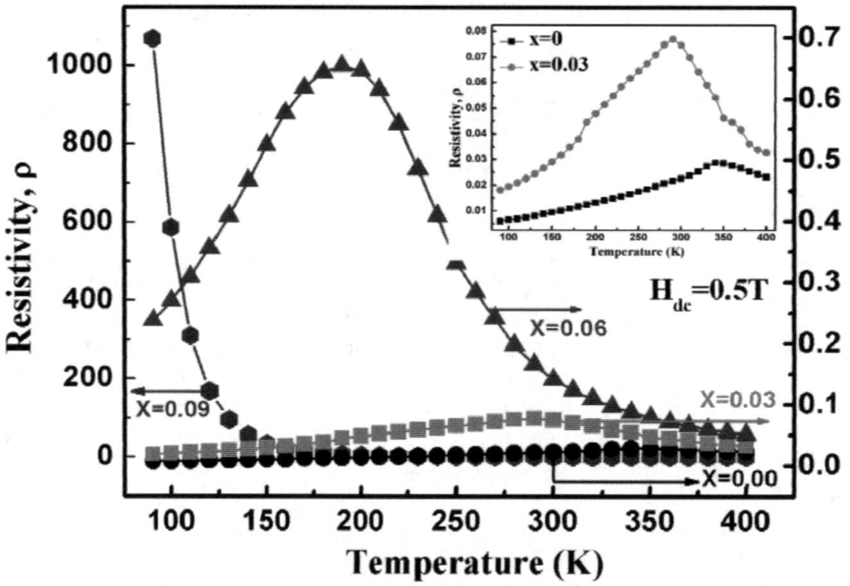

Figure 4. Temperature dependence of the resistivity at magnetic field of 0.5T for (1-*x*) $\mathbf{La_{0.7}Sr_{0.3}MnO_3}$/$x$$CuFe_2O_4$ combined samples. The inset shows the resistivity for the composites with x=0, 0.03.

Fig. 5 shows the temperature dependence of MR for all samples in the magnetic field of 0.5T and **1T.** In this paper, the MR is defined as MR (%) = [ρ(T, 0)—ρ(T, H)]/ ρ(T, 0)×100, where ρ(T, 0) and ρ(T, H) are resistivity at zero field and an applied field, respectively. It is found that the MR exhibits a maximum near the transition temperature and a rising part at low temperatures for all the samples. The MR compared with pure LSMO is enhanced at low temperatures for the composites. Besides, the maximum MR moves to lower temperatures with increment of CFO contents. The values of MR peak are 4.52, 5.12, 6.48 and 11.9% for x=0, 0.01, 0.03 and 0.06 at 0.5 T, respectively. **The MR% increases for all samples (Fig. 5b) as we increase the magnetic field from 0.5 T to 1 T.** There is one MR peaks in pure LSMO and composites with x=0.03, 0.06 have two peaks except for the sample with x=0.09. The composition x=0.03 shows the enhanced MR at room temperature. The low MR peak as the inter-grain magnetoresistance which was explained by the spin polarized tunneling at the grain boundaries and the high MR peak as the intrinsic CMR [1–3] were classified to compare the two MR peaks.

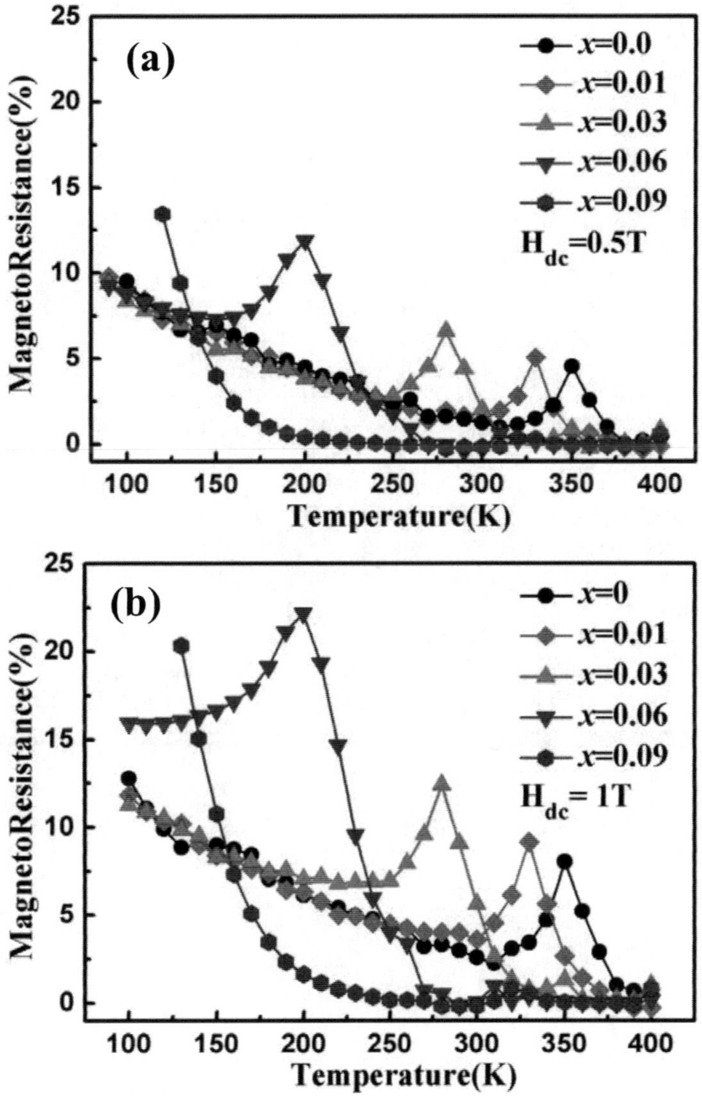

Figure 5. Temperature dependence of MR for the composite with x=0, 0.01, 0.03, 0.06 and 0.09, measured at (a) H_{dc}=0.5T and **(b) H_{dc} = 1T**

4. Conclusions

In summary, we have studied the magnetic and electrical transport properties of LSMO/$CuFe_3O_4$ composition system. The XRD and SEM results show that there is no reaction between LSMO and CFO grains, and CFO is distributed at the grain boundaries or surfaces of the LSMO grains. We have observed that with increasing CFO contents, resistivity increases but magnetization and transition temperature decreases. We can conclude that to form the composites using LSMO as matrix and CFO as dopants can enhance the LFMR in wide temperature range, which is encouraging as regards the practical application of MR materials.

References

[1] Jin S, Tiefel T H, McCormack M, Fastnachr P A, Ramesh R, Chen L H 1994 *Science* **264** 413.
[2] Xiong G C, Li Q, H.K. Ju, Mao S N, Senapati L, Xik X X, Greene R L, Venkatesan T 1995 *Appl. Phys. Lett.* **66** 1427.
[3] Ziese M 2002 *Rep. Prog. Phys.* **65** 143.
[4] Casanove M J, Roucau C, Baules P, Majimel J, Ousset J C, Magnoux D and Bobo J F 2002 *Appl. Surf. Sci.* **188** 19.
[5] Hwang H Y, Cheong S W, Ong N P, Batlogg B 1996 *Phys. Rev. Lett.* **77** 2041.
[6] Zener C 1951 *Phys. Rev.* **82** 403.
[7] Haghiri-Gosnet A M, Renard J P 2003 *J. Phys., D. Appl. Phys.* **36** R127.
[8] Gupta A, Sun J Z 1999 *J. Magn. Magn. Mater.* **200** 24.
[9] Li X W, Gupta A , Xiao G, Gong G G 1997 *Appl. Phys. Lett.* **71** 1124.
[10] Hueso L E, Rivas J, Rivadulla F, Lopez-Quintela M A 2001 *J. Non-Cryst. Solids* **287** 324.
[11] Eshraghi M, Salamati H, Kameli P 2007 *J. Alloy Compd.* **437** 22.
[12] Das D, Chowdhury P, Das R N, Srivastava C M, Nigam A K, Bahadur D 2002 *J. Magn. Magn. Mater.* **238** 178.
[13] Balcells L I, Carrillo A E, Martinez B, Fontcuberta J 1999 *Appl. Phys. Lett.* **74** 4014.
[14] Kameli P, Salamati H , Eshraghi M , Mohammadizadeh M R 2005 *J. Appl. Phys.* **98** 043908.
[15] Gupta S, Ranjit R, Mitra C, Raychaudhuri P, Pinto R 2001 *Appl. Phys. Lett.* **78** 3.
[16] Rama N 2006 *J. Magn. Magn. Mater.* **303** 342.
[17] Petrov D K, Krusin-Elbaum L, Suan J Z, Feild C, Duncombe P R 1999 *Appl. Phys. Lett.* **75** 7.

Acknowledgements

This work was supported by the Korea Research Foundation (KRF) grant funded by the Korean Government (MEST) (No.2009-0077268). This research was supported by the MKE(The Ministry of Knowledge Economy), Korea, under the ITRC(Information Technology Research Center) support program supervised by the NIPA(National IT Industry Promotion Agency) (NIPA-2010-C1090-1021-0015).

Journal of Physics: Conference Series **266** (2011) 012107 doi:10.1088/1742-6596/266/1/012107

Bias-voltage-dependence of magnetoresistance for epitaxial Fe/MgO/Co_2MnSn tunnel junctions

M A Tanaka[1], T Hori[1], S Hori[1], K Kondou[2], S Kasai[3], T Ono[2], K Mibu[1]
[1] Nagoya Institute of Technology, Nagoya, Aichi 466-8555, Japan
[2] Institute for Chemical Research, Kyoto University, Uji, Kyoto 611-0011, Japan
[3] Magnetic Materials Center, National Institute for Materials Science, Tsukuba, Ibaraki 305-0047, Japan

E-mail: mtanaka@nitech.ac.jp

Abstract. Tunnel magnetoresistance (TMR) was investigated for fully epitaxial Fe/MgO/Co_2MnSn magnetic tunnel junctions (MTJs) fabricated on MgO (001) substrates. The bias-voltage-dependence of the TMR ratio showed that the sign of TMR ratio at 300 K becomes negative for the bias voltage between -800 mV and +200 mV, which may be caused by the impurity state or other imperfections at the interface. The sign change in TMR effect was also observed with changing the temperature at low fixed bias voltage, implying that the effect is not simply due to a bias-dependent change of the density of state at the Fermi level.

1. Introduction

The tunnel magnetoresistance (TMR) of magnetic tunnel junctions (MTJs) made of two ferromagnetic electrodes separated by an insulating layer has attracted much interest for its use in new magnetic devices [1-2]. Within the Jullière model, the magnitude of TMR is determined by the spin polarization of the density of states at the Fermi energy of the two ferromagnetic electrodes [3]. Half metallic materials, which exhibit perfect spin polarization at the Fermi energy, are expected to be ideal materials to achieve large TMR ratio. Recently, TMR ratio of above 300 % at RT [4] was achieved in MTJs using Co-based Heusler alloys [4-7] which are theoretically predicted as half metallic materials [8-9]. On the other hand, we have reported that Co_2MnSn Heusler alloy films with relatively uniform local magnetism can be prepared when the films are grown by atomically controlled alternate deposition at 400 °C [10].

In this paper, we report on bias-voltage-dependence of the magnetoresistance for fully epitaxial Fe/MgO/Co_2MnSn MTJs. The negative TMR effect was observed between -800 mV and +200 mV at 300 K. Moreover, it was found that, at low bias voltage, the sign of TMR changed as a function of temperature.

2. Experiments

Layered structures of MgO (5 nm)/Cr (30 nm)/Fe (20 nm)/MgO (2.4 nm)/Co_2MnSn (26.1 nm)/Co (10 nm)/Cr (5 nm) were prepared on MgO (001) substrates using an electron beam deposition system. The bottom ferromagnetic Fe layer was deposited at RT and annealed at 350 °C in order to improve crystallographic quality. After the deposition of the MgO barrier layer, the top ferromagnetic Co_2MnSn layer was prepared by depositing one atomic layer of Co, half an atomic layer of Mn and Sn alternately in a controlled manner at the substrate temperature of 400 °C. The interface atoms of the

Co_2MnSn layer on the MgO barrier were designed to be Mn and Sn [10]. The crystal structures of the layered structures were investigated using x-ray diffraction (XRD) with θ - 2θ and polar plots. MTJs were patterned into ellipse-shaped pillars of 9.0×4.5 μm^2 or 6.0×3.0 μm^2 using a conventional photolithography and Ar ion milling process. Magnetoresistance measurements were carried out using a standard dc four-probe method. The positive current is defined as the current with the electrons flowing from the top Co_2MnSn layer to the bottom Fe layer.

3. Results

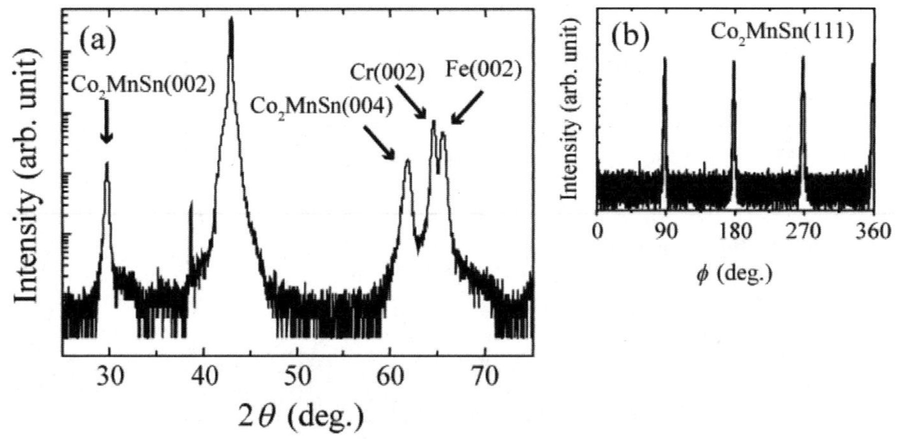

Figure 1. X-ray diffraction patterns of the MgO/Cr/Fe/MgO/Co_2MnSn/Co/Cr layered structure. (a) θ - 2θ plot with the scattering vector normal to the film plane, and (b) polar plot for the (111) peaks of the Co_2MnSn layer.

Figure 1(a) shows the θ - 2θ X-ray diffraction pattern of an MgO/Cr/Fe/MgO/Co_2MnSn/Co/Cr layered structure. The Co_2MnSn (002), Co_2MnSn (004) and Fe (002) peaks were observed, which indicates that the Co_2MnSn and Fe electrodes were grown with the (001) orientation epitaxially on the MgO (001) substrate. The polar plot for the (111) peaks of the Co_2MnSn layer is shown in figure 1(b). The pattern indicates that the Co_2MnSn layer has an $L2_1$ structure. The degree of $L2_1$ order of Co_2MnSn estimated from the intensity ratio of (111) and (002) was about 56%.

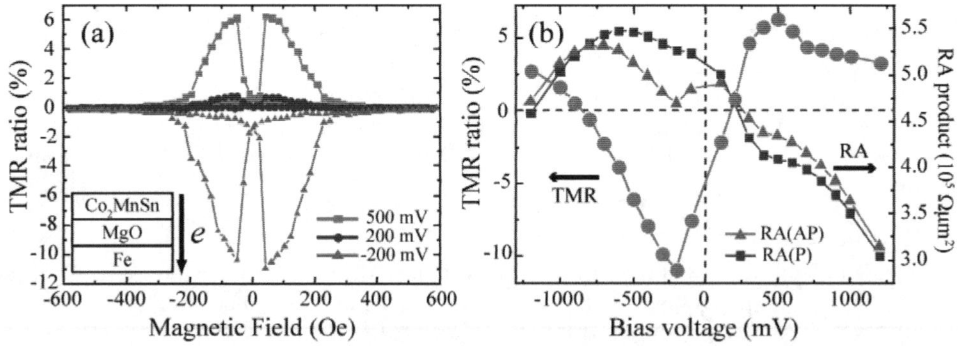

Figure 2. (a) Magnetoresistance curves for the epitaxial Fe/MgO/Co_2MnSn MTJ under various bias voltages at 300 K. (b) Bias-voltage-dependence of TMR ratio and RA products for anti-parallel (AP) and parallel (P) configurations at 300 K.

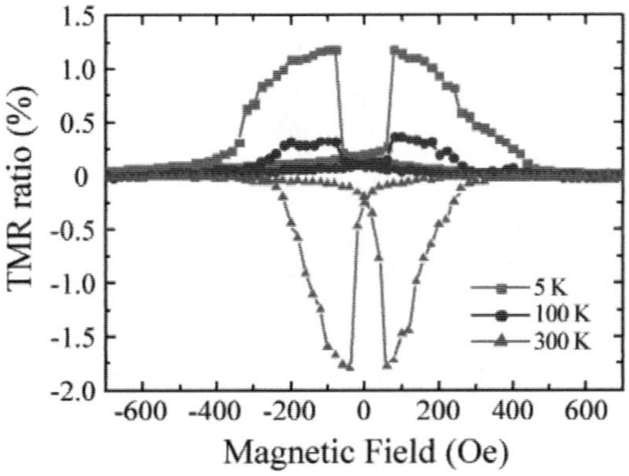

Figure 3. Magnetoresistance curves for the Fe/MgO/Co_2MnSn MTJ at a bias voltage of +100 mV at 5 K, 100 K and 300 K.

The magnetoresistance curves under various bias voltages at 300 K are shown in figure 2(a). At bias voltages of +500 mV and -200 mV, positive and negative TMR with the ratios of +6.3% and -10.9 % were observed, respectively. Figure 2(b) displays the bias-voltage-dependence of the TMR ratio and resistance area products (RA products) for anti-parallel (AP) and parallel (P) configurations. The bias-voltage-dependence of TMR exhibits an asymmetric character regarding the polarity of the voltage, and negative TMR ratios are realized in the bias voltage range between -800 mV and +200 mV. The RA products exhibit resistance drops at -200 mV and +400 mV for AP and P configurations, which correspond to the maximum and minimum TMR ratio, respectively. Figure 3 shows the magnetoresistance curves at a bias voltage of +100 mV at 5 K, 100 K and 300 K. A negative TMR effect was observed at 300 K, whereas a positive TMR effect appeared at 5 K. The sign change of the TMR effect with temperature was observed at bias voltage between -100 mV and +150 mV.

4. Discussion

The change in the sign of the TMR effect as a function of bias voltage at 300 K similar to that in figure 2(a) has been reported for some systems. Sharma *et al.* performed magnetoresistance measurements for Permalloy/Al_2O_3/Ti_2O_5/Permalloy MTJs and observed positive and negative TMR effects depending on the bias voltages [11]. They concluded that the phenomenon was caused by the difference of the spin polarization of two barrier/electrode interfaces and by the sign change of the spin polarization with the bias voltage in one side. The sign change of TMR effect with changing the bias voltages can also occur by the impurity states near the Fermi surface [12], Co anti-sites [13] and non-bonding states at the interface [14]. For comparison with the Fe/MgO/Co_2MnSn MTJs, Fe (30 nm)/MgO (2.4 nm)/CoFe (10 nm) MTJs were prepared. The magnetoresistance of the Fe/MgO/CoFe MTJs displayed large TMR ratio of about 150 % at a bias voltage of +100 mV and the TMR ratio simply decreased with increasing the bias voltage at 300 K, which means that the TMR behaviour of the Fe/MgO/Co_2MnSn MTJs may be due to the impurity states or other imperfections at the Co_2MnSn/MgO interface.

The temperature-dependent magnetoresistance curves in figure 3 indicate that the sign change of TMR ratio is not simply due to the difference of spin-dependent density of states (DOS) because the DOS-based spin polarization in barrier/electrode interfaces would be almost the same in the range between 5 K and 300 K. In other words, the temperature independent electron transport coexists with temperature dependent one in the TMR effect for the Fe/MgO/Co_2MnSn MTJs. The strong temperature dependence of the TMR ratio at low bias voltage may be due to the magnon-assisted inelastic tunnelling. Marukame *et al.* reported that the TMR ratio for the $Co_2MnGe/MgO/CoFe$ MTJs showed strong dependence on the temperature at a low bias voltage [15]. The TMR curves in their paper showed that the temperature-dependent sign change occurs in the negative bias voltage range between -300 mV and -200 mV, which is larger voltage than the present case. The result in the present paper implies that the mechanism is not simply due to the DOS-based spin-dependent tunnelling for the $Fe/MgO/Co_2MnSn$ MTJs.

5. Conclusion

We have investigated the tunnel magnetoresistance effect for fully epitaxial $Fe/MgO/Co_2MnSn$ MTJs with the Co_2MnSn layer prepared using atomically controlled alternate deposition. The TMR measurements at various bias voltages showed a negative TMR effect at the bias voltage between -800 mV and +200 mV at 300 K. At the low bias voltage, positive and negative TMR effects were observed at 5 K and 300 K, which may be caused by the increase of magnon scattering with increasing temperature.

Acknowledgments

The authors thank Prof. J. Inoue of Nagoya University for fruitful discussion. This work was supported by a Grant-in-Aid for Scientific Research in Priority Areas "Creation and Control of Spin Current" (No. 21019008) from the Ministry of Education, Culture, Sports, Science and Technology of Japan.

References

[1] Miyazaki T and Tezuka N 1995 *J. Magn. Magn. Mater.* **139** L231
[2] Moodera J S, Kinder L R, Wong T M and Meservey R 1995 *Phys. Rev. Lett.* **74** 3273
[3] Jullière M 1975 *Phys. Lett.* A **54** 225
[4] Wang W, Liu E, Kodzuka M, Sukegawa H, Wojcik M, Jedryka E, Wu G H, Inomata K, Mitani S and Hono K 2010 *Phys. Rev.* B **81** 140402(R)
[5] Inomata K, Okamura S, Goto R and Tezuka N 2003 *Jpn. J. Appl. Phys.* **42** L419
[6] Sakuraba Y, Hattori M, Oogane M, Kubota H, Ando Y, Kato H, Sakuma A and Miyazaki T 2006 *Appl. Phys. Lett.* **88** 192508
[7] Yamamoto M, Ishikawa T, Taira T, Li G-F, Matsuda K-I and Uemura T 2010 *J. Phys.: Condens. Matter.* **22** 164212
[8] Galanakis I, Dederichs P H and Papanikolaou N 2002 *Phys. Rev.* B **66** 134428
[9] Kurtulus Y, Dronskowski R, Samolyuk G D and Antropov V P 2005 *Phys. Rev.* B **71** 014425
[10] Mibu K, Gondo D, Tanaka M A, Kondou K, Kasai S and Ono T 2010 *J. Phys.: Conf. Series* **200** 062012
[11] Sharma M, Wang S X and Nickel J H 1999 *Phys. Rev. Lett.* **82** 616
[12] Inoue J, Nishimura N and Itoh H 2002 *Phys. Rev.* B **65** 104433
[13] Picozzi S, Continenza A and Freeman A J 2004 *Phys. Rev.* B **69** 094423
[14] Miura Y, Abe K and Shirai M 2010 *J. Phys.: Conf. Series* **200** 052016
[15] Marukame T, Ishikawa T, Matsuda K-I, Uemura T and Yamamoto M 2006 *J. Appl. Phys.* **99** 08A904

Journal of Physics: Conference Series **266** (2011) 012108 doi:10.1088/1742-6596/266/1/012108

Angle dependence of the magnetoresistance of CCP-CPP-GMR system

Tomohiro Taniguchi[1,2†] **and Hiroshi Imamura**[1*]

[1] Nanosystem Research Institute, National Institute of Advanced Industrial Science and Technology, 1-1-1 Umezono, Tsukuba, Ibaraki 305-8568, Japan

[2] Institute of Applied Physics, University of Tsukuba, 1-1-1 Tennou-dai, Tsukuba, Ibaraki 305-8573, Japan

E-mail: †tomohiro-taniguchi@aist.go.jp, *h-imamura@aist.go.jp

Abstract. Angle dependence of the magnetoresistance (MR) effect due to a magnetic domain wall confined in a current-confined-path current-perpendicular-to-plane giant magnetoresistance system was studied by solving the Boltzmann equation. We calculate the scattering rates by using the nonperturbative wave function of the conduction electrons, which enables us to invesigate the MR ratio of a sufficiently thin domain wall. We find that the dependence of the MR ratio on the rotation angle of the domain wall φ is approximately described by $\cos\varphi$, which has a good agreement with the recent experimental results.

Magnetoresistance (MR) effect due to a magnetic domain wall confined in the current-confined-path (CCP) current-perpendicular-to-plane (CPP) giant magnetoresistnace (GMR) system has drawn enormous attention because of its potential application to spin-electronics devices such as a read sensor for high-density magnetic recording [1–4]. In these devices, a high MR ratio with a small resistance area product is required. Recently, Takagishi *et al.* found the MR ratio of 11 – 12 % at a resistance area product of 0.3 – 0.55 $\Omega\mu m^2$ in an $Fe_{50}Co_{50}/AlO/Fe_{50}Co_{50}$ CCP-CPP-GMR system with the magnetic domain walls of 1.2 nm thick [2]. They also found that the dependence of the MR ratio due to the domain wall on the rotation angle of the wall φ is approximately described by $\cos\varphi$.

In 1997, Levy and Zhang investigated the MR effect due to the domain wall scattering by using the same Hamiltonian that was used to explain GMR effect in a spin-valve [5]. By solving the Boltzmann equation, they found that the MR ratio is proportional to $1/d^2$, where d is the thickness of the domain wall. Their theory has obtained good agreements with experiments [6].

However, as pointed in our previous paper [7], the theory of Levy and Zhang cannot be applied to a sufficiently thin domain wall such as a CCP-CPP-GMR system because they use the perturbative wave function in the calculation of the scattering rate and they apply the diffusion approximation to the Boltzmann equation. The validity of the theory of Levy and Zhang is restricted to the region $d \gg \ell_J, \ell_{\rm mfp}$, where both the electrons' traveling length during the precession of their spins around the *sd*-exchange field J, $\ell_J = \pi\hbar v_{\rm F}/(4J)$, and the mean free path, $\ell_{\rm mfp}$, are on the order of a few nanometer [8], i.e., they are comparable to the thickness of the domain wall in the CCP-CPP-GMR system. If we use the theory of Levy and Zhang, the MR ratio of the CCP-CPP-GMR system is estimated to be more than 50 % with $d = 1.2$ nm and the spin polarization $\beta \sim 0.5 - 0.6$, which is typical value of the conventional ferromagnetic metals such as FeCo [9]. This value of the MR ratio is much higher than the experimental

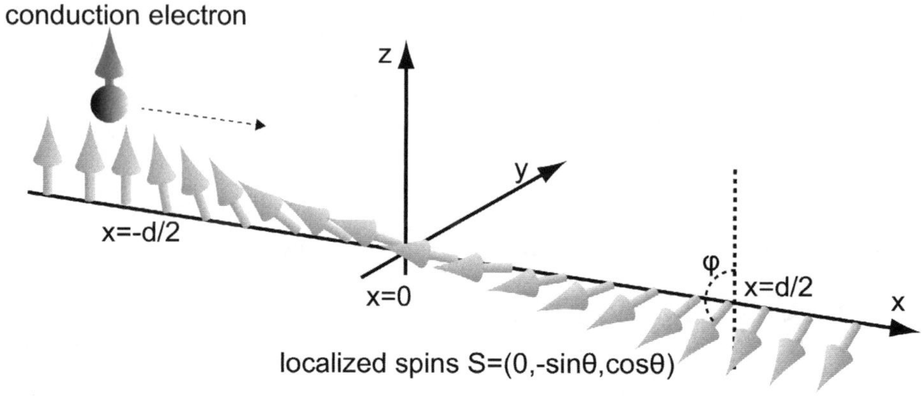

Figure 1. The schematic view of the electron transport in the CCP-CPP-GMR system. φ is the rotation angle of the domain wall.

values ($11-12$ % as shown in Refs. [2]). In Ref. [7], we revised the theory of Levy and Zhang to calculate the MR effect due to a magnetic domain wall by using the nonperturbative wave function in the calculation of the scattering rate and solving the Boltzmann equation without the diffusion approximation, and found that the MR ratio is about $10-20$ % for $d=1.2$ nm and $\beta \sim 0.5-0.6$, which is consistent with the experimental results in Refs. [1, 2].

In this paper, we study the dependence of the MR ratio of the magnetic domain wall on the rotation angle of the wall φ by developing the formalism presented in Ref. [7]. We found that the dependence of the MR ratio is approximately described by $\cos\varphi$, which agrees with the experimental results in Ref. [2].

We consider electron transport in a one-dimensional magnetic domain wall that lies over $-d/2 \leq x \leq d/2$. We assume that the spin-dependent transport of the conduction electrons is desribed by the Hamiltonian $\hat{H}_0 = -\hbar^2\nabla^2/(2m) - J\hat{\boldsymbol{\sigma}}\cdot\hat{\mathbf{S}}(\mathbf{r})$, where J is the sd-exchange coupling constant between the conduction (s-like) electrons and the localized (d-like) spins, $\hat{\boldsymbol{\sigma}}$ is the vector of the Pauli matrices and $\hat{\mathbf{S}} = (0, -\sin\theta, \cos\theta)$ is the unit vector along the direction of the localized spin (see Fig. 1). The angle θ is given by

$$\theta(x) = \frac{\varphi}{d}\left(x + \frac{d}{2}\right) , \tag{1}$$

where φ is the rotation angle of the domain wall. On the other hand, the spin-dependent impurity scattering is described by $\hat{V} = \sum_i [v - j\hat{\boldsymbol{\sigma}}\cdot\hat{\mathbf{S}}(\mathbf{r})]\delta(\mathbf{r}-\mathbf{R}_i)$, where $\mathbf{R}_i$ is the position of the i-the impurity, and v and j are the spin-independent and spin-dependent scattering potential, respectively.

The MR ratio is defined by $(\rho-\rho^{(0)})/\rho^{(0)}$, where $\rho = 1/(\sigma^+ + \sigma^-)$ and $\rho^{(0)}$ are the resistivity of the domain wall and that of the bulk ferromagnetic metal, respectively. The conductivity of the spin-s electrons ($s=\pm$) is defined as

$$\sigma^s = -\frac{e}{E}\int \frac{\mathrm{d}^3\mathbf{k}}{(2\pi)^3} v_x^s f^s(\mathbf{k}) , \tag{2}$$

where E is the strength of the applied electric field. The distribution function of the non-equilibrium conduction electrons $f^s(\mathbf{k})$ is obtained by solving the Boltzmann equation [7]

$$-ev_x^s E = -\frac{1}{\tau^s(k_x^s)} g^s(k_x^s) + \frac{m}{2\pi\hbar^3}\int_{-k_\mathrm{F}^s}^{k_\mathrm{F}^s} \mathrm{d}k'_x |V_{\mathbf{k}\mathbf{k}'}^{ss}|^2 g^s(k'_x) + \frac{m}{2\pi\hbar^3}\int_{-k_\mathrm{F}^{-s}}^{k_\mathrm{F}^{-s}} \mathrm{d}k'_x |V_{\mathbf{k}\mathbf{k}'}^{s-s}|^2 g^{-s}(k'_x) , \tag{3}$$

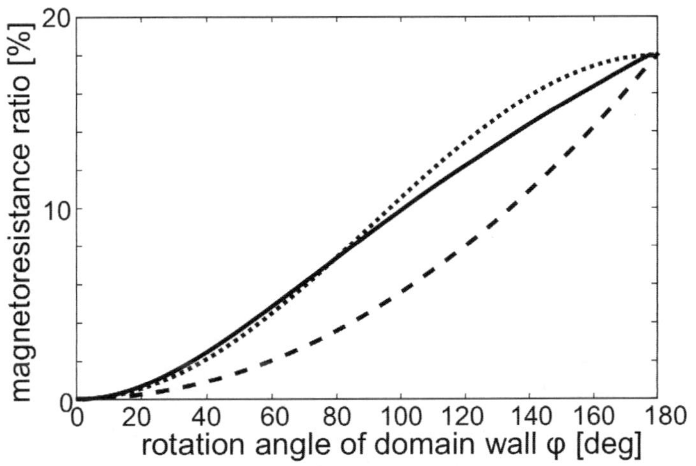

Figure 2. The dependence of the MR ratio of the CCP-CPP-GMR system on the rotation angle of the domain wall φ. The solid line is obtained from our formalism with $\beta = 0.6$ and $d = 1.2$ nm. The dotted line represents the fit of the experimental results of Ref. [2], i.e., MR ratio $\propto \cos\varphi$. The dashed line is proportional to φ^2.

where we neglect the effect of the spin accumulation, for simplicity. The function g^s is defined by $f^s(\mathbf{k}) = -\delta(\varepsilon(\mathbf{k}, s) - \varepsilon_F)g^s(\mathbf{k})$, where ε_F is the Fermi energy of the conduction electrons. The matrix elements of the scattering potential are given by

$$|V^{ss}_{\mathbf{kk}'}|^2 = c_i \left[(v - sj)\cos\frac{\phi(k^s_x)}{2}\cos\frac{\phi(k^{s\prime}_x)}{2} + (v + sj)\sin\frac{\phi(k^s_x)}{2}\sin\frac{\phi(k^{s\prime}_x)}{2}\right]^2 , \tag{4}$$

$$|V^{s-s}_{\mathbf{kk}'}|^2 = c_i \left[(-sv + j)\cos\frac{\phi(k^s_x)}{2}\sin\frac{\phi(k^{-s\prime}_x)}{2} + (sv + j)\sin\frac{\phi(k^s_x)}{2}\cos\frac{\phi(k^{-s\prime}_x)}{2}\right] , \tag{5}$$

where c_i is the impurity concentration. The factor $\phi(k^s_x)$ is defined by $\tan[\phi(k^s_x)/2] = (k^s_x\theta')/[k^2_J + \sqrt{(k^s_x\theta')^2 + k^4_J}]$, where $\theta' = \varphi/d$ and $k_J = \sqrt{2mJ}/\hbar$. The maximum value of k^s_x is given by $k^s_F = \sqrt{k^2_F + (\theta'/2)^2 + s\sqrt{(k_F\theta')^2 + k^4_J}}$. As mentioned in Ref. [7], the formalism presented above is reduced to that developed by Levy and Zhang [5] in the limit of $d \gg \ell_J$. The distribution function $f^s(\mathbf{k})$ is obtained by solving Eq. (3) numerically.

We calculate the MR ratio from Eq. (2) by varying the value of the angle φ from 0 to π. The values of the parameters we used are as follows. The Fermi energy ε_F and the sd-exchange coupling constant J are taken to be 5.0 eV and 0.5 eV, respectively [8, 10]. The resistivity of the bulk ferromagnetic metal $\rho^{(0)}$ is taken to be 150 Ωnm [9]. Using these parameters, ℓ_J is estimated to be 1.4 nm, and the mean free path $\ell_{\rm mfp} = (\ell^+_{\rm mfp} + \ell^-_{\rm mfp})/2$ is estimated to be 5.9 nm, where $\ell^s_{\rm mfp} = v^{s(0)}_F\tau^{s(0)}$, $v^{s(0)}_F = \hbar k^{s(0)}_F/m$, $k^{s(0)}_F = \sqrt{2m(\varepsilon_F + sJ)}/\hbar$, and $\tau^{s(0)} = \pi\hbar^3/[mc_i(v - sj)^2 k^{s(0)}_F]$.

The dependence of the MR ratio on φ with $d = 1.2$ nm and $\beta = 0.6$ shown by the solid line in Fig. 2 has good agreements with the fit of the experimental results of Takagishi *et al.* [2], i.e., the MR ratio is proportional to $\cos\varphi$ as shown by the dotted line in Fig 2. On the other hand, the MR ratio derived by Levy and Zhang [5] is given by

$$\text{MR ratio} = \frac{4}{5}\xi^2_{\rm eff}\left(\frac{\beta^2}{1-\beta^2}\right)\left(3 - \frac{5\sqrt{1-\beta^2}}{3}\right) , \tag{6}$$

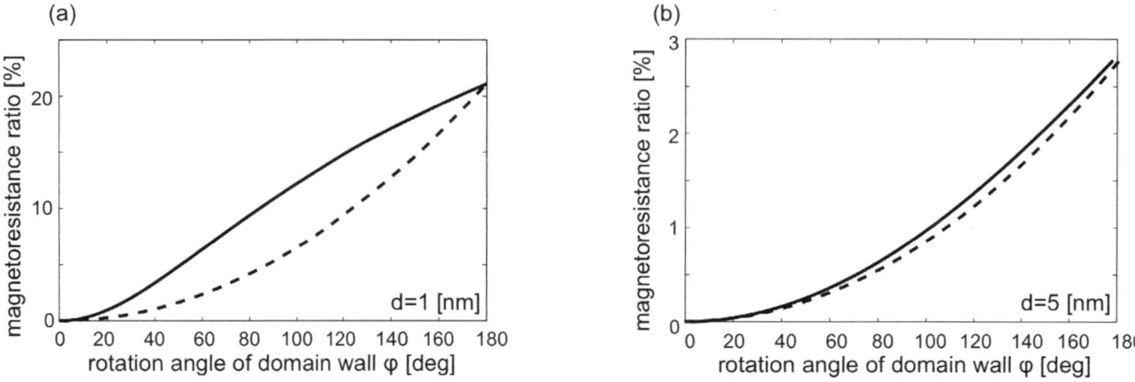

Figure 3. The dependence of the MR ratio on the rotation angle of the domain wall φ for (a) $d = 1.0$ nm $\ll \ell_J$ and (b) $d = 5.0$ nm $\gg \ell_J$, respectively. The dashed line is proportional to φ^2.

where $\xi_{\text{eff}} = (\varphi/\pi)\xi$ and $\xi = \ell_J/d$. i.e., the MR ratio is proportional to φ^2. The dashed line in Fig. 2 is proportional to φ^2, whose value at $\varphi = \pi$ is assumed to be equal to that of the solid line in Fig. 2. As shown in Fig. 2, the theory of Levy and Zhang has a poor agreement with the experiments. This is due to the fact that the domain wall in the CCP-CPP-GMR system [2] satisfies $d < \ell_J$.

In Ref. [2], Takagishi *et al.* discuss the possibility that the difference of the dependence of the MR ratio on φ between the experimental results, MR ratio $\propto \cos\varphi$, and the theoretical prediction of Levy and Zhang [5], MR ratio $\propto \varphi^2$, can be explained by taking into account the ballistic transport theory. Figures 3 (a) and (b) show the dependence of the MR ratio on φ in the ballisitic region ($d \ll \ell_{\text{mfp}}$) with $d = 1.0$ nm $\ll \ell_J$ and $d = 5.0$ nm $\gg \ell_J$, respectively. The dashed lines in Figs. 3 (a) and (b) are proportional to φ^2. β is taken to be 0.6. From these figures, it is interpreted that the difference of the dependence of the MR ratio on φ between the experiments ($\propto \cos\varphi$) and the theory of Levy and Zhang ($\propto \varphi^2$) is due to the failure of the theory of Levy and Zhang in the region $d \ll \ell_J$ because the angle dependence of the MR ratio with $d \ll \ell_J$ agrees with the experiments while that with $d \gg \ell_J$ agrees with the theory of Levy and Zhang. It is also interpreted that the ballistic transport properties cannot explain the difference because the angle dependence of the MR ratio of the domain wall in the ballistic region ($\ell_J \ll d \ll \ell_{\text{mfp}}$) agrees with the theory of Levy and Zhang ($\propto \varphi^2$) as shown in Fig. 3 (b). This means that to explain the experimental results of Ref. [2], the extension of the theory of Levy and Zhang to the ballistic region ($d \ll \ell_{\text{mfp}}$) is not sufficient and their theory should be extended to the region $d \ll \ell_J$. In Ref. [5], by assuming that $d \gg \ell_J$ the factor $\tan(\phi/2)$ is approximated to $(k_x/k_{\text{F}})\xi_{\text{eff}}$, which leads to that the MR ratio is proportional to $\xi_{\text{eff}}^2 \propto \varphi^2$. However, for a sufficiently thin domain wall $d \ll \ell_J$ [2], higer order terms of ξ_{eff} in $\tan(\phi/2)$ also contribute to the MR effect, and the dependence of the MR ratio on φ shifts from $\xi_{\text{eff}}^2 \propto \varphi^2$.

In conclusion, we studied the dependence of the MR ratio due to a magnetic domain wall confined in a CCP-CPP-GMR system on the rotation angle of the wall φ by solving the Boltzmann equation. We find that the dependence is approximately described by $\cos\varphi$, which has a good agreement with the experimental results of Ref. [2]. Our results interpret that the difference of the dependence of the MR ratio on φ between the experiments and the theory of Levy and Zhang [5] is due to the failure of the theory of Levy and Zhang in the region $d \ll \ell_J$.

Acknowledgments
The authors would like to acknowledge the valuable discussions they had with S. Kawasaki, N. Yokoshi, J. Sato, K. Matsushita, H. Tsukahara, H. Arai, and T. Kaneko. This work was supported by JSPS.

References

[1] Fuke H N, Hashimoto S, Takagishi M, Iwasaki H, Kawasaki S, Miyake K and Sahashi M 2007 *IEEE Trans. Magn.* **43** 2848
[2] Takagishi M, Fuke H N, Hashimoto S, Iwasaki H, Kawasaki S, Shinozaki R and Sahashi M *J. Appl. Phys.* **105** 07
[3] Sato J, Matsushi K and Imamura H 2008 *IEEE. Trans. Magn.* **44** 2608
[4] Matsushita K, Sato J and Imamura H 2008 *IEEE. Trans. Magn.* **44** 2616
[5] Levy P M and Zhang S 1997 *Phys. Rev. Lett.* **79** 5110
[6] Gregg J F, Allen W, Ounadjela K, Viret M, Hehn M, Thompson S M and Coey J M D 1996 *Phys. Rev. Lett.* **77** 1580
[7] Taniguchi T and Imamura H 2010 *Phys. Rev. B* **81** 012405
[8] Gurney B A, Speriosu V S, Nozieres J P, Lefakis H, Wilhoit D R and Need O U 1993 *Phys. Rev. Lett.* **71** 4023
[9] Bass J and Pratt J W P 2007 *J. Phys.: Condens. Matter* **19** 183201
[10] Kittel C 2004 *Introduction to Solid State Physics* 8th ed (Wiley)

Journal of Physics: Conference Series **266** (2011) 012109 doi:10.1088/1742-6596/266/1/012109

Magnetoresistance Effect through GaAs for Fe/MgO/GaAs/MgO/Fe Junctions

N. Tezuka[1], **F. Mitsuhashi, S. Sugimoto**
Department of Materials Science, Graduate School of Engineering, Tohoku University, Aobayama 6-6-02, Aoba-ku, Sendai, 980-8579, Japan

E-mail: tezuka@material.tohoku.ac.jp

Abstract. The magnetoresistance effect have been investigated for junctions consisting of Fe/MgO/n-GaAs/MgO/Fe. RHEED patterns and XRD pole profiles revealed that bottom electrodes, Fe, and inter layers, n-GaAs, grew epitaxially, while that upper electrodes, Fe, grew in a polycrystal form. The magnetoresistance curve correspond to the magnetization alignment between the lower and the upper Fe electrodes was obtained. The junctions exhibit maximum magnetoresistance ratio of 11.7% and 2.8% measured at 5 K and room temperature, respectively. The temperature and applied voltage dependence of resistance and magnetoresistance (MR) ratio were investigated. The MR ratio and resistance in parallel and antiparallel state for upper and lower Fe layer decreased with increasing temperature. It was found that with increasing voltage, MR ratio was maximum when the voltage was around several mV.

1. Introduction

Spintronics is a new field of electronics utilizing additional spin degree of freedom. Adding the characteristic of spin to conventional charge based electronics will improve the performance, such as non-volatility, power consumption, high density and so on. The realization of spintronics devices relies on the ability to inject, manipulate and detect spin polarized current in semiconductor. A number of efforts were reported to inject spins in a semiconductor from ferromagnetic material electrodes [1-5]. The spin injection efficiency through conventional ferromagnetic metal/semiconductor interface, however, is expected to be very small because of conductivity mismatch between a ferromagnetic electrode and a semiconductor in diffusive regime [6]. One approach for overcoming that problem is the insertion of a tunnelling barrier [7,8]. A large spin polarized tunnel current is anticipated for a (001) epitaxial growth of Fe electrode / MgO barrier [9].

The detection of spin polarized current in semiconductor is often done by using light emitting diodes [1-3] or a lateral non local geometry [4,5]. One of the other scheme is ferromagnet /insulator /semiconductor /insulator /ferromagnet heterostructure. The spin accumulation length in the semiconductor is typically several orders of magnitude larger than the spin diffusion length in the ferromagnetic layers[10]. For this structure, a spin polarized tunnel current from ferromagnetic electrode to the semiconductor produces an imbalance of the up and down electron numbers in the semiconductor conduction band or the up and down hole numbers in the valence band. Thus, a

1 tezuka@material.tohoku.ac.jp

magnetoresistance effect is predicted for this heterostructure. In our previous work, we have reported the structure and magnetoresistance effect of Fe/MgO/n-GaAs/MgO/Fe junctions measured at room temperature[11]. In this paper, we will describe temperature and applied voltage dependence of magnetoresistance effect for Fe/MgO/n-GaAs/MgO/Fe junctions.

2. Experimental procedure

The samples, comprise of MgO (001) sub. /Fe (30 nm)/MgO (2 nm)/n-GaAs (2.2 nm)/MgO (2 nm)/Fe (5 nm)/Co (10 nm)/Cr (2 nm)/Ta (3 nm), were grown by a molecular beam epitaxy (MBE) system. Prior to the growth of these layers, MgO substrates were annealed at 953 K. Fe bottom and Fe top layers were annealed for 1h at 573 K and 473 K, respectively. The epitaxial structure of bottom Fe layer, lower MgO insulator, and n-GaAs layer were checked by reflection high energy electron diffraction (RHEED) as shown in ref. 11. We also checked the structure of MgO sub./MgO buffer / Fe/MgO/n-GaAs by using X-ray diffraction. Pole profiles of MgO(220) , Fe(200) and GaAs (200) were measured. The epitaxial relationship Fe(001) [100]//MgO(001)[110]//GaAs(001)[100] was found, however, the angle between Fe [100] and GaAs [100] was about 10 degree. The upper MgO insulator and Fe electrode grew polycrystalline. The junctions with size of 100 μm^2 were patterned by typical microfabrication method using EB lithography and Ar ion milling. MR curves were measured by four prove method from 5 to 300 K and up to 500 mV.

3. Results and discussion

Figure 1 shows the typical magnetoresistance curves measured at from 5 to 300 K for a junction with n-GaAs thickness of 2.2 nm. In our previous work, magnetization hysteresis loop of the Fe/MgO/GaAs/MgO/Fe was also measured at room temperature. It revealed that the magnetic field for resistance change in the MR curve correspond to the coercivity of top and bottom Fe. It shows from 5

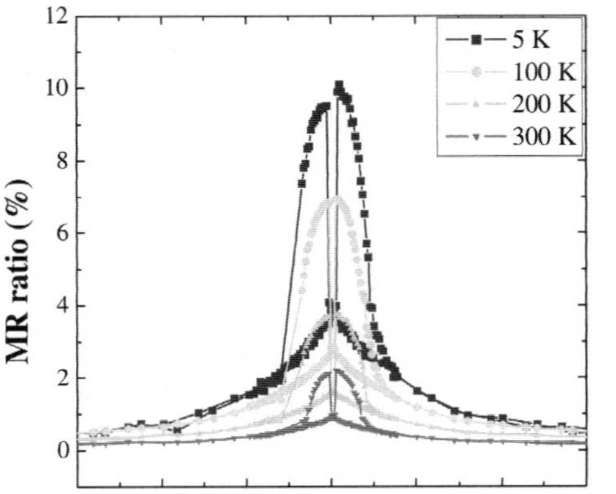

Figure 1. Magnetoresistance curves measured at temperatures of 300, 200, 100, and 5 K for a junction of Fe (30) /MgO (2) /n-GaAs(2.2)/MgO (2) /Fe (5) /Co (10)/ Cr (2) /Ta(3) (thickness in nm). The direction of magnetic field is parallel to the (100) easy axis of the Fe.

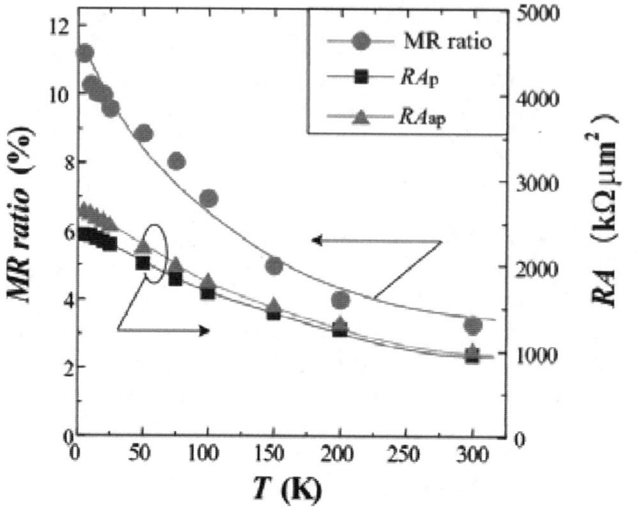

Figure 2. MR ratio and resistance area products (*RA*) in the parallel and antiparallel state as a function of measured temperature for a junction of Fe (30) /MgO (2) /n-GaAs(2.2)/MgO (2) /Fe (5) /Co (10)/ Cr (2) /Ta(3) (thickness in nm).

to 300 K that high and low resistance in the antiparallel and parallel magnetization state between top and bottom Fe electrode, respectively. A magnetoresistance ratio of 2.8 % is found measured at 300 K and increases to 11.7% at 5 K. The tunnelling spin polarization through GaAs estimated by Julliere's equation [12] was about 0.14 at 9 K and 0.035 at room temperature, assuming the spin polarization of Fe was 0.4 and a series circuit of Fe/MgO, GaAs and MgO/Fe.

The resistance area product in parallel state is about 1.0×10^3 k$\Omega\mu$m^2 at room temperature. We fabricated a single junction of Fe/MgO/Fe by using the same fabrication condition for Fe/MgO/GaAs/MgO/Fe. The resistance area product of this single junction is 14 k$\Omega\mu$m^2. From the simple tunnelling model, the resistance area product was estimated to be more than 5.0×10^5 k$\Omega\mu$m^2, assuming the tunnel barrier thickness of 4 nm. Thus , the measured resistance might not come from the single barrier of 4 nm but a series circuit of Fe/MgO, GaAs and MgO/Fe. Detail explanation can be found in Ref. 11.

Figure 2 shows the resistance area products in the antiparallel (RA_{ap}) and parallel (RA_p) configurations and magnetoresistance ratio as a function of measured temperature. RA_{ap} and RA_p decreases with increasing temperature, however, the changing ratio of RA_{ap} is large, compared to that of RA_p. MR ratio decreases with increasing temperature, especially below 10 K, the MR ratio decreased rapidly. RA_p was reported to be almost independent of temperature for an epitaxial Fe(001)/MgO(001)/Fe(001) junction. In this case, the lower Fe/MgO was grown epitaxially, while upper that was polycrystalline. Thus, the spin polarization in tunnelling current might be small.

Figure 3 (a) shows the applied voltage dependence of MR ratio for the junction measured at room temperature and 10 K. Asymmetric voltage dependence of MR ratio was clearly observed. From the RHEED pattern, it was found that lower MgO grew epitaxially, while upper MgO polycrystalline. This asymmetry might come from the difference of the crystal quality between lower and higher MgO insulator. With increasing voltage, MR ratio was maximum when the voltage was several mV. Next, we investigated the voltage dependence of MR ratio in small voltage region. Fig. 3 (b) and (c) show the MR ratio as a function of current density up to 2.0×10^4 A/m^2 measured at 9 K and room temperature, respectively. It is clearly shown that MR ratio becomes maximum around 0.7×10^4 A/m^2. The relationship between current density and spin polarization in semiconductor was calculated at room temperature by Smith and Silver[13]. It was predicted that spin polarization of semiconductor increases with increasing current density for ferromagnet/semiconductor, which was depend on the resistivity of the junction. In our sample, *RA* was about 1.0×10^{-2} Ωcm^2, assuming a series circuit of Fe/MgO/GaAs/MgO/Fe. The spin polarization of about 0.1 is in agreement with that estimated by this calculation.

4. Conclusion

Magnetoresistance effect was measured for junctions consisting of Fe/MgO/n-GaAs/MgO/Fe. The magnetoresistance curve correspond to the magnetization alignment between the bottom and the upper Fe electrodes was obtained. The junctions exhibit maximum magnetoresistance ratio of 11.7% and 2.8% measured at 5 and 300 K, respectively. The MR ratio and resistance in parallel and antiparallel state for upper and lower Fe layer decreased with increasing temperature. It was found that with increasing voltage, MR ratio was maximum when the voltage was around several mV. This voltage dependence of MR ratios could be explained by the spin current injection to semiconductor.

Acknowledgments
This work was partially supported by the Asahi Glass Foundation.

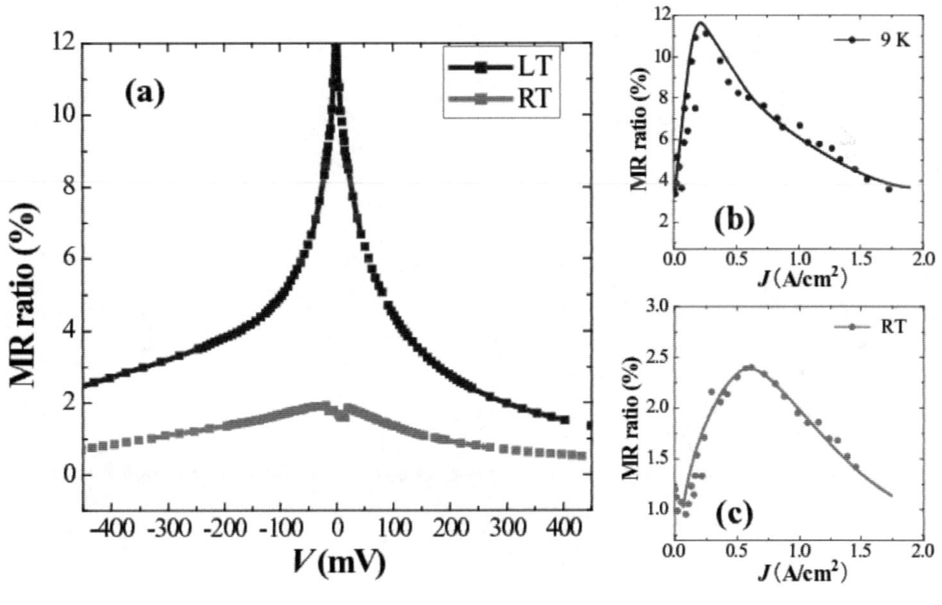

Figure 3. (**a**)MR ratio as a function of applied voltage measured at room temperature (RT) and 5 K (LT) for a junction of Fe (30) /MgO (2) /n-GaAs(2.2)/MgO (2) /Fe (5) /Co (10)/ Cr (2) /Ta(3) (thickness in nm). MR ratio as a function of current density measured at 9 K(b) and RT(c). The lines were described just guide to eyes.

Journal of Physics: Conference Series **266** (2011) 012109 doi:10.1088/1742-6596/266/1/012109

5. References

[1] Zhu H J, Ramsteiner M, Kostial H, Wassermeiner M, Schönherr H P, Ploog K H, 2001 *Phys. Rev. Lett.* **87** 016601.
[2] Hanbicki A T, Jonker B T, Itskos G, Kioseoglou G, Petrou A, 2002 *Appl. Phys. Lett.* **80** 1240.
[3] Motsnyi V F, Dorpe P V, Roy W V, Goovaerts E, Safarov V I, Borghs G, Boeck J D, 2002 *Appl. Phys. Lett.* **81** 265.

[4] Lou X, Adelmann C, Crooker S A, Garlid E S, Zhang J, Reddy K S M, Flexner S D, Palmstrøm C J, Crowell P A, 2007 *Nature Physics* **3**, 197.
[5] Lee T H, Koo H C, Kim K H, Kim H, Chang J, Han S H, Hong J, Lim S H, 2009 *J. agn Magn Mater.* **321** 3795.
[6] Schmid G, Ferrand D, Molenka,p L W, Filip A T, van Wee B J, 2000 *Phys. Rev.* B **62** R4790.
[7] Jiang X, Wang R, Shelby R M, Macfarlane R M, Bank S R, Harris J S, Parkin S S P, 2005 *Phys. Rev. Lett.* **87** 056601.
[8] Salies G, Wang R, Jiang X, Shelby R M, J S, Parkin S S P, Bank S R, Harris, 2005 *Appl. Phys. Lett.* **87** 262503.
[9] Buttler W H, Zhang X G, Shulthess MacLaren J M, 2001 *Phys. Rev.* **B** 63 054416.
[10] Kikkawa J M, Awshalom D D, 1999 *Nature* **397** 139.
[11] Mitsuhashi F, Tezuka N, Sugimoto S, 2009 *J.Japan Inst.Metals* **73** 251.
[12] Julliere M, 1975 Phys. Lett. **54A**, 225.
[13] Smith D L, Silver R N 2001 *Phys. Rev.* B **64** 045323.

Journal of Physics: Conference Series **266** (2011) 012110 doi:10.1088/1742-6596/266/1/012110

Current-induced domain wall motion in Co/Ni nano-wires with different Co and Ni thicknesses

K. Ueda[1], D. Chiba[1], T. Koyama[1], G. Yamada[1], H. Tanigawa[2] S. Fukami[2], T. Suzuki[2], N. Ohshima[2], N. Ishiwata[2], Y. Nakatani[3], and T. Ono[1*]

[1]Institute for Chemical Research, Kyoto University, Gokasho, Uji, Kyoto, 611-0011, Japan
[2]NEC Corporation, 1120 Shimokuzawa, Chuo-ku, Sagamihara, Kanagawa 252-5298, Japan.
[3]University of Electro-communications, Chofu, Tokyo, 182-8585, Japan

*E-mail: ono@scl.kyoto-u.ac.jp

Abstract. The authors have investigated magnetic domain wall motion induced by electric currents in ferromagnetic nano-wires made of Co/Ni multilayers. The thicknesses of Co and Ni layers were changed, whereas the numbers of layer stacks of Co and Ni were the same in all samples. The sample with thinner total Co/Ni thickness showed the lower threshold current density for the domain wall motion as an overall trend, which is qualitatively in agreement with the expectation by the theory based on the adiabatic spin-transfer model. The lowest threshold current density was 2.9×10^{11} A/m^2, obtained in the sample with the total Co/Ni thickness of 3.4 nm and the wire width of 110 nm.

1. Introduction

Current-induced domain wall (DW) motion in ferromagnetic nano-wires has been widely investigated[1-17)] because of its potential applications for magnetic random access memories[10-11,17)] or logic devices[18)] as well as novel physics relevant to the interaction between spin current and local magnetic moment[19-26)]. NiFe with in-plane magnetization has been actively investigated for this phenomenon[1,3-8,10,11)]. The threshold current density J_{th} for the DW displacement in this system, however, is determined by the strength of pinning field of a DW[10)]. This is unfavorable for the device application because the reduction of J_{th} and high thermal stability of DW are incompatible. In contrast, Co/Ni nano-wires with perpendicular magnetization[9,13-17)], which is of focus here, has much higher DW pinning field than that of in-plane magnetized system, despite the DW can be reproductively displaced by current. In addition, our recent experiments on our system shows the threshold current density J_{th} for the DW displacement can be controlled by the wire dimension[14)], which is reasonably a qualitative agreement with recent theoretical calculations based on the spin transfer model[21,27)]. These advantages of high pinning field and reduction of J_{th} in the perpendicularly magnetized system realize the compatibility of high thermal stability and low power consumption for devices. In this paper, we report the dependence of J_{th} on total thicknesses of Co and Ni layers in Co/Ni nano-wires to obtain the guideline for further reduction of it.

* To whom any correspondence should be addressed.

Journal of Physics: Conference Series **266** (2011) 012110 doi:10.1088/1742-6596/266/1/012110

2. Experiment

The film structures and their magnetic properties were summarized in Table 1. Multilayered Co/Ni films with the cap and buffer layers were deposited on silicon substrates by dc magnetron sputtering. The thickness of Co and Ni layers were varied as follows; Co(0.2)/[Ni(0.4)/Co(0.2)]$_4$ (t = 2.6 nm, sample #1), Co(0.2)/[Ni(0.6)/Co(0.2)]$_4$ (t = 3.4 nm, sample #2), Co(0.3)/[Ni(0.6)/Co(0.3)]$_4$ (t = 3.9 nm, sample #3), Co(0.3)[Ni(0.9)/Co(0.3)]$_4$ (t = 5.1 nm, sample #4), where t is the total thickness of the multilayer. The magnetic properties of the films were evaluated by magnetization measurements. For each samples, a series of devices with different wire width, w = 110 nm, 220 nm and 330 nm were fabricated by electron-beam lithography and ion-milling. Figure 1 shows the schematic illustration of the device structure. Two Au(100 nm)/Ti(5 nm) electrodes, labelled as A and B were formed at both ends of the wire. A Hall probe made of Ta(50 nm) was placed on the wire to detect the magnetization direction through the anomalous Hall effect. All measurements were performed at room temperature. Figure 2 shows the typical hysteresis loop obtained by the Hall measurement (w = 110 nm, sample #2), where the vertical axis indicates the offset Hall resistance ΔR_{Hall}. The square loop was observed, indicating the wire shows perpendicular magnetic anisotropy.

The experiments for the current-induced DW motion were performed as following procedure. First, the large positive magnetic field of +4 kOe was applied along perpendicular direction to make the wire single domain state. Next, the local magnetic field were applied by injection of the current through the electrode A in the orthogonal direction with respect to the wire. From the separate measurements (not shown), the distribution of the pinning-fields for the DW were confirmed to be 105 Oe to 303 Oe in all devices, which is one or two orders magnitude larger than that of NiFe nano-wires. After introducing DW, multiple current pulses with the duration of 10 ns were injected into the wire. At every injection of pulsed currents, Hall resistance was measured to check whether the DW was propagated under the Hall probe. In each device, the same procedures were performed for 20 times to determine the probability of the DW propagation from the initial position to the position under the Hall probe.

Table 1. Thickness and magnetic properties of four Co/Ni multilayers.

Sample	Co (nm)	Ni (nm)	total thickness t (nm)	M_s ($\times 10^5$ A/m)	K_u ($\times 10^5$ J/m^3)
#1	0.2	0.4	2.6	7.58	4.1
#2	0.2	0.6	3.4	6.68	4.2
#3	0.3	0.6	3.9	7.79	6.0
#4	0.3	0.9	5.1	6.84	3.9

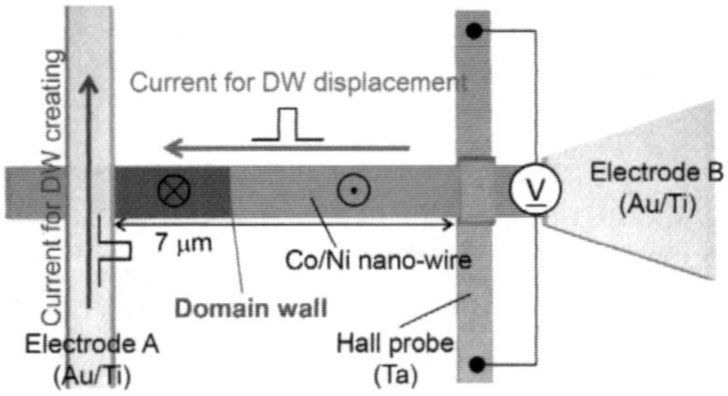

Figure 1. Schematic illustrations of device structure and measurement configuration.

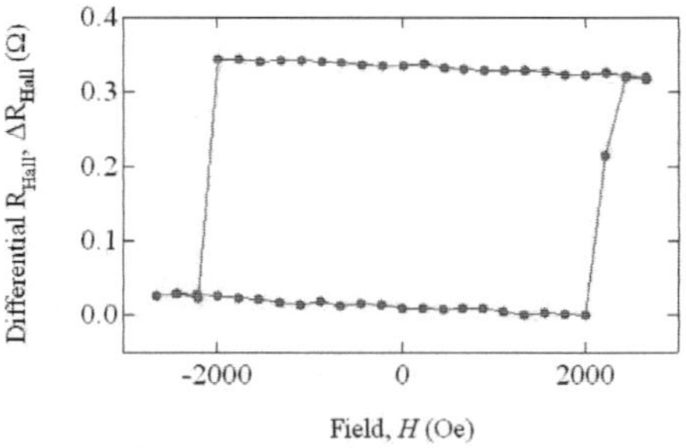

Figure 2. Result of Hall measurement in the sample #2 with w = 110 nm.

Journal of Physics: Conference Series **266** (2011) 012110 doi:10.1088/1742-6596/266/1/012110

3. Result and discussion

A typical current density J dependence of the propagation probability P of the three devices with different wire width w (sample #2) is shown in Fig. 3. One can see that P abruptly increases with J above threshold values in all devices with various w. With decreasing w, the threshold value was found to be lower, as previously reported by our group[14)]. This tendency is in agreement with the theoretical prediction based on the adiabatic spin transfer model in which the threshold current density J_{th} is dominated by the intrinsic DW pinning mechanism. If we defined J_{th} as J where the probability is 0.5, from which they were determined to be 4.5×10^{11} A/m^2, 4.4×10^{11} A/m^2, and 2.9×10^{11} A/m^2 for w = 330, 220, and 110 nm in this series of devices, respectively. J_{th} as a function of w in all samples is shown in Fig. 4. The same trend was obtained reproducibly in all samples with different Co and Ni thicknesses.

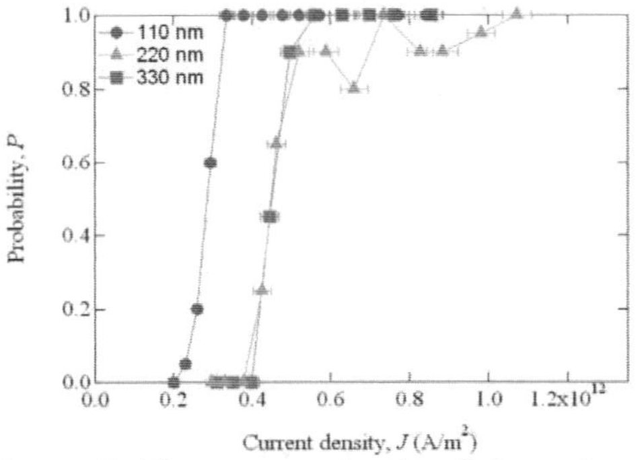

Figure 3. The current density J dependence of probability P for DW displacement in devices with various w made of sample #2.

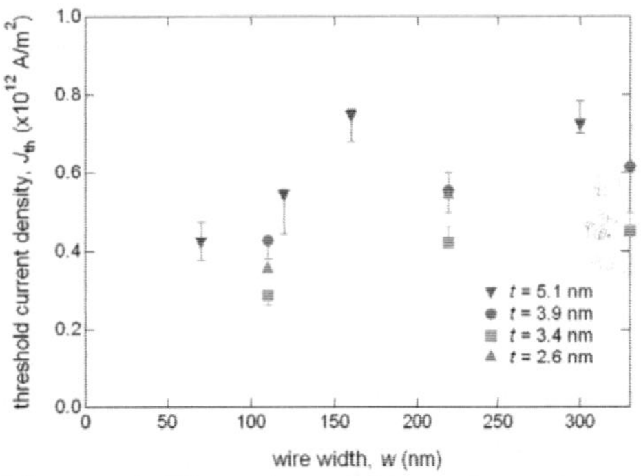

Figure 4. The wire width w dependence of the threshold current density J_{th} of all samples.

Finally, we discuss the thickness dependence. From Fig. 4, overall tendency shows that J_{th} reduces with decreasing the sample total thickness t. In the theory based on the adiabatic spin transfer model[21,22,27,28)], the current flowing in the wire exerts the torque on the local spins inside the DW and rotates them in the sample plane, resulting in the periodic change of DW structure between Bloch (stable state) and Néel wall (higher energy state). Within this scenario, the threshold current density is proportional to the hard axis anisotropy constant $K_{h.a.}$,which is the energy density of magnetic anisotropy associated with the rotation of the DW spins in the sample plane. $K_{h.a.}$ reduces with decreasing t because the difference between the in-plane demagnetizing factors decreases. Our experimental results are qualitatively in accordance with this scenario, although further comparison between the experiments and theory is needed because J_{th} also depends on the carrier spin polarization and the width of the DW.

4. Summary

In summary, we have described the DW motion induced by the application of electric current in the perpendicularly magnetized Co/Ni wire, and found that J_{th} for DW motion is tend to decrease with reducing Co/Ni thickness. The lowest current density of 2.9×10^{11} A/m^2 was obtained in the sample with the Co/Ni thickness of 3.4 nm and the wire width of 110 nm. The small threshold current density as well as high de-pinning field for the DW shows that our system is the favorable candidate for the realization of the memory and logic devices based on the electrical DW motion[10-11, 17-18)].

Acknowledgement

Journal of Physics: Conference Series **266** (2011) 012110 doi:10.1088/1742-6596/266/1/012110

We thank K. Kobayashi for useful discussion. This work was partly supported by the New Energy and Industrial Technology Development Organization (NEDO) Spintronics non-volatile devices project.

References

[1] A. Yamaguchi, T. Ono, S. Nasu, K. Miyake, K. Mibu, and T. Shinjo 2004 *Phys. Rev. Lett.* **92** 077205
[2] M. Yamanouchi, D. Chiba, F. Matsukura,and H. Ohno 2004 *Nature* **428** 539
[3] N. Vernier, D. A. Allwood, D. Atkinson, M. D. Cooke, and R. P. Cowburn 2004 *Europhys. Lett.* **65** 526
[4] E. Saitoh, H. Miyajima, T. Yamaoka, and G. Tatara 2004 *Nature* **432** 203
[5] M. Kläui, C. A. F. Vaz, J. A. C. Bland, W. Wernsdorfer, G. Faini, E. Cambril, L. J. Heyderman, F. Nolting, and U. Rüdiger 2005 *Phys. Rev. Lett.* **94** 106601
[6] G. S. D. Beach, C. Knutson, C. Nistor, M. Tsoi, and J. L. Erskine 2006 *Phys. Rev. Lett.* **97** 057203
[7] L. Thomas, M. Hayashi, X. Jiang, R. Moriya, C. Rettner, and S. S. P. Parkin 2006 *Nature* **443** 197
[8] Y. Togawa, T. Kimura, K. Harada, T. Akashi, T. Matsuda, A. Tonomura, and Y. Otani 2006 *Jpn. J. Appl. Phys.* **45** L683
[9] D. Ravelosona, S. Mangin, J. A. Katine, Eric. E. Fullerton and B. D. Terris 2007 *Appl. Phys. Lett.* **90** 072508
[10] S. S. P. Parkin, M. Hayashi, and L. Thomas 2008 *Science* **320** 190
[11] M. Hayashi, L. Thomas, R. Moriya, C. Rettner and S. S. P. Parkin 2008 *Science* **320** 209
[12] H. Tanigawa, K. Kondou, T. Koyama, K. Nakano, S. Kasai, N. Ohshima, S. Fukami, N. Ishiwata, and T. Ono 2008 *Appl. Phys. Express* **1** 011301
[13] T. Koyama, G. Yamada, H. Tanigawa, S. Kasai, N. Ohshima, S. Fukami, N. Ishiwata, Y. Nakatani, and T. Ono 2008 *Appl. Phys. Express* **1** 101303
[14] H. Tanigawa, T. Koyama, G. Yamada, D. Chiba, S. Kasai, S. Fukami,T. Suzuki, N. Ohshima, N. Ishiwata, Y. Nakatani, and T. Ono 2009 *Appl. Phys. Express* **2**, 053002
[15] T. A. Moore, I. M. Miron, G. Gaudin, G. Serret, S. Auffret, B. Rodmacq, A. Schuhl, S. Pizzini, J. Vogel, and M. Bonfim 2008 *Appl. Phys. Lett.* **93**, 262504
[16] C. Burrowes, A. P. Mihai, D. Ravelosona, J-.V-.Kim, C. Chappert, L. Vila, A. Marty, Y. Samson, F. Garcia-Sanchez, L. D. Buda-Prejbeanu, I. Tudosa, E. E. Fullerton and J.-P. Attene 2010 *Nature Physics* **6**, 17
[17] S. Fukami T. Suzuki, K. Nagahara, N. Ohshima, Y. Ozaki, S. Saito, R. Nebashi, N. Sakimura, H. Honjo, K. Mori, C. Igarashi, S. Miura, N. Ishiwata, and T. Sugibayashi, 2009 Symposium on VLSI Technology, Digest of Technical Papers: p.p. 230
[18] D. A. Allwood, G. Xiong, C. C. Faulkner, D. Atkinson, D. Petit, R. P. Cowburn 2005 *Science* **309** 1688
[19] L. Berger 1984 *J. Appl. Phys.* **55** 1954
[20] L. Berger 1996 *Phys. Rev.* B **73**, 014407
[21] G. Tatara, and H. Kohno 2004 *Phys. Rev. Lett.* **92** 086601
[22] Z. Li, and S. Zhang 2004 *Ph.ys. Rev. Lett.* **92** 207203
[23] S. Zhang, and Z. Li 2004 *Phys. Rev. Lett.* **93** 127204
[24] X. Waintal, and M. Viret 2004 *Europhys. Lett.* **65** 427
[25] A. Thiaville, Y. Nakatani, J. Miltat, and Y. Suzuki 2005 *Europhys. Lett.* **69** 990
[26] S. E. Barnes, and S. Maekawa 2005 *Phys. Rev. Lett.* **95** 107204
[27] S.-W. Jung, W. Kim, T.-D. Lee, K.-J. Lee, and H.-W Lee 2008 *Appl. Phys. Lett.* **92** 202508
[28] S. Fukami, T. Suzuki, N. Ohshima, K. Nagahara, and N. Ishiwata 2008 *J. Appl. Phys.* **103** 07E718

Growth of mixed phased films of γ-Fe_2O_3 and half metallic Fe_3O_4 using (100) MgO/Fe under layer for spin electronic devices

M L Wang[1,2], R H Yu[2], and S Nakagawa[1]
[1] Department of Physical Electronics, Tokyo Institute of Technology, 2-12-1 O-okayama, Meguro, Tokyo 152-8552, Japan
[2] Department of Materials Science and Engineering, Tsinghua University, Beijing 100084, China

E-mail: nakagawa@pe.titech.ac.jp

Abstract. Half metallic magnetite films using (100) MgO/Fe under layer deposited at room temperature were prepared using facing targets sputtering method. X-ray diffractometry and TEM analyses confirmed epitaxial growth of (100) oriented Fe_3O_4 layer on (100) oriented MgO under layer prepared on 5nm-thick Fe buffer layer on glass substrates. Resistivity analyses clarified that higher oxygen partial pressure caused higher oxidation of Fe ions in the films which tend to result in γ-Fe_2O_3. Substrate heating is effective to improve crystallinity and magnetic properties of Fe_3O_4 film. Oxygen partial pressure of 7.5%, substrate temperature of 150 °C and sputtering discharge current of 0.45A was suitable to prepare magnetite Fe_3O_4 film with (100) preferential orientation and good magnetic properties.

1. Introduction

Half metallic materials which have only minority spins in the density of states at Fermi level are desirable for spintronic devices because of their theoretical 100% spin polarization. Among these materials, Fe_3O_4 has gained an extraordinary interest due to its relatively high Curie temperature (858K) which is suitable for room temperature operation of spintronic devices[1][2]. Meanwhile the tunnelling barrier in magnetic tunnel junction (MTJ) has been developed from amorphous Al_2O_3 layer to (100) oriented MgO layer to achieve high TMR ratio [4]. The epitaxial multilayer that takes the advantages of both half-metallic Fe_3O_4 layer and (100) oriented MgO layer is expected to get a much greater TMR ratio [3]. Thus magnetite/MgO layers have potential applications in various magneto-resistive devices such as MTJs, spin filters and so on. In our study, we aim to prepare an epitaxial Fe_3O_4 (100) oriented film using MgO (100) oriented film as under layer (lattice misfit: ~ 0.3%) prepared on a Fe buffer layer on glass substrates.

2. Experiment

All the films (including Fe oxide single layers, MgO/Fe bilayers, Magnetite/MgO/Fe trilayers) have been prepared on glass substrates using facing targets sputtering (FTS) method. Background pressure was less than 10^{-6} Torr. Fe and Mg disks with diameter of 10 cm were used as targets. For MgO and magnetite layers, O_2 is added into the chamber to achieve reactive sputtering during the deposition.

---

1 nakagawa@pe.titech.ac.jp

The MgO/Fe under layers were deposited without substrate heating, while the third Fe oxide layers were grown at substrate temperature ranging from room temperature (R.T.) to 200 °C. Structural characterizations of thin films were performed by X-Ray Diffracometry (XRD) and Transmission Electron Microscopy (TEM). Analysis of magnetic and electrical properties was done by using Vibrating Sample Magnetometer (VSM) and four-point probe method, respectively.

3. Results and Discussion

3.1. Effect of Fe buffer layer and MgO under layer on Fe oxide layer

We tried to find the appropriate preparation conditions of MgO (100) oriented layers with Fe buffer layer on glass substrate at room temperature by changing different deposition parameters and the thickness of Fe layer. Figure 1 shows the out-of-plane XRD diagrams of MgO/Fe films for various thicknesses of Fe buffer layers. Values on the right side indicate the thickness of Fe layer. If there is no Fe buffer layer, MgO layer exhibits (220) preferential orientation. While the (220) orientation disappeared when 1 nm-thick Fe buffer layer was prepared but there is still no MgO (100) orientation. According to the diagram, MgO layer exhibits (100) orientation only for the cases of 5 nm and 30 nm-thick Fe buffer layers. MgO (100) oriented film can be grown epitaxially on Fe layer with (100) preferential orientation in the 5nm-thick and 30nm-thick Fe layers, since the Fe (200) peak can be observed in the films with 5nm-thick and 30nm-thick Fe layers. After the confirmation of MgO (100) orientation for the MgO/Fe under layer, we fabricated Fe oxide thin films on MgO/Fe(5 nm) under layers. The diffraction lines of Fe oxide/MgO/Fe multilayer was compared with that of single Fe oxide layer which was deposited at the same deposition condition and that of MgO/Fe bilayers as shown in figure 2. Although the Fe oxide single layer showed (311) orientation the multilayer exhibits strong (100) orientation. Thus the under layer is very effective to change the orientation of Fe oxide layer. The epitaxial growth of Fe oxide layer on MgO/Fe under layer is achievable.

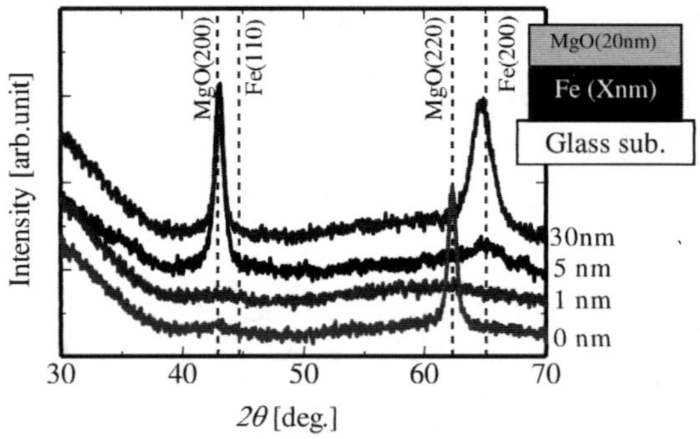

Figure 1. XRD diagrams of MgO/Fe films with different thickness (X nm) of Fe buffer layer.

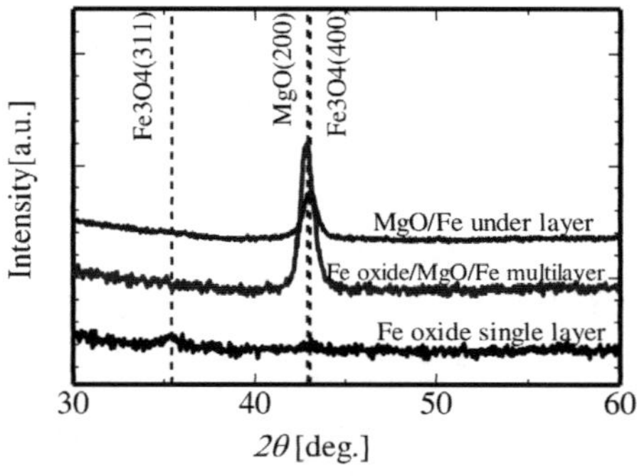

Figure 2. XRD diagrams of Fe oxide single layer, Fe oxide layer/MgO/Fe multilayer and MgO/Fe bilayer.

3.2. TEM observation

TEM analyses were performed on cross sectional specimens to study the structure and the morphology of the interface between under layer and Fe oxide layer. 100nm-thick Fe-oxide films were prepared at 0.57 Pa of Ar and O_2 mixture gas with 7.5% of O_2 flow rate. Figure 3 shows the electron diffraction (ED) patterns and the TEM photographs of the Fe_3O_4 films deposited at R.T. and 150 °C on MgO(20nm)/Fe(5nm) under layer deposited at room temperature. The direction of electron beam is perpendicular to the plane of paper. In ED observation, 10 μ m- area selected aperture slot was used. The regions observed in ED patterns were all most same as that in Bright field patterns. (a) - (d) and (e) - (h) correspond to the Fe_3O_4 films deposited at R.T. and 150 °C, respectively.

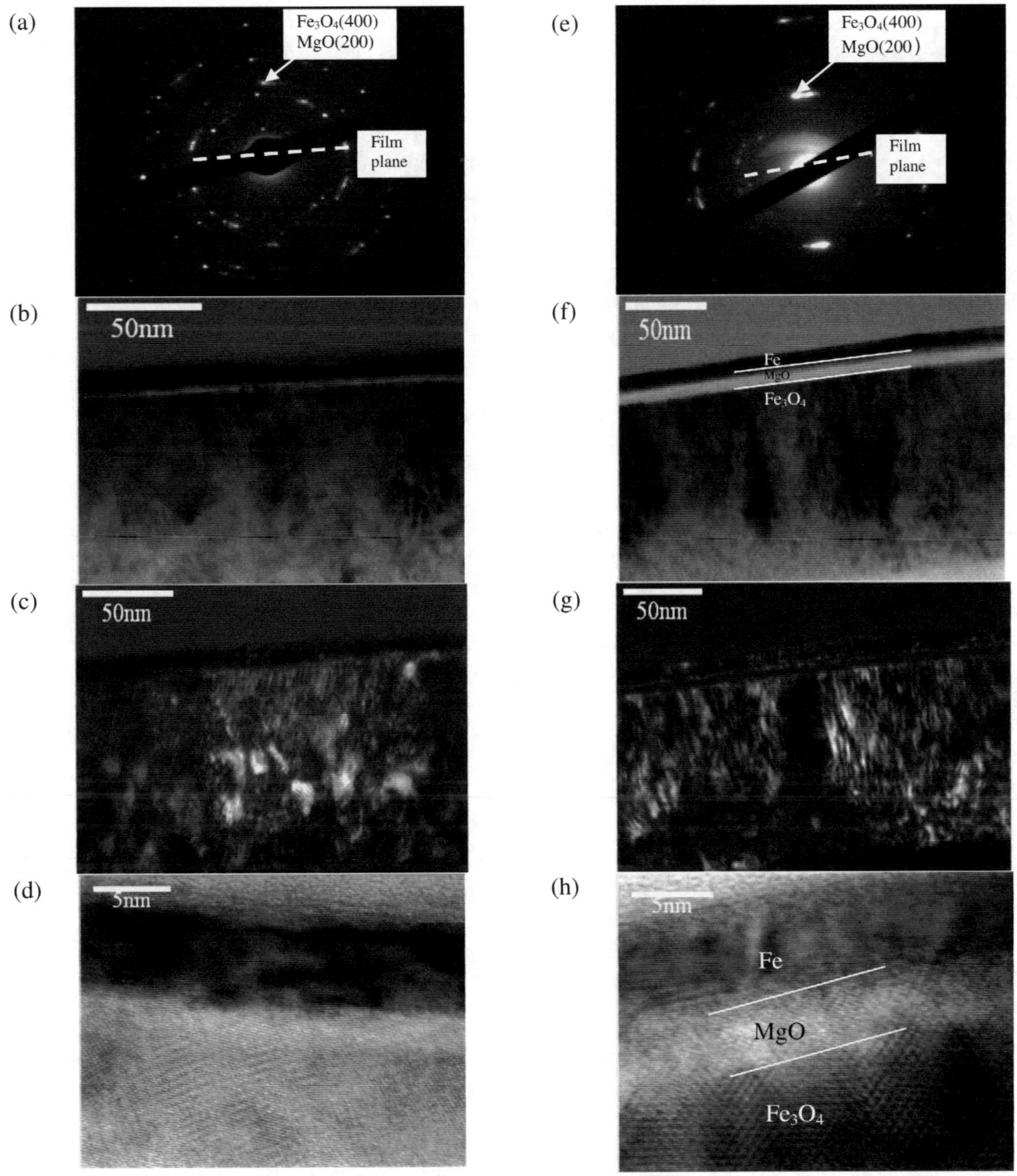

Figure 3. TEM images of 100nm-thick Fe_3O_4 layer on MgO/Fe under layer deposited at room temperature. (a) & (e):ED images; (b) & (f):Bright fields; (c) & (g): Dark fields; (d) & (h): HRTEMs. (a) - (d) and (e) - (h) correspond to the Fe_3O_4 films deposited at R.T. and 150 °C, respectively.

ED images, Bright field images, Dark field images corresponding to (400) selected area diffraction and HRTEM images of those two films correspond to (a) & (e), (b) & (f), (c) & (g) and (d) & (h), respectively. ED image of R.T.-deposited Fe_3O_4 film shows very fine diffraction spots in the perpendicular direction to the film plane (labelled in the figure), as shown in figure 3 (a) which implies that the grains in R.T.-deposited Fe_3O_4 film are small and the vertical growth of the crystallites is not good. The film deposited at 150 °C shows very obvious diffraction spots corresponding to Fe_3O_4 (400) and MgO (200) in the direction perpendicular to the film plane. Figure 3 (b) and (f) are bright field images and Figure (c) and (g) show dark field images corresponding to (400) selected diffraction spot of the films. Figure (f) and (g) also confirmed the better crystallinity of Fe_3O_4 layer deposited at substrate temperature of 150 °C. Figure (d) and (h) are the HRTEM images of the specimens. For the magnetite layer, substrate temperature of 150 °C is suitable to get good crystalline state. What is interesting, the thickness of MgO layer seems to be less than designed value (20nm), this maybe because the diffusion occurred during the process of TEM sample preparation. Ar ions with high energy sputtered to the very thin film of very small volume and then the temperature increased which may cause diffusion of Mg atoms.

3.3. Dependence of magnetic and electrical property on substrate temperature and O_2 partial pressure

Figure 4 shows the XRD diagrams of the Fe-oxide films prepared at different substrate temperature on MgO/Fe underlayers which are deposited at R.T. All the Fe-oxide films were prepared at 0.57 Pa of Ar and O_2 mixture gas with 7.5% of O_2 flow rate. In the case of R.T. deposition, the diffraction line of Fe_3O_4 is so weak. Increase of substrate temperature around 150 °C for the Fe-oxide film causes crystallization and (100) orientation of Fe_3O_4 crystallites. Further increase of substrate temperature above 200 °C causes random growth of crystallites which reveals appearance of (311) diffraction line. VSM measurements allow us to estimate the value of the magnetization of the magnetic oxide films. Magnetization measured along the film plane for Fe-oxide layers prepared at different substrate temperatures deposited on MgO/Fe under layer were demonstrated in figure 5. It is regarded that substrate heating is effective to improve saturation magnetization M_S and the coercive force H_c of films due to the better crystallization caused by the increase of the deposition temperature for the Fe-oxide layers. M_S values of 300, 210 and 175 emu/cc can be determined for the films grown at 150 °C, 200 °C and room temperature, respectively. The bulk value of Fe_3O_4 for M_S is 471emu/cc. The reason for the decrease of M_S may be ascribed to the existence of Anti-phase Boundaries (APBs) and γ-Fe_2O_3 phase [5].

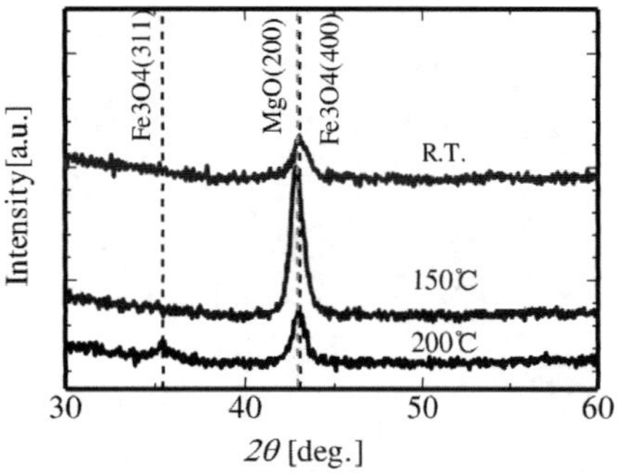

Figure 4. XRD diagrams of 100nm-thick Fe_3O_4 layers deposited at various substrate temperature on MgO/Fe under layer.

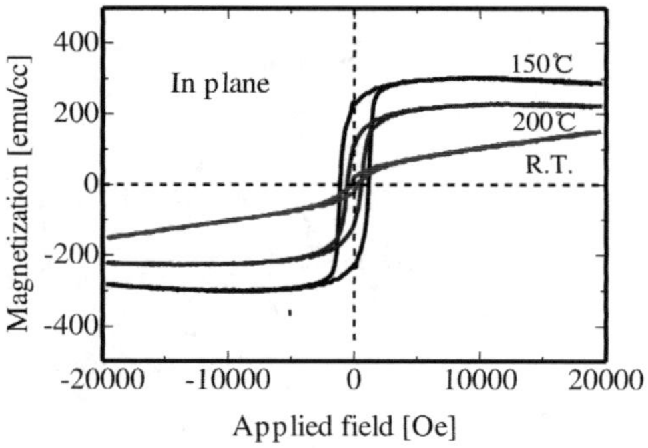

Figure 5. M-H loops of 100nm-thick Fe_3O_4 layers deposited at various substrate temperatures on MgO/Fe under layer.

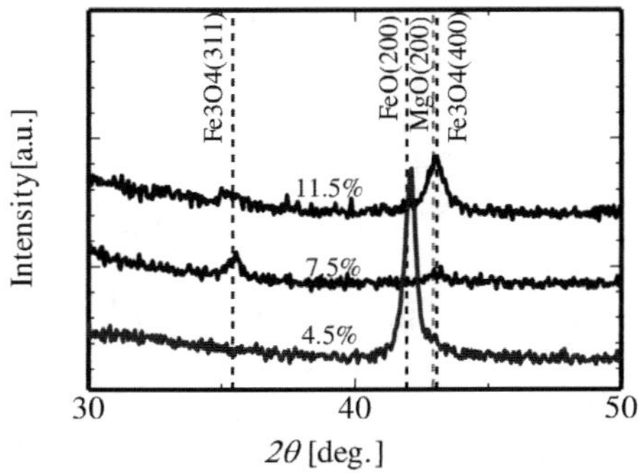

Figure 6. XRD diagram of Fe_3O_4 layer deposited on MgO/Fe under layer under different O_2 partial pressure ratio.

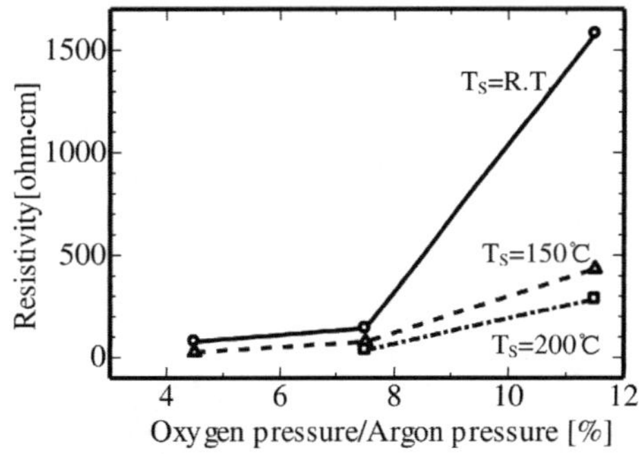

Figure 7. Resistivity of Fe oxide films as a function of O_2 partial pressure ratio for various deposition temperatures for Fe-oxide films.

Figure 6 shows XRD diagrams of Fe oxide/MgO/Fe multilayer with different O_2 partial pressure ratio. For the 4.5% of O_2 partial pressure ratio, Fe oxide layer has the dominant FeO phase. As the O_2 partial pressure ratio increases, it changed gradually to be Fe_3O_4 or γ-Fe_2O_3 phase. Since in our case we want to get (100) preferential oriented Fe_3O_4 and γ-Fe_2O_3 phase and they have almost the same spinel structure, these two phases with same (100) preferential orientation can't be distinguished easily through XRD. Figure 7 shows resistivity of Fe oxide films as a function of O_2 partial pressure ratio for various deposition temperatures for Fe-oxide films. The resistivity of Fe oxide films increases significantly with increasing O_2 partial pressure which indicate that higher O_2 partial pressure causes higher oxidation of Fe ions in the films which tend to result in higher resistivity. The film is oxidized transforming into maghemite γ-Fe_2O_3 film with higher resistivity. As the substrate temperature increases, the oxygen partial pressure at which the spinel ferrite is formed increases. At the same O_2 partial pressure, the lower substrate temperature causes higher oxidation. This may be because lower temperature can hold the O ions in the film while higher temperature can drive out O ions among the Fe ions. This tendency agrees with Ellingham diagram thermodynamics [6][7]. The bulk resistivity values of Fe_3O_4 and γ-Fe_2O_3 are about 10^{-2} and 10^{4} $\Omega\cdot$ cm, respectively. Thus it is probably that our film has a mixed phase of γ-Fe_2O_3 among Fe_3O_4 phase. In order to ensure the growth of Fe_3O_4, the next work is to get the temperature dependence of resistivity which can give the information of Verwey transition.

4. Conclusion

Fe_3O_4 (100) oriented film can be grown epitaxially on MgO (100) oriented under layer prepared on 5 nm-thick Fe buffer layer on glass substrates. Resistivity analyses confirmed that higher oxygen partial pressure causes higher oxidation of Fe ions in the films which tend to result in γ-Fe_2O_3. It is verified that substrate heating is effective to improve crystallinity and magnetic properties. However, high temperature above 200 °C causes diffusion of MgO which weaken Fe_3O_4 (100) orientation and growth indirectly. Experimental results showed that oxygen partial pressure of 7.5%, substrate temperature of 150 °C and sputtering discharge current of 0.45A was suitable to prepare magnetite Fe_3O_4 film with (100) preferential orientation, less density of γ-Fe_2O_3 mixed phase and good magnetic properties, which may be suitable for spintronic devices.

Acknowledgments

The authors would like to thank Prof. Y.Nakamura and Prof. J.Shi of Dept. of Metallurgy & Ceramics Science, Tokyo Institute of Technology for TEM observation.

References

[1] Jain. S, Dai. D. Y and Adeyeye. A. O 2004 *J. Appl. Phys.* **95** 7237
[2] Jain. S, Boothroyd, C. B and Adeyeye, A. O 2005 *J. Appl. Phys.* **97** 093713-1
[3] Sakuma. H, Sato. S, Gomimoto. R, Hiyama. S, and Ishii. K 2007 *IEEJ. Trans.* **2** 431-35
[4] Yuasa. S, Nagahama. T, Fukushima. A, Suzuki. Y and Ando. K 2004 *Nature Materials.* **3** 868
[5] A.R. Fert *et al* 2001 *Eur. Phys. J. B* **24** 43-49
[6] Pan. L. P, Zhang. G. M, Fan. C.F, Qiu. H, Wu. P, Wang. F. P and Zhang. Y 2004 *The Solid Films* **473** 63–69
[7] Sakuma. H, Ishii. K 2009 *J. Magn. Magn. Mater.* **321** 872–75

Journal of Physics: Conference Series **266** (2011) 012112 doi:10.1088/1742-6596/266/1/012112

Influence of composition on structure and magnetic properties of epitaxial Mn-Ga films

F Wu [1,*], S Mizukami [1], D Watanabe [1], H Naganuma [2], M Oogane [2], Y Ando [2], T Miyazaki[1]

[1] WPI-AIMR, Tohoku Univ. 2-1-1, Katahira, Aoba-ku, Sendai, Japan.
[2] Graduate School of Engineering, Tohoku Univ. Aobayama-05, Sendai, Japan.

E-mail: fengwu@wpi-aimr.tohoku.ac.jp

Abstract. In this study, influence of composition on structure and magnetic properties of epitaxial Mn-Ga films were investigated. The epitaxial Mn-Ga films with different composition were grown on metal (Pt and Cr) buffered MgO substrates by co-sputtering technique. By use of the Pt buffered MgO substrates, the $D0_{22}$ phases can be obtained in the Mn-Ga films with the composition ranging from $Mn_{65}Ga_{35}$ to $Mn_{75}Ga_{25}$. The crystalline quality and perpendicular magnetic anisotropic properties of the Mn-Ga films were deteriorated with increase of Mn content. The saturation magnetization decreased with the increase of Mn content, which is consistent with the reported results of bulk alloys.

1. Introduction

New materials with multiple properties, i.e. large perpendicular magnetic anisotropy (PMA), high spin polarization and low saturation magnetization (M_s), are desired to accelerate further progress of spintronics [1-2]. Mn-Ga alloy with $D0_{22}$ crystal structure has great possibility to be such material owing to its special crystalline and magnetic structure [3]-[6]. Recently, the $D0_{22}$-structured epitaxial $Mn_{70}Ga_{30}$ films have been reported to exhibit giant PMA (K_u^{eff}=1.2×10^7 erg/cc) and low M_s (250 emu/cc) simultaneously [6]. However, there is a strong influence of composition on the magnetic properties and electronic structure of the $D0_{22}$ Mn-Ga bulk alloy [4-5]. Investigation of the effects of the composition on the structure and magnetic properties of the Mn-Ga films is also very interesting for practical applications. In this work, the $D0_{22}$ epitaxial Mn-Ga films with wide composition range have been obtained by using suitable buffer, and the effects of the composition on its structural and magnetic properties were also studied.

2. Experimental details

Two different materials, namely, epitaxial Cr (40 nm) and Cr (40 nm) /Pt (10 nm) layers grown on MgO (001) single crystal substrates, were used as buffer layers for epitaxial growth of the Mn-Ga films. The experimental details for preparation of the Cr buffers have been described in literature [7]. The Pt layer was deposited at 350 °C on Cr buffered MgO substrate, and was confirmed to be epitaxially grown by pole figure measurement. The 30 nm thick Mn-Ga films were deposited at room temperature by co-sputtering technique in ultra-high vacuum magnetron sputtering system, the composition was controlled from $Mn_{65}Ga_{35}$ to $Mn_{75}Ga_{25}$ by changing sputtering power of $Mn_{60}Ga_{40}$ and Mn target, respectively. Then, the films were in-situ annealed at 400□. . Finally, the samples were deposited with 5 nm thick Ta capping layers to protect oxidation of surface. The composition of the sample was analyzed by inductively coupled plasma mass spectrometry (ICP). The structural properties were measured by X-ray diffractionmeter (XRD) with Cu $K\alpha$ radiation, and the magnetic properties were evaluated by superconducting quantum interference device (SQUID).

3. Results and Discussion

3.1 MgO/Cr/Mn-Ga

Figure 1 shows the XRD patterns of the Mn-Ga films grown on MgO/Cr with different composition. For the films with composition of $Mn_{65}Ga_{35}$ and $Mn_{68}Ga_{32}$, only (002) and (004) peaks from Mn-Ga can be observed, indicating that the Mn-Ga films are grown along *c*-axis. The films were confirmed to be epitaxially grown with the relationship as follows by pole figure measurements, MgO (001)<100>// Cr (001)<110>// Mn-Ga (001)<100>. The lattice constant of film (bulk) along a and c axis is 3.95 (3.91) Å and 7.04 (7.13) Å. The slight difference with the result of bulk alloy may be due to the existence of tensile strain existed in the Mn-Ga films, which results from lattice mismatch (about 5 %) between Cr and Mn-Ga because the lattice constant of Cr and Mn-Ga is 0.408 nm and 0.391 nm, respectively. The full width of half maximum (FWHM) of Mn-Ga (004) peak is only 2.5° from rocking curve measurement. It should be noted that the Mn-Ga (004) peaks become broad and shift to higher angle with increase of Mn concentration to $Mn_{72}Ga_{28}$. This means that the $Mn_{72}Ga_{28}$ film does not form $D0_{22}$ phase. The difficulty might be due to meta-stability of the $D0_{22}$ structure when the composition is approaching to stoichiometric composition ($Mn_{75}Ga_{25}$) in bulk alloy, as described in [5].

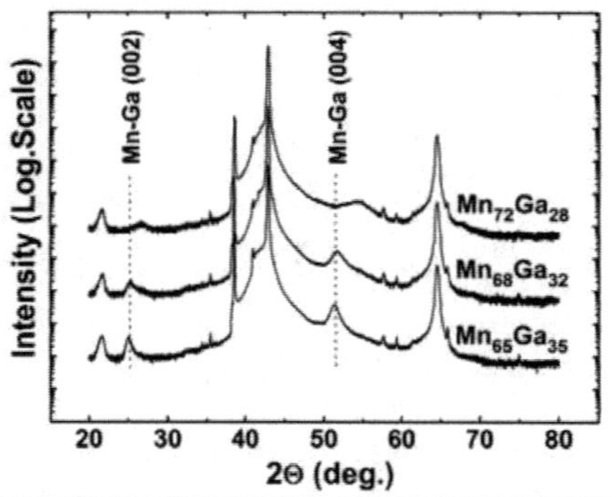

Figure 1. XRD patterns of the Mn-Ga films with different composition grown on MgO/Cr substrates

Figure 2 shows the M-H curves of the epitaxial Mn-Ga films with composition of $Mn_{65}Ga_{35}$ and $Mn_{68}Ga_{32}$. Both of them exhibit perpendicular magnetic anisotropy (PMA) property because the magnetization easy axis is perpendicular to film plane. The saturation magnetization and effective anisotropy energy (K_u^{eff}) is estimated to be around 350 emu/cc and 8×10^6 erg/cc. However, there is step in M-H loop of the film with composition of $Mn_{68}Ga_{32}$ in perpendicular direction, indicating other phases may be co-existed with the $D0_{22}$ phase although it cannot be detected by X-ray diffraction meter.

It can be concluded that preparation of $D0_{22}$ structured $Mn_{75}Ga_{25}$ films on MgO/Cr substrates is very difficult. There are at least two reasons, one is the relative large lattice mismatch between Cr and Mn-Ga, the other is the meta-stability of the stoichiometric $D0_{22}$ Mn-Ga.

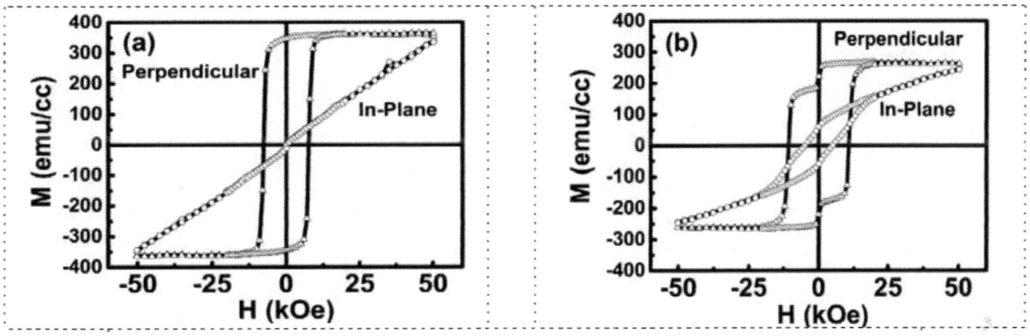

Figure 2. M-H curves of Mn-Ga films with different composition grown on MgO/Cr substrates, (a) $Mn_{65}Ga_{35}$, and (b) $Mn_{68}Ga_{32}$. Perpendicular means the applied magnetic field is perpendicular to the films' plane, and the In-plane means the field is parallel to the films' plane.

3.2 MgO/Cr/Pt/Mn-Ga

The reason of using Pt as buffer is that the lattice mismatch between Pt and Mn-Ga is almost zero because the lattice constant along *a*-axis of Mn-Ga and Pt is 3.91 Å and 3.93 Å, respectively.

Only (002) and (004) peaks from Mn-Ga can be observed for the films with composition ranging from $Mn_{65}Ga_{35}$ to $Mn_{75}Ga_{25}$, and peaks' position does not change with composition, as shown in Figure 3. The FWHM is only around 1 ° from rocking curve measurements of Mn-Ga (004) peaks, this value is much smaller than that of the film grown on MgO/Cr substrates, indicating that the crystalline quality was much improved. The epitaxial relationship is MgO (001)<100>// Cr (001)<110>// Pt (001)<100>// Mn-Ga (001)<100> according to pole figure results. The lattice constant along *a* and *c* axis is 3.91 Å and 7.20 Å, respectively, which is independent on composition. This independence is quite different from that of bulk alloy, and the reason is unclear.

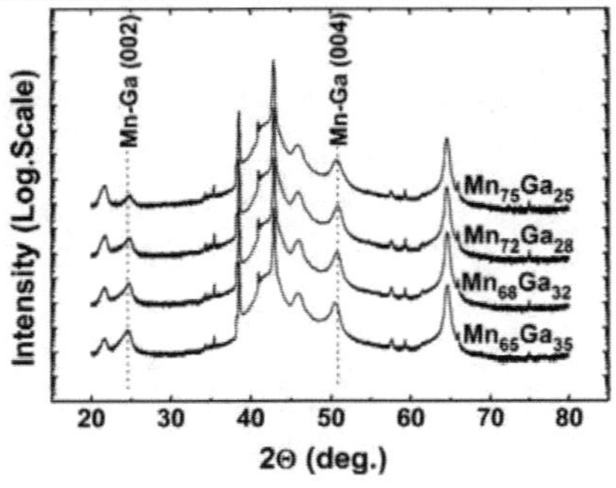

Figure 3. XRD patterns of the Mn-Ga films with different composition grown on MgO/Cr/Pt substrates

Figure 4 gives M-H curves of the films with different composition grown on MgO/Cr/Pt substrates. The perpendicular magnetic anisotropy can be observed in all the samples. However, the squareness of loop in perpendicular direction becomes poor with increase of Mn content, suggesting that the quality

Journal of Physics: Conference Series **266** (2011) 012112 doi:10.1088/1742-6596/266/1/012112

of PMA properties was deteriorated. Figure 5 shows intensity of Mn-Ga (004) peak as function of composition. As can be seen, the intensity of (004) peak decreases with increase of Mn content. That means the crystallization quality decrease with increasing Mn content. Therefore, the deteriorated crystalline quality leads to the poor perpendicular magnetic anisotropy property. Further work is needed to improve the quality of the epitaxial Mn-Ga films.

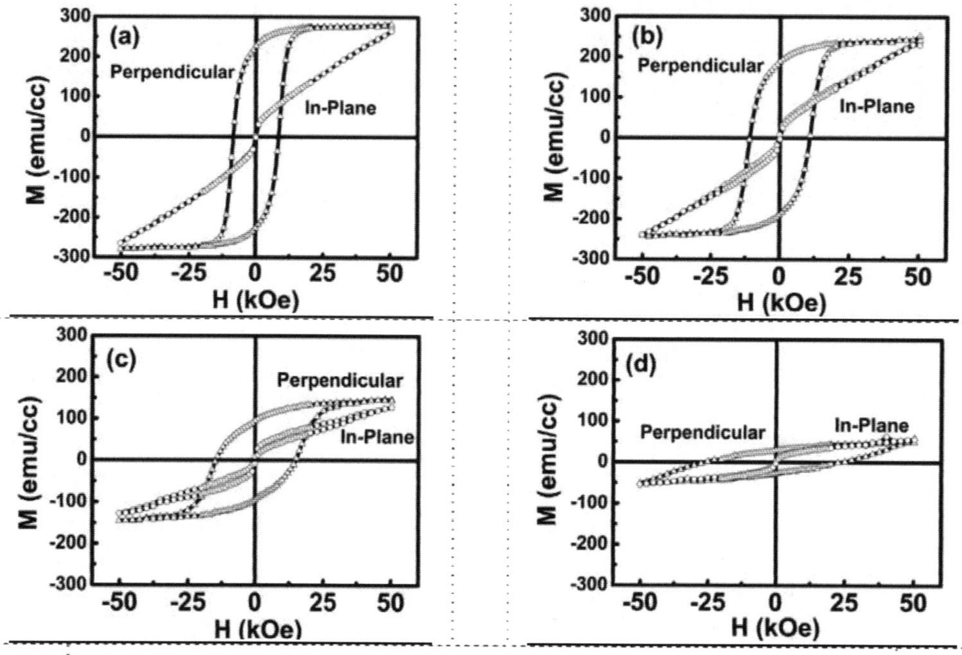

Figure 4. M-H curves of the Mn-Ga films with different composition grown on MgO/Cr/Pt substrates, (a) $Mn_{65}Ga_{35}$, (b) $Mn_{68}Ga_{32}$, (c) $Mn_{72}Ga_{28}$, and (d) $Mn_{75}Ga_{25}$. Perpendicular means the applied magnetic field is perpendicular to the films' plane, and the In-plane means the field is parallel to the films' plane.

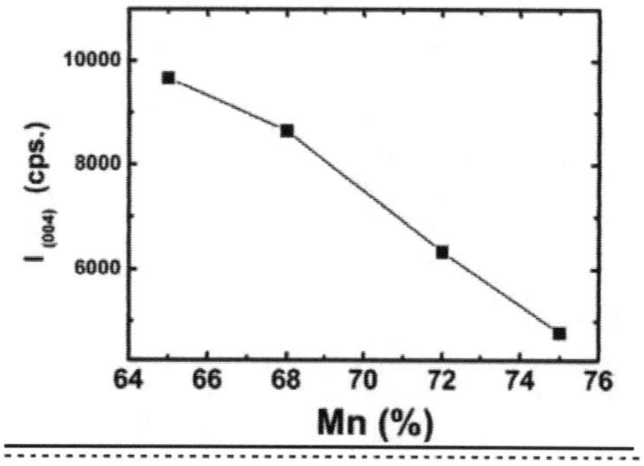

Figure 5. The intensity of (004) peak as function of Mn content.

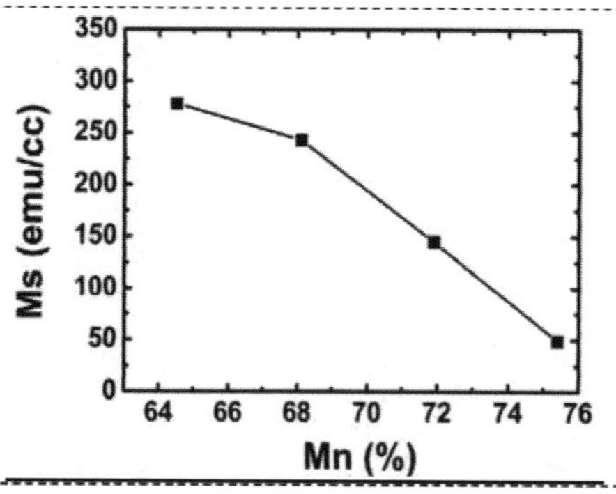

Figure 6. The dependence of saturation magnetization on Mn content.

Figure 6 summaries saturation magnetization of the films with different composition. The saturation magnetization decreases continuously from 270 emu/cc to 50 emu/cc when the composition approach to $Mn_{75}Ga_{25}$, which is roughly in agreement with the reports results for Mn-Ga bulk alloy including in figure 6 [4].

Conclusion

The $D0_{22}$ structured epitaxial Mn-Ga films with wide composition range can be prepared on Pt buffer. And the effects of composition on structure and magnetic properties were investigated. However, the perpendicular magnetic anisotropy property becomes poor with composition approaching to stoichiometry. Further studies are needed to overcome this difficulty.

Acknowledgment

This work was supported in part by Strategic Japanese-German cooperative program (ASPIMATT) from JST, Grant-in-Aid for Scientific Research, and Grant for Industrial Technology Research from NEDO.

Reference

[1] Mangin S, Ravelosona D, Katine J A, Carey M J, Terris B D and Fullerton E E 2006, *Nature Mater.* **5**, 210.
[2] Yagami K, Tulapurkar A A, Fukushima A and Suzuki Y 2004 *Appl. Phys. Lett.* **85**, 5634.
[3] Balke B, Fecher G H, Winterlik J and Felser C 2007 *Appl. Phys. Lett.* **90**, 152504.
[4] J. Winterlik, B. Balke, G. H. Fecher, and C. Felser 2008, *Phys. Rev. B.* **77**, 054406.
[5] Niida H, Hori T, Onodera H, Yamaguchi Yand Y. Nakagawa 1996 *J. Appl. Phys.* **79,** 5946.
[6] Wu F, Mizukami S, Watanabe D, Naganuma H, Oogane M, Ando Y and Miyazaki T 2009 *Appl. Phys. Lett.* **94**,122503.
[7] Sakuraba Y, Nakata J, Oogane M, Kubota H, Ando Y, Sakuma A and Miyazaki T 2005 *Jpn. J. Appl. Phys.* **44,** 6535.

Journal of Physics: Conference Series **266** (2011) 012113 doi:10.1088/1742-6596/266/1/012113

High-order standing spin wave modes in $Fe_{19}Ni_{81}$ micron wire observed by homodyne method

A Yamaguchi[1, 2*], K Motoi[1], H Miyajima[1], T Uchiyama[3] and Y Utsumi[4]
[1] Department of Physics, Keio University, Hiyoshi, Yokohama 223-8522, Japan
[2] PRESTO, JST, Honcho, Kawaguchi, Saitama 332-0012, Japan
[3] Department of Electrical Engineering and Computer, Nagoya University, Chikusaku, Nagoya 464-8603, Japan
[4] Laboratory of Advanced Science and Technology fro Industry, University of Hyogo, Koto, Ako, Hyogo 678-1205, Japan

E-mail: yamaguch@phys.keio.ac.jp

Abstract. The broadband spin dynamics of patterned ferromagnetic $Fe_{19}Ni_{81}$ microwire with thickness of 80 nm has been investigated experimentally using broadband rectifying method. The rectifying effect provides a highly sensitive method to detect the high-order perpendicular standing spin wave (PSSW) mode. Present analytical calculation reproduces the observed relation between resonance frequency and applied magnetic field. The effective thickness is explained by the pinning condition of magnetic moment at the surface of the wire.

1. Introduction

Spin dynamics in confined structure has been extensively studied by physics and engineering groups [1], since it is significant and important in both fundamental magnetism and device applications. In particular, the excitation of several spin wave modes accompanies non-linear dynamics in mesoscopic magnets whose dimension is comparable to the wavelength and frequency of dipole-exchange spin waves. Such spin dynamics in small magnetic elements has been explored in various ways such as Billouin light scattering [2], ferromagnetic resonance (FMR) [3, 4], time-resolve magneto optical Kerr effect [5]. In particular, a homodyne method of the microwave propagating in a small magnet provides highly sensitive detection and easy operation for spin dynamics in a small magnet [6 – 13].

The study of quantized and localized spin-wave modes of small magnet is very important because the external radio-frequency (rf) field couples to nonuniform or spin-wave mode of precession in addition to an uniform mode. The contribution to the magnetic energy of these modes arise from a combination of exchange and dipolar interactions.

In this study, we show the relationship between excited spin wave modes and rectifying spectra by comparing the field and frequency dependences of rf current rectification. We discuss them in term of the perpendicular standing spin wave (PSSW) mode. The mode depends on the exchange interaction in the ferromagnet and is quantized due to pinning of the moments at the surface [12 – 15].

2. Experimental procedure

In this study, we detect the high-order PSSW mode using the rectifying effect [12, 13]. A schematic experimental diagram and an optical microscope image of the wire are shown in figure 1(a) and 1(b), respectively. A single layered 80 nm thick $Fe_{19}Ni_{81}$ wire is fabricated onto a polished MgO substrate

Journal of Physics: Conference Series **266** (2011) 012113 doi:10.1088/1742-6596/266/1/012113

by means of ultra-high vacuum e-beam evaporation, e-beam lithography and lift-off method. The width and length of the wire are 5 μm and 150 μm, respectively. The electrodes made from Cr (5 nm thickness) and Au (85 nm thickness) form a coplanar waveguide structure [8, 13]. The centre conductive strip line is placed on the wire within the coplanar waveguide structure. A sinusoidal continuous wave rf current with power of +10 dBm was injected into the wire by a signal generator. We measure the rectifying voltage of the wire along the long axis.

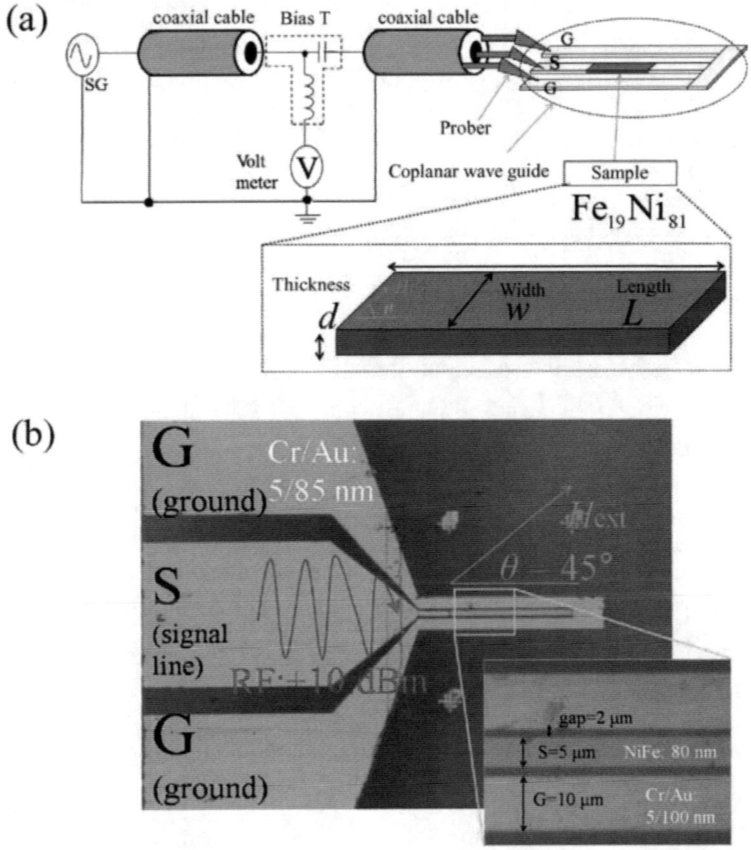

Figure 1. (a) Schematic circuit and experimental setup of the system consisting electrodes and magnetic wire. (b) Optical microscope image (top view).

3. Experimental result and discussion

The magnetic field is applied at angle $\theta = 45^{\circ}$ to the longitudinal axis of the wire in plane. Figure 2 shows the frequency dependence of the detected rectifying spectra at the external magnetic fields $H = 190$, 290 and 390 Oe. For clarity, the detected spectra are shown in the rf frequency range between (a) 0 and 15, (b) 15 and 20 and (c) 20 and 30 GHz.

Four resonance frequencies are observed as shown in figure 2. The lowest resonance frequency is derived from the uniform mode. The higher-order PSSW modes of the index n = 1, 2, and 3 are also clearly observed. The PSSW mode with the index n = 1 reconfirmed our previous experimental result [13]. The same behaviour of the PSSW mode was observed in the present study, while the present resonance frequency was not consistent with the previous one because each thickness was different [13].

The detected signal is derived from the time variation of the anisotropic magnetoresistance. As the sample resistance $R(t) = R_0 + \Delta R \cos^2(\omega t + \varphi)$ oscillates in response to the rf excitation current $I(t) = I_0 \cos(\omega t)$, a rectifying voltage is generated across the wire due to the phase difference φ between the current and the resistance, where R_0 is the resistance independent of the time t and $\Delta R \equiv R_{\parallel} - R_{\perp}$ is the resistance difference between the magnetizations parallel and perpendicular to the current. The details of the analytical model are presented by ref. 9, 10 and 11. The sign of the rectifying signal is determined by the phase difference φ that is derived from the spin wave modulation mode.

The magnetic field dependence of the resonance frequency of the wire is shown in figure 3. The frequency variation in the PSSW mode with applied field is qualitatively understood in a term of the rectangular cross-sectional wire. Accordingly to Kalinikos and Slavin [14], the dispersion relationship of the PSSW in a confined magnetic structure is given by

$$\omega_n^2 = \gamma^2 \left[H + \frac{2A}{M_S}\left(\frac{n\pi}{d_{\text{eff}}}\right)^2 \right] \left[H + \frac{2A}{M_S}\left(\frac{n\pi}{d_{\text{eff}}}\right)^2 + \frac{M_S}{\mu_0} \right], \quad (1)$$

where d_{eff} is the effective thickness for the PSSW mode wavelength including the boundary conditions [12, 14, 15], n the quantized number for the PSSW along the thickness direction, and μ_0 and γ the vacuum permeability and gyromagnetic ratio, respectively.

Figure 3 shows that the experimentally obtained lowest resonance frequency correlates well with calculation of the uniform mode. The comparison between the experimental result and the analytical PSSW eigenfrequency estimated from equation (1) is also shown in figure 3. The result implies that the above analytical relation explains our experimental results very well. Consequently, the effective thickness d_{eff} is estimated to be 64 nm of which are smaller than the actual thickness of the wire. This indicates that the surface is oxidized and contributes to the pinning of the magnetic moment [12, 15].

Fitting a mode spectral line shape to a linear combination of symmetric and antisymmetric Lorentzians gives the model full width at half maximum Δf_n (FWHM). The FWHM carries information about the mode damping parameter α_n. The damping parameters of the uniform and higher order PSSW modes are estimated to be $\alpha_{\text{uniform}} \approx 0.0094$ and $\alpha_n = 0.0177(n=1), 0.0177(n=2)$ and $0.0147(n=3)$, respectively. Compared to damping of the

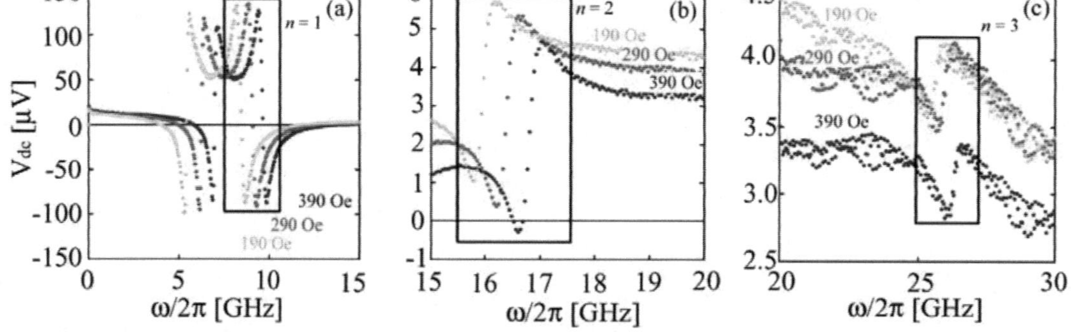

Figure 2. Typical observed rectifying spectra in the external static field of 190, 290 and 390 Oe at $\theta = 45^\circ$. The rectifying spectra are shown in the rf frequency range between (a) 0 and 15, (b) 15 and 20 and (c) 20 and 30 GHz.

uniform mode, the damping parameters of the PSSW modes are enhanced. However, the damping parameters between the higher order PSSW modes are almost same values. If two-magnon scattering process dominates the system, the damping parameters of the higher order PSSW modes are enhanced compared to damping of the $n = 1$ mode. The increase of damping of the PSSW mode with respect to the uniform mode is attributable to a random magnetic potential induced by the wire edge roughness or magnetization pinning centres. In particular, the surface pinning plays an important role on the enhancement of the damping of the PSSW modes. This is in agreement with Ref. 16 and the effective thickness estimation described the above.

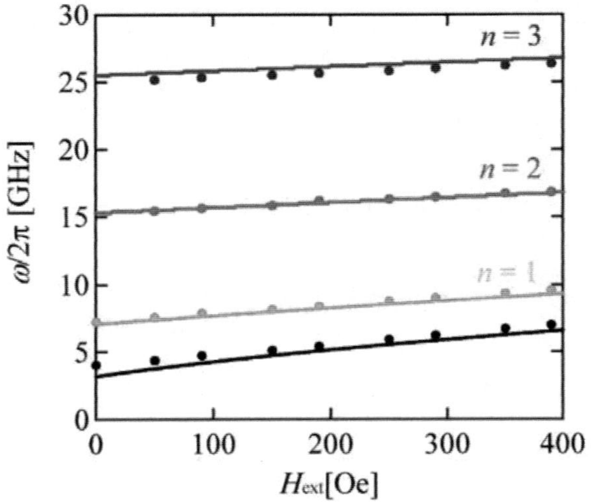

Figure 3. Magnetic field dependence of the resonance frequency observed by the rectifying spectra. The lines show the calculations as Kittel mode and equation (1).

4. Conclusions

In summary, highly sensitive measurements of the rectifying spectra in a micron-scale single-layered $Fe_{19}Ni_{81}$ strip were performed as a function of rf frequency and external magnetic field. The method provides an easy operation and highly sensitive detection for higher-order perpendicular standing spin wave modes. The observed characteristic for the resonance frequency and magnetic field is quantitatively reproduced by the analytical calculation. This highly sensitive detection of microscopic spin dynamics via the rectifying effect provides a powerful technique for studying the spin dynamics in a nano- or micron-scale confined magnetic structure.

Acknowlegements
The present study is partly supported by MEXT Grants-in-Aid for Scientific Research in a Priority Area, a JSPS Grant-in-Aid for Scientific Research and the JST PRESTO program.

References

[1] Hillebrands B and Ounadjela, Spin dynamics in Confined Magnetic Structures (Springer, Berlin, 2003), Vol. 1-3
[2] Bayer C, Jorzick J, Hillebrands B, Demokritov S O, Kouba R, Bozinoski R, Slavin A N, Guslienko K Y, Berkov D V, Gorn N L and Kostylev M P, 2005 *Phys. Rev*. B **72** 064427
[3] Berteaud A J and Pascard H, 1996 *J. Appl. Phys*. **37** 2035
[4] An S Y, Krivosik P, Kraemer M A, Olson H M, Nazarov A V, Patton C E, 2004 *J. Appl. Phys*. **96** 1572

[5] Berman A, Kruglyak V V, Hicken R J, Rowe J M, Kundrotaite A, Scott J and Rahman M, 2004 *Phys. Rev.* B **69** 174426
[6] Tulapurkar A A, Suzuki Y, Fukushima A, Kubota H, Maehara H, Tsunekawa K, Djayaparawira, Watanabe N, Yuasa S, 2005 *Nature* **438** 339
[7] Sankey J C, Braganca P M, Garcia A G F, Krivorotov I N, Buhrmann R A, Ralph D C, 2006 *Phys. Rev. Lett.* **96** 227601
[8] Yamaguchi A, Miyajima H, Ono T, Suzuki Y, Yuasa S, 2007 *Appl. Phys. Lett.* **90** 182507; Yamaguchi A, Miyajima H, Kasai S, Ono T, 2007 *Appl. Phys. Lett.* **90** 212505; Yamaguchi A, Miyajima H, Ono T, Suzuki Y, Yuasa S, 2007 *Appl. Phys. Lett.* **91** 132509
[9] Yamaguchi A, Motoi K, Hirohata A, Miyajima H, Miyashita Y, Sanada Y, 2008 *Phys. Rev.* **B 78** 104401
[10] Yamaguchi A, Motoi K, Hirohata A, Miyajima H, 2009 *Phys. Rev.* B **79** 224409
[11] Gui Y S, Mecking N, Zhou X, Williams Gwyn, Hu C –M, 2007 *Phys. Rev. Lett.* **98** 107602
[12] Gui Y S, Mecking N., Hu C –M, 2007 *Phys. Rev. Lett.* **98** 217603
[13] Yamaguchi A, Motoi K, Miyajima H, Nakatani Y, 2009 *J. Appl. Phys.* **105** 07D301
[14] Kalinikos B A and Slavin A N, 1986 *J. Phys.* C **19** 7013
[15] Guslienko K Y, Demokritov S O, Hillebrands B and Slavin A N, 2002 *Phys. Rev.* **66** 132402
[16] McMichael R D, Twisselmann D J and Kunz A, 2003 *Phys. Rev. Lett.* **90** 227601

Control of magnetic anisotropy field of (001) oriented $L1_0$-Fe(Pd_xPt_{1-x}) films for MRAM application

Satoru Yoshimura[1], Shogo Omiya[1], Genta Egawa[1], Hitoshi Saito[1], Jianmin Bai[2]
[1] Center for Geo-environmental Science, Faculty of Engineering & Resource Science, Akita University,
Tegatagakuen-machi 1-1, Akita, 010-8502, JAPAN
[2] Key Laboratory for Magnetism and Magnetic Materials of the Ministry of Education, Lanzhou University,
TianShui Road, S 222, LanZhou, GanSu, 730000, P.R. CHINA

E-mail:syoshi@gipc.akita-u.ac.jp

Abstract. $L1_0$-Fe(Pd,Pt) films which are expected to have tunable perpendicular magnetic anisotropy and lower ordering temperature compared with the $L1_0$-FePt films were fabricated and investigated in order to realize high-performance Magnetic Random Access Memory (MRAM) with spin-transfer magnetization switching method and Magnetic Domain-Wall Racetrack Memory with current-driven domain wall motion. The main results are as follows: (1) The long-range chemical order degree S for the $L1_0$-Fe(Pd,Pt) films with the optimized thermal treatment temperature fabricated on MgO(001) substrate was about 0.8. (2) The perpendicular coercive force and magnetic anisotropy field for the $L1_0$-FePd film with the thermal treatment temperature of 600 ℃ were 300 Oe and 14 kOe respectively, whereas the $L1_0$-FePt film with the thermal treatment temperature of 800 ℃ were 2000 Oe and 102 kOe, respectively. (3) The optimized ordering temperature to obtain $L1_0$ single phase decreased continuously from 800℃ to 600℃ with the increasing Pd content for the Fe(Pd,Pt) films. (4) The perpendicular magnetic anisotropy field for the $L1_0$-Fe(Pd,Pt) films decreased continuously from 102 kOe to 14 kOe with the increasing Pd content. It is found that the Fe(Pd,Pd) films which have tunable perpendicular magnetic anisotropy field is one of suitable ferromagnetic material for high-performance magnetic recording devices.

1. Introduction

Magnetic thin films with perpendicular magnetic anisotropy are suitable for high density Magnetic Random Access Memory (MRAM) with spin-transfer magnetization switching method [1] and Magnetic Domain-Wall Racetrack Memory with current-driven domain wall motion. The critical current, I_c, of the spin-transfer switching method and the current-driven domain wall motion is dependent on the magnetic properties of a storage layer of the magnetic tunnel junctions (MTJs) as follows.

$$I_c = \alpha e \gamma M_s A d\,[\,H_{ext} - 4\pi M_s + 2K_u / M_s + M_s / 2\mu_0\,] / \mu_B g$$

Here, α is the damping constant, γ is a gyromagnetic effect, M_s is the saturation magnetization, d is magnetic layer thickness, H_k is magnetic anisotropy field, K_u is magnetic anisotropy energy, and g is efficiency of the spin momentum transfer of the storage layer of the MTJ, A is magnetic cell size, and

H_{ext} is an applied field for the MTJ, respectively [2,3]. To realize high-performance MRAM and Racetrack Memory with low critical current, it is necessary to reduce the damping constant, and to control the perpendicular magnetic anisotropy and the saturation magnetization of the storage layer, independently.

The FePt and CoPt films with the structure of $L1_0$ are widely discussed for magnetic recording devices due to its very high perpendicular magnetic anisotropy, high Cure temperature, and high stability of crystalline structure. Therefore, the $L1_0$-FePt and $L1_0$-CoPt film is suitable material for magnetic recording device compared with multilayer with perpendicular magnetic anisotropy such as $[Co/Pd]_N$ and $[Co/Ni]_N$. However, very high perpendicular magnetic anisotropy field and very high ordering temperature of $L1_0$-FePt and $L1_0$-CoPt films make it difficult to realize the low critical current.

The MTJs with the storage layer of $L1_0$-CoPt film which was fabricated with low thermal treatment temperature and has relatively low perpendicular magnetic anisotropy are reported [4]. However, the spin polarization of CoPt film with low long-range chemical order degree is low, and consequently TMR ratio of the MTJs is relatively low [4]. Therefore, it is necessary to develop new magnetic thin films with highly ordered $L1_0$ structure, tunable perpendicular magnetic anisotropy, high Cure temperature, and relatively low ordering temperature.

Cu added $L1_0$-FePt thin films is one of possible materials for the magnetic recording devices due to its relatively low ordering temperature and low perpendicular magnetic anisotropy field [5], however addition of Cu decreases the Cure temperature and saturation magnetization. Consequently, Fe(Pt,Cu) thin films is not suitable material for high-performance magnetic recording devices.

On the other hand, $L1_0$-FePd thin films have relatively low perpendicular magnetic anisotropy field compared with $L1_0$-FePt thin films [6], and also have high Cure temperature. Moreover, FePd thin film can easily form ordered structure [7]. Here, Pd and Pt can easily form alloy with pure fcc structure.

In this study, we fabricated the Fe(Pd,Pt) thin films and investigated the magnetic properties and crystalline structures of the films to realize the control of perpendicular magnetic anisotropy and the reduction of ordering temperature.

2. Experimental procedure

The Fe(Pd,Pt) thin films with the thickness of 10 nm were prepared on MgO(001) substrates. All films were deposited at room temperature by using magnetron sputtering method. After the deposition of the films, thermal treatment was performed to form $L1_0$ ordered structure at a temperature of from 550 ℃ to 800 ℃ for 60 minutes. The Fe concentration of the films was fixed at 50 at%, the Pd concentration of the films was varied at 0, 18.5, 24.5, 40.0, 50 at%. In other words, we fabricated the $Fe_{0.5}(Pd_xPt_{1-x})_{0.5}$ films with the x of 0, 0.37, 0.49, 0.80, 1, respectively. The crystallographic orientation was analyzed by using an X-ray diffraction (XRD) system. The magnetization curves were measured by using a vibrating sample magnetometer (VSM). The film composition was analyzed by using an electron probe micro-analysis (EPMA) system.

3. Results and Discussions

Figure 1 shows the XRD (θ–2θ) patterns for the $Fe(Pd_{0.8}Pt_{0.2})$ films deposited on MgO substrate under different thermal treatment conditions. From all films, only diffraction peaks from c-plane and a-plane (the Fe(Pd,Pt)(001), (002), and (200) diffraction peaks) are observed. This indicates that all films epitaxially grow along

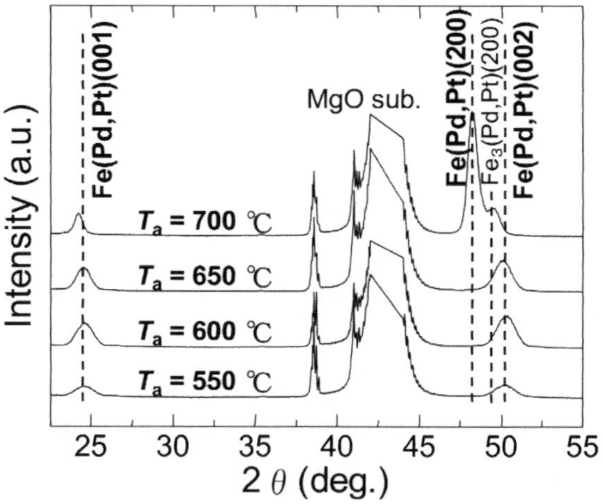

Figure 1. X-ray diffraction (XRD) (θ–2θ) patterns for the $Fe(Pd_{0.8}Pt_{0.2})$ films deposited on MgO substrates under various thermal treatment conditions.

(001) orientation onto MgO(100) substrates. With increasing thermal treatment temperature, T_a, up to 650 ℃, the intensity of the $L1_0$-Fe(Pd,Pt)(001) and (002) diffraction peaks increase, whereas the fcc-Fe(Pd,Pt)(200) diffraction peak is not observed. The long-range chemical order degree S for the Fe(Pd,Pt) films with the T_a of 600 - 650 ℃ is about 0.8. These results indicate that the Fe($Pd_{0.8}Pt_{0.2}$) films with the T_a of 600 - 650 ℃ show (001) orientation with $L1_0$ ordered structure. However, when the T_a is higher than 700 ℃, the $L1_0$-Fe(Pd,Pt)(001) and (002) diffraction peaks decrease and fcc-Fe(Pd,Pt)(200) and $L1_2$-Fe_3(Pd,Pt)(200) diffraction peak increase. This result indicates that the Fe($Pd_{0.8}Pt_{0.2}$) films with the T_a of more than 700 ℃ shows (100) orientation with fcc and $L1_2$ [8] structure. Here, in the case of $Fe_{50}Pd_{50}$, the transformation temperature from fct to fcc structure is about 800 ℃, whereas in the case of $Fe_{50}Pt_{50}$, the temperature is about 1300 ℃. This is a reason that FePd film forms fcc phase easily with high T_a.

Figure 2(a) shows the normalized intensity of $L1_0$ (001) diffraction peak of Fe(Pd_xPt_{1-x}) films as a function of thermal treatment temperature, T_a. The intensity is normalized by the maximum intensity in the same composition. Here, the x = 1 corresponds to FePd, and the x = 0 to FePt. In the case of x of more than 0.8, high (001) intensity ratio is obtained at relatively low T_a of 600℃. With the decreasing of x, the T_a at which Fe(Pd,Pt) films have high (001) intensity ratio increases. In the case of x = 0, high (001) intensity ratio is obtained at high T_a of 800℃. These results indicate that the ordering temperature of $L1_0$ structure decreases with the increasing of Pd content.

Figure 2(b) shows the peak intensity of $L1_0$ (001) normalized by that of fcc (200) as a function of T_a. The small intensity indicates the formation of fcc structure. In the case of x of more than 0.49, low (001) intensity ratio is obtained at T_a of higher than 700℃. With decreasing x, the T_a at which (001) intensity starts to decrease become low ratio increases. In the case of x = 0, decrease of (001) intensity ratio is not obtained at high T_a of 800℃. These results indicate that the transformation temperature into fcc decreases with increasing Pd content.

As mentioned above, optimized ordering temperature to obtain $L1_0$ single phase decreases with increasing Pd content for Fe(Pd,Pt) film, and the optimized ordering temperature for FePd film is 600 ℃ whereas the optimized ordering temperature for FePt film is 800 ℃. Here, the long-range chemical order degree S for the $L1_0$-Fe(Pd,Pt) film with the optimized T_a was about 0.8. However, the ordering temperature of FePd film is still high. Previously we studied the effect of plasma irradiation

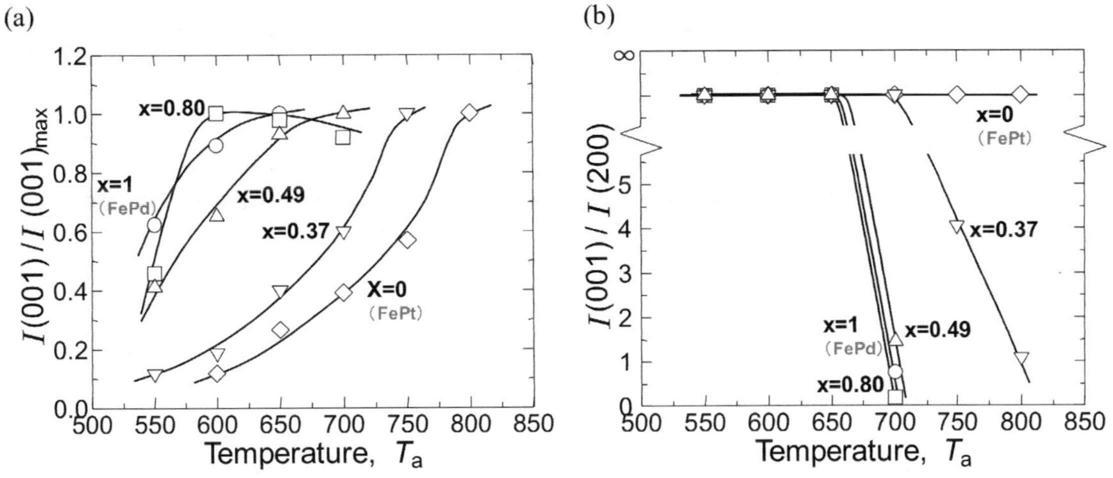

Figure 2. The normalized intensity of $L1_0$ (001) diffraction peak of Fe(Pd_xPt_{1-x}) films. The intensity normalized by a maximum intensity in the same composition. The optimized thermal treatment temperature to obtain $L1_0$ single phase (a), and normalized by that of fcc (200) intensity of each film. The small intensity indicates the formation of fcc structure (b), as a function of thermal treatment temperature, T_a.

during film deposition on the ordering temperature of $L2_1$-Fe_2(Mn,Cr)Si Heusler alloy [9,10]. The ordering temperature of the Heusler alloy was reduced by the plasma irradiation. We expect that the technique can also reduce the ordering temperature of the $L1_0$Fe(Pd,Pt) film.

Figure 3 shows the out-of-film-plane and in-film-plane hysteresis loops for the Fe($Pd_{0.8}Pt_{0.2}$) film deposited on a MgO substrate under the thermal treatment temperature, T_a of 600 ℃. The out-of-plane hysteresis loop shows very small coercive force, H_c of 300 Oe and large residual magnetization. On the other hand, the in-plane hysteresis loop shows large saturation magnetic field of the magnetization and very small residual magnetization. These indicate that the Fe($Pd_{0.8}Pt_{0.2}$) film has perpendicular magnetic anisotropy and its perpendicular magnetic anisotropy field, H_k is 14 kOe. Here, the perpendicular magnetic anisotropy field is determined by saturation magnetic field of in-plane hysteresis loop which is magnetically hard axis direction.

Figure 4 shows the dependence of the H_c and H_k on the composition of Fe(Pd_xPt_{1-x}) films deposited onto MgO(100) substrate with optimized T_a on the composition of Fe(Pd,Pt) films. The MFM images of FePt and FePd films demagnetized by DC field are also shown in this figure. The H_k is also determined by saturation magnetic field of in-plane hysteresis loop. The saturation magnetization, M_s for the Fe(Pd,Pt) is almost constant of 1200 emu/cm^3. The crystallographic orientation and the long-range chemical order degree S of these films are (001) and about 0.8, respectively. The H_k decreases gradually from over 100 kOe to less than 15 kOe with increasing Pd content, as a result, H_c of FePd film decreases from 2.0 kOe to 0.3 kOe.

In the MFM images, domain size of the FePd film is smaller than that of the FePt film. This difference corresponds to the H_k. However, the domain size of FePd film is in the scale of micrometer. This indicates that $L1_0$-FePd film is applicable for the nano-scale memory cell.

From these results, we can say, $L1_0$-Fe(Pd,Pt) films which can tune the uniaxial magnetic anisotropy field by changing Pd and Pt content with maintaining the saturation magnetization are suitable for high-performance magnetic recording devices.

4. Conclusions

We have investigated crystalline structure and the magnetic properties of the Fe(Pd,Pt) films with various Pd concentration prepared under various annealing temperature. Highly ordered $L1_0$-Fe(Pd,Pt) structure was obtained successfully. The ordering temperature of single phase $L1_0$ single phase $L1_0$-Fe(Pd,Pt) (the long-range chemical order degree S was about 0.8) decreased continuously from 800 ℃ to 600 ℃ with increasing the Pd concentration. The perpendicular magnetic anisotropy field (coercive

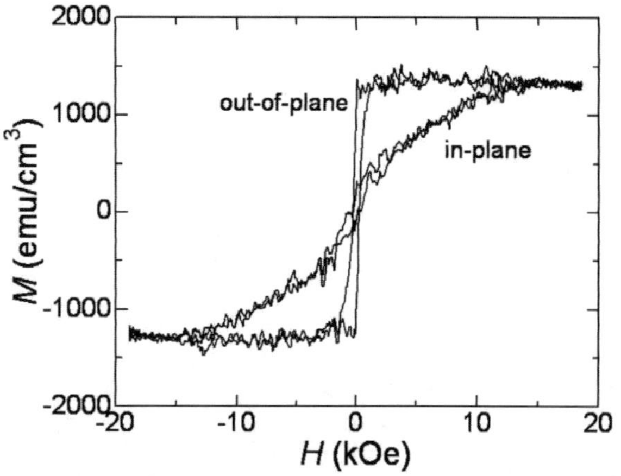

Figure 3. The out-of-film-plane and in-film-plane hysteresis loops for the Fe($Pd_{0.8}Pt_{0.2}$) film deposited on a MgO substrate under the thermal treatment temperature, T_a of 600 ℃.

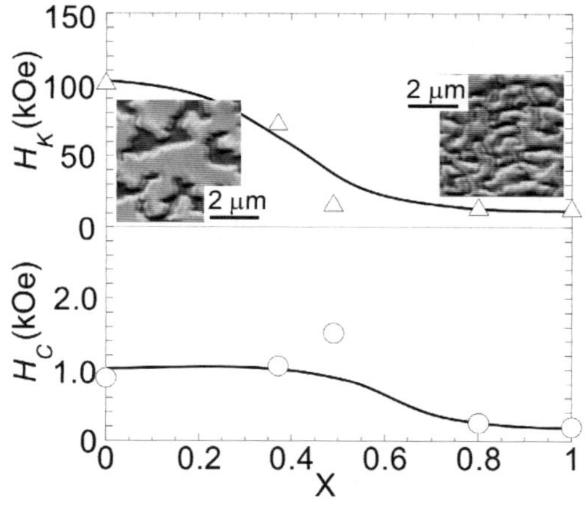

Figure 4. The H_c and H_k for Fe(Pd_xPt_{1-x}) films with the optimized thermal treatment condition to obtain $L1_0$ single phase, as a function of Pd content. The MFM images of FePt and FePd films are also shown.

force) for the films decreased continuously from 102 kOe to 14 kOe (1 kOe to 0.3 kOe) with increasing the Pd concentration, while saturation magnetization remained constant. We found that the Fe(Pd,Pt) films have tunable perpendicular magnetic anisotropy field, which is very useful to the high-performance magnetic recording devices.

5. References

[1] Yakushiji K, Yuasa S, Fukushima A, Kubota H, Nagahama T, Katayama T and Ando K 2008 *Magne Japan* vol 3 p 262
[2] Yagami K and Suzuki Y 2003 *Materia Japan* vol 42 p 640
[3] Yagami K, Tulapurker A A, Fukishima A and Suzuki Y 2004 *Appl. Phys. Lett.* vol 85 p 5634
[4] Kim G, Sakuraba Y, Oogane M, Ando Y and Miyazaki T 2008 *Appl. Phys. Lett.* vol 92 p 172502
[5] Maeda T, Kikitsu A, Kai T, Nagase T and Akiyama J 2005 *IEEE Trans. Magn.* vol 38 no 5 p 2796
[6] Clavero C, Garcia-Martin J M, Costa J L, Armelles G and Cebollada A 2006 *Phys. Rev. B* vol 73 p 174405
[7] Kovács A, Sato K and Hirotsu Y 2007 *J. Appl. Phys.* Vol 102 p 123512
[8] Kang S, Jia Z, Nikles D E and Harrell J W 2004 *J. Appl. Phys.* vol 95 p 6744
[9] YoshimuraS, Kobayashi H, Egawa G, Saito H, Bai J 2010 The 2nd International Symposium on Advanced Magnetic Materials and Applications (ISAMMA 2010) QD-01
[10] Yoshimura S, Kobayashi H, Egawa G, Saito H 2010 The 55th Conference on Magnetism and Magnetic Materials (MMM2010) CV-07

Journal of Physics: Conference Series **266** (2011) 012115 doi:10.1088/1742-6596/266/1/012115

Quantifying spin mixing conductance in *F*/Pt (*F* =Ni, Fe, and $Ni_{81}Fe_{19}$) bilayer film

T Yoshino[1], K Ando[1], K Harii[1], H Nakayama[1], Y Kajiwara[1] and E Saitoh[1, 2, 3]
[1] Institute for Materials Research, Tohoku University, Sendai, 980-8577, Japan
[2] PRESTO, Japan Science and Technology Agency, Sanbancho, Tokyo, 102-0075, Japan
[3] The Advanced Science Research Center, Japan Atomic Energy Agency, Tokai 319-1195, Japan

E-mail: ytatsuro@imr.tohoku.ac.jp

Abstract. The spin-mixing conductances in *F*/Pt (*F* = Ni, Fe, and $Ni_{81}Fe_{19}$) bilayer films were quantified from the peak-to-peak linewidth of ferromagnetic resonance (FMR) spectra based on the model of the spin pumping. When the Pt layer is attached to the *F* layer, we found the enhancement of the FMR linewidth due to the spin pumping. The experimental results show that the spin-mixing conductances in *F*/Pt (*F* = Ni, Fe, and $Ni_{81}Fe_{19}$) bilayer films have the same order of magnitude, showing that spin injection efficiency in the spin pumping is almost identical in these films.

1. Introduction

There has been a rapidly growing interest in the field of spintronics, which manipulates the electron's spin degree of freedom in addition to the charge degree of freedom [1, 2]. In the field of spintronics, methods for generating and detecting spin currents, a flow of electron spins in a solid, are essential for spin-based magnetic memories and computing devices [3-5].

In a ferromagnetic/paramagnetic bilayer film, the spin pumping induced by ferromagnetic resonance (FMR) is an efficient method to realize a pure spin-current injection into the paramagnetic layer from the ferromagnetic layer; when the magnetic field is applied to the film, the magnetization precession in the ferromagnetic metal pumps a spin current into the adjacent paramagnetic metal [6-12]. The pumped spin currents $\mathbf{J}_s$ induced by the spin pumping is described as

$$\mathbf{J}_s = \frac{\hbar}{4\pi}\left(g_r^{\uparrow\downarrow}\mathbf{m}\times\frac{d\mathbf{m}}{dt}\right). \tag{1}$$

Here, **m** is a unit vector of magnetization, $g_r^{\uparrow\downarrow}$ is real part of the spin-mixing conductance [13]. $g_r^{\uparrow\downarrow}$ is essential parameter to investigate the spin pumping experimentally. In this paper, we quantify $g_r^{\uparrow\downarrow}$ in *F*/Pt (*F* = Ni, Fe, and $Ni_{81}Fe_{19}$) bilayer films.

2. Experimental Procedure

Figure 1(a) shows a schematic illustration of the sample used in the present study. The samples are plane *F* films and *F*/Pt films where, *F* = Ni, Fe, and $Ni_{81}Fe_{19}$. The 10-nm thick Pt layer was sputtered in Ar gas atmosphere on a thermally oxidized Si substrate and then the 10-nm thick *F* layer was deposited with an electron beam evaporator. We measured the FMR spectra in all samples using the electron spin resonance spectroscopy.

During the measurement, the sample is placed at the center of a TE_{011} cavity at which the magnetic-field component of the microwave mode is maximized while the electric-field component of the microwave mode is minimized [14]. The microwave mode with a frequency of $f = 9.44$ GHz exists in the cavity, and the external magnetic field **H** along the film plane is applied to the sample. The microwave power (100 mW) is much lower than the saturation of the FMR absorption for the present sample. All the experiments were performed at room temperature.

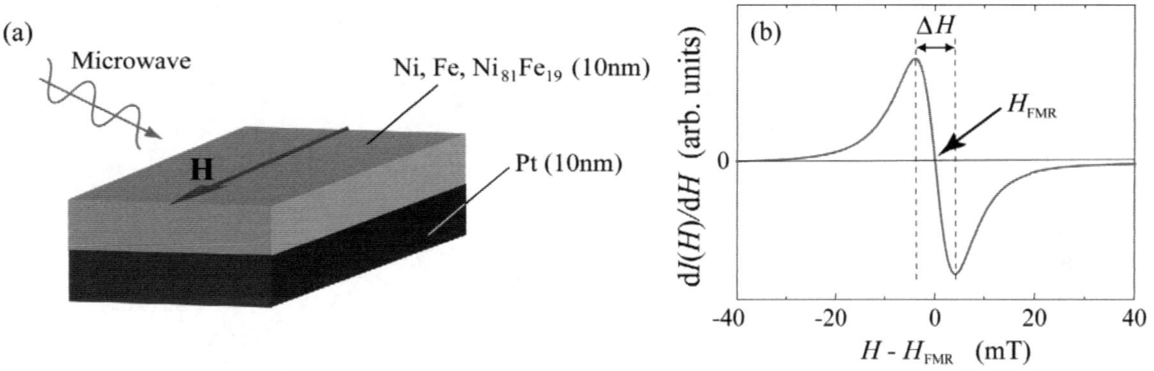

Figure 1. (a) A schematic illustration of the sample used in the present study. We used Ni, Fe, and $Ni_{81}Fe_{19}$ for the ferromagnetic layer. **H** is the external magnetic field. (b) FMR spectrum for the Fe film. I, H_{FMR}, and ΔH denote the microwave absorption intensity, the resonance field, and the peak-to-peak linewidth, respectively.

3. Results and Discussion

Figures 2(a), 2(b), and 2(c) show the FMR spectra $dI(H)/dH$ for the Ni, Fe, and $Ni_{81}Fe_{19}$ single layer films under the 100 mW microwave excitation, where I represents the microwave absorption intensity. Figures 2(d), 2(e), and 2(f) show the FMR spectra $dI(H)/dH$ for the F/Pt (F = Ni, Fe, and $Ni_{81}Fe_{19}$) bilayer films under the 100 mW microwave excitation.

From the FMR spectra, we can estimate the damping constant α using the relation $\alpha = (\sqrt{3}\gamma / 4\pi f)\Delta H$, where ΔH is defined as shown in figure 1(b) [4]. Here, γ represents the gyromagnetic ratio. The saturation magnetization $4\pi M_s$ can also be obtained from the resonance condition when **H** is applied along the film plane:

$$\left(\frac{\omega}{\gamma}\right)^2 = (H_{\rm FMR} + 4\pi M_s)H_{\rm FMR}, \qquad (2)$$

where $\omega = 2\pi f$ [15].

When attaching the Pt layer to the ferromagnetic layer, the FMR linewidth ΔH is enhanced, e.g., ΔH extracted from figures 2(a) and 2(d) are $\Delta H_{\rm Ni} = 26.3$ mT for Ni and $\Delta H_{\rm Ni/Pt} = 27.3$ mT for Ni/Pt, respectively. The additional contribution to α is attributed to the spin-current emission induced by the spin pumping. We define the additional damping constant as $\Delta\alpha$ as $\Delta\alpha = \alpha_{F/\rm Pt} - \alpha_F$, where $\alpha_{F/\rm Pt}$ and α_F are the damping constant for the F/Pt (F = Ni, Fe, and $Ni_{81}Fe_{19}$) bilayer films and the F films, respectively. Using $\Delta\alpha$, $g_r^{\uparrow\downarrow}$ is expressed by

$$g_r^{\uparrow\downarrow} = \frac{4\pi M_s d_{\rm FM}}{\gamma\hbar}\Delta\alpha, \qquad (3)$$

Journal of Physics: Conference Series **266** (2011) 012115 doi:10.1088/1742-6596/266/1/012115

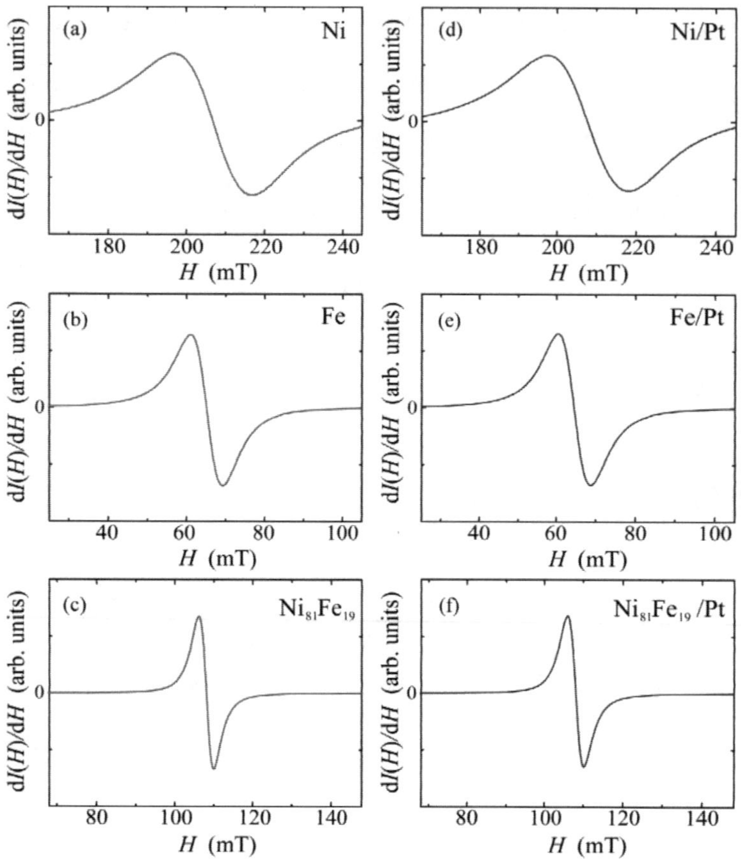

Figure 2. (a), (b), and (c) The FMR spectra d$I(H)$/dH for the Ni, Fe, and $Ni_{81}Fe_{19}$ single layer films under the 100 mW microwave excitation. (d), (e), and (f) The FMR spectra d$I(H)$/dH for the F/Pt (F = Ni, Fe, and $Ni_{81}Fe_{19}$) bilayer films under the 100 mW microwave excitation.

where d_{FM} is the thickness of the ferromagnetic layer [13]. Here, in the F/Pt (F = Ni, Fe, and $Ni_{81}Fe_{19}$) bilayer films, the Pt layer acts as a spin sink, since the thickness of the Pt layer is comparable to the spin diffusion length λ of Pt $\lambda = 7$ nm [13, 16].

In table 1, we calculate the values of $g_r^{\uparrow\downarrow}$ for the F/Pt (F = Ni, Fe, and $Ni_{81}Fe_{19}$) bilayer films obtained from figure 2 using equation (3). We also show the values of γ, $4\pi M_s$, and $\Delta\alpha$ for these films. Table 1 indicates that the spin-mixing conductances at the F/Pt (F = Ni, Fe, and $Ni_{81}Fe_{19}$) interface have the same order of magnitude, showing that spin injection efficiency in the spin pumping is almost identical in these films.

4. Conclusion

The spin-mixing conductances were quantified in F/Pt (F = Ni, Fe, and $Ni_{81}Fe_{19}$) bilayer films. From the FMR spectra using the model of the spin pumping, we found the spin-mixing conductances in F/Pt (F = Ni, Fe, and $Ni_{81}Fe_{19}$) bilayer films have the same order of magnitude. This shows that spin injection efficiency in the spin pumping is almost identical in these films.

Table 1. The values of γ, $4\pi M_s$, $\Delta\alpha$, and $g_r^{\uparrow\downarrow}$ for the *F*/Pt (*F* = Ni, Fe, and $Ni_{81}Fe_{19}$) bilayer films. The values of $g_r^{\uparrow\downarrow}$ are calculated from figure 2 using equation (3). The spin-mixing conductances at the *F*/Pt (*F* = Ni, Fe, and $Ni_{81}Fe_{19}$) interface have the same order of magnitude.

	γ ($T^{-1}s^{-1}$)	$4\pi M_s$ (T)	$\Delta\alpha$	$g_r^{\uparrow\downarrow}$ (m^{-2})
Ni	1.85×10^{11}	0.218	3×10^{-3}	3×10^{18}
Fe	1.84×10^{11}	1.60	3×10^{-4}	2×10^{18}
$Ni_{81}Fe_{19}$	1.85×10^{11}	0.787	8×10^{-4}	3×10^{18}

Acknowledgements
This work was supported by a Grant-in-Aid for Scientific Research in Priority Area "Creation and control of spin current" (19048028) from MEXT, Japan, a Grant-in-Aid for Scientific Research (A) from MEXT, Japan, a Grant for Industrial Technology Research from NEDO, Japan, and Fundamental Research Grant from TRF, Japan.

References

[1] Zutic I, Fabian J and Sarma S D 2004 *Rev. Mod. Phys.* **76** 323
[2] Maekawa S 2006 *Concepts in Spin Electronics* (Oxford Univ. Press, Oxford)
[3] Uchida K, Takahashi S, Harii K, Ieda J, Koshibae W, Ando K, Maekawa S and Saitoh E 2008 *Nature* **455** 778-81
[4] Ando K, Takahashi S, Harii K, Sasage K, Ieda J, Maekawa S and Saitoh E 2008 *Phys. Rev. Lett.* **101** 036601
[5] Kajiwara Y, Harii K, Takahashi S, Ohe J, Uchida K, Mizuguchi M, Umezawa H, Kawai H, Ando K, Takanashi K, Maekawa S and Saitoh E 2010 *Nature* **464** 262-66
[6] Saitoh E, Ueda M and Miyajima H 2006 *Appl. Phys. Lett.* **88** 182509
[7] Costache M V, Sladkov M, Watts S M, van der Wal C H and van Wees B J 2006 *Phys. Rev. Lett.* **97** 216603
[8] Ando K, Yoshino T and Saitoh E 2009 *Appl. Phys. Lett.* **94** 152509
[9] Ando K, Ieda J, Sasage K, Takahashi S, Maekawa S and Saitoh E 2009 *Appl. Phys. Lett.* **94** 262505
[10] Yoshino T, Kajiwara Y, Ando K, Nakayama H, Ota T, Uchida K and Saitoh E 2010 *J. Phys.: Conf. Ser.* **200** 062038
[11] Nakayama H, Ando K, Harii K, Kajiwara Y, Yoshino T, Uchida K, Ota T and Saitoh E 2010 *J. Phys.: Conf. Ser.* **200** 062014
[12] Nakayama H, Ando K, Harii K, Kajiwara Y, Yoshino T and Saitoh E 2010 *J. Magn. Soc. Jpn.* **34** 337-41
[13] Tserkovnyak Y, Brataas A, Bauer G E W and Halperin B I 2005 *Rev. Mod. Phys.* **77** 1375
[14] Inoue H Y, Harii K, Ando K, Sasage K and Saitoh E 2007 *J. Appl. Phys.* **102** 083915
[15] Ando K, Kajiwara Y, Takahashi S, Maekawa S, Takemoto K, Takatsu M and Saitoh E 2008 *Phys. Rev.* B **78** 014413
[16] Bass J and Pratt Jr W P 2007 *J. Phys.: Condens. Matter* **19** 183201

Journal of Physics: Conference Series **266** (2011) 012116 doi:10.1088/1742-6596/266/1/012116

Magnetic Properties of Fe(001) Thin Films on GaAs(001) Deposited by RF Magnetron Sputtering

Hirokazu Ikeya[1], Yutaka Takahashi[1,4], Nobuyuki Inaba[1], Fumiyoshi Kirino[2], Mitsuru Ohtake[3] and Masaaki Futamoto[3]

[1] Department of Electrical Engineering, Graduate School of Science and Engineering, Yamagata University, Yonezawa-shi, Yamagata, 992-8510, Japan

[2] Graduate School of Fine Arts, Tokyo National University of Fine Arts and Music, Tokyo, 110-8714, Japan

[3] Faculty of Science and Engineering, Chuo University, Tokyo, 112-8551, Japan

E-mail: takahasy@yz.yamagata-u.ac.jp

Abstract. Fe thin films, down to 6 nm thick, were prepared on GaAs(001) substrates by RF magnetron sputtering. The x-ray diffraction (XRD) analyses show that the epitaxial thin films of Fe(001) were grown with cube-on-cube orientation on GaAs(001). Magnetic properties were investigated by vibrating sample magnetometry (VSM) and ferromagnetic resonance (FMR) spectroscopy. The magnetization curves obtained by applying in-plane magnetic fields indicate that easy (hard) direction is along [100] ([110]) and the saturation magnetization is close to the bulk values. The in-plane magnetic anisotropy measured by FMR shows four-fold symmetry, as expected for bcc Fe. We did not observe the in-plane uniaxial magnetic anisotropy reported on the MBE-grown Fe films on GaAs substrates.

1. Introduction

The magnetic and transport properties of magnetic metal/semiconductor structures are of scientific and technological importance in view of developing magnetic/semiconductor hybrid (spintronic) devices. Fe thin films on GaAs substrates are intensively studied as a model structure since they are well-known materials in magnetic/semiconductor engineering with small lattice mismatch ($a_{GaAs} \approx 2 \times a_{Fe}$). Most of the Fe/GaAs structures have been deposited by molecular beam expitaxy [1, 2, 3, 4], and the epitaxial growth of Fe on GaAs has been reported. Another approach, i.e., sputtering is also important, as it is conventionally used for wide variety of magnetic film deposition. However, we can find a small number of previous works on the crystallographic [5, 6] and the magnetic [7] properties of Fe/GaAs structures fabricated by sputtering. In this report we present our study on the magnetic properties of Fe(001) thin films deposited on GaAs(001) substrates by RF magnetron sputtering. The crystallographic quality of Fe films was analyzed by XRD and the magnetic properties were studied by VSM and FMR.

2. Growth and structure

[4] To whom any correspondence should be addressed.

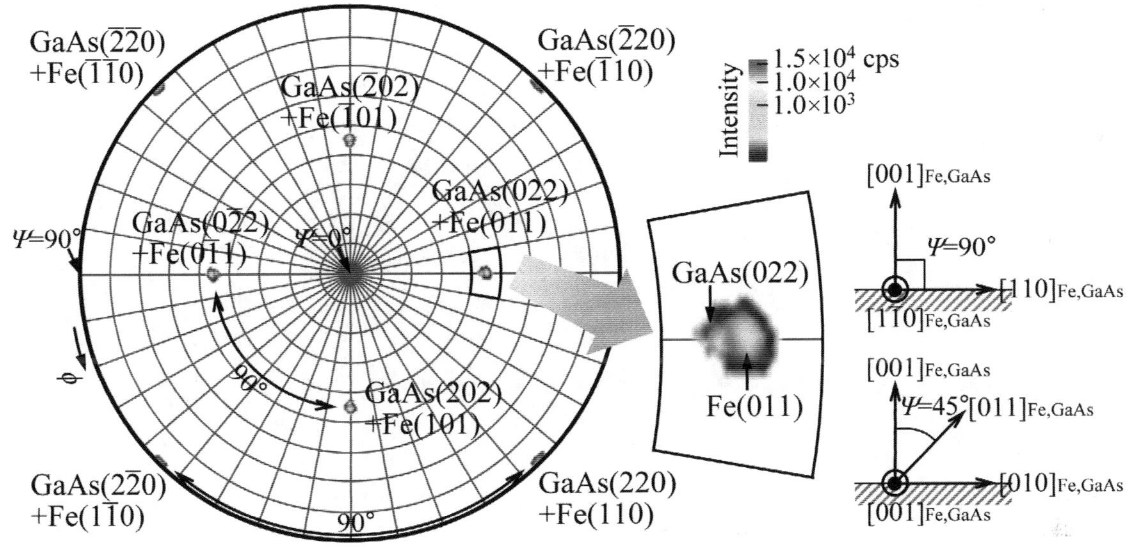

Figure 1. X-ray pole figure profile with a diffraction angle fixed at $2\theta = 44.4°$, corresponding to Fe{110}$_{bcc}$ and GaAs{220} poles of an Fe film grown on a GaAs(100) substrate. The inset shows the enlarged profile of Fe(011) and GaAs(022). The back-ground smaller than 20 cps was subtracted from the profile.

The Fe films of thickness 6 - 100 nm were prepared on epi-ready n-GaAs(001) wafers (Wafer Technology) at room temperature (RT) using a 3N Fe target by RF magnetron sputtering. The base pressure of the growth chamber was 0.8 - 1.0 x 10^{-4} Pa, and the Ar gas pressure during the deposition was kept at 0.8 Pa. The GaAs substrates were heated to 550 °C to remove a surface oxide layer prior to the deposition. The growth rate of Fe was 0.03 nm/s. The thicker film (100 nm) was used for XRD analyses in order to obtain the data with sufficient signal intensities. For magnetic properties by VSM and FMR, we show the data of thinner films (6, 12, 25 nm) below as we are interested in the magnetic properties of thin films.

The crystal structure of a 100-nm thick Fe film on GaAs(001) was analyzed by XRD using Cu $K\alpha$ lines. First the crystallographic quality and orientation of the Fe film were investigated by pole figure measurements. In Fig. 1, the pole figure profile with a diffraction angle fixed at $2\theta = 44.4°$ is plotted. Four-fold diffraction spots are observed at polar angle $\psi = 45°$ which correspond to the diffractions from GaAs{220} and Fe{110}$_{bcc}$. These two spots are very close because of the small difference in lattice constants (a_{GaAs} = 0.56530 nm and 2 x a_{Fe} = 0.57328 nm). The enlarged profile in the inset shows that the spot from Fe(011) is observed, separated from the strong spot of GaAs(022). No circular streaks were observed due to misaligned Fe{110} plane. This result indicates that the Fe film is epitaxially grown on GaAs substrate with the crystallographic orientation relationship with respect to a GaAs(001) substrate being Fe(001)[100]$_{bcc}$//GaAs(001)[100].

The lattice spacing of the Fe(001) film on the GaAs(001) substrate was investigated by out-of-plane θ-2θ scan and in-plane $\theta\chi$-$2\theta\chi$ scan. The out-of-plane diffraction spectrum is shown in Fig. 2 (a). (The diffraction was measured with off-set 0.4° to mitigate the intense signal from the GaAs substrate.) An Fe(002) peak is observed adjacent to a diffraction from GaAs(004). (Small peaks on the left shoulder are due to a Cu $K\beta$ line and a WL line from a contamination.) Other Fe lines, e.g., Fe(011) at 44.7°, were not detected. After peak separation the lattice spacing in the direction normal to the film surface was estimated to be 0.2920 nm, expanded by 1.8% relative to the bulk lattice spacing. The in-plane diffraction spectra with the scattering vector parallel to a GaAs[400] axis and a GaAs[220] axis are plotted in Figs. 2 (b) and (c), respectively. The diffractions from Fe(200) and

Journal of Physics: Conference Series **266** (2011) 012116 doi:10.1088/1742-6596/266/1/012116

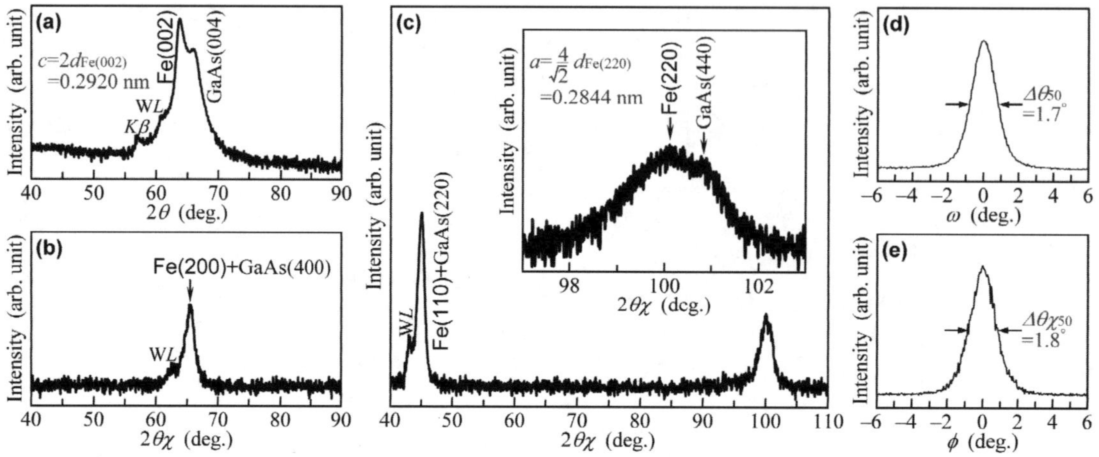

Figure 2. (a) Out-of-plane and (b), (c) in-plane XRD spectra of an Fe film grown on a GaAs(001) substrate. The in-plane XRD measurements were performed by making the scattering vector parallel to (b) the GaAs[400] or (c) the GaAs[220] direction. Rocking curves observed for (d) the $Fe(002)_{bcc}$ out-of-plane and (e) the $Fe(220)_{bcc}$ in-plane XRD peaks. The intensities in (a)–(c) are shown in a logarithmic scale and in (d) and (e) in a linear scale.

Fe(110) are not distinctly separated from GaAs peaks; it is likely that Fe(200) overlaps with GaAs(400) in Fig. 2 (b), and Fe(110) overlaps with GaAs(220) in Fig. 2 (c). However, as shown in the inset of Fig. 2 (c), the diffraction from Fe(220) is separated from GaAs(440). The in-plane latticespacing was estimated to be 0.2844 nm by peak separation. In the film plane the Fe thin film is compressed by 0.8% compared with the bulk lattice constant of Fe, but it is larger than the half of GaAs lattice constant, 0.28265 nm.

Out-of-plane and in-plane dispersions of crystalline orientations were investigated by measuring rocking curves of the Fe(002) out-of-plane XRD peak (Fig. 2 (d)), and the Fe(220) in-plane XRD peak (Fig. 2 (e)). The FWHM line width $\Delta\theta_{50}$ was 1.7° for Fe(002) plane, and $\Delta\theta\chi_{50}$ = 1.8° for Fe(220) plane. The observed width is larger than those reported previously for dc-sputtered Fe films on GaAs [5] or ion-beam sputtered films [6].

3. Magnetic properties

The magnetization curves of Fe thin films with various thicknesses were investigated by VSM at RT. The representative data with the in-plane field at $[100]_{Fe}$ and $[110]_{Fe}$ for a 25 nm film are shown in Fig. 3. The saturation magnetization (M_s) of 1.7 x 10^3 emu/cm^3, and the shape of loops with a $[100]_{Fe}$-direction ($[110]_{Fe}$-direction) being an easy direction (a hard direction), are identical to the well-known data for a bulk Fe single-crystal specimen [8]. For the thinner films, 6 and 12 nm in thickness, and for the thicker film, 100 nm in thickness as used for XRD analyses, the identical shape of loops was observed.

There are many studies on the magnetic anisotropy of thin films of Fe grown on GaAs(001) substrates by MBE. They reported the in-plane uniaxial magnetic anisotropy (UMA); a $[110]_{Fe}$-direction was an easy direction and $[1\text{-}10]_{Fe}$ was a hard direction [9, 10]. Its origin is still controversial. UMA decreases with film thickness, and it is likely that the anisotropy originates from the interfacial effect between Fe and GaAs. FMR spectroscopy is one of the important methods to study magnetic anisotropy. However, care should be taken to interpret the data. In order to obtain the data that could be unambiguously analyzed, the external magnetic field should be so high that sample magnetization was saturated (parallel to the external field) and a sample was of single domain [11]. We investigated the in-plane magnetic anisotropy of Fe thin films using a conventional FMR spectrometer at a microwave frequency 35 GHz (Q-band) and at RT. The external field was in the plane of films and its

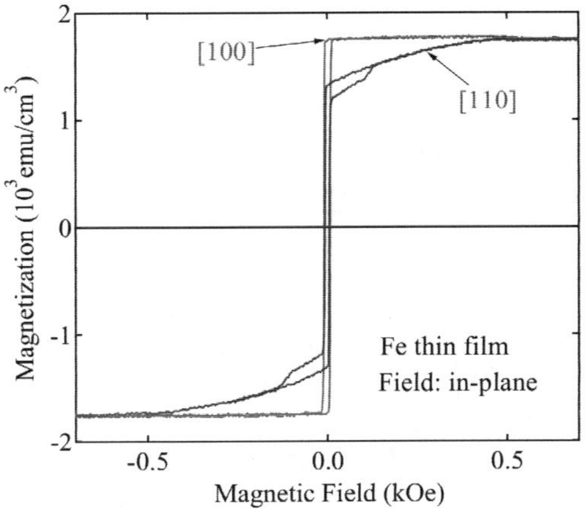

Figure 3. Magnetization loops of a 25 nm Fe film for $[100]_{Fe}$ and $[110]_{Fe}$ directions.

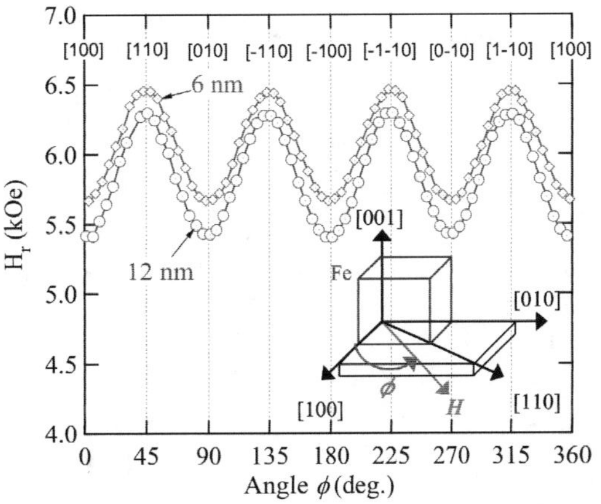

Figure 4. In-plane angle dependence of FMR resonance fields of Fe films with thickness 6 nm and 12 nm.

direction was varied from 0° ($[100]_{Fe}$) to 360°. (See the inset of Fig. 4) The in-plane angle dependence of resonance fields for 6-nm and 12-nm Fe films are plotted in Fig. 4. In our measurements the resonance field (H_r) was large enough to saturate the in-plane magnetization of samples.

In the MBE grown Fe thin films UMA becomes conspicuous in thinner samples, e.g., in an 8 nm Fe film in [10]. In the present case, from the angular dependence of H_r in Fig. 4, the four-fold in-plane anisotropy of films is clearly shown, as expected for cubic lattice of Fe. The in-plane four-fold anisotropy constant K_1 was estimated from H_r in FMR and M_s observed in VSM measurements [11, 12]. We obtained K_1 = 3.9 x 10^5 erg/cm^3 for the 12-nm Fe film. We did not observe uniaxial component even in the 6-nm film. This observation was also confirmed in our VSM measurements. The magnetization curve with the external field in $[110]_{Fe}$-direction is identical to that in $[1\text{-}10]_{Fe}$ (not shown). The different kinetics of crystal growth or the surface preparation of GaAs substrates in MBE and RF magnetron sputtering may result in the different in-plane magnetic anisotropy. Considering that the UMA is of interfacial origin [13], it is likely that the uniaxial component could be seen in sputter-deposited films if they were much thinner.

4. Conclusion

In conclusion, we have grown epitaxial Fe(001) thin films on GaAs(001) substrates by RF magnetron sputtering. The crystal structure was confirmed by XRD measurements, including pole figures, out-of-plane θ-2θ scan and in-plane $\theta\chi$-2$\theta\chi$ scan, and rocking curves. The magnetization curves measured by VSM show features common to bulk Fe single crystal. The in-plane magnetic anisotropy was investigated by Q-band FMR spectroscopy. The angle dependence of the resonance fields shows four-fold in-plane magnetic anisotropy. The uniaxial component observed in MBE-grown Fe thin films was not detected in our sputter-deposited samples.

Acknowledgements
This work was supported in part by Grant-in-Aid for Scientific Research (Grant Nos. 20560315 and 20560611) from the Japan Society for the Promotion of Science.

References
[1] Prinz G A 1994 Magnetic Metal Films on Semiconductor Substrates *Ultrathin Magnetic Structures II,* ed Heinrich B and Bland J A C (Berlin: Springer) p. 1
[2] Filipe A, Schuhl A, and Galtier P 1997 *Appl. Phys. Lett.* **70** 129
[3] Xu Y B, Kernohan E T M, Freeland D J, Ercole A, Tselepi M and Bland J A C 1998 *Phys. Rev.*

B **58** 890
[4] Schönherr H P, Nötzel R, Ma W and Ploog K H 2001 *J. of Appl. Phys.* **89** 169
[5] Bernstein S D, Wong T Y and Tustison R W 1992 *J. Appl. Phys.* **72** 4358
[6] Monteverde F, Michel A, Eymery J -P and Guérin Ph 2004 *J. of Crystal Growth* **267** 231
[7] Wada E, Itoh M and Taniyama T 2008 *J. Appl. Phys.* **103** 07A702
[8] Honda K and Kaya S 1926 *Sci. Repts. Tohoku Imp. Univ.* **15** 721, reproduced in Bozorth R M 1951 *Ferromagnetism* (D. Van Nostrand: New York)
[9] Krebs J J, Jonker B T and Prinz G A 1987 *J. of Appl. Phys.* **61** 2596
[10] Lu J, Meng H J, Deng JJ, Xu P F , Chen L, Zhao J H and Jia QJ 2009 *J. of Appl. Phys.* **106** 8757
[11] Kobayashi K, Inaba N, Fujita N, Sudo Y, Tanaka T, Ohtake M, Futamoto M and Kirino F 2009 *IEEE Trans. Magn.* **45** 2541
[12] Vonsovskii S V 1966 *Ferromagnetic Resonance* (Pergamon: Oxford)
[13] Tournerie N, Schieffer P, Lépine B, Lallaizon C, Turban P and Jézéquel G 2008 *Phys. Rev.* B **78** 134401

Magnetization reversal of nanodots with different magnetic anisotropy and magnetostatic energy

Y P Ivanov[1,2,3], K V Nefedev[1], A I Iljin[1], E V Pustovalov[1] and L A Chebotkevich[1,2]
[1] Institute of Physics and Informational Technologies, Far Eastern National University, Vladivostok, Sukhanova street 8, 690950, Russia
[2] Institute of Automatic and Control Processes, Far Eastern Branch of Russian Academy of Science, Vladivostok, Radio street 5, 690041, Russia

E-mail: yivanov@phys.dvgu.ru

Abstract. The arrays of circle and square nanodots were prepared by Ga^{+} ion-beam etching of thing cobalt films with induced uniaxial magnetic anisotropy. The size of researched nanodots was d=600, 450 and 250 nm that equal or less then radius of ferromagnetic correlation for continuous Co film. The distances l=2d and 3d between center of nanodots in the arrays were taken. The interpretation of experimental magnetic force microscopy data was done by means of micromagnetic simulation and MFM-experiment simulation. Each nanodot of arrays with d=600 nm has multidomain structure at demagnetisation state for l=3d and magnetic vortex for l=2d. For arrays with size of element d<600 nm and l=2d and 3d at same condition occurred only vortex state.

1. Introduction

Magnetic nanoarchitectures today are very perspective object of research from application point of view. It can be used at the development of new devices for information storage [1], sensors and other elements on the base of magnetic logic [2]. For implementation of these appliances it is necessary the well understanding of the reversal magnetization processes, the origins of specific magnetic state and the influence of magnetostatic interaction between elements in the nanosystem. Nanomagnets in arrays with sizes from some microns to hundreds nanometers can have deferent magnetic states such as multidomain, "S" and "C" states, states with nonuniform magnetization such as vortex and others [3]. The magnetic state of single nanodot in array can depend from external magnetic field, magnetostatic interaction between neighbors, geometry shape of the array and it's elements, from different types of magnetic anisotropy, properties magnetic material, etc. Cowburn et. al. [4] observed the transition from single domain to vortex state of nanodots in array of the permalloy nanodots at the variation of diameter to thickness ratio of nanodisk. The change of elliptic shape to disk in permalloy can attended by transition from single domain state to two-domain structure and vortex state at variation of eccentricity value [5]. The magnetic structure Landau-Lifshits with four 90^0 and one 180^0 boundaries was observed in thin film rectangular permalloy elements with size of some tens micron. The magnetocrystalline anisotropy [6] and induced magnetic anisotropy [7] play important role in the reversal magnetization process of nanomagnets additionally to shape anisotropy. Jubert and Allenspach [8] with help of theoretical analysis and numerical simulation found that these two type of

[3] E-mail: yivanov@phys.dvgu.ru

anisotropy in the absence of shape anisotropy and in dependence of the ratio of diameter to thickness can move line of transition between single domain state and vortex.

In this work the processes of magnetization reversal and formation of magnetic structure for the arrays of circle and square Co nanodots with different period between there and sizes of elements and with induced uniaxial magnetic anisotropy were investigated.

2. Experimental technique

The Co films with thickness 10 nm were deposited in ultrahigh vacuum P = 10^{-10} Torr on naturally oxidised monocrystals (100) Si from diffusion cells at room temperature. From above Co film was covered with 3 nm layer of Cu for the prevention of oxidation. Then the arrays of square and circle nanodots were prepared from Co films by focused beam of Ga^{+}. The estimated deep of etching was 50 nm. The raster images of circle and square holes onto scanning field was used by way of mask during etching. Ion current of gun at the etching varied from 2 pA to 20 pA, and diameter of ion beam was varied from 10 to 17 nm in depending of formed nanodot size. Minimal size of element on images was interval from 10 to 30 nm, that was more than size of ion beam. The space around arrays was "cleaned" off magnetic material for the influence exclusion of the entire Co-film on the experimental measurement. The ion current of cleaning beam was equal 500 pA and 1 nA. In this case was raster image with rectangle in centre which one closed nanodots array from ion beam. For controlling of formation process was used scanning electron microscope with increment from 7000 to 25000.

The structure of films and arrays was investigated by means of reflection high energy electron diffraction (RHEED), transmission electron microscopy (TEM), scanning electron microscopy (SEM) and atomic-force microscopy (AFM).

The magnetic hysteresis loops of nanodots array were measured by means of longitudinal magnetooptic Kerr effect (NanoMOKE-2). The method of magneto-force microscopy (MFM) was used to obtain the images of magnetic structure and hysteresis loops of individual nanodots. The micromagnetic simulation of magnetic moment distribution in nanodots was made by means of program package The Object Oriented MicroMagnetic Framework (OOMMF) [9], which one uses Landau-Lifshits-Hilbert formalism. The correctness of micromagnetic simulation was collaborated by simulation of MFM-experiment in frame of model magneto-hard tip by means of author program package.

3. Result and discussion

An RHEED patterns showed that Co films was polycrystalline. The mean size of grain is 6 nm from TEM measurements. The loops of magnetic hysteresis showed presence of pronounced uniaxial induced magnetic anisotropy, which one was formed by inclined incidence of molecular beam at deposition. The coercive force of Co films is H_c=18 Oe, the anisotropy field is H_c=40 Oe (K_u= $H_a M_s/2 \approx 3*10^4$ erg/cm^3, where M_s=1400 G – saturation magnetization, which one was measured by SQUID magnetometer at the precision 50 G).

In nanocrystalline ferromagnetic films the effect of magnetization vector oscillation appears on the score of the random orientation of local crystallographic anisotropy in grains and magnetostatic interaction. This one leads to appearance of thin magnetic structure (magnetization ripple). For such films it is possible to enter the ferromagnetic correlation radius R_f [10] which one can be determined from experiment as a period of short-wave magnetization ripple λ_{RS} on images of thin magnetic structure. The short-wave component of magnetization ripple in researched films was measured from MFM images of a domain structure $R_f=\lambda_{RS}\approx 600$ nm. Therefore the circle and square nanodots Co arrays with d=600; 450 and 250 nm (i.e. $d \leq R_f$) were prepaid from polycrystalline films with pronounced uniaxial in-plane magnetic anisotropy. 10x10 nanodots were in each array. Magnetic properties of one nanomagnet in array have strong dependence from period l (the distance between the centers of nanodots). As it was showed in [7] at $l=3d$ nanodisks practically have not interaction between each other, and the magnetostatic (dipole-dipole) interaction became essential at $l=2d$. In researched array period was $l=3d$ and $2d$.

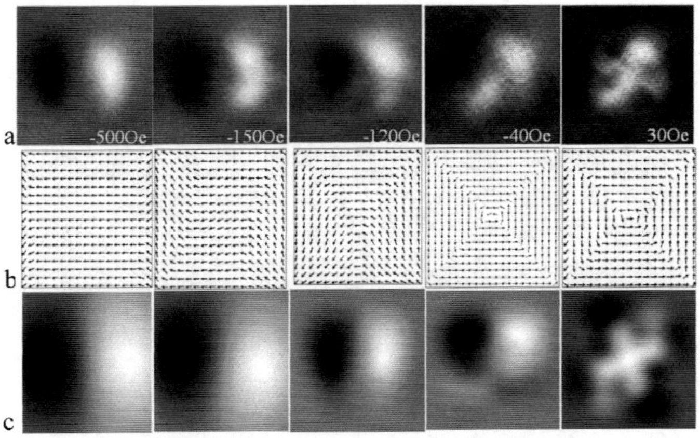

Figure 1. MFM-images of domain structure dynamics in the external magnetic field for square nanodots $d=R_f$ and $l=3d$ (a), results of micromagnetic simulation (b) and results of MFM-experiment simulation.

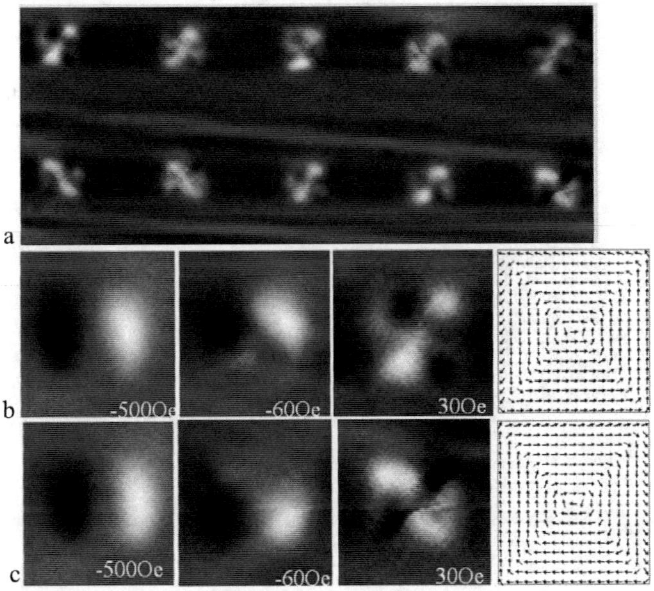

Figure 2. MFM-images of domain structure of square nanodots array part in field H=30 Oe (a), and initiation of transformation into 'C'-state of 4-domain state with contrary chirality of magnetization 2b and 2 c.

At 'figure 1' (a, b and c) the MFM-images, configuration of magnetization from micromagnetic simulation and results of MFM experiment simulation for square nanodots of array with $d=R_f$ and period $3d$ in dependence from external magnetic field are shown. The external magnetic field was oriented along induced easy magnetization axis. In the micromagnetic simulation following parameter were used: a saturation magnetization M_s=1400 G, a constant of exchange interaction $A=2*10^{-6}$ erg/cm, a constant of induced uniaxial magnetic anisotropy $3*10^{-4}$ erg/cm^3, and the size of unit cell 6 nm, which one corresponds to the average film grain size. In a simulation of MFM experiment the ratio between nanodot and tip sizes was the same as in physical experiment.

From 'figure 1' it is visible that the decreasing of external magnetic field from the negative saturation value to the positive, the square nanodot has sequential transitions from single domain state (H=-500 Oe) to 'S'-state (H=-150 Oe), 'C'-type state (H=-120 Oe), which one transforms to four domain configuration (H=30 Oe). The simulation results obtained that sizes of domains unequal. The size is more for domains where vector of magnetization is collinear to the direction of induced easy magnetization axis, than for domains where magnetization is orthogonal to easy axis ('figure 1b, H=30 Oe). There is light area with 'cross' type in the centre of square nanodot. Such distribution of the MFM-contrast can take place if there is small component of magnetization vector ($M_z/M_s \approx 0.01$), which one has orthogonal direction to sample plane. We supposed that presence of perpendicular component connect with cone like shape of nanodot, and it was confirmed by SEM and AFM.

On 'figure 2a' is MFM-image of 4-domain structure of array part in field H=30 Oe. The magnetization of nanodots has different 'chirality' 'figure 2b' and 'figure 2c' (for example, the highest and the bottom-most MFM-images of array) on dependence from magnetization direction in C-state.

The magnetic loop square nanodots array, which one was obtained by micromagnetic simulation 'figure 3b', has qualitatively and quantitatively corresponds to hysteresis loop of longitudinal magnetooptical Kerr effect 'figure 3a'.The jumps of magnetization on hysteresis curves correspond to the transitions 'S'→'C'→4-domain state and backwards. But the magnetic fields, in which ones the transition from one magnetic state to another takes place for single nanodot and for total array can be different. This one can be explained by asynchronous processes of array nanodots magnetic reversal 'figure 2a' and by presence of magnetic states with opposite chirality of magnetization.

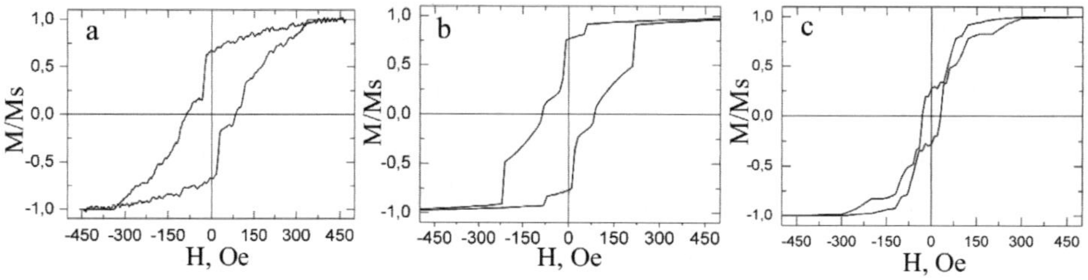

Figure 3. The magnetic hysteresis loops for array of square nanodots $d=R_f$ and with period $l=3d$ which ones was obtained: (a) –MOKE, (b) – micromagnetic simulation, (c) – from MFM-images for single nanodot.

The recent research [7] of circle nanodots arrays with $d=R_f$ and period $l=3d$ showed that two-domain configuration generate at demagnetization state.

The MFM experimental data, micromagnetic simulation and MFM experiment simulation for circle and square nanodot arrays with $d=R_f$, $l=2d$ and $d<R_f$ showed that magnetization reversal carry out as with the formation of the vortex states with the vector of magnetization which one is orthogonal to the nanodot plane in core 'figure 4.'

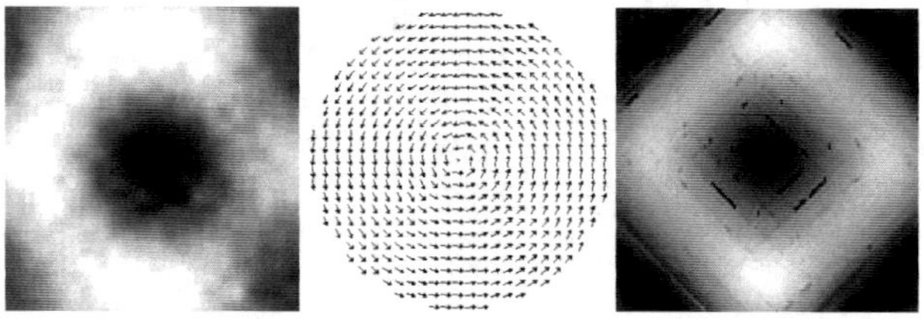

a b c

Figure 4. MFM image of vortex state of circle nanodot from array with $d=450$ nm at demagnetization state (a), micromagnetic simulation (b), MFM experiment simulation (c).

As shown the MOKE measurements the magnetic loops had the hysteresis. The hysteresis in such arrays was caused by existence of induced magnetic anisotropy and probably by the presence of uniform rotation of magnetization vector in nanodots at changing of external magnetic field. For arrays with $l=2d$ critical field of reversal magnetization and saturation field more than for arrays with $l=3d$, that explained by gaining of magnetostatic interaction at the decreasing of l. The pronounced core of vortex on MFM-images of circle and square elements of arrays with $d<R_f$, probably, caused by with cone-like shape of nanodot, which one depend on the fabrication method. SEM and AFM images testify to it: the shape of nanodots is truncated cone-like. For small size of nanodot this geometry stands more pronounced that one evidence by research of MFM-images, where the vortex core was more stable and it can be observed in more wide range of external magnetic field. By the indirect evidence of this can be the fact that decreasing of nanodots sizes leads to increasing of switching magnetic field value. It need to say that the analysis of MFM images let us to find only two from four possible types of vortex.

The Hamiltonian for polycrystalline nanodots is $E= E_{ex}+E_{an}+E_{mc}+E_z$, where E_{ex} – the exchange energy, E_{an} – energy of uniaxial induced magnetic anisotropy, E_{mc} – magnetostatic energy, E_z – energy of nanodots in the external magnetic field. The calculation of magnetostatic energy is very complicated task, but for case of uniform magnetization of disk

$$E_{mc} = \frac{1}{2} N M_S^2, \quad (1)$$

where N – demagnetization factor in-plane of disk. If ratio between thickness and diameter of disk more less then unity

$$N = \frac{t}{\pi d}\left[\ln(4\frac{d}{t}) - 0.5\right], \quad (2)$$

where t – thickness of disk [9].

On the 'figure 5' are dependence of saturation field H_s (the field of uniform magnetization of nanodots) and demagnetization field $H_N = NM_s$ in dependence from their sizes. It is visible that the behavior of both H_s and H_N is identical and independent from nanodots shape. The deviation of behavior of experimental curve from theoretical, especially for square shape nanodots, is caused rather by real shape of nanodots, which one of course is differ from ideal disk. Under the ideal nanodisk we understend the disk for which the ratio of a thickness to diameter is much less than unit and defects of form are not considered.

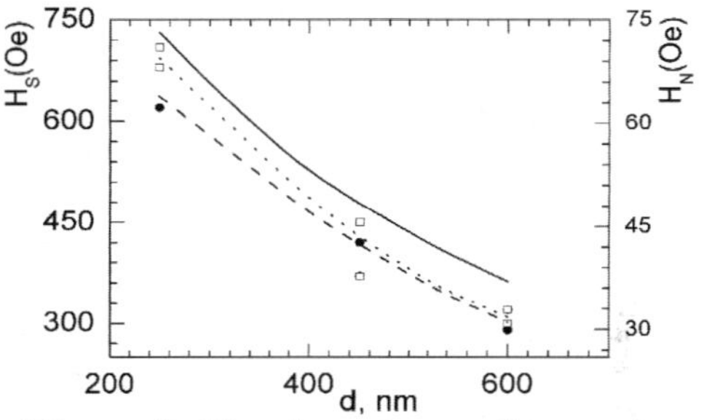

Figure 5. The dependence of saturation field H_s of square (empty squares and dotted curve) and circle (black circle and dashed straight line) nanodots and demagnetization field H_N of ideal nanodisk (solid curve) from size of nanodots.

The saturation field (the external magnetic field in which the sample is magnetized completely) develops of an internal field direct influencing orientation of the magnetic moments and demagnetization field. On 'figure 5' the dependence of a saturation field (it is received experimentally) for circle and square nanodot arrays and a demagnetization field for an ideal nanodisk (is received from the analytical formula) is resulted.

At the $d<R_f$ demagnetization energy of nanodot in array increases and stands more than energy of induced magnetic anisotropy, therefore at zero external magnetic field in given nanodots magnetic vortex was generated. The similar situation is in case of nanodots with $d=R_f$ at decreasing of array period from $l=3d$ to $l=2d$, as far as energy of dipole-dipole interaction between nanodots in array increases.

4. Conclusions

The arrays of circle and square ferromagnetic ultrathin nanodots were investigated. Source Co films have pronounced magnetic induced anisotropy to which one added the shape anisotropy (truncated cone) at ion beam etching. Vortex state of all nanodots in array is possible at $E_z=0$, if energy of induced iniaxial magnetic anisotropy less than energy of dipole-dipole interaction and magnetostatic energy single nanodot in arrays. The competition between energy of induced magnetic anisotropy and energy of shape anisotropy at decreasing of nanodot size can be used as control factor for generation definite magnetic state.

It is known from the literature, that in circle and square nanodots with diameter of less micron are observed either a vortex or single-domain state, and there is no accurate criterion of transition from these states in the multidomain state. Also there is a number of theoretical works where influence of magnetic anisotropy on transition from one magnetic state in another is calculated at change of geometrical parameters of nanodot. Our paper shows that even in the presence of the induced magnetic anisotropy the transition from a multidomain state in the vortex occurs at diameter of a nanodot (and round and square) equal to radius of ferromagnetic correlation of a continuous Co film, that is caused by increase magnetostatic energy (the reduction of the nanodot size or distances between them in the array). The nanodots which size is equal to radius ferromagnetic correlation reverse magnetization with formation of a multidomain state (for circle nanodot - a two-domain state, for square - four-domain). Both circle, and square nanodots which size is less than radius of ferromagnetic correlation, reverse magnetization through formation, displacement and annihilation of a magnetic vortex.

Also we have found that the used by us conditions of a preparation (FIB) of the truncated form of nanodot are causes the domination only two types of a magnetic vortex with different chirality of magnetization.

References

[1] Parkin S S P, Hayashi M, Thomas L 2008 *Science* **320** 190
[2] Cowburn R P 2006 *Science* **311** 183
[3] Rahm M, Schneider M, Biberger J, Pulwey R, Zweck J, Weiss D 2003 *Appl. Phys. Lett.* **82** 4110
[4] Cowburn R P, Koltsov D K, Adeyeye A O, Wellend M E, Tricker D M 1999 *Phys. Rev. Lett.* 83 1042
[5] Imre A, Csaba G, Ji L, Orlov A, Bernstein G H, Porod W 2006 *Science* **311** 205
[6] Hanson M, Kazakova O, Blomqvist P, Wappling R, Nilsson B 2002 *Phys. Rev.* B **66** 144419
[7] Ivanov Yu P, Iljin A I, Pustovalov E V, Chebotkevich L A , 2010 *Fizika Tverdogo Tela* **52** 1576
[8] Jubert P-O and Allenspach R *Phys. Rev.* B 2004 **70** 144402
[9] http://math.nist.gov/oommf/
[10] Loffler J F, Braun H-B, Wagner W *Phys. Rev. Lett.* 2000 **85** 1990

2nd Int. Symp. on Advanced Magnetic Materials and Applications (ISAMMA 2010) IOP Publishing
Journal of Physics: Conference Series **266** (2011) 012118 doi:10.1088/1742-6596/266/1/012118

Highly (100) oriented MgO growth on thin Mg layer in MTJ structure

K Jimbo[1] and S Nakagawa
[1] Department of Physical Electronics, Tokyo Institute of Technology, 2-12-1-S3-42 O-okayama, Meguro, Tokyo 152-8552, Japan

E-mail: jimbo.k@spin.pe.titech.ac.jp

Abstract. In order to apply Stress Assisted Magnetization Reversal (SAMR) method to perpendicular magnetoresistive random access memory (p-MRAM) with magnetic tunnel junction (MTJ) using MgO (001) oriented barrier layer, multilayer of Ta/ Terfenol-D/ Mg/ MgO and Ta/ Terfenol-D/ MgO were prepared. While the MgO layer, deposited directly on the Terfenol-D layer, did not show (100) orientatin, very thin metallic Mg layer, deposited prior to the MgO deposition, was effective to attain MgO (100) orientation. The crystalline orientation was very weak without Mg, however, the multilayer with Mg showed very strong MgO(100) peak and the MgO orientation was shifted depending on the Mg thickness.

1. Introduction

Perpendicular magnetoresistive random access memory (p-MRAM) which has various advantages including nonvolatility, large capacity and unlimited read/write endurance is a promising candidate for future memory devices. Especially, magnetic tunnel junction (MTJ) structure with MgO tunnelling barrier layer is used widely since the TMR ratio of Fe(001)/ MgO(001)/ Fe(001) structure is predicted to be more than 1,000% [1]. However, relatively high current density is required to induce magnetization reversal under the spin injection method in order to put MRAM into practical use. Previously, we have proposed a Stress Assisted Magnetization Reversal (SAMR) method in which a magnetization of free layer can be reversed less than its original coercivity under an application of in-plane stress thereby reducing power consumption of p-MRAM. We have demonstrated a validity of SAMR in spin valve structure with "Cu" spacer layer as shown in Figure 1. However, MTJ with MgO barrier layer described above will be widely installed in an MRAM cell structure. Thus, in this study, we show that a highly (100) orientation MgO thin film can be grown on Mg thin layer on amorphous rare earth - transition metal (RE-TM) alloy, Terfenol-D, layer. Terfenol-D is well known as an alloy of $Dy_{0.73}Tb_{0.27}Fe_2$ which has large magnetostriction constant in the order of 10^{-4}[3] and is suitable to attain stress induced magnetic anisotropy as well as perpendicular magnetic anisotropy[4].

2. Experimental

All films were grown on glass substrates by facing targets sputtering method. Plasma was induced by DC power supply under a current control mode. Mg and MgO were deposited in 0.13 Pa Ar gas, especially MgO was grown by reactive sputtering in 0.13 Pa Ar(98 %) and O_2(2 %), and the others were in 0.13 Pa of Kr gas. The crystal structures were analyzed with a X-Ray diffractometer(XRD).

Journal of Physics: Conference Series **266** (2011) 012118 doi:10.1088/1742-6596/266/1/012118

3. Result and discussion

Figure 1 shows anomalous Hall loops of Cu(15 nm)/ Terfenol-D(10 nm)/ Cu (3 nm)/ TbFeCo(30 nm)/ Cu(15 nm) multilayer observed under various tensile stress. The major magnetization loop under the tensile stress of 279 MPa indicates the disappearance of the steep transition of magnetization around 400 Oe, which was observed in the multilayer without tensile stress as shown in Fig.1(a). Figure 1 (b) shows minor loops on the multilayer in the range of ±2 kOe. Since the magnetization reversal of the Terfenol-D free layer occurs in this range, it can be infered that the easy axis of the magnetization of the Terfenol-D free layer is changed from perpendicular to in-plane with the application of the tensile stress due to the stress induced magnetic anisotropy. In order to apply the multilayer structure using RE-TM films with perpendicular magnetization to the MTJ structure, MgO tunnelling barrier layer should be sandwiched between the TbFeCo and Terfenol-D RE-TM layers.

Figure 2 shows XRD diagrams of Ta(5 nm)/ Terfenol-D(X nm)/ MgO(30 nm) films with various layer thickness of Terfenol-D film. The diffraction line of MgO(200) appears little, indicating that the MgO layer deposited directly on a Terfenol-D layer is not suitable to attain MTJ structure with (100) orientation. However, as shown in Figure 3 (a) and (b), strong MgO(200) peak can be observed by inserting a very thin metallic Mg thin layer between MgO and Terfenol-D layers. Mg(111) peak appears when Mg thickness is more than 1 nm(Figure 3(a)) or 4 nm(Figure 3(b)) and the intensity of

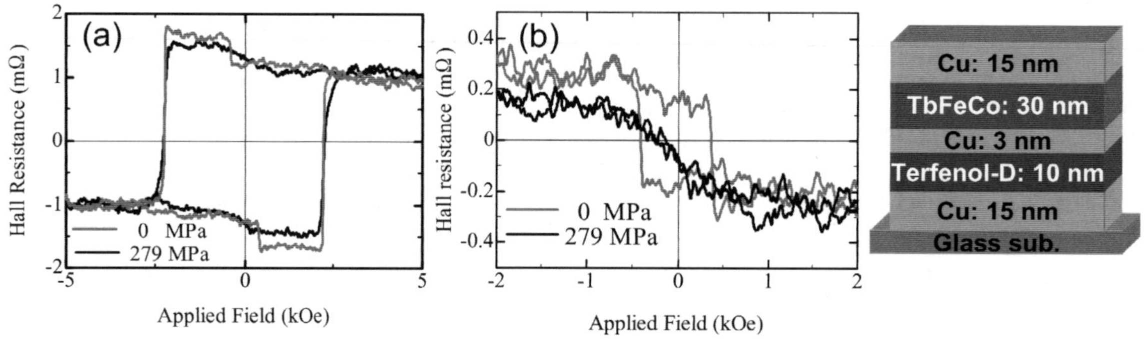

Figure 1. (a) is major loop and (b) is minor loop between ±2 kOe of Anomalous Hall loops of spin valve multilayer. (a) indicates only magnetization of Terfenol-D is shifted. (b) shows magnetization of free layer is changed from perpendicular to in-plane with stress.

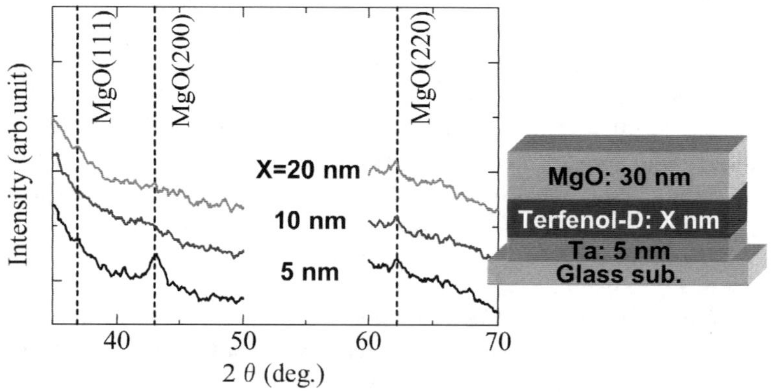

Figure 2. XRD curves of Ta/Terfenol-D(X nm)/MgO films

(111) diffraction peak is increased by increasing Mg thickness. Here, figure 3 (a) and (b) correspond to different deposition rates of Mg, (a) is 4.7 nm/min and (b) is 0.97 nm/min. These results indicate that MgO orientation in the Ta/Terfenol-D/Mg/MgO structure strongly depends on the metallic Mg layer thickness and deposition rate of Mg. Very thin metallic Mg layer is effective to attain (100) alignment of MgO, however, excess thickness of the metallic Mg layer causes (111) texture of the MgO structure.

MgO (111) alignment is composed of the alternation of depositions of the metallic Mg^{2+} layers and O^{2-} layers alternately. Since a close-packed plane of metallic Mg thick layer deposited on Terfenol-D is similar to Mg^{2+} layer of MgO (111), oxygen covers the Mg (111) layer to form MgO (111) texture as schematically illustrated in Figure 4. The MgO (111) texture will be grown on a thick Mg layer. The deposition of very thin Mg layer prior to the deposition process of the MgO layer may be effective to prevent the oxidation of the surface of Terfenol-D layer upon the introduction of oxygen during the deposition of the MgO layer.

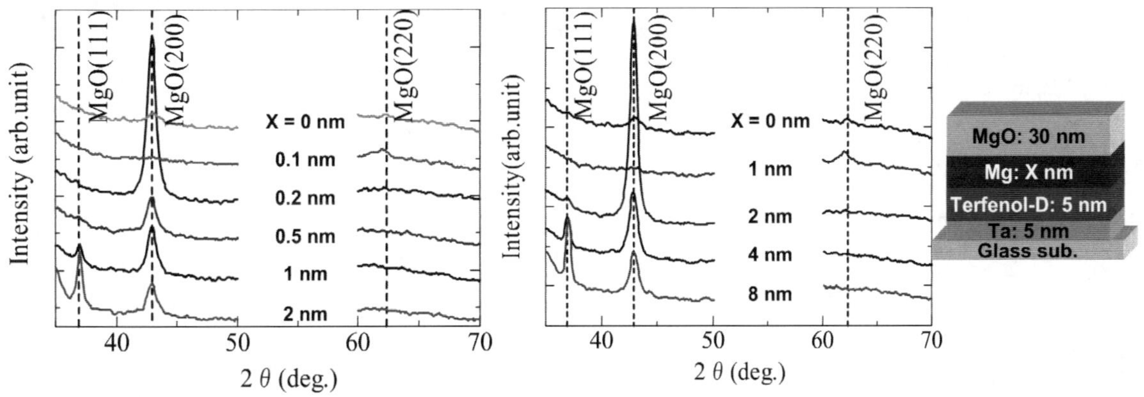

Figure 3. XRD diagrams of 30 nm-thick MgO films deposited on Mg layer with different thickness (X nm). Sputtering rates of Mg were (a) 4.7 nm/min and (b) 0.97 nm/min.

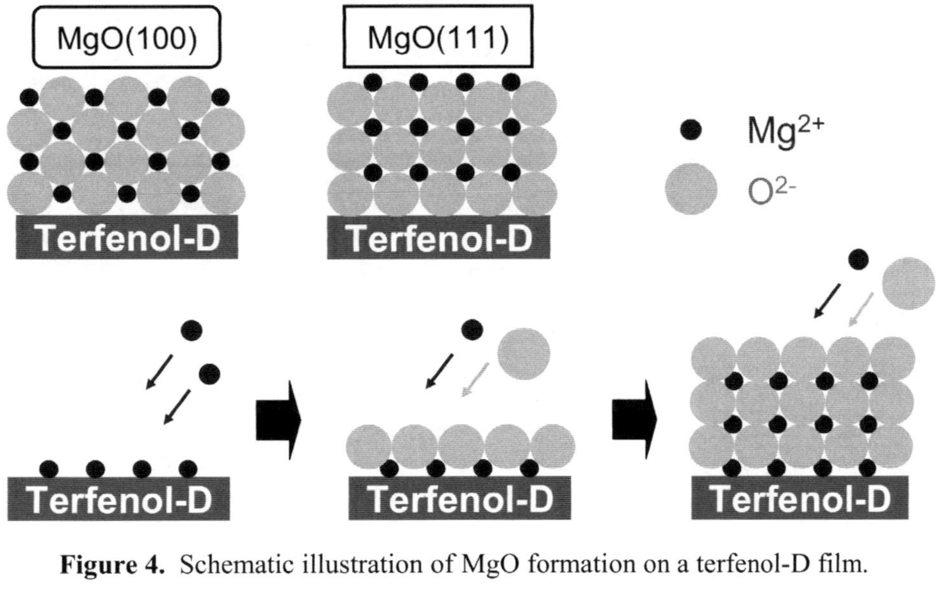

Figure 4. Schematic illustration of MgO formation on a terfenol-D film.

Journal of Physics: Conference Series **266** (2011) 012118 doi:10.1088/1742-6596/266/1/012118

4. Conclusion

A crystalline orientation of MgO deposited directly on a Terfenol-D amorphous alloy layer did not show (100) orientation while metallic Mg thin layer, inserted between Terfenol-D and MgO, caused a grain growth of MgO with high (100) orientation. Thicker Mg layer beneath the MgO layer caused the transition of the crystalline texture from (100) to (111) orientation. This result indicates that Mg film grown on Terfenol-D was required to attain (100) orientation of MgO layer deposited on it. Further, MgO had various thresholds at which the orientation was shifted from (100) to (111) under different Mg deposition rates. This result also showed that MgO crystalline orientation depended on Mg layer.

5. References

[1] Butler W H, Zhang X G, Schulthess T C and Maclaren J M 2001 *Phys. Rev.* B **63** 054416
[2] Nakagawa S, Yamada M, and Tokuriki N 2006 *IEEE Trans. on. Magn.* vol. **42** pp. 3773-3775
[3] Sandlund L, Fahlander M, Cedell T, Clark A E, Restorff A E and Wun-Fogle M 1994 *J.Appl. Phys.* vol. **75** No.10
[4] Saito N and Nakagawa S 2008 *IEEE Trans. on. Magn.* vol. **44** issue 11 pp. 2487-2490

Journal of Physics: Conference Series **266** (2011) 012119 doi:10.1088/1742-6596/266/1/012119

$L1_0$-ordered FeNi film grown on Cu-Ni binary buffer layer

T Kojima, M Mizuguchi and K Takanashi
Institute for Materials Research (IMR), Tohoku University, Sendai 980-8577, Japan

E-mail: takayuki.kojima@imr.tohoku.ac.jp

Abstract. An FeNi film with a metastable $L1_0$-ordered phase was fabricated by an alternate monatomic layer deposition. An Fe monolayer and a Ni monolayer were alternately grown on a $Cu_{0.6}Ni_{0.4}$ binary buffer layer. A (001) $L1_0$-FeNi superlattice peak, which indicates the formation of a $L1_0$ phase, was clearly observed by an X-ray diffraction. A long-range order parameter S of $L1_0$-FeNi was estimated to be 0.5 ± 0.1. It was found that the film has a uniaxial magnetic anisotropy along c-axis by magnetization measurement. The uniaxial magnetic anisotropy energy K_u was evaluated to be $(5.0 \pm 0.1) \times 10^6$ erg/cc.

1. Introduction

Perpendicularly magnetized films (PMF) have been investigated as materials for magnetic recording media. By using PMF, magnetic recording media with higher density can be developed. However, even devices with PMF have a limit of integration, because magnetizations easily fluctuate by external heats when a cell size of magnetic materials is too small. A uniaxial magnetic anisotropy enhances a thermal stability of a magnetization. Therefore, development of PMF with high uniaxial magnetic anisotropy energy K_u is required. $L1_0$-type ordered alloys have attracted as such materials. Particularly, $L1_0$-type alloys with noble metals such as FePt, FePd and CoPt have high K_u of the order of 10^7 erg/cc [1, 2, 3, 4]. On the other hand, it was reported that $L1_0$-FeNi also has high K_u of 1.3×10^7 erg/cc in a bulk sample [5]. Therefore, it is expected to realize a "noble metal free perpendicularly magnetized film" by the development of $L1_0$-FeNi films.

Existence of natural $L1_0$-FeNi has been only confirmed in meteorites [6, 7]. An order-disorder transition temperature of $L1_0$-FeNi phase is 320°C [8]. The ordering to $L1_0$ phase never occurs over this temperature, but diffusion rates of atoms are extremely slow below this temperature. It takes more than 10^4 years to finish one atomic jump at 300°C [9]. Thus, the formation of $L1_0$-FeNi is not easy by simple deposition processes. An Fe monolayer (ML) and a Ni ML are alternately stacked along the c-axis in $L1_0$-FeNi. Thus, $L1_0$-FeNi films may be fabricated by the alternately atomic deposition of Fe (1 ML) and Ni (1 ML). Actually, naturally non-existent materials such as $L1_0$-FeAu and hcp-CoRu were artificially created by the alternately monatomic layer deposition [10, 11].

Previously, T. Shima *et al.* fabricated $L1_0$-FeNi films [12]. The long-range order parameter S and the K_u were estimated to be 0.6 ± 0.2 and 6.3×10^6 erg/cc, respectively. However, the S estimated by X-ray diffraction (XRD) had large margin of errors, because $L1_0$-NiPt diffraction peaks overlapped $L1_0$-FeNi superlattice peaks.

In this paper, we focused on a $Cu_{0.6}Ni_{0.4}$ binary buffer layer. This material has no diffraction peak around $L1_0$-FeNi superlattice peaks in the XRD measurement, and it is expected this buffer layer should be non-magnetic. We have attempted to estimate the order parameter of $L1_0$-FeNi more precisely by using a $Cu_{0.6}Ni_{0.4}$ buffer layer.

Journal of Physics: Conference Series **266** (2011) 012119 doi:10.1088/1742-6596/266/1/012119

2. Experimental procedure

Films were grown by molecular beam epitaxy using two *e*-guns (Fe and Ni) and two Knudsen-Cells (Au and Cu) in ultrahigh vacuum (UHV) chamber (base pressure: 1×10^{-8} Pa). Polished MgO (001) substrate was annealed at 570°C for 30 min in the UHV chamber. Fe (1 nm) and Au (20 nm) layers were grown on the MgO substrate at 80°C. Then a $Cu_{0.6}Ni_{0.4}$ (93 nm) buffer layer was grown by co-deposition of Cu and Ni at 500°C. Finally, Fe (1 ML) and Ni (1 ML) were alternately grown at 187°C. The repetition time of Fe and Ni was 50 times. The deposition rate of Fe and Ni was 0.01 nm/s, which was controlled by a thickness monitor with quartz oscillators. Surface morphology and flatnesses of each layer were monitored by a reflection high-energy electron diffraction (RHEED). X-ray diffraction (XRD) with Cu-$K\alpha$ radiation was performed to characterize crystal structures and confirm the formation of $L1_0$-FeNi. Magnetic properties were investigated by magnetization measurement using the superconducting quantum interference device (SQUID) magnetometer at room temperature. A same stacking structure without only an $[Fe\ (1\ ML)\ /\ Ni\ (1ML)]_{50}$ layer was also fabricated as a reference sample.

3. Results and discussion

Figure 1 shows RHEED images of (a) $Cu_{0.6}Ni_{0.4}$ buffer layer and (b) $[Fe\ (1\ ML)\ /\ Ni\ (1ML)]_{50}$ surfaces. Sharp streak patterns were observed for all the surfaces. This confirms the formation of flat surface and an epitaxial growth of $Cu_{0.6}Ni_{0.4}$ buffer layer, Fe, and Ni layers. Streaks between fundamental streaks in Fig. 1 (a) indicate the formation of an ordered alloy or any surface reconstruction.

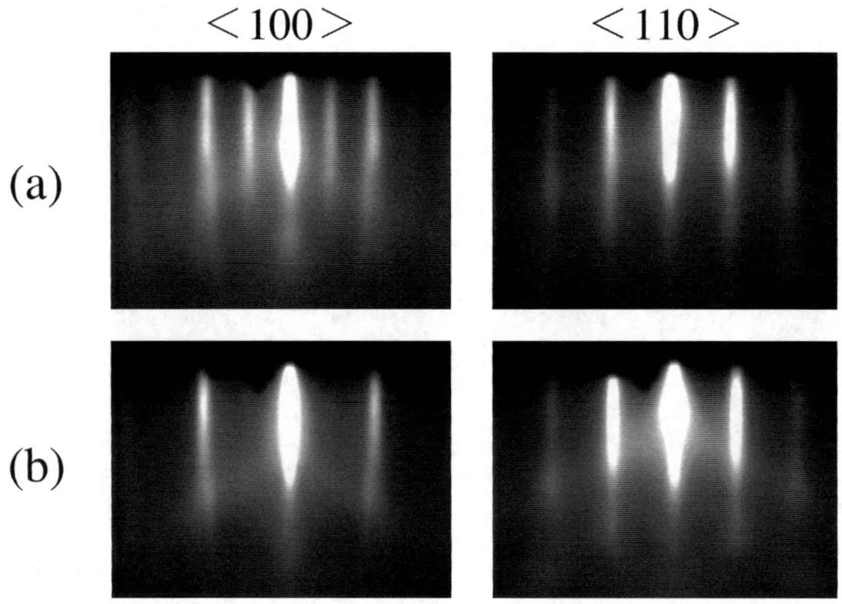

Figure 1. RHEED images of surfaces of (a) $Cu_{0.6}Ni_{0.4}$ buffer layer and (b) $[Fe\ (1\ ML)\ /\ Ni\ (1\ ML)]_{50}$. The incident beams were directed along the azimuth of MgO <100> and <110>.

Figure 2 shows XRD patterns of samples with and without an FeNi layer. (002) and (004) fundamental peaks of FeNi were observed around $2\theta = 51°$ and $2\theta = 119°$, respectively. In the pattern of sample without FeNi, there are three kinds of peaks from buffer layers. A peak around $2\theta = 49.5°$ is not (002) Au peak, because (002) Au peak should appear around $2\theta = 44.4°$. Thus, it is considered that the $Cu_{0.6}Ni_{0.4}$ layer and the Au layer were alloyed because the $Cu_{0.6}Ni_{0.4}$ was grown at high temperature (500°C). However, Au-Ni system is immiscible, therefore, there is a possibility that at least three phases were formed in the alloyed Au-Cu-Ni layer. *C*-axis lattice parameters (*c*) of the three phases in the buffer were estimated from their peak positions. They were 3.69 (lower angle), 3.57 (middle angle) and 3.51 (higher angle) Å. It is considered that the phase of c = 3.69 Å includes a large amount of Au compared with other phases, because the lattice parameters for Au, Cu and Ni are 4.08 Å, 3.62 Å and 3.52 Å, respectively. The phase of c = 3.51 Å might mainly consist of Ni. The phase of c = 3.57 Å is a phase with an intermediate composition, although it is not definitely identified.

Figure 2 (b) shows a magnified image of Fig. 2 (a) around 2θ = 25°. (001) $L1_0$-FeNi superlattice peak was clearly observed only for the pattern with FeNi layer. A long-range order parameter S of $L1_0$-FeNi was estimated to be 0.5 ± 0.1 from the ratio of integrated intensities of the (001) superlattice peak and the (002) fundamental peak.

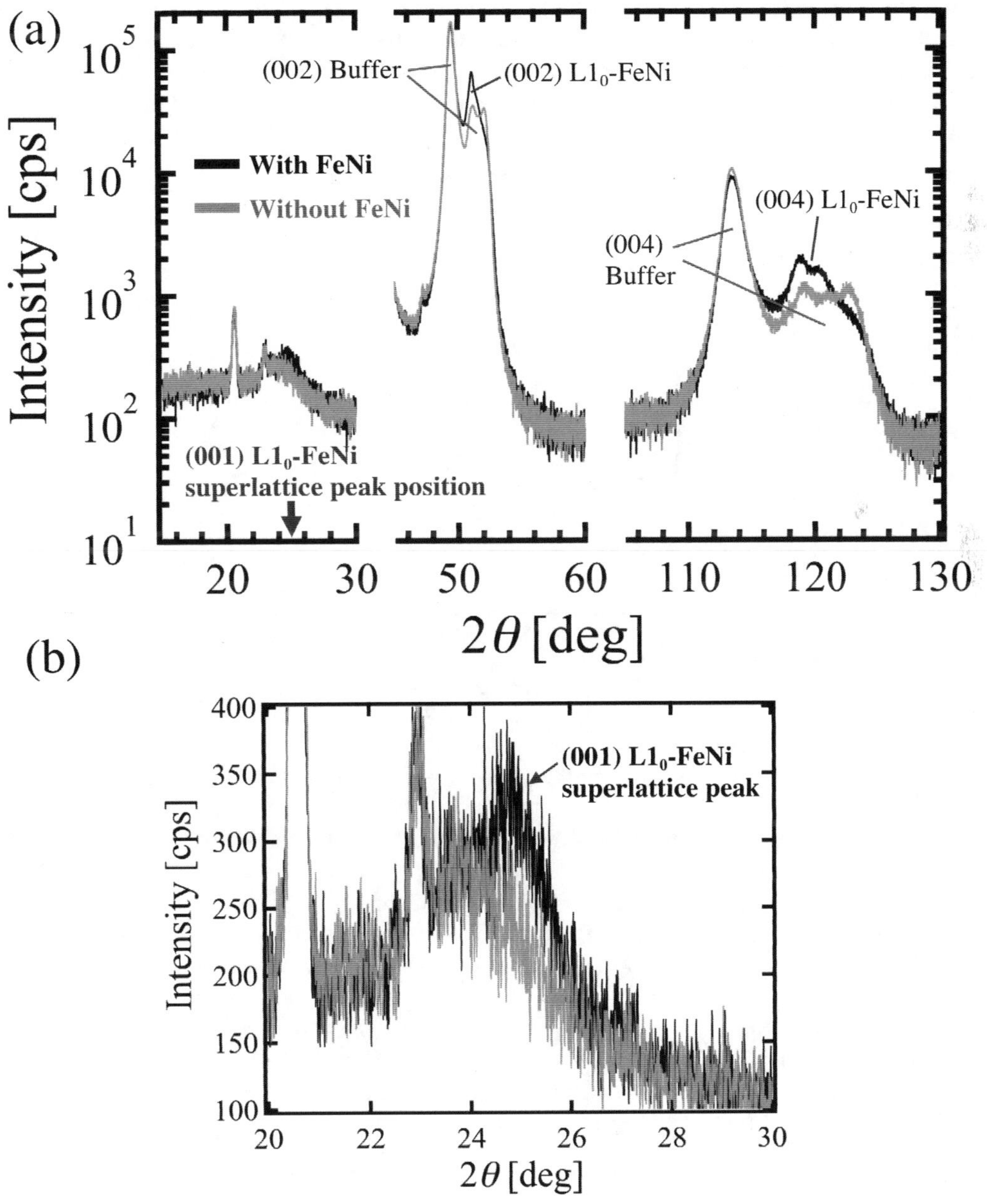

Figure 2. (a) XRD patterns of samples with and without [Fe (1 ML) / Ni (1 ML)]$_{50}$ layer. (b) Magnified view around 2θ = 25°.

Figure 3 shows magnetization curves of the $L1_0$-FeNi film when the magnetic field was applied in parallel and perpendicular to the film plane. Magnetization curves of the sample without FeNi layer were also measured. Unfortunately, it was found out that the sample without FeNi showed slight ferromagnetism (the saturation magnetization M_s of alloyed buffer layer was about 60 emu/cc). Thus, magnetization curves of the $L1_0$-FeNi film (Fig. 3) were obtained by subtracting magnetization curves of the sample without FeNi. The reason for the formation of ferromagnetic buffer layer even though $Cu_{0.6}Ni_{0.4}$ usually does not show ferromagnetism is considered as follows. The formation of at least three phases of the alloyed buffer layer was confirmed from the XRD measurements. Therefore, it is considered that any ferromagnetic phase different from $Cu_{0.6}Ni_{0.4}$ was formed, and this phase showed ferromagnetic behavior in the magnetization curve of the sample without FeNi layer. A K_u of the $L1_0$-FeNi film was evaluated by sum of an area enclosed between two magnetization curves (Fig.3) and the shape anisotropy $2\pi M_s^2$ (The value of area was minus because an easy magnetization axis was the in-plane direction). Uniaxial magnetic anisotropy of $K_u = (5.0 \pm 0.1) \times 10^6$ [erg/cc] was obtained ($M_s = 1125 \pm 5$ [emu/cc]). It is considered this value will be increased by further optimization of growth conditions.

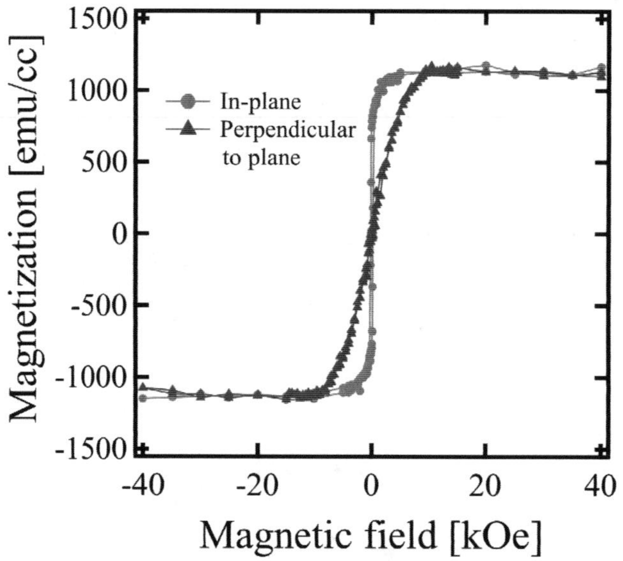

Figure 3. Magnetization curves of FeNi films when the magnetic field was applied in parallel and perpendicular to the film plane. Ferromagnetic components of buffer layers were subtracted.

4. Summary

An FeNi film with a metastable $L1_0$-ordered phase was fabricated by an alternate monatomic layer deposition of Fe (1 ML) and Ni (1 ML). Buffer layers and the FeNi layer were epitaxially grown on the MgO (001) substrate. A clear (001) $L1_0$-FeNi superlattice peak was successfully observed by an XRD measurement. The S of $L1_0$-FeNi was estimated to be 0.5 ± 0.1. It was revealed that the easy magnetization axis of FeNi film lay in parallel to the plane from magnetization curves obtained by subtracting magnetization of the buffer layer. The M_s and the K_u of FeNi film was estimated to be 1125 $\pm$ 5 emu/cc and $(5.0 \pm 0.1) \times 10^6$ erg/cc, respectively. There is a possibility that K_u will be further enhanced by an improvement of the long-range order parameter.

Acknowledgements
This work was partly supported by a Grant-in-Aid for Scientific Research (22246087) from Japan Society for the Promotion of Science.

References

[1] Lairson B M, Visokay M R, Sinclair R and Clemens B M 1993 *Appl. Phys. Lett.* **62** 639
[2] Gehanno V, Samson Y, Marty A, Gilles B and Chamberod A 1997 *J. Magn. Magn. Mater.* **172** 26
[3] Lairson B M, Visokay M R, Marinero E E, Sinclair R and Clemens B M 1993 *J. Appl. Phys.* **74** 1922
[4] Barmak K, Kim J, Lewis L H, Coffey K R, Toney M F, Kellock A J and Thiele J-U 2005 *J. Appl. Phys.* **98** 033904
[5] Paulevé J, Chamberod A, Krebs K and Bourret A 1968 *J. Appl. Phys.* **39** 989
[6] Albertsen J F, Jensen G B and Knudsen J M 1978 *Nature* **273** 453
[7] Mehta S, Novotny P M, Williams D B and Goldstein J I 1980 *Nature* **284** 151
[8] Paulevé J, Dautreppe D, Laugier J and Néel L 1962 *J. Phys. Radium* **23** 841
[9] Scorzelli R B 1997 *Hyp. Interact.* **110** 143
[10] Takanashi K, Mitani S, Sano M, Fujimori H, Nakajima H and Osawa A 1995 *Appl. Phys. Lett.* **67** 1016
[11] Himi K, Takanashi K, Mitani S, Yamaguchi M, Ping D H, Hono K and Fujimori H 2001 *Appl. Phys. Lett.* **78** 1436
[12] Shima T, Okamura M, Mitani S and Takanashi K 2007 *J. Magn. Magn. Mater.* **310** 2213

Effect of hydrogen plasma treatment on reduction process of iron oxide nanoparticles

T Maki[1, 2], H Kura[2], H Ishida[2], T Kaneko[2], R Hatakeyama[2], M Takahashi[2], and T Ogawa[2]
[1] Hitachi Metals, 1-2-1, Shibaura, Minato-ku, 105-8614 Tokyo, Japan
[2] Graduate School of Engineering, Department of Electronic Engineering, Tohoku University, 6-6-05, Aoba, Aramaki, Aoba-ku, Sendai, 980-8579 Miyagi, Japan

E-mail: Tomohito_Maki@hitachi-metals.co.jp

Abstract. Fe_3O_4 nanoparticles coated with oleic acid as a surfactant were reduced by hydrogen plasma generated by 13.56 MHz Radio-Frequency (RF). With increasing input RF power to the inductive coil, Fe_3O_4 nanoparticles could be transformed to almost single phase of Fe_3C and finally to α-Fe. The average grain sizes of Fe_3C and α-Fe nanoparticles are 19 - 23 nm and 52 - 59 nm, respectively. Saturation magnetizations of Fe_3C and α-Fe nanoparticles at 5 K were almost same as these of bulk Fe_3C and α-Fe. Field cooling procedure from room temperature to 5 K in the magnetic measurement revealed a formation of FeO phase from Fe_3O_4 by weak reduction. The desired ranges of electron temperature and electron density in order to obtain Fe_3C and α-Fe phases were determined by investigating plasma characteristics.

1. Introduction

Iron-based magnetic nanoparticles (NPs) such as iron oxide, carbide, nitride and metal have been studied for newly developing high performance magnetic materials. Various methods for fabricating iron-based NPs have been reported, which are mainly separated into two processes: reaction with gas phase (dry process)[1,2] and chemical synthesis in solution (wet process)[3,4]. In dry process, reaction gases are provided as stable molecules such as H_2 and CH_4, thus the reaction generally needs high temperature. High temperature treatment promotes sintering and grain growth of NPs, which brings about difficulty of controlling phases and nanostructures.

Low-temperature plasma treatment is well known as a useful method to improve surface property/activity of materials because of higher reactivity of exited elements than gas phase[5]. In addition, this method is expected to be suitable for treating NPs because it acts more effective to NPs having wide surface area than bulk materials. In this study, we apply hydrogen plasma treatment to the reduction of iron oxide NPs for developing high performance magnetic NPs with well-controlled phases and nanostructures.

2. Experimental procedures

Fe_3O_4 NPs were prepared by thermally decomposing iron pentacarbonyl ($Fe(CO)_5$) in the atmosphere. 5 mmol of $Fe(CO)_5$, 10 mmol of oreic acid (OA) and 30 ml of octyl ether were added into a flask. Then the solution was heated to 280 °C and kept for 1.5 h with stirring. After cooling down to the room temperature, the synthesized NPs were precipitated by adding acetone and separeted by

Journal of Physics: Conference Series **266** (2011) 012120 doi:10.1088/1742-6596/266/1/012120

centrifugation. The precipitation and the centrifugation processes were repeated four times and the NPs were re-dispersed in hexane without any precipitation. For spatially homogeneous irradiation of hydrogen plamsa to the NPs, the SiO_2 substrate was dipped into the hexane solution with Fe_3O_4 NPs and withdrawn at the rate of 4.5 mm/h. After evaporation of hexane at the interface between solution and atmosphere, NPs were adhered to the substrate. By withdrawing the substrate from the solution at a constant speed, NPs could be deposited on the substrate uniformly in thickness.

The NPs coated with OA on the substrate were inserted into the plasma treatment apparatus, which is shown schematically in figure 1. Hydrogen and argon gases were introduced into a quartz furnace tube with an inductive coil, and a steady gas flow was realized by pumping using a rotary pump. After pumping to a pressure under 1.0 Pa, hydrogen gas was introduced and the hydrogn plasma was generated by 13.56 MHz radio-frequency (RF). The input RF power for generating plasma was changed from 50 W to 450W and the hydrogen gas pressure and the treatment time were fixed at 6.0 Pa and 1 h, respectively. The sample position in the tube was at the center of the coil as shown in figure 1. After plasma treatment, the samples were cooled in an argon atmosphere and directly carried to a glove box in order to prevent re-oxidation.

Crystalline structure of NPs was characterized by an X-ray diffraction (XRD) and a transmission electron microscopy (TEM). Thermal analysis was performed by a thermogravimetry (TG). The weight of Fe in the NPs was estimated by using a X-ray fluorescence (XRF)[6]. The hysteresis loops of the samples were measured at 5 K with a superconducting quantum interference device (SQUID) after field cooled (FC) procedure from room temperature to 5 K under magnetic field of 4 T. In order to prevent re-oxidation during the magnetic measurement, samples were sealed by laminated films in the glove box. Plasma parameters were measured using a Langmuir probe that was installed at the center of the coil, which is the same position of samples.

3. Results and discussion

Figure 2 shows XRD patterns of NPs with various input RF power, P. As-prepared NPs were single phase of Fe_3O_4. Fe_3C phase appeared with residual Fe_3O_4 phase at P = 100 W and Fe_3O_4 phase was disappeared for $P \geq 150$ W. When P = 150 - 200 W, the NPs were supposed to be almost single phase of Fe_3C. From the TG measurement, the amount of OA covering the Fe_3O_4 NPs was estimated as 9.2

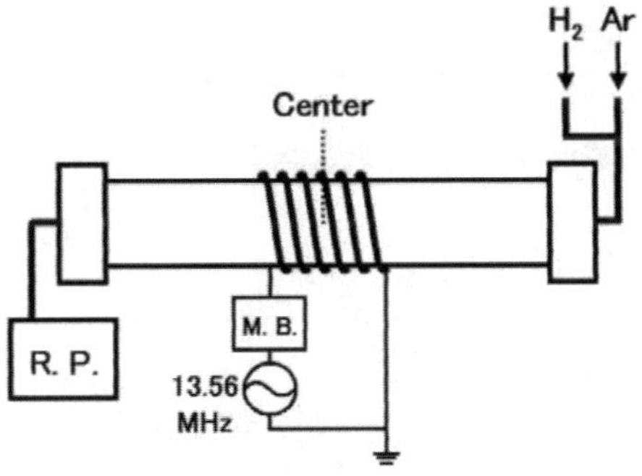

Figure 1. Schematic image of plasma treatment apparatus.

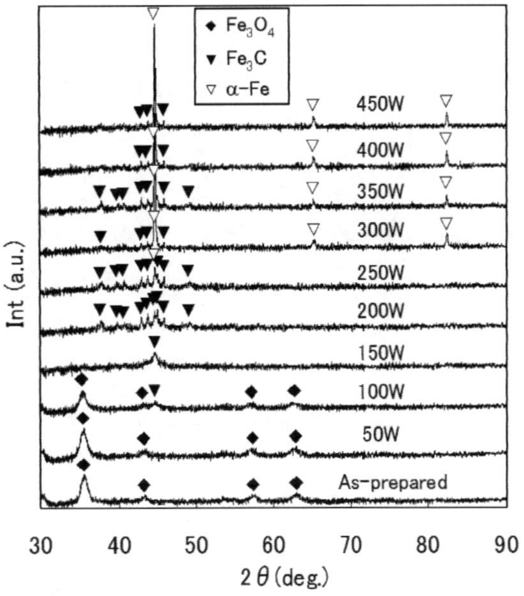

Figure 2. XRD patterns of NPs obtained with various input RF power, P.

wt%. Therefore, the molar ratio between Fe and C was calculated as 2 : 1, suggesting that an enough amount of C to carbonize the Fe existed in the NPs. From these results, within the intermediate range of input power, it was suggested that the decomposition of OA, reduction of Fe_3O_4, and carbonization to Fe_3C occurred almost at the same time. At $P = 250$ W, α-Fe began to appear with some additional Fe_3C. For $P \geq 300$ W, the peak intensity originating from α-Fe became dominant and that from Fe_3C negligible. This suggested that the hydrogen plasma excited by larger P could decompose Fe_3C to α-Fe.

Figure 3 shows the relationships between the grain size, D_S, of NPs and P. The D_S was estimated by the Scherrer's equation,

$$D_S = \frac{K\lambda}{\beta \cos\theta}, \quad (1)$$

where K is a shape factor, λ is the wavelength of the X-ray, β is the half-value width of the diffraction, and θ is the Bragg angle. In this paper, the value of K was 0.94[7]. For $P \leq 100$ W, the D_S calculated from Fe_3O_4(311) peak was around 10 nm. For P = 200 - 250 W, the D_S was increased to 19 nm - 23 nm, that was calculated from Fe_3C(210) peak. For $P \geq 300$ (W), the main phase became α-Fe and D_S further increased to 52 nm - 59 nm, which was calculated from α-Fe(110) peak. The average grain sizes of NPs, D_T, that were estimated from TEM images shown in figure 4, were also presented in figure 3. The D_T increased discontinuously with increasing P from 100 W to 350 W, and this tendency was in good agreement with that of D_S.

The structural transformation discussed above was also supported by the magnetic measurement. Figure 5 shows variation of the saturation magnetization, M_s per unit weight of the Fe in the NPs, the coercivity, H_{cJ}, and the exchange-anisotropy field, H_{ex} with the RF power P at 5 K. The procedure to determine H_{cJ} and H_{ex} were described elsewhere[8]. With increasing P up to 250 W, M_s increased from 118 $A{\cdot}m^2/kg_{Fe}$ to 160 $A{\cdot}m^2/kg_{Fe}$. This was mainly caused by the phase transition from Fe_3O_4 to Fe_3C. The saturation magnetization of the bulk Fe_3C is 137 - 141 $A{\cdot}m^2/kg$ at room temperature[9], and these values can be converted to 147 - 151 $A{\cdot}m^2/kg_{Fe}$ (per unit weight of Fe). M_s increased with increasing P from 300 W to 450 W and M_s reached 210 $A{\cdot}m^2/kg_{Fe}$ at $P = 450$ W. This value is close to the M_s of the bulk α-Fe, thus almost all of the NPs were transformed to α-Fe from Fe_3C. H_{cJ} took a maximum at P = 150 - 250 W. This could be attributed to the larger magnetocrystalline anisotropy field, H_a, of bulk

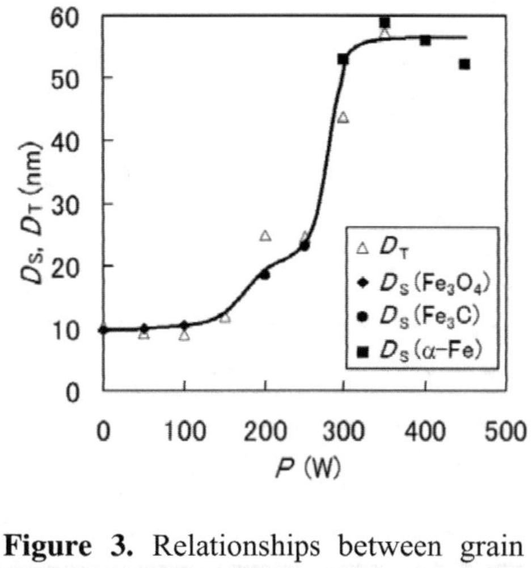

Figure 3. Relationships between grain sizes, D_S and D_T of NPs and the input RF power, P.

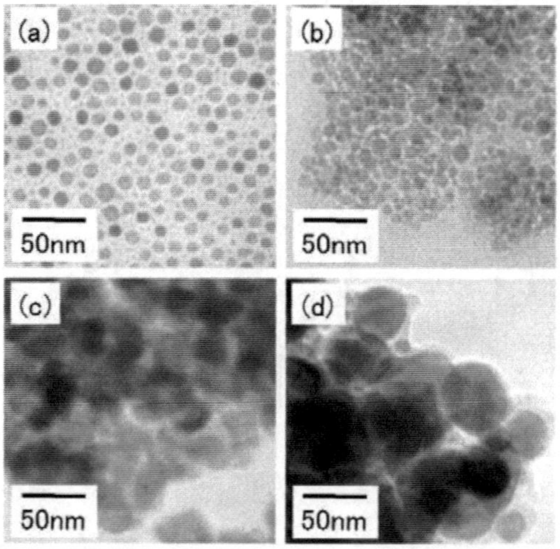

Figure 4. TEM images of NPs. (a) As-prepared Fe_3O_4 NPs. (b) After hydrogen plasma treatment with P = 100W. (c) P = 200W. (d) P = 350 W.

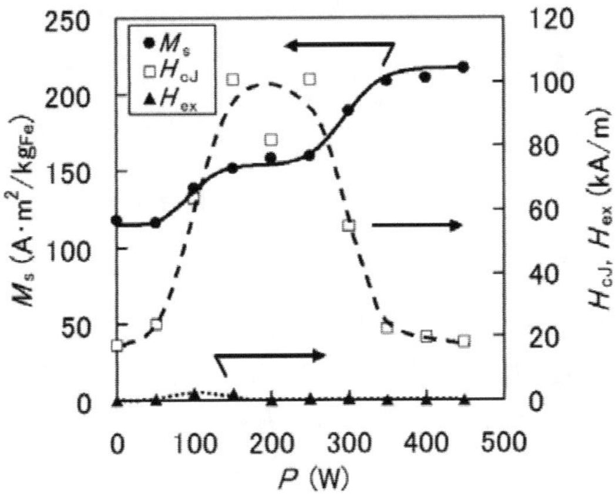

Figure 5. Variation of the saturation magnetization, M_s per weight of the Fe in the NPs, the coercivity, H_{cJ}, and the exchange-anisotropy field, H_{ex} with the input RF power, P at 5 K.

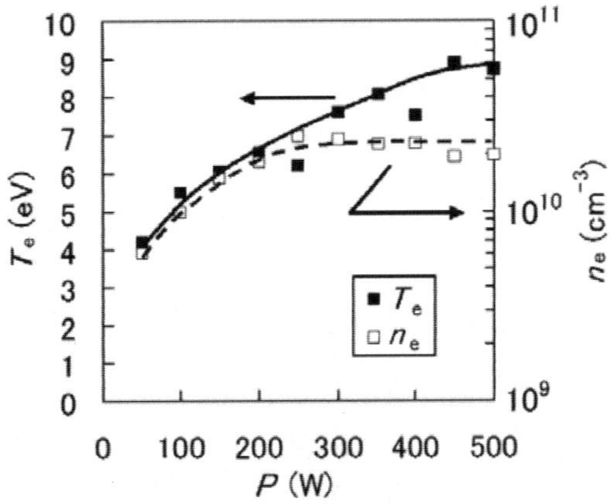

Figure 6. Electron temperature, T_e and electron density, n_e as a function of input RF power, P.

Fe_3C (H_a = 324 - 333 kA/m[9, 10]), compared to that of the bulk Fe_3O_4 ($|H_a|$ = 38 kA/m[11]) and the bulk α-Fe (H_a = 45 kA/m[12]). Furthermore, H_{ex} slightly increased at P = 100 - 150 W from those of as-made Fe_3O_4 NPs shown at P = 0. This would be a consequence of exchange-coupling between ferromagnet and antiferromagnet, thus the NPs were suggested to contain antiferromagnet FeO phase that was formed by a weak reduction of Fe_3O_4 phase.

Figure 6 shows electron temperature, T_e, and electron density, n_e, as a function of P. T_e increased from 4.2 eV to 8.9 eV with increasing P. n_e also increased from 6.1×10^9 cm^{-3} to 2.9×10^{10} cm^{-3} with increasing P in the range of $P \leq 250$ W, while n_e became constant when $P \geq 300$ W. From these results, in the case of hydrogen plasma treatment for 1 h, the proper values of the plasma parameters for the transformation of 10 nm Fe_3O_4 NPs into 20 nm Fe_3C NPs were determined to be T_e = 6 eV and n_e = 2×10^{10} cm^{-3}. Similarly, it is necessary to apply plasma with $T_e \geq 7.5$ eV and $n_e = 2 \times 10^{10}$ cm^{-3} in order to produce the transformation of 10 nm Fe_3O_4 NPs into 50 - 60 nm α-Fe NPs. T_e represents kinetic energy of electrons that collide with sample and heat it up. On the other hand, n_e represents the amount of excited ions, which is involved in activity of plasma. Therefore, when T_e and n_e have large values in the hydrogen plasma treatment, the reactivity between the hydrogen and the NPs become high by two effects: increased temperature of the NPs and high activity of the hydrogen plasma. It is suggested that this high reactivity induced the reduction and/or the carbonization of the NPs.

4. Conclusion

Fe_3O_4 NPs coated with OA as a surfactant was reduced by hydrogen plasma. By changing input RF power to the inductive coil, Fe_3C NPs that have 20 nm of average grain size and α-Fe NPs that have 50 - 60 nm of average grain size were obtained with almost single phase, respectively. The desired values of T_e and n_e have been also determined. These results suggest that the hydrogen plasma treatment can be useful to control the phase formation in Fe-based NPs.

5. References

[1] Lőffler J F, Scherrer P, Meier P, Doudin B, Ansermet J -P, Wagner W 1998 Phys. Rev. B 57 2915

[2] Prados C, Multigner M, Hernardo A, Sanchez J C, Fernandez A, Conde C F, Conde A 1999 *J. Appl. Phys.* **85** 6118

[3] Ogawa T, Takahashi Y, Yang H, Kimura K, Sakurai M, Takahashi M 2006 *Nanotechnology* 17 5539
[4] Yang H, Ito F, Hasegawa D, Ogawa T, Takahashi M 2007 *J. Appl. Phys.* **101** 09J112
[5] Moritani A, Yamada T, Kitamura T, Katayama H, Noda Y, Kanayama N 2000 *Thin Solid Films* **374** 298
[6] Matsuura H, Seto K, Yang H, Kawano K, Takahashi M, Ogawa T 2008 *IEEE Trans. Magn.* **44** 2804
[7] Scherrer P and Gottingen N G W 1918 *Math-Pys. Kl.* **2** 96
[8] Takahashi M and Tsunoda M 2002 *J. Phys. D* **35** 2365
[9] Hofer L J E and Cohn E M 1959 *J. Amer. Chem. Soc.* **81** 1576
[10] Takahashi M and Miyazaki T 1979 *Jpn. J. Appl. Phys.* **18** 743
[11] Rosencwaig A, Tabor W J, Hagedorn F B, Uitert L G V 1971 *Phys. Rev. Lett.* **26** 775
[12] Dillinger J F and Bozorth R M 1935 *Physics* **6** 279

Structural instability of FeCo ultrathin films grown on MgO(100)

S Mitani[1] and T Niizeki
National Institute for Materials Science
1-2-1 Sengen, Tsukuba 305-0047, Japan

E-mail: MITANI.Seiji@nims.go.jp

Abstract. Structures and their stability of FeCo ultrathin films grown on MgO(100) were investigated. Atomic force microscopy observations revealed dramatic structural changes from a continuous ultrathin layer to coarse islands by low temperature annealing. The results suggest that the thin layered structures of Fe based alloys as grown on MgO(100) are not so stable as expected from the relatively small lattice mismatch between Fe based alloys and MgO. This possibly explains annealing temperature dependence of tunnel magnetoresistance observed in a Fe/MgO/Fe(100) junction.

1. Introduction

Structures and their stability of thin films are of fundamental importance in science and technologies and are not fully understood in a wide range of materials although a great number of studies have been performed on thin film growth to date. Fe and Fe-based alloy thin films grown epitaxially on MgO(100) attract attention since MgO barrier magnetic tunnel junctions (MTJs) with Fe-based alloy electrodes play a significant role in current spintronic devices [1]. In this study, structures and their stability of Fe-based alloy "ultrathin" films grown on MgO(100) were investigated. The reason for choosing ultrathin films is that the effects of interfaces and/or epitaxial strain should appear clearly in thinner films. Annealing experiments were performed to see the stability (dynamic changes) of the film structures against (due to) epitaxial strain etc.

2. Experimental Procedure

Samples were prepared on a MgO(100) substrate by using an ultrahigh vacuum electron beam evaporator. The stacking structure is as follows: MgO(100) subs./MgO buffer 5nm/Fe 5nm/FeCo 50nm/MgO 2nm/FeCo 1.0 and 1.4nm. The Fe/FeCo bottom layers were annealed prior to the deposition of MgO and top FeCo layers. The evaporation sources used for MgO, Fe and FeCo layers were MgO (> 99.9at.%), Fe (> 99.95at.%) and $Fe_{50}Co_{50}$ (> 99.9at.%) blocks, respectively. In-situ reflection high-energy electron diffraction (RHEED) and ex-situ atomic force microscopy (AFM) were used for structural characterization. MTJs of MgO(100) subs./MgO buffer 10nm/Fe 60nm/MgO 2.1nm/Fe 5nm/IrMn 15 nm/Pt 7nm were also prepared to check the effect of post-annealing of tunnel magnetoresistance (TMR) in MTJs with a similar stacking structure. For the MTJs, top Fe/IrMn/Pt layers were deposited by sputtering.

[1] To whom any correspondence should be addressed.

Journal of Physics: Conference Series **266** (2011) 012121 doi:10.1088/1742-6596/266/1/012121

3. Results and discussion

Figure 1 shows RHEED patterns and AFM images for the sample with a 1.0 nm thick top FeCo layer. The streak RHEED pattern for the FeCo bottom layer annealed at 500°C shows that the layer is epitaxially grown and its surface is rather flat. The 2 nm thick MgO layer was also grown epitaxially. The RHEED pattern revealed that epitaxial strain is relaxed in the MgO lattice [2] and its crystallinity is poor, compared with MgO substrates. The RHEED patterns for 1.0 nm FeCo were spotty, indicating that island growth occurred on MgO(100). No significant changes appeared in the RHEED patterns through the post-annealing process in the temperature range from 200°C to 400°C. The weak streaks observed for the annealed states presumably come from the formation of facets of the FeCo islands. The AFM images for the FeCo islands at the annealed states correspond to the interpretation of the RHEED patterns. Considering the convolution effect of the AFM observations, the sizes of islands were roughly estimated to be 10 nm or smaller and 10 – 20 nm for 200°C- and 400°C-annealed states, respectively.

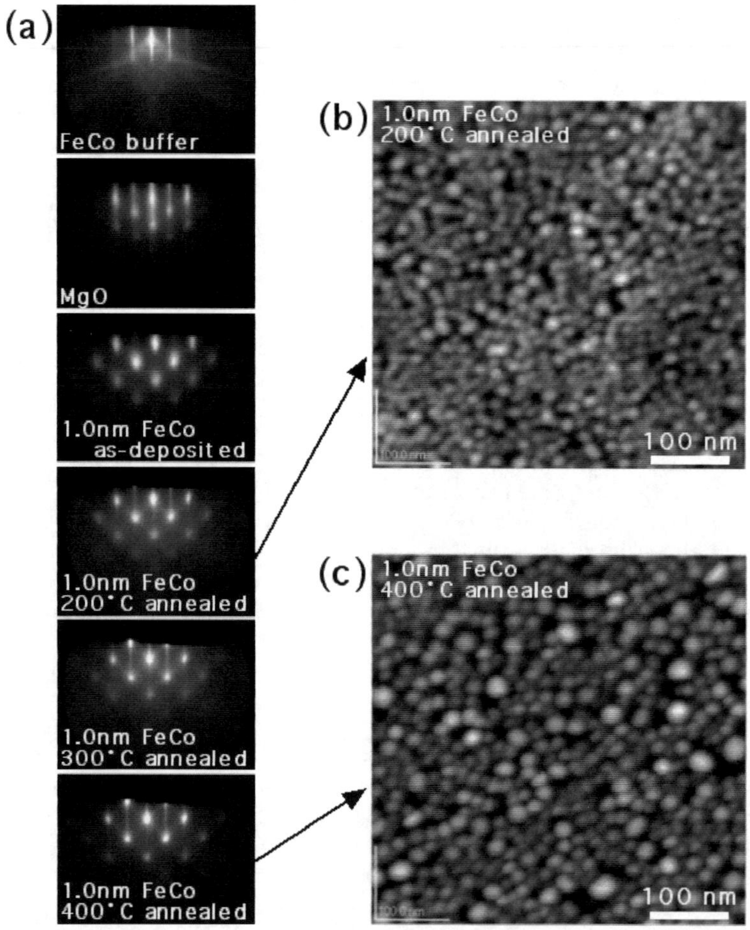

Fig. 1. (a) RHEED patterns for each layer of a sample with 1.0 nm thick FeCo, including the results of an annealing process, AFM images for the FeCo layer annealed (b) at 200°C and (c) at 400°C.

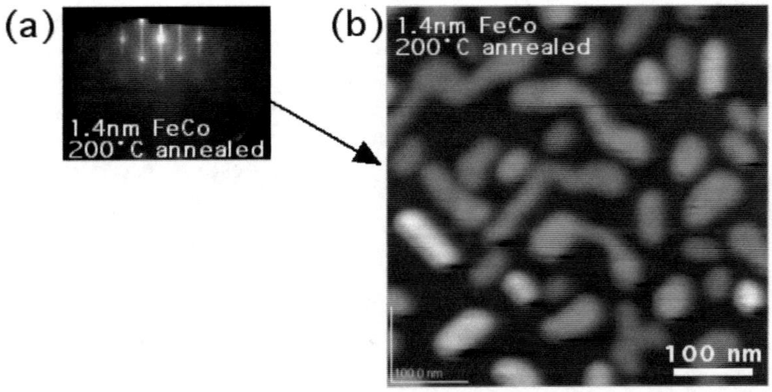

Fig. 2. (a) RHEED pattern and (b) AFM image for 200°C-annealed 1.4 nm thick FeCo grown on a MgO(100) layer.

Figure 2 shows a RHEED pattern and a AFM image for the sample with a 1.4 nm thick FeCo after annealing at 200°C for 10 min. Although the RHEED pattern is similar to that for the 1.0 nm thick FeCo annealed at 200°C, the result of the real space observation (Fig. 2(b)) is completely different. Namely, a number of large islands of a 100 nm scale appeared for the 1.4 nm thick FeCo after low temperature annealing. It has been confirmed by a previous study [3] and measurements of electrical resistivity and magnetization curves [4] that continuous layers of Fe and FeCo alloys are formed in the as-deposited state when the nominal thickness is larger than around 1.2 nm, while AFM is not useful to determine the critical thickness between continuous and discontinuous films due to the appearance of nanoparticle-like images even for continuous layers. Thus, the clear contrast in the AFM images (Figs. 1(c) and 2(b)) show that a dramatic structural change from a continuous layer to coarse islands occurred by low temperature annealing for the 1.4 nm thick FeCo layer. It is noted that a coarse-island structure of Fe on MgO, similar to that in Fig. 2 (b), was also observed by transmission electron microscopy in a study on double tunnel junctions [5] and that the experiments using pure Fe instead of FeCo in this study showed almost the same results as the FeCo case.

The lattice mismatches between most of body-centered-cubic Fe-based alloys and MgO are relatively small (2-3 %), so that epitaxial growth of Fe-based alloy films on MgO(100) is easily achieved as is supported by many experiments. In this sense, it could be expected that thin layered structures of Fe-based alloys are stable on MgO(100) surfaces. However, the present study suggests that as-deposited Fe-based alloy layers grown on MgO(100) is not so stable as expected from the small lattice mismatches, probably due to strain and defects introduced in the growth process. In addition, the present results do not meet a simple idea that continuous layers are in general more stable than nanoparticles. Namely, in thin film growth processes, one may consider that the system becomes stable at the moment when the film structure changes from nanoparticles to a continuous layer with increasing the nominal thickness, but the present results presumably show that the one's consideration is not in general correct.

The instability observed for the thin Fe-based alloy layers grown on MgO(100) could affect atomistic and/or layered structures even for thicker layers in actual spintronic devices. Concerning this idea, post-annealing temperature dependence of tunnel magnetoresistance was investigated for a Fe 60nm/MgO 2.1nm/Fe 5nm MTJ, in which no annealing was performed for the 5-nm Fe top electrode in the growth process. Fig. 3 shows magnetoresistance (MR) ratios and resistance-area products (RA) of the MTJ as a function of post-annealing temperature. At a relatively low temperature of 225°C, MR ratio takes a peak value, suggesting that low temperature annealing is sufficient to modify the MTJ

Journal of Physics: Conference Series **266** (2011) 012121 doi:10.1088/1742-6596/266/1/012121

structure for obtaining larger TMR. Further annealing above 225°C much affected the structure of the MTJ, resulting in degradation of TMR.

Considering the magnitude of lattice mismatch and/or the strength of chemical bonding to MgO, stability of Fe-based alloy thin layers should depend on the constitute elements, and therefore the results and discussion would be limited within simple Fe-based alloys such as Fe and FeCo. If the constitute elements are not only Fe and its neighbours, behaviour in annealing experiments is quantitatively different. Higher optimal annealing temperatures for TMR were reported in MTJs with Co_2FeAl [6] and CoFeB [7,8].

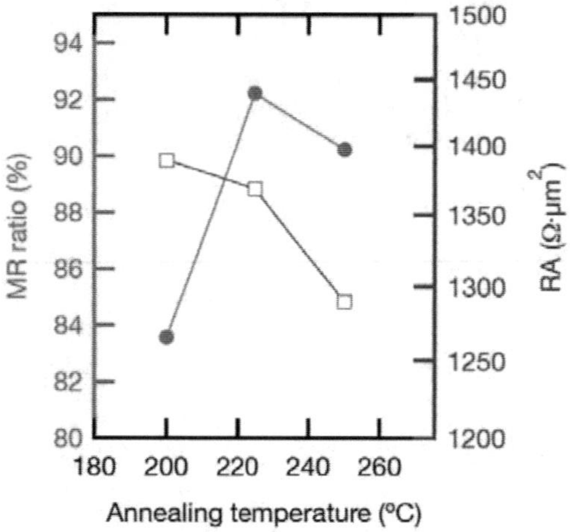

Fig. 3. Post-annealing temperature dependence of MR ratio and RA for a Fe/MgO/Fe tunnel junction.

4. Summary

Structural evolution of FeCo ultrathin films grown on MgO(100) were investigated by using RHEED and AFM. It was revealed that dramatic structural changes occurred by low temperature annealing, showing that the thin continuous layered structures of Fe based alloys are not so stable as expected from the small lattice mismatch. This possibly explains annealing temperature dependence of TMR in a Fe-based MTJ.

Acknowledgments

The authors thank Dr. S. Kasai and Dr. H. Sukegawa, NIMS and Prof. K. Takanashi, IMR Tohoku University for discussion and experimental support. A part of this study is supported by the Grant-in-Aid for Scientific Research (20360002) from MEXT.

[1] Yuasa S and Djayaprawira D D 2007 *J. Phys. D: Appl. Phys.* **40** R337
[2] Klaua M, Ullmann D, Barthel J, Wulfhekel W, Kirschner J, Urban R, Monchesky T L, Enders A, Cochran J F and Heinrich B 2001 *Phys. Rev.* B **64** 134411
[3] Ernult F, Mitani S, Takanashi K, Takahashi Y K and Hono K 2006 *Phase Trans.* **79** 717
[4] Mitani S, et al., unpublished.
[5] Nozaki T, Tezuka N and Inomata K 2006 *Phys. Rev. Lett.* **96** 027208
[6] Wang W, Sukegawa H, Shan R, Mitani S and Inomata K 2009 *Appl. Phys. Lett.* **95** 182502
[7] Parkin S S P, Kaiser C, Panchula A, Rice P M, Hughes B, Samant M and Yang S H 2004 Nature Mater. 3 862
[8] Ikeda S, Hayakawa J, Ashizawa Y, Lee Y M, Miura K, Hasegawa H, Tsunoda M, Matsukura F and Ohno H 2008 Appl. Phys. Lett. 93 082508

Journal of Physics: Conference Series **266** (2011) 012122 doi:10.1088/1742-6596/266/1/012122

Microstructure of hcp–Ni($1\bar{1}00$)/bcc–Cr(211) Bi-layer Film Grown on MgO(110) Substrate

Mitsuru Ohtake, Yoichi Sato, Jumpei Higuchi and Masaaki Futamoto

Faculty of Science and Engineering, Chuo University, 1-13-27 Kasuga, Bunkyo-ku, Tokyo 112-8551, Japan

E-mail: ohtake@futamoto.elect.chuo-u.ac.jp

Abstract. A Ni epitaxial film with hcp($1\bar{1}00$) single-crystal structure is obtained on a Cr(211) underlayer hetero-epitaxially grown on an MgO(110) single-crystal substrate in an early stage of film growth by ultra high vacuum rf magnetron sputtering. Reflection high energy electron diffraction (RHEED) observation indicates that the metastable hcp–Ni crystal tends to transform into more stable fcc crystal with increasing the film thickness. The crystallographic orientation relationship between the hcp and the fcc crystals is determined by RHEED and ϕ–scan X-ray diffraction. High-resolution transmission electron microscopy shows that the hcp and the fcc crystals exist separately on the Cr underlayer. The lattice constants of hcp–Ni crystal are determined from X-ray diffraction analysis as a=0.249 nm, c=0.415 nm, and c/a=1.67.

1. Introduction

3*d* ferromagnetic transition metal thin films have been widely studied for applications such as magnetic recording media, magnetic heads, etc. Metastable 3*d* transition metal films have also recently attracted much attention for practical applications like tunneling magnetoresistance devices [1]. The magnetic properties are greatly affected by their crystal structures. Ni is a typical soft magnetic material with fcc structure. hcp–Ni crystal is metastable and does not appear in the bulk phase diagram. A theoretical calculation predicts that the magnetic anisotropy of bulk hcp–Ni crystal is very similar to that of bulk hcp–Co crystal [2]. Preparation of hcp–Ni crystal has been reported in the forms of nano-particles and thin films [3]–[7]. However, parts of the hcp crystal tend to transform into more stable fcc structure during sample preparation processes, and these samples generally include large volume of fcc phase.

Recently, epitaxial thin film growth has made it possible to prepare metastable structures. In our previous studies, hcp–Ni [8]–[10] and hcp–permalloy [11] epitaxial films have been prepared on Au(100) underlayers at temperatures below 300 and 400 °C, respectively. The films consisted of two types of hcp($11\bar{2}0$) variants whose orientations were rotated around the film normal by 90° each other. However, the hcp crystal also transformed into fcc structure. X-ray pole figure analysis showed that the crystallographic orientation relationships between the hcp and the fcc crystals were hcp($11\bar{2}0$)[$1\bar{1}00$] || fcc($01\bar{1}$)[211] and fcc($01\bar{1}$)[$\bar{2}\bar{1}\bar{1}$] [12]. In order to investigate the hcp–Ni film growth and the transformation from hcp to fcc phase in detail, it is desirable to employ an hcp–Ni single-crystal film. bcc–Cr(211) has been frequently used as an underlayer to prepare an hcp–Co($1\bar{1}00$) single-crystal film with the c–axis parallel to the substrate surface [13]. The bcc–Cr(211) is expected also to promote the growth of hcp–Ni single-crystal film. In the present study, a Cr(211) underlayer hetero-epitaxially grown on an MgO(110) single-crystal substrate is employed to prepare an hcp–Ni single-crystal film. The film growth and the detailed film microstructure are investigated.

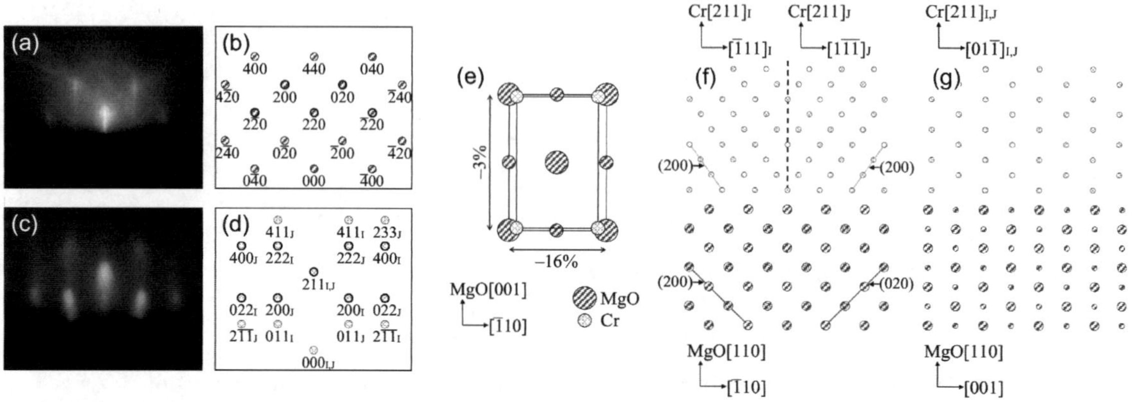

Figure 1. (a,c) RHEED patterns and (b,d) the RHEED spot maps of (a,b) an MgO(110) substrate and (c,d) a Cr underlayer grown on the substrate at 300 °C. The incident electron beam is parallel to the MgO[001] direction. (e,f,g) Epitaxial orientation relationship and the lattice mismatches of Cr(211) || MgO(110). The schematic models are viewed along (e) the MgO[110], (f) the MgO[001], and (g) the MgO[$\bar{1}$10] directions. The lattice mismatches are calculated from the lattice constants of the bulk MgO and the bulk Cr crystals. The dotted line in (f) shows a variant boundary. The symbols, I and J, respectively correspond to the orientation relationships of Type I and J explained in the text.

2. Experimental procedure

Thin films were prepared on polished MgO(110) substrates by using an rf magnetron sputtering system, where the base pressures were lower than 4×10^{-7} Pa. Substrates were heated at 600 °C for 1 hour in the ultra high vacuum chamber to obtain clean surfaces. The surface structure was checked by reflection high energy electron diffraction (RHEED). The RHEED pattern observed for the substrate exhibited *Kikuchi* patterns [figures 1(a) and (b)], indicating that the surface was clean and smooth. The surface roughness R_a examined by atomic force microscopy (AFM) was 0.66 nm. Pure Ni (99.9%) and Cr (99.9%) targets were employed. The rf powers for Ni and Cr targets were respectively fixed at 50 and 40 W and the Ar gas pressure was kept constant at 0.7 Pa, where the deposition rate was 0.02 nm/s for both materials.

The film layer structure was Ni(40 nm)/Cr(10 nm)/MgO(110). A 10-nm-thick Cr underlayer was deposited on the substrate at 300 °C. The substrate temperature was used to promote epitaxial growth of bcc–Cr(211) on MgO(110). The epitaxial orientation relationships between the substrate and the underlayer was determined by RHEED observation [figures 1(c) and (d)] as follows,

$$\mathrm{Cr}(211)[\bar{1}11] \parallel \mathrm{MgO}(110)[\bar{1}10] \quad \text{(Type I)},$$
$$\mathrm{Cr}(211)[1\bar{1}\bar{1}] \parallel \mathrm{MgO}(110)[\bar{1}10] \quad \text{(Type J)}.$$

The Cr(211) underlayer consists of two types of variants whose orientations are rotated around the film normal by 180° each other, as shown in figures 1(e)–(g). A 40-nm-thick Ni film was deposited on the Cr underlayer at room temperature.

The surface structure during Ni rf–sputter deposition process was studied by RHEED. The structural properties were investigated by out-of-plane ($2\theta/\omega$–scan), in-plane ($2\theta\chi/\phi$–scan), and ϕ–scan X-ray diffraction (XRD) with Cu–$K\alpha$ radiation (λ=0.15418 nm). The surface morphology and the cross-sectional microstructure were respectively observed by AFM and transmission electron microscopy (TEM). The TEM sample was first thinned mechanically and then ion-milled to be transparent for the electron beam accelerated at 300 kV.

3. Results and discussion

A Ni epitaxial thin film was obtained on a bcc–Cr(211) underlayer. Figures 2(a) and (b) show the RHEED patterns observed in the Ni rf–sputter deposition process. A clear diffraction pattern of hcp(1$\bar{1}$00) reflection shown in the RHEED spot map of figure 2(e) is observed in an early stage of Ni film growth, as shown in figure 2(a). An hcp–Ni(1$\bar{1}$00) single-crystal film is obtained on the underlayer. The epitaxial orientation relationship between the underlayer and the hcp–Ni crystal is determined by RHEED observation as follows,

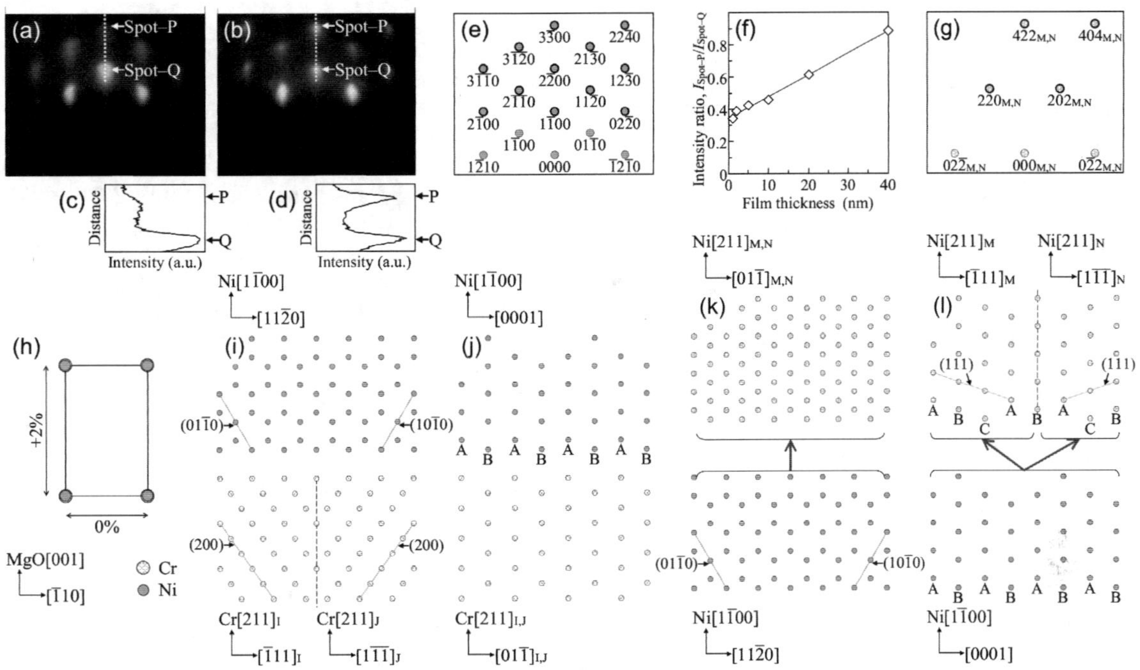

Figure 2. (a,b) RHEED patterns observed for a Ni film grown on a Cr(211) underlayer. The film thicknesses are (a) 2 nm (b) 40 nm. (c,d) RHEED intensity profiles along the dotted lines in (a,b), respectively. (e) RHEED spot map of hcp(1$\bar{1}$00). (f) Film thickness dependence on the RHEED spot peak intensity ratio of Spot–P to Spot–Q shown in (a,b). (g) RHEED spot map of fcc(211). The incident electron beam is parallel to the MgO[001] direction. (h,i,j) Epitaxial orientation relationship and the lattice mismatches of Ni(1$\bar{1}$00) || Cr(211). The lattice mismatches are calculated from the lattice constants of the bulk Cr crystal and the hcp–Ni film prepared in the present study. (k,l) Crystallographic orientation relationship between the hcp–Ni(1$\bar{1}$00) crystal and the fcc–Ni(211) crystal transformed from the hcp–Ni crystal. The schematic models are viewed along (h) the MgO[110], (i,k) the MgO[001], and (j,l) the MgO[$\bar{1}$10] directions. The dotted lines in (i,l) show variant boundaries. The symbols, I, J, M, and N, respectively correspond to the orientation relationships of Type I, J, M, and N explained in the text. The symbols, A, B, and C, show the stacking sequence of closed packed plane.

$$\text{Ni}(1\bar{1}00)[0001] \parallel \text{Cr}(211)[01\bar{1}]_{\text{Type I}} \text{ and } \text{Cr}(211)[0\bar{1}1]_{\text{Type J}} \parallel \text{MgO}(110)[001].$$

With increasing the film thickness, the RHEED spot intensities of 3$\bar{3}$00, 2$\bar{1}\bar{1}$0, 11$\bar{2}$0, and 22$\bar{4}$0 become stronger, whereas those of 2$\bar{2}$00, 3$\bar{1}\bar{1}$0, 3$\bar{1}\bar{2}$0, 21$\bar{3}$0, and 12$\bar{3}$0 weaken. The RHEED data suggest that another pattern overlaps with the hcp(1$\bar{1}$00) pattern. The pattern of increasing spot intensity is analyzed to be due to an overlap of two fcc(211) reflections, as shown by the spots, M and N, in the spot map of figure 2(g). The hcp structure in the Ni film is apparently transforming into more stable fcc structure. The transformed fcc–Ni crystal is formed on the Cr(211) underlayer in the following orientation relationships [figures 2(k) and (l)],

$$\text{Ni}(211)[\bar{1}11] \parallel \text{Ni}(1\bar{1}00)[0001] \parallel \text{MgO}(110)[001] \ (\text{Type M}),$$
$$\text{Ni}(211)[11\bar{1}] \parallel \text{Ni}(1\bar{1}00)[0001] \parallel \text{MgO}(110)[001] \ (\text{Type N}).$$

Figure 2(f) shows the film thickness dependence on the RHEED spot peak intensity ratio of 3$\bar{3}$00 (Spot–P) to 2$\bar{2}$00 (Spot–Q) estimated from the RHEED intensity profiles, observed, for example, as figures 2(c) and (d). The ratio gradually increases with increasing the film thickness. The increase in the intensity ratio shows that the ratio of fcc to hcp crystal exposed to the film surface increases with increasing the film thickness.

In order to confirm the orientation relationship between the hcp and the transformed fcc crystals, ϕ–scan XRD analysis was performed. Figure 3 shows the ϕ–scan XRD spectra of MgO{200}, Cr{200}, hcp–Ni{10$\bar{1}$0}, and fcc–Ni{111} poles. For the spectra of MgO{200} and hcp–Ni{10$\bar{1}$0} poles, two-fold symmetrical diffraction peaks are recognized. For the spectra of Cr{200} and fcc–Ni{111} poles, two diffraction peaks which originate from the Type I and the Type J bcc–Cr(211) variants and the Type M and the Type N fcc–Ni(211) variants are observed with 180° separation, respectively. The orientation relationship between the hcp–Ni and the fcc–Ni crystals expected from the RHEED study is confirmed by XRD analysis.

Figures 4(a)–(d) show the out-of-plane and the in-plane XRD spectra of the Ni/Cr/MgO specimen. hcp–Ni(1$\bar{1}$00) out-of-plane, hcp–Ni(0002) and/or fcc–Ni($\bar{1}$11) in-plane, hcp–Ni(11$\bar{2}$0) and/or fcc–Ni(02$\bar{2}$) in-plane, and hcp–Ni(11$\bar{2}$2) in-plane XRD reflections are clearly recognized. The lattice constant, a, of the hcp–Ni crystal is determined from the Ni(1$\bar{1}$00) XRD peak to be $2d_{(1\bar{1}00)}/(3)^{1/2}$ = 0.249 nm. The lattice constant, c, is calculated from the lattice constant, a, and the lattice spacing, $d_{\mathrm{Ni}(11\bar{2}2)}$, to be 0.415 nm, by considering the structure shown in figure 4(e). The c/a ratio is thus 1.67.

Figure 5(a) shows the AFM image of the Ni/Cr/MgO specimen. Small islands with rounded tops of 10–30 nm in diameter are observed. Figure 5(b) shows the cross-sectional TEM image observed along the MgO[$\bar{1}$10] direction. Sharp boundaries are recognizable between the MgO substrate and the Cr underlayer and between the underlayer and the Ni film. Figure 6(a) shows the high-resolution cross-sectional TEM image around the Ni/Cr interface. hcp–Ni and fcc–Ni crystals exist separately on the Cr underlayer as clearly shown by the high-resolution TEM images and the 2D FFT maps depicted in figures 6(b)–(d).

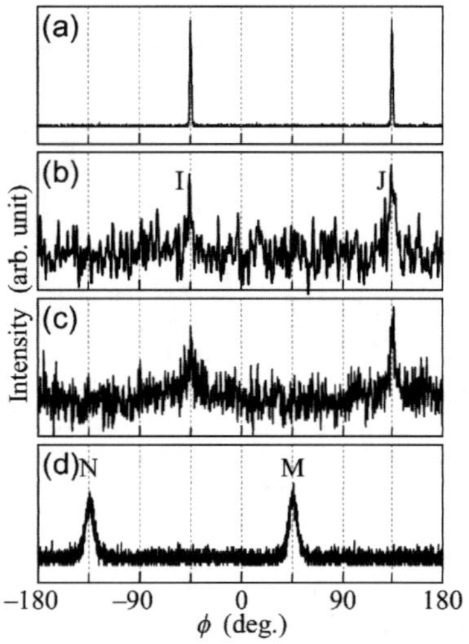

Figure 3. ϕ–scan XRD spectra of (a) MgO{200}, (b) Cr{200}, (c) hcp–Ni{10$\bar{1}$0}, and (d) fcc–Ni{111} poles obtained from a Ni/Cr/MgO specimen. The intensity is shown in a logarithmic scale. The symbols, I, J, M and N, respectively correspond to the orientation relationships of Type I, J, M, and N explained in the text.

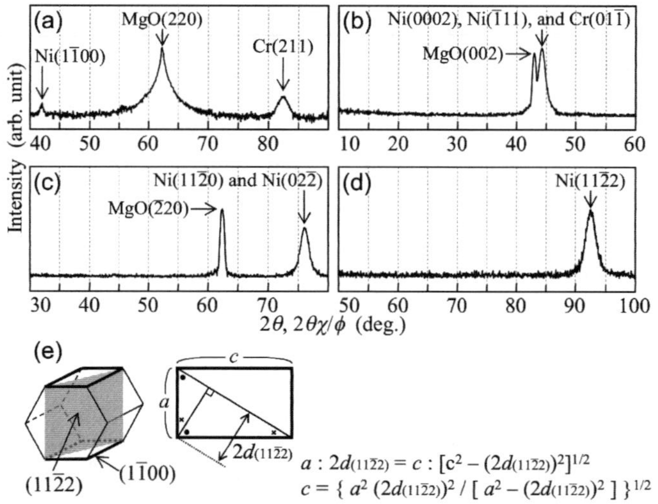

Figure 4. (a) Out-of-plane and (b,c,d) in-plane XRD spectra of a Ni/Cr/MgO(110) specimen. The in-plane XRD is measured by making the scattering vector parallel to (b) the MgO[002], (c) the MgO[2$\bar{2}$0], and (d) the Ni[11$\bar{2}$2] directions, and the incident X-ray angel is fixed at 0.4°. The intensity is shown in a logarithmic scale. (e) Schematic model of relationship between Ni(1$\bar{1}$00) and Ni(11$\bar{2}$2).

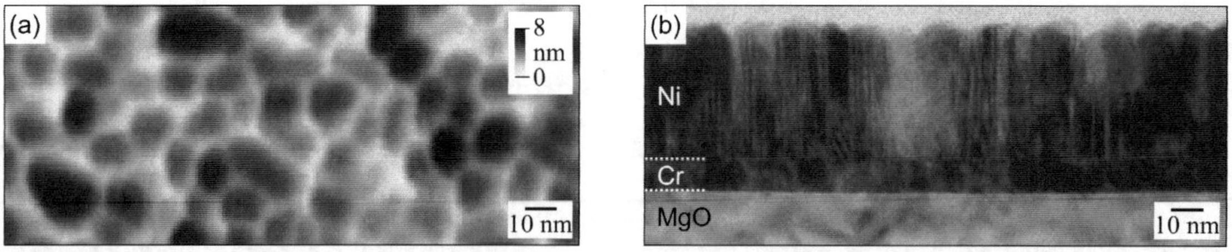

Figure 5. (a) AFM and (b) cross-sectional TEM images of a Ni/Cr/MgO specimen observed along the MgO[110] and the MgO[$\bar{1}$10] directions, respectively.

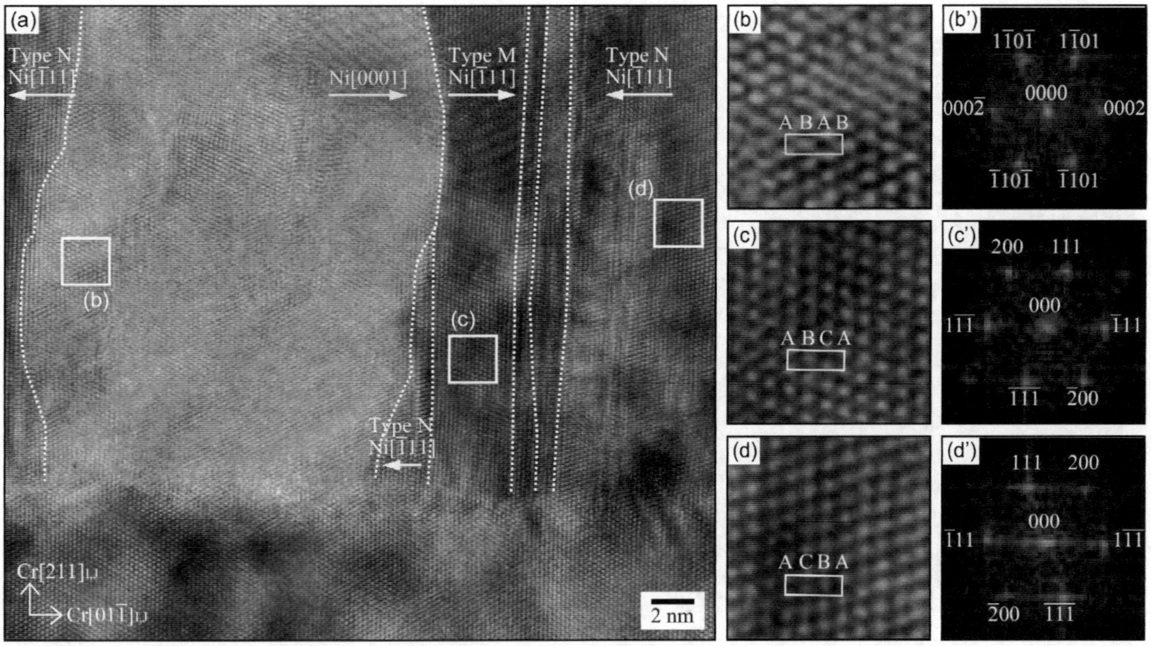

Figure 6. (a) Cross-sectional high-resolution TEM image around the Ni/Cr interface. The image is observed along the MgO[$\bar{1}$10] direction. The dotted lines in (a) show variant boundaries. (b,c,d) Enlarged views and (b',c'.d') 2D FFT maps of the areas surrounded by white square lines in (a).

4. Conclusion

A Ni epitaxial film with hcp(1$\bar{1}$00) orientation is obtained on a Cr(211) underlayer formed on an MgO(110) substrate in a early stage of film growth. With increasing the film thickness, Ni crystals with (211) orientations with fcc structure start to coexist with the hcp structure and the fcc/hcp ratio increases. This is due to a phase transformation from metastable hcp to more stable fcc structure. The resulting film consists of a mixture of hcp and fcc crystals. The crystallographic orientation relationships between the hcp and the transformed fcc crystals are Ni(211)[$\bar{1}$11] and Ni(211)[11$\bar{1}$] || Ni(1$\bar{1}$00)[0001]. The lattice constants of hcp–Ni crystal are determined as a=0.249 nm, c=0.415 nm, and c/a=1.67.

Acknowledgements

A part of this work was supported by NEDO-Japan and STFJ (Science and Technology Foundation of Japan). M. Ohtake is a Research Fellow of the Japan Society for the Promotion of Science.

References

[1] Yuasa S, Fukushima A, Kubota H, Suzuki Y and Ando K 2006 *Appl. Phys. Lett.* **89** 042505
[2] Papaconstantopoulos D A, Fry J L and Brener N E 1989 *Phys. Rev.* B **39** 2526
[3] Hemenger P and Weik H 1965 *Acta Crystallogr.* **19** 690
[4] Carturan G, Cocco G, Enzo S, Ganzerla R and Lenarda M 1988 *Mater. Lett.* **7** 47
[5] González V R *et al.* 2007 *J. Solid State. Chem.* **180** 22
[6] Tian W *et al.* 2005 *Appl. Phys. Lett.* 86 131915
[7] Tanaka T, Ohtake M, Kirino F and Futamoto M 2009 *IEEE Trans. Magn.* **45** 2515
[8] Ohtake M, Tanaka T, Kirino F and Futamoto M 2010 *J. Phys. Conf. Ser.* **200** 072072
[9] Ohtake M, Tanaka T, Kirino F and Futamoto M 2010 *J. Appl. Phys.* **107** 09E310
[10] Ohtake M, Tanaka T, Kirino F and Futamoto M 2010 *J. Magn. Soc. Jpn.* **34** 267
[11] Tanaka T, Ohtake M, Kirino F and Futamoto M 2010 *IEEE Trans. Magn.* **46** 1947
[12] Sato Y, Ohtake M, Higuchi J, Tanaka T, Kirino F, Futamoto M 2010 Structural Analysis of hcp–Ni and hcp–NiFe Thin Films Epitaxially Grown on Au(100) Single-Crystals *Abstr. ISAMMA 2010 (Sendai, Japan, 12–16 July 2010)* PF–03
[13] Nakamura A and Futamoto M 1993 *Jpn. J. Appl. Phys.* **32** L1410

Journal of Physics: Conference Series **266** (2011) 012123 doi:10.1088/1742-6596/266/1/012123

Origin of high-coercivity in nanocomposites by single-precursor sol-gel method

J Poltierova Vejpravova[1] P Brazda[2] A Mantlikova[1] D Niznansky[3] J Rohovec[4]
[1]Charles University Prague, Faculty of Mathematics and Physics, Department of Condensed Matter Physics, Ke Karlovu 5, 121 16 – Prague 2, Czech Republic
[2]Institute of Inorganic Chemistry, Academy of Sciences of the CR, 250 68 - Rez near Prague, Czech Republic
[3]Charles University Prague, Faculty of Sciences, Department of Inorganic Chemistry, Hlavova 2, 121 16 – Prague 2, Czech Republic
[4]Institute of Geophysics, Academy of Sciences of the CR, Rozvojova 269, 16500 – Prague 6, Czech Republic

E-mail: jana@mag.mff.cuni.cz

Abstract. We have fabricated high-coercivity nanocomposites constituted of Fe_2O_3 or $CoFe_2O_4$ nanocrystals embedded in amorphous silica matrix by a single-molecule precursor sol-gel method. Our approach enables smooth variation of the Fe(Co) : Si ratio up to 40 weight % of the magnetic phase, resulting in high density nanocomposite. Final crystallite size and phase composition was adjusted by subsequent heat treatment. The Fe_2O_3/SiO_2 samples annealed at 1000 and 1100 °C revealed coercivity, $H_c \sim 2$ T at 300 K due to formation of the metastable ε-Fe_2O_3 phase. The $CoFe_2O_4/SiO_2$ composites exhibit enhancement of the H_c value with increasing annealing temperature. Origin of the large coercivity in the two types of the magnetically hard nanocomposites will be discussed.

1. Introduction

Magnetic nanomaterials have been intensively studied as promising candidates for high-density magnetic storage media, ferrofluid technologies, and biomedical applications [1-3]. Their functionality is often conditioned by large coercivity (H_c) values. The most promising candidate is the $L1_0$ FePt phase, which exhibits $H_c \sim 2.2$ T, however, the nanocrystalls must be protected against rapid oxidation, that introduces additional technological treatment [4]. Our aim was to create a smart method for preparation of high-concentration nanocomposites and provide air-resistant materials with sufficient magnetic characteristics for high-density magnetic storage. In this paper, we report on magnetic properties of the two series of oxide-based nanocomposites: Fe_2O_3/SiO_2 and $CoFe_2O_4/SiO_2$.

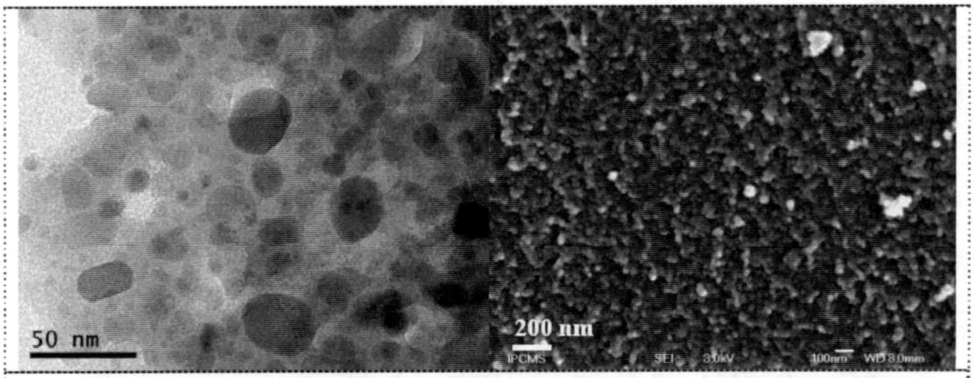

Figure 1. Typical TEM (left) and SEM (right) images of the high-density composite (Fe1000 sample).

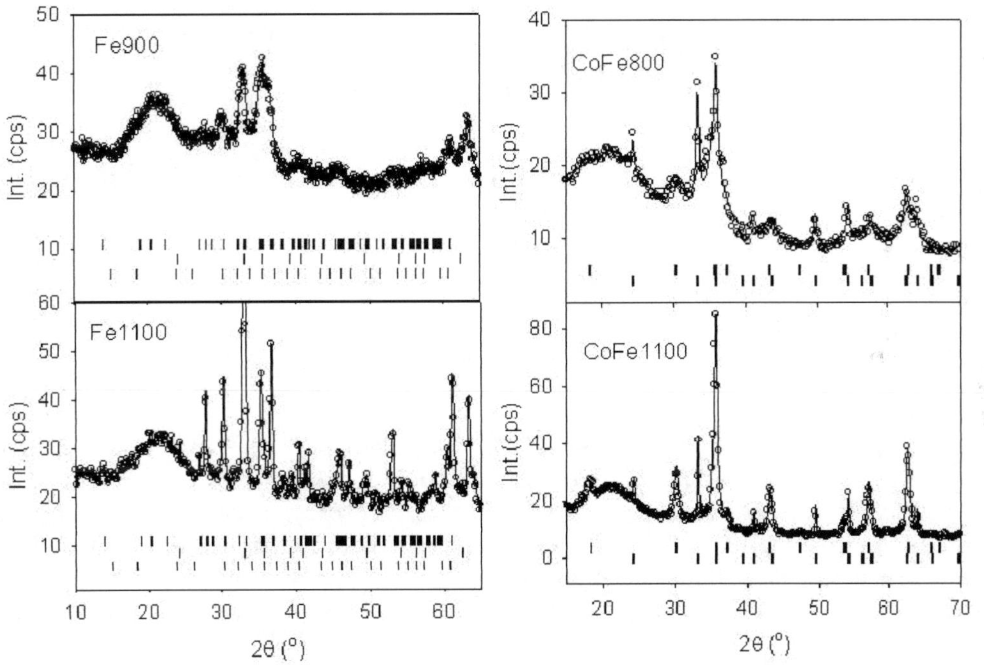

Figure 2. X-ray diffraction patterns. The solid lines correspond to the fit. Vertical lines mark positions of the Bragg reflections of the phases included in the Rietveld analysis. From top row to bottom row: epsilon – maghemite - hematite for Fe900 and Fe1100, and $CoFe_2O_4$ - hematite for CoFe800 and CoFe1100, respectively. The broad peak at around 20 ° is originated by the amorphous SiO_2.

2. Experimental details – synthesis, characterization and magnetic measurements

2.1. Synthesis

The samples were obtained by smart sol-gel method, where a single molecule serves as a source of the silica matrix and the nanocrystals. Details of the precursor molecule synthesis are described in [5]. For purpose of this work, the Fe : Si molar ratio was fixed to 1 : 2. Wet gels were obtained by hydrolysis of the precursor molecule. The gel was left for ageing at 50°C for two days, and then it was

Journal of Physics: Conference Series **266** (2011) 012123 doi:10.1088/1742-6596/266/1/012123

progressively heated up to 300°C in vacuum. The obtained xerogels were divided into several batches and each was annealed up to 800, 900, 1000 and 1100 °C, respectively, on air for 1 hour. Annealing times above 1 hour did not revealed significant changes in sample constitution. The nanocomposites were labeled according to the final treatment temperature as Fe800 – Fe1100 for the Fe_2O_3/SiO_2 and CoFe800 – CoFe1100 for the $CoFe_2O_4/SiO_2$.

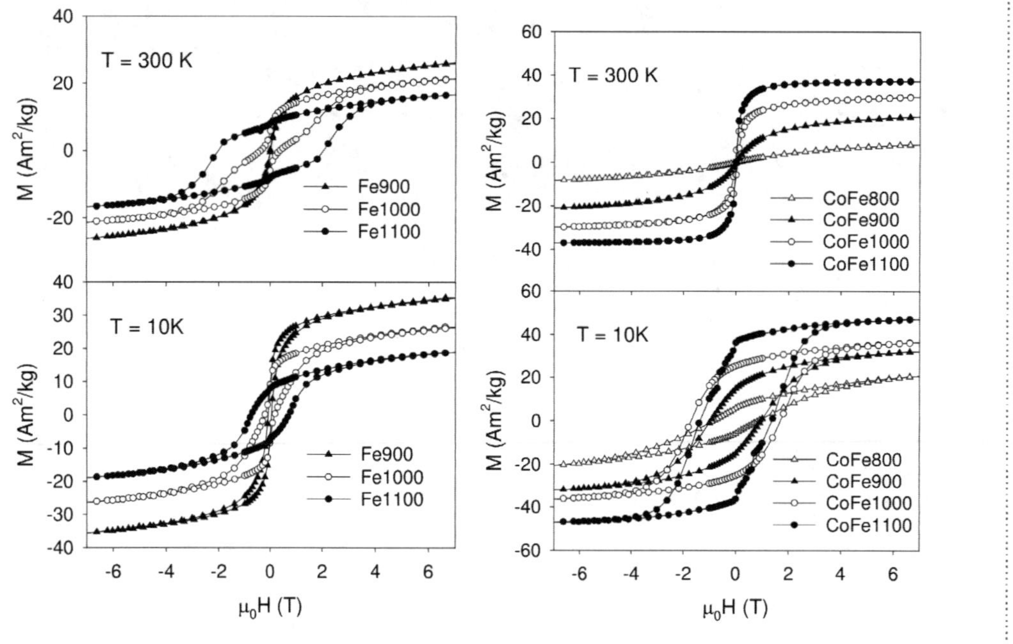

Figure 3. Hysteresis loops of the two types of composites at 300 K and 10 K.

2.2. Characterization

Final products were characterized by scanning electron microscopy (SEM), transmission electron microscopy (TEM), energy dispersive X-ray analysis (EDX) and X-ray diffraction (XRD). The SEM analysis revealed morphology of the powders, while the TEM observations served for determination of the size, shape and packing of the nanocrystals in the silica matrix. The XRD data were treated by Rietveld analysis in order to refine lattice parameters, atomic positions and apparent crystallite size corresponding to the volume of the coherently scattering domain of a nanocrystal [6]. Only spherical shape of the crystallites was considered in the model.

2.3. Magnetic measurements

Magnetic properties of the nanocomposites were measured using SQUID magnetometer (MPMS by Quantum Design) in the temperature range 2 – 400 K and magnetic fields up to 7 T. To avoid rotation of the crystallites when applying magnetic field, the powders were fixed by glue in gelatine capsules. The specific magnetization data were determined relative to the mass of the magnetic oxide phase, obtained from the EDX analysis.

3. Results and discussion

3.1. Synthesis and characterization

3.1.1. Fe_2O_3/SiO_2 nanocomposites

The typical constitution of the nanocomposites as observed by the TEM and SEM methods is demonstrated in Figure 1. The composites are formed of well-isolated crystallites embedded in the amorphous silica. The average crystallite size, determined from the XRD profile analysis increases with increasing annealing temperature; it ranges from 5 nm to 29 nm for the Fe900 – Fe1100 samples, respectively, as presented in Table 1. The molar ratio of the Fe, Si and O, obtained by the EDX analysis corresponds to the nominal stoichiometry. The key observation is the influence of the annealing temperature on the phase composition of the samples. While the Fe800 contains only the maghemite phase with defect spinel structure, the sample Fe900 revealed two phases, the maghemite and the orthorhombic ε-phase [7,8]. The nanocrystals in the Fe1000 and Fe1100 are mostly single-phase ε- Fe_2O_3; the Fe1100 contains below 5% of the rhomboedric hematite phase and both also samples include small amount of a very rare variant of the Fe_2O_3, the β- phase [9]. The resulting lattice parameters and the average crystallite size for the Fe900 - 1100 samples is summarized in the Table 1.

3.1.2. $CoFe_2O_4/SiO_2$ nanocomposites

The morphology and constitution of the $CoFe_2O_4/SiO_2$ nanocomposites is almost identical to the Fe_2O_3/SiO_2 samples, shown in the Figure 1. The crystallites appear to be spherical, as expected when considering the cubic symmetry of the ferrite (inverse spinel) structure. The average crystallite size, determined from the XRD profile analysis also increases with increasing annealing temperature from 3 nm to 13 nm in the CoFe800 – CoFe1100 samples, respectively, while the lattice parameter *a* remains intact (see Table 1). The required elemental ratio was confirmed by the EDX method. The lower particle size in comparison to the Fe_2O_3/SiO_2 composites can be explained as follows. There is limited intermixing of the cations (Co^{2+}, Fe^{3+}) into the silica matrix, which influences its viscosity and hence formation of the pores and crystallite growth during annealing. Also the degree of compaction of the starting xerogels can be different for the two types of composites.

3.2. Magnetic measurements

3.2.1. Fe_2O_3/SiO_2 *nanocomposites*

The typical hysteresis loops of the composites measured at the room temperature and at 10 K are presented in Figure 3. It is evident, that the hysteresis parameters significantly differ with respect to the phase composition of the samples, as summarized in the Table 1. The Fe900 reflects a typical trend expected for superparamagnetic maghemite nanoparticles with weak inter-particle interaction; the 10 K – loop shows moderate coercivity H_c = 0.07 T, which decreases down to 0.01 T at 300 K. Considering the fact, that the 5 nm maghemite nanoparticles should be in paramagnetic state at the room temperature [10], the low coercivity observed can be attributed to inter-particle interaction of dipolar origin. In contrast, the Fe1100 sample exhibit H_c = 0.7 T at 10 K, but enhancement of the value at 300 K up to 2.1 T. The observation can be explained by the typical features of the ε- Fe_2O_3 phase, which exhibits order-to-order magnetic phase transition at 150 K from the high-temperature high-coercivity magnetic phase to the low-coercivity ground-state magnetic phase [8]. The Fe1000 sample follows the expected behavior of the hybrid between the Fe900 and Fe1100 samples.

3.2.2. $CoFe_2O_4/SiO_2$ *nanocomposites*

The hysteretic behavior of the $CoFe_2O_4/SiO_2$ nanocomposites is demonstrated in the Figure 3 and the important parameters (coercivity, H_c and remanence, M_r) are summarized in the Table 1. As the particle size increases, the H_c and the M_r values increase when comparing the CoFe800 – CoFe1100 samples at the same temperature. Also the H_c and M_r decrease with increasing temperature as expected for an ensemble of nanosized particles, which usually exhibit superparamagnetic phenomena [11,12]. Due to rather high density of the composites, additional inter-particle interactions are expected to occur and hence the values of the H_c can be arbitrary enhanced due to them. When comparing the values to those obtained on $CoFe_2O_4/SiO_2$ nanocomposites with lower density (10 weight % of the magnetic oxide), prepared by conventional sol-gel approach, they are very similar. Therefore, the dipolar interactions are of equal importance regardless the concentration of the nanocrystals in the matrix.

Journal of Physics: Conference Series **266** (2011) 012123 doi:10.1088/1742-6596/266/1/012123

4. Conclusions

We have prepared high-concentration nanocomposites constituted of magnetic oxide nanocrystals (Fe_2O_3 and $CoFe_2O_4$) embedded in amorphous silica matrix using a smart sol-gel method. The single-molecule precursor technology can be easily utilized in production of corresponding thin films and coatings down to nm thickness and enables smooth variation of the magnetic phase-to-silica ratio. The final particle size and hence the magnetic properties of the composites can be easily controlled by annealing of the dry gels at temperatures from 800 °C to 1100 °C. The most important is the appearance of the 2 T-coercivity at room temperature in the Fe_2O_3/SiO_2 nanocomposite annealed at 1100 °C due to presence of the ε-phase, which designates it as a suitable candidate for magnetic storage and bio-sensing devices.

5. Acknowledgments

This work was supported by the Czech Science Foundation, project no. 202/08/P006 and Grant Agency of the AS CR, project no. 40010065.

Table 1. Summary of the lattice parameters (*a*, *b*, *c*) and average particle size (*d*) obtained from the Rietveld analysis. Only results obtained for the majority phase are presented (ε-Fe_2O_3 and $CoFe_2O_4$ in the Fe and FeCo series, respectively). The coercivity values (H_c) at 2 K and 300 K are also shown.

	d(nm)	*a* (pm)	*b* (pm)	*c* (pm)	H_c (T) *T=10 K*	H_c (T) *T=300 K*
Fe900	5±2	512±2	880±3	938±3	0.07	0.01
Fe1000	11.6±2.5	508.5±0.2	876.1±0.2	943.1±0.1	0.20	0.40
Fe1100	29.0±9.1	508.6±0.1	876.2±0.2	943.2±0.2	0.70	2.10
CoFe800	3.0±0.8	837.0±0.3	-	-	0.85	0.008
CoFe900	3.9±0.7	837.2±0.4	-	-	0.95	0.014
CoFe1000	13.2±0.6	837.9±0.3	-	-	1.83	0.044
CoFe1100	12.0±1.2	837.8±0.3	-	-	1.53	0.090

6. References

[1] Tartaj P, Morales M P, Veintemillas-Verdaguer S, Gonzales-Carreno T and Serna C Synthesis, properties, and biomedical applications of magnetic nanoparticles, in Handbook of Magnetic Materials, edited by Buschow K H J, Elsevier 2006
[2] Tartaj P, *Current Nanosci* 2006 **2** 43
[3] Weller D and Moser A 1999 *IEEE Trans Magn* **35** 4423
[4] Sun S, Murray C B, Weller D, Folks L and Moser A 2000 *Science* **287** 1989
[5] Brazda P, Niznansky D, Rehspringer J L and Poltierova Vejpravova J 2009 *J. Sol Gel Sci* **83** 826
[6] *Microstructure Analysis from Diffraction*, edited by Snyder R L, Bunge H J, and Fiala J, International Union of Crystallography, 1999
[7] Jin J, Hashimoto K and Ohkoshi S 2005 *J. Mater. Chem.* **15** 1067
[8] Kurmoo M, Rehspringer J L, Hutlová A, D'Orléans C, Vilminot S, Estournès C and Nižňanský D 2005 *Chem. Mater.* **17** 1106
[9] Zbořil R, Mashlan M and Petridis D 2002 *Chem. Mater.* **14** 969
[10] Fiorani D, Testa A M, Lucari F, D'Orazio F and Romero H 2002 *Physica B* **320** 122
[11] Cao X and Gu L 2005 *Nanotechnology* **16** 180
[12] Hutlová A, Nižňanský D, Rehspringer J L, , Estournès and Kurmoo M 2003 *Adv. Mater.* **15** 1622

Journal of Physics: Conference Series **266** (2011) 012124 doi:10.1088/1742-6596/266/1/012124

Crystal Structure and Morphology Dependence of the Phase of Mollusc Shell: A Case Study of XRD, SEM and ESR

W Siriprom[1] **J Kaewkhao**[2*] K. Phachana[3] and P Limsuwan[4]

[1]Faculty of Resources and Environment, Kasetsart University, Sriracha Campus, Chonburi,Thailand, 20230
[2]Center of Excellence in Glass Technology and Materials Science (CEGM), Nakhon Pathom Rajabhat University, Thailand, 73000
[3]Department of Chemistry, Faculty of Science, Burapha University, Chonburi, Thailand, 20131
[4]Department of Physics, King Mongkut's University of Technology Thonburi, Bangkok, Thailand, 10140

E-mail: mink110@hotmail.com

Abstract. The aim of this study is to investigate the physical properties in the *Paphia undulate* and *Amusium pleuronectes* shells collected from the coastal area of Chonburi province, Thailand. The crystal structure of the shells was studied by x-ray diffraction (XRD). The XRD patterns spectra reveal that *the Paphia undulate* shells and *Amusium pleuronectes* shells are made of a pure aragonite phase, and a mixture phase of aragonite and calcite, respectively, the identification and quantitative analysis were performed by using the Rietveld method. In this study, we also used the scanning electron microscope (SEM) to study the Morphology of the *Paphia undulate* shells and *Amusium pleuronectes* shells. The results on SEM micrographs agree well with those of XRD. In addition, the ions in the shell samples were also studied by Electron Spin Resonance Spectroscopy (ESR). The ESR spectra show that the samples of both shells were consisting of Mn^{2+} ions. The pattern of the ESR spectrum is thus strongly influenced by the environment of the paramagnetic ions, and then it can be used to identification of crystal structure of mollusc shell

1. Introduction

Mollusc shells are fascinating organic-material bio-composites with high mechanical performances. The marine mollusk exoskeletons are expected to contain carbonaceous matter as they form themselves towards the carbonate exoskeletons by absorbing much of atmosphere emissions **[1]**. The shells are found to have two common mineral forms of $CaCO_3$; calcite and aragonite. The crystal structures of calcite and aragonite and their inter-conversion have been studied **[2]**. Aragonite shells and aragonite mineral deposits are known to undergo a slow metamorphosis into calcite. The major and minor parts of organic materials are intercrystalline and intracrystalline, respectively **[2-4]**.

Sea shells are found as two common mineral forms of $CaCO_3$, calcite and aragonite. Sea shells are composed of 97-99% $CaCO_3$ (calcite, aragonite, or vaterite) with lesser amounts of $MgCO_3$, $(Al,Fe)_2O_3$, SiO_2, $Ca_3P_2O_8$, $CaSO_4$, protein, and mucopolysaccharides. In addition to these major and minor constituents, trace amounts of Sn, Mo, Mn, Cd, Ti, B, Pb, Au, Ag, Ni, Co, Bi, Cu, Sr, and Rb. As have also been found in varying amount **[1]**.

The purpose of this study was to obtain information of the shell structures of marine mussel shells. The information on microstructure is one of the most significant relative data for clarification of evolutionary trends of shell structure, and assessment the potentiality of using the *Amusium pleuronectes* shell and *Paphia undulate* shell as alternative biomaterials. It is of great importance to understand the formation and stability conditions for all crystal modifications of calcium carbonate. A better understanding of this phenomenon will aid in the development of new approaches, especially for the formation of complex metamorphic for uses as composite materials, bone replacement, device fabrication in microelectronics and optoelectronics.

2. Materials and Method

The *Paphia undulate* and *Amusium pleuronectes* samples were obtained from the coastal area of Chonburi province, the eastern part of Thailand. First of all the soft parts inside the marine mussel samples were removed after opening the bivalve of the shell. The bivalve shell was brushed and washed with distilled water. Afterward the bivalve samples were cleaned by soaking in dilute HCl solution for several minutes to remove all organic materials, followed by washing thoroughly with distilled water. The cleaned sample was ground into fine powder to desired sizes by sieves and labels. The powder samples are denoted as M-CL and M-SC where the labeling numbers indicate *Paphia undulate* (Carpet Clam) and *Amusium pleuronectes* (Radoated Scallop), respectively.

The crystal structure was characterized by a powder X-ray diffractometer. Powder diffraction data were recorded at room temperature using a powder diffractomter, Bruker AXS D8 Advance [CuKα (Ni filtered) with scintillation detector; 2θ range from 20-80° and step size of 0.02°]. The diffraction data were analyzed by the Reitveld method using the computer program FULLPROF SUITE-2000. Therefore, this method was used to estimate the mineral percentage in the sample. Another that, all shell samples were cut and polished for Scanning Electron Microscopy. The morphology were characterized by Field-emission scanning electron microscopy (FE-SEM , Hitachi S-4700, Japan)

The Metal ions in the exoskeletons were studied by ESR spectrometer. All ESR spectra of the powder samples were measured on Bruker E500 CW ESR spectrometer operating at X-band microwave frequency of 9.8 GHz using 0.5 mT field modulation amplitude with a time constant of 0.03 s.

3. Results and Discussion

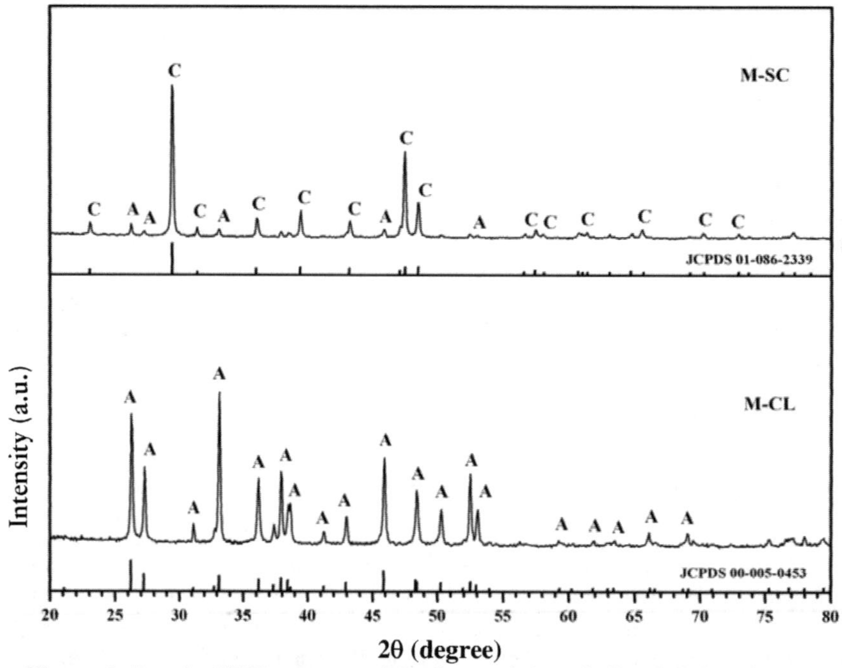

Figure 1. Powder XRD patterns of *Paphia undulate* shell (M-CL) and *Amusium pleuronectes* shell (M-SC) samples

The crystal structure of *M-CL* shell samples was made of aragonite phase, a common form of $CaCO_3$ mineral. For *M-SC* shell has a mixture phase of aragonite and calcite as shown in Figure 1. The mineral phases of the bivalve samples all species were identified from the diffractograms by reference to the Joint Committee on Powder Diffraction Standard (JCPDS) numbers 00-005-0453, and 01-086-2339 corresponding to aragonite phase and calcite phase respectively.

The powder X-ray diffraction pattern of M-CL in Figure 2 shows that the crystal structure of *M-CL* shell sample was made of a pure aragonite phase, a common form of calcium carbonate ($CaCO_3$) mineral. The X-ray diffraction patterns of *M-SC* shell sample in Figure 3 exhibits peak characteristics of calcite and aragonite. For all refinements, lattice parameters and phase fractions of M-SC sample are in good agreement. The percentage of calcite and aragonite is 94 % and 6%, respectively.

Figure 4 shows the SEM micrographs of the *M-CL* shell. It is seen that the *M-CL* shells composed of aragonite plates with the porous scattered all over the shell sample, a rod-like shape of $CaCO_3$ is obtained in these SEM images, which is the typical shape of aragonite. For *M-SC* shell samples, the columnar structures of mixture phase of aragonite and calcite can be observed in Figure 5. It is clear that amount of aragonite and calcite material adheres to the SEM micrographs of the shell.

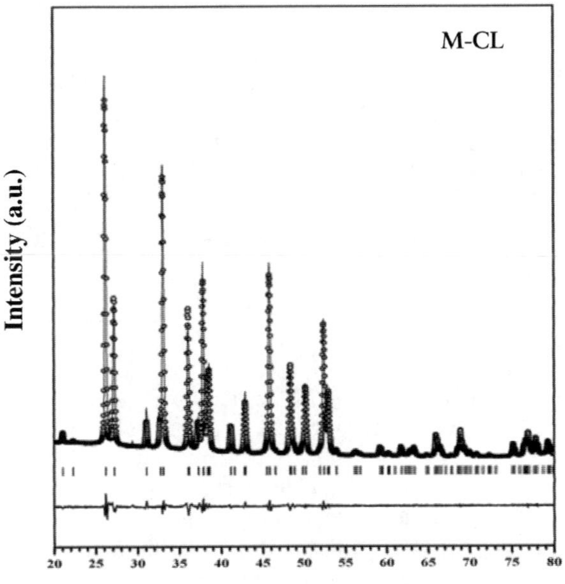

Figure 2. Difference X-ray diffraction profiles for the Rietveld refinement of *Paphia undulate* shell (M-CL) sample. Experimental (○mark), calculated (solid line) and difference (bottom) powder diffraction profile. Rwp = 8.17, $\chi 2 = 1.37$.

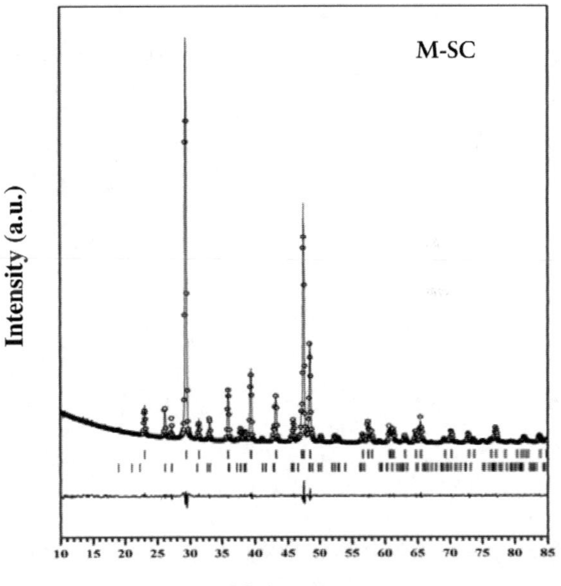

Figure 3. Difference X-ray diffraction profiles for the Rietveld refinement of *Amusium pleuronectes* (M-SC) shell sample. Experimental (○mark), calculated (solid line) and difference (bottom) powder diffraction profile. Rwp = 7.31, $\chi 2 = 1.31$.

Figure 4. The figure and microstructure of *Paphia undulate* (M-CL).

Figure 5. The figure and microstructure of *Amusium pleuronectes* (M-SC).

The ESR spectra of the *M-CL* shell samples are shown in Figure 6. The six hyperfine lines are obtained due to a trace of the Mn^{2+} ions present in the carbonate of aragonite structure. The ground state $^6S_{5/2}$ in the absence of an external magnetic field splits due to spin-spin interaction, into three Kramer's doublets with separations of 4D and 2D. In the presence of an applied magnetic field along the symmetry axis, these Kramer's doublets split further and the energy between successive levels, correct to first order, is $g\beta B+4D$, $g\beta B+2D$, $g\beta B$, $g\beta B-2D$, $g\beta B-4D$, with the result that they give rise to five equally spaced resonance. Since the nuclear spin of ^{55}Mn is 5/2 each of these five lines split further to a sextet and hence one can observe 30 line pattern per molecule per unit cell.

The broad absorption lines in *M-CL* shell could be due to low symmetry environment. The ESR spectra of *M-SC* shell shown in the Figure 7. The ESR spectra of *M-SC* shell, the Mn^{2+} ions present in the carbonate deviate from the higher symmetry environment. Each spectrum consists mainly of six strong Mn^{2+} lines are more intense. The spectrum is obviously attributed to the $m_S = -1/2 \leftrightarrow +1/2$ sextet fine structure transition of Mn^{2+} ions (electron spin $S = 5/2$ and nuclear spin $I = 5/2$) according to the selection rule $\Delta m_S = \pm 1$, $\Delta m_I = 0$. The weaker pairs of peaks between the main peaks are so called the forbidden transitions in which both electron and nuclear spin state change, each Mn^{2+} peak splits into a doublet, which results from the formation of calcite phase. This in equivalence gives rise to a doublet splitting of each fine-structure transition. The ESR spectra results are in good agreement with those of XRD and SEM as previously described.

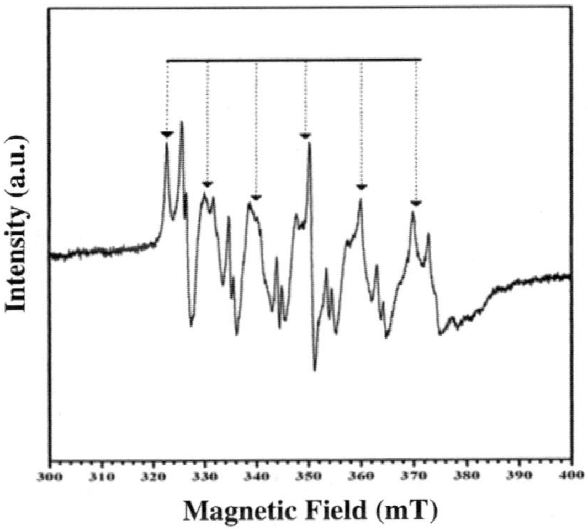

Figure 6. X-band ESR spectra of *Paphia undulate* (M-CL)shell samples in the magnetic field range 300-400 mT .

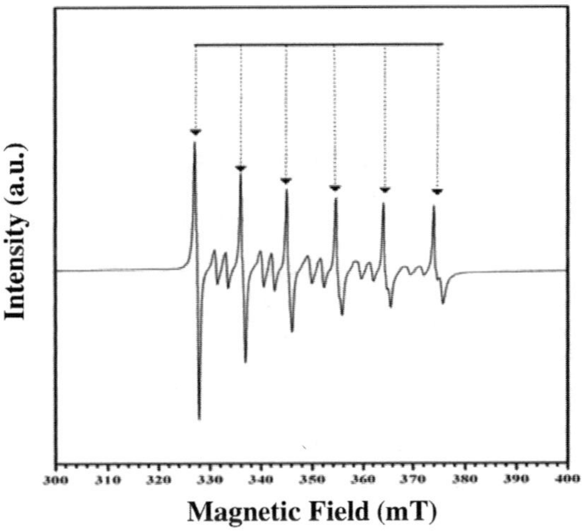

Figure 7. X-band ESR spectra of *Amusium pleuronectes* (M-SC) shell samples in the magnetic field range 300-400 mT.

4. Conclusion

The *M-CL* shell and *M-SC* shell samples obtained from the coastal area of Chonburi province, the eastern part of Thailand consists mainly of $CaCO_3$. The aragonite phase is not the only type of calcium carbonate crystal to make up the shell structure in mussel. The calcified part of the shell consists of two types of calcium carbonate crystals, either pure aragonite and mixture phase between aragonite and calcite. Moreover, the ESR spectroscopy can be profitably used to examine information concerning structural organization. Then biomineralized can provide unusual metal ion environments, related to heavy metals control in coastal areas involve monitoring a series of metal in water, which to study crystal structure on transition metal ions. The information from ESR spectroscopy microstructure is one of the most significant relative data for clarification of evolutionary trends of shell structure and assessment the potentiality of using the mussel shell as biomonitors of heavy metal pollution.

Acknowledgement

This work supported by Science, Technology and Innovation (STI), Burapha University. The authors would like to thanks Center of Excellence in Glass Technology and Material Science (CEGM), NPRU for facilities.

5. References

[1] Narasimhulu K V and Rao J L 2000. *Spectrochim. Acta. Part A* 56 1345
[2] Vongsavat V Winotai P and Meejoo S 2006. *Nucl Instrum Meth. Part B* **243** 167
[3] Rao J L Narasimhulu K V Gopal N O Linga Raju C H and Reddy B C V 2003 *Spectrochim. Acta. Part A* **59** 2955
[4] Rizzuti A Leonelli C 2008 *Powder Technol* **186** 255

2nd Int. Symp. on Advanced Magnetic Materials and Applications (ISAMMA 2010) IOP Publishing
Journal of Physics: Conference Series **266** (2011) 012125 doi:10.1088/1742-6596/266/1/012125

Hysteresis loop design by geometry of garnet film element with single domain wall

V A Skidanov[1], **P M Vetoshko and A L Stempkovskiy**
Institute for Design Problems in Microelectronics RAS
Sovetskaja street, 3, Moscow 124681, Russia

E-mail: skidanov@ippm.ru

Abstract. Numerical modeling and experimental investigation of magnetostatic stable states of two-domain structure in Bi-substituted uniaxial garnet film elements was made. Single domain walls (DW) between two opposite normally magnetized parts in isolated rectangular strip and strip-like bridge are found to exhibit different behavior. DW inside strip (bridge) suffers increasing repulsion (attraction) from nearest edge when shifted from element center. DW position center position is stable in isolated strip but bridge is magnetized spontaneously to one of two saturated states in zero external field. Isolated strip magnetization process occurs reversibly while bridge magnetization reversal occurs by coercive manner. Strip susceptibility and bridge coercive field are entirely defined by magnetostatic barrier created by element boundary stray field in case of constant DW length during magnetization reversal. Variation of strip and bridge boundary shape along DW trajectory gives the opportunity to create additional controllable potential profile due to DW surface energy modulation by DW length. Garnet elements with high Faraday rotation and low light switching field were developed for fine magnetic sensing and optical data processing applications.

1. Introduction

Bi-substituted epitaxial garnet films $(BiLu)_3$ $(FeGa)_5$ O_{12} with uniaxial anisotropy and high Faraday rotation are expected for years to be used for optical data processing devices. In most cases garnet film elements in magnetooptical matrix are found in monodomain state which is known to be rather coercive. Magnetization reversal requires external field which is comparable with uniaxial garnet anisotropy field and reaches values and spreads up to few hundred Oersteds which are unacceptable for practical applications [1].

But if garnet magnetization M is small enough to maintain monodomain state of the film or element size is more than domain wall (DW) width but less than domain dimension element can be divided into two or more stable domains. This case it can exhibit entirely different behaviour in magnetization process. Simultaneously transition to monodomain state can be closed by high energy barrier. Such elements can be fabricated by lithography and etching of uniaxial garnet film with partial mutual compensation of magnetizations of tetrahedral and octahedral iron sublattices.

It is known that compensation of sublattices magnetization isn't followed by Faraday rotation compensation [2,3]. So such garnet elements could be used for various types of magnetooptical devices.

---

1 Vladimir Skidanov, Institute for Design Problems in Microelectronics RAS,
Sovetskaja st. 3, Moscow 124681, Russia, phone: +7-499-735-3872, E-mail: skidanov@ippm.ru

The purpose of present work is numerical modelling and experimental magnetooptical investigation of magnetization process of two-domain configurations in uniaxial garnet film elements and estimation of magnetization reversal field possible control and reduction to value suitable for applications.

2. Magnetization by constant length DW motion

2.1. Numerical modeling

Numerical modeling of magnetostatic stable states of two-domain structure in rectangular elements etched in uniaxial garnet film with arbitrary thickness h was made. DW is proposed straight and connecting opposite element's boundaries. DW position is defined this case only by balance of external magnetic field and own stray field created by garnet element boundary. Stray field was calculated at domain wall center in dependence upon DW position inside element. Element edges were considered vertical in relation of substrate plane. Stray field distribution of such an element is calculated as magnetic field created by electric surface currents Mh flowing along element contour and $2Mh$ flowing along DW [4]. Domain wall surface energy is high enough to prevent DW bending.

Common view of stray field distributions is shown in figure 1.

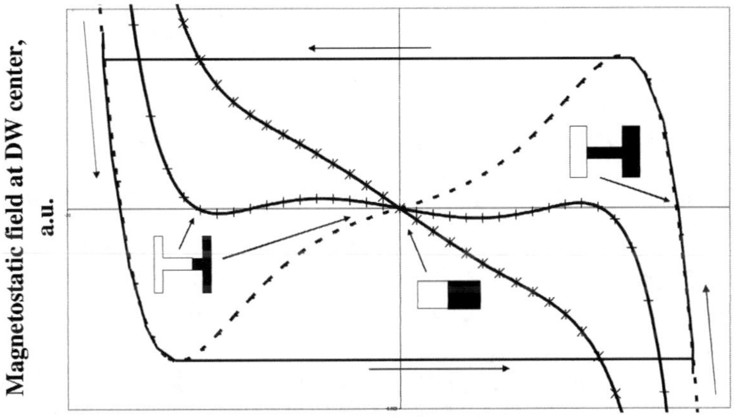

Figure 1. Stray field distributions scheme in uniaxial garnet film elements with constant DW length during magnetization reversal: isolated strip, bridge between two separated half-planes, bridge connecting two parallel strips. DW stable positions in zero external field are marked by arrows. Dashed line represents DW unstable positions.

Isolated strip and strip-like bridge between two opposite normally magnetized infinite half-planes are found to exhibit different behavior. Domain wall (DW) inside strip (bridge) suffers increasing repulsion (attraction) from nearest edge when shifted from center. DW position is strictly stable in isolated strip but bridge is magnetized spontaneously to one of two saturated states in zero external field. Isolated strip magnetization process occurs reversibly while bridge magnetization occurs by coercive manner. Coercive force is entirely defined by magnetostatic barrier created by garnet film boundary. Strip susceptibility and bridge coercive field are found to be functions of strip length, width and film magnetization M and thickness h.

Hybrid element made as a bridge between two narrow strips is found to possess an indefinite DW equilibrium position which is highly sensitive to external field. DW repulsion by outer hybrid element vertical boundaries can be compensated in wide region inside bridge by attraction DW by inner element vertical boundaries.

It should be noticed calculation results in figure 1 are represented with 90 degrees rotation in comparison to usual hysteresis loop presentation due to calculation algorithm peculiarities.

2.2. Experimental results

Bi-substituted uniaxial garnet films with magnetization M in the range 25 - 50 Gauss were grown on GGG substrate with (111) orientation. Various types of elements were etched chemically in garnet films. Matrix of quadratic elements (a) and bridges between parallel strips (b) are shown in figure 2.

Single DW in element was created by garnet film cooling from temperature exceeding Curie point to room temperature in external magnetic field gradient ~ 0.01 Oe/mcm with microscope remnant and Earth magnetic field compensation. Magnetization process was investigated in quasi-static regime.

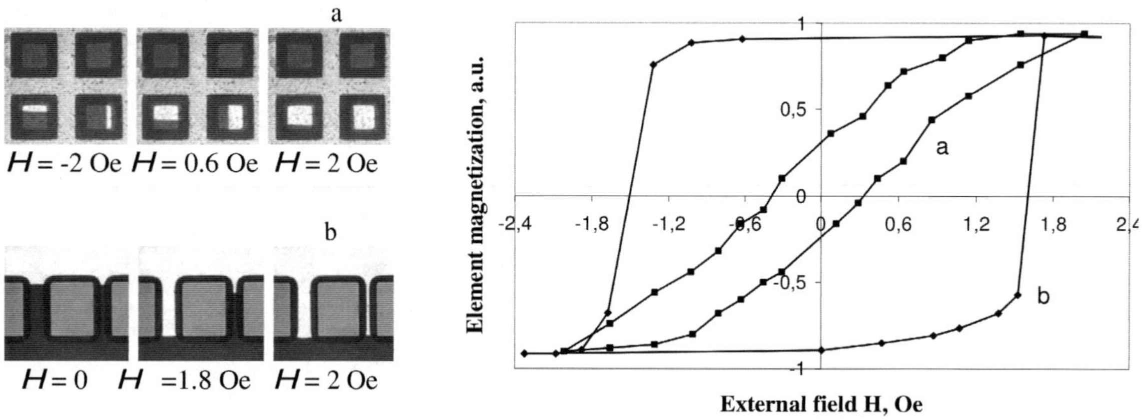

Figure 2. Experimental microphotographs of DW positions at critical points and magnetization reversal loops: a - for quadratic elements 30x30 mcm in matrix, b - for rectangular bridges 40x120 mcm and 10x120 mcm. Garnet magnetization M = 50 Gauss, thickness h = 3 mcm.

Magnetization reversal curves shown in figure 2 confirm qualitatively theoretical predictions. Quadratic elements in matrix in figure 2a are magnetized by almost reversible manner with coercive field ~ 0.4 Oe which equals to own coercive force of garnet film. Positions of DWs in different elements and different orientations are reproducible so that the spread of reversal parameters doesn't exceed garnet coercive field.

Bridge between two half-planes with opposite magnetization directions demonstrates almost rectangular hysteresis loop with reproducible coercive field defined by magnetostatic stray field of garnet film boundaries.

Magnetization reversal by constant length DW motion occurs at very small external field in the range (1 – 10) Oe. Switching field turned out two orders less than for monodomain state and few times less than predicted.

Unlike assumption for theoretical calculation garnet elements edges have very smooth wedge ~ 10 degrees to substrate plane. Epitaxial garnet film wedge is seen as dark element contour in microphotographs in figure 2. Wedge is formed due to great difference between etching speed values along normal to (111) crystal plane and along other crystal directions. Since wedge width exceeds significantly garnet film thickness element boundary stray field is reduced in comparison to calculated value. It is useful effect from one sight since it leads to switching field reduction. From other sight garnet film wedge restricts space resolution in microelectronics applications. In addition wide wedge reduces stability margin of two-domain element state. This effect is especially harmful for rectangular strip due to irreversible collapse of compressed domain at element edge.

3. Magnetization by variable length DW motion

3.1. Numerical modeling

Variation of DW length during magnetization gives an additional opportunity for hysteresis loop design due to high surface energy of domain wall which is proportional to DW length. So gradient of element width creates constant effective field barrier for DW motion in external field. DW equilibrium position coincides with element width minimum. Effective field barrier value is proportional to inclination angle tangent. Shapes of calculated magnetization curves corresponding to concave and convex elements are shown schematically in figure 3. Calculation was made for straight DW and negligible own coercive force of garnet film.

Hysteresis type of magnetization reversal is obvious in both concave and convex elements. But concave element shows distinct barrier position in demagnetized state. Such construction can be used as a threshold element in optical data processing.

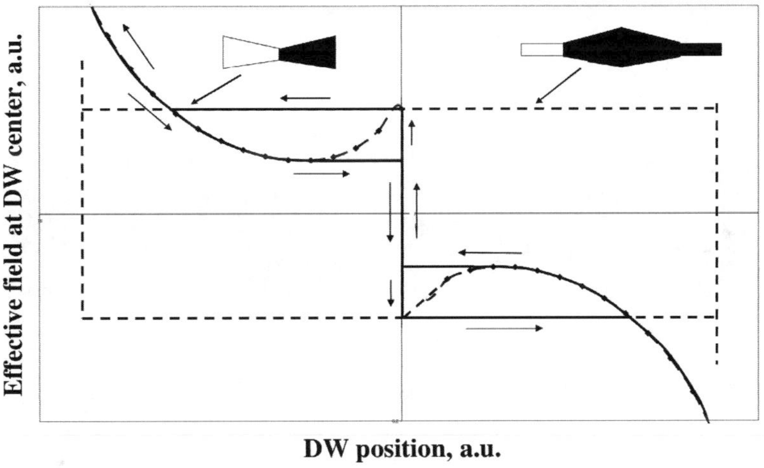

Figure 3. Scheme of effective field distributions in uniaxial garnet film elements with variable DW length during magnetization reversal. Concave (solid line) and convex (dashed line) elements have equal inclination angles.

3.2. Experimental results

Microphotographs of DW stable positions during concave element magnetization reversal are presented in figure 4. Magnetization curve seems qualitatively similar to calculated shape. Loop slope and narrow hysteresis at low field region are cased by DW bending and garnet film coercive force accordingly. Hysteresis at magnetization curve edge is caused by magnetostatic barrier formed by inclined element boundaries adjacent to element narrowest section. Magnetostatic barrier profile is asymmetrical as it is shown in scheme in figure 3 by dashed parts of magnetization curve for concave element. Local hysteresis loop width is defined by garnet magnetization and inclination angle while critical field value depends upon DW energy as well.

Magnetization reversal of convex rhomb-like elements occurs by rectangular hysteresis loop in accordance with calculated curve. Microphotographs of magnetization states in rhomb-like elements at zero external field and critical external fields H_{c1} and H_{c2} just before and after DW leaps are shown in figure 5. Ratio of critical field values coincides with ratio of elements inclination tangents in accordance with theoretical predictions.

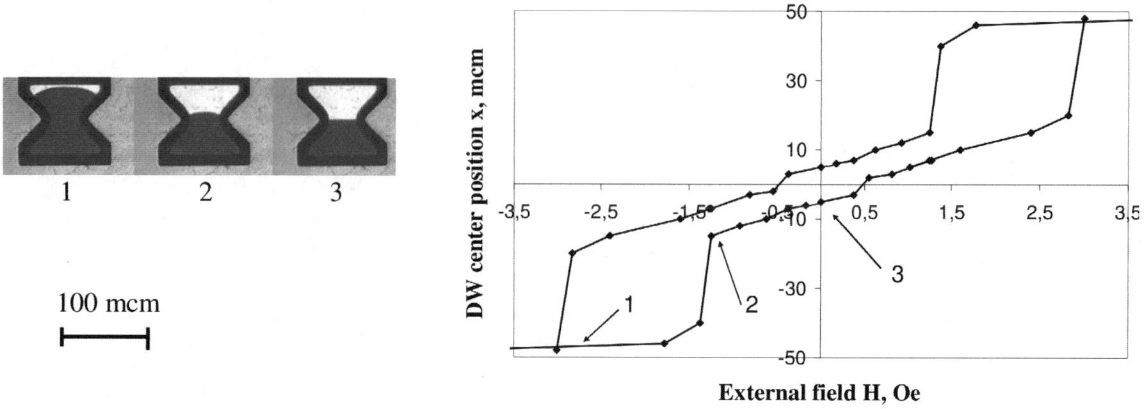

Figure 4. Experimental microphotographs of DW stable positions at critical points and magnetization reversal loop in concave garnet element. $M = 50$ G, $h = 3$ mcm.

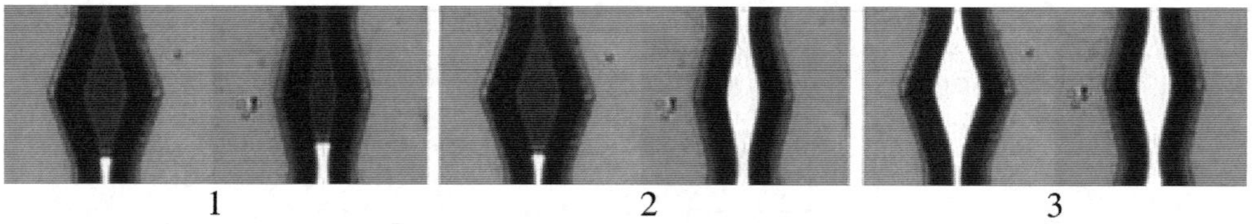

Figure 5. Experimental microphotographs of magnetization states in convex rhomb-like garnet elements near critical external field values: 1 – $H = 0$ Oe, 2 – $H_{c1} = 4$ Oe, 3 – $H_{c2} = 5.5$ Oe. Initial white domain width is 10 mcm. Wedge width is 15 mcm. $M = 50$ G, $h = 3$ mcm.

4. Conclusion

Two-domain stable state of uniaxial garnet film element with low magnetization exhibits hysteretic properties which can be effectively controlled by element shape. Switching field values are defined by garnet magnetization and element shape and dimensions and can be reduced by two orders of magnitude in comparison with monodomain element switching field. Nonlinear curvature of element contour will ensure additional possibility for design useful properties of magnetooptical elements due to smooth variation of domain wall surface energy.

To improve switching parameters and element space resolution significant reduction of garnet film wedge formed during chemical etching is required.

Due to high Faraday rotation and low stable switching field values uniaxial garnet film elements in two-domain states are promising objects for application in optical data processing and fine magnetic sensing.

5. References

[1] Zvezdin A K and Kotov V A 1997 *Modern Magnetooptics and Magnetooptical Materials* (Bristol and Philadelphia: IOP Publishing)
[2] Aichele T, Lorenz A, Hergt R, Gornert P 2003 *Crystal Research and Technology* **38** 575
[3] Sugano S and Kojima N 2000 *Magnetooptics* (Berlin: Springer)
[4] O'Dell T H 1974 *Magnetic Bubbles* (Bristol: Macmillan Press Ltd)

Journal of Physics: Conference Series **266** (2011) 012126 doi:10.1088/1742-6596/266/1/012126

Irradiation Induced Modifications in Magnetic Property of Mn/n-Si Structure

M K Srivastava, P S Pandey and P C Srivastava*
Department of Physics, Banaras Hindu University, Varanasi-221 005, India

*Corresponding Author: Tel: +91 9839234757; Fax: +91 542 2368390
E-mail Address: pcsrivastava50@gmail.com (P. C. Srivastava)

Abstract:
Mn films of ~50 nm were deposited on n-Si (100) substrates by electron beam evaporation technique. The Mn/n-Si structures were irradiated from swift heavy ions with a fluence of $1x10^{13}$ ions/cm^2. The irradiated and unirradiated structures have been characterized from atomic force microscopy, X-ray diffractometry and vibrating sample magnetometer facilities. It has been found that surfacial/interfacial granular magnetic silicide phases (of Mn_xSi_y) are formed before and after the irradiation with irradiation induced modifications of surface morphology and magnetic property. The surface roughness has been found to decrease on the irradiation from the atomic force microscopy data. From the X-ray diffraction data, it has been found that after the irradiation MnSi is formed in addition to Mn_5Si_2 as compared to unirradiated ones. The magnetization characteristics show that the magnetic parameters such as coercivity, saturation magnetization, remanance and squareness have decreased on the irradiation for in-plane orientation whereas coercivity and remanance is increased after the irradiation for out of plane orientation. The increase of remanance shows the presence of strong exchange coupling in the structure after the irradiation.

1.Introduction:

Recently, magnetic thin films on semiconductor substrates have attracted considerable attention due to its significant role in spintronics. Utilizing the spin of the electron in semiconductor devices holds great potential to provide high-speed device structures [1]. Spin injection from a magnetic metal to a semiconductor is one of the significant key processes for spintronics [2]. For such applications, magnetic metal/Si heterojunctions have attracted much attention and would be highly desirable to grow epitaxial ferromagnetic films on Si substrates. There have been some reports on this issue, concerning the epitaxial growth of Fe on GaAs, and the injection of spin-polarized electrons from a metal into a semiconductor [3, 4]. However, such systems are also susceptible to intermixing which causes the formation of nano cluster magnetic phases whose magnetic properties are of fundamental and technological significance.

Manganese is a magnetic element and reacts easily with Si to form manganese silicide. Among the manganese silicides, attention has been focused on the MnSi, Mn_5Si_2 and Mn_5Si_3. MnSi [5] is ferromagnetic with a curie temperature of 30K, Mn_5Si_3 [6] is antiferromagnetic, Mn_5Si_2 [6] is known to be paramagnetic and $MnSi_{1.73}$ is a higher manganese silicide (HMS) of $MnSi_{2-x}$ (x: 0.25 ~ 0.27) which is a narrow band gap semiconductor. Magnetic properties of these grown or deposited silicides are strongly dependent on microstructure and hence the magnetic properties can be tailored by modifying the microstructure. Swift heavy ion irradiation is an excellent tool for tailoring [7] the microstructure as a result of irradiation induced thermal spike phenomenon [8]. Thermal spike phenomenon of the irradiation results in the recrystallization and growth phenomena in the irradiated structures to modify the microstructures [9]. Ion beam irradiation is also known to modify the extrinsic magnetic properties, such as magnetic anisotropy, coercivity and magnetic exchange coupling [10-14].

In order to achieve the growth of these silicides on silicon substrate and to investigate the magnetic properties of these magnetic silicides interfacing with silicon, the Mn/n-Si interfaces have been prepared and irradiated from swift heavy ions with a fluence of $1x10^{13}$ ions/cm^2. The structures have been characterized from atomic force microscopy (AFM), X-ray diffractometry (XRD) and vibrating sample magnetometer (VSM) to investigate the magnetic property in relation to the microstructures.

2.Experimental details:

n-Si (100) wafers of 8-10 Ω-cm resistivity have been used for the fabrication of Mn/n-Si structure. n-Si substrates have been degreased by ultrasonic cleaning in trichloroethylene (TCE) solution and then etched by diluted HF for 15 sec. and then dried in a clean air chamber for the metallization. A thin layer (of ~ 50 nm) of Mn (99.99 % purity) was deposited on the n-Si substrates in a 1 cm x 1 cm area. The above fabricated Mn/n-Si interface structures were irradiated at room temperature from the Mn layer side (fronting the beam) by 100 MeV Fe^{7+} ions with a fluence of ~$1x10^{13}$ ions/cm^2, at Inter University Accelerator Centre, New Delhi from 15 UD pelletron facility. The ion beam was made incident 30° off-normal upon the samples during the irradiation for the wider exposure of the sample surface. The size of the beam spot was around 2.5 mm x 2.5 mm which was constantly scanned on the sample surface during the irradiation. The typical beam current was varied between 2 to 3 pnA (particle nano ampere).

3.Results and discussion:

Figure 1 shows the AFM micrograph (with x-y, 0.2 μm/div. and z, 100 nm/div.) of unirradiated Mn/n-Si structure of the top side of manganese film on silicon substrate. The observed feature in the micrograph shows the grainy microstructure where the grains are observed to be overlapping with each other. Figure 2 shows the AFM feature (with x-y, 0.2 μm/div. and z, 100 nm/div.) for the irradiated structure which also shows nice ellipsoidal shaped granular particles. The grains are well resolved without overlapping which could be formed due to the irradiation induced melting and recrystallization phenomena [9]. Moreover, the grains seems to be embedded in silicon substrate.

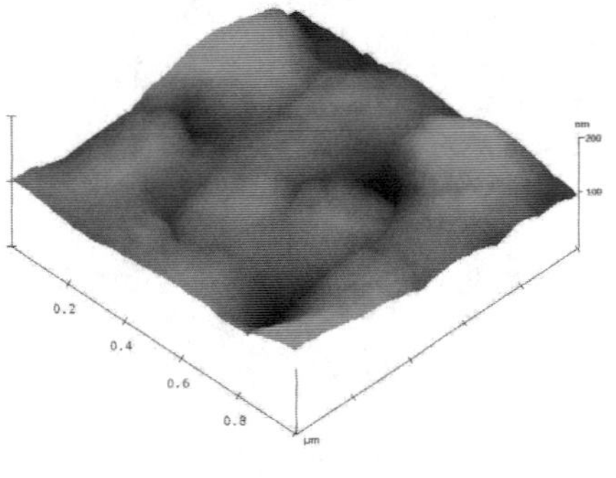

Figure 1 AFM micrograph of unirradiated structure.

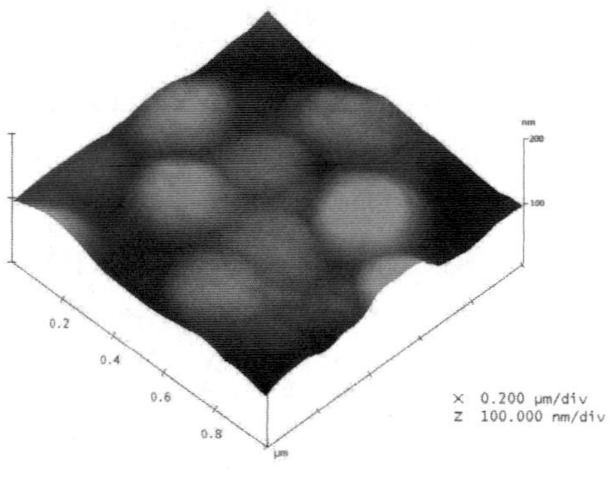

Figure 2 AFM micrograph of irradiated structure.

AFM image clearly shows the decrease in roughness after the irradiation. It is found to decrease from 15.3 nm to 7.8 nm after the irradiation as compared to unirradiated ones. Moreover, roughness is a measure of the texture of a surface. It is quantified by the vertical deviations of a

Journal of Physics: Conference Series **266** (2011) 012126 doi:10.1088/1742-6596/266/1/012126

real surface from its ideal form. If these deviations are large, the surface is rough; if they are small the surface is smooth. The AFM features of the irradiated ones (figure 2) clearly shows that grains are well shaped, resolved and there is no protrusion in horizontal or vertical direction on the scale of 1μm x 1μm. It clearly appears from figures 1 and 2, that the surface roughness is reduced after the irradiation (in figure 2 as compared to figure1). The smoothening mechanism on the surface during ion irradiation include volume diffusion, viscous flow, sputter redeposition and sputter removal affected by shadowing processes [15].

In order to identify the formed silicide phases, XRD data have also been recorded. For XRD, Cu alpha 1 (λ = 1.54439 Å) is used as an x-ray target. XRD pattern for the unirradiated and irradiated structures are shown in figure 3 and figure 4, respectively. The data of figure 3 shows two distinct peaks of Mn_5Si_2 whereas the data of figure 4 show that there are five distinct peaks of Mn_5Si_2/MnSi. It is further observed that the intensity of Mn_5Si_2 (213) has increased and the intensity of Mn_5Si_2 (331) has decreased after the irradiation as compared to unirradiated ones. Moreover, it is also interesting and significant to observe the evolution of two new peaks of MnSi after the irradiation, i.e., MnSi (310) and MnSi (321). The evolution of new phases of manganese silicides can be understood due to the irradiation induced interfacial intermixing. The crystallite size estimated from Scherrer relation [16] shows a decrease of crystallite size after the irradiation. For one of the peak it is found to decrease from 310 nm to 161 nm after the irradiation. The observed features from AFM and XRD could be understood in the realm of irradiation induced strong interfacial intermixing to cause the microstructural modifications. Swift heavy ion induced interfacial modifications has been earlier shown by Bolse [17].

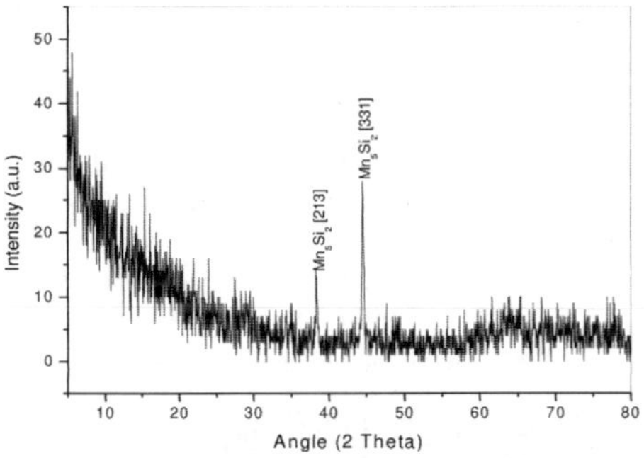

Figure 3 XRD pattern of unirradiated structure.

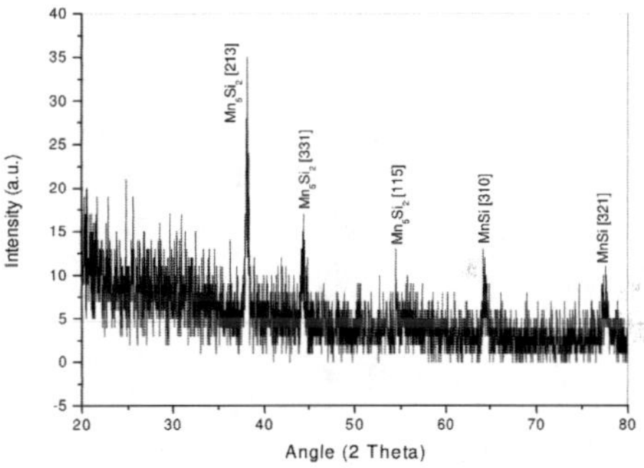

Figure 4 XRD pattern of irradiated structure.

The magnetization characteristics of the structure (i.e., Mn film deposited onto n-Si substrate) have been measured from vibrating sample magnetometer (VSM). The VSM measurements have been done at room temperature. Figures 5a, b, 6a and b show the M-H behavior for the unirradiated and irradiated ones for the applied magnetic field parallel and perpendicular to the plane of the structure, i.e., along and across the interface. The parameters of magnetization behavior are listed in table 1.

It is interesting to observe that in-plane seems to be the easy axis of magnetization whereas out of plane seems to be the hard axis of magnetization. It is also interesting to observe room temperature ferromagnetism in the unirradiated/irradiated structures of Mn/Si. The origin of ferromagnetism in the manganese silicide is known to be due to the small amount of carbon atoms incorporated into the paramagnetic Mn_5Si_2 [18]. In the case of silicon crystals, there is always presence of some carbon impurities which may play role to cause the observed room temperature ferromagnetism. The dissolved carbon in our silicon substrate has been found by our several EDAX and IR studies in various contexts [19]. The origin of ferromagnetism has been understood by Takeuchi et al. [20] as the effect of the anisotropic modifications of the local structure around the manganese sites by the carbon (C) atom. It is possible that the irradiation may change such modifications to change the magnetic property as observed for the irradiated structure.

Table 1 Magnetic parameters of Mn/n-Si structure

Magnetic parameters	In-plane orientation of Mn/Si		Out of plane orientation of Mn/Si	
	Unirradiated	Irradiated ($1x10^{13}$ ions/cm^2)	Unirradiated	Irradiated ($1x10^{13}$ ions/cm^2)
Hc (Oe)	80.67	69.45	189.69	237.13
Ms (emu)	6.6×10^{-4}	2.5×10^{-4}	5.2×10^{-4}	3.1×10^{-4}
Mr (emu)	7.9×10^{-5}	1.9×10^{-5}	-4.7×10^{-6}	6.9×10^{-6}
S= Mr/Ms	0.119	0.077	0.077	0.023

However, the saturation magnetization is reduced after the irradiation for both the orientations of the applied magnetic field. This can be understood by evolvement of the MnSi silicide phase, which is known

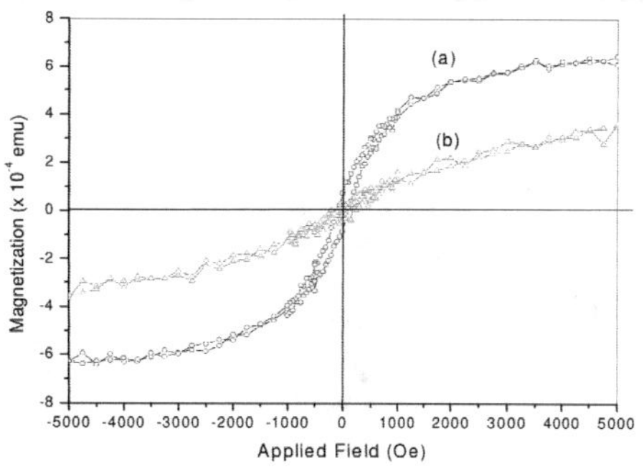

Figure 5 M-H characteristics of unirradiated structure for in-plane magnetic field (a), and for out of plane magnetic field (b).

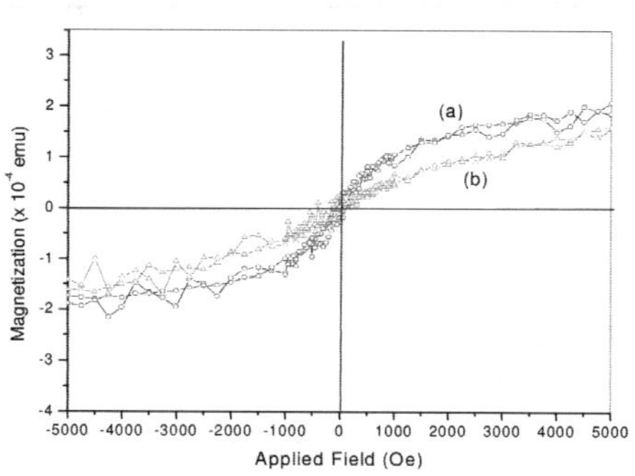

Figure 6 M-H characteristics of irradiated structure for in-plane magnetic field (a), and for out of plane magnetic field (b).

to be paramagnetic at room temperature. There is a possibility that due to the enhancement of paramagnetic phase there is a reduction in the saturation magnetization after the irradiation. Moreover, for in-plane orientation, other magnetic parameters such as coercivity, remanence and squareness are found to decrease after the irradiation, whereas, for out of plane magnetization, coercivity and remanence is found to increase on the irradiation. The increase in coercivity and remanence for the out of plane magnetization can be understood by considering the out of plane oriented grains of the magnetic silicide phases. The AFM features of the irradiated structures (figure 2) reveals such possibility. Moreover, the AFM features clearly show the shape anisotropy which shall result the observed differences for in plane and out of plane magnetizations.

4.Conclusion:

It is found that ion beam irradiation modifies the surface morphology and magnetic property of Mn on Si substrate. Surfacial/interfacial granular/ellipsoidal shaped silicide phases have been observed to grow on the irradiation with paramagnetic Mn_5Si_2 and ferromagnetic MnSi silicide phases. Surface roughness has been found to decrease on the irradiation. The microstructural changes on the irradiation have been understood in the realm of swift heavy ion induced interfacial intermixing processes. The irradiation induced changes of coercivity and remanence seems to be due to the oriented grains of MnSi silicide phase which are formed on the irradiation.

Acknowledgements:
The authors are thankful to Mr. D. Kabiraj of IUAC, New Delhi for his excellent co-operation during the irradiation experiments. Financial support from DST and UGC is gratefully acknowledged.

References:

[1] Wolf S A, Awschalom D D, Buhrman R A, Daughton J M, Molnar S von, Roukes M L, Chtchelkanova A Y and Treger D M, 2001 *Science* **294** 1488
[2] Prinz G A, 1998 *Science* **282** 1660
[3] LaBella V P, Bullock D W, Ding Z, Emery C, Venkatesan A, Oliver W F, Salamo G J, Thibado P M and Mortazavi M, 2001 *Science* **292** 1518
[4] Zhu H J, Ramsteiner M, Kostial H, Wassermeier M, Schönherr H –P and Ploog K H, 2001 *Phys. Rev. Lett.* **87** 016601
[5] Kwon D, Kim H K, Kim J H, Ihm Y E, Kim D, Kim H, Baek J S, Kim C S and Choo W K, 2004 *J. Magn. Magn.Mater.* **282** 240
[6] Takano F, Akinaga H, Ofuchi H, Kuroda S and Takita K, 2006 *J. Appl. Phys.* 99, 08J506
[7] Bolse W, 2006 *Nucl. Instrum. Methods B* **244** 8
[8] Wang Z G, Dufour C, Paumier E and Toulemonde M, 1994 *J. Phys.:Condens. Matter* **6** 6733
[9] Srivastava P C, Ganesan V and Sinha O P, 2004 *Nucl. Instrum. Methods Phys. Res. B* **222** 491
[10] Chappert C, Bernas H, Ferré J, Kottler V, Jamet J-P, Chen Y, Cambril E, Devolder T, Rousseaux F, Mathet V and Launois H, 1998 *Science* **280** 1919
[11] Veres T, Cai M, Cochrane R W and Roorda S, 2000 *J. Appl. Phys.* **87** 8504
[12] Cai M, Veres T, Roorda S, Cochrane R W, Abdouche R and Sutton M, 1997 *J. Appl. Phys.* **81** 5200
[13] Mewes T, Lopusnik R, Fassbender J, Hillebrands B, Jung M, Engel D, Ehresmann A and Schmoranzer H, 2000 *Appl. Phys. Lett.* **76** 1057
[14] Fassbender J, Ravelosona D and Samson Y, 2004 *J. Phys. D: Appl. Phys.* **37** R179
[15] Dash P, Mallick P, Rath H, Tripathi A, Prakash J, Avasthi D K, Mazumder S, Verma S, Satyam P V and Mishra N C 2009 *Applied Surface Science* **256** 558
[16] Eberhart J P, 1989 *Analyse Structurale et Chimiques des Materiaux (Paris: Bordas)* **46** 407
[17] Bolse W, 2003 *Radiation Measurements* **36** 597
[18] Wang W, Takano F, Akinaga H and Ofuchi H, 2007 *Phys. Rev. B* **75** 165323
[19] Tripathi J K 2006 *A Study on Magnetic Metal/Silicon Interface Devices; with and without Swift Heavy Ion Irradiation, Ph.D. Thesis: Department of Physics, Banaras Hindu University, Varanasi, India*
[20] Takeuchi T, Igarashi M, Hirayama Y, and Futamoto M, 1995 *J. Appl. Phys.* **78**, 2132

Journal of Physics: Conference Series **266** (2011) 012127 doi:10.1088/1742-6596/266/1/012127

High thermal stability of the ferrimagnetic moment in exchange biased FeO(core)/spinel(shell) nanocubes

H.T. Hai*, H. Kura, M. Takahashi and T. Ogawa
Graduate School of Engineering, Department of Electronic Engineering, Tohoku University, 6-6-05 Aza-Aoba, Aramaki, Aoba-ku, Sendai 980-8579, Japan

E-mail: phai@ecei.tohoku.ac.jp

Abstract. We demonstrate herein a basic approach to a new type of exchange-coupled iron oxide-based core/shell nanocubes consisting of an antiferromagnetic (AFM) core of ferrous oxide (wustite, FeO) surrounded by a ferrimagnetic (FM) shell of ferrite (spinel, γ-Fe_2O_3 or Fe_3O_4). By coupling with AFM core, effective blocking temperature of FM moments is strongly enhanced up to Néel temperature (T_N =198 K) of AFM core. This is assigned to contribution of an extra anisotropy induced by exchange bias at the interface between AFM and FM components in the nanocubes. Because of this special coupling structure, the high stability of FM moments can be achieved even at very small volume fraction of AFM with respect to FM matrix, thus significantly eliminate contribution of AFM to the magnetization of whole coupling system, suggesting a new hybrid structure for magnetic devices applications.

1. Introduction

Interest in magnetic nanoparticles has increased in the past few years by virtue of their potential for applications in fields such as ultrahigh-density recording media and medicine [1, 2]. Most applications reply on the magnetic order of the nanoparticles being stable with time evolution. However, with decreasing particle size, the magnetic anisotropy energy per particle responsible for holding the magnetic moment along magnetic anisotropy direction becomes comparable to the thermal energy. This brings about the thermal fluctuations of the magnetic moment, and the nanoparticles lose their stable magnetization state and show superparamagnetic. Thus, the demand for further miniaturization comes into conflict with the superparamagnetism caused by the reduction of the anisotropy energy per particles. It has been demonstrated that the magnetic coupling of ferromagnetic nanoparticles with an AFM matrix is a source of a large effective additional anisotropy [3-6]. This leads to a marked improvement in the thermal stability of the moments of the FM nanoparticles. This mechanism provides a way to artificially control the superparamagnetic limit in isolated nanoparticles. Herein, we show a simple approach to a new type of strongly exchange-coupled iron oxide-based core/shell nanocubes consisting of an AFM core of FeO, which is surrounded by a ferrimagnetic (FM) shell of spinel. The AFM/FM structure presented here is fundamentally interesting since it is inverted in respect to the classical and mostly studied FM/AFM paradigm based on a metallic particle covered by a thin oxide layer, Co/CoO [3]. Furthermore, the FeO and spinel components are highly compatible not only compositional but also structurally with a small lattice mismatch (~3%) [7], which is expected to provide a better structural perfection of the AFM constituent and a better control of its finite size. In this study, thermal stability of FM moment in 16 nm FeO (core)/spinel (shell) nanocubes with different FeO sizes was investigated. The advantage of this AFM/FM coupling system with respect to thermal stability of FM component as compared to single FM or classical FM/AFM coupling systems was also discussed.

2. Experiment

The 16 nm FeO (core)/spinel (shell) nanocubes were synthesized by the following method reported elsewhere [8]. Size of FeO core was atomically controlled by slow oxidation of the as-synthesized nanocubes at low temperature of 80°C in air ambient, and vigorous agitation condition and estimated using a simple calculation based on the saturation magnetizations of the nanocubes [8]. Herein, it was assumed that the nanocubes have perfect cubic cores and that the spinel phase in the nanocubes is γ-Fe_2O_3 and is preferentially distributed in the shells.

Transmission electron microscopy (TEM, Jeol-3010 at 200 kV) was employed to determine the morphology, particle size and size distribution of nanocubes. A superconducting interference device (SQUID; Quantum Design MPMS-5) with fields up to 40 kOe and temperatures from 5 K to 300 K was employed for magnetic measurements. Samples for the magnetic characterization were prepared by adding the hot dispersion of the nanocubes in tetracosane (melting point: 65°C) with 0.1 % volume fraction into a quartz tube (ø5 X 75 mm) and rapidly solidifying using liquid nitrogen. The tube was then sealed in a glovebox using Araldite and the diamagnetic background from the quartz tube was subtracted. Mass of Fe in the samples was determined by x-ray fluorescence (Rigaku RIX 2100) [8].

3. Results and discussion

Figure 1a shows low resolution TEM image of 16 nm as-synthesized FeO/spinel nanocubes. The nanocubes are highly monodispersed with small size deviation of 5 %. Size of FeO core as estimated from TEM image is about 10 nm. This is consistent with 3 nm penetration length of oxygen into FeO lattice from the nanocube surface at ordinary temperature as well as the value estimated using saturation magnetization (M_s) of the nanocubes. The high resolution TEM picture of an isolated 16 nm nanocube in figure 1b shows that it is highly crystallized, as indicated by clear atomic lattice planes. The d spacing determined from HR-TEM image is about 0.21 nm match well with that between <400> planes of spinel phase. Minimal lattice strain between FeO core and spinel shell could be observed, consistent with small lattice mismatch (~ 3%) of the constituents [7]. Consequently, structure matching of AFM core is expected to be well stabilized by spinel shell lattice. Figure 2a shows the temperature-dependent ZFC-FC susceptibility curves of the 16 nm as-synthesized core/shell nanocubes at different field-cool strengths in the range from 50 Oe to 1000 Oe. Both ZFC and FC curves exhibit the independent-applied field peaks at about 198 K, consistent with Néel temperature

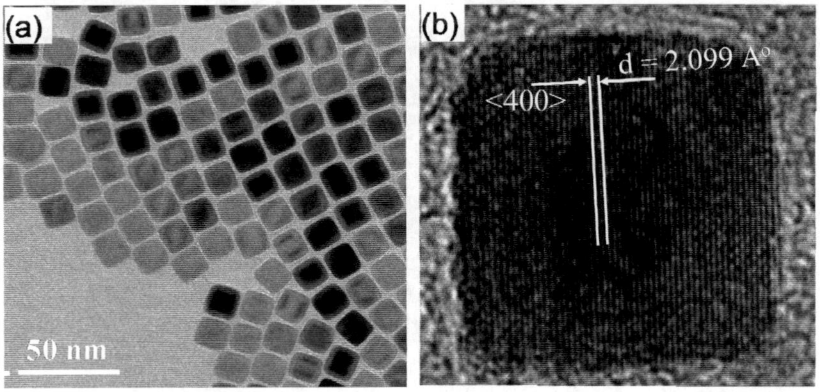

Figure 1 Low (a) and (b) high resolution TEM images of the as-synthesized 16 nm FeO(core)/spinel (shell) nanocubes.

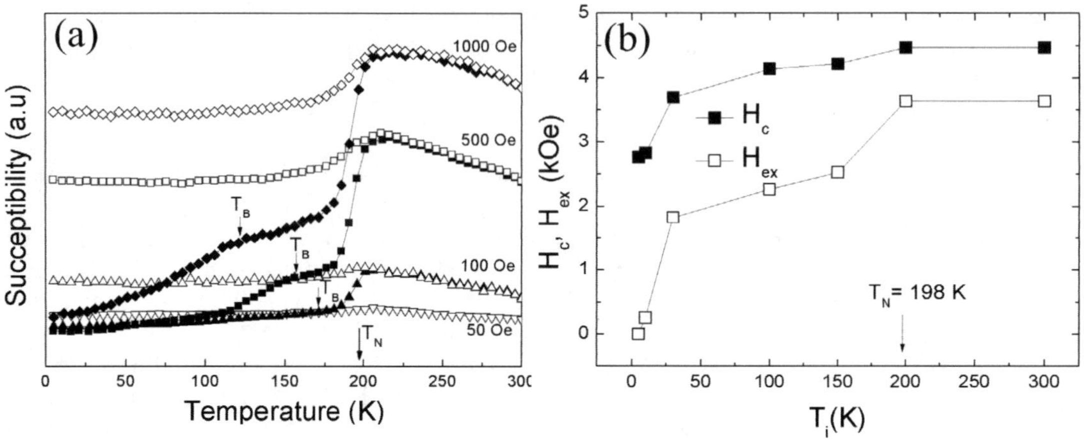

Figure 2 (a) Temperature-dependent ZFC-FC susceptibility curves measured at different field cool strengths of the as-synthesized 16 nm FeO(core)/spinel(shell) nanocubes. Herein, the solid and open symbols correspond to ZFC and FC data, respectively. (b) Loop-shift (H_{ex}) and coercivity (H_c) of the as-synthesized 16 nm FeO(core)/spinel(shell) measured at 5 K after the field cool of 10 kOe from T_i in the range from 5 K to 300 K.

(T_N) of FeO phase. Interestingly, in addition to the AFM transition peak, a broad shoulder in the low temperature range below AFM transition point in the ZFC curves appears and shifts gradually toward lower temperature as increasing field cool strength, revealing a feature of superparamagnetic transition (T_B) of FM component in the nanocubes. Moreover, the decrease of T_B from T_N of FeO with increasing field cool strength suggests that by coupling with AFM core the spinel shell remains FM up to T_N of FeO core. In this case, the nanoshell moments are prevented from thermal fluctuation over the magnetic anisotropy energy barrier for all temperature below T_N of FeO, and thus the nanocubes remain magnetically stable below T_N. In fact, the highly thermal stability of FM moments in FC mode may be understood by high alignment of FM spins and uncompensated spins of AFM at interface. It is really interesting because even in ZFC mode highly thermal stabilization of FM moments is also achieved. In order to get insight, the loop shift (H_{ex}) and coercivity (H_c) of the nanocubes were measured at 5 K after the FC of 10 kOe from T_i and shown in figure 2b. Even when the FC takes place from as low as T_i = 50 K, a large exchange bias (nearly a haft of the maximum value) still remains, meanwhile a large coercivity (nearly 62 % of the maximum value) can be achieved even the FC starts from T_i = 5 K (or zero-field cooling). This means that even in ZFC procedure, a significantly additional anisotropy is still provided to the FM component in the exchange-bias coupling system, thus stabilizing their magnetic moment. The same dependence of H_{ex} and H_c on the temperature point (T_i) at which the FC starts is also observed in the Co/CoO exchange coupling system [9], somewhat attributed to single domain nature of FM component in the nanoparticles, and still an open question needs to be further studied in the future.

As revealed, T_B of FM moment in an exchange-bias coupling system is equivalent with T_N of AFM constituent. Subsequently T_N is governed by its physical dimension due to the so-called "finite-size effect" [10]. Accordingly, study about thermal stability of FM moment in the nanocubes identifies with that about FeO core size-Néel temperature dependent.

Figure 3a shows temperature-dependent ZFC-FC susceptibility curves of the 16 nm nanocubes with different FeO core sizes of 10 nm, 7 nm, 5 nm, and 0 nm (or single phase spinel nanocubes). Herein, the existence of FeO in the nanocubes was proved through observation of exchange-bias while size of FeO core was estimated using M_s deduced from the hysteresis loops of the nanocubes (figure 3b) [8]. Indeed, the H_{ex} is reduced to zero from 3.6 kOe of 10 nm FeO core while M_s is increased up to 110 emu/g_{Fe} of the single phase spinel nanocubes as the spinel shell grows atomically on the expense of FeO core. The T_N as indicated in the figure 3a are 198 K, 160 K, 136 K,

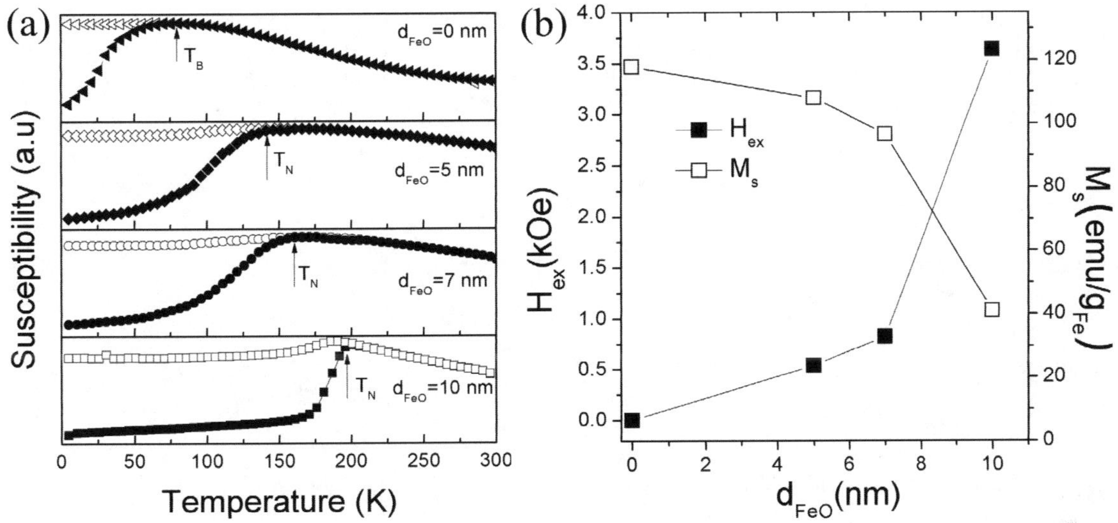

Figure 3 (a) Temperature-dependent ZFC-FC susceptibility curves measured at field of 50 Oe and (b) the dependence of H_{ex} and M_s on FeO core size of the 16nm FeO(core)/spinel(shell) nanocubes measured at 5 K after a field cool of 10 kOe from 300 K. In (a), the solid and open symbols correspond to ZFC and FC data, respectively.

and 80 K for 10 nm, 7 nm, 5 nm, and 0 nm FeO core in the nanocubes respectively. These were confirmed by the temperature-dependent ZFC-FC susceptibility curves measured at different field-cool strengths and can be understood by the finite-size effect in AFM material [10]. Herein, slow reduction of T_N with FeO core size may be attributed to the lattice perfection of FeO core stabilized by spinel shell constituent. This result suggests that by implant of a small AFM core with size of 5 nm, 7 nm and 10 nm, T_B of FM shell component in the 16 nm FeO/spinel core-shell nanocubes was strongly enhanced up to 136 K, 160 K, and 198 K (T_N of FeO phase) as compared to 80 K of the 16 nm single spinel nanocubes (or no FeO core). This is very interesting because the reduction of volume of the FM spinel shells (or enlargement of FeO core size) in 16 nm nanocubes should bring about a reduction of T_B according to the Néel-Brown relaxation model for superparamagnetic nanoparticles [11]. In other words, the exchange-bias coupling induced at the interface between FM and AFM phases in the nanocubes presented herein provides a new approach for artificially controlling superparamagnetic limit of nanoparticles in addition to the traditional one completely basing on size of FM nanoparticles. Moreover, by considering the volume factions of spinel constituent with respect to FeO one in the 16 nm FeO/spinel nanocubes, the inverted paradigm AFM/FM exchange-bias coupling system presented herein exhibits an anomalous advantage as compared to the classical FM/AFM one because of highly thermal stability of FM moments achieved with minimum amount of AFM implant, thus strongly reduces mass contribution of AFM to magnetization of whole coupling system.

4. Conclusions

In summary, we have demonstrated a new type of exchange-bias coupled iron oxide-based core/shell nanocubes consisting of an AFM core (FeO) surrounded by a FM shell (spinel). An extra anisotropy induced by exchange bias at the interface between AFM and FM components in the nanocubes strongly enhances the effective blocking temperature of FM moment up to Néel temperature of FeO core. Well structure stabilized FeO core by spinel shell due to their small lattice mismatch (~ 3 %) leads to a slow reduction of Néel temperature, and thus blocking temperature of FM moment in spinel shell with FeO core size. Whereas the 16 nm single spinel nanocubes lose their magnetic moment at 80 K, the spinel shell in the 16 nm nanocubes with FeO core size of 5 nm, 7 nm, and 10 nm remains ferrimagnetic up to about 136 K, 160 K, and 198 K respectively. This exchange-bias coupling system really provides and additional approach for artificially controlling superparamagnetic limit of nanoparticles and thus a new ideal for constructing new hybrid structure for magnetic device applications.

Acknowledgement

This work was supported by the Japan Society for the Promotion of Science (JSPS).

Reference

1. Kodama R H. 1999 *J. Mag. Mag. Mater.* **200** 359.

2. Sun S H, Murray C B, Weller D, FolkS L and Moser A. 2000 *Science* **187** 1989.

3. Meiklejohn W H and Bean C P. 1956 *Phys. Rev.* **102** 1413.

4. Nogues J and Schuller I. 1999 *J. Mag. Mag. Mater.* **192** 203.

5. Skumryev V, Stoyanov S, Zhang Y, Hadjipanayis G, Givord D and Nogues J. 2003 *Nature* **423** 850.

6. Yamamoto Y, Nakagawa H and Hori H. 2007 *J. Mag. Mag. Mater.* **310** 2384.

7. Cornell R M and Schwertmann. U., The Iron Oxides: Structure, Properties, Reactions, Occurrence and Uses (VCH, New York, 1996), pp. 28-29.

8. Hai H T, Yang H T, Kura H, Hasegawa D, Ogata Y, Takahashi M, OgawaT. 2010 *J. Colloids and Inter. Sci.* **346** 37.

9. Riveiro J M, Toro D, Andres J A, Gonzakez J P, Munoz J A and Goff J P. 2005 *App. Phys. Lett.* **86** 172503.

10. Ambrose T and Chien C L. 1996 *Phys. Rev. Lett.* **76** 1743.

11. Néel L. 1949, *Ann. Grophys* **5** 99.

2nd Int. Symp. on Advanced Magnetic Materials and Applications (ISAMMA 2010) IOP Publishing
Journal of Physics: Conference Series **266** (2011) 012128 doi:10.1088/1742-6596/266/1/012128

Effects of Si substitution on phase transformation and exchange bias in Ni-Mn-In ribbons

X G Zhao[1, 2], C C Hsieh[1], W C Chang[1], W Liu[2] and Z D Zhang[2]

[1]Department of Physics, National Chung Cheng University, Chiayi, Taiwan 62102
[2]Shenyang National Laboratory for Materials Science, Institute of Metal Research, and International Centre for Materials Physics, Chinese Academy of Sciences, 72 Wenhua Road, Shenyang 110016

E-mail: phywcc@ccu.edu.tw

Abstract The effects of partial Si substitution for In on phase transformation and exchange bias (EB) behaviour in as-spun and annealed $Ni_{48}Mn_{39}In_{13-x}Si_x$ (x = 0-4) ribbons have been investigated. It was observed that an increase in Si concentration strongly affects the magnetic state of the martensitic phase and the EB properties of the ribbons. Both martensitic transition temperature (T_M) and the Curie temperature (T_C) of the austenitic phase decrease with increasing Si concentration for $x \leq 3$, while the ribbon for x = 4 possess the modulated martensitic monoclinic structure at room temperature, but its magnetization is unlike (considerably smaller than) those of other ribbons. Both the exchange bias field (H_E) and the coercivity (H_C) are strongly dependent on Si content. The values of H_E and H_C vary from 216 Oe and 314 Oe to 434 Oe and 1682 Oe, respectively, for the as-spun ribbons at 10 K after cooling in a field of 20 kOe, as x is increased from 0 to 4, while they change from 258 Oe and 192 Oe to 456 Oe and 1488 Oe for the annealed ones. The EB phenomena are attributed to the exchange interactions that occurred in AFM-FM interfaces in Si-substituted Ni-Mn-In ribbons.

1. Introduction

The exchange bias (EB) phenomenon is usually described as a shift from the origin of the magnetic hysteresis loop of a material with ferromagnetic/antiferromagnetic (FM/AFM) interfaces, in which the Curie temperature (T_C) of FM phase is higher than the Néel temperature (T_N) of AFM phase, after it is cooled from above T_N to low temperatures in the presence of a magnetic field. Since it was discovered in Co particles surrounded by a layer of AFM CoO[1], extensive research has been conducted on thin films, bulk inhomogeneous materials and magnetic nanoparticles, because of its potential application in many aspects, including spin-valve and tunneling devices[2], permanent magnets[2,3], magnetic recording media and other devices[2]. In recent years, Ni-Mn-X (X = In, Sn, Sb) FM Heusler alloys have attracted considerable interest[4]. These alloys undergo martensitic transformation from FM austenite to low magnetic state martensite. This characteristic makes them exhibit several interesting physical properties, such as the shape memory effect [5], magnetocaloric effect [6], magnetoresistance[7], and the EB behaviour[8]. Recently, Pathak et al. reported that the effects of the substitution of In by Si, Ge and Al on phase transition temperature, magnetic entropy changes, electrical and exchange bias properties of the FM $Ni_{50}Mn_{35}In_{15}$ Heusler alloys [9-12]. In previous work, we reported the magnetocaloric effects properties associated with both the first- and the second-order phase transitions and EB behaviour of $Ni_{48}Mn_{39}In_{13}$ ribbons [13]. In this work, we report phase transformation and EB behaviour in the as-spun and annealed Si-doped Ni-Mn-In ribbons.

2. Experiment

The ingots with the nominal compositions of $Ni_{48}Mn_{39}In_{13-x}Si_x$ (x = 0-4) were prepared by arc melting pure elements under argon atmosphere in a water-cooled Cu crucible. The ingots were remelted several times and subsequently melt-spun using a single-roll melt-spinner at a wheel linear speed of 30m/s. Some of the as-spun ribbons were annealed at 1173 K for 20 min. and finally quenched in ice water. The structural analyses of the ribbons were carried out at room temperature with conventional powder X-ray diffraction (XRD) techniques using Cu-*Kα* radiation. Magnetization measurements were performed in the temperature interval of 10–350 K, using a superconducting quantum interference device magnetometer (SQUID, Quantum Design Inc.) in fields up to 20 kOe. The magnetic field was applied along the ribbon axis. Zero-field-cooled (ZFC) and field cooled (FC) thermomagnetic curves were recorded at 200 Oe with a temperature heating or cooling rate of 5 K/min.

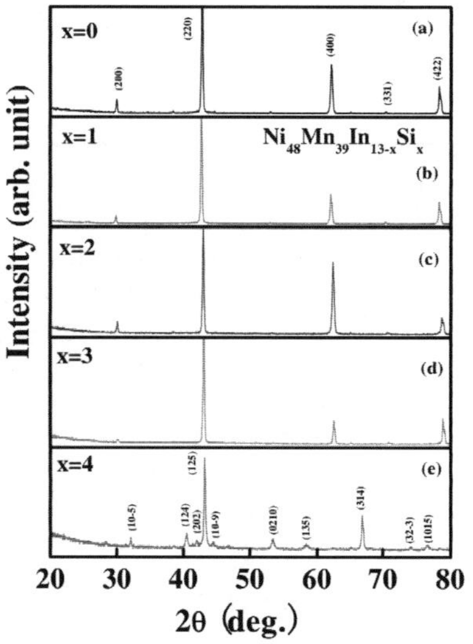

Figure 1. XRD patterns of the as-spun $Ni_{48}Mn_{39}In_{13-x}Si_x$ (x = 0-4) ribbons at room temperature

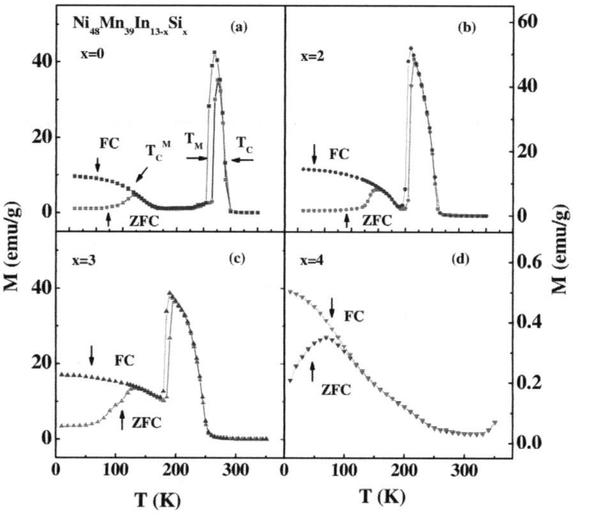

Figure 2. Temperature dependence of ZFC and FC magnetization curves *M(T)* for the annealed $Ni_{48}Mn_{39}In_{13-x}Si_x$ (x = 0, 2, 3, 4) ribbons, measured in a field of 200 Oe

3. Results and discussion

Figure 1 gives XRD patterns of the as-spun $Ni_{48}Mn_{39}In_{13-x}Si_x$ (x = 0-4) ribbons at room temperature. The lattice constants obtained from XRD data are summarized in Table 1. It can be seen that the ribbons for $x \leq 3$ crystallize in a partially ordered B2 phase instead of a highly ordered $L2_1$ phase, due to the near absence of superstructure peaks (111) and (311)[13]. In this case, the lattice constant is half that of the $L2_1$ phase, while the $Ni_{48}Mn_{39}In_9Si_4$ ribbon is in the martensitic phase with a ten-layered modulated (10M) structure. The unit cells are monoclinic with β=88.96° and 89.08° for as-spun and annealed ribbons, respectively. Due to the smaller atomic radius of Si than that of In, increasing Si concentration decreases lattice constants. After annealing, the lattice constants little change in the error bar range for the same Si concentration (As seen in table 1). However, the grain size increases upon annealing, according to the results of SEM observation (not shown here).

The temperature dependence of zero-field-cooled (ZFC) and field-cooled (FC) magnetizations of the annealed $Ni_{48}Mn_{39}In_{13-x}Si_x$ (x = 0, 2, 3, 4) ribbons, measured in a field of 200 Oe, is shown in figure 2. The magnetization curves *M(T)* for the as-spun and annealed ribbons revealed that all the ribbons show a splitting of ZFC and FC magnetizations at low temperatures, indicating irreversible behaviour.

Table 1. Lattice parameters of $Ni_{48}Mn_{39}In_{13-x}Si_x$ (x = 0-4) ribbons

	Crystal structure	As-spun Lattice parameter (Å)			Annealed Lattice parameter (Å)		
x		a	b	c	a	b	c
0	B2	3.000(1)			2.999(5)		
1	B2	2.994(5)			2.992(7)		
2	B2	2.989(2)			2.987(3)		
3	B2	2.982(5)			2.981(4)		
4	10M	4.411(3)	5.900(6)	21.258(3)	4.410(5)	5.899(5)	21.257(1)

From figure 2, it can be seen that as the temperature increase, a FM transition occurs above the splitting point. This temperature is defined as T_C^M , which represents the Curie temperature of the FM martensitic phase. An increase in Si concentration results in a decrease in both the martensitic structural transition temperature (T_M) and the Curie temperature (T_C) of austenitic phase for $x \le 3$, which is consistent with those of reference[9], whereas for x = 4, the ribbon is in martensitic state in the vicinity of room temperature, as seen in the figure 1(e). In Ni-Mn-In alloys, actually, the structural transition of martensitic phase is mainly controlled by In content[14] and the Si substitution for In only extends composition range of austenitic phase, resulting in decrease of T_M and T_C. As the Si content further increases (namely, the In content decreases), it makes the structure of the sample transforms martensitically above room temperature. For the ribbon with lower In concentration, long-range FM ordering weakens, so its magnetization is unlike (considerably smaller than) those of other ribbons due to the introduction of large amounts of AF phase at this concentration [14].

Figures 3(a) and 3(b) shows magnetic hysteresis loops of the as-spun and annealed $Ni_{48}Mn_{39}In_{13-x}Si_x$ (x = 0-4) ribbons at 10 K after FC at 20 kOe from 300 K. The measurements were actually carried out from -20 to 20 kOe. For clear visualization of the variation of the loop shift, only the hysteresis loop from -3 to 3 kOe is shown in the figure. The corresponding EB field H_E and coercivity H_C as a function of Si contents are shown in figures 3(c) and 3(d) for the $Ni_{48}Mn_{39}In_{13-x}Si_x$ ribbons. The values of H_E and H_C are calculated using $H_E = -(H_1 + H_2)/2$ and $H_C = |H_1 - H_2|/2$, where H_1 and H_2 are negative and positive fields at which the magnetization equals to zero. As it is evident from the figures, the hysteresis loop at 10 K is shifted significantly to the negative field axis at the region of measured temperature, which indicates that the exchange bias indeed exists in these ribbons. The values of H_E and H_C vary from 216 Oe and 314 Oe to 434 Oe and 1682 Oe for the as-spun ribbons, respectively, as x is increased from 0 to 4, while they change from 258 Oe and 192 Oe to 456 Oe and 1488 Oe for the annealed ones. Moreover, the EB properties are different in the as-spun and annealed ribbons. H_E was found to increase, while the coercivity H_C decreases, after annealing. It has been well-known that the origin of exchange bias in the compounds can be attributed to different interaction mechanisms.

One important mechanism is that the coupling of the ferromagnet to the antiferromagnet at the interface causes a unidirectional anisotropy [1-3], which may interpret our results. In fact, neutron diffraction technique study showed that antiferromagnetic order exists in Ni–Mn–Sn alloys [15].

In addition, as Mn is substituted for Sn, excess Mn atoms occupy Sn sites in Ni-Mn-Sn system. In such spatial configurations, Mn atoms can have Mn atoms as nearest neighbours along the [110] directions. The Mn-Mn spacing, in this case, is smaller than that of in the stoichiometric compound; therefore, the Mn in the Sn sites interacts antiferromagnetically with the Mn in the Mn sites inducing AF exchange interaction, leading to local noncollinear spin structures, which can pin the FM moments in different configurations at low temperature [16]. Similarly, for $Ni_{48}Mn_{39}In_{13-x}Si_x$ ribbons, the EB

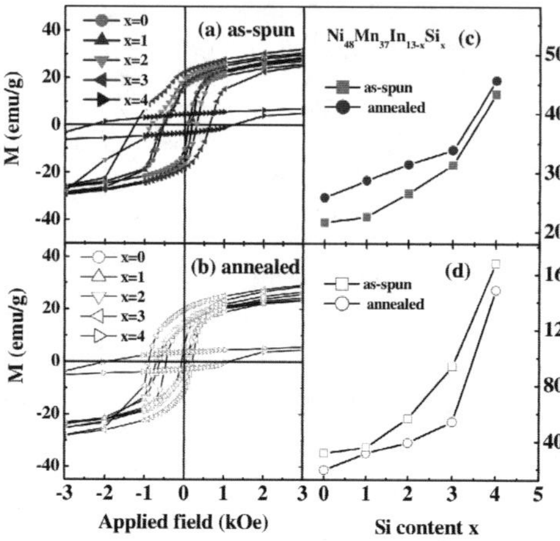

Figure 3. The magnetic hysteresis loops measured at 10 K after FC at 20 kOe of the $Ni_{48}Mn_{39}In_{13-x}Si_x$ (x = 0-4) ribbons, (a) as-spun, (b) annealed, (c)-(d) the corresponding H_E and H_C as a function of Si contents

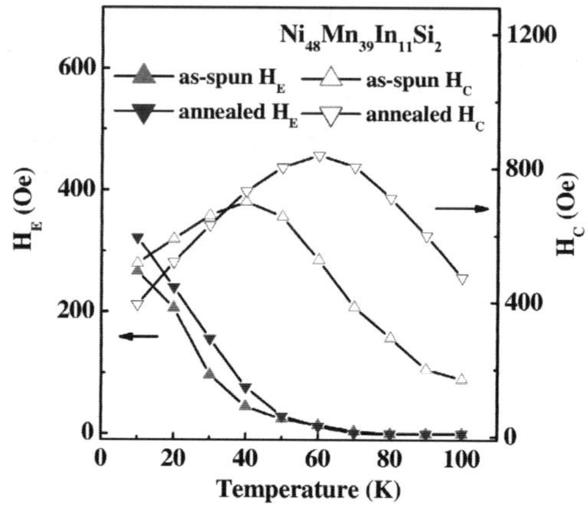

Figure 4. Variation of the H_E and H_C as a function of temperature from 10 K to 100 K for the as-spun and annealed $Ni_{48}Mn_{39}In_{10}$-Si_2 ribbons

phenomena can be qualitatively understood in terms of the coupling between FM and AFM spins at their interfaces in the martensitic state. The reasons are as follows: firstly, in off-stoichiometrtric Ni-Mn-In alloys, the rapid solidification can increase site disorder and excess Mn atoms in the In sites interact antiferromagnetically with Mn atoms in the Mn sites. Si atoms with smaller atomic radius substitute for In atoms, leading to the decrease of lattice constants and also the Mn-Mn distance between Mn atoms occupying in these Mn and In sites. This may effectively enhance the AFM interaction between the Mn atoms in the system, leading to the increment of H_E and H_C with the increase in Si contents. Secondly, the existence of internal stress in the as-spun ribbons with small grain sizes results in smaller H_E and larger H_C. However, after annealing, the relaxation of internal stress results in the decrease in the Mn-Mn distance for the same Si content. This strengthens the exchange coupling at the AFM/FM interfaces, giving rise to the increment of H_E and the decrease of H_C because of the growth of grains after annealing [13]. Thirdly, as the Si content further increases, namely, the In content decreases, the sample transforms from partially ordered B2 structure of the austenitic phase for x = 3 to 10M modulated monoclinic structure of martensitic phase for x = 4. Furthermore, taking the effect of martensitic transformation into account, the different martensitic structure may be formed with further increasing the Si content at a low temperature, resulting in different fractions of AFM and FM phases in the phase separated martensitic state. At lower In concentration, the relative large amounts of AFM phase is introduced at this concentration, which can strongly pin the FM moments in different configurations at low temperature, so an more large extra field is required to overcome the microscopic torque from different spin configurations, resulting in the H_E and H_C values increase enormously. The EB behavior for ribbons agrees in general well with for bulk Ni-Mn-In-Si alloys previously reported by A. K. Pathak et al. [11], the shifts of the hysteresis loops increase with increasing Si concentration. The value of H_E of the ribbon relatively larger than that of bulk compounds is presumably resulted from the forming of different volume fraction of martensitic phases at low temperature due to different preparation processes with their different cooling speed.

For exploring the temperature dependence of EB, as an example, figure 4 shows variation of the H_E and H_C as a function of temperature from 10 K to 100 K for the as-spun and annealed $Ni_{48}Mn_{39}In_{11}Si_2$ ribbons. The values of H_E almost linearly decrease with increasing temperature and becomes zero above 80 K. On the other hand, H_C initially increases with temperature and then decreases after reaching a maximum. Such temperature dependencies of H_E and H_C are typical for EB systems.

4. Conclusions

The effects of the partial substitution of In by Si on the magnetic phase transition and EB behavior in $Ni_{48}Mn_{39}In_{13-x}Si_x$ (x = 0-4) ribbons have been investigated. The substitution of a small amount of Si in In sites affects phase constitutions. The transition temperatures T_M and T_C decrease with an increase in Si concentration. Shift in the hysteresis loops of the ribbons occurs, when the samples were cooled down to 10 K in an applied magnetic field of 20 kOe. The values of H_E and H_C vary from 216 Oe and 314 Oe to 434 Oe and 1682 Oe for the as-spun ribbons, respectively, as x is increased from 0 to 4, while they change from 258 Oe and 192 Oe to 456 Oe and 1488 Oe for the annealed ones. The EB phenomena observed in the $Ni_{48}Mn_{39}In_{13-x}Si_x$ (x = 0-4) ribbons are attributed to the occurrence of AFM and FM interactions at low temperature of martensitic state. The behaviour is an addition to multifunctional properties of the $Ni_{48}Mn_{39}In_{13-x}Si_x$ Heusler alloy ribbons.

5. Acknowledgements

This work has been supported by the National Science Council, Taiwan under grant no. NSC-98-2112-M-194-005-MY3 and the National Natural Science Foundation of China under grant no. 30331030 and 50831006.

References

[1] Meiklejohn W H and Bean C P 1956 *Phys. Rev.* **102,** 1413
[2] Nogués J, Sort J, Langlais V, Skumryev V, Suriñach S, Muñoz J S, Baró M D, 2005 *Phys. Reports.* **422**, 65
[3] Nogués J and Schuller I K, 1999 *J. Magn. Magn. Mater.* **192**, 203
[4] Sutou Y, Imano Y, Koeda N, Omori T, Kainuma R, Ishida K and Oikawa K, 2004 *Appl. Phys. Lett.* **85**, 4358
[5] Kainuma R, Imano Y, Ito W, Sutou Y, Morito H, Okamoto S, Kitakami O, Oikawa K, Fujita A, Kanomata T and Ishida K, 2006 Nature **439,** 957
[6] Sharma V K, Chattopadhyay M K, Shaeb K H B, Chouhan A and Roy S B, 2006 *Appl. Phys. Lett.* **89**, 222509
[7] Koyama K, Okada H, Watanabe K, Kanomata T, Kainuma R, Ito W, Oikawa K and Ishida K, 2006 *Appl. Phys. Lett.* **89**, 182510
[8] Khan M, Dubenko I, Stadler S and Ali N, 2007 *Appl. Phys. Lett.* **91**, 072510
[9] Pathakl A K, Dubenko I, Stadler S and Ali N, 2008 *J. Phys. D: Appl. Phys.* **41,** 202004
[10] Pathakl A K, Dubenko I, Stadler S and Ali N, 2009 *J. Appl Phys.* **105**, 07B103
[11] Pathak A K, Dubenko I, Stadler S and Ali N, 2009 *IEEE Magn.* **45**, 3855
[12] Pathakl A K, Dubenko I, Mabon J C, Stadler S and Ali N, 2009 *J. Phys. D: Appl. Phys.* **42,**045004
[13] Zhao X G , Hsieh C C, Lai J H, Cheng X J, Chang W C, Cui W B, Liu W and Zhang Z D, 2010 *Scripta Mater.* **63**,250
[14] Krenke T, Acet M, Wassermann E F, Moya X, Mañosa L and Planes A, 2006 *Phys. Rev. B* **73**, 174413
[15] Brown P J, Grandy A P, Ishida K, Kainuma R, Kanomata T, Neumann K U, Oikawa K, Ouladdiaf B and Ziebeck K P A, 2006 *J. Phys.: Condens. Matter.* **18**, 2249
[16] Krenke T, Acet M, Wassermann E F, Moya X, Mañosa L and Planes A, 2005 *Phys. Rev. B* **72**, 014412

AUTHOR INDEX

AUTHOR INDEX

AUTHOR INDEX

AUTHOR INDEX